ISBN 978-0-331-66052-4
PIBN 11071606

1 MONTH OF
FREE
READING

at

www.ForgottenBooks.com

By purchasing this book you are eligible for one month membership to ForgottenBooks.com, giving you unlimited access to our entire collection of over 700,000 titles via our web site and mobile apps.

To claim your free month visit:

www.forgottenbooks.com/free1071606

English
Français
Deutsche
Italiano
Español
Português

www.forgottenbooks.com

Mythology Photography **Fiction**
Fishing Christianity **Art** Cooking
Essays Buddhism Freemasonry
Medicine **Biology** Music **Ancient
Egypt** Evolution Carpentry Physics
Dance Geology **Mathematics** Fitness
Shakespeare **Folklore** Yoga Marketing
Confidence Immortality Biographies
Poetry **Psychology** Witchcraft
Electronics Chemistry History **Law**
Accounting **Philosophy** Anthropology
Alchemy Drama Quantum Mechanics
Atheism Sexual Health **Ancient History**
Entrepreneurship Languages Sport
Paleontology Needlework Islam
Metaphysics Investment Archaeology
Parenting Statistics Criminology
Motivational

SMITHSONIAN INSTITUTION.
UNITED STATES NATIONAL MUSEUM.

BULLETIN

OF THE

NITED STATES NATIONAL MUSEUM

No. 47.

THE FISHES

OF

NORTH AND MIDDLE AMERICA:

ESCRIPTIVE CATALOGUE OF THE SPECIES OF FISH-LIKE VERTEBRATES FOUND I.
THE WATERS OF NORTH AMERICA, NORTH OF THE ISTHMUS OF PANAMA.

BY

DAVID STARR JORDAN, Ph. D.,
PRESIDENT OF THE LELAND STANFORD JUNIOR UNIVERSITY AND OF THE
CALIFORNIA ACADEMY OF SCIENCES,

AND

BARTON WARREN EVERMANN, Ph. D.,
ICHTHYOLOGIST OF THE UNITED STATES FISH COMMISSION.

PART III.

SMITHSONIAN INSTITUTION.

UNITED STATES NATIONAL MUSEUM.

———

THE FISHES

OF

NORTH AND MIDDLE AMERICA:

A DESCRIPTIVE CATALOGUE

OF THE

SPECIES OF FISH-LIKE VERTEBRATES FOUND IN THE
WATERS OF NORTH AMERICA, NORTH OF
THE ISTHMUS OF PANAMA.

BY

DAVID STARR JORDAN, Ph. D.,

PRESIDENT OF THE LELAND STANFORD JUNIOR UNIVERSITY AND OF THE
CALIFORNIA ACADEMY OF SCIENCES,

AND

BARTON WARREN EVERMANN, Ph. D.,

ICHTHYOLOGIST OF THE UNITED STATES FISH COMMISSION.

PART III.

WASHINGTON:

GOVERNMENT PRINTING OFFICE.

1898.

TABLE OF CONTENTS, PART III.

CLASS III. PISCES—Continued.
  ORDER BB. ACANTHOPTERI—Continued.
    Family CLXXXIV. Triglidæ—Continued.   Page.
      Group Gobioidea.. 2184
    Family CLXXXVII. Callionymidæ 2184
      Genus 799. Callionymus, Linnæus 2185
        2511. bairdi, Jordan 2185
        2512. himantophorus, Goode & Bean................................ 2186
        2513. calliurus, Eigenmann & Eigenmann.......................... 2187
        2514. pauciradiatus, Gill... 2188
    Family CLXXXVIII. Gobiidæ.. 2188
      Genus 800. Ioglossus, Bean 2192
        2515. calliurus, Bean 2193
      Genus 801. Philypnus, Cuvier & Valenciennes...................... 2194
        2516. dormitor (Lacépède).. 2194
        2517. lateralis, Gill... 2195
      Genus 802. Dormitator, Gill.:..................................... 2195
        2518. maculatus (Bloch).. 2196
      Genus 803. Guavina, Bleeker 2198
        2519. guavina, Cuvier & Valenciennes 2198
      Genus 804. Eleotris (Gronow) Bloch & Schneider 2199
        2520. amblyopsis (Cope).. 2199
        2521. abacurus, Jordan & Gilbert 2200
        2522. pisonis (Gmelin) 2200
        2523. perniger (Cope).. 2201
        2524. pictus (Kner & Steindachner) 2201
      Genus 805. Alexurus, Jordan 2202
        2525. armiger, Jordan & Richardson 2203
      Genus 806. Erotelis, Poey... 2203
        2526. smaragdus (Cuvier & Valenciennes).......................... 2204
      Genus 807. Gymneleotris, Bleeker 2204
        2527. seminudus (Günther).. 2204
      Genus 808. Chriolepis, Gilbert 2205
        2528. minutillus, Gilbert.... 2205
      Genus 809. Sicydium, Cuvier & Valenciennes....................... 2205
        2529. plumieri (Bloch)... 2206
        2530. antillarum, Ogilvie-Grant 2206
        2531. vincente, Jordan & Evermann 2207
        2531 (a). punctatum, Perugia 2867
      Genus 810. Cotylopus, Guichenot.................................. 2207
        Subgenus Sicya, Jordan & Evermann 2207
        2532. gymnogaster (Ogilvie-Grant) 2207
        2533. salvini (Ogilvie-Grant) 2208
      Genus 811. Evorthodus, Gill 2208
        2534. breviceps, Gill.. 2208
      Genus 812. Lophogobius, Gill...................................... 2209
        2535. cyprinoides (Pallas) 2209

CLASS III. PISCES—Continued.
  ORDER BB. ACANTHOPTERI—Continued.
    Family CLXXXVIII..Gobiidæ—Continued.   **Page.**
      Genus 813. Gobius (Artedi) Linnæue 2210
        Subgenus Gobius 2216
          2536. soporator, Cuvier & Valenciennes............................... 2216
        Subgenus Ctenogobius, Gill... 2218
          2537. nicholsii, Bean... 2218
          2538. eigenmanni, Garman..................................... 2218
          2539. glaucofrænum (Gill)..................................... 2219
          2540. manglicola, Jordan & Starks............................. 2220
          2541. stigmaturus, Goode & Bean............................... 2220
          2542. quadriporus, Cuvier & Valenciennes...................... 2221
          2543. shufeldti, Jordan & Eigenmann.......................... 2221
          2544. boleosoma, Jordan & Gilbert............................. 2221
          2545. fasciatus (Gill).. 2222
          2546. encæomus, Jordan & Gilbert.............................. 2223
          2547. stigmaticus (Poey)...................................... 2224
          2548· lyricus, Girard... 2224
          2549. garmani, Eigenmann & Eigenmann.......................... 2225
          2550. zebra, Gilbert..................................... 2226; 2867
        Subgenus Euctenogobius, Gill... 2226
          2551. poeyi, Steindachner..................................... 2226
          2552. badius (Gill)... 2227
        Subgenus Gobionellus, Girard... 2227
          2553. microdon, Gilbert....................................... 2227
          2554. smaragdus, Cuvier & Valenciennes........................ 2227
          2555. strigatus, O'Shaughnessy................................ 2228
          2556. sagittula (Günther)..................................... 2228
          2557. hastatus, Girard.. 2229
          2558. oceanicus, Pallas....................................... 2230
        Subgenus Lythrypnus, Jordan & Evermann 2230
          2559. dalli, Gilbert.. 2230
      Genus 814. Garmannia, Jordan & Evermann 2231
        Subgenus Garmannia... 2232
          2560. paradoxa (Günther)...................................... 2232
          2561. hemigymna (Eigenmann & Eigenmann) 2233
        Subgenus Enypnias, Jordan & Evermann 2233
          2562. seminuda (Günther)...................................... 2233
      Genus 815. Awaous, Steindachner.. 2234
          2563. flavus (Cuvier & Valenciennes).......................... 2235
          2564. nelsoni, Evermann....................................... 2235
          2565. taiasica (Lichtenstein) 2236
          2566. mexicanus (Günther)..................................... 2237
      Genus 816. Bollmannia, Jordan... 2237
          2567. ocellata, Gilbert....................................... 2238
          2568. chlamydes, Jordan....................................... 2238
          2569. macropoma, Gilbert 2239
          2570. stigmatura, Gilbert..................................... 2239
      Genus 817. Aboma, Jordan & Starks 2240
          2571. etheostoma, Jordan & Starks............................. 2240
          2572. lucretiæ (Eigenmann & Eigenmann) 2241
          2573. chiquita (Jenkins & Evermann).......................... 2241
      Genus 818. Microgobius, Poey 2242
          2574. gulosus (Girard).. 2243
          2575. eulepis, Eigenmann & Eigenmann 2244
          2576. thalassinus, Jordan & Gilbert.......................... 2245
          2577. signatus, Poey 2246
      Genus 819. Zalypnus, Jordan & Evermann................................ 2246

CLASS III. PISCES—Continued.
ORDER BB. ACANTHOPTERI—Continued.
Family CLXXXVIII. Gobiidæ—Continued.   Page.
    2578. cyclolepis (Gilbert).. 2246
    2579. emblematicus (Jordan & Gilbert) 2247
    Genus 820. Eucyclogobius, Gill... 2248
    2580. newberryi (Girard).. 2248
    Genus 821. Lepidogobius, Gill.. 2249
    2581. lepidus (Girard).. 2249
    Genus 822. Gillichthys, Cooper 2249
    2582. mirabilis, Cooper... 2250
    2583. detrusus, Gilbert & Scofield...................................... 2251
    Genus 823. Quietula, Jordan & Evermann 2251
    2584. y-cauda (Jenkins & Evermann) 2251
    Genus 824. Ilypnus, Jordan & Evermann 2253
    2585. gilberti (Eigenmann & Eigenmann)............................... 2253
    Genus 825. Clevelandia, Eigenmann & Eigenmann 2254
    2586. ios (Jordan & Gilbert).. 2254
    2587. rosæ, Jordan & Evermann 2255
    Genus 826. Evermannia, Jordan.. 2256
    2588. longipinnis (Steindachner)...................................... 2256
    2589. zosterura (Jordan & Gilbert)..................................... 2256
    Genus 827. Gobiosoma, Girard... 2257
    2590. histrio, Jordan 2258
    2591. molestum, Girard 2258
    2592. bosci (Lacépède)... 2259
    2593. crescentale, Gilbert 2259
    2594. multifasciatum, Steindachner 2260
    Genus 828. Barbulifer, Eigenmann & Eigenmann...................... 2260
    2595. ceuthœcus (Jordan & Gilbert)................................... 2260
    Genus 829. Typhlogobius, Steindachner................................ 2261
    2596. californiensis, Steindachner..................................... 2262
    Genus 830. Tyntlastes, Günther 2262
    2597. brevis (Günther) 2262
    2598. sagitta (Günther)... 2263
    Genus 831. Gobioides, Lacépède 2263; 2868
    2599. broussonnetii, Lacépède 2263
    2600. peruanus (Steindachner)... 2264
    Genus 832. Cayennia, Sauvage... 2265
    2601. guichenoti, Sauvage.. 2265
SUBORDER DISCOCEPHALI... 2265
    Family CLXXXIX. Echeneididæ.. 2265
    Genus 833. Phtheirichthys, Gill.. 2268
    2602. lineatus (Menzies)... 2268
    Genus 834. Echeneis (Artedi) Linnæus................................. 2268
    2603. naucrates, Linnæus.. 2269
    2604. naucrateoides, Zuieuw... 2270
    Genus 835. Remilegia, Gill... 2270
    2605. australis (Bennett).. 2270
    Genus 836. Remora, Gill... 2271
    Subgenus Remora... 2271
    2606. remora (Linnæus)... 2271
    Subgenus Remorina, Jordan & Evermann............................. 2272
    2607. albescens (Temminck & Schlegel) 2272
    Subgenus Remoropsis, Gill... 2272
    2608. brachyptera (Lowe).. 2272
    Genus 837. Rhombochirus, Gill... 2273
    2609. osteochir (Cuvier)... 2273
    Group Trachinoidea... 2273

CLASS III. PISCES—Continued.
ORDER BB. ACANTHOPTERI—Continued.   Page.
Family CXC. Malacanthidæ.................................... 2274
Genus 838. Malacanthus, Cuvier.......................... 2275
2610. plumieri (Bloch)............................... 2275
Genus 839. Caulolatilus, Gill........................... 2276
2611. princeps (Jenyns).............................. 2276
2612. microps, Goode & Bean......................... 2277
2613. cyanops, Poey................................. 2278
Genus 840. Lopholatilus, Goode & Bean 2278
2614. chamæleonticeps, Goode & Bean 2278
Family CXCI. Opisthognathidæ............................... 2279
Genus 841. Opisthognathus Cuvier...................... 2280
2615. lonchurum, Jordan & Gilbert.................. 2281
2616 punctatum, Peters............................ 2281
2617. macrognathum, Poey........................... 2281
2618. ommatum, Jenkins & Evermann................. 2282
Genus 842. Gnathypops, Gill............................. 2283
2619. scops, Jenkins & Evermann.................... 2283
2620. maxillosa (Poey)............................. 2284
2621. macrops (Poey)............................... 2284
2622. rhomalea (Jordan & Gilbert).................. 2285
2623. snyderi, Jordan & Evermann 2285
2624. mystacina, Jordan 2286
Genus 843. Lonchopisthus, Gill......................... 2286
2625. micrognathus (Poey).......................... 2287
Family CXCII. Bathymasteridæ 2287
Genus 844. Bathymaster................................. 2288
2626. signatus, Cope............................... 2288
Genus 845. Ronquilus, Jordan & Starks 2289
2627. jordani (Gilbert)............................ 2289
Genus 846. Rathbunella, Jordan & Evermann 2289
2628. hypoplecta (Gilbert)......................... 2290
Family CXCIII. Chiasmodontidæ.............................. 2291
Genus 847. Chiasmodon, Johnson......................... 2291
2629. niger, Johnson 2291
Genus 848. Pseudoscopelus, Lütken 2292
2630. scriptus, Lütken............................. 2292
Family CXCIV. Chænichthyidæ................................ 2293
Genus 849. Hypsicometes, Goode 2293
2631. gobioides, Goode............................. 2294
Family CXCV. Trichodontidæ 2295
Genus 850. Trichodon (Steller) Cuvier.................. 2295
2632. trichodon (Tilesius) 2295
Genus 851. Arctoscopus, Jordan & Evermann............. 2297
2633. japonicus (Steindachner)..................... 2297
Family CXCVI. Dactyloscopidæ............................... 2297
Genus 852. Gillellus, Gilbert......................... 2298
2634. semicinctus, Gilbert......................... 2298
2635. arenicola, Gilbert........................... 2299
2636. ornatus, Gilbert............................. 2299
Genus 853. Dactyloscopus, Gill 2300
Subgenus Dactyloscopus............................ 2301
2637. pectoralis, Gill............................. 2301
2638. tridigitatus, Gill........................... 2301
2639. poeyi, Gill.................................. 2302
2640. lunaticus, Gilbert........................... 2302
Subgenus Esloscopus, Jordan & Evermann........... 2303.
2641. zelotes, Jordan & Evermann 2303

CLASS III. PISCES—Continued.
ORDER BB. ACANTHOPTERI—Continued.
Family CXCVI. Dactyloscopidæ—Continued.   Page.
Genus 854. Dactylagnus, Gill 2304
    2642. mundus, Gill 2304
Genus 855. Myxodagnus, Gill.. 2305
    2643. opercularis, Gill 2305
Family CXCVII. Uranoscopidæ... 2305
Genus 856. Astroscopus, Brevoort 2306
    2644. y-græcum (Cuvier & Valenciennes)............................. 2307
    2645. zephyreus, Gilbert & Starks................................... 2309
    2646. guttatus (Abbott).. 2310
Genus 857. Kathetostoma, Günther 2311
    2647. averruncus, Jordan & Bollman.................................. 2311
    2648. albiguttum, Bean... 2312
SUBORDER HAPLODOCI 2313
Family CXCVIII. Batrachoididæ... 2313
Genus 858. Batrachoides, Lacépède 2314; 2868
    2649. surinamensis (Bloch & Schneider)............................. 2314
    2650. pacifici (Günther)... 2314
Genus 859. Opsanus, Rafinesque 2315
    2651. tau (Linnæus).. 2315
    2652. pardus (Goode & Bean).. 2316
Genus 860. Porichthys, Girard... 2317
    2653. porosissimus (Cuvier & Valenciennes)........................ 2319
    2654. notatus, Girard... 2321
    2655. margaritatus (Richardson) 2322
Genus 861. Thalassophryne, Günther....................................... 2323
    2656. maculosa, Günther 2324
    2657. reticulata, Günther.. 2325
Genus 862. Dæctor, Jordan & Evermann..................................... 2325
    2658. dowi (Jordan & Gilbert) 2325
SUBORDER XENOPTERYGII 2326
Family CXCIX. Gobiesocidæ... 2326
Genus 863. Caularchus, Gill 2327
    2659. mæandricus (Girard)... 2328
Genus 864. Bryssetæres, Jordan & Evermann............................... 2328
    2660. pinniger (Gilbert).. 2328
Genus 865. Gobiesox, Lacépède.. 2329
    Subgenus Bryssophilus, Jordan & Evermann.......................... 2330
    2661. papillifer, Gilbert 2330
    Subgenus Gobiesox.. 2331
    2662. gyrinus, Jordan & Evermann..................................... 2331
    2663. nigripinnis (Peters)... 2331
    •2664. cephalus, Lacépède.. 2332
    2665. tudes, Richardson.. 2333
    2666. strumosus, Cope.. 2333
    2667. virgatulus, Jordan & Gilbert................................. 2333
    2668. adustus, Jordan & Gilbert 2334
    2669. funebris, Gilbert.. 2334
    2670. pœcilophthalmus, Jenyns 2335
    2671. rhodospilus, Günther 2335
    2672. macrophthalmus, Günther 2335
    2673. cerasinus, Cope.. 2336
    Subgenus Sicyases, Müller & Troschel............................... 2336
    2674. erythrops, Jordan & Gilbert................................... 2336
    2675. rubiginosus (Poey)... 2337
    2676. carneus (Poey).. 2337
    2677. hæres, Jordan & Bollman....................................... 2337

CLASS III. PISCES—Continued.
  ORDER BB. ACANTHOPTERI—Con ued.   **Page.**
    Family CXC. Malacanthidæ. 2274
      Genus 838. Malacanthus, vier... 2275
        2610. plumieri (Bloch) 2275
      Genus 839. Caulolatilus, ₵ 2276
        2611. princeps (Jenyu 2276
        2612. microps, Goode Bean... 2277
        2613. cyanops, l'oey.. 2278
      Genus 840. Lopholatilus, ₵ode & Bean 2278
        2614. chamæleonticeps ₵oode & Bean 2278
    Family CXCI. Opisthognathia.. 2279
      Genus 841. Opisthognathu 'uvier 2280
        2615. lonchurum, Jord & Gilbert....................................... 2281
        2616 punctatum, Pete 2281
        2617. macrognathum, ley.. 2281
        2618. onmatum, Jenk & Evermann....................................... 2282
      Genus 842. Gnathypops, ₵l.. 2283
        2619. scops, Jenkins ₵▮ermann.. 2283
        2620. maxillosa (Poey) 2284
        2621. macrops (Poey). 2284
        2622. rhomalea (Jorda   Gilbert).. 2285
        2623. snyderi, Jordan ₵Evermann 2285
        2624. mystacina, Jorda... 2286
      Genus 843. Lonchopisthu₵ill... 2286
        2625. micrognathus (▮▮).. 2287
    Family CXCII. Bathymasterⁱᵉ 2287
      Genus 844. Bathymaster 2288
        2626. siguatus, Cope.. 2288
      Genus 845. Ronquilus, Jonn & Starks 2289
        2627. jordani (Gilbert). 2289
      Genus 846. Rathbunella, Jdan & Evermann 2289
        2628. hypoplecta (Gilbe)... 2290
    Family CXCIII. Chiasmodonlæ.. 2291
      Genus 847. Chiasmodon, Jnson... 2291
        2629. niger, Johnson 2291
      Genus 848. Pseudoscopelu₵Lütken 2292
        2630. scriptus, Lütke▮₵... 2292
    Family CXCIV. Chænichthyi 2293
      Genus 849. Hypsicometes oode 2293
        2631. gobioides, Goode 2294
    Family CXCV. Trichodontidæ 2295
      Genus 850. Trichodon (Ster) Cuvier..................................... 2295
        2632. trichodon (Tilesi) 2295
      Genus 851. Arctoscopus, Jdan & Evermann.............................. 2297
        2633. japonicus (Stein hner)... 2297
    Family CXCVI. Dactyloscopi 2297
      Genus 852. Gillellus, Gilbₑ... 2298
        2634. semicinctus, Gilb₵... 2298
        2635. arenicola, Gilber 2299
        2636. ornatus, Gilbert 2299
      Genus 853. Dactyloscopus ill 2300
        Subgenus Dactyloscopus.. 2301
        2637. pectoralis, Gill 2301
        2638. tridigitatus, Gil 2301
        2639. poeyi, Gill...... 2302
        2640. lunaticus, Gilbe 2302
        Subgenus Esloscopus, J an & Evermann.............................. 2303.
        2641. zelotes, Jordan ₰ vermann 2303

ORDER BB. ACANTHOPTERI—Continued.
Family CXCVI. Dactyloscopidæ—Continued.　Page.

.... 2304
.... 2304
.... 2305
.... 2305
.... 2305
.... 2306
.... 2307

2645. zophyreus, Gilbert & Starks.... 2309
2646. guttatus (Abbott)............. 2310
Genus 857. Kathetostoma, Günther 2311
2647. averruncus, Jordan & Bollman. 2311
.... 2312

SUBORDER HAPLODOCI 2313
.... 2313
.... 2314; 2368
.... 2314

2650. pacifici (Günther)...... 2314
Genus 859. Opsanus, Rafinesque . 2315
2651. tau (Linnæus)......... 2315
2652. pardus (Goode & Bean) 2316
Genus 860. Porichthys, Girard... 2317
2653. porosissimus (Cuvier & 2319
2321
.... 2322
Genus 861. Thalassophryne, Günther..... 2323
2324
.... 2325
Genus 862. Daector, Jordan & Evermann.. 2325
2658. dowi (Jordan & Gilbert) 2325
2326

Family CXCIX. Gobiesocidæ....... 2326
Genus 863. Caularchus, Gill 2327
2659. meandricus (Girard)............ 2328
Genus 864. Bryssetaeres, Jordan & 2328
2660. pinniger (Gilbert)..... 2328
Genus 865. Gobiesox, Lacépède. 2329
Subgenus Bryssophilus, Jordan 2330
2330
2331

2662. gyrinus, Jordan & Evermann.. ,.............. 2331
2331
2332
2333

2666. strumosus, Cope................ 2333
2667. virgatulus, Jordan & Gilbert....

2334
2669. funebris, Gilbert.... 2334
2335
2335
2335
.... 2336
Subgenus Sicyases, Müller & Troschel.. 2336
2674. erythrops, Jordan & Gilbert..... 2336
2337
.... 2337
.... 2337

CLASS III. PISCES—Continued.
ORDER BB. ACANTHOPTERI—Continued.   Page.
  Family CXC. Malacanthidæ............................... 2274
    Genus 838. Malacanthus, Cuvier........................... 2275
      2610. plumieri (Bloch)................................ 2275
    Genus 839. Caulolatilus, Gill............................ 2276
      2611. princeps (Jenyns)............................... 2276
      2612. microps, Goode & Bean.......................... 2277
      2613. cyanops, Poey................................... 2278
    Genus 840. Lopholatilus, Goode & Bean 2278
      2614. chamæleonticeps, Goode & Bean 2278
  Family CXCI. Opisthognathidæ............................. 2279
    Genus 841. Opisthognathus Cuvier 2280
      2615. lonchurum, Jordan & Gilbert 2281
      2616 punctatum, Peters................................ 2281
      2617. macrognathum, Poey............................. 2281
      2618. ommatum, Jenkins & Evermann.................... 2282
    Genus 842. Gnathypops, Gill............................. 2283
      2619. scops, Jenkins & Evermann...................... 2283
      2620. maxillosa (Poey) 2284
      2621. macrops (Poey)................................. 2284
      2622. rhomalea (Jordan & Gilbert).................... 2285
      2623. snyderi, Jordan & Evermann 2285
      2624. mystacina, Jordan 2286
    Genus 843. Lonchopisthus, Gill......................... 2286
      2625. micrognathus (Poey)............................ 2287
  Family CXCII. Bathymasteridæ 2287
    Genus 844. Bathymaster................................ 2288
      2626. signatus, Cope................................. 2288
    Genus 845. Ronquilus, Jordan & Starks 2289
      2627. jordani (Gilbert).............................. 2289
    Genus 846. Rathbunella, Jordan & Evermann 2289
      2628. hypoplecta (Gilbert)........................... 2290
  Family CXCIII. Chiasmodontidæ........................... 2291
    Genus 847. Chiasmodon, Johnson........................ 2291
      2629. niger, Johnson 2291
    Genus 848. Pseudoscopelus, Lütken 2292
      2630. scriptus, Lütken............................... 2292
  Family CXCIV. Chænichthyidæ............................. 2293
    Genus 849. Hypsicometes, Goode 2293
      2631. gobioides, Goode............................... 2294
  Family CXCV. Trichodontidæ 2295
    Genus 850. Trichodon (Steller) Cuvier.................. 2295
      2632. trichodon (Tilesius) 2295
    Genus 851. Arctoscopus, Jordan & Evermann............. 2297
      2633. japonicus (Steindachner)....................... 2297
  Family CXCVI. Dactyloscopidæ 2297
    Genus 852. Gillellus, Gilbert.......................... 2298
      2634. semicinctus, Gilbert........................... 2298
      2635. arenicola, Gilbert............................. 2299
      2636. ornatus, Gilbert............................... 2299
    Genus 853. Dactyloscopus, Gill 2300
      Subgenus Dactyloscopus............................... 2301
      2637. pectoralis, Gill............................... 2301
      2638. tridigitatus, Gill............................. 2301
      2639. poeyi, Gill.................................... 2302
      2640. lunaticus, Gilbert............................. 2302
      Subgenus Esloscopus, Jordan & Evermann.............. 2303.
      2641. zelotes, Jordan & Evermann 2303

CLASS III. PISCES—Continued.
ORDER BB. ACANTHOPTERI—Continued.
Family CXCVI. Dactyloscopidæ—Continued.   Page.
  Genus 854. Dactylagnus, Gill 2304
    2642. mundus, Gill 2304
  Genus 855. Myxodagnus, Gill.. 2305
    2643. opercularis, Gill.. 2305
Family CXCVII. Uranoscopidæ... 2305
  Genus 856. Astroscopus, Brevoort 2306
    2644. y-græcum (Cuvier & Valenciennes)................................... 2307
    2645. zephyreus, Gilbert & Starks.. 2309
    2646. guttatus (Abbott).. 2310
  Genus 857. Kathetostoma, Günther 2311
    2647. averruncus, Jordan & Bollman....................................... 2311
    2648. albiguttum, Bean... 2312
SUBORDER HAPLODOCI 2313
Family CXCVIII. Batrachoididæ... 2313
  Genus 858. Batrachoides, Lacépède...................................... 2314; 2868
    2649. surinamensis (Bloch & Schneider)................................... 2314
    2650. pacifici (Günther)... 2314
  Genus 859. Opsanus, Rafinesque 2315
    2651. tau (Linnæus).. 2315
    2652. pardus (Goode & Bean).. 2316
  Genus 860. Porichthys, Girard.. 2317
    2653. porosissimus (Cuvier & Valenciennes)............................... 2319
    2654. notatus, Girard.. 2321
    2655. margaritatus (Richardson) 2322
  Genus 861. Thalassophryne, Günther... 2323
    2656. maculosa, Günther 2324
    2657. reticulata, Günther... 2325
  Genus 862. Dæctor, Jordan & Evermann....................................... 2325
    2658. dowi (Jordan & Gilbert) 2325
SUBORDER XENOPTERYGII 2326
Family CXCIX. Gobiesocidæ.. 2326
  Genus 863. Caularchus, Gill 2327
    2659. mæandricus (Girard).. 2328
  Genus 864. Bryssetæres, Jordan & Evermann.................................. 2328
    2660. pinniger (Gilbert)... 2328
  Genus 865. Gobiesox, Lacépède... 2329
    Subgenus Bryssophilus, Jordan & Evermann................................. 2330
    2661. papillifer, Gilbert 2330
    Subgenus Gobiesox.. 2331
    2662. gyrinus, Jordan & Evermann... 2331
    2663. nigripinnis (Peters).. 2331
    ·2664· cephalus, Lacépède... 2332
    2665. tudes, Richardson 2333
    2666. strumosus, Cope.. 2333
    2667. virgatulus, Jordan & Gilbert....................................... 2333
    2668. adustus, Jordan & Gilbert 2334
    2669. funebris, Gilbert... 2334
    2670. pœcilophthalmus, Jenyns 2335
    2671. rhodospilus, Günther 2335
    2672. macrophthalmus, Günther 2335
    2673. cerasinus, Cope 2336
    Subgenus Sicyases, Müller & Troschel..................................... 2336
    2674. erythrops, Jordan & Gilbert.. 2336
    2675. rubiginosus (Poey)... 2337
    2676. carneus (Poey).. 2337
    2677. hæres, Jordan & Bollman.. 2337

CLASS III. PISCES—Continued.
  ORDER BB. ACANTHOPTERI—Continued.
    Family CXCIX. Gobiesocidæ—Continued.   Page.
      2678. punctulatus (Poey) 2338
      2679. fasciatus (Peters) 2338
    Genus 866. Rimicola, Jordan & Evermann............................... 2338
      2680. muscarum (Meek & Pierson).................................... 2338
      2681. eigenmanni (Gilbert)... 2339
    Genus 867. Arbaciosa, Jordan & Evermann........................... 2340
      2682. rhessodon (Rosa Smith)...............................`.......... 2340
      2683. humeralis (Gilbert)... 2341
      2684. rupestris (Poey)... 2341
      2685. zebra (Jordan & Gilbert).. 2341
      2686. eos (Jordan & Gilbert)..............................`............. 2343
    Group Blennioidea... 2343
  Family CC. Blenniidæ.. 2344
    Genus 868. Enneanectes, Jordan & Evermann 2349
      2687. carminalis (Jordan & Gilbert) 2350
    Genus 868(a). Dialommus, Gilbert 2368
      2687(a). fuscus, Gilbert ....`......................... 2368
    Genus 869. Heterostichus, Girard.. 2350
      2688. rostratus, Girard 2351
    Genus 870. Gibbonsia, Cooper... 2351
      2689. evides (Jordan & Gilbert)................................. 2352; 2869
      2690. elegans (Cooper) 2353
    Genus 871. Neoclinus, Girard... 2354
      Subgenus Neoclinus 2354
      2691. blanchardi, Girard.. 2354
      Subgenus Pterognathus, Girard... 2355
      2692. satiricus, Girard.......................`.......................... 2355
    Genus 872. Malacoctenus, Gill 2356
      2693. ocellatus (Steindachner) 2356; 2869
      2694. varius (Poey) 2357
      2695. macropus (Poey) 2357
      2696. lugubris (Poey) 2357
      2697. gillii (Steindachner)...................... 2358
      2698. bimaculatus (Steindachner)`...................... 2358
      2699. delalandi (Cuvier & Valenciennes).............................. 2358
      2700. versicolor (Poey) 2359
      2701. biguttatus (Cope)... 2360
    Genus 873. Labrisomus, Swainson 2360
      2702. herminier (Le Sueur) 2361
      2703. nuchipinnis (Quoy & Gaimard) ....`....... 2362
      2704. xanti, Gill... 2362
      2705. bucciferus, Poey 2363
      2706. microlepidotus, Poey... 2363
    Genus 874. Mnierpes, Jordan & Evermann 2364
      2707. macrocephalus (Günther) 2364
    Genus 875. Gobioclinus, Gill.. 2364
      2708. gobio (Cuvier & Valenciennes).................................. 2365
    Genus 876. Starksia, Jordan & Evermann 2365
      2709. cremnobates (Gilbert) 2365
    Genus 877. Cryptotrema, Gilbert 2366
      2710. corallinum, Gilbert... 2366
    Genus 878. Exerpes, Jordan & Evermann.....:.......................... 2367
      2711. asper (Jenkins & Evermann) 2367
    Genus 879. Auchenopterus, Günther 2369
      Subgenus Corallicola, Jordan & Evermann 2369
      2712. nigripinnis (Steindachner) 2369

CLASS III. PISCES—Continued.
ORDER BB. ACANTHOPTERI—Continued.
Family OC. Blenniidæ—Continued. Page.

2713. altivelis (Lockington) 2370
2714. marmoratus (Steindachner) 2371
Subgenus Auchenopterus 2371
2715. affinis (Steindachner) 2371
2716. monophthalmus, Günther 2372
2717. integripinnis (Rosa Smith) 2372
2718. fasciatus (Steindachner) 2373
2719. nox (Jordan & Gilbert) 2373
Genus 880. Paraclinus, Mocquard 2374
2720. chaperi, Mocquard 2374
Genus 881. Emmnion, Jordan 2375
2721. bristolæ, Jordan 2375
Genus 882. Atopoclinus, Vaillant... 2376
2722. ringens, Vaillant... 2376
Genus 883. Runula, Jordan & Bollman 2377
2723. azulea, Jordan & Bollman 2377
Genus 884. Blennius (Artedi) Linnæus 2377
Subgenus Lipophrys, Gill.. 2378
2724. carolinus (Cuvier & Valenciennes)................................ 2378
2725. fucorum, Cuvier & Valenciennes 2379
2726. stearnsi, Jordan & Gilbert 2379
2727. favosus, Goode & Bean.. 2380
2728. pilicornis, Cuvier & Valenciennes................................ 2380
2729. marmoreus, Poey 2381
2730. truncatus (Poey)... 2381
2731. vinctus, Poey.. 2382
2732. cristatus, Linnæus 2382
Genus 885. Scartella, Jordan 2384
2733. microstoma (Poey) 2384
Genus 886. Hypleurochilus, Gill 2385
2734. geminatus (Wood) 2385
Genus 887. Hypsoblennius, Gill.. 2386
Subgenus Hypsoblennius 2386
2735. gilberti (Jordan) 2386
2736. gentilis (Girard).. 2387
2737. striatus (Steindachner) 2388
2738. ionthas (Jordan & Gilbert)....................................... 2388
2739. hentz (Le Sueur)... 2390
Subgenus Blenniolus, Jordan & Evermann 2390
2740. brevipinnis (Günther).. 2390
Genus 888. Chasmodes, Cuvier & Valenciennes 2391
2741. jenkinsi (Jordan & Evermann)..................................... 2391
2742. quadrifasciatus (Wood)... 2392
2743. saburræ, Jordan & Gilbert 2392
2744. novemlineatus (Wood) 2393
2745. bosquianus (Lacépède).. 2394
Genus 889. Homesthes, Gilbert 2394
2746. caulopus, Gilbert.. 2394
Genus 890. Scartichthys, Jordan & Evermann 2395
2747. rubropunctatus (Cuvier & Valenciennes) 2396
Genus 891. Rupiscartes, Swainson 2396
2748. atlanticus (Cuvier & Valenciennes)............................... 2397
Genus 892. Entomacrodus, Gill... 2397
2749. chiostictus (Jordan & Gilbert) 2398
2750. margaritaceus (Poey) 2398
2751. decoratus, Poey 2399

CLASS III. PISCES—Continued.
  ORDER BB. ACANTHOPTERI—Continued.
    Family CC. Blenniidæ—Continued.   Page.
      2752. nigricans, Gill................................ 2399
      Genus 893. Salariichthys, Guichenot................. 2400
        2753. textilis (Quoy & Gaimard).................... 2400
      Genus 894. Ophioblennius, Gill...................... 2400
        2754. webbii (Valenciennes)........................ 2401
        2755. steindachneri, Jordan & Evermann............. 2401
      Genus 895. Emblemaria, Jordan & Gilbert............. 2401
        2756. atlantica, Jordan & Evermann................. 2402
        2757. nivipes, Jordan & Gilbert.................... 2402
        2758. oculocirris, Jordan......................... 2403
      Genus 896. Chænopsis, Gill......................... 2403
        2759. ocellatus, Poey............................. 2403
      Genus 897. Lucioblennius, Gilbert.................. 2404
        2760. alepidotus, Gilbert.......................... 2404
      Genus 898. Pholidichthys, Bleeker 2405
        2761. anguilliformis, Lockington.................. 2405
      Genus 899. Psednoblennius, Jenkins & Evermann...... 2406
        2762. hypacanthus, Jenkins & Evermann.............. 2406
      Genus 900. Stathmonotus, Bean...................... 2407
        2763. hemphillii, Bean............................. 2407
      Genus 901. Bryostemma, Jordan & Starks............. 2408
        2764. polyactocephalum (Pallas) 2408
        2765. nugator, Jordan & Williams.................. 2410
      Genus 902. Apodichthys, Girard..................... 2411
        2766. flavidus, Girard 2411
        2767. univittatus, Lockington 2412
      Genus 903. Xererpes, Jordan & Gilbert.............. 2413
        2768. fucorum (Jordan & Gilbert)................... 2413
      Genus 904. Ulvicola, Gilbert...................... 2413
        2769. sanctæ-rosæ, Gilbert & Starks 2413
      Genus 905. Pholis (Gronow) Scopoli 2414
        Subgenus Urocentrus, Kner......................... 2415
        2770. pictus (Kner) 2415
        Subgenus Rhodymenichthys, Jordan & Evermann. 2416
        2771. dolichogaster (Pallas)...................... 2416
        Subgenus Pholis................................... 2417
        2772. fasciatus (Bloch & Schneider)............... 2417
        2773. gunnellus (Linnæus)......................... 2419
        2774. ornatus (Girard) 2419
      Genus 906. Gunnellops, Bleeker 2420
        2775. roseus (Pallas) 2420
      Genus 907. Asternopteryx, Rüppell.................. 2420
        2776. gunnelliformis, Rüppell..................... 2420
      Genus 908. Anoplarchus, Gill....................... 2421
        2777. atropurpureus (Kittlitz) 2422; 2869
      Genus 908(a). Alectrias, Jordan & Evermann......... 2869
        2778. alectrolophus (Pallas).................. 2421; 2869
      Genus 909. Xiphistes, Jordan & Starks 2423
        2779. ulvæ, Jordan & Starks....................... 2423
        2780. chirus (Jordan & Gilbert)................... 2424
      Genus 910. Xiphidion, Girard...................... 2424
        2781. mucosum, Girard............................. 2425
        2782. rupestre (Jordan & Gilbert)................. 2426
      Genus 911. Cebedichthys, Ayres.................... 2426
        2783. violacens (Ayres) 2427
      Genus 912. Plagiogrammus, Bean...... 2427

CLASS III. PISCES—Continued.
ORDER BB. ACANTHOPTERI—Continued.
Family CC. Blenniidæ—Continued.   Page.
2784. hopkinsi, Bean 2428
Genus 913. Opisthocentrus, Kner 2428
2785. ocellatus (Tilesius) 2429
Genus 914. Pholidapus, Bean & Bean 2430
2786. dybowskii (Steindachner) 2430
Genus 915. Plectobranchus, Gilbert 2431
2787. evides, Gilbert 2432
Genus 916. Leptoclinus, Gill 2432
2788. maculatus (Fries) 2433
Genus 917. Poroclinus, Bean 2433
2789. rothrocki, Bean 2434
Genus 918. Lumpenus, Reinhardt 2435
Subgenus Anisarchus, Gill 2435
2790. medius (Reinhardt) 2435
Subgenus Lumpenus 2436
2791. anguillaris (Pallas) 2436
2792. mackayi (Gilbert) 2436
2793. fabricii (Cuvier & Valenciennes) 2437
2794. lampetræformis (Walbaum) 2438
Genus 919. Stichæus, Reinhardt 2439
2795. punctatus (Fabricius) 2439
Genus 920. Ulvaria, Jordan & Evermann 2440
2796. subbifurcata (Storer) 2440
Genus 921. Eumesogrammus, Gill 2441
2797. præcisus (Kröyer) 2441
Family CCI. Cryptacanthodidæ 2442
Genus 922. Delolepis, Bean 2442
2798. virgatus, Bean 2442
Genus 923. Cryptacanthodes, Storer 2443
2799. maculatus, Storer 2443
Genus 924. Lyconectes, Gilbert 2444
2800. aleutensis, Gilbert 2444
Family CCII. Anarhichadidæ 2445
Genus 925. Anarhichas (Artedi) Linnæus 2445
2801. latifrons, Steenstrup & Hallgrimsson 2446
2802. minor, Olafsen 2446
2803. lupus, Linnæus 2446
2804. lepturus, Bean 2447
2805. orientalis, Pallas 2447
Genus 926. Anarrhichthys, Ayres 2447
2806. ocellatus, Ayres 2448
Family CCIII. Cerdalidæ 2448
Genus 927. Cerdale, Jordan & Gilbert 2448
2807. ionthas, Jordan & Gilbert 2449
Genus 928. Microdesmus, Günther 2450
2808. dipus, Günther 2450
2809. retropinnis, Jordan & Gilbert 2450
Family CCIV. Ptilichthyidæ 2451
Genus 929. Ptilichthys, Bean 2452
2810. goodei, Bean 2452
Group Ophidioidea 2453
Family CCV. Scytalinidæ 2453
Genus 930. Scytalina, Jordan & Gilbert 2454
2811. cerdale, Jordan & Gilbert 2454
Family CCVI. Zoarcidæ 2455
Genus 931. Zoarces, Gill 2456

CLASS III. PISCES—Continued.
ORDER BB. ACANTHOPTERI—Continued.
Family CCVI. Zoarcidæ—Continued.   Page.
  Subgenus Macrozoarces, Gill.. 2457
    2812. anguillaris (Peck) 2457
  Genus 932. Embryx, Jordan & Evermann 2458
    2813. crassilabris (Gilbert) 2458
    2814. crotalinus (Gilbert).. 2458
  Genus 933. Lycodopsis, Collett... 2460
    2815. pacificus (Collett)... 2460
  Genus 934. Aprodon, Gilbert.. 2460
    2816. cortezianus, Gilbert 2461
  Genus 935. Lycodes, Reinhardt... 2461
    Subgenus Lycodes 2463
    2817. esmarkii, Collett... 2463
    2818. vahlii, Reinhardt 2463
    2819. concolor, Gill & Townsend 2463
    2820. zoarchus, Goode & Bean... 2464
    2821. reticulatus, Reinhardt... 2465
    2822. perspicillum, Kröyer 2465
    2823. frigidus, Collett 2465
    2824. terræ-novæ, Collett 2466
    2825. digitatus, Gill & Townsend..................................... 2466
    2826. palearis, Gilbert.. 2466
    2827. brevipes, Bean 2467
    Subgenus Lycias, Jordan & Evermann.................................. 2468
    2828. nebulosus, Kröyer.. 2468
    2829. seminudus, Reinhardt... 2468
  Genus 936. Lycodalepis, Bleeker....................................... 2468
    2830. polaris (Sabine) 2468
    2831. mucosus (Richardson)... 2470
  Genus 937. Lycenchelys, Gill.. 2470
    2832. verrillii (Goode & Bean)....................................... 2470
    2833. paxillus (Goode & Bean).. 2471
    2834. porifer (Gilbert)....................................... 2471; 2869
  Genus 938. Furcella, Jordan & Evermann................................ 2472
    2835. diaptera (Gilbert) 2472
  Genus 939. Lycodonus, Goode & Bean.................................... 2473
    2836. mirabilis, Goode & Bean.. 2474
  Genus 940. Lyconema, Gilbert 2474
    2837. barbatum, Gilbert.. 2474
  Genus 941. Bothrocara, Bean... 2475
    2838. pusilla (Bean)... 2476
    2839. mollis, Bean... 2476
  Genus 942. Gymnelis, Reinhardt.. 2477
    2840. viridis (Fabricius) 2477
    2841. stigma (Lay & Bennett)... 2477
  Genus 943. Lycocara, Gill... 2478
    2842. parrii (Ross).. 2478
  Genus 944. Melanostigma, Günther...................................... 2478
    2843. gelatinosum, Günther... 2479
    2844. pammelas, Gilbert.. 2479
Family CCVII. Derepodichthyidæ.. 2480
  Genus 945. Derepodichthys, Gilbert 2480
    2845. alepidotus, Gilbert 2480
Family CCVIII. Ophidiidæ.. 2481
  Genus 946. Lepophidium, Gill... 2482
    2846. marmoratum (Goode & Bean)...................................... 2482
    2847. emmelas (Gilbert).. 2483

CLASS III. PISCES—Continued.
ORDER BB. ACANTHOPTERI—Continued.
Family CCVIII. Ophidiidæ—Continued. Page.
  2848. stigmatistium (Gilbert) 2483
  2849. profundorum (Gill).. 2484
  2850. cervinum (Goode & Bean).. 2484
  2851. prorates (Jordan & Bollman)..................................... 2485
  2852. brevibarbe (Cuvier) 2485
  2853. pardale (Gilbert) 2486
  2854. microlepis (Gilbert) 2486
  Genus 947. Ophidion (Artedi) Linnæus............................... 2487
  2855. beani, Jordan & Gilbert.. 2487
  2856. holbrooki (Putnam) 2487
  2857. graellsi, Poey.. 2488
  Genus 948. Chilara, Jordan & Evermann............................. 2488
  2858. taylori (Girard) 2489
  Genus 949. Rissola, Jordan & Evermann............................. 2489
  2859. marginata (De Kay).. 2489
  Genus 950. Otophidium, Gill 2490
  2860. omostigma (Jordan & Gilbert) 2490
  2861. indefatigabile, Jordan & Bollman 2490
  2862. galeoides (Gilbert) 2491
Family CCIX. Lycodapodidæ... 2491
  Genus 951. Lycodapus, Gilbert...................................... 2492
  2863. dermatinus, Gilbert... 2492
  2864. fierasfer, Gilbert.. 2493
  2865. parviceps, Gilbert... 2493
  2866. extensus (Gilbert)... 2494
Family CCX. Fierasferidæ.. 2494
  Genus 952. Fierasfer, Cuvier 2495
  2867. affinis (Günther) 2495
  2868. arenicola, Jordan & Gilbert..................................... 2496
  2869. bermudensis (Jones) 2497
Family CCXI. Brotulidæ... 2498
  Genus 953. Brotula, Cuvier 2500
  2870. barbata (Bloch & Schneider) 2500
  Genus 954. Stygicola, Gill.. 2500
  2871. dentatus (Poey) 2500
  Genus 955. Lucifuga, Poey.. 2501
  2872. subterraneus, Poey 2501
  Genus 956. Brosmophycis, Gill...................................... 2502
  2873. marginatus (Ayres) 2502
  Genus 957. Ogilbia, Jordan & Evermann............................. 2502
  2874. ventralis (Gill) 2503
  2875. cayorum, Evermann & Kendall.................................. 2503
  Genus 958. Bythites, Reinhardt..................................... 2504
  2876. fuscus, Reinhardt 2504
  Genus 959. Catætyx, Günther....................................... 2504
  2877. rubirostris, Gilbert... 2505
  Genus 960. Dicromita, Goode & Bean................................ 2506
  2878. agassizii, Goode & Bean.. 2506
  Genus 961. Bassozetus, Gill.. 2507
  2879. normalis, Gill.. 2507
  2880. compressus (Günther) 2508
  2881. catena, Goode & Bean.. 2509
  2882. tænia (Günther).. 2510
  Genus 962. Mœbia, Goode & Bean................................... 2510
  2883. promelas (Gilbert)... 2511
  Genus 963. Neobythites, Goode & Bean.............................. 2512

CLASS III. PISCES—Continued.

ORDER BB. ACANTHOPTERI—Continued.

Family CCXI. Brotulidæ—Continued.   Page.

2884. gillii, Goode & Bean 2512
2885. marginatus, Goode & Bean 2513
Genus 964. Benthocometes, Goode & Bean 2513
2886. robustus, Goode & Bean 2514
Genus 965. Bassogigas, Gill. 2515
2887. gillii, Goode & Bean. 2515
2888. stelliferoides (Gilbert) 2516
Genus 966. Barathrodemus, Goode & Bean 2517
2889. manatinus, Goode & Bean 2517
Genus 967. Nematonus, Günther 2518
2890. pectoralis, Goode & Bean 2518
Genus 968. Porogadus, Goode & Bean 2519
2891. miles, Goode & Bean 2520
Genus 969. Penopus, Goode & Bean 2520
2892. macdonaldi, Goode & Bean 2521
Genus 970. Dicrolene, Goode & Bean 2522
2893. intronigra, Goode & Bean 2522
Genus 971. Mixonus, Günther 2523
2894. laticeps (Günther) 2523
Genus 972. Barathronus, Goode & Bean 2524
2895. bicolor, Goode & Bean 2524
Genus 973. Aphyonus, Günther 2525
2896. mollis, Goode & Bean 2525
Family CCXII. Bregmacerotidæ 2525
Genus 974. Bregmaceros, Thompson 2526
2897. macclellandii, Thompson 2526
2898. atlanticus, Goode & Bean 2527
SUBORDER ANACANTHINI 2528
Family CCXIII. Merlucciidæ 2529
Genus 975. Merluccius, Rafinesque 2529
2899. merluccius (Linnæus) 2530
2900. bilinearis (Mitchill) 2530
2901. productus (Ayres) 2531
Family CCXIV. Gadidæ 2531
Genus 976. Boreogadus, Günther 2533
2902. saida (Lepechin) 2533
Genus 977. Pollachius, Nilsson 2534
2903. virens (Linnæus) 2534
Genus 978. Theragra, Lucas 2535
2904. chalcogramma (Pallas) 2535
2905. fucensis (Jordan & Gilbert) 2536
Genus 979. Eleginus, Fischer 2537
2906. navaga (Kölreuter) 2537
Genus 980. Microgadus, Gill 2538
2907. proximus (Girard) 2539
2908. tomcod (Walbaum) 2540
Genus 981. Gadus (Artedi) Linnæus 2540
2909. callarias, Linnæus 2541
2910. macrocephalus, Tilesius 2541
2911. ogac, Richardson 2542
Genus 982. Melanogrammus, Gill 2542
2912. æglefinus, Linnæus 2542
Genus 983. Lepidion, Swainson 2543
2913. verecundum, Jordan & Cramer 2543
Genus 984. Antimora, Günther 2544
2914. viola (Goode & Bean) 2544

CLASS III. PISCES—Continued.
ORDER BB. ACANTHOPTERI—Continued.
Family CCXIV. Gadidæ—Continued. Page.

2915. microlepis, Bean.. 2545
Genus 985. Uraleptus, Costa.. 2545
2916. maraldi (Risso)... 2545
Genus 986. Lotella, Kaup... 2546
2917. maxillaris, Bean.. 2546
Genus 987. Physiculus, Kaup... 2547
2918. fulvus, Bean... 2547
2919. nematopus, Gilbert... 2548
2920. kaupi, Poey... 2548
2921. rastrelliger, Gilbert... 2549
Genus 988. Lota (Cuvier) Oken... 2550
2922. maculosa (Le Sueur).. 2550
Genus 989. Molva, Fleming... 2551
2923. molva (Linnæus)... 2551
Genus 990. Urophycis, Gill... 2552
Subgenus Urophycis.. 2553
2924. regius (Walbaum)... 2553
2925. cirratus (Goode & Bean).. 2553
2926. floridanus (Bean & Dresel)... 2554
Subgenus Emphycus, Jordan & Evermann.............................. 2554
2927. earlli (Bean)... 2554
2928. tenuis (Mitchill).. 2555
2929. chuss (Walbaum)... 2555
2930. chesteri (Goode & Bean).. 2556
Genus 991. Læmonema, Günther... 2556
2931. barbatulum, Goode & Bean.. 2556
2932. melanurum, Goode & Bean.. 2557
Genus 992. Gaidropsarus, Rafinesque................................... 2557
2933. ensis (Reinhardt)... 2558
2934. argentatus (Reinhardt).. 2559
2935. septentrionalis (Collett).. 2559
Genus 993. Enchelyopus, Bloch & Schneider.......................... 2560
2936. cimbrius (Linnæus).. 2560
Genus 994. Brosme (Cuvier) Oken.. 2561
2937. brosme (Müller).. 2561
Family CCXV. Macrouridæ.. 2561
Genus 995. Bathygadus, Günther.. 2563
2938. arcuatus, Goode & Bean.. 2564
2939. favosus, Goode & Bean... 2565
2940. macrops, Goode & Bean.. 2566
2941. longifilis, Goode & Bean... 2566
Genus 996. Steindachneria, Goode & Bean............................. 2567
2942. argentea, Goode & Bean... 2568
Genus 997. Trachyrinchus, Giorna....................................... 2568
2943. helolepis, Gilbert.. 2569
Genus 998. Malacocephalus, Günther.................................... 2569
2944. occidentalis, Goode & Bean....................................... 2570
Genus 999. Moseleya, Goode & Bean.................................... 2570
2945. cyclolepis (Gilbert)... 2570
Genus 1000. Nematonurus, Günther..................................... 2571
2946. goodei (Günther).. 2571
2947. suborbitalis (Gill & Townsend)................................... 2572
Genus 1001. Albatrossia, Jordan & Evermann........................ 2573
2948. pectoralis (Gilbert).. 2573
Genus 1002. Bogoslovius, Jordan & Evermann........................ 2574
2949. clarki, Jordan & Gilbert.. 2575

CLASS III. PISCES—Continued.
    ORDER BB. ACANTHOPTERI—Continued.
        Family CCXV. Macrouridæ—Continued.   **Page.**
            2950. firmisquamis (Gill & Townsend) 2575
            Genus 1003. Chalinura, Goode & Bean 2576
            2951. serrula, Bean 2576
            2952. filifera, Gilbert 2577
            2953. simula, Goode & Bean 2578
            Genus 1004. Coryhænoides, Gunner 2578
            2954. rupestris, Gunner 2579
            2955. carapinus, Goode & Bean 2579
            Genus 1005. Hymenocephalus, Giglioli 2580
            2956. cavernosus (Goode & Bean) 2580
            Genus 1006. Macrourus, Bloch 2581
            2957. berglax, Lacépède 2582
            2958. holotrachys, Günther 2582
            2959. bairdii, Goode & Bean 2583
            2960. lepturus, Gill & Townsend 2584
            2961. acrolepis, Bean 2585
            2962. stelgidolepis, Gilbert 2585
            2963. cinereus, Gilbert 2586
            Genus 1007. Cœlorbynchus, Giorna 2587
            2964. occa (Goode & Bean) 2588
            2965. carminatus (Goode) 2588
            2966. caribbæus (Goode & Bean) 2589
            2967. scaphopsis (Gilbert) 2590
            Genus 1008. Trachonurus, Günther 2591
            2968. sulcatus (Goode & Bean) 2591
            Genus 1009. Lionurus, Günther 2592
            2969. filicauda (Günther) 2592
            2970. liolepis, Gilbert 2593
    SUBORDER TÆNIOSOMI 2594
        Family CCXVI. Regalecidæ 2595
            Genus 1010. Regalecus, Brünnich 2595
            2971. glesne (Ascanius) 2596
        Family CCXVII. Trachypteridæ 2597
            Genus 1011. Trachypterus, Gouan 2599
            2972. rex-salmonorum, Jordan & Gilbert 2599
            2973. trachyurus, Poey 2600
        Family CCXVIII. Stylephoridæ 2601
            Genus 1012. Stylephorus, Shaw 2601
            2974. chordatus, Shaw 2601
    SUBORDER HETEROSOMATA 2602
        Family CCXIX. Pleuronectidæ 2602
            Genus 1013. Atherestbes, Jordan & Gilbert 2609
            2975. stomias (Jordan & Gilbert) 2609
            Genus 1014. Reinhardtius, Gill 2610
            2976. hippoglossoides (Walbaum) 2611
            Genus 1015. Hippoglossus, Cuvier 2611
            2977. hippoglossus (Linnæus) 2611
            Genus 1016. Lyopsetta, Jordan & Goss 2612
            2978. exilis (Jordan & Gilbert) 2612
            Genus 1017. Eopsetta, Jordan & Goss 2613
            2979. jordani (Lockington) 2613
            Genus 1018. Hippoglossoides, Gottsche 2614
            2980. platessoides (Fabricius) 2614
            2981. elassodon, Jordan & Gilbert 2615
            2982. robustus, Gill & Townsend 2616
            2983. hamiltoni, Jordan & Gilbert 2616

CLASS III. PISCES—Continued.
ORDER BB. ACANTHOPTERI—Continued.
Family CCXIX. Pleuronectidæ—Continued. Page.
Genus 1019. Psettichthys, Girard 2617
    2084. melanostictus, Girard 2618
Genus 1020. Verasper, Jordan & Gilbert 2618
    2985. moseri, Jordan & Gilbert 2619
Genus 1021. Hippoglossina, Steindachner 2620
    2986. stomata, Eigenmann & Eigenmann 2620
    2987. macrops, Steindachner 2621
    2988. bolimani, Gilbert 2621
Genus 1022. Lioglossina, Gilbert 2622
    2989. tetrophthalma, Gilbert 2622
Genus 1023. Xystreurys, Jordan & Gilbert 2623
    2990. liolepis, Jordan & Gilbert 2623
Genus 1024. Paralichthys, Girard 2624
    2991. californicus (Ayres) 2625
    2991(a). Paralichthys magdalenæ, Abbott 2872
    2992. æstuarius, Gilbert & Scofield 2626
    2993. brasiliensis (Ranzani) 2626
    2994. sinaloæ, Jordan & Abbott.............................. 2627, 2872
    2995. woolmani, Jordan & Williams 2628
    2996. dentatus (Linnæus) 2629
    2997. lethostigmus, Jordan & Gilbert 2630
    2998. squamilentus, Jordan & Gilbert 2631
    2999. albiguttus, Jordan & Gilbert 2631
    3000. oblongus (Mitchill) 2632
Genus 1025. Ramularia, Jordan & Evermann 2633
    3001. dendritica (Gilbert) 2633
Genus 1026. Ancylopsetta, Gill 2634
    3002. quadrocellata, Gill 2634
Genus 1027. Notosema, Goode & Bean 2635
    3003. dilectum, Goode & Bean 2635
Genus 1028. Gastropsetta, B. A. Bean 2636
    3004. frontalis, B. A. Bean 2636
Genus 1029. Pleuronichthys, Girard 2637
    3005. decurrens, Jordan & Gilbert 2637
    3006. verticalis, Jordan & Gilbert 2638
    3007. cœnosus, Girard 2638
Genus 1030. Hypsopsetta, Gill 2639
    3008. guttulata (Girard) 2639
Genus 1031. Parophrys, Girard 2640
    3009. vetulus, Girard 2640
Genus 1032. Inopsetta, Jordan & Goss 2641
    3010. ischyra (Jordan & Gilbert) 2641
Genus 1033. Isopsetta, Lockington 2642
    3011. isolepis (Lockington) 2642
Genus 1034. Lepidopsetta, Gill 2642
    3012. bilineata (Ayres) 2643
Genus 1035. Limanda, Gottsche 2644
    3013. ferruginea (Storer) 2644
    3014. aspera (Pallas) 2645
    3015. proboscidea, Gilbert 2645
    3016. beanii, Goode 2646
Genus 1036. Pseudopleuronectes, Bleeker 2646
    3017. americanus (Walbaum) 2647
    3018. pinnifasciatus (Kner) 2647
Genus 1037. Pleuronectes (Artedi) Linnæus 2648
    3019. quadrituberculatus, Pallas 2648

. CLASS III. PISCES—Continued.
  ORDER BB. ACANTHOPTERI—Continued.
  Family CCXIX. Pleuronectidæ –Continued.　　　　　　　　**Page.**
    Genus 1038. Liopsetta, Gill 2649
      3020. glacialis (Pallas).. 2649
      3021. putnami (Gill)... 2650
      3022. obscura (Herzenstein)... 2651
    Genus 1039. Platichthys, Girard 2651
      3023. stellatus (Pallas) 2652
    Genus 1040. Microstomus, Gottsche.. 2653
      3024. kitt (Walbaum) 2654
      3025. pacificus (Lockington) 2655
    Genus 1041. Embassichthys, Jordan & Evermann 2655
      3026. bathybius (Gilbert) 2655
    Genus 1042. Glyptocephalus, Gottsche..................................... 2656
      3027. cynoglossus (Linnæus).. 2657
      3028. zachirus, Lockington.. 2658
    Genus 1043. Lophopsetta, Gill 2659
      3029. maculata (Mitchill)... 2660
    Genus 1044. Platophrys, Swainson... 2660
      3030. spinosus (Poey) 2662
      3031. constellatus, Jordan.. 2663
      3032. ocellatus (Agassiz) 2663
      3033. maculifer (Poey) 2664
      3034. ellipticus (Poey)... 2665
      3035. lunatus (Linnæus) 2665
      3036. leopardinus (Günther)... 2666
    Genus 1045. Perissias, Jordan & Evermann................................ 2667
      3037. tæniopterus (Gilbert) 2667
    Genus 1046. Engyophrys, Jordan & Bollman 2668
      3038. sancti-laurentii, Jordan & Bollman 2668
    Genus 1047. Trichopsetta, Gill.. 2669
      3039. ventralis (Goode & Bean)... 2669
    Genus 1048. Syacium, Ranzani... 2670
      3040. papillosum (Linnæus) 2671
      3041. micrurum, Ranzani.. 2672
      3042. latifrons (Jordan & Gilbert)...................................... 2673
      3043. ovale (Günther) 2674
    Genus 1049. Cyclopsetta, Gill 2675
      3044. querna (Jordan & Bollman).. 2675
      3045. chittendeni, B. A. Bean 2676
      3046. fimbriata (Goode & Bean) 2676
    Genus 1050. Azevia, Jordan 2677
      3047. panamensis (Steindachner).. 2677
    Genus 1051. Citharichthys, Bleeker....................................... 2678
    Subgenus Orthopsetta, Gill... 2679
      3048. sordidus (Girard)... 2679
      3049. fragilis, Gilbert 2680
      3050. xanthostigmus, Gilbert 2680
      3051. stigmæus, Jordan & Gilbert....................................... 2681
    Subgenus Citharichthys 2682
      3052. dinoceros, Goode & Bean 2682
      3053. platophrys, Gilbert... 2683
      3054. arctifrons, Goode..... 2683
      3055. unicornis, Goode 2683
      3056. uhleri, Jordan... 2684
      3057. macrops, Dressel 2684
      3058. spilopterus, Günther 2685
      3059. gilberti, Jenkins & Evermann 2686

CONTENTS. **XXI**

CLASS III. PISCES—Continued.

ORDER BB. ACANTHOPTERI—Continued.

Family CCXIX. Plenronectidæ—Continued.   Page.
 Genus 1052. Etropus, Jordan & Gilbert........................... 2687
  3060. microstomus (Gill).................................... 2687
  3061. rimosus, Goode & Bean 2688
  3062. crossotus, Jordan & Gilbert......................... 2689
 Genus 1053. Monolene, Goode................................... 2690
  3063. sessilicauda, Goode 2691
  3064. atrimana, Goode & Bean 2692
Family CCXX. Soleidæ... 2692
 Genus 1054. Achirus. Lacépède................................. 2693
  Subgenus Baiostoma, Bean 2695
  3065. achirus (Linnæus) 2695
  3066. inscriptus, Gosse.................................... 2696
  3067. klunzingeri (Steindachner) 2697
  3068. lineatus (Linnæus)................................... 2697
  3069. mazatlanus (Steindachner).......................... 2698
  3070. fonsecensis (Günther)............................... 2699
  3071. fischeri (Steindachner) 2699
  3072. scutum (Günther)................................... 2700
  Subgenus Achirus 2700
  3073. fimbriatus (Günther)................................ 2700
  3074. fasciatus, Lacépède 2700
  3075. panamensis (Steindachner).......................... 2702
 Genus 1055. Apionichthys, Kaup............................... 2702
  3076. unicolor (Günther) 2702
 Genus 1056. Gymnachirus, Kaup............................... 2703
  3077. fasciatus, Günther 2703
 Genus 1057. Symphurus, Rafinesque........................... 2704
  Subgenus Symphurus.. 2705
  3078. piger (Goode & Bean) 2705
  3079. marginatus (Goode & Bean) 2706
  3080. atramentatus, Jordan & Bollman 2706
  3081. fasciolaris, Gilbert................................ 2707
  3082. elongatus (Günther) 2707
  3083. atricaudus (Jordan & Gilbert) 2707
  3084. leei, Jordan & Bollman 2708
  3085. plagusia (Bloch & Schneider) 2709
  3086. plagiusa (Linnæus) 2710
  3087. pusillus (Goode & Bean) 2710
  3088. diomedeanus (Goode & Bean) 2711
  3089. willlamsi, Jordan & Culver......................... 2711
  Subgenus Acedia, Jordan................................... 2712
  3090. nebulosus (Goode & Bean)........................... 2712
ORDER CC. PEDICULATI 2712
Family CCXXI. Lophiidæ 2713
 Genus 1058. Lophius (Artedi) Linnæus........................ 2713
  3091. piscatorius, Linnæus............................... 2713
 Genus 1059. Lophiomus, Gill.................................. 2714
  3092. setigerus (Vahl).................................... 2714
Family CCXXII. Antennariidæ 2715
 Genus 1060. Pterophryne, Gill................................ 2715
  3093. histrio (Linnæus) 2716
  3094. gibba (Mitchill) 2717
 Genus 1061. Antennarius, Lacépède............................ 2717
  3095. inops, Poey 2718
  3096. principis (Cuvier & Valenciennes) 2719
  3097. tenebrosus (Poey)................................... 2719

CLASS III. PISCES—Continued.
  ORDER CC. PEDICULATI—Continued.
    Family CCXXII. Antennariidæ—Continued.   Page.
      3098. reticularis, Gilbert 2719
      3099. strigatus, Gill.. 2720
      3100. sanguineus, Gill... 2721
      3101. ocellatus (Bloch & Schneider) 2721
      3102. scaber (Cuvier)... 2722
      3103. tigris, Poey 2723
      3104. nuttingii, Garman... 2723
      3105. multiocellatus (Cuvier & Valenciennes)........................ 2724
      3106. radiosus, Garman 2725
    Genus 1062. Chaunax, Lowe.. 2726
      3107. pictus, Lowe... 2726
      3108. nuttingii, Garman 2727
    Family CCXXIII. Ceratiidæ 2727
    Genus 1063. Ceratias, Kröyer... 2729
      3109. holbolli, Kröyer... 2729
    Genus 1064. Mancalias, Gill.. 2729
      3110. uranoscopus (Murray) 2729
      3111. shufeldti (Gill)... 2730
    Genus 1065. Cryptopsaras, Gill 2731
      3112. couesii, Gill.. 2731
    Genus 1066. Oneirodes, Lütken 2732
      3113. escrichtii, Lütken... 2732
    Genus 1067. Himantolophus, Reinhardt................................... 2732
      3114. grœnlandicus, Reinhardt...................................... 2733
    Genus 1068. Corynolophus, Gill 2733
      3115. reinhardti (Lütken) 2733
    Genus 1069. Liocetus, Günther.. 2733
      3116. murrayi (Günther).. 2733
    Genus 1070. Linophryne, Collett 2734
      3117. lucifer, Collett... 2734
    Genus 1071. Caulophryne, Goode & Bean.................................. 2734
      3118. jordani, Goode & Bean 2735
    Family CCXXIV. Ogcocephalidæ. 2735
    Genus 1072. Ogcocephalus, Fischer 2736
      3119. vespertilio (Linnæus) 2737
      3120. nasutus (Cuvier & Valenciennes) 2737
      3121. radiatus (Mitchill) 2738
    Genus 1073. Zalieutes, Jordan & Evermann............................... 2738
      3122. elater (Jordan & Gilbert)..................................... 2738
    Genus 1074. Halieutichthys, Poey....................................... 2739
      3123. aculeatus (Mitchill)... 2739
      3124. caribbæus, Garman.. 2741
    Genus 1075. Halieutæa, Cuvier & Valenciennes.......................... 2741
      3125. spongiosa, Gilbert... 2741
    Genus 1076. Halieutella, Goode & Bean 2742
      3126. lappa, Goode & Bean 2742
    Genus 1077. Dibranchus, Peters.. 2743
      3127. atlanticus, Peters 2743

LIST OF NEW NAMES.

The following is a list of the new generic, subgeneric, specific, and sub-specific names which appear as new in Part III of the present work:

	Page.
Sicydium vincente, Jordan & Evermann	2207
Enypnias, Jordan & Evermann	2231
Gnathypops snyderi, Jordan & Evermann	2285
Dactyloscopus zelotes, Jordan & Gilbert	2303
Dæctor, Jordan & Evermann	2325
Bryssophilus, Jordan & Evermann	2329
Gobiesox gyrinus, Jordan & Evermann	2331
Corallicola, Jordan & Evermann	2369
Blenniolus, Jordan & Evermann	2386
Homesthes, Gilbert	2394
Homesthes caulopus, Gilbert	2394
Scartichthys, Jordan & Evermann	2395
Ophioblennius steindachneri, Jordan & Evermann	2401
Emblemaria atlantica, Jordan & Evermann	2402
Enedrias, Jordan & Gilbert	2414
Embryx, Jordan & Evermann	2458
Lycias, Jordan & Evermann	2461
Emphycus, Jordan & Evermann	2552
Albatrossia, Jordan & Evermann	2573
Bogoslovius, Jordan & Evermann	2574
Hippoglossoides hamiltoni, Jordan & Gilbert	2616
Verasper, Jordan & Gilbert	2618
Verasper moseri, Jordan & Gilbert	2619
Ramularia, Jordan & Evermann	2633
Perissias, Jordan & Evermann	2667
Carcharhinus cerdale, Gilbert	2746
Carcharhinus velox, Gilbert	2747
Myliobatis asperrimus, Gilbert	2754
Aspistor, Jordan & Evermann	2763
Galeichthys xenauchen, Gilbert	2777
Tachysurus emmelane, Gilbert	2785
Aztecula, Jordan & Evermann	2799
Notropis chamberlaini, Evermann	2800
Notropis louisianæ, Evermann	2801
Pisoodonophis daspilotus, Gilbert	2803
Muræna clepsydra, Gilbert	2805
Stolephorus rastralis, Gilbert & Pierson	2811
Stolephorus mundeolus, Gilbert & Pierson	2812
Stolephorus naso, Gilbert & Pierson	2813
Stolephorus starksi, Gilbert & Pierson	2813
Cetengraulis engymen, Gilbert & Pierson	2815
Argyrosomus alascanus, Scofield	2817
Osmerus albatrossis, Jordan & Gilbert	2823
Bathylagus milleri, Jordan & Gilbert	2825

	Page.
Zaphotias, Goode & Bean	2826
Characodon garmani, Jordan & Evermann	2831
Siphostoma sinaloæ, Jordan & Starks	2838
Rhynchias, Gill	2841
Oligoplites mundus, Jordan & Starks	2844
Hemicaranx zelotes, Gilbert	2845
Ulocentra meadiæ, Jordan & Evermann	2852
Lobotes pacificus, Gilbert	2857
Porocottus bradfordi, Rutter	2862
Sigmistes Rutter	2863
Sigmistes caulias, Rutter	2863
Crystallichthys, Jordan & Gilbert	2864
Crystallichthys mirabilis, Jordan & Gilbert	2865
Allinectes, Jordan & Evermann	2866
Prognurus, Jordan & Evermann	2866
Prognurus cypselurus, Jordan & Gilbert	2866
Sicyosus, Jordan & Evermann	2867
Alectrias, Jordan & Evermann	2869
Furcimanus, Jordan & Evermann	2869
Salmo clarkii tahoensis, Jordan & Evermann	2870
Oligocottus snyderi, Greeley	2871
Flammeo, Jordan & Evermann	2871
Paralichthys magdalenæ, Abbott	2872
Paralichthys sinaloæ, Jordan & Abbott	2872

THE FISHES

OF

NORTH AND MIDDLE AMERICA.

BY DAVID STARR JORDAN AND BARTON WARREN EVERMANN.

PART III.

PREFATORY NOTE.

This volume is the third of a descriptive catalogue of the fishes and fish-like vertebrates of North and Middle America. For the sake of greater completeness the marine fishes of the Galapagos Islands and the South American coast north of the equator have been included, as all of these are sure, sooner or later, to be found within our limits. For the same reason the few species known from Kamchatka and the Kuril Islands are included as a part of the fauna of the Alaskan Sea.

The pagination and the numbering of the species, genera, and higher groups are continuous throughout the three parts.

Part I, *Branchiostomatidæ* to *Priacanthidæ* inclusive (pages 1 to 1240), was published October 3, 1896; Part II, *Lutianidæ* to *Cephalacanthidæ* inclusive (pages 1241 to 2183), was published October 3, 1898; and Part III, *Callionymidæ* to *Ogcocephalidæ* appears on November 26, 1898. Parts I, II, and III have each their own table of contents, while in Part IV (the Atlas) is given a table of contents complete for the entire work and corrected to include the Addenda.

The present part includes also an artificial key to the families of true fishes, an addendum containing species overlooked or described subsequently to the publication or casting of the part to which they belong, a glossary of scientific terms, and a general index complete for the entire work.

A fourth volume, or Atlas of plates, containing illustrations of one or more species of each of the more important genera, will follow within the year.

The preparation of the manuscript for this work was begun by the senior author in 1891. In 1893 the junior author became associated with him, and since then both have given to it such of their time and energy as could be spared from engrossing official duties to which systematic ichthyology bears no relation.

The insertion of the comma between generic and specific names and the authorities for them, as practiced in this publication, is in accordance with the views held by the authorities of the United States National Museum, and does not express the views of the authors of this work.

Class PISCES—Concluded.

Subclass TELEOSTOMI—Concluded.

Order BB. ACANTHOPTERI—Concluded.

Group GOBIOIDEA.

(THE GOBIES.)

Body elongate, variously scaled or naked; head usually large, armed or not, the suborbital ring without a bony stay for the preopercle; gill openings reduced, the membranes attached to the isthmus. Gills 4, a slit behind the last; pseudobranchiæ present. Ventral rays I, 4 or I, 5, inserted below pectoral, the fins close together or united or widely separated or otherwise peculiar; dorsal fins separate or united, the first of a few weak spines, sometimes wanting; anal rather long, usually with a single weak spine, similar to soft dorsal; caudal rounded. Usually no air bladder nor pyloric cœca. Vertebræ 24 to 35. Carnivorous bottom fishes, mostly of small size in warm regions, some marine, others of the fresh waters. Two families.

  a. Ventral fins widely separated; preopercle strongly armed; lateral line present.
  CALLIONYMIDÆ, CLXXXVII.
  aa. Ventral fins close together, usually united; preopercle with a weak spine or none; no lateral-line. GOBIIDÆ, CLXXXVIII.

Family CLXXXVII. CALLIONYMIDÆ.

(THE DRAGONETS.)

Body elongate, naked; head usually broad and depressed; the mouth narrow, the upper jaw very protractile; teeth very small, in jaws only; preopercle armed with a strong spine, which is usually branched. Eyes moderate, usually directed upward. Lateral line present, often duplicated. Dorsal fins 2, the anterior with 3 or 4 flexible spines; soft dorsal and anal short, the latter without distinct spine; ventrals I, 5, widely separated from each other; pectoral fins large. Gill openings small, the membranes broadly attached to the isthmus; gills 4, a slit behind the fourth; pseudobranchiæ present; no air bladder. Vertebræ usually $8 + 13 = 21$. Small fishes of the shores of warm seas, chiefly of the old world. Allied to the Gobies, but often resembling the *Cottidæ* in form. Genera 4, species about 30. (*Gobiidæ Callionymina*, Günther, Cat. Fishes, III, 138–152.)

  a. Ventrals entire, the outer ray not detached; head depressed; gill opening reduced to a very small foramen on upper surface of head; lateral line single.
  CALLIONYMUS, 799.

799. CALLIONYMUS, Linnæus.

Callionymus, LINNÆUS, Syst. Nat., Ed. x, 249, 1758 (*lyra*).

This genus includes Dragonets with the ventral fins entire, without detached ray, the gill opening reduced to a small foramen opening upward, and the lateral line single; head triangular, depressed; eyes directed upward; preopercular spine very large; sexual differences strongly marked. Species numerous, living on sea bottoms at some depth. (καλλις, beauty; ὀνομα, name.)

> *a.* Dorsal rays IV, 8 or 9; anal rays 8; some of the dorsal spines filamentous.
>> *b.* Preopercular spine very long, armed with about 9 hooks or spinules; caudal not filamentous. BAIRDI, 2511.
>> *bb.* Preopercular spine strong, bifurcate; caudal fin more or less produced or filamentous. HIMANTOPHORUS, 2512.
> *aa.* Dorsal rays III, 6 or IV, 6; anal rays 4.
>> *c.* Preopercular spine with 2 barbs, the anterior turned forward; body with white spots. CALLIURUS, 2513.
>> *cc.* Preopercular spine with 3 teeth above, ending in an acute point.
>>> PAUCIRADIATUS, 2514.

2511. CALLIONYMUS BAIRDI, Jordan.

Head $3\frac{1}{5}$; depth $9\frac{1}{4}$. D. IV, 9; A. 8. Body long and low, very slender, the head much depressed, the least depth of the caudal peduncle about equal to the diameter of the eye. Head triangular as viewed from above, its breadth $\frac{2}{3}$ its length, exclusive of the preopercular spine. Snout bluntish as seen from above, sharp in profile, its outline straight and moderately steep until above the eyes; profile behind the eyes considerably depressed. Snout $2\frac{2}{3}$ in head to gill opening; eye 4; mouth small, inferior, the maxillary reaching front of eye, as long as snout; lower lip conspicuous. Teeth slender, in villiform bands in both jaws, none on vomer. Interorbital area a simple narrow ridge. Bones of head behind eyes rugose; a low rough tubercle of bare bone above the temporal region on each side, somewhat behind each eye. Preopercular spine very long, as long as eye, its exterior ridge with a single antrorse spinule at its base, its posterior edge with 8 conspicuous hooks turned forward and inward, these growing progressively smaller from the second. Gill opening reduced to a pore at upper posterior angle of opercle, its width rather less than that of pupil. Dorsal spines strong, the first ending in a slender filament, the whole as long as head; second and third spines broken (probably each with a short filament in life, as a short filament is still present on the fourth spine); fourth spine well behind third (leaving room for another spine, although no trace of such spine is present); soft dorsal high, most of its rays slightly filamentous at tip, the longest about $\frac{2}{3}$ head; caudal subtruncate, not filamentous, about as long as head to base of preopercular spine; anal fin rather high, the length of its base 3 in body; pectorals about as long as ventrals, each as long as head without preopercular spine. Lateral line single. Color light grayish, mottled or spotted with yellowish and dark brown; cheeks with steel-bluish spots; first dorsal with dusky reticulations around pale gray spots; second dorsal and caudal with narrow dusky cross streaks; anal with its posterior half chiefly black, the anterior pale; ventrals black; pectorals pale. Type, a specimen 4½ inches long, in good condition, from the "spewings" of a Snapper or a Grouper (*Neomœnis aya* or *Epinephelus morio*), taken on the Snapper Banks, between Pensacola and Tampa; 1 other specimen known. ("I have named this species for Prof. Spencer F. Baird, to whom I have been indebted for aids of many kinds in connection with my studies of American fishes." Jordan.)

Callionymus bairdi, JORDAN, Proc. U. S. Nat. Mus. 1887, 501, Snapper Banks off Pensacola. (Type, No. 39300. Coll. Silas Stearns.)

2512. CALLIONYMUS HIMANTOPHORUS, Goode & Bean.

Head 3½; depth of head equal to length of its postorbital portion or to greatest depth of body. Greatest depth of body at the head and the anterior portion of the trunk. D. IV, 8; A. 8; P. 19; V. I, 5. Body slender, moderately elongate, fins all well developed, the tail tapering and with some of its rays produced into a filament. Caudal peduncle very slender, the least height of tail scarcely more than ¼ greatest height of body. Profile descending very rapidly at snout. Mouth small and the intermaxillary very protractile, but may be almost entirely concealed under the preorbitals. Intermaxillary reaching to front of orbit. Maxillary a roundish, slender bone, extending backward to end of intermaxillary. Mandible about as long as eye, extending to vertical through front of pupil. Teeth in villiform bands on intermaxillary and mandible. Interorbital space very narrow, less than ¼ length of eye, which is 1¼ times as long as snout and nearly ⅙ of total without caudal. A strong bifurcated spine at angle of the preoperculum extending backward slightly beyond the gill opening; length of this spine at its upper articulation ⅔ length of eye. Gill opening reduced to a small slit, placed at a distance behind eye about equaling length of eye and above median line of body. Skin naked. Lateral line abruptly arched over gill opening and connected across nape with its fellow of the opposite side. Spinous dorsal somewhat elevated in front, the first spine nearly twice as long as last, its length about ⅓ total length of caudal; sixth and seventh rays longest, their length nearly equaling that of base of fin; caudal consisting of 4 simple and 8 divided rays; of the divided rays the fifth and sixth are the longest, the lower portion of the fifth and the upper portion of the sixth being produced into a filament, making these rays as long as the distance from the tip of the intermaxillary to the fourth anal ray. It is worthy of remark that in another example of the same species and of about the same size as the type, the sixth of the divided rays alone contributes to form the filament; and in a young example, about ⅓ as large as the type, the first dorsal spine when laid back reaches to the end of soft dorsal. Some of the numerous examples of this species have none of the caudal rays much produced, even in large individuals. Anal fin beginning directly under third ray of soft dorsal, its rays increasing in length to the sixth, which is the longest and twice as long as the first, its length 5⅔ in total without caudal. All the rays simple except the last, which is divided. The pectoral beginning under middle of spinous dorsal and extending to below the fifth ray of the soft dorsal, its rays all simple. The ventral base overlapping lower extremity of pectoral base, its origin under the gill opening. The fourth and longest ray equaling ½ of total length without caudal. A small but distinct anal papilla. Color generally light brown, the back with numerous narrow streaks and blotches of slightly darker brown; a dark blotch on membrane between the third and fourth dorsal spines, in some cases occupying nearly all the membrane, in other cases more limited and nearly elliptical in shape; anal with a broad subvertical dark band, the tips of rays and a small area of the membrane behind each ray pale; the lower caudal lobe with a narrow submarginal dark band; ventral with 2 indistinct narrow dark bands on its outer half. From Blake Station XXX, off Barbados, in 209 fathoms; Station CLXXX, at 137 fathoms; Station XXXIII, off Santa Cruz, at 115 fathoms; Station 2 CCXVI, at 119 fathoms; Station CCXXX, at 84 fathoms. (ἵμας, whip; φορέω, bear.)

Callionymus himantophorus,[*] GOODE & BEAN, Ocean. Ichth., 296, pl. LXXVI, figs. 268, 268 a, b, 1896, off Barbados.

2513. CALLIONYMUS CALLIURUS, Eigenmann & Eigenmann.

Head 3½ to tip of opercular spine (5 in total); depth 7 (9). D. IV, 6; A. 4. Body flat below, the ventral surface bordered on each side with a fold of skin which is wider than the pupil; a single lateral line; diameter of eye equaling length of snout, 3½ in head; maxillary not extending to eye; preopercular spine with 2 barbs above, the anterior one larger and turned forward; gill opening a minute foramen opening upward. The last dorsal ray equaling length of head, and the first dorsal spine reaching its tip when the fin is depressed; ventral fins connected by a broad membrane to the middle of the outer pectoral region; pectoral fins as long as the head. Cheeks, opercles, connecting membrane of ventral fins and antepectoral region with milk-white spots; lower jaw black near the rictus; a series of black dots on branchiostegal membranes, 1 or 2 similar dots in front of pectorals, 2 on the cheek forming a series with the second branchiostegal spot; 4 black spots on the marginal membrane of the belly, other black spots above it; lower half of body with numerous dirty white spots; pectorals transparent, ventrals dusky; membrane of anal sprinkled with minute black points aggregated into black spots in places, and with opaque white spots; caudal transparent, having minute points, its upper half with opaque milk-white bars running obliquely downward and backward from ray to ray; lower half with interrupted longitudinal lines of opaque white, alternating with black spots; dorsal transparent, with white and dark dots most conspicuous between last rays; body marbled with light and darker. Key West, Florida; 1 specimen dredged in 5 fathoms. (κάλλος, beauty; οὐρά, tail.)

Callionymus calliurus, EIGENMANN & EIGENMANN, Proc. Cal. Ac. Sci. 1888, 76, South Beach, Key West. (Type, No. 26265. M. C. Z.)

---

* The species was listed by Eigenmann, Proc. Cal. Ac. Sci., 2d ser. 1, 78, as "*Callionymus agassizii,* Goode & Bean," a name only, accompanied by no description.

2514. CALLIONYMUS PAUCIRADIATUS, Gill.

"D. III, 6; A. 4. The preopercular spine is armed with three teeth above and terminates in an acute point." (Gill.) Matanzas, Cuba; an imperfectly described species, known only from the above note. (*pauci*, few; *radiatus*, rayed.)

Callionymus pauciradiatus, GILL, Ann. Lyc. Nat. Hist. N. Y., VIII, 1865, 143, Matanzas, Cuba.

Family CLXXXVIII. GOBIIDÆ.

(THE GOBIES.)

Body oblong or elongate, naked or covered with ctenoid or cycloid scales. Dentition various, the teeth generally small; premaxillaries protractile; suborbital without bony stay. Skin of head continuous with covering of eyes. Opercle unarmed; preopercle unarmed or with a short spine; pseudobranchiæ present. Gills 4, a slit behind the fourth; gill membranes united to the isthmus, the gill openings thus restricted to the sides. No lateral line. Dorsal fins separate or connected, the spinous dorsal least developed, of 2 to 8 flexible spines, rarely wanting; anal usually with a single weak spine, similar to the soft dorsal; ventral fins close together, separate or fully united, each composed of a short spine and 5 (rarely 4) soft rays, the inner rays longest; the ventral fins, when united, form a sucking disk, a cross fold of skin at their base completing the cup; caudal fin convex; anal papilla prominent. No pyloric cæca; usually no air bladder. Carnivorous fishes, mostly of small size, living on the bottoms near the shores in warm regions. Some inhabit fresh waters, and others live indiscriminately in either fresh or salt water. Many of them bury in the mud of estuaries. Few of them are large enough to be of much value as food. Genera about 80; species nearly 600. The species are for the most part easily recognized, but their arrangement in genera is a matter of extreme difficulty. Until the multitude of Asiatic forms are critically studied, any definition of the American genera must be tentative only. (*Gobiidæ*, part; groups *Gobiina, Amblyopina,* and *Trypauchenina,* Günther, Cat. Fishes, III, 1–138.)

ANALYSIS OF GENERA OF NORTH AMERICAN GOBIIDÆ.

a. Ventral fins separate; body scaly.
    OXYMETOPONTINÆ:
      b. Ventral rays I, 4.
          c. Forehead bluntly rounded, without sharp keel; tongue very slender, sharp; body elongate, compressed, covered with very small scales; head short, compressed, rather broad above; mouth oblique, the lower jaw projecting; teeth in few series, some of them canine-like; isthmus narrow. Dorsals separate, the first of 6 slender spines; soft dorsal and anal elongate; caudal lanceolate.   IOGLOSSUS, 800.
    ELEOTRIDINÆ:
      bb. Ventral rays I, 5.
          d. Vomer with a broad patch of villiform teeth; gill openings extending forward to below posterior angle of mouth, the isthmus thus very narrow;

teeth villiform, the outer scarcely enlarged; vertebræ 12 + 13 (*dormitor*); skull above with conspicuous elevated ridges, one of these bounding the orbit above, the orbital ridges connected posteriorly above by a strong cross ridge; a sharp longitudinal ridge on each side of the occipital, the two nearly parallel, the post-temporals being attached to the posterior ends. Insertions of post-temporals widely separated, the distance between them greater than the rather narrow interorbital width; the post-temporal bones little divergent; top of head depressed, both before and behind the cross ridge between eyes; a flattish triangular area between this and the little elevated supraoccipital region; preopercle without spines; lower pharyngeals with slender, depressible teeth, and without lamelliform appendages; scales of moderate size, ctenoid. PHILYPNUS, 801.

. Vomer without teeth; isthmus broad; gill openings scarcely extending forward below to posterior angle of preopercle; skull without crests.
   e. Body scaly, both anteriorly and posteriorly.
     f. Lower pharyngeal teeth stiff and blunt; the bones with an outer series of broad flexible lamelliform appendages, which are rudimentary gill filaments; body short and elevated; teeth slender, those in the outer row scarcely larger, and movable; top of head without raised crests, flattish, its surface uneven; post-temporal bones rather strongly diverging, the distance between their insertions about ½ the broad flattish interorbital space; no spine on preopercle or branchiostegals; scales large, ctenoid. Species herbivorous. DORMITATOR, 802.
     ff. Lower pharyngeals normal, subtriangular, the teeth stiff, villiform, no lamelliform appendages; scales of moderate or small size; body oblong or elongate.
      g. Body moderately robust, the depth 4 to 5½ times in the length to base of caudal; scales ctenoid; cranium without distinct median keel; a small supraoccipital crest.
       h. Post-temporal bones little divergent, not inserted close together, the distance between their insertions greater than the moderate interorbital space, or 3¼ in length of head; top of skull little gibbous; lower pharyngeals narrower than in *Eleotris;* preopercle without spine; scales very small, about 110 in a longitudinal series. Vertebræ 11 + 13; teeth moderate, the outer series on lower jaw enlarged. GUAVINA, 803.
       hh. Post-temporal bones very strongly divergent, their insertions close together, the distance between them about ⅔ the narrow interorbital space, and less than -⅓ length of head; top of skull somewhat elevated and declivous; interorbital area somewhat convex transversely; lower pharyngeals rather broad, the teeth bluntish; preopercle with partly concealed spine directed downwards and forward at its angle; scales moderate, 45 to 60 in a longitudinal series; vertebræ (*pisonis*) 11 + 15; teeth small. ELEOTRIS, 804.
     gg. Body very slender, elongate, the depth 8 to 9 times in length to base of caudal; scales very small, cycloid.
      i. Preopercle with a partly concealed antrorse hook at its angle; caudal with numerous accessory rays at base. ALEXURUS, 805.

    ii. Preopercle without spine; caudal without many accessory rays at base; post-temporal bones short, strongly divergent, the distance between their insertions about equal to the narrow interorbital space, or about ⅓ length of head; top of head with a strong median keel, which is highest on the occipital region; no supraoccipital crest; mouth very oblique; the teeth small.   EROTELIS, 806.

 ee. Body naked on the anterior part; head naked; lower jaw with 4 larger recurved teeth.   GYMNELEOTRIS, 807.

 eee. Body entirely naked.   CHRIOLEPIS, 808.

aa. Ventral fins united.

 j. Dorsal fins separate, free from caudal.

  SICYDIINÆ:

   k. Ventral disk short, adnate to belly; body subcylindrical, covered with ctenoid scales; lips very thick; upper teeth mostly small and movable, lower fixed; dorsal spines 6.

    l. Teeth simple; no canines in front of lower jaw.   SICYDIUM, 809.

    ll. Teeth trifid (or bifid); no canines in front of lower jaw.
                 COTYLOPUS, 810.

  GOBIINÆ:

   kk. Ventral disk free from the belly.

    m. Dorsal spines 4 to 8; eyes well developed.

     n. Teeth emarginate, uniserial, those of the lower jaw nearly horizontal; dorsal spines 6; scales large, ctenoid; gill openings moderate.   EVORTHODUS, 811.

    nn. Teeth simple.

     o. Body scaly, more or less.

      p. Maxillary normal, not prolonged behind the rictus; skull of the usual gobioid form, comparatively short and abruptly broadened behind the orbits; occiput depressed; supraoccipital and temporal ridges continuous.

       q. Dorsal spines 6; scales evidently ctenoid; head naked (the nape scaly as usual.)

        r. Interorbital area anteriorly elevated, with a large foramen-like depression in front of eye; body short, compressed, formed much as in *Dormitator;* nape with a fleshy crest; scales large. Vertebræ 11 + 15.   LOPHOGOBIUS, 812.

       rr. Interorbital area not elevated in front; body more elongate; no fleshy nuchal crest; isthmus broad.

        s. Inner edge of shoulder girdle without fleshy cirri or papillæ; cranium anteriorly short; interorbital space narrower, grooved, with a low median ridge or none; median crest on cranium low.

         t. Body scaly anteriorly and posteriorly (sometimes a naked strip on back or belly). Vertebræ 12 + 16 to 10 + 15. GOBIUS, 813.

         tt. Body entirely naked anteriorly, the posterior half scaled; scales moderate or small.
              GARMANNIA, 814.

ss. Inner edge of shoulder girdle with 2 or 3 conspicuous dermal flaps; preorbital region very long; premaxillary and maxillary strong; interorbital groove with a conspicuous median crest; scales rather small (45 to 70.) AWAOUS, 815.

qq. Dorsal spines 7 or 8 (very rarely 6, especially in *Eucyclogobius.*)

u. Scales large, ctenoid; shoulder girdle without dermal flaps.

v. Sides of head scaled; soft dorsal and anal rather short, of 11 to 14 rays each; deep-water species.
BOLLMANNIA, 816.

vv. Sides of head naked; soft dorsal and anal short, of 10 to 12 rays each; shore species. AROMA, 817.

uu. Scales very small, cycloid or nearly so.

w. Inner edge of shoulder girdle without fleshy processes; head naked; body more or less compressed; mouth very oblique; teeth strong; interorbital groove with or without a median ridge. Vertebræ 11 + 15 or 16; soft dorsal and anal long, of 15 to 17 rays each.

x. Body chiefly scaly, anteriorly as well as posteriorly.
MICROGOBIUS, 818.

xx. Body naked anteriorly, scaled posteriorly. ZALYPNUS, 819.

ww. Inner edge of shoulder girdle with 2 or 3 dermal flaps, or processes, as in *Awaous.*

y. Head naked, the interorbital groove with the median ridge high, not extending forward to orbit; body rather robust; soft dorsal and anal short; fresh-water species.
EUCYCLOGOBIUS, 820.

yy. Head scaled like the body; the interorbital groove with the median ridge little developed; soft dorsal and anal long; body elongate; marine species.
LEPIDOGOBIUS, 821.

pp. Maxillary much produced backward, extending beyond the gill opening in the adult; skull comparatively long, gradually (not abruptly) broadened behind orbits; median crest of cranium well developed; scales small, cycloid; head naked, occipital region narrowed forward; supraorbital and temporal crests not continuous.

z. Occiput depressed, with a blunt median keel.

    a'. Shoulder girdle without dermal flaps; dorsal spines 6; soft dorsal and anal short; mouth very large; isthmus broad; vertebræ 14 + 16 (*mirabilis*).

        GILLICHTHYS, 822.

    aa'. Shoulder girdle with 1 to 3 small dermal flaps on the inner edge; dorsal spines 5; soft dorsal and anal long.

        QUIETULA, 823.

zz. Occiput transversely rounded without median keel.

    b'. Shoulder girdle with 1 to 3 small dermal flaps on its inner edge; dorsal spines 5; soft dorsal and anal long. ILYPNUS, 824.

    bb'. Shoulder girdle without dermal flaps; dorsal spines 4 or 5; soft dorsal and anal long.   CLEVELANDIA, 825.

oo. Body and head entirely naked.

    c'. Dorsal spines 4; body long and slender; mouth large, the lower jaw projecting; no barbels; soft dorsal and anal long; male with ornate colors.

        EVERMANNIA, 826.

    cc'. Dorsal spines 7 (rarely 6).

    d'. Chin without barbels; mouth small, little oblique; body robust, soft dorsal and anal short.   GOBIOSOMA, 827.

    dd'. Chin with a fringe of short barbels; mouth terminal, oblique; soft dorsal and anal very short.   BARBULIFER, 828.

CRYSTALLOGOBIINÆ:

mm. Dorsal spines 2 (or 1); body wholly naked.

    e'. Eyes reduced to small rudiments; interorbital area forming a sharp median range; skull rather abruptly widened behind orbits; anterior portion of skull unusally long; no flaps on shoulder girdle; skull highest at nape, depressed above the eyes; soft dorsal and anal short.

        TYPHLOGOBIUS, 829.

GOBIOIDINÆ:

jj. Dorsal fin continuous, the soft part and the anal joined to base of caudal; eyes minute; body elongate; scales minute or wanting; mouth very oblique, the lower jaw projecting; gill openings moderate.

f'. Dorsal rays VI, 16 to 23; anal rays 17 to 23.

    g'. Teeth small, in a single series; scales present.   TYNTLASTES, 830.

    gg'. Teeth in a band, those of the outer series being very strong; scales present.

    h'. Body entirely scaled.   GOBIOIDES, 831.

    hh'. Anterior part of body naked.   CAYENNIA, 832.

800. IOGLOSSUS, Bean.

Ioglossus, BEAN, in Jordan & Gilbert, Proc. U. S. Nat. Mus. 1882, 297 (*calliurus*).

Body elongate, strongly compressed, of equal depth throughout, covered with very small, mostly cycloid, scales. Head short, compressed, not keeled above; mouth large, oblique, the lower jaw projecting; teeth in narrow bands or single series, some of them canine; no teeth on vomer or palatines; tongue very slender, sharp; opercles unarmed. Gill openings very wide, the membranes narrowly joined to isthmus on median line. No lateral line. Branchiostegals 5. Dorsals separate, the first of 6 very slender, flexible spines; the second elongate, similar to the anal; caudal long and pointed, free from dorsal and anal; ventrals close together, separate, each of 1 spine and 4 rays, their insertion below or behind pectorals; anal papilla present. A remarkable type, belonging to the *Oxymetopontinæ*, differing widely from our other Gobioid fishes. Gulf of Mexico, in rather deep water. (ἰός, arrow; γλῶσσα, tongue.)

2515. IOGLOSSUS CALLIURUS, Bean.

Head 5; depth 7 to 7½. D. VI-22 to 24; A. I, 21 to 23. Body very elongate, slender, much compressed, of equal depth throughout; head compressed, without osseous crest; mouth very oblique, the lower jaw strongly projecting; premaxillaries in front on the level with pupil; maxillary extending to opposite front of pupil, its length 2⅘ in head; upper jaw with a narrow band of about 2 series of conical cardiform teeth, those of the outer row much larger than the others, behind these 2 small conical curved canines; lower jaw with a single row of smaller teeth, behind which are about 4 short canines directed somewhat backward; the posterior pair strongly curved; no teeth on vomer or palatines. Tongue narrow, pointed. Eye large, nearly twice length of snout, 3⅓ in head, its diameter considerably more than depth of cheek, about ⅛ more than interorbital width; opercles unarmed. Pseudobranchiæ present. Gill openings wide, extending forward below, the membranes attached mesially to the very narrow isthmus, across which they do not form a fold. Gill rakers long and slender. Dorsal fins separated by a short interval, the first of very slender somewhat filamentous spines, the longest about as long as head; second dorsal little more than ½ as high as first, apparently nearly uniform, separated from the caudal by an interval nearly ½ length of head; caudal lanceolate, its middle rays filamentous, about ⅓ the length of rest of body; anal rather high, similar to soft dorsal; ventrals I, 4, inserted very slightly in advance of base of pectorals, the 2 fins very close together, but apparently quite separate and without basal fold of skin, the fin little longer than head, the inner rays filamentous; pectoral with broad base, about 1¼ in head. Anal papilla very short, midway between tip of snout and base of caudal. Body with very small, nonimbricate, embedded scales, these a little larger and imbricate on the tail; cheeks with embedded cycloid scales; scales very weakly ctenoid, most of them appearing cycloid; no lateral line. Color light olive, everywhere densely punctate; dorsals edged with black; middle of caudal reddish, with paler bluish edgings. Length 4½ inches. Here described from specimens from off Pensacola. Gulf of Mexico; known only from the Snapper Banks off Pensacola, in rather deep water. (κάλλος, beauty; οὐρά, tail.)

Ioglossus calliurus (BEAN MS.), in JORDAN & GILBERT, Proc. U. S. Nat. Mus. 1882, 297, Pensacola, Florida; BEAN, Proc. U. S. Nat. Mus. 1882, 419; JORDAN & GILBERT, Synopsis, 949, 1883; JORDAN, Proc. U. S. Nat. Mus. 1884, 437; JORDAN & EIGENMANN, Proc. U. S. Nat. Mus. 1886, 481.

801. PHILÝPNUS,* Cuvier & Valenciennes.

(GUAVINAS.)

Gobiomorus,† LACÉPÈDE, Hist. Nat. Foiss., II, 699, 1798, in part (*dormitor*, etc.); restricted to *dormitor* by JORDAN & GILBERT, Proc. U. S. Nat. Mus. 1882, 571; restricted to *Gobiomorus taiboa*, Lacépède (Valenciennes' *strigata*), by GILL, Proc. U. S. Nat. Mus. 1888, 79, in accordance with the law of exclusion.

Philypnus, CUVIER & VALENCIENNES, Hist. Nat. Poiss., XII, 255, 1837 (*dormitator*).

Lembus, GÜNTHER, Cat. Fishes, I, 505, 1859 (*maculatus*).

Body elongate, terete anteriorly, compressed behind. Head elongate, depressed above. Mouth large; lower jaw considerably projecting; teeth in jaw rather small, slender, recurved, the outer scarcely enlarged; teeth on vomer villiform, in a broad, crescent-shaped patch; gill openings extending forward to below posterior angle of mouth, the isthmus very narrow. Scales moderate, ctenoid, covering most of the head, 55 to 66 in a longitudinal series. Dorsal with 6 spines and 9 or 10 rays; anal rays I, 9 or 10; ventrals separate. No preopercular spine; insertion of post-temporals almost midway between occipital crest and edge of skull; parietals with a crest running from insertion of post-temporal forward to just behind eye, where they are connected by a thin, high, transverse crest; supraocular with a short, high crest, extending from above front of eye back to posterior edge of orbit, thence extending outward parallel with the transverse crest, leaving a deep groove between them; bony projections before and behind eye prominent. Vertebræ $12 + 13 = 25$; lower pharyngeals triangular, with slender teeth. Largest of the Gobies, some of the species reaching a length of 2 or 3 feet and valued as food. Tropical rivers. ($\phi i\lambda \upsilon \pi \nu o\varsigma$, slumber-loving; $\phi i\lambda o\varsigma$, loving; $\dot\upsilon \pi \nu o\varsigma$, sleep.)

a. Coloration rather obscure, the dark lateral band indistinct or wanting; scales 55 to 57. DORMITOR, 2516.

aa. Coloration bright, the black lateral band distinct; scales 52 to 55. LATERALIS, 2517.

2516. PHILYPNUS DORMITOR (Lacépède).

(SLEEPER; GUAVINA.)

Head 2⅜ to 2¹⁰⁄₁₁; depth 5 to 5¾. D. VI-10; A. I, 9; scales 55 to 57; eye 6⅓ to 7¼ in head; snout 3⅔; maxillary 2⅜. Body elongate, terete anteriorly, compressed behind. Head elongate, depressed above. Mouth large; maxillary reaching to middle of pupil. Lower jaw considerably projecting. Teeth on jaws slender, depressible. Interspace between dorsals slightly greater than interorbital width; dorsal spines slender, the second the longest, 2¼ in head; length of base of anal about 2¼ in head; ventrals reaching ⅔ of the distance to vent; tips of pectorals reaching ventral.

---

* The *Eleotrinæ* have been made the subject of a special paper (A Review of the American Eleotridinæ, in Proc. Ac. Nat. Sci. Phila. 1885, 66-80) by Eigenmann & Fordice. The *Gobiidæ* of America have been discussed in detail by Jordan & Eigenmann (Proc. U. S. Nat. Mus. 1886, 477-518) and later by Eigenmann & Eigenmann (Proc. Cal. Ac. Sci., 2d ser.. vol. 1, 1888, 51-78). In this paper are valuable notes on the specimens in the Museum of Comparative Zoology.

† For the reasons in favor of the use of the name *Gobiomorus* for *Valenciennea*, Bleeker, instead of using it for the present genus, see GILL, Proc. U. S. Nat. Mus. 1888, 69.

Dark brownish or olive, lighter below; an interrupted dark lateral band extending from base of pectoral to base of caudal (not always present); fins dusky, and with the exception of the anal and ventrals, all distinctly mottled; spinous dorsal margined with blackish; head often with dark spots. Streams of the West Indies and Atlantic shores of Central America, Mexico, and Surinam; everywhere common, reaching a length of 2 feet or more. Here described from Cuban specimens. (*dormitor,* sleeper.)

Guavina, PARRA, Descr. Dif. Piezas Hist. Nat. Cuba, tab. 39, fig. 1, 1787, Havana.

Gobiomorus dormitor, LACÉPÈDE, Hist. Nat. Poiss., II, 599, 1798, Martinique ; from a drawing by PLUMIER; EIGENMANN & EIGENMANN, Proc. Cal. Ac. Sci. 1888, 52.

Platycephalus dormitator, BLOCH, Ichth., 1801, Martinique ; after LACÉPÈDE.

Batrachus guavina, BLOCH & SCHNEIDER, Syst. Ichth., 44, 1801; based on *Guavina* of PARRA.

Eleotris longiceps, GÜNTHER, Proc. Zool. Soc. Lond. 1864, 151, Nicaragua; GÜNTHER, Fish. Centr. Amer., 440, 1869.

Electris dormitatrix, CUVIER, Règne Animal, Ed. II, vol. 2, 246, 1829, Antilles ; GÜNTHER, Cat. Fish., III, 119, 1861.

Gobiomorus dormitator, JORDAN & GILBERT, Proc. U. S. Nat. Mus. 1882, 572.

Philypnus dormitator, CUVIER & VALENCIENNES, Hist. Nat. Poiss., XII, 255, 1837; POEY, Mem. de Cuba, II, 381, 1860; GIRARD, U. S. and Mexican Boundary Survey, Zool., 27, pl. 12, fig. 13, 1859; JORDAN & GILBERT, Synopsis, 631, 1883.

2517. PHILYPNUS LATERALIS, Gill.

(ABOMA DE MAR.)

Head $2\frac{3}{10}$; depth $5\frac{1}{3}$. D. VI-10; A. I, 10; scales 52 to 55; eye 6 to $6\frac{1}{4}$ in head; snout $3\frac{1}{3}$ to $3\frac{3}{4}$; maxillary $2\frac{1}{2}$ to $2\frac{3}{4}$. Brownish, lighter or white below; a distinct dark brown or blackish band extending from base of pectoral to base of caudal; dorsals, pectoral and caudal dusky; ventrals and anal lighter; dorsals, caudal, and in some specimens the anal, distinctly blotched. The only constant difference between this species and *Philypnus dormitor* seems to be the brighter coloration of *lateralis*. Streams of Pacific Coast of Mexico and Central America, from Sonora to Panama, entering the sea; common, reaching a much larger size than any other of our Gobies. Here described from specimens from Rio Presidio, Mazatlan. (*lateralis,* pertaining to the side.)

Philypnus lateralis, GILL, Proc. Ac. Nat. Sci. Phila. 1860, 123, Cape San Lucas (Coll. Xantus); JORDAN & GILBERT, Proc. U. S. Nat. Mus. 1882, 377.

Eleotris lateralis, GÜNTHER, Cat., III, 122, 1861.

802. DORMITATOR, Gill.

(PUÑECAS.)

Prochilus, CUVIER, Règne Animal, Ed. 1, vol. II, 294, 1817 (*macrolepidota = maculatus*); name preoccupied.

Dormitator, GILL, Proc. Ac. Nat. Sci. Phila. 1862, 240 (*gundlachi*).

Body short, robust; head broad and flat above; mouth little oblique; maxillary reaching to anterior margin of orbit; lower jaw little projecting; no teeth on vomer; lower pharyngeal teeth stiff and blunt, the bones with an external series broad, flexible, lamelliform, these being rudimentary gill filaments; scales large, ctenoid, 30 to 33 in a longitudinal series; skull much as in *Eleotris;* D. VII–I, 8; A. I, 9 or 10; no spine on preopercle; post-temporals inserted midway between occipital crest and edge of skull; supraoccipital crest low. (*dormitator,* one who sleeps.)

2518. DORMITATOR MACULATUS (Bloch).

(GUAVINA MAPO; PAÑECA.)

Head 3½; depth about 3 in adult. D. VII–I, 8 or 9; A. I, 9 or 10; lateral line 33. Body short, robust; head broad and flat above; eye small, less than snout; caudal a little shorter than head; mouth little oblique; maxillary reaching to anterior margin of orbit; lower jaw little projecting; no teeth on vomer; interspace between dorsals equaling orbit; highest anal ray 1¾ in head; highest dorsal ray 1¼ in head; skull much as in *Eleotris,* but everywhere broader; no spine on preopercle; post-temporal inserted midway between occipital crest and edge of skull; supraoccipital crest low; scales large, becoming much smaller on belly, 25 series on median line from base of ventrals to vent; 18 series across breast from pectoral to pectoral; 18 on a median line from posterior border of orbit to dorsal. Dark brown, with lighter bluish spots; a faint dark stripe along sides; a conspicuous large dark blue spot edged with black above base of pectorals; a dark streak from eye to angle of mouth; 2 dark streaks on side of head; branchiostegal membrane blackish; dorsals barred with spots; anal dusky, barred with bluish, and with white margin; a dark bar on base of pectoral. Length 1 to 2 feet. Both coasts of America, ranging from South Carolina through the West Indies to Pará, Cape San Lucas, and Panama, in fresh or brackish water; everywhere abundant and used as food. Dr. Eigenmann observes:

There seem to be 2 forms of the adult—one with the profile gibbous, the dorsal outline forming a regular curve; the other having the profile depressed over the eyes, the anterior portion being subhorizontal. The specimens from Gurupa and the Rio Grande have the profile depressed; all the other specimens have a gibbous profile. A comparison in detail of the two forms is appended. Only extreme differences are given.

West Indian specimens 5 to 7½ inches.	Rio Grande specimens 5, 6½, and 7¼ inches.
Profile regularly curved from first dorsal spine to snout.	Profile depressed over eye, becoming horizontal anteriorly.
Head 3½ to 4; depth 3 to 3½; depth always greater than length of head.	Head 3; depth 3 to 3½; depth usually less than length of head.
Highest anal ray 1¼ to 1¾ in head.	Highest anal ray 1⅜ to 2 in head.
Distance from first dorsal spine to snout greater than distance from first dorsal spine to first anal ray.	Distance from first dorsal spine to snout equals distance from first dorsal spine to base of last anal ray.
Scales in median series 29 to 32.	Scales in median series 30 to 34.
Color usually dark brown, a black spot above base of pectoral, a short bar on base of pectoral.	Color gray, a jet-black spot above base of pectoral; a black bar at base of pectoral; a black line from eye to mouth; longitudinal black lines on cheeks and opercles; dark spots on back; some silvery scales on sides.

Among our specimens from Mazatlan are 3 markedly different forms which seem like distinct species. In view of the great variations to which this species is subject we do not, however, regard them as such, especially as none of the three corresponds exactly to the account above given of the 2 Atlantic forms.

I. Deep-bodied Specimens (*Dormitator latifrons*, Richardson).

Head 3; depth 3. D. VII–I, 8; A. I, 8 or 9; scales 30 to 33; eye $4\frac{3}{4}$ to $5\frac{1}{4}$ in head; snout $3\frac{1}{4}$ to $3\frac{1}{2}$ in head; interorbital width $2\frac{3}{5}$ in head; ventrals reaching $\frac{2}{3}$ the distance to vent, $1\frac{1}{2}$ to $1\frac{3}{5}$ in head; highest anal ray $1\frac{3}{4}$ to 2 in head. Body short, robust, the back elevated; head broad and flat above, the anterior profile from first dorsal spine to tip of snout oblique, descending abruptly; mouth oblique, maxillary reaching anterior margin of orbit; lower jaw little projecting. Color greenish, lighter below; body with cross bars of dark brown; fins dusky, the dorsals distinctly blotched with darker; a dark cross bar at base of pectorals; a dark-blue humeral blotch, becoming blackish in spirits; 3 or 4 dark cross bands extending from eye and below eye to posterior margin of preopercle; a dark band extending from below eye to below tip of maxillary. Two specimens from Rio Presidio, Mazatlan.

II. Common Form, at Mazatlan.

Head $3\frac{1}{5}$; depth $3\frac{2}{3}$ to $3\frac{3}{4}$. D. VII–I, 7; A. I, 8; scales 33 or 34; eye $4\frac{1}{4}$ to $4\frac{3}{4}$ in head; snout $3\frac{1}{2}$ to 4 in head; interorbital width 3 to $3\frac{1}{6}$ in head; ventrals reaching about $\frac{2}{3}$ the distance to vent, $1\frac{1}{2}$ in head; highest anal ray $1\frac{5}{6}$ to 2 in head. Body short, compressed, the back little elevated; head rather broad and slightly convex above, the anterior profile from first dorsal spine to tip of snout slightly convex; mouth oblique, maxillary reaching anterior margin of orbit; lower jaw little projecting. Color olive brown, with cross bars of darker brown, lighter below; fins dusky, the dorsals with about 3 darker cross bars; pectorals with a darker cross bar at base; a distinct dark-brown humeral spot slightly larger than eye; 3 or 4 dark cross bands extending from eye and below eye to posterior margin of preopercle; a distinct dark-brown bar extending from below eye to below tip of maxillary; a dark lateral band extending from base of pectoral to base of caudal. Many specimens from Mazatlan.

III. Large-headed Form.

Head $3\frac{1}{5}$; depth $3\frac{1}{2}$. D. VI–I, 8; A. I, 9; scales 32 or 33; eye $4\frac{1}{4}$ in head; snout $3\frac{1}{2}$ in head; interorbital width $2\frac{1}{4}$ in head; ventrals reaching $\frac{3}{4}$ the distance to vent, $1\frac{1}{4}$ in head; highest anal ray 2 in head. Body moderately compressed, the back little elevated; head very broad above, convex; the anterior profile from first dorsal spine to tip of snout oblique, gently descending; mouth oblique, maxillary reaching anterior margin of orbit; lower jaw little projecting. Color brownish, middle of back darker, lighter below; body with darker cross bands; ventrals yellowish; other fins dusky; dorsals with darker blotches; a dark crossbar at base of pectoral; a dark humeral spot; four cross bands extending from eye and below eye to posterior margin of preopercle; a dark band extending from below eye to below top of maxillary; a dark lateral band extending from base of pectoral to base of caudal. One specimen, from near Mazatlan. (*maculatus*, spotted.)

Sciæna maculata, BLOCH, Ichth., pl. 299, fig. 2, 1790, West Indies.

Eleotris mugiloides, CUVIER & VALENCIENNES, Hist. Nat. Poiss., XII, 226, 1837, Martinique; Surinam.

Eleotris sima,* CUVIER & VALENCIENNES, Hist. Nat. Poiss., XII, 232, 1837, Vera Cruz.

Eleotris latifrons, RICHARDSON, Voy. Sulphur, Fishes, 57, pl. 35, figs. 4 and 5, 1837, locality unknown, supposed to be from Pacific coast, Central America.

? *Eleotris grandisquama*,† CUVIER & VALENCIENNES, Hist. Nat. Poiss., XII, 229, 1837, America; locality unknown.

Eleotris somnolentus, GIRARD, Proc. Ac. Nat. Sci. Phila. 1858, 169, near mouth of Rio Grande.

Eleotris omocyaneus, POEY, Memorias, II, 269, 1860, Havana.

Dormitator microphthalmus, GILL, Proc. Ac. Nat. Sci. Phila. 1863, 170, Panama. (Coll. Capt. John M. Dow.)

Dormitator gundlachi, POEY, Synopsis, 396, 1868, Cuba.

Dormitator lineatus, GILL, Proc. Ac. Nat. Sci. Phila. 1863, 271, Savannah.

Dormitator maculatus, JORDAN & GILBERT, Synopsis, 632, 1883; JORDAN & EIGENMANN, l. c., 482; EIGENMANN & EIGENMANN, Proc. Cal. Ac. Sci., 2d series, vol. 1, 1888, 52.

803. GUAVINA, Bleeker.

Guavina, BLEEKER, Esquisse d'un Syst. Nat. Gobioid., 302, 1874 (*guavina*).

This genus is allied to *Eleotris*, differing in having the post-temporal bones little divergent, not inserted close together, the distance between their insertions greater than the moderate interorbital space, or 3⅓ in length of head; top of skull little gibbous; lower pharyngeals narrower than in *Eleotris*; preopercle without spine; scales very small, ctenoid, about 110 in a longitudinal series. Vertebræ 11 + 13; teeth moderate, the outer series on lower jaw enlarged. Fresh waters of the West Indies and Brazil. Two species known; *Guavina brasiliensis* (Sauvage) from Bahia, and the following. (*Guavina*, the Spanish name.)

2519. GUAVINA GUAVINA (Cuvier & Valenciennes).

(GUARUBACO; GUÁVINA.)

Head 3½; depth 4½ to 5¼. D. VI, or VII-I. 10; A. I, 9 or 10. Body stoutish, oblong; mouth oblique; maxillary reaching opposite middle of eye, its length 2½ to 3½ in head. Lower jaw little projecting; teeth in broad bands, the outer ones on lower jaw enlarged. Scales on head embedded; those on body very small, ctenoid on sides, cycloid on back and belly, 100 to 110 in a longitudinal series. Isthmus very broad. Pectorals reaching to middle of spinous dorsal. Highest anal ray 1⅓ in head. Post-temporals inserted twice as far from occipital crest as in *Eleotris pisonis*. Parietals ending

---

* Types, 2 specimens in poor order, from Vera Cruz, 0.09 mm. long. Snout a little more steep and convex than usual in *Dormitator maculatus*. Head 3½ in length; depth 3⅓. Eye 4½ in head. D VII, 9; A. 11; scales 31-11. Soft dorsal very high, with round black spots. Caudal and anal plain. This seems to be inseparable from *Dormitator maculatus*.

† We have the following note on the type of *Eleotris grandisquama*: Type specimen in fair condition, 0.14 mm. long, from "Amérique Méridionale!" Head slenderer than in *D. maculatus*, and much depressed, its depth at the eyes less than its width, which is less than that of body. Anterior profile almost concave. Caudal fin large; other fins moderate. Dorsal VI, 9; anal I, 9; scales about 29-11. A few dusky spots on dorsal and anal. According to Dr. Eigenmann, specimens of *Dormitator maculatus* from the Rio Grande agree fairly with this type, and it is not likely that it is different.

in a sharp point behind. Preopercular spine none, a broad, thin extension on the lower limb of preopercle taking its place. Lower pharyngeals triangular, normal, rather narrow; the teeth small. Vomer without teeth. Length 1 foot. East coast of tropical America, Cuba to Rio Janeiro, in fresh and brackish waters; very common. (*guavina*, Spanish name.)

Eleotris guavina, CUVIER & VALENCIENNES, Hist. Nat. Poiss., XII, 223, 1837, Martinique.
Guavina guavina, JORDAN & EIGENMANN, *l. c.*, 483.

804. ELEOTRIS (Gronow) Bloch & Schneider.

Eleotris, GRONOW, Zooph., 83, 1763 (nonbinomial).
Eleotris, BLOCH & SCHNEIDER, Syst. Ichth., 65, 1801 (*pisonis*).
Culius, BLEEKER, Esquisse d'un Syst. Nat. des Gobioid., 303, 1874 (*fuscus*).

Body long and low, compressed behind. Head long, low, flattened above, without spines or crests, almost everywhere scaly. Mouth large, oblique, lower jaw projecting. Lower pharyngeals rather broad, the teeth small,[*] bluntish. Preopercle with a small concealed spine below, its tip hooked forward. Branchiostegals unarmed. Eyes small, high, anterior; isthmus broad. Post-temporal bones very strongly divergent, their insertions close together, the distance between them about $\frac{2}{3}$ the narrow interorbital space, and less than $\frac{1}{4}$ length of head; top of skull somewhat elevated and declivous; interorbital area slightly convex transversely; dorsal fins well apart, the first of 6 or 7 flexible spines; ventrals separate. Scales moderate, ctenoid, 45 to 62 in a longitudinal series; vertebræ (*pisonis*) 11 + 15. Tropical seas, entering fresh waters. (ἠλεός, bewildered.)

> *a.* Teeth subequal, those of inner or outer series enlarged.
>> *b.* Cheek entirely scaled.
>>> *c.* Teeth of inner series of each jaw enlarged.
>>>> *d.* Scales in a median series 40 to 51, in a cross series 12 to 20.
>>>>> *e.* Eye large, 5 to 6 in head; scales 40 to 44—12 to 14.
>>>>> AMBLYOPSIS, 2520.
>>>>> *ee.* Eye small, 8 in head; scales 51-20.   ABACURUS, 2521.
>>>> *dd.* Scales in a median series 57 to 66; in a cross series 18 to 24.
>>>> PISONIS, 2522.
>> *bb.* Lower half of cheek naked; scales 61.   PERNIGER, 2523.
> *aa.* Teeth all equal; scales 60.   PICTUS, 2524.

2520. ELEOTRIS AMBLYOPSIS (Cope).

Head $3\frac{2}{3}$; depth $4\frac{1}{4}$. D. VI-9; A. I, 8; scales 46 (40 to 44–12 to 14 according to Eigenmann); eye $5\frac{1}{2}$ in head, 2 in interorbital width; preopercular spine strong, decurved; width of head $\frac{2}{3}$ in its length; chin prominent; premaxillary spines forming a projection in profile. Brown, a black spot above at base of pectoral; first dorsal and anal dusky; second dorsal and caudal delicately cross-barred with blackish; 3 black lines from orbit behind and below. Surinam. Described from 3 specimens each 3 inches long. (Cope.) Dr. Eigenmann mentions 15 other examples,

---

[*] The characters of the skeleton are taken from *Eleotris pisonis* and have not been verified on other species. The hooked preopercular spine supposed to characterize *Culius* is found on the typical species of *Eleotris*, as well as in *Alexurus*.

the longest 2¼ inches long, from Surinam, in the Museum of Comparative Zoology. (ἀμβλύς, blunt; ὄψις, face.)

Eleotris amblyopsis, COPE, Trans. Amer. Philos. Soc. 1870, 473, Surinam (Coll. Dr. Charles Hering); JORDAN & EIGENMANN, *l. c.,* 483, 1886; EIGENMANN & EIGENMANN, *l. c.,* 55.

2521. ELEOTRIS ABACURUS, Jordan & Gilbert.

Head 3; depth 4⅓. D. VI–9; A. I, 8; scales 51–20; eye 8 in head, 2⅕ in interorbital width; pectoral 1⅓; ventral 1⅓; highest dorsal ray 2; highest anal ray 2; caudal 1¼. Body slender, compressed, the head depressed, becoming very narrow anteriorly, its width ⅔ its length; a notable depression above orbits, the premaxillary processes protruding before it; lower jaw the longer; maxillary reaching vertical behind pupil, 2⅔ in head. Teeth in jaws in narrow villiform bands, becoming a single series on sides of lower jaw, those of the outer and inner series in each jaw somewhat enlarged, the largest being a single series in sides of lower jaw. Preopercular spine as usual in the genus. Scales smooth above and below, ctenoid on sides. Color in spirits, brown, lighter above and below; each scale on middle of sides with a dusky streak, these forming obscure lengthwise lines; back anteriorly with a few small black spots; under parts, including sides of head, very thickly punctulate with black; no dark stripes from orbit; lips black; a dark streak from snout through eye to upper angle of preopercle; 2 dusky streaks from eye downward and backward across cheek; a very conspicuous black blotch as large as eye in front of upper pectoral rays; pectorals and ventrals transparent, dusky; vertical fins all barred with light and dark in fine pattern. Coast of South Carolina. Known from a single specimen, 4 inches long, taken in the harbor of Charleston. This species agrees very well with Cope's account of *Culius amblyopsis,* but the scales are larger, the eye is smaller, and there is some difference in color, besides the remote habitat. (ἄβακος, checker; οὐρά, tail.)

Culius amblyopsis, JORDAN & GILBERT, Proc. U. S. Nat. Mus. 1882, 610; not of Cope.
Eleotris abacurus, JORDAN & GILBERT, Proc. Cal. Ac. Sci. 1896, 228, Charleston. (Coll. Dr. C. H. Gilbert. Type, No. 2009, L. S. Jr. Univ. Mus.)

2522. ELEOTRIS PISONIS (Gmelin).

(GUAVINA TÉTARD; SLEEPER.)

Head 3 to 3⅓ in body; depth 4⅓ to 5. D. VI–9; A. I, 8; scales 62; eye 5¾ to 8 in head; maxillary 2⅔; pectoral 1⅓; ventral 2; caudal 1¼. Body not much compressed; head somewhat depressed; mouth rather large, the maxillary reaching to below posterior margin of pupil; lower jaw much projecting, a knob at symphysis; wide bands of villiform teeth in jaws, none on vomer or palatines; interorbital region nearly twice as wide as the horizontal diameter of eye; top of head, cheeks, and opercles covered with small scales; a stout, concealed spine projecting downward on edge of preopercle. Origin of dorsal about midway between tip of snout and end of last dorsal rays; tips of first dorsal spines not reaching front of second dorsal when fin is depressed; origin of anal a little behind that of soft dorsal; pectorals reaching to posterior spine of first dorsal; ventrals inserted very slightly behind base of pectorals; caudal peduncle as wide as length of maxillary. Color brownish; fins with dark spots and wavy lines; ventrals dusky; 2 dark stripes behind the orbit. Here described from specimens, 6 or 7 inches long, collected in the Rio Almendares, Cuba, by Dr. Jordan. Streams of the West Indies, generally common from southern Florida to Rio Janeiro. Dr. Eigenmann enumerates many specimens from various localities in Brazil. (Named for Dr. William Piso, of the University of Leyden, associate of George Marcgraf and Prince Maurice of Nassau, in 1648, in the study of the natural history of Brazil.)

Amore pixuma, MARCGRAVE & PISO, Hist. Brasil., IV, 166, 1648, Brazil.
Eleotris capite plagioplateo, GRONOW, Mus. Ichth., II, 168, 1757; after MARCGRAVE.
Gobius pisonis, GMELIN, Syst. Nat., 1206, 1788; based on *Eleotris* of GRONOW.
Gobius amorea, WALBAUM, Artedi Piscium, III, 205, 1792; based on *Eleotris* of GRONOW.
Eleotris gyrinus, CUVIER & VALENCIENNES, Hist. Nat. Poiss., XII, 220, pl. 356, 1837, Martinique; San Domingo; Surinam.
*Eleotris (Culius) belizianus,** SAUVAGE, Bull. Soc. Philom. Paris 1879 (1880), 55, Belize (Coll. Morelet), Cayenne (Coll. Mélinon); EIGENMANN & FORDICE, Proc. Ac. Nat. Sci. Phila. 1885, 75; EIGENMANN & EIGENMANN, Proc. Cal. Ac. Sci. 1888, 55.
Eleotris pisonis, JORDAN & EIGENMANN, *l. c.*, 483; EIGENMANN & EIGENMANN, *l. c.*, 55.

2523. ELEOTRIS PERNIGER (Cope).

Head 4¼; depth 4¾. D. VI-I, 9; A. I, 9; scales 61; eye 3 in interorbital width; no vomerine teeth. A strong spine at posterior angle of preoperculum, directed downward. Premaxillary spines not prominent in profile; scaling of vertex extending to their extremities. Longitudinal diameter of orbit ⅙ length of head. Color black, abdomen brown, fins dusky; first dorsal with white extremity and 2 longitudinal black bars, 1 along the base; other fins with small black bars; [no] maxillary or caudal spot or ocellus. Length 5 inches. West Indies, south to Rio Janeiro. (Cope.) A specimen in our collection from Jamaica. It is close to *E. pisonis*, but the cheeks are not fully scaled. (*perniger*, very black.)

Culius perniger, COPE, Trans. Am. Philos. Soc. 1870, 473, St. Martins. (Coll. Dr. R. E. van Rijgersma.)
Eleotris perniger, EIGENMANN & EIGENMANN, *l. c.*, 55.

2524. ELEOTRIS PICTUS, Kner & Steindachner.

(GUAVINA.)

Head 3 to 3⅛; depth 6. D. VI-I, 7 or 8; A. I, 7 or 8; lateral line 60; 24 scales in an oblique series from front of soft dorsal downward and back-

---

*Eleotris belizianus is described as follows: Head 4 in total; depth 5. D. VI-I, 9; A. I, 8; scales 60; eye 5 in head. Preopercle with a spine turned downward; 16 rows of scales between soft dorsal and anal; scales of top of head a little smaller than those of body, extending forward nearly to front of eyes; cheeks scaly; scales ciliate. Interocular-space flattened, ½ broader than eye; snout depressed a little longer than eye; lower jaw prominent; outer teeth enlarged; maxillary reaching front of eye. Dorsals contiguous. Color brownish, faint dark streaks on the fins. Belize; Cayenne. (Sauvage.) Length 100 mm. Evidently not different from *E. pisonis*.

ward to anal; about 20 in a vertical series. Body elongate, depressed anteriorly; head especially very broad and flat; mouth large, broad, very oblique, the maxillary reaching nearly or quite to opposite posterior margin of eye, its length 2⅓ to 2⅔ in head; lower jaw considerably projecting. Teeth in jaws all equal, in broad bands, the outer not at all enlarged. Eye small, anterior, its length in adult 2 in interorbital width, which width is about 3 in head; a conspicuous knob at upper anterior and posterior angles of orbit; preopercular spine well developed, strong, compressed, directed downward and forward. Scales on head very small, mostly cycloid, covering cheeks and opercles and upper part of head to the eyes; scales on body smaller and smoother than in most other species, those on belly much smaller than those on sides; scales on back and belly cycloid, only those on sides distinctly ctenoid. Pectoral fins moderate, reaching to near end of base of first dorsal, 1⅜ in head; ventrals inserted just behind axil, reaching halfway to vent, about 2 in head. Interspace between dorsals about equal to diameter of eye. Soft dorsal and anal short and high, very similar, coterminous; last ray of anal a little longer than ½ length of head; caudal peduncle long, a little shorter than head. Caudal fin rounded, 1⅛ in head. Color* dark, dull olivaceous brown, paler below; younger individuals mottled below with bluish and speckled with dark brown; sides without longitudinal stripes; fins dusky, all of them finely mottled and speckled with darker, the dark markings on dorsal and anal forming undulated longitudinal stripes; on pectorals and ventrals forming dark bars. Distinguished from related species by the larger mouth with small, equal teeth, and the small, smoothish scales. Length about 18 inches. Streams of the Pacific Coast, from Sonora, south to Panama; abundant in Rio Presidio, at Mazatlan, where the types of *E. æquidens* were taken; not rare about Panama. (*pictus*, painted.)

Eleotris pictus, KNER & STEINDACHNER, Abh. Ah. Wiss. Wien 1864, 18, pl. 3, f. 1, Rio
  Bayano, near Panama; depth 6 to 7 in total length; scales 60.
Culius æquidens, JORDAN & GILBERT, Proc. U. S. Nat. Mus. 1881, 461, Rio Presidio, near
  Mazatlan. (Types, Nos. 28268 and 29240. Coll. Gilbert.)
Eleotris æquidens JORDAN & EIGENMANN, *l. c.*, 483.

805. ALEXURUS, Jordan.

Alexurus, JORDAN, Proc. Cal. Ac. Sci. 1895, 512 (*armiger*).

Body elongate, covered with small cycloid scales; preopercle with a small, concealed, hooked spine at its angle, as in *Eleotris;* caudal fin broad, its base with many procurrent rays. In other respects similar to *Eleotris.* One species known; marine. (ἀλέξω, to defend; οὐρά, tail, from the caudal fulcra.)

---

* A young example shows the following details of coloration in life: Blackish everywhere, sides with faint whitish streaks, along rows of scales a broad, blackish lateral band occupying whole of side, back and belly paler, traces of faint dark cross bands; caudal black, with a pale margin and some dark cross shades; pectorals, dorsals, and ventrals more or less barred with black; preopercular spine well developed; a whitish bar at base of caudal with a darker one before it.

2525. ALEXURUS ARMIGER, Jordan & Richardson.

Head $4\frac{2}{3}$; depth 8. D. VI-13; A. 11; V. I, 5; scales about 102-30; eye 8 in head; maxillary $2\frac{2}{3}$; mandible $2\frac{1}{2}$; snout $5\frac{2}{3}$; interorbital $4\frac{1}{3}$; pectoral $1\frac{1}{3}$; caudal equals head; ventral 2; last dorsal ray $1\frac{3}{5}$. Body long and low, compressed posteriorly, depressed in front. Head flattish and broad above, the cheeks moderately tumid. Eyes small, high up, separated by a broad, flattish, interorbital space; snout short; mouth moderate, very oblique, the maxillary ceasing below the center of pupil; lower jaw very heavy, oblique, projecting beyond upper, its outline horseshoe-shaped, obtuse in front. Teeth in rather broad bands, the outer enlarged below, but scarcely so above; none of them canine-like. Top of head with very small scales; cheeks and opercles with rudimentary scales above; preopercle with a concealed antrorse hook below, as in *Eleotris;* scales on body very small, perfectly smooth, partially embedded; scales on nape and throat minute. Gill membranes extending a little forward below, so that the branchiostegals are free from the isthmus. Insertion of dorsal twice as far from middle of base of caudal as from tip of snout; the fin low, its slender rays slightly filamentous; soft dorsal low, its last ray highest; anal similar, beginning under second dorsal ray; caudal long, bluntly pointed behind, with strongly procurrent base above and below, the base above $\frac{2}{3}$ length of head, formed of 14 short rays, that below a little shorter, of 12 rays, this procurrent portion forming an angle with the caudal proper where it joins it; pectorals and ventrals short, the ventrals inserted under pectorals. Color olive green, dusky above, paler below, but everywhere covered with fine black dots; both dorsals with the membranes pale, the rays each barred with black; caudal mesially blackish, all the rays barred or checkered in fine pattern; pectoral and anal pale, similarly speckled, base of pectoral dusky; ventral finely speckled. La Paz, Lower California; 1 specimen, $6\frac{1}{4}$ inches long, taken by Mr. James A. Richardson. (*armiger*, bearing arms, from the concealed spine.)

Alexurus armiger, JORDAN & RICHARDSON, Proc. Cal. Ac. Sci. 1895, 511, pl. 48, La Paz. (Type in L. S. Jr. Univ. Mus. Coll. James A. Richardson.)

806. EROTELIS, Poey.

(ESMERALDAS DE MAR.)

Erotelis, POEY, Memorias, II, 273, 1861 (*valenciennesi*=*smaragdus*).

Body very slender, elongate, covered with minute cycloid scales. Ventrals separate, the rays I, 5. No teeth on vomer. Lower pharyngeals subtriangular, the teeth stiff, villiform, none of them lamelliform. Posttemporal bones short, strongly divergent, the distance between their insertions about equal to the narrow interorbital space; top of head with a strong median keel, highest on the occipital region; no supraoccipital crest; no preopercular spine. Mouth very oblique. One species known; strictly marine. (Name an anagram of *Eleotris*.)

2526. EROTELIS SMARAGDUS (Cuvier & Valenciennes).

(ESMERALDA NEGRA; ESMERALDA DE MAR.)

Head 4⅓ to 5½; depth 8 to 12. D. VI–I, 10; A. I, 9; V. I, 5; scales 100.
Body very long and slender, compressed behind, the form much as in
Gobius oceanicus. Head depressed, flattish above, the eyes mostly superior, not ⅓ the width of the interorbital area, which has a knob near its
middle. Mouth very oblique, the lower jaw much projecting, the maxillary about reaching front of eyes; teeth rather small, in bands. Fins
rather high; dorsal spines slender, lower than the highest soft rays, which
are 1⅓ in head; caudal lanceolate, ⅓ longer than head; ventrals moderate, 2 in head. Scales very small, cycloid. Color very dark green, almost
black; the fins mostly bluish, the dorsal with brown lines; some dark
markings about eye and on base of pectoral above. Length 8 inches.
Coral shores among green algæ; known from Key West and Cuba; not
common; not entering rivers. Here described from Key West specimens.
(σμαραγδός, emerald.)

Eleotris smaragdus, CUVIER & VALENCIENNES, Hist. Nat. Poiss., XII, 231, 1837, Cuba; JOR-
DAN & GILBERT, Proc. U. S. Nat. Mus. 1884, 143.
Erotelis valenciennesi, PORY, Memorias, II, 273, 1861, Cuba.
Erotelis smaragdus, JORDAN & EIGENMANN, *l. c.,* 484.

807. GYMNELEOTRIS, Bleeker.

Gymneleotris, BLEEKER, Esquisse d'un Syst. Nat. des Gobioid., 304, 1874 *(seminuda).*

Body scaled only posteriorly, the anterior half and the head naked. Ventrals separate, I, 5. Vomer without teeth. Isthmus broad; skull without
crests. Lower jaw with 4 large recurved teeth. Otherwise essentially as
in *Eleotris,* the preopercle probably without spine. (γυμνός, naked;
Eleotris.)

2527. GYMNELEOTRIS SEMINUDUS (Günther).

Head 3¼. D. VII–11; A. 9. Head depressed, broader than high, flat
above. Snout rather obtuse, longer than eye, lower jaw somewhat prominent; cleft of mouth extending to below anterior margin of orbit. Teeth
in upper jaw in a narrow band, the lower having 4 somewhat larger and
recurved teeth in front, appearing to form a single series; palate toothless. None of the fin rays prolonged; pectoral not quite extending to
origin of second dorsal; ventral much shorter than pectoral, its inner ray
the longest, the others gradually decreasing in length outward; caudal
fin rounded. Head and trunk naked; tail covered with small scales.
Brown, with numerous well-defined white cross stripes on head as well as
on body; vertical fins black. Panama. (Günther); known from the type
only, a young example, 1¾ inches long; not seen by us. *(seminudus,* half-
naked.)

Eleotris seminuda, GÜNTHER, Proc. Zool. Soc. London 1864, 24, pl. 4, figs. 2, 2a, Panama;
GÜNTHER, Fish. Centr. Amer., 441, 1869.
Gymneleotris seminuda, JORDAN & EIGENMANN, *l. c.,* 484.

8o8. CHRIOLEPIS, Gilbert.

Chriolepis, GILBERT, Proc. U. S. Nat. Mus. 1891, 557 (*minutillus*).

This genus differs from *Gymneleotris*, Bleeker, in the total absence of scales, and the absence of enlarged canines in the front of the mandible. Head and body compressed, the former as deep as wide. Ventrals separate, near together, the inner rays longest, each with 1 spine and 5 soft rays. Teeth in a rather wide band in upper jaw, the outer series somewhat enlarged. Teeth in mandible in a single series, similar to outer row in upper jaw, none of them canine-like. Gill slits narrow; no dermal flaps on inner edge of shoulder girdle. Size small. (χρεία, want; λεπίς, scale.)

2528. CHRIOLEPIS MINUTILLUS, Gilbert.

Head 3⅓; depth 4⅓ in length. D. VII–12; A. 11. Mouth oblique, the maxillary reaching to below middle of orbit, 2⅓ in head; eyes high up, but with lateral range, separated by a narrow interorbital space less than diameter of pupil; diameter of orbit nearly twice length of snout, 3⅓ in head; dorsal spines high and slender, but not filamentous, the longest ½ length of head; soft dorsal rays higher, nearly ⅔ length of head; the anal lower; caudal short, broadly rounded, the depth of peduncle ½ length of head; length of pectoral equaling that of head without snout. Color uniform light brown on head and body, above and below; fins dusky, the anal blackish. A single specimen, 1 inch long, from *Albatross* Station 2825, off the east coast of Lower California. (*minutillus*, very small.)

Chriolepis minutillus, GILBERT, Proc. U. S. Nat. Mus. 1891, 558, Albatross Station 2825, Gulf of California, in 79 fathoms.

8o9. SICYDIUM, Cuvier & Valenciennes.

Sicydium, CUVIER & VALENCIENNES, Hist. Nat. Foiss., XII, 168, 1837 (*plumieri*).

Body subcylindrical, covered with rather small ctenoid scales; head oblong and broad, with cleft of mouth nearly horizontal; upper jaw prominent; snout obtusely rounded; lips very thick, the lower with a series of numerous slender horizontal teeth, of which sometimes only the extremities are visible; upper jaw with a single uniform series of numerous movable small teeth attached by ligament to edge of maxillary; behind this outer visible series lie numerous other parallel series of young teeth hidden in the gum, which succeed the former as they become worn out or broken; lower jaw with a series of widely set conical teeth; teeth all simple, slender, the distal half bent inward nearly at a right angle; eyes of moderate size; 2 dorsal fins, the anterior with 6 (5 or 7) flexible spines; caudal quite free; ventrals united into a short cup-shaped disk; gill openings of moderate width; 4 branchiostegals. Species few in the streams of the West Indies. (σικύδιον, diminutive of σικύα, a gourd, or gourd-shaped cupping glass, from the ventral disk.)

a. Body covered with small scales.
    b. Scales very small, about, 84.   PLUMIERI, 2529.
    bb. Scales moderate, about 68.   ANTILLARUM, 2530.
aa. Body nearly naked.   VINCENTE, 2531.

2529. SICYDIUM PLUMIERI (E h).

(SIRAJO.)

Head 4 to 4⅔; depth 4½; eye 6 to 7 in head. 2 t 3 in interorbital width. D. VI-I, 10; A. I, 10; scales 84.   Teeth in upper aw long, slender, bent inward at right angles, only the lips protrudin from the gums.   Front teeth of lower jaw not larger than those behind a single row of inconspicuous papillæ on the gum beneath the upper l a large median papilla above the maxillary suture; a median cleft in th upper lip.   Pectorals longer than head; third, fourth, and fifth dorsal armes produced into long ribbons, the fourth, which is the longest, being 2 to 3 times height of body.   Body usually covered with small scales, reduced in size on neck and belly; frequently almost naked, the scales presnt only on posterior part of body.   Caudal deeply emarginate.   Color oli or violet brown, with about 7 more or less distinct dark vertical bars · a dark bar at base of pectoral; dorsal with irregular dark marking anal fin with a dark marginal band, sometimes edged with white; aul-shaped figure on base of caudal fin, and a black bar on its posterior h.f.   Fresh waters of the West Indies.   (Named for Père Charles Plumier, w o discovered the species at Martinique.)

Gobius plumieri, BLOCH, Ichth., 125, pl. 178, fig. 3, 1786. lartinique; on a drawing by
    PLUMIER.
Sicydium siragus, POEY, Memorias. II, 278, 1861, Santiago d: uba.
Sicydium plumierii, CUVIER & VALENCIENNES, Hist. Nat. Pos., XII, 168, 1837; GILL, Proc.
    Ac. Nat. Sci. Phila. 1860, 101; GÜNTHER, Cat., III, 92, 18C OGILVIE-GRANT. Proc. Zool.
    Soc. Lond. 1884, 156, pl. 11, fig. 1; JORDAN & EIGENMANN ., 484; EIGENMANN & EIGEN-
    MANN, l. c., 56.

2530. SICYDIUM ANTILLARUM. Og ie-Grant.

Head 4⅔; depth 6; width of head ¾ length. I VI-I, 10; A. I, 10; scales 68.   Teeth in upper jaw long, slender, and bens iward over the gum at right angles.   A row of small lamelliform trans se papillæ on the gum beneath upper lip, with a larger median lamellurm papilla above maxillary suture; a slight median cleft in upper lip: iaxillæ at right angles to one another; horizontal teeth conspicuous. cales on body and tail subequal and larger than those on neck and bel.   Maxilla not extending to vertical from posterior margin of eye. the ameter of which is contained 6¼ times in length of head and twice in ir rorbital space.   Length of pectoral greater than that of head. The thir fourth, and fifth dorsal spines produced into long narrow ribands; the ourth, which is longest, nearly 3 times height of body; second dorsa: nsiderably higher than body.   Color uniform violet brown; dorsal fins ith irregular wavy dark markings; anal with a black and white mar al band; caudal with a dark band on upper margin.   One specimen, 4: ches long, from Barbados (Ogilvie-Grant); not seen by us.   (*antillar* of the Antilles.)

Sicydium antillarum, OGILVIE-GRANT, Proc. Zool. Soc. Lc: 1884, 157, Barbados.

2531. **SICYDIUM VINCENTE**, Jordan & E......new species.

Another species of *Sicydium* or of some rela...... is thus mentioned
by Dr. Eigenmann: "Mr. Samuel Garman col......eral hundred specimens of this species at Kingston, St. Vince......s of these specimens
are less than an inch in length, the longest ......; they differ considerably in coloration from the adult; most......ely naked, a few of
those examined having scales only on the ......r part of the body.
Caudal deeply emarginate. There are tra......out 7 dark vertical
bars; a black bar at base of pectoral; dorsa......eral series of black
spots; an H-shaped figure on base of caudal, ... bar on the posterior
half of caudal fin; belly and lower part of ...... in: everywhere else
with black points. The specimens collecte......Garman may be the
types of a new species. No large specimen......lected at the Island
of St. Vincent. Specimens 1½ inches in le......Hayti have the fins
plain and a series of blotches along the mi......he posterior part of
the body; the body, except the belly, is entir......ed with scales which
are plainly ctenoid." (Eigenmann.) (Nam......Vincent.)

Sicydium vincente, JORDAN & EVERMANN, Check-List F......96, St. Vincent Island;
     name only.

8ro. COTYLOPUS, G......

Cotylopus, GUICHENOT, in Maillard, Notes sur l' Isle ......on. x, Addendum 3, 1864
     (*acutipinnis*).
Sicya, JORDAN & EVERMANN, Check-List Fishes, 496, 18.....paster).

This genus is closely allied to *Sicydium*, agre......osely with the latter
in external characters and in the absence of......th in front of lower
jaw; it differs chiefly in the form of the up......which are curved,
tricuspid, and trident-shaped, the middle cu......permanent (*Cotylo-*
pus) or else worn away leaving the teeth ......ly bicuspid (*Sicya*).
(κοτύλη cup; πούς, foot.)

SICYA (*sucia*, a gourd, or gourd-shaped cup):
  a. Teeth in upper jaw curved, tricuspid, trident-...... lateral lobes long, the
    middle short and suspended between the ...... and soon wearing away
    leaving the teeth apparently bicuspid.
      b. Neck and belly naked: a double or triple row of ...... papillae on the gum be-
        neath the upper lip.   GYMNOGASTER, 2532.
      bb. Neck and belly covered with small scales; gum which the upper lip smooth.
               SALVINI, 2533.

Subgenus SICYA, Jordan & E......

2532. **COTYLOPUS GYMNOGASTE......** Grant).

Head 4⅓ to 5; depth 5⅓ to 6. D. VI-I, 1......10: scales 60 to 64;
eye 6 in head, twice in interorbital space. ...... the upper jaw tricuspid, the middle cusp, which is situated a......erior end of tooth, is
very soft and soon becomes worn away. A ......treble row of small
papillae on the gum beneath the upper lip, w......t a larger median
papilla; upper lip with a very slight median ......maxillae containing
an angle of about 75°; horizontal teeth more ......conspicuous. Scales
strongly ctenoid; neck and belly naked. Length ......ectoral greater than

2529. SICYDIUM PLUMIERI (Bloch).

(SIRAJO.)

Head 4 to 4⅔; depth 4½; eye 6 to 7 in head, 2 to 3 in interorbital width. D. VI–I, 10; A. I, 10; scales 84. Teeth in upper jaw long, slender, bent inward at right angles, only the lips protruding from the gums. Front teeth of lower jaw not larger than those behind; a single row of inconspicuous papillæ on the gum beneath the upper lip, a large median papilla above the maxillary suture; a median cleft in the upper lip. Pectorals longer than head; third, fourth, and fifth dorsal spines produced into long ribbons, the fourth, which is the longest, being 2 to 3 times height of body. Body usually covered with small scales, reduced in size on neck and belly; frequently almost naked, the scales present only on posterior part of body. Caudal deeply emarginate. Color olive or violet brown, with about 7 more or less distinct dark vertical bars; a dark bar at base of pectoral; dorsal with irregular dark markings; anal fin with a dark marginal band, sometimes edged with white; an H-shaped figure on base of caudal fin, and a black bar on its posterior half. Fresh waters of the West Indies. (Named for Père Charles Plumier, who discovered the species at Martinique.)

Gobius plumieri, BLOCH, Ichth., 125, pl. 178, fig. 3, 1786, Martinique; on a drawing by PLUMIER.

Sicydium siragus, POEY, Memorias, II, 278, 1861, Santiago de Cuba.

Sicydium plumierii, CUVIER & VALENCIENNES, Hist. Nat. Poiss., XII, 168, 1837; GILL, Proc. Ac. Nat. Sci. Phila. 1860, 101; GÜNTHER, Cat., III, 92, 1861; OGILVIE-GRANT, Proc. Zool. Soc. Lond. 1884, 156, pl. 11, fig. 1; JORDAN & EIGENMANN, *l. c.,* 484; EIGENMANN & EIGENMANN, *l. c.,* 56.

2530. SICYDIUM ANTILLARUM, Ogilvie-Grant.

Head 4⅔; depth 6; width of head ⅔ length. D. VI–I, 10; A. I, 10; scales 68. Teeth in upper jaw long, slender, and bent inward over the gum at right angles. A row of small lamelliform transverse papillæ on the gum beneath upper lip, with a larger median lamelliform papilla above maxillary suture; a slight median cleft in upper lip; maxillæ at right angles to one another; horizontal teeth conspicuous. Scales on body and tail subequal and larger than those on neck and belly. Maxilla not extending to vertical from posterior margin of eye, the diameter of which is contained 6½ times in length of head and twice in interorbital space. Length of pectoral greater than that of head. The third, fourth, and fifth dorsal spines produced into long narrow ribands; the fourth, which is longest, nearly 3 times height of body; second dorsal considerably higher than body. Color uniform violet brown; dorsal fins with irregular wavy dark markings; anal with a black and white marginal band; caudal with a dark band on upper margin. One specimen, 4¾ inches long, from Barbados (Ogilvie-Grant); not seen by us. (*antillarum* of the Antilles.)

Sicydium antillarum, OGILVIE-GRANT, Proc. Zool. Soc. Lond. 1884, 157, Barbados.

2531. SICYDIUM VINCENTE, Jordan & Evermann, new species.

Another species of *Sicydium* or of some related genus is thus mentioned by Dr. Eigenmann: "Mr. Samuel Garman collected several hundred specimens of this species at Kingston, St. Vincent. Most of these specimens are less than an inch in length, the longest 1¼ inches; they differ considerably in coloration from the adult; most are entirely naked, a few of those examined having scales only on the posterior part of the body. Caudal deeply emarginate. There are traces of about 7 dark vertical bars; a black bar at base of pectoral; dorsals with several series of black spots; an H-shaped figure on base of caudal, a black bar on the posterior half of caudal fin; belly and lower part of body plain; everywhere else with black points. The specimens collected by Mr. Garman may be the types of a new species. No large specimens were collected at the Island of St. Vincent. Specimens 1½ inches in length from Hayti have the fins plain and a series of blotches along the middle of the posterior part of the body; the body, except the belly, is entirely covered with scales which are plainly ctenoid." (Eigenmann.) (Named for St. Vincent.)

Sicydium vincente, JORDAN & EVERMANN, Check-List Fishes, 456, 1896, St. Vincent Island; name only.

810. COTYLOPUS, Guichenot.

Cotylopus, GUICHENOT, in Maillard, Notes sur l' Isle de la Réunion, II, Addendum 9, 1864 (*acutipinnis*).
Sicya, JORDAN & EVERMANN, Check-List Fishes, 456, 1896 (*gymnogaster*).

This genus is closely allied to *Sicydium*, agreeing closely with the latter in external characters and in the absence of larger teeth in front of lower jaw; it differs chiefly in the form of the upper teeth which are curved, tricuspid, and trident-shaped, the middle cusp either permanent (*Cotylopus*) or else worn away leaving the tooth apparently bicuspid (*Sicya*). (κοτύλη cup; πούς, foot.)

SICYA (σικύα, a gourd, or gourd-shaped cup):
- *a.* Teeth in upper jaw curved, tricuspid, trident-shaped, the lateral lobes long, the middle short and suspended between the outer lobes, and soon wearing away leaving the tooth apparently bicuspid.
  - *b.* Neck and belly naked; a double or triple row of small papillæ on the gum beneath the upper lip. GYMNOGASTER, 2532.
  - *bb.* Neck and belly covered with small scales; gum beneath the upper lip smooth. SALVINI, 2533.

Subgenus SICYA, Jordan & Evermann.

2532. COTYLOPUS GYMNOGASTER (Ogilvie–Grant).

Head 4½ to 5; depth 5½ to 6. D. VI-I, 10; A. I, 10; scales 60 to 64; eye 6 in head, twice in interorbital space. Teeth in the upper jaw tricuspid, the middle cusp, which is situated at the anterior end of tooth, is very soft and soon becomes worn away. A double or treble row of small papillæ on the gum beneath the upper lip, without a larger median papilla; upper lip with a very slight median notch; maxillæ containing an angle of about 75°; horizontal teeth more or less inconspicuous. Scales strongly ctenoid; neck and belly naked. Length of pectoral greater than that of head. Second, third, and fourth dorsal spines produced into filaments; the third, which is the longest, twice height of body; second dorsal higher than body. Color violet brown, yellowish in young specimens, shaded with indistinct transverse bands of darker; irregular brown spots on axis of pectoral, and a broad dark band from base of pectoral to root of caudal, both more or less indistinct in adult specimens; fins violet, clouded with darker. Length 4¼ inches. Streams about Mazatlan (Ogilvie-Grant); not seen by us. (γυμνός, naked; γαστήρ, belly.)

Sicydium gymnogaster, OGILVIE-GRANT, Proc. Zool. Soc. Lond. 1884, 158, pl. 11, fig. 2, and pl. 12, fig. 6, Mazatlan.
Sicyopterus gymnogaster, JORDAN & EIGENMANN, *l. c.*, 485.

2533. COTYLOPUS SALVINI (Ogilvie-Grant).

Head 4⅔; depth 6¼. D. VI–9 or 10; A. I, 10; scales 78; eye 5¼ in head, twice in interorbital space. Teeth in upper jaw tricuspid; the middle cusp, which is situated at anterior end of tooth, very soft and soon becomes worn away. Gum beneath upper lip smooth; a median papillose tubercle above maxillary suture; upper lip with a small median notch; maxillæ containing an angle of about 75°; horizontal teeth conspicuous. Scales ctenoid, those on neck and belly smaller than those on body and tail. Length of pectoral rather greater than that of head. Second and third dorsal spines subequal and produced into short filaments, 1½ times height of body; second dorsal not so high as body. Color olive brown; anal yellow, with a black and white band along margin; membrane of second dorsal clear, spotted with brown; caudal with a dark and yellow band round the extremity. Length 4⅔ inches. Streams near Panama; 1 specimen known. (Ogilvie-Grant.) (Named for Osbert Salvin, who collected largely in Central America for the British Museum.)

Sicydium salvini, OGILVIE-GRANT, Proc. Zool. Soc. Lond. 1884, 159, pl. 12, fig. 2, Panama.
Sicyopterus salvini, JORDAN & EIGENMANN, *l. c.*, 485.

811. EVORTHODUS, Gill.

Evorthodus, GILL, Proc. Ac. Nat. Sci. Phila. 1859, 195 (*breviceps*).

Body elongate, covered with ctenoid scales of moderate size. Head thick, short, naked. Isthmus moderate. Teeth in a single series, with the crown emarginate, those of the lower jaw horizontal; no canines. First dorsal of 6 spines; ventral fins united, not adherent to the belly, otherwise as in *Gobius*, so far as known. (εὖ, well; ὀρθός, straight; ὀδούς, tooth.)

2534. EVORTHODUS BREVICEPS, Gill.

Head 4½, about as deep as wide; depth 4¼. D. VI–I, 10; A. I, 11; eye 3. Teeth emarginate, uniserial, those of lower jaw nearly horizontal. Snout blunt, profile evenly decurved; caudal rounded, 3 in length of body; some of the dorsal rays filamentous. Color light brown, with irregular blackish blotches along sides; 2 black spots at base of caudal fin, 1 above the other, alternating with 1 more anterior on the peduncle; first dorsal with 2 bands parallel with its upper margin; second dorsal with 3 narrow longitudinal bands. (Gill.) Fresh waters of Trinidad and Surinam; not seen by us. (*brevis*, short; *-ceps*, head.)

Evorthodus breviceps, GILL, Proc. Ac. Nat. Sci. Phila. 1859, 195, Trinidad; JORDAN & EIGENMANN, *l. c.*, 486.

812. LOPHOGOBIUS, Gill.

(CRESTED GOBIES.)

Lophogobius, GILL, Proc. Ac. Nat. Sci. Phila. 1862, 240 (*cristagalli=cyprinoides*).

Dorsal spines 6; scales evidently ctenoid. Body short, compressed, form much as in *Dormitator*; nape with fleshy crest; scales large. Vertebræ 11 + 15. Interorbital area of cranium anteriorly elevated, with a large foramen-like depression in front of eye. One species, differing considerably in form from the other Gobies. The study of its skeleton shows no distinction of much importance unless the peculiar form of its interorbital area be regarded as such. (λόφος, crest; *Gobius*.)

2535. LOPHOGOBIUS CYPRINOIDES (Pallas).

Head $3\frac{2}{3}$; depth $3\frac{2}{3}$; greatest width $5\frac{1}{4}$ to $6\frac{1}{4}$. D. VI or VII–10 or 11; A. 9 or 10; scales 26 to 30; vertebræ 11 + 15; eye $3\frac{1}{2}$ to 4. Body short and deep, little compressed, formed much as in *Cyprinodon;* head naked, a prominent naked dermal crest extending from above middle of eye to near front of spinous dorsal; interorbital width slightly less than diameter of eye; profile convex; snout short, bluntish, about as long as eye; mouth very oblique, the gape slightly curved; front of upper lip on level of lower border of eye; lower jaw somewhat projecting; teeth in both jaws in bands, the outer series erect and somewhat enlarged, those of the inner series small; scales large, reduced on breast and nape; a few scales on upper part of opercle; median line before dorsal naked; dorsal spines produced in short filaments; last rays of soft dorsal reaching caudal; caudal rounded; pectorals lanceolate, reaching beyond insertion of anal, the upper rays not silk-like; skull very broad and short, with low, median crest, highest behind; double crests of temporal region joining at the upper posterior angles of the eyes and forming a bridge over the interorbital area, the crests ending abruptly above the anterior part of the orbit, forming a decided angle, the bridged interorbital leaving a large foramen in front of this angle. Color blackish green in life; spinous dorsal black; soft dorsal, ventrals, and anal dark, plain; pectorals lightish, plain; caudal finely mottled. Length 2 inches. West Indies, north to southern Florida; generally common in the streams and brackish waters of Cuba and other islands. Recently taken by Dr. Evermann in brackish water at Biscayne Bay, Florida. (κυπρῖνος, carp; εῖδος, resemblance.)

Gobius cyprinoides, PALLAS, Spicilegia, Zool., VIII, 17, pl. 1, fig. 5, 1770, Amboina; CUVIER & VALENCIENNES, Hist. Nat. Poiss., XII, 129, 1837; GÜNTHER, Cat. Fish., III, 8, 1861.
Gobius cristagalli, VALENCIENNES, in CUVIER & VALENCIENNES, Hist. Nat. Poiss., XII, 130, 1837, Havana; GUICHENOT, in Ramon de la Sagra, Hist. Cuba, 128, pl. 3, fig. 3, 1850.
Lophogobius cyprinoides, POEY, Repertorio, I, 335, 1867; POEY, Synopsis, 393, 1868; POEY, Enumeratio, 125, 1876; JORDAN & EIGENMANN, Proc. U. S. Nat. Mus. 1886, 487; EVERMANN & KENDALL, Bull. U. S. Fish Comm. 1897, 131, plate 9, fig. 13.

813. GOBIUS (Artedi) Linnæus.

(GOBIES.)

Gobius, ARTEDI, Genera, 28, 1738 (*Gobius ex nigricante varius*, etc., = *niger*).
Gobius, LINNÆUS, Syst. Nat., Ed. x, 262, 1758 (*niger*, etc.), and of authors generally.
Gobionellus, GIRARD, Proc. Ac. Nat. Sci. Phila. 1858, 168 (*hastatus* = *oceanicus*).
Ctenogobius, GILL, Fish. Trinidad, 374, 1858 (*fasciatus*).
Euctenogobius, GILL, Annals Lyc. Nat. Hist. New York 1859, 45 (*badius*).
Smaragdus, POEY, Memorias, II, 279, 1861 (*smaragdus*).
? Pomatoschistus, GILL, Proc. Ac. Nat. Sci. Phila. 1863, 263, footnote (*minutus*).
Coryphopterus, GILL, Proc. Ac. Nat. Sci. Phila. 1863, 263 (*glaucofrænum*).
? Deltentosteus, GILL, Proc. Ac. Nat. Sci. Phila. 1863, 263, footnote (*quadrimaculatus*).
? Gobiichthys, KLUNZINGER, Fisch. Rothen Meeres, 479, 1871 (*petersii*).
? Mesogobius, BLEEKER, Esquisse d'un Syst. Nat. Gobioid., 317, 1874 (*guavina*).
? Stenogobius, BLEEKER, *l. c.*, 317 (*gymnopomus*).
? Oligolepis, BLEEKER, *l. c.*, 318 (*melanostigma*).
? Gnatholepis, BLEEKER, *l. c.*, 318 (*anjerensis*).
? Callogobius, BLEEKER, *l. c.*, 318 (*hasselti*).
? Hypogymnogobius, BLEEKER, *l. c.*, 318 (*xanthozona*).
? Hemigobius, BLEEKER, *l. c.*, 318 (*melanurus*).
? Cephalogobius, BLEEKER, *l. c.*, 320 (*sublitus*).
? Acentrogobius, BLEEKER, *l. c.*, 321 (*chlorostigma*).
? Porogobius, BLEEKER, *l. c.*, 321 (*schlegeli*).
? Amblygobius, BLEEKER, *l. c.*, 322 (*sphinx*).
Zonogobius, BLEEKER, *l. c.*, 323 (*semifasciatus*).
? Odontogobius, BLEEKER, *l. c.*, 323 (*bynoënsis*).
? Stigmatogobius, BLEEKER, *l. c.*, 323 (*pleurostigma*).
? Oxyurichthys, BLEEKER, *l. c.*, 324 (*belosso*).
Lythrypnus, JORDAN & EVERMANN, Check-List Fishes, 458, 1896 (*dallii*).

Body oblong or elongate, compressed behind. Head oblong, more or less depressed. Eyes high, anterior, close together; opercles unarmed. Mouth moderate. Teeth on jaws only, conical, in several series, those in the outer row enlarged; no canines. Isthmus broad. Shoulder girdle without fleshy flaps or papillæ. Skull depressed, abruptly widened behind the eyes and without distinct median keel. Scales moderate, ctenoid, permanently covering the body; cheeks usually naked; belly generally scaly. Dorsal with 6 rather weak spines; pectorals well developed, the upper rays sometimes very slender and silky; ventrals completely united, not adnate to the belly; caudal fin usually obtuse. Species very numerous. The genus *Gobius*, as here understood, comprises a very large number of species more or less closely related to the European type of the genus, *Gobius niger*, and its American relative, *Gobius soporator*. An examination of skulls or skeletons of numerous European and American species shows a remarkable uniformity in most respects. The general form and structure of the cranium is the same in all, the only differences being very minor ones in the height of certain crests. *Gobius oceanicus* seems the most aberrant, but seems to be inseparable generically on account of intermediate forms. Probably several of the many genera indicated by Bleeker will prove valid, but only a thorough study of skeletons can establish them. It is not unlikely that *Ctenogobius*, to which group most of our species belong, may be separable from *Gobius*. (κωβιός; Latin, *Gobius* or *Gobio*, a name applied to the gudgeon (*Gobio gobio*) and

to other small fishes; allied to *Cobitis*, chub, etc. ·According to Aposto-
lides κωβιός and γωβιός are common names in modern Greek for all
species of the genus *Gobius*. Aristotle κωβιός, 610*b*, 4, 598*a* 11 16, 508*b*
16, 569*b* 23, 621*b* 13 19, 567*b* 11, 591*b* 13, 601*b* 22, 835*b* 14. The κωβιός has
many pyloric appendages above the stomach, spawns near the land on the
rocks, the bunches of eggs are flat and crumbling; it feeds on mud, seaweed, sea moss, etc.; lives near the land, gets fat in the rivers, and is
found in schools. The white κωβιός, found in the Euripus of Lesbos,
never leaves that lagoon for the open sea as the other fishes found there
do. Latin *Gobio* and *Cobio*, Plin. *Gobius*, Ovid., Hal. 12, 8. Martial 13,
88. Horace A. Hoffman.)

GOBIUS:
  a. Upper rays of pectoral fin silk-like; i. e., short and very slender and flexible, free for
  nearly their whole length.
    b. Body robust, compressed posteriorly; depth 4⅔ to 5½ in length; head broad, low,
    rounded in profile, its length 3¹⁄₁₀ to 3¾ in body; eye 4 to 5 in head; mouth
    large, little oblique; lips thick; teeth in both jaws in bands, the outer
    series enlarged; those on lower jaw subequal; scales large, strongly ctenoid,
    smaller on nape and belly; dorsal spines short, none filamentous. Color oliva-
    ceous, light or dark, varying from sand color to greenish black, every-
    where mottled and marbled with dark and paler; a faint dusky spot behind
    eye. D. VI-9 or 10; A. I, 7 to 9. Scales 36 to 41. SOPORATOR, 2536.
CTENOGOBIUS (κτεις, comb; *Gobius*):
  aa. Upper rays of pectoral normal, not silk-like, similar to the others.
    c. Scales large, 25 to 33.
      d. Color in life olivaceous, more or less spotted, never red.
        e. Dorsal soft rays 12 to 14; vertex and nape with a slight median fold
        of skin.
          f. Body compressed, its depth 5 in length; head 3½ to 3⅔; eye 3 to
          3¼ in head; vertex and nape with a slight median fold of
          skin; maxillary reaching about to front of pupil; lower jaw
          very slightly produced; teeth in bands, the outer slightly
          enlarged. Olivaceous; spinous dorsal black at tip; second
          dorsal finely checkered in adult. D. VI-14; A. I, 11. Scales
          25 or 26-10. NICHOLSII, 2537.
        ff. Body long, not much compressed; head 3½; eye 3 in head; no
        median fold on vertex and nape; a dark spot on first dorsal.
          EIGENMANNI, 2538.
      ee. Dorsal soft rays 10 to 12; no median fold of skin on vertex and nape.
        g. Caudal with 2 spots at its base; jaws unequal, the lower
        slightly produced; body robust, compressed behind, the
        depth 5 in total length; head 4½; eye longer than snout, 3½
        in head; maxillary reaching pupil; teeth in a band, the outer
        enlarged and distant, the inner enlarged and bent back-
        ward. Brownish; a faint blue spot on each scale; six spots
        along middle of back; similar spots on scapular region and
        middle of sides; 2 spots on base of caudal; a dark spot above
        opercle; blue dots on head; a straight blue line crossing
        cheek above and continued on opercle; dorsals faintly spot-
        ted. D. VI-10; A. 10. Scales 25-7. (Gill.)
          GLAUCOFRÆNUM, 2539.
      gg. Caudal plain or with but a single spot at its base.
        h. Dorsal spines low, the highest little longer than head.
          i. Region from nape to dorsal entirely scaled.
            j. Pores on preopercle not very conspicuous; no
            canine teeth.

k. Body very slender, compressed, the depth 5⅔ in length; caudal much longer than head; mouth rather large, the lower jaw projecting; teeth unequal, rather strong; yellowish, much spotted with darker. D. VI–12; A. 12. Scales 35.

MANGLICOLA, 2540.

kk. Body subfusiform, little compressed; depth 4½ in length; head blunt, 4 in length, rounded in profile; eye equal to snout, 4 in head. Mouth small, horizontal, the lower jaw included; maxillary 3 in head, reaching to below eye. Teeth small, in bands in both jaws, the outer enlarged, those of the upper jaw very slender. Scales large, ctenoid, those of nape and belly little reduced. Longest dorsal spine shorter than head. Caudal scarcely pointed, about as long as head. Color whitish gray, middle of sides with 4 or 5 dark blotches, from each of which a narrow dark bar extends downward and forward; a large black blotch above pectorals, obsolete in female; a small black spot at base of caudal; a dark mark below eye; vertical fins barred. D. VI–12; A. 11 or 12. Scales 33.

STIGMATURUS, 2541.

jj. Pores on preopercle very conspicuous; lower jaw with small canies. D. VI–J, 9; A. I, 9.

QUADRIPORUS, 2542.

ii. Region between nape and dorsal with a narrow naked median strip. Body moderately elongate, subfusiform, the depth 5½ in length. Head large, not so blunt as in *G. boleosoma*, its length 3⅔ to 3⅔ in length; anterior profile gently decurved; snout 3¼ to 3½ in head; eye 4; mouth large, slightly oblique; maxillary entending to front of pupil, 2½ in head. Teeth small, slender and curved, in moderate bands. Scales moderate, ctenoid, those in front much reduced in size; breast naked. Longest dorsal spine 1½ in head. Caudal as long as head, somewhat pointed. Olivaceous, mottled with gray; about 5 rounded dark blotches along middle of sides, the last forming a spot at base of caudal; no dark spot on side of nape; some dark marks on head; vertical fins barred. D. VI–12; A. 13. Scales 33 to 35.

SHUFELDTI, 2543.

iii. Region between nape and dorsal entirely naked.

l. Highest rays of second dorsal little more than ½ head, none of them reaching base of caudal.

m. Profile much decurved, skull rounded behind, without distinct median ridge; mouth horizontal. Body elongate, deepest below front of dorsal, tapering regularly backward, the greatest depth 5¼ in length. Head short, blunt, profile anteriorly abruptly decurved, cheek somewhat swollen. Length of head 3½ in body. Snout about equal eye, 3¾ in head. Mouth horizontal, maxillary reaching to below pupil (in male); lower jaw included. Teeth in each jaw in a band, the outer row of the upper jaw large, recurved. Scales large, ctenoid, somewhat reduced anteriorly. Nape, breast, and belly naked. Dorsal spines about ⅔ of head. Caudal pointed, 2¾ to 3½ in body. Color olivaceous, with numerous dark reticulations on the back; 5 black spots along the sides, the last forming a spot on base of caudal, sometimes with V-shaped dark bars extending from them to dorsal; breast and sides of belly with numerous dark specks in male; a dark line between eyes; a dark line from eye to middle of premaxillary, some dark spots below eye, sometimes forming bars, sometimes a stripe; a large oblique spot above pectorals, continued on opercle; a black spot at base of pectoral; dorsals and caudal barred, anal uniform dusky, ventrals and pectorals black in male, white in female. D. VI-11; A. 10 to 12. Scales 25 to 30.

BOLEOSOMA, 2544.

mm. Profile moderately decurved; eye longer than snout, 3¾ in head. Color yellowish, oblong dark blotches on middle of sides; dorsal and caudal barred. Head 4; depth 6. D. VI-12; A. 10.

FASCIATUS, 2545.

ll. Highest rays of second dorsal as long as head, the last reaching base of caudal. Body elongate, the back not arched; depth 6 in length; head 4, not compressed, the cheeks tumid. Profile abruptly decurved, the snout 3½ in head. Mouth large, nearly horizontal, the maxillary reaching posterior edge of eye in males, middle of eye in females. Teeth in narrow bands in each jaw, the outer somewhat enlarged, the outer in some (males?) much enlarged above and recurved, the enlarged teeth fixed, the others movable. Scales large, ctenoid, reduced anteriorly; belly naked. Dorsal spines little filamentous, the longest about equal to head; caudal 2¼ to 3 in body. Males dark olive, with 4 oblong dark blotches along middle of sides; a dark caudal spot; a black blotch larger than eye on each side of shoulder; dorsal spotted;

u. Scales 55 to 60; eye longer than snout, 4 in head; mouth slightly oblique, the jaws equal, the maxillary not reaching center of eye; teeth in a narrow band, the outer much enlarged and separated from the others by a narrow interspace. Second dorsal spine not equal to depth of body. Caudal 3½ in body. Scales on nape and axil very small, those on posterior part of body much larger. Light olive green; a series of brown spots along middle of tail; sides of head with dusky blotches, vertical fins dotted with black. D. VI-13; A. 14. Scales 58-20.   SAGITTULA, 2556.

uu. Scales very small (60 to 90); caudal more than twice as long as head in adult. Body compressed, extremely elongate, the depth 6 to 9 in length; head higher than wide, short, compressed, 4½ to 5 in length; mouth wide, oblique; maxillary in adult reaching to below posterior border of eye. Lower jaw very thin and flat; teeth in both jaws small, subequal, those in the upper jaw in a single series, those of the lower in a narrow band; outer teeth somewhat movable. Scales anteriorly small, cycloid, embedded, those behind larger and ctenoid; a few scales on upper anterior corner of opercle; dorsal fins high, some of the spines filamentous, longer than head. Caudal very long filamentous, 2 to 2⅔ in body. Light olive; fins dusky in male; a round, black spot on side, a little larger than eye, below spinous dorsal; first dorsal spine with 2 or 3 black spots; a small dusky spot at base of caudal; emerald spot on tongue conspicuous, fading in spirits. D. VI-14; A. 14 or 15.

    v. Head 5½ to 6 in length; scales 60 to 70; patch of scales on opercle obsolete.   HASTATUS, 2557.

    vv. Head 7 to 8 in length; scales about 90; patch of scales on opercle well developed.

                                               OCEANICUS, 2558.

LYTHRYPNUS (λύθρον, gore; ὕπνός, slumberer; a red sleeper):

oo. Soft dorsal and anal very long; D. VI-17; A. 14. Body short, compressed; mouth very oblique; jaws with distant canine-like teeth. Coral red, with bluish crossbands and markings. Scales 40.   DALLII, 2559.

Subgenus GOBIUS.

2536. GOBIUS SOPORATOR,* Cuvier & Valenciennes.

(SLEEPER; MAPO; CAIMAN.)

Head 3¹⁄₁₀ to 3⅔; depth 4⅔ to 5½; eye 4 to 5. D. VI-I, 9 or 10; A. I, 7 to 9; scales 35 to 41—13 to 15. Vert. 11+16. Body robust, compressed pos-

---

* The specimens examined are from Panama, Barbados, Pará, Itapuana, Cuba, Galapagos, Sambara, Bahia, Orange Key, Bahamas, Pernambuco, St. Thomas, Tortugas, Florida Keys, Martinique, Sao Matheas, Curuca, Rio de Janeiro, Rio Doce. "The color variations among examples of this species are very great, specimens from one locality varying from plain sand color, or gray, to greenish black; some dark brown specimens have light bars across the back; in others the scales have light centers forming horizontal series of light lines; sometimes there are light spots on sides of head and cheek; some specimens are conspicuously marbled with light and dark brown, and white spots occur in the centers of some of the scales on specimens of any ground color, these white spots being brighter on some of the scales than on the others, forming interrupted longitudinal lines. If any value could be placed upon the coloration, almost every specimen would be a distinct species. The color variation is irrespective of locality, some localities having all the above-described variations. The types of Poey's *mapo*, *lacertus*, and *brunneus* prove to be color varieties of *Gobius soporator*." (Eigenmann.)

teriorly; head broad, low, rounded in profile; mouth large, little oblique; lips thick; teeth on upper jaw in a broad band, those of outer series enlarged, the inner ones minute; teeth on lower jaw in a broad band, the outer row enlarged, but not quite as large as the outer series on upper jaw. Anterior half of trunk scaled, head naked; scales large, strongly ctenoid, smaller on nape and belly. Dorsal spines short, not filamentous; upper rays of pectoral fin silk-like, short, and very slender and flexible, free for nearly their whole length; caudal short. Skull posteriorly much as in *Lophogobius cyprinoides*, but the median crest reduced to a slight ridge. Lateral crests very high and closely approximated, rising obliquely outward; the inner crests meeting behind eye, the outer ones forming a very high border about the orbit. Interorbital very narrow and deep, with a median ridge. Coloration that of the rocks, usually granite gray or olivaceous, light or dark, varying from sand color to greenish black, everywhere mottled and marbled with darker and paler, often with brassy or greenish; a faint dusky spot behind eye; coloration varying indefinitely with the surroundings; pectorals, dorsals, and caudal generally mottled; anal and ventrals usually plain. Length 3 to 6 inches. Specimens from Pensacola show the following characters: Head 3⅓ (4 in total); depth 4 (5). D. VI-10; A. I, 9; scales 30 to 38; 12 rows of scales from first dorsal downward and backward to anal. Scales on nape extremely small, those on sides firm, ctenoid; first dorsal with an oblique median shade of blackish, the base in front and the distal part light orange; second dorsal dusky at the base, with some spots, its margin light orange; caudal reddish, with dusky cross lines or spots; anal and ventral dusky, yellowish at base in the female; pectoral olivaceous, yellowish at base, reddish at tip, 2 dark spots on base of pectorals. Form robust. Head rather blunt and heavy, the snout less abruptly decurved than in *G. lyricus*. Mouth moderate, the jaws equal, the maxillary reaching about to front of pupil, 2⅔ in head. Teeth in moderate bands, the outer series somewhat enlarged. Cheeks full, tumid. Eyes moderate, placed rather high, much broader than the interorbital space. Dorsal spines slender, the first longer than the other, but not filamentous, 1⅗ in head; caudal rounded, 1⅓ in head; upper rays of pectorals silk-like, the fin somewhat longer than the ventral, 1¼ in head. Color in life, very deep olive green, the back and sides obscurely barred and much marbled with different shades of olive green; cheeks with dark markings, forming reticulations around pale spots; whole under part of head blackish in the males, yellowish in the females. Tropical seas; universally distributed and almost everywhere common, lurking among stones or on sand in shallow water, or in rock pools, moving very quickly when disturbed; north on our coast to Carolina and Gulf of California. The commonest of all shore fishes in tropical America. Among our species it seems to be the one most nearly related to the European *Gobius niger*, and it may, therefore, be held to represent the subgenus *Gobius*, if our other species be placed in different subgenera. Perhaps all the others will ultimately be removed from *Gobius*.

Gobius soporator, CUVIER & VALENCIENNES, Hist. Nat. Poiss., XII, 56, 1837, Martinique; GÜNTHER, Cat. Fish., III, 26, 549, 1861; POEY, Enumeratio, 124, 1876; JORDAN & GILBERT, Synopsis, 634, 1883.

Gobius lineatus, JENYNS, Zool. Voy. Beagle, 95, pl. 19, fig. 2, 1842, Galapagos Archipelago. (Coll. Charles Darwin.)

Gobius catulus, GIRARD, Proc. Ac. Nat. Sci. Phila. 1858, 169, St. Joseph Island, Texas; GIRARD, U. S. and Mex. Bound. Survey, Zool., 26, pl. 12, figs. 9 and 10, 1859; JORDAN & EIGENMANN, *l. c.*, 493.

Gobius mapo, POEY, Memorias, II, 277, 1861, Cuba; POEY, Synopsis, 392, 1868.

Gobius lacertus, POEY, Memorias, II, 278, 1861, Cuba; POEY, Synopsis, 392, 1868; POEY, Enumeratio, 125, 1876.

Gobius andrei, SAUVAGE, Bull. Soc. Philom., Ser. 7, IV, 44, 1880, Rio Guayas, Ecuador. (Coll. André.)

Gobius carolinensis, GILL, Proc. Ac. Nat. Sci. Phila. 1863, 268, Charleston, South Carolina; JORDAN & GILBERT, Synopsis, 634, 1883.

Gobius brunneus, POEY, Synopsis, 393, 1868, Havana; name preoccupied.

Evorthodus catulus, JORDAN & GILBERT, Synopsis, 632, 1883.

According to Dr. Eigenmann, *Gobius albopunctatus* of the Western Pacific can not be separated from *Gobius soporator*. In this case several other synonyms should be added.

Subgenus CTENOGOBIUS, Gill.

2537. GOBIUS NICHOLSII, Bean.

Head 3⅛ to 3⅗; depth 5 to 5¼. D. VI-I, 12 to 14; A. I, 11; scales 25 or 26–10. Body compressed; width of head about twice in its length. Mouth oblique, the maxillary reaching to front of pupil, 2½ to 3 in head; lower jaw very slightly produced. Teeth present on both jaws; the outer series of long, conical teeth, placed at a considerable distance apart; the enlarged teeth on lower jaw not extending on the sides; the inner series of a band of small teeth. Interorbital space very narrow, equaling pupil. Snout 4 to 4⅓ in head. Eyes large, placed high, 3 to 3¼ in head. Caudal peduncle 2¾ to 3⅙ in head. Scales large, caducous, ctenoid; lacking on head, nape, and fins. Dorsal spines slender, flexible; base of first dorsal 1⅗ to 2 in head; soft dorsal and anal similar; base of anal 1⅕ to 1⅓ in base of soft dorsal, and 1¼ to 1⅓ in head; ventrals 1¼ to 1⅗ in head, inserted below or slightly behind origin of pectorals; pectorals reaching a considerable distance beyond ventrals, 1$\frac{1}{12}$ to 1¼ in head; caudal rounded, not equaling head. Color in spirits, light yellowish brown, with traces of darker, lighter below; ventrals usually dusky; spinous dorsal narrowly margined with black; second dorsal finely checkered in adult; other fins yellowish, not distinctly marked. Length 2 to 3½ inches. Coast of British Columbia; not rare. Here described from 5 specimens from *Albatross* Station 2944, numbered 66 in the L. S. Jr. Univ. Museum. (Named for Capt. Henry E. Nichols, U. S. N., its discoverer.)

Gobius nicholsii, BEAN, Proc. U. S. Nat. Mus. 1881, 469, Departure Bay, British Columbia; JORDAN & GILBERT, Synopsis, 946, 1883; JORDAN & EIGENMANN, *l. c.*, 494.

2538. GOBIUS EIGENMANNI, Garman.

D. VII–12; A. 13; P. 19; scales 27–7. Body rather stout, body cavity more than ½ the length from snout to base of caudal. Head $\frac{2}{9}$ of the total length or $\frac{2}{7}$ of the distance to the caudal base, blunt and rounded anteriorly, very narrow between the eyes, slightly compressed. Eyes large, ⅓ of the head, very close together. Snout short, little more than ¼ as long as the eye. Mouth wide; maxillary reaching a vertical from the middle of the eye, moderately oblique. First dorsal higher, anterior 3 rays prolonged in the filaments, third ray longest and reaching to the eighth ray of the second dorsal; origin of anal fin midway from edge of preopercle to base of caudal; pectorals nearly as long as the head; caudal as long as head, pointed. Scales large, thin, deciduous, 27 in a longitudinal series, 2 rows above the lateral line. Yellowish, with a few punctulations of black near the bases of the caudal rays, with a light-edged black spot on the outer halves of the fourth to the sixth rays of the first dorsal and with a black streak around the mouth immediately above the maxillary. The long body, the large eye, the dorsal spot, and the streak above the mouth serve to distinguish this species from its nearest allies of the same locality. Off Key West, in 60 fathoms. ("The specific name is given in honor of the distinguished ichthyologists who have added so much to our knowledge of the American Gobiidæ, C. H. and R. S. Eigenmann.")

Gobius eigenmanni, GARMAN, Bull. Lab. Nat. Sci. State Univ. Iowa, vol. IV, No. 1, 88, 1896; off Key West in 60 fathoms. (Coll. Iowa Univ. Bahama Expedition.)

2539. GOBIUS GLAUCOFRÆNUM (Gill).

Head 3⅓; depth 4¼. D. VI-10; A. I, 9; P. 18; lateral transverse 7. Body robust, compressed; head naked; mouth oblique, the lower jaw slightly projecting, the maxillaries extending to below pupil; teeth long, in many series, the outer curved; scales ctenoid, large. Pectoral fin with the upper rays little branched, not silk-like; cheeks scarcely tumid; caudal and pectoral longer than ventrals, about as long as head. Tawny, with a faint blue spot in the center of each scale, and with 6 spots, each formed by aggregation of dark dots, on the ridge of the back between the second dorsal spine and the axil of the soft dorsal fin; another row of similar but fainter spots runs from the scapular region, and a third row along the middle of the sides; head tawny, with dark spots and blue dots; a straight blue line across the cheek; dorsal fins with faint blue spots. Length 1¼ inches. Florida Keys; said to have come from the coast of Washington, but this is probably an error, as the species has not since been taken there, while 1 apparently identical has been taken at Tortugas.* (*glaucus*, glaucous; *frænum*, bridle.)

---

* Dr. Eigenmann thus describes the specimens from the Tortugas examined by him: *Gobius glaucofrænum* (Gill). Head 3⅓ (4⅔ in total); depth 4½ (5⅓). D. VI-10; A. 10; scales in a median series, 23, in a transverse series, 8; eye as long as snout, 3½ in head, jaws equal, maxillary barely reaching pupil. Teeth in bands in both jaws, those of the outer row of lower jaw enlarged. Dorsal spines scarcely filamentous, the third highest and equaling depth of body. Posterior dorsal rays highest, as high as spines; anal similar to soft dorsal; pectoral long and narrow, longer than head, 3½ in body; ventral reaching past vent; scales large, thin, finely toothed, reduced on breast; nape naked. Color in spirits, light yellowish brown; a light spot on each scale, the spots especially conspicuous near shoulder; 6 dark spots on middle of back; fainter but similar spots along middle of sides; a conspicuous dark spot above opercle; a wavy light line extending forward from it through lower rim of eye to snout; a straight pale-blue bar extending parallel to it across preopercle and cheek to corner of mouth; a narrow faint bar below it; a triangular dark spot at corner of mouth; cheeks and preopercle purplish chocolate; opercle and snout plain yellowish; 2 brown spots at base of caudal; the smaller specimens differing from this in having the markings more distinct. Length of 4 specimens examined, 1⅜, 1₁⁷₂, 1⅝, 1¼ inches. (Eigenmann.)

Coryphopterus glaucofrænum, GILL, Proc. Ac. Nat. Sci. Phila. 1863, 263, Coast of Washing.
ton (evidently an error).
Gobius glaucofrænum, JORDAN & GILBERT, Synopsis, 635, 1883; JORDAN & EIGENMANN,
l. c., 494; EIGENMANN & EIGENMANN, Proc. Cal. Ac. Sci. 1888, 59.

2540. GOBIUS MANGLICOLA, Jordan & Starks.

Head $4\frac{1}{4}$; depth $5\frac{2}{3}$. D. VI–12; A. 12; scales about 35, not to be exactly
counted; caudal lanceolate, $2\frac{2}{3}$ in body; pectoral about equal to head;
dorsal spine slender, not filamentous, $1\frac{2}{3}$ in head; eyes large, close together,
the range partly vertical, the narrow interorbital deeply furrowed; no
flaps on shoulder girdle; scales moderate, ctenoid anteriorly, becoming
smooth behind; median keel on head slight; head naked. Body long,
compressed, the head depressed, the cheeks tumid; snout bluntly truncate; mouth large, the maxillary reaching the middle of eye, not produced backward, truncated behind, somewhat oblique, the lower jaw a
little the longer; lower jaw flat; teeth strong, the outer in both jaws enlarged; cranium without median crest, abruptly widened behind eyes.
Color light olive, mottled with darker; 6 oblong blotches of blackish on
sides as in *Gobius boleosoma*, the last at base of caudal; dorsals and caudal
finely checkered and barred with dark brownish orange and blackish;
anal mottled; a dark shoulder spot; a dark bar before eye and 1 below
eye; ventrals dusky, the edge pale. One specimen, $1\frac{1}{4}$ inches long. Mazatlan; found in the mud of the Astillero among the roots of mangrove
bushes (*Rhizophora mangle*), (whence the name *mangle; colo*, I inhabit).

Gobius manglicola, JORDAN & STARKS, Proc. Cal. Ac. Sci. 1895, 496, Mazatlan. (Coll. Hopkins Expedition to Mazatlan. Type, 3095, L. S. Jr. Univ. Mus.)

2541. GOBIUS STIGMATURUS, Goode & Bean.

Head 4; depth $4\frac{1}{2}$; eye 4, about equal to snout. D. VI–12; A. 11 or 12;
scales 33. Body subfusiform, little compressed; head blunt, the profile
rounded. Mouth small, horizontal, the lower jaw included; maxillary 3
in head, reaching eye. Teeth small, in bands in both jaws, the outer enlarged, those of the upper jaw very slender. Region from nape to dorsal
entirely scaled, the scales large, ctenoid, those on nape and belly little
reduced. Dorsal spines short, the longest shorter than the head; caudal
fin scarcely pointed, about as long as head. Grayish white, middle of
sides with 4 or 5 dark blotches, from each of which a dark bar extends
downward and forward; a large black blotch above pectoral, obsolete in
the female; a small black spot at base of caudal, and a dark mark below
the eye; vertical fins barred. Two specimens taken in a shallow bay at
Key West are thus described: Very pale olive, everywhere freckled and
spotted; lower part of sides silvery, crossed by faint and narrow cross
streaks of light brown; sides with about 5 faint dark blotches; a dark
blotch below eye and 1 on opercle; a round black spot at base of caudal;
bars on verticle fins light olive. Numerous other specimens are less
freckled in coloration, and have a more diffuse caudal spot as well as a
vague dark spot at the shoulder. The dusky marks on the sides are larger.

We find no other differences, and refer all of them to *G. stigmaturus*. The relations of *G. boleosoma*, *G. stigmaturus*, and *G. encœomus* are certainly very intimate. Florida Keys, not very common, our specimens from Key West. (στίγμα, spot; ὀυρά, tail.)

Gobius stigmaturus, GOODE & BEAN, Proc. U. S. Nat. Mus. 1882, 418, no type locality given, but specimens probably from Florida Keys; JORDAN & GILBERT, Synopsis, 946, 1883; JORDAN & GILBERT, Proc. U. S. Nat. Mus. 1884, 140; JORDAN & EIGENMANN, *l. c.*, 495.

2542. GOBIUS QUADRIPORUS, Cuvier & Valenciennes.

D. VI–I, 9; A. I, 9; scales as in *Gobius caninus*. The 2 pores on the vertical arm of preopercle very open; 2 smaller ones above them; teeth of outer series small; 2 small canines on each side of lower jaw; dorsal spines not prolonged as filaments. Color yellowish, with lighter lines which follow the rows of scales; brown spots on dorsal; 2 lines on cheek. Surinam. (Cuvier & Valenciennes.) Not seen by us. (*quatuor*, four; *porus*, pore.)

Gobius quadriporus, CUVIER & VALENCIENNES, Hist. Nat. Poiss., XII, 87, 1837, Surinam; EIGENMANN & EIGENMANN, Proc. Cal. Ac. Sci. 1888, 61.

2543. GOBIUS SHUFELDTI, Jordan & Eigenmann.

Head 3⅔ to 3⅞; depth 5⅓; eye 4; snout 3⅓ to 3⅔. D. VI–12; A. 13; scales 33 to 35. Body moderately elongate, subfusiform; head less blunt than in *Gobius boleosoma*, the anterior profile gently decurved; mouth large, slightly oblique; maxillary extending to front of pupil, 2⅓ in head. Teeth small, slender, and curved, in moderate bands; scales covering anterior half of trunk; head and breast naked; scales moderate, ctenoid, those in front much reduced. Longest dorsal spine 1⅓ in head; caudal fin as long as head, somewhat pointed. Olivaceous, mottled with gray; about 5 round dark blotches along middle of side, the last at base of caudal; no dark spot on side of nape; some dark marks on head; vertical fins barred. Gulf coast of the United States, known as yet only from fresh waters about New Orleans. (Named for Dr. Robert Wilson Shufeldt, U. S. A., who collected the types.)

Gobius shufeldti, JORDAN & EIGENMANN, Proc. U. S. Nat. Mus. 1886, 495, New Orleans. (Type, No. 35202.)

2544. GOBIUS BOLEOSOMA, Jordan & Gilbert.

Head 4 (5 in total); depth 4¼ to 5½. D. VI–12; A. I, 10 to 12; scales 25 to 30. Body slender, subfusiform, little compressed; head moderate, not very blunt, the anterior profile somewhat evenly decurved, the snout not very short, scarcely shorter than the large eye; mouth not very large, horizontal, the lower jaw included, the maxillary extending slightly beyond front of pupil, its length about 3 in head; teeth small, slender, in narrow bands, those of the outer series longer than the others; eyes placed high, about 4 in head; interorbital space not wider than pupil; scales moderate, ctenoid, those on nape and belly not much reduced in size; gill opening not continued forward above opercle; first dorsal with the spines slender but rather firm, none of them filamentous, the longest about ⅔ head; second dorsal and anal rather large; caudal long, pointed, slightly longer than head; pectorals large, slightly longer than head, none of the upper rays silk-like; ventrals slightly shorter than head, inserted below axil of pectorals; skull rounded behind, no ridges nor crests; crests at side minute; interorbital very narrow. Color in life: Male, deep olive green, mottled with darker; middle of side with 4 or 5 vague darker blotches; a jet-black spot above gill opening, on side of back; head mottled, dusky below; usually a dark bar below eye; dorsals tipped with bright yellowish, each crossed by numerous narrow, somewhat oblique, interrupted bars or series of spots, these being of a rich reddish brown color; caudal barred with black, its upper edge tinged with orange; anal nearly plain, with a slight orange tinge; ventrals bluish black, their edges whitish. Female, paler and duller in color, more mottled, the black spot above gill opening obsolete or nearly so; a dark spot at base of caudal; upper fins barred, as in the male; lower fins mostly pale, tinged with orange. Many specimens of this species, the largest about 2 inches in length, were obtained in the Laguna Grande at Pensacola. It lurks in sea wrack on muddy bottoms in very shallow water (6 to 12 inches). In form, size, coloration, and movements this little fish bears a remarkable resemblance to the percoid, *Boleosoma olmstedi.* Gulf of Mexico, Pensacola to Key West; common in shallow sandy bays, lurking in sea wrack at the depth of a foot (whence the name βολίς, dart; σῶμα, body).

Gobius boleosoma, JORDAN & GILBERT, Proc. U. S. Nat. Mus. 1882, 295, Laguna Grande, Pensacola; ibid, Synopsis, 946, 1883; JORDAN & EIGENMANN, *l. c.*, 495.

2545. GOBIUS FASCIATUS (Gill).

Head 4 in length (4⅔ in total); depth 6 (7). D. VI–12; A. 10; scales 30–7. Body slender, elongate; head somewhat pointed; profile rounded, not as much as in *Gobius garmani* and *Gobius boleosoma;* eye large, slightly longer than snout, 3¾ in head; interorbital area scarcely wider than pupil; mouth slightly oblique, maxillary extending to below anterior margin of pupil, 3⅓ in head; lower jaw thin and flat; teeth strong, recurved, in a band in each jaw, the teeth of the outer series of the upper jaw enlarged, several times as large as those of the inner series. Scales finely ctenoid (fallen off anteriorly in specimen examined); antedorsal region and breast naked. Dorsal spines slender, filamentous near tip, not reaching second dorsal, 1½ in length of head; second dorsal of moderate height; caudal (tips broken) about 5 in length, 1⅔ in length of head; ventral not reaching vent, 1½ in head; pectorals pointed, equaling the head in length. Color yellowish, marbled with darker above; 4 oblong dark blotches along middle of sides; a darker spot at base of caudal; narrow dark stripes across nape; a faint dark stripe along upper margin of opercle, through lower margin of eye to snout; another extending from angle of mouth to edge of preopercle, then extending down along the margin of the preopercle and ending in a dark blotch on the lower part of the cheek; a dark spot on opercle; first dorsal with 2 curved bars; caudal with 3 rather broad dark bars; anal dusky; connecting membrane of ventral white, its first rays blackish, outer rays yellowish; lower parts yellowish. West Indies; not seen by us. This description by Eigenmann, from a specimen 1¾ inches long, No. 13231, M. C. Z., collected in Hayti by Dr. Weinland. (*fasciatus*, branded.)

Ctenogobius fasciatus, GILL, Fishes Trinidad, 378, 1858, Trinidad.
Gobius fasciatus, GÜNTHER, Cat., III, 34, 1861; JORDAN & EIGENMANN, Proc. U. S. Nat. Mus. 1886, 495; EIGENMANN & EIGENMANN, Proc. Cal. Ac. Sci. 1888, 62.

2546. GOBIUS ENCÆOMUS, Jordan & Gilbert.

Head 4; depth 6; snout 3¼. D. VI-11; A. 12; scales 27 to 33. Body very elongate, much tapering backward; head compressed, the cheeks high and vertical; snout very short, compressed, obtusely rounded vertically. Mouth nearly horizontal, low, large, the maxillary 2 in head, nearly reaching vertical from posterior margin of orbit. Teeth in very narrow bands in both jaws, those of the outer series in the upper jaw much enlarged and recurved in some specimens; eyes inserted high, the interorbital space very narrow, about as wide as pupil; diameter of orbit much greater than snout, nearly ¼ of head. Gill opening 2½ in head, the isthmus wide. Dorsals contiguous, the membrane of spinous dorsal reaching nearly to base of soft dorsal; dorsal spines high, of nearly uniform length, the last reaching well beyond origin of soft dorsal when depressed; the longest spine about ½ length of head; soft dorsal and anal long and high, the posterior rays of both fins reaching at least to base of caudal when depressed; caudal lanceolate, the middle rays produced, 2⅔ in body; ventrals reaching vent, somewhat longer than pectorals, which about equal length of head; ventral sheath well developed, its length ⅖ that of fin. Body wholly covered with large, strongly ctenoid scales, which are much reduced in size anteriorly; head, antedorsal region, and breast naked. In female specimens the mouth is evidently smaller, and the caudal less elongate. Colors in life: Male, light olivaceous, mottled above with darker olive brown; a series of about 4 obscure oblong dark blotches along middle of sides; a dark spot at base of caudal; each side of nape with an intense blue-black spot larger than eye; an obscure dusky streak from eye forward to mouth; a small dusky spot sometimes present on upper portion of base of pectorals; both dorsals translucent, with a series of bright reddish-brown spots as large as pupil; upper lobe of caudal light reddish, the lower lobe blue black; anal and ventrals dusky bluish, pectorals slightly dusky, with a narrow, bright pink border behind. Female, without bright markings; body light olive, with 5 oblong dark blotches on sides, the last on base of caudal; from each of the 3 middle blotches a V-shaped bar runs to the back (these visible also in males); back somewhat mottled with dusky; a black blotch on scapula; a small one on opercle; a dark bar from eye forward to mouth. Vertical fins with dusky streaks, these appearing on caudal in the form of cross bars; ventrals light, with 2 lengthwise dark streaks; pectorals plain. South Carolina to Key West, in sandy bays; scarce. Length 2 inches. (ἐγκαίω, brand; ὦμος, shoulder.)

Gobius encœomus,* JORDAN & GILBERT, Proc. U. S. Nat. Mus. 1882, 611, Charleston, South
  Carolina (Type, No. 29673, 3 specimens. Coll. C. H. Gilbert); JORDAN & GILBERT,
  Synopsis, 945, 1883; JORDAN & GILBERT, Proc. U. S. Nat. Mus. 1884, 142; JORDAN &
  EIGENMANN, *l. c.,* 496.

2547. GOBIUS STIGMATICUS (Poey).

Head 4; depth 5 to 6; eye 3⅓. D. VI–12; A. 12 or 13; scales 27. Body
a little deeper and less compressed than in *Gobius encœomus.* Anterior
profile moderately decurved; back slightly arched; skull flattish behind,
much broader than in *G. boleosoma,* with an evident median ridge; mouth
oblique, large, lower jaw thin and flat, maxillary reaching to below pupil.
Teeth above uniserial, some of them enlarged and recurved; lower teeth
in a narrow band, males sometimes with the hindmost of the outer series
a strong, exserted, recurved canine (present in Poey's type). Anterior half
of body scaled except region between nape and dorsal, which is naked;
breast naked. Longest dorsal spine ⅔ head, sometimes elongate; caudal
3⅓ in body. Light greenish, sides of male with 5 or 6 narrow, straight,
whitish or yellowish cross bars, regularly placed; 4 dark bars on head, 3
below the eye and 1 on opercle; a small dark spot behind and above
opercle; ventral fins barred; female with a row of irregular dark spots
connected by a dusky streak, the pale cross bars obsolete. Coast of North
Carolina, Florida Keys, the West Indies, southward to Rio Janeiro;
common at Havana. Subject to considerable variation. Brazilian specimens said by Eigenmann to be darker, the bars on cheek conspicuous;
third dorsal spine often much elongate, reaching fifth dorsal ray, last
soft ray sometimes reaching caudal. (*stigmaticus,* spotty.)

Smaragdus stigmaticus, POEY, Memorias, II, 281, 1861, Cuba.
Gobionellus stigmaticus, POEY, Synopsis, 394, 1868; POEY, Enumeratio, 126, 1876; JORDAN
  & GILBERT, Synopsis, 947, 1883.
Gobius stigmaticus, JORDAN, Proc. U. S. Nat. Mus. 1886, 49; JORDAN & EIGENMANN, *l. c.,*
  496.

2548. GOBIUS LYRICUS, Girard.

Head 4⅓; depth 4⅔. D. VI–11; A. I, 10; scales 27. Body rather elongate, moderately compressed; head rather short, the profile very obtuse,
descending abruptly from before the front of the eye to the snout; eyes
small, placed high, about as long as snout, and about 4⅓ in head; mouth
nearly horizontal, much below level of eye, the maxillary extending to
beyond pupil, 2⅔ in head; jaws subequal; teeth strong, in 1 series in each
jaw; in the lower jaw about 4 shortish, canine-like teeth behind the other
teeth; anterior teeth of lower jaw small, of upper jaw rather large; gill

---

* One small specimen, taken with the seine in a shallow bay, at Key West, is described
as follows:
  Light green, with 5 diffuse spots of darker green on sides, the posterior one most conspicuous; pectorals, both dorsals, and caudal edged above with pale orange; ventrals
mostly black, edged with paler; anal dark; a conspicuous dusky shoulder spot; maxillary reaching to below middle of eye; caudal about ⅓ longer than head. Lateral line
about 30. This little specimen appears to be identical with that described by us from
Charleston under the name *Gobius encœomus.* The species is allied to *G. stigmaturus,*
but has a much slenderer body. The number of scales in a lateral series is less than 37,
the number originally stated by us. There are about 33 in this specimen. (Jordan &
Gilbert.)

opening not continued forward above opercle; first dorsal with 2 or 3 spines filamentous, the longest reaching past the middle of the second dorsal, which is of moderate height and similar to the anal; caudal long and pointed, ⅟₄ longer than the head; pectoral as long as head, about reaching front of anal; upper rays of pectorals not silk-like; ventrals somewhat shorter than head, their insertion below front of pectorals; scales large, rough, those on nape, pectoral region, and belly reduced in size; head naked. Color in life, dark olive, with 4 or 5 irregular confluent blackish cross bands, besides dark blotches and irregular markings; head marbled with darker, the jaws, opercles, and branchiostegals blackish; first dorsal mostly dusky translucent, somewhat barred; second dorsal and anal plain dusky; caudal dark blue, with 2 longitudinal stripes of bright red; pectoral finely barred or reticulated with blackish and pale; head and belly yellowish. Female specimens duller and paler. Gulf of Mexico, from Galveston to Cuba and the Lesser * Antilles; rather common. (*lyricus*, pertaining to a lyre, apparently an allusion to the dorsal spines.)

Gobius lyricus, GIRARD, Proc. Ac. Nat. Sci. Phila. 1858, 169, Brazos Santiago, Texas; GIRARD, U. S. and Mex. Bound. Surv., 25, pl. 12, figs. 4 and 5, 1859; GÜNTHER, Cat., III, 550, 1861; JORDAN & EIGENMANN, *l. c.*, 496; EIGENMANN & EIGENMANN, *l. c.*, 63.

Smaragdus costalesi, POEY, Memorias, II, 280, 1861, Havana. (Type, No. 13109, M. C. Z. Coll. Felipe Poey.)

Gobius wurdemanni, GIRARD, Proc. Ac. Nat. Sci. Phila. 1858, 169, Brazos Santiago; · probably the female; JORDAN & GILBERT, Synopsis, 634.†

Euctengobius lyricus, JORDAN & GILBERT, Synopsis, 633, 1883.

2549. GOBIUS GARMANI, Eigenmann & Eigenmann.

Head 4 in length (5⅓ in total); depth 4 (5½). D. VI–11; A. 11; scales 30–7. Body robust, head short and blunt; profile in front of eye abruptly decurved, rounded much as in *Gobius boleosoma;* mouth inferior, horizontal; lower jaw included; maxillary extending to below pupil, 2⅓ in head; lips thin; teeth short and thick, in a single series in each jaw. Dorsals contiguous; dorsal spines filamentous, the second and third longer than the rest, reaching past first third of second dorsal; last dorsal rays reaching base of caudal; pectorals equaling head in length; ventral short and broad, 5 in body; caudal rather long and pointed, 3 in body. Scales large, slightly reduced and cycloid on nape. Color yellowish, marbled with brown; a series of irregular blotches along the sides; a light spot at base

---

* A specimen from St. Kitts is thus described by Eigenmann: "Depth 5 in length; head 4. The second and third dorsal spines extend to base of caudal; dorsal scarcely less than length of head, the last rays reaching past base of caudal; the caudal fin is ⅔ longer than the head, 2½ in body. Color light brown, faintly marked with darker; the first dorsal with minute dark points, the lower fourth of the spines with simple dark spots, above which are jet-black spots ocellated with white; the second dorsal fin dusky, darker posteriorly, the basal portion of the last half of the fin evenly black, the anterior 4 rays marked with dark points similar to the spots on the lower parts of the spines of the first dorsal; caudal dusky, with 2 light bars; anal plain, darker than body; ventral fins blackish, edged with white; pectorals blackish, with many series of white spots on the membrane, and short, white bars at base; branchiostegal membrane black, with a light margin.

† *Gobius wurdemanni*, Girard. Appearance of *Gobius lyricus*. Reddish brown, obscurely barred with dusky. Head larger; caudal shorter; ventrals shorter; anal lower; scales smaller than in *G. lyricus;* teeth very slender, much smaller than in *G. lyricus;* third dorsal spine filamentous. D. VI–11; A. 12. Brazos Santiago, Texas. (Girard.) (Named for Dr. Gustav Würdemann, its collector.)

of caudal, partly or wholly surrounded by a broad ring of dark brown; head slate color, white below; 3 dark bars extending forward and downward from eye to mouth; a triangular spot on opercle; dorsals, caudal, and pectorals finely barred with black; a chocolate bar on base of ventral; anal margined with white; an irregular black bar on shoulder and upper half of pectoral; everywhere more or less blotched with darker, the blotches at times forming numerous bars across the back. Dominica, Fort de France, Martinique, St. Kitts. (Eigenmann & Eigenmann.) Not seen by us. Apparently very close to *Gobius lyricus*, if not the same. (Named for its discoverer, Prof. Samuel Garman.)

Gobius garmani, EIGENMANN & EIGEMANN, Bull. Cal. Ac. Sci. 1888, 61, Dominica, Fort de France, Martinique, St. Kitts. (Coll. Samuel Garman.)

2550. GOBIUS ZEBRA, Gilbert.

Head 3; •depth 4⅓; eye 3¼ in head. D. VI–11 or 12; A. 9. Body not elongate, the snout short, the mouth oblique, with maxillary reaching below middle of orbit. Mouth small, the maxillary 2⅓ in head. Interorbital space very narrow. Teeth in upper jaw in a narrow band or double series, the outer row enlarged and spaced; lower jaw apparently with a single series, similar to the outer row in the upper jaw. Scales cycloid, large, wanting on nape and a narow strip along base of spinous dorsal. Color cherry red, head and sides with 15 blue cross bars, a little narrower than interspaces, encircling the body posteriorly, lacking for a short distance on belly and under side of head; on upper side of head and nape these bars run obliquely forward and downward, but elsewhere vertical; on middle of each interspace a very narrow blue line, becoming indistinct on lower part of sides; on cheeks the blue bars are connected by narrow cross lines, forming blue reticulations surrounding round spots of the ground color. Length of types ⅓ inch. Two specimens from *Albatross* Station 2989, west coast of Mexico, in 36 fathoms. (*zebra*, zebra, from the stripes.)

Gobius zebra, GILBERT, Proc. U. S. Nat. Mus. 1890, 73, Albatross Station 2989, west Coast of Mexico.

Subgenus EUCTENOGOBIUS, Gill.

2551. GOBIUS POEYI, Steindachner.

Head broad and flattish; depth 6½ in total length; eye 4¼, 1⅓ in interorbital width, longer than snout; snout short and decurved. D. VI–9; A. 9; scales 40. Maxillary extending to below middle of eye. Some of the dorsal spines produced and filamentous, the third 1½ times depth of body; caudal short, rounded. Two rows of ill-defined blotches on upper half of body; dorsals and caudal sharply barred, anal and ventrals dusky (male). A small round dark spot at base of caudal. (Steindachner.) Barbados; not seen by us. (Named for Prof. Felipe Poey.)

Gobius poeyi, STEINDACHNER, Ichthyol. Notizen, VI, 44, 1867, Barbados; JORDAN & EIGENMANN, l. c., 497.

2552. GOBIUS BADIUS (Gill).

Head 6 in total; depth 7. D. VI–I, 10; A. I, 10; scales 50–18. Anterior profile very oblique; a line of pores above each eye; 2 on upper ascending margin of preopercle; eye 4 in head; interorbital space 3⅓ in eye; caudal 5 in total length; pectoral 6. Color dark bay with a posteriorly straight heavy dot in the center of each scale on back and sides above; head plumbeous, with 2 livid blue bands from eye to upper jaw. (Gill). About mouth of Amazon; not seen by us. (*badius*, bay color, dark red.)

Euctenogobius badius, GILL, Ann. Lyc. Nat. Hist. N. Y., VII, 1857, 47, Amazon.
Gobius bosci, SAUVAGE, Bull. Soc. Philom. Paris, IV, 44, 1880.
Gobius badius, EIGENMANN & EIGENMANN, *l. c.*, 65.

Subgenus GOBIONELLUS, Giraid.

2553. GOBIUS MICRODON, Gilbert.

Head 4⅛; depth 5. D. VI–13; A. 14; scales 62. Head and body compressed, everywhere deeper than wide. Mouth at lower profile of snout, nearly horizontal, the lower jaw extremely weak, broadly rounded anteriorly; maxillary reaching vertical from hinder margin of pupil, nearly ⅓ length of head. Teeth minute scarcely perceptible without the use of a lens, those in upper jaw in a single series. Mandible with a close set outer series of teeth, separated by an interval from an inner narrow band of still smaller teeth. Interorbital space narrow, less than diameter of pupil. Isthmus wide, the gill slits extending little below base of pectorals. Scales minute and cycloid anteriorly and on belly, becoming larger posteriorly; on sides they are everywhere ctenoid behind the middle of spinous dorsal; belly wholly scaled; nape scaled forward nearly to orbits, but with a narrow median naked streak running back to front of dorsal; breast and sides of head naked. Dorsal fins not connected. First 4 spines filamentous, the longest longer than head, reaching when depressed to base of third ray of soft dorsal. Soft dorsal and anal similar, not high, the last rays not extending beyond the base of caudal; caudal lanceolate, much longer than head; pectorals and ventrals about equal, reaching vent. Color nearly uniform light olive, with minute darker punctulations which sometimes form darker margins to the scales; an oblique dusky streak on opercle; 3 or 4 oblique obscure dark cross bars on spinous dorsal, and 4 or 5 on tail; ventrals with white pigment. Length 2 inches. San Juan Lagoon, west coast of Mexico. (Gilbert.) (μικρός, small; ὀδούς, tooth.)

Gobius microdon, GILBERT, Proc. U. S. Nat. Mus. 1891, 554, San Juan Lagoon, north of Rio Ahomé, Mexico. (Coll. Gilbert.)

2554. GOBIUS SMARAGDUS, Cuvier & Valenciennes.

(ESMERALDA.)

Head 4; depth 5¼ to 5⅘; eye 4 to 5. D. VI–11 or 12; A. 11 or 12; scales 39 to 42. Body moderately elongate, compressed; head not compressed; the cheeks tumid; the snout short and abruptly decurved; mouth large, little oblique; lower jaw slightly inferior; maxillary reaching to below pupil or to posterior marg f orbit, $2\frac{1}{3}$ to $2\frac{1}{4}$ in head; outer row of teeth on upper jaw enlarged; th arrow band of teeth back of this row separated from it by a space; t h on lower jaw in a band, subequal. Scales cycloid anteriorly, becomin rger and ctenoid posteriorly. Caudal $2\frac{1}{4}$ to $2\frac{1}{4}$ in body. Male, light oliv with dark-olive blotches; body and head with many conspicuous round cr o olored spots, each surrounded by a dusky ring, these smaller than pu; and most distinct on head; snout with dusky streaks; dorsal and caudal nly barred; pectoral crossed with dark wavy lines, dusky at base; anal ventrals dusky; a small dark spot at base of caudal; a shining deep-gr spot inside the mouth in life. Female, plain olivaceous, nearly or quite i aculate. West Indies, south to Rio Janeiro, north to St. Augustine, Fl (Dr. Oliver P. Hay), and to Charleston (C. H. Gilbert); specimens be us from Marco Island, Florida (J. A. Henshall). (ὁμάραγδος, eme , from the bright-green spot on the tongue.)

Gobius smaragdus, CUVIER & VAL IENNES, Hist. Nat. Poiss., XII, 120, 1837, Cuba ; JORDAN & EIGENMANN, *l. c.*, 497.
Smaragdus valenciennei, POEY, M orias, II, 280. 1861. Cuba.
Gobionellus smaragdus, POEY, Sy nis, 394, 1868 ; POEY, Enumeratio. 126, 1876.

2555. GOBI STRIGATUS, O'Shaughnessy.

Head $3\frac{1}{4}$; depth 5; eye 3 horter than the rounded snout. D. VI-12; A. 11 or 12; scales 53-13. dy elongate, compressed posteriorly; head little compressed; maxillar reaching to below middle of eye; teeth small, the outer a little enlarged; rsal spines all shorter than head, not filamentous. Head naked; an ior half of body covered with ctenoid scales, those on nape much reduce in size. Two violet stripes from mouth to eye, 8 or 9 violet bars on , 3 or 4 on caudal; second dorsal spotted. (Steindachner.) Coast of S nam. (*strigatus*, striped.)

Gobius strigatus, O'SHAUGHNESSY un. Mag. Nat. Hist., series 4, XV, 1875, 145, Surinam.
Gobius kraussi, STEINDACHNER, Io . Beiträge, VIII, 16, 1879, Surinam ; JORDAN & EIGENMANN, *l. c.*, 497.

2556. GO US SAGITTULA (Günther).

Head $4\frac{1}{4}$ to 5 in length to se of caudal; depth 6 to 8; eye $4\frac{1}{4}$ to $5\frac{1}{4}$. D. VI-13 or 14; A. 13; scales al t 66 in longitudinal, 15 in transverse series, counted just below space veen the two dorsals. Body slender, tapering pretty regularly from m lle of first dorsal to caudal, most compressed posteriorly, depth about u rm from head to origin of second dorsal. Head short, depressed, an road; mouth large, nearly horizontal, the maxillary in adults $2\frac{3}{4}$ in ead, reaching beyond middle of eye; distance between maxillaries their posterior ends greater than their length; eye about $\frac{2}{7}$ the bon nterorbital space. Teeth in a narrow band in each jaw, those in low jaw uniform, the outer series in upper jaw considerably enlarged and parated by an interspace from the inner band. Pseudobranchiæ we leveloped. Gill rakers short and flexible. Longest dorsal spine about ead; distance between dorsals less than diameter of eye; pectorals   n head, their tips reaching past middle of spinous dorsal; ventrals a equaling pectorals, reaching more than halfway to origin of anal; equal and opposite to the second dorsal, but slightly lower; caudal greatly elongate, more than $\frac{1}{2}$ head and body in largest specimens, 2¼ in smaller ones, its relative length increasing with age. Head scaleless, predorsal region with small scales; body covered with close-set ctenoid scales, small and greatly crowded anteriorly, toward the caudal fin growing gradually larger and more strongly ctenoid. General color light yellowish, palest below, upper parts darker; middle of sides with 5 elongate black blotches, most distinct in the young; the first under first dorsal, second under origin of second dorsal, the third, which is sometimes almost double, at about middle of second dorsal, the fourth near its posterior end, and the last at base of caudal; a large black spot upon each shoulder just above origin of pectoral fin; head plain; lips and maxillary dark; opercle with a dark blotch; basal portion of dorsal fins with dark lines formed of spots; anal unmarked; pectorals with cross lines formed of dots; ventrals plain; caudal crossed by numerous narrow dark bars. Reaching a length of 8 inches. Gulf of California and neighboring waters south to Panama; very common in lagoons and mouths of rivers. The types of *sagittula* are evidently the young, those of *longicauda* the adults of the same species. (diminutive of *sagitta*, arrow.)

Buctenogobius sagittula, GÜNTHER, Proc. Zool. Soc. Lond 1861. 3, West coast Central America, young individuals; GÜNTHER, Cat. Fishes i. 555, 1861; GÜNTHER, Fishes of Centr. Amer., 389, 1869.
Gobius longicauda, JENKINS & EVERMANN, Proc. U. S. Nat Mus 1888, 146, adult examples, Guaymas. (Coll. Evermann & Jenkins. Type, No.83
Gobius sagittula, JORDAN & EIGENMANN, l. s., 497.

2357. GOBIUS HASTATUS, Gir

(EMERALD FISH; SHARP-TAILED GOBY)

Head 4½ to 5; depth 6 to 7⅓. D. VI-14; A. 14 or 15; scales 60; vertebræ 11+15. Body compressed, extremely elongate; depth nearly equal throughout; head short, compressed, deeper than wide; mouth wide, oblique, the jaws equal; maxillary in adult reaching to below posterior border of eye; lower jaw very thin and flat; teeth in each jaw small, subequal, those in the upper jaw in a single series, those in the lower jaw in a narrow band; outer teeth somewhat movable; scales anteriorly small, cycloid, and embedded, those behind larger and ctenoid; the scales larger than in *Gobius oceanicus*; a few scales on upper anterior corner of opercle, but without the large patch seen in *G. oceanicus*; dorsal fins high, some of the spines filamentous and longer than the head; caudal very long and filamentous, 2 to 2⅓ in body; pectoral slightly longer than head or than ventrals, none of its rays silk-like. A single specimen from Ceylon belongs to this species, which appears to be characterized by a longer head (5 in length, 7 in total), by the much larger scales (6 in a lateral line), by the obsolescence of the patch of scales on opercle and by slightly different coloration. This may be really only the extreme of variation of *G. oceanicus*, with which species most authors have hitherto confounded it. The two need detailed comparison. Coast of Texas. (*hastatus*, spear-like.)

Gobionellus hastatus, GIRARD. Proc. Ac. Nat. Sci. Phila 1858. 168. St. Josephs Island, Texas; GIRARD, U. S. and Mex. Bound. Surv., 25, pl. 12 gs. 7 and 8, 1859.
Gobius lanceolatus, GÜNTHER, Cat., III, 50, 1861, and of authors; not of BLOCH.

pupil or to posterior margin of orbit, 2¼ to 2¼ in head; outer row of teeth on upper jaw enlarged; the narrow band of teeth back of this row separated from it by a space; teeth on lower jaw in a band, subequal. Scales cycloid anteriorly, becoming larger and ctenoid posteriorly. Caudal 2¼ to 2¼ in body. Male, light olive, with dark-olive blotches; body and head with many conspicuous round cream-colored spots, each surrounded by a dusky ring, these smaller than pupil and most distinct on head; snout with dusky streaks; dorsal and caudal plainly barred; pectoral crossed with dark wavy lines, dusky at base; anal and ventrals dusky; a small dark spot at base of caudal; a shining deep-green spot inside the mouth in life. Female, plain olivaceous, nearly or quite immaculate. West Indies, south to Rio Janeiro, north to St. Augustine, Florida (Dr. Oliver P. Hay), and to Charleston (C. H. Gilbert); specimens before us from Marco Island, Florida (J. A. Henshall). (σμάραγδος, emerald, from the bright-green spot on the tongue.)

Gobius smaragdus, CUVIER & VALENCIENNES, Hist. Nat. Poiss., XII, 120, 1837, Cuba ; JORDAN & EIGENMANN, *l. c.*, 497.
Smaragdus valenciennei, POEY, Memorias, II, 280, 1861, Cuba.
Gobionellus smaragdus, POEY, Synopsis, 394, 1868; POEY, Enumeratio, 126, 1876.

2555. GOBIUS STRIGATUS, O'Shaughnessy.

Head 3¾; depth 5; eye 3½, shorter than the rounded snout. D. VI-12; A. 11 or 12; scales 53–13. Body elongate, compressed posteriorly; head little compressed; maxillary reaching to below middle of eye; teeth small, the outer a little enlarged; dorsal spines all shorter than head, not filamentous. Head naked; anterior half of body covered with ctenoid scales, those on nape much reduced in size. Two violet stripes from mouth to eye, 8 or 9 violet bars on side, 3 or 4 on caudal; second dorsal spotted. (Steindachner.) Coast of Surinam. (*strigatus*, striped.)

Gobius strigatus, O'SHAUGHNESSY, Ann. Mag. Nat. Hist., series 4, XV, 1875, 145, Surinam.
Gobius kraussi, STEINDACHNER, Ichth. Beiträge, VIII, 16, 1879, Surinam ; JORDAN & EIGENMANN, *l. c.*, 497.

2556. GOBIUS SAGITTULA (Günther).

Head 4½ to 5 in length to base of caudal; depth 6 to 8; eye 4¾ to 5¼. D. VI-13 or 14; A. 13; scales about 66 in longitudinal, 15 in transverse series, counted just below space between the two dorsals. Body slender, tapering pretty regularly from middle of first dorsal to caudal, most compressed posteriorly, depth about uniform from head to origin of second dorsal. Head short, depressed, and broad; mouth large, nearly horizontal, the maxillary in adults 2⅔ in head, reaching beyond middle of eye; distance between maxillaries at their posterior ends greater than their length; eye about ⁴⁄₇ the bony interorbital space. Teeth in a narrow band in each jaw, those in lower jaw uniform, the outer series in upper jaw considerably enlarged and separated by an interspace from the inner band. Pseudobranchiæ well developed. Gill rakers short and flexible. Longest dorsal spine about ⅔ head; distance between dorsals less than diameter of eye; pectorals 1¼ in head, their tips reaching past middle of spinous dorsal; ventrals about equaling pectorals, reaching more than halfway to origin of anal; anal equal and opposite to the second dorsal, but slightly lower; caudal fin greatly elongate, more than ½ head and body in largest specimens, 2⅓ in smaller ones, its relative length increasing with age. Head scaleless, predorsal region with small scales; body covered with close-set ctenoid scales, small and greatly crowded anteriorly, toward the caudal fin growing gradually larger and more strongly ctenoid. General color light yellowish, palest below, upper parts darker; middle of sides with 5 elongate black blotches, most distinct in the young; the first under first dorsal, second under origin of second dorsal, the third, which is sometimes almost double, at about middle of second dorsal, the fourth near its posterior end, and the last at base of caudal; a large black spot upon each shoulder just above origin of pectoral fin; head plain; lips and maxillary dark; opercle with a dark blotch; basal portion of dorsal fins with dark lines formed of spots; anal unmarked; pectorals with cross lines formed of dots; ventrals plain; caudal crossed by numerous narrow dark bars. Reaching a length of 8 inches. Gulf of California and neighboring waters south to Panama; very common in lagoons and mouths of rivers. The types of *sagittula* are evidently the young, those of *longicauda* the adults of the same species. (diminutive of *sagitta*, arrow.)

Euctenogobius sagittula, GÜNTHER, Proc. Zool. Soc. London 1861, 3, West coast Central America, young individuals; GÜNTHER, Cat. Fishes, III, 555, 1861; GÜNTHER, Fishes of Centr. Amer., 389, 1869.

Gobius longicauda, JENKINS & EVERMANN, Proc. U. S. Nat. Mus. 1888, 146, adult examples, Guaymas. (Coll. Evermann & Jenkins. Type, No. 39636.)

Gobius sagittula, JORDAN & EIGENMANN, *l. c.*, 497.

2557. GOBIUS HASTATUS, Girard.

(EMERALD FISH; SHARP-TAILED GOBY.)

Head 4⅓ to 5; depth 6 to 7⅔. D. VI–14; A. 14 or 15; scales 60; vertebræ 11+15. Body compressed, extremely elongate; depth nearly equal throughout; head short, compressed, deeper than wide; mouth wide, oblique, the jaws equal; maxillary in adult reaching to below posterior border of eye; lower jaw very thin and flat; teeth in each jaw small, subequal, those in the upper jaw in a single series, those in the lower jaw in a narrow band; outer teeth somewhat movable; scales anteriorly small, cycloid, and embedded, those behind larger and ctenoid; the scales larger than in *Gobius oceanicus;* a few scales on upper anterior corner of opercle, but without the large patch seen in *G. occanicus;* dorsal fins high, some of the spines filamentous and longer than the head; caudal very long and filamentous, 2 to 2⅔ in body; pectoral slightly longer than head or than ventrals, none of its rays silk-like. A single specimen from Ceylon belongs to this species, which appears to be characterized by a longer head (5 in length, 7 in total), by the much larger scales (60 in a lateral line), by the obsolescence of the patch of scales on opercles, and by slightly different coloration. This may be really only the extreme of variation of *G. oceanicus*, with which species most authors have hitherto confounded it. The two need detailed comparison. Coast of Texas. (*hastatus*, spear-like.)

Gobionellus hastatus, GIRARD, Proc. Ac. Nat. Sci. Phila. 1858, 168, St. Josephs Island, Texas; GIRARD, U. S. and Mex. Bound. Surv., 25, pl. 12, figs. 7 and 8, 1859.

Gobius lanceolatus, GÜNTHER, Cat., III, 50, 1861, and of authors; not of BLOCH.

2558. GOBIUS OCEANICUS, Pallas.

(ESMERALDA; ENDORMI ÉMERAUDE; BACALHAO SABARA.)

Head $4\frac{1}{8}$ to 6; depth $6\frac{1}{6}$ to $8\frac{1}{2}$; eye 4 to 5 in head; ventral 6 to $6\frac{1}{2}$; pectorals $5\frac{7}{8}$ to $6\frac{1}{2}$. D. VI–14. A. I, 14 or 15; scales about 65. Body extremely elongate; head very short; upper part of opercle scaled, head otherwise naked. Scales on body very small, becoming much larger behind. All the dorsal spines more or less filamentous; caudal fin nearly half length of rest of body. Skull behind eye broad and short, its length $1\frac{1}{2}$ in width, no decided ridges nor crests; lateral crests large and stout behind, minute forward; interorbital area narrow, deeply grooved, with a median ridge. Color in spirits, reddish olive; a distinct, round, blackish blotch below spinous dorsal, twice as large as orbit; an indistinct dusky shade along middle of sides, terminating in a dusky blotch on base of caudal; middle of sides with a series of marks, formed by very veiny lines widely diverging backward; a similar narrow line from eye to maxillary, and 1 from eye backward to upper angle of preopercle; evident traces of the emerald spot at base of tongue; 2 small dark spots on first dorsal spine; spinous dorsal dusky, with a light and dusky streak at base; soft dorsal dusky, a light (bluish in life) area behind each ray; anterior rays barred with light and dark; anal and ventrals whitish (probably blue in life), the ventrals without dark markings; pectorals dusky, the base lighter, and with some indistinct dusky bars; a dusky half bar on the upper part of the axil; base of tongue tuberculate, and shining with bright blue and green reflections like a precious stone (hence the names *smaragdus*, *esmeralda*, etc.), this color fading in spirits. Vertebræ elongate, $11+15=26$. Length a foot. South Atlantic and Gulf coasts of the United States and southward through the West Indies; not rare, perhaps intergrading with the preceding. Here described from a specimen 11 inches long, taken by Dr. Gilbert in Charleston Harbor. (*oceanicus*, ocean.)

Gobius cauda longissima acuminata, GRONOW, Zooph., 82, No. 277, pl. 4, fig. 4, 1763, locality unknown.

Gobius oceanicus, PALLAS, Spicilegia, VIII, 4, 1769, locality unknown; after GRONOW; JORDAN & EIGENMANN, *l. c*, 497.

Gobius lanceolatus, BLOCH, Fische Deutschlands, II, 8, pl. 38, fig. 1, 1783, Martinique, figure probably from PLUMIER; CUVIER & VALENCIENNES, Hist. Nat. Poiss., XII, 114, 1837; POEY, Synopsis, 393, 1868.

Gobius bacalaus, CUVIER & VALENCIENNES, Hist. Nat. Poiss., XII, 119, 1837, Surinam (Coll. Le Valliant); Cayenne (Coll. Richard); Cuba (Coll. Poey).

Gobionellus oceanicus, JORDAN & GILBERT, Proc. U. S. Nat. Mus. 1882, 613; JORDAN & GILBERT, Synopsis, 636, 1883.

Subgenus LYTHRYPNUS, Jordan & Evermann.

2559. GOBIUS DALLI, Gilbert.

Head $3\frac{3}{8}$; depth $4\frac{1}{3}$. D. VI–17; A. 14; scales 40. Body short, compressed, resembling *Microgobius*. Head high, mouth moderate, very oblique; upper pectoral rays normal; scales ctenoid, of moderate size; anterior dorsal spines much produced. Mouth very oblique, the maxillary reaching vertical from front of pupil, 2⅓ in length of head. Snout short, ⅔ diameter of orbit, which is 3 in head. Jaws with an outer series of long, distant, canine-like teeth, and an inner series or a narrow band of minute teeth. Dorsal spines 6, the 2 anterior greatly elongate, not free, in our largest specimen extending beyond middle of soft dorsal; membrane from last dorsal spine reaching to, or nearly to, base of first soft ray; soft dorsal rather high, the fin long; caudal rounded, less than length of head; ventrals free from belly, fully united; pectorals short, the upper rays not free nor silk-like. Scales of moderate size, ctenoid, covering entire trunk, with possible exception of the nape; the scales are readily caducous, and are lacking on nape and frequently on anterior third of body in our specimens. Color light coral red, anteriorly with 4 to 6 narrow blue bands not reaching ventral outline, the posterior ones growing narrower and fainter; a blue streak upward and backward from each orbit, the 2 uniting on occiput; a transverse interorbital bar, a continuation of which encircles the orbit anteriorly; below orbit, a blue bar consisting of 2 portions, 1 running downward and obliquely backward, the other upward and backward; in the largest specimen a blue streak runs from occiput along profile to front of dorsal; the first blue bar runs from nape obliquely downward and forward, ending on opercle; the second vertically downward from front of spinous dorsal, the third under middle of spinous dorsal, the remaining bars under soft dorsal; fins unmarked. Several small specimens, the largest 1 inch long, from *Albatross* Station 3001, in 33 fathoms. A single slightly larger example dredged by Dr. W. H. Dall, in about 35 fathoms, off Catalina Harbor, California. Probably the type of a distinct genus distinguished by the many-rayed fins and the form of the body and head. (Named for its discoverer, William Healey Dall.)

Gobius dalli, GILBERT, Proc. U. S. Nat. Mus. 1890, 73, Albatross Station 3001, Lower California (Coll. *Albatross*); Catalina Harbor (Coll. W. H. Dall).

814. GARMANNIA, Jordan & Evermann.

(HALF-NAKED GOBIES.)

Garmannia, JORDAN & EVERMANN, Proc. Cal. Ac. Sci. 1895, 495, pl. 49 (*paradoxus*).
Enypnias, JORDAN & EVERMANN, new subgenus (*seminudus*).

Anterior half of body naked; posterior half covered with moderate or small scales; teeth rather strong, unequal, usually 2 small curved canines in front. Very small gobies. Otherwise essentially as in *Gobius*. ("Named for Mr. Samuel Garman, the accomplished ichthyologist of the Museum of Comparative Zoology at Cambridge, Mass., in recognition of his important contributions to ichthyology.")

GARMANNIA:
a. Scales moderate.
b. Scales ctenoid, 13 or 14 series developed; first dorsal spine filamentous; D. VI-11; A. 9. Body rather robust, the depth about 4⅔ in length; the head 3½; lower jaw with 2 curved canines. PARADOXA, 2560.
bb. Scales smaller, 17 series developed; depth 4⅔ in length. D. VI-10; A. 8; first dorsal spine not filamentous; lower jaw with small canines.
HEMIGYMNA, 2561.

ENYPNIAS (ἐνύπνιος, in one's sleep):

, *aa.* Scales excessively minute; body slender, the depth 6 in length. D. VI-15; A. 10;
dorsal spines not filamentous; lower jaw with 2 small curved canines in front.
SEMINUDA, 2562.

Subgenus GARMANNIA.

2560. GARMANNIA PARADOXA (Günther).

Head about 3½ (4¼ in total); depth about 4⅗ (5¾ with caudal). D. VI-11;
A. 9; scales 14. Head nearly as broad as high, its width being rather
more than ½ of its length. Eyes rather close together, of moderate size.
Snout obtuse, rounded, as long as the eye; cleft of the mouth slightly
oblique, with the jaws equal in length, and with maxillary extending to
below middle of the eye. Teeth in villiform bands; 2 curved canine
teeth on each side of the lower jaw. Head and trunk entirely naked to
between second dorsal and anal, the remainder covered with ctenoid scales
of moderate size, 9 or 10 of them in 1 of the anterior transverse series.
First dorsal spine elongate, filiform, sometimes extending to the base of
the caudal; caudal rounded, shorter than head; none of pectoral rays
silk-like; ventral terminating at a great distance from vent. Blackish
in spirits; caudal and ventral fins black, dorsal filament whitish. (Günther.) Panama to Mazatlan; scarce. Our single specimen from the estuary
at Mazatlan differs somewhat from Dr. Günther's account. It is thus
described: Head 3½; depth 4¼. D. VI-11; A. 9; scales 12; eye 4 in
head; snout 4¼; pectoral 1⅓ in head; dorsal spine 1⅓. Form of *Gobiosoma bosci*. Body compressed; head broad and depressed, with tumid
cheeks; snout not very blunt, short, oblique-truncate; eyes rather large,
high, the maxillary not produced, extending to their posterior margin;
mouth large, oblique; lower jaw heavy, slightly projecting; teeth strong;
gill openings narrow, not wider than base of pectoral. First dorsal rather
high, the first spine filamentous, reaching past soft dorsal; other fins low.
Head and anterior half of body to front of soft dorsal naked; scattering
scales coming in above, 12 rows of imbricated slightly ctenoid scales
along median line of caudal peduncle and forward to middle of soft dorsal,
the scaled area about as long as head, the upper parts better scaled than
lower. No flaps on shoulder girdle. Olivaceous, with 7 or 8 dark cross
shades, 2 on head, 1 across gill openings, 1 behind pectoral, and a broad 1
below soft dorsal; dorsals dusky, the filamentous ray pink; lower half
of soft dorsal yellowish, upper dusky; lower fins black; caudal dusky; a
dark speck at angle of opercle; skin everywhere punctate with black; a
pale olive bar at base of caudal. Skull without median crest; interorbital
space not concave; head not very abruptly widened behind eyes. Pacific
coast of Mexico and Central America. One specimen, 1¼ inches long,
recently obtained on muddy bottoms among the mangroves lining the
estuary at Mazatlan. (*paradoxus*, paradox.)

Gobius paradoxus, GÜNTHER, Proc. Zool. Soc. Lond. 1861, 3, west coast Central America;
GÜNTHER, Cat., III, 549, 1861; JORDAN & EIGENMANN, Proc. U. S. Nat. Mus. 1886, 498.
Garmannia paradoxa, JORDAN, Proc. Cal. Ac. Sci. 1895, 497, pl. 59.

2561. GARMANNIA HEMIGYMNA (Eigenmann & Eigenmann).

Head $3\frac{2}{3}$ ($4\frac{1}{4}$ in total); depth $4\frac{2}{3}$. D. VI-10; A. 8; scales smaller than in *Garmannia paradoxa*, 17-7. Body compressed, depressed anteriorly, the greatest depth in this specimen being at origin of anal and second dorsal fins. Head wider than deep, rounded; profile much decurved from eye to mouth as in *paradoxa*; eye perfectly round, smaller than in *paradoxa*, $1\frac{1}{3}$ in rounded snout, 5 in head; interorbital space scarcely wider than orbit; mouth somewhat oblique, larger than in *paradoxa*; maxillary reaching beyond posterior rim of orbit; lower jaw slightly shorter than upper; teeth in upper jaw in a band, the outer series remote, and the teeth several times as large as in the inner row, all more or less movable; teeth in lower jaw similar, a recurved canine on each side near the front. Scales very weakly ctenoid, covering only the sides of the posterior half of body, not extending quite to base of dorsal or anal fins even at their posterior insertion; the upper and lower edges of the caudal peduncle likewise free from scales, the scaly region, however, widest on peduncle and tapering forward to the central point opposite beginning of anal, where the scales are smallest. First spine of the dorsal not elongate as in *G. paradoxa*, $1\frac{1}{3}$ in head, the third, fourth, and fifth spines slightly exceeding the first in height, equaling the posterior rays of soft dorsal, which are little higher than the anterior rays of the soft dorsal; caudal rounded, about 4 in length of body, $1\frac{1}{3}$ in head; ventral not reaching vent, $1\frac{1}{3}$ in head; pectorals rounded, rather short and broad, $1\frac{1}{4}$ in head. Color light olivaceous, without distinct markings, everywhere with minute dark punctulations; 8 faint cross bars from dorsal to middle of sides, which, close under dorsal fins, are formed of 2 blackish dots; 8 black dots along lateral line, the last at base of caudal; fins all smutty, the pectoral lightest, white on its anterior half, 2 dusky spots at its base; opercle ashy; a light bar at base of caudal; iris blackish blue, a short straight streak of same color from eye to upper lip; an irregular bluish mark on cheeks formed of punctulations closely crowded. West Indies, exact locality unknown; taken with the dredge. ($\dot{\eta}\mu\iota$, half; $\gamma\nu\mu\nu\acute{o}\varsigma$, naked.)

Gobius hemigymnus, EIGENMANN & EIGENMANN, Proc. Cal. Ac. Sci. 1888, 66, dredged in the West Indies.

Subgenus ENYPNIAS, Jordan & Evermann.

2562. GARMANNIA SEMINUDA (Günther).

Head 4; depth 6. D. VI-15; A. 10. Head and anterior portion of trunk naked; sides with exceedingly small scales, becoming somewhat larger posteriorly. Head with the cheeks swollen, depressed, broader than high, its width $\frac{2}{3}$ length. Eyes close together, directed upward, of moderate size; snout obtuse, as long as the eye; cleft of the mouth slightly oblique, with the jaws equal anteriorly, and with the maxillary extending to below the middle of the eye. Teeth in villiform bands, the anterior of the lower jaw slightly enlarged; 2 small curved canine teeth on each side of lower jaw. Dorsal fins rather low, the hind part of the spinous dorsal scarcely lower than anterior; caudal rounded, as long as pectoral; none of pectoral

3030——63

rays silk-like; ventral rather short, terminating at a great distance from vent. Blackish; fins and sides of head dotted with black; ventrals black. (Günther.) Panama; not seen by us; probably the type of a distinct genus. (*seminudus*, half-naked.)

Gobius seminudus, GÜNTHER, Proc. Zool. Soc. London 1861, 3, west coast Central America; GÜNTHER, Cat., III, 554, 1861; JORDAN & EIGENMANN, Proc. U. S. Nat. Mus. 1886, 498.

815. AWAOUS, Steindachner.

Awaous,* STEINDACHNER, Verh. Mat. Phys. Naturw. 1860, 289; after *les Awaous* of CUVIER & VALENCIENNES (*ocullaris*, etc).

Chonophorus, POEY, Memorias, II, 274, 1861 (*buccelentus = taiasica*).

Awaous, BLEEKER, Esquisse d'un Syst. Nat. Gobioides, 320, 1874 (*ocellaris*); after *les Awaous* of CUVIER & VALENCIENNES.

Inner edge of shoulder girdle with 2 or more conspicuous dermal flaps; preorbital region very long; premaxillary and maxillary strong; lips thick; scales rather small, ctenoid, 40 to 80 in a longitudinal series; interorbital groove with a conspicuous median crest; otherwise essentially as in *Gobius*. The species reach a large size and are confined to the fresh waters of the tropics of America and the Hawaiian Islands. The Asiatic species of similar habit have much larger scales and seem to form a distinct genus, *Rhinogobius*, Gill. The physiognomy in each is peculiar, the snout being long and convex. (*Awaou*, a Hawaiian name.)

> *a.* Scales about 53, little crowded anteriorly, 21 before dorsal on nape; depth 5⅘ in length; head 4; eyes placed high, interorbital area equal to diameter of eye; mouth horizontal; maxillary extending to middle of eye, 2¼ in head, lower jaw more flat than in *A. taiasica;* teeth small, in narrow bands, those of the outer row above enlarged, some large teeth in band of lower jaw. D. VI-I, 12; A. I, 10. Uniform yellowish in spirits. FLAVUS, 2563.

> *aa.* Scales 60 to 70, crowded anteriorly, about 30 scales before the dorsal on nape; body compressed posteriorly, rather depressed anteriorly; greatest depth 5¼ in length; head 3¼ in length. Olivaceous, a series of irregular, roundish blotches along middle of sides; narrow dark streaks radiating from eye; a blackish streak running across upper margin of opercle and extending obliquely across base of upper pectoral rays; belly white; dorsal and caudal more or less distinctly barred with wavy blackish lines.

>> *b.* About 15 scales between second dorsal and base of anal. NELSONI, 2564.

>> *bb.* About 21 scales between second dorsal and base of anal. TAIASICA, 2565.

> *aaa.* Scales 76 to 82; 24 scales between second dorsal and anal; head as broad as high; depth of body 6⅘ in length; head 4; head flat above, snout elongate, upper profile oblique; eye ⅙ of head, equals interorbital area (in adult); mouth horizontal; lower jaw included; maxillary reaching to below anterior margin of eye; teeth of the outer series enlarged; canine teeth none; scales ctenoid, those on nape and anterior part of body very small; head naked; dorsal fins lower than body, none of the spines produced; caudal rounded, 7 in length of body. Yellowish olive; back and sides reticulated with blackish; head, dorsal, caudal, and pectoral fins dotted with blackish, the spots forming streaks on second dorsal; 6 cross series of dots on the caudal; an irregular small blackish spot on the upper part of the root of pectoral. D. VI-11; A. 11; scales about 80. MEXICANUS, 2566.

* The name "Les Awaous," given to this group by Valenciennes, was a French plural, not a generic appellation, and if used as the name of a genus must be dated from its use in that sense by Steindachner or Bleeker. The Hawaiian type of "*Awaous*" agrees with the American species (*Chonophorus*) in the character of the flaps on the shoulder girdle, as well as in general appearance. The Asiatic genus, *Rhinogobius*, Gill (*similis*), seems to be very close to *Chonophorus*, but the scales are larger, 28 in the lateral series.

2563. AWAOUS FLAVUS (Cuvier & Valenciennes).

Head 4; depth $5\frac{2}{3}$ to $6\frac{1}{3}$; eye equal to the interorbital width, placed high. D. VI–I, 12; A. I, 10; scales about 53 to 55, little crowded anteriorly, 21 before the dorsal. Mouth horizontal, maxillary extending to middle of eye, $2\frac{1}{3}$ in head; lower jaw flatter than in *Awaous taiasica;* teeth small, in narrow bands, those of the outer row enlarged; some large teeth in band of lower jaw. Yellowish, with a row of faint ocellated spots along middle of sides; dorsal and caudal faintly barred; lines radiating from eye, a line along opercle halfway to pectoral, sometimes uniform blue-black. Rivers of Surinam and Brazil, south to Bahia. (*flavus,* yellow.)

Gobius flavus, Cuvier & Valenciennes, Hist. Nat. Poiss., XII, 60, 1837, Surinam; Günther, Cat. Fish., III, 13, 1861.
Chonophorus flavus, Jordan & Eigenmann, *l. c.*, 500; Eigenmann & Eigenmann, *l. c.*, 67.

2564. AWAOUS NELSONI, Evermann.

Head $3\frac{1}{3}$; depth 6; eye $5\frac{1}{4}$ in head; snout 3; maxillary $2\frac{2}{3}$. D. VI–11; A. 11; scales about 63. Body long, compressed and tapering posteriorly; head large, quadrate, mouth nearly horizontal, lower jaw included; snout abruptly decurved; top of head flat, the interorbital with a slight median groove with a thin, raised edge on each side; maxillary reaching about to vertical of anterior edge of pupil; teeth in bands on jaws very small, the outer somewhat enlarged; pectoral rays normal, the longest $1\frac{1}{3}$ in head; ventrals completely united, the disk free from belly, $1\frac{1}{3}$ in head. Dorsal fins separated by a space about $\frac{2}{3}$ diameter of eye; dorsal spines slender, weak, about $1\frac{2}{3}$ in head; soft dorsal and anal similar, each free from caudal; caudal fin rather short and rounded, its middle rays about $1\frac{1}{4}$ in head. Gill membranes broadly united to the isthmus; eyes moderate, high up, the interorbital width equal to the eye's diameter. Scales ctenoid, very small and irregularly crowded anteriorly, much larger posteriorly, about 15 rows counting from origin of soft dorsal downward and backward to the anal fin; head naked, but with slight indication of a few minute embedded scales on opercles. Color grayish; head mottled and blotched with dark; side with 7 or 8 black blotches, the largest under middle of pectoral fin; dorsals pale, crossed by several lines of black spots; caudal pale, with about 6 or 7 dark cross bars; ventrals and anal pale; pectorals pale, dusted with dark specks and with a small dark blotch at base of upper rays. Close to *A. taiasica,* but with broader interorbital, longer snout and larger scales on posterior part of body. Length 4 inches. Known only from fresh water at Rosario, Sinaloa, where 8 specimens were obtained July 27, 1897, by Mr. E. W. Nelson. (Named for Mr. Edward William Nelson, the well-known ornithologist, in recognition of his work upon the fishes of Illinois in 1876.)

Awaous nelsoni, Evermann, Proc. Biol. Soc. Washington, vol. XII, 1898, 3, fresh-water pools at Rosario, Sinaloa, Mexico. (Type, No. 48836, U. S. Nat. Mus.; cotypes, No. 533 U. S. Fish Comm., 5793 L. S. Jr. Univ. Mus., and 48837 U. S. Nat. Mus.)

2565. AWAOUS TAIASICA (Lichtenstein).

(GUAVINA HOYERA; ABOMA DE RIO.)

Head 3¼; depth 5¼; eye small, less than interorbital width (in adult), 3 in snout (twice in young), and about 7 in head. D. VI–11; A. 11; scales 60 to 70, crowded anteriorly, about 30 before dorsal fin, 21 between second dorsal and anal. Body compressed posteriorly, rather depressed anteriorly; head broader than deep. Distance from eye to mouth 3½ in head, the preorbital being much enlarged; mouth large, horizontal, maxillary extending to below anterior part of orbit in adult male, shorter in young; lower jaw included. Teeth of the upper jaw in 2 series, those in anterior series much enlarged and recurved; teeth of lower jaw in a narrow band, the outer series scarcely enlarged. Inner edge of the shoulder girdle with 2 or 3 rather long papillæ. Body covered with ctenoid scales, much reduced in size anteriorly; nape closely scaled, breast scaly, head naked. Dorsal fins less than depth of body, the spines scarcely filamentous, not as long as the soft rays; caudal rounded, shorter than the head; ventrals very broad and short, 1½ to 1⅘ in head, the rays very much branched. Skull rounded behind, with a very short crest in its middle; lateral crests high and thin, converging into 1 opposite the insertion of suprascapula, inner crests not meeting behind eye, the outer ones extending around orbit. A low, blunt ridge between the posterior corners of orbit, becoming much higher forward, continued as the ethmoid and ending abruptly some distance in advance of orbit. Teeth in upper jaw in a few series, those of outer series many times larger than the others, which are minute; those of lower jaw all alike small, in a band. Olivaceous, with a series of irregular, roundish blotches along middle of side, and narrow dark streaks radiating from eye; a blackish streak running across upper margin of opercle and extending obliquely across base of upper pectoral ray; belly white; dorsal and caudal more or less distinctly barred with wavy blackish lines. Length a foot or more. Extremely variable in form and coloration, as is the case with most widely distributed fresh-water fishes. Fresh waters of the West Indies and both coasts of Mexico, south to Brazil; common in Cuba, in Sinaloa, and about La Paz in Lower California, thence southward to Panama. (*taiasica*, Brazilian name of some other goby.)

Amore guacu, MARCGRAVE, Hist. Brasil., 166, 1648, Brazil.
Gobius taiasica, LICHTENSTEIN, Berl. Abhandl., 273, 1822, Brazil; not *Tajasica* MARCGRAVE.
Gobius banana, CUVIER & VALENCIENNES, Hist. Nat. Poiss., XII, 103, 1837, San Domingo; GÜNTHER, Cat., III, 59, 1861.
Gobius martinicus, CUVIER & VALENCIENNES, Hist. Nat. Poiss., XII, 105, 1837, Martinique.
Chonophorus bucculentus, POEY, Memorias, II, 275, 1861, Cuba.
Rhinogobius contractus, * POEY, Memorias, II, 424, 1861, Cuba; POEY, Enumeratio, 125, 1875

---

* The following are the characters assigned to *Awaous contractus* (Poey): Head 4; depth 5½; D. VI–11; A. 11; eye 7 in head; maxillary ceasing ⅓ an eye's diameter before eye. Head smaller than in *A. taiasica*. Greenish brown; the cheeks with brown lines; body with brown points; dorsals brownish, with brown longitudinal bands more numerous on the second; caudal with 7 brown bands, made of lanceolate spots on the rays; pectorals speckled; ventrals and anal rose color. Cuba (Poey); probably not different from *A. taiasica*; said to differ in the small mouth, which probably varies with age and sex.

Gobius dolichocephalus, COPE, Trans. Amer. Philos. Soc. Phila. 1869, 403, near Orizaba, Mexico.
Euctenogobius latus, O'SHAUGHNESSY, Ann. Mag. Nat. Hist., Series 4, XV, 1875, 146, Bahia. (Coll. Dr. Wucherer.)
Chonophorus taiasica, JORDAN & EIGENMANN, *l. c.*, 500.

2566. AWAOUS MEXICANUS (Günther).

Head 4; depth 6⅗; eye 8. D. VI–11; A. 11; scales 76 to 82, 24 between second dorsal and anal. Head as broad as deep, flat above, snout elongate, upper profile oblique; mouth horizontal, lower jaw included, maxillary reaching to below anterior margin of eye. Teeth of the outer series enlarged; no canine teeth. Scales ctenoid, those on nape and anterior part of body very small; head naked. Dorsal fins lower than depth of body, none of the spines produced; caudal rounded, 7 in length of body. Yellowish olive; back and sides reticulated with blackish; head, dorsal, caudal, and pectoral fins dotted with blackish, the spots forming streaks on second dorsal; 6 cross series of dots on caudal; an irregular, small blackish spot on the upper part of the base of the pectoral. (Günther.) Fresh-water streams of the eastern slope of Mexico; known to us only from Dr. Günther's description.

Gobius mexicanus, GÜNTHER, Cat., III, 61, 1861, Mexico.
Chronophorus mexicanus, JORDAN & EIGENMANN, *l. c.*, 501.

816. BOLLMANNIA, Jordan.

Bollmannia, JORDAN, Proc. U. S. Nat. Mus. 1889, 164 (*chlamydes*).

This genius differs from *Lepidogobius* by having no fleshy processes on inner edge of shoulder girdle, the interorbital area of skull narrower and without trace of median keel, and by very large ctenoid scales. From *Gobius* proper it is distinguished by the presence of 7 dorsal spines and by the presence of large scales on the cheeks. Species inhabiting the depths of the Pacific; not found in shoal waters as is the case with most other gobies. ("I have named this species in honor of my late colleague, Mr. Charles Harvey Bollman, whose untimely death, while engaged in the exploration of the rivers of Georgia, took place while this paper was passing through the press."—Jordan.)

a. * A conspicuous black spot on posterior portion of spinous dorsal. Body deep, the least depth of caudal peduncle greater than diameter of orbit.
  b. Filamentous dorsal spines very long, reaching beyond middle of soft dorsal when depressed. Lower caudal rays black; dorsal spot conspicuously ocellated. Eye large, 3 to 3¼ in head.   OCELLATA, 2567.
  bb. Filamentous dorsal spines shorter. Lower caudal rays not black, and dorsal spot not ocellated. Eye smaller, 3¾ to 4 in head.   CHLAMYDES, 2568.
aa. No black spot on spinous dorsal. Body slender, the depth ⅕ the length. Least depth of caudal peduncle not greater than diameter of orbit.
  c. Head large, 3 to 3¼ in length. No black spot at base of caudal. Fins low.
      MACROPOMA, 2569.
  cc. Head smaller, 3⅖ in length. A black spot at base of caudal. Fins higher.
      STIGMATURA, 2570.

* This analysis of species is taken from Gilbert, Proc. U. S. Nat. Mus. 1891, 555.

2567. BOLLMANNIA OCELLATA, Gilbert.

Head 3¼ to 3⅔ in length; depth 4½. D. VII–14 or 15; A. 14; scales 27. Very close to *Bollmannia chlamydes*, differing from the latter constantly in the following respects: The eye is larger, 3 to 3½ in head (3¾ to 4 in *chlamydes*); the filamentous rays of spinous dorsal are much longer, reaching in adults, when laid back, to or nearly to end of base of soft dorsal, 1½ to 1⅔ times length of head; rarely the filamentous dorsal rays are little more elongate than in *chlamydes*. Teeth in a narrow band in each jaw, the outer series in upper jaw, and both outer and inner series in lower jaw enlarged, but not canine-like; maxillary not reaching vertical from middle of pupil, ½ length of head; interorbital width less than ½ diameter of pupil; opercle short, its length being less than the diameter of the eye; pectorals nearly as long as head, a trifle more than length of ventrals, which scarcely reach vent; caudal much longer than head in adults, 7 or 8 scales before dorsal. Black spot on posterior part of spinous dorsal jet black, conspicuously ocellated with white; a black streak along lower margin of caudal, including several of the lower rays, and running from base to tip of fin; no dusky bars visible on sides in any of the types; fins dusky; membranes uniting outer rays of ventrals white instead of black, as in *chlamydes;* no black spot at base of caudal; branchiostegal membrane with a medial black streak; anal blackish. Numerous specimens from the northern part of the Gulf of California, at *Albatross* Stations 3031 and 3035, in 30 and 33 fathoms. This species may vary into the typical *chlamydes*, but the material before us does not justify us in so identifying it. (Gilbert.) (*ocellatus*, ocellated.)

Bollmannia ocellata, GILBERT, Proc. U. S. Nat. Mus. 1891, 555, Gulf of California.

2568. BOLLMANNIA CHLAMYDES, Jordan.

Head 3½ (5 to 5½ in total); depth 4½ (6⅔ to 7). D. VII–15; A. 15; scales in a longitudinal series about 28, 8 or 9 in a cross series at vent. Body rather robust, compressed; head large and heavy, its profile evenly curved; mouth very large, oblique, the lower jaw projecting; maxillary reaching to opposite pupil, 2⅓ to 2⅔ in head; teeth small, sharp, in several series, the outer, especially in lower jaw, somewhat enlarged; eye longer than snout, 3¾ to 4 in head; interorbital area very narrow, concave, its least width about ⅓ of eye or almost equal to pupil; scales very large, ctenoid; little reduced on breast and nape; about 8 before dorsal, where they are little smaller than on body; top and sides of head with large scales; scales on cheek in 4 rows; 2 rows on upper part of opercle; the scales on head lost in some of the specimens; dorsal spines slender, filamentous, fifth longest, 1½ in head; first 2 in head, last 3½ to 4; first soft dorsal ray 2⅔ in head, the antepenultimate longest and about equal to head; first anal ray equal to snout, the antepenultimate 1¼ in head; middle caudal rays very long, somewhat more than ½ length of body; pectorals 1¼ in head; ventrals 1½. Color olivaceous, darkest above; scales with a few black dots, some of the posterior occasionally dark edged; sides with 8 or 10 obscure dusky vertical bars, which are narrower than the interspaces, and in some specimens wholly obsolete; snout bluish; opercles with a dark shade; lips, gular region, and anterior branchiostegals very dark in males; upper part of spinous dorsal darkest, with a few lighter dark-edged oval spots, a well-marked black blotch between last 2 spines; soft dorsal dusky, usually with about 3 well-developed rows of lighter, dark-edged oval spots; anal dusky, crossed by 2 narrow bluish streaks; some of the last rays occasionally with a few spots similar to those on dorsal; caudal, pectorals, and ventrals dusky, tinged with blue; ventrals edged with pale. Length 4¼ inches. West coast of Colombia. Many specimens of this abundant species were dredged at *Albatross* Stations 2800 in 7 fathoms and 2805 in 51½ fathoms. (χλαμύδης, cloaked.)

Bollmannia chlamydes, JORDAN, Proc. U. S. Nat. Mus. 1889, 164, Pacific Ocean, off coast of Colombia, Station 2800, 8° 51′ N., 79° 41′ 30″ W., and Station 2805, 7° 56′ N., 79° 41′ 30″ W. (Type, No. 41158, U. S. Nat. Mus. Coll. *Albatross.*)

2569. BOLLMANNIA MACROPOMA, Gilbert.

Head 3 to 3½; depth 5. D. VII–14; A. 14; scales 28. Characterized by its slender form, low fins, large opercle, and comparatively plain coloration. Caudal peduncle correspondingly slender, its least height equaling diameter of eye. Head very large and heavy; opercle conspicuously larger than in *B. ocellata,* agreeing in this respect more nearly with *B. chlamydes;* dentition as in other species of the genus; eye large, 3¼ to 3⅓ in the head. Dorsal spines slender, comparatively little produced, the longest usually not reaching the base of the first ray of second dorsal, and never beyond the base of the second or third ray; soft dorsal and anal low, the posterior rays usually not reaching the rudimentary caudal rays when depressed, about ⅓ length of head; pectoral long, extending beyond front of anal; the ventrals to or nearly to vent; middle caudal rays produced as usual, varying in length; scales 8 to 10 in front of dorsal. Color in spirits, light brownish, the sides with 3 vertical dusky bars; spinous dorsal dusky, but without distinct black spot; caudal slightly dusky, with rather large elliptical light spots, as in *B. chlamydes,* the lower rays not black and no black spot at its base; ventrals blackish, including anterior membrane; second dorsal and anal dusky, without evident light spots; branchiostegal membranes sometimes slightly dusky, but not black. Many specimens from the Gulf of California just north of La Paz Bay, at *Albatross* Station 2996, in 112 fathoms. (Gilbert.) (μακρός, large; πῶμα, opercle.)

Bollmannia macropoma, GILBERT, Proc. U. S. Nat. Mus. 1891, 556, Albatross Station 2996, near La Paz, Lower California.

2570. BOLLMANNIA STIGMATURA, Gilbert.

Head short, 3⅔ in length; depth 5; least depth of caudal peduncle slightly less than diameter of eye; eye large, 2⅔ in head. D. VII–15; A. 14; lateral line 28. Dorsal spines filamentous, longer than in *B. macropoma,* the longest reaching base of fifth to seventh ray of second dorsal; posterior rays of second dorsal and anal often reaching base of median rays when depressed; pectorals not reaching beyond front of anal. Color almost uniform light brownish; lips black, the fins only slightly dusky, the caudal with elliptical light spots; a roundish dusky spot at base of caudal; branchiostegal membranes not black. Many specimens from the northern part of the Gulf of California, at *Albatross* Stations 3016 and 3017, in 76 and 58 fathoms. This species agrees with *Bollmannia macropoma* in its elongate form, comparatively low fins, and in the absence of a black spot on the spinous dorsal. It differs conspicuously in the very short head and narrow opercle, and in the presence of a black spot at base of tail. The eye is also larger and the fins higher. None of the specimens shows dusky bars on the sides, a conspicuous feature in *B. macropoma.* (Gilbert.) (στίγμα, spot; ούρά, tail.)

Bollmannia stigmatura, GILBERT, Proc. U. S. Nat. Mus. 1891, 556, Gulf of California, Albatross Stations 3016, 3017.

817. ABOMA, Jordan & Starks.

Aboma, JORDAN & STARKS, Proc. Cal. Ac. Sci. 1895, 497 (*etheostoma*).

This genus, allied to *Microgobius*, is distinguished by the large, ctenoid scales, which cover the body; head naked, rather long, pointed in profile; the mouth moderate, not very oblique; teeth rather strong. Dorsal spines more than 6, none of them filamentous; soft dorsal and anal short; no flaps on shoulder girdle. Cranium with a slight median crest. (The name *Aboma* is used by the Mexicans in Sinaloa as synonymous with goby.)

> a. Scales very large, 26 or 27; profile not very steep, the snout rather pointed.
> > b. Sides with a jet-black lateral band; caudal with dark cross bars; maxillary 3 in head.   ETHEOSTOMA, 2571.
> > bb. Sides with 4 oblique dark cross bars; a large dark spot at base of caudal; mouth larger, the maxillary 2 in head.   LUCRETIÆ, 2572.
> aa. Scales smaller, about 37; profile very steep, the snout rounded; sides with numerous pale cross bands with darker spots.   CHIQUITA, 2573.

2571. ABOMA ETHEOSTOMA, Jordan & Starks.

Head 3⅓; depth 5. D. VIII-11; A. 10; scales 26; longest dorsal spine 1¼ in head; eye 3; snout 4; maxillary 3. Body long and low, moderately depressed and pointed forward. Scales large, ctenoid behind, none on head, those on nape and belly much reduced. Mouth moderate, terminal, moderately oblique; the maxillary reaching middle of pupil; jaws subequal, or the lower a little the longer; teeth rather strong. No flaps on shoulder girdle. Cranium with a slight median crest. Interorbital ridge not hollowed out; skull not abruptly widened behind. Color olivaceous, side with a very broad jet-black lateral band, 3 times interrupted by silvery; caudal white, with 4 < shaped bands, growing progressively fainter behind; pectoral mottled gray, with a jet-black oblique crescent toward its base surrounding a large yellow spot; side of head with 4 round gray spots separated by black, the largest below eye, with a black streak before it; first dorsal jet-black, second mottled, the produced spine with yellowish; ventral and anal pale. A single small specimen, 1⅓ inches long, found in the mud on a shallow bottom in the Astillero at Mazatlan. (*Etheostoma*, a darter, which this species strongly resembles.)

Aboma etheostoma, JORDAN & STARKS, Proc. Cal. Ac. Sci. 1895, 498, pl. 50, Mazatlan. (Coll. Hopkins Expedition to Mazatlan.)

2572. ABOMA LUCRETIÆ (Eigenmann & Eigenmann).

Head 3⅓ in length; depth 5⅓. D. VII–10; A. 12; scales 28–8. Body slightly compressed posteriorly; head little wider than high; eye placed high, its diameter equaling length of snout, 4⅓ in head; profile little decurved; mouth large, oblique; maxillary extending below posterior margin of orbit, 2 in head; intermaxillary anteriorly on a level with center of pupil; teeth all recurved, large, those of upper jaw in a narrow band; teeth of outer and inner series enlarged, those of lower jaw similar, largest in front. No dermal flaps on shoulder girdle. Scales large, very weakly ctenoid, becoming cycloid and very much crowded above and below pectoral; head, breast, and anterior part of nape naked. As seen through a lens, these regions seem to be covered with minute embedded scales; this effect is, no doubt, due to light reticulations on a darker ground. Dorsal spines slender, not filamentous; caudal pointed, 3 in length of body; ventrals 1⅓ in head; pectorals longer than head. Color light brownish, with 4 oblique dark cross bars as wide as interspaces; 4 narrower transverse bars on nape and back; a large dusky spot at base of caudal; upper half of base of pectoral black; a black spot on opercle, margined below and behind with silvery; fins dusky. Pearl Island, Gulf of Panama; only 1 specimen known. (Named for Mrs. Lucretia M. Smith of San Diego, mother of Mrs. Eigenmann.)

2573. ABOMA CHIQUITA (Jenkins & Evermann).

Head 3⅓ to 3½; depth 4⅓ to 4¾. D. VII–11; A. 10; eye 4¾ in head in adult, 4 in young; scales 37–17. Body rather stout, compressed; head short, somewhat depressed, widened behind orbits; snout short and narrowly rounded; profile in front of eye very steep, less so to occiput, and nearly straight from there to caudal fin; eyes moderate, well up; interorbital space very narrow, less than eye; greatest width of head equaling greatest depth of body. Top of head, opercles, and space in front of dorsal naked, rest of body covered with small, strongly ctenoid scales, which increase in size upon the caudal peduncle. Spinous dorsal with its first spine filamentous in adult, much longer than head and reaching middle of soft dorsal, this filament wanting in young; distance from snout to origin of spinous dorsal a little more than ⅓ distance to base of caudal; second dorsal but slightly separated from spinous, its origin about midway of total length of fish; anal of about the same shape and size as soft dorsal, but beginning a little behind it; pectorals tapering, about equaling head in length, their tips not reaching origin of anal, but to origin of soft dorsal; ventrals united, free from belly, inserted behind pectorals, but their tips not reaching tips of pectorals. Teeth apparently in a single series, small and weak. Ground color pale yellowish, thickly mottled with fine punctulations of dark; about 7 pretty well-defined larger spots of dark brown along middle of side; 8 or 9 faint cross bars of lighter, a number of small light spots scattered irregularly over the sides; head dark; dorsal, anal, and ventral fins covered with fine black points; in some specimens the dorsals and anal quite dark; pectorals plainer; caudal similar to ventrals; "the cranium is depressed and flattish behind the orbits, without distinct median keel on occiput or on interorbital area. The form of the head is as in typical *Gobius*, the occiput abruptly widened behind the eyes; the ridges also similar, the orbital ridge bounding the orbit behind as well as above the eye and joining the temporal ridge laterally." (Gilbert MS.) Length 1 to 2 inches. Gulf of California; abundant. The original description from young examples, here corrected in accordance with Dr. Gilbert's notes on many adults taken by him at La Paz. (Spanish, *chiquito;* a diminutive of *chico,* a little one.)

Gobius chiquita, JENKINS & EVERMANN, Proc. U. S. Nat. Mus. 1888,146, Guaymas, Sonora.
(Type, No. 39634. Coll. Jenkins & Evermann.)

818. MICROGOBIUS, Poey.

Microgobius, POEY, Enumeratio, 127, 1875 (*signatus*).

Dorsal spines 7 or 8; scales very small, cycloid or weakly ctenoid, the body scaled anteriorly as well as posteriorly, the head naked, the nape, belly, and breast usually so. Inner edge of shoulder girdle without fleshy processes; body more or less compressed; mouth large, very oblique; the lower jaw conspicuous, teeth strong; interorbital groove with or without a median ridge. Vertebræ 11+15 or 16. ($\mu\iota\varkappa\rho\acute{o}\varsigma$, small; *Gobius.*)

a. Scales about 42. Body elongate, moderately compressed, the depth 4 to 5 in length; head long and large, rather sharp in profile, 3 to 3½ in body; eye longer than snout, 4 in head; mouth large, very oblique, the lower jaw strongly projecting; maxillary 1½ to 2¼ in head, extending to opposite middle of eye, or much beyond front of orbit; teeth in few series, the outer very long and slender, curved, the lower longest, none canine-like; scales small, some of them with short, thick teeth, those of anterior part of body not well developed; dorsal spines more or less filamentous, the third and fourth or fourth and fifth sometimes with long filaments; caudal pointed, about as long as head. Grayish olive, with rather sharply-defined markings of darker brown overlaid with orange in life; head with a pale bluish or gilt stripe from maxillary backward across suborbital region to upper edge of gill opening;. another pale gilt streak from snout along lower part of eye, another from angle of mouth upward and backward; rest of head dark; opercle with an oblique blackish bar; top of head and nape with dark marbling surrounded by paler reticulations; back with a series of black cross blotches mostly separated on the median line; 2 narrower dark vertical bars behind pectoral; middle line of side posteriorly with longitudinally oblong black blotches; besides these, numerous other blotches not regularly arranged; first dorsal with 2 or 3 oblique black bands; second dorsal pale, with about 4 series of black dots; caudal spotted with black; pectoral yellowish; ventral black, its center yellowish (male); anal pale. D. VII-15; A. 16 or 17.   GULOSUS, 2574.

aa. Scales about 50; snout not pointed; depth 5½ in length; mouth large, the maxillary 2¼ in head; teeth strong. Color yellowish, much dotted, but without bars.
EULEPIS, 2575.

aaa. Scales 65 or more.

    b. Caudal fin more than $\frac{1}{4}$ ($\frac{2}{9}$) length of body. Scales very small, cycloid, deciduous. Body elongate, much compressed, highest in front of ventrals, tapering regularly to the very narrow, short caudal peduncle; greatest depth $4\frac{3}{4}$ in length; head $3\frac{1}{2}$. Head compressed, much higher than wide; snout very short, acute, preorbital not as wide as pupil; mouth terminal, very wide and oblique; jaws equal; maxillary reaching vertical from middle of orbit, 2 in head. Outer series of teeth enlarged. Eye 3 in head. Dorsals closely contiguous; spines very slender, the fifth slightly produced and filamentous; pectorals as long as head. Head and body translucent, overlaid by brilliant green luster, formed by minute, close-set green points; 3 conspicuous translucent bars wider than the interspaces, crossing body close behind head; head with 2 brilliant narrow blue and green lines running obliquely across cheek below eye; dorsal whitish, with 2 or 3 lengthwise series of large reddish-brown spots; spinous dorsal blackish at base, upper caudal rays marked with red, the lower portion of caudal and most of the anal fin blackish, anal whitish at base, the anterior rays tipped with white. In spirits, body dusted with dark points; 2 light cross bars toward head; lower part of caudal and anal black. D. VII–16; A. 15.   THALASSINUS, 2576.

    bb. Caudal fin less than $\frac{1}{4}$ length of body. Scales small, cycloid, embedded. Body very much compressed, more or less elongate, greatest depth at ventrals 4 (female) to $6\frac{1}{4}$ (male) in length; head $3\frac{1}{2}$ to 4. Head much compressed, much deeper than wide. Snout very short, acute, the anterior profile not decurved, not steep; preorbital not as wide as pupil; mouth very large, very oblique or almost vertical; maxillary extending to below pupil, 2 in head (in male, $2\frac{1}{4}$ in female). Lower jaw projecting, the teeth of the outer series enlarged, recurved. Eye $3\frac{1}{2}$ to 4 in head. Dorsals contiguous, spines very fine, produced in filaments, the third highest, a little longer than head; second dorsal and anal high. Head and nape naked. In the female the depth is greater, mouth less oblique, smaller; profile from spinous dorsal oblique. First dorsal spine highest, $3\frac{1}{2}$ in length. Ventrals much shorter than in males. Dark gray; female with a short bright blue bar bordered by blackish above pectorals; a blotch of sky blue and orange below eye; fins dusky, the ventrals pale in female, dusky in males. Males with the body plain bluish gray. D. VII–17 to 20; A. 18 to 21; scales 68 to 70.   SIGNATUS, 2577.

2574. MICROGOBIUS GULOSUS (Girard).

Head 3 to $3\frac{1}{2}$; depth 4 to 5; eye 4 in head, longer than snout. D. VII–15; A. 16 or 17; scales about 42; vertebræ $11+15$. Body elongate, moderately compressed; head long and large, rather sharp in profile; mouth large, very oblique, the lower jaw strongly projecting; maxillary $1\frac{1}{2}$ to $2\frac{1}{4}$ in head, extending to opposite middle of eye. Teeth in few series, the outer very long and slender, curved, the lower longest, none canine-like. Body entirely scaled, except the nape, belly, breast, and head, which are naked; scales small, some of them with short thick teeth, those on anterior part of body not well developed. Dorsal spines more or less filamentous, the third to fifth sometimes with long filaments; caudal pointed, as long as head. Ventrals as long as pectorals, which are $1\frac{1}{4}$ in head. Skull flattened behind, with a median ridge extending from eyes back to end of skull. Double crests bordering skull in front and on sides, the inner ones meeting in front of median crest. Interorbital very narrow and deeply grooved, with a median ridge. Frontal bones very thin and fragile.

tulations of dark; about 7 prey well-defined larger spots of dark brown along middle of side; 8 or 9 fat cross bars of lighter, a number of small light spots scattered irregulay over the sides; head dark; dorsal, anal, and ventral fins covered with fine black points; in some specimens the dorsals and anal quite dark; pectorals plainer; caudal similar to ventrals; "the cranium is depressed and flattish behind the orbits, without distinct median keel on occiput or on interorbital area. The form of the head is as in typical *Gobius*, the occiput abruptly widened behind the eyes; the ridges also similar, the orbital ridge bounding the orbit behind as well as above the eye and joining the temporal ridge laterally." (Gilbert MS.) Length 1 to 2 inches. Gulf California; abundant. The original description from young examples, here corrected in accordance with Dr. Gilbert's notes on many adts taken by him at La Paz. (Spanish, *chiquito;* a diminutive of *chico* little one.)

Gobius chiquita, JENKINS & EVERMANN Proc. U. S. Nat. Mus. 1888, 146, Guaymas, Sonora. (Type, No. 39634. Coll. Jenkins Evermann.)

818. MICROGOBIUS, Poey.

Microgobius, POEY, Enumeratio, 127 (signatus).

Dorsal spines 7 or 8; scales very small, cycloid or weakly ctenoid, the body scaled anteriorly as well as posteriorly, the head naked, the nape, belly, and breast usually so. Inner edge of shoulder girdle without fleshy processes; body more or less compressed; mouth large, very oblique; the lower jaw conspicuous, teeth long; interorbital groove with or without a median ridge. Vertebræ 11 + 5 or 16. ($\mu\iota\kappa\rho\acute{o}\varsigma$, small; *Gobius*.)

> *a.* Scales about 42. Body elongate, moderately compressed, the depth 4 to 5 in length; head long and large, rather deep in profile, 3 to 3½ in body; eye longer than snout, 4 in head; mouth large, very oblique, the lower jaw strongly projecting; maxillary 1½ to 2¼ in head, extending to opposite middle of eye, or much beyond front of orbit; teeth in few series, the outer very long and slender, curved, the lower longest, none canine-like; scales small, some of them with short, thick teeth, those of anterior part of body not developed; dorsal spines more or less filamentous, the third and fourth or fourth and fifth sometimes with long filaments; caudal pointed, about as long as head. Grayish olive, with rather sharply-defined markings of darker brown tinged with orange in life; head with a pale bluish or gilt stripe from maxillary backward across suborbital region to upper edge of gill opening; another pale streak from snout along lower part of eye, another from angle of mouth downward and backward; rest of head dark; opercle with an oblique blackish bar; top of head and nape with dark marbling surrounded by paler reticulations; back with a series of black cross blotches mostly separated on the median line; narrower dark vertical bars behind pectoral; middle line of side posteriorly with longitudinally oblong black blotches; besides these, numerous other blotches not regularly arranged; first dorsal with 2 or 3 oblique black bands; second dorsal pale, with about 4 series of black dots; caudal spotted with black; pectoral yellowish; ventral black, its center yellowish (male); anal pale. D. VII- A. 16 or 17.
>
> *aa.* Scales about 50; snout not pointed; depth 5½ in length; [...]
> 2¼ in head; teeth strong. Color yellowish, mu[...]

aaa. Scales 65 or more.

b. Caudal fin more than ⅓ (⅖) length of body. ...

... Body elongate, much compressed ... tapering regularly to the very narrow ... caudal ... depth 4½ in length; head 3½. Head ... and ... snout very short, ... head ... very wide and oblique; jaws equal; ... middle of orbit, 3 in head. Outer ... enlarged ... Dorsals closely contiguous; spinous ... the ... duced and filamentous; ... head. Head ... lucent, overlaid by brilliant ... points; 3 conspicuous ... ing body close behind head; head with ... lines running obliquely upward ... or 3 lengthwise series of large ... ish at base, upper caudal rays ... caudal and most of the anal fin ... rior rays tipped with white. ... 2 light cross bars toward body ...

D. VII–16; A. 15. ...

bb. Caudal fin less than ⅓ length of body. ... very much compressed, more ... 4 (female) to 6⅓ (male) in length; head ... much deeper than wide. Snout ... decurved, not ... ; ... very oblique or almost vert ... head (in male, ⅔ in female). ... series enlarged, recurved. ... very fine, ... in ... head; second dorsal and anal ... the depth is greater, ... dorsal oblique. First dorsal ... shorter than in males. ... bordered by blackish ... below eye; fins dusky, the ... Males with the body pink ... scales 68 to 70.

2574. NICROGOBIUS ...

Head 3 to 3½; depth 4 to 5; eye 4 in head ...
15; A. 16 or 17; scales about 42; ... ately compressed; head long and large, ... large, very oblique, the lower jaw ... in head, extending to opposite ... outer very long and slender, curved ... Body entirely scaled, except the ... naked; scales small, some of these ... rior part of body not well developed ... tous, the third to fifth sometimes ... long as head. Ventrals as long as ...

... dor- ... spots; ... red, the ... whitish ... light buff; ... with dark ... al and anal ... at 1½ inches ... tide pools in ... θάλασσα, the ...

...12, Charleston Har-

Teeth on each jaw in narrow bands, all alike. Coloration in life, light grayish olive, with rather sharply defined markings of darker brown; head with a pale bluish stripe from behind the angle of the mouth upward and forward parallel with the gape to below front of eye, then turning abruptly backward across suborbital region to upper edge of gill opening; another pale streak from snout along lower part of eye; between this and the first streak a dusky area; below the first-mentioned streak a dusky region on cheek; opercle with an oblique blackish bar; top of head with dark marblings surrounded by paler reticulations; back with a series of black cross blotches, mostly separated on the median line; 2 narrow vertical dark bars behind pectoral; middle line of side posteriorly with longitudinally oblong black blotches; besides these numerous other blotches not regularly arranged; first dorsal with 2 or 3 oblique black bands; second dorsal pale, with about 4 series of black dots; caudal spotted with black, pectoral yellowish, ventral black, its center yellowish; anal pale; lower side of head pale; jaws dusky. Coast of Florida to Texas, in sandy or weedy bays, common north to Indian River. A strongly marked species with no near relative among our other gobies. The specimens here described from Pensacola. (*gulosus*, large-mouthed.)

Gobius gulosus, GIRARD, Proc. Ac. Nat. Sci. Phila. 1858, 169, Indianola, Texas; GIRARD, U. S. and Mex. Bound. Surv., Zool., 26, 1859; JORDAN & GILBERT, Synopsis, 634, 1883.

Lepidogobius gulosus, JORDAN & GILBERT, Proc. U. S. Nat. Mus. 1882, 294; JORDAN & GILBERT, Synopsis, 945, 1883.

Microgobius gulosus, JORDAN & EIGENMANN, *l. c.*, 505.

2575. MICROGOBIUS EULEPIS, Eigenmann & Eigenmann.

Head 4 in length (5¼ in total); depth 5½ (7). D. VII–15; A. 16; scales 50–14. Body elongate, scarcely compressed; head slightly higher than wide, the depth 1⅓ in its length; eye large, longer than snout, 3⅓ in head; snout 5 in head, rather broad, not pointed as in *M. thalassinus;* preorbital narrower than pupil; mouth very oblique, maxillary not extending beyond anterior margin of pupil, 2¼ in head; teeth in upper jaw in a very narrow band, slightly enlarged in outer series, largest toward angle of mouth; teeth of lower jaw in a similar band, some of outer ones in front long and slender. Scales cycloid, rather large, crowded anteriorly, regularly arranged, not embedded as in *M. signatus*, not deciduous as in *M. thalassinus;* breast, nape, and region along spinous dorsal naked. First dorsal spine equidistant from tip of snout and first anal ray; longest dorsal spine 1⅓ in head; caudal fin about 4 in body; ventral not reaching vent, equaling length of head, the basal membrane ⅓ of its actual length; pectoral equaling length of head. Color yellow or very light brown, dotted with minute dark points above; scales along back with a dark margin; head and nape with minute points; spinous dorsal transparent, a marked black spot on upper part of membrane between fourth and fifth dorsal spines; other fins plain; a light vertical bar on posterior margin of preopercle;

no other bars or stripes anywhere. Fortress Monroe, Virginia; known from a specimen 1⅜ inches long. (εὖ, well; λεπίς, scaled.)

Microgobius eulepis, EIGENMANN & EIGENMANN, Proc. Cal. Ac. Sci. 1888, 69, Fortress Monroe, Virginia. (Type, No. 27123, M. C. Z. Coll. Mrs. C. N. Willard.)

2576. MICROGOBIUS THALASSINUS, Jordan & Gilbert.

Head 3½ in length; depth 4¾. D. VII–16; A. 15; eye 3 in head. Body elongate, much compressed, highest in front of ventrals, thence tapering regularly to a very narrow, short caudal peduncle; the body with a peculiar, translucent, fragile appearance, common also to *Z. emblematicus.* Head compressed, much higher than wide; snout very short, acute, the preorbital not as wide as pupil; mouth terminal, very wide and oblique, the jaws equal; maxillary reaching vertical from middle of orbit, ½ length of head; teeth in a narrow band in each jaw, the outer series enlarged, canine-like (under a microscope the band of small teeth behind the outer series seems evident, but the size of our specimens does not enable us to verify it with certainty); eyes placed high, separated by a narrow ridge, the diameter about ¼ length of head. Dorsals very closely contiguous; spines very slender, the fifth slightly produced and filamentous, reaching (in our specimens) to base of third soft ray when depressed; caudal lanceolate, very long and pointed, the middle rays produced, 2⅔ in body; pectorals as long as head; the upper rays not silk-like; ventrals with basal membranes well developed; the fin long, reaching to or slightly beyond front of anal, somewhat longer than head. Body covered with rather small cycloid scales; head naked; the scales very readily deciduous; as they have in our specimens mostly fallen off, the count can not be given. Head and body translucent, overlaid by brilliant green luster, which is formed by exceedingly minute close-set green points; the luster is intense toward the head, where it assumes a blue tint, and becomes hardly noticeable on caudal peduncle; 3 conspicuous translucent bars, wider than the interspaces, crossing body immediately behind head; head with 2 brilliant narrow blue or green lines running obliquely across cheek below eye; opercle with greenish luster; branchiostegal membrane white; dorsals whitish, with 2 or 3 lengthwise series of large reddish-brown spots; spinous dorsal blackish at base; upper caudal rays marked with red, the lower portion of caudal and the most of the anal fin blackish, anal whitish at base, the anterior rays tipped with brilliant white; ventrals light buff; pectorals translucent. In spirits, the body appears dusted with dark points; 2 light cross bars toward head; lower part of caudal and anal black. Coast of South Carolina; two specimens, the largest 1½ inches long (No. 29674, U. S. Nat. Mus.), were taken in muddy tide pools in Charleston Harbor. (θαλάσσινός, thalassinus, sea-green; θάλασσα, the sea.)

Gobius thalassinus, JORDAN & GILBERT, Proc. U. S. Nat. Mus. 1882, 612, Charleston Harbor, South Carolina. (Coll. C. H. Gilbert.)

Lepidogooius thalassinus, JORDAN & GILBERT, Synopsis, 947, 1883.

Microgobius thalassinus, JORDAN & EIGENMANN, *l. c.,* 505.

2577. MICROGOBIUS SIGNATUS, Poey.

Head $3\frac{1}{2}$ to 4; depth 4 (female) to $6\frac{1}{2}$ (male); eye $3\frac{1}{4}$ to 4. D. VII–17 to 20; A. 18 to 21; scales 68 to 70; vertebræ 14 + 15. Body very much compressed, more or less elongate; head much compressed, deeper than wide; snout very short, acute, the anterior profile not decurved, not steep; preorbital not as wide as pupil; mouth very large, almost vertical; maxillary extending to below pupil, 2 in head in male, $2\frac{1}{4}$ in female; lower jaw projecting; teeth of the outer series enlarged and recurved. Dorsals contiguous, spines very fine, produced in filaments, the third longest, a little longer than head; second dorsal and anal high. Scales as in *M. gulosus*. Skull rounded, very fragile; a median crest which is highest between eyes; lateral crests developed, the inner ones meeting above posterior part of eye; interorbital comparatively broad, the median crest ending above anterior part of the orbit. Teeth in each jaw in 2 or 3 series; outer series of the upper jaw enlarged and recurved, the inner ones minute; outer series of lower jaw smaller than those of upper jaw, the one nearest angle of mouth an enlarged canine. Dark gray; female with a short bright blue bar, bordered by blackish above pectoral; a blotch of sky blue and orange below eye; fins dusky, the ventrals pale in female; males with the body plain bluish gray. The sexual differences in this species are very strongly marked. West Indies, in salt water; common in Cuba; one of the smallest gobies, barely 2 inches long. Here described from Havana examples collected by Dr. Jordan. (*signatus*, marked.)

Microgobius signatus, POEY, Enumeratio, 127, pl. 5, fig. 3, 1875, Cuba (Type in M. C. Z. Coll. Poey); JORDAN, Proc. U. S. Nat. Mus. 1886, 49; JORDAN & EIGENMANN, *l. c.*, 505.

819. ZALYPNUS, Jordan & Evermann.

Zalypnus, JORDAN & EVERMANN, Check-List Fishes, 459, 1896 (*emblematicus*).

This genus differs from *Microgobius* in having the anterior half of the body naked. Soft dorsal and anal long, of 16 or 17 rays. Two species known. (ζάλη, surf; ὕπνος, slumber.)

   a. Scales 48; shoulder with a round black spot; none of the dorsal spines elongate.

                                             CYCLOLEPIS, 2578.

   aa. Scales 65; a silvery cross bar behind pectorals; some of the dorsal spines usually elongate.

                                               EMBLEMATICUS, 2579.

2578. ZALYPNUS CYCLOLEPIS (Gilbert).

D. VII–16; A. 17; scales 48. Body somewhat elongate, compressed, the mouth very large, narrow, and oblique; maxillary produced beyond the rictus for a distance equaling $\frac{2}{3}$ diameter of orbit, reaching vertical from posterior margin of pupil, $1\frac{2}{3}$ in head; snout short, 5 in head; eye larger, $3\frac{3}{4}$ in head; interorbital width $\frac{1}{2}$ orbit; teeth in upper jaw in 2 series, the outer enlarged and distant; in lower jaw apparently in a single series, similar to outer series of upper jaw, with 2 stronger canines anteriorly. Inner edge of shoulder girdle without fleshy prominences. Dorsal spines

7, none of them elongate, the membrane of last spine reaching base of first soft ray; soft anal rays of moderate height, 1½ in head, the tips of last rays reaching base of caudal, the fin similar to soft dorsal but lower; caudal long, apparently rounded posteriorly, longer than head (mutilated in our specimen); ventrals and pectorals reaching vent. Scales cycloid, small, absent on belly, nape, and on sides in front of fourth dorsal spine. Color in spirits, light olive, the fins dusky; a conspicuous round black spot on shoulder, ¼ size of eye, its posterior margin denser black. Resembling *Zalypnus emblematicus*, differing in its larger scales and different coloration. A single specimen, about 2 inches long, from Lower California, in 7 fathoms. (Gilbert.) ($\varkappa\acute{\nu}\varkappa\lambda o\varsigma$, circle, cycloid; $\lambda\varepsilon\pi\acute{\iota}\varsigma$, scale.)

Microgobius cyclolepis, GILBERT, Proc. U. S. Nat. Mus. 1891, 74, Albatross Station 3020, Lower California.

2579. ZALYPNUS EMBLEMATICUS (Jordan & Gilbert).

Head 3⅔; depth 5. D. VII–16; A. 17; scales about 65. Anterior part of body naked; teeth of upper jaw in one series; body elongate, compressed, heaviest forward; depth 5 in length; head 3⅔; snout short, rather broad, acute in profile; mouth terminal, very oblique; gape wide, its length nearly ¼ head; maxillary reaching to opposite middle of pupil; lower jaw projecting. Teeth in lower jaw partly in 2 series in front, forming a single row laterally; anterior teeth in both jaws strong, incurved. Eyes very large, about ⅓ of head; snout less than orbit. Scales extremely small, cycloid, scarcely increasing in size toward caudal peduncle; head and anterior part of body to front of dorsal fin naked; a narrow naked strip along base of anterior ⅓ of spinous dorsal. Dorsal spines very slender and weak, some of the middle ones usually prolonged, sometimes reaching nearly to the base of caudal, sometimes little elevated; second dorsal and anal similar to each other, the rays high, the last when depressed nearly reaching to the base of caudal; caudal pointed, a little longer than head. Light olivaceous; above thickly punctate with pale dots; sides very thickly covered with golden-green specks; back with 6 pairs of golden-green spots on each side of the dorsal fin, each nearly as large as pupil; sides of head and anterior half of body with wide streaks and bars alternately of purplish blue and golden bronze; those on cheek longitudinal; those on opercle extending obliquely upward and backward, those on body vertical; first dorsal dusky, second dorsal with about 3 series of light-blue spots; anal pale; caudal yellowish green below, dusky above, a very conspicuous narrow bright-red streak from the lower end of the base to the tip of the fifth or sixth ray from the bottom, thus crossing the rays obliquely; ventrals bluish. In spirits, plain light olive, with a silvery cross bar behind pectorals. Length 3½ inches. Panama; known only from the original types. ($\check{\varepsilon}\mu\beta\lambda\eta\mu\alpha$, a banner, from the high dorsal.)

Gobius emblematicus, JORDAN & GILBERT, Bull.U.S.Fish Comm.1881, 330, Bay of Panama.
Lepidogobius emblematicus, JORDAN & EIGENMANN, *l. c.*, 505.

820. EUCYCLOGOBIUS, Gill.

Eucyclogobius, GILL, Proc. Ac. Nat. Sci. Phila. 1862, 279 (*newberryi*).

This genus is allied to *Lepidogobius*, differing chiefly in the naked head and short, chubby body; shoulder girdle with a few dermal flaps; opercle adnate to shoulder girdle from the angle upward; dorsal spines 6 or 7; soft dorsal short; scales all cycloid; cranium depressed behind the parietal region, somewhat excavated, the supraoccipital crest rather high, not extending so far forward as the orbit. Species small, in fresh or brackish waters of California. ($\varepsilon\acute{v}$ well; $\varkappa\acute{v}\varkappa\lambda o\varsigma$, circle (cycloid); *Gobius*.)

2580. EUCYCLOGOBIUS NEWBERRYI (Girard).

Head $3\frac{2}{3}$ to $3\frac{3}{4}$; depth $4\frac{4}{7}$ to $5\frac{1}{4}$. D. VI or VII*-11; A. 10 or 11 (8 in one specimen, perhaps abnormal); scales about 60 to 70, too irregular for exact counting. Body moderately elongate, somewhat compressed, tapering posteriorly; head rounded above, its width $2\frac{1}{7}$ in its length; mouth large, oblique, the maxillary reaching to or beyond posterior margin of orbit, 2 to $2\frac{1}{8}$ in head; interorbital space wide, 4 to $4\frac{1}{4}$ in head; snout bluntish, broad, a little longer than interorbital width; eye small, 5 in head; teeth present on both jaws, slender, canine-like, arranged in series, the outer row enlarged; caudal peduncle 3 to $3\frac{1}{4}$ in head; gill slit about $2\frac{1}{4}$ in head, its upper edge opposite or slightly above uppermost ray of pectoral; scales minute, cycloid, inconspicuous, wanting on head, nape, and fins; shoulder girdle with 2 or 3 small dermal flaps; dorsals separated by a narrow space; dorsal spines very slender; base of spinous dorsal $2\frac{1}{4}$ to $2\frac{2}{3}$ in head; anal similar to soft dorsal, its base about $1\frac{1}{4}$ in head; caudal subtruncate, $1\frac{1}{4}$ to $1\frac{1}{2}$ in head; ventrals inserted under or slightly behind lower edge of base of pectorals, $1\frac{2}{3}$ to 2 in head; pectorals $1\frac{1}{2}$ to $1\frac{1}{4}$ in head. Dark olivaceous, mottled with darker; head with some dusky markings; the sides and back with irregular dark markings as in species of *Etheostominæ*; dorsals distinctly mottled; the first 3 or 4 dorsal spines margined with paler; caudal with faint, broad, wavy cross bars, a faint spot at its base; anal dusky; ventrals yellowish, dusky in males; pectorals plain. Length about 2 inches. Streams of California, in small clear brooks near the sea; locally common in San Luis Obispo Creek, where the specimens here described were taken; probably confined to fresh waters. (Named for Dr. John Strong Newberry of Columbia College, then also on the U. S. Geological Survey.)

Gobius newberryi, GIRARD, Proc. Ac. Nat. Sci. Phila. 1856, 136, **Tomales Bay** (Coll. E. Samuels); GIRARD, Jour. Bost. Soc. Nat. Hist. 1857, 530, pl. 25, figs. 5 to 8; GIRARD, Pac. R. R. Surv., X, 128, 1858.
Lepidogobius newberryi, JORDAN & GILBERT, Synopsis, 637, 1883; JORDAN & EIGENMANN, *l. c.*, 503.

---

* Of the nine specimens examined from San Luis Obispo Creek, five have 7 dorsal spines and the other four 6. Girard gives the fin rays as D. VIII-13; A. 12; but we have seen no specimens either with 8 spines or 13 rays. Six specimens from Wadell Creek, Santa Cruz County, California, show the following fin variation: D. VI in 4; D. VII in 1; D. V (?) in 1; D. rays 10 in 4; D. rays 9 in 2; A. 10 in 2; A. 8 in 1; A. 9 in 3.

821. LEPIDOGOBIUS, Gill.

Lepidogobius, GILL, Ann. Lyc. Nat. Hist. N. Y. 1859, 14 (*lepidus*).
Cyclogobius, STEINDACHNER, S. B. K. Ak. Wiss. Wien, XLII, 1860, 284 (*lepidus*).

This genus contains small gobies with the head and body covered with small cycloid scales; dorsal spines 7; inner edge of shoulder girdle with 2 or 3 dermal flaps; interorbital groove with the median ridge of skull little developed; body elongate, subterete; otherwise essentially as in *Gobius*, the skull nearly as in *Gillichthys*, with a median keel and not abruptly widened behind the eye. Pacific Ocean; not entering rivers. (λεπίς, scale; *Gobius*.)

2581. LEPIDOGOBIUS LEPIDUS (Girard).

Head $4\frac{1}{3}$, regularly conical; depth 7; eye 4, equal to snout, twice as long as deep. D. VII-16 to 18; A. 15; scales about 86. Body elongate, subfusiform, little compressed. Snout not obtuse in profile; interorbital space narrow, about equal to diameter of pupil. Mouth rather large, maxillary reaching to below posterior edge of pupil, $2\frac{1}{3}$ in head; teeth small, all similar, those of upper jaw in 2 or 3 series, those of lower jaw close set, in a broad band. Body covered with small cycloid scales which are very much reduced anteriorly, especially on the nape; cheeks, sides of head, and upper posterior part of opercles covered with small scales; top of head scaly to eye; breast scaled. Dorsal spines weak, the longest 2 in head; soft dorsal low, none of the rays reaching caudal; caudal long, somewhat pointed. Color very pale olive, with roundish blotches of rusty red on back and sides; vertical fins mottled with reddish; distal half of all fins and under side of head blackish, especially in the males. "This species is remarkable for numerous lines of papillæ on mandible, snout, and sides of head. The occipital region of the skull is somewhat more depressed than in *Gobius soporator*, and has much lower ridges. A low median carina is present and the low supraorbital ridges are continuous behind the eyes with the temporal crests." (Gilbert MS.) Pacific coast of North America, from Vancouver Island to Lower California; in rather deep water off San Francisco Bay; often seined in great numbers and sold in restaurants as "whitebait." (*lepidus*, pretty.)

Gobius gracilis, GIRARD, Proc. Ac. Nat. Sci. Phila. 1854, 134, San Francisco; preoccupied by *Gobius gracilis*, JENYNS.
Gobius lepidus, GIRARD, Pac. R. R. Surv., X. 127, pl. 25a, figs. 5 and 6, 1858; substitute for *gracilis;* GÜNTHER, Cat., III, 78, 1861.
Lepidogobius gracilis, GILL, Ann. Lyc. Nat. Hist. N. Y. 1859, 14; JORDAN & GILBERT, Synopsis, 637, 1883; JORDAN & EIGENMANN, *l. c.*, 502.

822. GILLICHTHYS, Cooper.

Gillichthys, COOPER, Proc. Cal. Ac. Sci. 1863, 109 (*mirabilis*).
Gillia, GÜNTHER, Zool. Record 1864, 157 (*mirabilis*); name preoccupied.
Saccostoma (GUICHENOT MS.) SUAVAGE, Bull. Soc. Philom. Paris 1882, 171 (*gulosum*); name preoccupied.

Body moderately elongate, compressed, covered with small, cycloid, embedded scales; belly and head naked. Scales of the young more or

less ciliated. Eyes small, almost superior. Gape wide, the maxillary in
the adult inordinately developed, prolonged backward to the base of the
pectorals, its posterior part a cartilaginous expansion, connected to an
expansion of the skin of the lower jaw, thus forming a channel backward
from the mouth, almost exactly as in *Neoclinus* and *Opisthognathus*,
genera otherwise very different. Teeth small, even, in broad bands.
Skull in adult with a strong median keel, not abruptly widened behind
the eye, triangular behind; young with the keel obsolete. Dorsal fins 2,
the second high, the first of 6 very weak spines, none of which is
exserted; soft dorsal and anal short; caudal less rounded; pectorals
large; isthmus broad. Singular little fishes, in brackish waters, burrowing in the mud; confined to the Pacific. (Named for Theodore Gill.)

a. Head moderately depressed; dorsal fins close together.   MIRABILIS, 2582.
aa. Head very broad and depressed; distance between dorsals ⅓ length of first dorsal.
DETRUSUS, 2583.

2582. GILLICHTHYS MIRABILIS, Cooper.

(LONG-JAWED GOBY.)

Head 3⅓; depth 5; eye 6 to 7; snout longer than eye, low, little
decurved. D. VI-12; A. 10; vertebræ 15+17. Body stout, somewhat compressed behind, broad and depressed anteriorly; head broader than deep,
its width 1¼, its depth 2 or more in its length; interorbital space greater
than eye. Mouth very large; maxillary variable, extending to base of
pectoral in adult, broadened behind; fold of lower lip extending its full
length. Teeth all alike, small, fixed, and in bands, the band of the lower
jaw broader than that of the upper. Scales small, cycloid, irregularly
placed, largest from front of dorsal backward, decreasing in size anteriorly; head, breast, belly, and ⅓ of nape naked. Dorsal spines not
filamentous, not as long as the soft rays which are little more than ⅓
depth of body; caudal broad, short, rounded; pectorals broad and
rounded, longer than ventrals, 2 in head. Skull not abruptly widened
behind eye, as in *Gobius*, being triangular posteriorly. No lateral ridges;
a strong median keel; a short transverse crest behind orbit. Interorbital
not deeply grooved, with a blunt median ridge. Orbit not bordered by
any prominent ridges. Teeth in both jaws, close set, in bands, all alike.
Dull olive, very finely marbled with darker; sides of head and maxillary
finely punctuate; fins olive; belly yellowish. Length 8 inches. Pacific
coast of North America, from San Francisco to Cerros Island; a most
remarkable little fish; very abundant in the mud flats in shallow water
along the California coast, burrowing in holes in the mud like a crawfish,
and readily taking the hook baited with flesh or worm when dropped
into the mouth of the burrow. (*mirabilis*, wonderful.)

Gillichthys mirabilis, COOPER, Proc. Cal. Ac. Sci. 1863, 109, San Diego Bay; LOCKINGTON,
    Amer. Nat. 1879; JORDAN & GILBERT, Synopsis, 636, 1883; JORDAN & EIGENMANN, *l. c.*,
    510; EVERMANN & JENKINS, Proc. U. S. Nat. Mus. 1891, 162.
Gobius townsendi, EIGENMANN & EIGENMANN, Proc. U. S. Nat. Mus. 1888, 463, San Diego;
    young.

2583. GILLICHTHYS DETRUSUS, Gilbert & Scofield.

Allied to *Gillichthys mirabilis*, Cooper, differing in the broader and more depressed head, the larger anal fin, and greater distance between the 2 dorsals. Head 3½; depth 5; eye 7 in head; snout 4; interorbital 5⅓. D. VI–13; A. 11 developed rays (10 in *G. mirabilis*); scales very fine anteriorly but becoming much larger posteriorly; about 75 scales from base of pectoral to caudal, and about 25 longitudinal rows between front of anal and front of second dorsal. The head is depressed, the frontals broad, the shortest distance across being contained in the head 8 times (11 times in *G. mirabilis*.) The postfrontals are small and project but very little, differing from *G. mirabilis*, where the postfrontals project into an elevated wing-like process. The width of the isthmus contained 3 times in the head; maxillary 1⅓ and mandible 1⅔ in head. Least depth of caudal peduncle 2⅔ in head. Distance between dorsals equal to ⅓ length of first dorsal; length of first dorsal 2¼ in head; second dorsal 1½; anal 2 in head; length of longest pectoral ray 1¾ in head. Color a very pale olive, some with dark punctulations about the head and fins. The pale coloration is probably due to their life in shallow water on bottom of pale sand. Gulf of California. The types and numerous other specimens, the longest about 5 inches long, were taken by Dr. C. H. Gilbert at Horseshoe Bend, near the mouth of the Colorado River, in Mexico, where they are quite abundant. These are numbered 3836 in L. S. Jr. Univ. Mus. (*detrusus*, depressed.)

Gillichthys detrusus, GILBERT & SCOFIELD, Proc. U. S. Nat. Mus., xx, 1897, 498, pl. 38, Horseshoe Bend, mouth of Colorado River. (Type, No. 48127. Coll. Gilbert & Alexander.)

823. QUIETULA, Jordan & Evermann.

Quietula, JORDAN & EVERMANN, Proc. Cal. Ac. Sci. 1895, 839 (*y-cauda*).

This genus is closely related to *Gillichthys*, from which it differs in the presence of 2 or 3 cutaneous flaps on the inner edge of the shoulder girdle. Maxillary elongate, as in *Gillichthys*; scales rather small, cycloid; cranium essentially as in *Gillichthys*. Small gobies living in the mud of lagoons and river mouths. (A diminutive, from *quies*, quiet.)

2584. QUIETULA Y-CAUDA (Jenkins & Evermann).

Head 3¼ (4); depth 7 (8); eye 3¼. D. V–14 or 15; A. 15; scales about 50–18; B. 5. Body moderately elongate, compressed, narrowing regularly from shoulder girdle to caudal fin; head not greatly depressed, broader than body, its length 4 in body; snout rounded, short, about equal to diameter of eye; interorbital space narrow, not greater than ½ diameter of eye; mouth rather large, its gape extending nearly to vertical of posterior margin of orbit; maxillary somewhat variable in length, but usually prolonged behind eye for a distance nearly equal to diameter of eye. Scales small, cycloid, about 50 in longitudinal series, 18 in transverse. Teeth in a narrow band on premaxillaries and mandible, short, blunt, and curved slightly backward, most closely set and most numerous on premaxillaries. Shoulder girdle with 2 or 3 small cutaneous flaps on its inner edge. Fins moderate; dorsal of 5 spines and 16 soft rays, the spines unconnected with the rayed portion, the space between them about equal to $\frac{1}{4}$ diameter of eye; the spines weak and flexible, their length $\frac{1}{4}$ that of head; soft dorsal beginning at a point a little nearer end of snout than tip of caudal and extending nearly to caudal, its height about equal to that of spinous portion, the first few rays slightly graduated; anal having 15 rays and beginning a little behind origin of soft dorsal, the rays about equaling those of dorsal in length; pectorals moderate, inserted a little below axis of the body, their length greater than depth of body, their tips reaching a vertical from posterior part of spinous dorsal; ventrals united, but not adnate to belly, inserted slightly in front of pectorals and their tips not quite reaching those of pectorals. Ground color light; head and body pretty uniformly covered with dark punctulations; an irregular dark bar across occiput; breast and belly pale; a row of 9 or 10 small dark blotches along middle of side, the one at base of caudal plainest and having a shape something like the Greek letter Υ; about 6 dark blotches along median line of back; peritoneum dark. Length about $1\frac{3}{4}$ inches. Pacific coast of North America, from Guaymas to Vancouver Island; excessively abundant from San Diego southward in mud flats; specimens recorded from Saanich Arm, Vancouver Island, San Diego, mouth of Colorado River, San Luis Gonzales Bay, St. Georges Bay, Concepcion Bay, Guaymas, and La Paz. It was at first confounded with the young of *Gillichthys mirabilis*, from which genus it differs in the presence of dermal flaps on the shoulder girdle.* (*cauda*, tail, which has a Y-like mark.)

Gillichthys y-cauda, JENKINS & EVERMANN, Proc. U. S. Nat. Mus. 1888, 147, Guaymas, Sonora. (Type, No. 39637. Coll. Jenkins & Evermann.)

Quietula y-cauda, JORDAN & STARKS, Proc. Cal. Ac. Sci. 1895, 839.

Gillichthys guaymasiœ,† JENKINS & EVERMANN, Proc. U. S. Nat. Mus. 1888, 148, Guaymas, Sonora; young specimens $2\frac{1}{4}$ inches long. (Type, No. 39637. Coll. Jenkins & Evermann.)

---

* "The cranium is similar to that of *Gillichthys mirabilis*, the occiput being depressed, wedge-shaped, narrowed anteriorly with a blunt median carina, the supraorbital and temporal ridges not continuous behind the eye. As in *Gillichthys mirabilis*, the supraorbital ridges end in wing-like expansions immediately behind the interorbital space." (Gilbert MS.)

† *Gillichthys guaymasiœ* is thus described: Head 3 ($3\frac{3}{4}$ in total); depth 6 (7). D. V–14; A. 13; eye 5. Body quite slender, elongate, but little compressed; head long, narrow, not much widened behind the eyes, not depressed, forming $\frac{1}{4}$ the length to base of caudal. Profile gently arched from snout to $\frac{1}{4}$ the distance to dorsal fin, from there nearly straight to dorsal, and then gently curved to caudal peduncle; ventral outline nearly straight; a considerable prominence on the snout made by the enlarged end of the turbinal bone. Eye somewhat above the median line, not quite equaling the snout in length; interorbital space narrow, $1\frac{1}{2}$ times in the eye. The maxillaries are much produced, in some specimens nearly reaching the gill openings, broadest at the middle, and tapering to a blunt point posteriorly; premaxillaries not protractile, but little movable at the symphysis, more than $\frac{1}{2}$ as long as the maxillaries. Gill rakers 2 above the angle, 10 below, short and blunt, the first 4 the largest, those on the second arch but little developed. Teeth well developed, in a single series, on mandible and premaxillaries, all slightly curved backward. Tongue not so broad as in *Gillichthys mirabilis*, Cooper; it is gently rounded at the tip, which is free for a much greater length than in *Gillichthys mirabilis*. Peritoneum black or blackish, and the intestine short, but little longer than the head, and not at all convoluted. Scales small, embedded, and scarcely perceptible except on sides; no pores appear to be developed. First dorsal of fine flexible spines, distance of origin from snout $2\frac{1}{4}$ length of body, and separated from the second dorsal by a distance but little greater than length of snout; second dorsal of 14 rays of nearly equal length, which equals the

824. ILYPNUS, Jordan & Evermann.

Ilypnus, JORDAN & EVERMANN, Check-List Fishes, 460, 1896 *(gilberti)*.

This genus is allied to *Clevelandia*, from which it differs chiefly in the presence of dermal flaps on the inner edge of the shoulder girdle; scales minute, embedded, cycloid; dorsal with 5 spines; occiput transversely rounded, without median keel; maxillary moderate. Small gobies, inhabiting mud flats. (ἰλύς, mud; ὕπνος, slumber.)

2585. ILYPNUS GILBERTI (Eigenmann & Eigenmann.)

Head 3 to 3⅓ (3⅗ to 4 in total); depth 5 to 5½ (6 to 7). D. V–15 to 17; A. 14 to 16; B. 5; vertebræ 15 + 19. Form elongate, compressed. Head long, subconical, about as high as wide, its width 2¼ in its length. Profile nearly straight from eyes to spinous dorsal, decidedly decurved in front of eyes. Eye entirely above the premaxillary level, 1 in snout, 4½ in head, ⅓ in interorbital. Mouth slightly oblique; maxillary extending to below middle of eye, lower jaw slightly included. Teeth villiform, in a broad band in each jaw, the outer series of lower jaw somewhat enlarged. One, rarely 2, dermal flaps on inner edge of shoulder girdle. Scales cycloid, embedded, very small; head, nape, and breast naked. Distance from tip of snout to insertion of spinous dorsal 2⅔ in length; highest dorsal spine about ⅔ length of head; soft dorsal rays lower; interdorsal area about ⅓ orbital diameter; tip of last dorsal ray not reaching base of caudal; caudal broad and rounded when expanded; anal similar to soft dorsal fin; ventral fins large, nearly reaching vent in specimens 1¾ inches long. Pectorals usually shorter than ventrals. Color in life, sand color; head and body with small rust-colored spots, which are dotted with black, the punctulations forming a more or less regular network; dorsal fins hyaline at base, bright rust-colored above, and rather broadly margined with white, everywhere black punctate except on margins; about 3 groups of black dots on each ray, giving a barred appearance to these fins; caudal margined with white, upper and lower parts of fin rust colored, median portion dark gray; about 5 wavy, rustlike, vertical bars; entire fin dotted with black except its margin; anal fin hyaline at base, sparsely dotted, its middle third jet-black, margined with white; pectorals and ventrals milky white, yellowish, sparingly black dotted and white edged; a large, conspicuous, metallic blue-black spot on opercle; top of head blackish; belly white or yellowish; chin and throat white, sometimes

distance from end of snout to middle of pupil; length of base of soft dorsal not quite equaling length of head; distance of posterior end from caudal fin equaling distance between the 2 dorsal fins. Origin of anal behind that of soft dorsal and a little posterior to middle of total length of fish; its base 1½ times in base of soft dorsal, or about 4 in length of fish to base of caudal fin; pectorals moderate, a little more than ½ length of head; ventrals inserted slightly behind the pectorals and about equaling them in length. Color in life whitish beneath, grayish or mottled above; 6 double white spots along the back, alternating with fine blackish areas; a white spot behind each eye on top of head; cheek with 2 dark bands extending obliquely backward and downward from eye; a number of dark splotches on opercles; about 7 dusky areas along the side, the last and most marked being upon the base of the caudal fin; dorsal fins finely marked lengthwise by about 4 series of small dark spots: caudal crossed by 5 or 6 wavy vertical bars of very fine dark spots or points; anal, pectorals, and ventrals plain. In alcohol these markings are less plain, especially the white and black areas upon the back. Length 2⅔ inches.

on premaxillaries. Shoulder girdle with 2 or 3 small cutaneous flaps on its inner edge. Fins moderate; dorsal of 5 spines and 16 soft rays, the spines unconnected with the rayed portion, the space between them about equal to $\frac{1}{2}$ diameter of eye; the spines weak and flexible, their length $\frac{1}{4}$ that of head; soft dorsal beginning at a point a little nearer end of snout than tip of caudal and extending nearly to caudal, its height about equal to that of spinous portion, the first few rays slightly graduated; anal having 15 rays and beginning a little behind origin of soft dorsal, the rays about equaling those of dorsal in length; pectorals moderate, inserted a little below axis of the body, their length greater than depth of body, their tips reaching a vertical from posterior part of spinous dorsal; ventrals united, but not adnate to belly, inserted slightly in front of pectorals and their tips not quite reaching those of pectorals. Ground color light; head and body pretty uniformly covered with dark punctulations; an irregular dark bar across occiput; breast and belly pale; a row of 9 or 10 small dark blotches along middle of side, the one at base of caudal plainest and having a shape something like the Greek letter Υ; about 6 dark blotches along median line of back; peritoneum dark. Length about 1$\frac{3}{4}$ inches. Pacific coast of North America, from Guaymas to Vancouver Island; excessively abundant from San Diego southward in mud flats; specimens recorded from Saanich Arm, Vancouver Island, San Diego, mouth of Colorado River, San Luis Gonzales Bay, St. Georges Bay, Concepcion Bay, Guaymas, and La Paz. It was at first confounded with the young of *Gillichthys mirabilis*, from which genus it differs in the presence of dermal flaps on the shoulder girdle.* (*cauda*, tail, which has a Y-like mark.)

Gillichthys y-cauda, JENKINS & EVERMANN, Proc. U. S. Nat. Mus. 1888, 147, Guaymas, Sonora. (Type, No. 39637. Coll. Jenkins & Evermann.)

Quietula y-cauda, JORDAN & STARKS, Proc. Cal. Ac. Sci. 1895, 839.

Gillichthys guaymasiœ,† JENKINS & EVERMANN, Proc. U. S. Nat. Mus. 1888, 148, Guaymas, Sonora; young specimens 2$\frac{3}{4}$ inches long. (Type, No. 39637. Coll. Jenkins & Evermann.)

---

* "The cranium is similar to that of *Gillichthys mirabilis*, the occiput being depressed, wedge-shaped, narrowed anteriorly with a blunt median carina, the supraorbital and temporal ridges not continuous behind the eye. As in *Gillichthys mirabilis*, the supraorbital ridges end in wing-like expansions immediately behind the interorbital space." (Gilbert MS.)

† *Gillichthys guaymasiœ* is thus described: Head 3 (3$\frac{2}{3}$ in total); depth 6 (7). D. V–14; A. 13; eye 5. Body quite slender, elongate, but little compressed; head long, narrow, not much widened behind the eyes, not depressed, forming $\frac{1}{4}$ the length to base of caudal. Profile gently arched from snout to $\frac{1}{2}$ the distance to dorsal fin, from there nearly straight to dorsal, and then gently curved to caudal peduncle; ventral outline nearly straight; a considerable prominence on the snout made by the enlarged end of the turbinal bone. Eye somewhat above the median line, not quite equaling the snout in length; interorbital space narrow, 1$\frac{1}{2}$ times in the eye. The maxillaries are much produced, in some specimens nearly reaching the gill openings, broadest at the middle, and tapering to a blunt point posteriorly; premaxillaries not protractile, but little movable at the symphysis, more than $\frac{1}{2}$ as long as the maxillaries. Gill rakers 2 above the angle, 10 below, short and blunt, the first 4 the largest, those on the second arch but little developed. Teeth well developed, in a single series, on mandible and premaxillaries, all slightly curved backward. Tongue not so broad as in *Gillichthys mirabilis*, Cooper; it is gently rounded at the tip, which is free for a much greater length than in *Gillichthys mirabilis*. Peritoneum black or blackish, and the intestine short, but little longer than the head, and not at all convoluted. Scales small, embedded, and scarcely perceptible except on sides; no pores appear to be developed. First dorsal of fine flexible spines, distance of origin from snout 2$\frac{3}{4}$ length of body, and separated from the second dorsal by a distance but little greater than length of snout; second dorsal of 14 rays of nearly equal length, which equals the

824. ILYPNUS, Jordan & Evermann.

Ilypnus, JORDAN & EVERMANN, Check-List Fishes, 460, 1896 (*gilberti*).

This genus is allied to *Clevelandia*, from which it differs chiefly in the presence of dermal flaps on the inner edge of the shoulder girdle; scales minute, embedded, cycloid; dorsal with 5 spines; occiput transversely rounded, without median keel; maxillary moderate. Small gobies, inhabiting mud flats. (ἰλύς, mud; ὕπνος, slumber.)

2585. ILYPNUS GILBERTI (Eigenmann & Eigenmann.)

Head 3 to 3⅓ (3⅔ to 4 in total); depth 5 to 5½ (6 to 7). D. V–15 to 17; A. 14 to 16; B. 5; vertebræ 15 + 19. Form elongate, compressed. Head long, subconical, about as high as wide, its width 2¼ in its length. Profile nearly straight from eyes to spinous dorsal, decidedly decurved in front of eyes. Eye entirely above the premaxillary level, 1 in snout, 4¼ in head, ¼ in interorbital. Mouth slightly oblique; maxillary extending to below middle of eye, lower jaw slightly included. Teeth villiform, in a broad band in each jaw, the outer series of lower jaw somewhat enlarged. One, rarely 2, dermal flaps on inner edge of shoulder girdle. Scales cycloid, embedded, very small; head, nape, and breast naked. Distance from tip of snout to insertion of spinous dorsal 2⅔ in length; highest dorsal spine about ⅔ length of head; soft dorsal rays lower; interdorsal area about ⅓ orbital diameter; tip of last dorsal ray not reaching base of caudal; caudal broad and rounded when expanded; anal similar to soft dorsal fin; ventral fins large, nearly reaching vent in specimens 1¾ inches long. Pectorals usually shorter than ventrals. Color in life, sand color; head and body with small rust-colored spots, which are dotted with black, the punctulations forming a more or less regular network; dorsal fins hyaline at base, bright rust-colored above, and rather broadly margined with white, everywhere black punctate except on margins; about 3 groups of black dots on each ray, giving a barred appearance to these fins; caudal margined with white, upper and lower parts of fin rust colored, median portion dark gray; about 5 wavy, rustlike, vertical bars; entire fin dotted with black except its margin; anal fin hyaline at base, sparsely dotted, its middle third jet-black, margined with white; pectorals and ventrals milky white, yellowish, sparingly black dotted and white edged; a large, conspicuous, metallic blue-black spot on opercle; top of head blackish; belly white or yellowish; chin and throat white, sometimes

distance from end of snout to middle of pupil; length of base of soft dorsal not quite equaling length of head; distance of posterior end from caudal fin equaling distance between the 2 dorsal fins. Origin of anal behind that of soft dorsal and a little posterior to middle of total length of fish; its base 1⅓ times in base of soft dorsal, or about 4 in length of fish to base of caudal fin; pectorals moderate, a little more than ½ length of head; ventrals inserted slightly behind the pectorals and about equaling them in length. Color in life whitish beneath, grayish or mottled above; 6 double white spots along the back, alternating with fine blackish areas; a white spot behind each eye on top of head; cheek with 2 dark bands extending obliquely backward and downward from eye; a number of dark splotches on opercles; about 7 dusky areas along the side, the last and most marked being upon the base of the caudal fin; dorsal fins finely marked lengthwise by about 4 series of small dark spots; caudal crossed by 5 or 6 wavy vertical bars of very fine dark spots or points; anal, pectorals, and ventrals plain. In alcohol these markings are less plain, especially the white and black areas upon the back. Length 2¼ inches.

punctate. Young lighter, showing the reticulations, but the other markings faint or undeveloped. Length about 2¼ inches. (Eigenmann.) "This species agrees with *Lepidogobius* in the presence of papillæ on the inner edge of shoulder girdle. It differs decidedly in the shape of the occipital region of the cranium, which is transversely evenly convex as in *Clevelandia;* not abruptly widened behind the orbits, not continuous laterally with the temporal ridge as in *Gobius, Lepidogobius,* etc. From *Clevelandia* and *Gillichthys, Lepidogobius gilberti* differs in the presence of papillæ on the shoulder girdle, and from *Gillichthys y-cauda* in the shape of the cranium." (Gilbert MS.) San Diego Bay and southward; found by Dr. Gilbert abundant at Magdalena Bay, at Concepcion Bay, and St. Georges Bay, in the Gulf of California. (Named for Charles Henry Gilbert, professor of Zoology in the Leland Stanford Junior University.)

Lepidogobius gilberti, EIGENMANN & EIGENMANN, Proc. U. S. Nat. Mus. 1888, 464, San Diego Bay. (Type, No. 40128, U. S. Nat. Mus. Coll. C. H. Eigenmann.)

825. CLEVELANDIA, Eigenmann & Eigenmann.

Clevelandia, EIGENMANN & EIGENMANN, Proc. Cal. Ac. Sci. 1888, 73 (*longipinnis,* EIGENMANN & EIGENMANN, = *rosæ*).

This genus is closely allied to *Gillichthys,* differing chiefly in the form of the skull, which is rounded above, strongly convex in transverse profile, perfectly smooth, without ridges or crests. Body long and slender; maxillary much produced, but not extending to the gill opening; mouth horizontal; dorsal spines 4 or 5, very weak; body covered with minute cycloid embedded scales; soft dorsal and anal long, each of 14 to 17 rays. (Named for Daniel Cleveland, esq., president of the San Diego Society of Natural History, a gentleman deeply interested in scientific matters.)

a. Caudal short, rounded; dorsal spines 5.   IOS, 2586.
aa. Caudal pointed, scarcely shorter than head; dorsal spines 4.   ROSÆ, 2587.

2586. CLEVELANDIA IOS (Jordan & Gilbert).

Head 3⅓ in length of body; depth 6. D. V-16; A. 14; eye 6⅓ in head; maxillary 1⅜; pectoral 1⅜; ventrals 1⅓; caudal 1⅓; base of soft dorsal 3 in length of body; base of anal 3⅓. Body long and slender, compressed, the back not elevated; caudal peduncle moderately wide; head long, profile steep to within a short distance of the front of the eye, thence horizontal; mouth very large, not very oblique, the maxillary projecting to opposite the middle of the cheek; jaws subequal; teeth in narrow villiform bands; eye small, longer than wide, set high in head; interorbital space narrow, about as wide as eye. Body covered with very small cycloid scales, too small to count; spinous dorsal well separated from soft dorsal, the spines slender; soft dorsal the higher, its origin a little nearer base of caudal fin than tip of snout; anal about equal to soft dorsal in height, its origin a little behind first dorsal ray, ending at about the same comparative place as soft dorsal; ventrals inserted slightly behind pectorals, reaching midway between their base and front of anal; caudal short, its end rounded.

punctate. Young lighter, showing the reticulations, but the other markings faint or undeveloped. Length about 2¼ inches. (Eigenmann.) "This species agrees with *Lepidogobius* in the presence of papillæ on the inner edge of shoulder girdle. It differs decidedly in the shape of the occipital region of the cranium, which is transversely evenly convex as in *Clevelandia;* not abruptly widened behind the orbits, not continuous laterally with the temporal ridge as in *Gobius, Lepidogobius,* etc. From *Clevelandia* and *Gillichthys, Lepidogobius gilberti* differs in the presence of papillæ on the shoulder girdle, and from *Gillichthys y-cauda* in the shape of the cranium." (Gilbert MS.) San Diego Bay and southward; found by Dr. Gilbert abundant at Magdalena Bay, at Concepcion Bay, and St. Georges Bay, in the Gulf of California. (Named for Charles Henry Gilbert, professor of Zoology in the Leland Stanford Junior University.)

Lepidogobius gilberti, EIGENMANN & EIGENMANN, Proc. U. S. Nat. Mus. 1888, 464, San Diego Bay. (Type, No. 40128, U. S. Nat. Mus. Coll. C. H. Eigenmann.)

825. CLEVELANDIA, Eigenmann & Eigenmann.

Clevelandia, EIGENMANN & EIGENMANN, Proc. Cal. Ac. Sci. 1888, 73 (*longipinnis,* EIGENMANN & EIGENMANN, = *rosæ*).

This genus is closely allied to *Gillichthys,* differing chiefly in the form of the skull, which is rounded above, strongly convex in transverse profile, perfectly smooth, without ridges or crests. Body long and slender; maxillary much produced, but not extending to the gill opening; mouth horizontal; dorsal spines 4 or 5, very weak; body covered with minute cycloid embedded scales; soft dorsal and anal long, each of 14 to 17 rays. (Named for Daniel Cleveland, esq., president of the San Diego Society of Natural History, a gentleman deeply interested in scientific matters.)

a. Caudal short, rounded; dorsal spines 5.   IOS, 2586.
aa. Caudal pointed, scarcely shorter than head; dorsal spines 4.   ROSÆ, 2587.

2586. CLEVELANDIA IOS (Jordan & Gilbert).

Head 3¼ in length of body; depth 6. D. V–16; A. 14; eye 6½ in head; maxillary 1⅘; pectoral 1⅔; ventrals 1⅘; caudal 1¼; base of soft dorsal 3 in length of body; base of anal 3¼. Body long and slender, compressed, the back not elevated; caudal peduncle moderately wide; head long, profile steep to within a short distance of the front of the eye, thence horizontal; mouth very large, not very oblique, the maxillary projecting to opposite the middle of the cheek; jaws subequal; teeth in narrow villiform bands; eye small, longer than wide, set high in head; interorbital space narrow, about as wide as eye. Body covered with very small cycloid scales, too small to count; spinous dorsal well separated from soft dorsal, the spines slender; soft dorsal the higher, its origin a little nearer base of caudal fin than tip of snout; anal about equal to soft dorsal in height, its origin a little behind first dorsal ray, ending at about the same comparative place as soft dorsal; ventrals inserted slightly behind pectorals, reaching midway between their base and front of anal; caudal short, its end rounded.

Color light olivaceous, the cheeks and sides with many dark points which form mottlings; snout dark; a dark spot on upper part of opercle; top of head black; dorsals light, with 3 or 4 dark lines running across the rays; some dark spots on base of anal; pectorals crossed with dark wavy lines; caudal with about 5 irregular cross bars. Puget Sound and neighboring waters. Here described from 2 specimens, each 2 inches in length, dredged off Port Orchard by Mr. Edwin C. Starks. The original description is imperfect and partly incorrect, the single type, from the stomach of *Hexagrammos asper*, being in bad condition. (*iós*, arrow.)

Gobiosoma ios, JORDAN & GILBERT, Proc. U. S. Nat. Mus. 1882, 437, Saanich Arm, Vancouver Island (Coll. Jordan & Gilbert); JORDAN & GILBERT, Synopsis, 948, 1883; JORDAN & EIGENMANN, *l. c.*, 509.

Clevelandia ios, JORDAN & STARKS, Proc. Cal. Ac. Sci. 1895, 839, pl. 100.

2587. CLEVELANDIA ROSÆ, Jordan & Evermann.

Head 4 (4¾ in total); depth 6⅔ (7). D. IV–16; A. 17; scales 70–18. Body very much elongate, slender; head long and slender, depressed anteriorly much as in *Lucius;* profile straight; eye moderate, slightly shorter than snout, 4⅓ in length of head; interorbital area about as wide as pupil; anteorbital area scarcely ⅓ diameter of eye; mouth large, maxillary extending much beyond orbit; lower jaw flat, slightly curved upward anteriorly; mouth very much as in *Lucius;* teeth all small, in narrow bands in each jaw; the outer ones of the upper jaw slightly larger than the others. Scales minute, slightly enlarged posteriorly; the margins plain, anterior part of the exposed area lengthwise striated; breast and antedorsal area naked. Distance from snout to insertion of first dorsal spine 2⅔ in body; the spines slender and short, 3 in head; interdorsal area equals snout and eye; dorsal rays slightly longer than spines, the last ray not extending halfway to caudal; caudal pointed, scarcely shorter than head; ventrals not reaching halfway to vent, 1⅗ in head; pectoral 1¼ in head; vent slightly behind middle of body. Color light brownish; numerous darker spots of aggregated points along nape and upper half of body; belly white; head slightly darker than body; posterior edge of opercle white; an oblique silvery bar on the lower half of opercle, and a light blotch at the upper corner of opercle; cheek with black points; some light areas below eye; lower surface of head and posterior part of maxillaries plain; 2 dark bars on spinous dorsal; second dorsal with 3 or 4 dark bars; a curved black bar at base of caudal; remainder of caudal irregularly barred with dark; other fins plain. Length 1⅞ inches. San Diego Bay (Eigenmann & Eigenmann); at first incorrectly identified by Mr. and Mrs. Eigenmann with *Evermannia longipinne* (Steindachner), a species similar in habit but wholly scaleless. (Named for Mrs. Rosa Smith Eigenmann.)

Clevelandia longipinnis, EIGENMANN & EIGENMANN, Proc. Cal. Ac. Sci. 1888, 73; not *Gobiosoma longipinne*, STEINDACHNER.

Clevelandia rosæ, JORDAN & EVERMANN, Proc. Cal. Ac. Sci. 1896, 229, San Diego. (Coll. R. S. Eigenmann.)

826. EVERMANNIA, Jordan.

Evermannia, JORDAN, Proc. Cal. Ac. Sci., IV, 1895, series 2, 592 (*zosterura*).

Body slender, compressed behind, entirely naked. Head long, slender. Snout rather pointed; mouth moderate, terminal, the maxillary more or less produced backward; teeth small and slender, the outer above slightly enlarged. Skull with a small median crest, not much widened behind. Interorbital space very narrow, channeled; no dermal flaps on shoulder girdle; first dorsal of 4 to 6 spines; second dorsal and anal long, of 14 or 15 rays. Caudal lanceolate. Ventrals formed as in *Gobius* and *Gobiosoma*. Size small, the sexes not colored alike. Species living in holes in sand and mud between tide marks. (Named for "my former student and later scientific associate, Dr. Barton Warren Evermann, now ichthyologist of the United States Fish Commission, in recognition of his work on the fishes of the Gulf of California."—Jordan.)

a. Head $3\frac{3}{4}$; depth $5\frac{1}{4}$; body and fins dotted. LONGIPINNIS, 2588.
aa. Head $3\frac{1}{3}$; depth $6\frac{2}{3}$; vertical fins in males banded with black and with white edgings. ZOSTERURA, 2589.

2588. EVERMANNIA LONGIPINNIS (Steindachner).

Head $3\frac{4}{5}$; depth 5 to $5\frac{1}{3}$. D. IV to VI-16 or 17; A. 16 or 17; snout slightly decurved in profile, $3\frac{1}{2}$ in head; eye 6, greater than interorbital width. Body very slender. Mouth somewhat oblique, the jaws equal; maxillary extending beyond middle of head to a distance behind eye equal to diameter of eye. Teeth in each jaw in 2 series laterally and 3 in front, those of the outer series somewhat enlarged. Fins low, the longest dorsal spine 2 in head; pectoral a little shorter than caudal, scarcely longer than ventrals. Caudal rounded, shorter than head, probably $4\frac{2}{3}$ in body. Body and head completely naked.* Brownish yellow; upper parts of head and body with small, irregularly placed brown spots and streaks; dorsals and caudal finely barred with dark specks. (Steindachner). Gulf of California; not seen by us; known from 3 specimens $37\frac{1}{2}$ mm. long. We refer this species provisionally to *Evermannia*, with which genus it agrees in external respects, although the mouth is much larger. It may be the type of a distinct genus. It differs from *Clevelandia* in the entire absence of scales. (*longus*, long; *pinna*, fin.)

Gobiosoma longipinne, STEINDACHNER, Ich. Beitr., VIII, 27, 1879, Las Animas Bay, Gulf of California; JORDAN & EIGENMANN, *l. c.*, 509.
Evermannia longipinnis, JORDAN, Proc. Cal. Ac. Sci. 1896, 229.

2589. EVERMANNIA ZOSTERURA (Jordan & Gilbert).

Head $3\frac{1}{4}$; depth 6. D. IV-15; A. 14; eye equals snout, 5 in head; P. $1\frac{2}{3}$; C. $1\frac{1}{3}$. Body compressed, profile convex; snout short, not very blunt; eyes high, the maxillary reaching to their posterior margin; mouth oblique, jaws equal; first spine of dorsal filamentous, reaching to middle

* At our request Dr. Steindachner has reexamined the types of this species. He still finds them "vollkommen schuppenlos."

of soft dorsal (male); body entirely naked. Body everywhere speckled with dots of dark brown. Male sometimes with traces of 8 olive cross bands; fins very ornate, the dorsal and anal yellowish at base, then a broad median band of jet black, then a broad white margin; middle of caudal yellow to the tip, with a black band above and below, and a white edge above and below this as in dorsal and anal; no bands on tail. Female with dorsal filament short, reaching about to first soft ray; dorsals and anal checkered with blackish; caudal faintly barred; all vertical fins with pale edgings, but without the black stripe of the males. Length 2 inches. Very common on sandy bottoms, everywhere about the estuary of Mazatlan, the numerous specimens here described being dug out of the sand. It is seldom found much, if any, below the mark of low tide. It is a very handsomely colored species, the male being more strikingly marked than any other of our gobies. (ζωστήρ, band; οὐρά, tail.)

Gobiosoma zosterurum, JORDAN & GILBERT, Proc. U. S. Nat. Mus. 1881, 361, Mazatlan (fin rays incorrect), (Type, No. 29245, U. S. Nat. Mus. Coll. C. H. Gilbert); JORDAN & EIGENMANN, *l. c.*, 509.
Evermannia zosterura, JORDAN, Proc. Cal. Ac. Sci., 2d ser., vol. IV, 1895, 498, pl. 51.

827. GOBIOSOMA, Girard.

(NAKED GOBIES.)

Gobiosoma, GIRARD, Proc. Ac. Nat. Sci. Phila. 1858, 169 (*alepidotus*).

Body entirely naked; mouth moderate, horizontal; snout blunt; teeth in several series, the outer row enlarged; no canines; dorsal spines normally 7, rarely 5 or 6; second dorsal and anal short; no barbels about head; shoulder girdle without flaps. Species chiefly American. (*Gobius:* σῶμα, body.)

 a. Coloration olivaceous, mottled with darker; no red nor blue.
  b. Maxillary extending to beyond pupil, 4¼ in head; color blackish, with sharply defined cross bars of whitish. Body rather short, the depth 5¼ in length; head 3¼; snout low, little obtuse; mouth large, rather oblique, the maxillary 2⅔ in head; teeth small, in few series above, in a band below, the outer enlarged; fins low; caudal 1¼ in head. Cross bands on body as wide as eye, not quite meeting below; a dark blotch on base of pectoral, a fainter one on base of caudal; fins nearly plain. D. VII–13; A. 12. HISTRIO, 2590.
  bb. Maxillary extending to below posterior part of orbit; coloration not sharply defined, the body usually with dark cross streaks.
   c. Body rather short, chubby, the depth about 4 in length; head about 3⅔; head rounded above; teeth in several series, slender, the outer ones somewhat elongate, none of the inner ones specially enlarged. Color olivaceous, with dark points; sides with narrow, alternating light and dark bars; a row of small linear dark spots along middle of sides; first dorsal with 3 oblique dark bars; second dorsal, caudal, and pectorals finely barred; base and edge of anal light, middle dark; breast with many well-defined spots; a dark line running forward and downward from eye to angle of mouth, another extending straight down; a black bar on edge of preopercle, a black spot on upper edge of opercle. D. VII–13; A. 10. MOLESTUM, 2591.

cc. Body more elongate, depth 5 to 6 in body; head very broad, flattish above, with tumid cheeks, its length 3½ in body; eye small, longer than snout, 5 in head; mouth large, little oblique, the jaws subequal.

  d. Soft dorsal with 14 rays; no crescent at base of caudal; maxillary extending to below posterior part of orbit (at least in male), 2½ in head; teeth in few series, the outer considerably enlarged; 2 teeth on each side of inner series of lower jaw especially large canines; dorsal spines slender, none filamentous; caudal rounded. Olivaceous, with darker cross shades of rounded spots; vertical fins dusky, faintly barred. Teeth of the female similar to those of the male but smaller; head narrower; more slender. D. VII-14; A. 10.   BOSCI, 2592.

  dd. Soft dorsal with 12 rays; a brown crescent at base of caudal.

                                               CRESCENTALE, 2593.

aa. Coloration not plain olivaceous; head with a red bar; anterior dorsal rays not produced in filaments; head and body compressed; greatest depth 5⅔ in total length, head about 4; angle of mouth little behind center of eye; eye 4 in head; teeth pointed, in several series, those of the outer series a little enlarged; caudal rounded. Head light yellow; a carmine-red bar extending along upper edge of head, from upper corner of gill opening to snout, where it joins its fellow, ending behind over the pectoral in a small indigo-blue spot; body with 16 or 17 light green, well-defined cross bars, separated by narrow white stripes; fins chiefly greenish. D. VII (VI)-11 or 12; A. 10.

                                            MULTIFASCIATUM, 2594.

2590. GOBIOSOMA HISTRIO, Jordan.

Head 3¼; depth 5⅔. D. VII-12 or 13; A. 11 or 12; maxillary 2¼ in head; caudal 1¼. Body rather short; snout depressed, little obtuse; mouth large, rather oblique, maxillary reaching to below posterior part of orbit; chin without barbels; many series of minute papillæ along mucous pores of head. Teeth small, in few series above, in a band below, the outer enlarged. Fins low. Cross bands on body whitish, as wide as eye, not quite meeting below; a dark blotch on base of pectoral, a fainter one on base of caudal; fins nearly plain. Length 2 inches. Gulf of California; known only from the Gulf of California at Guaymas (Emeric; Evermann & Jenkins) and La Paz (Gilbert). (*histrio*, a harlequin.)

Gobiosoma histrio, JORDAN, Proc. U. S. Nat. Mus. 1884, 260, Guaymas, Mexico (Coll. H. F. Emeric); JORDAN & EIGENMANN, *l. c.*, 508; EVERMANN & JENKINS, Proc. U. S. Nat. Mus. 1891, 162.

2591. GOBIOSOMA MOLESTUM, Girard.

Head about 3⅔; depth 4. D. VII-13; A. 10; vertebræ 12 + 15. Body rather short, maxillary extending to below posterior part of orbit. Teeth in several series, slender, the outer ones somewhat elongate, none of the inner ones specially enlarged. Skull flattish, with a slight median keel; lateral crests developed, lower and stronger than in *Gobius;* interorbital very narrow, bounded by 2 minute crests; bones of the skull very weak and fragile. Teeth in both jaws recurved, in 2 or 3 series. Olivaceous, with dark points; sides with narrow, alternating light and dark bars; a row of small dark spots along middle of side; first dorsal with 3 oblique dark bars; second dorsal, caudal, and pectorals finely barred; base and edge of anal light, middle dark; breast with many well-defined spots; a dark line running forward and downward from eye to angle of mouth, another extending straight downward from eye; a black bar on edge of preopercle, and a black spot on upper edge of opercle. A specimen taken at Key West is thus described: Pale olive, with darker cross bands formed of dark dots; a row of dark dots along middle of side; vertical fins all mottled and faintly barred with dark olive; pectorals and ventrals nearly plain. Length 2½ inches. Gulf coast of the United States; common in shallow waters along the coast from Key West to Texas and south to Bahia. (_molestus_, disturbed.)

Gobiosoma molestum, GIRARD, Proc. Ac. Nat. Sci. Phila. 1858, 169, Indianola, Texas; GIRARD, U. S. and Mex. Bound. Surv., 27, pl. 12, fig. 14. 1859; GÜNTHER, Cat., III, 556, 1861; JORDAN & GILBERT, Synopsis, 638, 1883; JORDAN & EIGENMANN, _l. c._, 508.
Gobiosoma alepidotum, JORDAN & GILBERT, Proc. U. S. Nat. Mus. 1882, 297, Laguna Grande, Pensacola. (Coll. Dr. Jordan.)

2592. GOBIOSOMA BOSCI (Lacépède).

(NAKED GOBY.)

Head 3⅛; depth 5 to 6. D. VII–14; A. 10; eye 5, longer than snout. Body more elongate; head very broad, flattish above, with tumid cheeks. Eye small. Mouth large, little oblique, jaws subequal, the maxillary extending to below posterior part of orbit (at least in male), 2⅓ in head. Teeth in few series, the outer considerably enlarged; 2 teeth on each side of inner series of lower jaw especially large canines. Dorsal spines slender, not filamentous; caudal rounded. Olivaceous, with darker cross shades of rounded spots; vertical fins dusky, faintly barred. Atlantic coast of the United States, Cape Cod to Florida; generally common, especially southward in shallow grassy bays. (Named for M. Bosc, French consul at Charleston in the last century; an ardent naturalist.)

Gobius bosci, LACÉPÈDE, Hist. Nat. Poiss., II, 555, pl. 16, fig. 1, 1798, Charleston, South Carolina. (Coll. M. Bosc.)
Gobius alepidotus, BLOCH & SCHNEIDER, Syst. Ichthyol., 547, 1801, after LACÉPÈDE; DE KAY, N. Y. Fauna: Fishes, 160, pl. 23, fig. 70, 1842.
Gobius viridipallidus, MITCHILL, Trans. Lit. and Philos. Soc. N. Y., I, 1814, 379, pl. 1, fig. 8, New York.
Gobiosoma bosci, JORDAN & GILBERT, Proc. U. S. Nat. Mus. 1882, 613; JORDAN & EIGENMANN, _l. c._, 508.
Gobiosoma alepidotum, GÜNTHER, Cat., III, 85, 1861; JORDAN & GILBERT, Synopsis, 638, 1883.

2593. GOBIOSOMA CRESCENTALE, Gilbert.

Head 3¾ in length; depth 6¼; eye 5⅓ in head; snout 5⅓. D. VII–12; A. 11. Body very slender, the head depressed, broad and flattened above, the head and body of nearly equal depth throughout. Mouth small, oblique, the maxillary not extending beyond the vertical from posterior border of orbit, 2¼ in head; eyes small, 1⅗ in the rather broad interorbital space. Teeth in bands in both jaws, the outer series enlarged, canine-like, and distant. Fins all small, the caudal short and rounded from a broad base, pectoral as long as head without snout; ventrals short, not

reaching ⅞ the distance from their base to vent; dorsal spines not filamentous; skin wholly naked. Color in spirits, lower half of head and
body uniform warm brown, the back much lighter, the two areas separated by a well-defined line along middle of sides; this line passing through
orbit and through the middle of the base of the pectoral fin; back light
grayish, with brownish reticulations, which tend to form 5 or 6 indistinct darker bars uniting with the darker area below the lateral line;
a conspicuous brown crescent at base of caudal and pectorals, broad
below, narrowing above, margined in front with whitish; anal brown at
base; dorsal and caudal with small brown spots forming faint cross series.
A single specimen known. Off coast of Lower California. (Gilbert.)
(*crescentalis*, pertaining to a crescent.)

Gobiosoma crescentalis, Gilbert, Proc. U. S. Nat. Mus. 1891, 557, off coast of Lower Cal-
    ifornia, at Albatross Station 2825, 24° 22′ N., 110° 19′ 15″ W., in 79 fathoms.

2594. GOBIOSOMA MULTIFASCIATUM, Steindachner.

Head about 4; depth 5⅔. D. VII–12 (Poey), VI–11 (Steindachner); eye 4
in head. Body and head compressed. Angle of mouth little behind center
of eye. Teeth pointed, in several series, those of outer series somewhat
enlarged. Dorsal rays not filamentous; caudal fin rounded. Head light
yellow; a carmine-red bar extending along upper edge of head, from
upper corner of gill opening to snout, where it joins its fellow, ending
behind over the pectoral in a small indigo-blue spot; body with 16 or 17
light-green, well-defined cross bars, separated by narrow white stripes.
(Steindachner.) West Indies; known from Cuba, St. Thomas, and the
Lesser Antilles; not seen by us. Its coloration is very different from that
of *Gobiosoma*, and it may belong to a distinct genus. (*multus*, many;
fasciatus, banded.)

*Gobius lineatus,** Poey, Memorias, ii, 424, 1861, Cuba; name preoccupied by *Gobius line-*
    aius, Jenyns.
Gobiosoma multifasciatum, Steindachner, Ichth. Beiträge, v, 183, 1870, Lesser Antilles;
    Jordan & Eigenmann, *l. c.*, 509; Eigenmann & Eigenmann, *l. c.*, 73.

828. BARBULIFER, Eigenmann & Eigenmann.

Barbulifer, Eigenmann & Eigenmann, Proc. Cal. Ac. Sci. 1888, 70 (*papillosus*).

A series of numerous minute barbels around the mouth and chin; otherwise as in *Gobiosoma*; body naked, the dorsal spines 7; second dorsal and
anal very short. (*barbula*, a small barbel; *fero*, I bear.)

2595. BARBULIFER CEUTHŒCUS (Jordan & Gilbert).

Head 3⅔; depth 7. D. VII–10; eye 4; A. 10. Body slender; head narrow and slender, depressed; snout not blunt; mouth terminal, oblique,
the maxillary reaching to below eye, 3 in head; eyes close together; chin

---

* *Gobius lineatus* is thus described: Head 3½; depth of body 6 in length. D. VII–12;
eye 6 in head. Body elongate, subcylindrical, maxillary extending almost to below
middle of eye; pectorals rounded; dorsals high. Yellowish green; the body with 20 vertical yellow bands; a red band extending from snout to point of opercle; fins yellowish.
Cuba. (Poey.) Type .43 mm. in length.

with a fringe of short barbles; vertical fins high, rays not filamentous. Upper half of head and body brown, finely speckled; 4 oblong, colorless areas along base of dorsals and a smaller one on back of caudal peduncle; lower parts abruptly pale; back with 5 or 6 blackish cross bars reaching to middle of sides, below which they extend as 5 or 6 short V-shaped projections; a brownish streak below eye; a small brown bar on base of pectoral, and a jet black bar at base of caudal. About Key West; scarce. (κεῦθος, a cavity; οἰκέω, to inhabit; the type specimen taken from the cavity of a sponge.)

Gobiosoma ceuthœcum, JORDAN & GILBERT, Proc. U.S.Nat.Mus. 1884, 29, Key West; young (Type in U. S. N. M.); JORDAN & EIGENMANN, *l. c.*, 508.

*Barbulifer papillosus,** EIGENMANN & EIGENMANN, Proc. Cal. Ac. Sci. 1888, 70, Key West, Florida; adult.

829. TYPHLOGOBIUS, Steindachner.

(BLIND GOBIES.)

Typhlogobius, STEINDACHNER, Ichth. Beiträge, VIII, 24, 1879 *(californiensis).*
Othonops, ROSA SMITH, Proc. U. S. Nat. Mus. 1881, 19 *(eos = californiensis).*

Body moderately elongate, compressed, covered with loose, smooth, naked skin. Head large, depressed, with tumid cheeks. Mouth large, the maxillary reaching to beyond the orbit; jaws equal, each with a narrow band of villiform teeth, the outer teeth slightly enlarged; lower jaw capable of little motion; snout rounded; no cirri. Eyes very small, reduced to mere vestages, covered by skin, and functional only in the young. Skull greatly modified, the brain case quadrate. Fins low; first dorsal of 2 flexible spines; second dorsal moderate; anal very short; caudal rounded; ventral disk as in *Gobius.* Gill openings rather narrow. One species known; singular blind gobies, living like slugs under rocks between tide marks. (τυφλός, blind; *Gobius.*)

* This species, which we suppose to be the adult of *Barbulifer ceuthœcus,* is thus described by Dr. Eigenmann: Head 3½ (4¾ in total); depth 4½ (5¾). D. VII-9; A. 9. Body short and robust, deepest below first dorsal spine; head blunt, profile straight from first dorsal spine to eye, much curved in front of eye; eye longer than snout, 3½ in head; interorbital area ⅔ diameter of eye; snout blunt; mouth small, oblique; maxillary 3 in head, reaching to below anterior margin of pupil; lips thick. About 21 barbels, in length ½ orbital diameter or longer, arranged as follows: A series of 7 cross the snout from one angle of the mouth to the opposite angle, the anterior 3 on the snout rather thick and colored (2 of them nasal), all the others yellowish, the barbel nearest each angle of the mouth longer than any of the others; on the lower jaw a barbel near each rictus, 2 on the chin, behind which are 2 pairs of barbels; posterior to these and below the rictus are 2 barbels on each side; 1 slender barbel on each side of preopercle below the posterior margin of the eye. Numerous rows of pores or papillæ on the head; 1 series extending straight downward on the anterior part of the opercle, from the upper end of which another series extends perpendicularly backward; other pores irregularly scattered on the opercle; a double series extending along edge of preopercle, the pores becoming larger and especially conspicuous below, meeting on the chin; 6 or 7 series radiating from eye, extending to snout, maxillary, and opercular series below; a row of pores nearly surrounding mouth, curving backward, encircling the nasal opening; 1 series about the eye posteriorly, otherwise none on top of head or nape; fins high and rounded; second dorsal higher than first, 1½ in head, caudal very broad and rounded, equal to the head in length; anal lower than soft dorsal; ventral reaching ⅔ to vent, 1½ in head; pectoral 1½ in head. Color yellow; upper half of body with a broad band of purplish spots; 6 diamond-shaped spots of darker cross the band, extending above and below it; nape, top of head, and upper part of cheek covered with dark points; opercle light yellow, cheeks darker; an oblique bar of black points on upper half of pectoral base, a curved bar of fainter spots on base of caudal; fins otherwise colorless and transparent. Length 1½ inches. (Eigenmann.)

2596. TYPHLOGOBIUS CALIFORNIENSIS, Steindachner.

(BLIND GOBY OF POINT LOMA; PINK-FISH.)

Head 3½; depth 5; eye 6; eye concealed, very small; D. II-12; A. 12. Vertebræ 17+13. Body subcylindrical, the males more compressed behind; head very broad behind, its greatest width ⅔ its length. Interorbital space a mere ridge; skin about mouth and eye very loose; a small papilla in front of nasal opening. Lower lip developed as a fold; another fold of skin behind it, bordered with fine cilia; behind this fold is a row of short, thick papillæ; edge of jaw rounded. Spinous dorsal remote from the soft dorsal in the male, but connected with it by a low membrane, this membrane absent in the female; soft dorsal much higher than the spinous; caudal broad, rounded; anal very short, inserted under sixth dorsal ray, and coterminous with dorsal; pectorals little longer than ventrals, 2 in head. Body naked; males with small tubercular plates irregularly placed. Skull highest at its posterior part, depressed forward; the bones all thick and strong. No lateral crests; a median keel which is lowest behind. Orbit not bounded by any ridges. Two keels diverge from the posterior end of the median keel to the insertion of the suprascapula. Premaxillaries and mandible very long. Teeth of the upper jaw all alike, long, close-set, in a broad band, those of the lower jaw in a narrow band, the inner ones apparently larger. Color uniform light pink. Length 2 inches. Coast of Lower California, from San Diego southward to Cerros Island; an extraordinary fish, found attached to the lower side of rocks in shallow water or surf; especially common at Point Loma.

Typhlogobius californiensis, STEINDACHNER, Ichth. Beiträge, VIII, 24, 1879, False Bay, San
   Diego, California (Coll. Prof. Essmark); JORDAN & GILBERT, Synopsis, 639, 1883;
   JORDAN & EIGENMANN, *l. c.*, 511.
Othonops eos, ROSA SMITH, Proc. U. S. Nat. Mus. 1881, 53, Point Loma, California.

830. TYNTLASTES, Günther.

Tyntlastes, GÜNTHER, Proc. Zool. Soc. London 1862, 193 (*sagitta*).

Body elongate, compressed, covered with small, imbricate, cycloid scales. Head elongate, quadrangular. Mouth wide, oblique, the lower jaw projecting; teeth small, in single series, none on vomer or palatines. Eyes very small, or rudimentary. Dorsal fin single, continuous, about 6 of its anterior rays simple; caudal fin pointed, more or less joined to the dorsal and anal; ventral fins united. Air bladder very small or absent. No pseudobranchiæ. Vertebræ 11+20. Pacific Ocean. (τυντλάστης, a mud-dabbler.)

a. Dorsal and anal each with 15 soft rays; head 4½ in length.   BREVIS, 2597.
aa. Soft dorsal and anal each with 21 unbranched or soft rays; head 5½ in length.
                                                                                    SAGITTA, 2598.

2597. TYNTLASTES BREVIS (Günther).

Head 4½; depth 8. D. VI, 15; * A. 15. Eyes minute. Jaws each with a

---

* The dorsal formula is apparently VIII, 14 in 2 half-digested specimens taken from the stomach of a *Centropomus* at Panama. (Gilbert.)

series of wide-set teeth. Caudal fin black. (Günther.) Panama; not seen by us. (*brevis*, short.)

Amblyopus brevis, GÜNTHER, Proc. Zool. Soc. London 1864, 151, Panama; GÜNTHER, Fishes Centr. Amer., 441, 1869.

Tyntlastes brevis JORDAN & EIGENMANN, *l. c.*, 512.

2598. TYNTLASTES SAGITTA (Günther).

Head $5\frac{2}{3}$; depth $9\frac{2}{3}$. D. VI, 21; A. 21. Body and head elongate, compressed. Maxillary reaching to behind eye; teeth subhorizontal, very small. Scales becoming larger posteriorly. Caudal arrow-shaped, about 4 in body; pectorals as long as ventrals, 2 in head. Grayish, sides and under parts silvery; an ovate gray spot before each dorsal ray; caudal grayish. (Günther.) Length $9\frac{1}{4}$ inches. Coast of Lower California; exact locality unknown. (*sagitta*, arrow.)

Amblyopus sagitta, GÜNTHER, Proc. Zool. Soc. London 1862, 193, "California," probably from Lower California.

Tyntlastes sagitta, JORDAN & GILBERT, Synopsis, 639, 1883; JORDAN & EIGENMANN, l. c., 512.

831. GOBIOIDES, Lacépède.

(BARRETOS.)

Gobioides, LACÉPÈDE, Hist. Nat. Foiss., II, 580, 1798 (*broussonnetii*).

Plecopodus, RAFINESQUE, Analyse de la Nature, 87, 1815 (*broussonnetii*); substitute for *Gobioides*, regarded as objectionable.

Ognichodes, SWAINSON, Nat. Hist. Class'n Animals, II, 183 and 278, 1839 (*broussonnetii*).

Body greatly elongate, compressed behind, the scales very minute; head small; eyes very small; mouth large, oblique, the lower jaw projecting; gill openings moderate. Teeth in a band, those in the outer series being very strong. Dorsal rays V to VII, 15 to 23; anal rays 16 to 23. Dorsal fin low, continuous, the spines similar to the soft rays, but more widely separated; the soft dorsal and the anal are joined to base of caudal; ventrals 45, united in a disk which is formed much as in *Gobius*. No air bladder; no pseudobranchiæ. From *Tænioides* (=*Amblyopus*) the genus *Gobioides* is distinguished by the absence of barbels, the presence of scales, and by the much smaller number of rays in its vertical fins. Brackish waters of the Tropics, reaching a considerable size. (*Gobius;* εῖδος, resemblance.)

a. Eye small, but evident; scales evident, larger behind.   BROUSSONNETII, 2599.
aa. Eye minute, not evident; scales minute.   PERUANUS, 2600.

2599. GOBIOIDES BROUSSONNETII, Lacépède.

Head $5\frac{1}{4}$ (young) to 7 (adult); caudal $3\frac{1}{2}$ to 5; eye small but evident, 7 to 10 in head; interorbital space 1 to $1\frac{2}{3}$ diameter of eye. D. VII, 16; A. I, 16. Body elongate, mouth oblique, maxillary extending beyond eye; teeth in bands, the outer series enlarged, shorter, and closer set than in *Gobioides peruanus;* scales twice as large as in *peruanus*, those on anterior part of body not imbricated, much smaller than those on posterior part, which are elongate oval in form. Violet bars extending downward and forward on the upper part of body; sometimes a violet spot with a lighter or darker dot at end of the bars; head marbled or spotted with dark violet or brown. (Steindachner.) Length 20 inches or more. West Indies to Brazil; common southward, ascending rivers; once taken near New Orleans (Bean & Bean). (Named for Dr. Augustin Broussonnet, professor in the University of Montpelier.)

Gobioides broussonnetii, LACÉPÈDE, Hist. Nat. Poiss., II, 580, 1798, probably from Surinam, "given by Holland to France."

Amblyopus brasiliensis, BLOCH & SCHNEIDER, Syst. Ichth., 69, 1801, Brazil; on drawing made by Prince Maurice; CUVIER & VALENCIENNES, Hist. Nat. Poiss., XII, 121, 1837.

Gobious oblongus, BLOCH & SCHNEIDER, Syst. Ichth., 548, 1801; based on LACÉPÈDE.

Gobioides barreto, POEY, Memorias, II, 282, 1861, Cuba; POEY, Synopsis, 394, 1868; POEY, Enumeratio, 125, 1876.

Amblyopus mexicanus, O'SHAUGHNESSY, Ann. Mag. Nat. Hist., series IV, vol. XV, 1875, 147, Mexico.*

Gobioides broussoneti, JORDAN & EIGENMANN, *l. c.*, 512; BEAN & BEAN,† Proc. U. S. Nat. Mus. 1895, 631.

2600. GOBIOIDES PERUANUS (Steindachner).

Head 5; depth 11. D. VII, 17; A. I, 16. Eye scarcely visible, much smaller than in *G. broussoneti;* scales very minute; snout $2\frac{1}{2}$ in postorbital part of head; interorbital 5 in head; lower jaw slightly projecting; maxillary $2\frac{3}{5}$ in head; a series of large slender teeth in each jaw, behind which, in each jaw, is a narrow band of fine teeth; caudal $4\frac{1}{3}$ in

* The following is Mr. O'Shaughnessy's description of *Amblyopus mexicanus:* D. VII, 15; A. I, 15. Depth 13 in total length. Body covered all over with scale-shaped crypts. Head naked. Dorsal $\frac{3}{4}$ height of body. Eye small, but distinct. Snout obtuse; lower jaw extending a little beyond upper. Teeth small, close set, the outer series much smaller and more closely set than in *G. broussonnetii.* Dorsal and anal connected with the caudal. Upper parts dark brown, with a series of white spots along the whole length of the side; lower parts of sides and belly white. One specimen in the British Museum, from Mexico, purchased. Length $20\frac{1}{4}$ inches. (O'Shaughnessy.)

This seems to differ from *G. broussonnetii* in color only.

† The following description is given by Bean & Bean of *Gobioides broussonnetii* (Lacépède): Head 7; depth 14. D. VI, 17; A. I, 16. The greatest depth of the head equals the length of the upper jaw, or about $\frac{1}{2}$ the length of head without snout. The body is compressed. Its greatest thickness is contained $1\frac{2}{3}$ times in its greatest depth. The teeth are in narrow bands in each jaw, some of those in the outer row enlarged, canine-like, and curved inward. All of the teeth are more or less curved inward and depressible. The vomer and palate are toothless. The mouth is oblique, the lower jaw projecting slightly beyond the upper. The maxilla extends well behind the eye, its length is slightly more than $\frac{1}{2}$ that of head without the snout. It is not much expanded posteriorly. Eyes very small, their diameter equaling $\frac{1}{2}$ length of snout, about equal to width of interorbital space. The snout scarcely equals more than $\frac{1}{8}$ of the head's length. Gill openings wide, the membranes wholly joined to the isthmus. Branchiostegals much curved, 4 in number. The dorsal begins at a distance from the nape equal to the postorbital part of the head, the origin being about over the end of the extended pectoral. The ventral reaches farther back than the pectoral, and is longer than that fin, its length equaling postorbital part of head. The distance of the vent from the tip of the snout equals somewhat more than 3 times the length of the head; it is under the interspace between the last spine and first ray of the dorsal, with a small genital papilla behind it. The caudal is very long and tapering, $1\frac{2}{3}$ times as long as the head. The dorsal spines are long and slender, the fifth nearly as long as the postorbital part of the head. The second dorsal ray is slightly longer. The anal rays are about as long as those of the dorsal. The scales are thin, not imbricated, except on the posterior part of the head, where they are long and elliptical in shape. The head and breast are naked. The colors have faded out in alcohol; the ground color appears to have been light brown, with darker blotches on the median line of the body under the spinous portion of the dorsal and the anterior part of the soft dorsal. (Bean & Bean.) Here described from a specimen obtained in the Gulf of Mexico by Mr. Robert S. Day, of New Orleans, Louisiana, and is No. 38220, U. S. Nat. Mus.

body, connected by membrane to dorsal and anal; sides with regular cross series of pores. Body with narrow angular cross bars; dorsal rays violet, the membrane yellowish. (Steindachner.) Shores of Ecuador and Peru, ascending rivers.

Amblyopus peruanus, STEINDACHNER, Fisch-Fauna des Cauca und Flüsse bei Guayaquil, 42, 1880, Guayaquil.

Gobioides peruanus, EIGENMANN & EIGENMANN, Proc. Cal. Ac. Sci., 2d ser., I, 1888, 75.

832. CAYENNIA, Sauvage.

Cayennia, SAUVAGE, Bull. Sci. Philom., ser. 7, IV, 1880, 57 *(guichenoti).*

Body much elongate; dorsals united, caudal free from dorsal and anal; ventrals united, not adhering to belly; teeth small, the outer enlarged; anterior part of body naked, posterior part covered with cycloid scales. Otherwise as in *Gobioides,* from which the genus may not be separable. (Name from Cayenne.)

2601. CAYENNIA GUICHENOTI, Sauvage.

Head 9; depth 17. D. VI, 17; A. I, 16; vertebræ about 36. Head deeper than wide; eye small, placed well forward; maxillary reaching to below posterior margin of eye; a low membrane connecting dorsal and caudal; caudal 7 in length; ventrals 1½ in head. Color brownish, marbled with black anteriorly. Cayenne (Sauvage); not seen by us. (Named for A. Guichenot, formerly ichthyologist of the Muséum d'Histoire Naturelle at Paris.)

Cayennia guichenoti, SAUVAGE, Bull. Soc. Philom., ser. 7, IV, 1880, 57 Cayenne; EIGENMANN & EIGENMANN, Proc. Cal. Ac. Sci. 1888, 76.

Suborder DISCOCEPHALI.

Bony fishes "with a suctorial transversely laminated oval disk on the upper surface of the head (homologous with a flat dorsal fin), thoracic ventral fins with external spines, a simple basis cranii, intermaxillary bones flattened, with the ascending processes deflected sideways, and with the supramaxillary bones attenuated backward, flattened, and appressed to the dorsal surface of the intermaxillaries; hypercoracoid (or scapula) perforated nearly in the center, and with 4 short actinosts (carpals)." (Gill.)

This remarkable group consists of a single family, *Echeneididæ.* (δίσκος, disk; κεφαλή, head).

Family CLXXXIX. ECHENEIDIDÆ.

(THE REMORAS.)

Body fusiform, elongate, covered with minute, cycloid scales. Mouth wide, with villiform teeth on jaws, vomer, palatines, and usually on tongue. Premaxillaries not protractile. Lower jaw projecting beyond upper. Spinous dorsal modified into a sucking disk, which is placed on

3030——65

the top of the head and neck, and is composed of a double series of transverse, movable, cartilaginous plates, serrated on their posterior or free edges. By means of this disk these fishes attach themselves to other fishes or to floating objects, and are carried for great distances in the sea. Opercles unarmed. Pectoral fins placed high; ventral fins present, thoracic and close together, I, 5; dorsal and anal fins long, without spines, opposite each other; caudal fin emarginate or rounded. Branchiostegals 7. Gills 4, a slit behind the fourth; gill rakers short; gill membranes not united, free from the isthmus. Pseudobranchiæ obsolete. Several pyloric appendages. No air bladder. No finlets. No caudal keel. Vertebræ more than $10 + 14$. Genera 4; species about 10, found in all seas, all having a very wide range. The species of this group are apparently descended from a fossil genus, *Opisthomyzon*,* Cope (*glaronensis*), characterized by the small posterior disk and slender body.

The following description of this family is given by Dr. Gill: Body elongated, subcylindrical, diminishing backward gradually from the head and into the slender caudal peduncle. Anus subcentral. Scales cycloid, very small, and not, or scarcely, imbricated. Lateral line nearly straight and very faint. Head above oblong and with a flattened straight upper surface, furnished with an adhesive oblong or elongated, laminated disk. The eyes are rather small, submedian, and overhung by the disk. Suborbital bones forming a slender infraorbital chain; the first or preorbital triangular and thick. Opercular apparatus normally developed and unarmed. Nostrils double, close together. Mouth terminal or, rather, superior, the lower jaw projecting, but with the cleft nearly horizontal and not extending laterally to the eyes. Teeth present on the jaws and palate. Branchial apertures ample and fissured forward. Branchiostegal rays 7 (or 8) on each side. The adhesive disk on the upper surface of the head is a modified first dorsal fin, and from the snout generally extends more or less posteriorly on the nape and back; it is oblong or elongated and of an oval or elliptical form, divided into equal halves by a longitudinal septum, and with more or less numerous transverse laminæ in each division, the laminæ being slightly erectile and depressible. Dorsal fin oblong or elongated on the posterior half of the body (including head), ending some distance from the caudal. Anal fin opposite and similar to the dorsal. Caudal fin rather small, variable in outline, but never deeply forked. Pectoral fins moderate, inserted high on the sides. Ventral fins thoracic, each with a spine and 5 branched rays. The vertebral column has vertebræ in slightly increased numbers, the abdominal vertebræ being about 12 to 14 and the caudal 15 or 16. The stomach is cæcal and the pyloric cæca are present in moderate numbers. The air bladder is obsolete.

---

* "A careful comparison of the proportions of all the parts of the skeleton of the fossil *Echeneis* with those of the living forms, such as *Echeneis naucrates* or *Echeneis remora*, shows that the fossil differs nearly equally from both, and that it was a more normally shaped fish than either of these forms. The head was narrower and less flattened, the preoperculum wider, but its two jaws had nearly the same length. The ribs, as also the neural and hæmal spines, were longer, the tail more forked, and the soft dorsal fin much longer. In fact, it was a more compressed type, probably a far better swimmer than its living congeners, as might be expected, if the smallness of the adhesive disk is taken into account." (Storms.) This form (*Echeneis glaronensis*, Wellstein) is made the type of the genus *Opisthomyzon*, Cope, the name referring to the posterior portion of the small disk. The vertebræ in *Opisthomyzon* are $10 + 13 = 23$.

Concerning the relations of this family, Dr. Gill has the following pertinent remarks:

"The family of *Scombéroïdes* was constituted by Cuvier for certain forms of known organization, among which were fishes evidently related to *Caranx*, but which had free dorsal spines. In the absence of knowledge of its structure, the genus *Elacate* was approximated to such-because it also had free dorsal spines. Dr. Günther conceived the idea of disintegrating this family, because; inter alias, the typical *Scomberoides* (family *Scombridæ*) had more than 24 vertebræ and others (family *Carangidæ*) had just 24. The assumption of Cuvier as to the relationship of *Elacate* was repeated, but inasmuch as it has 'more than 24 vertebræ' (it has $25 = 12 + 13$) it was severed from the free-spined *Carangidæ* and associated with the *Scombridæ*. *Elacate* has an elongated body, flattish head, and a colored longitudinal lateral band; *Echeneis* has also an elongated body, flattened head, and a longitudinal lateral band; therefore *Echeneis* was considered to be next allied to *Elacate* and to belong to the same family. The very numerous differences in structure between the two were entirely ignored, and the reference of the *Echeneis* to the *Scombridæ* is simply due to assumption piled on assumption. The collocation need not, therefore, longer detain us. The possession by *Echeneis* of the anterior oval cephalic disk in place of a spinous dorsal fin would alone necessitate the isolation of the genus as a peculiar family. But that difference is associated with almost innumerable other peculiarities of the skeleton and other parts, and in a logical system it must be removed far from the *Scombridæ*, and probably be endowed with subordinal distinction. In all essential respects it departs greatly from the type of structure manifested in the *Scombridæ* and rather approximates—but very distantly—the *Gobioidea* and *Blennioidea*. In those types we have in some a tendency to flattening of the head, of anterior development of the dorsal fin, a simple basis cranii, etc. Nevertheless, there is no close affinity nor even any tendency to the extreme modification of the spinous dorsal exhibited by *Echeneis*. In view of all these facts *Echeneis*, with its subdivisions, may be regarded as constituting not only a family but a suborder. * * * Who can consistently object to the proposition to segregate the *Echeneididæ* as a suborder of teleocephaleous fishes? Not those who consider that the development of 3 or 4 inarticulate rays (or even less) in the front of the dorsal fin is sufficient to ordinarily differentiate a given form from another with only 1 or 2 such. Certainly the difference between the constituents of a disk and any rays or spines is much greater than the mere development or atrophy of articulations. Not those who consider that the manner of depression of spines, whether directly over the following, or to the right or left alternately, are of ordinal importance; for such differences again are manifestly of less morphological significance than the factors of a suctorial disk. Nevertheless, there are doubtless many who will passively resist the proposition because of a conservative spirit, and who will vaguely recur to the development of the disk as being a 'teleological modification,' and as if it were not an actual fact and a development correlated with radical modifications of all parts of the skeleton at least. But whatever may be the closest relations of *Echeneis*, or the systematic value of its peculiarities, it is certain that it is not allied to *Elacate* any more than to others of the hosts of Scombroid, Percoid, and kindred fishes, and that it differs in toto from it, notwithstanding the claims that have been made otherwise. It is true that there is a striking resemblance, especially between the young—almost as great, for example, as that between the placental mouse and the marsupial *Antechinomys*—but the like is entirely superficial, and the scientific ichthyologist should be no more misled in the case than would the scientific therologist by the likeness of the marsupial and placental mammals."

    a. Body very slender, the vertebræ 14 + 16 = 30; ventrals narrowly adnate to abdomen; lower jaw produced in a flap; pectorals acute, with flexible rays.
        b. Laminæ 10 only.   PHTHEIRICHTHYS, 833.
        bb. Laminæ 20 to 28.   ECHENEIS, 834.
    aa. Body rather robust, the vertebræ 12 + 15 = 27; ventrals broadly adnate to abdomen; lower jaw not produced; pectorals rounded.
        c. Laminæ 24 to 27.   REMILEGIA, 835.
        cc. Laminæ 16 to 20.
            d. Pectoral rays soft and flexible.   REMORA, 836.
           dd. Pectoral rays stiff and ossified.   RHOMBOCHIRUS, 837.

833. PHTHEIRICHTHYS, Gill.

Phtheirichthys, GILL, Proc. Ac. Nat. Sci. Phila. 1862, 239 (*lineata*).

Disk with 10 laminæ; palatines with sharp teeth; teeth in pairs, uniform in all ages; otherwise as in *Echeneis*. A single species, found attached to spearfishes and Barracudas. (ψθείρ, a louse; ἰχθύς, fish.)

2602. PHTHEIRICHTHYS LINEATUS (Menzies).

Head 5; disk twice as long as broad, its length 4½ in body. D. X–33; A. 33. Lower jaw very narrow, much projecting. Body blackish, with 2 whitish lateral bands; all the fins white-margined. Tropical seas, ranging north to South Carolina and Pensacola; rather rare. (*lineatus*, striped.)

Echeneis lineata, MENZIES, Trans. Linn. Soc. London, I, 1791, 187, pl. 17, fig. 1, Pacific Ocean between the tropics; GÜNTHER. Cat., II, 382, 1860.
Echeneis tropica, EUPHRASEN, Nya Handl., XII, 317, 1791, Atlantic between the Tropics.
Echeneis apicalis, POEY, Memorias, II. 254, 1861, Cuba. (Coll. Poey.)
Echeneis sphyrænarum, POEY, Memorias, II, 255, 1861, Cuba, on Barracudas. (Coll. Poey.)
Phtheirichthys lineatus, JORDAN & GILBERT, Synopsis, 969, 1883.

834. ECHENEIS (Artedi) Linnæus.

Echeneis (ARTEDI) LINNÆUS, Syst. Nat., Ed. X, 260, 1758 (*naucrates*).
Leptecheneis, GILL, Proc. Ac. Nat. Sci. Phila. 1864, 60 (*naucrates*); the name *Echeneis* being transferred to *E. remora*, the only species known to Artedi.

Body comparatively elongate, the vertebræ 14 + 16 = 30; disk long, of 20 to 28 laminæ; pectoral pointed, its rays soft and flexible; soft dorsal and anal long, of 30 to 41 rays each; caudal lunate in the adult, convex in the young. Species of wide distribution, attaching themselves mainly to sea turtles and large fishes. (ἐχενηΐς, an ancient name, from ἔχω, to hold back; ναῦς, a ship.)

a. Disk of 22 to 26 laminæ (rarely 21 or 28), its length less than ¼ body. NAUCRATES, 2603.

aa. Disk of 20 or 21 laminæ, its length more than ¼ body. NAUCRATEOIDES, 2604.

2603. ECHENEIS NAUCRATES, Linnæus.

(SHARK-SUCKER; PEGA; PEGADOR; SUCKING-FISH.)

Head 5¼; depth 11 to 12. D. XXII to XXVIII (rarely XXI)–32 to 41; A. 31 to 38. Breadth between pectorals 7½; disk 4 to 5 in body; eye 5 in head; snout 2⅓; maxillary 3; from angle of mouth to tip of lower jaw 2⅔; pectoral 1⅗; ventrals 1½; middle caudal rays 1⅗; highest anal ray 2; highest dorsal ray 2⅓; width of disk 2½ in its length; base of dorsal 2½, anal 2½, in body. Body elongate, subterete, slender. Lower jaw strongly projecting, the tip flexible; maxillary reaching nostril; teeth uniform in the adult, the young with series of small slender teeth in advance of the others; gill rakers short and slender, about equal to pupil; vertical fins low. Anal rays higher than dorsal anteriorly; pectorals reaching very slightly past tips of ventrals; origin of ventral spine under middle of pectoral base; inner rays of ventral fins narrowly adnate to the abdomen; dorsal and anal commencing and ending opposite each other; caudal with the middle rays produced in the young, the fin becoming emarginate or lunate with age. Color brownish; belly dark, like the back, as usual in this family; sides with a broad stripe of darker edged with whitish extending through eye to snout; caudal black, its outer angles whitish; pectorals and ventrals black, sometimes bordered with pale; dorsal and anal broadly edged with white anteriorly; adult nearly uniform dark brown, not paler below. Warm seas, universally distributed; common north to Cape Cod and occasionally to San Francisco, attaching itself to turtles and to large fishes. This species is very common in the tropics, being found attached to sharks, groupers, or any other large fish, without regard to species. Few large sharks at Key West are without them. They are often caught with hook and line from the wharf, where they frequently forsake their host to take the bait. Lütken's remark that only *Remora remora* has been recorded from sharks is no longer true. Several writers have recognized 2 species of *Echeneis* proper—*naucrates*, with 22 to 26 laminæ, the disk 4 to 5 in body, and *naucrateoides* (= *albicauda* = *holbrooki* = *lineatus*), in which the disk is longer, 3⅔ to 4 in body, but composed of fewer, 20 or 21, laminæ. The latter form is rather common on our coast, the specimens from Key West above mentioned having 21. We doubt the existence of any permanent difference between the two, but provisionally retain *Echeneis naucrateoides* as a species distinct from *Echeneis naucrates* until more complete comparison can be made. (*naucrates*, a pilot; ναῦς, ship; κρατέω, to govern, guide.)

Echeneis neucrates (misprint for *naucrates*), LINNÆUS, Syst. Nat., Ed. x, 261, 1758, "in Pelago Indico;" GÜNTHER, Cat., II, 384, 1860; JORDAN & GILBERT, Synopsis, 416, 1883.
Echeneis albicauda, MITCHILL, Amer. Monthly Mag., II, 1817, 244, New York.
Echeneis lunata, BANCROFT, Proc. Comm. Zool. Soc., I, 1830, 134, Kingston, jamaica.
? *Echeneis vittata*, LOWE, Proc. Zool. Soc. Lond. 1839, 89, Madeira.
Echeneis fasciata, GRONOW, Ed. Gray, 92, 1854, Mediterranean Sea.
Leptecheneis naucrates, GILL, Proc. Ac. Nat. Sci. Phila. 1864, 60.

Echeneis vittata, RÜPPELL, Neue Wirb. Fische, 82, 1835, Red Sea.
Echeneis guaiacan, POEY, Memorias, II, 248, 1861, Cuba; young.   (Coll. Poey.)
Echeneis verticalis, POEY, Memorias, II, 253, 1861, Cuba; young.
Echeneis metallica, POEY, Memorias, II, 252, 1861, Cuba; D. XXIII, 40; A. 37; large speci-
    men, metallic green, the bands faint.   (Coll. Poey.)
Echeneis fusca, GRONOW, Cat. Fish., 92, 1854; after *E. naucrates*, L.

2604. ECHENEIS NAUCRATEOIDES, Zuiew.

Head 5; depth 11.   D. XX or XXI–32 to 35; A. 33 to 35.   Disk 3⅔ to 3¾ in
total, twice width of body between pectorals.   In all other respects essentially as in *Echeneis naucrates*, the disk longer, but composed of fewer
laminæ, the laminæ being farther apart.   Color of *Echeneis naucrates*.
Cape Cod to West Indies, common on our south Atlantic coast; specimens before us from Key West.   (*naucrates*, ναυκράτης, a pilot; εἶδος,
resemblance.)

Echeneis neucratoides, ZUIEW, Nova Acta Acad. Sci. Imp. Petropol., IV, 1789, 279, no locality.
Echeneis lineata. HOLBROOK, Ichth. S. C., 102, 1860, Charleston, South Carolina; not of
    MENZIES.
Echeneis holbrooki, GÜNTHER, Cat., II, 382, 1860, Jamaica; D. XIII, 35; A. 33.
Leptecheneis naucrateoides, GILL, *l. c.*, 61.

835. REMILEGIA, Gill.

Remilegia, GILL, Proc. Ac. Nat. Sci. Phila. 1864, 61 (*australis*).

This genus differs from *Remora* chiefly in the length of the sucking disk,
which has 24 to 27 laminæ; the soft dorsal and anal are proportionately
short.   (A metathesis for *remeligo*, the delayer or hinderer.)

2605. REMILEGIA AUSTRALIS (Bennett).

D. XXVII–22; A. 21 to 23.   The length of the disk is 2¼ in the total, the
width of the body between the pectorals 5⅔.   Caudal truncated; dorsal
and anal fins not continued to the caudal.   Color brown.   This species
has the general habit of *E. remora*, but may be readily distinguished from
all the others by the extraordinary size of the disk, which is elongate,
subelliptical, obtusely rounded anteriorly and posteriorly, and formed by
27 pairs of laminæ; it extends backward beyond the vertical from the
tip of the ventrals, and its length is 2¼ in the total.   The spines with
which the single lamina are armed are less conspicuous than in the other
species, and do not offer the same resistance to the touch.   There is a
large posterior portion of the disk which is not provided with laminæ,
but quite smooth.   The width of the disk, taken between the extremities
of the bony laminæ, is ⅓ of its length; the membranaceous margin is
bent upward.   The head and the body below the disk are depressed, and
their height is 9¼ in the total length, whilst the width between the pectorals is 5⅔ in it.   The body between the disk and the vertical fins is
quadrangular, tapering posteriorly.   The upper jaw is subtruncated, and
overreached by the lower, which is much narrower; both are armed with a
broad band of villiform teeth, and with an outer series of larger ones on

the sides; the vomerine and palatine bones have a continuous band of teeth, narrowest on the vomer; the tongue is hard, cartilaginous, and destitute of teeth. The cleft of the mouth reaches only to the vertical from the nostril; the eye is small. The pectoral is rounded and small, its length being $\frac{1}{3}$ of the total; the ventrals are slightly pointed, and, as in all the species of the genus, composed of 1 spine, hidden in the skin, and 4 soft rays; they are inserted immediately behind the vertical from the pectoral, which they equal in length; they can be received in a shallow groove on the abdomen. The distance between the dorsal and the disk is 3½ in the length of the latter; the dorsal is low, and enveloped in a thick membrane. The caudal is truncated when stretched out. The anal is very similar to the dorsal, and its origin and termination fall vertically below those of the latter. The scales are minute, and can be perceived only by the aid of a magnifier; they are embedded in pore-like cavities. (Günther: description of type of *Echeneis scutata*.) Tropical seas; rare; recorded by Dr. Lütken from 10° N., 39° W. (Coll. Capt. V. Hygom) from a dolphin; not seen by us. (*australis*, southern.)

Echeneis australis, BENNETT, Narr. Whaling Voyage, II, 273, pls. 24–26, 1840.
Echeneis scutata, GÜNTHER, Ann. Mag. Nat. Hist. 1860, 401, pl. 10, f. B, Ceylon (Coll. Dr. Sibbald); GÜNTHER, Cat. Fish., II, 381, 1860; LÜTKEN, Vid. Medd. Kjöbenh. 1875, 42.

836. REMORA, Gill.

(REMORAS.)

Remora, GILL, Proc. Ac. Nat. Sci. Phila. 1862. 239 (*remora*).
Echeneis, GILL, Proc. Ac. Nat. Sci. Phila. 1864, 60 (*remora*); not *Echeneis*, GILL, 1862, restricted to *naucrates*.
Remoropsis, GILL, Proc. Ac. Nat. Sci. Phila. 1864, 60 (*brachypterus*).
Remorina, JORDAN & EVERMANN, Check-List Fishes, 490, 1896 (*albescens*).

Body rather robust, the vertebræ $12 + 15 = 27$; disk shortish, of 13 to 18 laminæ; pectoral rounded, its rays soft and flexible; soft dorsal and anal moderate, of 20 to 30 rays; caudal subtruncate. Species attaching themselves to large fishes, especially to sharks. (*Remora*, an ancient name, "holding back.")

REMORA:
    a. Laminæ about 18; soft dorsal with 23 rays.   REMORA, 2606.
   aa. Laminæ 13 to 16.
      REMORINA:
        b. Dorsal rays XIII, 22.   ALBESCENS, 2607.
      REMOROPSIS (*Remora*; ὄψις, appearance):
       bb. Dorsal rays XIV, XVI, 29 to 32.   BRACHYPTERA, 2608.

Subgenus REMORA.

2606. REMORA REMORA (Linnæus).

(REMORA.)

Head 4; disk 2¾; width between pectorals 5¼. D. XVIII–23; A. 25; vertebræ $12 + 15$. Body comparatively robust, compressed behind. Pectoral fins rounded, short, and broad, their rays short and flexible; ventral fins

adnate to the abdomen for more than ⅓ the length of their inner edge.
Tip of lower jaw not produced into a flap; head broad, depressed; disk
longer than the dorsal or the anal fin; maxillary scarcely reaching front
of orbit. Caudal lunate; vertical fins rather high; pectoral ⅔ length of
head. Color blackish, nearly uniform above and below. Length 15 inches.
Warm seas, north to New York and San Francisco, where it is not rare;
usually found attached to large sharks; very common in the West Indies;
more robust than *Echeneis naucrates*, and reaching a smaller size.

Echeneis remora, LINNÆUS, Syst. Nat. Ed. x, 260, 1758, "in Pelago Indico;' GÜNTHER, Cat.,
    II, 378, 1860; LÜTKEN, Vid. Medd. Kjöbenh. 1875, 38; JORDAN & GILBERT, Synopsis, 417,
    1883.
Echeneis squalipeta, DALDORF, Skrivt. Naturh. Selsk., II, 1797, 157, Atlantic Ocean between
    the tropics; GÜNTHER, Cat., II, 377, 1860.
Echeneis jacobœa, LOWE, Proc. Zool. Soc. London 1839, 89, Madeira.
Echeneis remoroides, BLEEKER, Batoë, II, 70, Batoe.
Echeneis parva, GRONOW, Cat. Fish., Ed. Gray, 92, 1854, no locality; after *E. remora*, L.
Echeneis postica, POEY, Memorias, II, 255, 1861, Havana. (Coll. Poey.)
Remora jacobœa, GILL, Proc. Ac. Nat. Sci. Phila. 1862, 239.

Subgenus REMORINA, Jordan & Evermann.

2607. REMORA ALBESCENS (Temminck & Schlegel).

Length of disk 3⅕ to 3¼ in total length; width between pectorals 5 to 5⅓;
number of laminæ on disk 13 or 14. D. XIII–22; A. 22. Angle of mouth
in the vertical from the third lamina of the disk. Length of ventral fins
equal to the distance between root of pectoral and posterior margin of
eye. Color uniform grayish brown. (Günther.) Tropical Pacific, straying to America; a specimen taken at La Paz, Gulf of California (Streets),
and 1 in the Gulf of Mexico (Bean). (*albescens*, whitish.)

Echeneis albescens, TEMMINCK & SCHLEGEL, Fauna Japonica, Poiss., 272, pl. 120, fig. 3, 1842,
    Japan; GÜNTHER, Cat., II, 377, 1860; STREETS, Bull. U. S. Nat. Mus., VII, 54, 1877.
Remora albescens, JORDAN, Cat. Fishes, 66, 1885.

Subgenus REMOROPSIS, Gill.

2608. REMORA BRACHYPTERA (Lowe).

Head nearly 4; width between pectorals 6½. D. XIV to XVI–29 to 32;
A. 25 to 30. Body robust, the greatest depth nearly twice the length of
the short pectoral fins; disk shorter than base of dorsal, rather broad;
upper jaw angular. Caudal nearly truncate. Light brown, darker below,
fins paler. Warm seas, occasionally north to Cape Cod. (βραχύς, short;
πτερόν, fin.)

Echeneis brachyptera, LOWE, Proc. Zool. Soc. London 1839, 89, Madeira; GÜNTHER, Cat.,
    II, 378, 1860; JORDAN & GILBERT, Synopsis, 417, 1883.
Echeneis sexdecimlamellata, EYDOUX & GERVAIS, Voy. Favorite, v, 77, pl. 31, 1839, Indian
    Ocean?
Echeneis quatuordecimlaminatus, STORER, Rept. Fishes Mass., 155, 1839, Holmes Hole.
Echeneis pallida, TEMMINCK & SCHLEGEL, Fauna Japonica, Poiss., 271, pl. 120, fig. 2, 3, 1842,
    Japan.
Echeneis nieuhofii, BLEEKER, Sumatra, II, 279, Sumatra.
Remoropsis brachypterus, GILL, Proc. Ac. Nat. Sci. Phila. 1864, 60.

837. RHOMBOCHIRUS, Gill.

Rhombochirus, GILL, Proc. Ac. Nat. Sci. Phila. 1863, 88 (*osteochir*).

This genus agrees with *Remora* in every respect excepting the structure of the pectoral fins. These are short and broad, rhombic in outline, the rays all flat, broad and stiff, being partially ossified, although showing the usual articulation; upper rays of pectoral broader than the others. One species known. ($\rho \acute{o} \mu \beta o \varsigma$, rhomb; $\chi \varepsilon \acute{\iota} \rho$, hand.)

2609. RHOMBOCHIRUS OSTEOCHIR (Cuvier).

Head 4⅔ in length; disk 2¼; width between pectorals 5. D. XVIII–21 to 23; A. 20 or 21; P. 20. Mouth very small, maxillary not nearly reaching to the line of the orbit; outer series of teeth longer than the others. Disk very large, broader and rougher than in *Remora remora*, extending forward beyond the tip of the snout. Caudal fin emarginate, with rounded angles. Light brown; underside of head, ventral line, part of ventrals and a spot on pectorals pale. West Indies north to Cape Cod; parasitic on species of *Tetrapturus*; rather rare. ($\delta \sigma \tau \acute{\varepsilon} o \nu$, bone; $\chi \varepsilon \acute{\iota} \rho$, hand.)

Echeneis osteochir, CUVIER, Règne Animal, Ed. 2, vol. II, 348, 1829, no locality given; GÜNTHER, Cat., II, 381, 1860; JORDAN & GILBERT, Synopsis, 418, 1883.
Echeneis tetrapturorum, POEY, Memorias, II, 256, 1858, Cuba. (Coll. Poey.)

Group TRACHINOIDEA.

(THE TRACHINOID FISHES.)

A large group of transitional forms, some of them of doubtful relationships, showing affinities with the *Percoidea* on the one hand and with the *Batrachoididæ* and *Blennoidea* on the other. In general, the spinous dorsal is short or weak, the soft dorsal long and similar to the anal, and the squamation is less complete and less ctenoid than in the *Percoidea*. The skull is, in general, depressed, with the supraocular crest low, and the suborbital stay is wanting, although in some genera the suborbital bones are enlarged. The bones of the skull are not strongly armed, and the ventral fins are often inserted well forward, and they are sometimes reduced in size. The group is divided by Dr. Gill into *Percophidoidea*, *Trachinoidea*, and *Uranoscopoidea*. The two latter groups are natural and related, but, as Dr. Gill observes, "the *Percophidoidea* are undoubtedly a heterogeneous group and need a thorough revision." The relations of *Bathymaster*, *Trichodon*, and *Latilus* especially are uncertain. Several of the leading families of this group are confined to the South Temperate Zone, and none of the *Trachinidæ* occurs within our limits.

    a. Mouth horizontal or moderately oblique, the lips not fringed; eyes lateral; ventral rays I, 5, their insertion more or less before the pectorals; suborbitals moderate; gills 4, a slit behind the fourth.
        b. Snout subconic, not prolonged and spatulate; ventrals not widely separated.
            c. Body covered with scales; dorsal spines flexible.
                d. Lateral line complete; caudal fin forked; vertebræ 24 to 27.
                                    MALACANTHIDÆ, CXO.

  dd. Lateral line incomplete, running close to the back; caudal rounded or lanceolate; dorsal fin continuous.
   e. Vertebræ about 27; scales cycloid; maxillary more or less dilated behind, with a supplemental bone; middle rays of ventrals longest.   OPISTHOGNATHIDÆ, CXCI.
   ee. Vertebræ about 50; scales ctenoid; maxillary not dilated, without supplement bone; inner rays of ventrals longest.
                BATHYMASTERIDÆ, CXCII.
  cc. Body naked; snout short; mouth very large, the maxillary much produced behind; jaws with sharp canines; lateral line well developed; dorsals 2; caudal forked.   CHIASMODONTIDÆ, CXCIII.
 bb. Snout much prolonged and spatulate; ventrals widely separated; body scaly or naked; lateral line near the back; dorsal usually divided.
               CHÆNICHTHYIDÆ, CXCIV.
aa. Mouth vertical, the lips fringed.
 f. Eyes lateral; gills 4, a slit behind the last; preopercle armed; body naked, compressed; caudal lunate, on a slender peduncle; vertebræ about 48.
               TRICHODONTIDÆ, CXCV.
 ff. Eyes superior; gills more or less reduced, usually 3½, the last slit small or wanting; suborbitals more or less dilated; body scaly or naked.
  g. Lateral line well developed, concurrent with the back anteriorly; dorsal spines slender, not pungent; vertebræ about 25 to 30.
   h. Ventral rays I, 3.   DACTYLOSCOPIDÆ, CXCVI.
  gg. Lateral line obscure; dorsal spines few, more or less pungent, sometimes obsolete.   URANOSCOPIDÆ, CXCVII-

Family CXC. MALACANTHIDÆ.

(THE BLANQUILLOS.)

 Body more or less elongate, fusiform or compressed. Head subconical, the anterior profile usually convex; suborbital without bony stay; the bones not greatly developed; cranial bones not cavernous; opercular bones mostly unarmed. Mouth rather terminal, little oblique; teeth rather strong; no teeth on vomer or palatines; the premaxillary usually with a blunt posterior canine, somewhat as in the *Labridæ;* premaxillaries protractile; maxillary without supplemental bone, not slipping under the edge of the preorbital. Gills 4, a long slit behind the fourth; pseudobranchiæ well developed; gill membranes separate, or more or less united, often adherent to the isthmus; lower pharyngeals separate. Scales small, ctenoid; lateral line present, complete, more or less concurrent with the back; dorsal fin long and low, usually continuous, the spinous portion always much less developed than the soft portion, but never obsolete; anal fin very long, its spines feeble and few; caudal fin forked; tail diphycercal; ventrals thoracic or subjugular, I, 5, close together; pectoral fins not very broad, the rays all branched; vertebræ in normal or slightly increased number (24 to 30). Pyloric cœca few or none. Fishes of the temperate and tropical seas, some of them reaching a large size. Genera about 6; species about 8 to 10, mostly American. The relationships of the family are obscure, and it may be that the genera here associated are not really closely allied. (*Malacanthidæ*, Günther, Cat., III, 359, 1861; *Trachinidæ*, part, Günther, Cat., II, 225–264, 1860.)

MALACANTHINÆ:
a. Vertebræ 24; preopercle entire.
  b. Soft dorsal and anal extremely long, each with more than 40 rays; preopercle entire; form slender; scales very small. MALACANTHUS, 838.
aa. Vertebræ more than 24; preopercle more or less serrate.
CAULOLATILINÆ:
    c. Soft dorsal and anal moderate, each with 22 to 27 soft rays; preopercle serrate; scales rather small; form robust.
    d. Upper jaw with posterior canines; dorsal spines graduated.
                                                                CAULOLATILUS, 839.
    LATILINÆ:
    cc. Soft dorsal and anal short, each of 13 to 15 soft rays; preopercle denticulate; scales small; form robust.
    e. Nape with a large adipose appendage; a fleshy prolongation on each side of the labial fold, extending forward behind angle of mouth.
                                                                LOPHOLATILUS, 840.

838. MALACANTHUS, Cuvier.

(MATAJUELO BLANCO.)

Malacanthus, CUVIER, Règne Animal, Ed. 2, vol. II, 205, 1829 *(plumieri)*.

Body elongate, slightly compressed; cleft of mouth horizontal, with the jaws equal; eyes lateral; scales very small, minutely ciliated; one continuous dorsal, with the first 4 to 6 rays not articulated; dorsal and anal very long; pectoral rays all branched; jaws with villiform teeth; an outer series of stronger teeth, some of them canine-like, and with a canine at the posterior extremity of the intermaxillary; no teeth on the palate; preopercle entire; opercle with a spine; gill rakers little developed; vertebræ in small number, $10+14=24$. One species, a shore fish of tropical America. ($\mu\alpha\lambda\alpha\varkappa\acute{o}\varsigma$, soft; $\check{\alpha}\varkappa\alpha\nu\theta\alpha$, spine.)

2610. MALACANTHUS PLUMIERI (Bloch).

(MATAJUELO BLANCO.)

Head $3\frac{2}{3}$; depth $6\frac{1}{4}$. D. VI, 49; A. 48; scales 14–130–30; eye $5\frac{1}{2}$ in head; maxillary $2\frac{4}{5}$; snout $2\frac{1}{2}$; P. 2; longest dorsal rays 3, equal to anal ray; upper caudal lobe $1\frac{3}{4}$. Body elongate, little compressed. Head moderately long and pointed; eye placed high; interorbital flat, as wide as eye; profile of head obliquely straight from tip of snout to above nostril, where there is a slight angle formed, thence nearly horizontally straight to dorsal. Mouth large, maxillary reaching slightly past the vertical from posterior nostril; jaws equal; a band of villiform teeth in upper jaw growing broader anteriorly, and another row of small, even, conical teeth at the sides, and 6 well-developed canines in front, the 2 outer ones the largest; a canine on premaxillary at angle of mouth; villiform teeth in lower jaw not extending very far back; large recurved canines on side of jaw anteriorly, small conical teeth in front and on sides posteriorly, with a single large canine at angle of mouth; gill rakers rudimentary, about $5+7$. Top of head forward from above middle of eye, preorbital, and lower jaw, naked; fins withou scales. Dorsal and anal similar, long and low, continuous; pectoral reaching past tips of ventrals to front of anal; ventrals not reaching to vent, origin of ventral spine slightly behind base of pectoral; caudal forked, the lobes elongate, sometimes produced into a filament. Color in spirits, uniform, pale olive brown above, white below; fins light brownish; no distinct markings. Length 15 inches. West Indies, rather common; used as food. Here described from specimens from Havana. (Named for Père Plumier, of Martinique.)

Matejuelo blanco, PARRA, Dif. Piezas Hist. Nat. Cuba, 22, tab. 13, f. r. 1787, Cuba.

Coryphæna plumieri, BLOCH, Ichthyol., v, 119, pl. 175, 1787, Martinique; from a drawing by PLUMIER.

Malacanthus trachinus, VALENCIENNES, in CUVIER, Règne Animal, pl. 90, fig. 3.

Sparus oblongus, BLOCH & SCHNEIDER, Syst. Ichth., 283, 1801; after PARRA.

Malacanthus plumieri, CUVIER & VALENCIENNES, Hist. Nat. Poiss., XIII, 319, 380, 1839, specimens from San Domingo; GÜNTHER, Cat., III, 359, 1861.

839. CAULOLATILUS, Gill.

(BLANQUILLOS.)

Caulotatilus, GILL, Proc. Ac. Nat. Sci. Phila. 1862, 240 (no diagnosis), and GILL, Proc. Ac Nat. Sci. Phila. 1865, 66 (*chrysops*).

Dekaya, COOPER, Proc. Cal. Ac. Sci. 1864, 70 (*princeps*), not *Dekayia*, MILNE-EDWARDS & HAIME, 1851, a genus of corals.

Body elongate, subfusiform, not strongly compressed, heavy forward, tapering to a rather slender caudal peduncle; profile of head strongly arched; mouth moderate, little oblique, the jaws nearly equal; lips thick; maxillary narrow, not slipping under the preorbital; teeth in villiform bands, preceded by a row of stronger acute teeth; posterior teeth in each jaw canine-like, directed forward; posterior canines of upper jaw largest; no teeth on vomer or palatines; preopercle pectinate, the teeth nearly even; opercle with a blunt, flat spine; eyes large, lateral; gill membranes slightly connected, forming a fold across the isthmus, with which they are narrowly joined; branchiostegals 6; gill rakers short and stout; nostrils double, round, close together; scales small, firm, ctenoid; lateral line continuous, concurrent with the back; dorsal with 7 to 9 slender, pointed, graduated spines and 22 to 27 soft rays; anal similar to soft dorsal, with 1 or 2 small spines and more than 20 soft rays; caudal fin forked; ventral fins thoracic; no adipose appendage at the nape; vertebræ 12+15=27. Large fishes of the warm seas of America; valued as food. ($\varkappa\alpha\upsilon\lambda\acute{o}\varsigma$, stem; *Latilus;* being distinguished from *Latilus* by the many rays.)

a. Scales small, about 125 in the lateral line, about 50 in a transverse series.

   b. Eye large, 4½ in the head; depth 4 in length; scales 16-125-40.

                                                    PRINCEPS, 2611.

      bb. Eye small, 6 in head; depth 3⅓ in body; scales 13-120-35.   MICROPS, 2612.

aa. Scales larger, about 108 in the lateral line, about 25 in a transverse series; scales 12—108—25.   CYANOPS, 2613.

2611. CAULOLATILUS PRINCEPS (Jenyns).

(BLANQUILLO; WHITE-FISH.)

Head 3⅔; depth 4. D. IX, 24; A. II, 23; scales 16-125-40. Flesh of the occiput becoming thick with age, as in *Harpe.* Eye large, about ¼ the convex interorbital space, 4½ in head; maxillaries reaching front of eye;

teeth rather strong; preopercle finely, evenly, and acutely serrate behind, nearly entire below; preopercle, interopercle, and preorbital naked; cheeks and opercles scaly; top of head scaled on the median line to between the eyes; dorsal spines flexible; ventrals slightly behind the pectorals, the outer rays longest; caudal moderately forked, the upper lobe the longer; caudal peduncle short and slender, abruptly contracted; pectorals falcate, longer than caudal, ⅝ length of the head. Olivaceous, with bluish reflections; brownish above, greenish below; fins light greenish olive, tinged with bluish and orange, the colors always pale; dorsal and anal greenish, with a bluish band near the tip; axil dusky. Rocky islands of the Pacific coast from Monterey southward to the Galapagos; abundant about the Santa Barbara Islands; a food fish of considerable importance. Length 40 inches. We are unable to detect any differences by which the Californian form, *Caulolatilus anomalus*, can be separated from *Caulolatilus princeps*. (*princeps*, a leader.)

Latilus princeps, JENYNS, Zool. Beagle, Fishes, 52, pl. 11, 1840, Chatham Island, Galapagos Archipelago (Coll. Charles Darwin); GÜNTHER, Cat., II, 253, 1860.

Dekaya anomala, COOPER, Proc. Cal. Ac. Sci. 1864, 70, coast of Southern California.

Caulolatilus affinis, GILL, Proc. Ac. Nat. Sci. Phila. 1865, 68, Cape St. Lucas. (Coll. John Xantus.)

Caulolatilus princeps, GILL, *l. c.* 68.

Caulolatilus anomalus, GILL, *l. c.* 68; STREETS, Bull. U. S. Nat. Mus., VII, 48, 1877; JORDAN & GILBERT, Synopsis, 625, 1883.

2612. CAULOLATILUS MICROPS, Goode & Bean.

Head 3½; depth 3½. D. VII, 25; A. I, 23; scales 105 counting the oblique series, 120 counting the row above lateral line; transverse rows 12 + 30; eye 5¾ in head; snout 2; maxillary 2½; pectoral 1⅓; ventral 2¹⁄₁₀; highest dorsal spine 3½; highest anal rays 3⅓; upper caudal lobe 1½. Body rather robust; upper profile of head rather steep, evenly rounded from tip of snout to dorsal; nostrils small, midway between eye and tip of snout, separated by a distance equal to ½ diameter of pupil; mouth large, maxillary scarcely reaching to anterior margin of eye; lips thick; lower jaw included; jaws with small conical teeth, the outer row enlarged, canine-like, a large tooth on posterior end of maxillary at angle of mouth; preopercle finely and evenly serrate on its vertical limb; a broad flat spine on opercle; snout, preorbital and lower jaw naked; fins scaleless; dorsal and anal similar, long and low; pectorals reaching far past tips of ventrals to vent; origin of ventral spine about the length of 2 scales behind the vertical from pectoral base; caudal fin lunate when spread, its upper lobe slightly the longer. Color reddish, marked with yellow; a yellow band below the eye; a dark blotch in and above axil of pectoral; dorsal light at base, darker above, with many indistinct brownish spots. Gulf of Mexico, in rather deep water; not rare. Here described from a specimen from the Pensacola Snapper Banks, 26 inches in length. A rather doubtful species, perhaps not distinct from *C. cyanops* or *C. chrysops*. (μικρός, small; ὤψ, eye.)

Caulolatilus microps, GOODE & BEAN, Proc. U. S. Nat. Mus. 1878, 43, off Pensacola, Florida. (Coll. Silas Stearns.)

Caulolatilus chrysops, JORDAN & GILBERT, Synopsis, 626, 1883; not *Latilus chrysops*, CUVIER & VALENCIENNES.

2613. CAULOLATILUS CYANOPS,* Poey.

(BLANQUILLO.)

Head 4 in total length. D. VII, 24; A. I, 22 (scales 10–108–25. Bean). Profile convex before the eye, not ascending to the nape; no scales on the fins; soft rays little divided; caudal slightly lunate; first caudal vertebra spoon-like, its cavity receiving the air bladder; vertebræ 12 + 15; no pyloric cæca, stomach short, air bladder large. Color greenish above, a faint, broad, interrupted brown band above the lateral line; some small brown spots above and below it; region below the eye clear blue, not very different from the color of the belly; soft dorsal brown, paler at its base, edged with orange; spinous dorsal orange. (Poey.) Coast of Cuba; not seen by us. Both this and the preceding species may be identical with *Caulolatilus chrysops,* a species described from the coast of Brazil. (κυάνεος, blue; ὤψ, eye.)

?*Latilus chrysops,* CUVIER & VALENCIENNES, Hist. Nat. Poiss., IX, 496, 1833, Brazil (Coll. M. Gay); GÜNTHER, Cat., II, 253, 1860.
Caulolatilus cyanops, POEY, Repertorio, I, 312, 1867, Cuba. (Coll. Poey.)

840. LOPHOLATILUS, Goode & Bean.

(TILE-FISHES.)

Lopholatilus, GOODE & BEAN, Proc. U. S. Nat. Mus. 1879, 205 (*chamæleonticeps*).

Body stout, somewhat compressed; mouth moderate, maxillary reaching anterior margin of the orbit; opercle and preopercle scaly, the latter finely denticulate; upper jaw with outer series of stronger teeth, behind which is a band of villiform teeth; lower jaw with a few large canines, and an inner series of small conical teeth; vomer and palatines toothless; nape with a large adipose appendage; a fleshy prolongation upon each side of the labial fold, extending backward beyond the angle of the mouth; stomach small, siphonal, barely more than a loop in the very large intestine; alimentary canal short, less than total length of the body; air bladder simple, with thick muscular walls, strongly attached to the roof of the abdominal cavity by numerous root-like appendages, resembling somewhat that of *Pogonias.* Deep-sea fishes. (λόφος, crest; *Latilus.*)

2614. LOPHOLATILUS CHAMÆLEONTICEPS, Goode & Bean.

(TILE-FISH.)

Head 3; depth 3¼. D. VII, 15; A. II, 13; scales 8–93–30. Body stout, somewhat compressed, its greatest width equaling length of caudal peduncle; intermaxillaries supplied with a series of from 19 to 23 canine teeth, behind which is a band of villiform teeth, widest at the symphysis; mandible with about 12 large canines; eye rather small, its diameter 6¼ in

---

* The characters distinguishing *Caulolatilus chrysops* are thus given by Poey: Head 4½ in total length. D. VIII, 24; A. II, 22. Profile most gibbous behind the eye; a very bright gilded band below the eye, broader anteriorly; dorsal fin brown with irregular blue spots; axillary spot green. Coast of Brazil.

head, and about twice length of labial appendages; distance between posterior nostril and eye equal first anal spine, and $\frac{1}{4}$ distance from tip of snout to anterior nostril. Caudal fin emarginate, middle rays $1\frac{1}{2}$ in outer rays; vent under interval between fourth and fifth dorsal rays. Back bluish, with a green tinge, iridescent, changing through purplish blue and bluish gray to rosy white below, and milky white toward median line of belly; head rosy, iridescent, with red tints most abundant on forehead, blue under the eyes, cheeks fawn-colored; throat and under side of head pearly white, with an occasional tint of lemon yellow, most pronounced in front of ventrals and on anterior portion of ventral fins; back with numerous maculations of bright yellow or golden; anal purplish, with blue and rose tints, iridescent; margin of anal rich purplish blue, iridescent, like the most beautiful mother-of-pearl, this color pervading more or less the whole fin, which has large yellow maculations, the lower border rose-colored, like the belly, base of the fin also partaking of this general hue; dashes of milk white on base of anal between the rays; dorsal gray; in front of the seventh dorsal the upper third posterior to the upper two-thirds dark brown; spots of yellow, large, elongate, on or near the rays; adipose fin whitish brown or yellow, a large group of bright yellow confluent spots at the base; pectorals sepia-colored, with rosy and purplish iridescence. (Goode & Bean.) Deep waters of the western Atlantic, at times very abundant; now rare or almost extinct. "The tilefish was first observed in 1879 by fishermen fishing for cod on Nantucket Shoals. From its abundance it was thought to become of some economic importance. In March and April, 1882, vessels arriving at New York, Philadelphia, and Boston reported having passed large numbers of dead and dying fish, the majority of which were tilefish. Captain Collins estimated the area covered by dead and dying fish to be from 5,000 to 7,500 square statute miles, the number of fish to be 1,000,000,000. Several visits were made by the Fish Commission vessels to the grounds where these fishes were formerly abundant, but no specimen was obtained, and it was thought to have become extinct. In 1892 several specimens were taken by the *Grampus* in latitude 38° to 40° N., and longitude 71° to 73°. W. The wholesale destruction of the tilefish in 1882 is thought by Colonel McDonald to be due to climatic causes." (Goode & Bean.) (*chamæleon*, χαμαιλέων; - *ceps*, head.)

Lopholatilus chamæleonticeps, GOODE & BEAN, Proc. U. S. Nat. Mus. 1879, 205, Nantucket Shoals; JORDAN & GILBERT, Synopsis, 624, 1883; COLLINS, Rept. U. S. Fish Comm. (1882) 1884, 237; LUCAS, Rept. U. S. Nat. Mus. (Smithsonian Report) 1889, 647, with plate; GOODE & BEAN, Oceanic Ichthyology, 284, 1896.

Family CXCI. OPISTHOGNATHIDÆ.

(THE JAW-FISHES.)

Body oblong or elongate, low, moderately compressed, covered with small cycloid scales; lateral line present, straight, running close to the dorsal fin, not extending much behind middle of body. Head large, naked, the anterior profile decurved, no ridges, spines, or crests above. Mouth terminal, horizontal, its cleft usually very wide, the maxillary sometimes greatly dilated; supplemental maxillary present; premaxillaries protractile: jaws subequal, with conical or cardiform teeth; vomer usually with a few teeth; palatines toothless; opercles unarmed; no suborbital stay. Pseudobranchiæ present. Gill rakers rather long; gills 4, a slit behind the fourth; gill membranes somewhat united, free from the isthmus. Branchiostegals 6. Air bladder present. No pyloric cœca. Vertebræ large, about 27, in number. Dorsal fin long, continuous, its anterior half composed of slender, flexible spines, which pass gradually into soft rays; caudal distinct, rounded or lanceolate; tail not incurved, the last vertebra expanded (27 to 34); anal long and low, without distinct spines; ventrals separate, jugular, I, 5, the middle rays longest; pectorals fan-shaped. Three genera, of about 15 species; small fishes inhabiting rocky bottoms in tropical seas, many of them with bright markings. The American species are all rarities, living about rocks in deep or shallow water; nowhere abundant and none of the species well represented in collections. (Trachinidæ, genus *Opisthognathus.* Günther, Cat., ii, 256–261.)

a. Maxillary of great length, nearly as long as head, produced behind in a flexible lamina. .... **Opisthognathus, 241.**
aa. Maxillary normal, truncate behind, much shorter than head.
   b. Caudal moderate, rounded behind: body oblong, moderately compressed. .... **Gnathypops, 242.**
   bb. Caudal lanceolate, long and pointed: body elongate. .... **Lonchopisthus, 242.**

241. OPISTHOGNATHUS, Cuvier.

Opisthognathus, Cuvier, Règne Anim., Ed. 2, vol. ii, 249, 1829 (sconarus).

Maxillary prolonged backward in a long flexible lamina, which reaches about to base of pectoral. Characters of the genus otherwise included above. It has been suggested that the species of *Gnathypops* are females of analogous species of *Opisthognathus,* the long maxillary being a character of the male. This seems impossible, but deserves an investigation. The fact that *Gnathypops maxillosa* has but 27 vertebræ, while its long-jawed cognate. *Opisthognathus macrognathus,* is said by Poey to have 34 vertebræ, is opposed to this view, as is also the fact that the analogous species do not in other respects exactly correspond, as in *Gnathypops mystacina,* the scales are smaller than in *Opisthognathus lonchurus; Gnathypops rhomalea* has fewer fin rays than *Opisthognathus punctatus,* etc. But the parallelism of species in the two genera living in the same waters is remarkable. (ὄπισθε, behind; γνάθος, jaw.)

a. Scales moderate, about 67: D. X, 15; A. II, 15; body nearly plain olivaceous, the maxillary not distinctly striped within. .... **sconarus, 241.**
aa. Scales very small, 100 to 130 in longitudinal series: dorsal rays about XI, 15; A. II, 15: body and fins much variegated, the maxillary within with 2 ink-black stripes on a milk-white ground.
   b. Dorsal without large black spot in front: scales 130. .... **macrognathus, 241.**
   bb. Dorsal with a large black spot more or less ocellated.
      c. Scales about 100. .... **macrognathus, 241.**
      cc. Scales about 140. .... **scanurus, 241.**

2615. OPISTHOGNATHUS LONCHURUS, Jordan & Gilbert.

Head 3⅓; depth 4⅓. D. 25; A. 15; scales 67. Head moderate; snout very short, shorter than pupil; eye 3⅓ in head; maxillary 1⅓ in head, rather narrow; lower jaw included; vomer with 5 rather large teeth. Longest dorsal spine about as long as head, slightly higher than soft rays; caudal long, the middle rays scarcely shorter than head; longest anal rays 1⅓ in head; pectoral little more than ½ head. Scales moderate. Olivaceous; margin of upper lip with a narrow black stripe; caudal with 3 dusky bars; color of rest of body uniform. Gulf of Mexico, in deep water. Two specimens known, taken from the stomach of a Red Snapper at Pensacola, Florida. The species resembles *Gnathypops mystacinus*, found in the same waters, but the latter species has smaller scales. (λόγχη, lance; οὐρά, tail.)

Opisthognathus lonchurus, JORDAN & GILBERT, Proc. U. S. Nat. Mus. 1882, 290, Snapper Banks, off Pensacola, Florida (Type, No. 29671. Coll. Jordan & Stearns); JORDAN & GILBERT, Synopsis, 943, 1883.

2616. OPISTHOGNATHUS PUNCTATUS, Peters.

D. 28; A. 18. Body moderately elongate; scales very small, about 125 in lateral line. Dorsal spines continuous with the soft rays. No vomerine teeth. Maxillary very long, extending slightly beyond head. Head everywhere finely speckled with black, the body more coarsely and irregularly spotted; pectoral finely and closely speckled, its edge plain; ventral fin dusky, similarly marked; dorsal without large black blotch, finely spotted, the spots behind gradually forming the boundaries of white ocelli, the base of the fins having rings of white around black spots, the upper part with dark rings around pale spots; caudal with pale spots, its edge, like that of the dorsal, somewhat dusky, not black; anal with a broad, blackish edge, and with dark spots, those near the base of the fin largest; lining membrane of maxillary with the usual bands of white and inky black. Mazatlan. Only the type of this species is yet known, this description having been taken by us from the original specimen. It bears considerable resemblance to *Gnathypops rhomalea*, which is found in the same waters, differing in the generic character of the dilated maxillary. (*punctatus*, spotted.)

Opisthognathus punctatus, PETERS, Berliner Monatsberichte 1869, 708, Mazatlan; JORDAN, Proc. Ac. Nat. Sci. Phila. 1883, 290; JORDAN, Cat. Fish. N. A., 118, 1885.

2617. OPISTHOGNATHUS MACROGNATHUS, Poey.

Head 3⅓; depth 5. D. XI, 16; A. II, 16 or 17; P. 17; scales 100. Body moderately elongate, somewhat compressed. Head blunt anteriorly; snout very short, about as long as pupil; eye large, 4 in head; maxillary reaching slightly past edge of preopercle, but not to end of head, its length contained 3⅓ times in length of body. Teeth rather strong, wide set, forming 2 distinct series, directed backward, especially in the upper jaw; lateral teeth of lower jaw largest; a single vomerine tooth. Gill rakers long and slender, nearly 20 below angle. Scales very small.

3030——66

terminal, horizontal, its cleft usually very wide, the maxillary sometimes greatly dilated; supplemental maxillary present; premaxillaries protractile; jaws subequal, with conical or cardiform teeth; vomer usually with a few teeth; palatines toothless; opercles unarmed; no suborbital stay. Pseudobranchiæ present. Gill rakers rather long; gills 4, a slit behind the fourth; gill membranes somewhat united, free from the isthmus. Branchiostegals 6. Air bladder present. No pyloric cæca. Vertebræ large, about 27 in number. Dorsal fin long, continuous, its anterior half composed of slender, flexible spines, which pass gradually into soft rays; caudal distinct, rounded or lanceolate; tail not isocercal, the last vertebra expanded (27 to 34); anal long and low, without distinct spines; ventrals separate, jugular, I, 5, the middle rays longest; pectorals fan-shaped. Three genera, of about 15 species; small fishes inhabiting rocky bottoms in tropical seas, many of them with bright markings. The American species are all rarities, living about rocks in deep or shallow water; nowhere abundant and none of the species well represented in collections. (*Trachinidæ*, genus *Opisthognathus*, Günther, Cat., II, 254–256.)

a. Maxillary of great length, nearly as long as head, produced behind in a flexible
   lamina.   OPISTHOGNATHUS, 841.
aa. Maxillary normal, truncate behind, much shorter than head.
   b. Caudal moderate, rounded behind; body oblong, moderately compressed.
                                                        GNATHYPOPS, 842·
   bb. Caudal lanceolate, long and pointed; body elongate.   LONCHOPISTHUS, 843.

841. OPISTHOGNATHUS, Cuvier.

Opisthognathus, CUVIER, Règne Anim., Ed. 2, vol. II, 240, 1829 (*sonnerati*).

Maxillary prolonged backward in a long flexible lamina, which reaches about to base of pectoral. Characters of the genus otherwise included above. It has been suggested that the species of *Gnathypops* are females of analogous species of *Opisthognathus*, the long maxillary being a character of the male. This seems impossible, but deserves an investigation. The fact that *Gnathypops maxillosa* has but 27 vertebræ, while its long-jawed cognate, *Opisthognathus macrognathum*, is said by Poey to have 34 vertebræ, is opposed to this view, as is also the fact that the analogous species do not in other respects exactly correspond, as in *Gnathypops mystacina*, the scales are smaller than in *Opisthognathus lonchurum; Gnathypops rhomalea* has fewer fin rays than *Opisthognathus punctatum*, etc. But the parallelism of species in the two genera living in the same waters is remarkable. (ὄπισθε, behind; γνάθος, jaw.)

a. Scales moderate, about 67; D. X, 15; A. II, 13; body nearly plain olivaceous, the
   maxillary not distinctly striped within.   LONCHURUM, 2615.
aa. Scales very small, 100 to 150 in longitudinal series; dorsal rays about XI, 17; A.
   II, 16; body and fins much variegated, the maxillary within with 2 ink-black
   stripes on a milk-white ground.
   b. Dorsal without large black spot in front; scales 120.   PUNCTATUM, 2616.
   bb. Dorsal with a large black spot more or less ocellated.
      c. Scales about 100.   MACROGNATHUM, 2617.
      cc. Scales about 140.   OMMATUM, 2618.

2615. OPISTHOGNATHUS LONCHURUM, Jordan & Gilbert.

Head 3⅓; depth 4⅜. D. 25; A. 15; scales 67. Head moderate; snout very short, shorter than pupil; eye 3⅓ in head; maxillary 1⅓ in head, rather narrow; lower jaw included; vomer with 5 rather large teeth. Longest dorsal spine about as long as head, slightly higher than soft rays; caudal long, the middle rays scarcely shorter than head; longest anal rays 1⅓ in head; pectoral little more than ½ head. Scales moderate. Olivaceous; margin of upper lip with a narrow black stripe; caudal with 3 dusky bars; color of rest of body uniform. Gulf of Mexico, in deep water. Two specimens known, taken from the stomach of a Red Snapper at Pensacola, Florida. The species resembles *Gnathypops mystacinus*, found in the same waters, but the latter species has smaller scales. (λόγχη, lance; οὐρά, tail.)

Opisthognathus lonchurus, JORDAN & GILBERT, Proc. U. S. Nat. Mus. 1882, 290, Snapper Banks, off Pensacola, Florida (Type, No. 29671. Coll. Jordan & Stearns); JORDAN & GILBERT, Synopsis, 943, 1883.

2616. OPISTHOGNATHUS PUNCTATUM, Peters.

D. 28; A. 18. Body moderately elongate; scales very small, about 125 in lateral line. Dorsal spines continuous with the soft rays. No vomerine teeth. Maxillary very long, extending slightly beyond head. Head everywhere finely speckled with black, the body more coarsely and irregularly spotted; pectoral finely and closely speckled, its edge plain; ventral fin dusky, similarly marked; dorsal without large black blotch, finely spotted, the spots behind gradually forming the boundaries of white ocelli, the base of the fins having rings of white around black spots, the upper part with dark rings around pale spots; caudal with pale spots, its edge, like that of the dorsal, somewhat dusky, not black; anal with a broad, blackish edge, and with dark spots, those near the base of the fin largest; lining membrane of maxillary with the usual bands of white and inky black. Mazatlan. Only the type of this species is yet known, this description having been taken by us from the original specimen. It bears considerable resemblance to *Gnathypops rhomalea*, which is found in the same waters, differing in the generic character of the dilated maxillary. (*punctatus*, spotted.)

Opisthognathus punctatus, PETERS, Berliner Monatsberichte 1869, 708, Mazatlan; JORDAN, Proc. Ac. Nat. Sci. Phila. 1883, 290; JORDAN, Cat. Fish. N. A., 118, 1885.

2617. OPISTHOGNATHUS MACROGNATHUM, Poey.

Head 3⅔; depth 5. D. XI, 16; A. II, 16 or 17; P. 17; scales 100. Body moderately elongate, somewhat compressed. Head blunt anteriorly; snout very short, about as long as pupil; eye large, 4 in head; maxillary reaching slightly past edge of preopercle, but not to end of head, its length contained 3¾ times in length of body. Teeth rather strong, wide set, forming 2 distinct series, directed backward, especially in the upper jaw; lateral teeth of lower jaw largest; a single vomerine tooth. Gill rakers long and slender, nearly 20 below angle. Scales very small.

Dorsal fin low, continuous, the soft rays but little higher than the spines, which are slender and flexible, the longest $3\frac{1}{4}$ in head; caudal short, rounded, its length $5\frac{3}{4}$ in body; anal similar to soft dorsal; pectoral $\frac{1}{2}$ as long as head. Grayish olive, much variegated with yellowish and dark olive; about 6 irregular dusky bands on the body, which extend on the dorsal fin; whitish markings on body forming roundish spots, surrounded by reticulations of grayish olive; head marbled, its posterior part, as well as the sides of the back and pectoral base, with small blackish dots; membrane lining inside of maxillary with 2 curved inky-black bands on a white ground; angle of mouth with a black spot; lining of opercle black; fins all variegated like the body. Florida Keys to Cuba. Here described from the type of *O. scaphiurum*, from Garden Key, but *O. macrognathum* seems to be the same. ($\mu\alpha\kappa\rho\delta\varsigma$, long; $\gamma\nu\alpha'\theta\sigma\varsigma$, jaw.)

*Opisthognathus macrognathus,** POEY, Memorias, II, 284, July, 1860, Cuba. (Coll. Poey.)
Opisthognathus megastoma, GÜNTHER, Cat., II, 255, September, 1860, Gulf of Mexico. (Haslar Collection.)
Opisthognathus scaphiurus, GOODE & BEAN, Proc. U. S. Nat. Mus. 1882, 417, Garden Key, Florida (Type, No. 5936, U. S. Nat. Mus. Coll. Dr. Whitehurst); JORDAN & GILBERT, Synopsis, 943, 1883.

2618. OPISTHOGNATHUS OMMATUM, Jenkins & Evermann.

Head 3; width of head 5; depth 5; eye 3 in head. D. 28; A. 18; scales about 140. Body moderate, compressed, depth $4\frac{1}{7}$; width behind the head $8\frac{3}{10}$ in length of body. Head large, its breadth equaling its depth, being 5 in length of body. Scales small, embedded; head naked, lateral line extending past middle of dorsal fin. Mouth large. Maxillary long, $1\frac{1}{2}$ in head; postorbital portion $2\frac{3}{10}$ in head, not extending beyond head; snout short, its length less than $\frac{1}{2}$ diameter of eye; distance from tip of snout to end of maxillary $3\frac{3}{10}$ in length of body and $1\frac{1}{2}$ in head. Teeth in front part of each jaw in several series, on sides of jaws reduced to a single series, the outer series strong; a tooth on the vomer; gill membranes connected; the interorbital space very narrow, $11\frac{3}{4}$ in head. Distance from snout to origin of dorsal but little greater than length of head; space between dorsal and caudal fins $\frac{1}{3}$ greater than length of snout; no depression between spinous and soft rays of dorsal fin, the dorsal equaling the anal in height, its longest ray $1\frac{3}{4}$ times the eye; pectorals slightly longer than ventrals, being 2 in head; breadth of pectorals 3 in head; ventrals inserted slightly in front of pectorals; caudal rounded and narrow. Coloration: Body irregularly mottled with dark, head evenly blackish; dorsal fin blackish on the posterior portion, with 2 rows of 4 or 5 pale spots well separated; a large ocellated spot from the third to the sixth spines, including them, greater than diameter of eye; anal fin black, with a series of pale spots on the rays, the base pale; caudal black, with 2 pale

---

* Poey thus describes his specimens of *Opisthognathus macrognathum:* "Head $3\frac{1}{4}$ in total; depth $5\frac{1}{4}$; eye nearly 4, twice length of snout, 3 times interorbital width. D. XI, 16; A. II, 16. Vomer with 2 teeth; spines not pungent. Body covered with large yellowish points on a brown ground; 7 broad brown bands on sides, not reaching belly, but extending to middle of dorsal, which, like the anal, has yellow points; a large black ocellus between sixth and ninth spines of dorsal; maxillary with 2 ink-black bands on a milk-white ground; pectorals, ventrals, and caudal yellowish with black points. No pyloric cæca; vertebræ 10 + 24 = 34, the first 5 strong." (Poey.)

spots at the base and a row of spots across the middle; lining of maxillary with bands of black and white. Bay of Guaymas; 3 specimens known. (ὀμματός, eyed, from its ocellate dorsal.)

Opisthognathus ommata, JENKINS & EVERMANN, Proc. U. S. Nat. Mus. 1888, 153, Guaymas. (Type, No. 39640. Coll. Jenkins & Evermann.)

842. GNATHYPOPS, Gill.

Gnathypops, GILL, Proc. Ac. Nat. Sci. Phila. 1862, 241 (*maxillosus*).

This genus differs from *Opisthognathus* in having the maxillary of medium length and truncate behind, not extending to edge of opercle; caudal moderate, rounded behind. Species in form and habit agreeing closely with those of *Opisthognathus*. (γνάθος, jaw; ὕπο, below; ὤψ, eye.)

   a. Body and fins spotted with black and often with pale.
     b. Dorsal fin with a conspicuous dusky blotch in front.
       c. Scales very small, about 120; dorsal rays 26.   SCOPS, 2619.
       cc. Scales moderate, about 65; dorsal rays 15.   MAXILLOSA, 2620.
     bb. Dorsal fin without distinct blotch in front; scales small, about 100.
       d. Dorsal rays about XI, 16; dark spots on head and body few.
                                 MACROPS, 2621.
       dd. Dorsal rays about XI, 13; dark spots on head and body numerous.
        e. Dorsal fin distinctly notched; lateral line not reaching middle of dorsal.   RHOMALEA, 2622.
        ee. Dorsal fin not notched; lateral line reaching middle of dorsal.
                                   SNYDERI, 2623.
   aa. Body and fins nearly uniform olive; the spots few and spare; dorsal rays 24; scales 100; no black on membrane of maxillary.   MYSTACINA, 2624.

2619. GNATHYPOPS SCOPS, Jenkins & Evermann.

Head 3⅓; width of head 5¾, its depth 4½. D. 26 (X, 16); A. 19 (II, 17); scales 3–122–40. Scales small, none on head; lateral line extending to about middle of dorsal fin; mouth large; maxillary extending beyond eye a distance 4$\frac{1}{10}$ in head; snout 6½ in head; teeth in bands, outer series on upper jaw rather strong; a single tooth on vomer; gill membranes connected. Opercle ending in a long flap, which extends upward and backward, nearly meeting over the back in front of the dorsal fin. Eye large, 2½ in head; interorbital space narrow, 11 in head. No depression between the dorsal spines and the soft rays, which are scarcely distinguishable; height of dorsal equal to that of anal; ventrals inserted in front of pectorals; pectorals equal to ventrals in length, 7 in body; caudal rounded. Coloration, in alcohol: Body pale, covered with many dark spots about the size of 3 to 6 scales; top of head with smaller dark spots; sides of head with whitish spots; dorsal fin with a black ocellated spot equal to eye on the space between second and fifth spines; remainder of fin dark, with many white spots running into each other on some portions, so as to form irregular lines; base of anal pale, the outer edge black; caudal dark with 2 whitish spots at the base, and a row of 6 white spots across the middle on alternate rays; pectorals lighter, with small whitish specks; ventrals dusky; belly pale. Guaymas; 3 specimens known, respectively 115 cm., 10 cm., and 7 cm. in length to base of caudal. This species is the analogue of *Opisthognathus ommatum*, also from Guaymas. (*scops*, the screech owl; σκώψ, from σκοπέω, to look, in allusion to the large eyes.)

Gnathypops scops, Jenkins & Evermann, Proc. U. S. Nat. Mus. 1888, 152, Guaymas. (Type, No. 39641, U. S. Nat. Mus. Coll. Jenkins & Evermann.)

2620. GNATHYPOPS MAXILLOSA (Poey).

Head $3\frac{5}{6}$; depth $4\frac{1}{3}$. D. VIII, 17; A. II, 13; scales 65. Body moderately compressed; head not very large; maxillary truncate behind, extending behind eye for a distance for about $\frac{2}{3}$ diameter of eye, its length $1\frac{2}{3}$ in head; eye $3\frac{3}{8}$ in head. Teeth conical, curved, well separated, mostly in a single series; no teeth on vomer. Fins moderate; dorsal continuous, its spines slender; caudal short, its length $\frac{2}{3}$ head. Color grayish olive, with 7 irregular Λ-shaped bars of darker, everywhere much marbled and variegated; fins all similarly marked, the ventrals dusky, the dorsal with a dusky blotch in front. Cuba, north to Florida, from which locality the specimen here described was taken. Evidently very close to the Brazilian species *G. cuvieri*, the eye perhaps smaller. According to Poey, *Gnathypops maxillosa* has the eye 4 in head; D. VIII, 18; A. II, 15; 2 teeth on vomer; spinous dorsal lower than soft dorsal; body covered with large yellow spots on a ground color of clear brown; 6 brown cross bands reaching middle of dorsal, which is variegated with yellow and reddish, as is the anal; maxillary yellowish on its posterior border, the middle blackish; other spots on the jaws; ventrals, pectoral, and caudal yellowish, the ventrals finely spotted with brown, the caudal with 5 brown bands; base of pectorals with dark spots. Vertebræ $10 + 17 = 27$. (*maxillosus*, pertaining to the jaw.)

Opisthognathus maxillosus, Poey, Memorias, II, 286, 1860, Cuba. (Coll. Poey.)
Gnathypops maxillosus, Gill, Proc. Ac. Nat. Sci. Phila. 1862, 241; Poey, Synopsis, 400; Jordan & Gilbert, Synopsis, 942, 1883.

2621. GNATHYPOPS MACROPS (Poey).

Head $3\frac{1}{3}$ in total (with caudal?); depth about $4\frac{1}{2}$. D. XI, 16; A. II, 15 or 16; scales 100. Eye $3\frac{1}{2}$ in head; maxillary extending beyond eye $\frac{2}{3}$ of a diameter, $5\frac{1}{6}$ in head; vomer with 6 conical teeth. Color (faded in the type) reddish olive, with round, yellowish spots and vestiges of vertical bands; dorsal and anal plain, pectorals with brown bands; jaws not spotted with brown and white. Coast of Cuba (Poey); known from 1 specimen 132 mm. long; not seen by us. According to Poey, it may not be distinct from *G. maxillosa*, which in turn may possibly be the female of *Opisthognathus macrognathum*. This species may also be identical with the Brazilian species *Gnathypops cuvieri*,* but the latter has a dorsal ocellus and apparently larger scales. (μακρός, large; ὤψ, eye.)

Opisthognathus macrops, Poey, Memorias, II, 287, 1860, Cuba. (Coll. Poey.)

---

* *Gnathypops cuvieri* (Valenciennes). Head $3\frac{1}{3}$; depth $4\frac{1}{3}$; eye $3\frac{1}{3}$ in head. D. X, 18; A. II, 16; scales 70. Maxillary reaching beyond the vertical from posterior margin of orbit, $1\frac{1}{2}$ in head; eye $3\frac{1}{3}$ in head, dorsal fin not notched. Olivaceous; a large dark-blue ovate ocellate spot between the fourth and eighth dorsal spines; dorsal and anal mottled, the edge dusky posteriorly; caudal with 3 dark-bluish bands. Bahia (Valenciennes). (Named for Georges Dagobert Cuvier.)

Opisthognathus cuvieri, Cuvier & Valenciennes, Hist. Nat. Poiss., XI, 504, 1836, Bahia (Coll. Blanchet); Günther, Cat., II, 256, 1860.

2622. GNATHYPOPS RHOMALEA (Jordan & Gilbert).

Head 2$\frac{7}{8}$; depth 4. D. XI, 13; A. II, 13; scales 103 (pores fewer). Body rather robust, compressed; head very large, ovoid, thicker and deeper than body, with swollen cheeks, the occipital region high, the snout somewhat truncate, the intermediate profile forming a nearly even curve; greatest depth of head equal to its thickness and $\frac{2}{3}$ its length. Eye not very large, 6 in head, longer than snout, about equal to the width of the flattish interorbital space. Mouth large, the maxillary extending well beyond the eye, but not to the margin of the preopercle nor to the mandibulary joint, its posterior margin truncate; supplemental bone small, but distinct; length of maxillary from end of snout 1$\frac{3}{4}$ in head. Teeth moderate, in both jaws, in broad bands which become narrow on the sides; outer series of teeth somewhat enlarged, especially in upper jaw; 1 rather small, blunt tooth on middle of vomer. Gill membranes scarcely connected; gill rakers long and slender, about $\frac{3}{4}$ diameter of eye, 9 + 19; pseudobranchiæ situated in a cavity above the gill arches. Head naked; scales on body small, smooth, somewhat embedded; breast naked; lateral line ceasing opposite anterior third of second dorsal; 103 scales in a longitudinal series from head to caudal. Dorsal fin high; a rather deep notch separating the spines from the soft rays; the longest spines 3 in length of head, more than $\frac{1}{3}$ longer than the last spine, and scarcely lower than the soft rays. Insertion of dorsal opposite tip of the bony opercle, the opercular flap extending to opposite the third spine, last rays of dorsal and anal reaching past the base of caudal rays; caudal fin rounded, about $\frac{1}{2}$ length of head; anal higher than soft dorsal, its longest rays 2$\frac{1}{4}$ in head; ventrals large, close together, inserted in front of pectoral, 1$\frac{3}{5}$ in head; pectorals short and broad, 1$\frac{3}{5}$ in head. Color in spirits, olivaceous, slightly brownish above, scarcely paler below, everywhere more or less tinged and mottled with greenish; head everywhere thickly and closely covered with small rounded dark-brown spots, largest above and on cheeks, where they are about as large as pin heads; smaller on lips and opercles, most thickly set on the anterior part of the head; eye thickly spotted; spots similar to those on the head extending along upper part of back, forming a vague band, which grows narrower backward and disappears opposite front of second dorsal; front side of pectoral and first 3 or 4 dorsal spines with dark spots; dorsal dusky olive, with darker clouds, and with some dark spots, especially on the spinous part; caudal and anal plain dusky or faintly marbled with paler; ventrals blackish, greenish at base; pectorals dusky green. Gulf of California, in shallow water; 1 specimen known, the type (above described) 16 inches long. This is perhaps the largest species of the genus. ($\rho o\mu\alpha\lambda\epsilon o\varsigma$, robust.)

Opisthognathus rhomaleus, JORDAN & GILBERT, Proc. U. S. Nat. Mus. 1881, 276, Santa Maria Cove, Lower California. (Type, No. 29382. Coll. Lieut. Henry L. Nichols.)

2623. GNATHYPOPS SNYDERI, Jordan & Evermann, new species.

Head 3$\frac{1}{2}$; depth 4$\frac{1}{3}$. D. X, 14; A. II, 13; scales 93; 60 pores; 3 or 4 teeth on vomer; lateral line very distinct, extending to fourteenth ray of dorsal; no notch separating the dorsals, the spines and soft rays not separable, the last spine not much shorter than the longest, which is $2\frac{1}{4}$ head; longest anal ray 2, pectoral $2\frac{1}{4}$. Vertebræ $10 + 17 = 27$. Body olivaceous, with 5 broad faint dusky cross shades; head with many round black spots of varying sizes, some as large as pin heads, the largest below and between eyes, covering both jaws and the membrane of the maxillary, few on cheeks, most numerous on forehead; similar spots extending along side of back to end of lateral line; dorsal dusky, with 6 round dusky blotches at its base, corresponding to the dark shades on body; caudal, anal, and pectorals plain dusky olive; ventrals blackish. Gulf of California; known from 1 specimen collected by Dr. Gilbert in San Luis Gonzales Bay. The species is close to *G. rhomalea*, but has a different dorsal fin and lateral line. Type, No. 2014 L. S. Jr. Univ. Mus., about 8 inches long. (Named for John O. Snyder, curator of fishes in Leland Stanford Junior University.)

2624. GNATHYPOPS MYSTACINA, Jordan.

Head $3\frac{1}{12}$ in length ($3\frac{3}{5}$ to tip of caudal); depth $4\frac{4}{5}$ ($5\frac{5}{6}$). D. 23 or 24 (X, 14); A. II, 11; lateral line with about 54 tubes; 100 scales between gill opening and caudal. Head rather elongate, very blunt in profile; snout very short, not longer than pupil; eye large, about $3\frac{1}{3}$ in length; maxillary $1\frac{2}{3}$ in length of head, 5 in length to base of caudal, $6\frac{1}{4}$ in total length to tip of caudal; end of maxillary abruptly truncate, not ending in a flexible lamina, the supplemental bone well developed; lower jaw slightly included. Teeth in each jaw in a narrow band, the outer slender, enlarged; vomer with about 4 slender teeth; palatines toothless. Gill rakers long and slender. Gill membranes nearly separate, free from the isthmus. Scales very small; lateral line extending to below anterior part of soft dorsal, its length $\frac{3}{4}$ that of head. Dorsal spines not distinguishable from the soft rays, the rays apparently fewer than usual, none of them very high, the last ray $2\frac{1}{4}$ in head; caudal short, apparently truncate, $1\frac{1}{4}$ in head; anal rather low; pectorals 2 in head; ventrals $1\frac{3}{4}$. Color nearly plain olive green, without bands or spots on body or fins; vertical fins tipped with blackish; maxillary with a faint median blackish stripe; pectoral with 2 dusky cross shades; no black or white on lining membrane of jaws. Length $3\frac{1}{2}$ inches. Deep waters of Gulf of Mexico; the few specimens known from the stomachs of Red Snappers (*Neomænis aya*) from the Pensacola Snapper Banks. It resembles *Opisthognathus lonchurum*, but the scales are smaller. ($\mu\acute{v}\acute{\sigma}\tau\alpha\xi$, mustache, from the maxillary stripe).

Gnathypops mystacinus, JORDAN, Proc. U. S. Nat. Mus. 1884, 37, Snapper Banks off Pensacola. (Coll. Jordan & Stearns. Type, 34976, U. S. Nat. Mus.)

843. LONCHOPISTHUS, Gill.

Lonchopisthus, GILL, Proc. Ac. Nat. Sci. Phila. 1862, 241 (*micrognathus*).

This genus differs from *Gnathypops* in the slender, compressed body, the still smaller maxillary and the lanceolate caudal fin. The single species is very rare. ($\lambda\acute{o}\gamma\chi\eta$, lance; $\acute{o}\pi\iota\acute{\sigma}\theta\varepsilon$, behind, from the form of the caudal.)

2625. LONCHOPISTHUS MICROGNATHUS (Poey).

Head 5 in total; depth 6. D. X, 17; A. II, 16; scales 80; eye 3 in head. Body elongate, compressed, snout short; maxillary reaching ⅓ an eye's diameter behind the eye; no teeth on vomer; no second row of teeth in jaws; no scales on head except on cheek; lateral line almost touching profile of back. Caudal long and pointed as in *Gobius oceanicus.* Dark brown, paler below; 20 narrow vertical whitish bands from back to belly, the first 2 on cheek, the third on opercle; fins colored like body, except the pectorals, which are yellow, the edge orange. Vertebræ 10 + 18 = 28. Length 4 inches. Cuba; rare (Poey); only the types known, examined by us in the National Museum. (μικρός, small; γναθος, jaw.)

Opisthognathus micrognathus, POEY, Memorias, II, 287, 1860, Cuba. (Coll. Poey.)
Lonchopisthus micrognathus, GILL, Proc. Ac. Nat. Sci. Phila. 1862, 241.

Family CXCII. BATHYMASTERIDÆ.

(THE RONQUILS.)

Body rather elongate, moderately compressed, covered with small, ctenoid scales. Head rather large, subconic. Eyes large. Mouth moderate, nearly horizontal, the lower jaw slightly projecting; lips full; premaxillaries protractile, not extending to angle of the mouth; maxillary without supplemental bone, not slipping under the narrow preorbital. Teeth moderate, in a cardiform band in each jaw, the outer somewhat enlarged; bands of teeth on vomer and palatines. No barbels; no crests or spines on head. Branchiostegals 6. Gill membranes scarcely or broadly connected, free from the isthmus; gill rakers few, very short. Pseudobranchiæ large. Opercular bones unarmed. Mucous pores numerous on top and sides of head, sometimes provided with fringed flaps. Lateral line conspicuous, placed high, not quite reaching the caudal fin, its scales sometimes enlarged. Dorsal fin long, continuous, moderately high, a few of the foremost rays inarticulate, none of them pungent or spine-like; the posterior rays branched; anal fin long, similar to the dorsal; caudal convex; pectorals rather broad, their bases extending obliquely downward and backward, their rays all branched; ventrals slightly in front of pectorals, I, 5, close together, the inner rays longest. Skeleton well ossified. Pyloric cæca few (2 or 3). No anal papilla. Vertebræ in large numbers, about 14 + 35. Three species known, from the Northern Pacific; here referred to 3 genera. The relations of the group are uncertain; externally they resemble the *Opisthognathidæ,* but the relation can not be close, and the number of vertebræ is greatly increased.

a. Gill membranes not connected below.
  b. Head naked; scales in lateral line not enlarged; only first 3 or 4 rays in dorsal fin unbranched; pores of head with small flaps. BATHYMASTER, 844.
  bb. Head scaly on cheeks; scales in lateral line enlarged; anterior 20 to 30 rays of dorsal fin unbranched; pores of head mostly without flaps.
      RONQUILUS, 845.
aa. Gill membranes broadly connected; cheeks scaly; scales in lateral line enlarged; about 15 of anterior rays of dorsal simple. RATHBUNELLA, 846.

844. BATHYMASTER, Cope.

Bathymaster, COPE, Proc. Amer. Phil. Soc. 1873, 31 (*signatus*).

Head naked; pores of head large, many of them with dermal flaps; gill membranes scarcely connected; scales of lateral line similar to the others; dorsal fin with but 3 or 4 of its anterior rays unbranched; characters otherwise included above. ($\beta\alpha\theta\acute{v}\varsigma$, deep; $\mu\alpha\acute{\sigma}\tau\acute{\eta}\rho$, searcher.)

2626. BATHYMASTER SIGNATUS, Cope.

Head $3\frac{1}{2}$; depth 5. D. 47; A. 34; scales 6-95-19; eye $4\frac{1}{2}$ in head; maxillary $2\frac{1}{2}$; snout $4\frac{1}{2}$; pectoral $1\frac{1}{2}$; ventral $2\frac{1}{2}$; highest dorsal ray $2\frac{3}{4}$; highest anal ray $3\frac{1}{2}$; caudal $2\frac{3}{4}$. Body compressed, elongate, anterior profile convex from tip of snout to dorsal; mouth not very oblique, the maxillary reaching the vertical from posterior edge of orbit; snout about equal to eye; jaws equal, with bands of small conical teeth, outer row enlarged; lower jaw with a single row at the sides; well developed conical teeth on vomer and palatines. Branchiostegal membranes not united; margin of preopercle free, furnished with 5 or 6 conspicuous mucous pores; large pores on top and sides of head, each with a small flap; opercle ending in a flap behind; gill rakers moderately long and slender, $\frac{2}{3}$ eye, about 7 + 18; many mucous pores on top of head and under eye; head entirely naked; dorsal and pectoral with fine scales running about halfway up the fin; anal naked; a naked strip from nape to dorsal; pectoral broad and fan-shaped, its lower rays smaller, reaching to front of anal; origin of ventral spine about the diameter in front of the lower end of pectoral base; dorsal about uniform in height for nearly its entire length, higher than anal; dorsal and anal rays about reaching to base of caudal rays; first 3 or 4 rays of dorsal simple, the others branched; caudal truncate or slightly rounded. Color almost uniform warm brown with darker shades, the fins somewhat mottled with yellowish, the anal and ventrals blackish, other fins dusky; a conspicuous black ocellated blotch on front of dorsal, covering tips of 4 or 5 spines.* Shores of southern Alaska, from Unimak

---

* Concerning this species, Dr. Gilbert has the following note: "*Bathymaster signatus* is taken very abundantly in our series of shallow-water dredgings along the southern shore of the Alaskan Peninsula, and northward through Unimak Pass. The stations at which it was obtained are numbered 3211, 3212, 3213, 3214, 3215, 3217, 3220, 3222, and 3223, and the depth range from 34 to 56 fathoms. In addition, a very few small specimens were secured at Stations 3262, 3309, 3221 and 3333, north of the Aleutian Islands, in depths of 19 to 71 fathoms, but the species is evidently not abundant in Bering Sea. No examples were taken in any of the very numerous dredgings made in Bristol Bay. In life the sides are olive brown, and the upper parts show faint traces of 6 or 7 broad dusky cross bars, which correspond to or alternate with an equal number below the lateral line; the anal and ventral fins, the branchiostegal and gular membranes, the lower pectoral rays, and the snout blue black; anterior edge of orbit and front edge of preorbital light yellow; the pores on edge of preopercle, 2 pores above and behind maxillary, and 3 at upper edge of opercle, bright scarlet; a large black blotch on anterior dorsal rays; distal half of anterior portion of dorsal fin and the upper pectoral rays yellow. Outer ventral ray simple and inarticulate, followed by 5 branched rays. Only the first 2 dorsal rays spinous, being soft and flexible, but unjointed. The third and all following rays jointed and forked. All of the anal rays jointed. A specimen from *Albatross* Station 3211, 35 mm. in length to base of caudal, shows that the ventrals occupy very different positions in adults and in young. In the latter they are truly thoracic in position, and are inserted as much behind base of pectorals as they are located in advance of this point in adults. A specimen 65 mm. long is entirely similar to adults in this respect."

Pass to Sitka; not uncommon in water of moderate depths. Here described from a specimen collected by the *Albatross* (No. 2143, L. S. Jr. Univ. Mus.), Station 3214, 11 inches in length. Other specimens taken in rock pools at Sitka are dark green, almost black. (*signatus*, marked.)

Bathymaster signatus, COPE, Proc. Amer. Philos. Soc. 1873, 31, Sitka (Coll. Prof. George Davidson); GILBERT, Proc. U. S. Nat. Mus. 1888, 554.

845. RONQUILUS, Jordan & Starks.

Ronquilus, JORDAN & STARKS, Proc. Cal. Ac. Sci. 1895, 838 (*jordani*).

Cheeks scaly; scales of lateral line enlarged; anterior half of dorsal fin more or less composed of unbranched rays; mucous pores on head without conspicuous flaps; gill membranes separate. One species. (*Ronquil*, a Spanish name of the typical species, possibly from ρόγχος, one who grunts.)

2627. RONQUILUS JORDANI (Gilbert).

(RONQUIL.)

Head $4\frac{1}{5}$; depth $6\frac{3}{4}$. D. 41; A. 33; V. I, 5; P. 18; scales 92+6 (tubes), about 200 transverse. Body rather elongate, moderately compressed. Eye large, about as long as snout, 4 in head, its diameter much more than the interocular space; maxillary extending to below front of pupil; cheeks closely scaly; rest of head entirely naked; a narrow, naked area in front of dorsal, bounded by rows of mucous pores; skull with large mucous cavities behind the eyes, which are translucent in life; scales of lateral line enlarged, twice as far apart as the others. Dorsal fin inserted at a distance behind the occiput, less than the diameter of the eye; pectorals $\frac{5}{8}$ the length of the head; fourth ray of ventrals longest; vent much nearer snout than root of caudal; vertebræ 14+35=49; olivaceous, tinged with brown; about 8 round, faint-bluish blotches along the sides, each surrounded by rings of yellow spots; a yellow ring around the eye and a yellow band along the cheek; fins translucent, the anal with a yellowish strip and a deep-bluish or black edging; dorsal reddish or yellow, with a dusky blotch in front; ventrals dusky; pectorals with the lower rays blackish or dark blue, larger specimens nearly uniformly dark, the color varying with the surroundings. Bristol Bay to Puget Sound, about rocks, in water of moderate depth. Length 6 to 10 inches. Known from Seattle and from Wrangel and Bristol Bay, the latter specimen in 32 fathoms. (Named for its discoverer, David Starr Jordan.)

Bathymaster signatus, JORDAN & GILBERT, Synopsis, 623, 1883; not *B. signatus*, COPE.
Bathymaster jordani, GILBERT, Proc. U. S. Nat. Mus. 1888, 554, Elliott Bay at Seattle (Coll. Jordan) and Fort Wrangel, Alaska. (Coll. *Albatross*.)
Ronquilus jordani, JORDAN & STARKS, Proc. Cal. Ac. Sci. 1895, 838, pl. 99.

846. RATHBUNELLA, Jordan & Evermann.

Rathbunella, JORDAN & EVERMANN, Check-List Fishes, 463, 1896 (*hypoplectus*).

This genus differs from *Ronquilus* in having the gill membranes broadly united across the isthmus. The unbranched anterior rays form about $\frac{1}{2}$

of the dorsal fin. (Named for Mr. Richard Rathbun, then chief of the Division of Scientific Inquiry in the U. S. Fish Commission, in recognition of his many services to science.)

2628. RATHBUNELLA HYPOPLECTA (Gilbert).

Head 4⅓ in length; depth 7. D. 46; A. 33. Head and body compressed, elongate, the anterior profile of head compressed, declivous; mouth somewhat oblique, at lower side of snout, small, the maxillary reaching vertical from middle of pupil, 3¼ in head; snout very slightly shorter than orbit, 4⅓ in head; diameter of orbit 4 in head; teeth well developed, in broad bands on jaws, vomer and palatines, the vomer and palatine patches nearly continuous; branchiostegal membranes broadly united, free from isthmus, forming a fold whose depth exceeds ½ diameter of orbit. Margin of preopercle adnate behind, slightly free below, furnished with a series of 6 conspicuous mucous pores; head without spines, ridges or filaments; inner margin of shoulder girdle conspicuously notched above and below, but without hook; gill rakers tubercular, few in number; a well-marked slit behind last gill. Distance from nape to front of dorsal fin equals its distance from posterior border of eye; anterior 10 or 12 dorsal rays simple and apparently not articulate, but flexible and not spine-like; distance from front of anal to base of ventrals 2¼ in its distance from base of caudal, all but first ½ of dorsal rays, and all of anal rays forked at tip; dorsal not high, the longest rays ½ head; highest anal ray equals snout and ½ eye; last dorsal and anal rays entirely disconnected from caudal, leaving a free space on caudal peduncle ½ diameter of orbit; ventrals I, 5, in advance of base of pectorals, narrowly triangular, the inner rays longest; pectorals with curved base running backward and downward, the rays all branched, 18 in number, the width of base of fin 3¼ in head, the longest ray 1⅓ in head; caudal rounded, ⅔ length of head. Body covered with small, partially embedded, cycloid scales, including antedorsal region, belly, breast, and area in front of base of pectorals; cheeks covered with similar but smaller scales, the opercles and rest of head naked. Lateral line running high, parallel with back, on a series of enlarged scales, which are also partly embedded in the thick skin; the lateral line fails to reach base of caudal by a distance equaling ¼ of head, and is present on 82 scales. Color, dark olive-brown above, lighter below; a series of about 12 quadrate dark blotches below lateral line, connected more or less by dusky streaks with an alternating series along base of dorsal; no bright colors; dorsal, pectorals, ventrals and branchiostegal membranes dusky straw color; anal black, the rays white tipped; caudal blackish; peritoneum white. A single specimen, 8 inches long, from *Albatross* Station 2944, off Santa Barbara Islands, in 30 fathoms.

Bathymaster hypoplectus, GILBERT, Proc. U. S. Nat. Mus. 1890, 97, off Santa Barbara Islands, California, at Albatross Station 2944. (Coll. *Albatross*.)

Family CXCIII. CHIASMODONTIDÆ.

(The Black Swallowers.)

dy elongate, subcylindrical, or slightly tapering; head subconic.
naked; lateral line continuous, placed low; 2 dorsal fins, the first
r short, of slender spines, the second dorsal and anal long; ventrals
al, thoracic, inserted before pectorals, the rays I, 5; pectorals long and
w; mouth very deeply cleft, reaching beyond the eyes, with numer-
long, sharp, movable teeth, the anterior canines movable; teeth on
tines; upper jaw not protractile, the maxillary produced backward.
cular apparatus very oblique and reduced; no spines or cirri on head;
al fin forked. Genera 2; species 2; deep-sea fishes, notable for the
p teeth and for the extensible stomach. (*Chiasmodontidæ*, Gill, in Jor-
& Gilbert, Synopsis, 964, 1883.)

Jaws with some of the anterior canines extremely long and movable, the 2 anterior
   crossing each other; lower jaw projecting.   Chiasmodon, 847.
Jaws with slender, close-set teeth, some of them greatly produced; lower jaw not
   prominent.   Pseudoscopelus, 848.

847. CHIASMODON, Johnson.

(Black Swallowers.)

Chiasmodon, Johnson, Proc. Zool. Soc. London 1863, 406 (*niger*).
Chiasmodus, Günther, change of spelling.

dy elongate, compressed, and tapering posteriorly, naked; belly pend-
its walls membranaceous, capable of great dilation. Mouth very
); lower jaw longer than upper; each jaw with 2 series of large,
ted teeth, some of the anterior being very large and movable; vomer-
eeth none; palatines with teeth similar to those in the jaws. Gills 4.
seudobranchiæ. Gill openings very wide, the membranes joined to
isthmus for a short distance. Dorsal fins 2; anal single; ventrals
ted below pectorals, each of 5 soft rays. Tail truncate at base of
al. Caudal forked, free from dorsal and anal. Singular fishes of the
sea, remarkable for their ability to swallow fishes of many times
own size by means of the great distensibility of the walls of the
(χίασμα, a mark of the form of the letter χ; ὀδούς, tooth; the 2
rior canines crossing each other when depressed.)

2892. CHIASMODON NIGER, Johnson.

and 34. D. XI-28; A. 27; P. 13; V. 5. Head compressed, elongate, the
rn flat, its depth less than ¼ its length; maxillary reaching angle of
percle; both jaws armed with long, pointed, wide-set teeth, nearly
f which are movable; 2 anterior teeth of upper jaw very long, cross-
each other when depressed; 3 anterior pairs of teeth in lower jaw

of the dorsal fin. (Named for Mr. Richard Rathbun, then chief of the Division of Scientific Inquiry in the U. S. Fish Commission, in recognition of his many services to science.)

2628. RATHBUNELLA HYPOPLECTA (Gilbert).

Head 4⅔ in length; depth 7. D. 46; A. 33. Head and body compressed, elongate, the anterior profile of head compressed, declivous; mouth somewhat oblique, at lower side of snout, small, the maxillary reaching vertical from middle of pupil, 3⅕ in head; snout very slightly shorter than orbit, 4¼ in head; diameter of orbit 4 in head; teeth well developed, in broad bands on jaws, vomer and palatines, the vomer and palatine patches nearly continuous; branchiostegal membranes broadly united, free from isthmus, forming a fold whose depth exceeds ½ diameter of orbit. Margin of preopercle adnate behind, slightly free below, furnished with a series of 6 conspicuous mucous pores; head without spines, ridges or filaments; inner margin of shoulder girdle conspicuously notched above and below, but without hook; gill rakers tubercular, few in number; a well-marked slit behind last gill. Distance from nape to front of dorsal fin equals its distance from posterior border of eye; anterior 10 or 12 dorsal rays simple and apparently not articulate, but flexible and not spine-like; distance from front of anal to base of ventrals 2¼ in its distance from base of caudal, all but first ½ of dorsal rays, and all of anal rays forked at tip; dorsal not high, the longest rays ½ head; highest anal ray equals snout and ½ eye; last dorsal and anal rays entirely disconnected from caudal, leaving a free space on caudal peduncle ½ diameter of orbit; ventrals I, 5, in advance of base of pectorals, narrowly triangular, the inner rays longest; pectorals with curved base running backward and downward, the rays all branched, 18 in number, the width of base of fin 3¼ in head, the longest ray 1¼ in head; caudal rounded, ⅔ length of head. Body covered with small, partially embedded, cycloid scales, including antedorsal region, belly, breast, and area in front of base of pectorals; cheeks covered with similar but smaller scales, the opercles and rest of head naked. Lateral line running high, parallel with back, on a series of enlarged scales, which are also partly embedded in the thick skin; the lateral line fails to reach base of caudal by a distance equaling ¼ of head, and is present on 82 scales. Color, dark olive-brown above, lighter below; a series of about 12 quadrate dark blotches below lateral line, connected more or less by dusky streaks with an alternating series along base of dorsal; no bright colors; dorsal, pectorals, ventrals and branchiostegal membranes dusky straw color; anal black, the rays white tipped; caudal blackish; peritoneum white. A single specimen, 8 inches long, from *Albatross* Station 2944, off Santa Barbara Islands, in 30 fathoms.

Bathymaster hypoplectus, GILBERT, Proc. U. S. Nat. Mus. 1890, 97, off Santa Barbara Islands, California, at Albatross Station 2944. (Coll. *Albatross*.)

Family CXCIII. CHIASMODONTIDÆ.

(THE BLACK SWALLOWERS.)

Body elongate, subcylindrical, or slightly tapering; head subconic. Skin naked; lateral line continuous, placed low; 2 dorsal fins, the first rather short, of slender spines, the second dorsal and anal long; ventrals normal, thoracic, inserted before pectorals, the rays I, 5; pectorals long and narrow; mouth very deeply cleft, reaching beyond the eyes, with numerous long, sharp, movable teeth, the anterior canines movable; teeth on palatines; upper jaw not protractile, the maxillary produced backward. Opercular apparatus very oblique and reduced; no spines or cirri on head; caudal fin forked. Genera 2; species 2; deep-sea fishes, notable for the sharp teeth and for the extensible stomach. (*Chiasmodontidæ*, Gill, in Jordan & Gilbert, Synopsis, 964, 1883.)

> *a.* Jaws with some of the anterior canines extremely long and movable, the 2 anterior crossing each other; lower jaw projecting. CHIASMODON, 847.
> *aa.* Jaws with slender, close-set teeth, none of them greatly produced; lower jaw not prominent. PSEUDOSCOPELUS, 848.

847. CHIASMODON, Johnson.

(BLACK SWALLOWERS.)

Chiasmodon, JOHNSON, Proc. Zool. Soc. London 1863, 408 (*niger*).
Chiasmodus, GÜNTHER, change of spelling.

Body elongate, compressed, and tapering posteriorly, naked; belly pendent, its walls membranaceous, capable of great dilation. Mouth very large; lower jaw longer than upper; each jaw with 2 series of large, pointed teeth, some of the anterior being very large and movable; vomerine teeth none; palatines with teeth similar to those in the jaws. Gills 4. No pseudobranchiæ. Gill openings very wide, the membranes joined to the isthmus for a short distance. Dorsal fins 2; anal single; ventrals inserted below pectorals, each of 5 soft rays. Tail truncate at base of caudal. Caudal forked, free from dorsal and anal. Singular fishes of the deep sea, remarkable for their ability to swallow fishes of many times their own size by means of the great distensibility of the walls of the body. ($\chi i \alpha \sigma \mu \alpha$, a mark of the form of the letter χ; $\delta \delta o \dot{v} \varsigma$, tooth; the 2 anterior canines crossing each other when depressed.)

2629. CHIASMODON NIGER, Johnson.

Head 3¼. D. XI–28; A. 27; P. 13; V. 5. Head compressed, elongate, the crown flat, its depth less than ¼ its length; maxillary reaching angle of preopercle; both jaws armed with long, pointed, wide-set teeth, nearly all of which are movable; 2 anterior teeth of upper jaw very long, crossing each other when depressed; 3 anterior pairs of teeth in lower jaw likewise prolonged, the third pair the longest; palatines with a longer, fixed tooth in front. Eye moderate, above the anterior part of maxillary, 4¼ in head, shorter than snout, as wide as interorbital space. Lateral line in a longitudinal groove. First dorsal of slender rays, its base 2¼ in in that of second dorsal; anal commencing behind second dorsal, its anterior rays without connection with vertebral column; posterior rays of anal and dorsal very feeble; pectoral as long as head without snout; ventral ¼ as long as pectoral. Color entirely black. Length 12 inches. (Günther.) Deep waters of the Atlantic; a remarkable fish, the walls of the body inordinately extensible; taken at Madeira, in the mid-Atlantic, near the island of Dominica, and off the coast of Massachusetts.* (*niger*, black.)

Chiasmodon niger, JOHNSON, Proc. Zool. Soc. London 1863, 408, Madeira; JORDAN & GIL-
BERT, Synopsis, 964; GOODE & BEAN, Oceanic Ichthyology, 292, 1896.
Chiasmodus niger, GÜNTHER, Cat., v, 435, 1864; CARTER, Proc. Zool. Soc. 1866, 38; GÜNTHER,
Challenger Report, Deep Sea Fishes, XXII, 99, 1887.

848. PSEUDOSCOPELUS, Lütken.

Pseudoscopelus, LÜTKEN, Spolia Atlantica, Scopelini, 64, 1892 (*scriptus*).

Body perciform, scaleless, naked; mouth very large; eyes moderate; the slender maxillary reaching far beyond eye; jaws and palate with slender, close-set teeth; ventral fins short, subthoracic, of 1 spine and 5 rays; first dorsal short, of about 8 slender spines; posterior dorsal long, similar to the anal. Each jaw with a distinct line of pores, a median line of pores before ventrals, a cross line connecting ventrals, a series of pores from the vent passing around anal on each side. Lateral line well developed, running high. Head without spines. Gill openings very broad. Pectorals long; caudal short, forked. One species known, in deep water. (ψευδής, false; *Scopelus*.)

2630. PSEUDOSCOPELUS SCRIPTUS, Lütken.

Head 3½; depth 4¼. D. VIII-22; A. 22; V. I, 5. Body subfusiform, somewhat compressed. Head large, the snout short and pointed, 4¼ in head, the small eye, about 5. Jaws subequal, maxillary 1½ in head; cheek V-shaped, very oblique; bones of head not serrate. Form of head and mouth much as in *Engraulis* or *Scopelus*. Pectoral nearly as long as head, reaching past front of anal; soft dorsal higher than spinous, the anterior rays of soft dorsal and anal elevated. Pores as above described.

---

* The first specimen of this remarkable fish was obtained at Magdalena (Madeira), at a depth of 312 fathoms, in 1850, by Lowe, who, however, omitted to give a description of it. The species was rediscovered at the same locality by Johnson twelve years later. A third specimen was picked up from the surface, near the island of Dominica. A fourth example was obtained by the *Challenger* in mid-Atlantic, at Station 107, in 1,500 fathoms, on August 26, 1873. A fifth was obtained by the U. S. National Museum from Capt. Thomas F. Hodgdon of the Gloucester schooner *Bessie W. Somers*. It was found on Le Have Bank, floating on the surface, in June, 1880. (Goode & Bean.)

One specimen from Old Bahama Straits. (Lütken.) A singular fish of uncertain relationships, remarkable for the development of mucous pores. (*scriptus*, written.)

Pseudoscopelus scriptus, LÜTKEN, Spolia Atlantica, Scopelini, 64, 1892, Old Bahama Straits.

Family CXCIV. CHÆNICHTHYIDÆ.

Body rather elongated, gradually and regularly declining from the nape to the caudal fin; anteriorly subcylindrical or scarcely compressed. Skin naked or covered with small scales. Lateral line high on the sides and near the dorsal fin. Head moderate or large, with the snout prolonged, depressed, and spatuliform. Crown depressed, not relieved by crests or ridges. Preorbital bones large; suborbital chain very narrow, not articulated with the preopercle. Opercular bones all present, the interopercle and subopercle moderately developed. Mouth terminal, with the cleft lateral and large, extending to the vertical of the eye; upper jaw with its border formed almost entirely by the premaxillaries, whose posterior processes are very short; maxillaries with their articulations entirely posterior to the premaxillaries, slender and gradually enlarged toward their extremities. Teeth on the jaws; palate unarmed. Gill openings wide; gill membranes inferiorly deeply emarginated behind. Branchiostegals 6. Pseudobranchiæ developed. Dorsal fin with its spinous portion short, and usually distinct from the soft, the rays of the latter often simply articulated and not branched; anal fin a little shorter than the dorsal, its rays divided, the membrane notched behind each; caudal fin not forked; pectoral fins well developed, with their inferior rays divided; ventral fins jugular or subjugular, separated by a rhomboid area, each with a spine and 5 rays, the first of which is frequently thickened and entire. Cranium flattened behind, the crests little developed or obsolete. The spatuliform snout is principally formed by the elongated frontal bones. Stomach of moderate size and cæcal. Pyloric cæca in very small number. The chief distinctive characteristic of this family is doubtless the spatuliform extensions of the snout. This, combined with the extent of the fins, structure of the head, and general form, distinguish the group from all others. It appears to be most closely allied to the *Harpagiferidæ* and *Notothenidæ*. From the former it is separated by the form of the head, as well as by that of the body. From the latter, by the same features, and also by the naked skin. (Gill.) Genera 3 or 4, with about 6 species; inhabiting rather deep waters, mostly in the Tropics. (*Chænichthyoidæ*, Gill, Proc. Ac. Nat. Sci. Phila. 1861, 507.)

a. Body covered with cycloid, deciduous scales; maxillary with a flap; opercle with a dermal flap. HYPSICOMETES, 849.

849. HYPSICOMETES, Goode.

Hypsicometes, GOODE, Proc. U. S. Nat. Mus. 1880, 347 (*goboides*).

Body elongate, subcylindrical, tapering posteriorly. Head very large, much depressed, with snout elongate, spatulate; cleft of mouth very wide, horizontal, with lower jaw much the longer; the posterior margin of the maxillary wide, free, and with a long cutaneous flap. Eyes very large, close together, subvertical. Scales large, cycloid, deciduous; lateral line conspicuous and continuous, descending abruptly behind pectorals, its scales smaller than those of the body adjoining. Teeth acicular, in bands on the jaws, vomer, and palatines, the largest being upon the palatines, the vomer, and upon 2 pads on either side of the symphysis of the maxillaries. A sharp, short, strong scapular spine. Opercle with 3 feeble, sharp spines, each at the end of a strong feeble ridge; a long, skinny opercular flap extending far beyond the bony portion, and covered with scales. Branchiostegals 6. Gill membranes free from the isthmus, except far in front, where they are united to it, the left-hand flap overlapping the right at the point of junction. Pseudobranchiæ present. Gill rakers short. (ὕψι, below, i. e., in deep water; κωμήτης, dweller.)

2631. HYPSICOMETES GOBOIDES, Goode.

Head about 2½; depth 7½; orbit 4½ in head, or 1½ in snout. D. VI-15 to 17; A. 16 to 18; V. I, 5; P. 26; scales 65. Mouth very wide, horizontal, the maxillary, which is expanded spoon-like posteriorly, reaching considerably beyond vertical from anterior margin of orbit; eye considerably nearer tip of snout than end of flap, and equidistant between tip of snout and tip of uppermost spine of operculum; entire upper surface of head, cheeks, and opercula covered with scales, except upon bony portion of snout; first dorsal fin placed far forward, not far behind vertical from axil of pectoral; interspace between termination of first dorsal and beginning of second equal to diameter of the orbit, this fin composed of 6 spines, the first and second of which are longest, equal to distance from anterior margin of orbit to tip of lower jaw, and triangular in form; origin of second dorsal almost vertical from that of anal, and terminating a little in advance of the latter; second dorsal fin highest in front and low behind; length of caudal peduncle a little less than length of snout; caudal rounded; pectoral very broad at base, rounded, extending beyond vent and nearly to vertical from origin of anal; lower rays branched; ventrals far apart, horizontal, *Trigla*-like, composed of 1 flexible spine and 5 branched rays, their insertion far forward and far in advance of base of pectorals. Color grayish brown; lighter and yellowish below. Known only from a very small specimen, in which many of the important characters were not discernible. This specimen (No. 26007, U. S. Nat. Mus.) was taken by the *Fish Hawk* from Station 871, in 40° 02′ 54″ N. lat., 70° 23′ 40″ W. lon., at a depth of 115 fathoms, and is much contracted and distorted from immersion in strong alcohol. (Goode.) (*Gobius; εἶδος*, resemblance.)

Hypsicometes goboides, GOODE, Proc. U. S. Nat. Mus. 1880, 348, lat. 40°, 02′, 54″ N., lon. 70° 23′ 40″ W., in 115 fathoms (Coll. *Fish Hawk*); JORDAN & GILBERT, Synopsis, 808, 1883; GOODE & BEAN, Oceanic Ichthyology, 290, fig. 263, 1896.

Family CXCV. TRICHODONTIDÆ.

(THE SAND-FISHES.)

Body rather elongate, compressed, naked. Head short, flat on top, the sides vertical. Eyes large, high up, but not superior. Mouth large, almost vertical; lower jaw projecting, its tip entering the profile; lips fringed; premaxillaries protractile; maxillary very broad, without supplemental bone, not slipping under the very narrow preorbital. Teeth moderate, slender and sharp, but not setiform, in bands on jaws and vomer; palatines toothless; inner teeth of jaws depressible. Gill rakers short, slender; gill membranes narrowly united, free from the isthmus. Branchiostegals 5. Gills 4, a slit behind the fourth. Pseudobranchiæ large. Preopercle with 5 prominent spines, the 2 upper directed strongly upward, the 2 lower downward, the middle 1 downward and backward; no barbels; opercle small, strongly striate, unarmed; preorbital with spines; no suborbital stay. Lateral line obsolete. Dorsal fins separate, the first the larger, of numerous slender spines; anal fin elongate, without distinct spines, the rays of anterior third of the fin much shorter than the others, the beginning of the fin below middle of spinous dorsal; pectorals with a very broad, curved, procurrent base; a broad lunate area between pectoral and gill opening, nearly covered by the opercle; soft rays of dorsal, anal, and pectoral fins all simple; ventrals I, 5, close together, thoracic, but behind the pectorals, the middle rays longest; caudal lunate, with many accessory rays, on a slender peduncle. Vertebræ numerous, 48 in typical species. Two genera and 2 species known; from the North Pacific; living in sand near the shore. The fringed lips and other characters indicate the relationship of these fishes with the *Uranoscopidæ.* (*Trachinidæ*, genus *Trichodon*, Günther, Cat., II, 250.)

a. First dorsal long and rather low, of 14 or 15 spines.   TRICHODON, 850.
aa. First dorsal short and high, of 10 spines.   ARCTOSCOPUS, 851.

850. TRICHODON (Steller) Cuvier,

(SAND-FISHES.)

Trichodon, STELLER, in Tilesius, Mem. Acad. St. Petersburg, IV, 1811, 468 (*trichodon*).
Trichodon, CUVIER, Règne Animal, Ed. II, vol. 2, 149, 1829 (*trichodon*).

Characters of the genus included above, the first dorsal long and rather low, of 15 spines. One species. ($\theta \rho i \xi$, hair; $\delta \delta o \dot{v} \varsigma$, tooth.)

2632. TRICHODON TRICHODON (Tilesius).

(SAND-FISH.)

Head from tip of upper jaw, 3⅓; depth 3¼. D. XIII–I, 18; A. 28; P. 22; eye 4⅓ in head, snout 4½; maxillary 2; interorbital 3; pectoral 1⅓; ventral 1¾; height of spinous dorsal 3⅙. Body moderately elongate, compressed; dorsal outline slightly concave and sloping gently upward from snout to dorsal, thence turning at a very slight angle nearly straight to caudal;

ventral outline well rounded from chin to caudal peduncle, the curve much more gradual posteriorly; head and body everywhere covered with thin naked skin. Mouth large, superior, nearly vertical, the lower jaw projecting, its tips entering the profile; lips fringed; maxillary reaching to middle of pupil; teeth in 2 or 3 rows, small, sharp and recurved; teeth on vomer; palatines toothless. Eyes placed high, their diameter equal to length of snout; interorbital wide and flat, a third wider than eye; top of head smooth, sometimes rugose in younger individuals, covered with thin smooth skin; anterior nostril ending in a tube; preopercle with 5 spines, the 1 at angle largest, the 2 upper ones pointing upward and backward, the middle one pointing downward and backward, the 2 lower ones pointing downward and forward; opercle with radiating ridges; gill rakers short and slender, numerous. Origin of spinous dorsal behind base of pectoral, its distance from snout 3 in body, the spines not varying greatly in length, the last one connected by a membrane to the back; soft dorsal well separated from spinous, its rays about equal to spines in length, highest in front; anal long, its origin nearer to the snout than base of caudal by a distance equal to the length of the eye. Pectoral, when spread, broadly rounded behind, its lower rays rapidly decreasing in size below, reaching well past front of anal; ventrals inserted behind base of pectorals a distance equal to $\frac{2}{3}$ eye, their tips reaching to vent. Lateral line running high. Vertebræ $17 + 30 = 47$. Color silvery, light brown above; a dark brown streak following the lateral line, broken up into spots anteriorly; quadrangular, dark brown marks along the back at base of dorsals, chain-like markings in front of dorsal on nape; snout and tip of lower jaw dark; a dark line at lower part of eye; dorsals light, a dark streak along upper part of spinous dorsal; pectorals dusky; ventrals and anal colorless. Length 8 to 10 inches. North Pacific, on sandy shores, from Bering Sea to Monterey; very abundant northward; burying in the sand. Here described from a specimen, 8½ inches in length, from Herendeen Bay, Alaska (*Albatross* collection). Possibly detailed comparison may show a difference between California specimens and those from Bering Sea.

Trachinus trichodon, TILESIUS,* Mem. Acad. St. Petersburg, IV, 1811, pl. 15, fig. 8, 473, Kamchatka; PALLAS, Zoographia Rosso-Asiatica, III, 235, 1811.†

---

* The specific name *trichodon* should apparently date from Tilesius, 1811. Although Vol. IV, of the Mem. Acad. St. Petersburg bears the date 1813 it was for the year 1811, and it is evident that the plate containing the figure of this species was accessible to Pallas as early as 1811, for, in his "Zoographia," printed in 1811, though not published until 1831, Pallas refers to the plate of Tilesius in very definite terms. The fact that Pallas was, in 1811, thus able to refer definitely to Tilesius's plate of *Trachinus trichodon,* fixes the date of publication of that plate at least as early as 1811. That this plate appeared in the volume of Memoirs for 1811 (though the volume was not published until 1813), fixes 1811 as the date for the name. Though the "Zoographia" of Pallas was not formally published until 1831, it was printed in 1811, and Cuvier & Valenciennes evidently had a copy in 1829, as they refer to it.

† Tilesius confused matters greatly by using, in one and the same article, three different names or combinations of names for this fish. At the beginning of this article (p. 406) in a bald list of the species discussed in the paper, he has "*Drachinus trichodon.*" On page 466 he has "*Trachinus gasteropelecus,*" accompanied by a full description of the species. In a footnote on page 473, he has "*Trachinus trichodon*" together with a description which he says applies to the young, and finally his pl. 15, fig. 8, is marked "*Trachinus trichodon.*"

Drachinus trichodon, TILESIUS, Mem. Acad. St. Petersburg, IV, 1811, 406; name only.
Trachinus gasteropelecus, TILESIUS, *l. c.*, 466, 1811, Kamchatka.
Trichodon stelleri, CUVIER & VALENCIENNES, Hist. Nat. Poiss., III, 154, pl. 57, 1829; based on
  Trachinus trichodon PALLAS; GÜNTHER, Cat., II, 251, 1860; JORDAN & GILBERT, Synopsis, 627, 1883.
Trichodon lineatus, AYERS, Proc. Ac. Nat. Sci. Phila. 1860, 60, San Francisco; D. XV–18;
  A. 28; P. 23.

851. ARCTOSCOPUS, Jordan & Evermann.

Arctoscopus, JORDAN & EVERMANN, Check-List Fishes, 464, 1896 (*japonicus*).

This genus differs from *Trichodon* in the short, high, triangular spinous dorsal which is composed of 10 spines. ($\check{\alpha}\rho\kappa\tau o\varsigma$, northern; $\acute{o}\kappa o\pi\acute{o}\varsigma$, gazer; for *Uranoscopus*.)

2633. ARCTOSCOPUS JAPONICUS (Steindachner).

Head 3¾; depth 3¾. D. X or XI–13; A. 30 or 31; P. 25. Form of body and coloration of *Trichodon trichodon*. First dorsal high, triangular, the spines slender, separated by a long interval from the second dorsal. Preopercle with 5 sharp spines; the 2 spines on the preorbital very small. Pectoral well developed, all its rays simple, the lower a little thickened, the fin considerably longer than the head and reaching past the last spine of the dorsal; anal fin with its rays gradually longer posteriorly. Dentition as in *Trichodon trichodon*, but the mouth rather more oblique. Length 4½ inches. North Pacific; scarce. Recorded from Strietok, in the Sea of Japan, and Sitka, Alaska, by Steindachner, and by Jordan & Gilbert from Iturup Island (Kurils). (*japonicus*, from Japan.)

Trichodon japonicus, STEINDACHNER, Ichth. Beitr., X, 4, 1881, Strietok; Sitka; JORDAN, Cat.
  Fishes N. A., 117, 1885.
Arctoscopus japonicus, JORDAN & GILBERT, Rept. Fur Seal Investig., 1898.

Family CXCVI. DACTYLOSCOPIDÆ.

(THE SAND STAR-GAZERS.)

Body oblong, low, compressed posteriorly, covered with moderate, cycloid, imbricated scales; lateral line complete, anteriorly running along side of back, posteriorly median; head oblong, nearly plane above; eyes small, superior, well forward; suborbital bones enlarged, but without bony stay connecting with the preopercle; nostrils double; opercles fringed; mouth nearly vertical; premaxillaries protractile, not forming the entire edge of the upper jaw; lips fringed as in *Uranoscopidæ*; gill openings very broad, the membranes separated and free from the isthmus, pseudobranchiæ present or obsolete. Dorsal fin very long, continuous or divided, several of the anterior rays spinous; anal very long, commencing close behind the vent, which is near the breast; caudal diphycercal, free from dorsal and anal; pectorals variable, the base broad and procurrent; ventrals jugular, I, 3; vertebræ more than 10 + 14; pyloric cæca none. Genera 4; species about 10; small fishes living on sandy shores of tropical

America. This family is nearly related to *Uranoscopidæ*, of which group it seems to be a reduced or degenerate branch. Its relations with the Asiatic family *Leptoscopidæ* are most intimate, the incomplete ventrals and simple pectoral rays of *Dactyloscopidæ* being the chief distinctive features. (*Dactyloscopidæ*, Gill, Arrangm. Families Fishes, 1872.)

  a. Dorsal fin divided, the first dorsal composed of 3 spines inserted on the nape; head not cuboid; chin without flap; fringes of lips small.   GILLELLUS, 852.
  aa. Dorsal fin continuous.
    b. Dorsal fin commencing at the nape; pseudobranchiæ very small or obsolete; head cuboid.   DACTYLOSCOPUS, 853.
    bb. Dorsal fin commencing far behind the nape; pseudobranchiæ well developed.
      c. Head cuboid, formed as in *Dactyloscopus;* the mouth vertical.
              DACTYLAGNUS, 854.
      cc. Head elongate-conoid, the lower jaw projecting, with a fleshy flap at tip.
              MYXODAGNUS, 855.

852. GILLELLUS, Gilbert.

Gillellus, GILBERT, Proc. U. S. Nat. Mus. 1890, 98 (*semicinctus*).

A separate dorsal fin on the nape composed of 3 spines. Lateral line descending posteriorly, its dorsal and median portions about equal. Fringes of upper lip obsolete, those of lower lip little evident. Head not cuboid, the mouth moderately oblique, the lower jaw rounded in front and without symphyseal flap. The physiognomy is intermediate between *Dactyloscopus* and *Myxodagnus*, from each of which the genus is well separated by the characters of the dorsal fin and the lateral line. ("Named in honor of Dr. Theodore Gill, to whom we owe our knowledge of the previously described members of this most interesting group." Gilbert.)

  a. Tip of lower jaw projecting.
    b. Anterior portion of lateral line longer than posterior portion; the scales 25 to 28 + 3 + 15 to 18 = 43 to 49. D. III-IX, 28; A. II, 30 or 31.
              SEMICINCTUS, 2634.
    bb. Anterior portion of lateral line much shorter than posterior portion, 2½ times in the latter; scales 18 + 3 + 27 = 48. D. II-IX, 31; A. II, 35.
              ARENICOLA, 2635.
  aa. Tip of lower jaw scarcely projecting; anterior portion of lateral line 1¼ times in posterior. D. I-IX, 31; A. II, 34.   ORNATUS, 2636.

2634. GILLELLUS SEMICINCTUS, Gilbert.

Head 3⅔; depth 5¼. D. III-IX or X, 28; A. II, 30 or 31; scales 25 to 28–3-15 to 18 (43 to 49 scales in all). Body deep, tapering rapidly either way from front of dorsal. Mouth moderately oblique, the maxillary extending beyond orbit, 3 in head; tip of lower jaw projecting; teeth in a narrow band in front of jaws, becoming a single series laterally; none of the teeth enlarged. Opercular fringes well developed, 8 or 9 in number; fold of membrane between rami of lower jaw well developed; pseudobranchiæ apparently not developed; gill rakers obsolete. Dorsal fin beginning at a distance from occiput less than diameter of eye, the first 3 rays entirely detached from the rest of the fin, the first ray the highest, the second and third shortened; of the remaining part of the fin the first 9 or

10 rays are unarticulated and spinous; first 2 anal rays not articulated; caudal about 1¾ in head; pectorals 1¼. Lateral line running anteriorly along the very base of spinous dorsal, no scales intervening between it and base of fin; it descends to middle of sides posteriorly, the median portion of its length shorter than the dorsal portion. Color light oliva-ceous, the back with 6 broad cross bars of pink, narrowly margined behind and in front with blackish, terminating below on middle of sides; the lower of these bars frequently black; a black bar across caudal peduncle, and sometimes a black line at base of caudal; along median line of sides frequently a series of small black spots alternating with the cross bars; a similar series along median dorsal line; a large pink blotch covering occiput; a dusky bar across interorbital space, running downward and backward across cheek; silvery spots and blotches on cheeks and anterior portions of opercles; fins unmarked. Specimens have been obtained in the Gulf of California by the *Albatross*, at Stations 2827 and 2829, and by the *Grampus* in the Atlantic, at Stations 5108 and 5112, off the coast of Florida; no specific difference among them noticed, but the Atlantic form needs further study. (Gilbert.) (*semi*, half; *cinctus*, belted.)

Gillellus semicinctus, GILBERT, Proc. U. S. Nat. Mus. 1890. 98, Albatross Stations 2827, 2829, Gulf of California (Coll. *Albatross*); JORDAN, Proc. Cal. Ac. Sci. 1896, 229, pl. 32.

2635. GILLELLUS ARENICOLA, Gilbert.

Head 4¾ in length; depth 8¾. D. II-IX, 31; A. II, 35; scales 18-3-27. Body very slender and elongate, much as in *Myxodagnus*, the snout sharp, the mandible produced at symphysis and conspicuously projecting; labial fringes apparently obsolete; maxillary reaching vertical from middle of orbit; eye small, about equaling length of snout, 6 in head; opercular fringes nearly obsolete, 3 or 4 small ones at upper edge of opercle. Anterior dorsal inserted close behind occiput, composed of 3 rays, and separated by a short interspace from rest of fin; pectorals longer than head. Lateral line anteriorly running along base of dorsal, from which it is not separated by intervening scales, the anterior portion contained 2¼ times in the posterior median portion. Color light olivaceous, the head with grayish blotches and small pearly spots; 11 dark bars downward from back, the alternate ones narrower and fainter and not extending to middle of sides, as do the others; the margins of the larger bars darker than the median portion, the bars not continued onto dorsal fin; all the fins translucent. A single specimen 1½ inches long, from Cape San Lucas. (Gilbert.) (*arena*, sand; *colo*, I inhabit.)

Gillellus arenicola, GILBERT, Proc. U. S. Nat. Mus. 1890 99, Cape San Lucas. (Coll. Gilbert.)

2636. GILLELLUS ORNATUS, Gilbert.

Head 4½ in length; depth 8. D. III-IX, 31; A. II, 34; scales not counted. With the elongate form and general appearance of *Gillellus arenicola*, but differing in the subequal jaws and in the long anterior portion of the lateral line. Head conical, acute, very small; jaws nearly equal, the lower slightly longer than the upper, but not noticeably protruding. In this

respect the species resembles most strongly *G. semicinctus*, from which it varies widely in the general form and proportions. Snout extremely short, scarcely equaling diameter of the minute eye; diameter of orbit about 7 in head. Mouth oblique, the maxillary 4 in head, reaching nearly to vertical from posterior margin of orbit. Lips without fringes. Eyes separated by a narrow septum, the interorbital width being less than the diameter of the pupil. Opercular fringes few and small, flat, and not terminating evident ridges as in *Dactyloscopus*. Dorsal beginning well forward, its origin less than diameter of orbit behind the posterior line of occiput; anterior detached part of fin consisting apparently of 3 rays, the first of which is the longest, the second and third equal and short; fourth spine again longer; spines as usual slenderer than the rays, and showing no articulations, but with some difficulty discriminated from them; pectoral as long as head. Anterior part of lateral line running immediately along base of dorsal, without intervening scales, as in other members of this genus. It is much longer than in *G. arenicola* and is contained 1¼ times in the posterior median portion. There are 3 scales between the posterior part of the lateral line and the base of the dorsal. Color similar to that of *G. arenicola* and *G. semicinctus*, light olivaceous, unmarked below the middle of the sides, the back and upper half of sides with 8 brown bars which extend downward to lateral line; the upper part of each bar with a lighter central area, the light areas between the bars marked more or less with brown, which sometimes forms indistinct secondary bars; a blackish bar at base of caudal, and a faint streak below eye; a large pearly blotch on opercle. A single specimen, about 2 inches long, from *Albatross* Station 2828 in the Gulf of California. (Gilbert.) (*ornatus*, adorned.)

Gillellus ornatus, GILBERT, Proc. U. S. Nat. Mus. 1891, 558, Gulf of California. (Coll. Gilbert.)

853. DACTYLOSCOPUS,* Gill.

Dactyloscopus, GILL, Proc. Ac. Nat. Sci. Phila. 1859, 132 (*tridigitatus*).
Esloscopus, JORDAN & EVERMANN, Check-List Fishes, 465, 1896 (*zelotes*).

Body moderately elongate, covered with rather large, cycloid scales; head cuboid, oblong, and nearly flat above; eyes small; interorbital space broad; mouth nearly vertical; lower jaw not dilated beneath nor emarginate in front, without barbels; no intralabial filament; teeth villiform, on jaws only; pseudobranchiæ very small or obsolete. Dorsal commencing at the nape, with 6 or 12 slender spines, the soft rays numerous; anal

---

* This genus is thus defined by Dr. Gill: "Body elongate with the dorsal and abdominal outlines slowly converging to the caudal fin. Scales large, regularly imbricated. Lateral line straight, and running along the middle of the side. Head oblong, subcubical, and smooth. Preopercle entire, opercle radiately fringed behind. Mouth nearly vertical. Tongue thick, narrowed anteriorly, attached to the floor of the mouth. Labial velum without a barbel. Anus a short distance behind the base of the pectoral fins. Dorsal fin subequal, single, and very long, commencing above or before the anus and continued almost to the base of the caudal. Anal fin commencing behind the anus, and with the same form and termination as the dorsal. Caudal fin small and narrow, posteriorly subtruncated. Pectoral fins subangular. Ventral fins jugular, closely approximated, and each with 3 stout simple and articulated rays."

inserted behind dorsal; ventral rays I, 3. (δάκτυλος, finger; σκοπός, gazer, short for *Uranoscopus*.)

DACTYLOSCOPUS:
  a. Dorsal rays X to XII, 22 to 31; anal rays less than 35.
    b. Soft dorsal with 22 soft rays; anal with 26.   PECTORALIS, 2637.
    bb. Soft dorsal with 28 to 31 rays; anal with 32 or 33; scales about 45.
      c. Body rather slender, the depth about 6 in length (7 with caudal); opercular
          fringe of 15 filaments.   TRIDIGITATUS, 2638.
      cc. Body rather stout, the depth 5¼ in length (6 in total with caudal); oper-
          cular fringe of 18 filaments.
        d. Back not barred; head blotched and dotted.   POEYI, 2639.
        dd. Back with about 10 pale cross bars; head marked with whitish; a
          dark bar at base of caudal.   LUNATICUS, 2640.
ESLOSCOPUS (ἐσλός, good; σκοπός for *Uranoscopus*):
  aa. Dorsal rays VI, 38; anal rays II, 37; scales 6–51–5.   ZELOTES, 2641.

Subgenus DACTYLOSCOPUS.

2637. DACTYLOSCOPUS PECTORALIS, Gill.

Head about 5 in total length with caudal; depth about 7 (in total). D. XII, 22; A. II, 26; P. 12; V. I, 3. Width of head behind operculum 7 in total length with caudal; eye small, 10 in head; interorbital space equals diameter of eye; preoperculum broader at the angle than in *Dactyloscopus tridigitatus;* pores well developed; opercular fringe of 11 or 12 free filaments; origin of dorsal between ⅓ and ¼ length of fish from tip of snout; origin of anal under sixth or seventh dorsal ray, the first 12 dorsal and 2 anal rays simple. Pseudobranchiæ obsolete. Color light brownish yellow, with dark spots on the back, arranged in lines forming the outlines of about 6 quadrangular areas, from the angles of which irregular lines proceed downward, converging toward those departing from the angles of adjoining areas; more scattered and irregular spots and dots often present below the lateral line; head lighter, diffused with pink above. Each orbit with 4 diverging bands, 1 in front, a bifurcated one from the antero-inferior angle, and 2 from posterior border, a transverse sinuated nuchal line; upper angle of operculum whitish, bounded in front by a dark line or spot. (Gill.) Cape San Lucas; not seen by us. (*pectoralis*, pertaining to the breast.)

Dactyloscopus pectoralis, GILL, Proc. Ac. Nat. Sci. Phila. 1861, 267, Cape San Lucas. (Coll. John Xantus.)

2638. DACTYLOSCOPUS TRIDIGITATUS, Gill

Head 5 (in total) with caudal; depth 7. D. XII, 28; A. II, 32; P. 13; V. I, 3; scales 11 + 4 + 30 = 45. Body slender, much compressed posteriorly; opercular fringe of 15 separate filaments. Origin of dorsal fin over the lower angle of the base of the pectorals, or immediately before the margin of the operculum, its distance from snout to dorsal 5 in total length of body. Pseudobranchiæ very small (overlooked by Dr. Gill, but evident in living specimens). In life, pale sand color above, the lower part whitish; above 12 narrow cross bands of whitish on the back, not extending down far on the sides; head mottled above; fins all pale. West Indies, north to Key West; rather common in coral sand in shallow water about Key West. (*tres*, three; *digitus*, finger, from the 3 ventral rays.)

Dactyloscopus tridigitatus, GILL, Proc. Ac. Nat. Sci. Phila. 1859, 132, Barbados (Coll. Dr. Gill); GILL, *l. c.*, 1861, 264; GILL, *l. c.*, 1862, 505; GÜNTHER, Cat., III, 279, 1861; JORDAN & GILBERT, Synopsis, 753, 1883; JORDAN, Proc. U. S. Nat. Mus. 1884, 140.

2639. DACTYLOSCOPUS POEYI, Gill.

Head 5 in total length; depth 6¼ in total. D. XI, 31; A. II, 32. Body more robust than in *D. tridigitatus;* head plane above and obtusely angulated at the sides of the plane; thickness of the head behind the preoperculum exceeding ½ of its length; interorbital space ⅔ diameter of eye. Eye about 7 in head; preopercle as in *D. tridigitatus*, pores indistinct or obsolete; opercular fringe of about 18 filaments, the lowest of which are scarcely extended beyond the margin; origin of dorsal fin ⅙ distance from tip of snout; origin of anal fin under sixth dorsal ray; scales of moderate size and regularly imbricated. Color reddish brown, dotted with darker above the lateral line; head blotched and dotted with darker; opercles variegated; opercular bones nearly immaculate. (Gill.) Cuba; not seen by us. (Named for Prof. Felipe Poey.)

Dactyloscopus poeyi, GILL, Proc. Ac. Nat. Sci. Phila. 1861, 266, Cuba. (Coll. Felipe Poey.)

2640. DACTYLOSCOPUS LUNATICUS, Gilbert.

Head (to end of opercular fringes) 3⅗, from tip of lower jaw to base of fringes 4; depth greater than in related species, 5¼ in length. D. X or XI, 29 or 30; A. II, 32 or 33; scales about 11 + 4 + 30 = about 45. Head cuboid, narrowed forward, the vertex gently convex; width at occiput ½ length of head (to base of fringes on opercle). Mouth nearly vertical, maxillary 2⅔ in head. Labial fringes short but evident. A short nasal filament. Teeth in a rather broad cardiform band on front of upper jaw, becoming narrow laterally; in lower jaw a single series, or an irregular double series anteriorly; vomer and palatines toothless. Eyes small, very close together, the interorbital width about ½ their diameter, which equals length of snout, or about ⅛ head. Gill laminæ much reduced in size; a small round pore behind inner arch. Gill rakers obsolete; pseudobranchiæ small but evident. Opercular fringes composed of 18 filaments. Dorsal beginning at a distance behind occiput equaling diameter of orbit, its anterior rays but partly joined by membrane, the first 10 or 11 slender and not articulated, the last ray distant from base of caudal about a diameter of orbit; origin of anal under sixth dorsal spine, the 2 anterior rays not articulated; pectorals short, 1⅗ in head, containing 14 or 15 rays; caudal very small, with 10 developed rays, its length 2¾ in head. Lateral line running high in its anterior portion, declining on 3 or 4 scales, the posterior portion on middle of sides with 29 or 30 tubes; 4 scales between median portions of lateral line and base of dorsal. Color light olivaceous, a dark streak along back, 1 along middle of sides, and a fainter one along base of anal, formed by darker margins to the scales; median dorsal line with 10 or 11 more or less evident narrow pearly white cross bars; top of head and front of mandible colored like the back, the pearly blotches varying in size and shape, but symmetrically arranged, many of them narrowly edged with black; nasal tentacle white; white streaks on preopercle; caudal with a narrow black bar at base. Gulf of California. Three specimens, the longest 3 inches, from *Albatross* Stations 2797 and 3012, the latter in 22 fathoms. (Gilbert.) (*lunaticus*, moon-struck.)

Dactyloscopus lunaticus, GILBERT, Proc. U. S. Nat. Mus. 1890, 99, Gulf of California. (Coll. *Albatross*.)

Subgenus ESLOSCOPUS, Jordan & Evermann.

2641. DACTYLOSCOPUS ZELOTES, Jordan & Gilbert, new species.

Head 4¼ in length; depth 6¾. D. VI, 38; A. II, 37; V. 3; scales 6–51–5; B. 6. Head and body slender, compressed, the greatest width at occiput, ⅔ length of head; the greatest depth immediately behind insertion of anal fin, thence tapering to a very narrow tail. Head narrow, cuboid, compressed, the upper surface nearly plane, the cheeks vertical. Eyes very small, superior, with little lateral range; diameter of orbit about $\frac{1}{15}$ length of head; snout very short, about equaling orbit; anterior nostril in a short tube; gape subvertical, the lower jaw very heavy, projecting, as in *Uranoscopus;* premaxillaries protractile, the processes reaching far behind orbits; lips fringed; both jaws with bands of villiform teeth; no teeth on tongue, vomer, or palatines. Subopercle and interopercle very wide, flexible, striate, the latter overlapping throat and base of ventral fins, the former wholly covering base of pectoral fins; the striations of opercle terminate posteriorly in a wide, coarse, membranaceous fringe; branchiostegal membranes not united, free from the isthmus; pubic bones forming a sharp projection at throat; no pseudobranchiæ; gills small, a round pore behind the fourth. Dorsal beginning on the nape, its distance from snout about equaling depth of body, the first 6 rays shorter than those following and not connected by membrane; as no traces of articulation can be found, they are probably flexible spines, but are not clearly differentiated from those immediately following; origin of anal under fourth dorsal spine; caudal distinct, narrow, short; ventrals inserted under anterior margin of preopercle; ventrals 2 in head; pectorals 1¼. Scales large, with entire edges, wanting on head, breast, and region behind pectoral fins. Lateral line beginning at upper posterior angle of opercle, running parallel with the back on about 12 scales, then obliquely downward to middle of body. Color in spirits, light olivaceous, the edgings of the scales, some vermiculations on top of head, and the labial fringes clear brown; fins translucent, the caudal with a brown bar at base; eyes dark. Length 3¼ inches. Panama; 1 specimen known. The present description copied from the original in Proc. Nat. Mus. 1882, 628. (ζηλώτης, an imitator, from its resemblance to *Dactylagnus mundus*.)

Dactyloscopus, sp. nov., JORDAN & GILBERT, Proc. U. S. Nat. Mus. 1882, 628, Panama.
Dactyloscopus zelotes, JORDAN & GILBERT, new species (MS. 1882), Panama (Coll. Capt. Dow).

854. DACTYLAGNUS, Gill.

Dactylagnus, GILL, Proc. Ac. Nat. Sci. Phila. 1862, 505 (*mundus*).

Body moderately elongated, covered with rather large and uniform scales. Head cuboid, oblong, scarcely convex transversely above. Eyes small, directed obliquely upward, and situated near the snout on the upper surface of the head. Interorbital area moderate and channeled. Mouth very oblique or subvertical, the snout truncated in front; lower jaw transversely convex in front and with no barbel; teeth acute, in a narrow band along each jaw; palate smooth. Dorsal fin perfectly entire, commencing rather farther behind than the anal, and with its anterior portion armed with about 10 slender spines; anal fin longer than the dorsal. This genus closely resembles *Dactyloscopus* externally. It differs from the latter genus chiefly in the structure of the dorsal fin and the well-developed pseudobranchiæ. (δάκτυλος, finger; ἄγνος, *Agnus*, an old name of *Uranoscopus scaber*.)

2642. DACTYLAGNUS MUNDUS, Gill.

Head 4⅓; depth 6⅓. D. X, 31; A. II, 38; scales 2-48-10; eye 6 in head; maxillary 2⅝; snout equals eye; highest dorsal spine 3; highest anal ray 2¼; pectoral equals head; caudal 1⅗. Body elongate, compressed, tapering posteriorly; upper profile of head nearly horizontal, slightly convex; eyes superior, looking upward; interorbital narrow, concave; lower jaw strongly projecting, mouth nearly vertical; teeth small and conical, in narrow bands, widest in front; vomer and palatines toothless; lips furnished with labial fringes about as long as diameter of eye; nostril ending in a tube; preopercle entire; opercle fringed on its upper edge, a flap of skin downward from opercle covers the branchiostegals; pseudobranchiæ present; gill rakers not developed; head and belly naked; fins naked. Lateral line running near the back through 14 scales, deflected on 4, and thence continued along the middle through 36. Dorsal low, long, and continuous, distance from its origin to tip of snout 3¾ in body; anal similar, slightly higher and longer; posterior rays of dorsal and anal reaching to base of caudal rays; upper rays of pectoral the longest, reaching to the vertical from tenth anal ray, the lower rays short, graduated, tip of fin slightly curved up; origin of ventrals in front of pectorals, the inner rays the longest, reaching about to vent; caudal truncate, or very slightly rounded. Color in spirits, light brown above, white below, each scale on back with a dark brown spot; top of head with a few brown spots; fins colorless. Length 4¼ inches. Gulf of California. Here described from specimens collected by the *Albatross* at Carmen Island, Gulf of California; the type from Cape San Lucas. (*mundus*, neat.)

Dactylagnus mundus, GILL, Proc. Ac. Nat. Sci. Phila. 1862, 505, Cape San Lucas. (Coll. Xantus.)

855. MYXODAGNUS,* Gill.

Myxodagnus, GILL, Proc. Ac. Nat. Sci. Phila. 1861, 269 (*opercularis*).

This genus differs from *Dactyloscopus* in the form of the head, which is elongate-conoid, the lower jaw obtusely pointed and provided with a short flap in front. The pseudobranchiæ are well developed and the dorsal fin commences far behind the nape. One species known. (*Myxodes*, a genus of blennies, which this fish resembles in form; *Agnus*, ἄγνος, an old name of *Uranoscopus scaber*.)

2643. MYXODAGNUS OPERCULARIS, Gill.

Head 5 without lower jaw; depth 7. D. 36; A. II, 36; scales 2–44–9; pectorals equal head; ventrals 1¾; caudal 1¼. The body is deepest at front of dorsal fin, tapering regularly to the caudal fin. Head elongated, acutely conical; profile nearly straight, but slightly concave in front of the eyes; the crown is transversely arched and smooth; the frontal bones between the eyes are exceedingly narrow, so that the orbits appear separated by little more than a mere septum; eyes large, longitudinally elliptical; opercular pores obsolete; the postorbital or temporal ridge is nearly as long as the diameter of the orbit; the opercular fringe is composed of 6 or 7 short filaments; origin of dorsal above vent, the fin very low and continuous, its last rays not reaching to base of caudal rays; anal commencing slightly in front of dorsal, similar to it but higher, its last ray reaching to base of caudal rays; pectoral large and pointed, reaching to curve in lateral line; rays of ventral subequal, reaching about to vent; caudal truncate; scales moderate, finely striated concentrically and arranged in 11 rows on each side; the lateral line runs through 12 scales on the sides of the back, is then deflected through 3, and thence runs along the fifth row from the back through 36. Color light yellowish brown, rendered darker on the back by congregations of dark spots on the scales; there is a pearly patch behind and beneath the eye, and the operculum is also colored in the same manner. (Gill.) Cape San Lucas. Described from a specimen 2¼ inches in length. Not obtained by recent collectors. (*opercularis*, pertaining to the gill cover.)

Myxodagnus opercularis, GILL, Proc. Ac. Nat. Sci. Phila. 1861, 270, Cape San Lucas. (Coll. Xantus.)

Family CXCVII. URANOSCOPIDÆ.

(THE STAR-GAZERS.)

Head large, broad, partly covered with bony plates. Body elongate, conic, subcompressed, widest and usually deepest at the occiput. Body either naked or covered with very small, smooth, adherent scales, which

---

* This genus is thus defined by Dr. Gill: Body quite slender, the greatest height contained about 10 times in length. Head rather elongated and acutely conical, about twice as long as high; eyes large and elliptical, and very closely approximated; frontal bones extremely narrow. Mouth oblique; lower jaw projecting much beyond the upper and furnished with a short, compressed, and wide flap or barbel in front of the symphysis; villiform teeth present only on the jaws. Dorsal fin inserted behind the vertical of the anus, and furnished with simple and articulated rays; anal fin as long as or longer than the dorsal.

are arranged in very oblique series running downward and backward; the scales on the belly inconspicuous or obsolete. Lateral line little developed, running high. Eyes small, on anterior and upper portion of head, with vertical rings. Mouth vertical, with strong and prominent mandible; teeth moderate, on jaws, vomer, and palatines. Premaxillaries freely protractile; maxillary broad, without supplemental bones, not slipping under the preorbital. Gill openings wide, continued forward; gill membranes nearly separate, free from isthmus. Pseudobranchiæ present; 6 branchiostegals; 3½ gills, a slit behind the last; no anal papilla. Spinous dorsal very short or wanting; second dorsal long. Anal and pectorals large, the latter with broad oblique bases, the lower rays rapidly shortened, most of them branched; ventrals jugular, close together, I, 5, the spine very short, innermost ray longest; caudal not forked. Air bladder generally absent; pyloric cæca in moderate number. Vertebræ 24 to 26. Carnivorous fishes, living on the bottom of the shores of most warm regions. Genera 8; species 25.

URANOSCOPINÆ:

    a. Spinous dorsal of 4 or 5 well-developed spines; scales present.

        b. Head above not entirely covered with bone, the occipital plate ceasing much behind the orbits; from the middle line anteriorly a Y-shaped bony process extends forward, the tips of the fork between the eyes; a trapezoidal space on either side of the Y, covered by naked skin, bounded by the Y, the eyes, the suborbitals, and the occipital plate. A covered furrow behind and on the inner side of each eye terminating near front of orbits, its edges fringed. Head without spines; humeral spine obsolete; lips and nostrils fringed; no retractile tentacle in mouth. ASTROSCOPUS, 856.

KATHETOSTOMATINÆ:

    aa. Spinous dorsal obsolete; no scales; head above covered with bone except the groove of the premaxillary spine; the bony occipital plate coalescing with the orbital rims; humeral spine well developed; no distinct protuberances on top of head; no spine in front of humeral spine; 2 small forward-directed spines in front of eye; 3 small spines on lower margin of preopercle; upper lip scarcely fringed; no retractile tentacle in mouth; ventral fin not largely adnate to abdomen. KATHETOSTOMA, 857.

856. ASTROSCOPUS, Brevoort.

(ELECTRIC STAR-GAZERS.)

Astroscopus (BREVOORT) GILL, Proc. Ac. Nat. Sci. Phila. 1860, 20 (*anoplos;* young).
Agnus, GÜNTHER, Cat. Fishes, II, 229, 1860 (*anoplos*).
Upselonphorus, GILL, Proc. Ac. Nat. Sci. Phila. 1861, 113 (misprint for *Upsilonphorus*) (*y-græcum;* adult).

Body robust. Head above not entirely covered with bone, the occipital plate ceasing much behind the orbits; from the middle line anteriorly a Y-shaped bony process extends forward, the tips of the fork between the eyes; a trapezoidal space on either side of the Y, covered by naked skin, bounded by the Y, the eyes, the suborbitals, and the occipital plate. A covered furrow behind and on the inner side of each eye terminating near front of orbits, its edges fringed. Head without spines; humeral spine obsolete; lips and nostrils fringed; no retractile tentacle in mouth. Young individuals with top of head largely covered by bone. Head scaleless; back and sides covered with close-set scales; belly mostly naked. Humeral spine obsolete; no spine before the ventrals. First dorsal small, of 4 or 5 low, stout, pungent spines, connected by membrane to the second dorsal, which is rather high and long; pectorals and ventrals large. Species American, distinguished from the Old World genus, *Uranoscopus*,[*] chiefly by the unarmed head. ($\acute{\alpha}\acute{\sigma}\tau\rho\sigma\nu$, star; $\acute{\sigma}\varkappa\sigma\pi\acute{\epsilon}\omega$, to look.)

a. Naked space between forks of the Y on top of head long and narrow, but shorter than the vertical limb of the Y; no distinct spines before eye; sides with round pale spots, each with a dark ring.
    b. Dorsal spines 4, rather high; scales normal.   Y-GRÆCUM, 2644.
    bb. Dorsal spines 5, lower than in *y-græcum;* scales of sides cohering in oblique series.   ZEPHYREUS, 2645.
aa. Naked space between the forks of the Y short and broad, but longer than the very short vertical limb of the Y; 2 distinct spines directed forward before the eye; sides with small pale spots, not dark-edged.   GUTTATUS, 2646.

2644. ASTROSCOPUS Y-GRÆCUM (Cuvier & Valenciennes).

Head, without lower jaw, $2\frac{5}{6}$; depth $3\frac{1}{2}$. D. IV–I, 12; A. 13; scales 80; eye $12\frac{1}{4}$ in head; maxillary 2; pectoral $1\frac{1}{10}$; second dorsal spine 4; highest dorsal ray 2; highest anal ray 3; caudal $1\frac{1}{4}$. Body moderately elongate, very robust forward, greatest depth at occiput; anteriorly subcylindrical, posteriorly somewhat compressed. Head large and broad; mouth large, vertical, a fringe of barbels on each jaw, slightly longer than the diameter of the eye; tongue extremely large and fleshy, forming a pad under membrane of lower jaw which projects forward somewhat. Teeth conical, small and movable, in many bands in upper jaw, in lower jaw the teeth are larger and in fewer bands; teeth on vomer and palatines. Eyes very small but prominent, set on top of head; interorbital very wide, $3\frac{1}{2}$ times wider than the eye; bones on top of head coarsely granular; Y-shaped ridge on top of head conspicuous, on each side of which is a broad naked area; naked space between forks of Y on top of head long and narrow, but shorter than vertical limb of the Y which is very long; edges of nostrils fringed, anterior nostril round, separated from the eye by a high granular ridge; posterior nostril ending in a long curved furrow, which runs obliquely across the naked area behind eye, its posterior end not curved forward, its length $2\frac{1}{4}$ times the diameter of the eye; 2 or 3 small blunt spines in front of the eye; surface of the bones of opercle, preopercle, and humeral process coarsely granular; gill rakers not developed; pseudobranchiæ small. Head entirely scaleless; belly naked below a line drawn from fifth anal ray to upper end of pectoral base; fins without scales; scales very small and somewhat embedded. Width of pectoral at base less than $\frac{1}{4}$ length of the head, the upper rays longest, the lower rays very short, graduated from the lower side to the upper; fin somewhat pointed behind and curved up, its tip reaching to the vertical from base of sixth dorsal ray; the rays of ventrals very thick and swollen, the inner

---

[*] The following are the characters of *Uranoscopus:* Head with spines; humeral spine well developed; 1 strong spine on suboercle, 4 smaller ones on preopercle, all directed downward; 1 small spine directly above and in front of humeral spine; 4 low, stout protuberances on top of head pointing backward; naked space between eyes extending back to posterior part of orbits; upper lip and nostrils not fringed; retractile tentacle in mouth more or less developed. First dorsal with about 4 pungent spines; scales well developed.

rays the longest, reaching midway from their base to end of pectorals; origin of fin a distance of the width of pectoral in front of the lower edge of pectoral base; soft dorsal much higher than anal; posterior rays reaching slightly past the vertical from the base of the last anal ray; end of the last anal ray about reaching to base of caudal rays; caudal truncate or slightly rounded; a ridge of skin along middle of belly from the ventrals to vent. Dark brown above, paler below; upper parts densely covered with small rounded white spots, each surrounded by a black ring; lower jaw and labial fringes similarly spotted; spinous dorsal black, white posteriorly; soft dorsal brown anteriorly with a horizontal white and black band, then tipped with white; posteriorly with 2 vertical black stripes and a white one between them; caudal black, tipped with white, with 2 to 4 white longitudinal stripes, its upper and lower edges narrowly white; the anal white at base and tip, with a black median band, $\frac{1}{2}$ depth of fin, darkest posteriorly; pectorals brown, with a black band below, the lower edge white, the upper ray spotted; ventrals white with a black lengthwise streak. Old examples lose the black ring around the spots, and the edges of the spots are blended into the dark brown of the back; a dark stripe running from the upper angle of gill opening to caudal. South Atlantic coast from Cape Hatteras to the Caribbean Sea, in sandy bays, rather common in shallow water, varying much with age. Here described from a specimen, 15 inches in length, from Charleston, South Carolina. It is recorded from Charleston, Beaufort, Matanzas River, St. Johns River, Pensacola, Key West, and "the Caribbean Sea." According to Dr. James A. Henshall, the naked area on top of the head in *Astroscopus* is the seat of electric power. This interesting statement needs verification. (Named from the armature of the head, in the form of the Greek *Y.*)

Uranoscopus y-græcum, CUVIER & VALENCIENNES, Hist. Nat. Poiss., III, 308, 1829, origin unknown; GÜNTHER, Cat., II, 229, 1860.

*Uranoscopus anoplos,** CUVIER & VALENCIENNES, Hist. Nat. Poiss., VIII, 493, 1831 (young examples), Charleston, South Carolina.

Upsilonphorus y-græcum, GILL, Proc. Ac. Nat. Sci. Phila. 1861, 113; KIRSCH, *l. c.*, 263, 1889.

Astroscopus y-græcum, BEAN, Proc. U. S. Nat. Mus. 1879, 58; JORDAN & GILBERT, Synopsis, 628, 1883.

Agnus anoplus, GÜNTHER, Cat., II, 229, 1860.

Astroscopus anoplus, JORDAN & GILBERT, Synopsis, 629, 1883.

Astroscopus anoplos, KIRSCH, Proc. Ac. Nat. Sci. Phila. 1889, 262.

---

* The genus *Astroscopus* was based on small specimens which, in our present opinion are simply immature examples of the species *y-græcum*. The supposed genus is thus described in distinction from *Upsilonphorus*, which seems to us the adult of the same type: Head covered above with bone except a small region between and in front of the eyes, the bony occipital plate coalescing with the orbital rims; no spines on head; humeral spines obsolete; occipital region with bluntish projections; naked space between eyes extending back to near middle of orbits; lips and nostrils fringed; no retractile tentacle in mouth.
   The following characters are assigned by Dr. Kirsch to *Astroscopus anoplos*: Head 2½; depth 3¼. D. IV-I, 13; A. 13. Pectorals rather large, their longest ray equaling in length base of second dorsal and extending to front of that fin; ventrals equaling pectorals in length, and extending to front of that fin; the second dorsal equaling anal but its anterior insertion slightly posterior to that; anal rays reaching base of caudal; vent much nearer base of caudal than to tip of snout. Color dark brown above, yellowish below; lighter portions of body covered with small white specks; chin jet-black; all the fins whitish. Length 2 inches. (Specimen from Key West). Small individuals are found along the coast from Cape Hatteras to Florida wherever *A. y-græcum* is found. The adult differs mainly in the armature of the top of the head, a characteristic which is developed at different ages in different individuals.

2645. ASTROSCOPUS ZEPHYREUS, Gilbert & Starks.

Head, without lower jaw, $2\frac{3}{4}$; depth $3\frac{3}{5}$. D. V, 13; A. 14; scales 84; eye 12 in head; maxillary $2\frac{1}{4}$; pectoral $1\frac{1}{4}$; second dorsal spine 7; highest dorsal ray $2\frac{1}{4}$; highest anal ray $3\frac{1}{6}$; caudal $1\frac{1}{4}$. Body robust, widest at occiput, slightly compressed posteriorly; anteriorly subcylindrical. Head very large and broad, wider than the body; mouth large, vertical, a fringe of barbels curving over mouth on each jaw; length a little greater than the diameter of the eye; tongue very large and fleshy, forming a pad under the membrane of lower jaw, which projects forward somewhat; teeth conical, small and movable, in many bands in upper jaw; in lower jaw the teeth are larger and in 2 or 3 rows; vomer and palatines with teeth; eyes very small but prominent, set on top of head; interorbital very wide, 4 times as wide as the eye; bones on top of head coarsely granular; Y-shaped ridge on top of head conspicuous, on each side of which is a broad naked area, the form of these and other bones of the head exactly as in *A. y-græcum;* edges of nostrils closely fringed, anterior nostril round, the ridge between it and eye not very high or conspicuous; posterior nostril ending in a long curved furrow which runs obliquely across the naked area behind eyes; at its posterior end it turns sharply forward, its length $2\frac{3}{5}$ times the diameter of the eye; 2 very short blunt spines in front of the eye; surface of the opercle, preopercle, and humeral process granular, not so rough as in *Astroscopus y-græcum;* gill rakers not developed; pseudobranchiæ very small. Head entirely scaleless; belly naked below a line drawn from first anal ray to the middle of the pectoral base; fins without scales; scales small and nearly square, grown together side by side, forming series of oblique plates. Width of pectoral at base slightly less than $\frac{1}{2}$ length of head, the lower rays very short and graduated to the long upper rays, the fin pointed and slightly turned up, its tip reaching to the vertical from base of the third dorsal ray; the ventral rays thick and swollen, the inner rays the longest, its tip reaching about midway between its base and tips of pectorals; origin of fin in front of pectorals a distance equal to the width of pectoral base; soft dorsal somewhat higher than anal, its posterior rays reaching to the vertical from base of last anal ray; tip of last anal ray nearly reaching to the base of caudal rays; caudal truncate or slightly rounded; a fold of skin along middle line of belly from ventrals to vent. Color dark brown above, paler below; upper parts with many round white spots of various sizes, edged with rings of dark brown; spinous dorsal black, light posteriorly; soft dorsal light at base, the ends of the rays with black and white stripes; pectoral and anal dusky with light edge; caudal with longitudinal black and white stripes. Pacific coast of Mexico. One specimen, numbered 333, in the Leland Stanford Junior University Museum, collected by the *Albatross* at Magdalena Bay, Lower California. It is 12 inches in length. A distinct electric shock was given by this fish when alive, the electric organs being in the fleshy areas on top of head behind eyes. (Gilbert.) A second large specimen was sent from Mazatlan by Dr. George W. Rogers, having been taken by Ygnacio Moreno in January, 1896. (ζεφύριος, western; ζέφυρος, the west wind.)

Astroscopus zephyreus, GILBERT & STARKS, Proc U. S. Nat. Mus. 1896, 453, pl. 53, fig. 2, and pl. 54, Magdalena Bay, Lower California (Type No. 47743. Coll. *Albatross*).

2616. ASTROSCOPUS GUTTATUS (Abbott).

Depth 4 in length in young and 3½ in adult. D. IV or V-13 or 14; A. 13; V 1,5. Eye 5½ in interorbital space. Naked space between forks of Y on top of head short and broad, but longer than the vertical limb of the Y, which is very short; 2 distinct spinules directed forward before eye; white spots on body very small and irregular without dark rings; base of dorsals equalling in length the distance from front of first dorsal to tip of snout; base of first dorsal twice length of its longest spine; first spine equaling second in length, and 3 times length of last; length of middle caudal rays a little less than that of ventrals; pectorals slightly longer than ventrals, 3½ in total length, and extending to fifth anal ray. Color of upper parts of body and lower jaw bright chocolate; belly and throat white; darker portions covered with numerous circular spots much lighter than ground color; membrane of first dorsal black; second dorsal white with 3 irregular bands of dull black obliquely across it; the caudal with 3 parallel bands of blackish brown, the middle of which appears to be the continuation of a variable longitudinal band on the center of each side; the anal having a variable band of dull brown, darker upon the posterior termination. Length 12 inches. Atlantic coast of the United States, from Long Island to Virginia; apparently scarce. Recorded from Cape May; Tompkinsville, New York; Norfolk, Virginia; Somers Point, New Jersey, etc ; not known south of Cape Hatteras. In *Astroscopus guttatus* the pale spots are much smaller, less sharply defined, and occupy a smaller area than in *A. y-græcum;* the lower part of the head has 2 black blotches in each species; the second dorsal, anal, and ventrals are nearly or quite plain. The naked area behind each eye is (in *A. guttatus*) lunate, its length barely twice that of the snout; the bony Y-shaped plate is short and broad, concave on the median line, and forked for about ⅓ its length, the posterior undivided portion broader than long; the bony bridge across the occiput but little shorter than the part of the head which precedes it. In *A. y-græcum* the naked area is trapezoidal, longer than broad, and about 4 times the length of the snout; the Y is forked for more than ⅓ its length, its undivided part more than twice as long as broad, and not concave; the occipital plate is not ⅓ as long as the part of the head which precedes it. (*guttatus,* spotted, as with rain drops.)

Astroscopus guttatus ABBOTT, Proc. Ac. Nat. Sci. Phila. 1860, 365, Cape May, New Jersey.
Upsilonphorus guttatus BEAN, Proc. U. S. Nat. Mus. 1878, 60; KIRSCH. l. c., 264, 1888.

857. KATHETOSTOMA, Günther.

Kathetostoma, GÜNTHER, Cat. Fish., ii, 231, 1860 (*lœve*).

Body robust, formed as in *Astroscopus* and *Uranoscopus*. Scales none. One continuous dorsal without spines; ventrals jugular not adnate to the abdomen; pectoral rays branched; some bones of the head armed. Cavity of the gills without superior opening; 6 branchiostegals; pseudobranchiæ present. Air bladder none. Three species known, the type, *Kathetostoma lœve*, being from Australia. (καθετος, vertical; στόμα, mouth.)

a. Dorsal rays 13; anal 13, body shaded and dotted with blackish AVERRUNCUS, 2617.
aa. Dorsal rays 10; anal 13; body spotted with white. ALBIGUTTA, 2618.

2617. KATHETOSTOMA AVERRUNCUS, Jordan & Bollman.

Head 2⅝, 3⅓ with caudal; depth 3⅓. D. 13; A. 13. Body short and robust, its width behind base of pectorals equal to length of top of head. Head very large, its width at peropercle less than its length by ½ length of eye. Mouth large, vertical; maxillary 2 in head. Snout 1⅓ in eye. Eye rather small, 5 in head. Teeth of lower jaw largest, inner row of each jaw enlarged and movable; vomer and palatines with a few large, conical teeth. Lower jaw without tentacle. Interorbital space slightly concave, 1⅓ times length of eye. Premaxillary groove as broad as long, 1⅓ in eye, obtuse behind, extending backward just past middle of pupil. Distance between bases of humeral spines 1⅓ in top of head. Preorbital with 3 spines in front directed forward and downward. Preopercle with 3 spines below angle directed downward and forward. Two antrorse spines on mandible, and 2 on breast before ventrals. Bones of top of head coarsely granular, striate, no naked area above except premaxillary groove; 2 points on occipital region whence granular ridges radiate; opercles and orbital bones coarsely granular, but not striate. No trace of scales or of spinous dorsal. Base of dorsal equal to base of anal, 1⅓ in head, longest ray equal to depth of cheek; pectorals ⅔ eye, length greater than that of top of head; ventrals reaching more than halfway to vent, their length equal to that of top of head. A few small depressions resembling embedded scales on region before dorsal and above head. Color blackish brown, mottled with paler; lower parts pale, dusted with brown; lips and gular region black; dorsal dusty, with 5 indistinct, partly confluent, whitish spots along its base; anterior part of anal pale, posterior thickly dusted with blackish, tips of rays pale; pectorals blackish, faintly barred; axil dusted outside, inner part very pale; ventrals pale; caudal with 3 irregular oblique dark bars; floor of mouth pinkish; tongue dusted with dark specks. Length 4½ inches. Pacific Ocean, off coast of Colombia; a single specimen dredged at a depth of 7 fathoms; a most singular fish. (*averruncus*, a deity which wards off; from the mailed head.)

Kathetostoma averruncus, JORDAN & BOLLMAN, Proc. U. S. Nat. Mus. 1889, 168, off coast of Colombia, at Albatross Station 2801, 5° 57′ N., 79° 31′ 30″ W.; KIRSCH, l. c., 250, 1889; JORDAN, Proc. Cal. Ac. Sci. 1895, 228, pl. 31.

eyes. (Gilbert.) A second large specimen was sent from Mazatlan by Dr. George W. Rogers, having been taken by Ygnacio Moreno in January, 1896. (ζεφύριος, western; ζέφυρος, the west wind.)

Astroscopus zephyreus, GILBERT & STARKS, Proc. U. S. Nat. Mus. 1896, 453, pl. 53, fig. 2, and pl. 54, Magdalena Bay, Lower California (Type No. 47743. Coll. *Albatross*).

2646. ASTROSCOPUS GUTTATUS (Abbott).

Depth 4 in length in young and 3¼ in adult. D. IV or V–13 or 14; A. 13; V. I, 5. Eye 5¼ in interorbital space. Naked space between forks of Y on top of head short and broad, but longer than the vertical limb of the Y, which is very short; 2 distinct spinules directed forward before eye; white spots on body very small and irregular without dark rings; base of dorsals equaling in length the distance from front of first dorsal to tip of snout; base of first dorsal twice length of its longest spine; first spine equaling second in length, and 3 times length of last; length of middle caudal rays a little less than that of ventrals; pectorals slightly longer than ventrals, 3¼ in total length, and extending to fifth anal ray. Color of upper parts of body and lower jaw bright chocolate; belly and throat white; darker portions covered with numerous circular spots much lighter than ground color; membrane of first dorsal black; second dorsal white with 3 irregular bands of dull black obliquely across it; the caudal with 3 parallel bands of blackish brown, the middle of which appears to be the continuation of a variable longitudinal band on the center of each side; the anal having a variable band of dull brown, darker upon the posterior termination. Length 12 inches. Atlantic coast of the United States, from Long Island to Virginia; apparently scarce. Recorded from Cape May; Tompkinsville, New York; Norfolk, Virginia; Somers Point, New Jersey, etc.; not known south of Cape Hatteras. In *Astroscopus guttatus* the pale spots are much smaller, less sharply defined, and occupy a smaller area than in *A. y-græcum;* the lower part of the head has 2 black blotches in each species; the second dorsal, anal, and ventrals are nearly or quite plain. The naked area behind each eye is (in *A. guttatus*) lunate, its length barely twice that of the snout; the bony Y-shaped plate is short and broad, concave on the median line, and forked for about ½ its length, the posterior undivided portion broader than long; the bony bridge across the occiput but little shorter than the part of the head which precedes it. In *A. y-græcum* the naked area is trapezoidal, longer than broad, and about 4 times the length of the snout; the Y is forked for more than ½ its length, its undivided part more than twice as long as broad, and not concave; the occipital plate is not ½ as long as the part of the head which precedes it. (*guttatus*, spotted, as with rain drops.)

Astroscopus guttatus, ABBOTT, Proc. Ac. Nat. Sci. Phila. 1860, 365, Cape May, New Jersey.
Upsilonphorus guttatus, BEAN, Proc. U. S. Nat. Mus. 1879, 60; KIRSCH, *l. c.*, 264, 1889.

857. KATHETOSTOMA, Günther.

Kathetostoma, GÜNTHER, Cat. Fish., II, 231, 1860 (*læve*).

Body robust, formed as in *Astroscopus* and *Uranoscopus.* Scales none. One continuous dorsal without spines; ventrals jugular not adnate to the abdomen; pectoral rays branched; some bones of the head armed. Cavity of the gills without superior opening; 6 branchiostegals; pseudobranchiæ present. Air bladder none. Three species known, the type, *Kathetostoma læve,* being from Australia. (καθετος, vertical; στόμα, mouth.)

> *a.* Dorsal rays 13; anal 13; body shaded and dotted with blackish.   AVERRUNCUS, 2647.
> *aa.* Dorsal rays 10; anal 12; body spotted with white.   ALBIGUTTA, 2648.

2647. KATHETOSTOMA AVERRUNCUS, Jordan & Bollman.

Head $2\frac{3}{8}$, $3\frac{1}{8}$ with caudal; depth $3\frac{1}{4}$. D. 13; A. 13. Body short and robust, its width behind base of pectorals equal to length of top of head. Head very large, its width at peropercle less than its length by $\frac{1}{2}$ length of eye. Mouth large, vertical; maxillary 2 in head. Snout $1\frac{2}{3}$ in eye. Eye rather small, 5 in head. Teeth of lower jaw largest, inner row of each jaw enlarged and movable; vomer and palatines with a few large, conical teeth. Lower jaw without tentacle. Interorbital space lightly concave, $1\frac{1}{2}$ times length of eye. Premaxillary groove as broad as long, $1\frac{1}{2}$ in eye, obtuse behind, extending backward just past middle of pupil. Distance between bases of humeral spines $1\frac{1}{4}$ in top of head. Preorbital with 3 spines in front directed forward and downward. Preopercle with 3 spines below angle directed downward and forward. Two antrorse spines on mandible, and 2 on breast before ventrals. Bones of top of head coarsely granular, striate, no naked area above except premaxillary groove; 2 points on occipital region whence granular ridges radiate; opercles and orbital bones coarsely granular, but not striate. No trace of scales or of spinous dorsal. Base of dorsal equal to base of anal, $1\frac{2}{3}$ in head, longest ray equal to depth of cheek; pectorals $\frac{1}{4}$ eye, length greater than that of top of head; ventrals reaching more than halfway to vent, their length equal to that of top of head. A few small depressions resembling embedded scales on region before dorsal and above head. Color blackish brown, mottled with paler; lower parts pale, dusted with brown; lips and gular region black; dorsal dusty, with 5 indistinct, partly confluent, whitish spots along its base; anterior part of anal pale, posterior thickly dusted with blackish, tips of rays pale; pectorals blackish, faintly barred; axil dusted outside, inner part very pale; ventrals pale; caudal with 3 irregular oblique dark bars; floor of mouth pinkish; tongue dusted with dark specks. Length $4\frac{1}{2}$ inches. Pacific Ocean, off coast of Colombia; a single specimen dredged at a depth of 7 fathoms; a most singular fish. (*averruncus,* a deity which wards off; from the mailed head.)

Kathetostoma averruncus, JORDAN & BOLLMAN, Proc. U. S. Nat. Mus. 1889, 163, off coast of Colombia, at Albatross Station 2800, 8° 57′ N., 79° 31′ 30″ W.; KIRSCH, *l. c.,* 259, 1889; JORDAN, Proc. Cal. Ac. Sci. 1896, 229, pl. 31.

2648. KATHETOSTOMA ALBIGUTTA, Bean.

'Head 3; greatest width 3; depth 3⅓. D. 10; A. 12; interorbital space
4 in head, containing a deep groove, the length of which is slightly
greater than its width and nearly equaling length of eye. Mouth nearly
vertical when closed; intermaxillary slightly protractile, the length of
its tooth-bearing surface ⅖ length of head; maxillary very broadly
expanded behind, its greatest width about 3 in length, extending almost
to vertical from middle of eye; end of mandible not much farther back;
length of mandible 4⅔ in length; mandible having 2 blunt prominences
at its posterior end; the exposed portion of the maxillary traversed by
radiating striæ. The lower limb of preoperculum with 3 stout spines
along its lower border; length of humeral spine 3 in head; humerus very
strongly rugose on its upper border; 3 short spines on the anterior edge
of preorbital. Teeth in villiform bands in the intermaxillary and mandible, and on vomer; palatines in a very short band; a cavity between
head of vomer and the processes of the intermaxillary ending in a semicircular canal behind, which is separated from the anterior cavity by
a flap of skin. Gill openings very wide and only narrowly attached to
the isthmus, leaving a free posterior border. Pseudobranchiæ present,
small; a small, narrow slit behind the last gill, its length about ⅔ that
of eye; gill rakers tubercular, none on anterior arch. A pair of short
but stout spines in front of ventrals. The origin of dorsal a little nearer
to root of caudal than to tip of snout, midway between base of caudal
and middle of eye; length of dorsal base about 3 in length, the third
ray the longest, its length nearly ½ length of base of fin, the last ray
about as long as eye, and the first scarcely longer than this. The anal
origin directly under that of dorsal, the base of fin slightly longer than
that of dorsal; the seventh, eighth, and ninth anal rays about the longest, their length equaling about ½ that of middle caudal rays; the first
ray not much more than ¼ as long as the longest and the rays gradually increasing in size to the ninth; length of pectoral 3⅓ in body;
length of lowermost ray less than ¼ length of head; only the first ray
simple, the rest divided. Ventral origin under eye; the longest ray of
ventral slightly shorter than mandible. Caudal slightly rounded when
expanded, the middle rays as long as head without snout. The lateral
line beginning near the root of humeral spine, curving upward slightly
and running along back to end of dorsal, then curving downward to near
the middle of the caudal base; skin naked. Color, upper parts light
brown, the upper surface of the head minutely dotted with white; the
back with numerous roundish spots and oblong blotches of whitish; lower
parts pale; the dorsal with 2 or three dark blotches near the margin,
in some cases not much larger than eye, in others fully twice as long;
caudal with 9 black blotches, those on outer rays largest, differing in
size in different specimens, these blotches distributed over the greater
portion of the fin; anal pale, with the exception of a brownish blotch
on the membrane of the last 3 rays; pectoral with a brownish submarginal band on its outer half, this band sometimes broken up on the mem-

brane; ventrals pale. Length about 6¼ inches. Gulf of Mexico, in 27 to 88 fathoms; 1 specimen known. (*albus*, white; *gutta*, spot.)

Cathetostoma albigutta, BEAN, Proc. U. S. Nat. Mus. 1892, 121, Gulf of Mexico, at Albatross Station 2403, Lat. 28° 42′ 30″ N., Lon. 85° 29′ 00″ W. (Type, No. 39304, U. S. Nat. Mus.)

Suborder HAPLODOCI.

This group is distinguished mainly by the undivided post-temporal, the reduction in the number of gill arches to 3, and by the absence of peculiarities shown by related forms. One family. (ἁπλόος, simple; δοκος, a shaft or beam, from the form of the post-temporal.)

Family CXCVIII. BATRACHOIDIDÆ.

(THE TOAD-FISHES.)

Body more or less robust, depressed anteriorly, compressed behind; head large, depressed, its muciferous channels well developed; mouth very large, the teeth generally strong; premaxillaries protractile; gills 3, a slit behind the last; pseudobranchiæ none; gill openings restricted to the sides, the membranes broadly united to the isthmus; branchiostegals mostly 6; gill rakers present, moderate; suborbital without bony stay; post-temporal bone simple, undivided; scales small, cycloid, or wanting; dorsal fins 2, the first of 2 or 3 low, stout spines; soft dorsal very long; anal fin similar, but shorter; ventrals rather large, jugular, I, 2 or I, 3; pectorals very broad, the rays branched; pyloric cæca none; tail diphycercal, the caudal fin distinct, rounded; vertebræ in large number, 32 to 45. Carnivorous coast fishes, mostly of the warm seas, some of them ascending rivers; the young of some or all the species fasten themselves to rocks by means of an adhesive ventral disk, which soon disappears. In some species the spines of the head and dorsal fin are provided with poison glands. Genera 7; species about 15. (*Batrachidæ*, Günther, Cat., III, 166–177.)

 a. Dorsal spines 3; opercle developed as 2 strong diverging spines; subopercle rather strong, with 2 spines similar to those of opercle; no venom glands.

  b. Body scaly; branches of subopercular spine subequal and diverging; frontal region broad, flat, and slightly depressed, its median ridge rather prominent.   BATRACHOIDES, 858.

  bb. Body scaleless; branches of subopercular spine parallel, the lower branch much the shorter; vertebræ 10+22; frontal region not depressed, its median ridge prominent; axil with a large foramen. OPSANUS, 859.

 aa. Dorsal spines 2; opercle very small, its posterior part developed as a single strong spine; subopercle feebly developed, narrowed, and not ending in a spine; body scaleless.

  c. Spines solid, without venom glands; several lateral lines on sides of head and body, composed of pores and shining spots, some of these accompanied by cirri; canine teeth present; vertebræ 12+31; frontal region depressed, forming a triangular area below level of temporal region, its median ridge very low.   PORICHTHYS, 860.

  cc. Spines of dorsal fin and operculum hollow and connected with venom glands; lateral line on sides of body single; no canine teeth.

   d. Dorsal and anal free from caudal.   THALASSOPHRYNE, 861.

   dd. Dorsal and anal fully joined to caudal.   DÆCTOR, 862.

▪ ▪ 858. BATRACHOIDES, Lacépède.

Batrachoides, LACÉPÈDE, Hist. Nat. Poiss., III, 306, 1798 (*"tau,"* Lacépède*=*surinamensis*).
Batrachus (KLEIN), BLOCH & SCHNEIDER Syst. Ichth., 42, 1801 (*"tau,"* *didactylus*, *suri-namensis*, etc.; substitute for *Batrachoides*).
Batrictius, RAFINESQUE, Anal. Nat. 1815, 82 (substitute for *Batrachoides*).

Body robust, formed as in *Opsanus*. Dorsal spines 3; opercle developed as 2 strong diverging spines; subopercle strongly developed; branches of subopercular spine subequal and diverging; body covered with small ctenoid scales; frontal region broad, flat, and slightly depressed, its median ridge rather prominent. Mucous pores of sides not greatly developed. No poison glands. Shore fishes of warm regions. ($\beta\acute{a}\tau\rho\alpha\chi o\varsigma$, frog; $\epsilon\tilde{\iota}\delta o\varsigma$, resemblance.)

> *a.* Teeth small, about 14 on the vomer; anterior teeth of lower jaw in a band; lateral teeth of palatine enlarged and canine-like; irregularly arranged.
>
> SURINAMENSIS, 2649.
>
> *aa.* Teeth larger, about 8 on vomer; anterior teeth of lower jaw in 2 rows; 3 teeth on middle of palatines enlarged and canine-like, the middle one the smallest.
>
> PACIFICI, 2650.

2649. BATRACHOIDES SURINAMENSIS (Bloch & Schneider).

(SAPO.)

Head 3¼ in length of body; depth 6. D. III-29; A. 26. Teeth small, about 14 on vomer; anterior teeth on lower jaw in a band; lateral teeth on palatines enlarged and canine-like, irregularly arranged; pectoral without pores on its inner surface. Color grayish, darker on sides and head; base of soft dorsal pale, with a dark, irregular line above; upper part of fin lighter; caudal nearly black; anal fin light with some dark markings. Coasts of Guiana and Brazil; not rare on sandy shores; our specimen from Curaçao.

Batrachoides tau, LACÉPÈDE, Hist. Nat. Poiss., 306, pl. 12, fig. 1, 1798; not *Gadus tau*, LINNÆUS.
Batrachus surinamensis, BLOCH & SCHNEIDER, Syst. Ichth., 43, 1801, Surinam; from a specimen in the Museum of Vaillant in Paris; GÜNTHER, Cat., III, 173, 1861; MEEK & HALL, Proc. Ac. Nat. Sci. Phila. 1885, 61.

2650. BATRACHOIDES PACIFICI (Günther).

Head 3 in length; depth about 6. D. III-26; A. 22. Teeth rather large, about 8 on vomer; anterior teeth on lower jaw in 2 rows; lateral teeth on lower jaw gradually increasing to middle of jaw, behind which they become abruptly smaller and then gradually increase to end of jaw; 3 teeth on middle of palatines enlarged and canine-like, the middle one the smallest; pectoral with a row of pores on inner surface. Color olivaceous brown; some indistinct dark cross bands on body; dorsal with about 7 very irregular oblique dark bars, anal with about 5; pectorals and caudal

---

* "Il est revêtu d'écailles molles, petites, minces, rondes, brunes, bordées de blanc, et arrosées par une mucosité très abondante, comme celles de la lote et de la mustelle." (Lacépède.) Lacépède's specimen was therefore one of the scaly species, not an *Opsanus*. No species of the latter group seems to have been known to Lacépède or to Schneider.

dark, with few light cross bands. Panama; locally common, close to the preceding but with smaller teeth and fewer fin rays. The specimen examined by us collected by Dr. Gilbert.

Batrachus pacifici, GÜNTHER, Cat., III, 173, 1861, Panama; GÜNTHER, Fishes Centr. Amer., 435, 1869.

Batrachoides pacifici, GILL, Proc. Ac. Nat. Sci. Phila. 1863, 170; MEEK & HALL, Proc. Ac. Nat. Sci. Phila. 1885, 62.

859. OPSANUS,* Rafinèsque.

(TOAD-FISHES.)

Opsanus, RAFINESQUE, Amer. Monthly Mag. 1817, 203 *(cerapalus)*.
Batrachus, JORDAN & GILBERT, Synopsis, 751, 1883, and of authors; not of BLOCH & SCHNEIDER.

Body comparatively short and robust, scaleless; head large, depressed; jaws, vomer, and palatines each with a single series of strong blunt teeth; mandible with an additional external series at symphysis; teeth of upper jaw small; dentary bones forming an acute angle at symphysis; lips fleshy; upper angle of opercle with 2 diverging spines, more or less concealed in the skin; no poison glands; spinous dorsal of 3 stout, short spines, the second the longest; axil of pectoral with a large foramen;† lateral line obscure, its pores not conspicuous; young with a series of small, tufted cirri on back and sides; branchiostegals 6; vertebræ 12 + 22. Shore fishes, mostly of temperate regions; voracious creatures, living on the bottoms, feeding on mollusks and crustacea, and having great strength of jaw. (ὤψ, eye; ἄνω, upward; "the name means looking up." Rafinesque.)

a. Nostrils with fleshy tentacle between them. Color brownish or dusky greenish, mottled with darker and lighter, the dark on sides of body in large irregular blotches extending from base of dorsal to about ⅔ distance to base of anal, and more or less covered with small pale spots; belly and chin plain white or yellowish.
TAU, 2651.

aa. Nostrils without fleshy tentacle. Color whitish or gray, everywhere blotched or spotted with brownish yellow and black, the black spots on top of head smaller and more numerous than on rest of body; a large black blotch at base of spinous dorsal, running up on fin; 3 black blotches along base of soft dorsal, which do not extend ¼ the distance to base of anal; pectoral with black spots which do not form cross bands; ventrals with more dark markings than in *tau*; dorsal, anal, and caudal marked nearly as in *tau*. PARDUS, 2652.

2651. OPSANUS TAU (Linnæus).

(TOADFISH; SAPO; SLIMER; OYSTER-FISH.)

Head 2⅔; depth 4¼. D. III–26 to 28; A. 24. Body robust, naked, the head broad; mouth large, the very strong jaws closing with great force; teeth blunt, those on mandible small anteriorly, regularly increasing in

---

* The name *Batrachus* should not be used for this genus, as it was originally given merely as a substitute for *Batrachoides,* having properly the same type, *surinamensis,* wrongly supposed to be *tau* of Linnæus, a species unknown to Lacépède and Bloch & Schneider. No congener of *tau* was placed in *Batrachus* by Bloch & Schneider. Prior to any use of *Batrachus* as the generic name of the naked toadfishes, allied to *tau,* Rafinesque had given to one of the latter the generic name *Opsanus,* which can not be set aside for *Batrachus,* the latter being an unnecessary synonym of *Batrachoides.*
† The Brazilian genus, *Marcgravia (cryptocentra),* in which this foramen is wanting, has not been recorded from north of the equator.

size backward, those on vomer prominent; a broad flap above orbit; tip of maxillary and lower side of mandible with conspicuous cirri; a series of smaller cirri along margin of preopercle; subopercle ending in a long, sharp spine; orbit about equaling interorbital width or length of snout; pectoral with a large foramen in the axil. Dusky olive, with black markings confluent on the sides and forming irregular, indistinct bars; belly and under side of head lighter; sides often with many pale yellow or whitish spots; soft dorsal with 6 to 9 oblique light bands; anal with 5 to 9; caudal and pectoral fins with 5 to 7 light cross bands, these formed chiefly from light spots; ventrals with some dark markings. In specimens from shallow water or algæ, the brown becomes nearly black and more extended, the belly and chin spotted with darker, and top of the head has no distinct markings. The deeper-water specimens are lighter in coloration than those from near the surface, and those from the coral reefs (var. *beta,* Goode & Bean) are paler than those from the green algæ and sea wrack; otherwise no differences seem to exist. In young individuals the head is more narrow and rounded, and the lower branch of the subopercular spine proportionally larger than in the adult. Cape Cod to Cuba; very abundant among rocks and weeds close to the shore northward, in deeper water southward; the young clinging to rocks by a ventral sucking disk, which is soon lost. Length 15 inches. Not valued as food. (*tau,* T, the bones on the head when dried showing a T-shaped figure.)

Gadus tau, LINNÆUS, Syst. Nat., Ed. XII, 440, 1766, Carolina. (Coll. Dr. Garden.)
Cottus glaber, SCHÖPF, Schrift. Naturf. Freunde, VIII, 1788, 146, Long Island; D. 25; V. 3; A. 21; short cirri below mouth.
Cottus chætodon, BLOCH & SCHNEIDER, Syst. Ichth., 62, 1801, New York; after SCHÖPF.
Lophius bufo, MITCHILL, Trans. Lit. and Phil. Soc. 1815, 463, New York.
Opsanus cerapalus, RAFINESQUE, Amer. Monthly Mag., Jan., 1817, 204, south coast of Long Island. (Coll. C. S. Rafinesque.)
Batrachoides vernullas, LE SUEUR, Mém. Mus., V, 1819, 157, pl. 17, coast of Rhode Island.
Batrachoides variegatus, LE SUEUR, Jour. Ac. Nat. Sci. Phila., III, 1823, 399 and 401, Egg Harbor, New Jersey.
Batrachus celatus, DE KAY, New York Fauna: Fishes, 170, pl. 50, f. 161, 1842, New York.
Batrachus tau beta, GOODE & BEAN, Proc. U. S. Nat. Mus. 1882, 236, Gulf of Mexico.
Cottus glaber, WALBAUM, Artedi Piscum, III, 392, 1792; after SCHÖPF.
Batrachus tau, CUVIER & VALENCIENNES, Hist. Nat. Poiss., XII, 478, 1837; DE KAY, N. Y. Fauna: Fishes, 168, pl. 28, fig. 26, 1842; GÜNTHER, Cat., III, 167, 1861; JORDAN & GILBERT, Synopsis, 751, 1883; MEEK & HALL, Proc. Ac. Nat. Sci. Phila. 1885, 59.
Batrachus variegatus, CUVIER & VALENCIENNES, Hist. Nat. Poiss., XII, 484, 1837.

2652. OPSANUS PARDUS (Goode & Bean).

(SAPO.)

Head to end of opercular spine 3; depth 4. D. III–26; A. 22; maxillary 1⅔ in head; pectoral 2⅙; ventral 2; highest dorsal ray 2⅗; highest anal ray 3⅓; caudal 2. Body short and robust, compressed posteriorly; head large, somewhat depressed, wider than the body; eyes placed high, not so wide as the slightly concave interorbital space; mouth large, the maxillary reaching far beyond the eye, the lower jaw slightly projecting; a double row of small blunt teeth in upper jaw, not running very far back at the sides; lower jaw with a single row of much larger pebble-like teeth running well back and biting against a single row of similar teeth on palatines; a few teeth in front of jaw which bite against the premaxillary teeth; vomer with 1 or 2 irregular rows of large blunt teeth; head with many fleshy tentacles, 1 over each eye, a row around lower jaw, 1 on end of maxillary, and a row around preopercle; opercle ending in 2 diverging spines, the lower shorter; subopercle ending in a spine, its tip equal with the lower opercular spine, these spines not piercing the skin; gill rakers very short, scarcely developed. Body and fins covered with a soft smooth skin, which is exceedingly loose nearly to the ends of fin rays, and entirely covering the dorsal spines. Dorsal spines very short but stout; soft dorsal longer and higher than anal, but in other ways similar, reaching past base of caudal rays; pectoral short, as wide as long, round and fan-shaped behind, reaching to vertical from base of fourth dorsal ray; origin of ventral far in front of pectorals, the fins reaching to the vertical from the posterior edge of spinous dorsal; caudal well rounded, fan-shaped when spread. Color very pale yellowish brown, thickly covered with round spots of dark brown, those on head smaller; belly with numerous spots, the largest as large as eye; back with many oblong blotches, besides small round spots; fins blotched and banded. Gulf of Mexico, in deep water. This form has a very different coloration from *O. tau* and the texture of its skin and flesh is also less firm, but the technical differences are slight and it is rather a deep-water variety than a species. (*pardus*, leopard.)

Batrachus tau pardus, Goode & Bean, Proc. U. S. Nat. Mus. 1879, 336, Pensacola Snapper Banks; Jordan & Gilbert, Synopsis, 751, 1883; Meek & Hall, Proc. Ac. Nat. Sci. Phila. 1885, 60.

860. PORICHTHYS, Girard.

(Midshipmen.)

Porichthys, Girard, Proc. Ac. Nat. Sci. Phila. 1854, 141 (*notatus*).

Body rather elongate; head not very broad, depressed, the lower jaw projecting. Dorsal spines 2; pectoral broad, without foramen in axil; opercle very small, its posterior part developed as a strong, single spine; suboperculum feebly developed, narrowed and not ending in a spine; no scales on body; spines solid, without venom glands; several lateral lines on sides of head and body, composed of pores and shining spots, some of these accompanied by cirri; canine teeth present; vertebræ $12 + 31$; frontal region depressed, forming a triangular area below level of temporal region, its median ridge very low. Branchiostegals 6; interorbital area short, wide, and with shallow grooves. Air bladder more or less deeply divided into 2 lateral parts. Pyloric appendages none. Species American; remarkable for the very great development of mucous pores, some of which simulate the photophores of *Myctophum*, but are different in origin and not at all luminous. ($\pi \acute{o} \rho o \varsigma$, pore; $i \chi \theta \grave{v} \varsigma$, fish; in allusion to the extraordinary development of the mucous system.)

Note.—The following account of the distribution, structure, and development of the phosphorescent organs of *Porichthys* is furnished us by Prof. Charles Wilson Greene, who has made a careful study of these organs:

"*Porichthys* has numerous lines of conspicuous bright silvery spots distributed in rows over the surface of the body. These spots have been called phosphorescent organs, although no such function has yet been observed, the name arising out of a superficial resemblance. These so-called phosphorescent organs are arranged in rows over the body, and are definite and characteristic and quite constant in location in different individuals. They are accompanied by rows of epidermal sense organs, the two having an intimate relation in distribution over the surface of the fish. In surface view the shining organs have a bright silvery appearance, are more or less round in outline, size from a mere dot to 0.8 mm. in diameter, and surrounded or bordered on one side by an increased amount of pigment. The end buds present a round, transparent, or pellucid, and usually slightly raised, point. Each end bud is bordered by a pair of papillæ. There are about 20 well-defined lines as follows: The *lateral* row, from posterior upper border of pectoral straight alongside to upper third of base of caudal, 35 pairs with an end bud between each pair, upper series small or rudimentary, segmentally arranged and between myomeres. The *pleural* row, from middle of base of pectoral, curves backward and downward to a point above first anal ray then straight nearly to base of caudal, 43 to 62 organs. End buds below each organ to above middle of caudal, 31 organs. The *caudal* rows, end buds only, 2 longitudinal rows on upper and lower thirds of fin. The *anal* row, on either side base of anal fin from third anal ray to base of caudal. Phosphorescent organs in pairs, a pair for each anal ray, 1 end bud for each pair. The *gastric* row, from front around lower edge of pectoral and along side of belly to opposite anal papilla, 30 phosphorescent organs. The *gular* row, from isthmus along ventral side of ventral fin then outward to join gastric row, spur runs forward along external side of ventral fin, 27 organs. A parallel line of 50 end buds follows the gular row and posterior end of gastric. The *ventral* row with its fellow forms a parenthesis on the stomach from the side of the anus ¾ the distance to the ventral fin, 34 organs; no end buds. The *branchiostegal* row, from the isthmus outward over branchiostegal membrane and between first and second rays, no end buds. The *mandibular* row of phosphorescent organs extends around inner edge of ridge formed by the dentary bones; the row of end buds along the outer rim of the same ridge. The *opercular* rows, upper and lower, extend backward and upward across opercle. The *scapular* row, from above opercular spine straight back above pectoral fin, the curves in toward the base of the dorsal fin opposite the third dorsal ray. The *dorsal* row, along base of dorsal fin to base of caudal. This row and the scapular row consist of well developed end buds and rudimentary phosphorescent organs. The *occipital* and *frontal* rows, along the occipital and frontal regions, short rows of small and poorly developed organs. The *nasal*, from the posterior nasal tube to base of anterior tube. The *suborbital* and *postorbital*, from posterior nasal opening around under eye backward and downward to opercle. A *malar* row, from the suborbital down across the cheek. A *maxillary* across the posterior end of maxillary bone. The rows on the head consist of well-developed end buds with rudimentary and irregularly placed phosphorescent organs. The phosphorescent organs are embedded in the connective tissue dermis of the skin, and in section show a uniform general structure throughout the body. A typical organ from the anal or ventral rows consists of an outer spherical group of cells called a lens, resting in a deeper cup-like structure, the capsule, and this in turn in a cup of fibrillar connective tissue called the reflector. The lens consists of cells, polygonal in the center of the group and flattened or fusiform around the periphery. They have a large conspicuous nucleus and a dense, homogeneous, highly refracting cell body. The outlines of the cells are very distinct. In the cells of the capsule the nuclei stain readily, but the granular protoplasm with difficulty, and the cell boundaries are indistinct and usually obliterated. In some specimens connective tissue septa penetrate the capsule. Blood capillaries are always present. The reflector extends well up around the sides of the lens; it consists of fibrillar connective tissue which strongly reflects light. Much pigment is embedded in its meshes. No nerves have yet been traced to the organ. The developing phosphorescent organs do not appear in the embryo fish until it is 15 to 16 mm. long. Then a bud appears in the lower layer of the epidermis, which soon becomes constricted off as a spherical mass of shells lying in the subepidermal connective tissue. This mass later slightly elongates and gives rise by constriction to the lens and the capsule. The reflector is developed from the surrounding connective tissue, so also the pigment cells. Mature organs are not found until the fish reaches a length of over 20 mm. The end buds appear much earlier, 9 to 10 mm." (Charles Wilson Greene.)

    a. Abdomen with 4 longitudinal series of pores, each of which is accompanied by a shining silvery body; 4 rows of shining spots on sides of body; a white blotch below eye, with a black crescent below it.

        b. Teeth on palatines few (4 or 5), 1 to 3 of them developed as very strong canines, as large as canines on vomer; dorsal fin with distinct black blotches; back with dark saddles; third lateral line extending nearly to base of caudal.

                                      POROSISSIMUS, 2653.

        bb. Teeth on palatines numerous, none of them canine, and all much smaller than canines on vomer.

            c. Third lateral line ceasing at second third of anal; cross bands on back and dorsal fin very faint or wanting; dorsal fin with a faint dark edge; sides of head and shoulder without distinct spots; body rather elongate.   NOTATUS, 2654.

            cc. Third lateral line extending nearly to end of anal; cross bands on back and dorsal fin very distinct, appearing as roundish blotches, those on the dorsal fin along the margin; sides of head and humeral region much spotted with brown; body robust.   MARGARITATUS, 2655.

2653. PORICHTHYS POROSISSIMUS (Cuvier & Valenciennes).

(BAGRE SAPO.)

Head 3¾ (4¼ in total); depth 5⅔ (6). D. II–37; A. 34. Body rather elongate, tapering and compressed behind. Head depressed, ⅞ as broad as long and ⅓ wider than deep; lower jaw considerably projecting, maxillary reaching to well behind eye, its length 1¾ in head. Teeth in single series on jaws, vomer, and palatines, those of upper jaw very small, a few of the anterior and 2 or 3 of the lateral teeth somewhat enlarged, the latter strongly hooked forward; teeth in lower jaw strong, rather weaker than in *P. margaritatus*, those in the front of the jaw hooked strongly inward; the lateral teeth, which are larger, hooked backward and inward; 1 or 2 strong canines on each side of vomer, these curved backward and outward; teeth on palatines distant, few in number (usually 4 or 5); among these are 1 to 3 very strong canines (usually, but not always, much larger than canines on vomer), strongly curved forward and inward. In *P. margaritatus* and *P. notatus*, the palatine teeth are not especially enlarged, subequal and more numerous, the canines on the vomer being much larger than any of the other teeth. Gill openings extending from the upper edge of pectoral to just below lower edge. Pectoral without axilliary foramen; height of soft dorsal about 3 in head; length of caudal nearly 2; height of anal 3¼; length of pectorals 1⅔; of ventrals, 2⅔. Color in life, light brown above, the top of head much darker and clouded with dark brown; a row of about 10 bar-like dark blotches along middle of side, each larger than eye, those anteriorly deeper than long, the others longer than deep; each of these blotches usually more or less confluent with a saddle-like dark blotch across the back; a crescent-shaped pale translucent area below the eye; below this a larger blue-black area, irregularly crescent-shaped, covering the preorbital and suborbital region, bounded below and behind by a row of shining mucous pores; on it are about 4 large pores, and above and behind it, close behind and below eye, is a large shining pore bordered with black; cheek steel bluish; sides of body silvery, becoming golden below; lower part of head and belly bright golden; a dark stripe along base of dorsal; soft dorsal with 2 or 3 rows of small round dark olive spots, the upper row posteriorly becoming a dark edging to the fin; caudal, dull red, edged with dusky; anal very pale, edged with blackish; pectorals light orange, usually with some small dark spots above; ventrals orange, slightly darker anteriorly. Numerous series of pores on the body, those of the lateral line accompanied by shining golden bodies, as in other species of the genus. According to fishermen, these bodies are phosphorescent, shining at night; a statement which is probably true, although we have been unable to verify it; pores on sides of back not shining. Most of the pores, as in other species, accompanied by numerous small cirri or cilia; the arrangement of the lines of pores and shining bodies not very different from that found in *P. notatus.* It may be thus described in detail: A series of pores beginning at tip of snout, extending down around preorbital region, bounding the dark subocular blotch and joining almost at a right angle with a series of pores which extends downward from lower posterior corner of eye to angle of mouth. Another series diverges from the first in front of eye, passing close below eye, then upward above cheek, ending in a large pore behind preopercle. A curved series of pores extending backward along opercle, and another parallel with it along subopercle. Two obscure series from front of eye along top of head, becoming wide apart at the vertex, converging at the nape, then slightly diverging, converging in front of spinous dorsal, then again diverging to pass around the fin, each at last becoming straight at front of soft dorsal, extending close to its base to its last ray, there being about 2 pores to each ray. Just below this series, at front of soft dorsal on each side, begins a second series, with the pores wider apart and somewhat irregular, ceasing near the middle of the soft dorsal fin. The lateral line proper next begins above upper posterior angle of preopercle, whence a short branch passes directly upward. Opposite front of soft dorsal, the lateral line is interrupted for a distance a little more than diameter of eye. A short branch arises at this interruption and passes upward and backward at an angle from the end of the anterior part; thence the lateral line passes straight to base of caudal. The next series arises just behind axil of pectoral, then curves abruptly downward and backward, becoming straight opposite third ray of anal, thence proceeding to base of caudal, the pores small and close-set, anteriorly bead-like and shining, becoming dull toward the tail. Next comes a double series on each side of base of anal, the 2 series converging behind and finally coalescing. Another series begins at the middle of the base of the pectoral in front, curves downward, around the base of the fin, and, proceeding directly backward, ceases opposite vent. A series begins midway between gill opening and ventral and, extending straight backward, ceases opposite base of pectoral. Another begins, on each side, on lower side of head, directly below angle of mouth, the two diverging slightly between ventrals, then converging a little behind ventrals, then abruptly diverging, joining the series last mentioned, on each side, just in front of base of pectoral. A cross series of pores extends straight across belly, between vent and anal fin. At each end of this cross series a series of pores turns abruptly forward, the two meeting in an acute angle on the belly just in front of a vertical from base of pectorals. Finally, 3 parallel series on each side of lower parts of head meet in front, the two anterior in obtuse curves, the posterior in an acute angle. The anterior series along the mandible ends at the corner of the mouth. The next just behind the mandible ends just below the corner of the mouth. The next passes along the branchiostegal region, ending at the gill opening. Mandible with 2 large foramina. A series of dark-colored pores along each side of tongue. Length 8 inches. South Carolina to Texas, and southward to Argentina, on sandy shores; not very common, and found in rather deep water. Not rare about Galveston, -but unknown to fishermen at Pensacola. Here described from the types of *P. plectrodon*, the North American form. The types of *P. porosissimus* examined by us in Paris agree in dentition and other respects. Except for the remote locality there is no suggestion of differences. We are informed by Dr. Vaillant that the type specimen of *P. porosissimus* from St. Catherine, has 33 anal rays, that from Rio Janeiro 32, and that the number 27, given by Valenciennes for this species, represents an error in counting. According to Valenciennes, *P. porosissimus* has D. II-36; A. 27; each palatine bone with a row of small, pointed, unequal teeth; row of pores above anal reaching base of caudal. Color grayish brown above, silvery white below; dorsal and anal whitish, edged with brown; pectoral with longitudinal lines; ventrals brownish on the outer edge; caudal whitish at base, the rest brownish; some specimens with dark cross bands. _(porosissimus, most porous.)_

Batrachus porosissimus, CUVIER & VALENCIENNES, Hist. Nat. Poiss., XII, 501, 1837, Surinam (Coll. Leschenault & Doumerc), Cayenne (Coll. Poiteau), Rio Janeiro (Coll. Delalande), St. Catherine (Coll. Lesson & Garnot).
Porichthys plectrodon, JORDAN & GILBERT, Proc. U. S. Nat. Mus. 1882, 291, Galveston, Texas (Type, No. 30894. Coll. D. S. Jordan); JORDAN & GILBERT, Synopsis, 958, 1883.
Porichthys porosissimus, GÜNTHER, Cat., III, 176, 1861; JORDAN, Proc. Ac. Nat. Sci. Phila. 1883, 291; BERG, An. Mus. Nac. Buenos Aires, 1895, 70.
Porichthys porosissimus, JORDAN & GILBERT, Synopsis, 751, 1883; MEEK & HALL, Proc. Ac. Nat. Sci. Phila. 1885, 57.

2654. PORICHTHYS NOTATUS, Girard.

(SINGING FISH; MIDSHIPMAN; CABEZON; SAPO.)

Head 3⅔; depth. 6½. D. II-37; A. 33; V. I, 2; P. 18; eye 8 in head; maxillary 2; pectoral 1¾; ventral 2⅖; caudal 2¾. Head narrowed forward; opercle developed as a strong spine; maxillary reaching beyond orbit; lower jaw with a single row of about 10 large, recurved teeth, behind which is a patch of small teeth; sides of jaw with a single series of canines similar to those in front, but larger; upper jaw with an irregular series of small teeth; palatines with a single series of conical teeth; 2 large curved canines on vomer; head with several rows of fringed pores; 1 row along lower line of opercle and subopercle; another along upper edge of cheek, this branching behind and below the orbit, 1 branch running forward below the orbit and around the snout, the other vertically downward behind the maxillary; a series of fringes behind the lower lip; behind this a series of pores without fringes; a short straight series of pores on each side of vertex; a row of pores along the base of the dorsal fin, curving at front of dorsal, and terminating at upper angle of opercle; a row below this, not reaching base of pectoral; the third row not reaching base of caudal, but ceasing at second third of anal to about its twentieth ray, and is anteriorly strongly curved upward to base of pectoral; 2 concentric series on the abdomen, the outer extending forward between bases of ventrals. The so-called "shining pores" on the sides are not pores, but bright round pieces of shiny membrane, showing through a translucent skin; each of the spots has above it a pair of fringed flaps with a small pore between them; the rows of flaps along dorsal and anal similar, long and low, their last rays reaching base of caudal rays; pectoral broad, somewhat pointed behind, reaching to the vertical from the fifth anal ray; origin of ventrals in advance of pectorals, in distance equal to length of maxillary, their tips not reaching to pectoral base; caudal well rounded. Olive brown above, with coppery reflections, the belly brassy-yellow; sides with irregular broad vertical cross blotches, most distinct in the young; dorsal grayish, with oblique dark bars; vertical fins sometimes margined with black; pores of lateral line bead-like, shining silvery; a white space below eye, with a black crescent below it; head yellowish brown, with no dark spots on opercle and shoulder; peritoneum black. Length 15 inches. Pacific coast; very abundant from Lower California to Puget Sound; living under stones, near the shore northward, in deeper water southward. It makes a peculiar humming noise with its air bladder, hence the name singing fish. (*notatus*, spotted; noted.)

Porichthys notatus, GIRARD, Proc. Ac. Nat. Sci. Phila. 1854, 141, San Francisco; GIRARD, Pac. R. R. Surv., x, Fishes, 134, 1858.
Porichthys margaritatus, MEEK & HALL, Proc. Ac. Nat. Sci. Phila. 1885, 56; not of RICHARDSON.
Porichthys porosissimus, JORDAN & GILBERT, Synopsis, 751, 1883 (not of CUVIER & VALENCIENNES); GÜNTHER, Cat., III, 176, 1861 (in part).

2655. PORICHTHYS MARGARITATUS (Richardson).

Head $3\frac{1}{4}$ to $3\frac{3}{8}$; depth $4\frac{3}{4}$ to $5\frac{1}{4}$. D. II–37; A. 33. Similar to *Porichthys notatus*, differing chiefly in color. Top and sides of head and space above pectorals with numerous round dark brown spots and freckles, behind pectorals 6 to 8 vertical $\frac{1}{2}$ cross bars; dorsal not margined with black, but with 8 to 10 black submarginal spots; anal, with the exception of a few posterior rays, pale; caudal black at base and tip; pectorals with a few dots at base and on upper rays; a roundish white blotch below eye, below this a jet-black crescent. Palatine teeth small, 1 or 2 slightly enlarged. Series of shining spots arranged as in *P. notatus*, except that the third series extends almost to end of anal, to about its thirtieth ray. Pacific coast of tropical America. This species was obtained by the *Albatross* in large numbers off the west coast of Colombia, at Station 2795 at a depth of 33 fathoms, and at Station 2802 at a depth of 16 fathoms. The largest specimens are about 4½ inches long. In dentition it agrees with *Porichthys notatus*, but in color and arrangement of spots it resembles *P. porosissimus.* (*margaritatus*, bearing pearls; μάργαρος.) .

Batrachus margaritatus, RICHARDSON, Voyage Sulphur, Fishes, 67, 1845, Pacific coast of Central America ; coloration and arrangement of lines identical with *porosissimus*.
Porichthys nautopædium,* JORDAN & BOLLMAN, Proc. U. S. Nat. Mus. 1889, 171, Pacific Ocean, off coast of Colombia, Albatross Station, No, 2802, 8° 38' N., 78° 31' 30'' W., in 16 fathoms. (Type, No. 41145, U. S. Nat. Mus. Coll. *Albatross*.)

861. THALASSOPHRYNE, Günther.

(POISON TOAD-FISHES.)

Thalassophryne, GÜNTHER, Cat. Fishes, III,174, 1861 (*maculosa.*).

Body rather elongate, compressed; head moderate. Dorsal spines 2;† soft dorsal and anal rather short, free from caudal; opercle very small, its posterior part developed as a single strong spine; subopercle feebly developed, narrowed and not ending in a spine; no scales on body. Spines hollow, and connected with venom glands. Lateral line on sides of body single; jaws without canine teeth. Species all South American, some of them ascending rivers; all of them noted for their venomous spines.‡ (θάλασσα, the sea; φρύνη, toad.)

---

* *ναυτοπαίδιον*, sailor-boy, from the common name "midshipman," a name given in allusion to the "buttons" on the belly of the fish.
† In *Thalassothia*, Berg, a South American genus, likewise with poison glands, 4 dorsal spines are present.
‡ The poison organs of *Thalassophryne reticulata* are thus described by Dr. Günther: "In this species I first observed and closely examined the poison organ with which the fishes of this genus are provided. Its structure is as follows: (1) The opercular part: The operculum is very narrow, vertically styliform, and very mobile; it is armed behind with a spine, 8 lines long in a specimen of 10½ inches, and of the same form as the venom fang of a snake; it is, however, somewhat less curved, being only slightly bent upward; it has a longish slit at the outer side of its extremity, which leads into a canal perfectly closed, and running along the whole length of its interior; a bristle introduced into the canal reappears through another opening at the base of the spine, entering into a sac situated on the opercle and along the basal half of the spine: the sac is of an oblong-ovate shape, and about double the size of an oat grain. Though the specimen had been preserved in spirits for about 9 months, it still contained a whitish substance of the consistency of thick cream, which on the slightest pressure freely flowed from the opening in the extremity of the spine. On the other hand, the sac could be easily filled with air or fluid from the foramen of the spine. No gland could be discovered in the immediate neighborhood of the sac; but on a more careful inspection I found a minute tube floating free in the sac, whilst on the left-hand side there is only a small opening instead of the tube. The attempts to introduce a bristle into this opening for any distance failed, as it appears to lead into the interior of the basal portion of the operculum, to which the sac firmly adheres at this spot. (2) The dorsal part is composed of the 2 dorsal spines, each of which is 10 lines long. The whole arrangement is the same as in the opercular spines; their slit is at the front side of the point; each has a separate sac, which occupies the front of the basal portion; the contents were the same as in the opercular sacs, but in somewhat greater quantity. A strong branch of the lateral line ascends to the immediate neighborhood of their base. Thus we have 4 poison spines, each with a sac at its base; the walls of the sacs are thin, composed of a fibrous membrane, the interior of which is coated over with mucous. There are no secretory glands embedded between these membranes, and these sacs are probably merely the reservoirs in which the fluid secreted accumulates. The absence of a secretory organ in the immediate neighborhood of the reservoirs (an organ the size of which would be in accordance with the quantity of fluid secreted), the diversity of the osseous spines which have been modified into poison organs, and the actual communication indicated by the foramen in the sac, lead me to the opinion that the organ of secretion is either that system of muciferous channels which is found in nearly the whole class of fishes, and the secretion of which has poisonous qualities in a few of them, or at least an independent portion of it. This description was

MACILOSA, 2656.

RETICULATA, 2657.

BATUS MACILOSA, Günther.

(Günther.)

base of the pectoral obliquely downward and forward to the level of the inferior base of the pectoral. The 2 dorsal spines are slender, pungent, about ¼ the length of the head. Dorsal and anal fins terminate immediately before the root of the caudal, the length of which is ⅙ the total; pectoral obliquely rounded, extending to the origin of the anal; ventral rather short, not quite ¼ the length of the head, extending to the base of the pectoral. Skin perfectly smooth, with some very short tentacles at the lower jaw. Two short horizontal muciferous channels on the cheek and the lateral line are very distinct; they are not, as usually, composed of a series of distant pores, but the pores are confluent, forming 1 continuous groove of a white color. Other muciferous channels, as for instance along the base of the anal, are composed of separate indistinct pores. Color brown, marbled with darker; pectoral fin and sides of the body with some round black spots; chin and ventrals brownish; belly white. The general habit is that of a *Batrachus* [*Opsanus*]. One specimen, from Puerto Cabello, Caribbean Sea. (Günther.) (*maculosus*, spotted.)

Thalassophryne maculosa, GÜNTHER, Cat., III, 175, 1861, Puerto Cabello; GÜNTHER, Fishes of Centr. Amer., 436, pl. 68, fig. 1, 1869; MEEK & HALL, Proc. Ac. Nat. Sci. Phila. 1885, 54.

2657. THALASSOPHRYNE RETICULATA, Günther.

D. II-24; A. 24; V. I, 2; P. 16. The length of the head is ⅖ of the total length (without caudal). The teeth on the palate are in a single series, very short, obtuse, incisor-like. Pectoral very large, extending back to the sixth anal ray. Head, body, and fins brown, with a network of yellowish lines; vertical and pectoral fins with a white margin. In other respects this species agrees with *T. maculosa*. Length 13 inches. Panama; not rare. (*reticulatus*, netted.)

Thalassophryne reticulata, GÜNTHER, Proc. Zool. Soc. Lond. 1864, 150, 155, Panama; GÜNTHER, Fish. Centr. Amer., 437, pl. 68, fig. 2, 1869; JORDAN & GILBERT, Proc. U. S. Nat. Mus. 1882, 621; MEEK & HALL, Proc. Ac. Nat. Sci. Phila. 1885.

862a. DÆCTOR, Jordan & Evermann, new genus.

(POISON TOAD-FISHES.)

Dæctor, JORDAN & EVERMANN, new genus (*dowi*).

This genus differs from *Thalassophryne* in the more elongate body and the many-rayed soft dorsal and anal fins, the last rays of which are fully joined to the caudal. (*δαίκτωρ*, slayer; from *δαΐζω*, to slay.)

2658. DÆCTOR DOWI (Jordan & Gilbert).

Head 4 in length (4¾ with caudal); depth [illegible]. D. II-33; A. 30. Body comparatively elongate, compressed behind. Head low and rather narrow, its width 1⅓ in its length. Eye very small, the diameter not ½ the interorbital space, and about as long as snout, [illegible] in head. Interorbital width about 5½ in head. Opercular spine short, nearly 4 in head. Mouth oblique, the lower jaw much projecting. Maxillary 2 in head, extending

a. Dorsal and anal fins not joined to the caudal.

    b. Dorsal and anal fins rather short; D. II-19; A. 18; pectoral fins short, their tips reaching to origin of anal. Color brown, marbled with darker; pectoral fins and sides of body with some round black spots; chin and ventrals brownish; belly white.   MACULOSA, 2656.

    bb. Dorsal and anal fins longer; D. II-24; A. 24; pectoral fins longer, their tips reaching to sixth anal ray. Color of head, body, and fins brown, with a network of yellowish lines; dorsal, anal, caudal, and pectoral fins with white margins.   RETICULATA, 2657.

2656. THALASSOPHRYNE MACULOSA, Günther.

D.II-19; A.18; V.I, 2. The head is somewhat longer than broad, its length being contained 3⅓ in the total; it is moderately depressed. The snout is short, obtuse, with the cleft of the mouth ascending obliquely upward, and with the chin prominent. The maxillary extends to the vertical from the posterior margin of the orbit. The teeth are obtusely conical, standing in single series, except anteriorly in the lower jaw, where they form 2 series, and in the upper, where they are cardiform, in a narrow band. The eyes are directed upward and very small, their width being ⅓ of that of the bony bridge between the orbits. Gill covers with a single spine; it is long, slender, cylindrical, like one of the dorsal spines, and has the operculum for its base. Gill opening not very narrow; it extends from the upper

made from the first example; through the kindness of Captain Dow I received 2 other specimens, and in the hope of proving the connection of the poison bags with the lateral-line system, I asked Dr. Pettigrew, of the Royal College of Surgeons, a gentleman whose great skill has enriched that collection with a series of the most admirable anatomical preparations, to lend me his assistance in injecting the canals. The injection of the bags through the opening of the spine was easily accomplished; but we failed to drive the fluid beyond the bag, or to fill with it any other part of the system of muciferous channels. This, however, does not disprove the connection of the poison bags with that system, inasmuch as it became apparent that, if there be minute openings they are so contracted by the action of the spirit in which the specimens were preserved, as to be impassable to the fluid of injection. A great part of the lateral-line system consists of open canals; however, on some parts of the body, these canals are entirely covered by the skin; thus, for instance, the open lateral line ceases apparently in the suprascapular region, being continued in the parietal region. We could not discover any trace of an opening by which the open canal leads to the skin; yet we could distinctly trace the existence of the continuation of the canal by a depressed line, so that it is quite evident that such openings do exist, although they may be passable only in fresh specimens. Thus, likewise, the existence of openings in the bags, as I believed to have found in the first specimen dissected, may be proved by examination of fresh examples. The sacs are without an external muscular layer, and situated immediately below the loose, thick skin which envelops their spines to their extremity; the ejection of the poison into a living animal, therefore, can only be effected by the pressure to which the sac is subjected the moment the spine enters another body. Nobody will suppose that a complicated apparatus like the one described can be intended for conveying an innocuous substance; and therefore I have not hesitated to designate it as poisonous; and, Captain Dow informs me in a letter lately received, 'the natives of Panama seemed quite familiar with the existence of the spines and of the emission from them of a poison which, when introduced into a wound, caused fever, an effect somewhat similar to that produced by the sting of a scorpion; but in no case was a wound caused by one of them known to result seriously. The slightest pressure of the finger at the base of the spine caused the poison to jet a foot or more from the opening of the spine.' The greatest importance must be attached to this fact, inasmuch as it assists us in our inquiries into the nature of the functions of the muciferous system, the idea of its being a secretory organ having lately been superseded by the notion that it serves merely as a stratum for the distribution of peripheric nerves. Also the objection that the Stingrays and many Siluroid fishes are not poisonous, because they have no poison organ, can not be maintained, although the organs conveying their poison are neither so well adapted for this purpose nor in such a perfect connection with the secretory mucous system as in *Thalassophryne*. The poison organ serves merely as a weapon of defense. All the Batrachoids with obtuse teeth on the palate and in the lower jaw feed on Mollusca and Crustaceans." (Günther.)

base of the pectoral obliquely downward and forward to the level of the inferior base of the pectoral. The 2 dorsal spines are slender, pungent, about $\frac{1}{4}$ the length of the head. Dorsal and anal fins terminate immediately before the root of the caudal, the length of which is $\frac{1}{7}$ the total; pectoral obliquely rounded, extending to the origin of the anal; ventral rather short, not quite $\frac{1}{2}$ the length of the head, extending to the base of the pectoral. Skin perfectly smooth, with some very short tentacles at the lower jaw. Two short horizontal muciferous channels on the cheek and the lateral line are very distinct; they are nót, as usually, composed of a series of distant pores, but the pores are confluent, forming 1 continuous groove of a white color. Other muciferous channels, as for instance along the base of the anal, are composed of separate indistinct pores. Color brown, marbled with darker; pectoral fins and sides of the body with some round black spots; chin and ventrals brownish; belly white. The general habit is that of a *Batrachus* [*Opsanus*]. One specimen, from Puerto Cabello, Caribbean Sea. (Günther.) (*maculosus,* spotted.)

Thalassophryne maculosa, GÜNTHER, Cat., III, 175, 1861, Puerto Cabello; GÜNTHER, Fishes of Centr. Amer., 436, pl. 68, fig. 1, 1869; MEEK & HALL, Proc. Ac. Nat. Sci. Phila. 1885, 54.

2657. THALASSOPHRYNE RETICULATA, Günther.

D. II–24; A. 24; V. I, 2; P. 16. The length of the head is $\frac{2}{7}$ of the total length (without caudal). The teeth on the palate are in a single series, very short, obtuse, incisor-like. Pectoral very large, extending back to the sixth anal ray. Head, body, and fins brown, with a network of yellowish lines; vertical and pectoral fins with a white margin. In other respects this species agrees with *T. maculosa.* Length 13 inches. Panama; not rare. (*reticulatus,* netted.)

Thalassophryne reticulata, GÜNTHER, Proc. Zool. Soc. London 1864, 150, 155, Panama; GÜNTHER, Fish. Centr. Amer., 437, pl. 68, fig. 2, 1869; JORDAN & GILBERT, Proc. U. S. Nat. Mus. 1882, 62; MEEK & HALL, Proc. Ac. Nat. Sci. Phila. 1885 ?.

862. DÆCTOR, Jordan & Evermann, new genus.

(POISON TOAD-FISHES.)

Dæctor, JORDAN & EVERMANN, new genus (*dowi*).

This genus differs from *Thalassophryne* in the more elongate body and the many-rayed soft dorsal and anal fins, the last rays of which are fully joined to the caudal. ($\delta\alpha i\varkappa\tau\omega\rho$, slayer; from $\delta\alpha i\zeta\omega$, to slay.)

2658. DÆCTOR DOWI (Jordan & Gilbert).

Head 4 in length (4$\frac{3}{4}$ with caudal); depth 5$\frac{2}{3}$ (6$\frac{2}{3}$). D. II–33; A. 30. Body comparatively elongate, compressed behind. Head low and rather narrow, its width 1$\frac{1}{3}$ in its length. Eye very small, the diameter not $\frac{1}{4}$ the interorbital space, and about as long as snout, 8 in head. Interorbital width about 5$\frac{1}{2}$ in head. Opercular spine short, nearly 4 in head. Mouth oblique, the lower jaw much projecting. Maxillary 2 in head, extending to beyond eye. Teeth small, those on the palatine largest; teeth of upper jaw smaller than those of the lower; anterior teeth of the lower jaw in about 2 series. Pectoral fins long, $1\frac{1}{5}$ in head, reaching about to fifth anal ray; last rays of dorsal and anal fully joined to the caudal.   Color olivaceous, with darker blotches; first dorsal black; under parts pale; posterior portion of anal edged with dark.   Pacific coast of North America, from Punta Arenas to Panama; rare. (Named for Capt. John M. Dow, who obtained a fine specimen (now destroyed) from Panama.)

Thalassophryne dowi, JORDAN & GILBERT, Proc. U. S. Nat. Mus. 1887, 388, Punta Arenas (Type, No. 39085, U. S. Nat. Mus. Coll. Cornell University); JORDAN, Proc. Cal. Ac. Sci. 1896, 231, pl. 38.

Suborder XENOPTERYGII.

(THE CLING-FISHES.)

Breast with a broad sucking disk, between the wide-set ventral fins, this formed from the skin of the breast, not from the ventral fins themselves. Ventral rays I, 4 or I, 5; no scales; no spinous dorsal; no suborbital ring; palatine arcade materially modified; no air bladder; vertebræ in increased numbers; gill arches reduced. A well-marked group of small fishes, constituting a single family. ($\xi\acute{\epsilon}\nu o\varsigma$, strange; $\pi\tau\acute{\epsilon}\rho\upsilon\xi$, fin.)

Family CXCIX. GOBIESOCIDÆ.

(CLING-FISHES.)

Body rather elongate, tadpole-shaped, broad and depressed in front, covered by smooth, naked skin; mouth moderate; upper jaw protractile; teeth usually rather strong, the anterior conical or incisor-like; posterior canines sometimes present; suborbital ring wanting; no bony stay from suborbital across cheek; opercle reduced to a spine-like projection concealed in the skin, behind the angle of the large preopercle, this spine sometimes obsolete; pseudobranchiæ small or wanting, gills 3 or $2\frac{1}{2}$; gill membranes broadly united, free or united to the isthmus; dorsal fin on the posterior part of the body, opposite to the anal and similar to it, both fins without spines; ventral fins wide apart, each with 1 concealed spine and 4 or 5 soft rays. Between and behind the ventrals is a large sucking disk, the ventrals usually forming part of it. This sucking disk, which is wholly different in structure from that of *Cyclopterus* and *Liparis*, is thus described by Dr. Günther: "The whole disk is exceedingly large, subcircular, longer than broad, its length being (often) $\frac{1}{6}$ of the whole length of the fish. The central portion is formed merely by skin, which is separated from the pelvic or pubic bones by several layers of muscles. The peripheric portion is divided into an anterior and posterior part by a deep notch behind the ventrals. The anterior peripheric portion is formed by the ventral rays, the membrane between them and a broad fringe, which extends anteriorly from one ventral to the other. This fringe is a fold of the skin containing on one side the rudimentary ventral spine, but no cartilage. The posterior peripheric portion is suspended on each side on the coracoid, the upper bone of which is exceedingly broad, becoming a free, movable plate behind the pectoral. The lower bone of the coracoid is of a triangular form, and supports a very broad fold of the skin, extending from one side to the other, and containing a cartilage which runs through the whole of that fold. Fine processes of the cartilage are continued into the soft striated margin, in which the disk terminates posteriorly. The face of the disk is coated with a thick epidermis, like the sole of the foot in higher animals. The epidermis is divided into many polygonal plates. There are no such plates between the roots of the ventral fins." (Günther, Cat., III, 495.) No air bladder; intestines short; pyloric cæca few or none; skeleton firm; vertebræ 13 or 14+13 to 22 = 26 to 36. Carnivorous fishes of small size, chiefly of the warm seas, usually living among loose stones between the tide marks and clinging to them firmly by means of the adhesive disk. Their relations are obscure, but they are probably descended from allies or ancestors of the *Cottidæ* or *Batrachoididæ.* Genera about 15; species 50. The principal genus is *Gobiesox.* (*Gobesocidæ*, Günther, Cat., III, 489–515.)

GOBIESOCINÆ:
  a. Gill membranes free from the isthmus; gills 3; posterior part of sucking disk with no free anterior margin.
    b. Incisors of lower jaw with entire edges.
      c. Vertebræ about 32; anal fin long, nearly as long as dorsal.
                                                        CAULARCHUS, 863.
      cc. Vertebræ about 26; anal fin short.
        d. Dorsal fin very long, of about 17 rays, twice as long as the moderate anal, which has 8 or 9 rays; disk broad; upper teeth in several rows.   BRYSSETÆRES, 864.
        dd. Dorsal fin moderate or short, of 4 to 13 rays.
          e. Disk more or less broad, its length 2½ to 3 in body; dorsal and anal not very short, their rays 6 or more; body tapering rapidly backward; opercular spine strong. GOBIESOX, 865.
          ee. Disk very narrow, its width 4 to 5 in body; head short, 3½ to 4 in body; dorsal and anal very short and small; a patch of teeth in each jaw behind the large teeth; sucking disk small.   RIMICOLA, 866.
    bb. Incisors of lower jaw tricuspid or serrate; dorsal and anal fins short; vertebræ about 28.   ARBACIOSA, 867.

863. CAULARCHUS, Gill.

Caularchus, GILL, Proc. Ac. Nat. Sci. Phila. 1862, 330 (*mœandricus*).

This genus differs from *Gobiesox* chiefly in the numerous vertebræ, 32 in the only species known. The incisors are entire, the anal fin similar to the dorsal, each having 12 or 13 rays. The single species reaches a large size and is found farther north than any other of the group, a fact in accord with the increased number of vertebræ. (ϰαυλός, stem; ἀρχός, anus; from the many-rayed anal.)

2659. CAULARCHUS MÆANDRICUS (Girard).

(SUCK-FISH.)

Head $2\frac{3}{4}$; depth $6\frac{1}{4}$.   D. 13; A. 12; V. I, 4; vertebræ $13+19=32$; eye $7\frac{1}{4}$ in head; distance from vent to caudal $2\frac{3}{4}$ in length of body; sucking disk as broad as long, $3\frac{1}{2}$ in length. Head broad, nearly circular when viewed from above; interorbital width 3 in head; mouth wide, its width more than $\frac{1}{4}$ length of head; maxillary extending to below eye; outer teeth of upper jaw rather strong, close set, vertical, conical, or slightly compressed, a narrow band of small, conical teeth behind them; lower jaw with larger teeth, 6 or 8 of the anterior broad, incisor-like, with entire edges, placed nearly horizontally; lateral and posterior teeth small, as in upper jaw; nostrils ending in tubes; spine on opercle sharp, but not projecting through the skin; origin of dorsal fin a little in advance of vent, the fin much higher than the anal; vent midway between anal and posterior edge of disk; pectorals short and broad, not extending back past the margin of the ventral disk, the 3 lower rays forming part of disk; caudal rounded. Color light olive, everywhere reticulated with brownish orange; middle of upper lip black; a light bar between eyes and 1 across cheek; vertical fins dusky; caudal with 2 faint brownish bars near its base. Specimens from red algæ are light pink, mottled with darker, the pale band between eyes very distinct. Length 6 inches. Pacific coast of United States, from Vancouver Island to Point Concepcion; everywhere very abundant in rock pools; the largest species of *Gobiesocidæ*. (*mæandricus*, meandering, in allusion to the reticulated streaks.)

Lepadogaster reticulatus, GIRARD, Proc. Ac. Nat. Sci. Phila. 1854, 155, San Luis Obispo, California; name preoccupied.
Lepidogaster mæandricus, GIRARD, Pacific R. R. Surv., X, Fishes, 130, 1858, San Luis Obispo, California; substitute for *reticulatus*, preoccupied in *Lepadogaster;* GÜNTHER, Cat., III, 505, 1861.
Gobiesox reticulatus, JORDAN & GILBERT, Synopsis, 749, 1883.

864. BRYSSETÆRES, Jordan & Evermann.

Bryssetæres, JORDAN & EVERMANN, Proc. Cal. Ac. Sci. 1896, 230 (*pinniger*).

This genus differs from *Gobiesox* solely in the great development of the dorsal fin, which has 17 rays, the moderate anal having but 8 or 9; the vertebræ $10+16$, as usual in *Gobiesox*. One species known. ($\beta\rho\acute{\upsilon}\sigma\sigma o\varsigma$, sea-urchin; $\H{\epsilon}\tau\alpha\iota\rho o\varsigma$, comrade, the species living in rock pools with the sea-urchins.)

2660. BRYSSETÆRES PINNIGER (Gilbert).

Head $2\frac{3}{4}$ to $2\frac{7}{8}$; width of body $4\frac{1}{2}$; of head 3 in length. D. 16 or 17; A. 8 or 9. Interorbital width $3\frac{1}{4}$ in head; eye $\frac{1}{2}$ interorbital width; width of mouth $1\frac{3}{4}$ to $1\frac{3}{4}$ in head. Teeth in upper jaw conic, acute, in several series, the anterior row in front enlarged, unequal; in lower jaw the teeth mesially in 2 distinct series, those in middle of anterior row narrow, entire incisors, those laterally conic, canine-like. No evident opercular spine. Disk about as broad as long, its length about that of head. Front of dorsal varying in position, about midway between snout and base of caudal, its length about $\frac{1}{4}$ that of body. Vent nearly equidistant between disk and front of anal, the base of the latter $3\frac{1}{2}$ to $3\frac{3}{4}$ in body. Caudal $1\frac{1}{2}$ to $1\frac{3}{4}$ in head. Pectorals about $\frac{1}{2}$ of head, with a distinct fold at base. Color variable; anteriorly usually with reticulating dark lines surrounding yellowish spots; a narrow dark streak forward, 1 downward, and 1 backward from orbit; below dorsal fin about 6 dark bars running obliquely downward and backward, these sometimes in greater number, frequently more or less irregular and interconnected, often divided by vertical streaks or series of dots; body sometimes light in spirits, without distinctive markings; vertical fins usually dusky, narrowly margined with white, sometimes lighter with dark margins. Length $2\frac{1}{4}$ inches. (Gilbert.) Gulf of California, abundant; specimens known from Puerto Refugio (Angel Island), San Luis Gonzales Bay, and La Paz. Well distinguished by its long dorsal. (*pinniger*, fin-bearing.)

Gobiesox pinniger, GILBERT, Proc. U. S. Nat. Mus. 1890, 94, Puerto Refugio, Gulf of California. (Coll. *Albatross*.)

Bryssetæres pinniger, JORDAN, Proc. Cal. Ac. Sci. 1896, 230, pl. 34.

865. GOBIESOX, Lacépède.

(CLING-FISHES.)

Gobiesox, LACÉPÈDE, Hist. Nat. Poiss., II, 595, 1799 (*cephalus*).

Megaphalus, RAFINESQUE, Analyse de la Nature 1815, 86 (*cephalus*, substitute for *Gobiesox*, regarded as an objectionable compound).

Sicyases, MÜLLER & TROSCHEL, Archiv fur Naturgesch. 1843, 298 (*sanguineus;* small species, with upper teeth uniserial).

Tomicodon, BRISOUT DE BARNEVILLE, Rev. Zool., 144, 1846 (*chilensis = Sicyases*).

Sicyogaster, BRISOUT DE BARNEVILLE, Rev. Zool., 144, 1846 (*marmoratus = Gobiesox*).

Bryssophilus, JORDAN & EVERMANN, new subgenus (*papillifer*).

Body anteriorly very broad and depressed, posteriorly slender, covered with tough, smooth skin; opercle with a strong spine; head large, rounded in front; mouth terminal, crescent-shaped; lower jaw with a series of strong incisors in front, their edges rounded or truncate; upper jaw with a series of strong teeth, behind which are sometimes smaller teeth; no teeth on vomer or palatines; gills 3; gill membranes broadly united under the throat, not attached to the isthmus; sucking disk large, the posterior portion without anterior free margin. Dorsal and anal moderate, the dorsal rays 6 to 12, the anal rays 6 to 10. Vertebræ about 26, as far as known. Species numerous, all American; mostly tropical, clinging to rocks near the shore. (*Gobius; Esox;* the resemblances either to the goby or the pike being few or remote.)

BRYSSOPHILUS (βρύσσος, sea urchin; φιλέω, to love):

   a. Dorsal fin comparatively long, of about 13 rays; anal rays 9; disk broad; upper teeth in several rows; lower incisors narrow; papillæ below chin; color olivaceous.

                                                                 PAPILLIFER, 2661.

  aa. Dorsal fin moderate or short, its rays 6 to 11.

GOBIESOX:

c. Upper teeth in more than 1 series (character not verified in a few species); head broad.

 d. Coloration in life chiefly olivaceous, without red, sometimes banded with darker or paler.

  e. Dorsal rays 12; anal rays 7.   GYRINUS, 2662.

  ee. Dorsal rays 11; anal rays 6; fins black.   NIGRIPINNIS, 2663.

  eee. Dorsal rays 9 or 10; anal rays 6.   CEPHALUS, 2664.

  eeee. Dorsal rays 8; anal rays 6.   TUDES, 2665.

  eeeee. Dorsal rays 11; anal rays 10.   STRUMOSUS, 2666.

  eeeeee. Dorsal rays 10; anal rays 8.   VIRGATULUS, 2667.

  eeeeeee. Dorsal rays 9; anal rays 7.

   f. Width of head 3⅔ in length; color plain brown.

             ADUSTUS, 2668.

   ff. Width of head 5 in length; color blackish, with yellow vermiculations.   FUNEBRIS, 2669.

  eeeeeeee. Dorsal rays 7; anal rays 7; eyes variegated.

           POECILOPHTHALMUS, 2670.

 dd. Coloration in life chiefly bright red, or else with red spots or bands, the color not fading in spirits.

  g. Color red, with deep red spots. D. 6; A. 5. RHODOSPILUS, 2671.

  gg. Color uniform red, unspotted, the color not fading in spirits; dorsal rays 6 to 8; anal rays 6.

   h. Lower jaw with short incisors on each side, followed by canines.   MACROPHTHALMUS, 2672.

   hh. Lower jaw with 2 horizontal incisors on each side, the third horizontal tooth not incisor-like; no distinct canines.

           CERASINUS, 2673.

SICYASES (σικύα, a sucking cup made of a gourd):

 cc. Upper teeth in a single series (character not verified on some species); dorsal and anal short.

  i. Color chiefly red.

   j. Body with cross bands of deep red; iris red; dorsal rays 6; anal rays 5; head broad, the eyes very large. ERYTHROPS, 2674.

   jj. Body with dark cross bands and with spots of clear blue; body rather slender. D. 6 or 7; A. 6.   RUBIGINOSUS. 2675.

   jjj. Body plain, light red; form rather slender. CARNEUS, 2676.

  ii. Color olivaceous or brownish, not red.

   k. Dorsal rays 9; anal rays 6.

    l. Color olivaceous, without bands.   HÆRES, 2677.

    ll. Color greenish, with 3 dark cross bands and many dots.

          PUNCTULATUS, 2678.

   kk. Dorsal rays 7; anal rays 7; body with dark cross bands.

          FASCIATUS, 2679.

Subgenus BRYSSOPHILUS, Jordan & Evermann.

2661. GOBIESOX PAPILLIFER, Gilbert.

Head 2⅔; width of body 3⅔; width of head 2⅕. D. 13; A. 9. Width of mouth 1⅕ in head; interorbital width 3; eye ⅓ interorbital width; teeth in upper jaw conic, acute, very small, in 2 or more series, 2 of them slightly enlarged, canine-like; teeth in lower jaw in 2 series, the outer anteriorly, narrow entire incisors, with rounded tips, becoming conical laterally; opercular spine sharp, evident, though not projecting through the integument; lips and lower side of head anteriorly with fleshy papillæ; disk about as broad as long, its length 1⅕ in head; distance from front of dorsal to base of caudal 1¼ in its distance from tip of snout; vent exceptional in position, immediately in front of anal fin; base of anal 1¾ in head; caudal rather acute, 1⅞ in head; pectorals 2⅜ in head, a distinct fleshy fold at base. In spirits, uniform dark olivaceous, lower side of head and disk light; pectorals dusky; vertical fins with a black bar at base, then a white bar, followed by a wide, dusky area, and narrowly margined with white; caudal with all these marks except the black bar, having the posterior outlines curved, following margin of fin. Length 1½ inches. Magdalena Bay, Lower California. (Gilbert.) Possibly related to the genus *Caularchus.* (*papilla; fero,* I bear.)

Gobiesox papillifer, GILBERT, Proc. U. S. Nat. Mus. 1890, 96, Magdalena Bay, Lower California. (Coll. *Albatross.*)

Subgenus GOBIESOX.

2662. GOBIESOX GYRINUS, Jordan & Evermann, new species.

B. 6; D. 12; A. 7; V. I, 4; P. 20. A vertical fold of skin at base of pectoral; coracoid distinctly below level of upper margin of pectoral; teeth of upper jaw cardiform, lower jaw with very narrow but compressed incisors, which are as short as the other teeth. Lateral profile of head nearly semicircular; head much depressed, as long as broad, its length being ⅔ of the total; width of interorbital space somewhat less than ½ greatest width of head, or 3 times diameter of eye; cleft of mouth extending beyond anterior margin of eye; distance of origin of dorsal from caudal more than ½ of its distance from snout. Brownish, with scattered dark spots; a black blotch anteriorly on the dorsal fin. Length 3 inches. (Günther.) West Indies; not seen by us. A valid species, according to Dr. Günther, but apparently as yet without tenable specific name, as the original *Cyclopterus nudus,* Linnæus, must have been some other fish. (*gyrinus; γυρῖνος,* a tadpole.)

Lepadogaster nudus, BLOCH & SCHNEIDER, Syst. Ichth., 2, 1801; in part, description taken from *Cyclopterus nudus,* Linnæus, except the count of fin rays. D. 12; A. 6.
Cotylis nuda, MÜLLER & TROSCHEL, Hor. Ichth., III, 18, pl. 3, f. 2,
Gobiesox nudus, GÜNTHER, Cat. Fish., III, 502, 1861, Island of Cordova. (Coll. G. U. Skinner.)
Gobiesox gyrinus, JORDAN & EVERMANN, Check-List Fishes, 491, 1896, Cordova; after GÜNTHER; name only.

2663. GOBIESOX NIGRIPINNIS (Peters).

D. 11; A. 6; P. 22. "Nostrils, mouth, teeth, opercular spine, and fin rays as in *Cotylis stannii* (*Gobiesox cephalus*), but the dorsal fin longer. Light brown above (minutely dotted with black, if viewed by a magnifier); vertical fins black." Puerto Cabello (Peters); not seen by us; a doubtful species, perhaps identical with *G. cephalus* or *G. nudus.* (*niger,* black; *pinna,* fin.)

Cotylis nigripinnis, PETERS, Berl. Monatsber. 1859, 412, Puerto Cabello.
Gobiesox nigripinnis, GÜNTHER, Cat., III, 502, 1861; after PETERS; GÜNTHER, Fish. Centr. Amer., 390, 1869.

2664. GOBIESOX CEPHALUS, Lacépède.

(Tétard; Testar.)

D. 9 or 10; A. 6; C. 12; P. 19 or 20.   Head and anterior part of body very broad, much depressed; skin tough, naked, and smooth; head nearly as broad as long, with its profile semielliptical, the snout being very obtuse and rounded.   The upper surface of the head is quite flat, gently sloping downward in a straight line from the nape to the snout.   The greatest width of the interorbital space is ½ of that of the head, or 4 times the diameter of the eye.   The cleft of the mouth is horizontal, curved, wide, extending to below the center of the eye; the lips are thick, the lower being divided into 5 portions by 4 vertical grooves, the central portion being the smallest, the lateral ones the largest and hanging downward.   The upper jaw is slightly protractile, and there is a broad velum behind the teeth in each jaw.   A band of short conical teeth in the upper jaw; a single series in the lower, the anterior ones being slightly compressed incisors, and small like the lateral teeth, which are conical.   The eye is small, situated immediately below the upper profile of the head. Two nostrils, close together, opposite the upper angle of the orbit, their margins being slightly raised.   The lower angle of the opercular apparatus terminates posteriorly in an obtuse movable point enveloped in skin and directed backward.   The gill openings are somewhat narrow in consequence of the small degree of expansibility of the gill covers, but the gill membranes have the margin quite free, being united together under the throat, and not attached to the isthmus.   There are only 3 gills; the pseudobranchiæ are quite rudimentary, indicated by 2 or 3 short lamellæ. The distance of the origin of the dorsal fin from the caudal is nearly ⅓ of its distance from the snout, its first ray is much shorter than the others, and apparently without articulations.   The caudal rounded and of moderate length; the anal is only ½ as long as the dorsal, commencing below its middle and terminating in the same vertical.   The pectoral is broad and short, its lower ⅓ being longer than the upper; it is slightly connected with the ventral.   A vertical fold of the skin at the base of the pectoral; the coracoid is so high as to reach to the upper margin of the pectoral.   The adhesive apparatus as broad as long, its length being contained 3½ times in the total.   The vent and the porus urogenitalis are close together, situated midway between the margin of the ventral disk and the anal.   The anal papilla is small.   The color is brown (in spirits), whitish inferiorly. Length of adult, 7 inches.   (Günther.)   Caribbean Sea, said to be common; not seen by us.   The original *G. cephalus* seems nearer the next species, if the 2 are really different.   If that be the case the present species may stand as *Gobiesox stannii*.   But we have no material adequate to settle this question.   (*cephalus*, big-headed; κεφαλή, head.)

Gobiesox cephalus, Lacépède, Hist. Nat. Poiss., ii, 595, 1798, Martinique; on a drawing by Plumier; D. 8; A. 4 or 5; color plain reddish; anal inserted behind dorsal; head broad; eyes blue; Günther, Cat., iii, 499, 1861.
Lepadogaster testar, Bloch & Schneider, Syst. Ichth., 445, 1801, Martinique; after Plumier.
Cotylis stannii, Müller & Troschel, Hor. Ichthyol., iii, 18, taf. 3, fig. 3, 1845.

2665. GOBIESOX TUDES, Richardson.

Head $2\frac{1}{2}$; depth $4\frac{2}{3}$; width of head $2\frac{1}{4}$. D. 8; A. 6 in plate (5 in the description, the first short ray apparently not counted by Richardson). Head very broad, as broad as long, abruptly truncated anteriorly; mouth large, the maxillary reaching front of eye; lower jaw included; teeth entire; eye large, $4\frac{1}{4}$ in head, a little more than $\frac{1}{2}$ interorbital width, $1\frac{1}{4}$ in snout. Distance from front of dorsal to caudal about equal to length of head; insertion of dorsal before vent; the anal behind dorsal and much shorter than it; pectorals short. Color uniform, probably greenish, without spots or stripes. Length 5 inches. Locality "unknown, but supposed to be from China." (Richardson.) The species is, however, certainly not Chinese and is more likely to be from the West Indies. This species differs from *Gobiesox cephalus*, as described by Günther, in the larger eye and shorter dorsal. It is probably the same species. (*tudes*, hammer.)

Gobiesox tudes, RICHARDSON, Voy. Sulphur, Fish., 103, pl. 46, figs. 1–3, 1845, habitat unknown, erroneously supposed to be China.

2666. GOBIESOX STRUMOSUS, Cope.

D. 11; A. 10; C. 16; P. 21. Head extremely wide, its width $2\frac{3}{8}$ in total length; this width partly produced by a large fleshy mass extending from end of maxillary to end of interopercle; eye small; profile of head descending abruptly from posterior line of orbits. Superior dental series 12 on each side, externally, but the 3 median teeth conceal some series of which the second 3 external teeth are a continuation; inferior teeth 11 on each side; 4 median incisors horizontal and subequal; no marked canine. Bluish plumbeous, fins blackish. (Cope.) Hilton Head, South Carolina, and Indian River, Florida; 4 specimens recently taken at Titusville by Evermann & Bean; apparently distinguished from *G. virgatulus* by its longer anal. (*strumosus*, from struma, a scrofulous tumor, alluding to the swollen cheek.)

Gobiesox strumosus, COPE, Proc. Ac. Nat. Sci. Phila. 1870, 121, Hilton Head, South Carolina; JORDAN & GILBERT, Synopsis, 749, 1883; EVERMANN & BEAN, Fishes of Indian River, Florida, in Rept. U. S. Fish Comm. 1896, 248.

2667. GOBIESOX VIRGATULUS, Jordan & Gilbert.

Head $2\frac{1}{4}$ ($3\frac{2}{3}$ with caudal); width of head $3\frac{1}{4}$; depth 6 (7 in total). D. 10; A. 8 or 9; vertebræ $10+16=26$. Body rather slender, the head low and rather broad, broadly rounded anteriorly; eyes very small, about 4 to 6 in head, about $2\frac{1}{4}$ in interorbital width; interorbital space broad, slightly convex. Cheeks prominent; opercle ending in a sharp spine. Cleft of mouth extending to below front of orbit; lower jaw somewhat shorter than upper. Teeth of upper jaw in a narrow band of about 2 series; 4 teeth of outer series a little larger than the rest, somewhat canine-like; middle teeth of lower jaw incisor-like and partly horizontal, their edges entire or somewhat concave. Ventral disk considerably shorter than head. Distance from root of caudal to front of dorsal $2\frac{2}{3}$ in length. Pectoral short, about 2$\frac{3}{5}$ in head. Color in life olivaceous, with numerous paler spots and broad diffuse dark bars; the whole body covered with rather faint, wavy, longitudinal stripes or lines of a light orange-brown color, about as wide as the interspace, much as in some species of *Liparis*, these entirely disappearing in alcohol; skin everywhere with dark punctulations; caudal dusky, slightly barred with paler, its tip abruptly yellowish; dorsal and anal dusky, the darker parts corresponding to dark bars on the body, barred. A rather large species. Length 2 to 4 inches. Common among ballast rocks, from Pensacola Bay north to Charleston. Our specimens from Pensacola and Charleston. (*virgatulus*, narrowly striped.)

Gobiesox virgatulus, JORDAN & GILBERT, Proc. U. S. Nat. Mus. 1882, 293, Pensacola, Florida (Coll. Jordan & Stearns); JORDAN & GILBERT, Synopsis, 958, 1883; GOODE & BEAN, Proc. U. S. Nat. Mus. 1882, 236; JORDAN, Proc. U. S. Nat. Mus. 1884, 149.

2668. GOBIESOX ADUSTUS, Jordan & Gilbert.

Head 3; depth 5$\frac{1}{2}$. D. 9; A. 7. Head and body broad and flat, much depressed; width of head nearly equal to its length, 3$\frac{3}{8}$ in body. Incisors in middle of lower jaw entire, broad; those in upper jaw narrow, blunt, little compressed, entire, shorter than the lateral teeth; behind these 2 or 3 series of smaller teeth. Eyes rather large, separated by a broad interorbital space, which is $\frac{1}{3}$ length of head and about $\frac{1}{2}$ greater than diameter of eye. Opercular spine sharp. Pectoral short, about $\frac{1}{2}$ length of head; ventral disk as long as head; distance from base of caudal to front of dorsal equaling $\frac{3}{10}$ of the length; caudal rounded behind. Brown, banded with blackish on body, head marbled with darker brown; front of dorsal black, the fins dusky with darker points. Pacific Coast of Mexico. Three specimens, the largest about 2 inches long, were obtained in a tide pool at Mazatlan. (*adustus*, scorched; brown.)

Gobiesox adustus, JORDAN & GILBERT, Proc. U. S. Nat. Mus. 1881, 360, Mazatlan, Mexico; JORDAN & GILBERT, Proc. U. S. Nat. Mus. 1882, 627; JORDAN & GILBERT, Bull. U. S. Fish Comm. 1882, 108.

2669. GOBIESOX FUNEBRIS, Gilbert.

Body rather slender, its width 5 in length; width of head 3$\frac{1}{2}$ to nearly 4; head 2$\frac{3}{4}$ to 3 in length; depth $\frac{1}{2}$ head. D. 9; A. 6 or 7. Teeth in upper jaw conical, in several series, unequal but without canines; in lower jaw mesially in 2 series, the outer of narrow, entire incisors, truncate or rounded, without lateral canines. Interorbital space wide, 3 in head, the eye small, $\frac{2}{5}$ interorbital width. Mouth very wide, $\frac{1}{2}$ or more than $\frac{1}{2}$ length of head. Ventral disk wider than long, its length 1$\frac{1}{2}$ to 1$\frac{3}{8}$ in head. No evident opercular spine. Distance from front of dorsal to base of caudal 2$\frac{3}{8}$ to 3 in length anterior to dorsal; distance from vent to front of anal fin 1$\frac{1}{2}$ to 1$\frac{3}{8}$ in distance from vent to disk; base of dorsal from 1$\frac{1}{3}$ to 1$\frac{1}{4}$ in head; base of anal about $\frac{1}{2}$ head; caudal rounded, 1$\frac{2}{5}$ to 1$\frac{3}{4}$ in head; pectorals 3$\frac{1}{4}$ in head. Color varying from dark olive brown to black, everywhere covered with fine, yellowish vermiculations, usually arranged to form narrow lighter bars on the sides; 3 or 4 obscure dark streaks radiating from the eye; blackish below, the fins varying from blackish to straw color. Length 2¼ inches. (Gilbert.) Gulf of California; abundant at Puerto Refugio (Angel Island) and La Paz. (*funebris*, funereal, from the dark color.)

Gobiesox funebris, GILBERT, Proc. U. S. Nat. Mus. 1890, 95, Puerto Refugio, Gulf of California. (Coll. *Albatross*.)

2670. GOBIESOX PŒCILOPHTHALMUS, Jenyns.

Head 3, as wide as long. D. 7; A. 7. Opercular spine long and slender; teeth strong, somewhat crowded in front, the anterior in both jaws incisor-like; upper teeth conical, with smaller ones behind; 6 middle teeth of lower jaw incisor-like, projecting forward, their form not described (probably entire). Eyes large, close together, less than a diameter apart. General color olivaceous or brownish white, unmarked; iris golden, with pink and blue. Length 1⅗ inches. (Jenyns.) Chatham Island, Galapagos; only the single type known. (ποικίλος, variegated; ὀφθαλμός, eye.)

Gobiesox pœcilophthalmus, JENYNS, Voy. Beagle, Fishes, 141, pl. 27, figs. 2, 2a, 2b, 1842, Chatham Island (Coll. Darwin); GÜNTHER, Cat., III, 503, 1861.

2671. GOBIESOX RHODOSPILUS, Günther.

D. 6; A. 5; C. 8 or 9; P. 16. A vertical fold of skin along lower half of base of pectoral. Distance from front of dorsal to caudal 2⅔ in its distance from snout; anal before third dorsal ray. A very narrow band of short conical teeth in upper jaw, 1 lateral tooth larger than the others, recurved, canine-like; lower jaw with 1 series of teeth, the anterior narrow incisors, the outer distinctly canine, like the outer above. Rose-colored with rose-red transverse spots, each with an edge of deep-red dots. Panama. (Günther.) Not seen by us; known from 2 specimens, each 1¼ inches long. (ῥόδον, rose; σπίλος, spot.)

Gobiesox rhodospilus, GÜNTHER, Proc. Zool. Soc. Lond. 1864, 25, Panama (Coll. Captain Dow); GÜNTHER, Fish. Centr. Amer., 445, 1869.

2672. GOBIESOX MACROPHTHALMUS, Günther.

Eye 4½ in head. D. 8; A. 6; C. 12; P. 22. Head and anterior part of body very broad and much depressed, the head as broad as long, its profile semi-elliptical, the snout obtuse and rounded; top of head quite flat; interorbital width equal to eye. Mouth horizontal, curved, moderate, the cleft reaching beyond anterior margin of eye; an acute spine at lower angle of opercle; 4 short incisors on each side in lower jaw, separated from the conical lateral teeth by a larger canine-like tooth. Insertion of dorsal nearer caudal than snout; caudal rounded. Color, uniform reddish. (Günther.) Locality unknown; probably West Indies. (μακρός, large; ὀφθαλμός, eye.)

Dr. Eigenmann gives the following notes on a small specimen from St.

Thomas, which seems referable to *Gobiesox macrophthalmus*, differing in the slightly shorter dorsal and larger eye:

"Dorsal 6 or 7; anal about 7; head about 3; width of body 2¼; width of mouth 2 in head; interorbital width 4½, equal to snout. Eye large, 3½ in head. Teeth in the upper jaw conic, in more than 1 series in front, some in the outer row enlarged; teeth on the lower jaw in single series, about 4 blunt incisors on each side followed by the canines behind which the teeth are much smaller and conic. Width of disk 1½ in its length, 2½ in head; opercular spine strong; distance of origin of dorsal from caudal 2½ in its distance from tip of snout. Sides and back uniform bright red; eye black, iris bright red; lower surface yellow, dotted with bright red. One specimen 23 mm. long in the Museum of the University of Indiana from St. Thomas (Coll. Edward W. Brigham), much shrunken and fins hardened by strong alcohol." (Eigenmann, in lit.)

Gobiesox macrophthalmus, GÜNTHER, Cat., III. 502, 1861, locality unknown.

2573. GOBIESOX CERASINUS, Cope.

Head 3 in total with caudal. D. 6; A. 6; C. 12; P. 24; V. 4. Head very wide, ovate, as broad as long to upper base of pectoral. Eye large, 3½ in head, equal to frontal width. Ten teeth on each side of each jaw, none of the upper being incisors, the 2 median on each side larger than the others; 3 teeth on each side in lower jaw horizontal, the others vertical, 2 of the horizontal teeth incisors, the median one on each side of these much the larger; each horizontal tooth with a small one behind it; no canines. Profile regularly descending from supraoccipital; a long opercular spine. Dorsal beginning with last fourth of distance between tip of snout and base of caudal. Body and fins light crimson lake above, whitish below; no spots. One specimen, 2½ inches long, from St. Martin. (Cope.) (κεράσινος, *cerasinus*, cherry color.)

? *Cyclopterus nudus*, LINNÆUS, Syst. Nat., Ed. x, 260, 1758, "India;" from a specimen in Mus. Adolph Fred. (tab. 27), said to be 2 inches in length; the head broad with a sharp spine behind; dorsal rays 6; not *Lepadogaster nudus*, BLOCH & SCHNEIDER, Syst. Ichth., 2, 1801, who give "D. 12, A. 6." * the description otherwise that of Linnæus; not *Gobiesox nudus* of recent authors, which is a species (G. *gyrinus*, allied to G. *cingulatus*.

Gobiesox cerasinus, COPE, Trans. Am. Phil. Soc., XIV. 1871, 473, St. Martins West Indies. (Coll. Dr. R. E. Van Rijgersma. Type in Ac. Sci. Phila.)

Subgenus SICYASES,[†] Müller & Troschel.

2574. GOBIESOX ERYTHROPS, Jordan & Gilbert.

Head 2½; depth 6. D. 6; A. 5. Head scarcely longer than broad, proportionately very broad and depressed, its breadth 3 times in total.

"We do not know by what authority the number of fin rays given by Linnæus (D. 6) was altered to "D. 12, A. 6" by Schneider (Syst. Ichth.). The last-named rays agree with *nudus*, as described by Dr. Günther, that is, with our G. *gyrinus*. If the Linnæan type of *nudus* really had D. 6, it must have been *cerasinus* or *macrophthalmus* or some very similar species. The scanty Linnæan description agrees best with *cerasinus*. The name *nudus*, if used at all, must be taken for a species to which the Linnæan description may be applied. In our judgment the uncertainty is too great to justify the substitution of *nudus* for either *cerasinus* or *macrophthalmus*. It could be no other known species, however.
† This subgenus is composed of small species with the upper teeth in one series. This character should be verified on all our species, as perhaps none of them here

incisors in both jaws, entire and rather broad, the lateral teeth, as usual, pointed; no canines. Eyes very large, considerably wider than the narrow interorbital area, 3½ in head; interorbital area nearly 5 in head. Ventral disk a little longer than head, 2½ in body. Pectoral about ⅓ length of head. Distance from front of dorsal to caudal, 3½ in body. Caudal truncate with rounded edges. Light olivaceous; body with 3 or 4 bars of cherry red; head marbled with red; eyes intensely cherry red, their upper borders blackish; fins pale, the upper mottled with reddish; caudal barred with red. Two specimens, 1½ inches long, taken in a rock pool at Mazatlan; also recorded from the Tres Marias Islands. (ἐρυθρός, red; ὤψ, eye.)

Gobiesox erythrops JORDAN & GILBERT, Proc. U. S. Nat. Mus. 1881, 331, Mazatlan, Mexico (Type. No. 28136 Coll. Gilbert); JORDAN & GILBERT, Bull. U. S. Fish Comm. 1882, 110; JORDAN, Fishes Sinaloa, in Proc. Cal. Ac. Sci. 1895, 492.

2873. GOBIESOX RUBIGINOSUS (Poey).

D. 6 or 7; A. 6 or 4; P. 25. Head 3, including caudal, its greatest width twice its height; eye 2 in interorbital width. Body slender; head semioval, obtuse; distance from front of dorsal to caudal 1½ in length, including caudal; dorsal opposite anal, beginning at fourth seventh of total length; mouth terminal; teeth not examined by Poey. Color red, with 12 dark bands all with many scattered spots of clear blue on the body; eyes with a red iris. Length 22 mm. Cuba (Poey); not seen by us; locally common at Matanzas; perhaps a species of *Arbacioss*. (rubiginosus, ___)

Sicyeses rubiginosus POEY, Synopsis, 391, 1868, wharves of Palmascla, Matanzas.

2874. GOBIESOX CARNEUS (Poey).

Head rounded; body very slender; eyes large, as wide as interorbital space; mouth inferior; teeth not examined by Poey. Color pale red, with some white spots and bands. Length 22 mm. Otherwise essentially as in *Gobiesox rubiginosus*. (Poey.) Matanzas; a doubtful species. (carneus, flesh-colored.)

Sicyeses carneus POEY, Synopsis, 391, 1868, wharves of Palmascla, Matanzas, Cuba (Coll. ___

2875. GOBIESOX RARIES, Jordan & Bollman.

Head 2? (3½ in total); depth 6 (7½). D. 9; A. 6. Body rather slender; head low and broad, greatest breadth not quite equal to length, its anterior margin not so badly rounded as in *G. virgatulus*. Eyes very small, 1½ in interorbital space, 5 in head; interorbital bone appearing convex, least width 3½ in head and about equal to length of snout; cleft of mouth extending to beyond middle of eye; lower jaw included; teeth uniserial, those of upper jaw all canines, the first 3 on each side small, but becoming larger outward, next 3 or 4 much larger, rest smaller than those in front; anterior teeth of lower jaw entire incisors, which have on each side

Thomas, which seems referable to *Gobiesox macrophthalmus*, differing in the slightly shorter dorsal and larger eye:

"Dorsal 6 or 7; anal about 7; head about 3½; width of body 3½; width of mouth 2 in head; interorbital width 4½, equal to snout. Eye large, 2⅘ in head. Teeth in the upper jaw conic, in more than 1 series in front, some in the outer row enlarged; teeth on the lower jaw in a single series, about 4 blunt incisors on each side followed by the canines, behind which the teeth are much smaller and conic. Width of disk 1⅘ in its length, 2½ in head; opercular spine strong; distance of origin of dorsal from caudal 2⅕ in its distance from tip of snout. Sides and back uniform bright red; eye black, iris bright red; lower surface yellow, dotted with bright red. One specimen 23 mm. long in the Museum of the University of Indiana from St. Thomas (Coll. Edward W. Brigham), much shrunken and fins hardened by strong alcohol." (Eigenmann, in lit.)

Gobiesox macrophthalmus, GÜNTHER, Cat., III, 502, 1861, locality unknown.

2673. GOBIESOX CERASINUS, Cope.

Head 3 in total with caudal. D. 6; A. 6; C. 12; P. 24; V. 4. Head very wide, ovate, as broad as long to upper base of pectoral. Eye large, 3½ in head, equal to frontal width. Ten teeth on each side of each jaw, none of the upper being incisors, the 2 median on each side larger than the others; 3 teeth on each side in lower jaw horizontal, the others vertical, 2 of the horizontal teeth incisors, the median one on each side of these much the larger; each horizontal tooth with a small one behind it; no canines. Profile regularly descending from supraoccipital; a long subopercular spine. Dorsal beginning with last fourth of distance between tip of snout and base of caudal. Body and fins light crimson lake above, whitish below; no spots. One specimen, 2¼ inches long, from St. Martins. (Cope.)

($\varkappa\varepsilon\rho\acute{a}\delta\iota\nu o\varsigma$, *cerasinus*, cherry color.)

? *Cyclopterus nudus*, LINNÆUS, Syst. Nat., Ed. x, 260, 1758, "India;" from a specimen in Mus. Adolph Fred. (tab. 27), said to be 2 inches in length; the head broad with a sharp spine behind; dorsal rays 6; not *Lepadogaster nudus*, BLOCH & SCHNEIDER, Syst. Ichth., 2, 1801, who give "D. 12, A. 6,"[*] the description otherwise that of Linnæus: not *Gobiesox nudus* of recent authors, which is a species (*G. gyrinus*) allied to *G. virgatulus*.

Gobiesox cerasinus, COPE, Trans. Am. Phil. Soc., XIV, 1871, 473, St. Martins, West Indies. (Coll. Dr. R. E. Van Rijgersma. Type in Ac. Sci. Phila.)

Subgenus SICYASES,[†] Müller & Troschel.

2674. GOBIESOX ERYTHROPS, Jordan & Gilbert.

Head 2½; depth 6. D. 6; A. 5. Head scarcely longer than broad, proportionately very broad and depressed, its breadth 3 times in total.

---

[*] We do not know by what authority the number of fin rays given by Linnæus (D. 6) was altered to "D. 12, A. 6" by Schneider (Syst. Ichth.). The last-named figures agree with *nudus*, as described by Dr. Günther, that is, with our *G. gyrinus*. If the Linnæan type of *nudus* really had D. 6, it must have been *cerasinus* or *macrophthalmus* or some very similar species. The scanty Linnæan description agrees best with *cerasinus*. The name *nudus*, if used at all, must be taken for a species to which the Linnæan description may be applied. In our judgment the uncertainty is too great to justify the substitution of *nudus* for either *cerasinus* or *macrophthalmus*. It could be no other known species, however.

[†] This subgenus is composed of small species with the upper teeth in 1 series. This character should be verified on all our species, as perhaps none of them belongs to it.

Incisors in both jaws, entire and rather broad, the lateral teeth, as usual, pointed; no canines. Eyes very large, considerably wider than the narrow interorbital area, 3¼ in head; interorbital area nearly 5 in head. Ventral disk a little longer than head, 2⅔ in body. Pectoral about ⅓ length of head. Distance from front of dorsal to caudal, 3⅔ in body. Caudal truncate, with rounded edges. Light olivaceous; body with 3 or 4 bars of cherry red; head marbled with red; eyes intensely cherry red, their upper border blackish; fins pale, the upper mottled with reddish; caudal barred with red. Two specimens, 1¼ inches long, taken in a rock pool at Mazatlan; also recorded from the Tres Marias Islands. (ἐρυθρός, red; ὤψ, eye.)

Gobiesox erythrops, JORDAN & GILBERT, Proc. U. S. Nat. Mus. 1881, 320, Mazatlan, Mexico (Type, No. 29248. Coll. Gilbert); JORDAN & GILBERT, Bull. U. S. Fish Comm. 1882, 108; JORDAN, Fishes of Sinaloa, in Proc. Cal. Ac. Sci. 1895, 499.

2675. GOBIESOX RUBIGINOSUS (Poey).

D. 6 or 7; A. 6; V. 4; P. 25. Head 3, including caudal, its greatest width twice its height; eye 2 in interorbital width. Body slender; head semi-oval, obtuse; distance from front of dorsal to caudal 1⅓ in length, including caudal; dorsal opposite anal, beginning at fourth seventh of total length; mouth terminal; teeth not examined by Poey. Color red, with 12 dark bands and with many scattered spots of clear blue on the body; eyes with a red circle. Length 22 mm. Cuba (Poey); not seen by us; locally common at Matanzas; perhaps a species of *Arbaciosa.* (*rubiginosus,* reddish.)

Sicyases rubiginosus, POEY, Synopsis, 391, 1868, wharves of Palmasola, Matanzas, Cuba (Coll. Poey); POEY, Enumeratio, 124, 1875.

2676. GOBIESOX CARNEUS (Poey).

Head rounded; body very slender; eyes large, as wide as interorbital space; mouth inferior; teeth not examined by Poey. Color pale red, with some white specks and bands. Length 22 mm. Otherwise essentially as in *Gobiesox rubiginosus.* (Poey.) Matanzas; a doubtful species. (*carneus,* flesh-colored.)

Sicyases carneus, POEY, Synopsis, 392, 1868, wharves of Palmasola, Matanzas, Cuba (Coll. Poey); POEY, Enumeratio, 124, 1875.

2677. GOBIESOX HÆRES, Jordan & Bollman.

Head 2⅔ (3¼ in total); depth 6 (7½). D. 9; A. 6. Body rather slender; head low and broad, greatest breadth not quite equal to length, its anterior margin not so broadly rounded as in *G. virgatulus.* Eyes very small, 1⅓ in interorbital space, 5 in head; interorbital bone appearing convex, least width 3¼ in head and about equal to length of snout; cleft of mouth extending to beyond middle of eye; lower jaw included; teeth uniserial, those of upper jaw all canines, the first 3 on each side small, but becoming larger outward, next 3 or 4 much larger, rest smaller than those in front; anterior teeth of lower jaw entire incisors, which have on each side about 6 large graduated canines and behind these a few smaller ones; teeth of lower jaw slightly oblique. Distance from front of dorsal to root of caudal about 2⅖ in body (3½ in total). Pectorals moderate, 2 in head; ventral disk 1¼ in head. Color olivaceous, without any distinct bands; the occipital region and the caudal peduncle darker; body irregularly mottled with groups of darker spots; nape, preopercle, cheeks, and snout with numerous dark points; indistinct dark lines radiating from eye; lips dark; fins dusky; dorsal and anal with the first rays black; a pale spot near base of caudal; axil of pectoral dusky. Green Turtle Cay, Bahamas; a single specimen known, 2¼ inches in length. (*hæres*, one who clings.)

Gobiesox hæres, JORDAN & BOLLMAN, Proc. U. S. Nat. Mus. 1888, 552, Green Turtle Cay, Bahamas. (Coll. Dr. Charles L. Edwards.)

2678. GOBIESOX PUNCTULATUS (Poey).

Head very broad, 3 in total length with caudal. D. 9; A. 6. Color brown, covered with black points; 3 dark transverse bands; none on the head. Teeth not described, the incisors probably entire. Length 38 mm. Cuba (Poey); not seen by us. (*punctulatus*, speckled.)

Sicyases punctulatus, POEY, Enumeratio, 124, 1875, Havana.

2679. GOBIESOX FASCIATUS (Peters).

D. 7; A. 7; head and body with alternate dark green and yellowish cross bands. Commencement of dorsal before that of anal, its distance from caudal equal to length of its base. Type, 50 mm. long. (Peters); not seen by us; teeth not described. Puerto Cabello. (*fasciatus*, banded.)

Sicyases fasciatus, PETERS, Monatsber. Berl. Acad. 1859, 412, Puerto Cabello; GÜNTHER, Cat., III, 497, 1861; GÜNTHER, Fishes Centr. Amer., 390, 1869.

866. RIMICOLA, Jordan & Evermann.

Rimicola, JORDAN & EVERMANN, Proc. Cal. Ac. Sci. 1896, 231 (*muscarum*).

This genus differs from *Gobiesox* mainly in the very slender body and head. Head 3¼ to 4 in length, its width less than its length; dorsal and anal very short, of 4 to 6 rays each; incisors entire; a crescent-shaped patch of teeth in each jaw behind the large teeth; opercular spine weak or obsolete; sucking disk small. Species of small size; living below tide marks. (*rima*, a crevice; *colo*, I inhabit.)

> *a.* Dorsal rays 6; anal 5; color yellowish, with a brown lateral band and numerous brownish spots.   MUSCARUM, 2680.
> *aa.* Dorsal rays 4; anal 5; color uniform light green.   EIGENMANNI, 2681.

2680. RIMICOLA MUSCARUM (Meek & Pierson).

Head 3⅓ in length; depth 8¾; D. 6; A. 5. Body elongate, slender, depressed anteriorly, but very narrow, slightly compressed posteriorly, the greatest width of body immediately behind head, 7 in length. Head narrow, much depressed,—wider posteriorly. Eye small, its diameter 2½ in interorbital width, 5 in head. Maxillary reaching to the front of the eye, its length less than 3 in head. Teeth in upper jaw conical, acute, curved, forming a crescent-shaped patch, those of the anterior row enlarged; in the lower jaw an anterior row of about 5 broad, entire incisors placed nearly horizontally; behind these a crescent-shaped patch of teeth, similar to those in the upper jaw, becoming canine-like laterally. No evident opercular spine. Ventral disk longer than broad, its length $1\frac{1}{4}$ in head $6\frac{1}{4}$ in length; distance from vent to front of anal $2\frac{1}{4}$ in the distance from vent to disk; pectoral fin broad, short, $2\frac{1}{4}$ in head; dorsal and anal fins small, the anal slightly in advance; caudal fin rounded. Ground color, in alcohol, light yellowish, paler below; above everywhere sparsely covered with distinct brownish-red spots about as large as pupil; a lateral band of the same color begins on the front of the snout, where it joins the one on the opposite side, extends through the eye across the opercle to the caudal, becoming very indistinct posteriorly; this lateral stripe is in strong contrast with the uniform pale ventral surface. Coast of California. Two specimens were dredged in Monterey Bay at a depth of about 10 fathoms. One of these, the type, is $1\frac{1}{4}$ inches long. The second specimen ($1\frac{1}{10}$ inches long) has the dorsal spots confined to the top of the head and nuchal region and the lateral stripe disappearing slightly behind middle of body, and having the ventral surface marked posteriorly with brownish-red spots like the spots on the dorsal surface. (*muscarum*, of the flies, from the fly-speck markings.)

Gobiesox muscarum, .MEEK & PIERSON, Proc. Cal. Ac. Sci. 1895, 571, with colored plate, Monterey Bay. (Coll. S. E. Meek and Charles J. Pierson. Type in L. S. Jr. Univ. Mus.)

2681. RIMICOLA EIGENMANNI (Gilbert).

D. 4; A. 5. Head $3\frac{2}{3}$ in length; depth about $\frac{1}{2}$ head. Body very slender and narrow, the width of head $4\frac{1}{4}$ in length; width of body 6. Mouth wide, the distance between its angles $\frac{1}{2}$ length of head, the maxillary scarcely reaching vertical from front of orbit. Interorbital space wide, about $\frac{1}{4}$ head. Eye very small, about 3 in interorbital width. Teeth in upper jaw conic, acute, in several series, the anterior in upper jaw enlarged; teeth in lower jaw also in several series, those of front row narrow incisors, entire, with rounded or truncate edges; disk very small and narrow, its width about $\frac{2}{3}$ its length, the latter $1\frac{2}{3}$ in length of head. Fins all small, the base of dorsal $\frac{2}{3}$ length of head, less than free portion of caudal peduncle; distance from origin of dorsal to base of caudal $3\frac{1}{4}$ in length before dorsal; distance from vent to front of anal fin $1\frac{2}{3}$ in its distance from disk; caudal broadly rounded, its length $1\frac{1}{4}$ in head; pectoral somewhat pointed, about $\frac{1}{2}$ head; coracoid plate small, about $\frac{1}{4}$ height of pectoral and less than $\frac{1}{4}$ its length. Color uniform light olive green, without distinctive markings. Type, a single specimen, about 1 inch long, taken at Point Loma, near San Diego, California. Other specimens were taken some years since at San Cristobal Bay by Mr. Charles H. Townsend, and were referred to as *Gobiesox rhessodon* by Mrs. Eigenmann, Proc. U. S. Nat. Mus. 1884, page 553. (Named for Dr. Carl H. Eigenmann.)

Gobiesox eigenmanni, GILBERT, Proc. U. S. Nat. Mus. 1890, 96, Point Loma, near San Diego, California. (Coll. Gilbert.)

867. ARBACIOSA, Jordan & Evermann.

Arbaciosa, JORDAN & EVERMANN, Proc. Cal. Ac. Sci. 1896, 290 (*humeralis.*)

This genus differs from *Gobiesox* chiefly in the character of the incisor teeth of the lower jaw; these are strongly serrate, or tricuspid, making a ragged cutting edge. Size small; dorsal and anal comparatively short; head not very broad, the jaws contracted; vertebræ (in *Arbaciosa zebra*) 28. Some species provisionally referred to the section *Sicyases* of *Gobiesox* may prove to belong to *Arbaciosa;* small species, living in rock pools, among the sea urchins, by whose spines they are protected. This relation of *Arbaciosa zebra* with the Echinoid *Arbacia stellata* is especially constant. (*Arbacia*, a sea urchin.)

   a. Anal fin long, about 10 rays; dorsal rays 11; teeth above in 1 series; color olivaceous.
                                                RHESSODON, 2682.
  aa. Anal fin of 5 to 7 rays.
     b. Dorsal fin of 8 or 9 rays; teeth in single series; color brownish, with red bars
       and a large black humeral spot.   HUMERALIS, 2683.
    bb. Dorsal fin of 7 rays; color greenish, with pale spots and numerous pale cross
       bands; no red; body slender.   RUPESTRIS, 2684.
    bbb. Dorsal fin of 6 rays; color chiefly red.
     c. General color pinkish olivaceous, with some bright red; back with 5
       reddish-brown or blackish bars. Upper teeth in more than 1 series.
       Body comparatively slender, the depth nearly 8 in length.
                                         ZEBRA, 2685.
    cc. General color bright rosy red, black, with 1 to 3 faint dark bars. Upper
       teeth nearly uniserial. Body comparatively stout, the depth 5½ in
       length.   EOS, 2686.

2682. ARBACIOSA RHESSODON (Rosa Smith).

Head 3¼; depth 6¼. D. 11; A. 10; eye 4½ in head, ⅔ in interorbital space; ventral disk 1⅓ in head; pectoral 2; caudal 2⅕. Form much as in *Gobiesox mæandricus;* snout bluntly and evenly decurved; the greatest height of the body across the pectoral fins; head broader than body but less deep; maxillary extending to below the eye; incisors of lower jaw not much declined, each of them tricuspid, the central cusp longest; teeth of upper jaw conical, in an irregular series of 7 to 9; teeth in each jaw in single series; opercular spine sharp; distance from vent to caudal 2⅘ in length of body; dorsal a little longer than the anal, having its origin in advance of the anal and terminating opposite it; caudal rounded. Color dark olivaceous, usually with 3 broad yellowish cross bands above, the first across interorbital space and cheek, the second very wide, across back and front of dorsal fin, the third below middle of dorsal, some or all of these sometimes wanting; a dark bar at base of caudal; belly yellowish. Length 2¼ inches. San Diego to the northern part of the Gulf of California; locally abundant in rock pools. (ῥήσσω, to make ragged; ὀδούς, tooth.)

Gobiesox rhessodon, ROSA SMITH, Proc. U. S. Nat. Mus. 1881, 140, San Diego, California; JORDAN & GILBERT, Synopsis, 749, 1883; ROSA SMITH, Proc. U. S. Nat. Mus. 1883, 235.

2683. ARBACIOSA HUMERALIS (Gilbert).

Head 3 to $3\frac{1}{2}$; width of body $4\frac{1}{4}$; width of head $3\frac{3}{4}$; eye very small, 3 in interorbital width. D. 8 or 9; A. 7. Body of moderate width, the head not evenly rounded anteriorly, becoming contracted opposite eyes, the snout forming a quadrate projection beyond the profile, as seen from above. Teeth in a single series in each jaw, the anterior narrow incisors, trilobate at tip, the 2 posterior teeth on each side strong, conical canines, somewhat recurved; about 12 incisors in the upper jaw. Interorbital space very wide, about equaling width of mouth, $2\frac{1}{5}$ to $2\frac{1}{4}$ in length of head. Ventral disk about as wide as long, its length $1\frac{1}{4}$ in head. Opercular spine large and strong, but not exposed. Distance from front of dorsal to base of caudal 3 in length anterior to dorsal; base of dorsal $1\frac{2}{3}$ in head; base of anal about equals base of dorsal; distance from vent to front of anal half its distance from disk; caudal broadly rounded, $1\frac{3}{4}$ in head; pectoral $\frac{1}{3}$ head, without distinct fold of skin across it. Ground-color dark olive brownish, crossed by many carmine-red bars, these somewhat broken anteriorly and above, to form reticulating lines, posteriorly and on lower part of sides more regular and running obliquely downward and backward; a conspicuous round humeral spot, larger than eye, in life black with golden-green reflections; numerous streaks from eye backward across cheek and opercles. (Gilbert.) Gulf of California; abundant at Puerto Refugio (Angel Island); also known from La Paz. (*humeralis*, pertaining to the shoulder, *humerus*.)

Gobiesox humeralis, GILBERT, Proc. U. S. Nat. Mus. 1890, 95, Puerto Refugio, Gulf of California. (Coll. *Albatross*.)
Arbaciosa humeralis, JORDAN, Proc. Cal. Ac. Sci. 1896, 230, pl. 35.

2684. ARBACIOSA RUPESTRIS (Poey). •

Head 4 in total length with caudal; depth 6; eye 4 in head; snout less than eye. D. 7; A. 7. Forehead little decurved; eyes well separated; mouth small, with 1 row of compressed, close-set incisors with denticulated edges, 6 on each side in each jaw; snout truncate, as seen from above. Pectorals short, rounded. Dorsal and anal alike, opposed, highest in front. Caudal rounded. Color greenish ash, each side with 6 large oval spots, those behind touching; sides with about a dozen vertical bands of straw yellow or whitish, these bands sometimes interrupted, forming 2 series of points; 2 small similar bands from the eye, another toward tip of snout; a brown pale-edged band between eyes; some white spots on sides of head. Length $1\frac{1}{2}$ inches. Coral reefs of Cuba; not rare. (Poey.) not seen by us; said to be distinguished from other Cuban species by the slender body and narrow head. (*rupestris*, living among rocks.)

Gobiesox rupestris, POEY, Memorias, II, 283, 1861, Cuba.
Sicyases rupestris, POEY, Synopsis, 391, 1868; POEY, Enumeratio, 124, 1875.

2685. ARBACIOSA ZEBRA (Jordan & Gilbert).

Head $3\frac{3}{4}$ in length; depth nearly 8. D. 6 or 7; A. 5 or 6; vertebræ $11 + 17 = 28$. Body comparatively very long and narrow, the greatest width about $\frac{1}{5}$ the total length. Head narrow, depressed, its width about $4\frac{1}{2}$

tiues in length of body; eye small, its diameter about $\frac{1}{4}$ interorbital width; opercular spine well developed; ventral disk nearly as long as head; mouth rather small, anterior, maxillary reaching front of eye; incisors of lower jaw nearly horizontal, rather broad, 3-lobed at tip, the middle cusp the longest; upper teeth much smaller, the median ones compressed, blunt, close set, a little shorter than the lateral teeth and with dentate edges, 1 or 2 series of small teeth close behind them; anal beginning under middle of dorsal; the distance from insertion of dorsal to base of caudal contained $3\frac{2}{3}$ in length; pectoral $\frac{1}{2}$ as long as head; caudal truncate, with rounded angles. Back with 5 dark cross bars about as wide as the interspaces, 3 of them in front of dorsal fin, the 2 anterior much broader and more distinct than the others; these bars all distinct on back, fading on sides, which are often vaguely clouded with dark; the color of these dark bars varies from reddish brown to black, and that of the interspaces from olivaceous to light pink and bright rose red; top of head bright red, marbled with light slaty bluish; a black blotch on opercle, and 2 very distinct black cross spots, 1 on each side of median line, forming the front of first dorsal bar; cheek sometimes with 2 or 3 pale bluish streaks; dorsal, pectoral, and caudal more or less shaded with dusky; lower fins pale; usually a dark bar at base of caudal and 1 across middle of fin; shade of ground color extremely variable.* Very abundant in the rocky tide pools around Mazatlan, hiding everywhere under the numerous sea-urchins, especially *Arbacia stellata*, the protective coloration of both being that of the *Corallina*, which lines the rock pools. Length 2 to 3 inches. (*zebra*, from the banded coloration.)

Gobiesox zebra, JORDAN & GILBERT, Proc. U. S. Nat. Mus. 1881, 359, Mazatlan, Mexico (Type, No. 29250. Coll. Gilbert); JORDAN & GILBERT, Bull. U. S. Fish Comm. 1882, 108; JORDAN, Proc. Cal. Ac. Sci. 1895, 499.

---

* The following note on the variations is furnished by Miss Susan B. Bristol:
  " I find 4 specimens of this species which differ considerably from the typical form. These may represent a distinct species, but at present we are inclined to think that all these forms are modifications of one species, *Arbaciosa zebra*. The following is a description of a specimen $1\frac{1}{4}$ inches long, taken at Mazatlan (No. 4166 in the L. S. Jr. Univ. Museum): Head $3\frac{1}{2}$; depth 9. D.5 or 6; A.6. Body slender, much depressed, compressed posteriorly, the greatest width $4\frac{3}{4}$ in length. Head depressed, its width $1\frac{1}{4}$ in its length. Eye very small. about $1\frac{1}{2}$ in interorbital width. Snout rather rounded, $3\frac{1}{2}$ in head. Opercular spine present. Interorbital width $2\frac{3}{4}$ in head. Ventral disk $1\frac{1}{4}$ in head Mouth small, the lower jaw inferior; outer teeth in both jaws serrate. Anal beginning at end of the first $\frac{2}{3}$ of dorsal. Distance from front of dorsal to base of caudal $1\frac{1}{4}$ in head. Caudal terminate. Pectoral $2\frac{2}{3}$ in head. Color bright red, with very irregular yellow mottlings on back and sides, light yellow below; back with 4 irregular dark-red cross bars, the posterior 3 of which are wider than the interspaces; 3 of the cross bars in front of the dorsal fin, and the fourth on either side of the dorsal; 2 conspicuous black spots about $\frac{1}{4}$ as large as eye, 1 on either side of median line on back above the pectorals a short distance behind their origin; snout plain, dark red; pupil white; 2 yellow parallel stripes extending from eye backward and downward, the second ending at a point about $\frac{3}{4}$ the distance from tip of snout to end of opercle; dorsal, caudal, and anal dusky; ventrals and pectorals paler; a large red blotch at base of pectorals extending for a considerable distance on the fin. Another specimen from Mazatlan, bright red in color, about $\frac{3}{4}$ of an inch long (also in bottle No. 4166, L. S. Jr. Univ. Museum), differs from the preceding form in the following respects: In the greater depth, which is $6\frac{3}{4}$ in length, in the smaller ventral disk, which is $1\frac{1}{4}$ in head; in the more pointed snout; in the absence of the 2 black spots above pectorals; and in having the 4 dark red bands on the back more distinctly marked. Two specimens from Guaymas, Mexico, 1 and $1\frac{1}{4}$ inches long, No. 92 in the L. S. Jr. Univ. Museum, are chocolate brown in color, the shorter having on its back, including the bar at base of caudal, 7 dark brown cross bars and no dark spots above the pectoral, while in the longer there are no cross bars but a dark brown spot about $\frac{1}{4}$ as large as the eye is present above the pectoral; also, in the longer one, the dorsal begins at the end of first third of anal. The eye in the larger specimens of *zebra* is larger than in these 4 specimens. but some of the smaller specimens seem to be intermediate in this regard between the typical form and these forms."

2686. ARBACIOSA EOS (Jordan & Gilbert).

Head 3; depth 5½; eye moderate, 1⅓ in interorbital width, which is about 3⅓ in head. D. 6; A. 6. Body comparatively short, stout, and narrow; the head rather broad, but, like the body, much less depressed than in *G. erythrops;* width of head less than its length, or 3⅔ in body. Incisors serrate or tricuspid. Pectorals about 4 in head; ventral disk shorter than head. Distance from base of caudal to front of dorsal 3⅓ in total length; caudal truncate. Bright rosy red, sometimes dusky above with black points; back with 1 to 3 faint dark bars; 3 dark lines downward and backward from orbit, and usually 1 or 2 more on opercle; caudal usually with a reddish bar at base and a dusky one toward tip; fins otherwise nearly plain. Pacific coast of Mexico; abundant in rock pools about Mazatlan in company with *Arbaciosa zebra*, hiding under sea-urchins, especially with *Arbacia stellata*. Length 1¼ inches. ($\dot{\eta}\dot{\omega}\varsigma$, sunrise; from the red colors.)

Gobiesox eos, JORDAN & GILBERT, Proc. U. S. Nat. Mus. 1881, 360, Mazatlan, Mexico (Coll. Gilbert. Type, No. 29247); JORDAN & GILBERT, Bull. U. S. Fish Com. 1882, 108; JORDAN, Proc. Cal. Ac. Sci. 1895, 499.

Group BLENNIODEA.

(BLENNIOID FISHES.)

Body more or less elongate, naked or with scales, large or small; ventral fins small, more or less advanced in position, often wanting, the number of soft rays always less than 5; hypercoracoid perforate, the shoulder girdle normally formed; skull not armed with spines; suborbital not developed as a bony stay articulating with the preopercle; pseudobranchiæ present; dorsal fin long, its anterior half, and sometimes the whole fin, composed of spines; anal long; tail homocercal, the caudal usually rounded, rarely forked; vertebræ numerous, especially in the arctic species. A large group, with ill-defined boundaries, the more primitive forms showing affinities with the *Trachinoidea, Cirrhitidæ,* and other more typical fishes, the extremes very aberrant and passing directly into the *Ophidioidea,* and other forms lacking spines in the fins. We begin the series with the least modified of the type, the *Clininæ,* from ancestors of which group the others have doubtless descended.

    a. Caudal fin present, sometimes united to dorsal and anal; dorsal spines connected by membrane.
        b. Gill openings not reduced to horizontal slits below the pectoral fins.
           c. Teeth not developed as coarse molars.
               d. Mouth not vertical.   BLENNIIDÆ, cc.
               dd. Mouth nearly vertical; scales small or wanting; no lateral line; no ventral fins; dorsal composed entirely of slender spines; gill membranes attached to the isthmus; teeth strong.
                            CRYPTACANTHODIDÆ, cci.
           cc. Teeth developed as coarse molars on vomer, palatines, and sides of lower jaw; dorsal of flexible spines only; scales minute; gill membranes joined to the isthmus; no ventral fins; air bladder present; no lateral line.   ANARHICHADIDÆ, ccii.

bb. Gill openings reduced to separate, narrow, nearly horizontal slits below and in front of the pectoral fin; ventrals small; dorsal fin long and low, anteriorly of slender spines; vertical fins connected.   CERDALIDÆ, CCIII.
aa. Caudal fin none, the tail tapering to a point; no ventral fins; no lateral line; scales rudimentary; anterior part of dorsal of low free-hooked spines, the posterior part of many slender soft rays; teeth in jaws only, close set in 1 row.
PTILICHTHYIDÆ, CCIV.

Family CC. BLENNIIDÆ.

(THE BLENNIES.)

Body oblong or elongate, naked or covered with moderate or small scales which are ctenoid or cycloid; lateral line variously developed, often wanting, often duplicated; mouth large or small, the teeth various; gill membranes free from isthmus or more or less attached to it; pseudobranchiæ present; ventrals jugular or subthoracic, of 1 spine and 1 to 3 soft rays, often wanting; dorsal fin of spines anteriorly, with or without soft rays; anal fin long, similar to soft dorsal; caudal well developed. Vertebræ in moderate or large number, 30 to 80. Carnivorous fishes of moderate or small size, mostly living near the shore in the tropical and temperate or arctic seas; most of them are carnivorous, the *Clininæ*, so far as known, ovoviviparous, the rest mostly oviparous. Genera, about 80; species, about 400; chiefly of the rock pools and algæ; some species in the lakes of Italy. Dr. Gill divides the tropical Blennies into 3 families, *Clinidæ, Blenniidæ,* and *Chænopsidæ.* The first and second of these are fairly well defined. The third is now heterogeneous, and some of its members are intermediate between the other two. The arctic Blennies he again divides into *Xiphidiidæ, Cebedichthyidæ,* and *Stichæidæ,* but the first and last of these groups intergrade, the *Xiphidiinæ* are modified *Clininæ,* and there are other forms as well entitled to separate rank as *Cebedichthys.* It seems to us better to treat the group as a single family with many subfamilies. (*Blenniidæ,* Günther, Cat., III, 206–297.)

I. Tropical Blennies, with the vertebræ mostly in moderate number, usually fewer than 45; lateral line usually arched high above the pectoral, if present; dorsal fin with soft rays, at least 1 being present; anal spines little developed; ventrals well developed, usually I, 3.

a. Body scaly.
CLININÆ:
  b. Lateral line present, arched anteriorly over the pectoral, becoming posteriorly median in position, or else obsolete; species ovoviviparous.
    c. Scales ctenoid, very rough, 35 to 40 in lateral line; dorsal divided into 3 fins; no cirri above eye.   ENNEANECTES, 868.
    cc. Scales cycloid; dorsal fin not divided into 3 fins.
      d. Dorsal with 6 to 20 soft rays.
        e. Shoulder girdle with a small upturned hook on its inner edge.
          f. Scales along lateral line anteriorly not enlarged; snout sharp; first 5 spines of dorsal more or less modified.
            g. Caudal fin forked; air bladder present; scales minute; teeth in jaws in more than 1 series, on vomer and palatines; first 5 dorsal spines lengthened, and partly separated.   HETEROSTICHUS, 869.

    gg. Caudal fin truncate; air bladder wanting; scales minute; teeth in jaws in more than 1 series; teeth on vomer, none on palatines; first 5 dorsal spines lengthened and partly separated. GIBBONSIA, 870.

ee. Shoulder girdle without upturned hook on its inner edge above.

    h. Maxillary greatly developed, reaching much beyond eye; teeth on vomer and palatines; scales minute; soft dorsal long. NEOCLINUS, 871.

    hh. Maxillary normal, not greatly expanded.

        i. Anterior part of lateral line normally formed; usually a comb of filaments at the nape.

        j. Palatines without teeth; scales moderate or small, 38 to 110 in lateral line.

        k. Teeth in jaws in 1 row only; teeth usually on vomer, none on palatines; usually a comb of filaments at the nape.
        MALACOCTENUS, 872.

        kk. Teeth in jaws in more than 1 row, a band of villiform teeth behind the others; teeth on vomer, none on palatines.

        l. Body oblong, the depth $3\frac{1}{2}$ to $4\frac{1}{2}$ in length; a filament above the eye.
        LABRISOMUS, 873.

        ll. Body elongate, the depth about 6 in length; no filaments above the eye. MNIERPES, 874.

        jj. Palatines with teeth, those in jaws in more than 1 series; scales large, 30 to 37 in lateral line; no nuchal filaments.

        m. Head very broad, depressed; soft dorsal of about 20 rays. GOBIOCLINUS, 875.

        mm. Head moderate, not depressed; soft dorsal of about 8 rays. STARKSIA, 876.

    ii. Anterior part of lateral line running on a series of enlarged scales without visible pores; teeth in more than 1 series in jaws; teeth on vomer and front of palatines. CRYPTOTREMA, 877.

dd. Dorsal with 1 short soft ray only; scales large; teeth in jaws in more than 1 series; teeth on vomer, none on palatines.

    n. Dorsal fin more or less deeply notched behind the third spine.

        o. First 3 spines of dorsal very slender, close set, forming a separate ribbon-shaped fin, which is much higher than any of the spines in the second dorsal; anal spines rather high; body strongly compressed, the snout very sharp. EXERPES, 878.

        oo. First 3 dorsal spines stiff, wide set, not remote from rest of fin behind dorsal notch; anal spines short; body more elongate, the snout less acute.
        AUCHENOPTERUS, 879.

    nn. Dorsal fin continuous, not notched. PARACLINUS, 880.

EMMNIINÆ:

bb. Lateral line straight, close to the dorsal fin; scales small, cycloid; dorsal notched, its anterior half of slender spines; no cirri on head; ventrals thickish, inserted slightly before pectorals; teeth in bands, the outer enlarged.
        EMMNION, 881.

aa. Body scaleless; species oviparous, so far as known.

   p. Teeth comb-shaped, in a single row in each jaw, behind which are sometimes long canines; vomer and palatines usually toothless; lateral line usually single, with a strong arch anteriorly; dorsal fin long, continuous, or divided into 2 fins, the anterior portion composed of spines, which are stiff or flexible; anal fin long, usually with 1 or 2 small spines; ventrals well-developed, jugular, of 2 or 3 rays.

     q. Teeth all fixed, attached to the bone of the jaws and not movable.

     RUNULINÆ:

       r. Caudal fin lunate or forked; teeth compressed; spines and soft rays of dorsal indistinguishable.

         s. Ventral fins very long, each of a spine and a soft ray.

                   ATOPOCLINUS, 882.

        ss. Ventral fins not ½ length of head, each with about 2 soft rays; gill opening reduced to a small slit above pectoral.

                   RUNULA, 883.

  BLENNIINÆ:

       rr. Caudal fin rounded; teeth slender; gill membranes not reduced to a small slit.

        t. Teeth all fixed, attached to the bone of the jaws.

          v. Gill membranes free from the isthmus, or at least forming a distinct fold across it.

            w. Jaws one or both with a posterior fang-like canine, much longer than the anterior teeth.

                   BLENNIUS, 884.

            ww. Jaws without canines, the teeth all equal.

                   SCARTELLA, 885.

          vv. Gill membranes broadly united to the isthmus, the gill openings restricted to the sides.

            x. Jaws one or both with posterior fang-like canines.

                   HYPLEUROCHILUS, 886.

            xx. Jaws without posterior canines; the teeth equal.

              y. Three articulated ventral rays.

                z. Mouth small, the maxillary extending scarcely beyond front of eye; the head decurved in profile.

                   HYPSOBLENNIUS, 887.

                zz. Mouth large, the maxillary extending beyond vertical from middle of eye; the head rather pointed in profile.   CHASMODES, 888.

              yy. Four articulated ventral rays.

                   HOMESTHES, 889.

  SALARIINÆ:

       qq. Teeth of front of jaws all movable, implanted on the skin of the lips.

         a'. Vomer toothless

           b'. Jaws without posterior canines; dorsal fin deeply notched.

                   SCARTICHTHYS, 890.

           bb'. Jaws one or both with posterior fang-like canines.

            x. Dorsal fin continuous.   RUPISCARTES, 891.

            xx. Dorsal fin divided.   ENTOMACRODUS, 892.

         aa'. Vomer with a few teeth; posterior canines small.

                   SALARIICHTHYS, 893.

 pp. Teeth unequal, not comb-like; body oblong or elongate, more or less eel-shaped, naked, or rarely with rudimentary scales; supraocular flap sometimes present. Gill membranes united, free from the isthmus; dorsal fin very long, sometimes divided into 2 fins; formed of flexible spines, which often pass gradually into soft rays; anal fin long; ventral

fins thoracic or subjugular, usually, not much, if any, before the pectorals, composed of 2 soft rays each, the spine rudimentary; caudal well
developed, the dorsal and anal usually more or less joined to it at base.

OPHIOBLENNIINÆ:

    c'. Jaws each with 4 strong hooked canines in front; a hooked posterior
        canine below; a cirrus above eye and 1 above nostril; body scale-
        less; caudal fin forked; dorsal fin notched; body not eel-shaped;
        dorsal and anal free from caudal; ventrals small.

                                        OPHIOBLENNIUS, 894.

   cc'. Jaws with numerous teeth, not as above; caudal fin not forked.

    EMBLEMARIINÆ:

      d'. Body not eel-shaped; dorsal and anal not joined to caudal; no
        scales; no cirri; no lateral line; ventrals before pectorals;
        teeth on palatines; caudal fin rounded.

        e'. Dorsal fin very high, not notched, the spines passing grad-
          ually into the soft rays; jaws long, sharp at tip.

                                    EMBLEMARIA, 895.

   dd'. Body elongate or eel-shaped; the dorsal and anal low, joined to
      base of caudal.

    CHÆNOPSINÆ:

      f'. Ventrals subjugular, more or less before pectorals; pala-
        tines with teeth; jaws long and sharp.

      g'. Jaws with strong teeth, not as above described; dor-
        sal fin with its anterior half of flexible spines, the
        posterior half of soft rays, the former gradually
        passing into the latter; jaws long, pike-like;
        ventrals inserted slightly before pectorals; anal
        with 2 spines; a villiform band of teeth in each
        jaw behind anterior teeth.

        h'. Vomer toothless. Dorsal rays about XVIII, 38;
          anal II, 38.   CHÆNOPSIS, 896.

        hh'. Vomer with a few teeth. Dorsal rays XVIII,
          32; anal II, 30.   LUCIOBLENNIUS, 897.

   PHOLIDICHTHYINÆ:

      ff'. Ventrals subthoracic, inserted below pectorals; teeth in
        jaws uniserial; anal fin without spines.

      i'. Dorsal fin continuous, its spines indistinguishable
        from the soft rays.   PHOLIDICHTHYS, 898.

      ii'. Dorsal divided into 2 fins, the anterior portion of 3
        flexible spines behind the nape.

                          PSEDNOBLENNIUS, 899.

II. Blennies arctic or subartic; the vertebræ in large number, usually 50 or more; lateral line various, usually median; dorsal fin usually without soft rays; scales small,
cycloid, rarely wanting.

  j'. Gill openings not continued forward below, the membranes broadly united, some-
    times joined to the isthmus; ventral fins small or obsolete; scales small, cycloid.

    k'. Pectoral fins short or wanting, never pointed, and never more than ½ head;
      pyloric cæca usually, but not always, obsolete.

      l'. Body not covered with crosswise tubes at right angles to the lateral line.

        m'. Dorsal fin composed of spines only.

        STATHMONOTINÆ:

          n'. Body scaleless; ventrals moderately developed; anal spines 2;
            no lateral line; no pseudobranchiæ.   STATHMONOTUS, 900.

        nn'. Body covered with small smooth scales.

        CHIROLOPHINÆ:

          o'. Ventral fins well developed, of 1 spine and 3 rays; no
            anal spines; top of head with many cirri; a row of
            large pores above base of pectorals; gill membranes
            free from isthmus; no pyloric cæca.

p'. Lateral line obsolete, only the row of pores being present. BRYOSTEMMA, 901.

oo'. Ventral fins rudimentary or wanting, not more than 1 soft ray present; dorsal spines all short and rigid.

PHOLIDINÆ:

q'. Lateral line obsolete.

r'. Gill membranes broadly united, free from the isthmus; no pyloric cæca; carnivorous.

s'. Anal fin with a large sheathed spine; ventrals wanting.

t'. Anal spine very long, pen-shaped, its anterior surface channelled; pectoral fins moderate. APODICHTHYS, 902.

tt'. Anal spine moderate or small, not pen-shaped, its anterior edge convex, not channelled.

u'. Pectoral fins very small; anal spine moderate. XERERPES, 903.

uu'. Pectoral fins wholly wanting; anal spine small. ULVICOLA, 904.

ss'. Anal fin with 2 small spines or with none.

v'. Ventral fins reduced to a short spine, followed by a rudimentary ray.

w'. Caudal fin well developed. PHOLIS, 905.

ww'. Caudal fin very narrow, the dorsal and anal united around the tapering tail. GUNNELLOPS, 906.

vv'. Ventral fins entirely wanting; caudal as in *Pholis*. ASTERNOPTERYX, 907.

rr'. Gill membranes joined to the isthmus, sometimes forming a fold across it; no ventral fins; no anal spines; top of head with fleshy crests; pyloric cæca present; body naked anteriorly, with small scales posteriorly. ANOPLARCHUS, 908.

XIPHIDIINÆ:

qq'. Lateral lines several, each with many short cross branches; pyloric cæca present; gill membranes free from isthmus; ventrals none; anal spines 2 or 3, small; herbivorous.

w'. Pectorals small but well developed, much longer than eye. XIPHISTES, 909.

ww'. Pectorals minute, not longer than eye. XIPHIDION, 910.

CEBEDICHTHYINÆ:

mm'. Dorsal fin with its posterior half composed of soft rays; gill membranes broadly united, free from isthmus; ventrals wanting; lateral line single, high; pyloric cæca present; herbivorous. CEBEDICHTHYS, 911,

DICTYOSOMATINÆ:

ll'. Body covered with crosswise tubes at right angles with the lateral line and forming a network with it.

x'. Dorsal fin of spines only; teeth strong; ventral fins present, well developed; gill membranes broadly united, free from the isthmus. PLAGIOGRAMMUS, 912.

kk'. Pectoral fins long and rounded or pointed, nearly as long as head; dorsal fin high; gill membranes broadly united, free from the isthmus; no lateral line; species probably all herbivorous.

OPISTHOCENTRINÆ:

y'. Ventral fins wanting.

z'. Dorsal with its posterior spines rigid and sharp; head scaly. OPISTHOCENTRUS, 913.

zz'. Dorsal with its spines all flexible; head naked. PHOLIDAPUS, 914.

PLECTOBRANCHINÆ:

yy'. Ventral fins well developed; dorsal spines all pungent; body greatly elongate. PLECTOBRANCHUS, 915.

jj'. Gill openings continued forward below, the membranes separate or nearly so, scarcely joined to the isthmus; pectorals and ventrals well developed; dorsal spines slender, pungent, the fin without soft rays; herbivorous species.

LUMPENINÆ:

a''. Lateral line obsolete or obscure; body greatly elongate.

b''. Pectorals with the upper and middle rays shortened, shorter than lower; teeth on vomer and palatines. LEPTOCLINUS, 916.

bb''. Pectorals with the middle rays longest.

c''. Lateral line not wholly obsolete, a series of distant pores along sides; teeth on vomer and palatines. POROCLINUS, 917.

cc''. Lateral line obsolete, only a few small pores being traceable; no teeth on vomer; palatine teeth small or wanting. LUMPENUS, 918.

STICHÆINÆ:

aa''. Lateral line present, single, double, or triple; body moderately elongate; teeth on jaws, vomer, and palatines.

d''. Lateral line simple, one on each side of back. STICHÆUS, 919.

dd''. Lateral lines 2, or dividing into 2 on each side. ULVARIUS, 920.

ddd''. Lateral line forking, forming 3 on each side. EUMESOGRAMMUS, 921.

868. ENNEANECTES, Jordan & Evermann.

Enneanectes, JORDAN & EVERMANN, Proc. Cal. Ac. Sci. 1895, 501 (*carminalis*).

Body rather robust, covered with large, rough ctenoid scales; lateral line almost obsolete; mouth moderate, the jaws equal; no tentacle above the eye or on nape; no hook on shoulder girdle; eye large; dorsal fin divided into 3 fins, the first of 3 or 4 slender spines, the second of about 10, the soft dorsal of about 7 rays; caudal rounded; anal fin long; pectoral long, the lower rays simple and thickened. Small fish of the rock pools, closely allied to the Old World genus, *Tripterygion*, Risso, but distinguished by the chubby body, short fins, and large, rough scales. (ἐννέα, nine; νηκτηρ, swimmer, there being 9 fins.)

2687. ENNEANECTES CARMINALIS (Jordan & Gilbert).

Head $3\frac{2}{3}$; depth $4\frac{1}{2}$ to $5\frac{1}{3}$. D. III–XII, 9 (IV–X, 8 in the specimen before us); A. II, 11 (misprinted II, 17) scales 33 to 40. Body rather stout, heavy forward, rapidly tapering behind. Head short, the snout low and rather pointed, the profile straight and steep from the snout to opposite the front of the eyes, there forming an angle and extending backward nearly in a straight line; eyes very large, longer than snout, 3 in head, high up, and close together; mouth wide, the jaws subequal, the maxillary extending backward to front of pupil; teeth moderate, essentially as in species of *Labrisomus*, those of the outer series enlarged; no evident cirri on the head; scales on body of moderate size, ctenoid, the edges strongly pectinate; belly naked; lateral line extending to opposite last ray of soft dorsal, ascending anteriorly, but without convex curve; dorsals 3, the first and second contiguous, the second and third well separated; first dorsal of 3 spines, the first of which is the highest and about as long as diameter of eye; the second dorsal of higher and slenderer spines, the anterior the highest, the longest about equaling greatest depth of body; soft dorsal shorter and a little lower than second spinous dorsal; caudal small; anal long, beginning nearly under middle of spinous dorsal; pectoral long, longer than head, reaching much past front of anal; ventrals $\frac{3}{4}$ length of head. Color light brownish, with 4 dark-brown cross bars on sides, about as wide as the interspaces, which are marked with more or less reddish and with some lighter spots; belly pale; space behind pectoral dark; a dark bar downward and 1 forward from eye; first dorsal mottled with darker, second and third dorsals nearly plain; a narrow, dark bar at base of caudal and a broader one toward the tip, the fin sometimes entirely black; pectorals somewhat barred; lower fins plain. Mazatlan, in tide pools; the types, 4 specimens, each about $1\frac{1}{2}$ inches long. Another from the same locality, since figured by Dr. Jordan, differs somewhat in the count of the fin rays; but the very small size of the specimen prevents us from being entirely sure of its correctness. (*carmen*, a hetchel, from the rough scales.)

Tripterygium carminale, JORDAN & GILBERT, Proc. U. S. Nat. Mus. 1881, 362, Mazatlan.
    (Type, No. 28118. Coll. Gilbert.)
Enneanectes carminalis, JORDAN, Proc. Cal. Ac. Sci. 1895, 510, with plate of young example.

869. HETEROSTICHUS, Girard.

Heterostichus, GIRARD, Proc. Ac. Nat. Sci. Phila. 1854, 143 (*rostratus*).

Body rather elongate, compressed, covered with very small, smooth scales, those along lateral line not enlarged; head long and low, the snout conic, produced, very acute; premaxillaries protractile; mouth moderate, terminal; each jaw with a row of conical teeth, behind which anteriorly is a broad patch of villiform teeth; vomer and palatines with villiform teeth; gill rakers feeble; gill membranes broadly united, free from the isthmus; orbital cirri minute or wanting; cheeks scaly. Dorsal fin very long, the posterior rays soft, the 5 anterior spines wider apart than the rest and separated from them by a notch, the first and second spines longest, rather flexible, the other spines stiff; caudal fin forked; ventrals I, 3; pectorals moderate; lateral line simple, complete, abruptly curved behind pectorals; air bladder present, large. Size large. Close to *Gibbonsia*, from which the presence of the air bladder and the form of the caudal separate it. (ἕτερος, different; στῖχος, rank; in allusion to the differentiation of the anterior dorsal spines.)

2688. HETEROSTICHUS ROSTRATUS, Girard.

(KELPFISH.)

Head 3⅘ in body; depth 4⅛. D. V–XXXIII, 13; A. II, 34; eye 7 in head; maxillary 2½; pectoral 1⅘; ventral 2½; first dorsal spine 4⅕; highest ray of soft dorsal 2½; third anal ray 2⅚; caudal 2⅙. Body much compressed, deepest anteriorly; head slender, compressed and pointed; lower jaw strongly projecting, with thick lip; maxillary reaching pupil; width of interorbital a little greater than eye; orbital cirrus minute, usually entirely wanting; cheek and upper edge of opercle with small scales, rest of head naked. Origin of dorsal a little in front of the vertical from gill opening; pectoral under third dorsal spine, reaching to below the eleventh or twelfth; ventrals inserted in front of pectorals in distance equal to length of snout, their tips reaching about ¼ of their length beyond base of the pectoral; soft dorsal higher than spinous, ending slightly anterior to the anal; caudal furcate, the middle rays ⅘ length of outer. Color translucent, reddish brown, varying to blackish or olive, a series of large irregular light spots along sides below lateral line, continuous with a distinct light bar from eye to edge of opercle, bordered with black above, a similar spot on base of pectoral; an irregular line of large spots following outline of body under dorsal and above anal; a clear cut white streak from dorsal to tip of snout and continued on lower lip, the hue and pattern of color varying greatly; young examples most variegated; a translucent spot behind third dorsal spine, generally followed by similar spots for the whole length of the fin. San Francisco to San Diego. The largest of the Clinoid blennies, very abundant in the kelp, with which it agrees in coloration. Here described from a specimen, 16 inches in length, from San Francisco market. (*rostratus*, long-nosed.)

Heterostichus rostratus, GIRARD, Proc. Ac. Nat. Sci. Phila. 1854, 143, San Diego, California (Type, No. 284. Coll. A. Cassidy); GIRARD, Pac. R. R. Surv., X, Fishes, 26, pl. 13, 1858; GÜNTHER, Cat., 261, 1861; JORDAN & GILBERT, Synopsis, 764, 1883.

870. GIBBONSIA, Cooper.

Gibbonsia, COOPER, Proc. Cal. Ac. Nat. Sci., III, 1864, 109 (*elegans*).
Blakea, STEINDACHNER, Ichth. Beiträge, V, 148, 1876 (*elegans*).

Body less elongate and compressed, covered with minute cycloid scales, those along lateral line not enlarged; lateral line complete, abruptly decurved behind the pectoral; head somewhat pointed; snout unequal; conical teeth on jaws and vomer, the teeth mostly in single series, except in front, where there is a narrow villiform band; no conspicuous posterior canines; maxillary not produced backward from angle of mouth; a tentacle above eye, none at nape; gill membranes united, free from the isthmus. Shoulder girdle with an upturned hook on its inner edge as in *Clinus*. Dorsal fin long and low, chiefly composed of spines, 5 of the anterior spines different from the others, longer and set farther apart; anal fin low, with 2 spines; ventral fins jugular, of 1 spine and 2 or 3 rays; caudal fin truncate; branchiostegals 6; no air bladder; pyloric cæca absent. Viviparous. Pacific coast; bright-colored fishes, inhabiting rock pools among algæ. This genus is very close to *Clinus* (type *C. acuminatus*, Cuvier & Valenciennes), differing chiefly in the form of the dorsal fin and in the pointed snout. In *Clinus* the first 3 dorsal spines are shorter than the others. (Named for Dr. William Peters Gibbons, of Alameda, California, who was one of the early naturalists in the California Academy of Sciences.)

a. Dorsal rays about V-XXXI, 10; anal rays about II, 26; soft dorsal low; coloration comparatively plain, the soft dorsal without pellucid area.   EVIDES, 2689.
aa. Dorsal rays about V-XXVIII, 7; anal rays II, 24; soft dorsal high; coloration more or less highly variegated; soft dorsal with a large pellucid blotch posteriorly.   ELEGANS, 2690.

2689. GIBBONSIA EVIDES (Jordan & Gilbert).

(KELPFISH; SEÑORITA.)

Head 4⅘; depth 4½. D. V-XXX or XXXI, 10 or 11; A. II, 26 or 27. Body elongate, compressed; head small, rather pointed; mouth quite small, terminal, the maxillary about reaching pupil, 3¼ in head; lower jaw projecting, vomer with teeth; no teeth on palatines; posterior teeth not recurved; eye moderate, shorter than snout, 5 to 6 in head; a small supraocular flap, not higher than pupil; nasal cirrus very small; first spine of dorsal inserted over preopercle, its length more than ⅓ that of head, the second nearly equal; the third, fourth, and fifth progressively shorter; the sixth about as long as the fourth; the seventh longer; the rest nearly equal to the last, which is lower than the soft rays; the soft dorsal lower and more rounded than in *G. elegans*, the longest ray 2⅘ in head; pectorals moderate, not reaching vent; ventrals moderate; scales very small, smooth; head naked; no air bladder. Usual color of adult, translucent, reddish or orange, nearly plain or with oblong dark clouds below middle of sides anteriorly; often scattered blackish spots on sides, irregularly placed, forming a broken lateral band, most distinct in the young; a large pellucid spot on the membrane behind third dorsal spine, sometimes some small ones behind it; pectorals nearly plain; dorsal and anal plain, reddish, with a broad dusky shade distally; soft dorsal without pellucid area; caudal plain; a dark streak backward from eye; young examples often variegated, with light and dark shades of red, brown, and white, sometimes with 6 to 8 dark cross bars, sometimes with 4 or 5 lengthwise stripes alternating with paler ones, the hue varying exceedingly and dependent on the surroundings, but never so extravagantly spotted as in *Gibbonsia elegans*. Length 9 inches. Coast of California south to Point Concepcion; abundant in the kelp, rarely in rock pools. Here described from specimens from Monterey. (εὐειδής, comely; εὐ, well; εἶδος, appearance.)

Blakea elegans, STEINDACHNER, Ichth. Beiträge, v, 148, 1876, specimens from San Francisco; not *Myxodes elegans,* COOPER.

Clinus evides,* JORDAN & GILBERT, Synopsis, 763, 1883; specimens from Monterey, exclusive of part of synonymy; name a substitute for *elegans,* preoccupied in *Clinus.*

2690. GIBBONSIA ELEGANS (Cooper).

(SPOTTED KELPFISH.)

Head 4⅛; depth 4¼. D. V–XXVIII, 7; A. II, 24.. Body rather strongly compressed; head short, rather pointed, mouth small, terminal, rather oblique, the maxillary barely reaching pupil, 3⅓ in head; lower jaw projecting; teeth as in *Gibbonsia evides;* eye rather large, 4½ in head, as long as snout; a small fringed supraocular flap, as long as pupil; a slender nasal cirrus; first dorsal spine 2⅔ in head; fins as in *G. evides,* the soft dorsal shorter, higher, and less rounded, its longest ray 2¼ in head; pectorals and ventrals moderate, about as in *G. evides;* caudal fan-shaped on a slender peduncle; scales small and smooth; head naked. Color brown or red, agreeing with rocks or with *Corallina,* usually with eight irregular darker cross bars extending on the dorsal and anal, sometimes nearly plain brown; a dark spot probably always present behind head, and some, 1 or more, along lateral line posteriorly; spinous dorsal with a pellucid spot; usually many pale and dark spots and freckles on head and fins; pectoral and caudal usually barred, but plain in specimens taken in the kelp (*Macrocystis*), these latter much less variegated than tide-pool specimens; soft dorsal always with a large pellucid blotch posteriorly, this wanting or obscure in *G. evides.* Coast of southern California; abundant in rock pools lined with *Corallina* from Point Concepcion to Todos Santos; the specimens here described from Point Loma. Close to the preceding, but smaller and more brightly colored, the fin rays fewer. These differences, though small, seem to be constant; whether the 2 species overlap each other in geographical range is not known. (*elegans,* elegant.)

Myxodes elegans, COOPER, Proc. Cal. Ac. Sci., III, 1864, 109, San Diego and Santa Barbara.
Clinus ocellifer,† MOCQUARD, Bull. Soc. Philom. Paris 1886, 44, California.
Clinus evides, ROSA SMITH, Proc. U. S. Nat. Mus. 1883, 235, specimens from Todos Santos; not of JORDAN & GILBERT.

---

* The name *evides* may apparently be retained for this species, as the description of Jordan & Gilbert (Synopsis, 763) is based entirely on Monterey specimens, typical of this species. It was intended, however, as a substitute for the name *elegans,* already used in the genus *Clinus,* to which these species were then referred.

† The following is a translation of the description of *Clinus ocellifer* (Mocquard):
Head 4½; depth 4½. D. III–XXX, 8; A. II. 24; C. 13; P. 12; V. I, 3. Body strongly compressed, tapering rapidly behind. Eye a little longer than snout, 3¾ in head, twice interorbital space; lower jaw a little longer than maxillary, reaching front of eye; a little tentacle on anterior nostril, elongate, with 4 or 5 unequal branches; a tentacle over eye. Three first dorsal spines nearly double length of those which follow, and separated by an interval equal to that which separates the first spine from the third; last dorsal spines longer than those that precede and stronger than any of the others; dorsal and caudal well separated. Scales very small. Opposite fifth and sixth dorsal spines immediately below lateral line is a lens-shaped spot of brownish black with a dull border; a second ocellus a little before the posterior extremity of the spinous dorsal; this surrounded by a pale brown circle in 1 specimen; the sides also with 5 irregular bands of a paler brown than that of the spots; the anterior is a little behind the corresponding spot; the posterior opposite the third or fourth soft ray of the dorsal; the posterior spot at the upper extremity of the fourth band; other spots of the same color at the base of the dorsal on the right of the caudal; sides with a longitudinal series of small white spots, not surrounded by black circle; other spots on the anal in 6 transverse lines; larger spots on subopercle and about the ventrals. Teeth on the vomer, none on the palatines. Coast of California. Two specimens, 93 mm. long. (Mocquard.)

871. NEOCLINUS, Girard.

Neoclinus, GIRARD, U. S. Pac. R. R. Surv., x, Fish., 114, 1858 (*blanchardi*).
Pterognathus,* GIRARD, Proc. Ac. Nat. Sci. Phila. 1859, 57 (*satiricus*).

Body compressed, rather elongate, covered with minute cycloid scales; lateral line present, incomplete, high anteriorly; head naked, the cheeks tumid; upper jaw protractile; maxillary greatly produced backward, more than $\frac{2}{3}$ length of head, reaching far beyond the eye; both jaws, vomer, and palatines with stout, unequal, conical teeth in a single series, besides which, in the front of the jaws, are smaller teeth; nasal and supra-ocular region with fringed tentacles; gill membranes broadly united, free from the isthmus; gill rakers weak. Dorsal fin long, scarcely emarginate, its anterior $\frac{2}{3}$ composed of slender, flexible spines, which are similar to the soft rays, all of which are simple; anal long, its rays all simple; ventrals moderate, I, 3; caudal fin distinct; pectorals rather broad, rounded; no air bladder; no pyloric cæca. Pacific coast, in shallow water; remarkable for the great development of the maxillary, as in *Opisthognathus* and *Gillichthys*. (νέος, new; κλῖνος, *Clinus*.)

NEOCLINUS:
a. Maxillary long, but not reaching beyond head; membrane of jaws white,
BLANCHARDI, 2691.
PTEROGNATHUS (πτερόν, wing; γνάθος, jaw):
aa. Maxillary inordinately developed, reaching gill opening in the adult; maxillary flap blackish, edged with bright yellow. SATIRICUS, 2692.

Subgenus NEOCLINUS.

2691. NEOCLINUS BLANCHARDI, Girard.

Head 4; depth 5⅓. D. XXIV, 17; A. II, 30; eye 5 in head; maxillary variable, about 1½; pectoral 2; caudal 1¾ to 2. Upper profile of head convex, snout rather steep; jaws subequal; teeth on jaws, vomer, and palatines, subequal, canine-like; eye set high in head, equal to length of snout. Males with a long thick cirrus over front of middle of eye, twice as long as eye, its end multifid, 3 or 4 short, slender ones behind it over posterior half of eye; females with a much smaller cirrus in front, seldom as long as eye, the posterior ones similar to those of male; both with a multifid flap at anterior nostril; maxillary never reaching past preopercle (in specimens from 6 to 8 inches in length), not longer in males than in females. Head naked; scales on body very small, somewhat embedded; no scales on fins; origin of dorsal directly behind occiput, no notch between spinous and soft dorsals; the tips of last dorsal and anal rays reaching to base of caudal fin; pectorals broad, scarcely reaching to vent; about ⅓ the length of ventrals in front of base of pectoral. Color varying from dark red or

---

* " It is more than probable that had we been acquainted with this second species of *Neoclinus* first, we would have been misled as to its real generical characters, and framed a name in allusion to the condition of the upper jaw, such as *Pterognathus*, for example, which would have been most characteristic, for that upper jaw is as truly winged as the upper members of the flying squirrels. We can not help thinking that Cuvier himself would not have coined the name of *Opisthognathus* had he had before him the species which bears his name instead of that which he dedicated to Sonnerat. These two genera (*Opisthognathus* and *Neoclinus*) will furnish one of the best themes to ichthyological studies, as they exemplify the fact that specific characters may be developed to exaggeration, and become more conspicuous than the generic characters themselves." (Girard.)

plum color to olive green; sides mottled and spotted with darker; a dark spot, ocellated with yellow, generally present between first and second dorsal spines; dorsal blackish toward ends of rays; pectorals and anal white in female, slightly dusky in male; unexposed portion of lower lip entirely white; a yellow spot on base of caudal rays below and above. Coast of California, from Monterey to Santa Barbara; not rare; a remarkable fish. Here described from specimens from 6 to 8 inches in length, from Pacific Grove, California. We do not know what variation there may be in maxillary and barbels in larger or smaller specimens. This species differs from *N. satiricus* in having no second spot behind seventh spine of dorsal; in having that part of lower lip which is covered by the maxillary entirely white; barbels in male much longer; maxillary shorter; head slightly shorter; and in having the pectorals and anal lighter. (Named for its discoverer, Dr. S. B. Blanchard.)

Neoclinus blanchardi, GIRARD, U. S. Pac. R. R. Surv., x, Fish., 114, 1858, San Diego (Type, No. 691. Coll. Dr. S. B. Blanchard); GÜNTHER, Cat., III, 259, 1861; JORDAN & GILBERT, Synopsis, 761, 1883.

Subgenus PTEROGNATHUS, Girard.

2692. NEOCLINUS SATIRICUS, Girard.

Head $3\frac{1}{2}$ in body; depth 6. D. XVI, 17; A. 30; eye 5 in head; pectoral 2; caudal $2\frac{1}{2}$. Head bluntish, convex in profile; snout steep; jaws subequal; unequal, small canines on jaws, vomer, and palatines; eye about equal to length of snout, interorbital flattish, about $\frac{1}{4}$ eye in width; 3 or 4 small barbels above eye, seldom as long as eye, the anterior one sometimes absent on one or both sides; cirri not differentiated in the female; a multifid flap on anterior nostril; maxillary always reaching past edge of preopercle (in examples 6 to 9 inches in length), just past in females, longer than head in males. Head naked, scales on body small, partly embedded; no scales on fins. Origin of dorsal directly behind occiput; no notch between spinous and soft dorsals; pectorals in the larger examples reaching to vent; last rays of dorsal and anal reaching base of caudal fin; anterior half of ventrals in front of base of pectorals. Color in spirits, reddish brown or olive green, mottled and spotted with darker; a dark spot ocellated with yellow between first and second dorsal spines, a similar one between seventh and ninth; dorsal blackish, pectoral, anal and ventrals varying from dusky to black, in no case light in our specimens; a yellow spot sometimes present, below and above, on base of caudal rays; the membrane connecting maxillary with lower jaw blackish, broadly and abruptly edged with white (probably yellow in life). Coast of California, from Monterey to Santa Barbara; a rare and most interesting species. Here described from specimens, 6 to 9 inches in length, from Pacific Grove, California. Differing from *N. blanchardi* in length of maxillary; slightly larger head; males without long cirri; a second spot on dorsal; fins darker, and especially in having the membrane of lower lip blackish, edged with white. (*satiricus,* satirical.)

Neoclinus satiricus, GIRARD, Proc. Ac. Nat. Sci. Phila. 1859, 57, Monterey, California, in 30 fathoms (Coll. A. S. Taylor); GÜNTHER, Cat., III, 260, 1861; JORDAN & GILBERT, Synopsis, 761, 1883.

872. MALACOCTENUS, Gill.

Malacoctenus, GILL, Proc. Ac. Nat. Sci. Phila. 1860, 103 (*delalandi*).

This genus is very close to *Labrisomus*, differing in the dentition, the teeth in the jaws being in single series; vomer with a few teeth or with none, none on palatines. The form of the dorsal fin in some species is different, there being usually a notch behind the fourth dorsal spine as well as at front of soft dorsal. Most of the species are not well known, and perhaps more than 1 genus is here included. (μαλαϰός, soft; ϰτείς, comb, in reference to the comb of filaments at the nape in the typical species.)

a. Nape without filaments.
   b. Orbital tentacle present.
      c. D. XXI, 8; spinous dorsal not notched, the first rays shortest; body elon-
         gate; snout pointed; scales large, about 38.   OCELLATUS, 2693.
      cc. D. XX, 12; spinous dorsal weakly notched; body rather robust.
                                           VARIUS, 2694.
   bb. Orbital tentacle wanting; dorsal rays XXI, 11; spinous dorsal weakly notched;
      ventrals long.   MACROPUS, 2695.
aa. Nape with a single tentacle. D. XVIII, 9; a tentacle above eye.   LUGUBRIS, 2696.
aaa. Nape with a comb of slender tentacles; spinous dorsal more or less notched
    behind fourth or fifth spine.
   d. Orbital tentacle present. D. XVIII to XX, 11 or 12 vomer with teeth.
      e. Scales 43 or 44.
         f. Highest soft ray of dorsal 1½ in head; dorsal without ocelli.
                                     GILLII, 2697.
         ff. Soft rays of dorsal 1½ in head; dorsal fin with 2 large black ocelli;
           ventral fins long, as long as head.   BIMACULATUS, 2698.
      ee. Scales 55; ventrals moderate, shorter than head.   DELALANDI, 2699.
   dd. Orbital tentacle wanting; (no vomerine teeth?).   VERSICOLOR, 2700.
aaaa. Nuchal and other filaments undescribed; a black ocellus on front of dorsal.
    D. XX, 11; scales 46.   BIGUTTATUS, 2701.

2693. MALACOCTENUS OCELLATUS (Steindachner).

Head 4 to 4¼; depth 5 to 5⅔. D. XXI, 8; A. II, 8; scales 38; eye 4½ to 5 in head; snout 5⅔; interorbital width 10. Body elongate; the snout short; profile not steep; tentacle above eye very slender, none on nape. Maxillary ½ long as head, reaching posterior margin of eye. Teeth on jaws and vomer in 1 row, none on palatines. Dorsal with a notch between the spines and soft rays; spines all short, the longest not ½ head, the anterior shortest; the longest soft rays 1½ in head; dorsal slightly joined to base of caudal; ventral and caudal each 1½ in head; pectoral almost as long as head. Lateral line complete, strongly arched anteriorly. Color brownish; 8 pairs of narrow dark-brown cross bands on the body, most distinct above, sometimes broken up into cross spots; first membrane of the dorsal fin with black spot behind, sometimes a similar one, oval and indigo, behind eye; numerous sky-blue spots bordered with darker on sides of head and part of body; anal pale violet, edged with white, sometimes spotted; caudal gray, with darker spots in cross rows. Bahama Islands. Length 2 inches. (Steindachner.) Not seen by us. Perhaps not a member of this genus; the large scales, entire spinous dorsal, and short soft dorsal, indicating affinities with *Starksia*, which has, however, a different dentition. It may prove to be the type of a distinct genus. (*ocellatus*, with eye-like spots.)

Clinus ocellatus, STEINDACHNER, Ichth. Beitr., v, 182, 1876, Bahama Islands.

2694. MALACOCTENUS VARIUS (Poey).

Head 3⅔ in total length; depth 4½. D. XX, 12; A. 18; C. 14; pectoral 1⅔ in head; eye 3⅔ in head, equal to snout. Mouth small; maxillary reaching opposite front of eye; profile prolonged; nostrils small, not tubular. Teeth firm, in 1 row, the points sharp and incurved; no teeth on vomer; a tentacle over eye, none at the nape; head naked; body scaly; lateral line short; dorsal beginning over middle of opercle, the spinous part forming a sinuous curve; the spines firm; the first higher than the 4 which follow; the last low; the next to the last lower than the last; soft dorsal higher than the spines; ventral rays apparently 2, the last one deeply divided. Color clear yellowish; the body spotted with black; an isolated spot at the end of the dorsal fin; vertical fins with all the rays dotted with black; pectoral pale, without specks. Length 52 mm. Cuba. (Poey.) Not seen by us. (*varius*, variegated.)

Myxodes varius, POEY, Enumeratio, 132, pl. 5, f. 2, 1875, Havana. (Coll. Rafael Arango.)

2695. MALACOCTENUS MACROPUS (Poey).

Head 4¼ in total length with caudal; depth 5½. D. XXI, 11; A. I, 20; P. 17; eye ¼ longer than snout, 3¼ in head. Maxillary reaching front of eye. Teeth in 1 series, acute, not close-set; none on the vomer or palatines; no cilia over the eye nor on the nape; lateral line almost complete; ventrals as long as the depth of the trunk; first dorsal spine longest, the others forming a weak curve. Color uniform metallic coppery brown. Cuba (Poey); one specimen 35 mm. long. The type of this species examined by us in the Mus. Comp. Zool. It has scales 35; no hook on the shoulder girdle, and apparently no teeth on vomer or palatines. (μακρός, long; πούς, foot.)

Myxodes macropus, POEY, Synopsis, III, 99, 1868, Havana. (Coll. Poey.)

2696. MALACOCTENUS LUGUBRIS (Poey).

D. XVIII, 9; A. 20. Tentacle over eye; a filiform appendage on the side of the neck. Ventral very long, extending much beyond the vent. Dorsal fin with 2 depressions, the soft part short and very high, the first spine moderate. Color dark brown, with oblique vertical bands and brown points scattered over the head and trunk; a black spot at the base of the first 3 dorsal membranes; ventral entirely white. Cuba. One specimen 55 mm. long. (Poey.) Not seen by us. (*lugubris*, dismal, from the dark color.)

Myxodes lugubris, POEY, Enumeratio, 131, 1875, Cuba. (Coll. Poey.)

2697. MALACOCTENUS GILLII (Steindachner).

Head 4½; depth 4½; eye 4 in head; snout 3¼; interorbital 6 in head. D. XIX, 11; A. II, 17; P. 14; scales 43. Head pointed, conic anteriorly; snout longer than eye; a rather high tentacle above the eye, slender and split to the base, numerous others on the side of the nape; teeth not described. First three dorsal spines wider apart than others, first longest; eighth to tenth spines highest, ¼ head; highest soft ray 1½ in head; dorsal deeply notched. Body greenish gray, with brown spots or faint cross bands; head and dorsal marbled with darker; ventrals white, the longest ray a little longer than head, reaching anal; anal edged with dark; pectoral as long as head. Barbados. Two specimens, the larger 2 inches long. (Steindachner.) This species may be a *Labrisomus*. (Named for Dr. Theodore Gill.)

Clinus gillii, STEINDACHNER, Ichth. Notizen, VI, 46, 1867, Barbados.

2698. MALACOCTENUS BIMACULATUS (Steindachner).

Head 4⅜; depth 4½. D. XX, 10; A. II, 19; ventral 3; scales 44. Near to *M. delalandi,* but the body deeper (said to be 5¼ in the latter species, which is not the case). Profile to snout steep; eye a little shorter than snout, 3¼ in head; jaws equal, each like the vomer with ⅓ row of teeth; maxillary reaching about to front of pupil. Interorbital space narrow, more than ½ width of eye; a very slender, rather long, bifid tentacle above eye; tentacles on the nape, upper 1 almost as long as tentacle above eye. Upper margin of dorsal weakly notched between first and fifth spines, more deeply between spines and soft rays, the former as in *M. delalandi;* longest soft rays 1½ in head; longest spines 2 in head; first 4 spines more widely separated than the others; pectoral and ventral as long as head; caudal a little shorter. Body brown, with dark-brown bands and numerous blackish spots, only the cross bands on the head strongly marked; tips of the anal rays whitish; above these a bluish violet streak; pectoral with 2 milk-white spots at base; a large black isolated spot at the base of the first 4 dorsal spines, a second on the last 4 spines, extending on the body; anal and caudal thickly spotted with brown. Small rocky islands to the north of Cuba. (Steindachner.) Not seen by us; evidently close to *Malacoctenus delalandi,* but the scales larger. (*bis,* two; *maculatus,* spotted.)

Clinus bimaculatus, STEINDACHNER, Ichth. Beitr., V, 180, 1876, small, rocky islands north of Cuba.

2699. MALACOCTENUS DELALANDI (Cuvier & Valenciennes).

Head 3½; depth 3½ to 4½. D. IV–XVI, 11; A. II, 18; scales 55. Form rather stout, compressed; snout not very short, rather pointed, the profile gibbous above the eyes, thence declining straight to the tip of the snout; mouth rather small, the maxillary reaching front of eye; teeth in a single series in each jaw; vomer with a few teeth, none on palatines; eye large, 3½ in head, as long as snout; small slender cirri above the eyes, and a fringe of moderately long filaments at the nape rather longer than the orbital cirri. Outline of spinous dorsal emarginate; first spine a little longer than eye, the second, third, and fourth progressively shortened, the fifth again longer; the eighth to eleventh spines longest, thence gradually decreasing to the next to the last, which is much shorter than the last; soft dorsal rays considerably higher than the spines, the longest about $\frac{1}{2}$ length of head; anal long, not very high, the membrane deeply notched between all but the last 6 rays, which are the highest. Pectorals $\frac{5}{6}$ length of head; ventrals as long as from snout to edge of preopercle. Belly naked anteriorly; the scales small, cycloid; lateral line complete. Color olivaceous, darker above, much mottled and speckled with clear dark brown; sides with 5 distinct irregular dark-brown bars, extending from base of dorsal to level of lower margin of pectoral, their lower edges connected by a vague undulating longitudinal band; a blackish blotch on occipital region, and black blotches on cheeks, opercles, and before base of pectoral; opercle with several narrow pinkish streaks; head below with narrow streaks formed by series of dark-brown spots; an interrupted brown bar across lower jaw; belly unspotted; ventrals pale; other fins all barred with narrow series of dark-brown dots; anal somewhat dusky. Coast of Brazil and the west coast of Mexico; common. Here described from the types of *Clinus zonifer*. This is the most abundant denizen of the rock pools around Mazatlan, with the single exception of *Gobius soporator*, reaching a length of 3 to 5 inches. We are unable to separate *M. zonifer* from Mazatlan from Bahia examples of *M. delalandi*, and take our account from specimens of the former. (Named for Delalande, who collected for Cuvier in Brazil.)

Clinus delalandii, CUVIER & VALENCIENNES, Hist. Nat. Poiss., XI, 378, 1836, Brazil (Coll. Delalande); GÜNTHER, Cat., III, 264, 1861.
Clinus zonifer, JORDAN & GILBERT, Proc. U. S. Nat. Mus. 1881, 361, Mazatlan. (Coll. C. H. Gilbert.)
Clinus philipii, LOCKINGTON, Proc. Ac. Nat. Sci. Phila. 1881, 114; not of STEINDACHNER.
Labrisomus delalandi, JORDAN, Proc. U. S. Nat. Mus. 1888, 333.

2700. MALACOCTENUS VERSICOLOR (Poey).

Head $3\frac{1}{2}$; depth $3\frac{3}{4}$. D. XVIII, 12; A. 20. Body compressed; head moderate; snout prolonged; pectoral $1\frac{1}{4}$ in head; ventral $1\frac{3}{4}$; eye large, as long as snout, $3\frac{1}{2}$ in head; nostrils not tubular; mouth small; maxillary not reaching so far as eye; teeth firm, in 1 row, those above much larger and slightly curved backward; 12 teeth above and 9 below on each side [no teeth on vomer]; no tentacle over eye; no anal papilla; a comb of filiform tentacles on each side of the neck; head naked; body scaly; lateral line short; dorsal with 2 depressions, the first spine higher than the 4 which follow, the depressions much more marked than in *M. varius*; pectoral reaching beyond front of anal. Color yellowish brown; head, trunk, and fins varied with vertical brown bands and large brown spots; ventrals yellowish. On the figure the 5 bands behind the anal cross the body and extend on the vertical fins. Cuba. One specimen known, 53 mm. in length. (Poey.) Apparently very close to *M. delalandi*, but lacking the orbital tentacle, and, according to Poey, vomerine teeth also. (*versicolor*, variegated.)

Myxodes versicolor,[*] POEY, Enumeratio, 131, pl. 5, f. 1, 1875, Cuba. (Coll. Poey.)

---

[*] This species and its affines were referred by Poey to *Myxodes*, a South American genus allied to *Clinus* and *Gibbonsia*, but differing from the latter in its uniserial teeth.

2701. MALACOCTENUS BIGUTTATUS (Cope).

Dorsal XIX–I, 11; anal II, 16; The first dorsal spines the longest, last spine longer than penultimate; length of head without opercular flap, $3\frac{3}{4}$ times in length (exclusive of caudal fin); eye a little less than $\frac{1}{4}$ length of head, $\frac{3}{5}$ greater than interorbital width; pectoral fin reaching to fifth anal; scales large, 4–46–10 [cirri and teeth not described]. Pale reddish brown, humeral red-veined; rufous specks on anterior part of sides; 7 subquadrate brown blotches from nape to caudal fin, continued with interruptions as lateral bands, the fourth near end of spinous dorsal black; a black spot at base of membrane between first to third dorsal spines; 2 small brown spots behind orbit, the posterior on operculum. Length 2.25 inches. This species is well distinguished from *Labrisomus nuchipinnis* by the large scales, form of dorsal fin, coloration, etc. From New Providence, Bahamas; Dr. H. C. Wood's collection. Also a very small specimen from Dr. Rijgersma, St. Martins. (Cope.) Not seen by us. (*bis*, two; *guttatus*, spotted.)

Labrisomus biguttatus, COPE, Trans. Am. Philos. Soc. Phila. 1873, 473, New Providence, Bahama Islands. (Coll. Dr. H. C. Wood.)

873. LABRISOMUS,* Swainson.

Labrisomus, SWAINSON, Nat. Hist. Class'n Fishes, II, 277, 1839 (*pectinifer*).
Lepisoma, DE KAY, New York Fauna: Fishes, 41, 1842 (*cirrhosum*).
Labrosomus, GILL, amended spelling.
? *Blennioclinus*, GILL, Proc. Ac. Nat. Sci. Phila. 1860, 103 (*brachycephalus*).
? *Auchenionchus* (misprinted *Anchenionchus*), GILL, Proc. Ac. Nat. Sci. Phila. 1860, 103 (*variolosus*).
? *Calliclinus*, GILL, Proc. Ac. Nat. Sci. Phila. 1860, 103 (*geniguttatus*).
? *Ophthalmolophus*, GILL, Proc. Ac. Nat. Sci. Phila. 1860, 104 (*latipinnis*).

---

* Concerning this genus and its affines, Dr. Gill remarks:
"The name *Labrosomus* (or *Labrisomus*) was first published in 1839, in the second volume of the 'Natural History of Fishes, Amphibians, and Reptiles.' At the seventy-fifth page of that volume, Swainson has divided the Cuvieran genus *Clinus* into 5 genera: *Clinus*, of which the *Clinus acuminatus*, Cuvier, is taken as the type; *Labrisomus* with *Clinus pectinifer*, Valenciennes, as type; *Tripterygion*, Risso, *Clinitrachus*, Reese, which is typified by *Blennius variabilis* of Rafinesque, and *Blennophis*, of which the *Clinus anguillaris*, Valenciennes is the only true species. Of these genera, *Clinus* Swainson, and *Clinitrachus* Swainson, are distinguished by false or illusive characters, and cannot be regarded as distinct. The others are valid, but their characters require revision. The only claim to distinction of the genus *Labrosomus* given by Swainson, is founded on the strong, conic, and pointed row of front teeth, behind which are villiform ones; a thicker body than in *Clinus*, and the 'dorsal fin distinctly emarginate toward the caudal.' The genus resting on these characters alone is composed of very incongruous elements. To it are referred, at page 277 of the second volume, the following species, all of which are described as species of *Clinus* by Valenciennes: *Labrosomus gobio, L. pectinifer, L. capillatus, L. delalandii, L. linearis, L. variolosus, L. peruvianus, L. microcirrhis, L. ? geniguttatus, L. elegans, L. ? littoreus* and *L. latipinnis.* Of these species, not more than 3 can, with propriety, be regarded as congeners, if the *Labrosomus pectinifer* is taken as the type. These are *Labrosomus pectinifer, L. capillatus,* and perhaps *L. delalandii.* The latter is more probably the representative of a distinct genus. That genus is distinguished from *Labrosomus* by the smaller mouth, the presence of only 2 rays to the ventral fins, and perhaps by the undulating margin of the spinous portion of the dorsal fin. It may be named *Malacoctenus,* in allusion to the pectiniform row of filaments. This genus is the nearest ally of *Labrosomus.* All the others are very distinct. *Labrisomus* Swainson, is the type of a distinct genus, whose characters consist of a broad, depressed head, with a very short muzzle, large approximated eyes, superciliary and nasal tentacles, 2 ventral rays, and a comparatively short spinous dorsal. The genus may be called *Gobioclinus.* The only species, *Gobioclinus gobio,* is found in the West Indies, and has but 18 dorsal spines. *Labrisomus linearis* Swainson is synonymous with *Clinus brachycephalus,* Valenciennes. This, also, is the type of a

Body oblong, robust; head naked, short, compressed above; mouth rather large, with a row of stout, bluntish teeth in front of each jaw, behind which is a band of smaller teeth, broadest in lower jaw; teeth on vomer, no teeth on palatines; a tentacle above the eye; sides of neck with a tuft or series of fine filaments; dorsal fin continuous, with numerous slender spines and many soft rays, the spines not very unequal; pectorals long; lateral line continuous; scales moderate or small, cycloid; shoulder girdle without upturned hook-like process on its inner edge. Intestinal canal short, shorter than body. The limits of this genus are not well defined, and most of the nominal genera above named will probably be found worthy of recognition. This genus differs from *Clinus* chiefly in the absence of the upturned spine-like processes on the inner edge of the shoulder girdle. This process is found on *Clinus acuminatus*, the type of the genus *Clinus*. (*Labrus*; σῶμα, body.)

> *a.* Scales moderate, about 70 in lateral line (so far as known); soft dorsal with 11 to 13 rays.
> > *b.* Dorsal spines 16; anal rays 20; tentacles on nape.   HERMINIER, 2702.
> > *bb.* Dorsal spines 18; no teeth on palatines; first ray of dorsal not longest; orbital tentacle well developed; nape with a conspicuous comb of fringes.
> > > *d.* Vomer with a cluster of small teeth.   NUCHIPINNIS, 2703.
> > > *dd.* Vomer with 3 to 5 large blunt teeth arranged in the form of a ∧.
> > > > XANTI, 2704.
> > *bbb.* Dorsal spines 20; teeth on palatines (?); first dorsal spine longest.
> > > BUCCIFERUS, 2705.
> *aa.* Scales very small, about 110; a comb of fringes at nape; first dorsal spines low; head with yellow spots.   MICROLEPIDOTUS, 2706.

2702. LABRISOMUS HERMINIER (Le Sueur).

D. XVI, 11; A. 20; C. 14; P. 16; V. 3. Body slender, compressed. Cilia on nostrils, above the eye, and on the nape; lips thick, concealing conical teeth, behind each band of smaller teeth; teeth also on the palate and on the base of the gill arches. Scales rather large. Lateral line curved from the pectoral, becoming straight thence to the tail. Color reddish brown with numerous spots; a black spot at front of spinous dorsal.

---

distinct genus distinguished by its abbreviated and blenniform head, the profile being very convex; by the villiform teeth, the absence of superciliary tentacles, the spinous portion of the dorsal long, and the presence of only 2 rays to the ventral fins. The name *Blennioclinus* is conferred on it; for the species, the specific name of Valenciennes must be retained. *Labrisomus variolosus* is distinguished by a large thick head, with lateral eyes, short superciliary tentacles, and a small nuchal one. The mouth is large; the teeth of the jaws in an outer row strong and conical, behind which are villiform ones; those of the vomer and palate are villiform, in 3 patches, 1 on the vomer and 1 on each palatine bone. The spinous portion of the dorsal is long, and the ventrals have each 3 rays. The species thus characterized is the type of a new genus which may be named *Anchenionchus* (misprint for *Auchenionchus*). *Labrisomus microcirrhis*, *L. elegans* and *L. peruvianus* are nearly related to *Anchenionchus*, and are from the same zoological province. *Labrosomus ? geniguttatus* is distinguished from *Anchenionchus* by the more approximated eyes and by the disposition of the vomero-palatine teeth, as well as the small size of the anterior row of maxillary teeth. The dorsal is moderately long, and each of the ventrals has 3 rays. The mouth is comparatively small, and there are superciliary, nasal, and nuchal tentacles. For this species the generic name *Calliclinus* is proposed. *Labrisomus latipinnis* is related to *Blennioclinus*, but is distinguished from the species of that genus by the presence of superciliary tentacles. The generic name of *Ophthalmolophus* may be retained for it." (Proc. Ac. Nat. Sci. Phila. 1860, 102, 103.)

St. Bartholomew, West Indies; known from one specimen taken among madreporic rocks. (Le Sueur.) Not recognized by any recent author; perhaps not distinct from *L. nuchipinnis.*

Blennius herminier, LE SUEUR, Jour. Ac. Nat. Sci. Phila., IV, 1824, 361, St. Bartholomew.
Clinus hermineri, CUVIER & VALENCIENNES, Hist. Nat. Poiss., XI, 380, 1836.

2703. LABRISOMUS NUCHIPINNIS (Quoy & Gaimard).

Head 3½; depth 3¼. D. XVIII, 12; A. II, 17; scales 70. Body oblong, rather robust; head naked, thick, short, not very obtuse anteriorly, compressed above; mouth rather large, the maxillaries not prolonged backward, extending to opposite the posterior part of eye, 2¼ in head; teeth on vomer and palatines; front teeth of jaws conic, strong, behind them a band of villiform teeth, broadest in lower jaw; vomer with a patch of smallish teeth; eyes large; interorbital space very narrow; each side of neck with a long series of hair-like filaments, nearly as long as eye; orbital tentacle short and broad, multifid; nostril with a tufted barbel; lower jaw slightly projecting, its posterior teeth sometimes recurved; pectorals a little shorter than head, reaching vent. Dorsal spines rather slender, the 3 anterior spines scarcely shorter than the others, all the spines lower than the soft rays; dorsal fin commencing near the nape, the spinous portion long; soft rays higher than the spines; caudal small; pectorals rather large; ventrals moderate; gill-membranes broadly united, free from the isthmus; lateral line complete, high anteriorly, then abruptly decurved; membranes of vertical fins scaly; scales not very small, cycloid. Reddish brown, sometimes with vertical bands; a black spot on opercle, which is often edged with white; cheeks and fins reticulate or dotted. Length 6 to 8 inches. West Indies, north to Florida Keys, south to Brazil; generally common in rock pools; also recorded from the Canary Islands. (*nucha,* nape; *pinna,* fin.)

Clinus nuchipinnis, QUOY & GAIMARD, Voy. Uranie et Physicienne, Zool., 255, 1824, Brazil (Coll. M. Freycinet & M. Gay); GÜNTHER, Cat., III, 262, 1861; JORDAN & GILBERT, Synopsis, 762, 1883.
Clinus pectinifer, CUVIER & VALENCIENNES, Hist. Nat. Poiss., XI, 374, 1836, Bahia.
Lepisoma cirrhosum, DE KAY, N. Y. Fauna: Fishes, 41, 1842, Florida.
Clinus canariensis, VALENCIENNES, in WEBB & BERTHELOT, Poiss. Iles Canaries, 60, 17, f. 3, Canary Islands.
Clinus capillatus, CUVIER & VALENCIENNES, Hist. Nat. Poiss., XI, 377, 1836, Martinique.
Labrosomus pectinifer, GILL, Proc. Ac. Nat. Sci. Phila. 1860, 105.
Labrisomus capillatus, GILL, Proc. Ac. Nat. Sci. Phila. 1860, 107.

2704. LABRISOMUS XANTI, Gill.

Head 3¼ in body; depth 3½. D. XVIII, 12; A. II, 18; scales 10–64 (pores)–12 (from front of straight portion of lateral line to anal); eye 4¼ in head, maxillary 2; highest dorsal spine 2⅔; pectoral 1½; caudal 1¾. Body not greatly elongate, compressed, anterior profile well rounded from snout to nape; mouth rather large, the maxillary reaching to below middle of eye; teeth small, canine-like, growing gradually larger toward front of upper jaw; side teeth on lower jaw very small, abruptly-enlarged on front half of jaw; teeth on vomer Λ-shaped, in a single row, the ones at the angles enlarged, 1 or 2 small ones between them at the sides; small multifid dermal flaps at nape, over eye, and above nostril; interorbital concave at the middle, $\frac{2}{3}$ the diameter of eye; gill rakers small and short, $3 + 6$ in number. First dorsal spine inserted behind eye a distance equal to diameter of eye, about $\frac{1}{6}$ shorter than longest spine; soft dorsal the higher; origin of anal midway between snout and base of caudal, not running as far back as dorsal; pectoral reaching a little past front of anal; ventrals long and slender, inserted a little in front of pectorals, their ends not reaching vent; caudal rounded. Color in spirits, brownish gray, with about 6 wide irregular cross bars which are darker toward their edges, 2 black streaks running downward and backward from eye; cheeks and opercles with many small light blue spots; spinous dorsal mottled and spotted with darker, other fins with small irregular dark lines running across the rays; ventrals dusky; tentacles on head black. Described from a specimen 5 inches in length from La Paz, Lower California. Pacific coast of Tropical America from Gulf of California to Panama; common in rock pools; representing on the Pacific coast the scarcely different *L. nuchipinnis.* (Named for John Xantus.)

Labrosomus xanti, GILL, Proc. Ac. Nat. Sci. Phila. 1860, 107, Cerro Blanco (Type, Nos. 2334, 2335, 2478. Coll. J. Xantus); JORDAN & GILBERT, Proc. U. S. Nat. Mus. 1882. 368.

2705. LABRISOMUS BUCCIFERUS, Poey.

Head $3\frac{4}{5}$ in total length with caudal; depth 5. D. XX, 11; A. II, 19; eye 4 in head, a little longer than snout; anterior nostril with a little tube; lower jaw longer; forehead convex, the snout short; mouth large, reaching beyond middle of eye; a few filaments on nape and 1 above preopercle. Teeth cardiform, the outer ones large; teeth on vomer (and palatines). Body scaly. Lateral line complete. First ray of dorsal longest, the others forming a convex curve up to the 19, which is shortest; pectoral moderate, of 12 rays, the lower thickened. Color brownish yellow, with vertical brown points extending on fins; a series of pale points along sides; the head gray, cheek dark brown. One specimen 55 mm. long. Cuba. (Poey.) Not seen by us. (*bucca,* cheek; *fero,* I bear).

Labrisomus bucciferus, POEY, Synopsis, 399, 1868, Cuba. (Coll. Poey.)

2706. LABRISOMUS MICROLEPIDOTUS, Poey.

Length of head equal to depth; pectoral $1\frac{2}{3}$ in head; eye $1\frac{2}{3}$ in snout, $4\frac{1}{2}$ in head. Maxillary reaching to base of middle of eye, $2\frac{1}{2}$ in head. Mouth oblique, with strong teeth; the lower jaw the longer; small fringe of tentacles at anterior nostril above eye, and a comb of fringes at the nape. First 3 dorsal spines subequal, considerably lower than the second 3, which become progressively longer. Scales in lateral line about 110. Head brown, with small yellow spots scattered over its lower part and on the gill membranes. Pectoral and caudal with some black points. Cuba.

(Poey.) Known from an imperfect description, with a drawing of the head of a specimen 180 mm. long. (μικρός, small; λεπιδωτός, scaly.)

Labrosomus microlepidotus, POEY, Anal. Soc. Esp. Hist. Nat., XIX, 1880, 246, 1, 8, f. 2, Cuba. (Coll. Poey.)

874. MNIERPES, Jordan & Evermann.

Mnierpes, JORDAN & EVERMANN, Check-List Fishes, 468, 1896 *(macrocephalus).*

This genus is close to *Labrisomus,* from which it differs chiefly in the very elongate body and in the absence of an orbital tentacle. The dorsal spines are more numerous, and probably the vertebræ also. The lips are thick and there is no trace of hook on the shoulder girdle. A band of fillitorm teeth in the jaws behind the anterior series; teeth on vomer, none on palatines. (μνίον, moss; ἕρπης, creeper.)

2707. MNIERPES MACROCEPHALUS (Günther).

Head 4¼; depth 6 to 6¼ (7¼ in total). D. XXII, 12; A. II, 24; C. 13; P. 13; V. I, 3; scales about 70. The head is depressed, rather short, nearly as broad as long; crown of the head broad and flat; interorbital space concave, narrower than the orbit. Snout very short, obtuse, rounded; the maxillary not extending to behind the posterior margin of the orbit; lips thick. Teeth in jaws forming a band with an outer series of stronger ones; vomerine teeth in a narrow band; palatine teeth none. No orbital tentacles, those at the nostril and on the neck very small. Gill openings wide, the gill membranes being united at the throat. Head naked; scales on the body not very small, cycloid. Dorsal fin commencing at occiput, and terminating near base of caudal, the spines flexible, and much lower than the soft rays; the 3 anterior ones rather more remote from one another than the following; none of the rays of this or of the other fins branched; caudal rounded; anal higher posteriorly than ante_ riorly, about as high as the spinous dorsal; pectorals rounded, with the middle rays longest, shorter than the head; ventrals jugular, ¼ as long as the pectoral, with the spine and the outer ray enveloped in a common thick membrane. Dark grayish olive; head and fins blackish; head, base of the pectoral, anterior part of the body, and dorsal dotted with white. Pacific coast of Central America. (Günther.) Known from a few specimens from Panama. Those examined by us (Mus. Comp. Zool.) have the sides much freckled and mottled with pale. (μακρός, long; κεφαλή, head.)

Clinus macrocephalus, GÜNTHER, Cat., III, 267, 1861, Pacific coast of Central America (Coll. Capt. John M. Dow); GÜNTHER, Fish. Centr. Amer., 442, pl. 69, fig. 2, 1869. *Labrosomus macrocephalus,* JORDAN, Proc. U. S. Nat. Mus. 1885, 389.

875. GOBIOCLINUS, Gill.

Gobioclinus, GILL, Proc. Ac. Nat. Sci. Phila. 1860, 102 *(gobio).*

Body robust; head broad, depressed, with a very short muzzle. Eyes large, approximated, close together; palatine teeth present; a tentacle above eye; no nuchal filaments. Scales very large, about 30 in the lateral line. Spinous dorsal of 18 spines. This genus seems to differ from *Labrisomus* in the large scales, differently formed head, and in the absence of nuchal filaments. (*Gobio*, the gudgeon; *Clinus*.)

2708. GOBIOCLINUS GOBIO (Cuvier & Valenciennes).

Head 3½ in total length; depth 4½. D. XVIII, 19; A. II, 17; C. 15; P. 14; V. 2; scales 30–10. Head nearly as broad as long, its height a third less. Eye large, 2¼ in head, twice interorbital space; a very small tentacle over the eye, another on the nostril. Profile rounded between the eyes, descending vertically to the snout, which is very short. Cheeks inflated; the skull a little rough. Mouth reaching to opposite middle of eye, somewhat black; teeth small, conic, and pointed; upper jaw with 26 equal teeth, the lower with 16, the last 2 larger and more curved; teeth on vomer and palatines, simple, in 2 irregular rows; gill membranes united, free from isthmus. Body posteriorly compressed. Dorsal slightly notched between spines and soft rays of anal; pectorals equal to ventrals, 5 in total length; caudal obtuse, 6 in total length. Lateral line disappearing opposite tip of ventral. Color greenish, with traces of cloudy brownish; the cross bands a deep brown, pointed at base of caudal. Lesser Antilles. Known from several specimens, one 2 inches in length. (Cuvier & Valenciennes.) Not seen by us; apparently a strongly marked species. (*Gobio*, the gudgeon, from its resemblance to *Cottus gobio*, the miller's thumb.)

Clinus gobio, CUVIER & VALENCIENNES, Hist. Nat. Poiss., XI, 395, 1836, Lesser Antilles. (Coll. Plée.)
Gobioclinus gobio, GILL, Proc. Ac. Nat. Sci. Phila. 1860, 102.

876. STARKSIA, Jordan & Evermann.

Starksia, JORDAN & EVERMANN, Proc. Cal. Ac. Sci. 1896, 231 (*cremnobates*).

This genus is related to *Labrisomus*, differing in the large scales, the presence of palatine teeth, the short soft dorsal fin, and the absence of the comb of nuchal filaments. (Named for Mr. Edwin Chapin Starks. in recognition of his work on the fishes of Western America.)

2709. STARKSIA CREMNOBATES (Gilbert).

Head 3½ in length; depth 4½. D. XXI or XXII, 8; A. II, 19; scales 37. In appearance resembling very strongly the species of the genus *Auchenopterus*. Body slender, snout sharp, the jaws equal; mouth wide, oblique, the maxillary reaching vertical from posterior margin of orbit, 2¼ in head. Teeth small, villiform, forming a band in front of upper jaw, the outer series enlarged; in lower jaw a single series laterally, becoming double in front; similar teeth on vomer and palatines. Eye longer than snout, 4 in head; interorbital width less than diameter of pupil; opercle terminating in an evenly convex process behind, without spinous points; gill membranes broadly united, free from isthmus; no hook on inner edge of shoulder girdle; nostrils with a flap; a single slender filament above eye and 1 or more on each side of the nape. A slight notch between first and third dorsal spines and another between the eighteenth and twenty-first spines; the spines are low and strong, the highest equaling the snout and $\frac{1}{4}$ eye; soft rays higher, the longest equaling $\frac{1}{4}$ head; caudal short, rounded, entirely free from dorsal and anal; anal similar to soft dorsal, the first 2 rays spinous; ventrals inserted well in advance of pectorals, each consisting of 1 spine and 2 soft rays, which are joined only at base; pectorals pointed, the lower rays the longest, $1\frac{1}{4}$ in head. Scales large, cycloid, the lateral line running high in front, descending to middle of sides immediately behind pectorals, thence running straight to tail. In the types, which are probably immature, the pores are not developed on posterior part of body. Color in spirits, uniform light olivaceous, a small dusky spot behind orbit and 1 below and behind it; opercle dusky. In 1 specimen the rays of soft dorsal, anal, and caudal are finely barred with dusky. (Gilbert.) Length $1\frac{1}{2}$ inches. Gulf of California. Two specimens known, from *Albatross* Station 3001, in 71 fathoms. (*Cremnobates;* κρημνοβάτης, one that haunts rocks; a synonym of *Auchenopterus.*)

Labrosomus cremnobates, GILBERT, Proc. U. S. Nat. Mus. 1890, 100, Gulf of California. (Coll. *Albatross*).
Starksia cremnobates, JORDAN, Proc. Cal. Ac. Sci. 1896, 231.

877. CRYPTOTREMA, Gilbert.

Cryptotrema, GILBERT, Proc. U. S. Nat. Mus. 1890, 101 (*corallinum*).

This genus differs from *Labrisomus* chiefly in the absence of nuchal filaments and in the modified anterior portion of the lateral line, which runs on a series of enlarged scales having no externally visible pores. (κρυπτός, concealed; τρῆμα, pore.)

2710. CRYPTOTREMA CORALLINUM, Gilbert.

Head $3\frac{4}{5}$ to 4 in length; depth $5\frac{1}{4}$. D. XXVII, 12; A. II, 27. Body elongate, regularly tapering backward to caudal peduncle, whose depth equals length of snout, which is sharp; mouth nearly horizontal; maxillary reaching middle of eye or beyond, $2\frac{1}{4}$ to $2\frac{1}{2}$ in head; teeth strong, but none of them enlarged, in a single series in jaws laterally, becoming double anteriorly; teeth on vomer and in a small distinct patch on front of palatines; eyes large, the interorbital space flat, nearly $\frac{1}{4}$ diameter of orbit; orbit slightly exceeding length of snout, $3\frac{1}{4}$ in head; branchiostegal membranes broadly united, free from isthmus, the posterior edge on vertical from preopercular margin; anterior nostril in a short tube, a slender flap arising from its posterior margin; a pair of simple slender filaments arising from the upper edge of each orbit, 1 on each side of nape, none others on head; gill rakers very short and weak; shoulder girdle without hook on its inner edge. Scales rather large, cycloid, the head alone naked; lateral line in its upper anterior portion without externally visible tubes, its position shown by a series of enlarged scales twice the size of the others; on these the tubes are wholly on the under side, each opening anteriorly by a single pore under the edge of the preceding scale; anteriorly the lateral line runs near the back and parallel with it, becoming suddenly declined behind middle of trunk, thence running on middle of side; the oblique portion of lateral line rests on about 7 scales, and the externally visible tubes of lateral line begin at this point; posterior portion of lateral line contained 1⅓ to 1¾ in dorsal portion; scales of lateral line, 45 in dorsal portion, 7 in oblique portion, and 18 in posterior portion. A slight notch behind fourth dorsal spine, the second and third spines slightly longer than those following, the first little longer than the fourth, the longest spine about 2¼ in head; first 2 anal rays spinous, but weak and flexible; last dorsal and anal rays not joined by membrane to caudal peduncle, the depth of the latter equaling the length of its free portion; ventrals long and narrow, nearly reaching vent in males, consisting of 1 spine and 3 simple rays; pectorals with some of the lower rays longest, 1⅓ in head; all of pectoral rays simple, 14 in number; caudal fin truncate, 1¾ to 1⅓ in head. Length 5 inches. Color dusky olive above, with irregular narrow longitudinal streaks of bright coral red, and 7 round black blotches above middle of sides; reticulating red lines and spots on top and sides of head and snout; branchiostegal membranes dusky in males; 2 red streaks on base of pectorals; dorsal somewhat dusky, marked with lines of red spots; caudal with 3 rather faint cross bars; pectorals, ventrals, and anal largely black in males, pale in females; the red shades persistent in alcohol. Santa Barbara Islands. Three specimens from *Albatross* Station 2945, in 30 fathoms. (Gilbert.) (*Corallina*, a calcareous alga, among which it lives.)

Cryptotrema corallinum, GILBERT, Proc. U. S. Nat. Mus. 1890, 101, off Santa Barbara Islands. (Coll. C. H. Gilbert.)

878. EXERPES, Jordan & Evermann.

Exerpes, JORDAN & EVERMANN, Proc. Cal. Ac. Sci. 1896, 232 (*asper*).

Body slender, much compressed; the snout long, sharp in profile; first dorsal ribbon-shaped, the 3 slender spines close together, inserted at the nape, much in advance of the rest of the fin; ventrals very long and slender. Otherwise as in *Auchenopterus*, the scales large, and but 1 soft ray in the dorsal fin. (ἔξω, without; ἕρπης, creeper.)

2711. EXERPES ASPER (Jenkins & Evermann).

Head 3 (3⅔ in total); depth 5¼ (6¼); eye 4⅓ in head; scales 6–43–7, about 40 pores. D. III–XXV, 1; A. II, 20. Body compressed; head narrow, pointed; snout long, lower jaw slightly the longer; mouth a little oblique, cleft moderate, maxillary not reaching nearly to vertical at front of orbit. Teeth in 1 well-defined outer series and a broken inner one, those in the outer series strongest and of pretty uniform size, short and broad; vomerine teeth in a single patch; no palatine teeth. No tentacles of any kind about the head. Profile nearly straight from snout to origin of first dorsal, but very slightly arched from there to base of caudal fin. Scales rather large, cycloid, about 6 rows between origin of second dorsal and lateral line just behind its angle, and about 7 from there to middle of ventral surface; 9 rows from origin of second dorsal to upper limb of opercle; entire head, opercles, and fins naked. Lateral line beginning at upper limb of opercle on a level with the pupil, almost exactly under the middle of the first dorsal fin, and a little more than $\frac{1}{4}$ the distance from top of nape to the under side of the throat, arching gently for 7 or 8 scales, leaving but 1 row of scales between it and the first spines of the second dorsal; on the ninth, tenth, and eleventh scales it bears slightly downward until 2 rows are left between it and the dorsal, then a sharp turn is made which puts it 4 scales further down, and from there it pursues a nearly direct line to middle of base of caudal fin. Dorsal fins separate, the first of 3 slender, very close-set, flexible spines, their length about twice in that of head, the fin ribbon-shaped; second dorsal separated from first by a distance somewhat greater than diameter of eye, and composed of 25 rather stout, sharp spines and 1 terminal soft ray; the first 3 are graduated, the first being contained $1\frac{1}{2}$ times in distance between the 2 fins, the second is about $\frac{1}{4}$ longer, and the third still a little longer; the remaining 22 are of approximately equal length, about equaling distance from origin of first dorsal to that of second; the 1 soft ray somewhat shorter than spines, well separated from caudal by a space equal to that between dorsals; pectorals inserted under middle of space separating dorsals, composed of 14 rays, equaling eye and snout in length, and reaching slightly past origin of anal; ventral of 2 rays inserted directly under origin of first dorsal and considerably in front of pectorals, which they somewhat exceed in length, in some specimens reaching vent; anal fin beginning slightly in front of posterior end of pectorals, a little lower than second dorsal and reaching a trifle nearer to caudal fin; first spine longer and more slender than the first regular dorsal spine, while the second equals the third dorsal in length. Caudal rounded, equaling in length the greatest depth of fish. Coloration in alcohol, pale, pretty regularly covered with very fine dark punctulations, thickest on back, palest below; a large dark opercular blotch, 2 similar postocular blotches, and usually a darkish bar extends downward from eye; upper half of preorbital region dark, outer margin of jaws dark; breast and under parts of head pale, top of head and nape dark; first dorsal quite dark, almost black; second dorsal pale, obscurely mottled with brown, which is disposed in about 5 indistinct areas; a large black ocellus upon the twelfth and thirteenth spines of second dorsal, and a similar one upon the twenty-third and twenty-fourth spines; each ocellus is surrounded by a narrow circle of white or pale orange. In the 6 specimens before us there is a slight variation as to the exact position of the 2 ocelli; in 1 example the second ocellus extends back upon the twenty-fifth spine also, but in every case the twelfth and thirteenth and the twenty-third and twenty-fourth are the spines which most evidently locate the spots; pectorals and ventrals plain; anal paler than dorsal, sparsely covered with fine dark points, so grouped as to form 3 or 4 darker areas. Length $2\frac{1}{4}$ inches. Gulf of California. Known from 6 specimens taken from masses of kelp hauled out by the seine from the bay of Guaymas. (Jenkins & Evermann.) (*asper*, rough.)

Auchenopterus asper, JENKINS & EVERMANN, Proc. U. S. Nat. Mus. 1888, 154, Guaymas, Mexico. (Type, No. 39643. Coll. Jenkins & Evermann.)

879. AUCHENOPTERUS, Günther.

Auchenopterus, GÜNTHER, Cat., III, 275, 1861 (*monophthalmus*).
Cremnobates, GÜNTHER, Proc. Zool. Soc. Lond. 1861, 374 (*monophthalmus*). Substitute for *Auchenopterus*, regarded as preoccupied on account of its similarity to *Auchenipterus*, a genus of *Siluridæ*.
Corallicola, JORDAN & EVERMANN, new subgenus (*marmoratus*).

Body moderately elongate, compressed, covered with rather large, cycloid scales; head shortish, naked, the snout rather pointed; cheeks full; mouth moderate, with a band of conical teeth in the jaws and about 1 series on the vomer, none on the palatines; lower jaw prominent; gill membranes united, free from the isthmus; upper surface of head with tentacles. Dorsal fin composed of stiff spines, with but a single soft ray, which is lower than the spines; first 3 spines more or less separated from the others, stiff and rather wider set, sometimes higher than the others; anal fin low, with 2 short spines; ventrals jugular, well developed; pectorals broad; lateral line complete, strongly curved anteriorly. Warm seas. This genus differs from *Cristiceps* in having but 1 soft ray in the dorsal fin, and in the large scales. ($a\mathring{v}\chi\mathring{\eta}\nu$, nape; $\pi\tau\varepsilon\rho\acute{o}\nu$, fin.)

CORALLICOLA (*Corallus*, coral; *colo*, I inhabit):
    a. First 3 or 4 spines of dorsal forming a separate fin, being much higher than any of the spines in the posterior part of the fin; snout rather acute.
        b. Scales 33; dorsal with 1 ocellus, anal with none; a black cross bar at base of caudal; a yellow spot behind eye; snout pointed.   NIGRIPINNIS, 2712.
        bb. Scales 37 or 38.
            c. First dorsal spine longer than second; dorsal with 2 ocelli; anal blackish; D. IV-XXIV, 1.   ALTIVELIS, 2713.
            cc. First dorsal spine shorter than second; snout slender, very acute; caudal pale; dorsal with 2 ocelli, anal with 1; D. III-XXII, 1.   MARMORATUS, 2714.
AUCHENOPTERUS:
    aa. First 3 spines of dorsal scarcely forming a separate fin, none of them higher than the posterior spines; snout not very acute; anal without ocellus.
        d. Caudal fin pale, usually with a dark bar at its base; a notch between third and fourth dorsal spines.
            e. Dorsal spines about 31.
                f. Scales 34 to 36; membrane of third spine joining fourth at its base; dorsal and anal plain dusky.   AFFINIS, 2715.
                ff. Scales 38.
                    g. Membrane of third spine joining fourth slightly above its base.   MONOPHTHALMUS, 2716.
                    gg. Membrane of third spine joining fourth spine much above its base.   INTEGRIPINNIS, 2717.
            ee. Dorsal spines about 28; membrane of third spine joining fourth above its base; scales 38; body with distinct cross bars; dorsal with 1 ocellus.   FASCIATUS, 2718.
        dd. Caudal fin black; body chiefly black; head mottled with whitish; membrane of third dorsal fin joining fourth near its summit, the fin not notched; dorsal spines 30; dorsal with 2 ocelli.   NOX, 2719.

Subgenus CORALLICOLA, Jordan & Evermann.

2712. AUCHENOPTERUS NIGRIPINNIS (Steindachner).

Head 4; depth 5⅓. D. XXVIII, 1; A. II, 27; scales 33; eye 4⅓ in head; snout 4⅓, equal to interorbital space; snout pointed. Three first dorsal spines higher than the others and further apart. A tentacle over eye.

Scales of body much largest anteriorly; lateral line arched. A deep black spot with a white ring between the twenty-second and twenty-fourth spines; anal edged with white; black cross band at base of caudal with silvery point at upper base of pectoral; a diffuse yellowish spot below and behind eye. Barbados. One specimen 1 inch and 7 lines long. (Steindachner); not seen by us. (*niger*, black; *pinnis*, fin.)

Clinus nigripinnis, STEINDACHNER, Ich. Notizen, VI, 46, 1867, Barbados.

2713. AUCHENOPTERUS ALTIVELIS (Lockington).

D. IV–XXIV, 1; A. 21; P. 13; C. 13; V. 2; scales 37. Body compressed, greatest depth a little behind pectoral axil; greatest thickness at gill covers; dorsal and abdominal profiles of similar curvature, decreasing regularly to the caudal fin; profile of occiput and superorbital regions convex; snout somewhat produced, its upper outline slightly concave. Head 4 in total length; greatest depth a little less than length of head; caudal peduncle about ¼ of the greatest depth. Eye round, lateral, with a slight direction upward, its diameter less than the length of the snout; interorbital area nearly equal in width to the diameter of the eye, concave transversely, upper orbital borders slightly raised. A short nasal tentacle slightly anterior to the front margin of the eye; a large fimbriated tentacle on each side of the first dorsal ray. Cleft of mouth oblique, the lower jaw the longer; the posterior convex extremity of the club-shaped maxillary about vertical with the center of the pupil. Teeth of the outer row regular, sharp, incurved, the largest in front, gradually decreasing along the lateral portions of the jaws, and not extending much past the middle of their length; a narrow band of small teeth in the rear of the outer row; vomerine teeth present. Branchiostegals 6; gill openings continuous, membranes not attached to the isthmus. Distance from first ray of dorsal to posterior margin of eye equal to length of snout; first 2 rays of dorsal much developed, the first slightly the longer, and nearly equal in height to the distance of its base from the tip of the upper jaw; third ray about ⅓ the length of the first; fourth very short; succeeding rays to the twenty-sixth longer than the third, the last 3 somewhat decreasing. Anal commencing under eleventh dorsal ray, coterminous with, and equal in height to, the dorsal. Caudal with 13 simple jointed rays, the longest in the center, posterior margin convex. Pectorals narrow, lanceolate, the fifth and sixth rays longest and ⅔ the length of the head. Ventrals inserted in advance of the pectorals. Lateral line with 37 simple pores, parallel with dorsal outline to opposite the origin of the anal, where it is deflected almost perpendicularly downward to the middle of the side of the body, along which it continues to its termination. Scales rather large, about 10 in a transverse row in the central part of the body, their posterior margin membranaceous; no scales on fins; a line of pores around the margin of the orbit, another along the posterior margin of the preoperculum, connected with each other and with the lateral line by a line from the center of the binder border of the eye. Color in alcohol, bright pink above, becoming dusky below; underside of head light olivaceous, lower lip blackish; dorsal pink, dusky on its margin, a black spot on the fourth ray, and another on its hinder part upon the twenty-fourth and twenty-fifth rays, the latter spot extending on to the body; membrane of anal black; occipital tentacles black. La Paz, Lower California. A single specimen, $1\frac{9}{10}$ inches long, dredged at a depth of 22 fathoms. (Lockington.) (*altus*, high; *velum*, sail.)

Cremnobates altivelis, LOCKINGTON, Proc. Ac. Nat. Sci. Phila. 1881, 116, La Paz, Lower California. (Coll. W. J. Fisher.)

2714. AUCHENOPTERUS MARMORATUS (Steindachner).

Head $3\frac{2}{3}$ to $3\frac{4}{5}$ in body; depth $3\frac{1}{2}$ to $3\frac{5}{8}$. D. III–XXII, 1; A. II, 19; scales 2-36-9 (28 or 29 anteriorly); eye 4 to 5 in head; first dorsal $1\frac{1}{4}$; pectoral $1\frac{1}{4}$. Body comparatively deep, compressed, the back somewhat arched; head pointed; mouth large, the maxillary extending to behind the eye, 2 in head; opercle with a sharp spine; jaws equal; teeth pointed, in narrow bands, the outer larger; vomerine teeth in 1 row; supraocular tentacle small, about as large as nuchal tentacle; no nasal tentacle. Pectoral a little shorter than head; dorsals separate, the first dorsal higher than second dorsal, the spines of which are about $\frac{1}{4}$ head. Color in life of varying shades of olive gray or sand color, with a series of whitish blotches on head and along sides; markings on dorsal and anal whitish; 2 dark-blue ocelli on dorsal and 1 on anal, these edged with orange and interiorly with black; ventrals, pectorals, and caudal whitish, barred with clear orange red; first dorsal black at tip; a curved blackish line at base of caudal; lower side of head yellowish brown, with whitish bands; specimens from coral reefs more spotted. Florida Keys to Cuba; common in the eelgrass at Key West. Our specimens, 2 to $2\frac{1}{4}$ inches long, taken at Key West and Havana. (*marmoratus*, marbled.)

Cremnobates marmoratus, STEINDACHNER, Ichth. Beiträge, v, 174, pl. 12, f. 6, 1876, a small rocky island north of Cuba; JORDAN & GILBERT, Synopsis, 962, 1883; JORDAN & GILBERT, Proc. U. S. Nat. Mus. 1884, 142.

Subgenus AUCHENOPTERUS.

2715. AUCHENOPTERUS AFFINIS (Steindachner).

Head 4; depth $4\frac{2}{3}$. D. III–XXVII, 1; A. II, 19; V. I, 2; scales 33 to 35. Form of *A. integripinnis;* maxillary reaching to below posterior margin of eye; a fringed tentacle above eye and 1 on each side of occiput. First dorsal low, its longest (second) ray shorter than the highest of second dorsal; membrane of third spine joining the fourth spine just above its base; last ray of second dorsal joined by membrane to base of caudal. Dark brown, paler than in *A. nox*, but darker and more uniform than in *Auchenopterus fasciatus;* lower side of head pearly gray, thickly speckled with darker; sides with 5 very faint darker cross bands; dorsal and anal dusky, the latter with a pale edge; between the eighteenth and twenty-second dorsal spines a large dark spot ocellated with yellowish; caudal yellowish white, with darker cross streaks, a blackish band at its base; pectoral dusky at base, its posterior half yellowish, with darker cross streaks; ventral similar; a wedge-shaped, whitish band extending backward from eye to opercle. West Indies; recorded from Key West and St. Thomas. Here described from specimens from Key West. (*affinis*, related,—to *A. monophthalmus.*)

Cremnobates affinis, STEINDACHNER, Ichth. Beiträge, V, 178, 1876, St. Thomas; JORDAN, Proc. U. S. Nat. Mus. 1884, 142; JORDAN, Cat. Fishes N. A., 121, 1885.

2716. AUCHENOPTERUS MONOPHTHALMUS, Günther.

Head 3⅔; depth 4. D. III–XXVI, 1; A. II, 18; scales 2-32-9; eye 5 in head; maxillary 1⅖; pectoral 1⅕; caudal 1¼. Body compressed, deepest at middle of pectorals; head moderately pointed, the upper profile slightly and evenly convex; mouth large, maxillary reaching past eye; jaws subequal; teeth villiform, in bands on jaws, vomer, and palatines; interorbital space flat, as wide as eye; a multifid dermal flap over posterior edge of eye, and a smaller one on each side of nape; head naked; body with rather large, regular scales; fins naked. Origin of dorsal over edge of preopercle, the first 3 spines separated from rest of fin by a rather deep notch, the membrane from third spine joining fourth spine at about its middle; spines of posterior part of dorsal the highest; front of anal midway between tip of snout and base of caudal, tips of last rays reaching slightly beyond base of caudal and tips of last dorsal rays; pectorals reaching front of anal; ventrals long and slender, inserted in front of base of pectorals a distance equal to 1⅓ eye; caudal rounded. Color light grayish red or brown, with about 6 cross bars of darker brown, running up on dorsal; between the bars are scattered milky white irregular spots; a black spot, ocellated with white, on front of dorsal, a similar spot near posterior end, sometimes duplicated; narrow cross bars on anal; a dark bar on base of caudal, and a dark blotch on base of pectoral. Here described from specimens, a couple of inches in length, from La Paz, Lower California. Gulf of California to Panama, abundant in rock pools, creeping about among *Corallina;* close to *A. integripinnis*, but the first dorsal higher and more separate from rest of fin. (μόνος, one; ὀφθαλμός, eye, from the dorsal ocellus.)

Auchenopterus monophthalmus, GÜNTHER, Cat. Fish., III, 275, 1861, Panama; JORDAN, Proc. Cal. Ac. Sci. 1895, 501.
Cremnobates monophthalmus, GÜNTHER, Proc. Zool. Soc. Lond. 1861, 374; GÜNTHER, Fish. Centr. Amer., 442, pl. 69, fig. 1, 1869.

2717. AUCHENOPTERUS INTEGRIPINNIS (Rosa Smith).

Head 3½; depth 4½; eye 4 in head. D. III–XXVII, 1; A. II, 20; scales 2-36-9; pectoral 1⅕; caudal 1¼. Head stout, broad, conical; mouth little oblique, maxillary reaching posterior margin of eye; eyes large; nasal, supraocular and nuchal regions with fringed cirri, those at the nape flaplike. First and second dorsal spines low, a little higher than the third, which, in turn, is higher than the fourth and separated from it by an interspace, the membrane between the third and fourth spines not deeply emarginate, membrane from third spine attached to the lower ⅔ of fourth; anterior spines not forming a separate fin; highest anterior spine not higher than the highest of the posterior part of fin. Color dark brown, variegated with different shades of brown and reddish; about 5 indistinct dark cross bars; a distinct ocellated black spot on posterior part of dorsal fin; caudal fin abruptly translucent, speckled, a black bar at its base; base of pectorals violet, bordered with black, the rest of the fin checkered; ventrals barred. Length 2⅓ inches. Coast of California and southward to Todos Santos; abundant in rock pools among *Corallina.* Here described from a specimen, 1¼ inches in length, from San Cristobal, Lower California. (*integer,* entire; *pinna,* fin.)

Cremnobates integripinnis, ROSA SMITH, Proc. U. S. Nat. Mus. 1880, 147, La Jolla, near San Diego (Coll. Rosa Smith); JORDAN & GILBERT, Synopsis, 764, 1883.

2718. AUCHENOPTERUS FASCIATUS (Steindachner).

Head 4; depth 4¼. D. III–XXIV, 1; A. II, 18; scales 37. Body rather slender, a little deeper than in *A. integripinnis,* the snout less acute than in *A. marmoratus.* First dorsal spine rather higher than second and lower than the spines of posterior part of fin; membrane of third spine joining second dorsal at a point above its base, the two parts of the fin therefore separated only by an emargination. Tentacle above eye slender, small; cirri on side of occiput bluish. In life, light pinkish brown, much mottled, and with traces of 6 to 8 faint darker bars; head and its cirri above whitish; 3 blackish spots behind eye, radiating from it, the lower one largest; preopercle with 3 dark dots; dorsal pale, with 9 blackish blotches, in the next to the last of which is a large blue-black ocellus, edged with orange; anal with 5 dark blotches and no ocellus; a blackish bar across base of caudal; rest of caudal and pale part of anal with dark dots; ventrals whitish, barred with black; pectoral similar, its base with a whitish area, which has a brown center, below which is a small black spot. Length 2 inches. Florida Straits; north to Key West. Here described from specimens from Key West. (*fasciatus,* banded.)

Cremnobates fasciatus, STEINDACHNER, Ichth. Beiträge, V, 176, 1876, Florida Straits; JORDAN, Proc. U. S. Nat. Mus. 1884, 142; JORDAN, Cat. Fishes N. A., 121, 1885.

2719. AUCHENOPTERUS NOX (Jordan & Gilbert).

Head 3⅔; depth 3⅝. D. III–XXVII; A. II, 18; lateral line with 34 tubes. Snout not very acute, the upper and lower profiles of head nearly evenly convex; mouth large, maxillary reaching slightly beyond eye, ⅓ length of head; eye large, equaling length of snout, greater than interorbital width, 4 in head (to end of opercular spine); interorbital width 4¾ in head; nasal, supraorbital, and occipital tentacles present, those on snout and above the orbits simple, slender filaments, the latter about as long as diameter of orbit, 1 of them divided to the base, the other simple; the tentacle on each side of nape a compressed slip of skin higher than wide, the margin uneven, but not fringed. Anterior dorsal spines not much elevated, not higher than some of the posterior spines; the first and second spines about equal, 2½ in head; the third spine shorter, about equal in length to the fourth, from which it is separated by a wide membrane, which is, however, not at all notched; the spines thence increase in length toward the last; caudal 1¼ in head; pectorals reaching anal, nearly equaling length of head; ventrals not reaching vent, 1½ in head. Scales large, 4 series above lateral line and 4 below. Color, body and fins uniform blackish brown; a few small silvery-white specks on dorsal region, mostly along base of dorsal fin; head and base of pectoral fin with light pink areas and mottlings; snout pink above; nape with a pink cross bar; a dark streak upward and backward from eye to nape; a light streak from eye backward to opercle and 1 backward and downward; lower jaw mottled with light and dark; a small round, black spot near base of dorsal between twenty-third and twenty-fifth spines, and 1 between twenty-eighth and thirtieth, both very faintly ocellated with lighter; slight whitish tips on ventrals and lower edge of caudal. Key West; known from a single specimen, 1¾ inches long, taken with the seine in algæ on a rocky bottom at Key West. Its congeners, *A. marmoratus*, *A. fasciatus*, and *A. affinis*, were found in the same waters, *A. marmoratus* being much the most abundant of the 4, and reaching the largest size. (*nox*, night.)

Cremnobates nox, JORDAN & GILBERT, Proc. U. S. Nat. Mus. 1884, 30, Key West. (Coll. Jordan.)

880. PARACLINUS, Mocquard.

Acanthoclinus, MOCQUARD, Bull. Soc. Philom. Paris 1885, 18 (*chaperi*); name preoccupied.
Paraclinus, MOCQUARD, Bull. Soc. Philom. Paris 1886, 11 (*chaperi*).

Body elongate, compressed, covered with cycloid scales; mouth large, each jaw armed with an external row of conical teeth, with some teeth behind; teeth on the palate; dorsal very long, continuous, composed entirely of spines, anal with 2 spines; ventrals jugular, with few rays; tentacles on head; gill opening very broad; 6 branchiostegals; lateral line interrupted. Evidently very close to *Auchenopterus*, from which it may be distinguished by the continuous dorsal fin, a character which needs verification. (παρά, near; *Clinus*.)

2720. PARACLINUS CHAPERI, Mocquard.

Head 4¼; depth 4¼. D. XXXI; A. II, 19; P. 13; V. 2; scales 35. Body elongate, very strongly compressed; eye large, equal to snout or interorbital width; lower jaw slightly the longer; mouth oblique, reaching front of eye; outer row of teeth strong, canine-like, slenderer and more close set above, below diminishing rapidly in length, the bands of small teeth limited to front of each jaw, a curved group of teeth on palate; dorsal beginning over preopercle, not notched, composed entirely of stout spines; anal equally long; ventrals very narrow, of 2 soft rays, well separated, the inner slightly longer than outer; head with 3 pairs of tentacles, 1 at the nape, filiform, small, ½ as long as eye; the second below the orbit, broadened at base, separated into 3 or 4 branches, progressively longer from the inner outward, longer than eye; nuchal filament a little in front of dorsal, in form, oblong, entire, laminated, a little broader at its free edge, ¼ as long as eye; scales large, cycloid; lateral line interrupted before front of anal, anterior part rounding over eye with only 2 rows of scales between it and the dorsal, posterior part median; gill membranes broadly united, free from isthmus. Body brownish yellow, fins brown, the base and the caudal darker. Bay of Guanta, near Barcelona, in Venezuela; 1 specimen, 33 mm. long to base of caudal. (Mocquard.) Not seen by us. (Named for its collector, M. Chaper.)

Acanthoclinus chaperi, MOCQUARD, Bull. Soc. Philom. Paris 1885, 19, Bay of Guanta, Venezuela.

Paraclinus chaperi, MOCQUARD, Bull. Soc. Philom. Paris 1886, 41.

881. EMMNION, Jordan.

Emmnion, JORDAN, in Gilbert. Proc. U. S. Nat. Mus. 1896, 454 (*bristolæ*).

Body elongate, covered with caducous, cycloid scales of small size; lateral line straight, ending near base of last dorsal ray. Head moderate, decurved anteriorly, without cirri; mouth moderate; teeth in jaws in bands, the outer enlarged; no teeth on vomer or palatines; dorsal notched, its anterior ⅔ of flexible spines of moderate height; ventrals I, 3, the rays thickish, the fin inserted slightly before pectorals; caudal free. Galapagos Islands. (ἐν, in; μνίον, sea moss, or alga.)

2721. EMMNION BRISTOLÆ, Jordan.

Head 5⅞; depth 7¼. D. XXV, 13; A. I, 27; P. 13; V. I, 3; Br. 5; scales 3-50-11, the count not certain. Body slender, moderately compressed, the dorsal profile forming a nearly straight line from occiput to first dorsal ray, from thence descending very gently to base of caudal; ventral profile about straight. Head broad, slightly convex above, its width 1¼ in its length; anterior profile from first dorsal spine to a point above eye straight, thence abruptly descending to tip of snout; mouth horizontal, the lower jaw included; maxillary reaching nearly to posterior margin of eye, about 2¼ in head. Teeth present on both jaws, canine-like; upper jaw with 8 enlarged teeth in front, about 2 or 3 series of much smaller teeth behind these, only 1 series of which extends into posterior region of mouth; lower jaw with a series of teeth in front and on sides which is greatly enlarged in front; a patch of very small teeth behind the enlarged front teeth; no teeth on vomer or palatines. Premaxillary very protractile; snout blunt, 4⅛ in head; eyes large, round, placed close together, 3¼ in head; interorbital region very narrow, less than pupil; nostrils equal. Caudal peduncle 2⅜ in head; branchiostegal membranes deeply united, free from isthmus; gills 4, a small slit behind the fourth; no cirri above eyes, nor filaments on nape; head naked, body covered with cycloid scales, those on nape much smaller; belly naked. The scales on the body are apparently caducous as all have fallen, but the points are very distinct and they seem to have been embedded on their anterior edge, as the sac-like fold of skin is prominent. Lateral line simple, straight, running from upper edge of gill opening to last ray of dorsal when it is lost, not reaching the caudal; it is placed very high, and gradually approaches the dorsal fin, from which it is separated only by a very small distance. Dorsal extending from a point a short distance behind occiput nearly to base of caudal, emarginate; last spine shortest, about 2¼ in first soft ray, which is 2⅓ in head; the longest spines about 3 in head, all the spines slender and flexible. Anal extending from behind vent nearly to base of caudal; similar to soft dorsal, its rays lower. Ventrals well developed with broad base, the rays thickish, inserted very slightly in front of base of pectorals, 1¼ in head, reaching ⅔ the distance to vent. Caudal subtruncate. Pectorals reaching past vent, about as long as head. Dorsal and anal free from caudal. Color in spirits, dark dull reddish-brown, lighter below; head very dark; dorsals, pectorals, and caudal blackish, pectorals and caudal with lighter blotches; anal and ventrals dusky, anal margined with darker. Length about 3 inches. Galapagos Islands; one specimen known, evidently a rock-pool species. (Named for Miss Susan Brown Bristol, of the department of zoology in Leland Stanford Junior University.)

Emmnion bristolæ, JORDAN, in GILBERT, Proc. U. S. Nat. Mus. 1896, 454, pl. 55, fig. 1, Galapagos Islands. (Coll. *Albatross*.)

882. ATOPOCLINUS, Vaillant.

Atopoclinus, VAILLANT, Bull. Sci. Philom. Paris, serie 8, tome VI, 1894, 73 (*ringens*).

Body elongate, subcylindrical, without visible scales. Head obtuse, the snout short, rounded; mouth inferior, transverse, with compressed trenchant teeth in each jaw, those above at least in a single row, solidly fastened to the skeleton; teeth on vomer and palatines uncertain. Dorsal continuous, extending the whole length of the back, from the nape to the caudal peduncle, its rays mostly simple, only the posterior articulate; anal occupying nearly ½ the length, touching the caudal, which is, nevertheless, distinct; caudal deeply forked; ventrals distinctly jugular, very long, of a spine and a ray; no tentacles; gill membranes apparently rounded at the isthmus. Gulf of California; a singular genus evidently closely allied to *Runula*. (ἄτοπος, strange; *Clinus*.)

2722. ATOPOCLINUS RINGENS, Vaillant.

Head 5; depth 7. D. 24; A. 18; P. 15; V. I, 1. Eye large, 7 in head; interorbital space broad, 3 in head. Caudal a little longer than head. Color clear chamois brown, the belly pale; a brown band before the snout, across the eye to the caudal, on which it extends; a silvery stripe bordering this band above, and below for part of its length. Gulf of California. (Vaillant); known from 1 specimen badly shriveled, 39 mm. in length. (*ringens*, gaping.)

Atopoclinus ringens, VAILLANT, Bull. Sci. Philom. Paris, serie 8, tome VI, No. 2, February 25, 1894, 74, Gulf of California. (Coll. Léon Diguet.)

883. RUNULA, Jordan & Bollman.

Runula, JORDAN & BOLLMAN, Proc. U. S. Nat. Mus. 1889, 171 (*azalea*).

Body slender, its back not elevated; mouth small, inferior, destitute of canines; teeth fixed, upper largest; dorsal fin continuous, its spines and soft rays indistinguishable, most of them articulate; caudal fin lunate; gill openings reduced to a vertical slit in front of pectoral; scales none. This genus is remotely allied to the East Indian genus *Petroskirtes*, but has the mouth and dentition different, and the caudal fin, unlike that of most blennioid fishes, is forked. (Diminutive of *runa*, a dart or javelin.)

2723. RUNULA AZALEA, Jordan & Bollman.

Head 4⅘; depth 6⅛. D. 42; A. 26 or 27; V. I, 2. Body moderately elongate, not much compressed; head rather long, its upper outlines convex; snout short and very blunt; mouth entirely inferior, transverse, each jaw provided with long, slender, close-set curved teeth; no evident posterior canines; eye moderate, equal to snout and nearly equal to interorbital width, 4 in head; no tentacles on head; gill membranes fully united to the isthmus, the gill opening reduced to a vertical slit, its lower edge opposite middle of base of pectoral; no scales; lateral line very high, concurrent with the back; dorsal fin very low, continuous; the feeble spines and soft rays indistinguishable, the fin beginning at occiput; anal similar to soft dorsal; caudal lunate behind, well separated from dorsal and anal; pectorals small, rounded, about 1⅘ in head; ventrals short, before pectorals, about 2 in head. Color reddish brown, silvery below, about 5 dusky cross shades; a dusky lateral streak; a black spot surrounded by paler at base of caudal; dorsal with about 6 black crossbars; anal with 4; other fins pale; lower half of head abruptly pale. Galapagos Archipelago. The type, 2 inches long, taken at Indefatigable Island; 3 more specimens have since been obtained from the same island. (ἀζαλέος, parched, from the brown color.)

Runula azalea, JORDAN & BOLLMAN, Proc. U. S. Nat. Mus. 1889, 171, Indefatigable Island, Galapagos Archipelago (Coll. *Albatross*); JORDAN, Proc. Cal. Ac. Sci. 1896, 233, pl. 37.

884. BLENNIUS (Artedi) Linnæus.

(BLENNIES.)

Blennius, ARTEDI, Genera Piscium, 27, 1738.
Blennius, LINNÆUS, Syst. Nat., Ed. x, 256, 1758 (*galerita*).
Salaria, FORSKÅL, Descr. Anim., 22, 1777 (*basiliscus*).
Pholis, FLEMING, Brit. Anim., 207, 1828 (*lœvis=pholis*); not *Pholis* SCOPOLI, 1777.
Adonis, GRONOW, Cat. Fish., Ed. Gray, 93, 1754 (*pavoninus = ocellaris*).
Lipophrys, GILL, American Naturalist, June, 1896, 498 (*pholis*).

Body oblong, compressed, naked; head short, the profile usually bluntly rounded; mouth small, horizontal, with a single series of long, slender, curved, close-set teeth in each jaw, besides which, in the lower jaw at least, is a rather short and stout fang-like canine tooth on each side;

3030——72

premaxillaries not protractile; gill openings wide, extending forward below, the membranes free from the isthmus, or at least forming a broad fold across it. Dorsal fin entire, or more or less emarginate, the spines slender; pectorals moderate; ventrals well developed, I, 3; no pyloric cæca; lateral line developed anteriorly. Species numerous, lurking under rocks and algæ in most warm seas; some species in the lakes of northern Italy. The European species in general are larger in size than ours, with higher fins. (*Blennius*, the ancient name, from βλέννα, slime.)

LIPOPHRYS (λείπω, to disappear; ὀφρύς, eye-brow):
   a. Supraorbital cirrus wanting; snout not very blunt in profile.
     b. Posterior canine present in each jaw; dorsal slightly emarginate; D. XII, 18.
                                      CAROLINUS, 2724.

BLENNIUS:
  aa. Supraorbital cirrus present; profile of snout more or less blunt.
    c. Canines strong, present in both jaws; no nuchal cirri.
      d. Dorsal rays XI or XII, 17 or 18.
        e. Supraorbital cirrus bifid; dorsal free from caudal.
          f. Supraorbital cirrus as long as head; dorsal emarginate: sides
             spotted; D. XI, 17.   FUCORUM, 2725.
          ff. Supraorbital cirrus as long as eye and snout; dorsal continuous;
             color olivaceous, with dark bars; D. XI, 18.
                                      STEARNSI, 2726.
        ee. Supraorbital cirrus bifid, nearly as long as head; last ray of dorsal
          joined to caudal; sides with a network of blue lines; D. XII. 18.
                                      FAVOSUS, 2727.
      dd. Dorsal rays XII, 21 or 22; supraorbital cirrus long, fringed; dorsal free
        from caudal; cheeks with network of lines; body nearly plain.
                                  PILICORNIS, 2728.
    cc. Canines short and stoutish, present in lower jaw only (undescribed in *truncatus*
      and in *marmoreus*.)
      g. Nape without cirrus; snout abruptly decurved; body robust, marbled; D.
        XII, 20.   ·   MARMOREUS, 2729.
      gg. Nape with a cirrus on each side.
        h. Dorsal and anal free from caudal. Nape with a filiform bifid tentacle
          on each side; teeth undescribed; supraorbital tentacle simple;
          color olive, with bright spots.
                                  TRUNCATUS, 2730.
        hh. Dorsal and anal with the last ray largely joined by membrane to
          caudal; nape with a small cirrus; posterior canines strong,
          in lower jaw only; dorsal not notched; color uniform brown;
          D. XII, 13.   VINCTUS, 2731.
    ggg. Nape with a comb of many close-set cirri on a fleshy crest; lower
      jaw only with short posterior canines; dorsal fin continuous, free
      from caudal; D. XII, 16 or 17.   CRISTATUS, 2732.

Subgenus LIPOPHRYS, Gill.

2724. BLENNIUS CAROLINUS (Cuvier & Valenciennes).

D. XII, 18; A. 17. Body rather long and slender, more elongate than in the European species, *Blennius pholis*, more compressed, the head longer; maxillary extending to opposite middle of eye; teeth $\frac{14}{14}$, with strong canines on both jaws; gill membranes free from isthmus; no trace of tentacles above eye; dorsal spines slender, a little lower than the soft rays, the fin little emarginate; dorsal and anal not joined to the caudal. Greenish, with 4 or 5 irregular dark spots or shades along the back; dorsal with a large black spot in front; anal brown-edged. South Carolina. Only the original type in the museum at Paris known; from this the present description was taken. No later collector has recognized the species and it may not be American.

Pholis carolinus, CUVIER & VALENCIENNES, Hist. Nat. Poiss, XI, 276, 1836, Carolina. (Coll. M. Bosc.)
Blennius carolinus, JORDAN & GILBERT, Synopsis, 760, 1883.

2725. BLENNIUS FUCORUM, Cuvier & Valenciennes.

Head 5 in total length. D. XI, 17; A. 18. Orbitial cirri nearly as long as head, bifid at tip, and fringed at the base. Dorsal fin slightly emarginate, free from the caudal, the spines rather stiff. Head very short and steep, its profile nearly vertical; 24 teeth in each jaw; each jaw with very strong canines; gill membranes free from the isthmus posteriorly. Olive green, becoming darker above, with numerous brown spots on the cheeks and sides of the body; below reddish; dorsal with a large black spot in front, behind which are smaller spots; spinous dorsal edged with paler. (Cuvier & Valenciennes.) Open ocean in floating *Fucus*; the type from near the Azores; recorded by De Kay from the open sea, off New York, in floating seaweed. (*fucorum*, of the seaweed, *Fucus*.)

Blennius fucorum, CUVIER & VALENCIENNES, Hist. Nat. Poiss., XI, 263, 1836, 240 miles south of the Azores (Coll. Claude Gay); GÜNTHER, Cat., III, 217, 1861; DE KAY, N. Y. Fauna: Fishes, 149, pl. 22, fig. 66, 1842; JORDAN & GILBERT, Synopsis, 710, 1883.
Blennius oceanicus,* CUVIER & VALENCIENNES, Hist. Nat. Poiss., XI, 265, 1836, open sea, 29° N., 50° W.; on a drawing by CLAUDE GAY.

2726. BLENNIUS STEARNSI, Jordan & Gilbert.

Head $3\frac{4}{5}$ ($4\frac{3}{5}$ in total); depth $4\frac{2}{3}$ ($5\frac{2}{3}$); eye $4\frac{1}{3}$; snout $4\frac{1}{3}$. D. XI, 18; A. II, 21. Body much elongate, compressed, tapering regularly behind; anterior profile moderately decurved; snout short and blunt; mouth large, oblique, the jaws even; maxillary reaching slightly beyond middle of orbit, $2\frac{1}{4}$ times in head; teeth in the front of the jaws only, occupying on each side a space equal to $\frac{1}{2}$ length of maxillary; teeth $\frac{2\ 6}{2\ 4}$, the lateral one on each side much enlarged and canine-like, rather short but strongly curved; canine in upper jaw equaling about $\frac{1}{2}$ diameter of pupil; interorbital space very narrow, not as wide as pupil; upper posterior rim of orbit with a long slender filament, forked at base, its length equaling distance from tip of snout to posterior rim of orbit; no filaments at the nape; gill membranes somewhat united to the isthmus in front, but forming a broad fold across it posteriorly, the gill openings of the two sides therefore continuous below. Dorsals rather high; no notch between the spines and soft portion, the membrane of last ray not reaching base of caudal; spines of

* Very near *Blennius fucorum*, the profile more oblique, the cirri shorter, the spinous dorsal lower, the caudal more truncate; anal shorter. Color brown with brown spots on body and fins; sides clear green; belly silvery. Length 2 inches. (Cuvier & Valenciennes.)

nearly uniform height, all very slender and flexible, the tips almost filamentous; highest spine $\frac{1}{4}$ length of head; highest soft ray $1\frac{3}{4}$ in head;
anal lower than dorsal, its longest ray very slightly less than $\frac{1}{4}$ length of
head; length of caudal peduncle more than $\frac{1}{4}$ its height, about equaling
the diameter of orbit; caudal about equaling pectoral, $1\frac{1}{8}$ in head; ventrals long, the inner ray much the longest, $1\frac{1}{8}$ in head, not quite reaching
vent. Color light greenish olive, somewhat mottled; sides with irregular
dark bars formed of spots, these extending on the fin; skin everywhere
finely punctate; dorsal dark olive, the spinous part darker at tip; anal
blackish, with paler edge; ventrals dusky; pectorals and caudal olive.
Gulf of Mexico, in deep water. Three specimens known, the largest 3
inches long, taken from the stomach of a Red Snapper, at Pensacola.
(Named for Silas Stearns.)

Blennius stearnsi, JORDAN & GILBERT, Proc. U. S. Nat. Mus. 1882, 300, Pensacola Snapper
Banks. (Type, No. 29669, U. S. Nat. Mus. Coll. Jordan & Stearns.)

2727. BLENNIUS FAVOSUS, Goode & Bean.

Head $3\frac{2}{3}$; depth $4\frac{2}{3}$. D. XII, 18; A. II, 20. Body comparatively elongate and compressed; anterior profile moderately decurved; head nearly
$\frac{1}{4}$ longer than deep; snout very short and blunt; mouth large, horizontal;
jaws even; maxillary reaching posterior margin of orbit, its length $2\frac{1}{4}$ in
head. Each jaw with a long, curved, posterior canine; the canines of
lower jaw largest. Preorbital $\frac{2}{3}$ diameter of eye, which is $3\frac{2}{3}$ in head, and
equals more than twice interorbital width. An extremely long and slen-
der supraocular cirrus, trifid to the base, the longest branch nearly as long
as the head; no nuchal cirri. Gill membranes forming a rather narrow fold
across the isthmus. Dorsal low, continuous, the spines very slender and
flexible, the longest $\frac{1}{4}$ as long as the head; the longest soft ray $\frac{3}{4}$ as long
as head; the last ray slightly joined to base of caudal; caudal $\frac{3}{4}$ as long
as head; anal rather high; pectorals $\frac{4}{5}$ as long as head; only the straight
part of lateral line developed. Color faded, brownish, finely reticulated,
a series of obscure bluish blotches along the sides; front and sides of head
marked with very distinct blue, reticulating lines surrounding honey-
comb-like hexagonal interspaces; top of head with many small blue spots;
dorsal with black dots and streaks; a black spot bordered with whitish
between the first and second dorsal spines; anal with oblique blue streaks,
the fin margined with dusky, tips of rays whitish; base of pectorals with
blue reticulations. The whole body was probably reticulated with blue
in life. Gulf of Mexico. Known from 2 specimens collected at Garden
Key, Florida, by Gustav Würdemann; they are $3\frac{2}{3}$ inches and 3 inches
long, respectively. (*favosus*, honeycombed.)

Blennius favosus, GOODE & BEAN, Proc. U. S. Nat. Mus. 1882, 416, Garden Key, Florida
(Type, No. 2629, U. S. Nat. Mus. Coll. Gustav Würdemann); JORDAN & GILBERT,
Synopsis, 961, 1883.

2728. BLENNIUS PILICORNIS, Cuvier & Valenciennes.

Head $4\frac{3}{4}$ with caudal; depth $5\frac{1}{2}$. D. XII, 21 or 22; A. 23 or 24. Snout
obtuse, the upper profile very oblique. A strong curved canine in each
jaw. Orbital tentacle filiform, with several smaller ones at base. Inter-

orbital space flat, its width $\frac{1}{4}$ vertical diameter of eye; no groove or crest on the neck. Dorsal slightly notched, the spines flexible; caudal separate. Brown, dorsal and caudal spotted with darker. Length 5 to 6 inches. (Günther.) Coast of Brazil north to the West Indies, recorded from Rio Janeiro, Bahia, and the Tortugas, and off the coast of Florida. Mr. Garman gives the following color note on Tortugas specimens, collected by Prof. C. C. Nutting: Small, hexagonal reticulations on cheeks, resembling scales; anal darker toward ends of rays, the tips white; dorsal darker in outer half; basal part of dorsal and anal pale, sides with a few scattered black dots; median rays of caudal longer, the outer margin dark; caudal, pectorals, and ventrals paler than dorsal. (_pilicornis_, with downy horns.)

Blennius pilicornis, CUVIER & VALENCIENNES, Hist. Nat. Poiss., XI, 254, 1836, Rio Janeiro (Coll. Delalande and Gay); CASTELNAU, Anim. Nouv., etc., Amer. Sud, 25, 1855; GARMAN, Bull. Iowa Lab. Nat. Sci. 1896, 89.
Blennius filicornis, GÜNTHER, Cat., III, 216, 1861. (Coll. M. Parzudaki.)

2729. BLENNIUS MARMOREUS, Poey.

Head $4\frac{1}{4}$ in total length with caudal; depth 5. D. XII, 20; A. 16; P. 13. Eye very high, near the profile, twice length of snout. Snout round, falling off abruptly, but less so than in _Blennius truncatus_; posterior nostrils with a distinct tube; superciliary tentacle divided into 3 branches; no cilia at the nape. Teeth undescribed. Gill membranes not described. Pectoral and caudal round; dorsal low, the median spines highest, the soft rays a little higher, the difference slight. Color yellowish brown, darker medially, paler below; under the lens covered with small dots; fins below yellowish. This species differs from _Scartella microstomia_ in the stout trunk, the more blunt head, the cilia on the head and in the tube of the nostril. Cuba; 1 specimen 2 inches long. (Poey); not seen by us; perhaps not a _Blennius_ as here understood. (_marmoreus_, marbled.)

Blennius marmoreus, POEY, Enumeratio, 130, 1875, Cuba. (Coll. Rafael Arango.)

2730. BLENNIUS TRUNCATUS (Poey).

Head $5\frac{1}{4}$ in total length with caudal; depth $5\frac{1}{4}$. D. XII, 19; A. I, 20. Eyes placed very high, profile before them vertical, suggesting the forehead of a bull without horns; mouth small, maxillary reaching below posterior border of eye; anterior nostril divided into 5 at tip; 2 filiform tentacles with a common base on each side of nape; a simple tentacle behind eye; some pores on the head, which is compressed; teeth undescribed; gill membranes undescribed; gill membranes united and free from isthmus; dorsal notched medially; caudal truncate, with 2 faint angles; lateral line long, reaching beyond the point of the pectorals. Color olive, with some bright spots on trunk; the vertical fins darker. Cuba; 1 specimen $3\frac{1}{4}$ inches long. (Poey); not seen by us; perhaps not a species of _Blennius_. (_truncatus_, cut off short.)

Blennius truncatus, POEY, Memorias, II, 424, 1861, Cuba. (Coll. Poey.)

2731. BLENNIUS VINCTUS, Poey.

Head $3\frac{1}{2}$ to base of caudal; depth 4. D. XII, 13; A. I, 8; V. 3. Eye high, 4 in head, as long as snout. Anterior nostril in a short tube. Jaws equal; 4 pores on the side of the lower jaw; 1 on the opercle; 4 on the suborbital; 4 below eye. A long tentacle above eye; another very small one on the nape. Maxillary reaching to below front of pupil. Teeth large, not pointed, compressed, in 1 series of 10 to 12 on each side of each jaw, feeble, somewhat moveable; gill membranes united, free from isthmus. Dorsal elevated backward, connected by a membrane to the first third of the caudal, as is also the anal, twenty second ray highest, its height $\frac{1}{2}$ depth of body and double length of the dorsal ray above tip of pectoral; anal similar, $\frac{1}{8}$ also of the rays of the dorsal and anal simple; the spines flexible, differing from the others in not being articulate; pectoral pointed, its middle rays longest, and also more robust, all simple; ventral not very short; caudal rounded. Lateral line forming a curve anteriorly. Color uniform brown. Cuba. (*vinctus*, bound.)

Blennius vinctus, POEY, Repertorio, 243, 1867, Havana. (Coll. Felipe Poey. Type,* No. 12647, Mus. Comp. Zool.)

2732. BLENNIUS CRISTATUS, Linnæus.

Head 4; depth 4. D. XI, 16; A. 19; maxillary 3. Body moderately elongate, compressed; the head very blunt and deep, almost as deep as long, its anterior profile straight or slightly concave, and nearly vertical. Mouth moderate, the maxillary reaching to past front of eye; lower jaw with 2 short stoutish posterior canines, scarcely longer than the front teeth; upper jaw without canines. Teeth about $\frac{32}{28}$. Preorbital deep, its depth equal to diameter of eye and contained $4\frac{1}{4}$ times in length of head. Interorbital space flat, narrow, $\frac{3}{8}$ width of eye. Supraocular cirri small, fringed, their length about equal to that of pupil. Nape with a longitudinal dermal crest reaching to front of dorsal, provided with a series of about 20 filaments, the longest about as long as the eye. Gill membranes forming a broad fold across the isthmus, as in all species of *Blennius*. Dorsal nearly continuous, the last spine a little lower than the first soft ray, not very high, beginning on the nape in front of the vertical of the preopercle, the spines all slender and flexible, the longest $\frac{3}{8}$ as long as the head, the longest soft ray $\frac{4}{7}$ as long as the head; caudal free from dorsal and anal, $\frac{4}{5}$ as long as head; anal moderate, $\frac{4}{9}$ length of head; pectoral somewhat shorter than head; ventral a little more than $\frac{1}{2}$ length of head. Lateral line forming the usual arch above pectoral, and continued backward on median line to base of caudal, becoming indistinct posteriorly.

*On the type of *Blennius vinctus* we have the following notes: "No. 12647, M. C. Z. Cuba. (Poey.) One and a half inches long, in poor condition. Head ca $3\frac{1}{2}$; depth ca 4. D. XII, 13; A. II, 13. Dorsal joined to caudal as far as tips of the rays, which are high. Dorsal spines high and stiff, the fin not notched, the soft rays higher. A thick scale-like fringed cirrus above each eye, nearly as long as eye, which is small. Gill membrane free. Head blunt. Maxillary to front of pupil. Lower jaw with very strong canines; upper jaw with none. No nuchal cirri."

Color faded, apparently olivaceous, with about 6 dark cross bars, which extend on the dorsal fin; anal and posterior ½ of body with numerous round, whitish, stellate spots, probably bluish in life; bluish streaks from eye across the cheeks; anal edged with dusky; the other fins vaguely marked. Length 2½ to 4 inches. Tropical parts of the Atlantic, among rocks, widely diffused and variable. The above description from the type of *Blennius asterias*, from Garden Key, Florida. We have the following notes on numerous specimens from Abrolhos Islands, off the Coast of Brazil (Coll. *Albatross*): D. XII, 15. Nape with a fringed crest of 10 to 18 filaments. A small trifid tentacle above eye; posterior canines in lower jaw only, short and small; gill membranes broadly united, nearly free from the isthmus. Dorsal slightly notched; nasal tentacle present. Color excessively variable, mostly grayish, with 5 or 6 cross blotches on the back, extending to form quadrate blotches on the side; body mottled; fins also mottled; the anal dark, with a pale edge. Some specimens highly variegated, the caudal banded and with black and white spots; pale streaks from the eye across the cheek; dark bars on sides, extending on dorsal. Most specimens have the region above anal with numerous round whitish spots and some dark ones. These spots sometimes nearly obsolete, most evident on the paler specimens.

The following notes are taken from a specimen, No. 4635, M. C. Z., from Para, Brazil (Coll. Agassiz and Bourgeot): Head 4; depth 4¼. D. XII, 14; A. I, 16. Maxillary to front of eye, about equal to eye. Gill membranes free. Lower jaw with a very small canine, not twice the length of the upper teeth. Orbital cirrus quite small; a row of cirri along the nape, longer than the orbital cirrus. Head not very blunt, the anterior profile forming an angle above eye, thence straight and steep. Dorsal spines rather low and flexible, the fin scarcely notched. Color nearly lost; dark marblings on sides and on dorsal fin. This species is evidently the *Blennius crinitus* of Günther and the *B. asterias* of Goode & Bean, probably the *nuchifilis* of Cuvier and Valenciennes, and in all probability the *cristatus* of Linnæus, also. These nominal species are from various localities in the Atlantic. If our specimens are all alike, all these forms most likely belong to 1 species. For this *cristatus* is the oldest name. The very small canines show considerable divergence from the type of *Blennius*, approaching *Scartella*. (Eu.) (*cristatus*, crested.)

Blennius crista setacea longitudinale inter oculos, GRONOW, Museum, I, No. 75; D. 26; A. 16; locality unknown. (Coll. Vosmaer.)

Blennius cristatus, LINNÆUS, Syst. Nat., Ed. X, I, 256, 1758, Indies, after GRONOW; GÜNTHER, Cat. Fish., III, 223, 1861; JORDAN, Proc. U. S. Nat. Mus. 1890, 329.

Blennius crinitus, CUVIER & VALENCIENNES, Hist. Nat. Poiss., XI, 237, 1836, La Rochelle, France (Coll. D'Orbigny); GÜNTHER. Cat., III, 224, 1861.

Blennius nuchifilis, CUVIER & VALENCIENNES, Hist. Nat. Poiss., XI, 253, 1836, Isle of Ascension. (Coll. Quoy & Gaimard.)

Blennius asterias, GOODE & BEAN, Proc. U. S. Nat. Mus. 1882, 416, Garden Key, Florida (Type, No. 2620. Coll. G. Würdemann); Garden Key, Florida (Type, No. 2625. Coll. Dr. Whitehurst); Tortugas (Type, No. 6596. Coll. Dr. J. B. Holder); JORDAN & GILBERT, Synopsis, 961, 1883.

Adonis cristatus, GRONOW, Cat. Fish., Ed. Gray, 95, 1854.

885. SCARTELLA, Jordan.

Scartella, JORDAN, Proc. U. S. Nat. Mus. 1886, 50 (*microstoma*).

This genus differs from *Blennius* only in the entire absence of the posterior canine. The relations of this genus with such species of *Blennius* as *Blennius cristatus* are very close. It may be that the groups should be reunited, or that several species here referred to *Blennius* should be placed in *Scartella*. (σκάρτης, one who leaps.)

2733. SCARTELLA MICROSTOMA (Poey).

Head 4 in length (5 with caudal); depth 3⅝ (4⅔); eye 3¼. D. XI, 14; A. 15 or 16. Body rather stout, compressed posteriorly; head short, the anterior profile straight and very steep, almost vertical from tip of snout to above eye, where a sharp angle is formed with the straight line of the back. Eye large, longer than snout. Mouth moderate, the maxillary reaching to below front of pupil, its length 3⅓ in head. Teeth uniform; no posterior canines in either jaw. A small tufted or multifid cirrus over each eye, its length less than diameter of pupil; a row of about 3 short, slender cirri along each side of nape. Gill membranes broadly united, free from isthmus. Lateral line extending about to end of pectoral, each pore with a short, simple branch above and below, directed outward and backward; some conspicuous pores radiating from the eye. Dorsal fin low, subcontinuous, the spines rather slender, lower than the soft rays, the middle spines not much higher than the last; longest rays of dorsal about ¼ as long as head; caudal free from dorsal and anal, a little shorter than head; anal low; pectorals slightly longer than head; ventrals 1⅓ in head. The fins are somewhat shriveled, so that the count of the rays is made with difficulty and may not be perfectly exact. Color very dark olive brown, paler below; head and anterior half of body plain, posterior half sprinkled with sharply defined dots of a vivid sky-blue color, becoming white in alcohol; about 6 obscure round darker blotches in a longitudinal series along sides posteriorly; fins dusky olive, mottled with darker, the caudal obscurely barred, the anal with a pale edge; spinous dorsal nearly black. Length 3¼ inches. Cuba. Here described from a specimen taken by Dr. Jordan in Havana. We have also the following notes on Poey's type in the museum at Cambridge: D. XI, 15; A. 17. Dorsal and anal free from caudal. Body rather robust, the head blunt. Last tooth in each jaw a shade longer than its neighbor, but not canine-like. Gill membrane free from isthmus. Dorsal spines low, rather stiff, the fin deeply notched. Color much mottled, with some white spots on posterior half of body; a black ocellus behind first dorsal spine; 5 dark bars along back.

The following is Poey's description: Head 4⅓ in total length with caudal; depth 5¾. D. XII, 15; A. I, 17; P. 14. Snout short; profile falling abruptly; mouth small; eye 3 in head, twice interorbital space. Teeth 15 on each side in each jaw. Gill membrane broadly united, free from isthmus. Lateral line disappearing on middle of back; a row of 6 filaments arranged in pairs on each side of the nape. Membranaceous tentacles over the eye; dorsal somewhat notched, pectoral strongly developed at base. Color brown, with 5 or 6 darker points which form on the back and reach.base of the dorsal; pearly spots along sides and some below of the same color; caudal with 3 brown points. Cuba. One specimen, 46 mm. long. (Poey.) (μικρός, small; ϭτόμα, mouth.)

Blennius microstomus, POEY, Memorias, II, 288, 1861, Cuba. (Coll. Poey.)
Scartella microstoma, JORDAN, Proc. U. S. Nat. Mus. 1886, 50.

886. HYPLEUROCHILUS, Gill.

Hypleurochilus, GILL, Proc. Ac. Nat. Sci. Phila. 1861, 168 (*geminatus*).

This genus differs from *Blennius* in the restriction of the gill-openings to the sides, the gill-membranes being broadly and fully joined to the isthmus; canines well developed. (*v*, upsilon; πλευρον, side; χεῖλος, lip; in allusion to the V-shaped lateral lips.)

2734. HYPLEUROCHILUS GEMINATUS (Wood).

Head 3¼ to 3¾; depth 3½ to 4. D. XI, 15 to XIII, 14; A. II, 18. Head not very blunt, the anterior profile straight, oblique; male (*multifilis*) with the supraocular cirrus very large, each with 4 smaller ones at base; supraocular cirrus in female (*geminatus*) low, shorter than eye, branched at tip; interorbital space concave, not ½ diameter of eye; a slight transverse groove behind eye; canines in both jaws, very strong, hooked backward, the lower considerably stronger than upper; gill openings extending downward to opposite or slightly below lower edge of pectoral. Dorsal fin not emarginate, the spines slender, but rather stiff, lower than the soft rays; pectorals shortish, ventrals rather long. Olive brown, faintly barred with darker; sides plain, or with several pairs of spots of a reddish-brown color, arranged pretty regularly in a double row; vertical fins edged with darker, especially the anal; dorsal black in front. Length 2½ inches. South Atlantic and Gulf coast of the United States, in shallow water; abundant in empty shells and clusters of tunicates. The sexes quite unlike, the male (*multifilis*) distinguished by the high suborbital crest. (*geminatus*, twin.)

Blennius geminatus, WOOD, Journ. Ac. Nat. Sci. Phila., IV, 1824, 278, Charleston, South Carolina, female (Coll. Prof. Bache); CUVIER & VALENCIENNES, Hist. Nat. Poiss., XI, 265, 1836.
Blennius multifilis, GIRARD, Proc. Ac. Nat. Sci. Phila. 1858, 169, St. Josephs Island, Texas, male (Coll. Gustav Würdemann); GIRARD, U. S. and Mex. Bound. Surv., Zool., pl. 12, fig. 6, 27, 1859; GÜNTHER, Cat., III, 562, 1861
Hypleurochilus multifilis, GILL, Proc. Ac. Nat. Sci. Phila. 1861, 168; JORDAN & GILBERT, Synopsis, 758, 1883.
Hypleurochilus geminatus, JORDAN & GILBERT, Synopsis, 759, 1883.
?*Blennius geminatus,* GÜNTHER, Cat., III, 288, 1861.

887. HYPSOBLENNIUS,* Gill.

Hypsoblennius, GILL, Cat. Fish. East Coast U. S., 20, 1861 (*hentz;* no diagnosis).
Isesthes, JORDAN & GILBERT, Synopsis, 757, 1883 (*gentilis*).
Blenniolus, JORDAN & EVERMANN, new subgenus (*brevipinnis*).

This genus differs from *Blennius* in the absence of canine teeth and in the restriction of the gill openings to the sides, the gill membranes being fully united to the isthmus as far upward as the base of the pectorals; ventral with 1 short, strong spine and 3 simple, articulated rays. The known species are American. (*ὕψι*, high; *Blennius*.)

HYPSOBLENNIUS:
  a. Dorsal fin continuous, its margin entire or slightly notched.
    b. Dorsal rays XI to XIII, 17 to 19. Pacific species.
      c. Orbital cirrus multifid; spines of dorsal stiff; sides blotched or freckled.
                                                         GILBERTI, 2735.
      cc. Orbital cirrus simple or fringed.
        d. Spines of dorsal slender and flexible; sides with round dark spots; anal rays 21.   GENTILIS, 2736.
        dd. Spines of dorsal rather stiff; sides with irregular dark cross bands rather than spots; anal rays 19.   STRIATUS, 2737.
    bb. Dorsal rays XII, 14 or XII, 15. Atlantic species.
      e. Orbital cirrus simple, large or small; body everywhere with dark spots; dorsal spines rather low, stiffish.   IONTHAS, 2738.
      ee. Orbital tentacle forked at tip, long in males; dorsal spines stiff; body spotted.   HENTZ, 2739.
BLENNIOLUS, (diminutive form, from *Blennius*):
  aa. Dorsal fin deeply notched, very short, its rays XI, 12 or XII, 12; orbital tentacle slender, fringed; a dark lateral shade.   BREVIPINNIS, 2740.

Subgenus HYPSOBLENNIUS, Gill.

2735. HYPSOBLENNIUS GILBERTI (Jordan).

Head 4 in length (4⅔ with caudal); depth 4 (4⅔). D. XII, 19; A. II, 21. Body comparatively robust, deep, and compressed. Head large, rounded, the anterior profile less blunt than in *H. gentilis* and less rounded, nearly straight from tip of snout to above eye, thence again nearly straight to front of dorsal. Length of snout about equal to diameter of eye, 4¼ in head. Mouth rather small, terminal, the maxillary reaching to opposite middle of eye, 2⅔ in head. Teeth subequal, with no trace of posterior canines. Superciliary tentacle large, multifid, much branched from near the base, the principal division 3⅔ in head. Gill openings larger than in *H. gentilis*, extending downward to the level of lower edge of pectoral, the length of the slit 1¾ in head. Lateral line developed beyond the straight part, its posterior portion curved downward. Dorsal fin continuous, with a slight but distinct depression between the spinous and soft parts, the spines somewhat curved, but stiff and strong, the longest spine about 2¼ in head; longest soft rays 2 in head. Caudal fin free from dorsal and anal, 1¼ in head; ventrals 1¼ in head; pectorals about as long as head. Males,

---

* The recent identification of *Blennius hentz* with *Isesthes punctatus* enables us to understand the undefined genus *Hypsoblennius*, and to substitute it for the later *Isesthes*. Our judgment is opposed to the recognition of such unexplained "typonyms," but we defer to the custom of the American Ornithologists' Union.

as usual in this genus, with the anal spines partly detached, and provided with fleshy tips. Coloration olivaceous, the body and fins everywhere profusely mottled and reticulated with darker; obscure dark shades extending downward from eye across, or partly across, lower side of head; head without distinct spots or other sharply defined markings, except faint streaks radiating from eye; no pale bars on side of head in either sex; some yellowish markings on anterior part of dorsal. Length 5 inches. California, from Point Concepcion southward to Todos Santos or beyond; common among rocks in the kelp; our specimens from Santa Barbara and Point Loma. (Named for Charles Henry Gilbert.)

Isesthes gilberti, JORDAN, Proc. U. S. Nat. Mus. 1882, 349, Santa Barbara, California (Type, Nos. 26916 and 26917. Coll. Jordan & Gilbert); ROSA SMITH, Proc. U. S. Nat. Mus. 1883, 235, specimens from Todos Santos Bay; D. XI or XII, 16 to 21; A. 19 to 21; head $4\frac{1}{3}$; depth $4\frac{1}{3}$.

2736. HYPSOBLENNIUS GENTILIS (Girard).

Head $3\frac{2}{3}$ in length ($4\frac{1}{2}$ with caudal); depth 4 ($4\frac{4}{5}$). D. XIII, 17; A. II, 19. Body rather robust, deep and compressed, the head large, very bluntly and evenly rounded in profile, more obtuse and more evenly curved than in *H. gilberti*, the snout shorter, about equal to eye, $4\frac{1}{2}$ in head. Mouth rather small, terminal, the maxillary reaching to opposite middle of eye, its length 3 in head. Teeth subequal, the hindmost on each side of upper jaw shorter than the others, and a little apart from them but not forming "a small canine," as stated by Girard. Superciliary tentacle long and simple in the male, its edge fringed with short branchlets, its length about 3 in head; tentacles much smaller in the female, where they are scarcely visible. Gill opening extending downward not quite to lower edge of pectoral, its length (vertical) $2\frac{1}{6}$ in head. Lateral line with only the straight anterior portion developed, not curved downward posteriorly. Dorsal fins continuous, with scarcely a trace of emargination between the spinous and soft parts. Dorsal spines comparatively low and flexible, much less strong than in *H. gilberti*, the longest spines 3 in head; longest soft rays $1\frac{2}{3}$. Caudal free from dorsal and anal, $1\frac{2}{5}$ in head; ventrals $1\frac{2}{3}$ in head; pectorals $1\frac{1}{3}$. Coloration in spirits, brown, the whole body closely mottled and blotched with darker brown, so that the light ground color forms, especially anteriorly, light reticulations around darker spots; on the head the dark spots are small and close together, smallest anteriorly, the lower parts of the head being immaculate, extending from the curve of the preopercle downward, across the interopercle and branchiostegals, in a sharply defined white bar (said to be golden yellow in life), edged with black; behind this and parallel with it across subopercle and isthmus is a similar bar, these bars present only in the males; a few pale spots or bars in front of these; back with about 6 dusky cross shades, below each of these is an oblong dark blotch, the anterior placed along the lateral line, altogether forming an interrupted dark stripe; a similar dark stripe near the median line of the body, interrupted by some pale blotches. Fins all blotched and spotted by light and dark colors, but without distinct markings (a bluish spot in front of dorsal in life);

ventrals and anal nearly blackish in males, the base of the anal with a pale streak. Females more distinctly blotched, with a black spot in front of dorsal and white spots on middle of sides; head lacking the pale bars and black spots, but much mottled with brown and whitish; a very distinct blackish blotch on front of spinal dorsal; pectoral and caudal pale, a dark blotch on base of pectoral. Length about 4 inches. Monterey to Cape San Lucas; common southward in rock pools. Here described from specimens from Angel Island, Gulf of California, from Cape San Lucas, and from Monterey and San Diego. (*gentilis*, related.)

Blennius gentilis, GIRARD, Proc. Ac. Nat. Sci. Phil. 1854, 149, Monterey, California. Types, Nos. 690 and 785 (Coll. A. Cassidy; No. 489, Lieut. Trowbridge); GIRARD, Pac. R. R. Surv., X, Fishes, 113, pl. 25a, fig. 4, 1858; GÜNTHER, Cat., III, 217, 1861.

Isesthes gentilis, STEINDACHNER, Ichth. Beiträge, V, 150, 1876; JORDAN, Proc. U. S. Nat. Mus. 1882, 350; JORDAN & GILBERT, Synopsis, 956, 1883; JORDAN, Proc. U. S. Nat. Mus. 1882, 349.

2737. HYPSOBLENNIUS STRIATUS (Steindachner).

Head 4 to 4½; depth 4¾ to 5. D. XI or XII, 17; A. 19; P. 15; V. I, 3. Snout steep, and slightly concave in older examples; interorbital narrow, equal to ½ eye; origin of dorsal a little before the edge of preopercle; second and third dorsal spines equal to the distance from tip of snout to edge of preopercle; dorsal and anal free from caudal; pectoral reaching nearly to front of anal. Color yellowish below, sides brownish, irregular dark-brown cross bars on back and sides; toward the caudal are rows of spots, 4 or 5 wider cross bars of dark brown or violet; a dark blotch from the third to the fifth dorsal spine, behind which are irregular longitudinal dark stripes; anal edged with white, behind which runs a violet line; pectoral and caudal spotted; a dark oval spot behind eye; a brown line from first dorsal spine to eye. Panama (Steindachner), where specimens were also taken by Dr. Gilbert, none of these showing posterior canines, although Steindachner notes the presence of a small canine in 1 specimen. (*striatus*, striped.)

Blennius striatus, STEINDACHNER, Ichth. Beiträge, V, 15, 1876, with plates, Panama.
Isesthes striatus, JORDAN & GILBERT, Bull. U. S. Fish Comm. 1882, 111.

2738. HYPSOBLENNIUS IONTHAS (Jordan & Gilbert).

Head 3¾ to 4 (4¼ to 4¾ in total); depth 3⅙ to 3½ (3¾ to 4¾). D. XII, 13, or XII, 14; A. II, 13, or II, 14. Body rather deep, moderately compressed, the back little elevated. Head short, blunt, but less so than in *H. punctatus;* the profile prominent above the eye, thence descending abruptly but not vertically to the tip of the snout; length of snout 3⅛ in head. Mouth small, low, its cleft largely anterior, the short maxillary scarcely reaching past the front of the eye, 4 in head. Eyes large, placed high, 5 in head, the interorbital space about ½ their diameter. Female (*ionthas*) with the orbital cirrus low, scarcely larger than nasal cirrus, which is about equal to diameter of pupil. Teeth moderate, equal; no posterior canines. Gill opening extending

downward to a point varying from a little above to a little below middle
of base of pectoral, the height of the slit 3 in head. Lateral line not
reaching tip of pectoral. Dorsal fin continuous, the spines low and rather
stiff, slenderer than in *H. punctatus*, the longest spines a little lower than
the soft rays, which are about 1⅓ in head. Caudal free from anal, slightly
connected with dorsal; a little shorter than head; pectoral about as long
as head; ventrals shorter than head. Color of female clear olive green,
with only traces of darker bars; body everywhere densely freckled with
small round blackish spots, smaller than the pupil; on the sides and lower
part of head these spots are reduced to close-set dots; 2 dark lines,
separated by a golden area, downward from eye; a vertical curved blackish patch behind eye, in front of which is a golden area; vertical fins
olive green, dorsal and caudal usually mottled with dusky; paired fins
dusky olive; lower parts of head tinged with golden, sometimes with
dusky cross bars; cirri green.

The male (*scrutator*) is thus described: Head 4 (4⅔ in total); depth 3¾ (4¼).
D. XII, 14 or 15; A. II, 15 or 16. Body rather deep, compressed, the back
not elevated. Head short, very blunt, almost as deep as long, the profile
abruptly descending before eye, the snout about ⅓ length of head. Mouth
very small, anterior, the maxillary extending to opposite front of eye, 3⅓
in head; teeth subequal, without canines. Orbital cirri very long, reaching when depressed about to the front of dorsal, their length more than ⅓
head in adult, somewhat shorter in young; a short branch near its middle.
Nasal barbel minute. Eye large, much broader than the concave interorbital space, about 4⅓ in head. Lower edge of gill opening a little below
middle of base of pectoral, the depth of the slit 2½ to 3 in head. Dorsal fin
scarcely emarginate, the spines rather stiff, lower than the soft rays, the
longest spine 2 in head. Caudal slightly connected at base with dorsal, 1⅓
in head; pectoral about as long as head, reaching past front of anal;
ventrals 1⅔ in head. Lateral line extending to base of eighth spine, not to
tip of pectoral. Color in life, deep olive green, almost immaculate, or with
faint traces of darker vertical bars; a golden blotch behind eye, behind
which is a dusky crescent; 2 dark bars downward from eye, separated by
a yellowish area; fins all dusky greenish, nearly or quite immaculate;
front of spinous dorsal blackish. South Carolina to Texas, in rock pools;
numerous specimens, the largest about 2¼ inches long, were obtained with
hook and line from the wharves at Pensacola. (*ίονθάς*, freckled.)

Isesthes ionthas, JORDAN & GILBERT, Proc. U. S. Nat. Mus. 1882, 299, Pensacola, Florida
(Type, No. 30856, U. S. Nat. Mus. Coll. Jordan & Stearns), female; JORDAN & GILBERT,
Synopsis, 960, 1883.

Isesthes scrutator,* JORDAN & GILBERT, Proc. U. S. Nat. Mus. 1882, 300, Pensacola (Type,
No. 30850. (Coll. Jordan & Stearns); Galveston (Coll. Dr. August Galny); JORDAN &
GILBERT, Synopsis, 960, 1883.

* The form called *scrutator* agrees very closely with *Hypsoblennius ionthas* in all respects
except the great length of the orbital cirrus and the different coloration of the body. In
both the golden blotch and dark crescent behind the eye are distinct, as also the 2 dark bars
separated by a yellow one below the eye. Renewed comparison strengthens our impression that *Hypsoblennius scrutator* is the male of *Hypsoblennius ionthas*.

2739. HYPSOBLENNIUS HENTZ (Le Sueur).

Head $3\frac{2}{5}$; depth 3. D. XII, 15; A. 18; pectoral $1\frac{1}{4}$ in head; ventral $1\frac{3}{4}$; gill slit $2\frac{1}{4}$; eye $4\frac{1}{2}$; maxillary $2\frac{2}{3}$. Orbital tentacle very slender, once forked, 3 in head. Body rather deep; head large, obtuse; interorbital space concave, $\frac{1}{2}$ the diameter of orbit; orbital cirrus as long as dorsal spines, bifid at tip, branched below; a minute nasal cirrus; no canines; gill openings extending to about lower fourth of base of pectoral, thus narrower than in most related species. Dorsal fin high, little notched, the soft part highest, the spines stiff, $2\frac{2}{3}$ in head. Tip of each dorsal spine with a filiform, articulated, ray-like appendage. Color in spirits, olivaceous, back and sides of head and body everywhere covered with brown spots, very irregular in size and shape; on posterior part of body the spots are larger, and show a tendency to form vertical bars; cheeks dark; lower side of head with traces of 3 cross bars; spinous dorsal with an elliptical black spot on membrane of first 3 spines; soft dorsal and caudal obscurely barred; anal, ventrals, and lower rays of pectorals dusky; pectorals olivaceous, spotted with brown. Coasts of North and South Carolina, south to Indian River, Florida; locally common. (Named for its collector, Dr. Nicholas Marcellus Hentz, "the father of American Araneology.")

Blennius punctatus, WOOD, Journ. Ac. Nat. Sci. Phila., IV, 1825, 278, Charleston, South Carolina (Coll. Prof. Bache); CUVIER & VALENCIENNES, Hist. Nat. Poiss., XI, 267, 1836; GÜNTHER, Cat., III, 228, 1861; not *Blennius punctatus*, Fabricius, 1780, which is a *Stichæus*.

Blennius hentz,* LE SUEUR, Journ. Ac. Nat. Sci. Phila., IV, 1825, 363, Charleston, South Carolina. (Coll. Dr. Hentz.)

Hypsoblennius hentzi, GILL, Cat. Fish. East Coast N. A., 1861 (*nomen nudum*).

Hypleurochilus punctatus, GILL, Cat. Fish. East Coast N. A., 20, 1873.

Isesthes punctatus, JORDAN & GILBERT, Synopsis, 758, 1883; JORDAN & GILBERT, Proc. U. S. Nat. Mus. 1883, 616.

Isesthes hentzi, JORDAN & GILBERT, Synopsis, 960, 1883.

Subgenus BLENNIOLUS, Jordan & Evermann.

2740. HYPSOBLENNIUS BREVIPINNIS (Günther).

Head $3\frac{1}{5}$; depth 4. D. XII, 12; A. II, 14; pectoral $1\frac{2}{3}$ in head; ventral $1\frac{3}{4}$; gill slit $2\frac{1}{5}$; eye $3\frac{1}{3}$ in head; snout $2\frac{3}{4}$ in head; maxillary $2\frac{1}{2}$. Orbital tentacle slender, less than eye. Body rather deep, compressed, back not elevated; anterior profile from first dorsal spine to above eye almost horizontal or slightly decurved, from thence to tip of snout abruptly decurved; head large, its width not quite 2 in its length; interorbital space narrow, grooved, about equaling pupil; eyes large, placed high and close together. Mouth small, low, the maxillary reaching to pupil; teeth subequal, pectinate; no canines; dorsal fin continuous, deeply emarginate, the spines lower than the soft rays, the longest spine about $2\frac{1}{2}$ in head; caudal free

* The following is the substance of the account of "*Blennius hentz:*" Depth $3\frac{1}{3}$ (in total). D. XI, 14; A. 16. Body little elongate; snout very short, but not vertically truncate; eyes above angle of mouth, placed high; gill slit extending from level of base of pectoral fin to height of eye; teeth equal; dorsal slightly depressed in the middle; pectorals large; a short cirrus above each eye and a smaller one over each nostril. Light bluish ash, mixed with rufous, with numerous irregular black and rufous spots; dorsal black, with whitish spots; soft dorsal with 5 dark bands; ventrals blackish, with pale bands; caudal with 3 or 4 dark bands. Charleston Harbor, South Carolina. (Le Sueur.)

from anal and dorsal; lateral line not reaching soft dorsal. Olive brown, lighter below; back and upper half of sides irregularly marked with about 6 distinct dark-brown cross bars, these uniting at their lower edges and forming a continuous line from head to base of caudal; the bars nearly confluent on the back at base of dorsal fin; a dark lateral band nearly as wide as eye from opercle to base of caudal, containing 5 or 6 light-yellowish spots corresponding to the pale interspaces along the back; fins dusky, anal margined with black; head with a dark spot behind each eye, and 2 smaller blotches in the median line, 1 immediately behind the eyes, the other a short distance in front of dorsal. Pacific coast of Mexico, from Mazatlan to Panama; rather common. The specimens here described from Mazatlan. (*brevis*, short; *pinna*, a fin.)

Blennius brevipinnis, GÜNTHER, Cat. Fish., III, 226, 1861, Pacific coast Central America (Coll. Capt. John M. Dow); one specimen wrongly attributed to Hawaiian Islands.

888. CHASMODES, Cuvier & Valenciennes.

Chasmodes, CUVIER & VALENCIENNES, Hist. Nat. Poiss., XI, 295, 1836 (*bosquianus*).
Blenitrachus, SWAINSON, Class'n Fishes, etc., II, 78, 274, 1839 (*quadrifasciatus*).

Body oblong, compressed, naked; head triangular in profile, the snout somewhat pointed; mouth large, with lateral cleft, the maxillary usually, but not always, extending to beyond eye; premaxillaries not protractile; teeth in a single series, long and slender, comb-like, confined to the front of each jaw; no canines; cirri very small or wanting; gill openings very small, their lower edge above the middle of the base of the pectorals; lateral line incomplete. Fins as in *Blennius*. American. The species with smaller mouth approach *Hypsoblennius*, which genus is not far separated from *Chasmodes*. (χασμώδη.ς, yawning.)

a. Dorsal and anal free from caudal.
    b. Anal rays 18 or 19; body not banded.   JENKINSI, 2741.
    bb. Anal rays 15; body with 4 dark cross bands.   QUADRIFASCIATUS, 2742.
aa. Dorsal joined to base of caudal.
        c. Mouth moderate, the maxillary not extending to posterior border of eye, 2¼
           in head.   SABURRÆ, 2743.
        cc. Mouth large, maxillary reaching posterior border of eye.
                                  NOVEMLINEATUS, 2744.
        ccc. Mouth very large, the maxillary extending to beyond eye.
                                    BOSQUIANUS, 2745.

2741. CHASMODES JENKINSI (Jordan & Evermann).

Head 3⅓ (4 in total); depth 4 (5). D. XII, 17; A. 18 or 19; eye 4 to 5 in head. Body more robust than in related species, resembling *Hypsoblennius;* head large, gently rounded in profile, the snout steep, 4 in head; interorbital space narrow, grooved; orbital tentacle (male) much as in *Hypsoblennius gilberti*, about 3 in head, branched, the branches usually 4; mouth much larger than in *Hypsoblennius*, the maxillary 2¾ to 3 in head, reaching to below posterior margin of eye; teeth even, comb-like; gill opening 2 in head, extending downward nearly to lower edge of pectoral, much larger than in *Chasmodes saburræ*. Dorsal little notched, the spines slender, 2¼ in head, the rays a little higher; anal lower, the rays 3½ to 4 in head; pectorals reaching anal, 1⅓ in head; ventrals 2¼; dorsal and anal free from caudal. Color in life, according to Evermann & Jenkins, yellowish; 5 quadrate spots of darker extending from dorsal to a line drawn from middle of eye to lower base of caudal, the anterior one above tip of pectoral; median line of side with a more or less distinct series of small spots; a short dark vertical line behind the eye; a dark blotch in front of origin of dorsal fin and another on humeral region; underside of head with 2 ill-defined dark bands; dorsal fin more or less speckled with black, the anal with a narrow white border above which is a broader band of deep brown. Six specimens, the largest about 3 inches long, were obtained at Guaymas, Sonora, by Drs. Evermann & Jenkins, in 1887. One of these, (No. 412, L. S. Jr. Univ. Mus.), examined by us, is the type of the present description. The large mouth distinguishes this species at once from *Hypsoblennius striatus*, with which it has been identified. The species is intermediate between typical *Chasmodes* and *Hypsoblennius*, and its discovery may make it necessary to merge the latter in *Chasmodes*. (Named for Dr. Oliver Peebles Jenkins.)

Hypsoblennius striatus, EVERMANN & JENKINS, Proc. U. S. Nat. Mus. 1891, 163; not of STEINDACHNER.
Chasmodes jenkinsi, JORDAN & EVERMANN, Proc. Cal. Ac. Sci. 1896, 232, pl. 39, Guaymas. (Coll. Evermann & Jenkins.)

2742. CHASMODES QUADRIFASCIATUS (Wood).

D. 27; A. 15. Form of *Chasmodes bosquianus:* Lower jaw slightly longer than the upper. Dorsal and anal free from caudal; anal fin highest anteriorly. Body with 4 distinct brownish bands, a fifth broader and less marked on the neck; 4 round yellowish spots along base of anal; head spotted with blackish. (Wood.) *Habitat* uncertain, probably South Atlantic coast of the United States; not recognized by recent collectors; very likely based on the female of *C. bosquianus*, with the caudal torn from the other vertical fins. (*quadri-*, four; *fasciatus*, banded).

Pholis quadrifasciatus, WOOD, Journ. Ac. Nat. Sci. Phila., IV, 1825, 282; locality unknown, probably South Carolina. (Coll. Rubens Peale.)
Chasmodes quadrifasciatus, CUVIER & VALENCIENNES, Hist. Nat. Poiss., XI, 298, 1836; GÜNTHER, Cat. Fish., III, 229, 1861; JORDAN & GILBERT, Synopsis, 757, 1883.

2743. CHASMODES SABURRÆ, Jordan & Gilbert.

Head 3½ to 3¾; depth 3¼ to 3⅔. D. XI or XII, 17 to 19; A. II, 18 or 19. Body rather deep and compressed, less elongate than in *C. bosquianus;* the back somewhat arched. Head comparatively short, much shorter than in *C. bosquianus*, not ⅓ longer than deep; profile forming a nearly even curve from the base of the dorsal to the tip of the snout; mouth notably smaller than in *C. bosquianus;* maxillary not reaching posterior margin of eye, its length 2¼ in head; teeth occupying about ⅓ of lower jaw; height of gill slit 3⅔ in head, its lower ray opposite third ray of pectoral. A minute cirrus, shorter than pupil, above each eye and each nostril. Dorsal continuous, with slender rays, the last one joined to the caudal. First two rays of anal short, thick, and fleshy in the males. Male deep olive, with dark cross shades; numerous pale spots on the sides which form undulating lines converging backwards; dark stripes downward and forward from eye; top of head and upper part of dorsal fin usually with fine black spots; spinous dorsal with a median orange longitudinal band; other fins mostly dusky olive. Some specimens with the outer part of both dorsals and the top of head dusted with black spots, others with these spots obsolete; soft dorsal and caudal light orange, barred with light greenish; anal dull orange, with an obscure blackish median band, the exserted tips of the rays abruptly whitish; pectorals dusky olive, strongly tinged with orange; ventrals blackish, orange at tip. Female with about 8 blackish cross bands extending on the dorsal fin; the body everywhere with pale spots; fins all sharply barred with blackish and olive. Pensacola Bay, Florida; common about the wharves and ballast rocks in shallow water; taken with seines and pinhooks. Allied to *Chasmodes bosquianus*, but with the mouth smaller, the form less elongate. (*saburra*, ballast.)

Chasmodes saburræ, JORDAN & GILBERT, Proc. U. S. Nat. Mus. 1882, 298, Pensacola, Florida (Type, No. 30824. Coll. Jordan & Stearns); JORDAN & GILBERT, Synopsis, 958, 1883.

2744. CHASMODES NOVEMLINEATUS (Wood).

Head $3\frac{2}{3}$; depth $3\frac{2}{3}$; eye $4\frac{1}{3}$; snout $3\frac{1}{4}$; maxillary reaching posterior border of eye. D. XI, 18; A, III, 17. Head and shoulders heavy, the body lance-shaped, tapering gradually to tail; snout short, blunt, profile nearly vertical to eye, thence gently rounded; mouth rather large, somewhat oblique, the maxillary reaching posterior border of eye; dorsal and anal high, longest dorsal rays 2 in head; anal considerably lower; pectoral nearly as long as head; ventrals $1\frac{3}{4}$ in head. Color, side with 6 broad, dark, vertical bars, the anterior 4 extending on the dorsal fin, these bars separated by irregular narrow pale spaces; entire side profusely covered with small white spots; a small black spot at base of caudal; head mottled with light and dark; 2 small dark spots on under side of lower jaw; just behind these and extending downward from the angles of the mouth are 2 other larger, blacker spots, while behind these, extending downward and backward from middle of cheek, is an irregular black line; whole head with numerous fine dark punctulations; dorsal and anal variously spotted or barred with light and dark; spinous dorsal with a large dark area at top of anterior spines; caudal faintly barred; pectorals and ventrals more plainly barred. Length 2 inches. South Atlantic coast of the United States, South Carolina to Florida; abundant in Indian River, Florida, where numerous specimens were taken in January, 1896, by Evermann & Bean. (*novem*, nine; *lineatus*, lined.)

Pholis novemlineatus, WOOD,* Journ. Ac. Nat. Sci. Phila., IV, 1825, 280, Charleston Harbor, South Carolina.

---

* The following is the substance of Wood's original description of this species: "Body with 9 whitish longitudinal bands; dorsal fin with an irregular blackish spot between the first and second rays; remainder of the fin clouded with dusky brown. Head descending somewhat abruptly, tuberculated anteriorly; nostrils with a small appendage; head, lips,

Chasmodes saburræ, EVERMANN & BEAN, Fishes of Indian River, Florida, in Rept. U. S.
Fish Comm. 1896, 247; not of JORDAN & GILBERT.
Chasmodes novemlineatus, GÜNTHER, Cat., III, 229, 1861.

2745. CHASMODES BOSQUIANUS (Lacépède).

Head 3½; depth 3¼. D. XI, 19; A. II, 19. Orbital tentacle very minute
or wanting; maxillary extending to rather beyond eye; interocular space
very narrow, not concave. Dorsal fin not emarginate, the spines slender.
Dorsal joined to base of caudal; anal free. Color (in male) olive green,
with about 9 horizontal narrow blue lines, these somewhat irregular and
interrupted, converging backward; opercular membrane and a broad
stripe through middle of spinous dorsal deep orange yellow; anal fin dark,
the rays with white membranaceous tips; female dark olive green, reticulated with narrow, pale green lines, and with several broad dark bars,
which are more distinct posteriorly; vertical fins similarly marked; head
finely dotted with black; a dusky spot at base of caudal in both sexes.
New York to Florida; common southward in shallow water. (Named for
M. Bosc, who collected at Charleston for Lacépède.)

Blennius bosquianus, LACÉPÈDE, Nat. Hist. Poiss., II, 493, 1800 (female), South Carolina.
(Coll. Bosc.)
?Pholis quadrifasciatus, WOOD, Journ. Ac. Nat. Sci. Phila., IV, 1824, 282, locality unknown,
probably South Carolina. (Coll. Rubens Peale.)
Chasmodes boscianus, GÜNTHER, Cat , III, 229, 1861.
Chasmodes bosquianus, JORDAN & GILBERT, Synopsis, 756, 1883.

889. HOMESTHES, Gilbert, new genus.

Homesthes, GILBERT, new genus (*caulopus*).

Differing from *Hypsoblennius* chiefly in the presence of 4 articulated ventral rays instead of 3, as usual in *Blenniinæ.* We have examined the ventrals of *Hypsoblennius striatus, punctatus, ionthas, gentilis,* and *gilberti,* and
have found them to consist constantly of 1 short, strong spine and 3 simple
articulated rays. In *Homesthes caulopus* there is 1 strong, short spine and
4 well-developed simple jointed rays. (ὁμός, uniform; ἐσθίω, to eat.)

2746. HOMESTHES CAULOPUS, Gilbert, new species.

Head 3⅔ in length; depth at base of ventrals 4, at middle of abdomen
3⅜; least depth of caudal peduncle ⅓ length of head; snout 4; eye 4 to
4⅕. D. XII, 15 or 16; A. II, 17; P. 14. Longest dorsal spine 2⅔; last dorsal

opercula, etc., and base of the pectoral fins, finely spotted with bluish black, the spots
being larger on the front and opercula; branchial opening extremely small, extending
⅓ of the length of the external curve of the operculum; mouth descending little; gape
moderate; sides of the head fleshy; body compressed; rib spaces evident; sides with
9 longitudinal whitish lines, some of which are interrupted; behind the eye and under
the dorsal fin are 2 irregular whitish patches; dorsal fin commencing before the pectoral
fins; between the first and second rays is an irregular blackish spot, several of the following rays are also spotted, the color of the spots becoming lighter as they recede
toward the tail, where they mingle with the dusky color of the fin and are lost; fin rising posteriorly, and joining the caudal fin at about ⅓ the distance from its extremity;
anal fin commencing under the termination of the pectoral fin, and extending nearly to
the tail; caudal fin rounded; ventral fins 2-rayed; pectoral fins rather large, the base
thick and fleshy, finely spotted with bluish black; anus small, tubercle small; color
brownish, fins dusky. D. 30; C. 12⅜; A. 20; V. 2; P. 13. Length 3¼ inches; depth, exclusive of the dorsal fin, hardly 1 inch."

spine $3\frac{4}{5}$; longest (tenth) dorsal ray 2; longest (fifteenth) anal ray $2\frac{1}{2}$; ventrals $1\frac{5}{6}$; longest pectoral ray $1\frac{3}{5}$ to $1\frac{2}{3}$; caudal $1\frac{1}{2}$. Robust, moderately compressed, with wide heavy head and short, bluntly rounded snout, the anterior profile of which is nearly vertical. In shape and general appearance much resembling *Hypsoblennius gilberti*. Mouth very wide, horizontal, short, the maxillaries reaching vertical from hinder edge of pupil, 3 to $3\frac{1}{5}$ in head. Teeth, as usual in this group, the posterior not enlarged or canine-like. Nostrils with slightly elevated margins, scarcely tubular, the hinder edge of anterior nostril produced into a conspicuous laciniate flap, about $\frac{2}{3}$ as long as the diameter of orbit. A similar but larger orbital cirrus, divided nearly to the base into 6 or 8 slender filaments. Interorbital space deeply grooved, without median ridge, opening posteriorly into the deep transverse groove which separates the orbital region from the somewhat swollen occiput, its width $1\frac{1}{3}$ eye. The mucous canals of head give off transverse branches which open by numerous pores. These thickly beset the snout, subocular region, top of head, preopercle, and upper portion of opercle. Width of gill slit equaling or slightly exceeding $\frac{1}{2}$ length of head, confined to area above lower base of pectorals. First dorsal spine over margin of preopercle; spinous dorsal low, of nearly uniform height, much lower than second dorsal, the spines rather strong at base, with weak reflexed tips; membrane of last dorsal ray joined to extreme base of rudimentary caudal rays; anal low, rising slightly posteriorly, leaving a short free interval between its last ray and caudal. Lateral line strongly developed anteriorly for a distance equaling length of head; from that point it is only faintly visible, declining abruptly to middle of sides, along which it may be traced to base of caudal; the anterior portion gives off numerous pairs of short transverse lines, each of which ends in a pore; no pores or lines are visible posteriorly. Blackish, without sharp markings, the sides with irregular light blotches, some of which are subcircular in outline and contain 1 or more black central specks; the light markings near the back elongate and vertically placed, faintly outlining dark bars of the ground color; a vertical black blotch on cheek behind eye; lower parts lighter; no distinct bars on head; fins all blackish, the anal, the ventrals, the lower caudal and pectoral rays deeper black; anal and caudal margined with white, some of the dorsal rays narrowly tipped with white; tentacles whitish. Two specimens, 4 and $4\frac{1}{2}$ inches long, from Panama Bay. (Gilbert.) (καυλός, stem; πούς, foot.)

Homesthes caulopus, GILBERT MS., Fishes of Panama, Panama. (Coll. Gilbert. Type, No. 5623, L. S. Jr. Univ. Mus.)

890. SCARTICHTHYS, Jordan & Evermann, new genus.

Scartes, JORDAN & EVERMANN, Check-List Fishes, 471, 1896 (*rubropunctatus*); preoccupied by *Scartes*, Swainson, a genus of mammals.
Scartichthys, JORDAN & EVERMANN, new genus (*rubripunctatus*).

Body elongate, slowly declining to the caudal. Head obliquely compressed, oblong, the profile more or less vertical. Eyes lateral, closely approximated, situated at the angle of the profile with the postocular region. Gill apertures continuous under the throat, gill membrane free from isthmus. Branchiostegals 6. Mouth moderate, the contour of the upper jaw semicircular; upper jaw protruding beyond the lower; lips moderate, uniform and free, concealing the teeth. Teeth labial and movable, very slender and recurved, contiguous and uniserial; no posterior canines. Dorsal fin divided; anal similar to soft dorsal; caudal obtusely rounded; pectorals moderate, angularly rounded; ventrals approximated, each with 3 simple rays, the internal of which is smallest. This genus is very close to the Old World genus, *Salarias*,* Cuvier, which differs in having the dorsal fin continuous, as in *Rupiscartes*. (σκάρτης, one who leaps; ἰχθύς, fish.)

2747. SCARTICHTHYS RUBROPUNCTATUS (Cuvier & Valenciennes).

Head 4; depth 4 (5 with caudal); D. XI–16; A. 20; eye 4⅓ in head; teeth less flexible than in *Rupiscartes atlanticus;* no canine teeth; the forehead not projecting beyond the mouth; a very small tentacle on the neck, a longer fringed one above the orbit; dorsal fin deeply notched, not extending on to the caudal. Color brown, marbled with black, and dotted with reddish; a black spot on the anterior part of the dorsal; throat with 2 or 3 brownish cross bands; a jet-black spot behind eye, with a narrow edge posteriorly. (Günther.) Coast of Peru and Chile, north to Panama. Specimens examined by us collected by Prof. Frank H. Bradley at Pearl Islands, near Panama, and at Callao. Length 3 inches. (*ruber*, red; *punctatus*, spotted.)

Salarias rubropunctatus, CUVIER & VALENCIENNES, Hist. Nat. Poiss., XI, 348, 1836, Juan Fernandez (Coll. Claude Gay); GÜNTHER, Cat., III, 249, 1861; JORDAN & GILBERT, Proc. U. S. Nat. Mus. 1882, 628; not of KNEE, Novara-Fische, 198.

891. RUPISCARTES, Swainson.

Alticus,† COMMERSON, in LACÉPÈDE, Hist. Nat. Poiss., II, 458, 1800 (*saliens*).
Alticus,‡ CUVIER & VALENCIENNES, Hist. Nat. Poiss., XI, 337, 1836 (*alticus*).
Rupiscartes, SWAINSON, Nat. Hist. Class'n Anim., II, 275, 1839 (*alticus*).

This genus, as here understood, differs from *Salarias* only in the presence in 1 or both jaws of posterior canines. Dorsal fin continuous, without deep notch. Vertebræ 12 + 22 = 34 (*atlanticus*). (*rupes*, rock; σκάρτης, one who leaps; *Rupiscartes tridactylus* (*alticus*), "said to jump on the sea rocks like a lizard." Swainson.)

---

* *Salarias*, CUVIER, Règne Anim., Ed. 2, II, 175, 1829 (*quadripinnis*). *Erpichthys*, SWAINSON, Nat. Hist. Class'n. Anim. II, 275, 1839 (*quadripinnis*, etc). (σαλάρια, a modern Greek name of *Blennius basilicus*.)
† We do not think that the name *Alticus* can be substituted for *Rupiscartes*, because Lacépède does not adopt this genus of Commerson, but merges it in *Blennius*, quoting Commerson's account as a footnote. This is as follows: "*Alticus saltatorius*, pinna spuria in capitis vertice; seu pinnula longitudinali pone oculos cartilaginea; seu alticus desultor, occipite cristato, ore circulare deorsum patulo." Apparently this quotation of a generic description not approved, does not give priority to the latter.
‡ This genus *Alticus* is not adopted by Cuvier & Valenciennes. Valenciennes speaks of "un petit *Salarias* que nous paraît être celui-là même sur lequel Commerson avait établi son genre *Alticus*." But a genus is not established until it is accepted by some authority as well as defined.

2748. RUPISCARTES ATLANTICUS (Cuvier & Valenciennes).

Head 4 to $4\frac{1}{2}$; depth $3\frac{1}{2}$ to $3\frac{3}{4}$. D. XII, 21 or XIII, 20; A. 24 or 25; vertebræ $12 + 22 = 34$; eye 4 to $4\frac{1}{2}$ in head. Body rather high, compressed. Head short, very blunt, its width about 2 in the length; anterior profile from first dorsal spine to above eye straight or slightly convex; from thence to tip of snout abruptly decurved, in some specimens nearly vertical. Mouth inferior, lower jaw included; maxillary about reaching posterior border of eye. Teeth small, pectinate, the lower canines exceedingly large and entering the cavity in the palate. Supraorbital tentacle well developed, slender; a group of 5 or 6 short tentacles on either side of head in front of nostrils and on either side of neck, these shorter than pupil. Dorsal fin not emarginate, extending from a point above middle of operculum to base of caudal; anal lower than soft dorsal, $1\frac{1}{2}$ to 2 in dorsal; pectorals reaching past vent, about equaling head; ventrals about 2 in head. "The intestinal tract is more than 3 times as long as the entire body. The structure of the skeleton is very similar to that of the Blennies; the jaw bones, however, are still shorter, and the intermaxillary and mandibulary are deeply concave anteriorly. There are 12 abdominal and 22 caudal vertebræ, the former portion being only $\frac{1}{4}$ as long as the caudal." (Günther.) Some specimens, apparently males, with the anterior profile vertical and very high; fins high; caudal lanceolate, the black median rays much exceeding the outer pale ones. Females with the anterior profile a nearly even curve, the caudal lunate, its median black rays shorter than the outer pale ones. Body liver brown, paler below, with usually 5 or 6 darker cross bars extending on the dorsal; a black spot behind eye in all; fins mostly blackish, an orange area on upper edge of caudal; a yellow one tinged reddish below; eye red posteriorly. Length 6 to 8 inches. Tropical America, on both coasts, very abundant in rock pools, north to West Indies and to Todos Santos. Here described from specimens from Mazatlan.

Punaru, MARCGRAVE, Hist. Brazil, 165, 1648, Brazil.
Salarias atlanticus, CUVIER & VALENCIENNES, Hist. Nat. Poiss., XI, 321, 1836, Madeira
    (Coll. Richardson), Antilles (Coll. Plée); GÜNTHER, Cat. Fish., III, 242, 1861.
Rupiscartes atlanticus, JORDAN, Proc. U. S. Nat. Mus. 1888, 333.

892. ENTOMACRODUS,* Gill.

Salarias, SWAINSON, Nat. Hist. Class'n Fishes, II, 274, 1839 (*vermicularis;* not of CUVIER).
Entomacrodus, GILL, Proc. Ac. Nat. Sci. Phila. 1859, 168 (*nigricans*).

This genus has large posterior canines as in *Rupiscartes*, but the dorsal fin is divided into 2 fins as in *Scartichthys*. ($\dot{\varepsilon}\nu$, in; $\tau\acute{o}\mu o\varsigma$, cutting; $\dot{\alpha}\varkappa\rho\acute{o}\varsigma$, sharp; $\dot{o}\delta o\acute{v}\varsigma$, tooth.)

   a. Orbital cirrus present; dorsal rays XII or XIII-15; canines small.
      b. Cirrus above eye divided; anal rays 15.   CHIOSTICTUS, 2749.
      bb. Cirrus above eye simple or nearly so; anal rays 18; body with pearly spots.
                                        MARGARITACEUS, 2750.

---

* This genus is equivalent to *Salarias* of Swainson, but the generic name *Salarias* was based on *Salarias quadripinnis*, before either of the species referred to it by Swainson was made known.

aa. Orbital cirrus wanting; no cirri at nape.
  c. Dorsal rays XII–19; anal 15; body rather slender, the depth about 5 in length;
      body with bands and spots.   DECORATUS, 2751.
  cc. Dorsal rays XI–15; anal 17; body very slender, the depth about 6 in length;
      color blackish, nearly plain.   NIGRICANS, 2752.

2749. ENTOMACRODUS CHIOSTICTUS (Jordan & Gilbert).

Head 4⅛ in length; depth 5¼. D. XII–15; A. 15; eye 3⅓ to 4¼ in head,
varying with age. Body moderately elongate, compressed, the head
short, blunt, almost globular, about as broad as deep, and a little longer
than broad. Mouth inferior, with little lateral cleft, the lower jaw
included; width of cleft of mouth ⅔ length of head. Teeth small, weak,
finely pectinate; canine teeth small, not so long as diameter of pupil.
Supraorbital cirrus divided into 4, its height ⅔ that of eye; a few minute
slips at the nape. Interorbital space channeled, narrower than eye.
Maxillary extending to behind middle of eye. No crest on top of head.
First dorsal low and even, its spines rather slender, the last spines short,
scarcely connected by membrane with the soft rays; soft dorsal well
separated from caudal; caudal subtruncate, with rounded angles; anal
lower than soft dorsal, with a little longer base; pectorals a little longer
than head; ventrals about ⅓ as long. Color in life, olive brown above,
lighter below; 5 broad, dark bars from dorsal fin to middle of sides,
each terminating above on the fin, and below on sides in a pair of black
spots; sometimes only the dark spots are distinguishable, the bars being
obscure; sides below spinous dorsal with numerous black specks, and
with numerous oblong spots of bright silvery; sometimes a silvery streak
from upper portion of base of pectorals to base of caudal; a broad salmon-
colored streak on each side of ventral line; sometimes the space between
the silvery lateral band and the base of the anal is darker, the vertical
bars again appearing as pairs of black, vertical blotches; head yellowish
olive, darker above, and reticulating with narrow brown lines, these
appearing as parallel bars on the upper lip, and radiating from the median
line on the upper side of the head; vertical fins light grayish, with black
spots, which appear as wavy bars on the caudal fin; pectorals and ventrals pale, the former with a yellowish shade at base; orbital tentacles
bright red. Pacific coast of Mexico. Known from 4 specimens (the
largest 2¼ inches in length), taken in a deep rock pool at Mazatlan. Two
others taken by the *Albatross* from Clarion Island. ($\chi\iota\acute{\omega}\nu$, snow; $\sigma\tau\iota\varkappa\tau\acute{o}\varsigma$,
spotted.)

Salarias chiostictus, JORDAN & GILBERT, Synopsis, 363, 1883, Mazatlan, Mexico. (Coll.
  Jordan & Gilbert.)

2750. ENTOMACRODUS MARGARITACEUS (Poey).

Head 5 in total length with caudal; depth 6½. D. XII–14; A. I, 14; eye
4½ in head, well forward. Body large, snout abruptly decurved; mouth
very low, maxillary reaching anterior nostril, which has a little tentacle;
(canines small); small tentacle over eye; gill membranes broadly connected,
free from isthmus; dorsal deeply emarginate, almost divided; anal beginning under middle of body without caudal, and anal papillæ and caudal

rounded; ventrals short; lateral line present anteriorly, no tentacles on nape. Color brown, with 2 vertical bands of a dusky silvery; a central point in each band shining bright. One specimen, 2¼ inches long. Cuba. (Poey.) Perhaps a *Salarias*.

We have the following notes on a specimen, possibly the type of this species, sent by Poey to the museum at Cambridge: Head 4⅔; depth 5. D. XII–15; A. 18. Body slender. Interorbital concave. Head short, blunt, almost round; a small cirrus over the eye, none on nape. Canines present, small. Body with about 6 dark cross bars besides pearly spots and various markings. Dorsal divided nearly to base. Closely resembles *Salariichthys textilis.* (*margarita*, μαργαρίτης, pearl.)

Salarias margaritaceus, POEY, Memorias, II, 289, 1861, Cuba. (Coll. Poey.)

2751. ENTOMACRODUS DECORATUS, Poey.

Head 5 in total length with caudal; depth 5. D. XII–19; A. 15; P. 14. Eye very high; anterior nostril prolonged in a tube; nape following a straight line to the posterior nostril, profile thin, following a straight and oblique line to mouth, which is very low and short, the maxillary reaching posterior nostril. Lower jaw shorter. Teeth movable, numerous, incurved, close set, in 1 row. (Canines not described.) No cilia on head. Dorsals of equal length, the soft rays more elevated; anal similar to second dorsal; pectoral broad, its lower rays thickened; caudal rounded. Color brownish yellow; the body with darker cross bands, which begin below the middle of the first dorsal, alternating with narrower spaces of the ground color; along the middle and edges of the bands vertical rows of sky-blue spots; in the pale interspaces below the lateral line, which is much curved, a white spot; 3 pale spots placed obliquely below the eye; rays of dorsal and caudal dotted with black. One specimen, 2 inches long. Cuba. (Poey.) Not seen by us; perhaps a *Salarias*. (*decoratus*, decorated.)

Entomacrodus decoratus, POEY, Synopsis, 398, 1868, Cuba. (Coll. Poey.)

2752. ENTOMACRODUS NIGRICANS, Gill.

The elongated body, from the snout to the end of the caudal fin, is between 7 and 8 times longer than it is high at the pectorals. Its height at the caudal is about ⅟₁₃ of the same length. The head is subquadrate, and forms ⅔ of the total length. Its greatest height equals ⅔ of its length. Its sides decline obliquely outward and downward. The first dorsal commences near the nape, and 2 of its rays are in advance of the pectorals. The second dorsal commences immediately behind the first, and nearly over the fourth ray of the anal, it ceases some distance from the base of the caudal. The anal is more uniform in height than the dorsal, and ceases before it does. The caudal forms less than ⅓ of the total length. D. XI–15; A. 17; P. 15; V. 3. The general color of the body and fins is blackish. West Indies. A single specimen was caught in shallow water, at the island of Barbados, near Bridgetown. (Gill.) Not seen by us. (*nigricans*, blackish.)

Entomacrodus nigricans, GILL, Proc. Ac. Nat. Sci. Phila. 1859, 168, Barbados. (Coll. Dr. Gill.)

893. SALARIICHTHYS, Guichenot.

Salariichthys, GUICHENOT, Mém. Soc. Sci. Nat. Cherbourg, XIII, 1867, 96 *(textilis)*.

This genus differs from *Entomacrodus* in the presence of teeth on the vomer; dorsal deeply notched; cirri present over eye and on nape; posterior canines small. (*Salarias; ἰχθύς*, fish.)   •

2753. SALARIICHTHYS TEXTILIS (Quoy & Gaimard).

D. XII, 16; A. 18. A few bluntish teeth on vomer; tentacles very small, fringed over nostril and eye, simple on neck; canines quite short; depth $4\frac{3}{5}$; head $4\frac{3}{5}$; pectoral short, little longer than head; gill membranes broadly united, free from isthmus; dorsal notched almost to base, free from caudal; orbital filament $\frac{1}{4}$ eye. Olive, with 13 silvery cross streaks, not $\frac{1}{4}$ as wide as the dark interspaces, some of the cross streaks Y-shaped; both dorsals with cross markings, the second with 12 or 13 streaks of dark obliquely upward and backward, alternately with similar pale streaks; cross bars on sides bent in middle, extending up and back and down and back from middle line parallel with muscular impressions; sides with some obscure pale dots; caudal barred with 7 dark bars; anal darkest mesially; lower side of head with dark streaks radiating from the isthmus; bars at chin Y-shaped, upper part of head with darker markings; pectoral nearly plain; a dusky area at base below which is a dusky spot; marblings at base of dorsal. West Indies, from Bermudas to Brazil. Here described from a specimen from Abrolhos Islands (Coll. *Albatross*). This specimen agrees fairly with the account given by Jenyns, but Jenyns describes 5 bars on the tail. It also agrees fairly with the account of the Bermuda specimens given by Goode. It is evidently the *Salarias vomerinus* of Cuvier & Valenciennes, and probably their *textilis* also; but their description of the latter does not apply very well to the coloration of our specimen. (*textilis*, woven.)

Salarias textilis, QUOY & GAIMARD MS., CUVIER & VALENCIENNES, Hist. Nat. Poiss., XI, 307, 1836, Ascension Island (Coll. Quoy & Gaimard); GÜNTHER, Cat., III, 248, 1861; GOODE, Bull. U. S. Nat. Mus., V, 29, 1876.

Salarius vomerinus, CUVIER & VALENCIENNES, Hist. Nat. Poiss., XI, 349, 1836, Bahia. (Coll. Blanchet.)

Salariichthys textilis, JORDAN, Proc. U. S. Nat. Mus. 1890, 329.

894. OPHIOBLENNIUS, Gill.

Blennophis, VALENCIENNES, in WEBB & BERTHELOT, Poiss. Îles Canar., 60, 1844 *(webbii;* not *Blennophis,* CUVIER & VALENCIENNES, a genus of *Clininæ*).
Ophioblennius, GILL, Proc. Ac. Nat. Sci. Phila. 1860, 103 *(webbii;* substitute for *Blennophis*).

Body oblong, strongly compressed, scaleless; snout short, high, abruptly decurved anteriorly; symphysis of lower jaw of 4 hooked canines, the outer strongest and bent backward, almost forming a right angle; sides of lower jaw with 2 or 3 still larger canines, the hindermost very large and bent backward; upper jaw with 4 slender canines in front, followed by a long row of shorter, slender, movable teeth, which are set close together; nasal tentacle digitate; a low, simple tentacle above eye; gill openings wide. Dorsal fin long, the spines slender, separated by a slight notch from the soft rays; caudal lunate or forked, free from dorsal and anal; ventrals small, I, 2; lateral line incomplete; pectorals large. A strongly marked genus, perhaps more nearly allied to *Blennius* than to *Emblemaria* or *Chœnopsis*. (ὄφις, snake; *Blennius*, in allusion to the fang-like teeth.)

> *a.* D. X, 20; A. 20; depth 5½ in length.   WEBBII, 2754.
> *aa.* D. XI, 22; A. II, 23; depth 4¼ in length.   STEINDACHNERI, 2755.

2754. OPHIOBLENNIUS WEBBII (Valenciennes).

Head 5; depth 5½. D. X, 20; A. 20; P. 16. A slender tentacle above eye in front, and a much broader one, divided into 4 to the base, above the nostril. Snout obtuse, nearly vertical at tip; eye large; 4 teeth at end of upper jaw, strongly pointed, curved backward like hooks; lower jaw with 4 teeth at tip, the two middle ones like upper teeth, the two outer hidden and turned backward; a little recurved tooth on side of lower jaw; caudal fin forked; dorsal somewhat notched at the last spine; lateral line ending near middle of body. Olive green, light or dark; dorsal and anal dusky violet, the base pale; back and sides often with fine points; a dark spot behind eye; the silvery swim bladder showing through sides of belly. (Steindachner). Tropical Atlantic; known only from the Canaries and Barbados; not seen by us. (Named for P. B. Webb, one of the explorers of the Canary Islands.)

Blennophis webbii, VALENCIENNES, in WEBB & BERTHELOT, Îles Canar., Poiss., 60, pl. 20, f. 1, 1844, Fortaventura, Canary Islands (Coll. Webb); ''caught in myriads at Puerto de Cabras in August, eaten as Anchovias'' (Webb); GÜNTHER, Cat., III, 259, 1861; STEINDACHNER, Ichth. Notizen, VI, 48, 1867.
Ophioblennius webbi, JORDAN & GILBERT, Synopsis, 756, 1883.

2755. OPHIOBLENNIUS STEINDACHNERI, Jordan & Evermann, new species.

Head 4 to 4½; depth 4 to 4¼. D. XI, 22; A. II, 23; V. I, 2; P. 15. Head much compressed; eye 3¼ in head; snout 4¼. Dorsal beginning above gill opening, ending just before caudal, its soft rays somewhat higher than the spines, the highest spine 1¾ in head; caudal and pectorals each about as long as head; ventrals 1½ in head. Dark golden brown, sometimes with a broad cross band of dusky violet on back and dorsal fin; caudal with 2 dark longitudinal stripes; dorsal and anal purplish or orange; an intense, round, dark, ocellated spot behind eye. (Steindachner.) West coast of Mexico; not seen by us; recorded from near Mazatlan and the Tres Marias Islands. (Named for Dr. Franz Steindachner.)

Blennophis (Ophioblennius) webbi, STEINDACHNER, Ich. Beitr., VIII, 41, 1879, 5 specimens 70 mm. long, from Navidad near Mazatlan and the Tres Marias Islands.
Ophioblennius steindachneri, JORDAN & EVERMANN, Check-List Fishes N. and M. A., 472, 1896, name only, Tres Marias Islands; after STEINDACHNER.

895. EMBLEMARIA, Jordan & Gilbert.

Emblemaria, JORDAN & GILBERT, Proc. U. S. Nat. Mus. 1883, 627 (*nivipes*).

Body slender, not eel-shaped, compressed, scaleless. Ventrals present, jugular, each of 1 spine and 2 soft rays. A single high dorsal fin beginning on the nape and extending to the caudal, with which it is not confluent; no notch between spinous and soft parts. Head cuboid, compressed, narrowed anteriorly. Symphysis of lower jaw forming a very acute angle. A single series of strong, blunt, conical teeth on each jaw, and on vomer and palatines. Vomer and palatine teeth larger, their series continuous, parallel to the series in upper jaw. No cirri at the nape; sometimes a cirrus on upper part of eyeball: Gill openings very wide, the membranes broadly united below, free from the isthmus. Lateral line obsolete. This genus bears some resemblance to *Blennius*, but the dentition is entirely different, approaching that of *Chænopsis*. Tropical America, in rather deep water. (*Emblema; ἔμβλημα*, a banner.)

  a. Eye without cirrus.
    b. Depth 5 in length; dorsal rays 33; ventrals not pure white.   ATLANTICA, 2756.
    bb. Depth 7 in length; dorsal rays 37; ventrals pure white.   NIVIPES, 2757.
  aa. Eye with a long cirrus on eyeball above pupil; ventrals dusky; maxillary not
      extending beyond eye.   OCULOCIRRIS, 2758.

2756. EMBLEMARIA ATLANTICA, Jordan & Evermann, new species.

Head 3⅔; depth 5. D. 35; A. 24; P. 15; V. 3. Body slender, compressed; head heavy; snout evenly decurved; mouth large, horizontal, reaching back of eye. Jaws with short, strong, incurved conical teeth. Fin rays long and filamentous, the longest dorsal rays as long as head; anal rays shorter. Coloration faded in the type, but traces of about 7 broad brown vertical bars as broad as eye and twice as broad as the pale interspaces, the dark bars extending upon dorsal fin; ventrals pale. Gulf of Mexico. Known from 1 specimen, 3½ inches long, taken from the stomach of *Neomœnis aya*, on the Snapper Banks off Pensacola, Florida; very close to *E. nivipes*, but more robust, with fewer dorsal rays.

Emblemaria atlantica, JORDAN & EVERMANN, Check-List Fishes, 472, 1896, name only, Snapper Banks off Pensacola, Florida. (Type, No. 33915. Coll. Silas Stearns.)

2757. EMBLEMARIA NIVIPES, Jordan & Gilbert.

Head 3¾ in length; depth 7. D. XXIII, 14; A. 25. Body everywhere equally compressed, posteriorly tapering; head wider than body, of about equal depth, with very short, subvertical, sharply compressed snout; eyes very large, approximated above, with some vertical range; orbital ridges sharply raised above, the interorbital region very narrow, channeled, about equaling diameter of pupil; eye 3¾ in head. Gape very wide, horizontal, low, reaching much beyond eye, the maxillary about ½ head, not produced beyond angle of mouth; intermaxillaries separated by a groove from the snout, this groove continuous for the entire length of the upper jaw, maxillary not evident, apparently adnate to the skin of the preorbital. First dorsal spine inserted over margin of preopercle; spines all very slender and flexible, the posterior but weakly differentiated from the soft rays, the anterior portion of fin very high, the spines filiform, not exserted beyond the membrane; the longest dorsal spine about ⅓ length of body, the last spine about ⅓ head; membranes of last rays of both dorsal and anal slightly joined to base of caudal. Front of anal nearer snout than base of caudal by a distance equaling $\frac{1}{3}$ length of head. Caudal $\frac{3}{5}$ length of head; ventrals and pectorals slightly less. Color in spirits, sides dark brown, with 8 to 10 lighter vertical bars of variable width; body lighter below; obscure cross bands on lower side of head; dorsal blackish anteriorly, whitish behind, with membrane at intervals of every second, third, or fourth ray dusky; caudal light at base, its tip blackish; anal dusky translucent; ventrals bright white, the basal portion dusky. Pearl Islands, near Panama. A specimen 2 inches long is the type of the species. Numerous smaller specimens were obtained at the same time. (*nix, nivis*, snow; *pes*, foot.)

Emblemaria nivipes, JORDAN & GILBERT, Proc. U. S. Nat. Mus. 1883, 627, Pearl Islands, near Panama. (Type, No. 29676. Coll. Prof. Frank H. Bradley.)

2758. EMBLEMARIA OCULOCIRRIS, Jordan.

Head $3\frac{3}{4}$; depth $6\frac{2}{3}$. D. about 35; A. 25. Upper part of eyeball above pupil (sclerotica) with a slender cirrus tipped with black, this nearly as long as eye; eye longer than snout, about $3\frac{3}{4}$ in head, the maxillary extending to below posterior part of pupil; snout sharper than in *Emblemaria nivipes*, $\frac{2}{3}$ eye; teeth small, rather sharp, directed backward; longest dorsal spines as long as head; pectorals $1\frac{1}{3}$ in head; ventrals $1\frac{2}{3}$, inserted before pectorals. Color in spirits, brown, with traces of about 9 blackish cross bars, which are separated on the back by whitish, quadrate interspaces; a white spot at nape; some dusky below eye; dorsal dusky, the pale bars of back extending on its base; anal dusky; ventrals blackish; caudal pale, its tip black; pectorals pale. Gulf of California. Known from 1 specimen, $1\frac{1}{2}$ inches long, from La Paz. It is shriveled and in poor condition. It seems to be very close to *Emblemaria nivipes*, but differs in the presence of an ocular cirrus, in the sharper snout, smaller mouth, and dusky ventrals. The teeth seem rather more slender, but can not be well examined. (*oculus*, eye; *cirrus*, filament.)

Emblemaria oculocirris, JORDAN, in GILBERT, Proc. U. S. Nat. Mus. 1896, 456, La Paz. (Type, No. 47749. Coll. *Albatross*.)

896. CHÆNOPSIS, Gill.

Chænopsis, GILL, Ann. Lyc. Nat. Hist. N. Y., VIII, 1865, 141 (*ocellatus*).

Body naked, eel-like. Head much elongate, quadrate behind, conic in front, profile straight; snout acute, jaws produced; no teeth on vomer, teeth in front of jaws strong, with villiform teeth behind them. Dorsal and anal long, continuous, confluent with the caudal. Dorsal rays about XVIII, 38; anal II, 38. Ventrals inserted slightly before pectorals. West Indies. (*χαίνω*, to yawn; *ὄψις*, face.)

2759. CHÆNOPSIS OCELLATUS, Poey.

D. XVIII, 38; A. II, 38; C. 15. Body naked, eel-like; anus submedian. Head much elongate, quadrate behind at the opercular region, conic in front, with the profile rectilinear and the snout acute; eyes moderate;

mouth large, with the cleft wide and nearly horizontal. Teeth subcylindrical, in a uniform row, behind which, in front, there is a broad band of villiform teeth; on the palatine bones, uniserial and obtusely subcylindrical like those of the jaws; the palatine rows are parallel; vomer edentulous. Gill membranes confluent below, free from the isthmus. Dorsal and anal long, confluent with caudal; ventrals slightly in advance of pectorals, with 2 or 3 rays. (Gill.) Matanzas, Cuba; 1 specimen, examined by us in the National Museum. (*ocellatus*, having eye-like spots.)

Chœnopsis ocellatus, Poey, in Gill, Ann. Lyc. Nat. Hist. N. Y., VIII, 1867, 143, Matanzas, Cuba. (Coll. Poey.)

897. LUCIOBLENNIUS, Gilbert.

Lucioblennius, Gilbert, Proc. U. S. Nat. Mus. 1890, 103 (*alepidotus*).

Body very elongate, wholly naked; gill membranes broadly united, free from isthmus; dorsal fin single, extending along the entire back, its anterior half spinous. Ventrals in front of pectorals, I, 2. First two anal rays spinous. Last rays of dorsal and anal joined to caudal. Teeth conic, not movable, in jaws and on vomer and palatines. Lateral line not described. A strange genus, evidently very close to *Chœnopsis*. (*Lucius*, pike; *Blennius*, blenny.)

2760. LUCIOBLENNIUS ALEPIDOTUS, Gilbert.

Head 3 in length; depth 3¼ in head. D. XVIII, 32; A. II, 30. Body much compressed, slender throughout, the head rather deeper and wider than body. Snout long, depressed, and rather wide, the anterior profile descending very gradually. Mouth nearly horizontal, the lower jaw protruding, the gape extending to much behind orbit, the entire physiognomy remarkably pike-like. Snout 4 in head; maxillary 1¾; eye 4¾ to 5. Teeth in a villiform band in upper jaw, the outer series slightly larger; in lower jaw in a single series laterally, widening into a patch anteriorly, the outer enlarged; a few teeth only on vomer; palatines with a long and rather broad patch similar to those in jaws. Dorsal fin beginning on the nape in advance of middle of opercle, the fin uniformly low, extending the whole length of back, the posterior ray joined by membrane with the caudal; the spines and rays are similar in appearance, flexible and simple, none of the soft rays branched; the spines are more slender, and show no joints, the articulations being present in small number on all the soft rays; the highest ray is less than diameter of orbit; anal and caudal rays similar to those of soft dorsal; caudal short, rounded; origin of anal midway between tip of snout and end of caudal fin, its first 2 rays spinous; ventrals under opercular margin, of 1 spine and 2 well-developed rays, nearly ¼ as long as head; pectorals narrow, of apparently unbranched rays, about ¼ as long as head. Color light olivaceous, with 11 vertical dark blotches on sides, most of which divide to form on middle of sides double vertical bars; top and sides of head with dark cloudings, and with numerous black specks of varying size; middle of sides and base

of dorsal with numerous pearly dots nearly as large as pupil; branchiostegal membrane black posteriorly; the lateral bars extended to base
of dorsal, the anterior ones usually forming conspicuous black blotches
which extend well up on the fin; other fins unmarked. Length 1½ inches.
Gulf of California; two specimens from *Albatross* Station 3005, in 21
fathoms. (ἀλεπιδωτός, scaleless.)

Lucioblennius alepidotus, GILBERT, Proc. U. S. Nat. Mus. 1890, 103, Lower California
(Coll. *Albatross*); JORDAN, Proc. Cal. Ac. Sci. 1896, 233, pl. 37.

898. PHOLIDICHTHYS, Bleeker.

Pholidichthys, BLEEKER, Boeroe, 406, 1856 (*leucotænia*).

Body elongate, tapering, naked; snout obtuse; no cirri; teeth unequal,
on jaws only; dorsal, anal, and caudal fins distinct, but connected by a
membrane; the dorsal formed of flexible spines; the soft rays, if present,
not distinguishable from them; ventrals inserted scarcely before the pectorals, of 2 rays. Lateral line and vertebræ undescribed. Tropical parts
of the Pacific. (*Pholis*; ἰχθύς, fish.)

2761. PHOLIDICHTHYS ANGUILLIFORMIS, Lockington.

Head 6⅔ in total length with caudal; depth 16. Body exceedingly
elongate, much compressed, naked; upper profile of head forming a continuous convex curve to the tip of the snout, which is about equal in
length to the eye. Eye lateral, round; interorbital space about ⅔ of the
diameter of the eye, convex transversely. Posterior extremity of maxillary vertical with the hinder margin of the eye. Tip of snout a little
below the level from the center of the eye; mouth moderately oblique,
lower jaw slightly the longer. Teeth of lower jaw in a close-set row, the
largest in front, diminishing along the sides; teeth of upper jaw similar,
but smaller; palate smooth. Vertical fins continuous, but distinct; dorsal entirely spinous; anal commencing a little behind the middle of the
entire length of the fish; ventrals 2-rayed, very slightly in advance of
the pectorals, which are about equal in length to the distance of their base
from the eye. Color in spirits, dark blackish brown mingled with white
upon top, sides, and lower parts of head; interorbital area and top of
snout white. Gulf of California; a single specimen dredged off San Jose
Island, Amortiguado Bay. Total length 1₁⁵₂ inches. Head ¼ inch. The
example is broken across, the branchiostegals are defective, the caudal fin
broken and some fin rays missing; so that the fin formula can not be
exactly given. The dorsal fin has above 60 rays. The body is much more
slender than that of *P. leucotænia*, Bleeker, and there is no trace of the
longitudinal bluish-white band of that species. (Lockington.) (*Anguilla*,
eel; *forma*, shape.)

Pholidichthys anguilliformis, LOCKINGTON, Proc. Ac. Nat. Sci. Phila. 1881, 118, San Jose
Island, Lower California. (Coll. W. J. Fisher.)

899. PSEDNOBLENNIUS, Jenkins & Evermann.

Psednoblennius, JENKINS & EVERMANN, Proc. U. S. Nat. Mus. 1888, 156 (*hypacanthus*).

Body compressed, elongate, naked; head short, blunt; no cirri; mouth large, the jaws subequal; teeth in a single series in each jaw, none on vomer or palatines; lateral line not developed. Dorsal fins 2, the first at the nape, of three flexible spines; second dorsal with a few slender spines which pass into the soft rays; anal much shorter than second dorsal, both fins joined to base of caudal; dorsal rays III–34; anal 27; ventral rays 2, the fin directly below pectorals. Apparently close to *Pholidichthys,* but with the dorsal divided and changing gradually from spines to soft rays. (ψεδνός, naked; *Blennius.*)

2762. PSEDNOBLENNIUS HYPACANTHUS, Jenkins & Evermann.

Head 4⅔ (5 in total); depth 7 (8); eye 4, equal to snout; B. 6. D. III–34; A. 27. Body greatly compressed, elongate; head short, snout blunt, about equal to eye; anteorbital profile very steep, gently rounded from front of eye to first dorsal, from there nearly straight to caudal; ventral line nearly straight. Body naked, no membranaceous appendages. Mouth large, horizontal, jaws subequal, extending to beyond middle of eye. Teeth in a single series in each jaw, well developed, pretty uniform in size, slightly projecting backward; vomer and palatines apparently smooth. Eye large, equal to twice interorbital space, high up. Dorsal fins 2, the first of 3 very slender, flexible spines, hard to distinguish from soft rays, but they do not appear to be at all jointed. This fin is inserted upon the nape immediately above the posterior edge of the preopercle, and a distance in front of second dorsal nearly equal to length of snout, its very soft spines equal distance from end of snout to posterior rim of orbit; second dorsal begins directly over origin of pectorals and extends to caudal, with which it is slightly connected; first few rays of second dorsal very weak, flexible spines, the last few pretty evidently soft, jointed rays, while the intermediate ones are not distinguishable as definite spines or soft rays—in short, there seems to be a gradual change from spines to soft rays from the anterior to the posterior part of the fin. This character, if we mistake not, is entirely unique. The fin is of nearly uniform height, the rays about equaling those of the first dorsal in length; anal similar to second dorsal in shape and height, but much shorter, its origin being much behind that of the second dorsal or nearly halfway from the snout to base of caudal; posteriorly it extends coterminously with the dorsal, and, like it, is slightly joined to the caudal fin; caudal fin apparently rounded, fan-shaped, but its shape can not be exactly made out, as some of its rays are broken off; pectorals inserted below axis of body, directly over ventrals, their length about ¾ that of head; ventrals of 2 rays, inserted under pectorals, about equal to pectorals in length; body entirely scaleless. Coloration in alcohol, pale, mottled with fine dark points so arranged as to inclose circular areas with fewer spots; a long dark blotch behind the axil, inclining downward and backward; head covered with similar punctulations; opercles dusky; chin with 2 dark cross lines, separated by 1 of white, extending onto upper jaw on each side; top of head with a purple spot; sides with a series of about 6 short black lines, the last broadest and plainest; base of caudal with a distinct black blotch; first dorsal quite dark, almost black; second dorsal with about 8 pretty well-defined dark blotches at its base, rest of fin with numerous dark spots of different sizes; anal with about 12 dark blotches extending somewhat regularly from the base slightly forward, these separated by plain unmarked spaces of a little greater width; caudal sparingly marked with dark points arranged in wavy cross bars; pectorals and ventrals unmarked. Gulf of California at Guaymas. A single specimen, 1¾ inches long, obtained from a shallow arm of the bay. (Jenkins & Evermann.) (ὑπό, below (imperfect); ἄκανθα, spine.)

Psednoblennius hypacanthus, Jenkins & Evermann, Proc. U. S. Nat. Mus. 1888, 156, Guaymas, Mexico. (Type, No. 39638. Coll. Jenkins & Evermann.)

900. STATHMONOTUS, Bean.

Stathmonotus, Bean, Proc. U. S. Nat. Mus. 1885, 191 *(hemphillii).*

Body moderately long and low, much compressed; head small, compressed, naked; mouth small, oblique; conical teeth in both jaws, in 2 series, the outer slightly enlarged and, in the upper jaw, somewhat recurved; a few teeth on the vomer. Gill membranes, as in *Pholis,* broadly united, free from the isthmus. Scales none. No lateral line. Dorsal fin long and low, beginning near the head, and consisting entirely of stiff, sharp spines, which are very short anteriorly and gradually increase in size posteriorly. Anal similar to dorsal, with 2 spines and many soft rays. Caudal short, rounded, scarcely separated from the dorsal and anal; pectorals small, much smaller than in *Pholis,* containing only a few rays; ventrals better developed than in *Pholis,* their position more anterior, consisting of a spine and 2 rays. Pseudobranchiæ absent. Branchiostegals 5. Coast of Florida. (σταθμή, a carpenter's rule; νῶτος, back.)

2763. STATHMONOTUS HEMPHILLII, Bean.

Head 7; depth 8 to 8½; D. LI; A. II, 27; V. I, 2; P. 5 or 6; eye 6 in head. Maxillary extending about to vertical through hind margin of eye; jaws subequal, or the lower projecting very slightly beyond upper; eyes small, separated by an interspace about equal to their own length, and very slightly greater than length of snout; pectoral very little more than ⅛ as long as head, and scarcely as long as ventral; dorsal beginning over posterior end of pectoral, its anterior spines very much shorter than the posterior ones; length of caudal about equal to length of postorbital part of head; vent slightly in advance of middle of total length to base of caudal, and about under the twentieth dorsal spine. Colors from the alcoholic specimen: A white line extending from tip of snout to caudal, divided into small segments by short cross bars, the first 2 on the head, and the last at origin of caudal; posteriorly, these short bars extend downward, terminating slightly below the base of the dorsal fin; several white blotches, simulating bars, on posterior half of anal fin; edge of

caudal white; sides and under surface of head with several whitish oblique bands forming V-shaped markings; a few roundish white blotches
on sides of head, the most conspicuous behind eye; general color darkish
brown, nearly black. Length about 2 inches. Key West; 2 specimens
known. (Bean.) (Named for the collector, Henry Hemphill.)

Stathmonotus hemphillii, BEAN, Proc. U. S. Nat. Mus. 1885, 191, pl. 13, Key West, Florida.
(Coll. Henry Hemphill. Type, No. 37193, U. S. Nat. Mus.)

901. BRYOSTEMMA, Jordan & Starks.

Bryostemma, JORDAN & STARKS, Proc. Cal. Ac. Sci. 1895, 841 (*polyactocephalum*).

Body moderately elongate, covered with small scales; snout short; no
teeth on vomer or palatines; teeth in jaws small; gill membranes united,
free from the isthmus; nostrils, orbital regions, and neck with dermal
flaps, the supraorbital flaps high. Dorsal fin long, of spines only; pectorals well developed, more than half length of head; ventrals well developed, jugular; caudal fin distinct. No air bladder or pyloric cæca. No
true lateral line; a short series of large pores above pectoral. North Pacific, representing *Chirolophis* of the Atlantic. This genus differs from the
European genus, *Chirolophis*, Swainson (*Blenniops*, Nilsson), in the absence
of a true lateral line. Dr. Boulenger informs us that a true median lateral line is developed in *Chirolophis ascanii*. (βρύον, moss; στέμμα,
crown.)

a. Dorsal with about 60 spines; anal with about 55 soft rays; a black spot on anterior
part of dorsal, but no ocelli posteriorily.   POLYACTOCEPHALUM, 2764.
aa. Dorsal with about 54 spines; anal with 40 soft rays; dorsal with several black
ocelli, most distinct posteriorly.   NUGATOR, 2765.

2764. BRYOSTEMMA POLYACTOCEPHALUM (Pallas).

Head 6½; depth 6. D. LXI; A. 55 (51 to 57); P. 14; V. I, 3; lateral series
with 9 to 15 pores. Body elongate, much compressed, covered with small,
smooth, embedded scales. Head very short, blunt in profile; mouth short,
terminal, the maxillary 3 in head; lower jaw heavy, projecting, its tip
with 2 small slender cirri, which are pale in color; teeth subequal, small,
bluntish, close set, in 1 row in each jaw; eyes 4 in head, near together;
the snout 4; supraorbital cirri 2⅓ in head; interorbital space flat; a flat
fringed cirrus over front of eye, these 2 joined at base, about 3 in head; a
small cirrus about ⅓ length of this over posterior part of each eye, these 5
to 6 in head; top of head and nape covered with series of erect cirri, the
longest nearly as long as eye; about 15 minute cirri along dorsal edge of
lateral line, 1 on each pore. Rows of pores running around eye, under preopercle, and along entire length of the short lateral line; lateral series of
pores ⅓ length of head; gill rakers not developed. Dorsal fin beginning over
pectoral and running to caudal; anterior rays fringed with fleshy cirri;
first ray, including cirri, 2 in length of head; anal beginning close behind vent and running to caudal, to which it is joined at base; distance
from tip of snout to vent nearly 3 in body; pectoral fin but little shorter
than head, its breadth at base not ⅓ its length. Color in spirits, pale

brownish, plain or mottled with darker, with about 13 dark blotches along dorsal and anal fins, more distinct on dorsal; a black spot on fourth to sixth dorsal spines very distinct; a faint one on anterior part of anal; a few dark markings about head and nape; cirri mostly pale. Bering Sea, south to Puget Sound and Yezo. Here described from a fine specimen, 6½ inches long, from Port Orchard, near Seattle, collected by Prof. O. B. Johnson. Other specimens before us from St. Paul (Pribilof Islands), from *Albatross* Stations 3213 and 3274, south and north of the Peninsula of Alaska, and from Petropaulski Harbor, Kamchatka. These specimens show a great deal of variation, and possibly represent 3 different species. It is more likely, however, that they represent extremes of variation. Young examples, collected by the *Albatross* in eastern Bering Sea, are more elongate and less compressed; body much mottled and vaguely barred; ventral fins checkered in fine pattern; head sand color; a black blotch on fourth to sixth dorsal spine; anterior dorsal spine little elevated and with few fringes; sides of head without cirri; anterior cirri joined almost to the tip, a little shorter than the posterior cirri, which are long and very slender. In 1 specimen of these, however, the cheeks are covered with densely matted cirri extending from the angle of the mouth to the dorsal. In these examples the anterior cirri are short and separate, about as long as the posterior cirri. The larger example, 75 cm. long, from Petropaulski, is evidently the typical *polyactocephalum*, and corresponds perfectly to Herzenstein's account of *B. japonicum*. It shows the following characters: Head 6⅓; depth 5½. D. LXI; A. 45; P. 14; V. I, 3; lateral series with 6 pores. Body a little deeper than in Puget Sound examples; head short, blunt in profile; mouth short, terminal, oblique, the maxillary 2⅔ in head; lower jaw heavy, projecting, its tip with 2 broad fringed flaps of a dark color; eyes 4 in head, close together, the interorbital space concave; a fringed cirrus above each eye in front, the 2 connected with each other only in the thickened skin at base; a similar cirrus over each eye behind; the posterior cirri ⅓ longer than the anterior ones, 2⅓ in head; top of head and nape with similar cirri, none of them longer than pupil; a few small cirri on cheeks and opercles; some along lateral series of pores, which is 2¼ in head; anterior rays of dorsal fringed with fleshy cirri, the first 2 in head; distance from snout to vent 2⅔ in body; pectorals nearly as long as head, the rays thickened in the adult, the base of the fin about ⅓ its length. Color very dark brown, with vague cross bands and many spots; dorsal and anal each with a broad black edge; other fins all black, the caudal barred. Perhaps the dark coloration and long cirri are characters of the adult male. ($\pi o \lambda \acute{v} \varsigma$, many; $\grave{\alpha} \varkappa \tau \acute{\iota} \varsigma$, ray; $\varkappa \varepsilon \varphi \alpha \lambda \acute{\eta}$, head.)

Blennius polyactocephalus, PALLAS, Zool. Rosso-Asiat., III, 179, 1811, Kamchatka.

Chirolophus japonicus, HERZENSTEIN, Mélanges Biologiques Soc. Sci. Petersb., XIII, 1890, 123, Yezo.

Chirolophis polyactocephalus, JORDAN & GILBERT, Synopsis, 765, 1883; BEAN in Nelson, Rept. Nat. Hist. Coll. Alaska, 305, pl. 15, f. 2, 1887.

Bryostemma polyactocephalum, JORDAN & STARKS, Proc. Cal. Ac. Sci. 1895, 841; JORDAN & GILBERT, Rept. Fur Seal Invest., 1898.

Head 5½; depth 5½; D. LIV; A. 41; V. I, 3; pores of lateral line 25. Body elongate, less compressed than in *Bryostemma polyactocephalum*, covered with small, smooth, embedded scales. Head short, very obtuse, almost truncate; top of head from nostrils to near front of dorsal covered with fleshy cirri, much smaller than in *B. polyactocephalum;* only 2 or 3 small ones extending on first dorsal spine; supraorbital cirrus short, 4 to 5 in head; 2 small cirri placed at the sides of snout with a larger median one behind them, forming a triangle; jaws equal; mouth horizontal, the angle extending to below pupil; eyes small, 4 in head; snout very short, almost vertically truncate, ⅔ in eye; teeth of both jaws subequal, short, bluntish, and close set. Lateral line short, 7¼ in length of body, concurrent with the dorsal outline of body. A line of pores begins in front of eye on a level with pupil, runs under eye and to a level with pupil again, then back to and along the entire length of the short lateral line. Gill rakers not developed; gill membranes free from isthmus. Vent ⅕ distance from tip of snout to tip of caudal; distance from origin of ventral to anus 4¼ in length of body   Pectoral fin 5¼ in body, as long as head. Dorsal fin beginning in front of the pectoral, highest along the posterior half, the longest spine 2¾ in head, the fin higher than anal; dorsal slightly joined to caudal; anal separated from caudal; caudal rounded, 1¾ in head; first dorsal spine 4¼ in head, its surface with 2 or 3 small cirri. Color in spirits of 1 specimen, probably male, dark brown, with 13 pale cross bars along back, extending on dorsal fin; along sides these become obsolete; on belly they become increased in number and broadened below; dorsal fin with 13 large, very distinct black ocelli with yellowish rings, 1 between each pair of the pale blotches; anal with about 7 small blackish spots at base on posterior part, the fin otherwise nearly plain; caudal faintly barred with light and dark; pectorals pale, with 2 dark pale-edged oblique bars before it; sides of head with irregular dark vertical bars, 1 of them forming an inverted Λ below eye, this and others extending across lower jaw; cirri mostly black. The other specimen, probably the female, has the body nearly plain brown, the dorsal with but 4 ocelli, the anterior 9 being replaced by dark bars on the fin; anal with dark oblique cross bars; pectorals barred with black; markings on head more sharply defined, coloration otherwise similar. This second specimen is 4¾ inches in length, the other 4. Puget Sound; the above account from the 2 original types from near Seattle. Three others since obtained near Channel Rocks, Port Orchard, show the following life coloration: Dark red above, orange brown below, belly cream color; sides below with cream-colored cross bars, wider than eye, running from the axis of body downward and fading into the general color below; a Λ-shaped mark downward from eye across branchiostegals to isthmus, a similar mark behind eye across edge of preopercle, this last sometimes broken up and chain-like; top of head dark; snout light; 2 oblique dark bars at base of pectoral; dorsal with 12 or 13 sharp dark brown spots as large as eye, edged with bright red, these arranged regularly along the whole length of fin; pectorals and caudal bright red with wavy, irregular, brown lines running across the rays; anal red, with dark brown bars as wide as the interspaces running obliquely downward and forward; ventrals light brown. (*nugator*, a fop.)

Bryostemma nugator, JORDAN & WILLIAMS, Proc. Cal. Ac. Sci. 1895, 843, pl. 101, Seattle, Washington. (Coll. Young Nat. Soc. Type, No. 3134, L. S. Jr. Univ.)

902. APODICHTHYS, Girard.

Apodichthys, GIRARD, Proc. Ac. Nat. Sci. Phila.1854, 150 (*flavidus*).

Body elongate, compressed, covered with very small scales; no lateral line; snout short; mouth moderate, oblique; teeth in the jaws moderate, stouter anteriorly; vomer with teeth; gill membranes united, free from the isthmus. · Dorsal fin long, low, even, of spines only; anal fin similar, preceded by a very large pen-shaped spine channeled along its anterior surface and hidden in a pouch of skin; caudal fin short, connected with dorsal and anal; no ventral fins; pectoral fins moderate; intestinal canal short, without pyloric cæca. Small, bright-colored fishes of the Pacific, living among rocks near shore. (ἄπους, without feet; ἰχθύς, fish; in allusion to the want of ventral fins.)

a. Color various, green, olive, or scarlet; sides of head without silvery band; depth 7 to 8 in length; head 9.   FLAVIDUS, 2766.
  b. General color olivaceous.   var. *flavidus*, 2766a.
  bb. General color scarlet.   var. *sanguineus*, 2766b.
  bbb. General color grass green.   var. *virescens*, 2766c.
aa. Color reddish; a bluish silvery stripe on side of head; depth 9 to 10 in length; head 7.   UNIVITTATUS, 2767.

2766. APODICHTHYS FLAVIDUS, Girard.

Head 9⅓; depth 7½. D. XCIII; A. I, 40. Head short; mouth very oblique; maxillary reaching pupil; upper jaw with a series of conical teeth, behind which is a patch of smaller teeth; sides of mandible with conical teeth in a single series, forming a patch in front; vomer with 3 conical teeth; palatines toothless; nape equidistant between front of dorsal and pupil. Anal spine very large, ⅔ length of head, shaped like a pen, deeply excavated on its anterior side, and very convex behind, very thin, flexible, and with sharp edges, entirely included in a pouch of skin; pectoral fins about ⅔ length of head. Color orange, varying with the surroundings to intense grass-green (var. *virescens*), yellowish brown (var. *flavidus*), crimson and dark purple (var. *sanguineus*); a few light round spots along axis of body posteriorly; a narrow black bar downward and backward from eye; a shorter, less distinct bar from upper margin of orbit backward to occiput; anal fin obliquely barred with brownish. Length 18 inches. Pacific coast, Vancouver Island to the Santa Barbara Islands; abundant; usually found below low tide mark. The following color notes are from specimens taken in Puget Sound belonging to the green form (var. *virescens*), the larger 10 inches in length, the smaller 3 inches. The large one is a bright grass-green, mottled with light gray; a series of blended white spots, as large as eye, along the axis of body from the pectoral fin to the middle of caudal peduncle; belly with many similar spots smaller in size and somewhat sharper in outline; a row of conspicuous black spots, irregular in size, shape, and position, along back at the base of dorsal spines; a black line as wide as pupil from nape to eye, a similar line from eye to posterior end of maxillary; a faint light streak across cheek posteriorly; cheek and base of pectoral dusted with fine dark points. The small one is bright green without distinct markings on body; a silvery bar, running posteriorly from tip of snout through eye, across cheek, to the middle of opercle; no bar downward from eye to maxillary, or from eye to nape as in the large one. (*flavidus*, yellowish.)

Apodichthys flavidus, GIRARD, Proc. Ac. Nat. Sci. 1854, 150, Presidio, San Francisco Bay (Coll. Dr. Kennerly. Type, No. 494, U. S. Nat. Mus.); GIRARD, Pac. R. R. Surv., X, Fishes, 117, 1858; GÜNTHER, Cat., 290, 1861; JORDAN & GILBERT, Synopsis, 769, 1883.

Apodichthys virescens, AYRES, Proc. Cal. Ac. Nat. Sci. 1855. 55, San Francisco; GIRARD, Pac. R. R. Surv., X, Fishes, 118, 1858.

Apodichthys inornatus, GILL, Proc. Ac. Nat. Sci. Phila. 1862, 279, Puget Sound, probably (Coll. Northwestern Boundary Commission); D. XC; A. 38.

Apodichthys sanguineus, GILL, Proc. Ac. Nat. Sci. Phila. 1862, 279, California. (Coll. Dr. Samuel Hubbard.)

2767. APODICHTHYS UNIVITTATUS, Lockington.

D. about XCV; A. about I, 40. Body elongate, much compressed, band-like, preserving almost same depth to about posterior fifth of body, thence tapering more rapidly to caudal fin. Head 7; depth nearly 10 times in total length; depth of caudal peduncle about $\frac{1}{4}$ of that of body; snout obtuse, about $\frac{2}{3}$ as long as diameter of eye, upper profile of head a continuous curve from snout to occiput. Interorbital area highly convex transversely, about equal in width to $\frac{1}{2}$ diameter of eye. Eye entirely lateral, round, contained entirely in anterior half of head; iris golden. Mouth small, posterior extremity of the maxillary reaching to anterior margin of eye. Teeth small. Branchiostegals 5. Dorsal continuous with, but distinct from, anal, arising vertically from tip of operculum, and composed of spines only. Anal preceded by a long, sharp, slender spine of V-shaped transverse section, hollow side anterior, length of spine equal to about $\frac{1}{4}$ depth of fish. Distance from anal spine to tip of operculum a little more than to tip of caudal. Caudal with numerous accessory rays, so that its sides are almost straight, posterior margin broken in the type, all rays simple. General color in spirits, light reddish, vertical fins rather bright, and top of head reddish brown; tip of snout brown; a silvery band (possibly bluish in life) from tip of snout, across lower part of eye, cheek, and opercles, terminating at about middle of length of operculum, this band bordered above by a narrower brown band. Lower California, probably from the gulf. A single specimen. Length 1.88 inches. The peculiar vitta upon each side of the head at once distinguishes this species from the other described forms. (Lockington.) Not seen by us. (*uni-vittatus*, having one band.)

Apodichthys univittatus, LOCKINGTON, Proc. Ac. Nat. Sci. Phila. 1881, 118, Gulf of California.

903. XERERPES, Jordan & Gilbert.

Xererpes, JORDAN & GILBERT, Proc. Cal. Ac. Sci. 1895, 846 (*fucorum*).

This genus differs from *Apodichthys* in the moderate size of the anal spine, which is rounded and not channeled on its anterior edge, and in the small size of its pectoral fins. The single known species lives in *Fucus* chiefly above low-tide mark and may often be shaken out of half-dry mats of seaweed on rocks well above the water. (ξερός, dry; ἕρπης, creeper.)

2768. XERERPES FUCORUM (Jordan & Gilbert).

Head 10; depth 9½. D. LXXXIII; A. 35. Form and dentition as in *Apodichthys flavidus*. Mouth very oblique, the maxillary reaching center of pupil; nape nearer front of dorsal than end of snout. Anal spine comparatively small, about ⅓ length of head, transversely very convex in front, and slightly concave or grooved behind, the pouch of skin at its base little developed; pectorals very small, shorter than eye; anal fin beginning nearer tip of caudal than tip of snout by about 3 times length of head. Bright olive green or deep red, the color varying with the surroundings; a row of dark spots along axis of body, these sometimes with light-bluish center, and connected by a very narrow dark streak; generally a dark streak downward from eye, but no other markings about head. Length 6 inches. Monterey to Puget Sound; abundant in rock pools and bunches of *Fucus*; remarkable for its active movements. It is found mostly in masses of *Fucus* attached to rocks between tide marks, and it is often found at low tide at a considerable distance from any water, kept damp by the masses of algæ. Sometimes a dozen of them can be shaken from a bunch of algæ attached to a dry rock. It is, like the species of *Xiphidion*, very active, moving over stones or sand, and showing less anxiety about the presence of its native element than any other fish known to us. (*fucorum*; of the *Fucus* or seaweed.)

Apodichthys fucorum, JORDAN & GILBERT, Proc. U. S. Nat. Mus. 1880, 139, Monterey (Coll. Jordan & Gilbert); JORDAN & GILBERT, Synopsis, 770, 1883.
Xererpes fucorum, JORDAN & STARKS, Proc. Cal. Ac. Sci. 1895, 846.

904. ULVICOLA, Gilbert & Starks.

Ulvicola, GILBERT & STARKS, Proc. U. S. Nat. Mus. 1896, 455 (*sanctæ-rosæ*).

This genus is allied to *Xererpes*, but differs in having the opercle above angle adnate to shoulder girdle, in the smaller size of the anal spine, and especially in the entire absence of pectoral fins. (*Ulva*, sea lettuce; *colo*, I inhabit.)

2769. ULVICOLA SANCTÆ-ROSÆ, Gilbert & Starks.

Head 10 in body; depth 13. D. XCVII; A. I, 40; eye 4½ in head; caudal 1½. Body elongate, as in *Apodichthys*, strongly compressed, upper profile of head slightly convex, no construction at nape; mouth very small, oblique, the maxillary reaching about to front of eye; teeth very small in a single row on jaws; vomer with teeth; interorbital a narrow, sharp ridge; snout about equal to length of eye; gill opening short, limited to the part below angle of opercle, above adnate to shoulder girdle. Origin of dorsal above upper end of gill opening, much nearer occiput than tip of snout; anal spine small, not channeled as in *Apodichthys flavidus*; origin of anal nearer base of caudal than tip of snout by a distance equal to twice length of head; pectorals and ventrals obsolete; caudal rather long, confluent with dorsal and anal. Color in spirits, light brown, slightly lighter under head and on belly; no markings. The type is a specimen $4\frac{1}{2}$ inches in length, collected by the *Albatross* at Santa Rosa Island, off Santa Barbara, January 6, 1889. (Type, No. 47579. Coll. *Albatross*.)

Ulvicola sanctæ-rosæ, GILBERT & STARKS, Proc. U. S. Nat. Mus. 1896, 455, pl. 55, fig. 2, Santa Rosa Island, California.

905. PHOLIS (Gronow) Scopoli.

(GUNNELS.)

Pholis, GRONOW, Zoophylaceum, 78, 1765 (not binomial).
Pholis, SCOPOLI, Introd. Hist. Nat., 456, 1777 (*gunnellus*).
Murænoides, LACÉPÈDE, Hist. Nat. Poiss., II, 324, 1800 (*sujef*).
Centronotus, BLOCH & SCHNEIDER, Syst. Ichth., 165, 1801 (*fasciatus*).
Dactyleptus, Rafinesque Anal. de la Nature 1815, 82; substitute for *Murænoides*.
Centronotus, CUVIER, Règne Animal, Ed. 2, II, 239, 1829 (*gunnellus*).
Ophisomus,[*] SWAINSON, Nat. Hist. Class'n. Anim., II, 277, 1839 (*gunnellus*).
Urocentrus, KNER, Sitzber. K. Akad. Wiss. Wien, LVIII, 1868, 51 (*pictus*).
Rhodymenichthys, JORDAN & EVERMANN, Check-List Fishes, 474, 1896 (*ruberrimus = dolichogaster*).

Body long and low, considerably compressed, somewhat band-shaped, the tail slowly tapering; head small, compressed, naked;[†] mouth rather small, oblique; jaws with rather small teeth in narrow bands or single series; vomer and palatines usually toothless; gill membranes broadly united, free from the isthmus; scales very small, smooth; no lateral line. Dorsal fin long and low, beginning near the head, composed entirely of stiff, sharp, subequal spines; anal similar in form, of 2 spines and many

---

[*] Substitute for *Gunnellus*, the latter being a barbarous word derived from "gunwale." "Nomina generica quæ ex Græca vel Latina lingua radicem non habent, regicienda sunt." This rule has never been generally adopted.
[†] In *Pholis nebulosus*, a Japanese species, the head is scaly. This species is the type of a distinct genus; which may be called

ENEDRIAS, Jordan & Gilbert, new genus.

Enedrias, JORDAN & GILBERT, new genus (*nebulosus*).
    This genus differs from *Pholis* in the scaly head. (ἐνέδρα, lurking place.)
Enedrias nebulosa (SCHLEGEL).
    Head $7\frac{1}{2}$ to 8; depth $8\frac{1}{2}$ to $9\frac{1}{2}$. D. LXXX; A. II, 39. Dorsal and anal somewhat connected to caudal; pectoral $2\frac{1}{2}$ to $2\frac{2}{3}$ in head. Head small. Body everywhere freckled with dark blotches; 12 dark triangular blotches along base of dorsal; a row of dusky blotches on middle of side posteriorly; 10 or 12 dark blotches on base of anal; caudal dusky, edged with pale, 2 pale cross streaks on top of head; pectoral pale. Northern Japan to Okhotsk Sea, Gulf of Strietok; our specimens from Hakodate. (*nebulosus*, clouded.)

Gunnellus nebulosus, SCHLEGEL. Fauna Japonica, Poiss., 138, 1850, Bay of Magi, Japan.
Centronotus nebulosus, STEINDACHNER, Ichth. Beitr., IX. 24, 1880.
Enedrias nebulosus, JORDAN & GILBERT, Rept. Fur Seal Invest., 1898, with plate.

soft rays; caudal fin short and small, more or less joined to dorsal and anal; pectorals short, rather shorter than head; ventrals very small, of 1 spine and a rudimentary ray; intestinal canal short, without cæca. Shore fishes of the Northern seas. ($\phi\omega\lambda\acute{\iota}\varsigma$, name of some fish said to shelter itself when lying in wait by producing a cloud of mucus; $\phi\omega\lambda\acute{\alpha}\varsigma$, one who lies in wait.)

UROCENTRUS ($o\grave{v}\rho\acute{a}$, tail; $\kappa\acute{\epsilon}\nu\tau\rho o\nu$, spine):

    a. Pectoral fin small, 3½ to 4 times in length of head; dorsal spines about 93; anal rays 48; body with 2 rows of dark blotches; fins nearly plain.   PICTUS, 2770.
    aa. Pectoral fin moderate, 2 to 2¼ times in length of head.

       RHODYMENICHTHYS (*Rhodymenia*, a large red alga; $\dot{\rho}\acute{o}\delta o\nu$, rose; $\dot{v}\mu\acute{\eta}\nu$, membrane; $\dot{\iota}\chi\theta\acute{v}\varsigma$, fish).

       b. Dorsal and anal joined to the caudal to the full height of the spines, without constriction at base of caudal; body greatly compressed, ribbon-like. Dorsal spines about 93; anal about 47; pectorals short, 2⅔ in head; no ocelli along base of dorsal.   DOLICHOGASTER, 2771.

   PHOLIS:

       bb. Dorsal and anal slightly connected with caudal, leaving a constriction of outline at base of caudal; body less compressed; dorsal fin with dark blotches or ocelli.

          c. Pectoral fins well developed, about ½ length of head. Dorsal spines about 88; anal rays about 42; pectoral 2½ in head; dorsal fin with dark quadrate blotches rather than ocelli; sides scarlet in adult, bounded with black.   FASCIATUS, 2772.

        cc. Dorsal spines about 80 (76 to 85); anal rays about 40; pectoral 2 in head; dorsal fin with small rounded black blotches.   GUNNELLUS, 2773.

        ccc. Dorsal spines about 77; anal rays about 35; pectoral 2 in head; dorsal fin with ocelli, or lunate, dark blotches.   ORNATUS, 2774.

Subgenus UROCENTRUS, Kner.

2770. PHOLIS PICTUS (Kner).

Head 9⅓ to 10½; depth 8 to 10. D. XCIII or XCIV; A. II, 46 to 48 (misprinted 40 in Kner's account). Eye as long as snout; mouth oblique, the upper jaw the longer, reaching to front of eye; pectorals very short, scarcely longer than eye, 3 to 4 in head; anal said to have an isolated channeled spine hidden in the skin, but our specimens show no peculiar structure. Color yellowish, with 2 lengthwise series of large oblong blackish blotches, the one along base of dorsal, but not on the fin, of 21 or 22 blotches, the other on lower part of sides, of about 25; a series of fainter blotches along base of anal; in other specimens the lower row becomes obscure, the upper more distinct, and the series above anal disappears; a black bar downward from eye, a whitish band behind it; opercles dusky. West side of Bering Sea; our specimens from Shana Bay, Iturup Island, Kuril Group.

As already shown by Steindachner, this is a typical *Pholis*, Kner having been in error in ascribing to it an isolated and channeled first anal spine. The ventral spines are bound down by the integument more closely than usual, but they are in other respects not peculiar. Each is accompanied by 2 short spinous rays concealed in the membrane, and difficult to detect.

The latter are stiff and pungent, and seem to be not articulated. The ventrals of *P. ornatus* show the same structure. Kner gives the anal formula as II, 40. This must be a misprint for II, 49, as the artist figures 51 rays in the fin, not differentiating the 2 anterior ones. (*pictus*, painted.)

Urocentrus pictus, KNEE, Sitzungsb. d. k. Akad. D. Wissench., LVIII, 1868, 51, taf. 7, fig. 21, Singapore; an error.

Centronotus.pictus, STEINDACHNER, Ichth. Beiträge, IX, 25, 1880.

Pholis pictus, JORDAN & GILBERT, Rept. Fur. Seal Invest., 1898.

Subgenus RHODYMENICHTHYS, Jordan & Evermann.

2771. PHOLIS DOLICHOGASTER * (Pallas).

(BUTTER-FISH.)

Head 9¼ in length; depth 8. D. XCII; Á. II, 44; pectoral 14; eye 5 in head; maxillary 2¾; pectoral 2½; caudal 2; ventral spines 1⅔ in eye. Body elongate, much compressed; head small, its upper profile convex; mouth moderate, very oblique, the maxillary reaching to below middle of eye; teeth rather large and blunt, arranged in a single row, the anterior one not enlarged; interorbital space narrow, without a sharp ridge, its width less than eye; snout equal in length to eye; distance from tip of snout to occiput 1⅔ in head. Head entirely naked; body covered with small, cycloid, inconspicuous scales. Origin of dorsal over upper end of gill slit, its distance from nape equal to distance from nape to front of eye, the spines toward the anterior end of the fin the highest; origin of anal a little nearer tip of caudal than snout; dorsal and anal confluent with caudal, the anal more broadly connected than dorsal; pectorals small, rounded behind; ventral spines inserted directly under base of pectorals, their length little greater than their distance apart; caudal short and broad, well rounded in outline. Bering Sea; recorded from the Kurils, and from Robben, Bering, and Medni islands, and from Kigiktowik Bay. The specimen above described was taken at Robben Island by Capt. J. G. Blair, then in command of the guardship *Leon.* It is 9 inches long and is uniform red in color, with a few pale dots. Another specimen, 18 cm. long, taken by Mr. Gerald E. H. Barrett-Hamilton at Bering Island, shows the following characters: The color is cherry red on the body and fins, lighter on belly, lower half of cheek and under side of head; lips blackish anteriorly, a narrow black streak running from them along snout to eye and from eye across cheek and opercles toward upper edge of pectoral base; this line separates the deep-red upper part of the head from the lighter area below;

---

* The following species is allied to *Pholis dolichogaster:*

Pholis taczanowskii (Steindachner).

Head 9: depth 10; D. LXXXII; A. II, 45; teeth bluntly conical; dorsal very low, joined to the caudal without constriction. Snout scarcely longer than eye, which is 5¼ in head. Pectoral 3 in head. Scales very small, the head naked. Clear, yellowish gray, finely dotted, fins grayish, the pectoral yellowish; a yellowish streak edged with darker from eye to axil. Gulf of Strietok. (Steindachner.) (A personal name.)

Centronotus taczanowskii, STEINDACHNER, Ichth. Beitr., IX, 24, pl. 3, fig. 1, 1880, Gulf of Strietok, Okhotsk Sea. (Coll. Prof. Dybowsky.)

sides of body with a number of minute scattered black spots; along middle of sides is a distant series of light spots as large as pupil, the margin of each with 2 to 4 black specks like those scattered over sides. The dorsal and anal are more widely joined to the caudal than in other species, the fins being higher posteriorly and without perceptible notch. The dorsal contains 93 spines, the anal 2 spines and 47 rays, the pectorals 15 rays. Head $9\frac{1}{4}$ in length; depth $7\frac{5}{6}$. Eye 5 in head; maxillary $3\frac{1}{4}$; pectorals $2\frac{1}{2}$; caudal $2\frac{1}{4}$; ventral spine $2\frac{1}{3}$ in eye. *Blennius dolichogaster*, Pallas, is undoubtedly identical with *Gunnellus ruberrimus*, Cuvier & Valenciennes. They agree in the very long dorsal and anal fins (D. XCIII, A. II, 50 in *dolichogaster*), and in the color. *P. dolichogaster* is described as having the color brownish olive, shaded with greenish and yellowish, spotted with green above the lateral line; belly yellow; anal, caudal, and pectorals yellowish; dorsal and anal dusky, with transverse pale bars. Compare with this, details of coloration recently published concerning *P. ruberrimus* by Bean & Bean (Proc. U. S. Nat. Mus. 1896, 248): "Color olive brown, with minute black spots; belly yellowish." In another specimen, "Across the spinous dorsal there are 20 narrow, nearly vertical pale streaks. Similar streaks to the number of 12 cross the anal." The species is evidently not always red in life. ($\delta o\lambda\iota\chi\acute{o}\varsigma$, long; $\gamma\alpha\acute{o}\tau\acute{\eta}\rho$, belly.)

Blennius dolichogaster, PALLAS, Zoogr. Rosso–Asiat., III, 175, 1811, Kamchatka. (Type in Mus. Berlin.)
Gunnellus ruberrimus, CUVIER & VALENCIENNES, Hist. Nat. Poiss., XIV, 440, 1839, Kuril Islands; after notes of PALLAS,* Zoogr. Rosso–Asiat., III, 178, 1811.
Gunellus dolichogaster, CUVIER & VALENCIENNES, Hist. Nat. Poiss., XI, 436, 1836.
Centronotus dolichogaster, GÜNTHER, Cat., 288, 1861.
Muraenoides dolichogaster, JORDAN & GILBERT, Synopsis, 768, 1883.
Pholis dolichogaster, JORDAN & GILBERT, Rept. Für Seal Invest., 1898.
Rhodymenichthys ruberrimus, JORDAN & EVERMANN, Check-List Fishes North and Middle America, 474, 1896.
Pholis ruberrimus, BEAN & BEAN, Proc. U. S. Nat. Mus. 1896, 248·

Subgenus PHOLIS.

2772. PHOLIS FASCIATUS (Bloch & Schneider).

Head 8 to $9\frac{1}{2}$; depth 7 to 9. D. LXXXVI to LXXXIX; A. II, 42 to 44; V. I, 1. Head scaleless; mouth decidedly oblique, the tip of lower jaw on level of middle of the eye; width of mouth nearly $\frac{1}{2}$ head. Eye equal to snout, a little more than interorbital width; ventral spine $\frac{2}{3}$ eye, $\frac{1}{7}$ length of mandible; caudal $\frac{1}{5}$ head; pectoral $2\frac{1}{4}$ in head; vertical fins slightly joined at base. Ground color yellowish gray in life, the sides of a brilliant scarlet; base of dorsal occupied by 10 or 11 oblong blotches of dark brown, which extend to the tips of the fins; these blotches each divided upon the fin by a median spot of the ground color; the areas of the ground color alternating with these blotches are finely speckled with brown, a large spot of brown usually occupying a median

* Bright red. Form of *taenia;* scales inconspicuous; ventrals each a single scarcely projecting spine; caudal broad, rounded, distinct. D. CXV. Kuril Islands. (Pallas.) *Muraenoides ruberrimus*, BEAN, in Nelson, Rept. Nat. Hist. Collections made in Alaska, 305, pl. XIV, fig. 1, 1887.

position upon the fin; middle and lower part of sides occupied by vermiculating brown lines on the ground color, these vermiculations arranged in more or less distinct cross bars, about 20 in number, reaching to or nearly to the midventral line, the posterior ones often continued faintly onto the anal fin; pectoral and caudal fins yellow, unmarked; a brown blotch across snout and tip of mandible, followed by a narrow yellowish bar descending to front of eye; interorbital space crossed by a broad brown bar with blackish margins, which become much narrower below and traverse the eye and the cheek; behind this a broader yellow bar, margined behind with a narrow brown line. In life the coloration is extremely brilliant, the pale markings being bright orange or scarlet. Bering Sea and Arctic Ocean, from Greenland to the Kurils; locally abundant; numerous fine large specimens taken from the stomachs of cormorants on St. Paul Island, Pribilof Group; others dredged in shallow waters. Our specimens from St. Paul, Bristol Bay, and Upernavik, Greenland. Three large specimens from St. Paul Island, the type locality of *P. maxillaris*, have been compared with a number of individuals of *P. fasciatus* from Upernavik, Greenland. We can appreciate no differences between the two. The size of the mouth and the length of the head are the same in specimens of equal length, and no difference exists in the development of the ventrals. The agreement seems to be perfect in the fin rays, relative proportion and coloration. Pallas's short account of *Blennius tænia* contains nothing distinctive except the number of fin rays and the statement that the body is banded. As both of these items agree with the present species, we may safely follow Bean & Bean in making the identification. In a specimen from St. Paul, 29 cm. long, the length of the maxillary is contained 2⅔ times in distance from tip of snout to origin of dorsal; the mandible equals the length of the pectoral. In a younger example, 15 cm. long, from Bristol Bay, the maxillary is contained 3½ in predorsal length; the mandible approximately equals length of pectoral. (*fasciatus*, banded.)

Centronotus fasciatus, BLOCH & SCHNEIDER, Syst. Ichth., 165, pl. 37. fig. 1, 1801, Tranquebar; an error? GÜNTHER, Cat., III, 287, 1861.

Gunnellus grœnlandicus, CUVIER, & VALENCIENNES, Hist. Nat. Poiss., XI, 442, 1836, Greenland, after BLOCH & SCHNEIDER; REINHARDT, Dansk. Vidensk. Selsk. Nat. og Mathem. Afh., VII, 122, 1838.

Gunnellus murænoides, VALENCIENNES, in CUVIER, Règne Animal, Poiss., pl. 78, fig. 2, 916; after BLOCH & SCHNEIDER.

*Blennius tænia,** PALLAS, Zoogr. Rosso-Asiat., III, 1811, 178, Kuril Islands.

Murænoides maxillaris, BEAN, Proc. U. S. Nat. Mus. 1881, 147, St. Paul Island, Alaska (Type, No. 23999. Coll. Henry W. Elliott); JORDAN & GILBERT, Synopsis, 768, 1883.

Gunnellus fasciatus, CUVIER & VALENCIENNES, Hist. Nat. Poiss., XI, 441, 1836.

Murænoides fasciatus, JORDAN & GILBERT, Synopsis, 767, 1883.

Murænoides tænia, JORDAN & GILBERT, Synopsis, 766, 1883.

Pholis fasciatus, GILBERT, Rept. U. S. Fish Comm. 1893, 449; JORDAN & GILBERT, Rept. Fur Seal Invest., 1898.

Pholis tænia is thus described: Body banded; teeth obtuse, subdistinct; head subtriangular, compressed; body ensiform, covered with minute embedded scales; vent median. Dorsal fin extending from near the head to the tail, the spines subequal; caudal subdistinct; pectorals small; ventrals represented by 2 recurved spines. Body banded. D. LXXXVII; A. 47. Kuril Islands. (Pallas.)

2773. PHOLIS GUNNELLUS (Linnæus).

(GUNNEL; BUTTER FISH.)

Head from 7 to 8 in body; depth 7 to 8; D. LXXVI to LXXXV; A. II, 38 to 44; V. I, 1; eye 5 in head; maxillary 3; P. 2; C. 1⅓. Head compressed, naked; mouth oblique, the maxillary reaching to front of pupil; teeth blunt, in a single row, somewhat enlarged anteriorly; interorbital a narrow ridge about ½ eye. Distance from origin of dorsal to nape equal to distance from nape to middle of eye; pectoral rather large, about 2 in head, inserted directly under front of dorsal. Color olive brown, sides with numerous obscure darker bars; base of dorsal with blackish spots, generally bordered with a narrow yellow line, a dark bar running downward and backward from eye; anal with dusky bars across the rays. This species differs from *Pholis ornatus* in the more numerous fin rays and in coloration; the spots on dorsal are black, edged with yellow; in *P. ornatus* they are yellow with a black bar before and behind, each partly encircling it; no black bordered light streak from eye to occiput. Length 12 inches. North Atlantic, from Labrador south to Woods Hole and Norway to France; abundant on rocky shores among algæ, both in America and Europe. Here described from specimens from Salem, Massachusetts. (Eu.) (*grunnellus*, English gunnel, said to be corrupted from gunwale.)

Blennius pinna dorsalis ocellis X nigris, LINNÆUS, Mus. Adolph-Fred., I, 69.

Blennius gunnellus, LINNÆUS, Syst. Nat., Ed. X, 257, 1758, Atlantic Ocean; after *Blennius pinna dorsalis*, etc.

Ophidion imberbe, LINNÆUS, Syst. Nat., Ed. X, 259, 1758, Europe; after *Oph. cirris careus*, ARTEDI.

Centronotus gunnellus, BLOCH & SCHNEIDER, Syst. Ichth., 167, 1801; GÜNTHER, Cat., III, 285, after RÜPPELL'S type.

Murænoides gunnellus, JORDAN & GILBERT, Synopsis, 767, 1883.

Blennius europæus, OLAFSEN, Reisei Island, I, 81, 1772, Iceland.

Blennius murænoides, SUJEE, Act. Petrop. II, 1779, 195, no locality, probably the Baltic; GMELIN, Syst. Nat., 1184, 1788.

Murænoides sujef, LACÉPÈDE, Hist. Nat. Poiss., II, 324, 1800; after SUJEF.

Ophidium mucronatum, MITCHILL, Trans. Lit. & Phil. Soc. N. Y., II, 1815, 361, pl. 1, f. 1, New York; earliest American name.

Gunellus vulgaris, FLEMING, British Anim., 207, 1828, England.

Murænoides guttatus, LACÉPÈDE, Hist. Nat. Poiss., II, 324, 1800; YARRELL, Brit. Fish., I, 269.

Gunellus ingens, H. R. STORER, Bost. Journ. Nat. Hist., VI, 1850, 261, pl. 8, f. 1, Labrador. (Coll. H. R. Storer.)

Gunellus macrocephalus, GIRARD, in H. R. STORER, Bost. Journ. Nat. Hist., VI, 1850, 263, Chelsea Beach, Massachusetts (Coll. Chas. Girard); D. H. STORER, Rept. Fish. Mass., 261, pl. 17, f. 3.

2774. PHOLIS ORNATUS (Girard).

Head 8; depth 8. D. LXXVII to LXXIX; A. II, 35 to 37. Head naked, very narrow above; nape nearly equidistant between origin of dorsal and front of orbit; origin of anal equidistant between base of caudal and base of pectoral; pectoral 2 in head. Coloration, usually olive green above, yellow or orange below,[*] but varying with the surroundings to brown

---

[*] A specimen from near Seattle varies much in color from all the others before us. It is purplish red, paler below; 2 conspicuous white spots bordered with white on front of dorsal; a pale streak bordered with black from eye to nape.

and cherry red; traces of about 20 darker bars along sides; a dark bar downward from eye; fins reddish; a V-shaped mark from eye to occiput, grayish, bordered by jet-black; the common form with about 14 red spots along base of dorsal, each with a curved black bar in front and behind, partly encircling it; others with about as many broad ⋀-shaped darker blotches, which extend on the fin, the first one or two blotches often shaped as in the former case; anal white, unmarked. Length 12 inches. San Francisco to Bering Sea; very common northward, its range extending to Kamchatka;* very common at Unalaska; always in shallow water. (*ornatus*, ornamented.)

Gunnellus ornatus, GIRARD, Proc. Ac. Nat. Sci. Phila. 1854, 149; GIRARD, Pac. R. R. Surv., x, Fishes, 116, pl. 25b, figs. 6 and 7, 1858 (Type, No. 490, Presidio, California, Coll. Lieut. Trowbridge; No. 491, Shoalwater Bay, Washington, Coll. Dr. J. G. Cooper; No. 492, Fort Steilacoom, Washington, Coll. Dr. Geo. Suckley).
Centronotus lœtus, COPE, Proc. Amer. Phil. Soc. Phila. 1873, 27, Sitka or Unalaska (Coll. George Davidson); A. II, 33.
Murænoides ornatus, JORDAN & GILBERT, Synopsis, 767, 1883.
Pholis ornatus, GILBERT, Rept. U. S. Fish Comm. 1893, 450.

906. GUNNELLOPS, Bleeker.

Gunnellops, BLEEKER, Versl. Ak. Amst., 2, VIII, 1874, 368 (*roseus*).

This genus is apparently distinguished from *Pholis* by the tapering tail, around which the vertical fins are confluent; palatine teeth present. (*Gunnellus*, Gunnel, an old name of *Pholis gunnellus*; ὤψ, appearance.)

2775. GUNNELLOPS ROSEUS (Pallas).

D. ca. C; A. ca. 90; P. 9; V. I.   Head obtuse, the lower jaw projecting; eyes large; body very long, compressed, tapering into a slender tail; pectorals small, ovate, hyaline; 2 spines in place of ventrals; dorsal extending from the nape to the end of the tail; anal joined to caudal.   Color intensely red. Kuril Islands. (Pallas.) Not seen by any recent collector. (*roseus*, rosy.)

Blennius roseus, PALLAS, Zoogr. Rosso-Asiat., III, 177, 1811, Kuril Islands.
Centronotus roseus, GÜNTHER, Cat., III, 290, 1861.
Gunnellops roseus, JORDAN & EVERMANN, Check-List Fishes N. and M. A., 474, 1896.

907. ASTERNOPTERYX, Rüppell.

Asternopteryx (RÜPPELL MS.) GÜNTHER, Cat. Fishes Brit. Mus., III, 288, 1861, name only; JORDAN & GILBERT, Synopsis, 769, 1883 (*gunelliformis*).

This genus is closely allied to *Pholis*, differing chiefly in the entire absence of ventral fins. From *Pholidapus* it is distinguished by the shorter pectorals and by the more broadly united gill membranes. Greenland. A single species known. (ἀ-, without; στέρνον, breast; πτέρυξ, fin.)

2776. ASTERNOPTERYX GUNELLIFORMIS, Rüppell.

Head 9; depth 8½.   D. LXXXVII (LXXXI, according to Günther); A. II, 40.   Head and body strongly compressed; head bluntish, snout short,

---

* We have specimens collected at Tareinsky Bay by Mr. Barrett-Hamilton.

jaws equal; maxillary reaching pupil, 3 in head; eye 5½; gill membranes broadly united, their outline not notched; no trace of ventral fins; pectoral large, 2 in head (3 according to Günther). Dorsal and anal joined to the caudal, the anal with a slight notch behind the last ray; dorsal spines short and all pungent. Color dark brown, clouded with darker; about 11 quadrate pale areas along dorsal fin extending on the sides, these areas each with a black central spot at tip and faintly marked with dark blotches; dorsal with dark spots; a dark band from eye downward, a pale band behind it; lips dark; anal fin bright orange; pectorals and gill membranes pale orange; caudal orange. Greenland. Here described from a fine specimen, 9¼ inches long, in the U. S. National Museum, from Omanak Fjord, Karsak, Noursoak Peninsula, taken in 1897 by Schuchert and White; only the original type in the Senckenburg Museum hitherto known. (*Gunellus; forma*, shape.)

Asternopteryx gunelliformis, RÜPPELL MS.; type (in Senckenburg Museum) from Greenland.

Centronotus gunelliformis, GÜNTHER, Cat., III, 288, 1861.

Murænoides gunelliformis, JORDAN & GILBERT, Synopsis, 769, 1883.

908. ANOPLARCHUS, Gill.

Anoplarchus, GILL, Proc. Ac. Nat. Sci. Phila. 1861, 261 (*atropurpureus*).

Body elongated, compressed, covered with very small, embedded scales which are obsolete or concealed anteriorly; lateral line obsolete. Head small, compressed; eyes small; mouth oblique; teeth in each jaw in a narrow band, the outer somewhat enlarged; narrow bands of teeth on vomer and palatines; gill membranes attached to the isthmus; sometimes with a free fold behind; branchiostegals 5. Dorsal fin not very low; no anal spines; ventrals wanting; caudal fin small, entire; pectoral fins moderate or small; pyloric cæca present, few. Pacific. (ἄνοπλος, unarmed; ἄρχος, anus; the anal fin being without spines.)

    a. Gill membranes narrowly joined to the isthmus, with a free fold behind; dorsal with about 63 spines.   ALECTROLOPHUS. 2777.

    aa. Gill membranes broadly joined to the isthmus, without free fold behind; dorsal with 54 to 57 spines.   ATROPURPUREUS, 2778.

2777. ANOPLARCHUS ALECTROLOPHUS (Pallas).

Head 6⅔ in length; depth 7⅜. D. LXII or LXIII; A. 43. Mouth oblique, maxillary reaching vertical behind pupil, 2¼ in head. Teeth in narrow bands on the jaws, the outer series in upper jaw somewhat enlarged; vomer and palatines with narrow bands of teeth; dentition similar to that in *A. atropurpureus*, which has been erroneously described as having the teeth in the jaws in single series and the vomer and palate toothless; gill membranes rather narrowly joined to the isthmus and with a free posterior edge slightly wider than pupil. *A. atropurpureus* has the gill opening somewhat more restricted and the gill membranes without free fold. Large pores on head arranged similarly in the two species. Spinous dorsal beginning slightly in advance of base of pectoral, its distance frȯm snout less than length of head; distance from origin of anal to tip of snout 2⅔ in length to base of caudal; pectoral short and broad, rounded, 2⅙ in head. Scales small, embedded, those on the anterior part of the body concealed by the thickened integument, as in *A. atropurpureus.* Coloration in our specimens nearly uniform dark olive, with obscure dusky mottlings on the side. In 1 specimen there is a light bar extending obliquely downward and backward from eye, with a dark bar above and below it, the 3 separated by narrow light gray lines; caudal narrowly cross-banded with light and dark as in *A. atropurpureus,* and the anal obliquely barred with the same. In the smallest specimen is a series of roundish spots about as large as eye along back just below dorsal fin; each spot seems to have a narrow dark margin, a light ring, a dusky ring, and a light center; a series of similar but smaller spots along middle of sides posteriorly; the colors were probably brighter and more varied in life. Western part of Bering Sea and Sea of Okhotsk. Here described from 3 small specimens, 3⅓ to 9 inches long, taken at Tareinsky Bay, Kamchatka, by Mr. Barrett-Hamilton; 2 other fine specimens since taken by Arthur W. Greeley in Monterey Bay; the only ones recorded since Pallas. They differ from specimens of *A. atropurpureus* in the higher crest, the more numerous fin rays, and in having the gill membranes with a distinct free margin. (ἀλέκτωρ, cock; λόφος, crest.)

Blennius alectrolophus, PALLAS, Zoogr. Rosso-Asiat., III, 174, 1811, Island of Talek, Gulf of Penshin, Okhotsk Sea.

Gunnellus alectrolophus, CUVIER & VALENCIENNES, Hist. Nat. Poiss., XI, 447, 1836.

Centronotus alectrolophus, GÜNTHER, Cat., III, 289, 1861.

Anoplarchus alectrolophus, JORDAN & GILBERT, Rept. Fur Seal Invest., 1898.

2778. ANOPLARCHUS ATROPURPUREUS (Kittlitz).

Head 6½ in body; depth 7. D. LV; A. 40; eye 5⅓ in head; maxillary 2⅓; pectoral 2½; caudal 1¾. Head with a fleshy crest, which rests on a ridge of bone, its height in older examples about equal to eye; mouth rather large, the maxillary reaching beyond the orbit. Dorsal and anal comparatively high, barely connected with the base of caudal; nape midway between origin of dorsal and pupil. Body naked anteriorly, scaled behind. Color grayish olive, varying to brown; everywhere above finely marked with blackish reticulations; along each side of back a series of small, irregular, sharply defined grayish spots; a series of small pale spots along lateral line; belly pale; crest and middle line of back rather pale; under parts of head yellowish; an oblique, wedge-shaped, pale streak extending downward and backward from the eye, bounded on each side by a sharp light-red line, and then by a dusky area; lower jaw mottled; dorsal olivaceous, speckled, a blackish spot on front; anal olive, tinged with red; pectorals dull orange, barred at base; caudal reddish, with narrow pale streaks, and a light bar at base; color sometimes nearly plain purplish, but more often grayish and mottled. Alaska to San Francisco; abundant northward; common in Bering Sea. Here described from specimens from Neah Bay, Straits of Fuca, Washington. We have also specimens from the Pribilof Islands. (*ater,* black; *purpureus,* purple.)

Ophidium atropurpureum, KITTLITZ, Denkwürd einer Reise Russ.-Amer., I, 225, 1858, Alaska.
Centronotus cristagalli, GÜNTHER, Cat., III, 289, 1861, Vancouver Island.
Anoplarchus purpurescens, GILL, Proc. Ac. Nat. Sci. Phila. 1861, 261, Washington Territory. (Coll. Dr. Kennerly.)
Anoplarchus cristagalli, GÜNTHER, Cat., III, 564, 1861.
Anoplarchus atropurpureus, JORDAN & GILBERT, Synopsis, 771, 1883; JORDAN & STARKS, Proc. Cal. Ac. Sci. 1895, 846.

909. XIPHISTES, Jordan & Starks.

Xiphistes, JORDAN & STARKS, Proc. Cal. Ac. Sci. 1895, 846 (*chirus*).

This genus is very close to *Xiphidion*, differing in the well-developed pectoral fins, which are longer than eye; lower lateral line not connected with abdominal line. (ξιφιστής, a sword belt.)

a. Anal spines 3; branches of upper lateral line extending on dorsal fin; color grass-green.   ULVÆ, 2779.
aa. Anal spines 2; branches of upper lateral line shorter; color brownish, marbled, and with red blotches.   CHIRUS, 2780.

2779. XIPHISTES ULVÆ, Jordan & Starks.

Head 8; depth 10. D. LXXIV; A. III, 48; eye 5 in head; maxillary 2¾; pectoral 3¼. Body eel-shaped, as in the related species; head short; mouth small, oblique, maxillary extending to below posterior margin of eye; jaws subequal, with canine teeth; 4 enlarged canines in front of lower jaw; teeth in upper jaw gradually enlarged from behind forward; eye moderate, equal to length of snout; interorbital space prominent, sharply convex, narrower than width of eye; nape not constricted. Five mucous canals radiating downward and backward from eye, not reaching to edge of preopercle, the branches running upward from upper lateral line ending on the membrane of dorsal, the lower lateral line not connected with the abdominal line. Lateral line otherwise as in *Xiphistes chirus*. Origin of dorsal at a distance behind nape equal to distance from nape to middle of eye, the fin posteriorly barely connecting with caudal; anal with 3 spines, its origin about a head's length nearer snout than base of caudal, connected with caudal posteriorly; pectorals equal in length to snout and ½ eye, slightly shorter than caudal; caudal rounded, fan-shaped. Color olive-green above, very bright green below; middle and lateral line posteriorly, with conspicuous white spots, ½ as large as pupil, each with a black spot before and behind it; a black streak from tip of snout, through eye, to nape, a streak starting from eye behind quickly fading out; dorsal darker than body, unmarked; the anterior third of anal green, without markings, behind this, faint cross bars of brown appear, growing broader and darker posteriorly; caudal olive green, with a light bar across base; pectorals green, without markings. One specimen obtained at Waadda Island, Neah Bay. It was found high on the rocks, among algæ, just below high water mark. Length 5 inches. This species is very closely related to *Xiphistes chirus;* it differs from it chiefly in having 3 anal spines, in the branches of the upper lateral line running higher, and in coloration. (*Ulva*, the green sea lettuce.)

Xiphidion ulvæ, JORDAN & STARKS, Fishes of Puget Sound, 847, 1895, **Waadda Island, Neah Bay.** (Type, No. 3132, L. S. Jr. Univ. Mus. Coll. E. C. Starks.)

2780. XIPHISTES CHĬRUS (Jordan & Gilbert).

Head 7; depth 9. D. LXX; A. II, 50. Head short; nape not constricted; mouth small; maxillary extending to middle of pupil; teeth strong, the anterior canine-like, bluntish; about 4 canines in lower jaw, 5 or 6 in the upper, similar to the teeth behind them, but somewhat larger. Abdominal lines meeting on the breast, but not connected with the lower lateral line. Dorsal fin beginning close behind pectoral; nape midway between middle of eye and front of dorsal; anal beginning about a head's length nearer snout than base of caudal; pectoral fin comparatively large, longer than the eye, its length about equal to distance between middle and lower lateral lines. Color olive brown, yellowish below; sides with marblings of different shades of brown, sometimes with short blackish vertical bars; some round black spots along the back and sides; a black spot behind opercles; numerous black spots on sides of head, forming in older individuals light and dark streaks, which radiate from eye across cheek and opercles, the pale streaks forming reticulations; dorsal with black spots and a series of bright reddish brown cross blotches; pectorals and caudal plain. Monterey to Alaska; smaller than the other species, and living in deeper water; abundant about Cape Flattery. (Jordan & Gilbert.) (χεῖρ, hand.)

Xiphister chirus, JORDAN & GILBERT, Proc. U. S. Nat. Mus. 1880, 135, **Point of Los Pinos, near Monterey, California** (Coll. Jordan & Gilbert); JORDAN & GILBERT, Synopsis, 772, 1883; JORDAN & STARKS, Fishes of Puget Sound, 846, 1895.

910. XIPHIDION, Girard.

Xiphidion, GIRARD, Pac. R. R. Surv., x, Fishes, 119, 1858 (*mucosum*); not *Xiphidium*, Serv., a genus of Grasshoppers.
Xiphister, JORDAN, Proc. U. S. Nat. Mus. 1879, 2⁄1 (*mucosum*); substitute for *Xiphidion*, regarded as preoccupied by *Xiphidium*.

Body elongate, eel-shaped, covered with small scales; lateral lines several: 1 along the median line of the side, 1 above this, and 1 below it; 1 on each side of the abdomen, the 2 meeting in front, and 1 from the occiput toward the base of the dorsal fin. Each of these has on each side series of short branches, placed at right angles to the main line, those on opposite sides alternating. Each of these branches has about 2 open mucous pores. Lower lateral line connected with the abdominal line. Head short, bluntish, scaleless; mouth moderate, oblique; jaws with rather strong teeth, the anterior canine-like; no teeth on vomer or palatines. Branchiostegals 6; gill membranes separate, free from the isthmus. A single long, low, uniform dorsal fin, consisting of spines only; anal fin similar in form, with small spines, indistinct or obsolete; caudal short, joined to dorsal and anal; no ventral fins; pectoral fins very small, shorter than eye. Intestinal canal moderately elongate, with 4 to 6 well-developed pyloric cæca. Herbivorous, feeding on algæ. Active fishes, inhabiting tide pools and crevices among rocks in the North Pacific. (ξιφίδιον, a small sword.)

> *a.* Distance from origin of dorsal to occiput less than that from occiput to tip of snout; streaks radiating from eye paler in the center, edged above and below with blackish.
> MUCOSUM, 2781.
> *aa.* Distance from origin of dorsal to occiput greater than that from occiput to snout; streaks radiating from eye black, abruptly margined with pale olive.
> RUPESTRE, 2782.

2781. XIPHIDION MUCOSUM, Girard.

Head 7 in body; depth 8⅓. D. LXXIV; A. 46; eye 7½ in head; maxillary 2⅓; caudal 2⅔; pectoral a little longer than eye. Lower jaw with a series of short, stout conical teeth; upper jaw with a narrow band of similar teeth; 2 strong canines in upper jaw, 4 in the lower. Lower lateral line sending a branch to the abdominal line; nape not constricted. Dorsal beginning anteriorly, distance from its origin to occiput less than that from occiput to tip of snout; origin of anal nearer snout than tip of caudal by about ⅓ length of head. Blackish green, pale on belly and sides of head, marked posteriorly with olive green in various pattern; a transverse light-greenish bar at base of caudal; 3 olive-brown streaks radiating backward from eye, paler in the center and edged above and below with blackish, outside of which is sometimes a streak of pale olive; these streaks all merge backward into the color of the head; middle streak broadly wedge-shaped, the third streak terminating before reaching margin of preopercle; old individuals often coarsely blotched with yellow. Length 18 inches. Monterey to Alaska; very abundant among rocks and algæ. Here described from specimens, 9 or 10 inches in length, from Neah Bay, Straits of Fuca, Washington. (*mucosus,* slimy.)

Xiphidion mucosum, GIRARD, Pac. R. R. Surv., x, Fishes, 119, 1858, South Farallones, California (Coll. Lieut. Trowbridge. Type, No. 493, U. S. Nat. Mus.); GÜNTHER, Cat., III, 291, 1861; JORDAN & STARKS, Fishes Puget Sound, 848, 1895.
Xiphidion cruoreum,* COPE, Proc. Amer. Phil. Soc. Phila. 1873, 27, Sitka (Coll. Prof. George Davidson); JORDAN & GILBERT, Proc. U. S. Nat. Mus. 1880, 137.
Xiphister mucosum, JORDAN & GILBERT, Synopsis, 772, 1883.

---

* The following is the original description of *Xiphidion cruoreum:* Head 8½ in total length; depth 9½; eye 7 in head, equal to length of pectoral fin. D. about 70; A. 48; Br. 5. Teeth, 2 canines above, 4 below, subequal. Dorsal spines not commencing near the head, the anterior buried in a soft fold of skin; caudal fin not distinct. Three lateral mucous canals extending entire length of caudal fin, which have numerous alternating transverse branches, those of the superior reaching base of dorsal, those of inferior reaching base of anal; each of the cross branches with several excretory pores, none on the main stem; a similar but short tube extending from near base of dorsal fin to supra-occipital region, and not branching anteriorly; the superior lateral canal descending to near the median, but not joining it, nor does the latter extend into the inferior; another tubular line on each side of abdomen, these uniting on jugular region by a continuation of the inferior lateral tube. Vent nearer end of muzzle than end of caudal fin, by length of head. Color maroon, more reddish below; a vertical, broad, reddish bar at base of tail, beyond which is a dark spot; 2 brown radii, black-edged, extending backward and downward from eye. Body covered with small scales, except on the jugular and abdominal regions, which are naked. Length 8 inches. This fish is not very different from *X. mucosum,* Girard. It differs in the smaller eye, the more remote origin of the dorsal fin from the head, the lack of anterior union of the mucous canals, and the coloration. (Cope.)

2782. XIPHIDION RUPESTRE (Jordan & Gilbert).

Head 7⅔ in body; depth 9. D. XVIII; A. 50; eye 6 in head; maxillary 2⅔; caudal 2⅔. Teeth essentially in *X. mucosum*. The lower lateral line sends a branch to the abdominal line; a constriction at the nape. Distance from origin of dorsal to the occiput greater than the distance from the occiput to the snout. Anal fin beginning much in advance of middle of body, the distance from the first ray to tip of caudal exceeding the distance to snout by nearly twice length of head; pectoral very short, its length less than diameter of eye. Reddish brown, uniform or variously shaded with lighter; a light olivaceous bar at base of caudal, extending on dorsal and anal, behind this a blackish area; tip of caudal usually pale; 3 long, well-defined stripes radiating backward from eye, these stripes uniform black, abruptly margined with very light olive; the central stripe proceeding straight backward from the eye breadth of cheek, at which point it is broadest; it is then narrowed and bent abruptly downward; both the middle and lower stripes reach the margin of preopercle. Length 12 inches. Smaller than the preceding, and equally abundant; among rocks and algæ, from Vancouver Island to Monterey. Here described from specimens, 6 or 7 inches in length, from Neah Bay, Straits of Fuca, Washington. (*rupestris*, living among rocks.)

Xiphister rupestris, JORDAN & GILBERT roc. U. S. Nat. Mus. 1880, 137, Monterey Bay,
　　California (Coll. Jordan & Gilbert) JORDAN & GILBERT, Synopsis, 773, 1883.
Xiphidion rupestre, JORDAN & STARKS, hes Puget Sound, 848, 1895.

911. CEBEDICHTHYS,[*] Ayres.

Cebedichthys, AYRES, Proc. Cal. Ac. Nat. Sc. I, 1855, 59 (*violaceus*).

Body comparatively short, compressed, covered with minute scales; lateral line distinct, running very high, with very short branches, each ending in a pore, as in *Xiphistes*, but the branches more oblique and less regular. Head short; crown with a conspicuous fleshy longitudinal crest in the adult; jaws subequal, with conical teeth; villiform teeth on vomer and palatines; gill membranes united, free from the isthmus. Dorsal fin continuous, long and low, the anterior part composed of sharp spines, which are rather lower than the soft rays; caudal fin rounded, connected with dorsal and anal; anal fin similar to soft dorsal, with 1 or 2 small spines; pectorals small; ventrals wanting. Intestinal canal elongate, with several pyloric cæca. Pacific Ocean. Herbivorous; similar in habits

---

[*] The following remarkable genus may be allied to *Cebedichthys:*
NEOZOARCES, Steindachner.
Neozoarces, STEINDACHNER, Ichth. Beitr., 26, 1880 (*pulcher*).

to the species of *Xiphidion*. (ξιφθος, the Sap... a kind of monkey; ιχθύς, fish; in allusion to the "peculiar monkey-li... physiognomy as seen from the front.)

2788. CEBEDICHTHYS VIOLA... (Ayres).

Head 6¼; depth 6. D. XXIII, 41; A. I, 41... maxillary extending to or beyond orbit. Dorsal scaly at base; vent n... snout than base of caudal; pectoral ⅓ length of head; nape mid... between dorsal and eye. Dull olive grayish, mottled with lighter, som... es reddish tinged; vertical fins all edged with reddish; cheek w... darker stripes, edged with paler, 1 downward and backward fro... eye, close behind angle of mouth; another above it to root of pect... another running upward and backward from the eye, and meeting its... w over the crest. Length 30 inches. San Francisco to Point Concepci... bundant; often brought into the markets. (*violaceus*, violet.)

Apodichthys violaceus, GIRARD, Proc. Ac. Nat. Sci. Ph... ..., 150, San Luis Obispo, California. (Coll. Dr. Kennerly.)

Cebedichthys cristagalli, AYRES, Proc. Cal. Ac. Nat. Sci., ... 55, 55, San Francisco.

Cebedichthys violaceus, GIRARD, Pac. R. R. Surv., x, Fish... pl. 34, figs. 4 and 5, 1858; JORDAN & GILBERT, Synopsis, 774, 1882.

912. PLAGIOGRAMMU... Bean.

Plagiogrammus, BEAN, Proc. U. S. Nat. Mus. 1893, 699 (...).

Body moderately elongate, compressed, c...ed with very small scales; lateral lines 2, viz, 1 beginning above an... ghtly in advance of the upper angle of the gill opening and extendi... along the upper part of the body, but not reaching the tail, and 1 be...ng in advance of the end of this and reaching to the caudal; numero... teral ridges on the sides, similar to those on *Dictyosoma* of Temmin... Schlegel; a series of subpentagonal plate-like bodies along the abd...al edge on each side between the ventral and the anal. Head ... rately long, naked, with pointed snout; mouth oblique and rather ...; jaws subequal, or the lower slightly projecting; jaws with stron... th in broad bands, the intermaxillaries with an outer series of enlarg... nine-like teeth; teeth on vomer and palate; a pair of large canines n...e symphysis in each jaw, the canines of the upper jaw fitting into an ...space behind the man ulary canines. A series of pores on the ram... the mandibula continuing around the preopercular edge; a seri... similar pores along the lower margin of the preorbital continued ... ard and upward toward the nape. Anterior nostril tubular; post... without tube. Maxillary broadly expanded posteriorly; lips well de...oped. Branchiostegals 5; gill membranes partly united, but free fr... he isthmus behind. Gill rakers minute, tubercular, in moderate num... A single long dorsal fin consisting of spines only, the spines long... n the posterior portion; anal fin lower than the dorsal, but simila... shape. Pectoral large, ...irely below median line. Ventrals well ...loped, in advance of pec... ...caudal rounded, distinct. Intestin... nal short, with 5 small ... (πλάγιος, oblique; γραμμή, ...)

spines, in the branches of the upper lateral line running higher, and in coloration. (*Ulva*, the green sea lettuce.)

Xiphidion ulvæ, JORDAN & STARKS, Fishes of Puget Sound, 847, 1895, **Waadda Island, Neah Bay**. (Type, No. 3132, L. S. Jr. Univ. Mus. Coll. E. C. Starks.)

2780. XIPHISTES CHIRUS (Jordan & Gilbert).

Head 7; depth 9. D. LXX; A. II, 50. Head short; nape not constricted; mouth small; maxillary extending to middle of pupil; teeth strong, the anterior canine-like, bluntish; about 4 canines in lower jaw, 5 or 6 in the upper, similar to the teeth behind them, but somewhat larger. Abdominal lines meeting on the breast, but not connected with the lower lateral line. Dorsal fin beginning close behind pectoral; nape midway between middle of eye and front of dorsal; anal beginning about a head's length nearer snout than base of caudal; pectoral fin comparatively large, longer than the eye, its length about equal to distance between middle and lower lateral lines. Color olive brown, yellowish below; sides with marblings of different shades of brown, sometimes with short blackish vertical bars; some round black spots along the back and sides; a black spot behind opercles; numerous black spots on sides of head, forming in older individuals light and dark streaks, which radiate from eye across cheek and opercles, the pale streaks forming reticulations; dorsal with black spots and a series of bright reddish brown cross blotches; pectorals and caudal plain. Monterey to Alaska; smaller than the other species, and living in deeper water; abundant about Cape Flattery. (Jordan & Gilbert.) ($\chi\varepsilon i\rho$, hand.)

Xiphister chirus, JORDAN & GILBERT, Proc. U. S. Nat. Mus. 1880, 135, **Point of Los Pinos, near Monterey, California** (Coll. Jordan & Gilbert); JORDAN & GILBERT, Synopsis, 772, 1883; JORDAN & STARKS, Fishes of Puget Sound, 846, 1895.

910. XIPHIDION, Girard.

Xiphidion, GIRARD, Pac. R. R. Surv., X, Fishes, 119, 1858 (*mucosum*); not *Xiphidium*, Serv., a genus of Grasshoppers.
Xiphister, JORDAN, Proc. U. S. Nat. Mus. 1879, 241 (*mucosum*); substitute for *Xiphidion*, regarded as preoccupied by *Xiphidium*.

Body elongate, eel-shaped, covered with small scales; lateral lines several: 1 along the median line of the side, 1 above this, and 1 below it; 1 on each side of the abdomen, the 2 meeting in front, and 1 from the occiput toward the base of the dorsal fin. Each of these has on each side series of short branches, placed at right angles to the main line, those on opposite sides alternating. Each of these branches has about 2 open mucous pores. Lower lateral line connected with the abdominal line. Head short, bluntish, scaleless; mouth moderate, oblique; jaws with rather strong teeth, the anterior canine-like; no teeth on vomer or palatines. Branchiostegals 6; gill membranes separate, free from the isthmus. A single long, low, uniform dorsal fin, consisting of spines only; anal fin similar in form, with small spines, indistinct or obsolete; caudal short, joined to dorsal and anal; no ventral fins; pectoral fins very small, shorter than eye. Intestinal canal moderately elongate, with 4 to 6 well-developed pyloric cæca. Herbivorous, feeding on algæ. Active fishes, inhabiting tide pools and crevices among rocks in the North Pacific. (ξιφίδιον, a small sword.)

> *a.* Distance from origin of dorsal to occiput less than that from occiput to tip of snout; streaks radiating from eye paler in the center, edged above and below with blackish.
> MUCOSUM, 2781.
> *aa.* Distance from origin of dorsal to occiput greater than that from occiput to snout; streaks radiating from eye black, abruptly margined with pale olive.
> RUPESTRE, 2782.

2781. XIPHIDION MUCOSUM, Girard.

Head 7 in body; depth 8¼. D. LXXIV; A. 46; eye 7½ in head; maxillary 2⅓; caudal 2⅔; pectoral a little longer than eye. Lower jaw with a series of short, stout conical teeth; upper jaw with a narrow band of similar teeth; 2 strong canines in upper jaw, 4 in the lower. Lower lateral line sending a branch to the abdominal line; nape not constricted. Dorsal beginning anteriorly, distance from its origin to occiput less than that from occiput to tip of snout; origin of anal nearer snout than tip of caudal by about ¼ length of head. Blackish green, pale on belly and sides of head, marked posteriorly with olive green in various pattern; a transverse light-greenish bar at base of caudal; 3 olive-brown streaks radiating backward from eye, paler in the center and edged above and below with blackish, outside of which is sometimes a streak of pale olive; these streaks all merge backward into the color of the head; middle streak broadly wedge-shaped, the third streak terminating before reaching margin of preopercle; old individuals often coarsely blotched with yellow. Length 18 inches. Monterey to Alaska; very abundant among rocks and algæ. Here described from specimens, 9 or 10 inches in length, from Neah Bay, Straits of Fuca, Washington. (*mucosus*, slimy.)

Xiphidion mucosum, GIRARD, Pac. R. R. Surv., x, Fishes, 119, 1858, South Farallones, California (Coll. Lieut. Trowbridge. Type, No. 493, U. S. Nat. Mus.); GÜNTHER, Cat., III, 291, 1861; JORDAN & STARKS, Fishes Puget Sound, 848, 1895.
Xiphidion cruoreum,* COPE, Proc. Amer. Phil. Soc. Phila. 1873, 27, Sitka (Coll. Prof. George Davidson); JORDAN & GILBERT, Proc. U. S. Nat. Mus. 1880, 137.
Xiphister mucosum, JORDAN & GILBERT, Synopsis, 772, 1883.

---

* The following is the original description of *Xiphidion cruoreum:* Head 8¼ in total length; depth 9½; eye 7 in head, equal to length of pectoral fin. D. about 70; A. 48; Br. 5. Teeth, 2 canines above, 4 below, subequal. Dorsal spines not commencing near the head, the anterior buried in a soft fold of skin; caudal fin not distinct. Three lateral mucous canals extending entire length of caudal fin, which have numerous alternating transverse branches, those of the superior reaching base of dorsal, those of inferior reaching base of anal; each of the cross branches with several excretory pores, none on the main stem; a similar but short tube extending from near base of dorsal fin to supra-occipital region, and not branching anteriorly; the superior lateral canal descending to near the median, but not joining it, nor does the latter extend into the inferior; another tubular line on each side of abdomen, these uniting on jugular region by a continuation of the inferior lateral tube. Vent nearer end of muzzle than end of caudal fin, by length of head. Color maroon, more reddish below; a vertical, broad, reddish bar at base of tail, beyond which is a dark spot; 2 brown radii, black-edged, extending backward and downward from eye. Body covered with small scales, except on the jugular and abdominal regions, which are naked. Length 8 inches. This fish is not very different from *X. mucosum,* Girard. It differs in the smaller eye, the more remote origin of the dorsal fin from the head, the lack of anterior union of the mucous canals, and the coloration. (Cope.)

2782. **XIPHIDION RUPESTRE** (Jordan & Gilbert).

Head 7⅓ in body; depth 9. D. LXVIII; A. 50; eye 6 in head; maxillary 2⅓; caudal 2⅔. Teeth essentially as in *X. mucosum.* The lower lateral line sends a branch to the abdominal line; a constriction at the nape. Distance from origin of dorsal to the occiput greater than the distance from the occiput to the snout. Anal fin beginning much in advance of middle of body, the distance from the first ray to tip of caudal exceeding the distance to snout by nearly twice length of head; pectoral very short, its length less than diameter of eye. Reddish brown, uniform or variously shaded with lighter; a light olivaceous bar at base of caudal, extending on dorsal and anal, behind this a blackish area; tip of caudal usually pale; 3 long, well-defined stripes radiating backward from eye, these stripes uniform black, abruptly margined with very light olive; the central stripe proceeding straight backward from the eye, ¼ breadth of cheek, at which point it is broadest; it is then narrowed and bent abruptly downward; both the middle and lower stripes reach the margin of preopercle. Length 12 inches. Smaller than the preceding, and equally abundant; among rocks and algæ, from Vancouver Island to Monterey. Here described from specimens, 6 or 7 inches in length, from Neah Bay, Straits of Fuca, Washington. (*rupestris,* living among rocks.)

Xiphister rupestris, JORDAN & GILBERT, Proc. U. S. Nat. Mus. 1880, 137, Monterey Bay, California (Coll. Jordan & Gilbert); JORDAN & GILBERT, Synopsis, 773, 1883.
Xiphidion rupestre, JORDAN & STARKS, Fishes Puget Sound, 848, 1895.

911. CEBEDICHTHYS,* Ayres.

Cebedichthys, AYRES, Proc. Cal. Ac. Nat. Sci., I, 1855, 59 (*violaceus*).

Body comparatively short, compressed, covered with minute scales; lateral line distinct, running very high, with very short branches, each ending in a pore, as in *Xiphistes,* but the branches more oblique and less regular. Head short; crown with a conspicuous fleshy longitudinal crest in the adult; jaws subequal, with conical teeth; villiform teeth on vomer and palatines; gill membranes united, free from the isthmus. Dorsal fin continuous, long and low, the anterior part composed of sharp spines, which are rather lower than the soft rays; caudal fin rounded, connected with dorsal and anal; anal fin similar to soft dorsal, with 1 or 2 small spines; pectorals small; ventrals wanting. Intestinal canal elongate, with several pyloric cæca. Pacific Ocean. Herbivorous; similar in habits

---

* The following remarkable genus may be allied to *Cebedichthys:*

NEOZOARCES, Steindachner.

Neozoarces, STEINDACHNER, Ichth. Beitr., IX, 26, 1880 (*pulcher*).

NEOZOARCES PULCHER, Steindachner.

Body elongate, tapering backward, the dorsal and anal united at the tail without distinct caudal. Scales small, embedded, no lateral line. Mouth very large, the maxillary extending far beyond eye; lower jaw slightly longer than upper; blunt, conical teeth in many rows on jaws, vomer, and palatines. A thick tentacle above nostril; gill membranes united, free from isthmus. Dorsal low, the anterior portion of short, stiffish spines; no anal spine; ventrals wanting; pectorals moderate; pseudobranchiæ present. Head 6; depth 9. D. XLI, 50; A. I, 75. Color highly variegated. Gulf of Strietok, Okhotsk Sea. (νέος, new; *Zoarces*; but it has little affinity with the latter genus.)

Neozoarces pulcher, STEINDACHNER, Ichth. Beitr., IX, 27, taf. 6, f. 2, 1880, Gulf of Strietok. (Coll. Professor Dybowski.)

to the species of *Xiphidion.* (κῆβος, the Sapajou, a kind of monkey; ἰχθύς, fish; in allusion to the "peculiar monkey-like" physiognomy as seen from the front.)

2783. CEBEDICHTHYS VIOLACEUS (Ayres).

Head 6½; depth 6. D. XXIII, 41; A. I, 41. Maxillary extending to or beyond orbit. Dorsal scaly at base; vent nearer snout than base of caudal; pectoral ⅔ length of head; nape midway between dorsal and eye. Dull olive grayish, mottled with lighter, sometimes reddish tinged; vertical fins all edged with reddish; cheek with 3 darker stripes, edged with paler, 1 downward and backward from the eye, close behind angle of mouth; another above it to root of pectoral; another running upward and backward from the eye, and meeting its fellow over the crest. Length 30 inches. San Francisco to Point Concepcion; abundant; often brought into the markets. (*violaceus,* violet.)

Apodichthys violaceus, GIRARD, Proc. Ac. Nat. Sci. Phila. 1854, 150, San Luis Obispo, California. (Coll. Dr. Kennerly.)

Cebedichthys cristagalli, AYRES, Proc. Cal. Ac. Nat. Sci., I, 1855, 58, San Francisco.

Cebedichthys violaceus, GIRARD, Pac. R. R. Surv., X, Fishes, 121, pl. 26, figs. 4 and 5, 1858; JORDAN & GILBERT, Synopsis, 774, 1883.

912. PLAGIOGRAMMUS, Bean.

Plagiogrammus, BEAN, Proc. U. S. Nat. Mus. 1893, 699 (*hopkinsi*).

Body moderately elongate, compressed, covered with very small scales; lateral lines 2, viz, 1 beginning above and slightly in advance of the upper angle of the gill opening and extending along the upper part of the body, but not reaching the tail, and 1 beginning in advance of the end of this and reaching to the caudal; numerous lateral ridges on the sides, similar to those on *Dictyosoma* of Temminck & Schlegel; a series of subpentagonal plate-like bodies along the abdominal edge on each side between the ventral and the anal. Head moderately long, naked, with pointed snout; mouth oblique and rather large; jaws subequal, or the lower slightly projecting; jaws with strong teeth in broad bands, the intermaxillaries with an outer series of enlarged canine-like teeth; teeth on vomer and palate; a pair of large canines near the symphysis in each jaw, the canines of the upper jaw fitting into an interspace behind the mandibulary canines. A series of pores on the ramus of the mandibula continuing around the preopercular edge; a series of similar pores along the lower margin of the preorbital continued backward and upward toward the nape. Anterior nostril tubular; posterior without tube. Maxillary broadly expanded posteriorly; lips well developed. Branchiostegals 5; gill membranes partly united, but free from the isthmus behind. Gill rakers minute, tubercular, in moderate number. A single long dorsal fin consisting of spines only, the spines longest in the posterior portion; anal fin lower than the dorsal, but similar in shape. Pectoral large, entirely below median line. Ventrals well developed, in advance of pectorals; caudal rounded, distinct. Intestinal canal short, with 5 small pyloric cæca. (πλάγιος, oblique; γραμμή, line.)

Head 4; depth 5½; eye 5. D. XLI; A. II, 29; V. I, 5; B. 5; scales about 95; ridges on side 32. Snout acute; anterior nostril tubular and nearer eye than tip of snout; posterior nostril close to upper anterior margin of eye; maxillary extending almost to vertical through hind margin of eye; intermaxillary long, slender, and reaching nearly as far back as maxillary; intermaxillary teeth in broad bands, with an outer series of 5 or 6 large canines, those near the symphysis largest; teeth in mandible in broad bands in front, followed by several enlarged canine-like teeth; a large canine on each side of symphysis, the interspace between the 2 mandibulary canines receiving the canines of the intermaxillary when the jaws are closed. A row of 8 pores along ramus of mandible and edge of preopercle; another series around lower margin of preorbital bone as described for the genus; about 8 gill rakers on first arch below angle. Distance of dorsal origin from snout nearly equal to length of head; spines lowest in front, the longest spine ⅞ length of head; longest rays of anal near end of fin and scarcely exceed length of eye; length of pectoral equaling that of postorbital part of head; ventrals close together; inner rays longest, ⅔ as long as head; caudal rounded, its length nearly ¼ that of head; vent under eleventh spine of dorsal. Upper lateral line beginning above and slightly in advance of upper angle of gill opening, curving very slightly over pectoral and extending to below twenty-fifth spine of dorsal, its distance from dorsal edge equal to diameter of eye and also equal to its distance from lower lateral line; lower lateral line beginning under sixteenth spine of dorsal and extending to caudal. On each side of the abdominal ridge, between the ventrals and the vent, are about 10 subpentagonal plate-like bodies, the largest about ⅓ as long as eye. Color dusky brown, the fins black. Monterey, California; a few specimens dredged among rocks. Little is known about the habits of the species, beyond the fact that in the aquarium it hides in rock crevices and seldom ventures from its hiding place. (I take pleasure in associating with this blenny the name of Mr. Timothy Hopkins, of Menlo Park, California, the founder of the Seaside Laboratory at Pacific Grove, Monterey Bay, in commemoration of his services in behalf of science. Bean.)

Plagiogrammus hopkinsi, BEAN, Proc. U. S. Nat. Mus. 1893, 699, Monterey Bay, California. (Type, No. 44721, U. S. Nat. Mus.)

913. OPISTHOCENTRUS, Kner.

Opisthocentrus, KNER, Sitzber. Akad. Wiss. Wien 1868, 49 (*quinquemaculatus*).
Blenniophidium, BOULENGER, Proc. Zool. Soc. Lond. 1892, 583 (*petropauli*).

Body moderately elongate, compressed, covered with very small cycloid scales. Mouth small, horizontal, protractile, with fleshy lips; small conical teeth in jaws and on vomer and palatines. No cirri. Gill membranes broadly connected, but free from isthmus; branchiostegals 4. Dorsal fin very long, extending from the nape to the caudal, with which it is subcontinuous; a few of the posterior rays are stiff spines, the rest being simple and not articulate, but flexible; anal fin extending from the anus, which is a little nearer the anterior than the posterior extremity, to the caudal, formed exclusively of soft rays; no ventrals. No lateral line. No prominent anal papillæ. Pyloric appendages present. A remarkable genus, allied to *Lumpenus*, or rather to *Plectobranchus*, distinguished by having only the posterior spines rigid. North Pacific. (ὄπισθε, behind; κέντρον, spine.)

2785. OPISTHOCENTRUS OCELLATUS (Tilesius).

Head 6¼; depth 6¼ (without caudal). D. LV to LXI, usually LIX; A. 36 to 39; 5 to 7 of the posterior dorsal spines rigid; Eye as long as snout, 4 in head, and a little more than interorbital width; maxillary extending to below anterior fourth of eye; some wide pores on the head; cheeks, opercles, and occiput closely scaled; strips of small scales on the branchiostegal membrane between the rays. Dorsal rays continuous and subequal in depth, the longest spine 2⅓ in head in females, 1⅙ in males; pectoral 1⅓ in head, about as long as caudal. Anus twice as far from caudal as from base of pectoral. Yellowish brown, with ill-defined darker marblings; a crescentic black line on the top of the head from eye to eye; a black line, obliquely directed forward, below the eye, and another, in opposite direction, from the eye to the opercle; 2 dark-brown streaks across the nape, the second crossing the origin of the dorsal fin and extending to the base of the pectoral; dorsal and caudal fins grayish olive, lighter at the base, the dorsal with 5 to 9 (usually 6) large black spots at regular intervals, these wanting in the males; pectorals and anal colorless. Numerous specimens are from Tareinsky Bay, Kamchatka; Petropaulski Harbor, and Shana Bay, Iturup Island. The number of dorsal ocelli varies from 5 to 9 in our specimens, 6 being the prevailing number. Of 24 specimens whose fins we have enumerated, 4 have 58 dorsal spines, 10 have 59, 5 have 60, and 5 have 61. In addition, 1 specimen has but 55 spines. The latter is the only male in the collection and is conspicuous by the absence of distinct dorsal ocelli and the great height of the vertical fins, the longest dorsal spine exceeding the length of the pectoral and contained 1⅙ times in head. In females the longest spine is 2⅓ in head. The anal contains 36 to 39 rays in all our specimens. The dorsal fin is composed exclusively of spines, the anterior flexible ones passing into the strong pungent ones near the posterior end. The stronger spines vary from 7 to 12 in number in our specimens. Our material answers the description of the type of *O. quinquemaculatus* which had 57 dorsal spines and 36 anal rays. It also agrees with specimens from Petropaulski, reported on by Bean & Bean (Proc. U. S. Nat. Mus. 1896, 391), with dorsal spines 58 in number. *Blenniophidium petropauli*, Boulenger, has but 52 dorsal spines, but it is otherwise not to be distinguished from *O. ocellatus*. Still more aberrant are 4 specimens from Gulf of Strietok, northern Japan, mentioned by Steindachner (Ichth. Beiträge, IX, 25), with but 50 to 53 spines and 32 to 34 anal rays. These may represent a distinct species. *Ophidium ocellatum* of Tilesius must be this species, but the count of fin rays is incorrect and may be taken from the rough figure. *Opisthocentrus tenuis* is probably also identical with *O. ocellatus,* though the writers did not think so until after examination of the present large material. Coast of Kamchatka, southwestward to Okhotsk Sea, generally common from Komandorski Islands to Yezo. (*ocellatus,* with eye-like spots.)

Ophidium ocellatum, TILESIUS, Mém. Ac. St. Petersb., II, 1811, 237, Kamchatka. D. 80; A. 50; evidently an error. The rude figure shows D. 73; A. 50, the spines low; the dorsal with 5 ocelli.

Centronotus (Opisthocentrus) quinquemaculatus, KNEE, Sitzber. Akad. Wiss. Wein 1868. 48, taf. 7, f. 20, "Pinang." Described from a young specimen 2 inches long, No. 6353, Mus. Wien.

Gunnellus apos, CUVIER & VALENCIENNES, Hist. Nat. Poiss., XIV, 426, 1839; after TILESIUS.

Centronotus apus, GÜNTHER, Cat., III, 288, 1861.

Blenniophidium petropauli, BOULENGER, Proc. Zool. Soc. London 1892, 584, with plate, Petropaulski (Coll. George Baden-Powell); D. 52; A. 37; 5 ocelli.

Opisthocentrus tenuis,* BEAN & BEAN, Proc. U. S. Nat. Mus. 1897, Volcano Bay, Port Morusan, Japan. (Coll. Col. Nicolai A. Grebnitski. Type, No. 47565, U. S. Nat. Mus.)

Opisthocentrus quinquemaculatus, STEINDACHNER, Ichth. Beitr., IX, 25, 1880; BEAN & BEAN, Proc. U. S. Nat. Mus. 1896, 381, 392.

Opisthocentrus ocellatus, JORDAN & GILBERT, Rept. Fur Seal Invest., 1898.

914. PHOLIDAPUS, Bean & Bean.

Pholidapus, BEAN & BEAN, Proc. U. S. Nat. Mus. 1896, 389 (*grebnitskii*).

Body moderately elongate, compressed, covered with very small, smooth scales. Mouth small, horizontal; bands of small teeth on jaws and vomer, none on palatines. Head naked; gill membranes broadly connected, free from the isthmus; dorsal very long, composed entirely of flexible spines; anal of soft rays; caudal short, rounded, separate; no ventral fins; no lateral line; pyloric cæca present. This genus is close to *Opisthocentrus,* but has no pungent spines, and the head is naked. Okhotsk Sea. (φολίς, *Pholis*; ἄπους, without feet, i. e., ventral fins.)

2786. PHOLIDAPUS DYBOWSKII (Steindachner).

Head 5¼ to 6⅔; depth 6 to 6¼. D. LXII or LXIII; A. II, 39. Eye 3¾ to to 4⅘ in head; snout a little longer than eye; lower jaw scarcely included; 1 or 2 strong conical teeth on each side behind the narrow premaxillary band of teeth; teeth on vomer, none on palatines; no cirri; large pores

---

* *Opisthocentrus tenuis* is thus described:
   D. 39, XV; A. 38. Length of fish to caudal base 5½ inches; length of head 1; depth of body ⅞; the greatest width of the body is contained 2¼ times in the length of the head. The diameter of the eye is nearly equal to the length of the snout and is contained 4½ times in the length of the head; the width of the interorbital space is almost equal to the long diameter of eye. The maxilla reaches to the vertical past front of eye. Teeth bluntly rounded, embedded in flesh; vomerine teeth present; palatines none. The origin of the dorsal fin is over the end of the gill cover, its first 39 rays are simple and flexible, the last 15 are strong spines and end slightly above the membrane in stiff points, the longest spine is almost ¼ as long as the head. The anal originates under the twentieth ray of the dorsal; its rays are divided and articulated; the longest ray is ⅓ as long as the head. The general color is brown with cross reticulations of black. Sides of head and body along base of anal, orange; anal, caudal, and pectorals light with dusky shadings; dorsal finely mottled with black and bearing 6 black spots on areas of white, the first of these spots being on the sixth ray and the last on the next to last spine; a black bar from front of eye downward, and another from posterior margin obliquely down and backward. This species differs from the typical form in its greater compression of the body and its increased number of dorsal spines. (Bean & Bean.)

about eye and on opercles; longest dorsal spines 2¼ to 3 in depth of body, last spines shorter and stiffer than others; dorsal and anal slightly joined to caudal; pectoral as long as caudal, about 1¼ in head. Head naked. Brown or grayish, with faint spots or marblings; 1 or 2, rarely 3, dark ocelli on the dorsal; 3 or 4 dark streaks radiating from eye, the uppermost joining its fellow. Length 10 to 15 inches. . Coast of northern Japan and sea of Okhotsk, north to the Kuril Islands. Our specimens, 5 in number, the largest 25 cm. long, from Shana Bay, Iturup Island. Steindachner's excellent and detailed description leaves nothing to be desired, and corresponds perfectly with our material except in the character of the scales. A careful examination of these under high magnification fails to show that they are "am hintern Rande mit kurzen Zähnchen bewaffnet." The posterior border is entire and the scales strongly marked with concentric striæ. Dorsal spines number 62, 63, 63, 64, 64. Dorsal ocelli present in all our specimens, 2 of them being faintly visible even in the youngest, 55 mm. long. *Pholidapus grebnitskii* seems to differ only in the shorter dorsal fin (57 spines). (Named for Professor Dybowski, its first collector.)

Centronotus dybowskii, STEINDACHNER, Ichth. Beiträge, IX, 22, 1880, Gulf of Strietok, northern Japan (Coll. Prof. Dybowski); JORDAN & GILBERT, Rept. Fur Seal Invest., 1898.

?*Pholidapus grebnitskii,** BEAN & BEAN, Proc. U. S. Nat. Mus. 1896, pl. 34, 390, Yezo, Japan. (Coll. Col. Nicolai A. Grebnitski.)

915. PLECTOBRANCHUS, Gilbert.

Plectobranchus, GILBERT, Proc. U. S. Nat. Mus. 1890, 102 (*evides*).

Teeth conic, on jaws, vomer, and palatines, some of them canine-like. Body scaly; lateral line obsolete, its course indicated by a lighter streak on middle of sides. Gill slits not continued far forward, the membranes

---

* *Pholidapus grebnitskii,* Bean & Bean, is thus described: The specimens are 141 mm. long, including caudal; 126 mm. to base of caudal. The head (22 mm.) is equal to the greatest depth of body. The eye is slightly longer than the snout and ¼ as long as the head. The interorbital space is narrow, ⅔ of the length of the eye. The naked head resembles that of *Pholis,* its length is contained about 5½ times in total length without the caudal. The mouth is small and very oblique; the mandible is slightly included and has a well developed lip. The maxilla is partly concealed under the preorbital bone; it does not quite reach to below the anterior margin of the pupil. The anterior nostril is midway between the eye and the tip of the intermaxilla. Seven mucous pores around the orbit; 3 on the preorbital bone. The pore in the origin of the semicircular dark band around the nape is continued backward by a series of 6 similar ones ending near the upper angle of the gill opening. A series of 10 or 11 pores beginning near the front of the chin on each side, extending backward and curving upward to the upper anterior edge of the operculum. The gill membranes are broadly united, but they are not joined to the isthmus. The dorsal origin is over the end of the head; the fin is low, and consists of spines, the longest and strongest in the posterior third being slightly longer than the eye. The distance of the vent from the tip of the snout contains the head length 2⅓ times. The anal is slightly lower than the dorsal, the rays longest posteriorly. The caudal is rounded, and is barely separated from the dorsal and anal. The pectoral base is broad, and the fin is ⅔ as long as the head. The intestine is slender, and is more than twice as long as the head. Stomach short, pear-shaped, with 6 slender pyloric cæca of unequal length, the longest about twice as long as the eye. The body is completely scaled, the scales very small, cycloid, closely imbricated, with numerous concentric striæ, and they extend halfway up the membrane connecting the dorsal spines. The general body color is brown, the sides sparsely and vaguely mottled. The pectorals are pale. A narrow, dark band extends from the middle of the eye downward and forward, a similar band running backward from the eye on the preopercle; an interrupted semicircular dark band from eye to eye across the nape. D. LVII; A. II, 39 or 40. (The species is named for Mr. N. Grebnitski, to whose industry and zeal the Museum is indebted for many valuable collections. Bean & Bean.)

broadly united, wholly free from isthmus. Dorsal of spines only. Anal
with 2 spines. Ventral with 1 spine and 3 well-developed rays. Lower
pectoral rays longest, as in *Leptoclinus*. North Pacific. (πλεκτός, enfolded;
βράγχος, gill.)

2787. PLECTOBRANCHUS EVIDES, Gilbert.

Head rather long. 4⅓ in length, extending well beyond origin of dorsal
fin; depth about 11. D. LVI; A. II, 34. Body very slender, the depth
nearly constant throughout. Caudal peduncle without free portion, its
depth 2⅓ in that of body. Upper jaw with a broad inner band of minute
teeth in front and on the sides, the outer series enlarged, 2 in the front of
the jaw distinctly canine-like; teeth in the lower jaw similar to the outer
series above, in a single series laterally, forming a patch in front of jaw,
where 2 of them are much enlarged canines, the largest teeth in the jaws;
vomer and palatines with bands of small but very evident teeth. Eyes
large, close together, the interorbital space ⅓ pupil. Orbit 3⅓ in head,
longer than snout. Posterior nostril with a short flap, the tube obsolete.
Mouth large, somewhat oblique, maxillary reaching middle of orbit, 2⅓ in
head. Top of head with very large pores, a series running backward from
each eye, the two joined by a cross series on occiput. Body covered with
very small cycloid scales, including belly, nape, breast, and cheeks, those
on breast and cheeks not imbricated; lateral line without visible pores.
Spinous dorsal beginning well forward, the distance from its origin to
nape less than from latter to posterior margin of orbit. Anterior spines
short, but fully united by membrane, the longest spine 3⅓ in head; membrane of last spine reaching base of upper caudal rays; origin of anal
very slightly in advance of middle of body; anal with 2 short, sharp
spines, the rays longer, their terminal ⅓ free from membrane; last anal
ray connected with base of lower caudal ray; ventrals well developed,
nearly ⅓ head; pectorals with lowermost rays abruptly lengthened, ⅔ head;
caudal short, rounded, little more than ⅓ head. Color dusky olive above,
lighter below; sides crossed by about 25 narrow white bars, narrower
than interspaces; 3 equidistant dark blotches near back, each double,
the two halves occupying contiguous interspaces between white bars;
branchiostegal membrane black; head without markings; pectorals white
at base, the distal half black, margined with white; ventrals white; dorsal with alternating oblique bars of white and blackish, 2 jet-black
roundish spots on its posterior portion; caudal whitish at base, then
dusky, margined with white, its upper ray jet-black; anal light at base,
becoming black at edge of membrane, the free tips of rays white. Coast
of Oregon. A single specimen, 4 inches long, from *Albatross* Station 3064,
in 46 fathoms. (εὐειδής, comely.)

Plectobranchus evides, GILBERT, Proc. U. S. Nat. Mus. 1890, 102, coast of Oregon, at
Albatross Station 3064. (Coll. *Albatross*.)

916. LEPTOCLINUS, Gill.

Ctenodon, NILSSON, Skandinav. Faun., IV, 190, 1853 (*maculatus*) (name three times * preoccupied).
Leptoclinus, GILL, Proc. Ac. Nat. Sci. Phila. 1864, 209 (*aculeatus*).

* *Cten*

Body much elongated; lateral line obsolete; teeth on jaws, vomer, and palatines; pectoral fins with the upper rays shortened; caudal fin subtruncate. Arctic seas. This genus is close to *Lumpenus*, differing mainly in the form of the pectoral. (λεπτός, slender; *Clinus*.)

2783. LEPTOCLINUS MACULATUS (Fries).

(LANGBARN.)

Head 5; depth 8. D. LX (LVIII to LX); A. 36 (35 to 38). Eye large, 3¼ in head; snout short and blunt, 4¼ in head, maxillary reaching past middle of eye, 2¼ in head. Teeth in jaws, vomer, and palatines; jaws each having 2 strong canines in front. Scales small, cycloid. First 3 or 4 dorsal spines short and free; longest dorsal spines as long as eye; caudal fin free from dorsal and anal; ventrals 3 in head; pectorals rather large, 1¼ in head. Color yellowish, irregularly marked with dark spots, a series of about 6 of these spots extending along sides close to base of dorsal fin; a series of smaller spots extending along center of sides from upper base of pectoral to caudal; dorsal irregularly covered with dark spots; caudal with 4 dark cross bands; anal, ventral, and pectorals plain yellowish. Bering Sea to Spitzbergen, south to Aleutian Islands and the coasts of Sweden and Norway. This description is taken from a specimen, 5¼ inches long, from Alaska, near Unimak Pass (*Albatross* Station 3309). A few young individuals of this species, hitherto known only from the North Atlantic, were taken in Unimak Pass and in Bristol Bay, in 29¼ to 70 fathoms. Three small specimens were also taken off Robben Reef, near the Kamchatka coast, in 28 fathoms, and one off Karluk, Kadiak Island. Having no Atlantic specimens of this species, we are unable to satisfy ourselves of the identity of the two, but no difference is evident from descriptions. The lateral line is much more distinct than in our specimens of *Lumpenus medius*, in which it can be made out with difficulty on scattered scales along middle of sides. (Eu.) (*maculatus*, spotted.)

Clinus maculatus, FRIES, Kgl. Vet. Ak. Handl. 48, 1837, Bohūslän, Sweden.
Lumpenus aculeatus, REINHARDT, Kong. Dansk. Vid. Selsk., vi, 1837, 100, no description.
Clinus aculeatus. REINHARDT, Dansk. Vidensk. Selsk. Natur. Afh., vii, 1838, 114, 122, 194, Spitsbergen.
Clinoden maculatus, NILSSON, Skand. Fauna, iv, 100, 1853.
Stichæus maculatus, GÜNTHER, Cat., iii, 281, 1861.
Lumpenus aculeatus, KRÖYER, Naturhist. Tidsskr., i, 377, 1862.
Stichæus aculeatus. GÜNTHER, Cat., iii, 282, 1861; COLLETT, Norske Nord-Havs Exp., 67, 1880.
Lumpenus maculatus, JORDAN & GILBERT, Synopsis, 777, 1883; Lilljeborg, Sveriges Och Norges Fisk., 560, 1891.
Leptoclinus maculatus, GILBERT, Rept. U. S. Fish Comm. 1893, 450.

917. POROCLINUS, Bean.

Poroclinus, BEAN, Proc. U. S. Nat. Mus. 1890, 40 (*rothrocki*).

Body elongate, moderately compressed, covered with small scales; lateral line obsolete. Head moderately long; snout short; eyes large; interorbital space narrow. Mouth small, lower jaw slightly included; teeth on vomer and palate; narrow bands of teeth in jaws, the outer series broadly united, wholly free from isthmus. Dorsal of spines only. Anal with 2 spines. Ventral with 1 spine and 3 well-developed rays. Lower pectoral rays longest, as in *Leptoclinus*. North Pacific. (πλεκτός, enfolded; βράγχος, gill.)

2787. PLECTOBRANCHUS EVIDES, Gilbert.

Head rather long, 4¾ in length, extending well beyond origin of dorsal fin; depth about 11. D. LVI; A. II, 34. Body very slender, the depth nearly constant throughout. Caudal peduncle without free portion, its depth 2⅓ in that of body. Upper jaw with a broad inner band of minute teeth in front and on the sides, the outer series enlarged, 2 in the front of the jaw distinctly canine-like; teeth in the lower jaw similar to the outer series above, in a single series laterally, forming a patch in front of jaw, where 2 of them are much enlarged canines, the largest teeth in the jaws; vomer and palatines with bands of small but very evident teeth. Eyes large, close together, the interorbital space ⅓ pupil. Orbit 3⅓ in head, longer than snout. Posterior nostril with a short flap, the tube obsolete. Mouth large, somewhat oblique, maxillary reaching middle of orbit, 2⅓ in head. Top of head with very large pores, a series running backward from each eye, the two joined by a cross series on occiput. Body covered with very small cycloid scales, including belly, nape, breast, and cheeks, those on breast and cheeks not imbricated; lateral line without visible pores. Spinous dorsal beginning well forward, the distance from its origin to nape less than from latter to posterior margin of orbit. Anterior spines short, but fully united by membrane, the longest spine 3½ in head; membrane of last spine reaching base of upper caudal rays; origin of anal very slightly in advance of middle of body; anal with 2 short, sharp spines, the rays longer, their terminal ⅓ free from membrane; last anal ray connected with base of lower caudal ray; ventrals well developed, nearly ⅓ head; pectorals with lowermost rays abruptly lengthened, ⅔ head; caudal short, rounded, little more than ⅓ head. Color dusky olive above, lighter below; sides crossed by about 25 narrow white bars, narrower than interspaces; 3 equidistant dark blotches near back, each double, the two halves occupying contiguous interspaces between white bars; branchiostegal membrane black; head without markings; pectorals white at base, the distal half black, margined with white; ventrals white; dorsal with alternating oblique bars of white and blackish, 2 jet-black roundish spots on its posterior portion; caudal whitish at base, then dusky, margined with white, its upper ray jet-black; anal light at base, becoming black at edge of membrane, the free tips of rays white. Coast of Oregon. A single specimen, 4 inches long, from *Albatross* Station 3064, in 46 fathoms. (εὐειδής, comely.)

Plectobranchus evides, GILBERT, Proc. U. S. Nat. Mus. 1890, 102, coast of Oregon, at Albatross Station 3064. (Coll. *Albatross*.)

916. LEPTOCLINUS, Gill.

Ctenodon, NILSSON, Skandinav. Faun., IV, 190, 1853 (*maculatus*) (name three times* preoccupied).
Leptoclinus, GILL, Proc. Ac. Nat. Sci. Phila. 1864, 209 (*aculeatus*).

---

* *Ctenodon*, Wagler, 1830, a lizard; Ehrenberg, 1838, a rotifer, and Swainson, 1839, a fish.

Body much elongated; lateral line obsolete; teeth on jaws, vomer, and palatines; pectoral fins with the upper rays shortened; caudal fin subtruncate. Arctic seas. This genus is close to *Lumpenus*, differing mainly in the form of the pectoral. (λεπτός, slender; *Clinus*.)

2788. LEPTOCLINUS MACULATUS (Fries).

(LANGBARN.)

Head 5; depth 8. D. LX (LVIII to LX); A. 36 (35 to 38). Eye large, 3¼ in head; snout short and blunt, 4⅘ in head, maxillary reaching past middle of eye, 2⅕ in head. Teeth in jaws, vomer, and palatines; jaws each having 2 strong canines in front. Scales small, cycloid. First 3 or 4 dorsal spines short and free; longest dorsal spines as long as eye; caudal fin free from dorsal and anal; ventrals 3 in head; pectorals rather large, 1¼ in head. Color yellowish, irregularly marked with dark spots, a series of about 6 of these spots extending along sides close to base of dorsal fin; a series of smaller spots extending along center of sides from upper base of pectoral to caudal; dorsal irregularly covered with dark spots; caudal with 4 dark cross bands; anal, ventral, and pectorals plain yellowish. Bering Sea to Spitzbergen, south to Aleutian Islands and the coasts of Sweden and Norway. This description is taken from a specimen, 5¼ inches long, from Alaska, near Unimak Pass (*Albatross* Station 3309). A few young individuals of this species, hitherto known only from the North Atlantic, were taken in Unimak Pass and in Bristol Bay, in 29½ to 70 fathoms. Three small specimens were also taken off Robben Reef, near the Kamchatka coast, in 28 fathoms, and one off Karluk, Kadiak Island. Having no Atlantic specimens of this species, we are unable to satisfy ourselves of the identity of the two, but no difference is evident from descriptions. The lateral line is much more distinct than in our specimens of *Lumpenus medius*, in which it can be made out with difficulty on scattered scales along middle of sides. (Eu.) (*maculatus*, spotted.)

Clinus maculatus, FRIES, Kgl. Vet. Ak. Handl. 49, 1837, Bohüslän, Sweden.
Lumpenus aculeatus, REINHARDT, Kong. Dansk. Vid. Selsk., VI, 1837, 190, no description.
Clinus aculeatus, REINHARDT, Dansk. Vidensk. Selsk. Natur. Af h., VII, 1838, 114, 122, 194, Spitzbergen.
Ctenodon maculatus, NILSSON, Skand. Fauna, IV, 190, 1853.
Stichæus maculatus, GÜNTHER, Cat., III, 281, 1861.
Lumpenus aculeatus, KRÖYER, Naturhist. Tidsskr., I, 377, 1862.
Stichæus aculeatus, GÜNTHER, Cat., III, 282, 1861; COLLETT, Norske Nord-Havs Exp., 67, 1880.
Lumpenus maculatus, JORDAN & GILBERT, Synopsis, 777, 1883; Lilljeborg, Sveriges Och Norges Fish., 500, 1891.
Leptoclinus maculatus, GILBERT, Rept. U. S. Fish Comm. 1893, 450.

917. POROCLINUS, Bean.

Poroclinus, BEAN, Proc. U. S. Nat. Mus. 1890, 40 (*rothrocki*).

Body elongate, moderately compressed, covered with small scales; lateral line obsolete. Head moderately long; snout short; eyes large; interorbital space narrow. Mouth small, lower jaw slightly included; teeth on vomer and palate; narrow bands of teeth in jaws, the outer series enlarged. Gill openings slightly prolonged forward below, narrowly attached to the isthmus anteriorly. Dorsal composed of many sharp, flexible spines, diminished in length anteriorly. Caudal long, pointed; anal with 3 spines and many rays; pectorals large, the middle rays longest; ventrals jugular, with 1 spine and 3 rays. Intestine short; pyloric cæca 1 or 2; no air bladder. Northern Seas. ($\pi \acute{o} \rho o \varsigma$, pore; *Clinus.*)

2789. POROCLINUS ROTHROCKI, Bean.

Head 6¼ in length; depth at nape 12. D. LVII to LX; A. III, 40 to 42. Body tapering uniformly backward. Vent placed anteriorly, its distance from snout 1⅜ to 1¾ in its distance from base of caudal. Snout compressed, slightly projecting, the lower jaw included; maxillary reaching vertical from front of pupil, 3¼ to 3½ in head. Teeth acute, in narrow bands in the jaws, a single well-marked series on vomer, and a patch on front of palatines, those on vomer and palatines fully as large as those on jaws, and equally developed in young and adults. Eyes large, close together, the interorbital space convex, its width about ½ pupil. Diameter of orbit equaling length of maxillary, about 3½ in head. Nostril tubes well developed, ½ diameter of pupil. Gill openings narrower than in other described members of this group, extending forward below the vertical from posterior part of cheek, where they are firmly joined to isthmus, across which they do not form a fold. Gill rakers obsolete. Dorsal beginning over end of opercular flap, its distance from nape equaling distance of latter from posterior margin of pupil; membrane of last spine slightly joined to base of caudal; anterior dorsal spines short, but well connected by membrane; anal with 3 distinct spines, shorter than the rays that follow, the second the longest, all as strong as dorsal spines, and fully connected by membrane, rays all branched at tip, membrane of last ray joined only slightly to base of caudal; caudal sharply pointed in all our specimens, the median rays longest, about as long as head; pectorals evenly rounded, the median rays longest, 14 or 15 in number, all branched; ventrals well developed, about ⅔ as long as head, consisting of 1 short, sharp spine and 3 rays, the spine not closely joined to rays. Lateral line indistinct, usually appearing obsolete, more evident toward head, consisting of a series of distant pores along median line; scales very small, cycloid, imbricated, covering body, inclosing abdomen, breast, and nape; cheeks scaled, the head otherwise naked, or sometimes with a small patch of scales on upper part of opercles. Color, sides with a series of 10 to 12 narrow white cross bars, the first in front of dorsal fin, the last under last dorsal spine, the bars about ⅓ interspaces; above lateral line scales conspicuously margined with darker, below lateral line they broaden out and become forked; upper caudal rays at base with an oval white ring inclosing a darker area, this mark more conspicuous in the young; belly and ventrals white, other fins dusky, but without definite markings. (Gilbert.) Bering Sea. Known from 2 specimens; the type, 7 inches long, was taken August 4, 1888, at *Albatross* Station 2852, north latitude 55° 15′, west longitude 159° 37′, at a depth of 58 fathoms, between Nagai and Big Koniushi Islands. The specimen here described from Unalaska. (Named for Dr. J. T. Rothrock, professor of botany, University of Pennsylvania.)

Poroclinus rothrocki, BEAN, Proc. U. S. Nat. Mus. 1890, 40, 55° 15′ N., 159° 37′ W., between Nagai Island and Koniushi Islands. (Coll. *Albatross*.)

918. LUMPENUS, Reinhardt.

(SNAKE BLENNIES.)

Lumpenus, REINHARDT, Dansk. Vidensk. Selsk. Natur., VI, 1837, 110](*lumpenus=fabricii*).
Leptogunnellus, AYRES, Proc. Cal. Ac. Nat. Sci., I, 1854, 26 (*gracilis*).
Centroblennius, GILL, Proc. Ac. Nat. Sci. Phila. 1864, 209 (*nubilus*).
Leptoblennius, GILL, Proc. Ac. Nat. Sci. Phila. 1864, 209 (*serpentinus*).
Anisarchus, GILL, Proc. Ac. Nat. Sci. Phila. 1864, 209 (*medius*).

Body greatly elongate, moderately compressed, covered with small scales; lateral line indistinct or obsolete. Head long; snout short; no cirri; eyes large, placed high; mouth moderate, with a single row of rather small conical teeth on each jaw; palatine teeth present or absent; gill openings prolonged forward below, very narrowly united anteriorly to the isthmus, not forming a free fold across it. Dorsal composed of numerous, sharp, flexible, rather high, spines; caudal fin long; anal many-rayed; pectorals large, more than ½ length of head, the middle rays longest; ventrals well developed, jugular, I, 3 or I, 4; intestinal canal long; pyloric cæca present; no air bladder. Chiefly herbivorous. Northern seas. (*Lumpen*, a Danish name of *Zoarces viviparus*, with which these fishes were at first confounded.)

ANISARCHUS (ἄνισος, unequal; ἄρχος, for anal):
  a. Anal fin very low in front, the rays gradually lengthened; dorsal spines 61; anal rays 42.   MEDIUS, 2790.
  aa. Anal fin not much lower in front than behind.
      LUMPENUS:
        b. Teeth on palatines more or less developed, at least in the adult; anal rays 40 to 46; dorsal spines 63 to 71.
          c. Dorsal spines 69 to 71.
              d. Anal rays 46; dorsal separate from caudal.   ANGUILLARIS, 2791.
              dd. Anal rays 41; dorsal slightly joined to caudal.   MACKAYI, 2792.
          cc. Dorsal spines about 63; anal rays 43.   FABRICII, 2793.
      LEPTOBLENNIUS (λεπτός, slender; *Blennius*).
        bb. Teeth on palatines wanting; dorsal spines 72 to 75; anal rays about 50.
                                                              LAMPETRÆFORMIS, 2794.

Subgenus ANISARCHUS, Gill.

2790. LUMPENUS MEDIUS (Reinhardt).

Head 6; depth 10. D. LXI; A. 42; V. I, 3. Lower jaw scarcely included, the maxillary reaching front of eye; teeth on palatines, none on vomer; ventrals slender, ⅓ length of head; lower rays of pectoral shorter than middle ones, the fin shorter than head. Dorsal and anal slightly joined to the truncate caudal; anterior half of anal with the rays shortened. (Collett.) Yellowish, nearly plain. Greenland to Norway and Spitzbergen and westward to Bering Sea and Kamchatka. Specimens from the coast of Kamchatka are not evidently different from the current figures and descriptions of Atlantic specimens. (Eu.) (*medius*, middle.)

Clinus medius, REINHARDT, Dansk. Vidensk. Afh:, VII, 1838, 194, Greenland.
Lumpenus medius, KRÖYER, Naturh. Tidsskr., I, 377, 1837; JORDAN & GILBERT, Synopsis, 777, 1883.
Stichæus medius, GÜNTHER, Cat., III, 281, 1861.
Anisarchus medius, GILL, Proc. Ac. Nat. Sci. Phila. 1864, 210.
Lumpenus medius, COLLETT, Norske Nord-Havs Exp., 62, 1880; JORDAN & GILBERT, Rept. Fur Seal Invest., 1898.

Subgenus LUMPENUS.

2791. LUMPENUS ANGUILLARIS (Pallas).

Head 8; depth 14. D. LXXI; A. 46 (45 to 50); V. I, 4; B. 7. Cheeks scaly; mouth somewhat oblique, the lower jaw included; maxillary reaching front of pupil; teeth on palatines, none on the vomer; a single series of rather long, conical, and not very closely-set teeth in each jaw. Gill openings prolonged forward a distance greater than length of snout; pyloric cæca 4, unequal. Fins all comparatively high, pectorals ⅜ length of head, the middle rays longest; ventrals ⅓ length of head; dorsal and anal distinct from the pointed caudal, which is nearly as long as head. Olive green above, pale below; sides marked above with dark olive brown; a series of more or less distinct oblong blotches of olive brown along middle of sides; dorsal barred or spotted; anal pale; opercle with a dark blotch; head dusky above. Length 18 inches. San Francisco to Alaska; very abundant northward to Sitka and Unalaska; originally recorded from Kamchatka. (*anguillaris*, eel-like.)

Blennius anguillaris, PALLAS, Zoogr. Rosso-Asiat., II, 176, 1811, Kamchatha and Aleutian Islands. (Coll. Billings and Merk.)
Septogunnellus gracilis, AYRES, Proc. Cal. Ac. Nat. Sci., I, 1855, 26, San Francisco.
Gunnellus anguillaris, CUVIER & VALENCIENNES, Hist. Nat. Poiss., XI, 434, 1836.
Lumpenus anguillaris, GIRARD, Pac. R. R. Surv., X, Fishes, 123, pl. 25b, figs. 1 to 3, 1858; STORER, Synopsis, 121, 1846; JORDAN & GILBERT, Synopsis, 777, 1883; JORDAN & STARKS, Fishes Puget Sound, 848, 1895.
Stichæus anguillaris, GÜNTHER, Cat., III, 282, 1861.

2792. LUMPENUS MACKAYI (Gilbert).

Head 6⅔; depth 13 or 14; eye 8 in head; snout 4. D. LXIX; A. II, 41. Very elongate. Head compressed and high, especially anteriorly, the upper profile of snout very convex, the upper jaw decidedly longer than the lower. Mouth nearly horizontal. Maxillary reaching vertical from front or middle of pupil, its length 3⅔ to 3¾ in head. Teeth small, in a narrow band in jaws; a single series of weak teeth on palatines; vomer toothless. Gill openings continued forward to below middle of cheeks, the membranes then narrowly joined to isthmus; gill rakers short and weak, about 10 on horizontal limb of arch. Eye small, its horizontal diameter ¼ longer than its vertical, slightly longer than interorbital width. Distance from snout to nape equaling length of postorbital part of head. Opercles large, continued to beyond base of pectorals. Dorsal beginning immediately above upper end of gill slit, the spines short, strong, and pungent, not flexible; some of the anterior spines short, but not free, the fin increasing in height to opposite front of anal, the longest spine equaling length of snout, the membrane of last spine joining base of upper rays of caudal; anal with 2 strong spines similar to those of the dorsal fin, the second twice length of first and $\frac{4}{5}$ that of highest dorsal spines; anal rays all forked, the posterior longest, equaling length of snout and eye, free from the caudal; caudal fin rounded in younger specimens, lanceolate in adults, becoming in the latter $\frac{3}{4}$ as long as head; ventrals short, of 1 short spine and 3 simple rays, the fin $\frac{1}{4}$ length of head; pectorals large, the middle rays longest, $\frac{2}{3}$ length of head. Scales small, smooth, elongate, imperfectly imbricated, partially embedded or altogether wanting on anterior part of back; cheeks scaled, head otherwise naked; faint traces of a lateral line sometimes visible on middle of sides anteriorly. Color in spirits, light olivaceous (light yellowish in life); a continuous jet-black streak from occiput along each side of dorsal to base of caudal, with 2 interrupted black streaks below it, the lowermost running on middle of side; top and sides of head darker, variously marked with anastomosing black lines and spots; opercles blackish; dorsal and caudal fins dusky translucent, without distinctive markings; anal and ventrals white; pectorals white or dusky; roof of mouth black; peritoneum black dorsally, white ventrally. Bering Sea. Several specimens were seined near the mouth of the Nushagak River, Alaska. (Gilbert.) (Named for Charles Lesley McKay, of Appleton, Wisconsin, a very able young ichthyologist, who was drowned at Nushagak, in Bristol Bay, in 1883.)

Lumpenus mackayi, GILBERT, Rept. U. S. Fish Comm. 1893 (1896), 450, pl. 32, mouth of Nushagak River, Bristol Bay. (Coll. Gilbert.)

2793. LUMPENUS FABRICII (Cuvier & Valenciennes).

Head 8 or 9; depth 11 to 15. D. LXIII to LXV; A. 41 to 43; V. I, 3; P. 15. Upper jaw scarcely longer than lower; teeth on palatines few and small, often really or apparently wanting, especially in the young; maxillary not reaching eye; vertical fins distinct; pectorals large, ovate. Color light brown, with large pale rounded blotches separated by brown shades; head yellowish; pectorals yellowish mottled, with a dusky spot at base. Arctic seas; recorded from Spitzbergen, Greenland, the Gulf of St. Lawrence, Wellington Sound, Bristol Bay, and other localities in Bering Sea (Petropaulski and Plover Bay, as *L. anguillaris*). We have specimens from Bristol Bay, Disco, Upernavik, and the Gulf of St. Lawrence. These are apparently identical, and they show clearly the identity of *L. nubilus* with *L. fabricii*.

The following notes are from specimens taken in Bristol Bay, in $4\frac{1}{4}$ to 14 fathoms: These specimens seem to agree in structural details with specimens of *Lumpenus fabricii* from the North Atlantic. The Pacific specimens are lighter in color, with the dusky mottlings confined to the dorsal region, and with a very distinct series of oblong brown blotches along lateral line, alternating with a lower series of small, faint, round spots.

Under parts immaculate; the mottlings along base of dorsal frequently uniting to form a series of oblong blotches alternating with those of lateral line; other specimens show no traces of dorsal blotches; dorsal fin translucent, faintly mottled with darker; caudal with brownish cross bars; pectoral with a round dusky shade at base; fins otherwise unmarked. Mandible with a single series of conical teeth, which widens at symphysis into an irregular double series or narrow patch; a similar series of conical teeth in premaxillaries, within which is a band of fine villiform teeth. A number of small specimens from Disco, Greenland, are entirely similar except for the darker coloration. This species is near *L. anguillaris*, but the latter has a larger mouth, larger teeth, and more numerous fin rays. (Named for Otho Fabricius, the first student of the fishes of Greenland.)

Blennius lumpenus, FABRICIUS, Fauna Grönlandica, 151, 1780, Greenland; not *Blennius lumpenus*, LINNÆUS, which is a species of *Gaidropsarus*, with 2 barbels at the chin.
Gunnellus fabricii, CUVIER & VALENCIENNES, Hist. Nat. Poiss., XI, 431, 1836, Greenland; after FABRICIUS; KRÖYER, Naturhist. Tidsskr., I, 377, 1837; GAIMARD, Voy. Scand., Zool., Poiss., pl. 14, fig. 1.
Lumpenus nubilus, RICHARDSON, Last Arctic Voyage, Fishes, 13, pl. 28, 1855, Wellington Sound. (Coll. Edward Belcher.)
Blennius (Clinus) lumpenus, RICHARDSON, Fauna Bor.-Amer., 90, 1836.
Clinus lumpenus, REINHARDT, Dans. Vidensk. Selsk. Nat. Afh., VII, 194, 1838.
Stichœus lumpenus, GÜNTHER, Cat., III, 280, 1861.
Stichœus nubilus, GÜNTHER, Cat., III, 564, 1861.
Centroblennius nubilus, GILL, Proc. Ac. Nat. Sci. Phila. 1864, 209.
Lumpenus fabricii, JORDAN & GILBERT, Synopsis, 778, 1883.
Leptoblennius nubilus, JORDAN & GILBERT, Synopsis, 778, 1883; GILBERT, Rept. U. S. Fish Comm. 1893 (1896), 451.

2794. LUMPENUS LAMPETRÆFORMIS (Walbaum).

(SNAKE BLENNY; TANGBROSME.)

Head 9; depth about 15. D. LXXIII (LXVIII to LXXIV); A. 50 (49 to 52); V. I, 3. Body elongate, head slender; lower jaw little shorter than upper; maxillary reaching front of eye. Vent well forward, near end of first third of body; pectoral convex, somewhat shorter than head; first 3 or 4 rays of dorsal short, little connected; caudal acuminate, free from dorsal and anal. Yellowish or greenish, with numerous (about 20) faint brown blotches of different sizes, some of them confluent and extending obliquely upward on dorsal; caudal with transverse dark shades. (Collett.) North Atlantic and Arctic on both shores, south to Sweden and Norway, east to Spitzbergen; rare south to Cape Cod, if *L. serpentinus* is the same. We can find no difference on a comparison of our notes with published figures and descriptions, except that Storer describes *serpentinus* as having the caudal plain yellowish. (Eu.) (*Lampetra*, lamprey; *forma*, form.)

Blennius capiti lœvi, etc., MOHR, Hist. Nat. Islandiæ, 85, taf. 4, 1786, Iceland; D. 72; A. 54.
Blennius lampetræformis, WALBAUM, Artedi Piscium, III, 184, 1792, Iceland; after MOHR.
Centronotus islandicus, BLOCH & SCHNEIDER, Syst. Ichth., 157, 1801, Iceland; after MOHR.
Clinus nebulosus, FRIES, Vet. Akad. Handl., 55, 1837, Bohuslän, Sweden.
Clinus mohri, KRÖYER, Naturh. Tidsskr., 1 R, 1837, 32, Iceland.
Blennius gracilis, STUVITZ, Nye Mag., Naturvid., I, 406, 1838, west coast of Norway.

Blennius serpentinus, STORER, Proc. Bost. Soc. Nat. Hist., III, 1848, 30, Massachusetts Bay, from the stomach of a codfish (Coll. Capt. Nathaniel E. Atwood); STORER, Hist. Fish. Mass. 169, pl. 18, f. 1, 1867.

Gunnellus islandicus, CUVIER & VALENCIENNES, Hist. Nat. Poiss., XI, 433, 1836.

Stichæus islandicus, GÜNTHER, Cat., III, 281, 1861.

Lumpenus lampetræformis, COLLETT, Norske Nord-Havs Exp., 71, 1880; JORDAN & GILBERT, Synopsis, 778, 1883.

Leptoblenninus serpentinus, JORDAN & GILBERT, Synopsis, 778, 1883.

919. STICHÆUS, Reinhardt.

Stichæus, REINHARDT, Dansk. Vidensk. Natur. og Math. Afhandl. 1837, 109 (*punctatus*).

Notogrammus, BEAN, Proc. U. S. Nat. Mus., IV, 1881, 147 (*rothrocki*); young.

Body moderately elongate, covered with small scales; teeth on jaws, vomer, and palatines. Lateral line present, single, running along side of back; pectorals and ventrals well developed. Dorsal moderately high, of spines only; gill openings continued forward below, the membranes scarcely united to the isthmus; pyloric cæca present. Arctic seas. (*στιχάω*, to set in rows.)

2795. STICHÆUS PUNCTATUS (Fabricius).

Head 4½; depth about 7. D. XLVIII or XLVIX; A. 32 to 35; eye twice interorbital width, 4½ in head; snout subconical, 4 in head. Maxillary about equal to snout, 3⅘ in head, reaching slightly beyond front of eye. Narrow bands of teeth in the jaws and present on vomer and palatines, the outer series in the upper jaw and the inner series in lower jaw enlarged. Scales small, cycloid; head and cheeks scaleless; longest dorsal spines slightly longer than snout. The membrane from last dorsal spine joining extreme base of upper caudal ray; anal wholly distinct; pectorals rather long, reaching vent, 1⅓ in head; ventrals 2½ in head. Numerous large pores scattered over top and sides of head. Lateral line rather close to back, running along the upper fourth of height of body and ending abruptly at about ⅔ the length of body. Color bright scarlet, the head marked below with 5 or 6 brown reticulations; a brown streak from snout through eye; fins irregularly marked by dark bars or spots; a narrow row of 5 large round black spots, each with a white band near its posterior margin, occurring at regular instances along dorsal fin; a row of about 8 large dark spots on anal. Arctic seas, from Greenland to northern Siberia, south to Bristol Bay and Newfoundland. Our description (from Dr. Gilbert) taken from a specimen, about 5 inches in length, from Karta Bay, Alaska. It agrees very closely with the account by Ensign H. G. Dresel, of 2 examples from Godhavn, Disco Island, Greenland. The Alaska species must be the same as the other. Dresel finds the depth 7⅘ in length. Dr. Gilbert further observes: A single specimen, 86 mm. long, was dredged in Bristol Bay, Alaska, Station 3239, depth 11½ fathoms. Several larger individuals were seined in Karta Bay, Prince of Wales Island, Alaska, July 12, 1889. The position of the lateral line in this species is incorrectly given as "median" by Jordan & Gilbert in the Synopsis, pp. 755 and 775. Cuvier and Valenciennes, in their description, drawn from the writings of Fabricius, state

that the lateral line runs along the upper fifth of the height of the body
and terminates at about the middle of the length. This correctly describes
its position in all our specimens, where it originates immediately above
the opercle, exhibiting at first rather a strong upward convex curve, then
running nearly parallel to the back, separated from base of dorsal fin by ⅓
height of body. It is very distinct throughout its course, and terminates
at about the middle of the length. The narrow brown streak bounding
the lateral line above, in *Notogrammus rothrocki*, is conspicuous in our
smallest specimen (86 mm.). Branchiostegal membranes very narrowly
joined anteriorly, forming a narrow free fold across the isthmus, from
which they are entirely distinct. Narrow bands of teeth in the jaws, and
distinctly present on vomer and palatines; the outer series in upper jaw
and the inner series in the lower jaw enlarged. D. XLVII or XLVIII;
A. I, 32 to 35. The membranes from last dorsal spine join extreme base
of upper caudal ray; anal wholly distinct. We have not the material for
a comparison of Pacific with Atlantic representatives of this species, and
the published descriptions of the latter lack detail. (*punctatus*, spotted.)

Blennius punctatus, FABRICIUS, Fauna Grönlandica, 153, 1780, Greenland; REINHARDT,
  Naturhist. Selsk. Skrift., II, pt. 2, pl. 10, fig. 3.
Notogrammus rothrocki, BEAN, Proc. U. S. Nat. Mus., IV, 1881, 146, Plover Bay and Cape
  Lisburne, Siberia; young. (Types, Nos. 27565, 27580, and 27573. Coll. Dr. Bean.)
Clinus punctatus, RICHARDSON, Fauna Bor.-Amer., III, 88, 1836.
Gunnellus punctatus, CUVIER & VALENCIENNES, Hist. Nat. Poiss., XI, 428, 1836.
Stichæus punctatus, KRÖYER, Naturhist. Tidsskr., I, 377, 1837; GAIMARD, Voy. en Scand. et
  Lapon., Zool., Poiss., pl. 20, fig. 2; GÜNTHER, Cat., III, 283, 1861; JORDAN & GILBERT,
  Synopsis, 775, 1883; DRESEL, Proc. U. S. Nat. Mus. 1884, 249; GILBERT, Rept. U. S.
  Fish Comm. 1893 (1896), 450.

920. ULVARIA, Jordan & Evermann.

Ulvaria, JORDAN & EVERMANN, Check-List Fishes N. and M. A., 475, 1896 (*subbifurcatus*).

This genus is very close to *Eumesogrammus*, from which it differs in the
absence of the lowermost or third lateral line, the median line being bifurcate. (*Ulva*, the sea lettuce, in which many Blennioid fishes live.)

2796. ULVARIA SUBBIFURCATA (Storer).

Head 4⅓; depth nearly 5. D. XLIV; A. 30. Mouth rather large; maxillary reaching to below orbit; back somewhat arched; ventral outline
nearly straight; eyes large; lateral lines 2 (the lowermost lateral line
wanting); median lateral line forked; upper branch of median lateral
line about ⅔ length of the head. Brownish, with several round paler
blotches above at the base of the dorsal fin; spaces between these blotches
darker, appearing like bars; a broad black bar crossing the opercle
obliquely from below the orbit, and 2 parallel dark bars running backwards from orbit; belly yellowish white; dorsal fin with numerous black
dots. North Atlantic, south to Cape Cod; very rare. (*subbifurcatus*, somewhat forked.)

Pholis subbifurcatus, STORER, Rep. Fish. Mass., 63, 1839, Nahant, Mass. (Coll. Dr. Thos.
  M. Brewer); DE KAY, N. Y. Fauna: Fishes, 150, 1842; STORER, Hist. Fish. Mass., 258,
  1867.
Eumesogrammus subbifurcatus, JORDAN & GILBERT, Synopsis, 775, 1883.

921. EUMESOGRAMMUS,* Gill.

Eumesogrammus, GILL, Proc. Ac. Nat. Sci. Phila. 1864, 210 (*præcisus*).

Body comparatively short, the back somewhat arched; mouth rather large, the jaws with villiform teeth; teeth on vomer and palatines. Scales small; lateral lines 3, without accessory branches; pectorals and ventrals well developed. Dorsal moderately high, of spines only, free or slightly connected with the rounded caudal; gill openings continued forward below, the membranes narrowly joined to the isthmus; pyloric cæca present. Northern seas. ($\varepsilon\dot{v}$ well; $\mu\dot{\varepsilon}\dot{\sigma}o\varsigma$, middle; $\gamma\rho\alpha\mu\mu\dot{\eta}$, line; the longest lateral line being the middle one.)

2797. EUMESOGRAMMUS PRÆCISUS (Kröyer).

Head 4; depth nearly 6. D. XLIX; A. 34; V. 3. Snout subconical; cleft of mouth slightly oblique; vomerine and palatine teeth present; 3 lateral lines on each side, the median one continued to the base of the caudal; ventral fin $\frac{1}{4}$ as long as the pectoral, which is much shorter than head; dorsal fin terminating just at root of caudal. An ovate, black, white-edged spot between the sixth and tenth dorsal spines. Coasts of Green-land. (Günther.) (*præcisus*, exact.)

Clinus præcisus, KRÖYER, Naturh. Tidsskr., I, 25, August, 1836,† Greenland.
Clinus unimaculatus, REINHARDT, Dansk. Vidensk. Selsk., VII, 114, Feb., 1837, Greenland.
Stichæus unimaculatus, REINHARDT, Dansk. Vidensk., 109, 1837; GÜNTHER, Cat., 283, 1861.
Eumesogrammus præcisus, JORDAN & GILBERT, Synopsis, 774, 1883.

* The 2 following species from the Okhotsk Sea seem to represent 2 new genera (*Ernogrammus* and *Ozorthe*) closely related to *Eumesogrammus*:

ERNOGRAMMUS ENNEAGRAMMUS (Kner).

Head 3¾; depth 6⅜. D. XLI; A. 33 or 34; P. 14 or 15. Eye 4 in head, as long as snout. Mouth large, nearly horizontal, the maxillary reaching middle of eye; lower jaw projecting; profile of snout nearly horizontal; fine pointed teeth in bands on jaws and across the vomer. Head naked; dorsal of high, slender spines; caudal separate, rounded; anal high; pectoral long, 1⅓ in head; ventrals ⅓ as long as pectorals; scales very small, smooth; lateral lines each with short oblique branches, each ending in a wide pore; 1 lateral line along base of dorsal from head to caudal, 1 along middle of side, 1 along base of anal to caudal, this forking at the vent and sending 2 parallel branches forward to the breast; brownish, 2 rows of small dark spots along middle lateral line; dorsal and anal with dark spots and a broad dark margin; pectorals with 3 black cross bands; a dark bar at base of caudal; 3 black bars from eye. Okhotsk Sea. Known from a specimen, 1¾ inches long, from Decastris Bay. (Kner.) ($\dot{\varepsilon}vv\dot{\varepsilon}\alpha$, nine; $\gamma\rho\alpha\mu\mu\dot{\eta}$, line.) *Ernogrammus*, new genus ($\dot{\varepsilon}\rho vo\varsigma$, branch), is distinguished from *Eumesogrammus* by the branching lateral line.
Stichæus enneagrammus, KNER, Sitzber. Akad. Wiss. Wien 1868, 16, taf. VI, f. 19 Decastris Bay. (No. 1401c Mus. Wien.)

OZORTHE HEXAGRAMMA (Schlegel).

Head 5¼; depth 5¼. D. XLIII; A. 24. Snout conical; mouth little oblique; the maxillary reaching front of eye; bands of fine teeth on vomer and palatines; a few large canine-like teeth in front; eye 5 in head; dorsal spines all stiff, the middle ones longest; dorsal joined to caudal by membrane; lateral lines 3, the upper partly interrupted, sending at right angles upward and downward lines which join the middle line; third lateral line along base of anal only. Scales small. Dorsal with large dark brown spots obliquely placed; 3 brown stripes across cheek; anal colored like dorsal; caudal pectoral and ventrals each with 3 dark cross bands. Northern coast of Japan to Okhotsk Sea. This description (after Kner) from a specimen from Decastris Bay (No. 5575, Mus. Wien.). This differs somewhat from the type of the species and may be different. ($\dot{\varepsilon}\xi$, six; $\gamma\rho\alpha\mu\mu\dot{\eta}$, line.) The new genus *Ozorthe* ($\dot{o}\gamma o\varsigma$, branch; $\dot{o}\rho\theta\eta$, right angle) is distinguished by the form of its lateral lines as above described.
Stichæus hexagrammus, SCHLEGEL, Fauna Japonica, Pisces, 136. pl. 3, f. 1., 1850. Bay of Simabara, Japan. Head 4½; depth 6½. D. XL; A. 29. GÜNTHER, Cat., III, 284, 1861; KNER, Sitzber. Akad. Wiss. Wien 1868, 45.

† These dates are thus given by Kröyer, as quoted by Dr. Gill, Proc. Ac. Nat. Sci. Phila. 1864, 210. We have been unable to verify them.

Family CCI. CRYPTACANTHODIDÆ.

(THE WRY-MOUTHS.)

Body very long and slender, compressed, naked or covered with small, cycloid scales; lateral line obsolete or composed of open pores without tubes; head oblong, cuboid, with vertical cheeks; conspicuous muciferous channels in mandible and preopercle; head flattish above, with deep rounded pits between and behind eyes; mouth large, very oblique; lower jaw very heavy, its tip projecting; premaxillary not protractile; jaws with rather sharp, conical teeth; larger teeth on the vomer and sometimes on palatines. Gill membranes joined to the isthmus, the gill openings prolonged forward below. Pyloric cæca 5. Pseudobranchiæ small. Dorsal fin long, composed entirely of spines, which are rather strong, but enveloped in the skin; dorsal and anal joined to the caudal; no ventral fins; pectorals short. Blennies of large size, of the Northern shores of America. Three species known, forming 3 genera. (*Blenniidæ*, genus *Cryptacanthodes*, Günther, Cat., III, 291, 1861.)

a. Body scaly; lateral line present, composed of open pores; isthmus narrow; teeth on palatines.　　　　　　　　　　　　　　　　　DELOLEPIS, 922.
　aa. Body naked; lateral line obsolete.
　　c. Palatines with teeth; isthmus narrow.　　　　　CRYPTACANTHODES, 923.
　　cc. Palatines toothless; isthmus rather broad.　　　　LYCONECTES, 924.

922. DELOLEPIS, Bean.

Delolepis, BEAN, Proc. U. S. Nat. Mus. 1882, 465 (*virgatus*).

Body anguilliform, moderately compressed posteriorly, covered with small, imbricated, cycloid scales; vent nearly median; a small anal papilla; lateral line continuous, straight, nearly median, composed of open pores, without prominent tubes. Head oblong, subquadrangular, naked, the muciferous channels well developed, the vertex shallow concave; snout short, obtuse; nostril tubular, close behind premaxillary; eyes small, high, separated by an interspace of moderate width, surrounded by a series of shallow pits; mouth wide, oblique, terminal, the lower jaw projecting beyond the upper; lips fleshy; premaxillaries slightly protractile, with 2 rows of small conical teeth; a few larger teeth at the symphysis; vomer and palatines with a few rather large teeth; tongue smooth, adherent; mandible with a few shallow pits, the series continued on the posterior border of preopercle; opercles unarmed. Gill membranes attached to a narrow isthmus; gill rakers very short; pseudobranchiæ present. Branchiostegals 6. Pectorals short, placed low, their bases vertical; ventrals none; dorsal beginning above gill opening, composed entirely of spines; anal with 2 spines and many split rays; dorsal and anal continuous with the caudal, which is rather long and pointed. Intestine short, with a few pyloric cæca. (δῆλος, visible; λεπίς, scale.)

2798. DELOPLEPIS VIRGATUS, Bean.

Head 6; depth 10. D. LXXVI; A. II, 46; P. 13; cæca 6. Width of head equal to greatest depth of body; interorbital area equal to snout, or ⅓ length of mandible; maxillary reaching a little behind eye, its length 3 in distance from snout to front of dorsal; eye 2 in snout, 11 in head. Beginning at a short distance behind origin of dorsal, small, oblong, cycloid scales, closely imbricated, cover a strip of the body along the lateral line; the scaled area gradually widens backward until, behind the vent, only a very narrow strip along bases of dorsal and anal is naked. Dorsal beginning over upper angle of gill opening; first spine ½ as long as the seventy-first or longest; caudal 11 in length; pectoral 3 in head. Brownish yellow; a brown stripe along lateral line, another along back, a third along base of anal. Length 30 inches. Coast of southern Alaska to Puget Sound; not rare about Seattle. (*virgatus*, striped.)

Delolepis virgatus, BEAN, Proc. U. S. Nat. Mus. 1881, 466, Kingcombe Inlet, British Columbia; Port Wrangel, Alaska (Coll. Capt. H. E. Nichols. Types, Nos. 29149 and 29150, U. S. Nat. Mus.); JORDAN & STARKS, Fishes Puget Sound, in Proc. Cal.Ac. Sci. 1895, 848.

923. CRYPTACANTHODES, Storer.

Cryptacanthodes, STORER, Rept. Fish. Mass., 28, 1839 (*maculatus*).

Body long and slender, compressed, naked, without lateral line; head cuboid, with vertical cheeks and conspicuous muciferous cavities; eyes small, placed high; mouth large, very oblique, the very heavy lower jaw prominent in front; jaws, vomer, and palatines with stoutish conical teeth, in few series. Gill openings prolonged forward below, narrowly attached to the isthmus. Dorsal fin of stoutish spines, hidden in the skin; dorsal and anal joined to caudal; pectorals short; ventrals wanting. ($\varkappa\rho\upsilon\pi\tau\acute{o}\varsigma$, hidden; $\dot{\alpha}\varkappa\alpha\nu\theta\acute{\omega}\delta\eta\varsigma$, spined.)

2799. CRYPTACANTHODES MACULATUS, Storer.

(WRY-MOUTH; GHOST-FISH.)

Head 6½; depth 13. D. LXXIII; A. 50. Eyes small, placed high, not so wide as the interorbital space, which has 2 ridges and 3 pits; orbital rim raised; 2 deep pits behind each eye at the temples; a deeper pit on the top of head between them; a raised ridge continued backward on each side of head behind orbital rim; maxillary extending to beyond eye; pseudobranchiæ small; pectorals short, 3 in head, their tips reaching beyond front of dorsal; vent a little in front of middle of body. Light brownish, with several series of smallish dark spots, arranged in more or less regular rows, from head to base of caudal; vertical fins closely spotted with darker; head above thickly speckled; body sometimes ("*inornatus*") entirely immaculate. Length 24 inches. Labrador to Long Island Sound; not very common; a few specimens have been taken at Woods Hole. The ghost-fish form (*inornatus*) occasionally seen, is doubtless an albino. (*maculatus*, spotted.)

Cryptacanthodes maculatus, STORER, Rept. Fish. Mass., 28, 1839, coast of Massachusetts; DE KAY, N. Y. Fauna: Fishes, 63, pl. 18, fig. 50, 1842; GÜNTHER, Cat., III, 291, 1861; JORDAN & GILBERT, Synopsis, 780, 1883.

Ophidium imberbe, PECK, Amer. Acad., 2d part, II, 1804, 46, pl. 4, New Hampshire; A. 49; P. 14; C. 22; not of LINNÆUS.

Fierasfer borealis, DEKAY, New York Fauna: Fishes, 316, 1842, New York; after PECK.

Cryptacanthodes inornatus, GILL, Proc. Ac. Nat. Sci. Phila. 1863, 332, Coast of Massachusetts; albino form.

924. LYCONECTES, Gilbert.

Lyconectes, GILBERT, Rept. U. S. Fish Comm. 1893 (1896), 452 (*aleutensis*).

Mouth subvertical; lower jaw projecting; premaxillary protractile. Teeth strong, conic, wide set, in more than 1 series. Mucous pits prominent on head. Gill opening narrow, ceasing opposite middle of base of pectorals, the membranes widely joined to isthmus. Dorsal and anal wholly joined to caudal, the latter extending well beyond them; dorsal fin composed of spines only; no ventral fins. Body naked; no lateral line. This genus differs from *Cryptacanthodes* principally in the absence of palatine teeth, agreeing with it in general appearance and in most details of structure. Alaska. (λύκος, wolf; νηκτήρ, swimmer.)

2800. LYCONECTES ALEUTENSIS, Gilbert.

Head 7⅓; depth 14¼. D. LXIX; A. 49; P. 13; caudal 18. Head square in cross section, the upper and lower surfaces plain, the cheeks vertical, the depth and width equal. Mouth still more oblique than in *Cryptacanthodes maculatus*, with much heavier mandible and less expanded maxillary, the exposed portion of the latter lying vertically, and not extending beyond vertical from middle of eye. Teeth all similar, few in number, those in premaxillary arranged in 2 series, the inner of which are smaller than the outer, from which they are separated by a wide interspace; teeth in mandible in a single series laterally, becoming a sparsely filled patch toward symphysis; 4 or 5 similar conical teeth on head of vomer; palatines toothless. A long nostril tube overhangs the upper lip. Upper lip separated by a fold from forehead, the upper jaw protractile. Eye extremely small, sunken in the socket, which it does not nearly fill, its diameter slightly less than ⅓ interorbital width; supraorbital rim not elevated, and containing no conspicuous projections; suborbital rim swollen, with an enlarged mucous channel; a conspicuous series of mucous pits along each mandible and the margin of preopercle; 2 series on top of head diverging backward from above the eyes; otherwise no pits or projections on head; a shallow triangular depression on occiput. Gill slit much less oblique than margin of preopercle, its length 1¼ times the distance between lower ends of gill slits, the latter reaching the vertical from middle of opercles. Dorsal fin of rather flexible spines, not concealed in heavy fin membranes; origin of dorsal immediately behind axil of pectorals. Hinder margin of occiput midway between front of dorsal and middle of eye. Origin of anal well in advance of middle of length, its distance from tip of snout contained 1⅔ times in its distance from base of caudal. Pectoral short, rounded, its base separated by a wide prepectoral area from gill slit, the width of area ¾ length of fin, the latter equaling distance from tip of snout to middle of eye. No ventrals. Body covered with lax naked skin, which also covers but does not obscure rays of anal fin; no pores to lateral line. Color in life, reddish on head, body, and fins, due to the blood vessels in the skin. Aleutian Islands. A single specimen, 180 mm. long, known. (Gilbert.).

Lyconectes aleutensis, GILBERT, Rept. U. S. Fish Comm. 1893 (1896), 452, pl. 34, fig. 3, Albatross Station 3312, north of Unalaska Island, in 45 fathoms. (Coll. *Albatross*.)

Family CCII. ANARHICHADIDÆ.

(The Wolf-Fishes.)

Body oblong or elongate, covered with rudimentary scales; no lateral line. Head scaleless, without cirri, its bones very thick and strong, the profile strongly decurved. Mouth very large, oblique, the jaws anteriorly with very strong conical canines; sides of lower jaw with very strong molar teeth, which shut against a series of very coarse molars on the palatines; vomer solid, armed with strong molar teeth, the dentition adapted for crushing sea-urchins and mollusks. Gill membranes broadly united to the isthmus; no pyloric cæca. Dorsal fin high, composed entirely of flexible spines; no ventral fins; pectoral fins broad, placed low. Large carnivorous fishes of the northern seas. Two genera and about 6 species known. (*Blenniidæ*, pt., Günther, Cat., iii, 208–211, 1861.)

ANARHICHADINÆ:
    a. Body moderately elongate, the tail not tapering to a point; dorsal and anal separate
        from the caudal.   ANARHICHAS, 925.
ANARRHICHTHYINÆ:
    aa. Body eel-shaped, excessively elongate; the dorsal and anal joined with the caudal at
        the end of the long and tapering tail.   ANARRHICHTHYS, 926.

925. ANARHICHAS (Artedi) Linnæus.

(Wolf-Fishes.)

Anarhichas (ARTEDI) LINNÆUS, Syst. Nat., Ed. x, 247, 1758 (*lupus*).

Body moderately elongate, covered with rudimentary scales; head scaleless, without cirri, compressed, narrowed above, the profile strongly decurved; mouth wide, oblique; premaxillary not protractile; jaws with very strong conical canines anteriorly; lateral teeth of lower jaw either molar or with pointed tubercles; upper jaw without lateral teeth; vomer extremely thick and solid, with 2 series of coarse molar teeth; palatines with 1 or 2 similar series. Gill membranes broadly joined to the isthmus; no lateral line. Dorsal fin rather high, composed entirely of flexible spines, which are enveloped in the skin; anal fin lower; caudal fin developed, free from dorsal and anal; no ventral fins; pectoral fins broad, placed low; air-bladder present; no pyloric cæca. Northern seas. (*Anarhichas* (or *Scansor*), the climber; an ancient name of *Anarhichas lupus;* from ἀναρριχάομαι, to climb or scramble up; the allusion not evident, the word spelled with a single *r* by Artedi and Linnæus.)

  a. Dorsal spines 60 to 70.
    b. Vomerine teeth not extending farther backward than the palatine teeth.
      c. Back and sides vaguely mottled, without spots or bands; vomerine teeth not
        extending nearly as far backward as palatine teeth.
                                 LATIFRONS, 2801.
      cc. Back and sides profusely covered with roundish black spots; vomerine
        teeth extending nearly as far backward as palatine teeth.
                                   MINOR, 2802.
    bb. Vomerine teeth extending much farther backward than the short band of palatine teeth; sides of body with 9 to 12 darker cross bars; nape and shoulder with dark spots.   LUPUS, 2803.

aa. Dorsal spines 80 to 85; body without bands or spots; vomerine teeth extending farther backward than palatine band.

  d. Head moderate, 4½ in length; caudal rays 20; upper canines 4. ·

  LEPTURUS, 2804.

  dd. Head very large; caudal rays 17; upper canines 6.

  ORIENTALIS, 2805.

2801. ANARHICHAS LATIFRONS, Steenstrup & Hallgrimosan.

Head 5; depth 4. D. LXVII; A. 45. Body more robust than *A. lupus*, the dorsal fin lower. Head broad, the profile not strongly decurved; teeth much smaller than in *A. lupus;* vomerine teeth not extending nearly as far back as the palatine series. Pectorals ⅔ length of head; dorsal fin not very high, beginning above the gill opening, the longest spine less than ⅓ head; caudal 2⅓ in head. Brown, obscurely spotted with darker; the sides without dark bars or black spots. (Collett.) North Atlantic on both coasts, chiefly north of the Arctic Circle, south to Banquereau on our coast.ʼ (Eu.) (*latus*, broad; *frons*, forehead.)

Anarrhichas latifrons, STEENSTRUP & HALLGRIMSSON, Förh. Skand. Naturf. 3 die Möte 1842, 647; BEAN, Proc. U. S. Nat. Mus., II, 1879, 218; COLLETT, Meddelsk. Norges Fiske 1879, 46; JORDAN & GILBERT, Synopsis, 782, 1883; GOODE & BEAN, Oceanic Ichthyology, 301, fig. 271, 1896.
Anarrhichas denticulatus, KRÖYER, Overs. Vidensk. Selsk. Kjöb. 1844, 140.

2802. ANARHICHAS MINOR, Olafsen.

Head 5⅓; depth 5⅓. D. LXXVIII; A. 46. Form of *Anarhichas lupus* or a little more slender; fins similarly formed, the dorsal a little lower, Vomerine teeth extending nearly or quite as far back as the palatines. Body pale olivaceous or yellowish; sides without vertical bars; round, black spots covering dorsal and caudal fins as well as back and sides down to the level of the pectoral; head spotted; belly immaculate. North Atlantic, on both coasts, chiefly north of the Arctic Circle, south to Eastport, Maine; Gloucester; and Norway. (Eu.) (*minor*, smaller.)

Anarrhichas minor, OLAFSEN, Reise i Island, 592, 1772, Iceland.
Anarrhichas pantherinus, ZUIEW, Nov. Act. Petrop. 1781, 271; BEAN, Proc. U. S. Nat. Mus., II, 1879, 217; JORDAN & GILBERT, Synopsis, 781, 1883; GOODE & BEAN, Oceanic Ichthyology, 301, fig. 270, 1896.
Anarrhichas karrak, BONNATERRE, Encycl. Ichth., 38, 1788, Iceland; after OLAFSEN.
Anarrhichas maculatus, BLOCH & SCHNEIDER, Syst. Ichth., 496, 1801, Iceland; after OLAFSÉN.
Anarrhichas leopardus, AGASSIZ in SPIX, Pisc. Brasil., tab. 51, 1829, "Atlantic Ocean."

2803. ANARHICHAS LUPUS, Linnæus.

(WOLF-FISH.)

Head 6; depth 5½. D. LXII; A. 42. Maxillary reaching beyond orbit; band of vomerine teeth extending much farther back than the short palatine band. Pectorals large, rounded, ⅔ length of head. Dorsal high, beginning over the gill opening, its longest rays about ½ length of head. Brownish or bluish gray; sides with numerous (9 to 12) very dark transverse bars, which are continued on the dorsal fin; besides these

numerous dark spots and reticulations, the spots most distinct below front
of dorsal; fins dark; caudal tipped with reddish. Length 3 to 4 feet.
North Atlantic, south to Cape Cod and France; rather common both in
America and Europe. A large voracious fish, not valued as food. The
American form, *vomerinus*, seems to be fully identical with the European.
(Eu.) (*Lupus*, a wolf.)

Anarhichas lupus, LINNÆUS, Syst. Nat., Ed. X, 247, 1758, no definite locality; after ARTEDI;
GÜNTHER, Cat., III, 208, 1861; JORDAN & GILBERT, Synopsis, 781, 1883; GOODE & BEAN,
Oceanic Ichthyology, 299, 1896.
Anarrhichas strigosus, GMELIN, Syst. Nat., I, 1144, 1788, British Sea.
Anarrhichas vomerinus, AGASSIZ in STORER, Hist. Fish. Mass., 265, pl. 18, fig. 1, 1867, Cusk
Rocks, between Boston and Cape Ann.

2804. ANARHICHAS LEPTURUS, Bean.

(ALASKA WOLF-FISH.)

Head 4½; depth 5; D. LXXXI; A. 52; C. 20 or 21. Head moderate;
maxillary ½ as long as head; nostril nearer eye than mouth. Four
large canines in the upper jaw and 5 in the lower, all of them strongly
recurved; behind the canines in each jaw are a few sharp, conical teeth,
also recurved; palatine teeth in 2 series, 4 in the outer and 5 in the inner
series, those in the outer series the longer; vomerine teeth in 2 series, the
vomerine patch beginning in advance of the palatine, and extending
farther back than the latter; head and fins scaleless; median line of body
and all of tail with small, widely separated scales. Dark brown, without
bands or spots; belly pale, clouded with very dark brown. (Bean.)
Coasts of Alaska, south to Vancouver Island; common about the Aleutian Islands, and perhaps identical with *Anarhichas orientalis*. (λεπτός,
slender; οὐρά, tail.)

Anarrhichas lepturus, BEAN, Proc. U. S. Nat. Mus., II, 1879, 212, St. Michaels, Alaska; JOR-
DAN & GILBERT, Synopsis, 782, 1883; GOODE & BEAN, Ocean. Ichth., 299, 1896.
? *Anarrhichas orientalis*, PALLAS, Zoogr. Rosso-Asiat., III, 77, 1811, Kamchatka.

2805. ANARHICHAS ORIENTALIS, Pallas.

This species, if correctly described, would differ from *Anarhichas lepturus* in the very large head, 2½ times in total length of body; in the
absence of scales; in having the nostril midway between eye and mouth,
and in having 6 canines in the upper jaw. Color plain brown. D. LXXXIV;
C. 17. Coast of Kamchatka. (Pallas.) As the first of these characters is
certainly erroneous, it is likely that the others are also, and that this
species is not distinct from *Anarhichas lepturus*. (*orientalis*, eastern.)

Anarhichas orientalis, PALLAS, Zoogr. Rosso-Asiat., III, 77, 1811, Kamchatka.
? *Anarrhichas lepturus*, BEAN, Proc. U. S. Nat. Mus., II, 1879, 212, St. Michaels.

926. ANARRHICHTHYS, Ayres.

Anarrhichthys, AYRES, Proc. Cal. Ac. Nat. Sci., I, 1855, 32 (*ocellatus*).

Body elongate, tapering backward into a very long and compressed
tail, around which the dorsal and anal are confluent with the caudal.
Scales rudimentary; no lateral line. Dorsal high, composed entirely of

flexible spines; pectoral fins broad, placed low; no ventral fins. Head very large, compressed, the snout rather short; mouth large; jaws with very strong, conical canines anteriorly; vomer and palatines each with about 2 rows of coarse molars, the palatine band shutting against similar teeth on the sides of the lower jaw. Gill membranes broadly united to the isthmus. No pyloric cæca. Large, eel-shaped fishes of the North Pacific, remarkable for the tremendous dentition, the head essentially as in *Anarhichas*, the body strikingly different. (*Anarhichas; ἰχθύς*, fish.)

2806. ANARRHICHTHYS OCELLATUS, Ayres.

(WOLF-EEL.)

Head 11; depth 15. D. CCL; A. 233; P. 19. Body elongate, formed as in an eel; the head and jaws very strong. Pectorals broad, more than ½ head; longest dorsal spine ½ head. Color dark greenish, the body and dorsal fin everywhere covered with round, ocellated black spots of various sizes, the light markings forming reticulations around the spots; head paler, with the reticulations in much finer pattern; anal pale-edged. Length 5 to 8 feet. Pacific coast, from Monterey north to Puget Sound; generally common. One of our most remarkable fishes; rarely used as food. It feeds chiefly on sea-urchins and sand dollars. (*ocellatus*, with eye-like spots.)

Anarrhichthys ocellatus, AYRES, Proc. Cal. Ac. Nat. Sci., I, 1855, 31, San Francisco; JORDAN & GILBERT, Synopsis, 782, 1893; JORDAN & STARKS, Fishes Puget Sound, 848, 1895.
Anarrhichthys felis, GIRARD, Proc. Ac. Nat. Sci. Phila. 1854, 150, San Francisco (Type, No. 511. Coll. W. O. Ayres), name only, no description; GIRARD, U. S. Pac. R. R. Surv., X, Fish., 125, pl. 25a, figs. 1 to 3, 1858; GÜNTHER, Cat., III, 211, 1861.

Family CCHI. CERDALIDÆ.

Body elongate, compressed, covered with small scales; no lateral line; head small; gill openings reduced to small slit-like openings more or less horizontal in position; dorsal fin very long and low, anteriorly of slender spines, which pass gradually into the soft rays; no free spines; no cirri; tail not isocercal; pseudobranchiæ well developed. Three species known, from the west coast of tropical America in rock pools near the shore. The presence of some spines in the dorsal separates them from the *Scytalinidæ*, while the small gill openings distinguish them from the *Blenniidæ*, to which they are more nearly allied.

 a. Ventral fins each with 2 rays; dorsal rays 41; body moderately elongate; greatest depth 10¾ in length; distance from insertion of dorsal to occiput equal to length of head.   CERDALE, 927.
 aa. Ventral fins each with 1 ray; dorsal rays 48 to 55; body very elongate, eel-like, its depth 15 to 18 in length.   MICRODISMUS, 928.

927. CERDALE, Jordan & Gilbert.

Cerdale, JORDAN & GILBERT, Bull. U. S. Fish Comm. 1881, 332 (*ionthas*).

This genus differs from *Microdesmus* in the presence of 2 rays in the ventral fin. Its body is much less elongate than in *Microdesmus*. The gill openings are reduced to small, nearly horizontal slits below and in front of the pectoral fins. (*κερδαλῆ*, the wary one, the fox-like.)

2807. CERDALE IONTHAS, Jordan & Gilbert.

Head 7⅔ in length; depth 10¾. D. 41; A. 36 to 38; C. 4–17–4; P. 12; V. 2. Body considerably elongate, compressed, of nearly equal depth throughout, the head tapering rapidly from occiput to snout; snout short, not obtuse, but the lower jaw heavy and blunt, much projecting beyond the premaxillaries; gape very short and oblique, the tip of the premaxillary not reaching ventral from orbit. Margin of upper jaw formed entirely by the premaxillaries, which are free laterally, but scarcely movable mesially. Maxillary not distinguishable, probably enveloped in the integument of the snout. Teeth rather strong, short, blunt, in a double series in each jaw, apparently wanting on the vomer and palatines. Lips developed laterally, where they form a fold around the angle of the mouth; lower lip adnate mesially, the upper reduced to an obsolete fold. Length of gape ⅓ length of head. Nostrils 2, distant, the anterior at the end of the snout, almost labial, the posterior above front of orbit, both circular. Eye very small, somewhat less than interorbital width or than length of snout. Distance from snout to past margin of orbit contained 2⅔ times in length of head. Pseudobranchiæ well developed. Gill openings very narrow, reduced to a short, nearly horizontal slit, extending forward from a point just below the lower base of the pectoral fin. Branchiostegals evident, apparently 4 in number. Vertical fins well developed; dorsal and anal both long, the membrane of the last ray of each joining the base of the rudimentary rays of the caudal. Distance from occiput to the origin of dorsal fin equal to the length of the head; rays of dorsal fin very slender, distinct, the membrane thin and transparent, the rays all, or nearly all, articulate, the anterior simple, the posterior bifid at tip. Vent slightly in advance of middle of length of body, the anal fin beginning immediately behind it; anal rays bifid at tip, excepting the first 2, which appear simple; tail not isocercal, truncate at base of caudal, most of the rays of the caudal springing from the expanding last vertebra; caudal fin rounded, ⅔ length of head, its rays much branched, more closely set than the rays of the dorsal and anal; rudimentary rays very numerous; ventral fins small, close together, inserted slightly in advance of the lower end of the pectoral, each fin composed of 2 rays, the inner prolonged beyond the outer, and bifid at tip, about as long as pectoral fin and ⅔ length of head; pectorals well developed, broad, the rays branched at tip. Head and body entirely covered with small scales, which are close set but hardly imbricate, not arranged in series; mandible, snout, and gill membrane scaly; scales on belly and breast smaller than the others and more thickly set; base of caudal and pectoral fins scaled. Coloration in life, body translucent light olive, immaculate below; back and sides very finely marked with clusters of fine dots, the ground color appearing as reticulations between the clusters, which are of irregular size and form; on the sides of the head these dots form bars, which radiate from the eye to the snout and lower side of the head. This species is known from 3 specimens, 2¼ to 3 inches in length, taken in a rock pool at Panama. (*ἰονθάς*, freckled.)

Cerdale ionthas, JORDAN & GILBERT, Bull. U. S. Fish Comm. 1881, 332, Panama. (Coll. Chas. H. Gilbert.)

928. MICRODESMUS, Günther.

Microdesmus, GÜNTHER, Proc. Zool. Soc. London 1864, 26 (*dipus*).

Body anguilliform, covered with rudimentary scales; head small, with short, obtuse snout and small mouth; lower jaw projecting; teeth minute, in jaws only; eyes very small; gill opening reduced to a very narrow, somewhat oblique slit, in front of lower part of pectorals; vertical fins well developed, the dorsal and anal joined to the caudal by a thin membrane; rays of dorsal mostly articulate, all but a few of the last simple; ventral fins very small, reduced to a single ray; pectorals moderate; vent normal, in middle of body. Pacific coast of tropical America. (μικρός, small; δεσμός, a band.)

a. Dorsal rays 55, the fin beginning less than a head's length behind occiput.

DIPUS, 2808.

aa. Dorsal rays 48, the fin beginning more than a head's length behind the occiput.

RETROPINNIS, 2809.

2808. MICRODESMUS DĪPUS, Günther.

Head about 11 in total length; depth about 18. D. 55; A. 34; C. 16; P. 12; V. 1. Head rather compressed, snout short, mouth very narrow, lower jaw very prominent. Eye minute, lateral, and in anterior third of head. Dorsal fin commencing at a distance from occiput which is somewhat less than length of head, nearly even, the rays very distinct, the interradial membrane being thin and transparent; anal fin commencing immediately behind vent. Caudal rays much more slender and more closely set than those of dorsal and anal; caudal fin rounded, $\frac{2}{3}$ length of head; pectorals as long as ventrals, and $\frac{1}{4}$ as long as head; ventrals close together, and inserted a little behind root of pectoral. Upper parts uniformly brownish olive. Panama. Known from a single specimen, 4$\frac{1}{4}$ inches long. (Günther.) (δίς, two; πούς, foot.)

Microdesmus dipus, GÜNTHER, Proc. Zool. Soc. London, January 26, 1864, 4, pl. 3, fig. 2, Central America (Coll. Capt. Dow); JORDAN, Cat., 126, 1885.

2809. MICRODESMUS RETROPINNIS, Jordan & Gilbert.

Head 14$\frac{1}{2}$ in length; greatest depth 15$\frac{3}{4}$. D. 48; A. 29; C. 3-17-3; P. 13; V. 1. Body very elongate, compressed, tapering somewhat from front of dorsal to caudal peduncle. Head very small, rapidly tapering forward from occiput; upper profile with a noticable depression behind the orbits, the outline thence to snout strongly convex. Mouth very small, somewhat oblique, the fleshy tip at symphysis of lower jaw projecting much beyond the premaxillaries; gape scarcely reaching vertical from orbit. Teeth small, apparently in a single series in each jaw only. Nostrils double, distant, the anterior near the end of snout, the posterior above anterior margin of orbit. Gill openings a very narrow, somewhat oblique slit, from front of lower third of pectoral fin downward and forward. Branchiostegals evident, 4 or 5 in number. Eye very small, lateral, situated near the upper profile of the head, its diameter nearly $\frac{1}{4}$ the length of the short snout. Vertical fins well developed; dorsal and anal connected with the caudal by a very delicate membrane. Distance from origin of dorsal fin to occiput 3 times the length of the head, its rays distinct, connected by thin transparent membrane, as are the rays of the anal; most of the rays simple and undivided (but mostly articulate), a few of the posterior only forked at tip; origin of anal fin nearly equidistant between gill rakers and tip of caudal, its rays mostly forked at tip; caudal rays much divided and more closely set than those of dorsal and anal, the fin somewhat pointed in outline, as long as the head; tail not isocercal, truncate at base of caudal fin; ventral fins very small, close together, inserted slightly behind base of pectorals; each fin reduced to a single undivided filament; pectoral fin small, pointed, the middle rays longest, much shorter than the ventrals, and ½ the length of the head. Vent considerably behind middle of total length of the fish (with caudal). Head and body covered with scattered rudimentary scales. Color in life, translucent light olive, with a series of irregular quadrate dark blotches along the back and a series along each side, these blotches formed of clusters of dark points. One specimen, nearly 4 inches in length, was taken in a rock pool at Panama; others since taken by Dr. Gilbert. This species differs from the description of the previously known *Microdesmus dipus*, Günther, in the posterior insertion of the dorsal and the posterior position of the vent, the smaller number of fin rays, the shorter head, longer ventrals, and mottled coloration. (*retro*, backward; *pinna*, fin.)

Microdesmus retropinnis, JORDAN & GILBERT, Bull. U. S. Fish Comm. 1881, 331, Panama. (Coll. C. H. Gilbert.)

Family CCIV. PTILICHTHYIDÆ.

(THE QUILL-FISHES.)

Body extremely elongate, serpentiform, little compressed, the tail tapering to a point. Skin with a few thin, loose, scattered scales; no lateral line. Head unarmed, rather small; upper jaw not protractile; snout short; mouth oblique; lower jaw projecting considerably beyond the upper, with a protruding fleshy appendage at tip. Maxillary reaching front of eye. Mandible little movable. Both jaws with fine, close-set, sharp teeth, in 1 row, the posterior teeth a little the largest; no evident teeth on vomer or palatines. Gill openings restricted to below the most convex part of the opercle, the membranes broadly united below, free from the isthmus. Gills 4, a slit behind the fourth. Pseudobranchiæ very small, almost obsolete. Gill rakers short and stout. Pectorals short; ventrals wanting; dorsal beginning close behind the nape, the anterior portion for about ½ the length of the body composed of very low, stiff, free spines, hooked backward, the posterior portion higher, of slender soft rays connected by thin membrane. No caudal fin, the tip of the tail free. Anal similar to the soft dorsal, without spines. Vent at considerable distance from the head. North Pacific. A single species known.

Concerning the relationships of this interesting group, Dr. Gilbert observes:

"The genus *Ptilichthys*, of which this species [*P. goodei*] is the sole representative, has been doubtfully referred by Dr. Bean to the *Mastacembelidæ*, a family of fresh-water fishes inhabiting the East Indies, characterized by having the shoulder girdle posteriorly placed and not articulating with the cranium (Order *Opisthomi*, Gill). The necessity for preserving intact the unique type of the species prevented Dr. Bean from making any anatomical examination of *Ptilichthys*, and it was reserved for Dr. Theodore Gill, in the Standard Natural History, III, 259, 1885, to express his disbelief in the relationships which have been suggested, and to make the fish the type of a peculiar family, the *Ptilichthyidæ*, to be placed provisionally among the Blennioid series. His adherence to this view is again expressed in his list of 'Families and Subfamilies of Fishes,' appearing as the Sixth Memoir of Volume VI, of the National Academy of Sciences. He has doubtless indicated the proper position of this peculiar fish as nearly as we are now able to determine it. An examination of its shoulder girdle shows it to be entirely normal. The post-temporal is not furcate, but is a very slender bony rod attaching to the epiotic region of the skull, and giving loose attachment posteriorly to the almost equally slender postero-temporal. The latter overlaps the upper end of the clavicle in the usual manner. A postclavicle was not detected. The coracoid portion consists of a roundish, oblong, perforated hypercoracoid meeting the hypocoracoid directly, without intervening cartilage. The curved line separating the two bones corresponds distally with the interspace between the first (upper) and second actinosts. The hypocoracoid is broad and short; its mesially directed (i. e., inferior) process joins at its tip the clavicle, but is elsewhere separated from the latter by the usual elongate membranaceous interspace. The actinosts are 4 in number, of large size, hourglass-shaped. The jaws are normal, the premaxillary alone occupying the front and sides of upper jaw and bearing the teeth, while the maxillary is a broad bone lying behind it, overlapped proximally by the maxillary process of the palatines. Both vomer and palatines seem to be toothless. The alimentary canal is almost perfectly straight, with the anterior portion entirely enveloped in the long, narrow liver. At the pylorus occurs a short and abrupt U-shaped flexure, scarcely noticeable on account of the closeness with which the sides are joined, and the fact that the width of the flexure is no greater than the cross diameter of the tube. Pyloric cæca are not evident. Air bladder is entirely wanting. The ovary is single, apparently without viaduct, and contains in our specimen eggs which are comparatively very large." (Gilbert.) (*Ptilichthyidæ*, Gill, Standard Nat. Hist., III, 259, 1885.)

929. PTILICHTHYS, Bean.

Ptilichthys, BEAN, Proc. U. S. Nat. Mus., IV, 1881, 157 (*goodei*).

Characters of the genus included above. ($\pi\tau\acute{\iota}\lambda o\nu$, quill; $\grave{\iota}\chi\theta\acute{\upsilon}s$, fish.)

2810. PTILICHTHYS GOODEI, Bean.

D. XC, 145; A. about 185; P. 12. Eye rather large, as long as snout, 5 in head; cheeks and opercles long; pectoral fin ¼ as long as head; soft dorsal and anal deeper than body posteriorly, anal a little lower than dorsal. Vent near end of anterior third of body; distance from vent to beginning of soft dorsal 3¼ times length of head; length of head twice its greatest depth, 5½ in distance to vent; appendage of mandible ¼ as long as eye; free tip of caudal ⅜ eye. Orange or yellowish, body with a blackish longitudinal stripe; anal darker in color than dorsal. Length about 12 inches. Aleutian Islands; rare; in water of moderate depth. Here described from the original type from Unalaska; 2 other specimens known, the one studied by Dr. Gilbert taken in the entrance to the harbor of Unalaska. (Named for Dr. George Brown Goode.)

Ptilichthys goodei, BEAN, Proc. U. S. Nat. Mus., IV, 1881, 157, Port Levachef, Unalaska (Col. Sylvanus Bailey. Type, No. 26619, U. S. Nat. Mus.); JORDAN & GILBERT, Synopsis, 369, 1883; GILBERT, Rept. U. S. Fish Comm. 1893 (1896), 453.

Group OPHIDIOIDEA.

(THE EEL-POUTS.)

This group, as a whole, agrees with the *Blennioidea* in all respects, except that no spines are developed in any of the fins, save sometimes in the posterior part of the dorsal. From the *Anacanthini*, with which the *Ophidioidea* agree in the jugular ventrals and in the absence of spines, they are separated by the form of the hypercoracoid, which is perforate, as in ordinary fishes. The group is a very large and varied one, widely distributed in all seas.

    a. Pseudobranchiæ well developed, very rarely small or obsolete.
        b. Ventral fins jugular, inserted much behind the eye, often wanting, never filamentous.
            c. Gill membranes broadly united, free from the isthmus; ventrals wanting; no scales. SCYTALINIDÆ, CCV.
            cc. Gill membranes united to the isthmus, the gill openings lateral.
                      ZOARCIDÆ, CCVI.
        bb. Ventral fins developed as slender filaments attached at the throat not far behind eye.
            e. Gill membranes broadly attached to the isthmus; no scales.
                      DEREPODICHTHYIDÆ, CCVII.
            ee. Gill membranes nearly separate, free from the isthmus; body scaly.
                      OPHIDIIDÆ, CCVIII.
    aa. Pseudobranchiæ absent (or rudimentary in some *Brotulidæ*).
        f. Ventral fins wanting; no scales.
            g. Vent normal, well behind pectorals. LYCODAPODIDÆ, CCIX.
            gg. Vent at the throat. FIERASFERIDÆ, CCX.
        ff. Ventral fins well developed; vent posterior, normal.
            h. Dorsal fin single, low; ventral fins short. BROTULIDÆ, CCXI.
            hh. Dorsal fins 2, the anterior, at the nape, of a single long ray; ventral fins elongate. BREGMACEROTIDÆ, CCXII.

Family CCV. SCYTALINIDÆ.

Body elongate, compressed, eel-shaped, naked. Head depressed, with tumid cheeks, like the head of a snake. Mouth moderate, horizontal, the lower jaw the longer; teeth in a single series in the jaws, vomer, and palatines; no barbels. Gills 4, a slit behind the fourth; pseudobranchiæ present. Gill membranes broadly connected, free from the isthmus. Dorsal fin long and low, beginning near middle of body, of slender rays embedded in the skin; anal similar to dorsal, both connected to the caudal fin; tail diphycercal; pectoral fins small, ventral fins wanting. Vent remote from the head, without papilla. Air bladder none; cæca none. Vertebræ numerous, small. The skeleton does not differ essentially from that of *Lycodopsis pacificus*, with which it has been compared. The skull is not at all depressed, the wide depressed form of the head of the fish is due to the fleshy cheeks. The frontal takes up the greater part of the top of the skull, the parietals are separated by the supraoccipital, which extends forward to the frontals. Opercles all present. Lower jaw large and strong. Post-temporal scarcely so firmly attached as in *Lycodes;* the clavicle long and slender. As here understood, this family consists of a single species, a shore fish of the Northern Pacific, living in the gravel between tide marks, and diving with great activity into the wet gravel when disturbed. Its relations are apparently with the *Zoarcidæ.* It is not certain that *Scytalina* has any special affinity with the *Congrogadidæ*, in which group it was at first placed by Jordan & Gilbert.

930. SCYTALINA, Jordan & Gilbert.

Scytalina, JORDAN & GILBERT, Proc. U. S. Nat. Mus. 1880, 266 (*cerdale*).
Scytaliscus, JORDAN & GILBERT, Proc. U. S. Nat. Mus. 1883, 111 (*cerdale*); substitute for *Scytalina* on account of the earlier *Scytalinus*, Erichson, a genus of Coleoptera.

Body very long and slender, covered with small scales. Head depressed, shaped like the head of a snake, with tumid cheeks and a distinct neck. Eyes small, superior. Mouth rather large, the lower jaw slightly projecting. Teeth conic, in single series on jaws, vomer, and palatines. Each jaw with 2 canines in front. No lateral line; pseudobranchiæ small. Gill rakers almost obsolete. Dorsal fin very low, its first ray near the middle of the body. Anal fin similar to dorsal, nearly as long. Tail diphycercal, the caudal well developed. (Diminutive of *Scytale*, from *σκυτάλη*, a viper.)

2811. SCYTALINA CERDALE, Jordan & Gilbert.

Head 8; depth 14. D. 41; A. 36. Head broader than body; body much deeper behind vent than anteriorly; snout depressed, rounded at tip; cheeks very long; opercle short; interorbital space rather broad, concave posteriorly; eyes very small, anterior and superior, 10 to 12 in head, 2 in snout, 3 to 4 in interorbital width; upper lip separated by a crease from the skin of the forehead; lower jaw scarcely projecting; edge of lower lip with pores, and small dermal flaps and fringes; maxillary extending somewhat beyond eye; anterior nostrils with small flaps. Lower jaw with a series of close-set, even, conical teeth, besides 2 divergent canines in front; upper jaw with similar teeth in several series in front, the canines smaller and closer together. Pectorals inserted high, little longer than eye; insertion of dorsal slightly in front of anal, a little in front of middle of body; rays of vertical fins low and weak, those of caudal most developed; dorsal and anal joined to caudal; vent close in front of anal, which is similar to dorsal. Flesh colored, with much mottling of purplish in fine pattern; belly nearly plain; caudal reddish-edged. Length 6 inches. Straits of Juan de Fuca; burrowing among rocks near tide mark. The 2 original types came from the shore of Waadda Island, near Cape Flattery, where the species lives in wet shingle and shows extraordinary activity in hiding among rocks when disturbed. In the same locality 25 additional specimens have been dug out of the gravel by Mr. E. C. Starks in 1895. The species is still unknown from any other locality. (κερδαλῆ, the wary one, the fox.)

Scytalina cerdale, JORDAN & GILBERT, Proc. U. S. Nat. Mus., III, 1880, 266, Waadda Island (Type No. 27400. Coll. Jordan & Gilbert); JORDAN & GILBERT, Synopsis, 791, 1883; JORDAN & STARKS, Fishes of Puget Sound, in Proc. Cal, Ac. Sci. 1895, 849, pl. 104.

Family CCVI. ZOARCIDÆ.

(THE EEL-POUTS.)

Body elongate, more or less eel-shaped, naked or covered with very small, embedded, cycloid scales; head large; mouth large, with conical teeth in jaws, and sometimes on vomer and palatines; bones of head unarmed. Gill membranes broadly united to the isthmus, the gill opening reduced to a vertical slit; pseudobranchiæ present; gills 4, a slit behind the fourth. Dorsal and anal fins very long, of soft rays only, or the dorsal with a few spines in its posterior portion; vertical fins sometimes confluent around the tail; pectorals small; ventrals jugular, very small or wanting, if present, inserted behind the eye. Lateral line obsolete or little developed, sometimes bent downward behind pectorals, sometimes sending a branch on median line backward. Gill rakers small; pyloric cæca rudimentary; vent not near head. Pseudobranchiæ present. Genera about 15; species 50. Bottom fishes, chiefly of the Arctic and Antarctic seas; some of them, at least, are viviparous, and some descend to considerable depths. Dr. Gill thus enumerates the skeletal characters of the *Zoarcidæ:*

Orbito-rostral portion of the cranium contracted and shorter than the posterior, the cranial cavity open in front, but bounded laterally by the expansion of the annectant parasphenoid and frontals, with the supraoccipital declivous and tectiform behind, the occipitals above inclined forward along the sides of the supraoccipital, and the exoccipital condyles distant, with the hypercoracoid foraminate about its center and the hypocoracoid with an inferior process convergent to the proscapula. These characters are formulated from the skeleton of *Zoarces anguillaris*. (Gill, Proc. Ac. Nat. Sci. Phila. 1884, 179.) *Zoarchidæ,** SWAINSON, Nat. Hist. Class. Fishes, II, 82, 184, and 283, 1839. *Lycodidæ*, GÜNTHER, Cat., IV, 319–326, 1862; genus *Zoarces*, GÜNTHER, Cat., III, 295. 1861.

ZOARCINÆ:

 I. Dorsal fin low behind, some of its posterior rays short and spine-like; ventrals small; scales present; teeth strong, conic, in jaws only; lateral line present, along middle of side.   ZOARCES, 931.

 II. Dorsal fin continuous, of soft rays only.

---

* The name *Zoarchidæ* or *Zoarcidæ* is prior to that of *Lycodidæ*.

LYCODINÆ:

  a. Ventral fins present.

    b. Vomer without teeth; body scaly.

      c. Palatines without teeth.

        d. Body very slender, the depth 12 to 16 times in length; lateral line short and faint, ventral in position.   EMBRYX, 932.

        dd. Body rather robust, the depth 8 to 9 in length; lateral line rather faint, lateral in position.   LYCODOPSIS, 933.

      cc. Palatines with teeth; lateral line distinct, running along middle of side.   APRODON, 934.

    bb. Vomer and palatines with teeth.

      e. Lower jaw without barbels.

        f. Dorsal fin without sculptured scutes at base.

          g. Body rather deep, the depth 6 to 8 times in the length.

            h. Body more or less scaly.   LYCODES, 935.

            hh. Body entirely naked, or with a few scales on tail only; none on body or fins.   LYCODALEPIS, 936.

          gg. Body more slender, the depth 12 to 20 in the length; lateral line lateral in position.

            i. Pectoral fin with rounded outlines, the lower rays not greatly produced.   LYCENCHELYS, 937.

            ii. Pectoral fin deeply notched, the lower rays much produced; lateral line ventral in position.   FURCIMANUS, 938.

        ff. Dorsal fin with the rays each provided with a sculptured scute or appendage at base; no lateral line; body elongate.   LYCODONUS, 939.

      ee. Lower jaw with many barbels; body slender, scaly.   LYCONEMA, 940.

  aa. Ventral fins entirely wanting.

    GYMNELINÆ:

      j. Teeth moderate, nearly uniform, on jaws, vomer, and palatines.

        k. Body scaly; vomer and palatines with teeth; body compressed, not very slender; skull cavernous.   BOTHROCARA, 941.

        kk. Body scaleless.

          l. Lower jaw not very prominent; body very slender; gill openings very narrow.   GYMNELIS, 942.

          ll. Lower jaw very prominent; body slender, tapering behind; scales undescribed.   LYCOCARA, 943.

    MELANOSTIGMATINÆ:

      jj. Teeth long, unequal, on jaws, vomer, and palatines; skin lax; gill openings reduced to a small foramen; body very slender; scales obsolete.   MELANOSTIGMA, 944.

931. ZOARCES, Gill.

(EEL-POUTS.)

Enchelyopus, KLEIN, Ichthyologia Missus, IV, 52, 1747; not as restricted by BLOCH & SCHNEIDER.

Zoarces, CUVIER, Règne Animal, Ed. 2, II, 240, 1829 (*viviparus*).

Zoarchus, SWAINSON, Nat. Hist. Class'n Fishes, II, 283, 1839 (*viviparus*).

Enchelyopus, GILL, Proc. Ac. Nat. Sci. Phila. 1863, 258 (*viviparus*); not of BLOCH & SCHNEIDER.

Macrozoarces, GILL, Proc. Ac. Nat. Sci. Phila. 1863, 258 (*anguillaris*).

Body elongate, compressed, tapering posteriorly; head oblong, heavy, narrowed above, the profile decurved; mouth large; teeth strong, conic, bluntish, in 2 series in the front of each jaw and 1 series on the sides;

teeth in outer series larger; no teeth ou vomer or palatines; dorsal fin very long, low, some of its posterior rays much lower than the others, developed as sharp spines; pectoral fins broad; ventrals jugular, of 3 or 4 soft rays; scales small, not imbricated, embedded in the skin; lateral line slender, lateral in position; size large; species viviparous. The American and Asiatic species (subgenus *Macrozoarces*) differ from the European type of *Zoarces*, Cuvier, in the increased number of fin rays and vertebræ. In *Zoarces viviparus* (Linnæus), the European eelpout, the dorsal rays are about 100, the anal about 85, and the number of vertebræ is proportionally diminished. (ζωαρκής, viviparous.)

Subgenus MACROZOARCES, Gill.

2812. ZOARCES ANGUILLARIS * (Peck).

(EEL-POUT; MUTTON-FISH; MOTHER OF EELS.)

Head 6; depth 7. D. 95, XVIII, 17; A. 105. Mouth moderate, lower jaw included; maxillary reaching beyond orbit; pectoral long, about $\frac{2}{3}$ length of head; ventrals $\frac{1}{4}$ head; highest ray of dorsal about equal to snout, the posterior spines about $\frac{1}{3}$ length of eye; first ray of dorsal above preopercle. Reddish brown, mottled with olive, the scales paler than the skin about them; dorsal fin marked with darker; a dark streak from eye across cheek and opercles. Length 20 inches. Delaware to Labrador; rather common north of Cape Cod. Two forms occur, distinguished by the size of the jaws. These have been regarded as distinct species, but the large-mouthed form (*ciliatus; labrosus*) is doubtless the male, as a similar variation occurs in *Lycodopsis pacificus*, and exists in some degree in species of *Lycodes*. (*anguillaris*, eel-like.)

? *Encheliopus*, GRONOW, Zoophyl., 77, No. 266, 1763, America (*unicolor*); dorsal and anal united with the caudal.

? *Blennius americanus*, BLOCH & SCHNEIDER, Syst. Ichth., 171, 1801, America; after Gronow.

Blennius anguillaris, PECK, Mem. Amer. Ac. Sci., II, 1804, 46, New Hampshire.

Blennius fimbriatus, MITCHILL, Trans. Lit. and Phil. Soc. N. Y. 1815, 374, pl. 1, fig. 6, New York.

Blennius ciliatus, MITCHILL, Trans. Lit. and Phil. Soc. N.Y. 1815, 374, pl. 1, fig. 7, New York.

Zoarces labrosus, CUVIER, Règne Anim., Ed. II, vol. 2, 240, 1829, America; CUVIER & VALENCIENNES, Hist. Nat. Poiss., XI, 466, 1836.

Zoarces gronovii, CUVIER & VALENCIENNES, Hist. Nat. Poiss., XI, 469, 1836; after Gronow.

? *Enchelyopus americanus*, GRONOW, Cat. Fishes, Ed. Gray, 101, 1854, American Ocean.

Zoarces fimbriatus, CUVIER & VALENCIENNES, Hist. Nat. Poiss., XI, 468, 1836.

Blennius labrosus, MITCHILL, Trans. Lit. and Phil. Soc. N. Y. 1815, 375.

Zoarces anguillaris, STORER, Fishes Mass., 66, 1839; STORER, Synopsis Fishes N. A., 375, 1845; GÜNTHER, Cat., III, 296, 1861; JORDAN & GILBERT, Synopsis, 784, 1883.

* Allied to *Zoarces anguillaris* is the following species from the Ochotsk Sea:

ZOARCES ELONGATUS, Kner.

Head 5$\frac{2}{3}$; depth 11$\frac{1}{4}$. D. 80, XII, 22. A. 90 or more. Lateral line extending somewhat beyond pectorals. Color brownish, no brown streak behind eye; dorsal with 12 to 14 large dark spots which extend on the back as faint bands, between which are smaller ones. Known from 1 specimen, 10$\frac{1}{2}$ inches long, from Decastris Bay, near the mouth of the Amur. (Kner). (*elongatus*, elongate.)

Zoarces elongatus, KNER, Sitzber. k. k. Akad. Wien 1868, 52, taf. 7, f. 2, Ochotsk Sea. (No. 1502, Wien Mus.)

932. EMBRYX, Jordan & Evermann.

Embryx, JORDAN & EVERMANN, new genus (*crotalinus*).

This genus differs from *Lycodopsis* in the very slender body, the depth being 12 to 16 times in the length, and especially in the ventral position of the lateral line which is faint and incomplete, only the anterior descending portion developed. Deep seas. (ἔν, in; βρύξ, abyss.)

  a. Ventrals nearly as long as eye; head 6¾ in length; no scales on head.

                                       CRASSILABRIS, 2813.

  aa. Ventrals shorter than pupil; head 5½ in length; head with some scales.

                                       CROTALINUS, 2814.

2813. EMBRYX CRASSILABRIS (Gilbert).

Head 6¾; depth 16; maxillary reaching vertical from front of pupil, 3 in head; exposed portion of eye 6; snout 4; width of snout 3. Body exceedingly slender. Occiput flat, forming a right angle with the descending cheeks, the snout short and wide, the upper lip conspicuously thickened and fleshy on the sides. Upper jaw with a single series of rather large, distant teeth; mandible with a broad patch of cardiform teeth anteriorly, which becomes abruptly constricted on middle of lateral portion of jaw, the inner series alone continued backward toward angle. Palate smooth. Head not conspicuously excavated with mucous canals; series of pores present on mandible and sides of head. Gill openings continued forward to below pectorals, and about to vertical from middle of opercle; the width of the isthmus ¼ the length of slit. Opercular flap with a wide membranaceous border, produced backward and largely covering base of pectorals. Gill rakers very little developed, about 12 movable rudiments on horizontal limb of arch. Origin of dorsal in front of middle of pectorals, slightly farther from occiput than is the latter from front of eye; distance from origin of anal to tip of snout 3⅓ in total length; ventrals nearly as long as eye, inserted under middle of opercle; pectorals with 14 or 15 rays, the upper portion of fin longest, the lower rays rapidly shortened, the longest rays ⅓ as long as head. Scales small, circular, covering nape, breast, and under side of pectorals, but absent on head. Lateral line single, inconspicuous, running below middle of sides, ventral in position, the pores not developed on the scales. Color light brownish above, dark below; lower side of head, margins of snout, gill membranes, part of opercles, and margins of vertical fins jet black; ventrals and posterior face of pectorals black; anterior face of pectorals light glaucous blue, margined with black; lining of mouth and gill cavity and peritoneum black. Pacific coast of southern California. A single specimen, 12 inches long, from *Albatross* Station 2839. (Gilbert.) (*crassus*, thick; *labrum*, lip.)

Lycodopsis crassilabris, GILBERT, Proc. U. S. Nat. Mus. 1890, 106, off southern California. (Type, No. 44280. Coll. *Albatross*.)

2814. EMBRYX CROTALINUS (Gilbert).

Head 5½; depth 12; maxillary reaching to behind middle of pupil, 2⅔ in head; eye 6; interorbital width 14; snout 4. Body very slender, with

much the appearance of *Lycenchelys paxillus,* the cheeks tumid, much projecting laterally, the greatest width of head more than ⅓ its length.
Snout short and broad, much depressed, the head scarcely constricted
opposite orbits. Eyes with little lateral range. In the single type specimen the upper jaw greatly overlaps the lower, the mandibular band of
teeth shutting entirely within those on premaxillaries. Teeth in upper
jaw in a single series, 2 or 3 small teeth sometimes present anteriorly,
giving traces of an inner series. In lower jaw the teeth are sparsely set
in a broad band anteriorly, becoming suddenly contracted to a single series
on middle of sides. None of the anterior mandibular teeth enlarged, 2 or
3 of posterior teeth on sides larger and hooked backward. No teeth on
vomer or palatines. Nostril in a short tube. Gill slits wide, reaching to
below pectorals, but not extending farther forward below than above.
Width of isthmus ⅓ length of slit. A series of 7 pores along mandible
and preopercle; a second series of 7 or 8 extending from snout along sides
of head above premaxillaries. Lateral line faint, descending, its position
ventral. Dorsal inserted over middle of pectorals, its origin as far from
occiput as is the latter from front of pupil; distance of front of anal
from snout equals ⅓ length of body; ventrals short, less than length of
pupil; pectorals with posterior margin obliquely truncate, the upper
rays longest, the lower growing regularly shorter, thickened at tips, the
rays 15 or 16, the longest 2⅔ in head. Scales small, embedded, covering body and most of vertical fins. A few very small, scattered scales
on nape, posterior part of occiput, and contiguous parts of cheeks
and opercles. Lateral line single, indistinct, running obliquely downward
to near base of anal, thence backward, not reaching base of caudal fin.
Color dark brown, black on opercles, sides of snout, fins, and lower parts
generally; a broad light bar across head behind eyes, extending down on
cheeks; some light mottling on mandible and gular membrane; lower
rays of pectorals margined with whitish; lining of mouth, gill cavity,
and peritoneum jet black. North Pacific. Two specimens known; the
type above described from Santa Barbara Islands, the second from *Albatross*
Station 3210, south of Saanak Islands, Alaska, depth 483 fathoms.

On this Dr. Gilbert has the following notes:

"The stomach contained remains of Crustacea. Colors in life, head and
body light brown, the lower parts darker; snout, suborbital region, and
a band across pectorals greenish gilt; no light bar on head. Depth 12¾ in
length; head 5⅓; maxillary 2⅓ in head; eye 7, equal to interorbital width.
Width of bone between orbits 17 in head. Snout 3⅓ in head. Teeth above
in a narrow band, reaching only about halfway of gape. In the mandible, teeth are absent on posterior ⅔ of gape. The gill slit extends a little
farther forward below than above. Ventrals as long as pupil. Longest
pectoral ray 2¾ in head. Head wholly scaled behind eyes. Lateral line
not evident."

(*crotalinus,* from *Crotalus,* κρόταλος, a rattlesnake.)

Lycodopsis crotalinus, GILBERT, Proc. U. S. Nat. Mus. 1890, 105, Albatross Station, 2980, off
Santa Barbara Islands. (Coll. *Albatross.*)

933. LYCODOPSIS, Collett.

Lycodopsis, COLLETT, Proc. Zool. Soc. London 1879, 381 *(pacificus).*
Leurynnis, LOCKINGTON, Proc. U. S. Nat. Mus. 1879, 326 *(paucidens).*

Body moderately elongate, the depth 8 to 9 times in length, covered with small, smooth, embedded scales. Lateral line rather faint, extending along middle of side. Head large; snout broad and long; interorbital space very narrow; mouth large, horizontal; teeth conical, those of the upper jaw in a single row; those of the lower in a band in front, the inner series enlarged, larger than the upper teeth; no teeth on vomer or palatines. Ventral fins very small; vertical fins continuous, without spines. Sexes more or less unlike, the mouth larger in the male. Pacific Ocean. (*Lycodes;* ὄψις, appearance.)

2815. LYCODOPSIS PACIFICUS (Collett).

Head 4⅓ (male) to 5⅓ (female); depth 8 (male) to 8⅓ (female). D. 100; A. 85. Female *(pacificus),* head comparatively short; orbital region not restricted, nor cheeks tumid; mouth comparatively small, the maxillary reaching center of pupil. Male *(paucidens),* with the head and mouth large, the snout very broad, the interorbital region constricted; maxillary reaching posterior edge of orbit. Head, nape, and axil of pectoral naked. Dorsal and anal fins enveloped in thick skin, which is covered with embedded scales like those on the body; pectoral ⅓ the length of head in female, ⅔ in male; ventrals ⅓ length of orbit; mandible ⅓ length of head in female, ⅔ in male; distance from snout to base of dorsal 4⅓ in length in female, 3¾ in male. Lateral line lateral in portion. Light reddish olive, becoming lighter below; vertical fins margined with black; the scales paler than skin, forming light spots; pectorals dusky. Length 12 to 18 inches. San Francisco to Puget Sound; rather common in water of moderate depth offshore. Sexes markedly different.

*Lycodes pacificus,** COLLETT, Proc. Zool. Soc. London 1879, 381, female, Japan. (Coll. Peters) the locality given probably an error.
Leurynnis paucidens,† LOCKINGTON, Proc. U. S. Nat. Mus. 1879, 326, off San Francisco, California, male (Type, No. 23502, U. S. Nat. Mus. Coll. W. N. Lockington); JORDAN & GILBERT, Synopsis, 785, 1883.
Lycodopsis paucidens, GILL, Proc. U. S. Nat. Mus. 1880, 248.
Lycodopsis pacificus, JORDAN & GILBERT, Synopsis, 785, 1883.

934. APRODON, Gilbert.

Aprodon, GILBERT, Proc. U. S. Nat. Mus. 1890, 106 *(cortezianus).*

This genus differs from *Lycodes* only in the dentition, the teeth being present in a single strong series on the palatines, but none on the vomer.

---

* In regard to the type specimen of *Lycodes pacificus,* Professor Collett writes us as follows (December 2, 1895):
"I got the specimen for describing from the Museum of Berlin from the hands of Professor Peters himself, and he told me that the specimen was from Japan. It| is not impossible that he was mistaken; but I can not have any opinion about that."
In view of the fact that the species is abundant off the California coast, whence Professor Peters had obtained collections, that it has not been found in Japan nor in Alaska, we have no doubt that the locality given by Professor Peters is erroneous, and that the fish really came from California.
† The examination of many specimens leaves no room for doubt that *L. pacificus* is the female and *L. paucidens* the male of the same species.

The genus is thus intermediate between *Lycodes* and *Lycodopsis*. (ἀ, without; πρό, before; ὀδούς, tooth.)

2816. APRODON CORTEZIANUS, Gilbert.

Head 4⅓ to 4½; depth 7½ to 9 in length; head high and narrow, snout broader, but long and very convex. Mouth large, maxillary reaching vertical from middle of orbit, 2⅓ in head; eye 4⅔; snout 3; depth of head 2. Teeth in premaxillaries strong, conical, in a single series; lower jaw with the teeth mainly in 2 series, an outer row of slightly enlarged teeth, and an inner row directed backward, a wide interspace between the two series with occasional scattered teeth only posteriorly; on sides of mandible a single series of teeth similar to those in upper jaw; vomer toothless; palatines with a single series of strong conical teeth. Head without conspicuous mucous pores; a strong ridge on middle of occiput anteriorly; gill slit wide, continued forward to vertical from preopercle, the width of isthmus 5 times in length of slit; gill rakers short, better developed than usual, 15 on horizontal limb of outer arch. The vertical limb of arches joined to gill cover by a fold of the lining membrane of the latter, as in *Macrourus*. Pseudobranchiæ well developed. Origin of dorsal but little behind base of pectorals; the hinder margin of occiput midway between dorsal and front or middle of eye; distance from snout to origin of anal 2⅗ in total length; ventrals inserted under front of opercles, their length about ⅔ of orbit; pectorals very large, broadly rounded, the upper portion of fin longest, the lower rays rapidly shortened, the lowermost with broad, fleshy tips; rays 20 or 21 in number; scales of the usual type, those on abdomen so deeply embedded as to be almost invisible; head, anterior half of nape, breast, and base of pectorals naked; pectorals and ventrals not scaled, other fins partly covered; lateral line little developed, running along middle of sides and tail. Color light brownish, lighter below; vertical fins broadly margined with black, becoming almost wholly black behind; pectorals light at base, black distally, with a conspicuous white edge; ventrals white; lining of mouth white, of gill cavity dusky; peritoneum black. Cortez Banks, near San Diego, California. The types, 6 specimens, the longest 15 inches, from *Albatross* Stations 2925 and 2948, in 339 and 266 fathoms. Dr. Gilbert also records 1 specimen from *Albatross* Station 3349, off the coast of northern California, depth 239 fathoms. (*cortezianus*, from Cortez Banks.)

Aprodon corteziana, GILBERT, Proc. U. S. Nat. Mus. 1890, 107, Cortez Banks, off San Diego. (Type, No. 46457. Coll. *Albatross*.)

935. LYCODES, Reinhardt.

Lycodes, REINHARDT, Kongl. Dansk. Vidensk. Selsk. Naturv., VII, 1838, 153 (*vahli*).
Lycias, JORDAN & EVERMANN, new subgenus (*seminudus*).

Body moderately elongate, more or less eel-shaped, tapering behind, the depth from 6 to 10 times in the length; head oblong; mouth nearly horizontal; lower jaw included; conical teeth on jaws, vomer, and palatines, those on jaws and palatines mostly in a single series. Dorsal fin beginning behind base of pectoral, without any spines; the rays all soft and articulate; pectorals moderate, inserted rather high, its outline rounded; ventral fins small, of 3 or 4 rays. Scales small and embedded, present on part or all of the body, the scaly area more extensive in the adult than in the young. Lateral line faint, sometimes obsolete, normally bent downward behind pectorals and following ventral outline, sometimes with an accessory branch following middle of side; the median branch usually wanting. No air bladder; no anal papilla; pyloric cæca 2 or none. Species numerous, chiefly of the northern seas, inhabiting considerable depths. In general, the male has the head and mouth larger than the female, and the lips thickened. (λυκώδης, wolfish.)

LYCODES:*
   I. Trunk more or less completely scaled.
      a. Dorsal rays about 115; anal rays 90 to 105.
         b. Head 4½ to 5 in length; depth 7 to 8.
            c. Nape wholly scaly.
               d. Lateral line double, with a median and a ventral branch; pectoral rays 22; body blackish with yellowish cross bands or series of spots.   ESMARKII, 2817.
               dd. Lateral line simple, ventral; body blackish, the young with 6 darker cross bands.   VAHLII, 2818.
            cc. Nape naked; lateral line obsolete; color plain brown, the fins edged with darker, pectoral rays 21; ventrals short.   CONCOLOR, 2819.
         bb. Head 5⅔ in length; depth 9; a naked area around dorsal; pectoral rays 19; lateral line ventral; color brownish mottled, the young barred; a black blotch at front of dorsal.   ZOARCHUS, 2820.
      aa. Dorsal rays 85 to 105; anal rays 68 to 93.
         e. Head large, 3⅔ to 4½ in length; ventrals about as long as eye; depth 8 to 9½ in length; body chiefly scaly, the fins naked.
            f. Body brownish, with a fine network of black lines on head and body, those on body in 5 groups; dorsal edged with black; lateral line probably developed anteriorly only, figured as median; pectorals broad, of about 23 rays.   RETICULATUS, 2821.
           ff. Body not covered with a network of black lines.
               g. Color pale, with dark bands and 2 ocellated spots on the forehead; pectoral rays about 17; lateral line figured as lateral.
                    PERSPICILLUM, 2822.
               gg. Color grayish, without bands or spots; pectoral rays about 20; lateral line single, ventral.   FRIGIDUS, 2823.
        ee. Head short, 5 to 5½ in length.
           h. Pectoral broad, of 23 or 24 rays; lateral line single, ventral; color plain.   TERRÆ-NOVÆ, 2824.
          hh. Pectorals narrow, of about 18 rays; ventral fins shorter than eye; lateral line obsolete, or nearly so.
             i. Dorsal rays 101 to 105; anal rays 81 to 90; dorsal and anal without dark markings; ventrals more than ½ length of eye; jaws with enlarged flaps of skin.
               j. Body in adult not barred, but with 4 dark longitudinal stripes.   DIGITATUS, 2825.
               jj. Body with 14 to 16 pale crossbars above, which disappear in the adult.   PALEARIS. 2826.
            ii. Dorsal rays 85; anal 74; ventrals minute, not ⅓ length of eye; flaps of jaws narrow or obsolete.   BREVIPES, 2827.

---

* The analytical key to the species here given is far from satisfactory. The species should be divided into groups distinguished by the development of the lateral line and the breadth of the pectoral; unfortunately the last-named character has been neglected in most of the current descriptions; we have examined all the species accessible to us.

Lycias (λύκος, wolf):

II. Trunk naked anteriorly, scaled only on the tail or posterior half.

 k. Dorsal fin scaled posteriorly; color brown, with faint yellow transverse bands on back.   NEBULOSUS, 2828.

 kk. Dorsal fin naked; color uniform pale grayish brown without spots or bands; pectoral rays 21; lateral line single, median.   SEMINUDUS, 2829.

Subgenus LYCODES.

2817. LYCODES ESMARKII, Collett.

Head $4\frac{1}{3}$; depth 8. D. 110 to 116; P. 22; A. 95; V. 4. Body behind front of dorsal scaled; vertical fins scaly; nape scaly; snout obtuse; maxillary not more than $\frac{1}{3}$ head; lateral line indistinct, divided, having a median branch besides the ventral series of pores, the median series faint, soon obsolete; pectorals 8 in length; vertebræ 25 + 87. Brownish black, with a whitish-yellow patch on the nape, and 5 to 8 transverse bands of the same color across the dorsal and posteriorly across the anal, these bands becoming broken into annular spots with age. North Atlantic; recorded from Finmark and Spitzbergen. American specimens from the Gulf Stream in about lat. 40°. (Collett.) (Named for Professor Lauritz Esmark, of Copenhagen.)

Lycodes esmarkii, COLLETT, Norges Fiske, 95, 1874, Varanger Fjord, Finmark (Coll. Lensmand Klerk and Prof. Esmark); COLLETT, Norske Nord-Havs Exp., Fiske, 84, pl. *3*, fig. 22, 1880.

Lycodes vahli, GOODE & BEAN, Proc. U. S. Nat. Mus. 1879, 209, not of REINHARDT; JORDAN & GILBERT, Synopsis, 786, 1883; GOODE & BEAN, Oceanic Ichthyology, 303, 1896.

2818. LYCODES VAHLII, Reinhardt.

Head $4\frac{1}{3}$; depth 8. D. 116; A. 93; V. 4. Head nearly twice as long as high; snout long, maxillary reaching to opposite middle of eye; distance of vent from ventrals nearly equal to length of head; ventral fins less than $\frac{1}{4}$ as long as pectorals; vertical fins scaly; body wholly scaly; lateral line distinct, ventral in position; vertebræ 25 + 87. Brownish yellow, with 6 blackish cross bands extending on the dorsal fin and confluent on the belly, the first cross band on and below the anterior dorsal rays, the second above the vent; adults nearly uniform blackish. Coast of Greenland. (Günther.) (Named for Martin Vahl, an early Danish naturalist.)

Lycodes vahlii, REINHARDT, Kon. Dan. Vidensk. Selsk. Nat. Math. Afh., VII, 1838, 153, tab. V, Greenland; GILL, Cat. Fishes East Coast N. A., 46, 1861; GÜNTHER, Cat., IV, 319, 1862; JORDAN & GILBERT, Synopsis, 786, 1883; GOODE & BEAN, Oceanic Ichthyology, 303, 1896.

2819. LYCODES CONCOLOR, Gill & Townsend.

Head 5 in total; depth about $7\frac{1}{2}$; eye $7\frac{3}{4}$ in head; snout 3; ventral fin 2 in eye; pectoral 2 in head. D. 118; A. 98; P. 21. Body rather elongate, covered with very small, entirely separated embedded scales which become more distinct anteriorly and extend in advance of the dorsal fin and scapular region, as well as on the vertical fins; lateral line obsolete; pectorals with scattered scales on external and internal surfaces near base; a specialized area of smaller scales behind base of pectoral and a naked area around upper axilla of pectoral; head moderate, entirely naked; nape naked. Upper teeth in a cardiform band in front, thinning out behind. Lips rather thin. Color nearly uniform, only relieved by the apparently lighter hue of the scales and the somewhat darker margins of the fins; the scales paler than the ground color, which is thus covered with whitish or silvery specks. Bering Sea. Only the type known, its length 22 inches, from which we have taken the above description. (*concolor*, uniformly colored.)

Lycodes concolor, GILL & TOWNSEND, Proc. Biol. Soc. Wash., XI, 1897 (Sept. 17, 1897), 233, Bering Sea, lat. 55° 19' N., long. 168° 11' W., Albatross Station 3608, (Aug. 12, 1895), in 276 fathoms. (Type, No. 48764, U. S. Nat. Mus. Coll. *Albatross*.)

2820. LYCODES ZOARCHUS, Goode & Bean.

Head nearly 5⅔ in total length; depth 9; eye 4 in head = snout. D. 116; A. 102; P. 19. Body covered with conspicuous embedded scales which extend behind the dorsal and anal, leaving only a narrow naked margin around these fins; head and pectorals naked. A lateral line begins slightly above the upper angle of the gill opening, rapidly curving downward and extending along the lower part of the body not far from base of anal fin; it can be traced above the anterior ⅔ of the anal. Interorbital distance, measured on the bone, 4 in eye; nostrils placed close to upper lip and as far from each other as from the eye; maxillary reaching to vertical through middle of eye; upper jaw 2¼ in head; mandible nearly ½ head; mandible with a conspicuous flap on each side, about as long as eye, beginning at a distance from the symphysis equal to ¼ length of eye; inner edge of mandible also with a slightly elevated ridge of skin. Length of intermaxillary series of teeth equal to ⅓ length of head; length of palatine series nearly equal to that of intermaxillary; vomerines in a round patch; mandibular teeth in 3 series; width of gill opening ⅔ length of head; ventrals in front of base of pectorals, their length 8 in head. Distance between lower angles of gill opening nearly ¼ length of head; origin of dorsal distant from the head a space equal to ¼ length of head, slightly behind middle of pectoral; pectoral, when extended, reaching to about vertical from sixth dorsal ray; longest ray of dorsal about ¼ length of head; anal origin under seventeenth ray of dorsal; vent under fifteenth ray of dorsal; longest pectoral ray contained about 9½ times in total length. Lateral line distinct, ventral in position, the median pores absent. Color grayish brown, lighter on the belly and under surface of the head; sides irregularly mottled with darker, a narrow dark edge at tip of first 4 dorsal rays. In a young example (No. 39299, U. S. Nat. Mus.) the mottlings on the sides are band-like, the bands not extending below the middle of the body entirely. This example is from lat. 44° 26' N., long. 57° 11' 15'' W., 190 fathoms. The type of the description is a specimen 366 mm. long, obtained by the *Albatross* in lat. 44° 46' 30'' N., 130 fathoms, off Nova Scotia. (*Zoarchus*, a synonym of *Zoarces*; from ζωαρχής, viviparous.)

Lycodes zoarchus, GOODE & BEAN, Oceanic Ichthyology, 308, 1896, off Nova Scotia, in 130 fathoms. (Type, No. 39298. Coll. *Albatross*.)

2821. LYCODES RETICULATUS, Reinhardt.

Head 4; depth about 8. D. 94; A. 75; V. 4. Body entirely scaly; lateral line faint, developed anteriorly (fide Günther's plate), probably becoming ventral; vertical fins naked. Head twice as long as high; snout long; maxillary extending to behind middle of eye; distance from vent to ventrals more than length of head; cæca 2. Brownish, with reticulated black lines on the head and body, those on the body disposed in 5 groups or cross bands, the 3 anterior of which emit 1 or 2 vertical streaks on the dorsal fin; dorsal dark edged. Length 14 inches. North Atlantic, from Greenland south to Narragansett Bay, in 17 to 140 fathoms; abundant also in northern Europe. (Eu.) (*reticulatus,* netted.)

Lycodes reticulatus, REINHARDT, Kong. Dansk. Vid. Afh., VII, 1838, 167, Greenland; GÜN-THER, Cat., IV, 320; GILL, *l. c.,* 260; COLLETT, Nord-Havs Exp., 84; JORDAN & GILBERT, Synopsis, 787, 1883; GOODE & BEAN, Oceanic Ichthyology, 305, 1896.
Lycodes rossi, MALMGREN, Om Spetsbergen Fiskfauna, 516, 1864, Spitzbergen.
Lycodes gracilis, SARS, Christ. Vid. Selsk. Forh. 1866, Dröbak.

2822. LYCODES PERSPICILLUM, Kröyer.

This species is distinguished by a light body color and dark bands, also 2 ocellated spots on the forehead, which have suggested the specific name. Still further separated from the previously known species of *Lycodes* by the smaller number of fin rays, larger eye, etc. (Kröyer.) Greenland and southward in deep water. Specimens were obtained by the *Albatross* from Station 2491, in 45° 24' 30'' N. lat., 58° 35' 15'' W. long., at a depth of 59 fathoms, and from Station 2456, in 47° 29' N. lat., 52° 18' W. long., at a depth of 86 fathoms.

The following is the substance of Dr. Günther's description:

Head 4 in total length; depth nearly 8. Head not quite twice as long as high; snout long; upper maxillary extending to below middle of eye. Distance of vent from ventrals nearly equal to length of head. Yellowish, with 9 or 10 brownish cross bands, edged with dark brown, and broader than the interspaces, the first occupying the upper parts of the head and inclosing a pair of roundish, yellowish spots situated behind the level of the eyes;. the second cross band is on and before the anterior dorsal rays.

(In the figure of Goode & Bean the lateral line is represented as median, which is probably not correct.) (*perspicillum,* eyebrow, from the spot above the eye.)

Lycodes perspicillum, KRÖYER, Dansk. Vidensk. Selsk. Afhandl., XI, 1845, 233, Greenland; GÜNTHER, Cat., IV, 320, 1862; GILL, Proc. Ac. Nat. Sci. Phila. 1863, 260.

2823. LYCODES FRIGIDUS, Collett.

Head 4 to 4$\frac{1}{2}$ in total length; depth 6$\frac{1}{2}$ (to 9$\frac{1}{2}$, young). D. 93 to 98 (including $\frac{1}{3}$ of the caudal, 99 to 104); A. 80 to 85 (including $\frac{1}{3}$ of the caudal 86 to 90); P. 20 to 21; V. 3. Head wide and flat. Scales with very conspicuous mucous cavities below, small, covering the entire body, but not the head, nor the base of the dorsal and anal fins. In the young the middle of the belly, the base of the fins, and the fins themselves are usually naked.

Teeth present on intermaxillary, mandible, palatines, and vomer; lateral line low, extending from upper end of gill opening in a curved direction down toward vent from which it runs close along anal to end of tail. (Goode & Bean.) Pectoral fin obliquely truncate at tip, appearing furcate when not spread open. North Atlantic and Arctic Ocean, from Spitzbergen south to the New England coast, where many specimens were taken in 516 to 1,423 fathoms. (Eu.) One of Collett's specimens from Hammerfest, examined by us.) (*frigidus,* frozen.)

Lycodes frigidus, COLLETT, Forh. Selsk. Christ. 1878, Nos. 14 and 15, Beeren Island and Spitzbergen; COLLETT, Norske Nord-Havs Exp., 96, pl. 3, f. 23, 24, 1880; GOODE & BEAN, Oceanic Ichthyology, 305, 1896.

2824. LYCODES TERRÆ-NOVÆ, Collett.

Head 5 to 5½; depth 8 to 11. D. 106 to 108; A. 89 to 93; P. 23 or 24. Body slender, head small; pectorals broad; maxillary reaching to middle of eye; band of palatine teeth very short, scarcely ½ length of maxillary band; body entirely scaly, head naked; lateral line ventral, extending along edge of belly, the median branch wanting; vent before middle of body. Color lost in type, the only specimen known. Banks of Newfoundland, in 155 fathoms. (Collett.) (*terra,* land; *novus,* new, from Newfoundland.)

Lycodes terræ-novæ, COLLETT, Campagnes Scientifiques, L'Hirondelle, x, 1896, 54, Bank of Newfoundland, Hirondelle Station 162, in 155 fathoms. (Coll. Albert, Prince of Monaco.)

2825. LYCODES DIGITATUS, Gill & Townsend.

Head 5 in total; depth about 8½; eye 6½ in head; snout 3; ventral fin 1¼ in eye; pectoral 1⅗ in head. D. 101; A. 81; P. 18. Body moderately elongate; covered with small, entirely separated embedded scales, which become nearer anteriorly and extend in advance of the dorsal fin as well as on the vertical fins; no specialized area of smaller scales behind base of pectorals; pectorals scaleless; head moderate, entirely naked; nape naked; upper jaw with outer row of close-set teeth, broader in front; teeth on vomer and palatines; lips rather thick. Color in alcohol, brownish yellow, suffused with reddish in front, variegated, darker anteriorly, with 4 dark longitudinal stripes most distinct about middle of body, fading out backward; fins light and without dark margins; head dark above and laterally light below. Bering Sea. Only the type known, from which we have taken this description, its length 18 inches; possibly the adult of *L. palearis,* but the pectoral fins are shorter than in the latter. (*digitatus,* fingered.)

Lycodes digitatus, GILL & TOWNSEND, Proc. Biol. Soc. Wash., XI, 1897 (Sept. 17, 1897), 232, Bering Sea, lat. 56° 14′ N., long. 164° 8′ W., at Albatross Station 3541, in 49 fathoms. (Type, No. 48765, U. S. Nat. Mus. Coll. *Albatross.*)

2826. LYCODES PALEARIS, Gilbert.

Head 5¼ in length; depth 9½ to 11 in length, 2⅛ in head; eye 5 to 6 in head, 1½ to 2 in snout. Dorsal with about 105 rays, counted to middle of caudal; anal about 90; pectoral 18; ventrals 1½ to 1⅛ in eye, twice as long as in *L. brevipes*; pectorals 1⅛ in head. Head naked; nape more or less naked, the scaleless area variable in extent, sometimes confined to its anterior third, sometimes reaching nearly to front of dorsal; body sparsely covered with embedded scales; axil naked; lateral line short, decurved, extending scarcely beyond middle of pectorals. Anal origin under eighteenth dorsal ray. Teeth present in jaws, vomer and palatines, those in premaxillaries laterally in a single series which widens anteriorly into a rather broad patch, the outer teeth somewhat enlarged, especially in front; all the premaxillary teeth shut outside on the mandibular series which are opposed to those on vomer and palatines; mandibular teeth arranged similarly to those in upper jaw, the lateral series somewhat enlarged, continuous with the inner edge of the symphyséal patch; vomerine teeth bluntly conic, 3 or 4 in number; palatines in a single series. Snout long, prominent, the upper jaw projecting beyond the lower for a distance equaling ⅔ of orbit; upper lip thin, much expanded laterally, continuous posteriorly with the lower lip which forms a wide free membranaceous lobe opposite middle of each mandible; anteriorly the lower lip becomes abruptly contracted and adnate to the jaw, leaving the symphyseal portion without free margin; inner edge of mandible with wide membranaceous borders, which increase in width anteriorly where they terminate in a pair of acutely pointed free flaps; these and the membranaceous margins very conspicuous in both young and old individuals. In *L. brevipes* they are very inconspicuous, becoming evident in adults only. General color brownish olive, growing lighter on the lower parts; dorsal with 14 to 16 white vertical bars, extending in young examples across back and sides and onto anal fin, in adults confined to the fins, and frequently indistinct or wanting; anterior dorsal angle frequently black, separated from remainder of fin by a curved white bar; dorsal and anal not black margined as in *L. brevipes;* in the latter, the white lateral bars are 9 to 12 in number, and are usually confined to upper half of body; there is also no black spot on anterior dorsal rays. This species is very close to *L. brevipes* Bean, differing constantly in the longer ventrals, the greater development of mandibular and labial folds, the more numerous white bars, and the smaller eye. Bering Sea. Three specimens, 113 to 166 mm. long, from *Albatross* Stations 3253 and 3254, in Bristol Bay, in 36 and 46 fathoms. (Gilbert.) (*paleœ*, the wattles of a cock.)

Lycodes palearis, GILBERT, Rept. U. S. Fish Comm. 1893 (1896), 454, Bristol Bay, Alaska. (Coll. *Albatross*.)

2827. LYCODES BREVIPES, Bean.

Head 5 in total length; depth 10. D. 85 to middle of caudal; A. 74; P. 21. Body covered with scales except immediately behind pectoral fins; head naked; dorsal and anal fins minutely scaled; diameter of eye equals the length of the snout, 4 in head; dorsal origin nearly over middle of pectoral; anal origin under eighteenth ray of dorsal; ventrals minute, scarcely more than ⅓ diameter of eye; pectorals 9 in length of the body; lateral line single, very faint, ventral in position, abruptly decurved and becoming obsolete over about the tenth anal ray. A narrow light band

across the nape and from 9 to 11 across the back extending downward
about to median line and becoming obscure in adults; dorsal and anal
each with a narrow dark margin. (Bean.) Aleutian Islands to Kadiak;
abundant; taken by us in large numbers off Karluk in 1897. (*brevis,*
short; *pes,* foot.)

Lycodes brevipes, BEAN, Proc. U. S. Nat. Mus. 1890, 38, between Unga and Nagai islands,
at Albatross Station 2848, in 110 fathoms. (Type, No. 45362. Coll. *Albatross.*)

Subgenus LYCIAS, Jordan & Evermann.

2828. LYCODES NEBULOSUS, Kröyer.

D. 87; A. 68; P. 19; V. 3.   Body naked anteriorly, the posterior part of
dorsal fin scaly; the anal naked or nearly so. Brown, with small, faint,
yellow, transverse bands across the back. Greenland. (Kröyer.) An
imperfectly described species, not recognized by any recent writer.
This species and the next should perhaps be placed in *Lycodalepis.* (*nebulosus,* clouded.)

Lycodes nebulosus, KRÖYER, Kong. Dan. Vidensk. Sel. 1844, 140, Greenland; GILL, Proc.
Ac. Nat. Sci. Phila. 1863, 261; JORDAN & GILBERT, Synopsis, 787, 1883.

2829. LYCODES SEMINUDUS, Reinhardt.

Head 3½; depth 7. D. 91; A. 74; P. 21. Body naked in front of vent,
scaly behind; fins naked. Head large. Distance of ventrals from vent
somewhat more than length of head; cæca 2. Color uniform pale grayish
brown, without spots or bands. North Atlantic, from Greenland to Spitzbergen; rare. (Collett.) (*semi-,* half; *nudus,* naked.)

Lycodes seminudus, REINHARDT, Kong. Dansk. Selsk., etc., 1838, 221, Omenak, Greenland;
GÜNTHER, Cat., IV, 320, 1862; JORDAN & GILBERT, Synopsis, 787, 1883; GOODE & BEAN,
Oceanic Ichthyology, 307, 1896.

936. LYCODALEPIS, Bleeker.

Lycodalepis, BLEEKER, Verl. Akad. Amst., Ed. 2, VIII, 1874, 369 (*mucosus*).

This genus differs from *Lycodes* in the absence of scales on trunk and
fins; scattered scales sometimes present on the tail only. (λυκώδης,
Lycodes; ἀλεπίς, without scales.)

   a. Color brownish, with many cross bands and streaks of cream color; head 4½ in length;
     depth 8; lateral line obsolete; tail sometimes with a few scales.   POLARIS, 2830.
   aa. Color blackish, with about 5 narrow pale cross bars on back; head 3½ in length;
     depth 8; lateral line double, a median and a ventral series of pores being faintly
     developed.   MUCOSUS, 2831.

2830. LYCODALEPIS POLARIS (Sabine).

Head 4½; depth 8. D. 85; A. 67; P. 18; V. 3; Br. 6. Head depressed,
its greatest width ¾ of its length; distance from tip of snout to nape nearly
equaling greatest width of head, 6 in length; upper jaw 1¾ to 2 in head,
extending to vertical of hind margin of orbit, larger in male than in the
female; a full series of teeth on premaxillaries, and in front of these a few
smaller teeth form an outer imperfect series; a toothless space at symphysis,

first tooth on each side of this larger than any of the rest; 1 complete series of teeth on mandible, and in front of it, about the symphysis, 2 irregular short series; a few teeth in a cluster on head of vomer; palatines with a short single series; teeth all slender and slightly recurved; long diameter of eye 9 in head. Pectoral $1\frac{3}{5}$ in head; ventral about as long as eye; longest dorsal ray $3\frac{2}{3}$ in head; vent in middle of total length, immediately behind third cross band; longest anal ray $4\frac{1}{3}$ in head; scattered scales present on posterior two-thirds of tail in 1 specimen (type of *L. coccineus*), wholly wanting in the others, typical of *L. turneri;* no scales on the fins; no trace of lateral line. Color light brown; abdomen grayish brown; lower parts of head cream; a band of cream on the anal from origin of rays to about their middle; a crescentic V-shaped band of same color, mottled with umber, crossing nape and continuing behind pectorals, extending backward to the first cross bar; a streak of cream more or less interrupted by umber, extending backward from eye across cheek almost to end of operculum; 10 bands of cream color, bordered with dark umber, from tips of dorsal rays extending on lower half of body, becoming wider and somewhat broken below middle of body; a very indistinct caudal tip of cream color. In young examples these markings are very distinct; in older ones they grow progressively more obscure, the oldest having scattering blotches of cream color instead of bands, the V-shaped nuchal band persisting longest. The type of *Lycodes coccineus* is described as brown, red below; pectorals reddish brown above, carmine below; 9 bluish-white bands on the dorsal; a few whitish blotches on sides and on head; anal brownish red; head white below; a whitish blotch as large as eye at upper angle of gill opening. Length 18 to 20 inches. Arctic Ocean, Bering Straits, and adjacent waters south to St. Michaels. Here described from the type of *Lycodes coccineus* and from a number of specimens from Point Barrow referred to *Lycodes turneri*. Evidently all belong to the same species, but 1 has a scaly tail while the others are wholly naked. In 2 large examples, supposed to be males, the head is very much depressed, broad and flat, and the maxillary is more than $\frac{1}{2}$ head. In the others the head is smaller, less flattened, with smaller mouth, the maxillary 2 in head. These are doubtless females and young. The species should probably stand as *Lycodalepis polaris.* (*polaris,* polar.)

?*Blennius polaris,** SABINE, Parry's Journal, Voyage 1819-20, Supplement, 212, North Georgia.

Lycodes turneri, BEAN, Proc. U. S. Nat. Mus. 1878, 464, St. Michaels, Alaska (Type, No. 21529. Coll. Dr. Lucien M. Turner); TURNER, Contr. Nat. Hist. Alaska, 93,.pl. 4, 1886.

Lycodes coccineus, BEAN, Proc. U. S. Nat. Mus., IV, 1881, 144, Big Diomede Island, Bering Strait (Coll. Dr. Bean. Type, No. 27748, 20 inches long, with scales on the tail); JORDAN & GILBERT, Synopsis, 787, 1883.

Blennius (Zoarches?) polaris, RICHARDSON, Fauna Bor.-Amer., III, 94, 1836.

Lycodes polaris, GÜNTHER, Cat., IV, 321, 1862.

Lycodalepis turneri, JORDAN & GILBERT, Synopsis, 788, 1883; SCOFIELD, in JORDAN & GILBERT, Fur Seal Invest., 1898.

Lycodalepis polaris, JORDAN & GILBERT, Synopsis, 788, 1883.

---

* *Blennius polaris* is thus described: Without any scales; length of the pectoral exceeding twice its breadth, having 15 rays. Yellowish, lighter on the belly, with 11 large saddle-like markings across the back, the middle of these markings being much lighter than their edges; the whole back and the sides marbled. (Sabine.) Coast of North Georgia.

2831. LYCODALEPIS MUCOSUS (Richardson).

Head $3\frac{1}{7}$; depth 8. D. (including $\frac{1}{2}$ of caudal) 90; A. (including $\frac{1}{2}$ of caudal) 71; P. 18; V. 3. Body robust, head very large; snout 3 in head; interorbital area 6 in head; nostrils much farther from eyes than from each other, their distance from eyes $4\frac{1}{8}$ in head; upper jaw $6\frac{1}{4}$ in total length; lower jaw $6\frac{3}{8}$; eyes small, close together, their long diameter 11 in the head; distance from tip of snout to base of pectoral fin $3\frac{1}{2}$ in total length; pectoral fin $6\frac{3}{8}$; length of ventrals equaling long diameter of eye. (Goode & Bean.) Lateral line (in specimens from Cumberland Gulf) very faint, but with both median and ventral branch. Blackish, with irregular white markings in the form of 5 faint and narrow bars across the back. Arctic America. (*mucosus*, slimy.)

Lycodes mucosus, RICHARDSON, Last Arctic Voyage, 362, pl. 26, 1855, Northumberland Sound; BEAN, Bull. U. S. Nat. Mus., No. 15, 112, 1879; GOODE & BEAN, Oceanic Ichthyology, 306, 1896.
Lycodalepis mucosus, JORDAN & GILBERT, Synopsis, 788, 1883.

937. LYCENCHELYS, Gill.

Lycenchelys, GILL, Proc. Ac. Nat. Sci. Phila. 1884, 110 (*murœna*).

This genus contains small and very slender species differing from *Lycodes* in the elongation of the body, the depth being from 10 to 20 times in the length. The lateral line is single and median in all known species. The genus is very close to *Lycodes*, but the position of the lateral line sufficiently defines it, especially in connection with the slender eel-like form. (λύκος, wolf; ἔγχελυς, eel.)

a. Lower half of pectoral not notably longer than upper; depth 12 to 16 in length.
  b. Dorsal rays 92; anal 88; color grayish, with irregular brown patches. VERRILLII, 2832.
  bb. Dorsal rays 118; anal 110; color brown, the head darker. PAXILLUS, 2833.
aa. Lower half of pectoral considerably longer than upper; head with large pores; depth 14 times in length; color dusky brown. PORIFER, 2834.

2832. LYCENCHELYS VERRILLII (Goode & Bean).

Head $5\frac{3}{4}$; depth about 13; eye 2 in snout. D. 92; A. 88; P. 15; V. 5. Body elongate; head much depressed. Distance of vent from ventrals slightly greater than head, its distance from snout about 3 in body; distance of dorsal fin from snout $\frac{1}{2}$ greater than head; distance of anal from snout twice head; dorsal and anal fins about equal in height, with even margins, not differentiated from caudal, the rays increasing somewhat in length posteriorly; distance of pectoral from snout about equal to head, twice length of pectoral; pectoral reaching vertical from base of second dorsal ray; distance of ventrals from snout less than head, their length less than $\frac{1}{2}$ that of pectorals. Head, body, and fins enveloped in tough, lax skin. Scales cycloid, circular, and ovate, with numerous concentric striæ, and about 18 lobes on margin, the whole perimeter being lobed; scales deeply embedded in the skin at distances from each other equal to their own diameters, most numerous on upper part of body and extending upon base of dorsal; very few scales upon lower half of body, none on anal fin. Upper jaw far overlapping the lower; gape reaching orbit. A series of 6 large pores on each side, extending backward from nostril toward angle of opercle, the fourth of the series under center of orbit; a similar series, 7 on each side, along line of lower jaw from its symphysis to angle of opercle, all slit-like, the others circular. Nostrils at extremities of fleshy tubes. Teeth in lower jaw in 2 rows, nearly uniform in size; teeth of upper jaw in a single series, somewhat enlarged near the symphysis; patches of smaller teeth behind; about 7 teeth on vomer; a single row on palatines; all the teeth curved. Gill opening narrow, the membranes attached to the isthmus. Color, body above lateral line light grayish brown with numerous minute circular dots marking the position of the scales; pearly white below lateral line; brown irregular patches upon sides, bisected by lateral line, the lower half color of dorsal, that above darker and with the white dots, these brown patches 7 to 10 in number; a brown spot on tip of tail; abdominal region livid blue. Coast of Massachusetts, in deep water; a dwarf species very small in size. (Named for Prof. Addison E. Verrill of Yale University.)

Lycodes verrillii, GOODE & BEAN, Amer. Journ. Sci. Arts, XIV, 1877, 474. off coast of New England in the Gulf Stream; JORDAN & GILBERT, Synopsis, 786, 1883.

Lycenchelys verrilli, JORDAN, Cat., 124, 1885; GOODE & BEAN, Oceanic Ichthyology, 309, figs. 277 and 277 A, 1896.

2833. LYCENCHELYS PAXILLUS (Goode & Bean).

Head 8; depth 16; eye 3½ to 4 in head, equal to snout, which is 4 times interorbital width. D. (with ½ of caudal) 118; A. 110; P. 16; V. 3. Body attenuate, head broad, flat above, with declivous profile; cheeks full and protuberant; teeth stout, recurved, and sharply pointed, in a single series in each jaw, except at the symphysis; a few teeth clustered at the head of the vomer; palatines with a single series; the tubular nostril much nearer tip of snout than eye. Lateral line median, faint and short (in specimens examined by us). Dorsal beginning over tip of pectoral; ventral little longer than pupil. Scales very small, present everywhere except on head and pectorals, nearly covering vertical fins. Light brown, the head somewhat darker. Gulf stream, lat. 35° to 41° N., in deep water, 263 to 904 fathoms. (Goode & Bean.) (*paxillus*, a peg.)

Lycodes paxillus, GOODE & BEAN, Proc. U. S. Nat. Mus. 1879, 44, between LaHave and Sable Island Banks (Type, No. 22177. Coll. Capt. J. W. Collins), a male in breeding form; JORDAN & GILBERT, Synopsis, 785, 1883.

Lycodes paxilloides, GOODE & BEAN, Bull. Mus. Comp. Zool., X, No. 5, 207, 1883, off New-foundland (Type in M. C. Z. Coll. *The Blake*); a normal, not sexually distorted individual.

Lycenchelys paxillus, JORDAN, Cat., 124, 1885; GOODE & BEAN, Oceanic Ichthyology, 311, figs. 279 and 282, 1896.

2834. LYCENCHELYS PORIFER (Gilbert).

Head 5⅔; depth 14. Body very slender. Head much contracted opposite orbits, the snout expanded, as in *Lycodopsis paucidens*. Mouth moderate, the maxillary reaching vertical from front of pupil, 3⅓ in head; eye 5⅔; snout 3⅔; interorbital width ⅓ eye. Teeth in front of premaxillaries in 2 series, merging into 1 laterally, the outer series anteriorly somewhat enlarged; teeth in front of mandible in a broad band, narrowing laterally to a single series, none of them enlarged; vomer and palatines with single series. Head with 2 series of large and very conspicuous elongate pores, 1 series on mandible and subopercle, the second parallel with it on level of snout. Gill openings wide, extending forward beyond preopercular margin, the width of isthmus less than ¼ length of slit. Distance from origin of dorsal to tip of snout 4¼ in length. Median dorsal rays simply forked near base, those posteriorly in both dorsal and anal repeatedly subdividing. Distance of anal from snout 2⅝ in length; pectorals rounded, the lower half of fin longer than the upper, the rays thickened, the fin containing 15 or 16 rays, its length less than ½ head; ventrals longer and slenderer than usual, each apparently composed of 2 rays closely joined, their length ¼ orbit, inserted unusually far forward, being in advance of preopercular margin. Scales very small, circular, partially embedded, covering body and vertical fins; head, antedorsal region, breast, and a strip connecting the two latter embracing base and axil of pectorals, naked; lateral line median. Color dusky brown, the fins, sides of head, and belly blackish; lining of mouth and gill cavity and peritoneum black. Off Lower California. A single specimen, 12 inches long, from *Albatross* Station 3009, in 857 fathoms. A transitional species approaching *Furcella*. (*porus*, pore; *fero*, I bear.)

Lycodes porifer, GILBERT, Proc. U. S. Nat. Mus. 1890, 104, off Lower California, in 857 fathoms. (Type, No. 44384. Coll. Dr. Gilbert.)

938. FURCIMANUS, Jordan & Evermann.

Furcimanus, new genus (*diapterus*); JORDAN & EVERMANN, Check-List Fishes, 480, 1896 (*diapterus*); preoccupied by Furcella, Lamarck, 1801, a genus of mollusca.

This genus differs from *Lycenchelys* in the forked pectorals, the upper and lower rays being much longer than the middle ones. The lateral line is single and ventral in position (not lateral as in *Lycenchelys*). (*furca*, a fork; *manus*, hand.)

2835. FURCIMANUS DIAPTERA (Gilbert).

Head 5¾ to 6; depth 12; eye large, usually longer than snout, 3 to 3⅓ in head; snout 3¼ to 3⅔; interorbital width about 10. Body slender. Mouth small, somewhat variable in length, the maxillary reaching vertical from between front and middle of pupil, 2¼ to 3 in head. Teeth in premaxillaries in a double row throughout, the 2 series well separated, rarely with 1 or 2 teeth intercalated, showing traces of a third row; the teeth of inner series small and directed obliquely inward; those of outer series anteriorly enlarged, becoming smaller on sides of jaw; on front of mandible the teeth are in a broad band, in which traces of 3 or 4 irregular series can be made out; none of these enlarged; laterally the teeth are arranged in a single series, those opposite middle of cleft considerably enlarged; a small patch of from 2 to 5 teeth on vomer; palatines with a single row much shorter than premaxillary patch. Nostril with a short inconspicuous tube. Mandible and preopercular border with deep pit-like excavations, which are not evident in fresh specimens; no evident mucous pores on the head.

Gill openings wide, extending below the base of the pectorals; the gill membranes joined to isthmus for a distance equaling $\frac{2}{3}$ length of slit; gill rakers very short, almost tubercular, but compressed and slightly movable, about 15 present on anterior limb of outer arch; a wide slit behind fourth gill. Ventrals short, inserted under middle of opercle. Pectorals deeply notched in both young and adults, the median rays much shorter than either upper or lower, the lobe produced by the elongate lower rays varying in length, being sometimes shorter than upper lobe, sometimes longer; the rays of lower lobe are thickened, and undoubtedly serve as a support to the fish when resting on the bottom, as has been observed in so many other forms; the pectorals contain 20 or 21 rays; in the structure of this fin the present species seems to differ from all previously described forms, with the exception of *L. esmarkii*, in which the notched condition of the fin does not persist in the adults. Scales small, embedded, covering entire body and vertical fins; the scales on nape are much reduced in size, and in 2 specimens (11¼ and 7¼ inches long) are continued onto occiput, which they entirely cover; in another specimen, 9 inches long, the occiput is naked, and in another, 5 inches long, the anterior part of nape is likewise naked; in the latter, as in other specimens, the dorsal and anal are well scaled. Lateral line single, wavy, ventral in position, extending from above gill slit obliquely downward to near base of anal, along which it is continued for a variable distance, not reaching base of caudal. Color dusky brownish, blue-black on belly and along anterior portion of base of anal; 8 or 9 narrow white bars on sides, most conspicuous in the young, in which they are continued up on dorsal fin and become forked below on middle of sides, forming Λ-shaped marks; in adults these bars become faint or wholly disappear; when present, they are not continued on dorsal, and are usually vertically divided by a streak of the ground color; in the small specimen there is a distinct black blotch on margin of anterior dorsal rays; in adults, the vertical fins are brownish on basal portion, their distal half black; pectorals and ventrals deep blue-black; mouth, gill cavity, and peritoneum dusky or black. Several specimens, from *Albatross* Stations 2892, 2896, 3067, and 3077, in depths from 82 to 376 fathoms, off the coasts of California and Oregon. (Gilbert.) A remarkable species. ($\delta\iota\acute{a}$, divided; $\pi\tau\epsilon\rho\acute{o}\nu$, fin.)

Lycodes·diapterus, GILBERT, Proc. U. S. Nat. Mus. 1891, 564, off the coast of Oregon, in 685 to 877 fathoms. (Type, No. 44385. Coll. Dr. Gilbert.)

939. LYCODONUS, Goode & Bean.

Lycodonus, GOODE & BEAN, Bull. Mus. Comp. Zool., x, No. 5, 208, 1883 (*mirabilis*).

Body elongate, formed as in *Lycenchelys* and *Lyciscus;* scales small, circular, embedded in the skin; lateral line very short or obsolete; jaws without fringes, lower jaw included; fin rays all articulated, each ray of dorsal and anal supported laterally by a pair of sculptured scutes; caudal distinct, not fully connate with dorsal and anal; ventrals present; gill opening narrow; teeth as in *Lycodes*. Deep water. (*Lycodes*, with a meaningless change of termination.)

3030——-78

.2836. LYCODONUS MIRABILIS, Goode & Bean.

Head 7 in total length; depth about 18. D. about 80; A. about 70; C. 9; P. 18; V. 3; scales as in *Lycodes*, the scales not extending out upon the fins; no scales on head and nape. Lateral line apparently obsolete posteriorly; not extending back of the extremity of the pectoral, its position median; eye high up, $2\frac{1}{2}$ in head, equal to postorbital portion of the head; the width of interorbital space less than diameter of pupil, $3\frac{1}{2}$ times in long diameter of eye; nostrils immediately in front of eye; maxillary extending to vertical through anterior margin of pupil; mandible, to a little behind vertical through posterior margin of the pupil; dorsal fin inserted slightly behind vertical through base of pectoral (the portion of the fin present in the mutilated specimen before us contains 80 articulated rays; the first 10 or 11 scutes do not support rays, but whether rays were originally present or not can not be ascertained); longest dorsal ray about equal to longest anal ray, its length about 3 in head; distance of vent from snout twice length of head; anal beginning immediately behind vent, of about 70 articulated rays; caudal rays extending beyond tips of ultimate dorsal and anal rays, about 9 in number; distance of ventral from snout equal to twice length of upper jaw; middle ventral ray longest, it being $\frac{1}{2}$ as long as postorbital part of head; length of pectoral equaling 3 times that of snout. Off the New England coast, in depths of 721 to 1,309 fathoms; a most remarkable little fish. (*mirabilis*, wonderful.)

Lycodonus mirabilis, GOODE & BEAN, Bull. Mus. Comp. Zool., x, No. 5, 208, 1883, New England Coast, lat. 38° 20′ 8″ N., long. 73° 23′ 20″ W., in 740 fathoms (Type in M. C. Z.); JORDAN, Cat. Fishes, 124, 1885; GOODE & BEAN, Oceanic Ichthyology, 312, 1896.

940. LYCONEMA, Gilbert.

Lyconema. GILBERT, Rept. U. S. Fish Comm. 1893 (1896), 471 (*barbatum*).

Generic characters as in *Lycodes*, but the lower jaw covered with a dense mass of slender filaments or barbels, between which can be seen the mucous pores of the mandible. In *Iluocœtes*, a related genus from the Antarctic, the mandible is provided with a series of hollow tubes, which are doubtless the produced margins of the pores. Alaska. ($\lambda\dot{\upsilon}\kappa o\varsigma$, wolf; $\nu\tilde{\eta}\mu\alpha$, thread.)

2837. LYCONEMA BARBATUM, Gilbert.

Head $6\frac{1}{2}$; depth $11\frac{1}{2}$; maxillary 3 in head; eye $3\frac{1}{2}$; snout $4\frac{1}{2}$. D. 103; A. 90 (each counted to middle of caudal); P. 15, its length $1\frac{9}{10}$ in head; ventrals very short, $\frac{1}{2}$ to $\frac{2}{3}$ diameter of orbit. A dense fringe of filaments covers the entire under surface of lower jaw, extending to behind angle of mouth; another series laterally on the throat, and a few scattering ones sometimes present on the branchiostegal membranes; upper jaw without barbels. Body slender; upper jaw overlapping the lower; mouth small, maxillary reaching vertical from front of pupil; teeth all conical, none of them much enlarged, those in lower jaw in a patch or irregular double series, narrowing to a single series laterally; in upper jaw, a single series, the teeth of which increase in size toward the middle line, the middle teeth being almost canine-like; behind the latter, a short inner series of small teeth directed backward; teeth on vomer and palatines in a single series. Gill slits continued forward to slightly beyond bases of ventrals, and to level of lower edge of base of pectorals; width between gill slits $\frac{1}{2}$ diameter of eye; pseudobranchiæ well developed; posterior line of occiput midway between origin of dorsal and front of pupil or front of eye; origin of anal fin at end of first third of length of body; pectorals broad, with the posterior edge emarginate, some of the upper and the lower rays longer than the intermediate ones. Scales showing traces of definite arrangement in series, widely separated anteriorly, becoming crowded toward end of tail, continued up on the vertical fins, but not on head, on anterior half of nape, nor on the pectoral fins; lateral line very faintly shown, and for only a short distance behind head, where its course is obliquely downward; the usual series of mucous pores present, but not conspicuous. In spirits this species has an olive-brown ground color, becoming white on underside of head and on abdomen; a series of 8 or 9 brown spots $\frac{1}{2}$ as large as eye, along middle of sides, those posteriorly continued downward onto base of anal, the last 2 or 3 reaching edge of fin and there developing into intense black blotches; a similar series of smaller spots corresponding in position to those just described occurs along the base of dorsal, these continued as faint bars on the fin, at the margin of which they develop into a black blotch, those posteriorly wider and more intense; an intermediate series of spots alternating with the 2 just described; an elliptical jet-black spot occupies the greater part of caudal fin, and is narrowly margined all around with white; peritoneum jet-black; the mouth and gill cavities white. Coast of Alaska, in rather deep water; known from 12 specimens, the longest 6$\frac{1}{4}$ inches; depth 204 fathoms. (*barbatus*, bearded.)

Lyconema barbatum, GILBERT, Rept. U. S. Fish Comm. 1893 (1896), 471, coast of Alaska, at Albatross Station 3129, lat. 36° 39′ 40″ N., long. 122° 01′ W., in 204 fathoms.

941. BOTHROCARA, Bean.

Bothrocara, BEAN, Proc. U. S. Nat. Mus. 1890, 38 (*mollis*).

Body elongate, compressed, semitranslucent, covered with small scales; small teeth in jaws and on vomer and palatines; mucous pores about head largely developed. No ventral fins; dorsal and anal joined to caudal. Deep-sea fishes, allied to *Lycodes*, but lacking ventrals. The species have been referred to the Antarctic genus *Maynea*, Cunningham. From the latter, however, *Bothrocara mollis* seems to be distinct, differing in the larger mouth, more cavernous head, and lower dorsal. In some regards *B. pusilla* is intermediate, and it may belong to *Maynea*. (βόϑρος, cavity; κάρα, head.)

a. Body elongate, with the head short, 6 in length; depth 9; mouth small; mucous cavities small; color light brown, the dorsal dark-edged. PUSILLA, 2838.

aa. Body deeper and more compressed; the large head 4$\frac{1}{4}$ in length (5$\frac{1}{2}$ in young); depth 6$\frac{2}{3}$; mouth large; mucous cavities large; color uniform brown, the vertical fins dark-edged. Size large. MOLLIS, 2839.

2838. BOTHROCARA PUSILLA (Bean).

Head 6 in the total length; depth 9. D. 95, including $\frac{1}{3}$ caudal; A. 81, including $\frac{1}{3}$ of caudal; P. 17; eye 3 in head; snout 4. Body elongate, little compressed; head short; mouth small; maxillary extending to below front of pupil; gill clefts narrow, the anterior end below margin of preopercle; width of isthmus rather less than $\frac{1}{3}$ of orbit; the low dorsal beginning nearly over axil of pectoral; pectoral nearly $\frac{2}{3}$ as long as head; vent as far from end of head as dorsal origin from tip of snout. Color light brown; dorsal and anal with a narrow dark margin. · Size small; length $6\frac{1}{2}$ inches. Eastern parts of Bering Sea, and about the Alaskan Peninsula, in rather deep water. Besides the original types Dr. Gilbert records a few specimens from north of Unalaska, at depths of 121 to 351 fathoms. (*pusillus,* weak.)

Maynea pusilla, BEAN, Proc. U. S. Nat. Mus. 1890, 39, off Nagai Island, lat. 55° 10′ N., lon. 160° 18′ W., in 110 fathoms (Type, No. 45360. Coll. *Albatross*); GILBERT, Rept. U. S. Fish Comm. 1893 (1895), 455.

2839. BOTHROCARA MOLLIS, Bean.

Head $4\frac{1}{4}$ in total length in adult, $5\frac{1}{8}$ to $5\frac{1}{4}$ in young; depth $6\frac{3}{4}$ in adult, 10 in young. D. 100 to 105 to middle of caudal; A. 89 to 95 to middle of caudal; eye 4 in head in adult, $3\frac{1}{4}$ in young. Body covered with embedded scales, which extend on dorsal and anal fins. Head naked, breast and nape scaly; snout blunt, the lower jaw included; maxillary reaching middle of pupil, $2\frac{1}{4}$ in head; large mucous cavities conspicuous along mandible, suborbital ring, and top of head; vomerine and palatine teeth present, the latter in a narrow band, obscure in the young. Pectoral 2 in head; origin of dorsal slightly behind base of pectoral, its distance from tip of snout 4 in total length; origin of anal under seventeenth dorsal ray; longest dorsal ray 5 in head; longest anal ray $8\frac{1}{4}$ in head. Gill openings wider than in *Bothrocara pusilla,* the anterior end of the cleft under posterior margin of eye, the width of the isthmus less than $\frac{1}{4}$ diameter of pupil. Color uniform brown, fins lighter; dorsal and anal margined with black, more prominent posteriorly where it covers the entire fins. North Pacific. Adult examples from southern California, 18 inches long, were described as *Maynea brunnea,* while a young individual, $5\frac{1}{4}$ inches long, from Queen Charlotte Islands, with the vomerine and palatine teeth not evident, was made the type of a distinct genus, as *Bothrocara mollis.* The two are identical and apparently belong to the same genus as *B. pusilla.* Similar specimens, 1 adult and 2 young, were dredged by us (*Albatross*) off Bogoslof Island in 664 fathoms. Dr. Gilbert records also specimens from near Unalaska, depth 316 fathoms. The teeth on the palatines are in a single series instead of a wide band, as stated in the original description. (*mollis,* soft.)

Bothrocara mollis, BEAN, Proc. U. S. Nat. Mus. 1890, 38, off Queen Charlotte Islands, in 876 fathoms (Type, No. 45359. Coll. *Albatross*); JORDAN & GILBERT, Rept. Fur Seal Invest., 1898.
Maynea brunnea, BEAN, Proc. U. S. Nat. Mus. 1890, 39, lat. 33° 8′ N., lon. 118° 40′ W., off San Clemente Island, southern California, in 414 fathoms. (Coll. *Albatross.*)

942. GYMNELIS, Reinhardt:

Gymnelis, REINHARDT, Dansk. Vidensk. Selsk. Afhandl., VII, 131, 1838 (*viride*).
Cepolophis, KAUP, in Archiv fur Naturgesch. 1856, 96 (*viridis*).

Body elongate, naked. Vertical fins without spines; ventral fins none.
Small conical teeth on the jaws, vomer, and palatines. Gill openings very
narrow. No air bladder; pyloric cæca none; no anal papilla. Size small.
Cold seas. Two or 3 species known: *G. pictus*, from the Antarctic, and
G. viridis, which ranges widely in Arctic waters, and with which the very
dubious *G. stigma* is probably identical. (γυμνός, naked; ἔγχελυς, eel.)

   a. Dorsal fin inserted close behind pectoral, its distance from it much less than diameter of eye; no ocellus on dorsal fin.   VIRIDIS, 2840.

   aa. Dorsal fin inserted an eye's diameter behind pectoral; a large black spot, ocellated with white, on dorsal fin above vent; other ocelli sometimes present.

                                           STIGMA, 2841.

2840. GYMNELIS VIRIDIS (Fabricius).

Head about 6⅓; depth about 13; eye 7 in head. D. 100; A. 80. Snout
subconical, longer than the eye; jaws equal; mouth oblique; maxillary
reaching beyond eye; teeth rather small, conical, in a single series on each
side, forming a patch anteriorly; distance from snout to vent 2⅓ times
length of head. Pectoral rounded, inserted low, its length less than ⅓ that
of head. Dorsal fin inserted close behind pectoral, its distance from it
much less than diameter of eye. Body pale, with faint dark cross shades;
dorsal clouded but without black spot; anal dusky. Arctic seas, Alaska
to Greenland and Nova Scotia; abundant in the Arctic waters south to
Unalaska and Bristol Bay, where specimens were taken in shallow water;
our specimens from Bristol Bay.

Ensign H. G. Dresel records 1 small specimen (No. 28636, U. S. Nat. Mus.),
badly preserved, obtained by Mr. Newton Pratt Scudder in Davis Straits,
July, 1879. Length 100 mm. D. ca. 97; A. ca. 80. In this specimen the
maxillary does not extend to the posterior margin of the eye, which is comparatively very large. Its diameter is longer than distance from tip of
snout to orbit, and is contained 4 times in head. Head 7 in total length;
depth 12. Pectoral 2 in head. (*viridis*, green.)

Ophidium viride, FABRICIUS, Faun. Grœn., 141, 1780, Greenland.
Ophidium unernak, LACÉPÈDE, Hist. Nat. Poiss., II, 280, 1800, Greenland; after FABRICIUS.
Gymnelis viridis, RICHARDSON, Last Arctic Voyage, 321, pl. 29, 1854.
Gymnelis viridis, REINHARDT, Dansk. Vidensk. Selsk. Afh., VII, 1838, 131; GÜNTHER, Cat.,
    IV, 323, 1862; KRÖYER, Poissons du Nord, Voy. en Scand. et Lap., pl. 15, a–f; COLLETT,
    Norske Nordh. Exped., Fiske, 123, pl. 4, fig. 32, 1880; JORDAN & GILBERT, Synopsis,
    789; GILBERT, Rept. U. S. Fish Comm. 1893 (1896), 455.
? *Gymnelis pictus*, GÜNTHER, Cat., IV, 324, 1862, no locality.

2841. GYMNELIS STIGMA (Lay & Bennett).

Head 6; depth 11. D. 90; A. 70. Form, size, and general appearance
of *G. viridis*, the dorsal inserted farther back, an eye's diameter behind
pectoral. A large, round black ocellus, ringed with white, on dorsal fin
above vent; 2 or 3 other ocelli sometimes present; head and nape with
small white spots; body with faint dark shades and bands. Otherwise

as in *G. viridis*, from which it may not be distinct; but the above characters appear in our specimens (from near the Pribilof Islands) and in Richardson's figure of *G. unimaculatus*. The white spots on the head were
mistaken for "very small scales" in the original description of *G. stigma*
from Dr. Collie's notes. This description is, in substance, as follows: No
trace of ventral fins; dorsal, caudal, and anal fins united into a transparent ridge; rays of branchial covering distinct; scales very small.
Color dilute brown, with void swathes and spots; a purplish spot near
beginning of dorsal fin. Snout obtuse; chin with a large gibbosity;
teeth small. Length about 5 inches. (Lay & Bennett.) Arctic regions,
Greenland to Bering Sea, with the preceding, and apparently equally
common. (ὅτίγμα, spot.)

Ophidium stigma, LAY & BENNETT, Zool. Beechey's Voy., 67, pl. 20, fig. 1, 1839, Kotzebue
    Sound. (Coll. Dr. Collie.)
Gymnelis viridis var. *unimaculatus*, RICHARDSON, Last Arctic Voyage, 367, 1854, Northum-
    berland Sound. (Coll. Edward Belcher.)
Gymnelis stigma, GÜNTHER, Cat., IV, 325, 1862; JORDAN & GILBERT, Synopsis, 789, 1883.

943. LYCOCARA, Gill.

Uronectes, GÜNTHER, Cat., IV, 325, 1862 (*parrii*); name preoccupied in Crustacea.
Lycocara, GILL, Proc. Ac. Nat. Sci. Phila. 1884, 180 (*parrii*).

Body ensiform, compressed; tail long and tapering; ventrals none; vent
not far distant from the head; numerous minute teeth in jaws and on
palate; lower jaw the longer; no barbel; scales and gill openings unknown.
One species, very imperfectly known, no specimens having been obtained
by any recent collector. (λύκος, wolf; κάρα, head.)

2842. LYCOCARA PARRII (Ross).

Head 4. D. 50; A. 45; P. 37. Head very obtuse, its length, depth, and
breadth equal; head broader than the body, flattened and grooved between the eyes, which are lateral and rather large; lower jaw the longer;
jaws and palate with minute teeth; greatest depth of body somewhat
more than length of head; neck much arched. Dorsal inserted just
behind head; pectoral extending beyond vent. Vent not far distant from
head. Color uniform. Baffins Bay. (Günther.) (Named for Capt. William
Edward Parry, the Arctic explorer.)

Ophidium parrii, ROSS, Parry's Third Voyage, App., 109, 1826, Baffins Bay.
Uronectes parrii, GÜNTHER, Cat., IV, 326, 1862; JORDAN & GILBERT, Synopsis, 789, 1883.

944. MELANOSTIGMA, Günther.

Melanostigma, GÜNTHER, Proc. Zool. Soc. Lond. 1881, 21 (*gelatinosum*).

This genus is distinguished from *Bothrocara* by the much more elongate
teeth, which in the jaws, as well as on the vomer and palatines, stand in
single series. Gill openings much smaller than in related forms, reduced
to a small foramen above the base of the pectoral. Skin loose and movable, as in *Liparis*, enveloping the vertical fins; pectorals very small;

ventrals none. Body tapering very rapidly backward; the tail very slender. Deep-sea fishes, of soft substance, allied to *Bothrocara*, but with
stronger teeth. ($\mu\acute{\epsilon}\lambda\alpha\varsigma$, black; $\acute{\sigma}\tau\acute{\iota}\gamma\mu\alpha$, spot.)

a. Maxillary reaching beyond front of pupil; color purplish gray, becoming black on
the tail. GELATINOSUM. 2843.

aa. Maxillary not reaching beyond vertical from front of pupil; color uniform deep
black. PAMMELAS, 2844.

2843. MELANOSTIGMA GELATINOSUM, Günther.

Body enveloped in a loose, delicate skin, as in *Liparis*. Head large,
deep, compressed, with obtuse snout. Eye large, 3½ in head, and longer
than snout. Cleft of mouth rather oblique, but lower jaw not projecting
beyond upper; lips not fleshy; gill opening reduced to a very narrow foramen above base of pectoral fin; origin of dorsal fin and root of pectoral
enveloped in loose skin of body; dorsal fin probably commencing above
middle of pectoral, low at first, but becoming considerably higher posteriorly; pectorals very narrow, consisting of a few rays only. Upper parts
tinged with a purplish-gray; sides marbled with same color, which toward
end of tail becomes more intense, almost black; inside of mouth, gill
openings, and vent black. Total length of the type specimen 5½ inches;
distance of the snout from the gill opening ⅞ inches, from the vent 1⅜
inches. (Günther.) Deep waters of the western Atlantic; originally
known from the Straits of Magellan, but since obtained at various localities from Cape Cod to West Indies, in 500 to 1,000 fathoms. The identity
of these specimens with the original types from South America may be
questionable. (*gelatinosus*, jelly-like.)

Melanostigma gelatinosum, GÜNTHER, Proc. Zool. Soc. London 1881, 21, Tilly Bay, Straits
of Magellan, in 24 fathoms (Coll. H. M. S. *Alert*, Dr. Coppinger); GÜNTHER, Challenger Report, XXII, 82, 1887; GOODE & BEAN, Bull. Mus. Comp. Zool., X, No. 5, 209, 1883;
JORDAN, Cat. Fish. N. A., 125, 1885; GOODE & BEAN, Oceanic Ichthyology, 314, 1896.

2844. MELANOSTIGMA PAMMELAS, Gilbert.

Head 8 in total length; depth 12½; pectoral narrow, its length 2⅔ in
head; eye large, 3½ in head; snout short and broad, 7 in head. Well distinguished from *M. gelatinosum* by the wider, blunter head, the smaller,
less oblique mouth, the uniform black coloration, and the arrangement of
the teeth in the jaws in 2 series. As in *M. gelatinosum*, the head and
body are enveloped in a loose, thin skin, which is thrown into folds in
alcoholic specimens, and entirely conceals anterior portion of dorsal and
anal fins. On dissection the dorsal is seen to have its origin close behind
the head, at a point over middle of pectoral fin; anal beginning immediately behind vent, the rays of both fins enveloped in a gelatinous, subcutaneous tissue. Head broad, with its greatest width equaling its
greatest depth; mouth broad, somewhat oblique, with equal jaws, the
maxillary reaching vertical from front of pupil; each jaw with teeth in
two distinct series in front, in a single series laterally in lower jaw, the
outer teeth in front enlarged, almost canine-like. Gill opening a small
pore above base of pectoral, its diameter about ½ that of eye. Color
intense black on head and abdomen, brownish black elsewhere. Length

of type 4½ inches.   Coast of southern Alaska.   Three other specimens are at hand from *Albatross* Station 3126 (lat. 36° 49′ 20″ N., long. 122° 12′ 30″ W.; depth 456 fathoms).   In the smallest, 2½ inches long, the head and abdomen are jet-black, but the rest of the body is only slightly dusky.   (Gilbert.) (πᾶς, all; μέλας, black.)

Melanostigma pammelas, GILBERT, Rept. U. S. Fish Comm. 1893 (1896), 472, pl. 35, coast of southern Alaska, at Albatross Station 3202, lat. 36° 46′ 10″ N., long. 121° 58′ 45″ W., in 382 fathoms.

Family CCVII. DEREPODICHTHYIDÆ.

Deep-sea fishes of slender body, scaleless, and without lateral line, somewhat resembling the *Zoarcidæ,* but with each ventral fin reduced to a slender, unbranched filament, the two very closely approximate, and springing from a common projecting base located far forward, below the eye. Gill opening a narrow, vertical slit.   Character otherwise given below. A single species known; apparently intermediate between the *Zoarcidæ* and the *Ophidiidæ.*

945. DEREPODICHTHYS, Gilbert.

Derepodichthys, GILBERT, Rept. U. S. Fish Comm. 1893 (1886), 456 (*alepidotus*).

Body slender; no scales; no lateral line; ventral fins reduced each to a slender, unbranched filament, the two very closely approximate, and springing from a common projecting base, which is located far forward below the eye, as *Ophidion.*   Gill opening a narrow, vertical slit, little wider than base of pectorals.   Teeth cardiform, curved, few in number, in narrow bands or irregular single series on jaws, vomer, and palatines. (δέρη, throat; πούς, foot; ἰχθύς, fish.)

2845. DEREPODICHTHYS ALEPIDOTUS, Gilbert.

Head 8¼ in total length; depth of head and body 2½ in head; width of head 2⅔ in head; distance from tip of snout to base of ventrals 2¼ in length of head.   Distance from tip of snout to front of dorsal 5 ¼ in total length, from tip of snout to vent 3⅔ in total.   Head and body very long and slender, the former resembling a *Lycodes* in appearance, being moderately compressed, with a flattish occiput and a gentle rounded decurved rostral profile. Mouth slightly oblique, quite at lower side of snout; the lower jaw shorter, fitting within the upper; maxillary and premaxillary entirely concealed within the thick skin of the upper lip, which is directly continuous with that of the forehead, the upper jaw being therefore nonprotractile; angle of mouth under front of pupil, its distance from tip of snout 2⅔ in head.   Teeth cardiform, curved, few in number, in narrow bands or irregular single series on jaws, vomer, and palatines.   Eye small, not filling the elongate orbit, the diameter of exposed portion of eyeball slightly less than ¾ length of snout, the latter 3¼ in head.   A series of large mucous pores on snout and lower part of cheeks; a second series on mandible; no pores on body.   Gill slit vertical, not continued forward, its lower end slightly above base of lower pectoral rays; length of slit ¼ length of head, slightly less than distance between slits. Pectorals long and slender, reaching halfway to vent, 1⅓ in head; dorsal and anal confluent with the caudal, concealed in the thick integument, so that the rays can not be counted. Color in spirits, light brownish, the dorsal and pectorals whitish, the anal with a dark margin which becomes black posteriorly; lips dusky; abdominal region blue black. Coast of British Columbia. A single specimen, 4¼ inches long, dredged off Queen Charlotte Island. (Gilbert.) (ἀλεπιδωτός, scaleless.)

Derepodichthys alepidotus, GILBERT, Rept. U. S. Fish Comm. 1893 (1896), 456, Queen Charlotte Island, at Albatross Station 3342, in 1,588 fathoms.

Family CCVIII. OPHIDIIDÆ.

(THE CUSK EELS.)

Body elongate, compressed, more or less eel-shaped, usually covered with very small scales, which are not imbricated, but placed in oblique series at right angles with each other; head large, lower jaw included; both jaws, and usually vomer and palatines also, with villiform or cardiform teeth; premaxillaries protractile; gill openings very wide, the gill membranes separate, anteriorly narrowly joined to the isthmus behind the ventrals; pseudobranchiæ small. Gills 4, a slit behind the fourth; vent more or less posterior. Vertical fins low, without spines, confluent around the tail; tail isocercal; ventral fins at the throat, each developed as a long, forked barbel. Air bladder and pyloric cæca present. To this Dr. Gill adds the following characters, shared more or less by related families: "Orbito-rostral portion of cranium contracted and shorter than the posterior, the cranial cavity closed in part by the expansion and junction of the parasphenoid and frontals, the supraoccipital horizontal and cariniform posteriorly, the exoccipitals expanded backward and upward behind the supraoccipital, the exoccipital condyles contiguous, and with the hypercoracoid (scapula, Parker) fenestrate (or foraminate) about its center, and the hypercoracoid with its inferior process divergent from the proscapula." Genera 7, species about 25. Carnivorous fishes; found in most warm seas, some of them descending to considerable depths, the group especially well represented in tropical America. (*Ophidiidæ*, group *Ophidiina*, Günther, Cat., IV, 376–380, 1862.)

a. Head scaly, at least above; body covered with scales imbricated in quincunx; snout usually with a spine at tip; opercle with or without spinous tip; air bladder, so far as known, ovate, without posterior foramen. LEPOPHIDIUM, 946.
aa. Head scaleless; scales of body rudimentary, scarcely embedded.
 b. Air bladder oblong-ovate, not contracted behind, and without posterior foramen.
  c. Opercle ending in a flat point behind, without spine. OPHIDION, 947.
  cc. Opercle ending behind in a strong spine concealed in the skin.
   CHILARA, 948.
 bb. Air bladder short, thick, reniform or orbicular, with a large foramen behind.
  d. Opercle ending in a flat point, without spine. RISSOLA, 949.
  dd. Opercle ending behind in a spine concealed in the skin.
   OTOPHIDIUM, 950.

946. LEPOPHIDIUM, Gill.

Leptophidium, GILL, Proc. Ac. Nat. Sci. Phila. 1863, 210 *(profundorum)*; name preoccupied
    in Serpents by *Leptophidium*, HALLOWELL, 1860.
Lepophidium, GILL, Amer. Nat., Feb., 1895, 16 *(profundorum)*.

Body much elongate, moderately compressed, with back and abdominal
regions arched, more compressed and slowly decreasing in height backward to an abruptly rounded point; scales regularly imbricated in quincunx oval, and with striæ radiating backward; head with imbricated
scales, extending to forehead; snout high, projecting forward, and
obtusely rounded, armed above with a short, nearly concealed spine
directed forward and somewhat downward, obsolete in 1 species; mouth
moderate, oblique; teeth of jaws villiform, immersed in a mucous membrane, separated by an interval from the longer ones in the outer row,
which are pointed and usually movable; vomer and palatines with teeth.
Deep waters of America on both coasts. Perhaps a fuller knowledge of
the species of this genus will lead to its subdivision. (λέπος, scale; *Ophidium*, from the squamation.)

a. Snout without decurved hook or spine; gill rakers 8; head 5 in length; depth 7;
    pectorals 10; body marbled, the vertical fins edged with black.
                                                                MARMORATUM, 2846.
aa. Snout with a decurved hook or spine at tip, sometimes more or less concealed in
    the skin.
    b. Gill rakers 7 to 9 in number.
        c. Head large, 3¾ to 4½ in length.
            d. Body stoutish, the depth 6 in length; scales 125; no black blotch on
                front of dorsal.   EMMELAS, 2847.
            dd. Body slender, the depth 9¼ in length; dorsal with a black blotch in
                front; scales 180.   STIGMATISTIUM, 2848.
        cc. Head moderate, 6 in length; depth 10; vertical fins black-edged.
            e. Anterior teeth in jaws movable; pectoral 11 in body; body without
                white spots.   PROFUNDORUM, 2849.
            ee. Anterior teeth in jaws not movable; pectoral 13 to 14 in body; body
                with whitish spots.   CERVINUM, 2850.
    bb. Gill rakers 4 in number.
        f. Scales moderate, 175 to 200 in lateral line.
            g. Body without dark cross bars; dorsal and anal margined with black;
                air bladder oblong.
                h. Head 4½ in length; depth 8; pectoral 10¼. Pacific species.
                                                                PRORATES, 2851.
                hh. Atlantic species imperfectly described. BREVIBARBE, 2852.
            gg. Body with dark cross bars; dorsal spotted with black; anal wholly
                black; head 5¼ in length; depth 8⅓.   PARDALE, 2853.
        ff. Scales minute, about 250 in lateral line; head 4⅔; depth 7¼ to 8; color
            nearly plain, the fins dark edged.   MICROLEPIS, 2854.

2846. LEPOPHIDIUM MARMORATUM (Goode & Bean).

Head 5: depth 7½; eye 4 in head; snout about 5. Body somewhat
elongate, stoutish anteriorly, gradually tapering; head thickish; interorbital area broad, convex, its width nearly equal to length of snout,
which is blunt, spineless; eye circular, somewhat exceeding length of
snout. Maxillary extending to vertical through posterior margin of orbit,

the mandible far beyond, its length equal to that of postorbital portion of head. Teeth on vomer and in jaws in villiform bands, the outer series in the latter slightly enlarged. Pseudobranchiæ present; gill rakers short, 8 below angle of first arch, the longest less than $\frac{1}{2}$ diameter of eye. Branchiostegals 7. Ventrals as long as postorbital part of head. Dorsal origin at distance from snout contained $4\frac{1}{4}$ in total length, with 28 rays in a space equal to length of head, counting from the origin of the fin; anal origin separated from snout by distance $2\frac{3}{4}$ in total length; length of pectoral 2 in head, or 10 in total. Scales closely imbricated, ornamented with delicate concentric striæ; lateral line apparently complete, located about $\frac{1}{4}$ distance from dorsal to ventral outline. Color yellowish gray, marbled along the upper half of head and body with olive brown; dorsal and anal fins with black margins. Gulf Stream, in 213 fathoms. (Goode & Bean.) (*marmoratus*, marbled.)

Leptophidium marmoratum, GOODE & BEAN, Proc. U. S. Nat. Mus. 1885, 423, lat. 23° 10′ 39″ N., long. 82° 20′ 21″ W., in 213 fathoms (Type, No. 37237, U. S. Nat. Mus. Coll. *Albatross*); GOODE & BEAN, Oceanic Ichthyology, 348, 1896.

2847. LEPOPHIDIUM EMMELAS (Gilbert).

Head $3\frac{3}{4}$ to 4; depth $5\frac{2}{3}$ to 6; eye $4\frac{3}{4}$ in head; snout $4\frac{3}{4}$; interorbital width 7; vertebræ 13 + 44 = 57; maxillary $2\frac{1}{2}$ to $2\frac{1}{4}$ in head; ventral filament $2\frac{3}{4}$; pectoral 2 in head; scales 8-125-18 or 20 before dorsal. Body deep, compressed. Maxillary reaching slightly beyond orbit. Jaws slender and weak, the teeth in very narrow bands, the outer not enlarged. Rostral ridge very sharp, bearing a flat spine at its base directed upward and backward, terminating in a very slender sharp spine anteriorly. Opercle ending in a weak spinous point behind. Gill rakers short and slender, the longest $\frac{1}{2}$ pupil, 8 or 9 movable ones developed. Skull and all bones of head very thin and papery. Dorsal beginning over base of pectorals, the nape midway between its origin and middle of orbit. Distance from snout to origin of anal $1\frac{1}{3}$ in distance from latter to end of tail. Scales large, covering cheeks, opercles and top of head forward to middle of interorbital space. Color brownish, much dusted with minute specks; fins blackish, the vertical fins with an indistinct narrow whitish margin; inside of mouth dusky; the roof of mouth, lining of gill cavity, and peritoneum jet-black. Coast of Lower California. Many specimens, the longest 9 inches, from *Albatross* Stations 3007 and 3008, in 362 and 306 fathoms. (Gilbert.) ($\dot{\epsilon}\nu$, within; $\mu\dot{\epsilon}\lambda\alpha\varsigma$, black.)

Leptophidium emmelas, GILBERT, Proc. U. S. Nat. Mus. 1890, 110, coast of Lower California. (Coll. *Albatross*.)

2848. LEPOPHIDIUM STIGMATISTIUM (Gilbert).

Head $4\frac{1}{4}$ in length; depth $9\frac{1}{4}$; eye $4\frac{1}{4}$ in head; snout 6; interorbital width $6\frac{1}{2}$; maxillary reaching slightly beyond posterior border of eyes, $2\frac{1}{4}$ in head. A strong rostral spine. Outer teeth scarcely enlarged, evidently so only in front of upper jaw. Gill rakers long and slender, strongly curved forward at tip, the longest equaling $\frac{1}{2}$ eye; 7 well-developed gillrakers present. Opercle ending in a rounded process, a broad soft flap projecting beyond it. Dorsal inserted behind middle of pectorals, the nape equidistant from front of dorsal and base of rostral spine; pectorals 2⅔ in head; longest ventral filament 2¾ in head; scales small, about as in *L. prorates*, 180 transverse series, 28 in front of dorsal, continued forward on top of head to front of pupil; cheeks and opercles scaly. Color dusky olivaceous, lighter below; dorsal with a large black blotch on anterior rays, the margin obscurely dusky; anal broadly margined with jet-black; caudal with median rays black at base, the outer rays and the margin light; lining of gill cavity jet-black; inside of mouth white; peritoneum bright silvery. A single specimen 10 inches long. Coast of Lower California. (Gilbert.) Much resembling *L. prorates*, differing in dentition, in gill rakers, and in color. (στίγμα, brand; ίστίον, sail.)

Leptophidium stigmatistium, GILBERT, Proc. U. S. Nat. Mus. 1890, 109, off Lower California, at Albatross Station 2996, in 112 fathoms.

2849. LEPOPHIDIUM PROFUNDORUM (Gill).

Head 6; depth 10. Body very slender; scales regularly arranged in quincunx order, those on head extending to forehead, opercles, and cheeks; snout high, projecting, armed with a concealed spinous hook; teeth villiform, separated by an interval from an outer row of longer, slender, movable teeth; eye longer than snout, 3½ in head; lateral line obsolete behind; vent toward end of first third of length; ventral fins short; gill rakers 8. Light rufous; vertical fins margined with black. Gulf Stream, off the coast of Florida. .(Gill.) One specimen known. (*profundorum*, of the depths.)

Leptophidium profundorum, GILL, Proc. Ac. Nat. Sci. Phila. 1863, 211, Gulf Stream, off the Coast of Florida (Coll. Commodore Rodgers); GOODE & BEAN, Oceanic Ichthyology, 347, 1896.

Ophidium profundorum, JORDAN & GILBERT, Synopsis, 793, 1883.

2850. LEPOPHIDIUM CERVINUM (Goode & Bean).

Head about 6¼; depth about 10½; eye 4 in head; ventrals 3 in head. Body elongate, slender; head slender, somewhat compressed; interorbital area broad, convex, its width equal to length of snout, and 5⅔ in head; snout sharp, conical, armed with a short but sharp spine, and somewhat overhanging mouth; eye much exceeding length of snout; maxillary extending nearly to vertical through posterior margin of orbit, 2⅔ in head; mandible extending behind same vertical, its length equal to that of head without postorbital portion. Jaws, vomer, and palatines with narrow bands of villiform teeth, some of which are noticeably enlarged (not movable). Pseudobranchiæ present. Gill rakers short, 8 below angle of first arch, 4 of which are rudimentary, the longest 5 in diameter of eye. (In *L. profundorum* the gill rakers are slenderer and longer, though about equally numerous on the first arch.) Scales in about 11 rows from the origin of the dorsal to the median line of the body. Dorsal origin far back, at a distance from the snout 4¾ in total length; at a distance from the eye equal to the head's length. (In *L. profundorum* this distance is ⅔ of the head's length and the first ray of the dorsal is nearly over the middle of the extended pectoral; in *L. cervinum*, over its tip, or nearly so.) Distance of anal origin with snout 3 in total length. Length of pectoral 2 in head's length and 13 to 14 in that of body (10 in *L. marmoratum*, 11 in *L. profundorum*). Scales ornamented with radiating striæ, densely covering all parts of the fish except snout, under surface of head, and the fins; lateral line continued almost to end of tail. Color brownish yellow, with numerous subcircular spots of white, with diameter $\frac{1}{2}$ that of eye, along the upper half of body; vertical fins with narrow black margin. Gulf Stream. (Goode & Bean.) A specimen from off Sand Key Light, Florida, recorded by Mr. Garman. (*cervinus*, deer-like, from the fauncolor.)

Leptophidium cervinum, GOODE & BEAN, Proc. U. S. Nat. Mus. 1885, 422, lat. 40° 1' N., long. 69° 56' W., depth 76 fathoms (Type, No. 28764. Coll. *Fish Hawk*); GOODE & BEAN, Oceanic Ichthyology, 346, 1896.

Lepophidium cervinum, GARMAN, Bull. Iowa Lab. Nat. Hist. 1896, 91.

2851. LEPOPHIDIUM PRORATES (Jordan & Bollman).

Head $4\frac{1}{5}$ to $4\frac{2}{3}$ ($4\frac{3}{5}$ to $4\frac{1}{4}$ in total); depth $7\frac{1}{5}$ to $8\frac{1}{9}$ ($7\frac{3}{5}$ to $8\frac{1}{2}$); eye $4\frac{1}{2}$ in head; snout 5; maxillary $2\frac{1}{4}$; interorbital $1\frac{3}{4}$ in eye; pectoral $2\frac{1}{5}$ in head; inner ventral filament shortest, the longer $2\frac{2}{3}$ in head. Body moderately elongate, compressed, considerably stouter than in *L. profundorum*. Mouth large, maxillary reaching about $\frac{1}{2}$ pupil's length beyond posterior border of eye. Outer teeth slightly enlarged, a little movable, those of upper jaw largest. Gill rakers rather long and slender, $\frac{1}{4}$ length of eye, 4 developed. Tip of snout with a strong spine directed forward and slightly downward; opercle without spine, ending in a flat projection covered by skin. Dorsal beginning over middle of pectorals, longest ray 4 in head. Scales regularly imbricated, but very small, about 225 in a longitudinal series; scales on top of head extending forward to base of ethmoid spine; sides of head covered with small scales; lateral line not reaching end of tail. Air bladder oblong-lanceolate. Color olivaceous, paler below; scales rather profusely dotted with black; a pale shade across opercles; lower jaw, gular region, and anterior branchiostegals dusted; dorsal and anal margined with black, the band on anal the broader; pectorals pale. Specimens of this species were obtained at Panama and at *Albatross* Station 2801, south of Panama. Length of type 10 inches. ($\pi\rho\omega\rho\acute{a}\tau\eta\varsigma$, prow-bearing, from the rostral spine.)

Leptophidium prorates, JORDAN & BOLLMAN, Proc. U. S. Nat. Mus. 1889, 172, Panama. (Type, No 41149, U. S. Nat. Mus. Coll. *Albatross*.)

2852. LEPOPHIDIUM BREVIBARBE (Cuvier).

A short decurved spine at tip of snout; teeth strong; occiput and opercles scaly. Vertical fins edged with black. (Kaup.) Air bladder oblong ovate, without contracted portion and without posterior foramen; no single anterior bone replaced by cartilage. (Müller.) West Indies and Brazil; a scarcely known species; apparently close to *L. prorates*, but very insufficiently described. (*brevis*, short; *barba*, beard.)

Ophidion brevibarbe, CUVIER, Règne Animal, Ed. 2, vol. II, 358, 1829, Brazil; MÜLLER, Abhandl. Berl. Acad. 1843, 153, pl. 4, f. 4; KAUP, Apodes, 154, pl. 16, f. 1; GÜNTHER, Cat. Fish., IV, 379, 1862.

2853. LEPOPHIDIUM PARDALE (Gilbert).

Head 5½ in length; depth 8½; eye 3⅗ in head; snout 4⅗; interorbital 1½ in eye. Body very slender, with a short head and small mouth; maxillary scarcely reaching vertical from posterior border of orbit, its length 2⅙ in head; outer teeth very little enlarged, not movable; teeth present on jaws, vomer, and palatines. Gill rakers slender, the longest ¼ eye, 4 developed. Tip of snout with a strong, concealed spine, as in *L. prorates*. Opercle ending in a short spine. Nape midway between front of dorsal and front of pupil; dorsal beginning over middle of pectorals, which are 2¼ in head; ventral filaments very short; the inner the longer, 4⅗ in head. Scales very small, about 200 in a longitudinal series, extending forward on top of head to middle of interorbital space; cheeks and opercles scaly. Light olive, a series of 8 black bars downward from back, scarcely reaching lateral line, sometimes continuous with their fellows of the other side, and alternating with smaller black spots on dorsal outline; below the smaller spots a series of round spots nearly as large as eye along middle line of sides; sides and lower parts of head and body dusted with rather coarse black specks; dorsal light, the margin with 10 elongate black blotches, usually longer than the interspaces; caudal dusky at base, its distal half white; anal wholly black; peritoneum and lining of gill cavity white. Lower California. A single specimen, length 7½ inches, from *Albatross* Station 3014, in 29 fathoms. (Gilbert.) (*πάρδαλις*, leopard.)

Leptophidium pardale, GILBERT, Proc. U. S. Nat. Mus. 1890, 108, off Lower California. (Type, No. 44382. Coll. Dr. Gilbert.)

2854. LEPOPHIDIUM MICROLEPIS (Gilbert).

Head 4⅗ in length; depth 7½ to 8; eye 4¼ to 5 in head; snout 5; interorbital width 6½; maxillary extending beyond orbit, 2¼ to 2½. Rostral spine very strong, as in *L. prorates*. Outer teeth enlarged, not at all movable, those in upper jaw largest. Four gill rakers developed, the longest 3½ in eye. Opercle ending in a short concealed spinous point. Dorsal inserted in front of middle of pectorals, the distance from nape to front of dorsal usually less than from nape to middle of eye; longest ventral filament 3⅓ to 3½ in length of head; pectorals 2¼ to 2½ in head. Scales exceedingly small, regularly imbricated, in about 250 transverse series, 35 transverse series between nape and dorsal (about 175 transverse rows in *L. prorates*, 25 series between nape and dorsal). Top of head scaly as far as front of eyes. Cheeks and opercles scaly. Color as in *L. prorates*, the lining of peritoneum and gill cavity silvery white, the former with little or no black specking. Closely related to *L. prorates*, differing principally in the much smaller scales. Gulf of California. Many specimens, the longest 14 inches, from *Albatross* Stations 3015 and 3016, in 145 and 76 fathoms. (Gilbert.) (*μικρός*, small; *λεπίς*, scale.)

Leptophidium microlepis, GILBERT, Proc. U. S. Nat. Mus. 1890, 109, Gulf of California. (Coll. Dr. Gilbert.)

947. OPHIDION (Artedi) Linnæus.

(CUSK EELS.)

Ophidion (ARTEDI) LINNÆUS, Syst. Nat., Ed. x, 259, 1758 (*barbatum*).
Ophidium, LINNÆUS, Syst. Nat., Ed. XII, 431, 1766, and of most recent authors; changed spelling.

Body moderately elongate, compressed; scales small, usually not imbricated, but arranged in short, oblique series, often placed at right angles with each other, much as in *Anguilla*. Head naked; teeth villiform, those of the outer series more or less enlarged, none of them movable; teeth on vomer and palatines bluntish, some of them enlarged. Vent well behind pectorals. Opercle without distinct spine; sometimes (*O. barbatum*) a distinct spine at tip of snout. Air bladder oblong-ovate, tapering behind, without foramen. Shore species mostly European. (*Ophidium*, an ancient name, from ὀφίδιον, a small snake.)

> a. Gill rakers 4.
>> b. Head 4⅔ in length; depth 7; fins not dark edged. BEANI, 2855.
>> bb. Head 6 in length; inner ray of ventral 1½ in length of outer, which is shorter than head; fins dark-edged. HOLBROOKI, 2856.
> aa. Gill rakers 6 or 7; head 5½ to 6 in length; depth 8 to 10; color silvery, unspotted; fins not dark-edged; ventrals nearly as long as head. GRAELLSI, 2857.

2855. OPHIDION BEANI, Jordan & Gilbert.

Head 4⅔ in length; depth about 7. Head small, the profile not very obtuse; snout 4⅔ in head; eye 3¼, more than twice the narrow interorbital space; mouth oblique, the maxillary reaching to posterior border of pupil, 2 in head; lower jaw slightly included; teeth small, in narrow bands in the jaws, the outer series in upper jaw somewhat enlarged; vomerine and palatine teeth small, subequal; head naked; snout spineless; opercle without spine; no evident pseudobranchiæ; gill rakers rather long and strong, 4 below angle of arch; occiput nearly midway between origin of dorsal and front of eye. Air bladder long and slender, occupying nearly the whole length of abdominal cavity, tapering backward. Very light olive, somewhat punctate above, slightly silvery below; fins without trace of dark edging (but being mutilated they may have been dark-edged in life). Gulf of Mexico. Two specimens, 1 of which is in good condition and about 4 inches long, were taken from the stomach of a red snapper, at Pensacola. (Named for Dr. Tarleton Hoffman Bean.)

Ophidium grællsi, JORDAN & GILBERT, Proc. U. S. Nat. Mus. 1882, 301; JORDAN & GILBERT, Synopsis, 963, 1883; not of POEY.
Ophidion beani, JORDAN & GILBERT, Proc. U. S. Nat. Mus. 1883, 43, Snapper Banks off Pensacola (Coll. Jordan & Stearns. Type, No. 30868, U. S. Nat. Mus.); JORDAN, Cat. Fishes N. A., 126, 1885.

2856. OPHIDION HOLBROOKI (Putnam).

Head 6 in total length. Inner barbel nearly ⅔ length of the outer; outer barbel equal to distance from center of eye to point of operculum; maxillary reaching to posterior border of eye. Length of eye equal to distance from its posterior margin to ridge of preoperculum. Dorsal and anal witlr a black margin. Gill rakers 4; air bladder long, pointed, with a foramen. (Putnam.) Length 6 inches. Gulf of Mexico; recorded from Key West, Florida; not seen by us. (Named for Dr. John Edwards Holbrook, the distinguished ichthyologist of Charleston.)

?Ophidion josephi, GIRARD, U. S. and Mex. Bound. Surv., Ichth., 29, 1859, St. Joseph Island, Texas; JORDAN & GILBERT, Synopsis 793, 1883; quite as likely to be *Rissola marginata*.

Ophidium holbrooki, PUTNAM, Proc. Bost. Soc. Nat. Hist. 1874, 342, Key West, Florida; JORDAN & GILBERT, Synopsis, 793, 1883.

2857. OPHIDION GRAELLSI, Poey.

Head 5¾ in body; depth 10; eye 3 in head; pectoral 2⅘; ventral scarcely as long as head. Body elongate, compressed; mouth large, the maxillary reaching to posterior margin of pupil; small teeth on jaws, vomer, and palatines; eye very large, greater than length of snout; interorbital space ⅔ of eye, a sharp ridge along its middle to tip of snout, where it ends in a sharp spine; opercles unarmed; about 6 gill rakers developed on lower part of gill arch, apparently none above; pseudobranchiæ small, if present. Air bladder, injured in specimen examined, apparently lanceolate; dorsal and anal low, confluent with caudal, which ends in a point; pectorals small, their ends scarcely reaching midway from their base to front of anal; ventrals with 2 filamentous rays, the outer scarcely as long as head, the inner ½ as long. Color in spirits, reddish brown, with silvery reflections on sides; head silvery, upper part of eye black; fins the color of the body, with no dark edgings. Coasts of Cuba; rare. Here described from a specimen from Havana, Cuba, 2¼ inches in length, sent by Professor Poey. Poey has also sent a drawing of his original type, a much larger specimen, which he describes as follows:

Head 5½; depth 8; snout rounded; eye large, 4½ in head; maxillary reaching posterior border of eye; teeth small, slender, with a villiform band behind them; teeth on vomer and palatines; scales small; head scaly, except on snout; lateral line high; branchiostegals 7; dorsal beginning over second third of pectoral, joining anal behind; about 100 rays in each fin; vent a little behind first third of length. Yellowish brown, silvery on side of head; no black on fins. Air bladder distinct; no pyloric cæca. Intestine with 2 short turns. Cuba. (Poey.) Air bladder, gill rakers, and ventral not described. Length 230 mm. Rare; not reaching a foot in length. (Named for Mariano de la P. Graëlls, director of the Botanic Garden at Madrid, "comme témoignage de mon estime pour ses travaux scientifiques, et pour la zèle qu'il déploit . . . pour l'acquisition des objets et l'acclimatation des espèces.")

Ophidion graellsi, POEY, Memorias, II, 425, 1860, Havana (Coll. Poey); POEY, Synopsis, 402, 1867.

948. CHILARA, Jordan & Evermann.

Chilara, JORDAN & EVERMANN, Check-List Fish. N. and M. A., 482, 1896 (*taylori*).

This genus contains a single robust species which differs from *Ophidion* only in the presence of a stout concealed spine at tip of opercle; the air bladder is oblong-ovale, the head naked and the snout without spine. ($χιλάρι$, the modern Greek name of the species of *Ophidion* and *Rissola*.)

2858. CHĪLARA TAYLORI (Girard).

Head 6; depth 8; head large, little compressed, naked; top of head with conspicuous mucous pores; dorsal fin beginning over the pectorals; outer ray of ventral little more than ¼ length of head, inner about ⅓; air bladder ovate, not contracted; 7 gill rakers below the angle of the arch; pseudobranchiæ developed; no spine on the end of the snout; opercle with a flat spine concealed in its membranes; outer teeth in both jaws considerably enlarged, the upper largest. Color light olive; head and upper parts covered with conspicuous round dark, olive-brown spots; chin dusky; vertical fins edged with black. Length 12 inches. Coast of California, from Monterey to San Diego; not rare in waters of moderate depth. (Named for A. S. Taylor, its discoverer.)

Ophidium taylori, GIRARD, Pac. R. R. Surv., X, Fishes, 138, 1858, Monterey, California (Type, No. 867. Coll. A. S. Taylor); JORDAN & GILBERT, Synopsis, 793, 1883.

949. RISSOLA, Jordan & Evermann.

Rissola, JORDAN & EVERMANN, Check-List Fish. N. and M. A., 483, 1896 (*marginatum*).

This genus contains species agreeing with *Ophidion* in general characters, but with the air bladder short, broad, spherical or kidney-shaped, with a posterior foramen. Species chiefly of the Mediterranean. (Named for Anastase Risso, apothecary at Nice, author of the Ichthyologie de Nice, 1810, and Histoire Naturelle de l'Europe Méridionale, 1826, two of the very best of local faunal works, the foundation of our knowledge of the fishes of the Mediterranean.)

2859. RISSOLA MARGINATA (DeKay).

Head 6½; depth 7½; eye 4 in head; maxillary reaching posterior margin of orbit; air bladder short and broad, with a foramen on the under side; upper ray of ventral about equaling length of head; inner ray ½ length of outer; gill rakers 4; color nearly plain brownish; dorsal and anal fins margined with black. Coast of the United States, from New York south to Pensacola and the coast of Texas; not very common; very similar to the Mediterranean species *Rissola rochii* (Müller), but probably distinct. (*marginatus*, margined.)

Ophidium marginatum, DE KAY, N.Y. Fauna: Fish., 315, 1842, New York Harbor; PUTNAM, Proc. Bost. Soc. Nat. Hist. 1874, 342; JORDAN & GILBERT, Synopsis, 792, 1883.

?*Ophidium josephi*, * GIRARD, U. S. and Mex. Bound. Surv., Zool., 29, 1859, Saint Joseph Island, Texas; JORDAN & GILBERT, Synopsis, 793, 1893.

---

* The scanty description of *Ophidion josephi* agrees fairly with either *Rissola marginata* or *Ophidion holbrooki*, and may be either. The following is the substance of Girard's account:
"Head 6 in length; eye moderate, 4 in head; maxillary extending to opposite its posterior margin; origin of dorsal at some distance behind base of pectorals. Body shorter and pectorals more elongate than in *O. taylori*. Pale olive, sprinkled all over with brownish specks; belly and sides of head plain; vertical fins edged with black."

950. OTOPHIDIUM, Gill.

Otophidium, GILL, in JORDAN, Cat. Fish. N. A., 126, 1885 (*omostigma*).

This genus differs from *Ophidion*, in the form of the air bladder, which is short, thick, and with a large foramen (not examined in *O. galeoides*). The opercle ends in a concealed spine as in *Chilara*. Species American, so far as known. (*οὖς ὠτός*, ear; *Ophidium.*)

a. Gill rakers 4.
b. Head long, $4\frac{1}{3}$ to $4\frac{1}{2}$ in length; depth $5\frac{1}{2}$ to 6.
c. Scapular region with a jet-black spot; pseudobranchiæ little developed; ventrals $\frac{1}{2}$ length of head; maxillary $1\frac{7}{8}$ in head. OMOSTIGMUM, 2860.
cc. Scapular region without jet-black spot; pseudobranchiæ well developed; ventrals with the inner ray longest, $\frac{1}{2}$ head; body with dark cross bands. INDEFATIGABILE, 2861.
bb. Head moderate, $5\frac{1}{2}$ in body; depth 6; a pale spot before dorsal; pale spots along lateral line; ventral $\frac{1}{2}$ head. GALEOIDES, 2862.

2860. OTOPHIDIUM OMOSTIGMUM (Jordan & Gilbert).

Head $4\frac{1}{8}$ in length; depth about 6. Body comparatively short, highest at occiput, thence tapering rapidly to tip of tail; upper profile of head very convex; snout blunt; mouth horizontal, the lower jaw included; maxillary not quite reaching posterior border of orbit; teeth in jaws uniform, strongly incurved, in rather broad bands; a single series of small teeth on vomer, those on palatines minute; maxillary $1\frac{7}{8}$ in head; eye large, 3 in head, much larger than snout, equaling twice interorbital width; opercle terminating in a strong, compressed spine, the length of which is about $\frac{3}{8}$ diameter of pupil; gill rakers very small, 4 below on anterior arch. Longest ventral filament $\frac{1}{2}$ length of head; the shorter $\frac{3}{4}$ length of longer. Distance from origin of dorsal to tip of snout $3\frac{1}{8}$ in total length; distance from origin of anal to snout $2\frac{1}{5}$ in total length. Scales minute, embedded. Pseudobranchiæ probably present (type reexamined by us). Air bladder short, thick, with a large posterior foramen. Color light olive green, silvery on belly, cheeks, and lower side of head; sides above with a few irregular, large, scattered, dark blotches, about 9 of these along base of dorsal fin; an intensely black, round blotch on scapular region, rather larger than pupil; dorsal with black blotches; anal largely black; upper half of eye black, lower half bright silvery. Gulf of Mexico. A single specimen, $3\frac{1}{2}$ inches long, taken from the stomach of a red snapper, at the Snapper Banks off Pensacola. (*ὦμος*, shoulder; *στίγμα*, spot.)

Genypterus omostigma, JORDAN & GILBERT, Proc. U. S. Nat. Mus. 1882, 301, Pensacola Snapper Banks (Coll. Jordan & Stearns. Type, 29670, U. S. Nat. Mus.); JORDAN & GILBERT, Synopsis, 963, 1883.
Otophidium omostigma, GOODE & BEAN, Oceanic Ichthyology, 345, fig. 305, 1896.

2861. OTOPHIDIUM INDEFATIGABILE, Jordan & Bollman.

Head $4\frac{3}{5}$ ($4\frac{1}{2}$ in total); depth $5\frac{3}{8}$ ($5\frac{6}{7}$); eye large, 3 in head; snout 4. Body rather short, compressed, width of nape $2\frac{1}{2}$ in head. Mouth large; maxillary reaching to opposite posterior margin of pupil, $1\frac{3}{5}$ in head; outer row of teeth of each jaw very slightly enlarged. Interorbital space 2 in eye; interorbital area with a thin crest under the skin, this ending in 2 compressed spines, 1 turning forward, the other backward over front of eye, these spines concealed by the skin. Gill rakers short and thick, less than $\frac{1}{4}$ pupil, 4 developed. Dorsal beginning at end of anterior third of pectorals, longest ray $3\frac{1}{2}$ in head; pectorals 2 in head; inner ventral filament. longest, 2 in head. Air bladder short and thick, with a foramen. Scales very small, more or less imbricated on body; head naked. Opercle with a sharp, partly concealed spine. Pseudobranchiæ present. Color pale yellowish brown, silvery on belly and sides of head; back with about 12 irregular dark cross bands, the alternate ones being narrower and broken up into spots, 2 before dorsal; a few scattered spots about as large as pupil on sides, these most distinct about the shoulder; dorsal pale, first rays black, and with 3 or 4 other black blotches on upper part; anal black, margined with white; pectorals pale, axil dusky; caudal and posterior part of anal pale; chin pale. A single specimen obtained at Indefatigable Island, Galapagos Archipelago. Length 4 inches. (*indefatigabilis,* tireless.)

Otophidium indefatigabile, JORDAN & BOLLMAN, Proc. U. S. Nat. Mus. 1889, 172, Indefatigable Island, in the Galapagos Archipelago. (Type, No. 44393. Coll. *Albatross.*)

2862. OTOPHIDIUM GALEOIDES (Gilbert).

Head $5\frac{1}{4}$ in length; depth 6. D. 125. Maxillary reaching beyond pupil, $2\frac{1}{4}$ in head; snout $4\frac{2}{3}$; eye $3\frac{2}{3}$. Gill rakers short and broad, 4 of them developed. Opercle ending in a sharp concealed spine. Outer teeth little enlarged. Dorsal beginning over middle of pectorals, the nape equidistant between front of dorsal and tip of snout. Caudal very short and bluntly rounded, as in *Chilara taylori,* the rays not projecting beyond dorsal and anal; pectorals $1\frac{2}{3}$ in head; ventral filament $\frac{1}{4}$ head. Scales as in *C. taylori,* not at all imbricated, arranged with their long axes frequently at right angles to each other; head naked. Color light olive, without bars, a narrow dusky streak along base of dorsal, and a round light spot at origin of dorsal; a series of small olive-brown spots along lateral line, with a few scattering spots below it but none above; nape and head without spots; vertical fins translucent; dorsal with a large black blotch on tip of anterior rays, the fin behind this narrowly edged with black, which does not surround the caudal; anal with much silvery-white pigment anteriorly on distal portion, becoming dusky behind; pectorals translucent, edged with white below; peritoneum, buccal, and gill cavities white. Closely related to *Otophidium indefatigabile,* differing in color, and in the much shorter head, smaller mouth, less imbricated scales, the more posterior insertion of dorsal, and the absence of spines on head. Air bladder not examined. Pseudobranchiæ present. Gulf of California. One specimen, $5\frac{1}{4}$ inches long, from *Albatross* Station 3025, in $9\frac{1}{4}$ fathoms. (Gilbert.) ($\gamma\alpha\lambda\tilde{\eta}$, shark; $\varepsilon\tilde{\iota}\delta o\varsigma$, appearance.)

Otophidion galeoides, GILBERT, Proc. U. S. Nat. Mus. 1890, 110, Gulf of California, lat. 31° 21′ 15″ N., long. 113° 59′ W. (Type, No. 44381. Coll. *Albatross.*)

Family CCIX. LYCODAPODIDÆ.

Deep-sea fishes allied to the *Fierasferidæ,* differing chiefly in the normal position of the vent, which is remote from the head, and just before the anal fin; gill openings large, the membranes united anteriorly only, free from the isthmus, as in *Fierasfer.* ·Pseudobranchiæ wanting; no scales; no lateral line; no ventral fins.   One genus with 4 known species, from the North Pacific.

951. LYCODAPUS, Gilbert.

Lycodapus, GILBERT, Proc. U. S. Nat. Mus. 1890, 107 (*fierasfer*).

Body naked.   Ventrals wanting.   Vertical fins united around the tail. Gill openings wide, continued forward under the throat; the gill membranes anteriorly narrowly united, loosely joined to the isthmus by a fold of lax skin.   Branchiostegals 6.   No pseudobranchiæ.   Gills 4, a wide slit behind inner arch.   Gill rakers developed.   Teeth present in jaws and on vomer and palatines, none of them enlarged.   Vent remote from the throat. *(Lycodes; ἄπους, footless.)*

> *a.* Body slender, the depth 8 to 11 in length.
> > *b.* Head rather large, 4¾ to 5½ in length.
> > > *c.* Head, body, and fins with very many mucous pores; dorsal rays 70; anal 60.
> > > > DERMATINUS, 2863.
> > > *cc.* Head, body, and fins with very few mucous pores; dorsal rays 82; anal 70.
> > > > FIERASFER, 2864.
> > *bb.* Head small, 7⅔ in length; gill openings not extending above base of pectorals.
> > > PARVICEPS, 2865.
> *aa.* Body very slender, the depth about 15 in length; dorsal rays about 100.
> > EXTENSUS, 2866.

2863. LYCODAPUS DERMATINUS, Gilbert.

Head 4¾; depth 1⅔ in head; eye 5 in head; snout 4; maxillary 2¼.   D. 70; A. 60.   Very similar to *L. fierasfer,* but the head, body and fins covered with a thick loose skin which contains numerous pores, or openings for the mucous canals.   One series of these runs along middle of sides and forms the lateral line; it rises anteriorly above the gill opening, and is continued forward on top of head, the two meeting between eyes; a second series runs between eye and upper lip, and curves around on middle of cheek, running upward to behind eye; one series runs along a fold bordering mandible, 1 along preopercular margin, and 1 on opercle.   In *L. fierasfer* a few pores are visible on mandible, and 1 or 2 can frequently be made out on preopercular margins.   The skin is very thin and delicate, and the fin rays are very evident through the membrane.   The general proportions and the dentition of the type are essentially as in *L. fierasfer,* but the vomerine teeth are long and hooked backward. Mandible heavier than in *L. fierasfer.*   Origin of dorsal vertically above axil of pectorals.   Length of head and trunk ⅓ total length.   Teeth in narrow bands in the jaws, a single series on vomer and palatines.   Gill membranes very narrowly joined below and free from the isthmus, as in *L. fierasfer.*   Pectorals much longer than in *L. fierasfer.*   General color in spirits light brownish yellow, made somewhat dusky by the pigment spots in the skin; body, and especially the fins, darker posteriorly.   Aleutian

Islands, in deep water. Only the type known, an example 4¾ inches long. (δέρμα, skin; *dermatinus*, skinny.)

Lycodapus dermatinus, GILBERT, Rept. U. S. Fish Comm. 1893 (1896), 471, pl. 35, Aleutian Islands, lat. 37° 54′ 10″ N., long. 123° 30′ W., at Albatross Station 3162, in 552 fathoms.

2864. LYCODAPUS FIERASFER, Gilbert.

Head 5½; depth 10; eye 4½ in head; snout 3⅓; maxillary 2 to 2¼. D. 82; A. 70. Body compressed, elongate, tapering rather rapidly backward, the tail not produced to a filament; head flat above, the cheeks deep, vertical, the mouth very oblique, with the lower jaw slightly the longer and nearly entering the upper profile; skull very thin and papery, translucent; jaws weak; gape of mouth wide, the maxillary reaching vertical from behind front of pupil; teeth all small, in a very narrow band in jaws, in a single series on vomer and palatines; interorbital width ⅔ of eye; snout broad, depressed, spatulate, its tip prominent, turned upward, the upper profile thus longitudinally concave; an evident median ridge on snout and interorbital space; gill slits continued forward. below to vertical from middle of eye, the membranes united for a distance equaling diameter of pupil; gill rakers short, less than diameter of pupil, strongly toothed, about 10 on horizontal limb of arch; head without conspicuous mucous pores or cavities. Dorsal beginning well forward, its distance from occiput slightly less than that from occiput to nostril. Dorsal and anal rays slender, all articulated, branched only at tips; caudal not distinct, the rays springing from end of tail not projecting beyond the others; origin of anal immediately behind vent, its distance from snout nearly equaling ⅓ total length; pectorals narrow, varying in length, about 2¼ in head. Body and fins invested in a rather lax transparent skin, without traces of scales. Color, body translucent, dusted with black specks; abdomen blackish; lips, inside of mouth, lining of gill cavity, and peritoneum jet-black; iris silvery. (Gilbert.) North Pacific. The types, several specimens, the longest 5¼ inches, from *Albatross* Stations 2980, 3010, 3072, off Lower California, in 610 to 1,005 fathoms. Also taken near Unalaska in 109 fathoms. (*Fierasfer*, the pearlfish.)

Lycodapus fierasfer, GILBERT, Proc. U. S. Nat. Mus. 1890, 108, off Lower California, in 610 to 1,005 fathoms (Coll. Dr. Gilbert); JORDAN, Proc. Cal. Ac. Sci. 1896, 234, pl. 23.

2865. LYCODAPUS PARVICEPS, Gilbert.

Head 7⅔; depth 11; eye 4½ in head; snout 3⅓; least interorbital width 5; maxillary 2⅙; pectoral 2¾. D. 100; A. about 85 (both counted to middle of caudal); P. 9; no ventrals. Upper profile of head nearly straight, not longitudinally concave as in *L. fierasfer;* head deeper and narrower, the snout less spatulate; skin thicker. A conspicuous series of pores on mandible and along preopercular margin; gill slit very oblique, extending anteriorly as far as vertical from eye, the membranes then narrowly united, free from the isthmus except at extreme front; gill slit superiorly much more restricted than in *L. fierasfer,* not extending above base of pectorals, while in the latter it extends above them for ⅔ diameter of eye. Mouth oblique, maxillary reaching vertical from middle of eye; jaws even at tip, the mandible slightly included laterally; mandibular teeth in a moderate band anteriorly, the inner series enlarged, narrowing posteriorly to a single row; premaxillary teeth of uniform size, in a 'narrow band throughout; vomer with 4 canine-like teeth; palatine teeth small, in a single close-set series. Distance from origin of dorsal to occiput slightly less than that from occiput to posterior nostril; head and trunk contained $3\frac{1}{2}$ in tail. Body brownish in spirits, fins whitish, translucent; everywhere dusted with black specks; tail and fins distinctly blackish posteriorly; orbit blackish above; gill cavity silvery, blackish anteriorly; mouth blackish, except anteriorly; peritoneum black, the color not showing through the abdominal wall. Similar to *L. fierasfer*, differing in the much smaller head, longer, slenderer body, the thicker skin with more evident mucous pores, and in the more restricted gill openings. Aleutian Islands, in moderately deep water. Only the type known, a specimen about 5 inches long. (Gilbert.) (*parvus*, small; -*ceps*, head.)

ﯨ*Lycodapus parviceps*, GILBERT, Rept. U. S. Fish Comm. 1893 (1896), 455, north of Unalaska Island at Albatross Station 3324, in 109 fathoms. (Coll. Dr. Gilbert.)

2866. LYCODAPUS EXTENSUS (Gilbert).

Head $6\frac{3}{4}$; depth $15\frac{1}{2}$; eye $4\frac{1}{4}$ in head; snout $3\frac{2}{3}$; interorbital width $1\frac{1}{4}$ in eye; pectoral $2\frac{2}{3}$ in head. D. 96 (the extreme end of the tail wanting). Gill openings as in *L. fierasfer*, extending well above base of ventrals. Skin thin, the mucous pores inconspicuous, evident on mandible and along margin of preopercle. Upper profile of head longitudinally concave, shaped as in *L. fierasfer*, but slenderer, its depth greater than that of body. Mouth oblique, the maxillary reaching vertical from middle of eye, $2\frac{1}{4}$ in head. Teeth in narrow bands in each jaw, tapering laterally to single series; vomerine teeth more numerous than in *L. parviceps* or *L. fierasfer*, small, not canine-like, in a single series; palatine teeth wanting, as in some individuals of *L. fierasfer*. Occiput midway between front of dorsal and anterior nostril; pectorals slenderer and longer than in *L. fierasfer*. Head and trunk contained $2\frac{2}{3}$ times in tail. Color light brownish, the black peritoneum visible through the skin of the abdomen; mouth and gill cavity largely dusky; a narrow dark-brown streak along base of dorsal and anal, occupying, toward tip of tail, the entire height of both fins. An extremely slender elongate form, with head smaller than *L. fierasfer*, but otherwise resembling that species more than *L. parviceps*. Aleutian Islands, in rather deep water. Only the type, a specimen 4 inches long, known. (Gilbert.) (*extensus*, stretched out.)

Lycodalepis extensus, GILBERT, Rept. U. S. Fish Comm. 1893 (1896), 455, north of Unalaska, at Albatross Station 3324, in 109 fathoms. (Coll. Dr. Gilbert.)

Family CCX. FIERASFERIDÆ.

(THE PEARL-FISHES.)

Body elongate, compressed, tapering into a long and slender tail; no scales; teeth cardiform, on jaws, vomer, and palatines; canine teeth often present; no barbels; lower jaw included; vent at the throat; gill membraneś somewhat united, free from the isthmus; no pseudobranchiæ; no pyloric cæca; vertical fins very low, confluent, without spines; no ventral fins; pectoral fins present or absent. Small shore fishes of tropical seas, often living in shells of mollusks, echinoderms, etc., being especially ofteñ commensal with the pearl oyster and with the larger *Holothuria*. Genera 3; species 12. (*Ophidiidæ*, group *Fierasferina*, Günther, Cat., IV, 381–384, 1862.)

a. Pectoral fins present; no distinct caudal fin; gill membranes connected anteriorly only. FIERASFER, 952.

952. FIERASFER, Cuvier.

Fierasfer, CUVIÈR, Règne Anim., Ed. 1, II, 239, 1817 (*imberbe=acus*).
Echiodon, THOMPSON, Proc. Zool. Soc. London 1837, 55 (*drummondi*).
Diaphasia, LOWE, Proc. Zool. Soc. London 1843, 92 (*acus*).
Oxybeles, RICHARDSON, Voy. Erebus and Terror, Fishes, 74, 1844–48 (*homei*).
Porobronchus, KAUP, Ann. Mag. Nat. Hist. 1860, 272 (larva of *Fierasfer acus*).
Carapus, * GILL, Proc. Ac. Nat. Sci. Phila. 1864, 152 (after RAFINESQUE, 1810; not type).
Vexillifer, GASCO, Bull. Assoc. Nat. Med. Napoli 1870, 59 (larva of *Fierasfer acus*).
Lefroyia, JONES, Zoologist, IX, 1874, 3838 (*bermudensis*).

Gill membranes little connected, leaving the isthmus bare. No distinct caudal fin; pectoral fins developed. The species of this genus are not well known, and their characters and nomenclatures are uncertain. It is not unlikely that the American species are all reducible to one, *Fierasfer affinis* or *dubius*, but our scanty material will not justify us in taking this view. (*Fierasfer*, the ancient name, from φιερός, sleek and shining.)

a. Vomer with canine teeth; pectoral about ½ length of head.
  b. Front teeth of upper jaw enlarged; head 7 to 8 in length; depth 11½ to 15 times in length of body. AFFINIS, 2867.
  bb. Front teeth of upper jaw not enlarged; head 6½ in length; depth about 10½ times in length of body. ARENICOLA, 2868.
aa. Vomer with small teeth, scarcely canine-like; pectoral about 2½ in head; head 7 to 8½ in body. BERMUDENSIS, 2869.

2867. FIERASFER AFFINIS† (Günther).

(PEARL-FISH.)

Head 7½; depth of head 15. Maxillary extending slightly beyond orbit; lower teeth larger than the upper, except 2 to 4 front teeth of upper jaw, which are about equal to lower teeth; vomer with 3 to 6 teeth, 2 or 3 of

* The name *Carapus*, Rafinesque, has been substituted for *Fierasfer* by Gill and Poey. This change seems to us not justifiable, as it is certainly not desirable. The name *Carapus* first appears in Rafinesque's Indice d'Iltiologia Siciliana, 57, 1810. No type is mentioned by Rafinesque, but the diagnosis is taken from that of Lacépède's second subgenus under *Gymnotus*, which contains the three species, *carapo, fierasfer*, and *longirostratus*. Of these species, *carapo* is the original Linnæan type (Ed. X) of the genus *Gymnotus*. *Carapus* should therefore be regarded as a synonym of *Gymnotus*. The Brazilian name *carapo* evidently suggested the word *Carapus*, although Dr. Gill derives the name from κάρα, head; ἄπους, footless, an ex post facto distinction from *Ophedion*. In a list of Sicilian fishes, on page 37 of Rafinesque's Indice, published somewhat later, the name *Carapus acus* appears for *Fierasfer acus*. This reference of a species of *Fierasfer* to *Gymnotus* or *Carapus* was due to Rafinesque's ignorance of its relations.

† In the Museum of Comparative Zoology is "one valve of a pearl oyster, in which a specimen of *Fierasfer dubius* is beautifully inclosed in a pearly covering, deposited on it by the oyster." (Putnam.)

these canine-like. Pectoral ⅓ head; vent under base of pectoral. Dorsal fin low, but distinct; anal much more developed than dorsal, its longest rays about in the middle of the fish. Air bladder long, slightly constricted behind. Gill membranes not covering isthmus. Color in spirits, uniform light brown, with a short silvery band along the sides of the abdomen made by confluent spots. (Putnam, description of *F. dubius.*) Panama; especially common among the Pearl Islands, chiefly in shells of pearl oysters. This species should probably stand as *Fierasfer affinis.*

The following notes are from numerous specimens, 3 to 4 inches long, from Pearl Islands, collected by Prof. Bradley, these also being types of *Fierasfer dubius:* Head 6¾ to 7½; eye 4¼ to 5 in head. Teeth in upper jaw small, acute, in a rather narrow band; sometimes a few in the front of the jaw inconspicuously enlarged; those in lower jaw and on palatines conic, blunt, in somewhat wider bands, the outer series of lower jaw enlarged, canine-like; vomer with a narrowly oblong patch of small, blunt teeth, surrounding a median series of 3 to 6 conspicuously enlarged, retrorsely curved canines, which are usually much the largest teeth in the mouth. Two specimens from *Albatross* Station 3021, Lower California, agree in general with the above account: Head 7½; depth 11½; eye 4; 2 upper teeth on each side somewhat enlarged, about as large as lateral teeth on mandibles; vomerine canines larger. Professor Putnam refers also to *Fierasfer dubius* specimens from Key Biscayne, Florida (Coll. Theodore Lyman); Tortugas (Coll. Gustav Würdemann); Cape Florida (Coll. Würdemann), and New Providence, Bahama (Coll. F. G. Shaw). These specimens apparently belong rather to *Fierasfer bermudensis,* if that species be different. (*affinis,* related, to *Fierasfer acus.*)

? *Fierasfer affinis,*[*] GÜNTHER, Cat., IV, 381, 1862, no locality given.

Fierasfer dubius, PUTNAM, Proc. Bost. Soc. Nat. Hist. 1874, 344, Pearl Islands (Coll. Prof. Frank H. Bradley); JORDAN & GILBERT, Proc. U. S. Nat. Mus. 1882, 629; JORDAN & GILBERT, Synopsis, 791, 1883.

2868. FIERASFER ARENICOLA, Jordan & Gilbert.

Head 6½ in length; depth 10½; eye 5 in head; snout 5. Body with nape slightly elevated, thence tapering regularly to the tail. Snout blunt, rounded, - protruding; mouth subinferior, nearly horizontal, large, the lower jaw included; gape wide, the maxillary ½ length of head, extending beyond vertical from orbit; teeth in upper jaw very small, acute, in a narrow band, none of them enlarged; those in lower jaw and on vomer blunt, conic, in a wide band; those in outer series acute; a few on each side of mandible and 2 or 3 anteriorly on vomer, enlarged, canine-like. Gill openings very wide, the branchiostegal membranes little united, leaving nearly all of isthmus uncovered; the membranes

---

* *Fierasfer affinis,* Günther, is thus described:
"The length of the head is ⅙ of the total; its greatest width is rather less than ½ of its length. Gill openings rather wide, the united gill membranes leaving the greater portion of the isthmus uncovered. Teeth cardiform; a pair in front of the upper jaw, a series on the side of the lower, and several others on the vomer larger than the rest. Dorsal fin low but very distinct. The length of the pectoral nearly ½ that of the head. (This species is) similar to *F. acus,* but with a very different dentition." (Günther.) Described from a specimen 8 inches long, from unknown locality. This description, so far as it goes, agrees with *Fierasfer dubius,* but the specimen may not be American.

united as far back only as vertical from end of maxillary; opercle adherent above the upper angle, which is produced in a point extending above the base of pectorals; below the angle the opercular margin runs very obliquely forward. Eye large, greater than interorbital width. Origin of dorsal fin distant from nape by the length of the head, the fin a very inconspicuous fold anteriorly, becoming higher posteriorly, where the rays are evident; anal well developed along entire length, beginning immediately behind vent and running to tail, its rays visible; caudal exceedingly short; pectorals very well developed, more than ⅓ length of head; vent just in front of base of pectorals. Head and body perfectly translucent; a faint silvery luster on middle of sides anteriorly; a few inconspicuous small light yellowish spots along middle of sides (disappearing in alcohol); tip of tail dusky; upper margin of orbit black. Pacific coast of Mexico. A single specimen, 3¼ inches long, was found buried in the sand at low tide on the beach at Mazatlan. This specimen may be identical with *Fierasfer dubius*, but it is more robust than Putnam's types, with longer head and without enlarged teeth in upper jaw. It may be regarded as distinct, pending investigation. (*arena*, sand; *colo*, I inhabit.)

Fierasfer arenicola, JORDAN & GILBERT, Proc. U. S. Nat. Mus. 1881, 363, Mazatlan. (Type, No. 29244. Coll. C. H. Gilbert.)

2869. FIERASFER BERMUDENSIS (Jones).

Head 8½ in length; eye 4, longer than snout; mouth large, the maxillary reaching beyond orbit; pectoral 2½ in head. Teeth small, acute, uniserial, 3 in a line on the vomer; palatine teeth small.* Color pale brownish, a bluish streak crossing the nape between the opercles, 4 pale points on the back. Vertebræ 100. Length 140 mm. West Indies. This description (by Poey) from a specimen taken in the stomach of a holothurian at Havana. Others are recorded from Key West and St. Thomas, the latter from an oyster; not seen by us; doubtfully distinct from *Fierasfer dubius* or *affinis*, but the vomerine teeth said to be smaller.

?*Carapus affinis*, POEY, Synopsis, 402, 1867; not *Fierasfer affinis*, GÜNTHER.
Lefroyia bermudensis, JONES, Zoologist, IX, 1874, 3838, Bermuda.† (Coll. General Lefroy.)

---

* In another specimen, according to Poey, the teeth are villiform, with an enlarged series outside, the lower teeth largest, the ninth, tenth, and eleventh largest; teeth on vomer small, acute, in a row; palatine teeth bluntish.

† *Fierasfer bermudensis* (Jones) was thus originally described:

"Total length rather more than 4½ inches. Greatest depth at the vertical of the pectorals 3½ lines. The length of the head is slightly more than ⅓ of the total length. The greatest width of the head rather less than ½ of its length. Body naked, attenuate, compressed. Facial outline rugose. Eye moderate; horizontal diameter of eyecup 1¼ lines; vertical diameter 1¼ lines. Gape of the mouth ovoid. Lower jaw shorter and received within the upper. Cardiform teeth of irregular size in both jaws, vomer, and palatines, those of the latter largest. Branchiostegals 7, inflated, united below. Vent thoracic. Pectorals originating at the upper angle of the operculum, 3 lines in extent, and composed of very delicate soft rays. Dorsal indistinct, commencing in a groove about the vertical of the twentieth anal ray, continuous to caudal extreme, where, in conjunction with the anal, it forms a small filamentous tip. Anal prominent, commencing immediately behind the vent in advance of the vertical of the upper angle of the operculum, and extending to the caudal extreme. About its center it is equal in depth to that of the body at same position. Owing to the delicate texture of the fins it is impossible to ascertain certainly the number of rays, but those of the anal exceed 140. Color, when dried out of spirits, golden yellow; the body transparent, showing the vertebræ within; a condition. according to Lefroy, equally observable in life.

"I propose to publish it as *Lefroyia bermudensis*, in compliment to the gallant officer to whom I am indebted for the specimen." (J. Matthew Jones.)
This species is probably identical with the one called *affinis* by Poey.

Family CCXI. BROTULIDÆ.

(THE BROTULOID FISHES.)

Body elongated, compressed, regularly tapering behind, the tail generally subtruncate at base of caudal fin, not isocercal; vent submedian; scales cycloid and minute, embedded in the lax skin, which more or less envelops the fins, sometimes wanting; gill openings very large, the membranes mostly free from the isthmus; vertical fins united or contiguous at base of caudal; dorsal fin commencing not far from nape; caudal narrow or pointed; ventral fins small, few-rayed, attached to the humeral arch and more or less in advance of pectoral. Pyloric cœca few (1 or 2), rarely obsolete or in increased number (12); maxillaries generally enlarged behind and produced toward their upper angle. (Gill.) Pseudobranchiæ small or wanting, hypercoracoid with the usual foramen, as in Blennioid fishes. These fishes are closely related to the *Zoarcidæ*. In spite of various external resemblances to the *Gadidæ*, their affinities are rather with the Blennioid forms than with the latter. Genera about 45, species about 100; largely of the depths of the sea; 2 species degenerated into blind cave fishes. We have not had material for any elaborate study of these fishes and follow closely the arrangement given by Goode & Bean. (*Brotuloidæ*, Gill, Proc. Ac. Nat. Sci. Phila. 1863, 252, and 1884, 175.)

BROTULINÆ:
    a. Snout and lower jaw each with well-developed barbels; vertical fins united; teeth on vomer and palatines.
        b. Ventrals each reduced to a bifid filament.   BROTULA, 953.
    aa. Snout and lower jaw without barbels.
        LUCIFUGINÆ:
        c. Species blind, dwelling in fresh-water streams in caves; barbels replaced by cilia.
            d. Palatines with strong teeth; teeth in lower jaw strong.   STYGICOLA, 954.
            dd. Palatines toothless; teeth in jaws villiform.   LUCIFUGA, 955.
        cc. Species marine, the eyes usually well developed.
            BROSMOPHYCINÆ:
            e. Caudal fin differentiated, on a distinct caudal peduncle.
                f. Snout and lower jaw with small cilia; head naked, or nearly so.
                        BROSMOPHYCIS, 956.
                ff. Snout and lower jaw without cilia; head more or less scaly.
                        OGILBIA, 957.
            ee. Caudal fin not differentiated, without distinct peduncle.
            BYTHITINÆ:
                g. Ventrals inserted on the isthmus, not far from the humeral symphysis.
                h. Pectorals normal, simple; eyes present.
                    i. Lateral line present posteriorly, but broken in the middle; palatines with teeth; ventrals a pair of filaments each of 2 closely united rays.   BYTHITES, 958.
                ii. Lateral line obsolete posteriorly.
                    j. Ventrals each of a single ray.
                      k. Lateral line distinct on front of body.
                        l. Preopercle without spines; head scaly (except snout); opercle with a single spine; vent median.   CATÆTYX, 959.

ll. Preopercle with 3 or 4 spines, opercle with a
single one; head partially naked.

DICROMITA, 960.

kk. Lateral line obsolescent, almost, or quite invisible; opercle with a feeble spine; head
smooth; eyes small.

m. Ventral consisting of a single ray.

BASSOZETUS, 961.

mm. Ventral bifid. MŒBIA, 962.

jj. Ventrals each of a pair of rays.

n. Caudal fin exserted, but confluent with anal and
dorsal.

o. Head scaly.

p. Preopercle with small spines at its
angle, opercle with 1 spine.

NEOBYTHITES, 963.

pp. Preopercle unarmed.

q. Opercle with 2 spines; ventrals
close together.

BENTHOCOMETES, 964.

qq. Opercle with 1 strong spine;
ventrals far apart.

BASSOGIGAS, 965.

nn. Caudal not confluent with vertical fins, but
without distinct peduncle; teeth on jaws,
vomer, and palatines in villiform bands;
preopercle unarmed; head scaly.

r. Opercle with a flat spine; snout much produced and dilated; lateral line very
indistinct (or absent?).

BARATHRODEMUS, 966.

rr. Opercle a triangular flap, unarmed; lower
pectoral rays prolonged, the lowest
filamentous. NEMATONUS, 967.

iii. Lateral line represented by 3 rows of pores—dorsal, lateral, and ventral; head with spines.

s. Ventrals of 2 distinct rays; opercular spine moderate, straight. POROGADUS, 968.

ss. Ventrals each of 2 united rays, opercular spines
strong, curved. PENOPUS, 969.

hh. Pectorals with the lower rays differentiated.

t. Preopercle armed with 3 spines; opercle armed with 1
spine; lateral line obsolete posteriorly; ventrals bifid.

DICROLENE, 970.

tt. Preopercle unarmed; a single spine on opercle; lateral
line absent (?); ventrals each a pair of filaments,
closely united throughout. MIXONUS, 971.

APHYONINÆ:

gg. Ventrals inserted on humeral symphysis; lateral line obsolete (in
almost every case); ventrals each of a single filament; body
naked; notochord persistent.

u. Eye visible through the skin; a few fang-like teeth on vomer
and mandible. BARATHRONUS, 972.

uu. Eye not visible; no teeth on maxillary or palatines; teeth on
vomer rudimentary, those on mandible small.

APHYONUS, 973.

953. BROTULA, Cuvier.

(BRÓTULAS.)

Brotula, CUVIER, Règne, Anim., Ed. 2, II, 296, 1829 (*barbata*).

Body elongate, compressed, covered with minute smooth scales; eye moderate; mouth medium, with villiform teeth on jaws, vomer, and palatines; lower jaw included; each jaw with 3 barbels on each side. Dorsal fin long and low, the dorsal and anal joined to the caudal. Ventral fins each reduced to a single filament of 1 ray. Eight branchiostegals. Air bladder large, with 2 horns posteriorly. One pyloric cæca. Vertebræ $16 + 39 = 55$. Tropical seas, in water of moderate depth. (*Brótula*, Spanish name of *Brotula barbata*.)

2870. BROTULA BARBATA (Bloch & Schneider).

(BRÓTULA.)

Head $4\frac{1}{2}$; depth about 5. D. 123; A. 93; V. 1. Upper jaw the longer. Ventral fin $\frac{1}{2}$ as long as head. Dorsal commencing behind vertical from root of pectoral; vertical fins covered with thick skin. Color nearly uniform brown. Length 12 to 18 inches. West Indies; rare; in water of moderate depth. One specimen obtained by us in the market of Havana. (*barbatus*, bearded.)

Brótula, PARRA, Dif. Piezas Hist. Nat., 70, lam. 31, fig. 2, 1780, Havana.
Enchelyopus barbatus, BLOCH & SCHNEIDER, Syst. Ichth., 52, 1801; after PARRA.
Brotula barbata, CUVIER, Règne Anim., Ed. 2, II, 296, 1829; POEY, Memorias, II, 102, lam. 9, fig. 2, 1860; GÜNTHER, Cat., IV, 371, 1862.

954. STYGICOLA, Gill.

Stygicola, GILL, Proc. Ac. Nat. Sci. Phila. 1863, 252 (*dentatus*).

This genus differs from *Lucifuga* in the presence of palatine teeth. The teeth in the jaws are larger. As in *Lucifuga*, the single known species inhabits cave streams in Cuba. (στύξ, *Styx*, the river of the lower regions; *colo*, I inhabit.)

2871. STYGICOLA DENTATUS (Poey).

Head $2\frac{2}{3}$; depth $3\frac{3}{4}$. D. 90; A. 70; P. 17; V. 1. Vertebræ $11 + 37 = 48$. Eyes usually wanting, occasionally represented by a rudiment; head elevated at the nape, the general form less slender than in *Lucifuga*, the belly more prominent; no scales on the nape; strong teeth, well separated, on the palatines as well as the vomer; teeth in the jaws larger than in *Lucifuga*; posterior with a large apophysis. Color translucent violet, with darker areas on nape and throat. Caves of the province of San Antonio, in southern Cuba. Largest specimen 5 inches long. (Poey.) (*dentatus*, toothed.)

Lucifuga dentatus, POEY, Memorias, II, 102, 1860, Cave of Cajío (Coll. Noda), Cave of Castle La Industria (Coll. Dubrocá), Cave of Ashton (Coll. Fabre); GÜNTHER, Cat., IV, 373, 1862.
Stygicola dentata, GILL, Proc. Ac. Nat. Sci. Phila. 1863, 252.

955. LUCIFUGA, Poey.

(CUBAN BLINDFISH.)

Lucifuga, POEY, Memorias, II, 95, 1860 (*subterraneus*).

Body moderately elongate, translucent pinkish, covered with minute scales. Eye rudimentary, covered by the skin; bands of villiform teeth in the jaws and vomer, none on the palatines; nostrils 2 on each side; no barbels; head with small tactile cirri; no spines on head; gills 4; no pseudobranchiæ; gill opening large, extending forward nearly to the symphysis, the gill membranes not united; branchiostegals 7 or 8; vertical fins low, united around the tail; ventrals each reduced to a short thin filament; male with an anal papilla, no pyloric cæca; air bladder large, rounded behind, joined to the base of the skull. Cave streams of Cuba; the eyes having undergone a degeneration similar to that seen in *Amblyopsis*. These fishes have no relation to the blind cave fishes of North America, but are derived from marine types, their ancestors being evidently allies of *Ogilbia* and *Brotula*. It is known that blindfishes are found also in caves of the islands of Jamaica, but no specimens have been seen by naturalists. (*lux*, light; *fugo*, I flee.)

2872. LUCIFUGA SUBTERRANEUS, Poey.

(PEZ CIEGO.)

Head 2⅔; depth 3⅔. Branchiostegals 7. D. 70; A. 70; P. 51; V. 1; C. 9. Vertebræ 11 + 36 = 47. Body elongate, compressed, tapering, pointed; head low at nape, much depressed anteriorly, broad, covered with soft, white, wrinkled skin, with microscopic cirri, having firm and conical tubes; no barbels on lips or chin; skin of head with many pores; scales not ciliate, present on body and top of head and on opercles; lateral line median, marked by a series of microscopic cirri like those on head, these wanting posteriorly; eyes wanting; nasal openings double; mouth large; lower jaw shorter; lips fleshy; maxillary broad at tip, ⅔ length of head; teeth in jaws very short and sharp, in a band; vomerine teeth larger; no palatine teeth; pharyngeal teeth slender; tongue smooth; gill openings large; males with an anal papilla; fin rays simple, flexible, jointed but not branched; dorsal beginning at a point about ¼ nearer tip of snout than tip of caudal; anal smaller, beginning farther back, the 2 fins fully joined to the pointed caudal; pectoral fin short, falcate, nearly ¼ head; ventral in front of pectoral, a slender ray not ½ length of maxillary. Color transparent rosy, head reddish, becoming darker in alcohol. No pyloric cæca; intestines short; air bladder large. Described from 12 specimens, the longest about 4½ inches; found in caves of the jurisdiction of San Antonio, in the southern part of Cuba. (Poey.) (*sub*, under; *terra*, earth.)

Lucifuga subterraneus, POEY, Memorias, II, 96, 1860, San Antonio, Cuba (Coll. D. Tranquilino); Sandalio de Noda (Coll. D. Juan Antonio Fabre); first coll. from Cajío Cave, 1831 (Noda); second, Cave at La Industria (Coll. Dubrocá); third, Ashton Cave, San Andreas (Coll. Fabre); fourth, Cave of the Dragon (Coll. Fabre); fifth, Cave at the Castle of Concord (Coll. Layunta).

956. BROSMOPHYCIS, Gill.

Brosmophycis, GILL, Proc. Ac. Nat. Sci. Phila. 1861, 168 (*marginatus*).
Halias, AYRES, Proc. Cal. Ac. Sci. 1861, 52 (*marginatus*); preoccupied.

Body elongate, moderately compressed; head unarmed; snout not long; teeth sharp, curved, in bands on jaws, vomer, and palatines; small cilia above snout and on anterior part of lower jaw. Body covered with thin cycloid scales; scales on head rudimentary or wanting. Caudal fin differentiated, entirely separated from the dorsal and anal; caudal peduncle slender. California. This genus is very close to *Ogilbia*, differing in the. ciliated lips. Its species reaches a larger size. (*Brosmius; Phycis.*)

2873. BROSMOPHYCIS MARGINATUS (Ayres).

Head 4½ in body; depth 6½. D. 92; A. 70; eye 7¼ in head; snout 4½; maxillary 2; pectoral 1⅔; caudal 3; body elongate, moderately compressed; snout blunt; profile of head straight from snout; snout scarcely overhanging mouth; jaws subequal, the teeth conical, sharp, and slightly curved back, in bands on jaws, vomer, and palatines; maxillary reaching ¼ the eye's diameter beyond eye; snout and lower jaw thickly covered with small cilia; head naked with the exception of small scales above; 2 large pores at tip of chin, a few large ones around preopercle and preorbital, 1 around gill opening, behind which is a pocket in the skin; about 3 short gill rakers developed below the angle of first arch, with many rough plates, not differentiated from those on the other arches. Dorsal and anal long and low, the rays embedded in the skin; tips of last rays each beyond the base of the caudal about ⅓ the length of caudal rays; distance of front of dorsal from snout 3½ in length of body; origin of anal a little nearer base of caudal than tip of snout; pectoral reaching about half way from its base to front of anal; ventrals developed as long filaments; caudal slender and rounded behind. Color bright reddish brown; fins edged with bright rose-red. Coast of California, in water of moderate depth; rare. Here described from a specimen, 12 inches in length, collected off San Francisco by Mr. W. G. W. Harford. (*marginatus*, edged.)

Brosmius marginatus, AYRES, Proc. Cal. Ac. Nat. Sci., I, 1854, 13, San Francisco (Coll. W. O. Ayres); GIRARD, Pac. R. R. Surv., x, Fishes, 141, 1858.
Brosmophycis marginatus, GILL, Proc. Ac. Nat. Sci. Phila. 1861, 168; GILL, Proc. Ac. Nat. Sci. Phila. 1862, 280.
Halias marginatus, AYRES, Proc. Cal. Ac. Nat. Sci., pt. 2, 1861, 52.
Dinematichthys marginatus, GÜNTHER, Cat. Fishes Brit. Mus., IV, 375; JORDAN & GILBERT, Synopsis, 796, 1883.

957. OGILBIA, Jordan & Evermann.

Ogilbia, JORDAN & EVERMANN, in EVERMANN & KENDALL, Bull. U. S. Fish Comm. 1897 (February 9, 1898), 132 (*cayorum*).

Body moderately elongate, covered with minute, smooth, embedded scales; sides of head with similar scales; lateral line inconspicuous; opercle with a very small spine, preopercle unarmed; no strong hook on maxillary; no barbels nor cilia; teeth in jaws in bands, similar teeth on vomer and palatines; caudal free from the dorsal and anal; lower lip without cirri; dorsal and anal rays covered by the skin, ventrals each reduced to a filament of 2 rays; anal papilla of the male without horny claspers. Small fishes of the tropical shores of America, living in rock pools and shallows among algæ. This genus is closely allied to the East Indian genus *Dinematichthys*, differing in the absence of anal papilla and claspers and in the shorter vertical fins. (Named for J. Douglas Ogilby, the accomplished naturalist of the museum of Sydney, in recognition of his excellent work on the fishes of Australia.)

a. Snout very short, about 7 in head; eye small, 10 or 11 in head; scales small, obscure, snout very short, 6½ in head. VENTRALIS, 2874.
aa. Snout longer, about 4 in head; eye about 8½ in head; scales larger, distinct. CAYORUM, 2875.

2874. OGILBIA VENTRALIS (Gill).

Head 4½ in body; depth 5½. D. 64; A. 50; scales about 100; eye 10 or 11 in head; maxillary 2; pectoral 1⅗; caudal 1⅚. Body elongate, moderately compressed; snout blunt, the profile behind snout nearly straight to occiput; mouth large, the maxillary extending 2 or 3 times the eye's diameter behind eye; teeth small, in bands on jaws, vomer, and palatines; eye very small, nearer snout than posterior end of maxillary; no cilia on snout and chin; body apparently naked to the unaided eye; but body and top of the head covered with small scales, which can be seen by the aid of a lens. Origin of dorsal distant from tip of snout by a space contained 3¼ times in body; front of anal about midway between tip of snout and base of caudal; tips of last dorsal and anal rays reaching about to the middle of caudal rays, but not connected; pectorals scarcely reaching midway between their base and the front of anal; ventrals filamentous; caudal slender and rounded behind. Color in spirits, light brown above, lighter below; fins all colorless; without distinct marking anywhere. Gulf of California; not rare in rock pools; several specimens, 2 to 4 inches in length taken by us at Mazatlan. Here described from a specimen, 2 inches in length, from La Paz Harbor, Lower California. (*ventralis*, pertaining to the belly.)

Brosmophycis ventralis, GILL, Proc. Ac. Sci. Phila. 1863, 253, Cape San Lucas. (Coll. Xantus.)
Dinematichthys ventralis, JORDAN, Proc. Cal. Ac. Sci. 1895, 502, pl. 54.

2875. OGILBIA CAYORUM, Evermann & Kendall.

Head 4; depth 4½; eye 8½; snout 4. D. about 68; A. about 50; scales about 14-87-13; maxillary 1⅔; pectoral 1⅘; ventral 1½; caudal 2⅓. Body moderately elongate, compressed; head moderate, snout blunt; mouth large, jaws subequal, maxillary extending beyond vertical of eye a distance nearly equal to length of snout; eye very small, high up, situated in anterior third of head; nostril small, close to eye; teeth small, in bands on jaws, vomer, and palatines; back elevated, strongly arched from snout to origin of dorsal fin, thence descending in a nearly straight line to base of caudal; ventral outline comparatively straight, slightly concave at front of anal. Dorsal and anal long and low, distinct from caudal, the posterior rays longest, about 3⅓ in head, base of each scaled; distance from tip of snout to origin of dorsal about 3 in length of body; origin of anal under about twenty-second dorsal ray, equidistant between tip of snout and base of caudal; scales very small, embedded, but showing distinctly under a lens; cheek and opercles partially covered with minute, embedded scales; top of head naked; opercle with a large, flat, flexible spine on level with eye. No barbels, cilia, nor tubercles; 2 large mucous pores at symphysis of lower jaw and 2 on preorbitals near anterior edge on each side; a row of 5 or 6 pores on lower jaw and edge of preopercle. Color uniform pale olivaceous or light brown, finely punctate with minute brown specks. Key West. Only the type known, an example, 2⅓ inches long, seined on a shoal covered with algæ at Key West. (Cayo Hueso, or Bone Key, the original Spanish name for the Island of Key West, whence the name *cayorum*, of the keys.)

Ogilbia cayorum, EVERMANN & KENDALL, Bull. U. S. Fish Comm. 1897 (Feb. 9, 1898), 132, pl. 9, fig. 14, Key West, Florida. (Type, No. 48792. Coll. Evermann & Kendall.)

958. BYTHITES, Reinhardt.

Bythites, REINHARDT, Dansk. Vidensk. Selsk. Afhandl., VII, 1838, 178 (*fuscus*).

Body elongate, covered with minute scales. Head large, thick; mouth large; jaws equal; no barbel; bands of teeth in the jaws and on vomer and palatines. Branchiostegals 8; gill membranes united, free from the isthmus; eyes moderate. Lateral line interrupted. Vertical fins united; ventral fins reduced to simple filaments, each composed of 2 rays closely united. Air bladder large; 2 pyloric cæca. A thick, conical, anal papilla (in the male). Greenland. ($\beta v\theta i\tau\eta\varsigma$, an animal of the depths, from $\beta v\theta\iota o\varsigma$, the deep.)

2876. BYTHITES FUSCUS, Reinhardt.

Head about 4; depth 4½. Body somewhat compressed, lipariform; snout obtuse, naked, with minute cirri. Mandible long, curved, extending far behind vertical from posterior margin of orbit; eye small; scales moderate on body; lateral line complete, but interrupted over vent, the two parts slightly overlapping the same vertical; vertical fins confluent, enveloped in thick skin; pectorals broad, lanceolate, with broad base; ventrals filiform, reaching behind origin of pectoral, as long as pectoral and ⅔ as long as head; a conspicuous anal papilla in the male. The only known specimen, now in the museum at Copenhagen, was obtained in Greenland half a century ago. (Goode & Bean.) (*fuscus*, dusky.)

Bythites fuscus, REINHARDT, Dansk. Vidensk. Selsk. Afh., VII, 1838, 178, Greenland; GÜNTHER, Cat., IV, 375, 1863; JORDAN & GILBERT, Synopsis, 795, 1883; GOODE & BEAN, Oceanic Ichthyology, 316, 1896.

959. CATÆTYX, Günther.

Catætyx, GÜNTHER, Challenger Report, XXII, 104, 1887 (*messieri*).

Body compressed, elongate, covered with very small and thin scales; lateral line indistinct, interrupted. Head oblong, with somewhat pointed snout, covered with very small scales, only the anterior part of the snout naked; bones of the head rather firm, but with the muciferous system well developed, the canals having wide openings along the infraorbital, and on the lower limb of the preoperculum; eye rather small; nostrils far apart, the posterior in front of the eye and the anterior at the extremity of the snout; operculum with a spine behind; no other armature on the head; snout not swollen, but the upper jaw slightly overlapping the lower; barbels none; mouth wide; bands of villiform teeth in the jaws, on the vomer, and the palatine bones; a series of larger teeth along the sides of the lower jaw; tail not much attenuated; vertical fins confluent; ventrals close together, reduced to a pair of fine, simple filaments, and inserted somewhat behind the isthmus, below the middle of the operculum. Gills 4, with short, broad gill rakers and well developed laminæ; pseudobranchiæ none; branchiostegals 8; pyloric appendages. Deep seas. Two species known. (*καταί*, at the bottom; *τύξις, τυγχάνω*, find.)

2877. CATÆTYX RUBRIROSTRIS, Gilbert.

Depth of body below origin of dorsal equals ⅓ distance from end of snout to vent, 7 in length; head 4; distance from snout to origin of dorsal 3⅓; from snout to vent 2⅞; maxillary extending beyond eye, 2⅔ in head; eye equaling snout, 5⅓; interorbital width 7; width of snout 3⅓. Teeth in upper jaw in a narrow band, minute, compressed, narrowly triangular, none of them enlarged; in the lower jaw a still narrower band of similar teeth, the posterior row slightly enlarged and increasing a little in size on sides of jaw, where it is accompanied by a single series only of the smaller teeth; this lateral series is continued backward far beyond premaxillary band; teeth on vomer and palatines similar to those in sides of lower jaw, the former in a V-shaped patch, the latter in a long and very narrow band. Anterior nostril in a short tube at tip of snout, the posterior large, without tube, immediately in front of eye; system of mucous pores well developed but not conspicuous, the pores collapsing on account of the thinness of the skin covering head; large mucous tube below eye, extending around front of snout and opening by slit-like pores along edge of snout and lower margin of infraorbital flap, opening posteriorly by a vertical slit ½ as long as pupil, immediately above end of maxillary; another series of pores along mandible and at edge of expanded limb of preopercle; no other evident pores. Angle of preopercle much expanded, its width equaling diameter of pupil; a sharp, strong spine arising from anterior portion of opercle, the structure of the gill flap apparently like that of *Bassogigas stelliferoides;* a short, sharp spine directed backward immediately behind posterior nostril; no other spines on head; gill openings wide, continuing forward to below posterior margin of orbit, the membranes wholly free from the isthmus; gill rakers short but not very broad, about ⅓ length of pupil, only 2 or 3 developed immediately in front of angle of arch. Dorsal beginning over or slightly behind middle of pectorals, the distance from its origin to occiput equaling or somewhat exceeding distance of latter from tip of snout; dorsal and anal fully united to caudal. The caudal has a base of appreciable width, bearing about 12 close-set rays, which extend much beyond tips of last dorsal and

3030——80

anal rays; origin of anal nearer snout than base of ·caudal; ventrals slender, each consisting of a single ray, inserted very near together, under anterior portion of opercle, their length about equaling that of maxillary; pectorals with about 23 rays, evenly rounded behind, their length 1⅔ in head. Scales very small, cycloid, regularly imbricated, in about 135 transverse series; nape and belly scaled, as is also the head, excepting snout, mandible, suborbital, and sometimes interorbital areas. Color dusky olive, the ventrals white, the other fins black, at least on distal portions; opercles, gill membranes, sides and top of snout, and posterior portion of abdomen blue black; snout flushed with dark ruby red in life; lining membrane of mouth and gill cavity, and peritoneum jet-black. Closely related to *C. messieri*, differing in the shorter, broader snout, the wider preopercle, the more anterior origin of anal, and apparently in the gill rakers and pores on head. Off coast of California. Four specimens known, the longest 4¼ inches in length. (Gilbert.) (*ruber*, red; *rostrum*, snout.)

Catætyx rubrirostris, GILBERT, Proc. U. S. Nat. Mus. 1890, 111, off coast of California, at Albatross Stations 2909, 2925, and 2936, in 205 to 359 fathoms. (Type, No. 44379.)

960. DICROMITA, ·Goode & Bean.

Dicromita, GOODE & BEAN, Oceanic Ichthyology, 319, 1896 (*agassizii*).

Brotulids resembling in form and general appearance *Catætyx* and *Diap. lacanthopoma*, having the lateral line obsolete, or interrupted posteriorly; ventrals a pair of simple, fine filaments, and with teeth upon the palatines. It has, however, 3 or 4 small spines upon the preoperculum, as well as a sharp spine upon the upper angle of the operculum; and the lateral line, though indistinct, is traceable for ½ or ¾ the length of the body, which, like the upper part of the head, is covered with small, deciduous scales, the opercular region being apparently scaleless, and the bones of the suborbital region almost uncovered, with conspicuous sinuses, which show through the transparent texture of the surface. Head oblong; snout somewhat produced, depressed, and turgid, resembling, though in a less degree, that of *Barathrodemus*. Eye moderate, conspicuous. Mouth wide; teeth villiform, in bands on the jaws and palatines, and very minute upon the vomer, which has a roughened, knob-like enlargement at its angle. Vent premedian. Ventral fins confluent; ventrals rooted very close together, each reduced to a fine, flexible, simple filament, planted somewhat behind the isthmus and below the middle of the operculum. Gills 4, with well-developed laminæ and rather long, slender gill rakers. Branchiostegals 8. Pseudobranchiæ apparently absent. (δίκρος, forked; μίτος, thread.)

2878. DICROMITA AGASSIZII, Goode & Bean.

Body elongate, much compressed, its height about ⅙ of its total length, its width about ½ its greatest height; head slightly greater than height of body, about twice its own width; mouth very large, the maxillary curved and much dilated at its extremity, reaching far behind the vertical from the posterior margin of orbit; jaws nearly equal, the snout considerably produced and dilated, its length equal to diameter of eye and ¼ length of head. Teeth very fine, villiform, in bands on jaws and palatines, and also present on vomer, though very small, especially upon the rounded, globular process of the angle. Lateral line very indistinct, interrupted, but extending behind the vent at least ⅓ of the way to tip of tail. Dorsal origin nearly in vertical from the axil of the pectoral; ventrals very slender, villiform, closely approximate at their roots, and less than ½ as long as the head. Color brownish. A specimen was obtained by the *Blake* off Granada, Station XCIII, at a depth of 291 fathoms. The collateral type was obtained by the *Albatross* at Station 2374, in lat. 29° 11′ 30″ N., long. 85° 29′ W., at a depth of 26 fathoms. (Goode & Bean.) (Named for Prof. Alexander Agassiz.)

Dicromita agassizii, GOODE & BEAN, Oceanic Ichthyólogy, 319, fig. 285, 1896, off Granada, in 291 fathoms (Coll. the *Blake*); Lat. 29° 11′ 30″ N.,⏀Long. 85° 29′ W., in 26 fathoms. (Coll. *Albatross.*) (Type in M. C. Z.)

961. BASSOZETUS, Gill.

Bathynectes, GÜNTHER, Ann. and Mag. Nat. Hist., II, 1878, 20 (*compressus*); name preoccupied in Crustacea.
Bassozetus, GILL, Proc. U. S. Nat. Mus., VI, 1883, 59 (*normalis*).
Bathyonus, GOODE & BEAN, Proc. U. S. Nat. Mus., VIII, 1886, 603 (*catena*).

Body compressed, with long tapering tail, covered with deciduous thin scales of moderate size. Bones of the head very soft and cavernous, the upper opercular spine very feeble, ridge-like; no other armature of the head. Head scaly, except the snout, which is obtusely rounded off, with the jaws equal or nearly equal in front. Mouth very wide; bands of villiform teeth in the jaws, on the vomer, and palatine bones. Barbels none. Eye small; anterior nostril about midway between the posterior and the extremity of the snout. Vertical fins confluent; ventrals close together, reduced to a pair of simple filaments, and inserted below the rounded angle of the preoperculum. Gills 4, with short gill laminæ, but with long stiff gill rakers on the first branchial arch. Pseudobranchiæ none. Branchiostegals 8. Pyloric appendages none. ($\beta\acute{a}\acute{o}\acute{o}\omega\nu$, for $\beta a\vartheta\acute{\upsilon}\varsigma$, deep; $\zeta\eta\tau\acute{\epsilon}\omega$, seek.)

  a. Body moderately elongate, the depth 9 to 10 in length; dorsal rays 116; anal 92 to 96.
    b. Head 6 in length.   NORMALIS, 2879.
    bb. Head about 7½ in length.   COMPRESSUS, 2880.
  aa. Body more elongate, the depth 12½ in length; head 8¾; head with conspicuous, chainlike rows of pores.   CATENA, 2881.
  aaa. Body excessively attenuate, the depth more than 16 times in length; dorsal rays 138; anal 115; distance from snout to vent nearly 4 times in body.
                                          TÆNIA, 2882.

2879. BASSOZETUS NORMALIS, Gill.

Head 6 in total length; eye 4 in snout. D. 116; A. 96. Body much compressed, its width in the region of vent not more than ⅓ of its height, which at the same point is about ⅑ of the total length; greatest height of body, over the origin of the pectorals, about ⅔ the distance from base of pectorals to vent, the vent being about twice as distant from base of caudal rays as from snout. Head moderately compressed, flat above; snout obtuse rounded, turgid; lower jaw considerably included. Bones of head not completely ossified, very cavernous in the alcoholic specimen, the head showing many deep sinuosities and depressions. Eye very small, situated about midway between the tip of the snout and the vertical from the posterior end of maxillary. Teeth all small and short, densely set, forming narrow, villiform bands; vomerine band open V-shaped. Dorsal fin beginning far in advance of origin of pectoral and above upper angle of gill opening, rays longest in region of vent; anal beginning immediately behind vent, its rays not quite so long as those of dorsal; pectoral with broad base, short, not exceeding much more than halfway to vertical from vent, its length considerably less than that of postorbital portion of head. Ventral rays very slender, villiform, reaching almost to vent, far beyond pectoral, their length almost equal to that of head. Scales moderate, very deciduous, extending upon cheeks and on top of head almost to tip of snout; no evidence of a lateral line. Color light, the head and abdomen blackish; inside of mouth purplish brown. The *Blake* secured specimens from Station CCIV, in lat. 24° 33′ N., at a depth of 1,920 fathoms, and from Station LXXXIV, off Dominica, in 1,131 fathoms. The *Albatross* also obtained examples (No. 49416, U. S. Nat. Mus.) from Station 2380, in Lat. 28° 02′ 30″ N., Long. 87° 43′ 45″ W., at a depth of 1,430 fathoms; (No. 33306, U. S. Nat. Mus.) from Station 2042, in lat. 39° 33′ N., Long. 68° 26′ 45″ W., at a depth of 1,555 fathoms. (Goode & Bean.) West Indies, Gulf of Mexico, and to lat. 40° N., in region of the Gulf Stream. (*normalis*, normal.)

Bassozetus normalis, GILL, Proc. U. S. Nat. Mus. 1883, 259, Lat. 39° 33′ N., Long. 68° 26′ 45′ W., in 1,555 fathoms (Type, No. 33306. Coll. *Albatross*); GOODE & BEAN, Oceanic Ichthyology, 322, fig. 287, 1896.

2880. BASSOZETUS COMPRESSUS (Günther).

D. 116; A. 92; P. 23; V. 1. The greatest depth of the body is above the end of the gill cover and about ½ length of trunk; vent twice as distant from extremity of tail as from snout, consequently the tail is more moderately attenuated. Head compressed like the body, and about ⅔ length of trunk; superficial bones form large muciferous cavities which, when full, must give to the head a much more evenly rounded appearance than in the preserved state, when the supporting bony ridges project more or less from under the skin. Snout slightly swollen, but the jaws nearly even in front, the wide mouth slightly ascending forward; maxillary with the form usual in these Gadoid fishes, dilated behind, and extending far behind the eye. Eye very small, ½ length of snout, and ₁₁ that of head, placed high up on the side, and not possessing an orbital fold of integument; interorbital space rather convex and equal in width to 3 diameters of eye. Teeth all very small, short, densely set, and forming villiform bands, the broadest on maxillary bone and quite uncovered on the sides, no labial folds being developed; palatine band broader than the mandibulary, and the vomerine band V-shaped, each arm being bent with the convexity inward. Gill opening and cavity very wide and of an intense black; gill rakers much longer than the laminæ, 15 in number on the anterior arch, besides some rudimentary ones above. Dorsal fin commencing above upper end of gill opening, with short rays partly hidden in the skin, becoming longer in middle of fin, but remaining of moderate length; anal shorter; pectoral with a rather narrow base, quite free, and composed of feeble rays, its length only $\frac{1}{2}$ that of head; ventral rays very feeble, reaching somewhat beyond the root of pectoral. In the specimens examined only very few of the thin, cycloid scales have been preserved; they are of moderate size, there being about 16 in a transverse series running from the vent to the dorsal fin; the lateral line, if it was developed, can no longer be traced. Blackish, with the fins, head, and abdomen black. Specimens of this very fine and truly bathybial fish were obtained at great depths on the southeast of New Guinea, off the Philippine Islands, and in the mid-Atlantic; the exact localities being 75 miles east-southeast of Raine Island, Station 184, depth 1,400 fathoms; two specimens, 17 and $4\frac{1}{2}$ inches long. Philippine Islands, Station 205, depth 1,050 fathoms; one specimen, $5\frac{1}{2}$ inches long. Mid-Atlantic, Station 107, depth 1,500 fathoms; one specimen, $5\frac{1}{2}$ inches long. The young are extremely similar to the old, but have a larger eye, which is $\frac{1}{3}$ of length of head. The specimen from Station 205 (Philippine Islands) has longer ventral filaments, extending nearly to the vent. (Günther.) (*compressus*, compressed.)

Bathynectes compressus, GÜNTHER, Ann. Mag. Nat. Hist., II, 1878, 20, Challenger Station 107, mid-Atlantic, in 1,500 fathoms. (Coll. *Challenger*.)
Bathyonus compressus, GÜNTHER, Challenger Report, XXII, 109, 1887.
Bassozetus compressus, GOODE & BEAN, Oceanic Ichthyology, 322, 1896.

2881. **BASSOZETUS CATENA**, Goode & Bean.

Head $8\frac{2}{3}$; depth $1\frac{1}{2}$ in head or $12\frac{1}{2}$ in body; eye 5; snout 5; interorbital width 5. Body very elongate, much compressed, and tapering into a slender, whip-like tail. Head without spines, very cavernous, not much compressed, higher than body. Interorbital area somewhat convex. The muciferous channel upon the infraorbital ring shows in its course several wide subcircular sinuses, closely approximated; a similar row upon the posterior edge of the preoperculum and continued forward upon the under surface of the mandible; the vertex also has a semicircle of similar sinuses. Maxillary extending beyond vertical through posterior margin of orbit, its length equal to that of postorbital part of head; mandible $\frac{2}{3}$ as long as head and equal in length to height of body; jaws, vomer, and palate with bands of villiform teeth, the vomerine band V-shaped. Nostrils in front of middle of eye, separated by a slight interspace, the anterior nearer to its mate than to tip of snout. Branchiostegals 8; pseudobranchiæ absent. Gill rakers long and numerous, the longest slightly exceeding diameter of eye, 15 developed below angle of first arch, besides several rudiments; dorsal origin slightly behind that of pectoral, its distance from tip of snout about $7\frac{1}{3}$ in total, rays well developed; in the anterior $\frac{1}{3}$ of the fin, in a space equal to length of head, were counted 20 rays, the longest of which is $\frac{2}{5}$ as long as head; anal origin under twenty-first dorsal ray, its rays shorter than those of dorsal; pectoral extending to vertical from eighteenth ray of dorsal, $\frac{4}{5}$ as long as head; ventrals composed each

2510 Bulletin 47, United States National Museum.

of a simple filament, the origin slightly in advance of vertical through pectoral origin, the length ⅔ that of head, not reaching nearly to vent, the distance of which from origin of ventrals is slightly greater than length of head. Color brownish yellow; head and abdomen blackish. Gulf of Mexico, in great depths. Only the type known, 237 mm. long. (*catena*, chain, from the arrangement of the mucous cavities on the head.)

Bassozetus catena, GOODE & BEAN, Proc. U. S. Nat. Mus. 1885, 603, Lat. 28° 00′ 15″ N., Long. 87° 42′ W., in 1,467 fathoms (Type, No. 37341. Coll. *Albatross*); GÜNTHER, Challenger Report, XXII, 111, 1887; GOODE & BEAN, Oceanic Ichthyology, 323, fig. 286, 1896.

2882. BASSOZETUS TÆNIA (Günther).

D. 138; A. 115; P. 30; V. 1. The greatest depth of the body is below the origin of the dorsal fin and about ⅓ of the length of the trunk, the vent being not quite thrice as distant from the extremity of the tail as from the snout; therefore, the whole of the fish, and especially the tail, is much attenuated. Head not compressed, low and long, forming ⅔ length of trunk. Structure of the bones of the head as in *B. compressus*. Snout rather swollen and broad, the upper jaw but slightly overlapping the lower; maxillary extending far behind the eye, which is very small, ⅓ length of snout, about ⅟₁₄ that of head, and ¼ width of interorbital space. Teeth very small and short, densely set, forming narrow, villiform bands; vomerine bands open Λ-shaped. Gill cavity deep black; gill rakers long and slender, 16 in number, with some rudimentary ones in front and behind. Dorsal fin commencing above upper end of gill opening, with short rays partly hidden in the skin, the rays becoming longer on the anterior third of tail, but remaining of moderate length, the anal rays still shorter; pectoral with a broad base, quite free, and composed of rather feeble rays, its length equal to that of postorbital portion of head; ventral rays very feeble, reaching nearly to the middle of the pectoral. The scales must have been extremely thin and rather small; there were probably about 20 in a transverse series running from the vent to the dorsal fin. The lateral line can not be made out. Light colored (possibly pink in life), with the head and abdomen black. Only 1 specimen known of this eminently bathybial fish, obtained in mid-Atlantic (*Challenger* Station 104) at a depth of 2,500 fathoms. Its total length is 10 inches. (ταινία, ribbon.)

Bathyonus tænia, GÜNTHER, Challenger Report, XXII, 110, 1887, pl. 23, fig. A, mid-Atlantic, Station 104, at a depth of 2,500 fathoms; GOODE & BEAN, Oceanic Ichth., 323, 1896.

962. MŒBIA, Goode & Bean.

Mœbia, GOODE & BEAN, Oceanic Ichthyology, 331, 1896 (*gracilis*).

Brotulids resembling *Bassozetus* in general form, excepting that the tail is prolonged in a very slender filament, the dorsal and anal rays being extremely short posteriorly, but positively confluent with the caudal rays, which are much longer and much exserted; ventrals each bifid, instead of a single ray, as in *Bassozetus*. Head very cavernous, the sinuses large and conspicuous on the infraorbital ring, on the mandible, and the preoperculum. A single, short, feeble spine on the shoulder, but none upon the operculum or preoperculum, though certain projections seem to show above the eye, doubtless due to the shrinkage of the integument upon the underlying projections of the bone. Mouth very wide, the extremity of the maxillary much dilated; posterior nostrils very wide and separated from the eye by a small, spinous projection of bone; teeth in narrow bands, that on the vomer V-shaped, with the 2 arms straight. A few large scales in a row starting from the upper angle of the gill opening and terminating over the axle of the pectoral. Gill rakers on outer arch rather numerous, long and slender. Pseudobranchiæ represented by 2 minute globules. Deep sea; 2 species known. ("Named in honor of Prof. Karl Möbius, director of the Royal Zoological Museum in Berlin, who has added much to our knowledge of marine life by his noble work, Die Fauna der Kielerbucht, and by numerous other writings.")

2883. MŒBIA PROMELAS (Gilbert).

Head 2 in trunk; depth 3. Body $3\frac{1}{5}$ to $3\frac{1}{4}$ in tail. Tail produced into a filament, the caudal basis extremely narrow, supporting 5 long slender rays which are firmly bound together. Mouth terminal, large, the maxillary much dilated at tip, reaching well behind the eye, $1\frac{3}{5}$ in head; lower jaw included, the tip slightly produced. Teeth in villiform bands on mandible, premaxillary, vomer, and palatines, the band on mandible very narrow, that on vomer with the diverging arms much incurved, the anterior angle rounded. Tongue toothless, some of the basibranchials forming a sharply elevated dentigerous crest. Gill laminæ extremely narrow, the gill rakers of outer arch very long and slender, 1 (with 4 rudiments) above angle, 15 below. Infraorbital chain with 6 mucous sinuses, the mandible with 5, preopercle with 5, and a number on top of head; these are all bridged over with very delicate membrane which is easily ruptured. A row of low, strong spinous points directed posteriorly on the ridge running backward from the eye; no other spines on head, though a number of short spinous points are made evident when the skin is removed; opercular spine rather weak. A distinct membranaceous flap runs along the projecting edge of shoulder girdle, connecting pectorals with upper end of gill flap; pectorals slender, equaling postorbital part of head; ventrals each of a bifid filament, the two branches joined at the base for a very short distance, variable in length, reaching to or nearly to tips of pectorals, usually contained about $1\frac{1}{2}$ times in head; dorsal beginning a trifle behind base of pectorals, its distance from occiput equaling distance of latter from front of eye. Scales very small, apparently covering a part of top of head; 3 series of large pores on sides; 1 from upper end of gill slit backward parallel with dorsal outline; a second along middle of sides; the third beginning halfway between base of pectorals and ventral outline, extending backward on belly and along base of anal fin, these lines all somewhat indistinct, and it can not be determined how far they extend backward. Color light brown; head (except occiput), mouth, gill cavity, and abdomen jet-black; fins dusky. This species closely resembles *Mœbia gracilis* (Günther), from New Guinea, dif-

of a simple filament, the origin slightly in advance of vertical through pectoral origin, the length ⅔ that of head, not reaching nearly to vent, the distance of which from origin of ventrals is slightly greater than length of head. Color brownish yellow; head and abdomen blackish. Gulf of Mexico, in great depths. Only the type known, 237 mm. long. (*catena*, chain, from the arrangement of the mucous cavities on the head.)

Bassozetus catena, GOODE & BEAN, Proc. U. S. Nat. Mus. 1885, 603, Lat. 28° 00′ 15″ N., Long. 87° 42′ W., in 1,467 fathoms (Type, No. 37341. Coll. *Albatross*); GÜNTHER, Challenger Report, XXII, 111, 1887; GOODE & BEAN, Oceanic Ichthyology, 323, fig. 286, 1896.

2882. BASSOZETUS TÆNIA (Günther).

D. 138; A. 115; P. 30; V. 1. The greatest depth of the body is below the origin of the dorsal fin and about ¼ of the length of the trunk, the vent being not quite thrice as distant from the extremity of the tail as from the snout; therefore, the whole of the fish, and especially the tail, is much attenuated. Head not compressed, low and long, forming ⅘ length of trunk. Structure of the bones of the head as in *B. compressus*. Snout rather swollen and broad, the upper jaw but slightly overlapping the lower; maxillary extending far behind the eye, which is very small, ¼ length of snout, about 1/14 that of head, and ¼ width of interorbital space. Teeth very small and short, densely set, forming narrow, villiform bands; vomerine bands open Λ-shaped. Gill cavity deep black; gill rakers long and slender, 16 in number, with some rudimentary ones in front and behind. Dorsal fin commencing above upper end of gill opening, with short rays partly hidden in the skin, the rays becoming longer on the anterior third of tail, but remaining of moderate length, the anal rays still shorter; pectoral with a broad base, quite free, and composed of rather feeble rays, its length equal to that of postorbital portion of head; ventral rays very feeble, reaching nearly to the middle of the pectoral. The scales must have been extremely thin and rather small; there were probably about 20 in a transverse series running from the vent to the dorsal fin. The lateral line can not be made out. Light colored (possibly pink in life), with the head and abdomen black. Only 1 specimen known of this eminently bathybial fish, obtained in mid-Atlantic (*Challenger* Station 104) at a depth of 2,500 fathoms. Its total length is 10 inches. (ταινία, ribbon.)

Bathyonus tænia, GÜNTHER, Challenger Report, XXII, 110, 1887, pl. 23, fig. A, mid-Atlantic, Station 104, at a depth of 2,500 fathoms; GOODE & BEAN, Oceanic Ichth., 323, 1896.

962. MŒBIA, Goode & Bean.

Mœbia, GOODE & BEAN, Oceanic Ichthyology, 331, 1896 (*gracilis*).

Brotulids resembling *Bassozetus* in general form, excepting that the tail is prolonged in a very slender filament, the dorsal and anal rays being extremely short posteriorly, but positively confluent with the caudal rays, which are much longer and much exserted; ventrals each bifid, instead of a single ray, as in *Bassozetus*. Head very cavernous, the sinuses large and conspicuous on the infraorbital ring, on the mandible, and the preoperculum. A single, short, feeble spine on the shoulder, but none upon the operculum or preoperculum, though certain projections seem to show above the eye, doubtless due to the shrinkage of the integument upon the underlying projections of the bone. Mouth very wide, the extremity of the maxillary much dilated; posterior nostrils very wide and separated from the eye by a small, spinous projection of bone; teeth in narrow bands, that on the vomer V-shaped, with the 2 arms straight. A few large scales in a row starting from the upper angle of the gill opening and terminating over the axle of the pectoral. Gill rakers on outer arch rather numerous, long and slender. Pseudobranchiæ represented by 2 minute globules. Deep sea; 2 species known. ("Named in honor of Prof. Karl Möbius, director of the Royal Zoological Museum in Berlin, who has added much to our knowledge of marine life by his noble work, Die Fauna der Kielerbucht, and by numerous other writings.")

2883. MŒBIA PROMELAS (Gilbert).

Head 2 in trunk; depth 3. Body $3\frac{1}{2}$ to $3\frac{1}{4}$ in tail. Tail produced into a filament, the caudal basis extremely narrow, supporting 5 long slender rays which are firmly bound together. Mouth terminal, large, the maxillary much dilated at tip, reaching well behind the eye, $1\frac{2}{3}$ in head; lower jaw included, the tip slightly produced. Teeth in villiform bands on mandible, premaxillary, vomer, and palatines, the band on mandible very narrow, that on vomer with the diverging arms much incurved, the anterior angle rounded. Tongue toothless, some of the basibranchials forming a sharply elevated dentigerous crest. Gill laminæ extremely narrow, the gill rakers of outer arch very long and slender, 1 (with 4 rudiments) above angle, 15 below. Infraorbital chain with 6 mucous sinuses, the mandible with 5, preopercle with 5, and a number on top of head; these are all bridged over with very delicate membrane which is easily ruptured. A row of low, strong spinous points directed posteriorly on the ridge running backward from the eye; no other spines on head, though a number of short spinous points are made evident when the skin is removed; opercular spine rather weak.. A distinct membranaceous flap runs along the projecting edge of shoulder girdle, connecting pectorals with upper end of gill flap; pectorals slender, equaling postorbital part of head; ventrals each of a bifid filament, the two branches joined at the base for a very short distance, variable in length, reaching to or nearly to tips of pectorals, usually contained about $1\frac{1}{2}$ times in head; dorsal beginning a trifle behind base of pectorals, its distance from occiput equaling distance of latter from front of eye. Scales very small, apparently covering a part of top of head; 3 series of large pores on sides; 1 from upper end of gill slit backward parallel with dorsal outline; a second along middle of sides; the third beginning halfway between base of pectorals and ventral outline, extending backward on belly and along base of anal fin, these lines all somewhat indistinct, and it can not be determined how far they extend backward. Color light brown; head (except occiput), mouth, gill cavity, and abdomen jet-black; fins dusky. This species closely resembles *Mœbia gracilis* (Günther), from New Guinea, differing in the following respects: Depth 3 in trunk (in *gracilis* 3½); eye 6 in head (in *gracilis* 5½); vomerine patch of teeth with the two arms incurved (U-shaped in *gracilis*); dorsal fin beginning behind the pectoral (over root of pectoral in *gracilis*); an additional series of large scales (lateral line) along middle of sides, and another along ventral outline; trunk 2¼ in tail (2⅜ in *gracilis*); ventrals shorter, not reaching past tips of pectorals. Five specimens, the largest 9½ inches long, from *Albatross* Station 3010, at a depth of 1,005 fathoms, in the Gulf of California. (Gilbert.) (πρό, before; μέλας, black.)

Porogadus promelas, GILBERT, Proc. U. S. Nat. Mus. 1891, 547, Gulf of California, in 1,005 fathoms. (Coll. *Albatross*.)

963. NEOBYTHITES, Goode & Bean.

Neobythites, GOODE & BEAN, Proc. U. S. Nat. Mus. 1885, 600 (*gilli*).

Brotulids having the body elongate, compressed, covered with small scales, and the head also scaled; lateral line incomplete, obsolete posteriorly. Eye moderate; snout moderate, rounded, slightly produced, the lower jaw slightly included; no barbel. Teeth villiform, in narrow bands in jaws and palatines; vomerine teeth in a V-shaped patch; 2 weak spines at angle of preoperculum, and a stronger one at angle of operculum. Gill openings wide, the membranes deeply cleft and not attached to the isthmus; vertical fins united; ventrals reduced each to a bifid ray. Branchiostegals 8. Pseudobranchiæ present, but small. Air bladder present. (νέος, new; *Bythites*.)

a. Scales about 88 in longitudinal series; depth 4⅔ in length.   GILLII, 2884.

aa. Scales about 123 in longitudinal series; depth 5¾ in length; dorsal rays 101.

MARGINATUS, 2885.

2884. NEOBYTHITES GILLII, Goode & Bean.

Body compressed, its height contained 4⅔ times in total length, and less than length of head; interorbital area convex, its width equal to diameter of circular eye, 3⅔ in length of head, and 1½ in length of snout in young. Head compressed, deeper than broad, with wide sinuses, its length contained 4¼ times in that of the body; snout obtusely rounded, slightly produced; mouth large, the maxillary extending considerably behind the vertical through posterior margin of eye, expanded posteriorly; mandible still longer, its length about 2¼ times in height of body; interorbital space convex. Teeth in villiform bands in jaws and on palatines; vomerine patch subcircular, with angles extended posteriorly. Gill rakers moderately long and slender, somewhat numerous, the longest about ⅔ diameter of eye, 11 developed and 3 rudiments below the angle. Pseudobranchiæ absent; gill opening wide, the membrane deeply cleft, free from the isthmus behind. A single long, flat spine attached to posterior portion of operculum, high up, extending back to its edge; a small hidden spine at lower angle of preoperculum. Nostrils small, the anterior one in a very short tube, almost upon tip of snout; posterior nostril slightly larger, not tubular, immediately in front of middle of eye. Scales mod-crate, upon head and body, in 88 vertical rows, 7 rows between dorsal origin and lateral line, which becomes obsolete in its posterior half, 16 or 17 from vent forward to lateral line; dorsal origin behind that of ventral and pectoral, its distance from snout contained 4 times in total length, its rays moderately long; anal origin under eighteenth dorsal ray, its distance from snout contained 2¼ times in body length, rays rather slenderer than those in the dorsal; caudal rays 6 or 7 in number, their length contained 9 times in total length, not differentiated from those of the adjacent fins; pectoral origin well forward, its base somewhat concealed by the flap of the operculum, its length about equal to ⅔ that of head; ventrals each a bifid ray, the inner filament the longer, inserted slightly in advance of the base of the pectoral, not far from humeral symphysis, and reaching nearly to vent, its length nearly equal to height of body; distance from origin of ventral to vent slightly greater than height of body; color light yellow, with silvery reflections, with cloudings of brown above lateral line and numerous black chromatophores; a series of irregular brown blotches above the lateral line, with 1 or 2 much darker, extending upon the dorsal fin. In many specimens the color is uniform yellow, with simply the dark ocelli showing. (Goode & Bean.) Atlantic, in rather deep water, from Gulf Stream to the coast of Brazil. (Named for Dr. Theodore Gill.)

Neobythites gillii, GOODE & BEAN, Proc. U. S. Nat. Mus. 1885, 601, Lat. 28° 36' N., Long. 85° 33' W., in III fathoms (Type, No. 37340. Coll. *Albatross*); GÜNTHER, Challenger Report, XXII, 103, 1887; GOODE & BEAN, Oceanic Ichth., 325, fig. 289, 1896.
Neobythites ocellatus, GÜNTHER, Challenger Report, XXII, 103, pl. 21, fig. B, 1887, off Pernambuco, in 350 fathoms.

2885. NEOBYTHITES MARGINATUS, Goode & Bean.

Head 4⅔ in total length; depth 5⅓. D. 101; scales 7–123–29. Body compressed, somewhat elongate; interorbital area convex, its width greater than the diameter of the circular eye. Mouth large, the maxillary extending considerably behind vertical through posterior margin of orbit, its length 2 in head; mandible slightly more than ⅔ height of body. Teeth as in *N. gillii*. Gill rakers slightly longer than ½ the diameter of eye, 7 and 3 rudiments below the angle of the anterior arch. Pseudobranchiæ absent. A long flat spine upon the upper edge of the operculum, extending back nearly to its margin; 2 short, flat spines upon the angle of the preoperculum. Nostrils as in *N. gillii*. Scales small, very closely imbricated, the lateral line obsolete in its posterior half. Distance of dorsal origin from snout 4 times in total length; anal origin under fourteenth dorsal ray, at a distance from the snout 2⅔ times in total length. Caudal of about 8 or 9 rays, very closely placed, about 10½ times in total length; pectoral placed much as in *Benthocometes*, its length about 2¼ times that of the head, extending to vertical through the vent; ventral a bifid ray inserted in advance of base of pectoral, not reaching to the vent, its length considerably less than height of body; distance of ventral origin from vent slightly more than height of body. Color light yellowish brown, an obscure narrow band of darker brown commencing on the snout, inter-

rupted by the eye, and extending backward ⅔ distance to tail; another beginning on the snout, extending over eye and back as far as first described, interrupted posteriorly; dorsal fin milky white at base in its anterior third; above this a blackish band extending whole length of fin; a narrow white margin above. The type is from *Blake* Station LXXIX, off Barbados, in 209 fathoms. (*marginatus*, edged.)

Neobythites marginatus, GOODE & BEAN, Bull. Mus. Comp. Zool., x, No. 5,162, 1883 off Barbados, in 209 fathoms (Coll. *Blake*); GOODE & BEAN, Oceanic Ichthyology, 326, fig. 290, 1896.

964. BENTHOCOMETES, Goode & Bean.

Benthocometes, GOODE & BEAN, Oceanic Ichthyology, 327, 1896 (*robustus*).

Brotulids, similar in appearance and structure to *Neobythites* and *Bassogigas*, distinguished by 2 short flat spines upon the anterior portion of the operculum, placed at some distance from each other, and by the absence of spines on the preoperculum. The lateral line is complete, and extends without interruption to the posterior fourth of the body, where it becomes obsolete. The vomerine teeth are bunched in a circular patch instead of being arranged in triangular form. The head is comparatively short, with the jaws in front nearly equal; the snout not produced, but obtuse, rounded, and almost declivous in its outline. Deep sea. Two species known. (βένθος, the depths; κωμήτης, inhabitant.)

2886. BENTHOCOMETES ROBUSTUS, Goode & Bean.

Body rather short and deep, its greatest height nearly 4⅔ in total length and about equal to length of head; interorbital area convex, its width greater than diameter of the circular eye, and 1½ times length of snout; head about 4 times diameter of eye; mouth moderate, the maxillary extending to vertical through posterior margin of eye, the mandible a little beyond, its length equal to that of postorbital part of head. Teeth in villiform bands in jaws and on palatines; vomerine teeth bunched in a circular patch. Gill rakers moderate, the longest a little more than twice in diameter of eye, 4 above angle of first arch, 11 below. Pseudobranchiæ rudimentary. Gill opening wide, the membrane deeply cleft behind, free from the isthmus. A pair of short flat spines upon the anterior portion of the operculum. Nostrils small, the anterior as close to the snout as the posterior ones are to the eyes; no apparent cirri. Scales minute; lateral line obsolete on the last fourth of body. Dorsal origin behind that of ventral and pectoral, its distance from snout 3⅔ times in body; height of dorsal fin moderate, the longest ray about 3 times in head; anal origin under eighteenth ray of dorsal, the height of fin about equaling that of dorsal; vertical fins not connate with the caudal, which consists of 12 or 13 very slender rays, its length nearly equal to ¼ head; pectoral with a broad base, close to gill opening, its length nearly ⅔ that of head; ventral a single bifid ray, inserted in advance of vertical through base of pectorals, and not far from humeral symphysis, reaching nearly halfway to vent, the distance of which from the origin of the ventral is equal to length of head. Color yellowish brown. The type of this species, a specimen 88 mm. long, was taken by the *Blake* from Station XCIV, off Moro Castle, Cuba, at a depth of from 250 to 400 fathoms. A collateral type specimen (No. 29057) was obtained by the *Fish Hawk* from Station 1043 in Lat. 38° 39' N., Long. 73° 11' W., at a depth of 130 fathoms. (Goode & Bean.) West Indies, to lat. 39° N., in Gulf stream. (*robustus,* stout.)

Neobythites robustus, GOODE & BEAN, Bull. Mus. Comp. Zool., x, No. 5, 161, 1883, off Moro Castle, Cuba, in from 250 to 400 fathoms. (Type in M. C. Z. Coll. *Blake.*)
Benthocometes robustus, GOODE & BEAN, Oceanic Ichthyology, 327, fig. 288, 1896.

965. BASSOGIGAS, Gill.

Bassogigas, GILL MS. in GOODE & BEAN, Oceanic Ichthyology, 328, 1896 (*gillii*).

Brotulids having the body elongate, compressed, covered with a thick, heavy skin, which upon the head covers and obscures all the angles of the skull; scales small, covering body and head completely; lateral line indistinct for the greater part of the course, but apparently extending at least ⅔ of the way from the operculum to the tail; eye moderate; vertical fins completely united; ventrals a pair of bifid filaments inserted behind the humeral symphysis and remote at their bases, short, rather stout; snout without barbels, slightly produced, the lower jaw being barely included; villiform teeth in the jaws, on the vomer and palatines; vomerine patch V-shaped, but with its arms broadly expanded and thicker at the angle, so that it is almost triangular; operculum with a long, sharp spine; preoperculum unarmed; branchiostegals 8; air bladder present; pseudobranchiæ small. Deep sea. (βάσσων, for βαθύς, deep; γίγας, giant.)

a. Dorsal fin with 83 rays; anal 67.   GILLII, 2887.
aa. Dorsal fin with 95 rays; anal 82.   STELLIFEROIDES, 2888.

2887. BASSOGIGAS GILLII, Goode & Bean.

D. 83; C. 6; A. 67. Head rather short and broad, with snout slightly overlapping the lower jaw; diameter of the eye scarcely ⅛ of the length of the snout and about $\frac{1}{12}$ that of head; maxillary extending far behind eye, the vertical from the anterior margin of orbit nearly bisecting it, its length ⅓ that of head, and its posterior margin ending in a broad triangular dilation; teeth normal; anterior and posterior nostrils separated by a space greater than diameter of eye; preoperculum with a square, rounded angle; no armature; operculum with a strong, sharp spine above, the tip of which projects slightly beyond the opercular flap; distance of vent from root of pectoral slightly more than length of head, as far removed from this point as is the anterior nostril; scales moderate, covering the entire head; lateral line somewhat conspicuous, obsolete in its posterior third. Dorsal and anal fins enveloped in thick scaly skin; origin of dorsal in advance of middle of pectoral; pectorals rounded, broad, and very short; less than ⅓ as long as head and extending about ⅓ distance from origin to vertical from vent; ventrals inserted somewhat behind angle of preoperculum, extending to vertical from axil of pectoral, and about ¼ of distance from origin to vent; each ventral filament bifid, the inner part being the longer. Color uniform grayish brown; fins darker. The type of this species was obtained by the *Albatross* from Station 2684, off Cape Henlopen, Delaware, in Lat. 39° 35' N., Long. 70° 51' W., at a depth of 1,106 fathoms. (Goode & Bean.) (Named for Dr. Theodore Gill.)

Bassogigas gillii, GOODE & BEAN, Oceanic Ichthyology, 328, fig. 291, 1896, off Cape Henlopen, Delaware, in 1,106 fathoms. (Type, No. 39417. Coll. *Albatross.*)

2888. BASSOGIGAS STELLIFEROIDES (Gilbert).

Head 4 to 4½ in length; depth 5 to 5¼. D. 95; A. 82; scales 110. Physiognomy strikingly like that of the Sciænoid genus *Stellifer.* Mouth large, oblique, the lower jaw included, maxillary reaching well beyond orbit, ½ length of head. Teeth uniform, small, in narrow bands, those on vomer in a ⋂-shaped patch; a well-developed band on palatines; tongue smooth, a well-developed dentigerous crest on median line behind it; no barbel at symphysis. Snout short, bluntly rounded, about equaling diameter of orbit, slightly overhanging mouth, 5 in head; interorbital width 4; upper limb of preopercle extending obliquely downward and backward, largely adnate, the angle produced into a free membranaceous flap which entirely conceals the narrow interopercle, and bears no spines. The structure of the gill flap does not appear to have been correctly interpreted. The opercle is strong, but of small extent, forking at its base, 1 branch continued straight backward as a strong spine, the second a narrow flat process downward and somewhat backward, parallel with and little distant from margin of preopercle. Filling the deep notch between these 2 processes, and forming the greater portion of the gill flap, is the thin membranaceous subopercle. Branchiostegal rays 7. Gill rakers long and slender, the longest ¾ diameter of orbit, 7 above angle, 13 and about 5 rudiments below. Nape midway between front of dorsal and front of eye; dorsal and anal similar, uniform, low, joined to base of caudal, the latter truncate, projecting well beyond them; ventrals inserted under angle of preopercle, each of a single ray forked to the very base, the 2 branches united by membrane for a distance equaling ⅔ orbit, the inner filaments being longest, ¼ longer than head, and extending well beyond front of anal; pectorals long and narrow, 1¼ in head; a narrow membranaceous flap connecting base of pectorals with upper angle of opercular flap. Scales small, well imbricated, entirely investing body and head, including gular membranes and part of gill membranes; lateral line nearly complete, lacking for about ½ length of body, running high, parallel with dorsal outline. Color silvery gray, dusted with coarse black specks, darker along dorsal outline; dorsal and anal with a narrow light streak at base, otherwise dusky, becoming black posteriorly, and with a narrow white margin; caudal black, with a broad white terminal bar; pectorals and ventrals white, with few black specks; peritoneum silvery white; mouth white anteriorly, its posterior portion and gill cavity jet-black. Pacific Ocean, off coast of Lower California. Many specimens from *Albatross* Station 2996, in 112 fathoms. Length 7 inches. (*Stellifer,* a genus of *Sciænidæ;* εἶδος, resemblance.)

Neobythites stelliferoides, GILBERT, Proc. U. S. Nat. Mus. 1891, 112, off Lower California. (Type, No. 44383. Coll. Dr. Gilbert.)

966. BARATHRODEMUS, Goode & Bean.

Barathrodemus, GOODE & BEAN, Bull. Mus. Comp. Zool., x, No. 5, 200, 1883 (*manatinus*).

Body brotuliform, much compressed; head compressed; mouth moderate. Head unarmed, except for a short flattened spine at upper angle of opercle. Snout long, projecting far beyond premaxillaries, its tip much swollen; jaws subequal in front. Teeth minute, in villiform bands on jaws, vomer, and palatines. No barbels. Anterior nostrils on the outer angles of the dilated snout, circular, each surrounded by a cluster of mucous tubes. Posterior nostrils above front of eye. Gill openings wide, the membranes not united. Gill rakers rather few. Body and head covered with small, thin, scarcely imbricated scales. Dorsal and anal long. Caudal fin separate, long, and slender. Ventrals close together, far in front of pectorals, each reduced to a single bifid ray. Deep-sea fishes. (*βάραθρον*, a gulf or deep abyss; *δῆμος*, people.)

2889. BARATHRODEMUS MANATINUS, Goode & Bean.

Head about 6 in total length; depth 7½. D. 106; A. 86; C. 2+5+2; P. 18 to 20; V. 1/1; scales about 175. Body much compressed. Dorsal and anal outline approaching at an equal angle the horizontal axis. Scales small, about 175 rows between the branchial opening and the tail, and about 34 rows, counting upward and forward obliquely from the origin of the anal to the dorsal line; lateral line apparently absent. Head considerably compressed, with rounded upper surface, its width contained 2½ times in its length, its greatest height equaling ⅘ its length. Snout slightly longer than the horizontal diameter of the eye, and projecting beyond tip of upper jaw a distance equal to vertical diameter of eye, much dilated and swollen, the anterior pair of nostrils being situated at the most salient angles; snout in general form resembling that of a manatee, whence the specific name. Mouth moderate, its cleft extending to the vertical from the center of the orbit; length of upper jaw equal to twice horizontal diameter of eye, and contained 2½ times in length of head; posterior portion of maxillary considerably expanded; maxillary largely included within a skinny sheath; when the mouth is closed the lower jaw is entirely included within the upper. Vomer and palatine with bands of teeth more than twice as broad as the bands of the intermaxillaries and on the mandible. Eye elliptical in form, its vertical diameter ⅘ of its horizontal, the latter being equal to distance from tip of snout to posterior nostril, and contained 5½ times in length of head; distance of eye from dorsal outline equal to ½ its horizontal diameter, and to ⅕ height of head in a perpendicular through center of eye; interorbital space rounded, its width equal to horizontal diameter of eye. Dorsal fin inserted in the vertical above insertion of pectoral, at a distance from end of snout equal to that of insertion of pectoral; anal inserted under twenty-first to twenty-third dorsal ray, and at a distance from snout about equal to ⅓ body length; height of dorsal and anal fins about equal to ½ height of body at insertion of anal, their bases extending almost to insertion of caudal; caudal composed of 9 rays, the 5 medial ones almost equal in length, though the tip of the tail is slightly rounded, about equal to height of body midway between branchial opening and base of tail; ventrals inserted almost under middle of operculum, in length about equal to ½ length of head; pectorals inserted under origin of dorsal, and at a distance behind branchial opening equal to ⅔ vertical diameter of eye, its length equal to greatest height of the body. Color grayish brown; abdominal region black. (Goode & Bean.) Gulf stream, north of the Bermudas, in 647 to 1,395 fathoms. (*Manati*, like the manatee or sea cow.)

Barthrodemus manatinus, GOODE & BEAN, Bull. Mus. Comp. Zool., x, No. 5, 200, 1883, Lat. 33° 35′ 20″ N., Long. 76° W., in 647 fathoms (Type in M. C. Z. Coll. *Blake*); JORDAN, Cat., 127, 1885; GÜNTHER, Challenger Report, XXII, 100, 1887; GOODE & BEAN, Oceanic Ichthyology, 332, fig. 294, 1896.

967. NEMATONUS, Günther.

Nematonus, GÜNTHER, Challenger Report, XXII, 114, 1887 (*pectoralis*).

Body compressed, with long tapering tail. Bones of head soft, muciferous channels moderately developed, and with the integument very thin or absent on the upper portion and snout. Operculum cartilaginous and flat; a broad process near its upper angle corresponding to the opercular spine in some of the related genera, the head otherwise unarmed, though irregular by reason of the cranial bones. Snout much depressed, broad, rounded; jaws equal in front; mouth very wide; bands of villiform teeth in jaws, on vomer and palatines. Barbel none. Eyes small. Vertical fins confluent; ventrals a pair of bifid filaments close together, on the isthmus, close to the humeral symphysis. . Gills 4, with very short laminæ and rather short, incurved, acicular gill rakers on the first arch, and much shorter, less numerous, spatulate ones on the 3 other arches. Pseudobranchiæ rudimentary. No traces of a lateral line, though the body is covered with scales of considerable size, almost as large as the eye, and the cheek with others still larger. *Nematonus* differs from *Porogadus* not only in the absence of spines upon the head, as Günther has indicated, but in the much less ossified opercular apparatus, in the shorter and thicker head, in the absence of the 3 series of pores simulating lateral lines, and in the tendency to prolongation in the lower rays of the pectoral, which increase from the uppermost to the lowermost in *Nematonus*, while *Porogadus* has a lanceolate fin, and also in the extreme exsertion of the caudal rays. ($\nu\tilde{\eta}\mu\alpha$, thread; *Onus*, the rockling.)

2890. NEMATONUS PECTORALIS, Goode & Bean.

D. 93; A. 73; P. 17; V. 2. Body moderately elongate, much compressed, the tail much shorter and more robust than in *Bassozetus catena*, its height equaling 1⅕ times length of head and ⅓ that of body. Head stoutish, not much compressed, lower than body, its length contained 5⅕ times in the body; snout compressed, broad at its tip, its length exceeding diameter of the circular eye; interorbital area slightly convex, its width slightly exceeding twice diameter of eye, 3 times in head. Maxillary extending far behind eye, its length less than that of preorbital portion of head; mandible as long as postorbital portion of head; jaws, vomer, and palatines with narrow bands of villiform teeth, normally arranged. Branchiostegals 8. Gill lamellæ very short; gill rakers long and numerous, 18 on first arch below the angle, 5 above, 4 of which are rudimentary. Pseudobranchiæ present, but very rudimentary. Anterior nostrils on the top of the snout and near the median line of the head, near its tip, separated by a space about equal to diameter of eye; posterior nostrils in front ᴠf eye. Muciferous pores large, arranged much as in *B. catena.* Dorsal origin in the same vertical with that of pectorals, its distance from tip of snout contained 5 times in total and equaling twice length of maxillary. Rays well developed in anterior third, the longest ⅔ of head; anal origin under twentieth dorsal ray, its rays nearly as long as those of dorsal; pectoral with its penultimate ray produced, extending to thirteenth ray of anal, nearly twice as long as head; ventrals originating in advance of vertical through pectorals, and each a bifid filament; distance of ventral origin from tip of snout equaling length of ventral and about ¼ as long as head; distance of ventral origin from vent considerably greater than length of head; distance from tip of ventral to vent equal to ½ the length of the head. Number of scales in transverse series from vent to dorsal about 23; from the upper angle of the gill opening to the vertical through origin of anal 32. Color brownish yellow; head and abdomen blackish. The type (No. 37342, U. S. Nat. Mus.) was taken at *Albatross* Station 2380, Lat. 28° 02′ 30″ N., Long. 87° 43′ 45″ W. in 1,430 fathoms. It is 183 mm. long to the caudal base, 215 with caudal. Another young specimen, 70 mm. long, was taken at *Blake* Station XCV, off Dominica, in 330 fathoms. (Goode & Bean.) (*pectoralis*, pertaining to the breast.)

Nematonus pectoralis, GOODE & BEAN, Proc. U. S. Nat. Mus. 1885, 604, Lat. 28° 02′ 30″ N., Long. 87° 43′ 45″ W., in 1,430 fathoms (Type, No. 37342. Coll. *Albatross*); GÜNTHER, Challenger Report, XXII, 114, 1887; GOODE & BEAN, Oceanic Ichth., 333, fig. 295, 1896.

968. POROGADUS, Goode & Bean.

Porogadus, GOODE & BEAN, Proc. U. S. Nat. Mus. 1885, 682 (*miles*).

Body brotuliform, much compressed; head with numerous spines on interorbital space, 2 pairs on the shoulders, 1 at angle of operculum and a double series on angle of preoperculum; head with numerous mucous pores, as in *Bassozetus;* mouth large; snout moderate, not projecting much beyond the upper jaw; jaws nearly equal in front; teeth in villiform bands in jaws and on vomer and palatines; barbel none; gill openings wide, membranes narrowly united, not attached to the isthmus; gills 4; gill laminæ short; gill rakers moderate, numerous; pseudobranchiæ absent; caudal fin of few rays, on a very narrow base, not prolonged, scarcely differentiated from the vertical fins; dorsal and anal fins well developed; pectorals simple, moderate; each ventral a single bifid ray close to the humeral symphysis; branchiostegals 8; scales small; lateral line apparently triple, or replaced by 3 series of pores—1 close to ventral outline, 1 median, and another along base of dorsal. (πόρος, pore; *Gadus*, the codfish.)

2891. POROGADUS MILES, Goode & Bean.

Head 6¼; depth 10; eye 5¾ in head; body much compressed, elongate,
tapering to a very slender tail; head long, moderately compressed, subconical, the profile gradually ascending in nearly a straight line from tip
of snout to origin of dorsal; interorbital space slightly convex, spiny, its
width 4⅔ times in length of head, and slightly greater than diameter of
eye; opercles and head generally covered with numerous and strong
spines, as described in the generic diagnosis; mouth very large and wide;
maxillary extending far behind eye and much expanded at its tip, its
length more than ½ that of head; length of mandible equal to greatest
height of body; jaws, vomer, and palatines with narrow bands of villiform teeth, none of which is enlarged; gill rakers 15 on anterior arch
below the angle, 3 rudimentary ones above. Anterior pair of nostrils
nearly on top of snout and somewhat nearer its tip than to eye, separated
by a narrow space and placed immediately in front of middle of eye;
behind each posterior nostril a strong spine projecting outward and
upward; pores of the head arranged much as in *Bassozetus*; scales minute;
lateral line not to be clearly made out; 3 rows of minute pores on each
side of dorsal, median, and ventral, beginning near head and extending
well toward extremity of tail. Dorsal origin slightly behind vertical
through pectoral base, its distance from snout nearly 6 times in length of
body, its rays moderately long, the longest about as long as snout, and
very numerous; anal origin in vertical from twenty-second or twenty-
third dorsal ray, its distance from snout 3⅓ times in length of body, its
rays about as long as those of dorsal; pectoral imperfect, its length in the
type equaling ½ that of head; ventrals a bifid filament, placed close to
the humeral symphysis, well in advance of pectoral, its length equal to
height of body; distance from origin of ventrals to vent nearly equal to
length of head; ventral not reaching vent by a distance equal to length of
snout. Color blackish brown. The type (No. 35625, U. S. Nat. Mus.) is 153
mm. in length, from *Albatross* Station 2230, lat. 38° 27′ N., long. 73° 02′ W.,
at a depth of 1,168 fathoms. (Goode & Bean.) (*miles*, a soldier.)

Porogadus miles, Goode & Bean, Proc. U. S. Nat. Mus. 1885, 602, Lat. 38° 27′ N., Long.
73° 02′ W., in 1,168 fathoms; Goode & Bean, Oceanic Ichthyology, 334, fig. 292, 1896.

969. PENOPUS, Goode & Bean.

Penopus, Goode & Bean, Oceanic Ichthyology, 335, 1896 (*macdonaldi*).

Body stout in front, tapering behind; tail not greatly exceeding the
length of the rest of the fish; head scaly, thick, its top surface flat, with
depressed and moderately projecting snout; a pair of minute postnasal
spines; a strong and much curved spine on the operculum; several weak
spines on the angle of the preoperculum, and several at the posterior
angle of the suboperculum; mouth moderately large, the lower jaw
included; several narrow slit-like pores along the margin of the preorbital and suborbital; 2 minute pores on under surface of mandible
near its symphysis, and not far behind them 2 long slit-like pores; the

anterior nostril in a long slit, the posterior larger, oblong in shape, and ½ concealed by a fold of skin; eye small; the teeth appear only in minute asperities, the intermaxillary band much wider in front than behind; mandibulary band narrow throughout; vomerine band very narrow, V-shaped; palatines in a long, broad band; gill openings wide, deeply cleft in front, narrowly joined to the isthmus; branchiostegals 8; no pseudobranchiæ; gill rakers long and slender, not numerous; gill laminæ moderately long, a long slit behind the fourth gill; scales very small; lateral lines 3; caudal fin consisting of few rays, well differentiated from the dorsal and anal; dorsal beginning not far behind head; ventrals slightly in advance of the pectorals and composed of 2 rays, united by membrane, which forms a margin around them; pectoral normal, several of its upper rays simple; vent not much in advance of middle of total length. This genus agrees with *Porogadus* in nearly every respect except in the scarcity of spines on the head and in the structure of the ventrals. *Porogadus* has the ventrals composed of 2 distinct rays which are separated throughout their entire length, but in *Penopus* the 2 rays are inclosed in a membrane which connects them and forms a margin around them. In *Porogadus*, also, the suboperculum has a smooth margin and the opercular spine is weaker than in *Penopus*, and is not curved. Deep seas. (πήνη, thread; πούς, foot.)

2892. PENOPUS MACDONALDI, Goode & Bean.

D. 137; A. 102. Greatest height of body equaling length of postorbital part of head and about ⅑ of total without caudal; greatest width of body anteriorly about ⅔ of its greatest height; head stout, its greatest width equaling ¾ of its greatest depth and more than ⅓ of its length; width of interorbital space about ⅓ length of head; eye very small, its length less than ⅓ width of interorbital space; distance from eye to tip of snout equaling length of intermaxillary; distance of anterior nostril from tip of snout equaling length of eye; distance of posterior nostril from eye slightly less than its distance from tip of snout; maxillary expanded behind and reaching somewhat behind eye, its length equaling that of snout; mandible extending much behind eye, its length equal to postorbital part of head. Dorsal beginning over middlle of pectoral, its rays well developed, those in middle of fin longer than anterior ones; anal beginning under twenty-seventh ray of dorsal, middle rays longest; pectoral nearly ½ length of head and about equal to distance of its tip from vent; ventral about ⅓ distance of its origin from origin of anal. Lateral lines 3, the uppermost beginning at the upper angle of the gill opening, quickly approaching top of body near base of dorsal and merging into dorsal base about middle of tail; median lateral line beginning a little behind head and extending almost to root of caudal, becoming very faint posteriorly; lowermost lateral line with its origin under and not far from base of pectoral, extending along lower side of tail and merging into base of anal fin somewhat beyond middle of length of tail. Color yellowish brown; operculum, opercular flap and branchiostegal membrane, pectoral, and ventral dusky. Only a single specimen, 315 mm. long, known; obtained by the *Albatross* September 18, 1886, at Station 2716, Lat. 38°

29' 30'' N., Long. 70° 57' W., in 1,631 fathoms. (Goode & Bean.) (Named for Hon. Marshall McDonald.)

Penopus macdonaldi, GOODE & BEAN, Oceanic Ichthology, 336, fig. 293, 1896, Lat. 38° 29' 30'' N., Long. 70° 57' W., in 1,631 fathoms. (Type, No. 39433.)

970. DICROLENE, Goode & Bean.

Dicrolene, GOODE & BEAN, Bull. Mus. Comp. Zool., x, No. 5, 202, 1883 (*intronigra*).

Brotulids with body moderately compressed; head somewhat compressed; mouth large; tip of maxillary much dilated; eye large, placed close to dorsal profile. Head with supraorbital spines; several strong spines on preopercle and 1 long spine at upper angle of opercle. Snout short, not projecting beyond the upper jaw; jaws subequal. Teeth in narrow villiform bands in each jaw, on head of vomer, and on palatines. No barbel. Gill openings wide, membranes not united; gills 4; gill laminæ of moderate length; gill rakers rather long, not numerous; pseudobranchiæ absent. Caudal not confluent with dorsal and anal, but without a distinct peduncle. Dorsal and anal fins long; pectoral rays in 2 groups, several of the lower ones being separated and much produced; ventrals a pair of bifid rays, close together on the isthmus. Branchiostegals 8. Body and head covered with small scales; lateral line close to base of dorsal fin, apparently becoming obsolete on posterior third of body. Stomach siphonal; pyloric cæca few and rudimentary; intestine shorter than body. Deep sea; a single species known. (δικρόος, forked; ὠλένη, limb.)

2893. DICROLENE INTRONIGRA, Goode & Bean.

Head 5; eye large, 4 in head; interorbital width 4. D. 100; A. about 85; C. 6 or 7; V. 1; P. 19+7 or 8; scales 110 to 120. Body moderately compressed, its dorsal and ventral outlines approaching at an equal angle the horizontal axis, and tapering to a narrow point. Head somewhat compressed, with flattish upper surface, which is encroached upon by the upper margin of orbit; a strong spine at posterior upper margin of orbit, pointing backward and upward; a long, sharp spine at upper angle of opercle, its exposed portion 2 in eye; 3 equidistant spines on lower posterior border of preopercle, much weaker than that on opercle. Large muciferous cavities in bones of head; a row of large cavities extending backward from upper angle of orbit, and continuous with those on lateral line. Mouth large, its cleft considerably more than ½ head, the maxillary extending beyond eye and with scales upon its expanded tip. Distance from snout to origin of dorsal fin ⅔ total length; anal inserted under twenty-fifth or twenty-sixth dorsal ray; height of dorsal and anal fins each about equal to eye; length of caudal fin 2 in distance from snout to dorsal; ventrals about equal to upper jaw; pectorals inserted close to branchial aperture, the 8 lower rays free and much prolonged, the longest and most anterior being about 3 in body, and more than 3 times as long as the contiguous posterior ray of the normally constructed portion of the fin, which is, however, about equal to the last free rays. West Indies, Gulf of

Mexico, and Gulf Stream in various localities, and off coast of Soudan and on the bank d'Arguin, in deep water. (*intro*, within; *niger*, black.)

Dicrolene intronigra, GOODE & BEAN, Bull. Mus. Comp. Zool., x, No. 5, 202, 1883, Gulf Stream, Lat. 39° 59′ 45″ N., Long. 68° 54′ W. (Coll. *Blake*); GÜNTHER, Challenger Report, XXII, 107, 1887; VAILLANT, Exp. Sci. Travailleur et Talisman, 258, pl. 23, fig. 2, 1888; GOODE & BEAN, Oceanic Ichthyology, 338, fig. 297 A and B, 1896.

971. MIXONUS, Günther.

Mixonus, GÜNTHER, Challenger Report, XXII, 108, 1887 (*laticeps*).

Lower pectoral rays free, not united by membrane with, but inserted on the same base as, the upper part of the fin; they are but slightly stronger than the other rays and prolonged. Body elongate, compressed, covered with small, very thin and deciduous scales. Head slightly compressed, broad and flat above, depressed in front, naked (with the exception of the parts between the mandibles, and, perhaps, of the cheeks). Bones thin, with muciferous system moderately developed; only 1 small spine above on the operculum; preoperculum without spine. Eye small. Vertical fins united, but the narrow caudal projecting beyond the short dorsal and anal rays. Ventrals each reduced to a filament, which consists of 2 rays firmly bound together in their whole length; they are inserted behind the humeral symphysis and close together. Snout broad, rounded, scarcely overlapping the lower jaw. Mouth very wide; villiform teeth in the jaws, on the vomer, and palatine bones. Gill laminæ short; gill rakers long, not very closely set. Pseudobranchiæ none. (μίξις, mixture, half; *Onus*, a synonym of *Gaidropsarus*, the rockling.)

2894. MIXONUS LATICEPS (Günther).

Head 2; depth 3; eye 8 in head; snout 4. P. 17. Greatest depth of body below origin of dorsal fin; distance of vent from snout ⅔ its distance from extremity of spinal column. Crown of head remarkably convex, covered with an extremely thin and transparent skin, which, perhaps, in older examples is scaly; interorbital space less convex, and equaling in width the length of snout including the eye; eye small, above middle of length of the maxillary; posterior nostrils wide, open, in front of the eye. Distance of vent from ventrals exceeds length of head; origin of dorsal fin above root of pectorals, its rays of moderate length, but longer than those of anal; pectoral with a rather narrow base, as long as head without snout, its rays feeble, 3 or 4 lower ones a little stouter, detached, and prolonged; ventral filaments not reaching as far backward as pectoral. Gill rakers 10, much longer than the laminæ. Whitish, with the abdomen and gill apparatus black. Mid-Atlantic, in profound depths. One specimen, 5½ inches long, was obtained in mid-Atlantic (*Challenger* Station 104), at the enormous depth of 2,500 fathoms. The second (type of *Sirembo guntheri*) was taken off Cape Verde, in 3,200 meters. (*latus*, broad; *-ceps*, head.)

Bathynectes laticeps, GÜNTHER, Ann. Mag. Nat. Hist. 1878, 20, mid-Atlantic, in 2,500 fathoms. (Coll. *Challenger*.)

Sirembo guntheri, VAILLANT, Exp. Sci. Trav. et Talisman, 268, pl. XXIV, fig. 5, 1889, off the Cape Verde Islands, at a depth of 3,200 meters.

Mixonus laticeps, GÜNTHER, Challenger Report, XXII, 108, pl. 25, fig. 8, 1887; GOODE & BEAN, Oceanic Ichthyology, 339, fig. 296 A, 1896.

972. BARATHRONUS, Goode & Bean.

Barathronus, GOODE & BEAN, Bull. Mus. Comp. Zool., x, No. 5, 164, 1883 *(bicolor).*

Brotulids having the head stout, body and tail compressed, covered closely by skin; scaleless; vent far behind pectoral, included in a cleft; mouth wide, oblique, the lower jaw projecting; intermaxillary teeth rudimentary; several fang-like teeth on the head of the vomer, none on palatines, a few rather large, recurved, separated teeth in the mandible; nostrils close together and small; eye visible through the skin, partly upon the top of the head, with or without dark pigment in the iris; barbel none; gill rakers very numerous and slender, and rather long; gill laminæ well developed on all the arches; no pseudobranchiæ; head full of muciferous channels; gill membranes not united, but covered by a fold of skin; ventrals reduced to single simple rays, placed in advance of the pectorals and close to the humeral symphysis; dorsal and anal placed far back; caudal scarcely differentiated, composed of rather numerous, very slender rays upon a somewhat narrow base. ($\beta\acute{\alpha}\rho\alpha\theta\rho\sigma\nu$, the abyss; $\breve{\sigma}\nu\sigma\varsigma$, *Onus,* the rockling.)

2895. BARATHRONUS BICOLOR, Goode & Bean.

Head $5\frac{1}{2}$ in total, its width $\frac{2}{3}$ its length; depth $6\frac{1}{2}$; orbit $4\frac{1}{4}$ in head; interorbital width $4\frac{3}{4}$. D. about 70; A. 57. Body much compressed; eye concealed by the skin; maxillary extending slightly beyond the perpendicular through posterior margin of orbit, almost entirely concealed under the preorbital, and much expanded at tip, where its width is rather greater than that of eye. Intermaxillary very thin, broad, and slightly protractile; vomer very close to intermaxillary symphysis, its head somewhat raised and bearing 3 fang-like teeth (2 of which are off one side and 1 on the other in the type), separated by a moderately wide interspace; mandible with 5 enlarged, separate, recurved teeth upon each side, which increase in size posteriorly, its upper edge, posteriorly, produced above the level of the tooth-bearing surface, and received under the expanded maxillary; longest gill raker about as long as eye. Dorsal origin distant from the snout about $\frac{1}{4}$ total length; dorsal rays well developed, numerous, long, and slender, the longest about 3 times in length of head; anal originating in vertical from fourteenth dorsal ray, equidistant from eye and base of caudal, longest rays about as long as those in the dorsal; pectoral with a fleshy base, its length a little less than height of body. Ventral well in advance of pectoral, close to humeral symphysis, the rays being placed very close together at their origin, the length of the fin contained about 9 times in the total length, about 3 times in the distance from its origin to the vent. Caudal with about 10 rays, its length about 8 times in total length. Color yellowish white, with a broad vertical band of black from origin of ventral nearly to vent; another similar and narrower band above it upon each side. The type, 120 mm. long, from *Blake* Station LXXI, off Guadaloupe, at a depth of 769 fathoms. (Goode & Bean.) *(bicolor,* two-colored.)

Barathronus bicolor, GOODE & BEAN, Bull. Mus. Comp. Zool., x, No. 5, 164, 1883, off Guadaloupe, in 769 fathoms (Coll. *Blake*); GOODE & BEAN, Oceanic Ichthyology, 341, fig. 298, 1896.

973. APHYONUS, Günther.

Aphyonus, GÜNTHER, Ann. Mag. Nat. Hist. 1878, 22 (*gelatinosus*).

Head, body, and tapering tail strongly compressed, enveloped in a thin, scaleless, loose skin. Vent far behind the pectoral, at nearly the middle of the total length. Snout swollen, projecting beyond the mouth, which is wide. No teeth in the upper jaw; small conical teeth in the lower, pluriserial in front and uniserial on the side. Vomer with a few rudimentary teeth; palatine teeth none. Nostrils close together, small. No externally visible eye. Barbel none. Ventrals reduced to simple filaments, placed close together and near to the humeral symphysis. Gill membranes not united. Four branchial arches, the posterior without gill laminæ, the anterior with very short gill rakers and with rather short gill laminæ. Head covered with a system of wide muciferous channels and sinuses, the dermal bones being almost membranaceous, while the others are in a semicartilaginous condition. Notochord persistent, but with a superficial indication of the vertebral segments, as in some Leptocephaline forms. (Günther.) (ἀφύη, anchovy, a small translucent fish; *Onus*, the rockling.)

2896. APHYONUS MOLLIS, Goode & Bean.

Body much compressed, its greatest height 6 in its total length. Head thicker than body, its height slightly greater. Length of head about 4¼ in total, width over ½ its length. Snout 3⅓ in length of head. Eye not externally visible. Diameter of orbit, as seen through the skin, about ¼ length of head. Maxillary extending to vertical through posterior margin of orbit, the mandible somewhat farther back, its length nearly equal to height of body. A few weak teeth on vomer, palatines, and mandible, and very rudimentary ones in maxillary, not visible to the eye, but appreciable to the touch. Gill laminæ on the fourth and rudimentary gill rakers, 8 rudiments and 4 developed below the angle. Dorsal origin almost over posterior edge of operculum, its distance from the snout ¼ of total length, dorsal rays more than 110, well developed, the longest 3 in head; anal origin slightly nearer base of caudal than to the tip of snout, its rays shorter than those in the dorsal; pectoral with a fleshy base, its origin somewhat behind that of the dorsal, its length equal to width of head; ventral origin in advance of that of pectoral, close to humeral symphysis, the fin a single simple ray, whose length equals that of the pectoral, its tip not reaching vent by a space equal to height of head. Skin not loose. Texture of body rather firm, not transparent; whitish. Gulf of Mexico, in deep water. This species is closely allied to *Aphyonus gelatinosus*. (Goode & Bean.) (*mollis*, soft.)

Aphyonus mollis, GOODE & BEAN, Bull. Mus. Comp. Zool., x, No. 5, 163, 1883, Lat. 24° 36′ N., Long. 84° 5′ W., in 955 fathoms (Coll. *Blake*); GOODE & BEAN, Oceanic Ichthyology, 342, fig. 299, 1896.

Family CCXII. BREGMACEROTIDÆ.

Body stout, with robust caudal portion, truncate or convex behind, almost without procurrent caudal rays above or below; vent before middle of body; suborbitals moderate; no barbels, spines, nor cirri on head; mouth terminal, with minute teeth on jaws and vomer, none on palatines; ventrals jugular, extremely long, few-rayed, the rays dilated and separate nearly to base. Dorsal fins 2, the first an elongate, slender occipital ray; second dorsal on posterior half of body, of soft rays, depressed medially, so that it forms 2 lobes; no spines in fins. Anal nearly similar to the soft dorsal and similarly depressed in the middle; dorsal and anal depressible in a groove of scales. Hypercoracoid perforate; no pseudobranchiæ; gill openings wide, the membranes free from the isthmus. A single genus with 2 or 3 species found in the open sea, probably near the surface; widely distributed. The presence of the hypercoracoid foramen shows that this family is allied to the *Brotulidæ* rather than to the *Gadidæ*. From the *Brotulidæ* it is mainly distinguished by the development of its dorsal and ventral fins.

974. BREGMACEROS, Thompson.

Bregmaceros, THOMPSON, in Charlesworth's Mag. Nat. Hist., IV, 1840, 184 *(macclellandii)*.
Calloptilum, RICHARDSON, Voy. Sulph.. Fish., 94, pl. 46, figs. 4–7, 1843 *(mirum)*.
Asthenurus, TICKELL, Journ. Asiat. Soc. Bengal 1865, 32 *(atripinnis)*.

Characters of the genus included above. ($\beta\rho\acute{\epsilon}\gamma\mu\alpha$, the upper part of the head, the nape; $\varkappa\acute{\epsilon}\rho\alpha\varsigma$, horn.)

a. Scales in transverse series 14; scales in lateral series 58 to 64.

MACCLELLANDII, 2897.

aa. Scales in transverse series 10; in lateral series 65; anterior lobes of dorsal and anal lower than in *B. macclellandii.*   ATLANTICUS, 2898.

2897. BREGMACEROS MACCLELLANDII, Thompson.

Head $5\frac{2}{3}$; depth $6\frac{2}{3}$. D. about I, 18–X–22 ($16+X-15$); A. about 18, X, 22 (22, X, 20); V. 4 or 5; scales 58–14 (64–14). Body moderately elongate, compressed, the form somewhat as in *Ophidion*, the back not elevated. Head short and small, moderately compressed; bones of head thin, without serrature or spine; eye moderate, 3 in head; interorbital space ridged, about as broad as eye; snout blunt, rather shorter than eye; mouth very oblique, the jaws subequal; maxillary reaching to beyond middle of eye, $2\frac{1}{3}$ in head; lower jaw flattish, curved upward; teeth in both jaws moderate, slender, close set, recurved, apparently in a single series. Tongue conspicuous; no teeth evident on vomer or palatines; branchiostegals 7 or 8; gill membranes separate, free from the isthmus; no evident pseudobranchiæ; gill rakers obsolete; no barbels about jaws. Body with rather large, thin, caducous scales (nearly all of them fallen in the typical specimens so that they can not be counted). Dorsal fin beginning with a single long and very slender spine on occiput, this nearly $\frac{1}{3}$ longer than head. Behind this, for a distance about equal to its length, the rudimentary rays, if present, do not rise above the sheath on each side. Nearly opposite the vent begins the dorsal proper, the distance of its first ray from snout being about $\frac{2}{3}$ length of body; about 12 rays are moderately elevated, about $\frac{3}{4}$ length of head. The others are gradually shorter and more slender, becoming too small to count, until just before caudal, where the fin becomes conspicuous again, this posterior lobe not ⅓ so high as the anterior. Anal opposite dorsal and similar to it, the first ray close behind vent; caudal free from dorsal and anal, the caudal peduncle truncate at its base. Ventrals of 3 long rays, with a fourth at the inner base of the third; this fourth is probably a rudiment of 2. The ventrals are jugular in position, the rays very long and filamentous, the longest about ⅓ the body, reaching to the middle of anal fin. Pectorals inserted high, somewhat shorter than head. Vent slightly behind end of anterior ⅓ of total length. Color brown above, sides and below silvery; back and base of anal closely dotted with dusky; dorsal mostly dusky; caudal pale, dusky at base, with a narrow white cross bar; lower fins pale; the dark marking on front of back assume something of the form of lengthwise streaks. Tropical Pacific; Bay of Bengal; Philippine Islands; coast of China, etc., east to the coast of Central America, living near the surface in the open sea. Here described from the types of *Bregmaceros bathymaster*, two specimens, 1¼ and 2 inches in length, dredged at *Albatross* Station 2804, south of Panama, in 47 fathoms depth. Two others, 4 inches long, found later off the coast of Panama. A recomparison of these latter specimens with Günther's * detailed account of *B. macclellandii* shows no difference whatever, and we regard *B. bathymaster* as identical with the latter. Günther counts the scales 64–14; we find 58–14. In our largest specimens the ventrals reach middle of anal. (Named for Dr. John McClelland, of the Bengal Medical Service, who first studied the fishes of the Ganges.)

Bregmaceros macclellandii, THOMPSON, in Charlesworth's Mag. Nat. Hist., IV, 1840, 184, mouth of the Ganges; GÜNTHER, Cat., IV, 368, 1862.
Calloptilum mirum, RICHARDSON, Voyage Sulphur, Fish., 95, pl. 46, figs. 4–7, 1843.
Asthenurus atripinnis, TICKELL, Journ. Asiat. Soc. Bengal 1865, 32, with plate, Bay of Bengal.
Bregmaceros bathymaster, JORDAN & BOLLMAN, Proc. U. S. Nat. Mus. 1889, Lat. 8° 13′ 30″ N., Long. 79° 37′ 45″ W., southwest of Panama. (Type, No. 41137. Coll. *Albatross*.)
Bregmaceros atripinnis, DAY, Proc. Zool. Soc. Lond. 1869, 522, Bay of Bengal; types, same specimens described by TICKELL.

2898. BREGMACEROS ATLANTICUS, Goode & Bean.

Head 5¼; depth 7⅔ in total length. D. I-15, X, 16; A. 15 or 16 + X (7 or 8) + 21 or 22; scales 65–10. Length 46 mm. Body compressed, moderately elongate. Interorbital area convex, its width greater than eye, which is 4 in head; jaws even in front; maxillary reaching to vertical through middle of eye; mandible to vertical through posterior margin of eye; teeth in intermaxillary minute, apparently in a single series; mandibu-

---

* The following is Dr. Günther's account of *Bregmaceros macclellandii,* taken from specimens from the China Sea:
"B. 7; D. I, 16+X+15; A. 22+X+20; V. 5 or 6; scales 64–14. Occipital ray very slender, longer than head; dorsal and anal fins depressible in a groove formed by the scales along the bases of these fins; anterior portions of dorsal and anal elevated, connected with the posterior lower portion by a series of very short extremely feeble rays. Vent at end of anterior third of total length. Three outer rays of ventral fins dilated, compressed, simple, much elongate, reaching to or nearly to middle of anal; the second and third rays sometimes united at base. Silvery, minutely dotted with brown."

lary teeth biserial, the inner teeth enlarged. Cephalic appendage reaching nearly to base of first dorsal, its length 4½ in total. Distance of dorsal from snout 2½ in total, that of the anal the same; the dorsal and anal fins received in a groove formed by the scales along their bases; anterior portion of second dorsal and second anal less elevated than in *B. macclellandii.* The differentiations between the developed and undeveloped rays of the anal are so slight that the limits of the so-called anterior and posterior sections of the fin can not be determined. Length of the longest anal ray about 2 in body length. Specimens were obtained by the *Blake* at the following stations: XCIX, off Granada, 90 fathoms; CXIII, off Neris, 305 fathoms; CLXXXV, Lat. 25° 33′ N., Long. 84° 21′ W., 101 fathoms. (Goode & Bean.) This species seems doubtfully distinct from *B. macclellandii.* (*atlanticus*, of the Atlantic.)

Bregmaceros atlanticus, GOODE & BEAN, Bull. Mus. Comp. Zool., XII, No. 5, 165, 1886, West Indies, off Granada and Neris (Coll. *Blake*); GOODE & BEAN, Oceanic Ichthyology, 388, fig. 331, 1896.

Suborder ANACANTHINI.

(THE JUGULAR FISHES.)

Vertical fins very long, destitute of true spines; tail isocercal, the posterior vertebræ progressively smaller; ventrals jugular, without spines; hypercoracoid without perforation or foramen; no pseudobranchiæ. The osteological characters of this group, called by him *Gadoidea,* are thus given by Dr. Gill:

"Jugulares with the orbito-rostral portion of the cranium longer than the posterior portion, the cranial cavity widely open in front; the supra-occipital well developed, horizontal and cariniform behind, with the exoccipitals contracted forward and overhung by the supraoccipital, the exoccipital condyles distant and feebly developed, with the hypercoracoid entire, the hypocoracoid with its inferior process convergent toward the proscapula, and the fenestra between the hypercoracoid and hypocoracoid." (Gill, Proc. Ac. Nat. Sci. Phila. 1884, 170.)

A large and important group, chiefly confined to the cold depths of the ocean and the northern seas. From all other typical fishes they are separated by the entire hypercoracoid. ($\dot{\alpha}\nu$- privative, without; $\ddot{\alpha}\kappa\alpha\nu\theta\alpha$, spine.)

    a. Caudal fin present; tail not greatly elongate; body tapering or coniform behind, with many procurrent caudal rays above and below; suborbitals moderate.
        b. Frontal bones paired, with a triangular excavated area above, the divergent frontal crests continuous from the forked occipital crest; ribs wide, approximated, channeled below or with inflected sides; no barbels.
                                    MERLUCCIIDÆ, CCXIII.
        bb. Frontal bones normal, not forming a triangular excavated area above; ribs normal; chin with a barbel (rarely obsolete).   GADIDÆ, CCXIV.
    aa. Caudal fin wanting; tail very long, tapering behind; suborbitals very broad.
                                        MACROURIDÆ, CCXV.

Family CCXIII. MERLUCCIIDÆ.

(THE HAKES.)

Body moderately elongate, covered with small, smooth, deciduous scales; posterior part of body coniform and with the caudal rays procurrent forward; vent submedian. Head elongate, depressed, pike-like; suborbital bones moderate; mouth terminal, with strong teeth; no barbels; ventrals subjugular; dorsal fins 2, a short anterior and long posterior one, a long anal corresponding to the second dorsal; ribs wide, approximated, and channeled below or with inflected sides; frontal bones paired, excavated, with divergent crests continuous from the forked occipital crest. A single genus, with about 4 species; large cod-like fishes, of voracious habit, inhabiting moderate depths, and distinguished from the *Gadidæ* mainly by the structure of the frontal bones and the ribs. (*Merlucciidæ*, Gill, Proc. Ac. Nat. Sci. Phila. 1884, 772.)

975. MERLUCCIUS, Rafinesque.

(HAKES.)

Merluccius, RAFINESQUE, Caratteri di Alcuni Nuovi Generi, etc., 26, 1810 (*merluccius*).
Onus, RAFINESQUE, Indice d'Ittiol. Sicil., 12, 1810 (*riali* = *merluccius*); substitute for *Merluccius.*
Merlangus, RAFINESQUE, Indice d'Ittiol. Sicil., 30, 1810 (*riali*); substitute for *Onus.*
Merlus, GUICHENOT, in Gay, Hist. Nat. Chili, Zool., II, 328, 1847 (*gayi*).
Stomodon, MITCHILL, Rept. Fish. N. Y. 1814. 7 (*bilinearis*).
Homalopomus, GIRARD, Proc. Ac. Nat. Sci. Phila. 1856, 132 (*trowbridgei*).
Epicopus, GÜNTHER, Cat. Fish. Brit. Mus., II, 248, 1860 (*gayi*).

Body elongate, covered with small, deciduous scales. Head slender, conical, the snout long, depressed; a well-defined, oblong, triangular excavation at the forehead, bounded by the ridges on the separated frontal bones, these ridges converging backward into the low occipital crest; eye rather large; edge of preopercle free; preopercle with a channel behind its crest, crossed by short radiating ridges; mouth large, oblique; maxillary extending to opposite the eye; lower jaw longer; no barbels; jaws with slender teeth, of various sizes, in about 2 series, those of the inner row longer and movable; vomer with similar teeth; palatines toothless. Branchiostegals 7. Gill rakers long; gill membranes not united. Dorsal fins 2, well separated, the first short, the second long, with a deep emargination; anal emarginate, similar to second dorsal; ventral fins well developed, with about 7 rays; vertebræ peculiarly modified, the neural spines well developed and wedged into one another; frontal bone double and the skull otherwise peculiar in several respects. Species several, very similar in appearance; ill-favored fishes of soft flesh and fragile fins, inhabiting water of some depth. Large voracious fishes, little valued as food. (*Merluccius*, the ancient name, meaning sea pike.)

a. Scales moderate, about 110 in lateral line; teeth very strong. D. 10–36; A. 36.
MERLUCCIUS, 2899.
aa. Scales small, 135 to 150 in lateral line; teeth moderate. D. 11 to 13–41; A. 41.
    b. Ventrals long, about 1⅔ in head. BILINEARIS, 2900.
    bb. Ventrals short, about 2½ in head. PRODUCTUS, 2901.

2899. MERLUCCIUS MERLUCCIUS (Linnæus).

(EUROPEAN HAKE.)

Head large, 3⅓; depth 6½. D. 10-36; A. 36; vertebræ 25+26; scales 150.
Ventrals a little more than ⅓ head; teeth very long. Dusky above, silvery below; dorsal, caudal, and distal part of pectoral blackish; inside of
opercle black; inside of mouth black posteriorly, pale in front; peritoneum
black. Coasts of Europe, generally abundant, south to Madeira and Italy,
straying to Greenland.* Here described from specimens taken at Genoa.
The identity of the Greenland Hake with *M. merluccius†* is perhaps
uncertain. (Eu.) (*merluccius*, ancient name; *mare*, sea; *Lucius*, pike.)

Gadus merluccius, LINNÆUS, Syst. Nat., Ed. x, 254, 1758, Europe; after authors.
Merluccius smiridus, RAFINESQUE, Caratteri, etc., 26, 1810; JORDAN & GILBERT, Synopsis,
    809, 1883; LILLJEBORG, Sveriges Fiske, II, 121, 1891.
Gadus ruber, LACÉPÈDE, Hist. Nat. Poiss., v, 673, 1803, Scotland; Dieppe; on notes by M.
    NÖEL; young.
Gadus merlus, RISSO, Ichth. Nice, 122, 1810, Nice.
Onus riali, RAFINESQUE, Indice d'Ittiol. Sicil., 26, 1810; substitute for *merluccius*.
Merlucius vulgaris, FLEMING, Brit. Anim., 195, 1828; GÜNTHER, Cat., IV, 344, 1862.
Merluccius esculentus, RISSO, Eur. Mérid., III, 1826, 220, Nice.
? Merluccius ambiguus, LOWE, Proc. Zool. Soc. Lond. 1840, 37, Madeira.
Merluccius sinuatus, SWAINSON, in Lowe, Proc. Zool. Soc. 1840, 38.
Merlucius lanatus, GRONOW, Cat. Fish., Ed. Gray, 130, 1854, Mediterranean.
Epicopus gayi, GÜNTHER, Cat., II, 248, 1860, no locality; not *M. gayi*, GUICHENOT, which is
    the Chilian Hake.
Merluccius linnœi, MALM, Götheborgs och Bohusläns Fauna, 489, 1877.

2900. MERLUCCIUS BILINEARIS (Mitchill).

(SILVER HAKE; NEW ENGLAND HAKE; WHITING.)

Head 3⅔; depth 6½. D. 13-41; A. 40; scales 100 to 110. Top of head
with W-shaped ridges very conspicuous; eye shorter than snout and less
than interorbital width; maxillary reaching posterior border of pupil;
teeth not very large, smaller than in the European species, *Merluccius
merluccius*. Scales larger than in other species; pectorals and ventrals
long, the latter reaching ¾ distance to vent, their length about ⅜ that
of head. Grayish, darker above; dull silvery below; axil and edge of
pectoral somewhat blackish; inside of opercle dusky silvery; inside of
mouth dusky bluish; peritoneum nearly black. Coasts of New England
and northward to Straits of Belle Isle; south, in deep water, to the Bahamas; rather common; used as food; breeding in deep water, though often
taken near shore, northward. This species resembles the European Hake,
Merluccius merluccius, but the latter has smaller scales, about 150, and
larger teeth. (*bilinearis*, two-lined.)

* The Iceland Hake has been described as *Merluccius argentatus* (Faber). According
to Faber, it has large teeth, the mouth white within, and the rays D. 15-43; A. 51; the fins
deeply notched. It is perhaps a valid species, and, if so, it doubtless occurs in Greenland.
(*argentatus*, silvered.)
Gadus merluccius (argentatus), FABER, Fische Islands, 90, 1829, Iceland.
Merluccius argentatus, GÜNTHER, Cat., IV, 346, 1862.

† "Dans l'Amérique du Nord, on cite ce poisson de Grœnland, mais l'exactitude de cette
indication parait douteuse." (Collett, Comp. Sci. Hirondelle, 1896, 58.)

Stomodon bilinearis, MITCHILL, Rept. Fish. N. Y., 7, 1814, New York.
Gadus albidus, MITCHILL, Journ. Ac. Nat. Sci. Phila., I, 1817, 409, New York.
Merlucius albidus, STORER, Hist. Fish. Mass., 363.
Merlucius bilinearis, GOODE & BEAN, Bull. Essex Inst., XI, 9, 1879, JORDAN & GILBERT, Syn-
· opsis, 809, 1883; GOODE & BEAN, Oceanic Ichthyology, 386, fig. 330, 1896.

2901. MERLUCCIUS PRODUCTUS (Ayres).

Head 3¾; depth 7. D. 11–41; A. 43; V. 7; scales 136. Head with the
W-shaped ridges less strongly marked; maxillary reaching center of
pupil; eye large; pectorals long and narrow, reaching vent; ventrals
much smaller than in *M. bilinearis*, reaching halfway to vent, their length
about ⅔ that of head; caudal somewhat forked. Scales quite small, deciduous. Teeth moderate. Silver gray; head dusted with coarse black
dots; inside of mouth and opercle jet-black; peritoneum silvery, with
black specks. Length 3 feet. Pacific coast of America, from Santa Catalina Island northward to Puget Sound; everywhere abundant at moderate
depths; used as food. (*productus*, drawn out.)

Merlangus productus, AYRES, Proc. Cal. Ac. Nat. Sci. 1855, 64, San Francisco.
Homalopomus trowbridgii, GIRARD, Proc. Ac. Nat. Sci. Phila. 1856, 132, Astoria, Oregon.
(Coll. Lieut. W. P. Trowbridge.)
Gadus productus, GÜNTHER, Cat., IV, 338, 1862.
Merluccius productus, GILL, Proc. Ac. Nat. Sci. Phila. 1863, 247; JORDAN & GILBERT, Syn-
opsis, 809, 1883.

Family CCXIV. GADIDÆ.

(THE CODFISHES.)

Body more or less elongate, the caudal region moderate, coniform
behind, and with the caudal rays procurrent above and below; vent submedian; suborbital bones moderate; scales small, cycloid; mouth large,
terminal; chin with a barbel, more or less developed. Gill openings very
wide; gill membranes separated or somewhat united, commonly free from
the isthmus; no spines, the fin rays all articulated. Dorsal fin extending
almost the length of the back, forming 1, 2, or 3 fins; anal fin long, single
or divided; caudal fin distinct, or confluent with the dorsal and anal;
ventral fins jugular, but attached to the pubic bone, each of 1 to 8
branched rays. Gills 4, a slit behind the fourth. No pseudobranchiæ.
Edge of preopercle usually covered by skin of head. Pyloric cæca usually
numerous, but sometimes few or none. Air bladder generally well developed. Genera about 25, species about 140; an important family, many of
its members being highly valued as food. They inhabit chiefly the northern seas, sometimes venturing into the oceanic abysses. One genus (*Lota*)
is confined to the fresh waters. (*Gadidæ*, Günther, Cat., IV, 326–369.)

GADINÆ:
  a. Anal divided into 2 separate fins; dorsal fin divided into 3.
    b. Lower jaw distinctly projecting; barbel small or obsolete; caudal concave
      behind.
      c. Teeth in upper jaw slender, wide, set in 1 or 2 series; caudal forked.
                                                                    BOREOGADUS, 976.
      cc. Teeth in upper jaw in a villiform band, the outer somewhat larger; caudal
        lunate.

　　d. Subopercle and postclavicle normal, both thin and flat, not enlarged
　　　　and ivory-like.　　　　　　　　　　　　　　.　POLLACHIUS, 977.
　　dd. Subopercle and postclavicle enlarged, the bone dense and smooth,
　　　　like ivory.　　　　　　　　　　　　　　　THERAGRA, 978.
　bb. Lower jaw included; barbel well developed; caudal not concave behind.
　　e. Hypocoracoid not swollen and ivory-like; lateral line pale; supraoccipital
　　　　crest moderate.
　　　f. Transverse processes of vertebræ thickened, swollen, and ivory-like
　　　　　at tip; small codfishes of the Arctic.　　　　ELEGINUS, 979.
　　　ff. Transverse processes of vertebræ not swollen at tip.
　　　　g. Vent in front of second dorsal; size very small.
　　　　　　　　　　　　　　　　　　　　　　MICROGADUS, 980.
　　　　gg. Vent below second dorsal; typical codfishes of large size.
　　　　　　　　　　　　　　　　　　　　　　　　GADUS, 981.
　　ee. Hypocoracoid much swollen and ivory-like; lateral line black; mouth
　　　　small, the maxillary not reaching to opposite eye; supraoccipital crest
　　　　very high.　　　　　　　　　　　　-　MELANOGRAMMUS, 982.
aa. Anal fin forming a continuous fin or sometimes deeply notched; dorsal not divided
　　into 3 fins.　　　　　　　　　　　　　　　　.
　h. Dorsal fin divided into 2 fins.
　　i. Anterior dorsal composed of distinct rays, similar to those in second
　　　　dorsal.
　　　j. Ventral fins rather broad, each of about 6 rays.
　　　MORINÆ:
　　　　k. Anal fin with a deep notch.
　　　　　l. Snout not much depressed, its edge without keel; tail slender.
　　　　　　　　　　　　　　　　　　　　　　LEPIDION, 983.
　　　　　ll. Snout flat, depressed, keeled on the edge; tail attenuate.
　　　　　　　　　　　　　　　　　　　　　　ANTIMORA, 984.
　　　LOTINÆ:
　　　　kk. Anal fin not notched; mouth terminal.
　　　　m. Vomer toothless.
　　　　　n. Teeth in jaws unequal, outer series enlarged.　　　　.
　　　　　　o. Barbel obsolete.　　　　　URALEPTUS, 985.
　　　　　　oo. Barbel well developed.　　　　LOTELLA, 986.
　　　　・ *nn.* Teeth in jaws all villiform; barbel developed.
　　　　　　　　　　　　　　　　　　　PHYSICULUS, 987.
　　　　mm. Vomer with teeth; head not compressed.
　　　　　　p. Vomer and mandible without canines.　Fresh-water
　　　　　　　species.　　　　　　　　　　LOTA, 988.
　　　　　　pp. Vomer and mandible armed with canines.　Deep-
　　　　　　　water species.　　　　　　　MOLVA, 989.
　PHYCINÆ:
　　jj. Ventral rays very slender, each of 1 or 2 rays.
　　　q. Ventrals each of 2 or 3 slender rays.　　UROPHYCIS, 990.
　　　qq. Ventrals each of a single bifid ray.　　LÆMONEMA, 991.
GAIDROPSARINÆ:
　ii. Anterior dorsal formed of a single slender ray, followed by a band of
　　fringes; ventrals each of 5 to 7 rays.
　　r. Barbels 3; snout with 2 barbels, 1 at each nostril, none at tip; chin
　　　with 1 barbel.　　　　　　　　　　GAIDROPSARUS, 992.
　　rr. Barbels 4; snout with 3 barbels, 1 at tip of snout and 1 on each
　　　nostril; chin with 1 barbel, head high and compressed; no canines.
　　　　　　　　　　　　　　　　　　　ENCHELYOPUS, 993.
BROSMINÆ:
　hh. Dorsal fin continuous, undivided; ventrals several-rayed; teeth on jaws,
　　vomer, and palatines; mouth large; frontal bone.　・　BROSME, 994.

976. BOREOGADUS, Günther.

Boreogadus, GÜNTHER, Cat. Fish. Brit. Mus., IV, 336, 1862 (*fabricii*).

This genus is closely allied to *Pollachius*, the body more slender, the caudal fin more deeply forked, and the teeth in both jaws slender, sharp, wide set, in 1 or 2 series. Small codfishes of the Arctic. (βόρειος, northern; *Gadus*.)

2902. BOREOGADUS SAIDA (Lepechin).

Head $3\frac{1}{2}$; depth $5\frac{1}{2}$; eye 4 in head; snout $3\frac{1}{5}$; interorbital space $4\frac{2}{3}$; gill rakers 9 to 13 + 30 to 32. D. 13–14–20; A. 16–21. Body slender, little compressed; head long, rather pointed, the lower jaw projecting; barbel minute; maxillary reaching middle of pupil; mandible 2 in head; teeth in upper jaw in 1 series, except in front, when the row is double; teeth in lower jaw uniserial; teeth nearly uniform in size, sharp, and wide set; teeth on vomer few, similar to those in jaws. Gill rakers numerous, long and slender, the longest $\frac{1}{4}$ eye; vent slightly before second dorsal; caudal peduncle slender, rounded, its depth scarcely more than $\frac{1}{4}$ eye. Pectorals reaching vent, $1\frac{1}{2}$ in head; ventrals $1\frac{1}{4}$, the second ray exserted for $\frac{2}{3}$ its length; first dorsal highest; front of second dorsal midway between tip of snout and base of caudal. Caudal forked for a distance equal to $\frac{1}{2}$ eye, the tips rounded. Color plain brownish, silvery below, the body with fine black points, most numerous above; dorsals and caudals dusky, the rays blackish distally, their edge narrowly white; anal similarly colored, pale at base; pectorals uniform dusky, pale-edged; ventrals somewhat dusky; peritoneum blackish. Length 6 to 8 inches. Arctic seas of Asia and America, from Greenland to Siberia; generally common in the far North, but rare in Bering Sea and south of Greenland. Here described (by Mr. Norman B. Scofield) from specimens from Davis Straits and Melville Bay, Greenland, the largest $6\frac{3}{4}$ inches long, and from specimens taken by Mr. Scofield at Point Barrow, Port Clarence, and Herschel Island. There is no difference between Greenland and Alaskan specimens. The range of fin rays is D. 12 to 15–12 to 15–18 to 22; A. 15 to 18–20 to 22. Concerning its habits Mr. Scofield observes:

"This fish appears to be quite abundant north of Bering Straits. It was especially brought to our notice by its habit of hiding in small holes in the floating ice, from which it was dislodged by our steamer striking and turning over the blocks of ice. This floating ice was usually in 7 fathoms of water and 1 or 2 miles from the coast. At Herschel Island we took it with the seine in shallow water along the beach. Lucien H. Turner reports it from St Michaels, where he took it through the ice in February, and was told by the natives that it appeared there only in winter. According to Richardson it spawns in Greenland in February, laying its eggs in the seaweeds along the shore under the ice." According to Richardson, in Northumberland Sound, "when hotly pursued by the Beluga or white whale, it has been observed, in its endeavors to escape, to leap by hundreds on the ice." (Eu.) (*saida*, Russian name.)

Gadus saida, LEPECHIN, Nov. Comm. Ac. Sci. Petrop. 1774, 512, White Sea; PALLAS,.
    Zoogr. Rosso-Asiat., III, 199, 1811; GÜNTHER, Cat., IV, 337, 1862; COLLETT, Norske Nord-
    Havs Exped , 126, 1880; JORDAN & GILBERT, Synopsis, 307, 1883.
Merlangus polaris, SABINE, Supp. Parry's Voyage, CCXI, 1824, Baffins Bay; RICHARDSON,
    Last Arctic Voyage, 27, 1824.
Gadus fabricii, RICHARDSON, Fauna Bor.-Amer., III, 245, 1836, northern bays of Green-
    land; after *Gadus æglifinus* of Fabricius.
Gadus agilis, REINHARDT, Danske Vid. Selsk. Afh., VII, 126, 1838, Greenland.
Gadus glacialis, PETERS, Nord Pol. Expd., II, 172, 1874.
Pollachius polaris, GILL, Cat. Fish. East Coast N. A., 218, 1861.
Boreogadus polaris, GILL, Proc. Ac. Nat. Sci. Phila. 1863, 233.
Boreogadus saida, BEAN, Bull. U. S. Nat. Mus., IV, 108, 1879; SCOFIELD, in Jordan & Gil-
    bert, Rept. Fur Seal Invest., 1898.

977. POLLACHIUS, Nilsson.

(POLLACKS.)

Pollachius, NILSSON, in Bonaparte, Catalogo Metodico Pesci Europ., 45, 1846 (*pollachius*).

Body rather elongate, covered with minute scales; mouth moderate or
large, the lower jaw projecting; barbel very small or obsolete; villiform
teeth on vomer, none on palatines; teeth in jaws equal or the outer
slightly enlarged; gill membranes more or less united; subopercle and
postclavicle not enlarged and not ivory-like; dorsal fins 3; anal 2; caudal
lunate; vent under first dorsal. Large fishes of the northern seas.
(Polog or Pollack, the English vernacular name, latinized as *Pollachius*,
as though derived from πολλαχῆ, many fashioned.)

2903. POLLACHIUS VIRENS (Linnæus).

(POLLACK; COAL-FISH; GREEN COD.)

Head 4; depth 4¼. D. 13-22-20; A. 25-20; scales about 150; vertebræ 54.
Body rather elongate, compressed; snout sharp and conic; mouth rather
small, oblique; maxillary reaching beyond front of orbit; lower jaw
slightly the longer; teeth in the upper jaw nearly equal, the outer series
not being especially enlarged; barbel rudimentary or obsolete; gill membranes considerably united, free from isthmus; vent under first dorsal;
caudal fin lunate; pectorals short, scarcely reaching anal; ventrals short,
their origin in front of base of pectoral a distance about equal to diameter of eye. Greenish brown above; sides and below somewhat silvery;
lateral line pale; fins mostly pale; sometimes a black spot in the axil.
North Atlantic; common northward on both coasts, south to Cape Cod
and France. (Eu.) (*virens*, green.)

Gadus virens, LINNÆUS, Syst. Nat., Ed. X, 253, 1758, Seas of Europe; after *Gadus triptery-
    gius imberbis* of the Fauna Suecica; GÜNTHER, Cat., IV, 339; JORDAN & GILBERT,
    Synopsis, 807, 1883.
Gadus carbonarius, LINNÆUS, Syst. Nat., Ed. X, 254, 1758, seas of Europe; after *Gadus
    dorso tripterygius imberbi* of ARTEDI.

Gadus colinus, LACÉPÈDE, Hist. Nat. Poiss., II, 416, 1800, England, etc.; after LE COLIN of Danberton.
Gadus virens, LACÉPÈDE, Hist. Nat. Poiss., II, 417, 1800.
Merlangus purpureus, MITCHILL, Trans. Lit. and Phil. Soc., I, 1815, 370, New York.
Merlangus leptocephalus, DE KAY, New York Fauna: Fishes, 288, pl. 45, fig. 146, 1842, Long Island.
Merlangus purpureus, STORER, Rept. Fish. Mass., 130, 1839.
Pollachius carbonarius, GILL, Proc. Ac. Nat. Sci. Phila. 1863, 233.

978. THERAGRA, Lucas, a new genus.

(ALASKAN POLLACKS.)

Theragra, LUCAS, in Jordan & Gilbert, Rept. Fur Seal Invest. 1896 (1898) (*chalcogrammus*).

This genus is closely allied to *Pollachius*, differing in the following respects: Suboperculum thick, smooth and dense instead of being thin and squamous as in *Pollachius;* the postclavicle is also similar in structure while its proximal portion is subcircular in *Theragra* and rhomboidal in *Pollachius;* this ivory-like character of the suboperculum and postclavicle is so marked that it serves to distinguish these bones at a glance, being entirely different from what is found in the corresponding bones of other gadoids. The Alaskan Pollack farther differs from the Atlantic Pollack in having 19 precaudal vertebræ and 33 caudal, instead of 23 precaudals and 32 caudals; the bodies of the vertebræ are also slightly longer and more deeply sculptured in the Alaskan fish and the spinous process of the anterior dorsals less elevated. The vertebral differences between the 2 genera are merely differences of degree and of specific value only, but the differences between the subopercula and postclavicula are different in kind, distinguishing the Alaskan Pollack not only from the Atlantic Pollack, but from other gadoids. (ϑήρ, beast; ἄγρα, prey or food; the Alaskan Pollack being a chief food of the fur seal, *Callorhinus.*)

a. Dorsal rays about 13 or 14–17–18 or 19; anal rays 20–20; side with 2 interrupted dark longitudinal bands. CHALCOGRAMMA, 2904.
aa. Dorsal rays 10 or 11–13 to 15–16; anal rays 16 to 19–16 to 19; sides plain dusky; body less elongate, the snout blunter, the fins lower. FUCENSIS, 2905.

2904. THERAGRA CHALCOGRAMMA (Pallas).

(ALASKA POLLACK.)

Head 4; depth 6. D. 12–14–18; A. 20–20. Eye 5 in head; snout 3⅓; maxillary 2½; snout conic, sharp, rounded in profile; mouth oblique; maxillary reaching middle of pupil; chin with a minute barbel; teeth small, those of the outer row above slightly enlarged; eye large, wider than the flat interorbital space, 4 in head. Gill membranes somewhat united, the posterior outline deeply emarginate; vent under interspace between first and second dorsal; first dorsal higher than the others, the second lowest; ventrals filamentous, reaching nearly to vent; pectorals long, reaching past front of anal, 1½ in head; caudal somewhat concave. Olivaceous above, sides silvery, with 2 interrupted stripes of dark, brassy, olive along sides, these irregular on their edges, about ¼ width of eye, with uneven edges; a trace of a third similar stripe below anteriorly, the stripes very irregular; back mottled. Dorsal plain dark olive; pectoral quite dark; lower fins ashy; caudal ashy olive. Bering Sea and neighboring waters, probably south to Sitka and the Kurils. Our specimens from Unalaska, Robben Reef, Komandorski and Pribilof islands and Bristol Bay. Excessively common throughout Bering Sea, swimming near the surface, and furnishing the greater part of the food of the fur seal. This animal rarely catches the true codfish, which swims nearer the bottom. Length 3 feet. (χαλκός, brass; γραμμή, line.)

Gadus chalcogrammus, PALLAS, Zoogr. Rosso-Asiat., III, 198, 1811, Kamchatka; GÜNTHER, Cat., IV, 340, 1862; JORDAN & GILBERT, Synopsis, 807, 1883.
Gadus periscopus, COPE, Proc. Am. Philos. Soc. Phila. 1873, 30, Unalaska (Coll. George Davidson).
Pollachius chalcogrammus, JORDAN, Cat. Fish. N. A. (130) 918, 1885.
Theragra chalcogramma, JORDAN & GILBERT, Rept. Fur Seal Invest., 1898.

2905. THERAGRA FUCENSIS (Jordan & Gilbert).

(WALL-EYED POLLACK; PUGET SOUND POLLACK.)

Head 3⅔ in body; depth 5¼. D. 10-13-16 to 11-15-16; A. 16-19 to 19-19; eye 4½ in head; maxillary 2⅔; pectoral 1⅜; longest caudal ray 2. Body elongate, not greatly compressed; mouth large, the maxillary reaching to below middle of eye; jaws with minute, sharp, curved teeth, the outer series enlarged; teeth on vomer, palatines toothless; lower jaw projecting, a very small barbel under its tip; interorbital space wide, very slightly and evenly convex, wider than the diameter of eye; nostrils much nearer eye than tip of snout, the posterior much the larger; head almost entirely covered with small scales; gill rakers numerous, the longest as long as pupil, about 5 + 27 in number. Distance of origin of first dorsal from snout 3⅓ in body; first rays of first dorsal reaching far past the ends of last rays where fin is depressed; first rays of other dorsals and anals scarcely reaching the base of last rays; caudal slightly forked or subtruncate when spread, the lobes subequal; end of pectoral reaching to front of anal; ventrals inserted in front of base of pectoral in distance a little more than diameter of eye, ending in a filamentous point. Color nearly plain sooty, with no distinct lateral bands, and with generally only a trace of a pale lateral streak along the side; on the head some diffuse dark spots; fins all dusky. The band of teeth in the premaxillary is wider than in *Theragra chalcogramma*, and the band is widened at the anterior end; the body is shorter; eye smaller; color darker; fins not so high; caudal not so deeply forked. Pacific coast, from Vancouver Island to Monterey, abundant in Puget Sound; probably northward to Kadiak, replacing *T. chalcogramma* to the southward. This form may intergrade with *Theragra chalcogramma*, though the original types seem well separated. Little is known of its range to the northward. Scofield and Seale took a specimen in Chignik Bay in northern Alaska, which seems as near *T. fucensis* as *T. chalcogramma*. Its rays * are D. 11–16–17; A. 18–17; ventrals reaching $\frac{2}{3}$ distance to vent; interorbital space wider than eye; coloration dark. But its body is as slender as in *T. chalcogramma*. (*fucencis*, from the straits of Juan de Fuca.)

Pollachius chalcogrammus fucencis, JORDAN & GILBERT, Proc. U. S. Nat. Mus. 1893, 315,
Puget Sound at Tacoma. (Type, No. 44455. Coll. David H. Hume.)

979. ELEGINUS, Fischer.

Eleginus, FISCHER, Mem. Soc. Nat. Moscow, V, 4, 2d Ed., 252–257, 1813 (*navaga*); not
Eleginus of later authors.
Tilesia, SWAINSON, Nat. Hist. Class'n Fishes, II, 300, 1839 (*gracilis*); name preoccupied.
Pleurogadus, BEAN, in JORDAN, Cat. Fish. N. A., 130, 1885 (*gracilis*); substitute for *Tilesia*
preoccupied.

This genus differs from the other codfishes in the structure of the transverse processes of the vertebræ, which are club-shaped, narrow at base, but expanding distally into a rounded hollow bulb at their tips. Skeleton otherwise essentially as in *Microgadus*, the skull similar. Small codfishes of the Arctic seas. (ἐλεγῖνος, a social fish mentioned by Aristotle.)

2906. ELEGINUS NAVAGA (Kölreuter).

(WACHNA COD.)

Head $3\frac{4}{5}$ in length of body; depth 6; eye $5\frac{4}{5}$ in head; snout 3; interorbital space $4\frac{1}{2}$; gill rakers 20 or 21; barbel small, equal to pupil; dorsal 13–18–18; anal 22–20; scales small, 157 transverse rows above lateral line from gill opening to first rudimentary caudal rays. Body slender and rounded with a rather long head; snout viewed from above rounded, but running to a rather sharp point when viewed from the side; lower jaw included, the fleshy snout projecting beyond the maxillary, its length slightly greater than that of the snout; tip of maxillary on a vertical with the front of the pupil; articulation of mandible with quadrate bone on a vertical running midway between pupil and posterior edge of eye; teeth

---

* The following is the count of fin rays in 13 specimens of *Theragra* of the two species:

Dorsal.	Anal.	Locality.
T. chalcogramma:		
13—15—20	19—20	Kamchatka.
13—15—20	19—21	Unalaska.
14—19—23	24—22	Pribilof Islands.
14—16—21	21—23	Do.
14—17—18	23—21	St. Paul Island.
13—15—19	21—22	Do.
14—17—19	21—20	Do.
14—17—18	22—20	Kamchatka.
11—16—17	19—17	Chignik Bay.
T. fucensis:		
10—15—17	18—16	Puget Sound.
11—15—16	19—18	Do.
10—14—16	16—19	Do.
12—13—17	19—19	Do.

all slender and curved backward, those in upper jaw in several irregular rows, the outer row regular and with slightly larger teeth; teeth in lower jaw in a single row except in front where they are in a double row; teeth on vomer few and about the size of the smaller teeth in the upper jaw; gill rakers moderate, the longest not quite equal to diameter of pupil; caudal peduncle compressed, its depth equal to diameter of eye; vent under front of second dorsal; pectoral fin not reaching vent, its length 1¼ times in head; ventrals reaching halfway to vent, the second ray moderately produced; first dorsal highest; distance between second and third dorsals twice distance between first and second; caudal fin very slightly concave; third ray of second dorsal midway between tip of snout and base of middle caudal rays. Color somewhat mottled, grayish brown above, light silvery below; the 3 dorsals and caudal dusky and edged with white; pectorals uniform dusky; ventrals but slightly dusted with black; anal with a few punctulations at their anterior ends; peritoneum pale. Arctic shores of Asia and North America, south to Bering Sea, locally abundant. It reaches the length of about a foot. Here described (by Norman B. Scofield) from numerous specimens, the largest 11 inches long, taken at Port Clarence by Scofield and Seale, and at Petropaulski by the *Albatross* (Fur Seal Invest. of 1896). The range of the fin rays is D. 12 to 15—18 to 21—18 to 21; A. 20 to 23—20 to 23. Mr. Scofield has prepared a skeleton of this species for comparison with that of *Microgradus proximus* from San Francisco. There is very little difference in the skulls. There is no difference in the neural spines of the vertebræ. The transverse processes of the vertebræ in *Microgadus proximus* are flattened and plate-like, while in *Eleginus navaga* they are club-shaped, narrow at base where they leave the centrum, but expanding into a rounded hollow bulb at the distal end. This character defines the genus *Eleginus*. (*navaga*, a Russian name.)

Gadus navaga, KÖLREUTER, Nov. Comm. Ac. Petrop., XIV, 1770, 484, pl. 12, coast of northern Russia; PALLAS, Zoogr. Rosso-Asiat., III, 196, 1811.
Gadus gracilis, TILESIUS, Mém. Ac. Imp. Petersb., II, 1810, 354, Kamchatka; JORDAN & GILBERT, Synopsis, 804, 1883.
Gadus wachna, PALLAS, Zoogr. Rosso-Asiat., III, 182, 1811, Kamchatka.
Tilesia gracilis, SWAINSON, Nat. Hist. Fish., II, 300, 1839; BEAN, Proc. U. S. Nat. Mus. 1881, 243.
Pleurogadus gracilis, BEAN, in JORDAN, Cat. Fish. N. A., 130, 1885.
Eleginus navaga, GILL, Proc. U. S. Nat. Mus. 1890, 303.

980. MICROGADUS, Gill.

(TOMCODS.)

Microgadus, GILL, Proc. Ac. Nat. Sci. Phila. 1865, 69 (*proximus*).

Very small codfishes allied to *Gadus*, but with the vent placed before the second dorsal and with a different structure of the cranium. The following is Professor Gill's account of the skull of *Microgadus proximus*, the italicised portions indicating the differences from *Gadus*:

The cranium is proportionally broader toward the front and less flattened, while the brain case is flattened below, *decidedly swollen* on each side of a depressed *sphenoidal groove*, and has an ovate cardiform shape; the *paraoccipital* or epiotic is not produced into an angle behind, but is obtusely rounded, and its posterior or *outwardly descending ridge blunt;* the opisthotic is well developed, oblong, and with its reentering angle *high up*, and, on a line with it, the surface is divided into 2 parts—a *narrow* and a flattened one, and a lower expanded one, much swollen; the alisphenoid or prootic is *oblong*, acutely emarginate in front, swollen from the region of the high anterior sinus, and above a little produced forward; the great *frontal* is a little longer than broad, with supraoccipital crest *continued forward* on the bone, and near the front expanded upward, and with the *expanded portion* behind dividing into narrow *lateral wings;* the lateral testiform ridges of the frontal are 'continued forward and *curved outward* toward the antero-lateral angles; the anterior frontals are *mostly covered in front* by the great frontal, and are much *developed* in the direction of the antero-lateral angles, the inferior expanded axillary portion being very narrow; the nasal has a rounded ridge in front, continued well below, and its posterior crest is *laminar* and trenchant.

Species American; valued as food. (μιχρός, small; γαδος, *Gadus*.)

a. Second anal with 21 or 22 rays; snout rather long; body semitranslucent; first anal and ventrals pale; body scarcely blotched with blackish.   PROXIMUS, 2907.

aa. Second anal with 16 to 20 rays; snout shorter; body opaque; first anal and ventrals dusky; body blotched above with blackish.   TOMCOD, 2908.

2907. MICROGADUS PROXIMUS (Girard).

(CALIFORNIA TOMCOD.)

Head 3½ in body; depth 5. D. 14–18–18 to 21; A. 21 or 22–21 or 22; V. 6–7; eye 5 in head; maxillary 2½; pectoral 2; highest dorsal spine 2; middle caudal rays equal to snout. Head long, convex above, somewhat compressed, with vertical sides; eye moderate; mouth rather large; maxillary reaching to below pupil; barbel small; teeth in each jaw in a band, the outer row a little enlarged. Gill membranes a little connected, free from the isthmus. First dorsal highest, somewhat falcate; first anal longer and higher than second; pectorals moderate, reaching anal; ventrals filamentous, scarcely reaching anal; caudal slightly emarginate or subtruncate when fin is spread. Lateral line very distinct, wavy, high anteriorly, slightly interrupted posteriorly. Vent below first dorsal. Color olivaceous above, pale, or slightly translucent white below; dorsal fins dusky, paler at base; first anal and ventrals uncolored; second anal dusted with dark points. Monterey to Unalaska; abundant; a food-fish of considerable importance, the flesh delicate but without much flavor. Here described from a specimen, 8 inches in length, from Alaska, *Albatross* Station 3213. It reaches the length of about a foot. (*proximus*, near, to *Microgadus tomcod*.)

Gadus proximus, GIRARD, Proc. Ac. Nat. Sci. Phila. 1854, 141, San Francisco; GIRARD, U. S. Pac. R. R. Surv., X, Fishes, 142, 1858; JORDAN & GILBERT, Synopsis, 805, 1883.

Morrhua californica, AYRES, Proc. Cal. Ac. Nat. Sci. 1854, 9, San Francisco.

Gadus californicus, GÜNTHER, Cat., IV, 332, 1862.

Microgadus proximus, GILL, Proc. Ac. Nat. Sci. Phila. 1865, 69.

2908. MICROGADUS TOMCOD (Walbaum).

(TOMCOD; FROSTFISH.)

Head 4 in body; depth 5. D. 13 to 15–15 to 19–16 to 18; A. 17 to 21–16 to 20; eye 5 in head; maxillary 2⅔; pectoral 1⅔; middle caudal rays 2¼; first dorsal rays 1⅜. Snout rounded, less produced than in *Microgadus proximus;* mouth short; maxillary 2½ in head, reaching pupil; eye large, 3⅔ in head; barbel small; pectorals reaching vent; ventrals filamentous, not reaching vent; vent under interval between first and second dorsals. Color olive brown, distinctly blotched and spotted with darker, lighter on the belly; more opaque than in *M. proximus;* back and sides profusely punctulate; dorsals and caudal blotched with darker; anals coarsely punctulate anteriorly, colorless posteriorly; ventrals and pectorals dusky. Virginia to Labrador; very common northward, and valued as a food-fish. Here described from a specimen, 9 inches in length, from Boston, Massachusetts. Length 1 foot. (*tomcod*, a vernacular name.)

Tomcod, SCHÖPF, Schrift. Naturf. Freunde, VIII, 140, 1780, New York.
Gadus tomcod, WALBAUM, Artedi Piscium, III, 133, 1792, after SCHÖPF; JORDAN & GILBERT, Synopsis, 806, 1883.
Gadus frost, WALBAUM, Artedi Piscium, III, 134, 1792, North America; after Frost-fish of ' Pennant.
Gadus tomcodus, MITCHILL, Trans. Lit. and Phil. Soc., I, 1815, 368, New York; GÜNTHER, Cat., IV, 331, 1862.
Gadus pruinosus, MITCHILL, Trans. Lit. and Phil. Soc., I, 1815, 368, New York.
Gadus tomcodus fuscus, MITCHILL, Trans. Lit. and Philos. Soc., I, 1815, 369, New York.
Gadus tomcodus luteus, MITCHILL, Trans. Lit. and Philos. Soc., I, 1815, 369, New York.
Gadus tomcodus mixtus, MITCHILL, Trans. Lit. and Philos. Soc., I, 1815, 369, New York.
Gadus polymorphus, MITCHILL, Trans. Lit. and Philos. Soc., I, 1815, 369, New York.
Morrhua americana, STORER, Rept. Fish. Mass., 120, 1839, coast of Massachusetts.

981. GADUS (Artedi) Linnæus.

(CODFISHES.)

Gadus, LINNÆUS, Syst. Nat., Ed. X, 251, 1758 (*morhua*); after ARTEDI.
Morrhua, OKEN, Isis 1817, 1182 (*morhua;* on *les Morrhues* of CUVIER).
Cepphus, SWAINSON, Nat. Hist. Class'n Fishes, II, 300, 1839 (*macrocephalus*).

Body moderately elongate, compressed and tapering behind. Scales very small; lateral line present, pale. Head narrowed anteriorly; mouth moderate, the maxillary reaching past front of eye; chin with a barbel; teeth in jaws cardiform, subequal; vomer with teeth; none on the palatines; cranium without the expanded crests seen in *Melanogrammus;* no part of the skeleton expanded and ivory-like. Dorsal fins 3, well separated; anal fins 2; ventral fins well developed, of about 7 rays. Species of the Northern Seas; highly valued as food. (*Gadus*, the Latin name, akin to the English word cod.)

a. Eye moderate, about ⅓ snout in adult; axil without dusky spot.
   b. Air bladder large. Atlantic codfish.   CALLARIAS, 2909.
   bb. Air bladder small. Pacific codfish.   MACROCEPHALUS, 2910.
aa. Eye large, more than ⅓ length of snout; axil with a dusky spot; caudal peduncle slender.   OGAC, 2911.

2909. GADUS CALLARIAS,* Linnæus.

(COMMON CODFISH.)

Head 3¼ to 4¼; depth about 4. D. 14–21–19; A. 20–18. Head large, but varying much in size; maxillary about reaching middle of orbit; occipital keel not greatly developed; teeth strong, cardiform, in narrow bands, those of the outer row in the upper jaw and of the inner row in the lower jaw somewhat enlarged. Eye moderate, about ¼ length of snout. First dorsal little elevated, its height about ¼ length of head; vent under front of second dorsal; caudal slightly emarginate; pectorals ¼ length of head. Greenish or brownish, subject to many variations, sometimes yellowish or reddish; back and sides with numerous rounded brownish spots; lateral line pale; fins dark. North Atlantic, south to Virginia, and France; one of the most important of food-fishes. (Eu.) (*Callarias*, an old name of the codfish.)

Gadus callarias, LINNÆUS, Syst. Nat., Ed. x, 252, 1758, young examples, Baltic Sea and oceans of Europe, after *Gadus*, etc., *cauda integra* of the Fauna Suecica; CUVIER, Règne Animal, Ed. 2, vol. II, 332, 1829; JORDAN & GILBERT, Synopsis, 804, 1883.

Gadus morhua, LINNÆUS, Syst. Nat., Ed. x, 252, 1758, seas of Europe; after *Gadus*, etc., *cauda subæquali* of the Fauna Suecica; RICHARDSON, Fauna Bor.-Amer., 242, 1836.

Gadus barbatus, LINNÆUS, Syst. Nat., Ed. x, I, 252, 1758.

Gadus vertagus, WALBAUM, Artedi Pisc., III, 143, 1792; after *Jägershen*, KLEIN, Hist. Nat. Pisc., v, 7, pl. 2, fig. 1, 1749.

? *Gadus heteroglossus*, WALBAUM, *l. c.*, 144; after *Hornbogen* of KLEIN.

Gadus arenosus, MITCHILL, Trans. Lit. and Philos. Soc., I, 1815, 368, New York.

Gadus rupestris, MITCHILL, Trans. Lit. and Philos. Soc., I, 1815, 368, New York.

? *Gadus nanus*, FABER, Fische Islands, 113, Iceland.

Morrhua americana, STORER, Hist. Fish Mass., 343, 1867.

Gadus morhua, GÜNTHER, Cat., IV, 328, 1862; GOODE & BEAN, Oceanic Ichthyology, 354.

2910. GADUS MACROCEPHALUS, Tilesius.

(ALASKA CODFISH.)

Head 3 in body; depth 4¾. D. 13–18–16. A. 21–17; eye 6 in head; maxillary 2½; highest dorsal ray 3; pectoral 2½; middle caudal rays 4. Head large, the snout blunt; mouth large, the maxillary reaching to below front of pupil, snout projecting beyond mouth, lower jaw included; teeth strong, cardiform, in narrow bands on jaws and vomer; interorbital wide, 1½ times wider than diameter of eye, very slightly convex. Gill rakers moderate, about equal to pupil in length, 3 + 17 in number. Pectoral reaching to below end of first dorsal; ends of first dorsal rays reaching second dorsal when fin is depressed; ventrals inserted in front of base of pectorals in distance equal to diameter of eye; veins under front of second dorsal; caudal subtruncate. Color brownish, lighter below, back and sides with numerous brownish spots; fins, with the exception of first anal and ventrals, dusky. This species is very abundant in Bering Sea,

---

* We retain the name *Gadus callarias*, Linnæus for the codfish, instead of the commonly used name *Gadus morhua*, applied by Linnæus to the same species, because the name *Gadus callarias* stands first on the page on which it occurs. To accord priority to the name standing first is essential to fixity, and not the less so if the competing names are of the same actual date, published by the same author. It is not justice nor elegance, but fixity, which the rules of nomenclature aim to secure.

on both shores, in 15 to 130 fathoms, forming an important article of commerce. Its range southward extends to the offshore banks of Oregon. In external respects we recognize no distinction between this species and the common eastern codfish, except that the head seems larger. Here described from a specimen 20 inches long, taken in the Straits of Fuca by the *Albatross.* Concerning this species Dr. Gilbert observes:

It has been frequently pointed out, and is well known to fishermen that the Pacific codfish has a smaller air bladder or sound than the Atlantic cod. Pending an examination of this question, which we are not now in a position to make, we propose to recognize the Pacific fish as a distinct species. (μακρός, long; κεφαλή, head.)

Gadus macrocephalus, TILESIUS, Mém. Acad. Sci. St. Petersb., II, 1810, 360, Kamchatka; GÜNTHER, Cat., IV, 330, 1862.
Gadus pygmœus, PALLAS, Zoogr. Rosso-Asiat., III, 1811, Kamchatka.
Gadus auratus, COPE, Proc. Am. Philos. Soc., 1873, 30, Unalaska.

2911. GADUS OGAC, Richardson.

(GREENLAND CODFISH.)

Head 3¼. D. 14 or 15–18 to 20–17 to 20; A. 20 to 22–18 to 19; V–6. This species resembles the common cod (*Gadus callarias*), but differs from it as follows: It has a more slender caudal peduncle, larger eye, greater interorbital width, longer barbel, more advanced position of ventral fins, and a longer pectoral fin. Color dark, blackish brown above, lighter below, with yellowish marblings; the tip of the dorsal, anal, and caudal fins black; ventral and pectorals dark brown or black, a dusky spot on the axil; barbel black. Coast of Greenland; not seen by us. The above notes from specimens collected at Godhavn, Greenland, examined by Ensign Dresel. (*ogac,* a native name.)

Gadus ogac, RICHARDSON, Fauna Bor.-Amer., 246, 1836, Greenland; REINHARDT, Vid. Selsk. Naturvid. Math. Afh. 1838; DRESEL, Proc. U. S. Nat. Mus. 1884, 246.
Gadus ogat, KRÖYER, Voy. Scand. et Lap., pl. 19.

982. MELANOGRAMMUS, Gill.

(HADDOCKS.)

Melanogrammus, GILL, Proc. Ac. Nat. Sci. Phila. 1862, 280 (*œglefinus*).
Æglefinus, MALM, Götheborgs och Bohusläns Fauna, 481, 1877 (*œglefinus*).

This genus is distinguished from *Gadus* by its smaller mouth, the produced first dorsal fin, black lateral line, and especially by the great enlargement of the hypocoracoid, which is dense and ivory-like. The lateral line is always black, and the supraoccipital and other crests on the head are largely developed. · Food fishes of large size. (μέλας, black: γραμμή, line.)

2912. MELANOGRAMMUS ÆGLEFINUS, Linnæus.

(HADDOCK.)

Head 3⅔; depth 4½. D. 15–24–21; A. 23–21. Snout long and narrow, overlapping the small mouth; maxillary barely reaching front of orbit;

teeth subequal, large, in a cardiform band in upper jaw; in a single series on lower jaw and on vomer; occiput carinated; a ridge extending backward from each orbit; eye very large, $\frac{2}{3}$ length of snout, 4 in head. Anterior rays of first dorsal elevated, $\frac{2}{3}$ length of head, the fin pointed, higher than second and third dorsals; caudal lunate; vent below front of second dorsal. The skull in this species is more depressed than in *Gadus callarias*, broader, and thinner in texture; occipital crest exceedingly high, much higher than in *Gadus*, the wing-like projections at its base anteriorly spreading widely, raised above the surface of the skull. Dark gray above, whitish below; lateral line black; a large dark blotch above the pectorals; dorsals and caudal dusky. North Atlantic, on both coasts, south to France and North Carolina; in deeper water to Cape Hatteras; an important food-fish, reaching a considerable size. (Eu.) (*æglefinus*, an old name of the haddock, from the French Aiglefin or Aigrefin, according to Bellon; perhaps from *aigre faim*, extremely hungry, voracious.)

Gadus æglefinus, LINNÆUS, Syst. Nat., Ed. x, 251, 1758, seas of Europe, after *Gadus*, etc., *cauda biloba*, of the Fauna Suecica; JORDAN & GILBERT, Synopsis, 803, 1883.
Morrhua æglefinus, FLEMING, British Animals, 191, 1828.
Morrhua punctatus, FLEMING, British Animals, 192, 1828.
Melanogrammus æglefinus, GILL, Proc. Ac. Nat. Sci. Phila. 1862, 280; *ibid.* 1863, 237; GOODE & BEAN, Oceanic Ichthyology, 354, 1896.
Æglefinus linnæi, MALM, Götheborgs och Bohusläns Fauna, 481, 1877.

983. LEPIDION, Swainson

Lepidion,* SWAINSON, Nat. Hist. Class'n Anim., I, 318, 1838, and II, 300, 1839 (*lepidion*).
Haloporphyrus, GÜNTHER, Cat. Fish. Brit. Mus., IV, 358, 1862 (*lepidion*).

Body elongate, covered with small scales; head not greatly depressed, higher than broad; the snout subconical, obtusely rounded; tail tapering behind; jaws with bands of villiform teeth; a roundish patch of teeth on vomer; no teeth on palatines; chin with a barbel; branchiostegals 7. Caudal fin separate; 2 dorsal fins and 1 anal; the first dorsal short; ventrals narrow, of 6 rays. Deep waters. The American species distinguished from the *Lepidion lepidion* (Risso), of the Mediterranean, by its non-filamentous first dorsal. (λεπίδιον, diminutive of λεπίς, scale:—small-scaled.)

2913. LEPIDION VERECUNDUM, Jordan & Cramer.

Head 3$\frac{1}{3}$; depth 4$\frac{1}{4}$. D. VIII–40; A. 37; V. apparently 4 (some rays broken on each side); scales about 75, not to be exactly counted. Body robust, compressed, tapering from the large head to the very slender, attenuate tail, which is not so broad as pupil; head large, not greatly compressed, not keeled above, its sides scaly; lower jaw with some scales; interorbital space depressed, 5$\frac{3}{4}$ in head; eye very large (in young), 2$\frac{2}{3}$ in head; snout short, depressed, not pointed, and with lateral keel, 5$\frac{3}{4}$ in head; preorbital very narrow; mouth rather large, oblique, the maxillary reaching to below front of pupil, 2$\frac{2}{3}$ in head; lower jaw slightly longer, its tip with a stiffish pointed projection representing the barbel; teeth small, in bands, a few

---

* *Lepidion* is sufficiently distinct from *Lepidia*, Savigny, 1817.

on vomer. No spines on snout or opercles. Gill membranes somewhat united, free from isthmus. Gill rakers slender, rather long, 10 to 12 on lower part of arch. Scales very small, mostly lost posteriorly (in our specimen) and not to be exactly counted; lateral line not evident. First dorsal rather low and long, none of its rays produced, the longest about ½ head; ventrals filamentous, ¼ head; pectorals about ⅓ head; caudal 2¼ head; anal deeply notched behind the middle, its posterior lobe highest. Color uniform purplish black, the fins paler. One young individual, 2¼ inches long, from *Albatross* Station 2993, off the Revillagigedo Islands. (*verecundus*, modest.)

Lepidion verecundum, JORDAN & CRAMER, Proc. U. S. Nat. Mus. 1896, 456, Revillagigedo
   Islands, at Albatross Station 2993. (Coll. *Albatross*.)

984. ANTIMORA, Günther.

Antimora, GÜNTHER, Ann. Mag. Nat. Hist. 1876, 2 (*rostrata*).

This group differs from *Lepidion* in the form of the snout, the backward position of the vent, the imperfect division of the anal, in which latter respect it approaches *Mora*. In *Lepidion* the snout is subconical, obtusely rounded; in *Antimora* it forms a flat, triangular lamina, sharply keeled at the sides, resembling the snout of *Macrourus*. Body elongate, compressed, tapering into à slender tail. Scales very small. Head entirely scaly, even to the gill membranes. Snout depressed, thin and flat, projecting beyond the mouth; mouth rather large; chin with a barbel; jaws with bands of villiform teeth; a small roundish patch of teeth on vomer, none on palatines. Dorsal fins 2, the first short, its anterior ray produced into a long filament; anal fin deeply notched, almost separated into 2 fins; ventral fins with 6 rays, 1 of them filamentous; caudal truncate. Branchiostegals 7. Deep-water fishes. (ἀντί, opposite; *Mora*, a related genus.)

  a. Head rather small, 4½ in length; scales 115.   VIOLA, 2914.
  aa. Head rather large, about 3½ in length; scales 130.   MICROLEPIS, 2915.

2914. ANTIMORA VIOLA (Goode & Bean).

Head 4½ in body; depth 5. D. 4–53; A. 40; V. 6; scales 11–115–27. Snout broad, pointed at tip, much depressed, forming a roof-like projection above mouth; a conspicuous keel extending backward from tip of snout along the suborbital to the posterior margin of the eye. Mouth U-shaped, wholly inferior; maxillary nearly reaching posterior margin of orbit; interorbital space flat, as wide as the large eye, the orbital ridges somewhat elevated; barbel about ¼ diameter of orbit. First dorsal with its first ray much produced, longer than head; anal fin deeply notched near its middle. Caudal peduncle as long as eye, its depth more than ½ its length; longest ray of ventrals reaching about halfway to vent; pectoral 1¼ in head. Color deep violet or blue black; inside of mouth and opercles blue black. Banks of Newfoundland and southward, in deep water. (Goode & Bean.) (*viola*, violet.)

Haloporphyrus viola, GOODE & BEAN, Proc. U. S. Nat. Mus., I, 1878, 256, La Have Bank, 400
   to 500 fathoms ; JORDAN & GILBERT, Synopsis, 800, 1883.
Antimora viola, GOODE & BEAN, Oceanic Ichthyology, 372, fig. 324, 1896.

2915. ANTIMORA MICROLEPIS, Bean.

Head about 4 in total length with caudal; depth 5⅔ without caudal; eye 4 in head, nearly equal to snout. D. 4 or 5-51; A. 41; barbel very slender, 2 in eye; gill rakers short, slender, 4 + 11. Maxillary reaching to nearly below posterior edge of eye; longest ray of first dorsal about ⅓ as long as head; anal deeply emarginate, beginning under twentieth ray of second dorsal; second ventral ray 1⅙ in head. Scales very small, about 9 rows between origin of second dorsal and lateral line, and about 130 in lateral line. Color olivaceous, deeper on opercles and branchiostegal membranes and on inside of mouth. Off Queen Charlotte Islands. Several specimens taken by the *Albatross* at different stations in Bering Sea, at depths of 350 and 351 fathoms, and off the coasts of the Queen Charlotte Islands and California, at depths of 1,588 and 455 fathoms. One large specimen, from off Bogoslof Island, has the filamentous ray of first dorsal ⅞ length of head, and the eye is shorter than the snout. ($\mu\iota\varkappa\rho\delta\varsigma$, small; $\lambda\varepsilon\pi\iota\varsigma$, scale.)

Antimora microlepis, BEAN, Proc. U. S. Nat. Mus. 1890, 38, off Cape St. James, Queen Charlotte Island, at Albatross Station 2860, in 876 fathoms (Type, No. 45361); GILL, BERT, Rept. U. S. Fish Comm. 1893 (1896), 456 and 473; JORDAN & GILBERT, Rept. Fur Seal Invest., 1898.

985. URALEPTUS, Costa.

Uraleptus, COSTA, Archiv für Naturgesh. 1858, 87 (*maraldi*).
Gadella, LOWE, Proc. Zool. Soc. Lond. 1843, 91 (*gracilis*).

Body elongate, compressed, and tapering posteriorly, covered with small scales. A separate caudal; 2 dorsal fins and 1 anal; ventral fins narrow, with flat base, composed of 6 rays. Upper and lower jaw with an outer series of strong curved teeth; vomerine and palatine teeth none; chin without barbel. Branchiostegals 7. Deep waters of the Atlantic. ($o\dot{\upsilon}\rho\dot{\alpha}$, tail; $\lambda\varepsilon\pi\tau\delta\varsigma$, slender.)

2916. URALEPTUS MARALDI (Risso).

Head 4; depth 6. Head rather thick, its greatest width equal to its height, which is somewhat more than ⅓ its length; cleft of mouth oblique, wide, the maxillary extending to below posterior margin of orbit; lower jaw received within the upper, but both nearly equal in length anteriorly, each armed with a series of rather large, curved, widely set teeth; another series of small teeth within the outer in the upper jaw. Snout rather broad, obtusely rounder, scarcely longer than eye, which is 4½ in head; interorbital space emarginate on each side of upper part of orbit, its width somewhat more than diameter of eye. Nape broad, scarcely elevated, with a spine on each side pointing outwards and covered by skin. Operculum small, with a slender horizontal spine posteriorly, the part below the spine being emarginate; gill membranes united below the throat by a rather narrow cutaneous bridge, not attached to the isthmus; gill openings wide; gills 4, a slit behind the fourth; pseudobranchiæ glandular. Trunk rather low; tail tapering into a very narrow band;

first dorsal fin commencing behind vertical from base of pectoral, somewhat higher than long, and not higher than second; second dorsal commencing immediately behind the first, its rays increasing somewhat in length posteriorly, one of the longest being ⅓ as long as head, the whole fin naked; caudal fin slender, slightly rounded, entirely free from dorsal and anal, and nearly ⅓ as long as head; anal fin commencing at some distance behind the vent, which is situated below the origin of the first dorsal, very similar to the second dorsal; pectoral inserted somewhat below middle of body, its length equaling distance between front margin of eye and end of operculum; ventrals narrow, slender, the outer ray produced into a filament shorter than the pectoral. Scales extending over the whole head, the chin and the thin lips being naked. (Günther.) Tropical Atlantic. This form, originally described from Nice, has since been found at Madeira by Johnson, and at Naples and Catania by Giglioli. The *Blake* obtained a poor specimen, apparently of this.form, at station LXXXI, off the Island of Nevis, in the West Indies. (Goode & Bean.) (Eu.) (A personal name for one of "quelques hommes que les talens, le mérite, la gloire ou l'amitié m'ont désignés.")

Gadus maraldi, RISSO, Ichth. Nice, 123, pl. 6, fig. 13, 1810, Nice.
Merlucius attenuatus, COCCO.
Gadella gracilis, LOWE, Proc. Zool. Soc. Lond. 1843, 91, Madeira. (Type in University of Cambridge.)
Merlucius maraldi, RISSO, Eur. Mérid., III, 220, 1826.
Uraleptus maraldi, GÜNTHER, Cat., IV, 349, 1862; GÜNTHER, Challenger Report, XXII, 87, 1887; GOODE & BEAN, Oceanic Ichthyology, 367, fig. 320, 1896.

986. LOTELLA, Kaup.

Lotella, KAUP, Archiv fur Naturgesch. 1858, 88 (*schlegeli*).

This genus differs from *Physiculus* chiefly in the presence in both jaws of an outer row of large teeth. Deep sea. (Name, a diminutive of *Lota*.)

2917. LOTELLA MAXILLARIS, Bean.

Head about 4⅓; depth 5. D. 5–55; A. 44; V. 10; scales about 7 or 8–115–14 or 15. Snout short; eye 3 in head; maxillary reaching vertical through anterior margin of pupil, its length equaling that of postorbital part of head. Teeth in narrow bands in jaws, the outer series enlarged; vomer and palate apparently without teeth. Vent situated about under eighth ray of second dorsal; distance of first dorsal from tip of snout 4 times in total length including caudal; ventrals extending to about vertical from origin of second dorsal, not reaching nearly to vent; longest ray of first dorsal a little more than ⅓ as long as head; none of the rays of second dorsal or of anal as long as first ray of first dorsal; longest ray of second dorsal not much exceeding ⅓ of height of body; longest ray of anal about ½ length of ventral; origin of anal about under tenth ray of second dorsal; ventrals situated about under beginning of posterior third of head, their length ⅓ that of second dorsal base; origin of pectoral somewhat in advance of that of first dorsal, the fin imperfect, but its length probably slightly exceeding that of ventral; caudal rounded. Color very light brown; the margins of the dorsal and anal, in their posterior portions, blackish. (Goode & Bean.) Gulf stream, Lat. 40° N., in 396 fathoms. (*maxillaris*, pertaining to the upper jaw.)

Lotella maxillaris, Bean, Proc. U. S. Nat. Mus. 1884, 241, Lat. 39° 55' N., Long. 70° 28' W. in 396 fathoms (Type, No. 29832. Coll. *Fish Hawk*); Goode & Bean, Oceanic Ichthy. ology, 368, 1896.

987. PHYSICULUS, Kaup.

Physiculus, Kaup, Archiv für Naturgesch. 1858, 88 (*dalwigkii*).

Body elongate, covered with small scales; head entirely scaly; snout broad, obtusely rounded, projecting beyond the mouth; mouth of moderate size; chin with a barbel; jaws with bands of villiform teeth; vomer and palatines toothless. Dorsals 2; anal fin single, not notched; ventral fin with 5 rays, the outer ray filamentous; caudal rounded, slender, free; branchiostegals 7. (φυκίς, an ancient name of some fish living in the *Fucus*, φῦκος, probably a species of *Gobius*.)

    a. Scales moderate, about 62 in a longitudinal series; gill rakers few; dorsal rays 10–49; anal 54; ventral reaching fourth anal ray.   FULVUS, 2918.
  aa. Scales very small, about 100 in a longitudinal series.
        b. Gill rakers few, about 11 below arch; head 4 in length; depth 5; ventrals filamentous at tip, as long as head, reaching tenth ray of anal.
                                                          NEMATOPUS, 2919.
      bb. Gill rakers undescribed, probably few; head 4 in length; depth 5; ventrals shorter than head, reaching front of anal.   KAUPI, 2920.
     bbb. Gill rakers very numerous, 7+18; head 3½ in length; depth 4¾; ventrals reaching seventh anal ray.   RASTRELLIGER, 2921.

2918. PHYSICULUS FULVUS, Bean.

Head about 4; depth 4⅜. D. 10–49; A. 54; V. 7; scales 6–61 to 62–16. Head broad and depressed; snout short; eye 3½ in head; the length of the upper jaw 2½ in head, about equal to space between ventrals and anal origin; maxillary not quite reaching vertical through hind margin of eye; barbel 6 in head. Teeth in narrow bands in jaw; no outer series of enlarged teeth, but a few in the middle of the bands in each jaw are slightly larger than the others; all of the teeth, however, inconspicuous; vomer and palate smooth. Vent situated about under third ray of first dorsal; distance of first dorsal from tip of snout equaling 3 times length of its base, its longest ray twice length of snout, and slightly exceeding length of longest of second dorsal; length of second dorsal base 3 times length of pectoral, which is nearly 5½ times in total without caudal. Origin of anal about in a vertical let fall from base of fifth ray of first dorsal; distance of ventral from tip of snout about 5¼ times in body. Tip of ventral when extended backward reaching base of fourth anal ray; length of middle caudal ray 3 in head. Lateral line very indistinct, situated rather high, following pretty closely the contour of back. Gill rakers moderately short and not numerous. General color a light yellowish brown; under surface of head, the abdomen, margins of dorsal and anal fins, lips, and axil of pectoral very dark brown; a dark brown blotch on the suboperculum; inside of mouth and gill membranes white. (Bean.)

Caribbean Sea, north to 40° in region of Gulf Stream, reaching a depth of 955 fathoms. (*fulvus*, brownish yellow.)

Physiculus fulvus, BEAN, Proc. U. S. Nat. Mus. 1884, 240, Lat. 40° 1′ N., Long. 69° 56′ W. in 79 fathoms (Type, No. 28766. Coll. *Fish Hawk*); GOODE & BEAN, Oceanic Ichthy-ology, 366, 1896.

2919. PHYSICULUS NEMATOPUS, Gilbert.

Head 4 in length; depth 5. D. 7 to 9–56 to 61; A. 59 to 64; scales 90 to 105. Length of caudal peduncle to base of median caudal rays 5¼ in head. Snout very broadly rounded, its width twice its length, which is 4⅔ in head; eye 3¼; interorbital 4; maxillary 2, reaching slightly beyond vertical from posterior margin of orbit. None of the teeth enlarged; palate smooth. Branchiostegal membranes more narrowly joined than in the *P. rastrelliger*, but wholly free from isthmus. Gill rakers short and slender, 11 movable ones on horizontal limb of arch. Origin of first dorsal over base of pectorals, its distance from tip of snout 3⅔ in length; base of first dorsal equaling snout and ½ eye, its highest ray 2¼ in head; free portion of caudal peduncle ⅓ diameter of orbit; notch of dorsal and anal fins not conspicuous, the posterior dorsal rays little longer than those which precede, 2¼ in head; caudal 2¼; pectorals 1½; ventrals with broad base and 7 rays, the outer 2 filamentous, the second the longest, reaching base of tenth to twelfth anal rays, and as long as head; distance between bases of ventrals equals interorbital width; scales small, regularly imbricated, becoming minute on snout, which they completely invest, as well as mandible and gular membranes; lateral line present on anterior half of body only, 8 scales above it anteriorly. Color light olive brown, sprinkled with dark specks, the sides of head and trunk with silvery luster; snout, mandible, and gular membrane dusky; abdominal area, branchiostegal membranes, base of ventrals, axillary blotch, and front of anal, purplish black; posterior edge of gill membranes and opercular flap white; dorsals dusky, with an inconspicuous darker margin, which becomes more marked posteriorly; anal darker, margined with black; caudal blackish; pectorals and filamentous portion of ventral white. Inside of mouth and gill cavity white; peritoneum silvery, rendered black on sides by clusters of spots. Coast of southern California. Many specimens, the largest 7 inches long, from *Albatross* Stations 2997, 3011, 3015, and 3016, in 71 to 221 fathoms. (Gilbert.) (νῆμα, thread; πούς, foot.)

Physiculus nematopus, GILBERT, Proc. U. S. Nat. Mus. 1890, 114, coast of southern Cali-fornia. (Types, No. 46486 and 46555. Coll. Dr. Gilbert.)

2920. PHYSICULUS KAUPI, Poey.

Head 3½; depth 4. D. 10–60; A. 60; P. 30; V. 8; C. 17; scales 12 to 15—over 100. Body and head short, swollen; tail regularly narrowed; vent below base of pectoral; eye high, equal to snout, 4½ in head; nostrils with valves; snout blunt; lower jaw the shorter; maxillary reaching slightly beyond eye; each jaw with a band of cardiform teeth, none on palate and tongue; maxillary sloping under skin of cheek; barbel a little longer than eye; opercular bones covered with skin, without spines; no spines at nape; gill membranes somewhat united, free from isthmus; gills 4, a slit behind fourth; no pseudobranchiæ; lateral line parallel with the back to beyond middle of body, then turning down suddenly, continuing to base of caudal. Scales small, cycloid, not easily counted; head scaly, even to the lips; vertical fins with small scales; ventral filamentous, equaling $\frac{2}{3}$ length of head, all the rays except the first short; ventral with 2 filamentous rays, which reach to front of anal and are about $1\frac{1}{4}$ in head; first dorsal as high as long, beginning behind base of pectoral, its longest rays about $2\frac{1}{4}$ in head; soft dorsal and anal low, free from the small rounded caudal; pectoral falcate, $1\frac{1}{4}$ in head. Color yellowish brown, bluish on belly; second dorsal and anal edged with darker brown. Type, 1 specimen, 250 mm. long. (Poey.) Deep waters of the Atlantic.

This species has constantly (as far as is shown in our specimens) a broader base to the ventral fins than *Physiculus dalwigkii*, and they are formed of 7 rays, of which the largest may or may not reach the anal fin; the fin rays vary within proportionate limits; they are, D. 9 or 10—60 to 66; A. 60 to 70; there are 13 scales between the anterior dorsal and lateral line; the caudal peduncle is shorter and less slender than in the Madieran form, but otherwise the species are so similar as to scarcely deserve specific separation. Poey obtained a specimen at Cuba, and Melliss 2 at St. Helena. These differ in no respect from 5 examples, 11 to 16 inches in length, found by the *Challenger* off Inosima in 345 fathoms. (Günther, Challenger Report, XXII, 88, pl. XVII, fig. A, 1887.) (Named for Dr. J. J. Kaup, author of a work on the Apodal fishes.)

Physiculus kaupi, POEY, Repertorio, I, 186, 1865, Matanzas. (Coll. Don Cirilo Dulzaides.)
? *Physiculus japonicus*, HILGENDORF, Sitz. Naturf. Freunde, Berlin, 1879, 80, Japan.

2921. PHYSICULUS RASTRELLIGER, Gilbert.

Head $3\frac{1}{4}$ in length; depth $4\frac{3}{4}$. D. 8 or 9-53 to 61; A. 57; scales 100 to 110. Length of caudal peduncle to base of median caudal rays, $2\frac{1}{4}$ in head. Snout short and broadly rounded, $4\frac{1}{4}$ in head; eye $3\frac{3}{4}$; interorbital width $4\frac{1}{8}$ to $4\frac{3}{5}$; maxillary $2\frac{1}{4}$, extending to vertical from posterior margin of pupil. Teeth in rather broad bands, none of them enlarged; width of patch on premaxillaries $\frac{1}{2}$ pupil; vomer and palatines toothless. Branchiostegal membranes broadly united, joined to the isthmus anteriorly, the width of the free fold more than $\frac{1}{2}$ pupil. Gill rakers numerous, slender, moderately long, the longest $\frac{1}{4}$ diameter of orbit, about 7 above angle, 17 to 19 below, the anterior ones short but movable. Origin of first dorsal slightly in advance of base of pectorals, its distance from tip of snout $3\frac{3}{4}$ in length; base of first dorsal equaling length of snout, its longest ray $2\frac{2}{5}$ in head; free portion of caudal peduncle equaling diameter of eye; second dorsal notched, the median rays $\frac{3}{4}$ the height of the highest anterior rays, the posterior highest, equaling first dorsal and longest caudal rays; anal similar to soft dorsal, but lower; ventrals under middle of opercle, the distance between their bases little less than interorbital width, equaling distance from vent to anal fin; ventrals with 7 rays, the outer 2 produced, the second the longest, reaching base of seventh or eighth anal ray; pectorals with broad base, covered with lax membrane, containing 26 to 28 rays, their length 1¼ to 1⅔ in head. Scales small, comparatively little reduced on top of head, a broad ring encircling snout in front of eyes naked, a very narrow patch of scales between this laterally and premaxillaries; scales in 100 to 110 transverse rows, 8 or 9 between lateral line and front of dorsal; lateral line wanting on posterior part of body, in the latter part of its course present on occasional scales only. Color uniform grayish olive on sides, each scale, or at least its marginal ½, closely covered with minute dark specks; gular and branchiostegal membranes, ventral region, and axil of pectorals blue black; basal portion of vertical fins light bluish, margined with blackish; pectorals dusky; ventrals blue black at base, the distal portion white; lining membrane of mouth and gill cavity white; peritoneum silvery, but in places so filled with black specks as to appear black. Coast of southern California. Many specimens, the longest 8 inches, from *Albatross* Stations 3045 and 2987, in 184 and 171 fathoms. (Gilbert.) (*rastrelliger*, bearing small gill rakers; *rastrum*, rake; *gero*, I bear.)

Physiculus rastrelliger, GILBERT, Proc. U. S. Nat. Mus. 1890, 113, coast of southern California. (Type No. 48266. Coll. *Albatross*.)

988. LOTA (Cuvier) Oken.

(BURBOTS.)

Les Lottes, CUVIER, Règne Anim., Ed. I, vol. 2, 215, 1817 (*lota*).
Lota, OKEN, Isis 1817, 1182 (*lota*).

Body long and low, compressed behind. Head small, depressed, rather broad; anterior nostrils each with a small barbel; chin with a long barbel; snout and lower parts of head naked; mouth moderate, the lower jaw included; each jaw with broad bands of equal, villiform teeth; vomer with a broad, crescent-shaped band of similar teeth; no teeth on palatines. Gill openings wide, the membrane somewhat connected, free from the isthmus. Scales very small, embedded; vertical fins scaly. Dorsal fins 2, the first short, the second long, similar to the anal; caudal rounded, its outer rays procurrent; ventrals of several rays. One or 2 species, living in fresh waters of northern regions. (*Lota*, the ancient name used by Rondelet, in French, *la Lotte*.)

2922. LOTA MACULOSA (Le Sueur).

(BURBOT; LAKE LAWYER; LING.)

Head 4½ in body; depth 5½. D. 13–76; A. 68; ventral 7; eye 7 in head; pectoral 1⅔ in head; maxillary 2⅔; middle caudal rays 2⅓. Body elongate, not much compressed anteriorly; head slightly depressed; mouth large, the maxillary reaching to posterior margin of eye; teeth villiform, in bands on jaws and vomer; barbel longer than the small eye; interorbital broad, nearly twice diameter of eye; gill rakers very short, about 3 + 6 in number; anterior nostrils with barbels; body covered with small embedded scales; pectorals scarcely reaching to below front of dorsal; ending of ventrals filamentous; caudal rounded; vertebræ 21 + 38 = 59; cæca 30. Dark olive, thickly marbled and reticulated with blackish;

yellowish or dusky beneath; young often sharply marked, the adult becoming dull grayish; vertical fins with dusky margins. Length 2 feet. Lakes and sluggish streams. New England and Great Lake region, north to the Arctic seas and west to the headwaters of the Missouri, the Frazer River basin, and Bering Straits; abundant northward; rare in the Ohio River and the Upper Mississippi; a rather coarse and tasteless fish, seldom used as food. Here described from a specimen, 18 inches long, from Lake Michigan at Michigan City, Indiana. The American Burbot is very close to the common species of northern Europe and Asia, *Lota lota* (Linnæus) $=$ *Lota vulgaris*, Cuvier $=$ *Lota communis*, Rapp, and may prove wholly identical with the latter. In *Lota lota* the pectorals reach beyond front of dorsal, being $1\frac{1}{5}$ in head. (*maculosus*, spotted.)

Gadus maculosus, LE SUEUR, Jour. Ac. Nat. Sci. Phila., I, 1817, 83, Lake Erie.
Molva maculosa, LE SUEUR, Mém. Mus., V, 1819, pl. 16.
Lota maculosa, DE KAY, New York Fauna: Fishes, 284, pl. 52, fig. 168, 1842.
Gadus compressus, LE SUEUR, Jour. Ac. Nat. Sci. Phila., I, 1817, 84, Connecticut River.
Lota compressa, DE KAY, New York, Fauna: Fishes, 285, pl. 78, figs. 244, 245, 1842.
Gadus lacustris, MITCHILL, Amer. Monthly Mag., II, 1818, 244, Sebago Pond, Maine (Coll. Henry A. S. Dearborn).
Molva huntia, LE SUEUR, Mém. Mus., V, 1819, 161, Connecticut River.
Lota inornata, DE KAY, New York Fauna: Fishes, 283, pl. 45, fig. 145, 1842, Hudson River, Lansingburgh, N. Y.
Lota brosmiana, STORER, Boston Journ. Nat. Hist., IV, 1839, pl. 5, fig. 1, New Hampshire.

989. MOLVA, Fleming.

(LINGS.)

Molva, FLEMING, British Animals, 192, 1828 (*vulgaris*).
Molva, NILSSON, Skandinav. Fauna, IV, 573, 1832 (*molva*).

Body elongate, covered with very small scales. Chin with a barbel; lower jaw included; bands of teeth on jaws and vomer; lower jaw with large canines which are arrow-shaped and movable; vomer with a curved series of canines mixed with small teeth, these mostly fixed; no teeth on palatines. Gill membranes broadly united. Two dorsal fins, both well developed; 1 anal fin; ventrals with several rays. Northern seas. (An old name of the salt-water ling.)

2923. MOLVA MOLVA (Linnæus).

Head 5; depth 7 or 8. D. 13 to 16–63 to 70; A. 57 to 66 (vertebræ $27+37=64$). Upper jaw the longer, the maxillary reaching to below middle of orbit. Teeth cardiform in the jaws, with an inner row of rather widely separated and larger ones on mandible; a semicircular band on vomer, among which a few larger ones are interspersed. First dorsal inserted over the latter half of pectoral, its greatest height $\frac{2}{3}$ that of body below it; pectoral about $\frac{1}{4}$ as long as head; anal insertion in vertical over seventh or eighth ray of second dorsal. Barbel longer than eye, which is about equal to width of interorbital space. Scales small, covering head and fins. Color black gray, lighter on the sides and beneath; vertical fins edged with white; a dark blotch at the posterior end of the first dorsal, and a more distinct one on the end of the second dorsal. Arctic parts of the

Atlantic, south in deep water. This fish, the "ling" of Europe, is found from Spitzbergen to the Gulf of Gascony, where specimens have been taken very exceptionally at Arcachon and San Juan de Luz. It is very rare, however, south of the British Channel, and most abundant along the coast of northern Europe, especially in the German Ocean and off Norway. It is rare about Iceland, Greenland, and the Faroe Islands, and has never been found in the Baltic. It is said to have been found in the deep water off Newfoundland, but we have been unable to find the specific record. Collett states that on the Norwegian coast young examples rarely occur in less depth than 100 fathoms, and according to Lilljeborg the largest are caught in from 80 to 150 fathoms. (Goode & Bean.) (*molva*, an ancient name.)

Gadus molva, LINNÆUS, Syst. Nat., Ed. x, 254, 1858, seas of Europe; after *Gadus dorso dipterygia*, ARTEDI.

Molva vulgaris, FLEMING, British Animals, 192, 1828; GÜNTHER, Cat., IV, 361, 1862; GOODE & BEAN, Oceanic Ichthyology, 364, fig. 317, 1896.

Gadus raptor, NILSSON, Prodromus, 46, Sweden.

Molva linnœi, MALM, Götheborgs och Bohusläns Fauna, 491, 1877.

990. UROPHYCIS, Gill.

(CODLINGS.)

Phycis, BLOCH & SCHNEIDER, Syst. Ichth., 56, 1801 (*tinca = blennioides*); not *Phycis*, Fabricius, 1798, a genus of Lepidoptera.

Phycis, RAFINESQUE, Amer. Monthly Mag. 1818, 243 (*marginata*).

Urophycis, GILL, Proc. Ac. Nat. Sci. Phila. 1863, 240 (*regius*).

Emphycus, JORDAN & EVERMANN, new subgenus (*tenuis*).

Body rather elongate; head subconic; mouth rather large, the maxillary reaching to below eye; lower jaw included; chin with a small barbel; jaws and vomer with broad bands of subequal, pointed teeth; palatines toothless. Dorsal fins 2, the first sometimes produced at tip; second dorsal long, similar to the anal; ventrals wide apart, filamentous, each of 3 slender rays, closely jointed, appearing like one befid filament. Gill membranes somewhat connected, narrowly joined to the isthmus. (οὐρά, tail; *Phycis*.)

UROPHYCIS:
   a. First dorsal fin not elevated, none of its rays filamentous.
      b. Scales moderate, 90 to 95 in a longitudinal series.
         c. Dorsal rays 8–43; anal 45; sides with some pale spots.   REGIUS, 2924.
         cc. Dorsal rays 10–66; anal rays 57; barbel minute.   CIRRATUS, 2925.
      bb. Scales small, 120 to 155 in a longitudinal series.
         d. Dorsal rays 13–57; anal about 50; scales 120; sides with some pale spots.
                                      FLORIDANUS, 2926.
         dd. Dorsal rays 10–62; anal about 53; scales 155.   EARLLI, 2927.
EMPHYCUS (ἐν-φύκος, in the seaweed):
   aa. First dorsal fin elevated, 1 or more of its rays filamentous.
      e. Scales about 140; dorsal rays 9–57; anal 48; ventrals reaching vent.
                                        TENIUS, 2928.
      ee. Scales about 110; dorsal rays 9–57; anal 50; ventrals reaching beyond vent, not longer than head.   CHUSS, 2929.
      eee. Scales about 90; dorsal rays 9–56; anal 56; second dorsal filamentous; ventrals very long, nearly 3 times length of head.   CHESTERI, 2930.

Subgenus UROPHYCIS.

2924. UROPHYCIS REGIUS (Walbaum).

Head 4⅓ in body; depth 5. D. 8–43; A. 40; scales about 90; eye 4¼ in head; maxillary 2; pectoral 1⅔; caudal 1⅔. Body rather elongate, compressed; mouth large, the maxillary reaching slightly past posterior margin of eye; lower jaw included; cardiform teeth on jaws and vomer; interorbital flattish, about equal in width to the diameter of eye; gill rakers short, 3 + 12 in number; origin of dorsal over base of pectorals; pectorals slender, barely reaching to front of anal; ventrals filamentous, composed of 2 rays each with the inner ray the larger, inserted in front of base of pectoral in distance equal to 1⅓ diameter of eye, their ends reaching beyond front of anal; front of anal nearer snout than base of caudal, by nearly a head's length; caudal subtruncate. Pale brownish tinged with yellowish, the lateral line dark brown, interrupted by white spots; inside of mouth white; first dorsal largely black, this color surrounded by white; second dorsal olivaceous, with irregular round dark spots; caudal, anal, and pectorals dusky; ventrals and lower edge of pectorals white; 2 vertical series of round dark spots on the sides of the head. North Atlantic, south to Cape Fear; ranging from shallow water to a depth of 167 fathoms. Here described from a specimen, 8 inches in length, from Charleston, South Carolina. The species is said to exhibit electric powers in life. (*regius*, royal.)

Blennius, sp., Schöpf, Schrift. Naturf. Freunde, Berlin, VIII, 1780, 142, New York.
Blennius regius, Walbaum, Artedi, Pisc, III, 186, 1792; after Schöpf.
Enchelyopus regalis, Bloch & Schneider, Syst. Ichth., 53, 1801, after Schöpf.
Gadus punctatus, Mitchill, Trans. Lit. and Phil. Soc. N. Y., I, 1815, 372, New York.
Urophycis regius, Gill, Proc. Ac. Nat. Sci. Phila. 1863, 240.
Phycis regius, Goode & Bean, Oceanic Ichthyology, 337, 1896.
Phycis regalis, Günther, Cat., IV, 355, 1862.
Phycis punctatus, De Kay, N. Y. Fauna: Fishes, 292, 1842.

2925. UROPHYCIS CIRRATUS, Goode & Bean.

Head 4; depth 5. D. 10–66; A. 57; scales 6–93–20. Body moderately stout; eye large, about 4 in head; interorbital space 2 in eye. Maxillary not reaching posterior margin of orbit in large specimens, but in smaller ones it extends fully to that vertical; mandible extending far beyond posterior margin of eye, its length about equal to postorbital part of head; barbel minute in all examples examined, its length usually about ⅓ that of eye. Teeth in villiform bands in both jaws, the intermaxillary bands being wider than those of mandible; vomerines in a narrow villiform band. Gill rakers 2 + 12, the largest club-shaped at end, the longest 4 in eye. Gill membranes attached to isthmus, but with a narrow, free posterior border. Length of pectorals about ⅓ distance from ventral to anal origin, reaching to about the twenty-sixth row of scales; ventral reaching in some specimens slightly beyond origin of anal; in 1 individual almost to middle of anal fin; none of the dorsal rays filamentous, the longest from 2⅓ to 3 times in head; base of first dorsal about equal to length of eye in most specimens; in smaller examples somewhat greater, about 3 in head;

3030——83

vent under sixteenth ray of second dorsal. Color light brown; lower
parts minutely dotted; dorsals with narrow dark margins; caudal with
a broad dark margin; anal with a narrow dark margin in its posterior
third; roof of mouth and interior of gill cavity dark brown. Deep water
of the Gulf of Mexico. (Goode & Bean.) (*cirratus*, bearing cirri.)

Phycis cirratus, GOODE & BEAN, Oceanic Ichthyology, 358, 1896, Gulf of Mexico at Lat.
    29° 03′ 15″ N., Long. 88° 16′ W. (Type, No. 39059. Coll. *Albatross*.)

2926. UROPHYCIS FLORIDANUS, Bean & Dresel.

Head 4 in body; depth 6. D. 13–57; A. 49; scales about 120; eye 6 in
head; maxillary 2; height of first dorsal 2½; middle caudal rays 2. Body
rather elongate, compressed, head subconic; mouth large, the maxillary
reaching to below posterior margin of orbit; upper jaw and snout somewhat projecting beyond lower; small cardiform teeth, in narrow bands
on jaws and vomer; barbel very slender, small; interorbital space wide,
slightly convex, nearly twice as wide as eye; gill rakers small, slender,
2 + 11 in number. Origin of dorsal a little behind the vertical from base
of pectoral; first dorsal high, slightly falcate; second dorsal a little
higher in its anterior end, higher than anal; origin of anal about midway
between tip of snout and base of caudal; pectoral slender, reaching an
eye's diameter beyond front of second dorsal; ventrals inserted twice
diameter of eye in front of pectorals, 2-rayed, the inner ray the longer,
not reaching to vent in larger examples, reaching to front of anal in small
ones; caudal long and rounded. Color in spirits, reddish brown, light
below, a small black spot above eye, a vertical series of 3 or 4 behind eye,
and 2 on opercle, these spots less than ½ pupil, distinct and clear cut; a
dark streak from preorbital across cheek to edge of opercle, lateral line
black, interrupted at short intervals by white spots; fins dusky, with the
exception of pectorals and ventrals, dark toward the ends of the rays.
Gulf of Mexico, in rather shallow water, coming to shore in abundance
about Pensacola in cold weather. Here described from a specimen, 7¼
inches in length, from Pensacola, Florida. (*floridanus*, from Florida.)

Phycis floridanus, BEAN & DRESEL, Proc. Biol. Soc. Wash. 1884, 100, Pensacola, Florida
(Coll. Silas Stearns); JORDAN, Cat. Fish. N. A., 129, 1885.

2927. UROPHYCIS EARLLI, Bean.

Head 3½ in body; depth 5. D. 10–60; A. 53; scales 155; eye 6 in head;
maxillary 2. Body moderately elongate, not much compressed anteriorly;
mouth large, the maxillary reaching to below or very slightly past posterior margin of eye; snout and upper jaw projecting beyond lower jaw;
teeth strong, cardiform in a narrow band on vomer and lower jaw, in a
rather wide band in upper; interorbital wide, convex, about 1½ times eye;
gill rakers short and blunt, about 2+9. Origin of dorsal slightly behind
the vertical from base of pectoral; origin of anal about midway between
snout and base of caudal; ventrals 2-rayed, the inner a little the longer,
not reaching to vent. Color brown, with some light spots on the second
dorsal fin and on the sides; anal and both dorsals margined with brown.

Atlantic coast of United States, southward in water of moderate depth; not common. Here described from a specimen, 17 inches in length, from Charleston, South Carolina. (Named for R. Edward Earll, then assistant to the United States Fish Commission.)

Phycis earlli, BEAN, Proc. U. S. Nat. Mus., III, 1880, 69, Charleston, S. C. (Coll. R. E. Earll. Type, Nos. 25207, 25208, and 25209); JORDAN & GILBERT, Synopsis, 798, 1883.

Subgenus EMPHYCUS, Jordan & Evermann.

2928. UROPHYCIS TENUIS (Mitchill).

(CODLING; WHITE HAKE; SQUIRREL-HAKE.)

Head 4¼; depth 5⅓. D. 9-57; A. 48; scales 138. Snout longer than eye, narrower and more pointed than in *P. chuss*. Eye large, usually wider than interorbital space; maxillary reaching beyond pupil. Filamentous dorsal ray about ⅔ length of head; ventral fins about reaching vent. Scales very small. Brownish, lighter and yellowish below; fins very dark. Banks of Newfoundland to Cape Hatteras; abundant northward in rather deep water, reaching a depth of 304 fathoms. The species resembles *Phycis chuss*, differing chiefly in the smaller scales. (*tenuis*, slender.)

Gadus tenuis, MITCHILL, Trans. Lit. and Phil. Soc. N. Y. 1815, 372, New York.
Phycis dekayi, KAUP, Archiv Natur. 1858, 89, North America.
?*Phycis rostratus*, GÜNTHER, Cat. Fish. Brit. Mus., IV, 353, 1862, no locality; D. 9-59 to 62; A. 49 to 50; scales ca. 150; ventrals immaculate, reaching front of anal.
Phycis tenuis, DE KAY, N. Y. Fauna: Fishes, 293, 1842; GILL, Proc. Ac. Nat. Sci. Phila, 1863, 238; JORDAN & GILBERT, Synopsis, 799, 1883; GOODE & BEAN, Oceanic Ichthyology, 359, fig. 312, 1896.

2929. UROPHYCIS CHUSS (Walbaum).

(CODLING; SQUIRREL-HAKE.)

Head 4½; depth 5. D. 9-57; A. 50; scales 110. Body rather slender; head depressed; eye large, about equal to interorbital width; maxillary reaching posterior margin of pupil; filamentous dorsal ray about ⅔ length of body, when perfect; pectorals ⅔ length of head; ventral fins extending beyond the vent; scales comparatively large. Brownish above, sides lighter and tinged with yellowish; thickly punctulate with darker; below pale; inside of mouth white; vertical fins somewhat dusky; anal fin margined with pale; lateral line not dark. Atlantic coast, from Gulf of St. Lawrence to Virginia; common northward; reaching a depth of 300 fathoms. (*chuss*, a vernacular name now obsolete, apparently derived from cusk.)

Chuss, SCHÖPF, Schrift. Naturf. Freunde, Berlin, VIII, 1780, 143, New York.
Blennius chuss, WALBAUM, Artedi Pisc., 186, 1792; after SCHÖPF.
Enchelyopus americanus, BLOCH & SCHNEIDER, Syst. Ichth., 53, 1801; after SCHÖPF.
Gadus longipes, MITCHILL, Trans. Lit. and Phil. Soc. N. Y., 1, 372, pl. 1, fig. 4, 1815, New York.
Phycis marginatus, RAFINESQUE, Amer. Month. Mag., Jan., 1818, 205, Point Judith, Rhode Island. D. 10-60; A. 40; ventral reaching anal; tail black-edged.
Phycis americanus, STORER, Rept. Fish. Mass., 138, 1839; GÜNTHER, Cat., IV, 353, 1862.
Phycis chuss, GILL, Proc. Ac. Nat. Sci. Phila. 1863, 237; JORDAN & GILBERT, Synopsis, 799, 1883; GOODE & BEAN, Oceanic Ichthyology, 359, fig. 311, 1896.

2930. UROPHYCIS CHESTERI, Goode & Bean.

Head 4⅓; depth 5; orbit 3⅓ in head; maxillary 2; barbel about 3 in orbit. D. 9 or 10-55 to 57; A. 56; C. 5, 18 to 21, 5; P. 17 or 18; V. 3; scales 7-90 or 91-28.   Vent situated under the twelfth ray of second dorsal, and equidistant from tip of snout and end of second dorsal; distance of dorsal fin from snout equal to twice length of mandible; third ray of first dorsal extremely elongate, extending to a point (thirty-third ray of second dorsal) ⅔ of distance from snout to tip of caudal, its length more than twice that of head, and more than 4 times as long as the rays immediately preceding and following it; anal fin inserted immediately behind vent, its distance from root of ventrals equal to that of dorsal from snout; as in other species of the genus, ventral of 3 rays, the first 2 much prolonged, the first contained 3 times in length of body, the second almost 3 times as long as head, reaching to fortieth anal ray or ¾ of distance from snout to tip of caudal, the third shorter than diameter of orbit; pectoral 4 times as long as operculum.   Scales large and thin, easily wrinkling with the folding of the thick, loose skin, particularly in the median line of sides of body. Lateral line much broken on posterior half of body.   (Goode & Bean.) Atlantic coast of United States, in 100 to 500 fathoms, with *Macrourus bairdi*, the most abundant fish on the continental slope, swarming everywhere below the 100-fathom line.   (Named for Capt. Hubbard C. Chester.)

Phycis chesteri. GOODE & BEAN, Proc. U. S. Nat. Mus. 1878, 256, off Cape Ann, in 140 fathoms (Coll. Captain Chester); JORDAN & GILBERT, Synopsis, 800, 1883; GOODE & BEAN, Oceanic Ichthyology, 360, fig. 313, 1896.

991. LÆMONEMA, Günther.

Læmonema, GÜNTHER, Cat. Fish. Brit. Mus., IV, 356, 1862 (*yarrellii*).

Body of moderate length, covered with small scales; fins naked.   A separate caudal; 2 dorsal fins and 1 anal, the anterior dorsal composed of 5 rays; ventrals reduced to a single long ray, bifid at its end.   Bands of villiform teeth in the jaws; a small group of vomerine teeth; none on the palatine bones.   Chin with a barbel.   Branchiostegals 7.   Deep sea.   (Λαιμός, throat; νῆμα, thread.)

  a. Scales 13-140-31; barbel ½ eye; dorsal and anal with narrow black edgings.

  BARBATULUM, 2931.

  aa. Scales 16-160-38; barbel ⅔ eye; a large, triangular, black blotch on tail and adjacent parts of vertical fins.   MELANURUM, 2932.

2931. LÆMONEMA BARBATULUM, Goode & Bean.

Head 4⅔; depth 4½; orbit 3 in head; upper jaw more than 2; barbel about 2 in eye.   D. 5-63; A. 59; P. 19; V. 2; scales 13-140-31.   Vent situated under sixth or seventh ray of second dorsal.   Distance of first dorsal from snout 4 in body; base of first dorsal ⅓ as long as middle caudal rays, that of second slightly more than 3 times length of head; first dorsal composed of 5 rays, the first of which is elongate, 3 times as long as middle caudal rays, extending to base of twenty-fourth ray of second dorsal; anal fin inserted at a distance from tip of snout equal to twice length of head, its distance from insertion of ventrals being equal to length of head; length of ventrals equal to that of pectorals, their tip not extending to vent. Scales small, very thin, deciduous, crowded anteriorly; lateral line not well defined on posterior part of body. Color similar to that of the various species of *Phycis;* the dorsal and anal fins with narrow black margins. The length of the first dorsal ray is very variable, being shorter in younger individuals. This species differs from *L. yarrellii* by its much smaller scales, and from *L. robustum* by the greater number of rays in the dorsal and anal fins, and its much shorter ventrals. (Goode & Bean.) Gulf Stream, reaching a depth of 312 fathoms. (*barbatulus,* having small barbels.)

Læmonema barbatula, GOODE & BEAN, Bull. Mus. Com. Zool., x, 204, 1883, Gulf Stream, Lat. 32° 43′ N., Long. 77° 20′ W., in 230 fathoms, and Lat. 28° 35′ N., Long. 73° 13′ W.; GOODE & BEAN, Oceanic Ichthyology, 362, figs. 315 and 315A, 1896.

2932. LEMONEMA MELANURUM, Goode & Bean.

Head about 4⅔; depth 4⅔; eye 3 in head; snout 4; interorbital width 6. D. 6–57; A. 55; P. 25; V. 2; Br. 7; scales 16–160–38. Maxillary extending to below middle of eye; intermaxillary nearly ½ length of head; mandible slightly more than twice length of snout. Teeth in intermaxillary and mandible in villiform bands; vomerine teeth in a small circular patch on middle of head of bone. Barbel about as long as snout. Distance of first dorsal from tip of snout about 4 in snout; length of first ray of dorsal equaling that of head without snout; last ray of dorsal scarcely more than ¼ as long as first; ventral consisting of a single bifid ray, its distance from tip of snout equal to length of head, its length nearly equal to that of dorsal or the pectoral when extended, not reaching vent by a distance equal to length of snout; pectoral equaling that of longest dorsal ray, and also equaling head without snout; second dorsal higher anteriorly, and posteriorly much higher than in middle; longest anterior ray ½ length of ventral; longest posterior ray ½ length of head. Vent under eighth ray of second dorsal. Gill rakers 5 + 15, the longest ¼ as long as snout. Color very light brown, the dorsals and anal with a narrow dark margin; a conspicuous, large, triangular, dark blotch on last rays of dorsal and anal, and a dark blotch occupying almost the whole of caudal, leaving a margin of whitish around it. (Goode & Bean.) Caribbean Sea, north to New York; reaching a depth of. 1,467 fathoms. (μέλας, black; οὐρά, tail.)

Læmonema melanurum, GOODE & BEAN, Oceanic Ichthyology, 363, fig. 316, 1896, Gulf Stream, Lat. 30° 44′ N., Long. 79° 26′ W., in 440 fathoms. (Type, No. 38270. Coll. *Albatross.*)

992. GAIDROPSARUS, Rafinesque.

(THREE-BEARDED ROCKLINGS.)

Gaidropsarus, RAFINESQUE, Indice d'Ittiol. Siciliana, 1810 (*mustellaris* = *mediterraneus*); description from a rough figure of RONDELET.

Les Mustèles, CUVIER, Règne Anim., Ed. I, vol. 2, 215, 1817 (*tricirrhatus* = *mediterraneus*).

?*Mustela,* OKEN, Isis. 1817 (for *les Mustèles;* not *Mustela,* a genus of mammals).

Onos, RISSO, Hist. Eur. Mérid., III, 214, 1826 *(mustella = mediterraneus).*
Mustela, STARK, Elem. Nat. Hist., I, 425, 1828 (after *les Mustèles*).
Motella, CUVIER, Règne Anim. Ed. 2, vol. II, 334, 1829 *(vulgaris = tricirratus).*
Onus, GÜNTHER, corrected spelling.

Body rather elongate, covered with minute scales; head not compressed, the upper jaw the longer; snout with 2 conspicuous barbels, the chin with 1; teeth on jaws and vomer in bands, palatines toothless; dorsals 2, the anterior of a single long ray followed by a series of short fringe-like rays concealed in a groove; second dorsal and anal long, similar to each other; caudal rounded or lanceolate; ventral rays 5 to 7. Small fishes of the northern seas, descending to deep water. We here regard the 5-bearded Rocklings (*Ciliata*, Couch, 1832) = *Couchia*, Thompson, 1856 = *Molvella*, Kaup, 1858, as a distinct genus, distinguished by the 5 barbels at the tip of the snout. (γαϊδραψάρα, a modern Greek name used by Rondelet for a species of this group.)

The name γαϊδροψάρον is now applied in Athens to the Pollack-like fish, *Micromesistius poutassou* (Risso).

According to Prof. Horace A. Hoffman "the name γαϊδουροψάρον is modern, meaning donkey fish. Γαϊδουρος = γάδαρος = ass, donkey. The ancients called a certain fish ὄνος, ass. Dorio, in Athenæus, VII, 99, says some persons call the ὄνος (i. e., the fish ὄνος) γάδος. Epicharmus, in his Marriage of Hebe, says: 'Wide-gaping χάνναι and monstrous-bellied ὄνοι.' (See Aristotle 599b 33, 601a 1, 620b 29, frag. 307, 1530a.) According to Aristotle the ὄνος has a mouth opening wide (literally, breaking back), like the γαλεοί. It leads a solitary life, is the only fish which has its heart in its belly, has stones in its brain like millstones in form, and is the only fish which lies torpid in the warmest days under the reign of the dog star, Sirius, the other fishes going into this torpid state in the wintriest days. The ὄνος, βάτος, ψῆττα, and ῥίνη bury themselves in the sand, and after they make themselves invisible they wave the things in their mouths which fishermen call little rods or little wands (ῥαβδία). (Hoffman & Jordan, Fishes of Athens, Proc. Ac. Nat. Sci. Phila. 1887, 146.)

 a. First ray of first dorsal long, as long as head; head small, 5¾ in length; teeth rather feeble, uniform. D. 59; A. 45; P. 25. Color uniform brick red.

                        ENSIS, 2933.

 aa. First ray of first dorsal short, about as long as snout.

  b. Pectoral rays 22 to 24; upper jaw without cirri or rudimentary barbels along the premaxillary; maxillary reaching posterior border of eye; head 5½ in length. D. 56; A. 45. Color reddish.   ARGENTATUS, 2934.

  bb. Pectoral rays 16; upper jaw with short cirri or barbels along the premaxillary; maxillary reaching far beyond eye. D. 50; A. 42. Head 4 in length. Color brownish.   SEPTENTRIONALIS, 2935.

2933. GAIDROPSARUS ENSIS (Reinhardt).

Head 5¾; depth 4¼. D. 59; A. 44 to 46; P. 22 to 27; V. 8. Body unusually deep, being greatest at the vent; head small; eye rather large, nearly as long as snout, equaling interorbital area, and in anterior half of head; posterior margin of orbit nearly equidistant between tip of snout and posterior margin of operculum. Mouth normal; supramaxillary ending under posterior margin of pupil. Teeth in a narrow band in each jaw, some of those at least in outer row of upper jaw slightly enlarged and brownish colored; teeth of vomer forming a short curved band in 2 rows. Nasal barbel about equaling diameter of eye. Chin barbel small and not much exceeding ⅓ diameter of eye. Foremost ray of first dorsal springing from back above opercular margin; second dorsal fin low in front, but rising rapidly to seventh or eighth ray, behind which it is nearly uniform for a long distance, and the highest at posterior portion; anal fin much lower than second dorsal; caudal slightly emarginate, almost truncate behind, its median rays about ⅔ as long as head; pectorals nearly ¾ as long as head, produced toward the upper angles, the third ray being longest; ventral fins with their bases mostly in advance of pectorals, the longest ray filamentous and nearly equaling pectoral. Lateral line obsolescent. (Goode & Bean.) Atlantic coast of North America, from Greenland to Cape Hatteras; in deep waters, reaching a depth in the Gulf Stream of 1,081 fathoms. (*ensis*, sword.)

Motella ensis, REINHARDT, Dansk. Vidensk. Selsk. Afhandl., VII, 15, 1838, Greenland.
Onos rufus, GILL, Proc. U. S. Nat. Mus. 1883, 259, Gulf Stream; GILL, Proc. Ac. Nat. Sci. Phila. 1884, 172; JORDAN, Cat. Fish. N. A., 128, 1885.
Onos ensis, GILL, Proc. Ac. Nat. Sci. Phila. 1863, 241; GILL, Cat. Fish. E. Coast U. S., 18, 1873; JORDAN & GILBERT, Synopsis, 797, 1883; JORDAN, Cat. Fish. N. A., 128, 1885; GOODE & BEAN, Oceanic Ichthyology, 381, fig. 327, 1896.

2934. GAIDROPSARUS ARGENTATUS (Reinhardt).

Head 5; depth 5¾. D. 54 to 59; A. 45 or 46. Body elongate; head small; teeth in several rows, 1 row more enlarged than the others; maxillary reaching posterior border of eye; eye large, 5⅓ in head; interorbital space scarcely exceeding the eye. Lateral line with about 27 enlarged pores along its entire length. First ray of first dorsal short, little longer than snout; vent near middle of length. Reddish gray, changing to bluish on the head and abdomen; tips of dorsal, anal, and caudal red, also the barbels and first ray of first dorsal; cavity of mouth pale. Coasts of Greenland (Collett), south to Faroë and Bear Islands; not seen by us. There can be no doubt that *Motella argentata* is the young of the species later called *Motella reinhardti*. (Eu.) (*argentatus*, silvered.)

Motella argentata, REINHARDT, Dansk. Vidensk. Selsk. Afh., VII, 128, 1838; Greenland; young.
Motella reinhardti, KRÖYER MS., 1852; COLLETT, Forh. Vid. Selsk. Chr., No. 14, 83, 1878, Greenland.
Couchia argentata, GÜNTHER, Cat., IV, 365, 1862.
Ciliata argentata, GILL, Proc. Ac. Nat. Sci. Phila. 1863, 241; GILL, Cat. Fish. E. Coast U. S., 18, 1873.
Onos reinhardti, GILL, Proc. Ac. Nat. Sci. Phila. 1863, 241; GILL, Cat. Fish. E. Coast U. S., 18, 1873; COLLETT, Norske Nord-Havs Exp., 131, 1880; JORDAN & GILBERT, Synopsis, 797, 1883; JORDAN, Cat. Fish. N. A., 128, 1885; GOODE & BEAN, Oceanic Ichthyology, 383, 1896.

2935. GAIDROPSARUS SEPTENTRIONALIS (Collett).

Head 4; depth 5⅔. D. 50; A. 42; P. 16. Three barbels, 2 at the nostrils, 1 at the chin, besides a row of about 8 shorter rudimentary barbels along the edge of the upper lip; eye small, ⅓ length of snout; cleft of mouth extending far beyond eye, its length nearly equal to that of postorbital part of head; teeth rather small, unequal; outer teeth of upper jaw and some of the inner teeth of lower enlarged; first ray of first dorsal short, about as long as snout; vent midway between tip of snout and last anal ray; lateral line with about 20 large pores. Grayish brown, paler below; cavity of mouth white. Coast of Norway; 1 specimen known from Greenland. (Collett.) (Eu.) (*septentrionalis*, northern.)

Motella septentrionalis, COLLETT, Ann. Mag. Nat. Hist. 1874, 15, 82, Lofoten, Norway.
Onos septentrionalis, COLLETT, Norske Nord-Havs Exped., 139, 1880; JORDAN, Cat. Fish. N. A., 128, 1885.

993. ENCHELYOPUS, Bloch & Schneider.

(FOUR-BEARDED ROCKLINGS)

Enchelyopus, BLOCH & SCHNEIDER, Syst. Ichth., 50, 1801 (*cimbrius;* the first species mentioned and the one left as type after elimination of the genera, defined prior to *Rhinonemus*).
Rhinonemus, GILL, Proc. Ac. Nat. Sci. Phila. 1883, 241 (*cimbrius*).

Barbels 4, 1 at each nostril, 1 at tip of snout, and 1 at the chin; head high and compressed anteriorly; teeth in narrow bands, some of them enlarged; otherwise essentially as in *Gaidropsarus.* North Atlantic. (ἐγχελυωπός, resembling an eel; "*facie anguillaris.*")

2936. ENCHELYOPUS CIMBRIUS (Linnæus).

(FOUR-BEARDED ROCKLING.)

Head 5½; depth 9. D. 45 to 50; A. 41 or 42; V. 5. Body slender, tapering from the shoulders back; caudal peduncle narrow, 4 in head; snout moderate, blunt, rounded, not depressed, a little shorter than the eye; eye large, subcircular, 4 in head; interorbital space narrow, equal to vertical diameter of eye, 6 in head; teeth villiform, those in the upper jaw unequal, small, with about 8 enlarged in front, those of the lower jaw long and slender, of equal length, a few somewhat enlarged in front; maxillary reaching beyond posterior border of eye, a barbel at each nostril, 1 on tip of snout and 1 on chin, stitch-like; lateral line with about 35 enlarged pores along its entire length; first (free) ray of dorsal nearly as long as head; ventral ½ head; caudal acute. Light olivaceous (salmon-red); first dorsal ray and posterior end of dorsal and anal abruptly black, as is lower half of caudal; pectorals and ventrals pale; sides of head somewhat silvery; cavity of mouth dark bluish. North Atlantic, on both coasts, south in deep water to the Gulf Stream; common in Massachusetts Bay; our specimens from Woods Hole; the young ("mackerel midges") silvery, unlike the adult in appearance. (Eu.) (*cimbrius,* welsh.)

Gadus cimbrius, LINNÆUS, Syst. Nat., Ed. 12, I, 440, 1766, Atlantic Ocean; Scania (Coll. Dr. Strussenfelt).
Motella caudacuta, STORER, Proc. Bost. Soc. Nat. Hist., III, 1848, 5, Cape Cod, Provincetown, Mass. (Coll. Herman M. Smith); STORER, Amer. Ac. Sci., 411, 1867; STORER, Hist. Fish. Mass., 183, 1867.
Rhinonemus caudacuta, GILL, Proc. Ac. Nat. Sci. Phila. 1863, 241; GOODE & BEAN, Amer. Journ. Sci. and Arts 1877, 476; JORDAN, Cat. Fish. N. A., 128, 1885.
Motella cimbria, NILSSON, Prod. Ich. Scand., 48, 1832; BELL, Cnn. Nat. and Geol., IV, 209, 1859.

Onos cimbrius, GOODE & BEAN, Proc. U. S. Nat. Mus. 1878, 349; GOODE & BEAN, Bull. Essex Inst., XI, 1879; JORDAN & GILBERT, Synopsis, 797, 1883; GOODE & BEAN, Bull. Mus. Comp. Zool., x, No. 5, 217, 1883.
Rhinonemus cimbrius, JORDAN, Cat. Fish N. A., 128, 1885; GOODE & BEAN, Oceanic Ichthyology, 384, fig. 328, 1896.
Enchelyopus cimbricus (misprint for *cimbrius*), BLOCH & SCHNEIDER, Syst. Ich., 50, pl. 9, 1801.

994. BROSME (Cuvier) Oken.

(CUSKS.)

Les Brosmes, CUVIER, Règne Animal, Ed. I, vol. 2, 216, 1817 (*brosme*).
Brosme, OKEN, Isis, 1817, 1182; after CUVIER.
Brosmius, CUVIER, Règne Animal Ed. 2, vol. II, 334, 1829 (*brosme*).

Body moderately elongate, covered with very small scales. Mouth rather large, with teeth in the jaws, vomer, and palatines, some of those on the vomer and palatines enlarged; chin with a barbel; branchiostegals 7. Dorsal fin single, continuous, not elevated, not notched; anal fin similar, but shorter; caudal fin rounded; ventral fin several-rayed. Northern seas. (From the Danish vernacular name, *brosme*.)

2937. BROSME BROSME (Müller).

(CUSK.)

D. 98; A. 71; P. 24; V. 5. Body cylindrical, posteriorly compressed; head flattened above. Mouth large, oblique, maxillary reaching beyond orbit; lower jaw included; several rows of sharp teeth on jaws, vomer, and palatines; barbel about 5 in head; interorbital greater than the diameter of eye. Origin of dorsal above anterior half of pectoral; pectoral round, 2⅓ in head; caudal rounded behind. Brownish above, the sides yellowish, sometimes mottled with brown; young uniform dark slate color, or with transverse yellow bands; vertical fins bordered with blackish, and with a white edge. (Storer.) North Atlantic, south to Cape Cod and Denmark; rare southward on our coasts. (*brosme*, a Danish name.)

Gadus brosme, MÜLLER, Prodr. Zool. Dan., 41, 1776, Denmark; FABRICIUS Fauna Grœnlandica, 140, 1780.
Gadus lubb, EUPHRASEN, Vet. Akad. Handl. 1794, 223, tab. 8.
Gadus torsk, BONNATERRE, Encycl. Meth., 51, 1788, Söndmöre, Norway; after Strom.
Brosmius vulgaris, FLEMING, British Anim., 194, 1828.
Brosmius flavesny, LE SUEUR, Mém. Mus., v, 1819, 158, Banks of Newfoundland; chin with 2 barbels; lower jaw longest.
Brosmius flavescens, GÜNTHER, Cat., IV, 369, 1862; STORER, Hist. Fish. Mass., 368, 1867.
Enchelyopus brosme, BLOCH & SCHNEIDER, Syst. Ichth., 51, 1801.
Brosmius brosme, GÜNTHER Cat., IV, 369, 1862; JORDAN & GILBERT, Synopsis, 802, 1883; GOODE & BEAN, Oceanic Ichthyology, 385, fig. 329, 1896.
Blennius torsk, LACÉPÈDE, Hist. Nat. Poiss., II, 508, 1800.

Family CCXV. MACROURIDÆ.

(THE GRENADIERS.)

Body elongate, tapering into a very long compressed tail, which ends in a point; scales moderate, usually keeled or spinous, sometimes smooth. Suborbital bones enlarged, sometimes cavernous. Teeth villiform or cardiform, in bands, on the jaws only; tip of lower jaw with a barbel; premaxillary protractile. Dorsal fins 2, the first short and high, of stiff, spine-like branched rays; the second dorsal very long, usually of very low feeble rays, continued to the end of the tail; anal fin similar to the second dorsal, but usually much higher; no caudal fin; ventrals small, subjugular, each of about 8 rays. Branchiostegals 6 or 7. Lateral line present. Gills 3½ or 4, a slit behind the fourth. Gill rakers small; gill membranes free or narrowly united to the isthmus, usually more or less connected; pseudobranchiæ wanting or rudimentary; pyloric cæca numerous; air bladder present. Genera 18; species about 50, chiefly of the northern seas, all in deep water. They differ from the codfishes chiefly in the elongate and degenerate condition of the posterior part of the body. Dr. Gill succinctly defines the group as "Gadoidea with an elongated tail tapering backward and destitute of a caudal fin, postpectoral anus, enlarged suborbital bones, inferior mouth, subbrachial ventrals, a distinct anterior dorsal, and a long second dorsal and anal converging on end of tail." We here follow Goode & Bean in the general arrangement of the genera of *Macrouridæ.* Some of these can, however, be only provisionally adopted, as the characters of dentition, form of mouth, and character of the second dorsal spine or ray, are subject to much intergradation. These characters seem much more distinct on paper than they are in fact. Still, most of the genera here adopted will ultimately prove valid. (*Macruridæ,* Günther, Cat., IV, 390–398, 1862.)

   a. First branchial arch free, without fold of membrane across it; mouth large; second
     dorsal well developed, higher than the anal.
   BATHYGADINÆ:
       b. Gills 3½; snout short and blunt, the jaws even in front; teeth in villiform bands,
         sometimes obsolete; bones of head soft and cavernous; scales smooth; first
         dorsal low, its spine not produced.   BATHYGADUS, 995.
   TRACHYRINCHINÆ:
       bb. Gills 4; snout rather long.
           c. Teeth in upper jaw in 2 series, the outer enlarged, those in lower jaw
             in 1 series; mouth subterminal; barbel obsolete; nape without scale-
             less fossæ; vomer with teeth; bones of head soft and cavernous; tail
             very long, flagelliform; anal fin with an elevated anterior lobe.
                                                          STEINDACHNERIA, 996.
           cc. Teeth in both jaws in villiform bands; barbel developed; mouth inferior;
             a naked fossa on each side of nape; a row of armed scales along base
             of dorsal anteriorly; opercle very small; anal not elevated in front.
                                                          TRACHYRINCUS, 997.
   MACROURINÆ:
   aa. First branchial arch with a fold of membrane across its terminal portion; gills 4, a
     slit behind the fourth; barbel well developed.
       d. Teeth not all in villiform bands, those of lower jaw in 1 series; mouth rather
         large, with more or less of lateral cleft.
           e. Upper jaw without villiform band behind the enlarged anterior teeth, the
             inner teeth, if present chiefly uniserial, not in villiform bands.
               f. Dorsal fins widely separated, the interspace greater than base of first.
                   g. First dorsal with the spine not serrate, its insertion over pectoral
                     or nearly so; pectoral placed high, opposite upper angle of
                     gill cleft; scales small, bristly; bones of head cavernous;
                     ventrals short and weak.   MALACOCEPHALUS, 998.
                   gg. First dorsal with its spine more or less strongly serrate; pectoral
                     inserted below upper angle of gill cleft.

> *h.* Scales nearly smooth, with weak ridges which are not
> > spinigerous.   MOSELEYA, 999.
> *hh.* Scales rough, with strong ridges.   NEMATONURUS, 1000.
> > *ff.* Dorsal fins near together, the interspace less than base of first; scales
> > > rough.
> > > *i.* Dorsal spine weak, unarmed or very nearly so; pectorals moder-
> > > > ate.   ALBATROSSIA, 1001.
> > > *ii.* Dorsal spine very strongly serrate; pectorals very long.
> > > > BOGOSLOVIUS, 1002.
> > *ee.* Upper jaw with a distinct villiform band behind the outer series of
> > > enlarged teeth; dorsal spine serrate; dorsal fins not widely separated.
> > > > CHALINURA, 1003.
> *dd.* Teeth in villiform bands above and below, the outer scarcely enlarged and
> > not separated from the rest; the lower band sometimes becoming a single
> > series laterally; scales rough.
> > *j.* Mouth wide, with considerable lateral cleft.
> > > *k.* Dorsal spine finely barbed; skull rather firm; dorsals moderately
> > > > separated.   CORYPHÆNOIDES, 1004.
> > > *kk.* Dorsal spine entirely smooth; bones of skull very thin and papery;
> > > > dorsals well separated.   HYMENOCEPHALUS, 1005.
> > *jj.* Mouth inferior, small, with little lateral cleft; a more or less distinct
> > > ridge across the suborbital region.
> > > *l.* Scales spinous, very rough.
> > > > *m.* Scales distinct, regularly imbricated.
> > > > > *n.* Long dorsal spine serrate in front; mouth subinferior,
> > > > > below the short snout.   MACROURUS, 1006.
> > > > > *nn.* Long dorsal spine smooth; mouth wholly inferior, below
> > > > > the long sturgeon-like snout.   CŒLORHYNCHUS, 1007.
> > > > *mm.* Scales indistinct, scarcely imbricated; the whole body rough-
> > > > villous; dorsal spine smooth.   TRACHONURUS, 1008.
> > > *ll.* Scales all thin and smooth, dorsal fin slightly serrulate.
> > > > LIONURUS, 1009.

995. BATHYGADUS, Günther.

Bathygadus, GÜNTHER, Ann. Mag. Nat. Hist. 1878, 23 (*cottoides*).

Head large, fleshy, without prominent ridges, spiny armature or external depressions; nape elevated, hump-like. Snout broad, obtuse, not
produced; mouth terminal very large, with small villiform teeth or none;
suborbital ridge very low, not joined to the angle of the preoperculum.
Maxillary entirely received within a groove under the prefrontal and suborbital bones, its tips narrowed and blade-like; premaxillaries protractile
downward, separated anteriorly, rib-shaped, compressed vertically, very
broad and without true teeth; provided posteriorly with a short flange,
which is received under the maxillary; mandible received within the
intermaxillary bones, without true teeth, but with minute asperities, similar to those in the upper jaw; vomer and palatines toothless. Barbel
sometimes present. No pseudobranchiæ. Gill rakers numerous, moderate, lanceolate, with minute denticulations along their inner edge.
Branchiostegal membrane free from the isthmus, deeply cleft. Branchiostegals 7, very stiff. Gill opening very wide; gills 3½; anterior gill arch
free. Operculum with a blunt, spine-like prominence at its angle. Ventrals below the pectorals, many-rayed, the anterior rays produced; dorsal consisting for the most part of branched rays, higher than the anal, the

first dorsal low, without differentiated spine. Scales cycloid, unarmed; lateral line strongly arched over the pectoral. Deep seas. This genus differs from *Macrourus* and its allies in the structure of both the first and last gill arches. It is perhaps the most primitive of the family and as such is nearest allied to the *Gadidæ*. (βαθύς, deep; *Gadus*, codfish.)

   a. Pectoral and ventral fins moderate, not much, if any, longer than head.
     b. Jaws without teeth; pectoral fin broad, of 25 rays; depth 5⅔ in length.
                                         ARCUATUS, 2938.
     bb. Jaws with small teeth; pectoral fin narrow; depth 6 to 6½ in length.
       c. Eye moderate, 5 in head; pectoral rays 14, the fin ½ as long as head.
                                         FAVOSUS, 2939.
       cc. Eye very large, 2¾ in head; pectoral as long as head without snout.
                                     MACROPS, 2940.
   aa. Pectoral and ventral fins much produced, much longer than head, each reaching about halfway to tip of caudal; depth 7½ in length; pectoral rays 13.
                                     LONGIFILIS, 2941.

2938. BATHYGADUS ARCUATUS, Goode & Bean.

Head 5 in total length; depth 5⅔; eye 4½ in head; snout 4½. D. II, 9 or 10–135; A. 120; P. 25; V. 8; scales 8–140–13 or 14 (counting backward from vent to lateral line), 22 counting forward. Body shaped much as in *Chalinura simula*, but the nape still more convex; back gibbous, the dorsal outline rising rapidly from interorbital region to origin of first dorsal, whence it descends gradually to end of tail. Scales moderate, cycloid, subovate, without armature, those of abdominal region and those above pectorals the largest; lateral line strongly arched over the pectorals, length of the arched portion contained about 3¼ times in straight portion, greatest height of arch about ¼ its chord; scales covering all parts of head except jaws and chin. Interorbital area flat, its width 6 in head; postorbital portion of head about 2½ times diameter of eye; operculum terminating in a flat obtuse spine, its length, including the flap, about equal to diameter of eye; preoperculum entire, with a prominent ridge in advance of its posterior edge; snout very broad, obtuse, the intermaxillaries extending beyond it, its width at nostrils equal to about twice length of eye; posterior extremities of intermaxillary processes elevated, producing a decided hump upon top of snout; ridge formed by prefrontal and suborbital bones terminating very slightly behind posterior margin of orbit, and not connected with angle of preoperculum. Nostrils immediately in front of lower part of eye, not tubular, the anterior one very small, porelike, only about ¼ as large as posterior one; distance of anterior nostril from tip of snout about ⅘ length of eye. Length of barbel 6⅔ in length of body, and equal to length of head without snout, more than 3 times as long as eye. No true teeth, the intermaxillaries and mandible being broad plates, covered with minute asperities; a naked space at the symphysis of intermaxillaries; distance of first dorsal from snout nearly 3½ times length of its base, the first spine minute, the second (in the type) somewhat mutilated, its length nearly 3 in length of head, not stouter than the branched rays, and entirely smooth; second dorsal fin separated from first by a very short interspace, equal to about ⅕ of length of eye, its rays long, subequal, the first slightly the longest, its length equal to that of base of first dorsal; anal much lower than dorsal, the longest rays being in front, its third ray about ¼ as long as first ray of second dorsal; this fin inserted under the seventh ray of second dorsal; about 3 of the terminal anal rays might be considered caudal rays; pectoral inserted slightly in advance of ventral, which is in about the same vertical with the origin of the first dorsal, second ray of pectoral slightly produced; length of pectoral equal to that of head without snout; ventral insertion distant from tip of snout a distance equal to that of first dorsal from snout, the first and second rays filamentous, the latter slightly the longer, and extending to the fifteenth or eighteenth ray of anal fin. Color brown; vertical fins bluish or black; peritoneum black; inside of gill covers and roof of mouth bluish. (Goode & Bean.) West Indies and Gulf of Mexico. Three specimens known; the type from near Martinique. (*arcuatus*, arched.)

Bathygadus arcuatus, GOODE & BEAN, Bull. Mus. Comp. Zool., XII, No. 5, 158, 1883, off Martinique, in 334 fathoms (Coll. *Blake*); GOODE & BEAN, Oceanic Ichthyology, 421, 1896.

2939. BATHYGADUS FAVOSUS, Goode & Bean.

Head 5⅓ in total length; depth about 6; eye 5 in head; snout about 4. D. II, 9-125; A. 110; V. 9; P. 14; B. 7; scales 10-135-16. Body heavy, stout, the profile descending gradually and in a slight curve from first dorsal to snout. Scales small, deciduous, cycloid, without armature; interorbital area slightly convex, its greatest width about 3 in head; the postorbital part of head 2⅓ times as long as eye; snout broad, oblique, its width at the nostrils a little more than that of interorbital area; nostrils close to and in front of middle of eye, the posterior somewhat the larger; no barbel. Teeth in both jaws in villiform bands, a naked space at symphysis of intermaxillaries; intermaxillary bands more than twice as wide as those of mandible; vomer and palatines toothless. Gill rakers 20 + 25, the longest on anterior arch slightly more than ¼ eye; pseudobranchiæ present, very rudimentary in some individuals, in others wanting or present only on one side; first dorsal distant from snout a distance slightly more than length of head, length of its base about equal to width of snout at nostrils, the fin consisting of 2 spines, the first minute, and 9 branched rays; length of longest dorsal spine, which is armed, 2 in head; second dorsal beginning immediately behind first, the membrane being continuous; anterior rays longest, apparently about ¼ length of head; anal lower than second dorsal, its distance from snout about equal to ⅓ of total length; pectoral inserted under anterior rays of first dorsal and very slightly in advance of origin of ventral, its length more than ¼ that of head; distance of ventral from snout 5 times in total length; this fin inserted nearly under base of pectoral; the first ray somewhat produced, its tip reaching to fourth ray of anal fin. Color bluish brown, darkest upon head and abdomen. West Indies. The type specimen, 350 mm. in length, was obtained by the *Blake* from Station LXXX, off Martinique, at a depth of 472 fathoms. (Goode & Bean.) (*favosus*, like honeycomb.)

Bathygadus favosus, GOODE & BEAN. Bull. Mus. Comp. Zool., XII, No. 5, 160, 1883, off Martinique in 472 fathoms (Coll. *Blake*); GOODE & BEAN, Oceanic Ichthyology, 420, fig. 352, 1896.

2940. BATHYGADUS MACROPS, Goode & Bean.

Head 5¼ in total length; depth 6½; eye 2¼ in head; snout 5. D. II,
8—about 125; V. 8. Body somewhat compressed; scales small, deciduous, about 25 rows in an oblique line from the vent to the dorsal fin,
24 from the upper angle of operculum to the vertical through origin of
the anal; interorbital area nearly flat, its width 4 in head; postorbital part of head somewhat longer than diameter of eye; snout broad,
obtuse; nostrils close to eye, the posterior nearly twice as large as
anterior one; maxillary extending to vertical through posterior margin
of orbit, its length equal to that of head without its postorbital portion;
length of mandible 3 times that of snout; intermaxillaries and mandible
provided with narrow bands of villiform teeth, those of the mandible much
shorter. A minute barbel, about ⅓ as long as snout. Vomer and palate
toothless. Gill rakers lanceolate, elongate, 7 + 26, the longest 7 in head;
pseudobranchiæ absent; distance of first dorsal from snout nearly 5 times
in total length, second or longest ray in the typical specimen twice
length of snout; second dorsal almost continuous with the first, its
anterior rays the longest, about 4 times in length of head; anal inserted
under fourteenth ray of second dorsal, its rays all very short; in a distance equal to length of head, counting back from insertion, there are 33
rays; pectoral inserted under first branched ray of first dorsal, its length
in the most nearly perfect specimens equaling length of head without
snout; ventral origin very slightly behind origin of pectoral under third
branched ray of dorsal, reaching nearly to vent when laid back, its length
equaling 3 times that of the snout. Branchiostegals 7. Color yellowish
gray, lighter below. (Goode & Bean.) In deep waters of the Gulf of
Mexico off the United States coast, in 321 to 347 fathoms. ($\mu\alpha\kappa\rho\delta\varsigma$, large;
$\tilde{\omega}\psi$, eye.)

Bathygadus macrops, GOODE & BEAN, Proc. U. S. Nat. Mus. 1885, 598, Gulf of Mexico,
    Lat. 28° 34′ N., Long. 86° 48′ W., in 335 fathoms (Type, No. 37339. Coll. *Albatross*);
    GÜNTHER, Challenger Report, XXII, 156, 1887; GOODE & BEAN, Oceanic Ichthyology,
    423, 1896.

2941. BATHYGADUS LONGIFILIS, Goode & Bean.

Head about 5¾ in total length; depth 7½; eye 4 in head; snout 4. D. II,
8 or 9—about 140; P. 13; V. 8; scales about 142. Body more compressed
than in *B. macrops*; scales small, cycloid, deciduous, about 25 rows from
the vent upward and forward to the dorsal fin, interorbital area flattened,
its greatest width 3⅘ times in total length of head; postorbital portion of
head twice as long as eye; snout and nostrils normal; maxillary reaching
somewhat beyond posterior margin of orbit, its length twice in distance
from snout to origin of first dorsal; length of mandible 2¼ times in snout;
barbel slender, long, its length equal to 1½ times orbital diameter. Teeth
in narrow villiform bands in each jaw, none on vomer or palatine bones;
gill rakers very long and slender, numerous, 17 + 35, the longest nearly 6
in head; pseudobranchiæ absent; first dorsal of 2 stout spines, the first
minute, the second elongate, and 8 or 9 branched rays, its distance from
snout 5¼ in total; second or longest simple ray nearly 8 times length of

snout, and reaching to or beyond the thirtieth ray of the second dorsal; second dorsal almost continuous with the first, its anterior rays longest and not diminishing rapidly in size toward tail; anal inserted under ninth ray of second dorsal, its rays much shorter than those of dorsal, and situated about same distance apart; pectorals inserted under anterior portion of first dorsal, first ray much ˙produced, extending more than halfway from its insertion to tip of tail; ventral origin slightly behind origin of pectoral, under third branched ray of dorsal, its first ray much enlarged, extending more than halfway from its insertion to tip of caudal, its length 2⅓ times in total length; branchiostegals 7. Color yellowish gray, abdomen bluish. This form is closely allied to *B. multifilis*, described by Günther from off the Philippines (Challenger Report, XXII, 155, pl. 42, fig. B, 1887), which, however, appears to have a smaller eye, less elongate filaments, and ventrals inserted in advance of the first dorsal, while the anal appears to be further back, under the twelfth or thirteenth ray of second dorsal. Both species are provided with long, slender barbels; in other respects they are closer to *B. cottoides*, the typical species, than to *B. macrops*. (Goode & Bean.) Deep waters of the Gulf of Mexico, in 525 to 739 fathoms. (*longus*, long; *filum*, thread.)

Bathygadus longifilis, GOODE & BEAN, Proc. U. S. Nat. Mus. 1885, 599, Gulf of Mexico. Lat. 28° 47′ 30″ N., Long. 87° 27′ W., in 724 fathoms (Type, No. 37338. Coll. *Albatross*); GÜNTHER, Challenger Report, XXII, 157, 1887; ALCOCK, Ann. Mag. Nat. Hist. 1890, 302; ALCOCK, *l. c.* 1891, 123; GOODE & BEAN, Oceanic Ichthyology, 422, 1896.
Hymenocephalus longifilis, VAILLANT, Exp. Sci. Trav. et Tails, 218, pl. 23, fig. 1, 1888.

996. STEINDACHNERIA, GOODE & BEAN.

Steindachneria, GOODE & BEAN, in AGASSIZ, Three Cruises of the *Blake*, II, 26, 1888 (no type; short diagnosis*); not *Steindachneria*, EIGENMANN, Nematognathi, Occasional Papers, I, Cal. Ac. Sci. 1890, 100 and 202, a genus of Siluroid fishes.
Steindachneria, GOODE & BEAN, Oceanic Ichthyology, 419, 1896 (*argentea*).
Steindachnerella,† EIGENMANN, American Naturalist, February, 1897, 159 (*argentea*).

Body compressed, with tapering tail. Mouth large, terminal. Dorsal fins continuous, both elevated anteriorly; anal divided, the anterior portion elevated, the posterior low. Teeth in each jaw biserial, the outer much enlarged, vomerine teeth present. Bones of head soft and cavernous. Eye large. Gill membranes connected anteriorly, free from the isthmus. Gill rakers slender, rather numerous; vent in anterior third of length. No pseudobranchiæ. Branchiostegals 7. No barbel. Pectorals and ventrals both below first dorsal. Scales thin, cycloid, deciduous. Deep seas. ("This remarkable genus is named in honor of Dr. Franz Steindachner, Custos of the Imperial Zoological Museum of Vienna," one of the ablest naturalists of the century.)

.* "*Steindachneria*; a Macruroid with a high differentiated first anal spine." (Goode & Bean.)
†As the original diagnosis of the Macrourid genus *Steindachneria*, although very short, is correct and sufficient for identification, the name in question should be retained for it rather than *Steindachnerella*, and the Silurid genus *Steindachneria*, Eigenmann should receive a new name.

2942. STEINDACHNERIA ARGENTEA, Goode & Bean.

Head 5⅓ in total; depth 7½, at anal origin 8; eye 3⅓ in head; snout about 5½; interorbital width 5½; maxillary 2; premaxillary 2; mandible 1⅔; gill rakers 4 or 5+19; D. VIII, 123+; A. 10+113; P. 15; V. 8. Head and body compressed; tail tapering to a very fine point. Scales small, deciduous, cycloid, 6 rows between lateral line and origin of soft dorsal. Nostrils nearer eye than end of snout, the anterior nostril nearly circular, the posterior much longer and slightly concave; no barbel. Maxillary dilated at the extremity and somewhat produced downward into an obtuse point, reaching nearly to a vertical at posterior margin of orbit, and concealed by the preorbital; premaxillaries slightly protractile, much attenuated posteriorly; mandible reaching slightly behind eye. Premaxillary and mandibular teeth biserial, those of the outer series enlarged and rather widely set, some of the enlarged teeth slightly sagittate at tip; vomerine teeth well developed; upper pharyngeal teeth in 2 broad, well-developed patches. Gill rakers slender, the longest about 2 in eye. Distance from snout to first dorsal about ⅕ total length, the first spine elongate, filiform, and reaching fourteenth ray of second dorsal; base of first dorsal about 1 in head; longest ray of second dorsal about 2⅓ in head, the rays diminishing in size rapidly, the last minute; origin of anal under sixth ray of second dorsal, not far behind the vent, the anterior elevated portion consisting of 10 rays, all of which except the first are divided, the second ray longest, twice length of eye, the tenth ray only about ⅓ length of second, and separated by a small membrane from rest of fin which consists of very minute rays. Vent under fourth ray of second dorsal. Origin of ventrals under base of pectorals and about under third spine of first dorsal; first ventral ray filamentous, reaching origin of anal; pectoral reaching to below fifteenth ray of second dorsal. Gulf of Mexico. Only the type known. Length 233 mm. (*argenteus*, silvery.)

Steindachneria argentea, GOODE & BEAN, Oceanic Ichthyology, 419, fig. 351, 1896, off delta of Mississippi River, Lat. 39° 14′ 30″ N., Long. 88° 09′ 30″ W., in 68 fathoms. (Type, No. 37350. Coll. *Albatross*.)

997. TRACHYRINCUS, Giorna.

Trachyrincus, GIORNA, Mem. Accad. Imp. Turin, XVI, 1803, 178 (no type mentioned).
Lepidoleprus, RISSO, Ichth. Nice, 197, 1810 (*trachyrincus*).
Oxycephas, RAFINESQUE, Caratteri, 31, 1810 (*scabrus=trachyrincus*).
Lepidosoma, SWAINSON, Nat. Hist. Class'n Fish., II, 261, 1839 (*trachyrhynchus*).
Trachyrhynchus, GÜNTHER, Challenger Report, XXII, 152, 1887; corrected spelling.

Snout produced in a long depressed process which is sharply pointed in front, with a sharp lateral edge, which is continued in a straight line across the suborbital region. Mouth inferior, horseshoe-shaped, placed like the mouth of a sturgeon. Teeth in both jaws in villiform bands; chin with a barbel; a scaleless fossa on each side of nape. Second dorsal well developed. Scales moderate, spinigerous; a series of larger scales, each armed with a projecting ridge, along each side of base of dorsal and anal anteriorly. Opercle small. Gill membranes scarcely united; gills 4; first gill arch free, with short, styliform gill rakers. Deep seas. This genus and its allies differ from *Macrourus* in the important character of the structure of the first gill arch. (τραχύς, rough; ρύγχος, snout; hence properly, but not originally, spelled *Trachyrrhynchus*.)

2943. TRACHYRINCUS HELOLEPIS, Gilbert.

Head 3⅛ in total; depth 7; eye large, 4 in head, = interorbital width; snout 2¼, its greatest width 1⅞ in its length. D. 11. Snout depressed, flat, narrowly triangular, tapering to a sharp point, its lateral ridges continuous backward over suborbital chain and across cheek. Interorbital space wide and flat. Ethmoidal ridge not prominent. Mouth wholly inferior, U-shaped, overpassed by the snout by a. distance contained 3⅛ in head. Barbel slender, short, less than ⅓ diameter of orbit. Teeth finely villiform, in very broad bands in each jaw, none of them enlarged. Maxillary reaching to or almost to vertical from hinder margin of orbit, 3⅓ in head. Opercle very small, triangular, its length behind preopercular margin scarcely more than ⅓ diameter of orbit; outer gill arch not adnate to the opercle, its lower limb with 17 short gill rakers, which are not tubercular. Distance of dorsal fin from nape 3⅓ in head, the 2 dorsal fins closely approximated; second dorsal ray not spine-like, soft and flexible, and not longer than the succeeding rays, its length ⅔ the diameter of orbit. Vent located immediately in front of origin of anal fin, its distance from ventrals 1½ in head. Ventrals short, inserted well in advance of base of pectorals, the outer ray little produced, its length 1⅓ in diameter of orbit. Scales all with their margins embedded, and therefore appearing non-imbricated, the central portion of each projecting, tubercle-like, and bearing a single strong central spine, with sometimes 2 or 3 smaller ones; belly and breast sometimes covered with much smaller scales similarly armed; no naked area between bases of ventrals; enlarged plates along bases of dorsals and anal bearing each a strong compressed backwardly-curved spine, usually without distinct serrations; from the base of the central spine radiate lines of short spinous points; dorsal series of plates continued forward to the nape, the predorsal portion of the included groove covered with scales; ventral series scarcely extending beyond vent, but extending farther posteriorly than do the dorsal plates; scales on top of head with a median serrated ridge; temporal fossæ small but evident, naked. Color apparently dark brown; gill cavity and peritoneum black. Pacific Ocean, off the coast of Central America, in deep water. Only the type known, a specimen 18 inches long. (ἧλος, tubercle; λεπίς, scale.)

Trachyrhynchus helolepis, GILBERT, Proc. U. S. Nat. Mus. 1891, 562, Pacific Coast of Central America in deep water. (Type, No. 48205.)

998. MALACOCEPHALUS, Günther.

Malacocephalus, GÜNTHER, Cat. Fish. Brit. Mus., IV, 396, 1862 (*lævis*).

Intermaxillary teeth biserial, mandibulary teeth uniserial. Mouth lateral; snouth short, obtuse. Head without prominent ridges, with wide muciferous cavities. Dorsal fin over origin of pectorals, its longest spine

3030——84

smooth; dorsal fins widely separated. Pectorals short, placed high, opposite upper angle of gill cleft. Scales small, bristly. Origin of lateral line at upper angle of gill cleft. (μαλακός, soft; κεφαλή, head.)

2944. MALACOCEPHALUS OCCIDENTALIS, Goode & Bean.

Eye 2¼ in head; barbel slightly longer than eye; snout 4 in head; interorbital space 4. Agreeing with Günther's description of *M. lævis*, but differing in the position of the vent, the ventrals, and the anal fin; the last commencing at a distance behind the vent equal to length of snout; distance of vent from origin of ventrals less than its distance from origin of anal; ventrals originate under middle of first dorsal; origin of pectorals under that of first dorsal, the pectorals as long as head without postorbital flap; ventrals reaching to or slightly beyond origin of anal. Gill rakers rudimentary, $x+11$. Second dorsal spine nearly equal to length of head; first branched dorsal ray about as long as head. Atlantic Ocean, off Cape Hatteras, and Caribbean Sea. Length 8½ inches; a doubtful species, perhaps identical with *M. lævis*. (*occidentalis*, western.)

Malacocephalus occidentalis, GOODE & BEAN, Proc. U. S. Nat. Mus. 1885, 597, off Cape Hatteras, at Albatross Station 2310, Lat. 35° 44' N., Long. 79° 51' W., in 132 fathoms. (Type, No. 37336.)

999. MOSELEYA, Goode & Bean.

Moseleya, GOODE & BEAN, Oceanic Ichthyology, 417, 1896 (*longifilis*).

This genus is near *Nematonurus*, having the mouth small, the upper teeth in 1 or 2 series, the dorsal spine weakly serrate, and the dorsal fins well separated. The chief difference lies in the scales, which are feebly ridged and nearly or quite smooth. The typical species, *M. longifilis* (Günther), is from off the coast of Japan. ("Named in honor of Prof. Henry N. Moseley, F. R. S., of Oxford University, whose contributions to natural history while naturalist of H. M. S. *Challenger* we desire to commemorate.")

2945. MOSELEYA CYCLOLEPIS (Gilbert).

Dorsal II-8 or 9; ventral 12; eye 4¼ in head; snout 3¾; maxillary 2⅔. Head smooth, compressed, without conspicuous ridges; median and lateral rostral ridges terminating in slightly projecting points, the median process, a short portion of the median ridge, and the edge of the membrane connecting median with lateral processes, with spinous scales and points. Snout projecting beyond the premaxillaries for ⅔ its length. Eye small, less than snout, very slightly exceeding interorbital space; mouth small, wholly inferior, maxillary reaching vertical from posterior margin of pupil. Premaxillary teeth in 2 series, the outer similar to those in mandible, not enlarged or canine-like, the inner series smaller, directed obliquely backward; a single series of teeth in mandible, not widening into a patch at symphysis. Barbel thick at base, ⅔ length of snout. Preopercle incurved above the angle, the lower limb expanded, the marginal region striate. First dorsal inserted behind axil of pectoral (second spine broken in both specimens examined), the basal portion smooth, a single sharp barb showing that the spine is serrate; base of first dorsal equals length of snout; interspace between dorsals exceeding length of first dorsal base by $\frac{1}{5}$ to $\frac{2}{3}$ length of latter. Vent immediately in advance of origin of anal, under middle of interspace between dorsals; dorsal low and inconspicuous and the anal higher, as usual in this group; pectorals very slender, $1\frac{9}{10}$ in length of head; outer ventral ray filamentous, reaching third or fourth anal ray. Scales mostly lost, the few remaining on head either entirely smooth or bearing a single median keel with 1 or 2 low spinous points; those on body without spines, either entirely smooth or showing traces of a low median keel; 6 scales in an oblique series between lateral line and middle of base of dorsal. Color dark brown, the anterior portion of back and sides with small scattered black spots; opercles, lower side of head including gill membranes and ventral area black, as are also the mouth and gill cavity and the peritoneum. A species with the general appearance, including the protruding snout, the inferior mouth and comparatively weak dentition of *Nematonurus armatus* and *N. affinis*, but with the dorsals less widely separated, the vent anterior in position, and the scales unarmed, as in *Moseleya longifilis*. (Gilbert.) Coast of British Columbia. Two specimens, the longest 150 mm., from Station 3342, off Queen Charlotte Islands, depth 1,588 fathoms. ($\varkappa \acute{\nu} \varkappa \lambda o\varsigma$, circle; $\lambda \varepsilon \pi \acute{\iota}\varsigma$, scale.)

Nematonurus cyclolepis, GILBERT, Rept. U. S. Fish Comm. 1893 (1896), 458, off Queen Charlotte Islands, at Albatross Station 3342, in 1,588 fathoms.

1000. NEMATONURUS, Günther.

Nematonurus, GÜNTHER, Challenger Report, Deep-Sea Fishes, XXII, 124, 150, 1887 (*armatus*).

Body rather robust, covered with rough, strongly-ridged scales. Head short; mouth small or moderate, more or less inferior; teeth in upper jaw rather strong, in 1 series or nearly so; lower teeth uniserial; mucous cavities small; pectoral fin inserted low, below upper angle of gill cleft; ventrals well developed, the outer ray filamentous; long ray of dorsal serrated; space between dorsals long, much greater than length of first dorsal. Deep seas. A well-marked genus, distinguished by its rough, firm scales and the wide space between dorsals. ($\nu\tilde{\eta}\mu\alpha$, thread; $o\dot{\nu}\rho\acute{\alpha}$, tail.)

a. Depth 6½ in length; scales without distinct median keel.   GOODEI, 2946.
aa. Depth 5¾ in length; scales with the median keel prominent; suborbital narrow, with well-marked mucous partitions.   SUBORBITALIS, 2947.

2946. NEMATONURUS GOODEI* (Günther).

Head 5⅔; depth 6½; eye 5 in head; snout 4¼; interorbital width 4¼; postorbital part of head 8½; first dorsal II, 8 or 9; second dorsal 105; A. 110; P. 20; V. 10; scales 7–150–18, small, strong, free portions covered by series of small vitreous spines arranged in about 6 rows; no specialization of the central row, though the median spine at margin of scale projects

---

* By some inadvertence this species is recorded by Goode & Bean as a *Hymenocephalus* (Oceanic Ichth., 407). On p. 408 it is said to be a *Nematonurus*. It has obviously no affinity with *Hymenocephalus*, and is, in fact, an ally of *Nematonurus armatus*.

most strongly and is longest. Width of interorbital area a little greater than horizontal diameter of orbit and length of operculum; snout triangular, depressed, its tip in axis of body nearly on a level with lower margin of eye, its lower surface forming an angle with the body axis, about equal to that formed with same by its upper profile; superior ridge pronounced anteriorly, but ending in advance of concavity in interorbital space; lateral ridges prominent, continuing posteriorly to eye, with strong angular projections in front of nostrils; no ridges continued from supraorbital region; nostrils rather close to eye; barbel shorter than eye; tip of lower jaw under anterior nostril; cleft of mouth under posterior margin of orbit; under surface of head naked, with the exception of a few minute, spiny tubercles on under surface of mandible; suborbital ridge very slightly developed; the intermaxillary a long bone, nearly as long as the maxillary; mouth large; teeth on intermaxillary in a double series, those of the outer series much larger than the inner; teeth in mandible uniserial. Dorsal spine strongly serrated; distance of first dorsal from snout equal to nearly 4 times length of its base, its distance from anterior margin of orbit equal to length of head; first spine minute, second strongly serrated, nearly $\frac{2}{3}$ length of head, when laid down is far from reaching origin of second dorsal; when the fin is erect its superior margin is nearly at right angles to plane of back and slightly convex; distance between dorsals twice length of base of first, the second beginning in the perpendicular from fifth ray of anal; anal about 3 times as high as second dorsal; vent under thirtieth scale of lateral line directly in advance of the anal and at a distance from ventral considerably greater than length of that fin; distance of pectoral from snout slightly more than length of head, its length less than that of dorsal spine, slightly more than $\frac{1}{2}$ its distance from the snout, its insertion (upper axil) in middle line of body; insertion of ventral under that of pectoral, slightly in advance of that of dorsal, its first ray not greatly prolonged, about $\frac{1}{4}$ length of distance of fin from snout; branchiostegal membrane narrowly attached to the isthmus, leaving no free margin behind; gill rakers very small tubercles, only 10 below angle on first arch. Color dark reddish brown, spines upon the scales with a metallic luster; young with 3 stellate bosses upon snout, 1 at tip, 1 at some distance upon each side. Length of specimen described 322 millimeters. (Goode & Bean.) Gulf Stream, from Cape Cod to Havana; generally abundant. (Named for George Brown Goode.)

Macrurus asper, GOODE & BEAN, Bull. Mus. Comp. Zool., x, No. 5, 196, 1883, Gulf Stream south of New England, Lat. 41° 24′ 25″ N., Long. 65° 35′ 30″ W., in 1,242 fathoms; name preoccupied by *Macrurus asper*, GÜNTHER; JORDAN, Cat., 131, 1885.

Macrurus goodei, GÜNTHER, Challenger Report, XXII, 136, 1887; substitute for *Macrurus asper*.

Hymenocephalus goodei, GOODE & BEAN, Oceanic Ichthyology, 407, fig. 340, 1896.

2947. NEMATONURUS SUBORBITALIS (Gill & Townsend).

Head $5\frac{1}{2}$; depth $5\frac{3}{4}$; eye 5 in head; snout $4\frac{3}{4}$; maxillary $2\frac{3}{4}$. D. 12–85; A. 102; P. 19; V. 11. Mouth wholly inferior; scales closely adherent and rather large, mostly short and roundish, with considerable exposed surfaces, having radiating ridges beset with weak spines; head a little more than $\frac{1}{5}$ of the entire length; snout projecting but little; median and lateral tubercles faintly developed; infraorbital narrow, divided into 2 well marked areas, an upper wider, distinguised by the glassy tubercular scales, and the narrow lower, almost skinny and scaleless; the ridge independently, is little marked; teeth biserial in the upper jaw, robust in the outer row, very weak in the inner; uniserial in lower jaw and scarcely incurved; dorsal spine strongly serrate, $1\frac{1}{2}$ in head; pectoral $1\frac{2}{3}$ in head; ventrals $1\frac{3}{4}$ in head, with short filaments, reaching vent; interspace between dorsals $\frac{1}{2}$ greater than base of first. Bering Sea. Only the type, 20 inches long, known, the above description taken from it by us. (*suborbitalis*, pertaining to the region below the eye.)

Macrurus (Nematonurus) suborbitalis, GILL & TOWNSEND, Proc. Biol. Soc. Wash., XI, 1897 (Sept. 17, 1897), 234, Bering Sea, southwest of Pribilof Islands, Albatross Station 3603, in 1,771 fathoms. (Type, No. 48773, U. S. Nat. Mus. Coll. *Albatross.*)

1001. ALBATROSSIA, Jordan & Evermann, new genus.

Albatrossia, JORDAN & EVERMANN, new genus (*pectoralis*).

This genus has the form and appearance of *Chalinura*, with the dentition of *Nematonurus*, and the dorsal spines of *Malacocephalus* and *Optonurus;* teeth in the upper jaw strong, in an irregular double series, the outer enlarged; the inner series growing double with age; lower teeth uniserial or nearly so; scales small, rather firm, rough; dorsal spine weak, smooth or very slightly serrate; dorsal fins close together; ventrals well developed; pectorals moderate. Size large. (Named for the good ship *Albatross*, in remembrance of her splendid contributions to our knowledge of the life of the deep seas.)

2948. ALBATROSSIA PECTORALIS (Gilbert).

Head 6 in total; depth $1\frac{3}{5}$ in head; eye $4\frac{1}{2}$ to 5 in head, $1\frac{1}{4}$ in snout. D. X–128; A. 121; V. 7; P. 17; mouth wide, lateral, the short snout projecting beyond premaxillaries for a distance about equaling $\frac{1}{2}$ diameter of orbit; suborbital ridge and lateral ridge on snout inconspicuous; a strong median ridge on snout and a pair of parallel ridges forward from above nostrils; maxillary reaching well behind vertical from posterior margin of orbit, $2\frac{1}{4}$ in head; teeth in 2 somewhat irregular series in front of premaxillaries, the outer series enlarged, the inner directed obliquely inward, the two series merging into one laterally; mandible with a single row, similar to inner series of upper jaw; barbel short, $\frac{2}{3}$ to $\frac{4}{7}$ diameter of orbit; angle of preopercle bluntly rounded, not produced; outer gill arch adnate, as usual in *Macrourus*, 7 short tubercular gill rakers present on its free portion; first dorsal spine slender and weak, with 1 or 2 small retrorse prickles near its middle; distance between dorsals equal to $\frac{2}{3}$ base of first; vent immediately in front of anal origin, its distance from base of ventrals slightly more than $\frac{1}{4}$ head; pectorals long and narrow, reaching vertical from ninth or tenth ray of second dorsal, more than $\frac{1}{4}$ length of head; outer ventral ray produced into a long slender filament, reaching $\frac{5}{8}$ the distance from its base to front of anal; scales rather small, 10 or 11 in a series between lateral line and origin of second dorsal or middle of first dorsal; scales on sides very thin and flexible, readily deciduous, each furnished with low diverging ridges, usually 3 in number, bearing few minute spinules, and projecting but little beyond the margins of the scales; entire head, including snout and mandibles, invested with much smaller scales irregularly imbricated, those on the opercles marked similarly to those on sides, the others usually each with a single median ridge terminating in a spinous point; no naked spots or pits on head or between ventral fins; a small narrow area behind and below axil of pectorals. Color light grayish, darker on belly and head; mouth, gill cavity, and peritoneum black; lateral line black; dorsals and ventrals dusky; anal lighter, edged with blackish; pectorals black. Bering Sea to Oregon. Specimens have been taken at *Albatross* Stations 3071, 3074, and 3075, in depths of 685 to 877 fathoms, off the coast of Oregon, and from near Bogoslof Island in Bering Sea in 664 fathoms. It is a large, firm-fleshed species, easily recognized. (*pectoralis*, pertaining to the pectoral.)

Macrurus (Malacocephalus) pectoralis, GILBERT, Proc. U. S. Nat. Mus. 1891, 563, off the coast of Oregon. (Coll. Dr. Gilbert.)

*Macrurus (Nematonurus) magnus,** GILL & TOWNSEND, Proc. Biol. Soc. Wash., XI, 1897 (Sept. 17, 1897), 234, Bering Sea, southwest of Pribilof Islands. (Types, No. 48770 and 48771, U. S. Nat. Mus. Coll. *Albatross*.)

Albatrossia pectoralis, JORDAN & GILBERT, Report Fur Seal Invest., 1898.

1002. BOGOSLOVIUS, Jordan & Evermann, new genus.

Bogoslovius, JORDAN & EVERMANN, new genus (*clarki*).

This genus is close to *Chalinura*, from which it is distinguished by its dentition, having the teeth in the upper jaw in 2 series, the outer slender and sharp, slightly arrow-shaped; those of the inner small, close set, replacing the villiform band of *Chalinura*. Scales excessively rough; ventral filament produced; dorsal spine filamentous, sharply serrate; dorsal fins close together; pectorals inserted below upper angle of gill opening. Deep seas. (Named for the volcanic island, St. John Bogoslof, in Bering Sea, near which the typical species was dredged.)

a. Ventrals much longer than head, reaching far beyond front of anal.   CLARKI, 2949.

aa. Ventrals shorter than head, scarcely reaching front of anal.   FIRMISQUAMIS, 2950.

---

* We have examined the type and cotypes of *Macrurus (Nematonurus) magnus*, Gill & Townsend, and find them to agree fully with *Albatrossia pectoralis* (Gilbert). The type may be redescribed as follows: Head 5¼; depth 7½; eye 4⅓ in head; snout 4⅔ to 4⅔; maxillary 2⅜; pectoral 2 in head; ventral with short filament, 2¼ in head. Mouth large, with lateral cleft. Dorsals well separated, the interspace not ⅝ base of first dorsal; long dorsal spine smooth, or with 1 or 2 roughnesses near its tip, its length 3⅔ in head; second dorsal low; pectoral inserted low, below angle of opercle. Scales moderately large, readily deciduous, decidedly oblong or long, with a small exposed surface which is beset with about 5 radiating ridges with conspicuous spinigerous ridges on dorsal surface, but not armed at tip; head regularly conical; snout rather long, projecting ⅓ its length beyond mandible; tubercles feebly developed, plain, and continuous from 3 parallel ridges; infraorbital flat, with the crest rather nearer the orbit than its lower margin; its entire surface scaly; teeth in the upper jaw biserial or triserial in front, the outer series strongly hooked, the inner series considerably smaller and well separated from the outer series; an irregular series between in the type specimen; teeth in lower jaw uniserial or irregularly biserial. Three specimens, the largest (type of *M. magnus*) 43 inches long.

2949. BOGOSLOVIUS CLARKI, Jordan & Gilbert.

Eye 4¼ in head; maxillary 2½. D. II, 12- ; P. 19; V. 10. Snout short, slightly exceeding diameter of eye, 3$\frac{1}{16}$ in head; median and nasal ridges very little projecting anteriorly, without radiating spines; tip of snout very little projecting beyond the mouth, for a distance not exceeding ¼ the interspace between ends of median and nasal ridges. Suborbital ridge inconspicuous, scarcely extending beyond the eye; mucous pores on head prominent. Mouth large, oblique, the lower jaw included, the maxillary nearly reaching vertical from posterior edge of orbit. Outer premaxillary teeth slender, sharp, unequal, rather distant, not very strong, slightly widened and arrow-shaped near tip, becoming very small toward angle of mouth; within this, and well separated from it, a close-set series of short teeth directed inward. Mandibular teeth slender, unequal, in a single series corresponding to outer series in upper jaw, slightly widening at symphysis, which is not prominent. Barbel very short, less than ½ diameter of pupil. Eye of moderate size, equaling distance from tip of snout to middle of anterior nostril, 1$\frac{1}{10}$ in interorbital width. Preopercle broadly rounded, the angle little produced backward, leaving a strip of interopercle exposed along its entire length. Gill membranes joined to the isthmus, with a narrow free edge. Gill rakers very short and thick, 3 + 12 in number, including rudiments. Dorsal beginning above base of pectorals, the second spine long, filamentous at tip, 1⅔ in head, its anterior margin sharply serrate, except in basal third; base of first dorsal 2⅘ in head; interspace between dorsals very short, usually less than diameter of pupil. Pectorals very long and slender, equaling or exceeding length of head behind snout; insertion of pectorals below upper angle of gill opening. Outer ventral ray excessively produced, twice or more than twice length of head in uninjured adults, reaching base of fiftieth anal ray or beyond. Vent immediately before anal origin. Scales in a strip along the back firm and very rough, none others preserved in our specimens; scales with 3 to 5 sharp, radiating ridges, each ridge with several sharply projecting spines, the posterior of which project beyond the margin of the scale. Color very light gray, the vertical fins blackish posteriorly; mouth and gill cavity and peritoneum jet-black. Bering Sea. Known from 4 specimens, 24 to 41 cm. long, from *Albatross* Station 3634, off Bogoslof Island, in 664 fathoms. (Named for George Archibald Clark, secretary of the Fur Seal Commission for 1896 and 1897, in recognition of his researches on the mammalia of Bering Sea.)

Bogoslovius clarki, JORDAN & GILBERT, Report Fur Seal Invest., 1898, Bering Sea off Bogoslof Island, in 664 fathoms.

2950. BOGOSLOVIUS FIRMISQUAMIS (Gill & Townsend).

Head 5 in total; depth 6½; eye 4⅔ in head; snout 3⅔; second dorsal spine 1⅔ in head; pectoral 2; ventral 1½; maxillary 2½. D. II, 10-126; A. 105; P. 20; V. 8. Scales firmly affixed, oblong or rather short, and with considerable exposed surfaces, which have subequal radiating ridges beset with numerous acute spinelets, the ridges varying from 3 to 8 in number; head regularly convex in profile: rostral tubercles obsolete and infraorbital ridge rounded; barbel greater than pupil; teeth biserial or partly triserial above; second dorsal spine with short retrorse serræ, the lower fifth smooth; base of first dorsal 3½ in head; interspace between dorsal fins ⅓ base of first dorsal, greater than diameter of pupil. This species is distinguishable from most American *Macrouri* by the very firm scales, and from *B. clarki* by the much shorter ventral. Bering Sea. Only the type, 31 inches long, known. (*firmus*, firm; *squama*, scale.)

Macrurus firmisquamis, GILL & TOWNSEND, Proc. Biol. Soc. Wash., XI, 1897 (Sept. 17, 1897), 234, Bering Sea, southwest of Pribilof Islands. (Type, No. 48772, U. S. Nat. Mus. Coll. *Albatross*.)

1003. CHALINURA, Goode & Bean.

Chalinura, GOODE & BEAN, Bull. Mus. Comp. Zool., X, No. 5, 198, 1883 (*simula*).
Chalinurus,* GÜNTHER, Challenger Report, XXII, 124, 144, 1887; change in spelling.

Scales cycloid, fluted longitudinally, with slightly radiating striæ. Snout long, broad, truncate, not much produced. Mouth lateral, subterminal, very large. Head without prominent ridges, except the subocular ones and those upon the snout. Suborbital ridge not reaching angle of preopercle. Teeth in the upper jaw in a villiform band, with an outer series much enlarged, those of the lower jaw uniserial, large. No teeth on vomer or palatines; small pseudobranchiæ present. Gill rakers spiny, strong, depressible, in double series on anterior arch. Ventrals below the pectorals; chin with a barbel. Dorsal spine serrate; soft dorsal much lower than anal. Deep sea fishes. Species numerous. This genus is allied to *Macrourus*, differing in the dentition; the genus *Optonurus*, with dorsal spine unarmed, is very close to *Chalinura*. (χαλινός, a strap or thong; οὐρά, tail.)

a. Snout long, longer than eye, which is 5 in head; pectoral 1¾ in head; dorsal spine 1¼ in head; scales 130.   SERRULA, 2951.
aa. Snout moderate, about as long as eye, which is 4 in head; dorsal, pectoral, and ventral produced, the pectoral 1¼ in head, the dorsal spine and ventral filament each about as long as head.   FILIFERA, 2952.
aaa. Snout very short, as long as eye, which is 5 in head; ventrals very long.
                                            SIMULA, 2953.

2951. CHALINURA SERRULA, Bean.

Head 5¼ in total length. D. II, 9–76 (?); scales 7 or 8–130–17; Br. 6. Cheeks and opercles scaly; snout with a median serrated keel on the nose; diameter of eye less than length of snout, 5 in head; maxillary reaching vertical from posterior margin of eye, its length 2¼ in head; mandible about 2 in head, a row of 5 pores on its under surface and 6 pores on the edge of the suborbitals; branchiostegal membrane narrowly free from the isthmus, the first gill opening restricted as in *Macrourus;* gill rakers small tubercles, 11 below the angle of the first arch, and only 1 or 2 above the angle; length of pectoral equals postorbital part of head; ventrals

---

* Goode & Bean rightly protest against the wanton "action of the English ichthyologists in changing the form of the generic name" *Chalinura*. *Chalinura* is perfectly correct, and should be used even if it were not so, as it is the original form, the only reason for changing it being that other generic names in the group end in *urus*.

about as long as head; longest dorsal spine strongly serrated and nearly equaling length of head without snout; dorsals separated by an interspace $\frac{2}{3}$ as long as head. Color brown; head, abdomen, and inside of mouth purple, the purple areas less marked in the type specimen, which is 12$\frac{1}{4}$ inches long. Coast of British Columbia, east of Prince of Wales Island, in 1,569 fathoms. (*serrula*, a fine saw.) .

Chalinura serrula, BEAN, Proc. U. S. Nat. Mus. 1890, 37, east of Prince of Wales Island, in 1,569 fathoms. (Coll. *Albatross*.)

2952. CHALINURA FILIFERA, Gilbert.

D. II, 12 to 14; P. 20 to 22; V. 9 or 10; eye 4 in head. Snout short, slightly exceeding diameter of eye, $3\frac{9}{10}$ in head, median ridge and nasal ridges terminating each in a much projecting point, furnishing each with a short rosette of radiating spines and ridges, outline between these points concave; tip of snout projecting beyond premaxillaries for a distance equaling that which separates the central rosette from 1 of the lateral ones; infraorbital ridges inconspicuous, not reaching angle of preopercle behind or bony portion in front. Mouth large, slightly oblique, with extensive lateral cleft, the maxillary reaching vertical from posterior margin of pupil, 2$\frac{3}{4}$ in head, equaling distance from tip of snout to middle of eye. Outer series of teeth in premaxillary strong, succeeding from a narrow band of smaller cardiform teeth; mandibular teeth similar to inner band of upper jaw, the band becoming slightly wider at the prominent symphysis. Barbel short, $\frac{1}{2}$ to $\frac{2}{3}$ length of snout. Eye large, the diameter of orbit slightly less than interorbital width on snout. Angle of preopercle produced backward, concealing all but the extreme posterior angle of interopercle, the margin appearing serrulate when divested of skin; gill membranes joined to isthmus, with a posterior free margin; gill rakers very short and heavy, 1 + 11. Dorsal beginning vertically above base of pectorals, the second spine extremely long and slender, smooth basally, the terminal half rather strongly toothed, becoming very slender toward tip and terminating in a long membranaceous filament. (In 1 specimen it exceeds length of head, in the others it equals $\frac{4}{5}$ that length.) Length of base of first dorsal equaling $\frac{1}{5}$ length of head; interspace between dorsals short, $\frac{2}{3}$ to $\frac{3}{4}$ length of snout. Pectorals very long and slender, equaling the head without the snout; outer ventral rays very long and filamentous, equaling length of head; vent immediately in advance of anal origin. Scales rather thin, those on back and sides with above 5 diverging ridges, each of which bears a number of short rigid spinules directed very obliquely backward, the posterior projecting but little beyond the margin of the scale; 8 or 9 scales in an oblique series between the middle of first dorsal and the lateral line. Dark brown; the fins, gill membranes, lips, nostrils, and underside of snout black; anterior part of mouth and lining of gill cavity purple; peritoneum blackish brown. Related to *C. serrula*, Bean, from the same region and depth, differing in the larger eye, shorter mental barbel, longer snout, longer pectoral fins, shorter interspace between dorsals and the longer dorsal fin. Coast of British Columbia; known from 3 specimens, 520 to 550 mm. long. (Gilbert.) (*filum*, thread; *fero*, I bear.)

Chalinura filifera, GILBERT, Rept. U. S. Fish Comm. 1893 (1896), 458, off Queen Charlotte Islands, at Albatross Station 3342, in 1,588 fathoms.

2953. CHALINURA SIMULA, Goode & Bean.

Head 5; depth 6; orbit 6 in head; snout 3; interorbital width greater than eye; postorbital part of head 3 times as long as eye; opercle 2 in upper jaw. D. II, 9–113; A. 118; P. 20; V. 9; Br. 6; scales 8–150–17 to 19. Body shaped much as in *Coryphænoides*, but rather stout; back more gibbous in profile, the dorsal outline rising quite rapidly from the interorbital region to origin of first dorsal, thence descending almost in a straight line to end of tail. Preopercle emarginate on its posterior limb. Snout broad, obtuse, scarcely projecting beyond the mouth, its width nearly as great at tip as its own length; median ridge very prominent, gibbous in outline when viewed laterally; lateral ridges starting almost at right angles with the median, and continued upon sides of head; no supraorbital ridges. Nostrils in front of middle of eye, and nearer its anterior margin than to tip of snout; barbel longer than eye; teeth in upper jaw in a broad villiform band, the outer series very much enlarged; lower jaw with teeth in a single series. Scales rather small, but with indications, particularly on the head, of radiating striæ. Origin of first dorsal from snout 4¼ in its base, or from anterior margin of orbit 1 in head; first dorsal spine very short, second rather stout, 1¼ in head, and with a simple serration anteriorly, the serræ closely appressed to the spine; second dorsal separated from the first by a distance equal to length of upper jaw; anal high, its average rays about 3 times as long as those of dorsal, inserted slightly behind perpendicular from last ray of first dorsal; pectoral inserted over base of ventral; origin of ventral from snout less than its longest ray, which is produced in a filament extending to base of eighteenth anal ray. (Goode & Bean.) West Indies and Gulf Stream, in deep water. (*simulus*, pug-nosed.)

Chalinura simula, GOODE & BEAN, Bull. Mus. Comp. Zool., x, No. 5, 199, 1883, Gulf Stream, at Blake Station 308, Lat. 41° 25' 45'' N., Long. 65° 35' 30'' W., in 1,242 fathoms; JORDAN, Cat., 132, 1885; GOODE & BEAN, Oceanic Ichthyology, 412, fig. 345, 1896.
Macrurus simulus, GÜNTHER, Challenger Report, XXII, 148, 1887.

1004. CORYPHÆNOIDES, Gunner.

Coryphænoides, GUNNER, Trondhj. Selsk. Skrift., III, 50, 1765 (*rupestris*).
Branchiostegus, RAFINESQUE, Analyse de la Nature 1810, 86 (substitute for *Coryphænoides*).

Snout short, obtuse, high, obliquely truncated, soft to the touch, except its bony center; mouth broad, terminal, its cleft lateral; head without prominent ridges, the membrane bones of the side of the head soft and papery; teeth villiform in both jaws, those in the outer series of upper jaws somewhat enlarged. Scales spinous, second or elongate dorsal ray finely serrated in front. Lower jaw with a barbel at tip. Deep Sea.

Close to *Macrourus*, differing in the larger terminal mouth. (*κορύφαινα*, *Coryphæna*; *εἶδος*, resemblance.)

a. Head 4 in length; gill rakers 4+15=19· RUPESTRIS, 2954.
aa. Head 6 in length; gill rakers 3+11=14. CARAPINUS, 2955.

2954. CORYPHÆNOIDES RUPESTRIS, Gunner.

D. 10; P. 19; V. 7; gill rakers 4+15=19. Head short, rather compressed; snout short, obliquely truncated in front; cleft of mouth wide, lateral, extending to beyond the center of eye; intermaxillary not much shorter than maxillary. Teeth in villiform bands in each jaw; barbel very small. Interorbital space convex, its width being considerably more than diameter of eye, which, in a specimen 3 feet long, is equal to the length of the snout and $\frac{1}{4}$ of that of the head. Scales equally rough over the whole of their surface, all the spinelets being directed backward; 7 or 8 scales in a transverse series between the dorsal fin and the lateral line; head entirely covered with small scales. Anterior dorsal spine armed with numerous small closely set barbs; outer ventral ray produced into a long filament. Distance between the vent and isthmus $\frac{2}{3}$ the length of the head. The gill membrane entirely free from the isthmus behind. Intermaxillary continues beyond its vertical process and extending almost as far back as the maxillary, these 2 bones being about equal in length; last third of intermaxillary toothless; intermaxillary teeth in a very narrow band, which is uniform in width, the outer teeth only slightly enlarged; mandible with villiform teeth in a broad bunch-like band at the symphysis and becoming uniserial behind. Eye nearly circular. Snout projecting slightly. Gill rakers longer and less tubercular in character than in *Macrourus berglax* and *M. acrolepis*. The suborbital ridge feebly developed and very abruptly curved upward and narrowed in front of the eye where it joins the nasal ridge. In *M. berglax* and *M. acrolepis* the suborbital ridge is very strong and is continued almost in a straight line toward the nasal ridge. (Goode & Bean.) Arctic seas and the north Atlantic, on both coasts south to the banks of Newfoundland and Norway, in deep water. (Eu.) (*rupestris*, living about rocks.)

Coryphænoides rupestris, GUNNER, Trondhjem Selsk. Skrift., III, 50, pl. 3, fig. 1, 1765, Norway; COLLETT, Norges Fiske, 131; JORDAN & GILBERT, Synopsis, 812, 1883; GOODE & BEAN, Oceanic Ichthyology, 402, 1896.
Lepidoleprus norvegicus, NILSSON, Prodr. Ichth. Scand., 51, 1832, Norway.
Coryphænoides norvegicus, GÜNTHER, Cat., IV, 396, 1862.
Macrourus stromii, REINHARDT, Dansk. Vidensk. Afhandl., VII, 129, 1828: GAIMARD, Voy. Skand., Poiss., pl. 11.
Macrurus rupestris, GÜNTHER, Challenger Report, XXII, 138, 1887.

2955. CORYPHÆNOIDES CARAPINUS, Goode & Bean.

Head 6. D. II, 8–100; A. 117; V. 10; eye 4 in head. Snout acute, projecting beyond the mouth, its tip at a distance from the mouth equal to or greater than diameter of eye. Bones of head very soft and flexible, its surface very irregular, there being a very prominent subocular ridge, a prominent ridge extending from tip of snout to middle of interorbital space, and a curved ridge extending from upper anterior margin of orbit over cavity containing nostrils to a prominent point at side of and slightly posterior to tip of snout; barbel ⅔ as long as eye. Interorbital space almost twice diameter of eye, equal to length of upper jaw; preoperculum crenulate; upper jaw extending to vertical through posterior margin of pupil, its length equaling ½ that of head without snout; mandible extending behind vertical through posterior margin of orbit, its length 3 times in distance from tip of snout to origin of first dorsal. Teeth in villiform bands on intermaxillary and mandible, the mandibulary series uniserial in about the second half of its length. First ray of dorsal very short, second compressed anteriorly and serrated, with slender teeth closely appressed and bent upward, its length equaling length of head and greater than height of body; this fin seated upon a hump-like elevation of the back, its base as long as snout; second dorsal beginning over tenth or twelfth anal ray, and at a distance from end of first dorsal equal to length of head without snout; vent located not far behind vertical from end of first dorsal. Scales 22 to 24 in a transverse series (the position of the lateral line can not be determined, but there appear to be 4 above it); scales oval, membranaceous, showing several parallel ridges composed of small spines. Gill membrane very deeply cleft and attached to the isthmus; gill rakers short and stout, about 11 below the angle on the first arch. (Goode & Bean.) Gulf Stream, in deep water. (*carapinus*, formed as in *Carapus*.)

Coryphænoides carapinus, GOODE & BEAN, Bull. Mus. Comp. Zool., X, No. 5, 195, 1883, Gulf Stream, Lat. 41° 24′ 45′′ N., Long. 65° 35′ 30′′ W., in 1,242 fathoms (Type in M. C. Z. Coll. *Blake*); GÜNTHER, Challenger Report, Deep-Sea Fishes, XXII, 139, 1887; GOODE & BEAN, Oceanic Ichthyology, 404, fig. 339, 1896.

1005. HYMENOCEPHALUS, Giglioli.

Hymenocephalus, GIGLIOLI, Pelagos, Genoa, 228, 1884 (*italicus*).
Mystaconurus, GÜNTHER, Challenger Report, Deep-Sea Fishes, XXII, 124, 1887 (*italicus*).

This genus is closely allied to *Coryphænoides*, differing in the smooth dorsal spine, and the membranaceous skull. First dorsal broad, placed far forward over base of pectoral; second dorsal and anal origins nearly opposite, and separated by a considerable space from the vertical from the end of first dorsal; vent far from ventrals. Head large, naked, soft, and cavernous; snout abrupt, perpendicular, or parabolic; mouth lateral, wide. Eye very large, orbital margin forming part of profile of head. Barbel long. Pectoral rather narrow·(10 to 16 rays). Scales thin, deciduous, with fine short spines. Under parts in advance of ventral wholly or partly naked. Deep seas. Remarkable for the papery structure of the bones of the head. (ὑμήν, membrane; κεφαλή, head.)

2956. HYMENOCEPHALUS CAVERNOSUS (Goode & Bean).

Head about 6 in total length; depth 7. D. II, 10–133; A. 27 rays, in a space equal to length of head. Body stoutish, the bones of head very soft and cavernous, spongy, in many places without muscular covering; interorbital area doubly concave, with a spinous medial ridge, its greatest width about 2⅔ in length of head; postorbital portion of head about ½ its length, 1⅓ as long as eye, which is circular, its diameter contained 2⅔ times in length of head. Snout broad, very obtuse, its width at nostril nearly equal to interorbital width, its length 4⅔ times in that of the head; nostrils normal. Teeth in each jaw in villiform bands, very small; a naked space at the symphysis of intermaxillaries; vomer and palatine toothless. Gill-rakers very short, minute, and rather numerous, about 18 below angle of anterior arch. Pseudobranchiæ absent. Barbel ⅔ as long as eye. First dorsal composed of 2 spines, the first minute, inserted at a distance from the snout equal to length of head, the second as long as head without snout, and 10 branched rays, its base equal to diameter of eye; second dorsal almost rudimentary, its rays remarkably short, about 133 in number, its distance from first dorsal ½ length of head; anal much higher than second dorsal, its distance from snout contained about 3½ times in total length; anterior anal rays longest, in length about ¾ diameter of eye; pectoral inserted under first branched ray of first dorsal, its length equal to twice that of eye and about ⅔ that of head. Scales (on type) mostly wanting, except a few on breast and nape, these being rough with small points, dentate behind. Ventral slightly behind the pectoral, its first ray filamentous, reaching to the base of the tenth anal ray, consisting of 11 rays. Color gray, with silvery tints on sides; abdomen and lips dark. (Goode & Bean.) Gulf of Mexico, in deep water. One young individual known. Length 162 mm. (*cavernosus*, cavernous.)

Bathygadus cavernosus, GOODE & BEAN, Proc. U. S. Nat. Mus. 1885, 598, Gulf of Mexico, at Albatross Station 2398, Lat. 28° 45′ N., Long. 86° 26′ W., in 227 fathoms (Type, No. 37337. Coll. *Albatross*); GÜNTHER, Challenger Report, XXII, 156, 1887.
Hymenocephalus cavernosus, GOODE & BEAN, Oceanic Ichthyology, 408, fig. 341, 1896.

1006. MACROURUS, Bloch.

Macrourus, BLOCH, Ichth., V, 152, 1787 (*rupestris*=*berglax*).
Macruroplus, BLEEKER, Versl. Med. Akad. Welenth. Amsterd., VIII, 1874, 369 (*serratus*).
Macrurus, GÜNTHER, Cat., IV, 392, 1862; corrected spelling.

Snout broadly conical, high, projecting beyond mouth; mouth moderate, its cleft horizontal, U-shaped, entirely inferior; teeth in both jaws in villiform bands, those of the outer series not enlarged; head with roughened bony ridges, one of which, on the suborbital and preorbital, simulates the suborbital stay of the Cottoids; eyes very large; scales imbricate, very rough, keeled. Dorsal spine long, serrated on the anterior edge. Deep water fishes. (μακρός, long; οὐρά, tail, hence correctly written *Macrurus*, but *Macrourus* is the original name as given by Bloch.)

a. Top of head with 4 to 6 distinct ridges; depth 6 to 7 in length; 5 scales between lateral line and dorsal.
  b. Anal rays 148; scales each with a strong ridge. BERGLAX, 2957.
  bb. Anal rays 121; scales each with 3 to 5 spinules, otherwise almost unarmed; ridges on top of head very rough. HOLOTRACHYS, 2958.
aa. Top of snout with indistinct ridges or with none.
  c. Pectoral fin moderate, 1½ to 2 in head.
    d. Body rather elongate, the depth 7 to 8 in length; bones of head rather firm; dorsal spine strongly serrated.

e. Head short, 6½ in head; pectoral more than ⅓ head; snout with bony
ridges above.   BAIRDII, 2959.
ee. Head 5½ in length; ventrals 5 in body; pectoral 1½ in head.
                                                        LEPTURUS, 2960.
eee. Head longer, 4¾ in length; pectoral ⅓ as long as head; eye as long
as snout, 4 in head; ventral 8 in body.   ACROLEPIS, 2961.
dd. Body rather robust, the depth 5½ in length; head without ridges above;
scales spinous, not ridged; dorsal spine 1¾ in head.
                                                    STELGIDOLEPIS, 2962.
cc. Pectoral fin elongate, about as long as head; head elevated, not ridged above,
the bones soft; eye large; second dorsal spine rough, nearly as long as
head; scales each with 7 to 9 ridges.   CINEREUS, 2963.

2957. MACROURUS BERGLAX, Lacépède.

D. 12–124; A. 148; P. 18 or 19; V. 8.   Short snout, subtrihedral, pointed
in front, much shorter than the large eye, which is ⅓ or ⅔ length of head
in the adult.   Intermaxillary very short, ½ length of maxillary, and not
continued beyond its expanded vertical process.   Eye oblong.   Whole
under surface of head below suborbital and nasal ridge naked; axil of
pectoral naked; space between ventrals scaled; body scales each with a
single strong median keel, made up of 5 to 8 spines directed backward;
some scales, particularly of head, have also 2 lateral keels; 6 longitudinal
series of scales between first dorsal fin and lateral line; first dorsal
spine indistinctly denticulated toward the point; length of pectoral
nearly or quite ⅓ length of head; longest spine of dorsal very finely serrated along its anterior margin, the serrations becoming obsolete near its
base.   Vent situated behind origin of second dorsal fin.   Gill rakers very
small, tubercular, 9 to 11 on the first arch; gill membranes broadly joined,
free from the isthmus behind.   This form, originally discovered on the
coast of Norway, has been found abundantly as far south as Georges Bank,
where the halibut fishermen catch it, or some closely allied form, on their
trawls.   The first specimen seen by American naturalists was picked up
floating at the surface off the mouth of New York Harbor.   The *Albatross*
obtained it from Station 2528, in Lat. 41° 47′ N., Long. 65° 37′ 30″ W., at a
depth of 677 fathoms.   Günther knew it from Finmark and Greenland, as
well as from New England.   He calls attention to remarkable individual
variations in the specimens examined by him.   (*berglax*, Norwegian name,
from *berg*, cliff; *lax*, salmon.)

Macrourus berglax, LACÉPÈDE, Hist. Nat. Poiss., III, 170, 1800, Greenland, Söndmöre; JOR-
DAN, Cat. Fish. N. A., 131, 1885.
Macrourus fabricii, SUNDEVALL, Vet. Akad. Handl. 1840, 6; COLLETT, Norges Fiske, 128,
1875; LILLJEBORG, Sverig. og. Norges Fiske, 242; GOODE & BEAN, Cat. Fish. Essex Co.
and Mass. Bay, 7, 1879; GÜNTHER, Challenger Report, XXII, 130, 1887.
Macrourus rupestris, GÜNTHER, Cat. Fish. Brit. Mus., IV, 390, 1862 (not of Gunner).
Macrurus berglax, GOODE & BEAN, Oceanic Ichthyology, 391, fig. 334, 1896.

2958. MACROURUS HOLOTRACHYS, Günther.

Head 4¾ in length; depth 6¾.   D. 12–115 to 125; A. 121; P. 20 or 21; V.
5; eye large, round, as long as snout, 2⅓ in head, much wider than interorbital space.   Snout triangular, each point with a tubercle, covered with

strong spines, this border continued as a strong ridge below eye, extending across opercle, this crest covered with coarse, spinous tubercles; mouth rather small, the maxillary reaching middle of eye; teeth very small, close set. Head with salient ridges above, covered with spinous scales; 1 ridge above eye, toward upper angle of gill opening, another ridge along the vertex, nearly parallel with this above it, besides a short temporal ridge; vent far back, under seventh ray of second dorsal. First dorsal not far behind eye, the long ray slightly serrulate; ventrals with a short filament. Scales each with a median crest of 3 to 5 spinules, otherwise almost unarmed. Five scales between lateral line and dorsal. (Collett.) Depths of the Atlantic. Known from 2 specimens, the type 9 inches long, from the mouth of Rio de la Plata, in 600 fathoms; the second, above described, about a foot long, from the banks of Newfoundland, in 1,267 fathoms. (ὅλος, wholly; τραχύς, rough.)

Macrurus holotrachys, GÜNTHER, Ann. Mag. Nat. Hist. II, 1878, 24 mouth of Rio de la Plata in 600 fathoms; GÜNTHER, Challenger Report, XXII, pl. 28, fig. B, 1887; COLLETT, Compagnes Scient. de l'Hirondelle, 1896, 83, pl. 2, fig. 6; GOODE & BEAN, Oceanic Ichthyology, 396, 1896.

2959. MACROURUS BAIRDII, Goode & Bean.

(COMMON RAT-TAIL.)

Head 6⅓ in total length; depth 8; greatest width 13. D. II, 11–137; A. 120; P. 15; V. 7; scales 6–152–19 or 20. Body much compressed posteriorly, tapering from first dorsal to tip of tail; scales irregularly polygonal, the free portions covered with transparent vitreous spines, arranged in from 10 to 12 irregular longitudinal rows. On head and upper part of body, in advance of the first dorsal, the median row of spines most prominent, and presenting the appearance of a low median keel. Lateral line nearly straight, formed by a smooth groove, which replaces 2 or 3 median rows of spines of each scale; greatest height at posterior margin of orbit greater than width at same point, 1¼ times in length of head; width of interorbital area equal to length of snout and length of maxillary; length of postorbital region about equal to horizontal diameter of orbit; length of operculum about ½ length of mandible. Snout sharp, a front view presenting 4 ridges radiating from tip at right angles to each other, the lower one being merely a fold in the skin of the under surface of the head, horizontal ridges continued into the ridges upon the suborbitals; ridge extending backward from tip of snout upon top of head lost in the interorbital space; branches of the horizontal ridges continued upon upper margins of orbits, and there disappearing. Nostrils immediately in front of orbit, the posterior pair much the longer. Mouth situated entirely on lower side of head; symphysis of lower jaw in vertical from anterior margin of orbit, and articulations of mandibles in vertical from posterior margin of orbit; width of cleft of mouth equal to distance between symphysis of maxillaries and line connecting their articulations; upper jaw protractile vertically. Teeth conical, somewhat recurved, of nearly uniform size, arranged in villiform bands; palate smooth. Distance of first dorsal from snout about 4 times the length of its base, and from anterior margin of orbit equal to length of head; first spine very short, not much longer than the teeth of the second spine; second spine in length twice horizontal diameter of orbit, stout, its anterior margin armed from base to tip with 15 teeth pointing upward, the uppermost slender; its length to tip of filament almost equal to distance from origin of second dorsal, this tip when laid back reaching almost to second dorsal; rays decreasing regularly in length so that, when the fin is upright, its shape approximates that of a right-angled triangle, the hypothenuse of which is the second dorsal spine, and its perpendicular side a line touching the tips of the rays; length of base of second dorsal less than that of the anal, its origin over the thirteenth scale of lateral line. Length of longest ray less than length of barbel; all rays very feeble; membrane scarcely perceptible; distance of anal from snout $3\frac{1}{2}$ times in its length at base, its origin under eighteenth scale of lateral line; length of first ray $\frac{1}{4}$ the length of tenth, and 3 times the length of last ray, the length of rays increasing to a point beneath anterior part of first dorsal, and thence gradually decreasing to tip of tail; distance of pectoral from snout 4 times width of interorbital area, its length twice length of mandible; insertion above the middle of depth of body, on a level with center of orbit, its third ray longest, its tip reaching to vertical from base of fourth ray; insertion of ventral behind pectoral and almost under that of first dorsal, its distance from snout slightly exceeding twice its length; tip of ventral filament reaching base of third anal ray. Ground color, light brownish gray; under parts silvery; belly darker, bluish; under surface of snout pink, as is also the first dorsal, except spines; spines of dorsal, ventral, and anterior anal rays blackish; throat, branchiostegal membrane, and isthmus rich deep violet; sclerotic coat green; eyes very dark blue. This species was the first deep-sea fish obtained by the Fish Commission or described by an American ichthyologist. It ranges in depths from 9 to 1,255 fathoms. This species is distinguished by Günther from his *Macrourus œqualis*, which it closely resembles, (1) by its longer snout, which is nearly equal to the diameter of the eye, and (2) by the smaller number of ventral rays (7). (Goode & Bean.) West Indies to Massachusetts Bay, usually in great depths; excessively abundant on the continental slope, with *Phycis chesteri*, far outnumbering all other deep-sea fishes in the region. (Named for Spencer Fullerton Baird.)

Macrourus bairdii, GOODE & BEAN, Amer. Journ. Sci. and Arts 1877, 471, Massachusetts Bay; GOODE, Proc. U. S. Nat. Mus. 1880, 337, 475; GÜNTHER, Challenger Report, XXII, 135, pl. 22, fig. B, 1887; GOODE & BEAN, Oceanic Ichthyology, 393, fig. 335, 1896.

2960. MACROURUS LEPTURUS, Gill & Townsend.

Head $5\frac{1}{2}$; depth 8; eye $4\frac{2}{3}$ in head; snout 4; maxillary $2\frac{2}{3}$. D. XIV–122; A. 116; P. 20; V. 8. Scales deciduous and moderate, oblong or oval with reduced exposed surfaces, those on the back or above the lateral line with a few, 3 to 5, ridges beset with spines, but those below mostly unarmed; head regularly conical; snout moderately extended; median tubercle very projecting, the lateral well developed, connected with the median by a well-defined ridge; infraorbital vertical, with the ridge linear and near the orbit; teeth cardiform in both jaws; the lower teeth beset the outer slope of the jaw. Ventral as long as head; pectoral $1\frac{3}{8}$ in head; dorsal spine serrate, $1\frac{1}{3}$ in head. Apparently close to *M. acrolepis*, but probably with shorter head, longer ventrals, and longer dorsal spine, the eye also larger. Length 22 to 26 inches. Bering Sea. Only 2 specimens known. (λεπτός, slender; οὐρά, tail.)

Macrurus lepturus, GILL & TOWNSEND, Proc. Biol. Soc. Wash., XI, 1897 (Sept. 17, 1897), 233, Bering Sea, southwest of Pribilof Islands, Albatross Station 3604, in 1,401 fathoms. (Type, No. 48767, U. S. Nat. Mus. Coll. *Albatross*.)

Macrurus dorsalis,* GILL & TOWNSEND, Proc. Biol. Soc. Wash., XI, 1897 (Sept. 17, 1897), 233, Bering Sea, southwest of Pribilof Islands, Albatross Station 3604, in 1,401 fathoms. (Type, No. 48768, U. S. Nat. Mus. Coll. *Albatross*.)

2961. MACROURUS ACROLEPIS, Bean.

Head about $4\frac{1}{2}$; depth at ventrals 7; eye $3\frac{1}{4}$ in head; snout 4; maxillary $2\frac{1}{5}$; mandible $2\frac{1}{3}$; pectoral about 2; ventral about $1\frac{2}{3}$. D. XI-111+; A. 94+; P. 20. Form of *M. berglax*, width of head $\frac{3}{4}$ its height; interorbital width $\frac{3}{4}$ eye; snout moderate, pointed. Origin of first dorsal from snout a distance 3 times length of upper jaw; base of first dorsal $3\frac{1}{2}$ in head, or about 3 times distance between dorsals; first dorsal spine very short, the second about $1\frac{3}{4}$ in head, serrate in front. Distance of anal from snout $2\frac{1}{2}$ times its length; distance of pectoral from snout slightly greater than head; distance of ventral origin from snout $\frac{1}{5}$ its length. Length 2 feet or more. Coasts of Vancouver Island, Washington and Oregon, in deep water, in 345 to 786 fathoms; common. A small specimen taken by us off Bogoslof Island. Our specimens have 11 rays in the first dorsal, not II, 11 or 13, as given by Bean. (ἄκρος, sharp; λεπίς, scale.)

Macrurus acrolepis, BEAN, Proc. U. S. Nat. Mus. 1883, 362, Straits of Juan de Fuca, near Neah Bay, Washington (Coll. James G. Swan, from the stomach of a fur seal); JORDAN, Cat. Fish. N. A., 131, 1885; GILBERT, Rept. U. S. Fish Comm. 1893 (1896), 457.

Macrourus acrolepis, JORDAN & GILBERT, Rept. Fur Seal Invest., 1898.

2962. MACROURUS STELGIDOLEPIS, Gilbert.

Head $4\frac{3}{5}$; depth $5\frac{1}{2}$; eye small, $3\frac{1}{2}$ to 4 in head; snout $4\frac{1}{2}$. D. II, 10 or 11; A. 130; scales 155; 5 or 6 scales between lateral line and base of first dorsal. Body deep, the lower profile rapidly rising along anterior portion of base of anal, the tail thus abruptly becoming slender. Head short and deep; snout heavy, little produced, acute at extreme tip; infraorbital ridge not prominent on sides of head or snout, not continued backward on preopercle. A pair of narrow, transverse naked strips on upper surface of snout near tip, separated on each side by a single scale from the naked

*The following is Gill & Townsend's description of *Macrourus dorsalis*:

"Dorsal 15-120; anal 122; pectoral 21; ventral 9. Scales deciduous and rather small, diversiform, with small exposed surfaces; near the dorsal they have about 5 radiating spinigerous ridges, but below the lateral line these ridges are fewer and unarmed; snout short, projecting a considerable length beyond the eye and a little beyond the supramaxillary; median tubercle very prominent; connecting ridge well defined; infraorbital nearly vertical, with the ridge linear and near the orbit; teeth cardiform."

To this we add the following, from our examination of their type: Head $5\frac{3}{8}$; depth 7; eye $4\frac{3}{5}$; snout 4; interorbital width slightly greater than eye; maxillary $2\frac{3}{5}$; ventral fin $\frac{1}{8}$ longer than head, $4\frac{1}{3}$ in body; pectoral $1\frac{1}{2}$ in head. Dorsal spine strongly serrate, its length equal to that of head.

nostril fossa; a double series of scales intervenes between the nostrils and the orbit; lower side of snout wholly naked anteriorly, partly scaled laterally. Mouth large, overhung by premaxillaries for a distance about ½ diameter of orbit; premaxillaries in advance of nostrils; maxillaries reaching vertical from posterior margin of pupil, 2⅜ in head; snout about equaling interorbital width; barbel long, ⅝ orbit. Teeth in cardiform bands of equal width in both jaws, narrowed laterally, but not to a single series; anterior series in upper jaw enlarged, in lower jaw all the teeth of equal size. Preopercle broadly rounded, the angle but moderately produced, a narrow strip of the interopercle visible for its entire length; outer gill arch partially joined to cover, as usual; gill rakers obsolete; gill membranes united, forming a wide free fold across isthmus posteriorly. Scales without ridges, their exposed surfaces thickly beset with spines which are usually without definite arrangement; the marginal spine longest, thence decreasing in length to the base, about 40 present on each scale on middle of sides; scales on head crowded, the spines shorter and not directed backward as on the body; a rosette of short spines on tip of snout; no naked area between ventrals; mandible and gill membranes partly scaled; no considerable naked area in axil of pectorals. Dorsal inserted over base of pectorals, the length of its base slightly less than ½ the interspace between base of dorsals; second dorsal·spine rather short and fragile, furnished anteriorly with a series of retrorse spinules, its length slightly exceeding ¼ that of head, its tip not reaching origin of second dorsal; origin of anal fin well in advance of second dorsal; the vent unusually far forward, its distance from base of ventrals 2 to 2½ in its distance from anal fin; ventrals less widely separated than in *M. scaphopsis*, the outer ray produced, extending beyond front of anal; ventrals with 10 rays; pectorals with 22 to 24 rays; longest pectoral ray equals ½ head. Color very dark brownish, lighter on tail; lower side of head, breast, and abdominal region, including front of anal and base of pectorals, blue black; roof of mouth, valvular flap of membrane behind bands of teeth, gill membranes, and upper posterior portion of opercular lining, black; mouth and gill cavity otherwise white; peritoneum bright silvery, with little black specking; fins dusky. (Gilbert.) Coast of southern California. Two specimens, the longest 12 inches in length, from *Albatross* Station 2960, in 267 fathoms. (στελγίς, a scraper; λεπίς, scale.)

Macrurus stelgidolepis, GILBERT, Proc. U. S. Nat. Mus. 1890, 116, coast of southern California, at Albatross Station 2960, in 267 fathoms.

2963. MACROURUS CINEREUS, Gilbert.

(POP-EYE.)

D. II, 10 or 11; ventral 9; 7 scales between lateral line and first dorsal. Eye 3¾ to 4 in head; snout about 4, high and blunt, but little overlapping the mouth, terminating in a pointed prolongation of the median ridge, which bears at its tip a bony tubercle furnished with radiating ridges; nasal ridges terminating in shorter and smaller, but similar, tubercles, the outline between them concave; tip of snout overpassing the premaxillaries for ⅔ its length; eye very large and protuberant; mouth of moderate size, the maxillary reaching vertical from hinder margin of orbit, equaling length of snout and $\frac{1}{2}$ of eye. Teeth finely villiform, in each jaw, the outer series not at all enlarged, the mandibular band narrow. Barbel short and slender, its length less than $\frac{1}{3}$ diameter of pupil; interorbital width $\frac{6}{7}$ diameter of orbit, equaling length of snout; preopercle greatly expanded, much overlapping the interorbital below, leaving exposed only the extreme posterior angle. Gill membranes narrowly joined, with a posterior fold, free from the isthmus; gill rakérs short, compressed, almost tubercular, 2 + 12. Origin of dorsal well behind base of pectorals; second dorsal spine long and filamentous, strongly spinous except on extreme base and tip; length of spine $\frac{5}{6}$ to $\frac{9}{7}$ head; base of first dorsal equaling diameter of orbit; interspace between dorsals $\frac{1}{2}$ to $\frac{2}{3}$ base of first dorsal; pectoral long and slender, equaling length of head behind anterior nostril opening, about as long as the filamentous outer ventral ray; vent immediately in front of anal origin. Scales on sides well imbricated, each with 7 to 9 parallel ridges which bear short sharp spines directed very obliquely backward; 7 scales between lateral line and base of first dorsal. Color uniform light grayish on body and fins, with the exception of the blackish pectorals and ventrals; sides of head silvery; mouth, gill cavity, and peritoneum brownish or purplish black; gill membranes and gular membrane dusky. (Gilbert.) Bering Sea; excessively abundant in the depths, where it outnumbers all other fishes. Numerous specimens from north of Unalaska Island, at *Albatross* Stations 3307 and 3329, in 1,033 and 399 fathoms; and the North Pacific, south of Unimak Island, *Albatross* Station 3340, in 695 fathoms. Our many specimens from near Bogoslof Island, in 664 fathoms. (*cinereus*, ashy gray.)

Macrourus cinereus, GILBERT, Rept. U. S. Fish Comm. 1893 (1896), 457, near Unalaska and Unimak Islands, in 399 to 1,033 fathoms; JORDAN & GILBERT, Report Fur Seal Invest., 1898.

1007. CŒLORHYNCHUS, Giorna.

Cœlorhynchus, GIORNA, Mém. Ac. Sci. Turin, XVI, 178, 1803 ("*Cœlorhynche la ville*").
Krohnius, COCCO, Lettera al Sig. Augusto Krohn, Pesci del Mare de Messina, I, 1844 (*filamentosus*; larva).
Paramacrurus, BLEEKER, Versl. Med. Ak. Wetensk. Amsterd. 1874, 103 (*australis*).
Oxymacrurus, BLEEKER, Versl. Med. Ak. Wetensk. Amsterd. 1874, 103 (*japonicus*).

This genus agrees with *Macrourus* in all essential respects, except that the small mouth is wholly below the long-pointed, sturgeon-like snout. Dorsal spine smooth in typical species, those with serrate spine having been lately separated under the generic name *Cœlocephalus*. (Gilbert & Cramer, Proc. U. S. Nat. Mus. 1896, 422) (*acipenserinus*). Species numerous. (κοῖλος, hollow; ῥυγχος, snout.)

a. Head large, $3\frac{1}{2}$ in length; depth 7; eye $\frac{1}{4}$ length of the long snout, 4 in head; dorsal spine moderate. OCCA, 2964.
aa. Head short, $4\frac{1}{2}$ to 5 in length.
  b. Body rather elongate, the depth 8 in length; eye as long as snout, 3 in head. CARMINATUS, 2965.
  bb. Body less elongate, the depth $6\frac{1}{2}$ to $6\frac{3}{4}$ in length.
    c. Dorsal spine long; anal rays about 110; scales 124. CARIBBÆUS, 2966.
    cc. Dorsal spine very short; anal rays 95; scales 98. SCAPHOPSIS, 2967.

2964. CŒLORHYNCHUS OCCA (Goode & Bean).

Head 3½ in total length; depth 7 in total length; snout exceedingly elongate, nearly twice as long as diameter of eye; a black flap between nostrils; angle of mouth nearly reaching vertical from posterior margin of the orbit; ridge of head very strong and continuous from snout to angle of preopercle, having, also, strong supraocular and occipital ridges; eye nearly round, its horizontal diameter 4 in head and equal to interorbital space; ventral originating under middle of first dorsal, and extending to fourth ray of anal; distance from ventral origin to vent 3½ in length of head; second spine of dorsal weak and smooth, its length equal to postorbital part of head, its base slightly less than distance between first and second dorsals; squamation excessively rough, each scale bearing about 5 large spines besides many smaller ones, the median spine of the large series being much the largest; 5 rows of scales between origin of dorsal and lateral line, 19 from vent forward to lateral line and 12 backward; barbel ¼ as long as snout. This species has scales similar to those of *Macrourus berglax*, there being a strong median keel formed by series of spines, of which the last is the largest; surface of each scale also with about 4 or more lateral ridges formed by series of short spines. In a much larger example (U. S. Nat. Mus. No. 37334), measuring 18 inches in length, the lateral series of keels have greatly increased in number, the individual spines having become more prominent, so that the median keel has become less conspicuous than in the type. In the larger specimen referred to, the nakedness of the under surface of the head is even more pronounced than in the smaller, in which the under surface of the head beneath the suborbital and nasal ridge is almost entirely naked. The intermaxillary has a very short bone similar in structure and dentition to that of *Macrourus berglax*, that is to say, the intermaxillary teeth are in a rather broad villiform band, and the outer teeth are not enlarged; mandibulary teeth in a similar broad villiform band; mouth entirely inferior and small. Gill membranes attached across the isthmus, very little emarginate, and not deeply cleft; in the large example the gill membrane is attached to the isthmus and not deeply cleft, but there is a very narrow free margin behind. The gill rakers are very short, tubercular, and few in number, certainly not more numerous than in *M. berglax ;* in the large example only 8 little tubercles can be seen on the first gill arch. Second spine of the dorsal in the type specimen is smooth, with the exception of 2 weak spines near its tip, but in the large example there is no trace of serrations on the dorsal spine. (Goode & Bean.) Length 450 mm. Gulf of Mexico and West Indies, in deep water. (*occa*, a harrow, from the rough scales.)

Macrurus occa, GOODE & BEAN, Proc. U. S. Nat. Mus. 1885, 595, Gulf of Mexico, Lat. 28° 34′ N., Long. 86° 48′ W., in 335 fathoms. (Type, No. 37334. Coll. *Albatross.*)

Cœlorhynchus occa, GOODE & BEAN, Oceanic Ichthyology, 400, figs. 332, 333, and 337, 1896.

2965. CŒLORHYNCHUS CARMINATUS (Goode).

Head about 5 in total length; depth 8; eye about 5 in head, equaling interorbital width; snout equaling eye or postorbital part of head; length of opercle about 2 in snout. Body less elongate than in *M. bairdii.* Snout long, sharp, depressed, triangular, the lower surface more nearly parallel with the axis of body than in *M. bairdii;* lateral ridges more pronounced, continued in a straight line under eye and upon preopercle; strong horizontal ridges running from supraorbital margins to gill openings, parallel with subocular ridges; nostrils immediately in front of orbit; barbel very short. Teeth small, conical, somewhat recurved, arranged in villiform bands. Origin of first dorsal to snout $4\frac{1}{2}$ times its base, its distance from anterior margin of orbit much less than length of head; first spine very short, hardly perceptible above the skin; second spine about 2 in head, slender and unarmed, when laid back its tip reaching to or beyond origin of second dorsal, the spines decreasing in length very gradually, the sixth being nearly as long as second, so that the fin is not so triangular as in *M. bairdii;* second dorsal beginning in a perpendicular from seventh anal ray; anal much higher than in *M. bairdii,* nearly equal to $\frac{1}{2}$ interorbital width, its origin under eighteenth scale of lateral line, its longest rays as long as interorbital width; distance of pectoral from snout equaling twice its own length, which about equals longest dorsal spine; origin of pectoral below middle of depth of body and below level of middle of orbit, its tip not reaching origin of anal; insertion of ventrals behind pectoral, slightly in advance of first dorsal, its distance from snout greater than twice its length, the long filament not reaching anal. Color silvery gray. Length 250 mm. This species is extremely close to the common Mediterranean species, *C. cœlorhynchus* (Risso), but the spines on the scales are a little larger. West Indies, Gulf of Mexico, and in the Gulf Stream in deep water; abundant; taken at many stations by the *Albatross,* the *Blake,* the *Fish Hawk,* and the *Challenger,* in 115 to 464 fathoms. (*carminatus,* from *carmen,* a wool card.)

Macrurus carminatus, GOODE, Proc. U. S. Nat. Mus., III, 1880, 346 and 475, Gulf Stream off Rhode Island, Lat. 40° 02′ 54″ N., Long. 70° 23′ 40″, at Fish Hawk Station 871, in 115 fathoms (Type, No. 26007); GOODE & BEAN, Bull. Mus. Comp. Zool., x, No. 5, 196, 1883.
Macrurus (Cœlorhynchus) carminatus, GÜNTHER, Challenger Report, Deep-Sea Fishes, XXII, 129, pl. 5, fig. 13, 1887.
Cœlorhynchus carminatus, GOODE & BEAN, Oceanic Ichthyology, 398, fig. 336, 1896.

2966. CŒLORHYNCHUS CARIBBÆUS (Goode & Bean).

Head $4\frac{1}{3}$ in total length; depth $6\frac{1}{4}$ in total length. 1 D. II, 8; 2 D. at least 110; A. 110+; scales 6-124-15 or 16. Body normal in shape; scales moderate, strong, densely covered with minute spines, without enlarged median keel; interorbital area flat, its greatest width about 5 times in length of head; postorbital portion of head about 3 in head, and just as long as eye, which is oval, and $1\frac{2}{3}$ as long as its vertical diameter. Snout long, thin, diaphanous, with acuminate point, its general form resembling that of *C. carminatus.* The nostrils close to the orbit, the posterior one much the larger. Teeth in each jaw in villiform bands, minute. Barbel slender and short, its length $\frac{1}{4}$ that of eye. Maxillary extending to vertical through middle of pupil; upper jaw about 3 in head; mandible $2\frac{1}{4}$; intermaxillary short. Outer series of teeth on intermaxillary and mandible not enlarged, the teeth not becoming uniserial. Gill membranes narrowly attached to the isthmus; gill rakers minute, tubercular, about 10 on first arch. Suborbital ridge very strong, continued almost in a straight line by the lateral ridge of the snout; under surface of head, except chin and branchiostegal region, densely covered with small, spiny tubercles; a naked space on underside of snout, occupying almost entire distance from front of mouth to tip of snout, widest anteriorly, the greatest width 5 in snout; intermaxillary protractile in a vertical direction; mouth distinctly inferior. Origin of second dorsal over seventh anal ray, about an eye's diameter behind first dorsal; length of anal rays about 4 in head; origin of pectoral in front of first dorsal, its length 2 in head, its tip reaching fifth anal ray. Color silvery gray, with yellowish and lavender tints. Length 290 mm. Caribbean Sea north to the Gulf of Mexico, in deep water. (*caribbæus*, of the Caribbean Sea.)

Macrurus caribbæus, Goode & Bean, Proc. U. S. Nat. Mus. 1885, 594, Gulf of Mexico, at
    Albatross Station 2377, Lat. 29° 07′ 30″ N., Long. 88° 08′ W., in 210 fathoms (Type,
    No. 37333); Günther, Challenger Report, Deep-Sea Fishes, xxii, 124, note 3, 1887.
Cœlorhynchus caribbæus, Goode & Bean, Oceanic Ichthyology, 401, fig. 338, 1896.

2967. CŒLORHYNCHUS SCAPHOPSIS (Gilbert).

Depth 6¾ in total length; head 4⅕; scales 98. D. II, 8; A. ca. 95. Snout flattened, acute, the conspicuous infraorbital ridge forming a strong ridge along its sides, the two meeting at tip in a salient point; an evident keel extending from tip of snout to middle of interorbital area; supraorbital ridge dividing anteriorly, 1 branch running down in front of nostril, the other separating nostril fossa from orbit; between the ridges the head is covered with a soft, yielding integument, which is semitranslucent. Lower side of snout wholly naked below, and with a large naked area above on each side of tip; snout projecting beyond mouth for a distance equaling length of maxillary. Mouth of moderate size, the maxillary reaching vertical from posterior margin of pupil, 3⅓ in head. Teeth villiform, in a broad band in upper jaw, in a narrower band below, not reduced to a single series laterally in either jaw, and none of the teeth enlarged. Eye large, elliptical, equaling length of snout, 3⅓ in head; interorbital width 4⅓. Barbel short, about ½ pupil; preopercular angle greatly produced backward, wholly concealing the interopercle, the strong infraorbital ridge failing to reach preopercular margin by only ⅓ diameter of pupil. Structure of gills as usual in this genus, the gill rakers obsolete; gill membranes broadly united, joined to isthmus, across which they form posteriorly a very narrow free fold. Besides the ridges already described on head, there are a pair on occiput, a pair from upper posterior margin of orbit to upper angle of gill opening, and a median ridge on nape reaching about halfway from occiput to dorsal. These ridges, as well as the interorbital space and the area between the occipital ridges, covered with scales compressed to a knife-like edge, which is provided with a single series of backward-directed spines; scales on infraorbital and rostral ridges bearing stellate spines or are similar to those on temporal region, sides of head, and body generally; scales on body large; 3 longitudinal series between lateral line and middle of first dorsal; each scale provided with a ridge bearing about 6 backward-directed spines, and from 2 to 4 pairs of lateral ridges also bearing spines, the lateral ridges sometimes extending the whole width of scales, sometimes confined to their basal portion; marginal spines longest; axil of pectoral naked, its base anteriorly with small cycloid scales; a naked, much depressed, elliptical area between bases of ventrals in all specimens; second dorsal spine smooth, weak, little exceeding length of soft rays, equaling length of snout and orbit; base of first dorsal 1½ in interspace between dorsals, which is 2⅔ in head; distance from front of anal to snout equaling ¼ total length; ventrals with the outer ray produced, about reaching front of anal; pectorals reaching beyond anal ¼ length of head; ventrals with 7 rays; pectorals with 15 to 17 rays. Color light olive brown, dusted with coarse black specks; axil of pectorals, belly, ventrals, and branchiostegal membranes blue black; lower side of head dusky; mouth anteriorly, including tongue and ½ of palate, white, its posterior part and most of lining of gill cavity jet-black; inner lining of cheeks abruptly white; lower part of iris silvery; peritoneum silvery, with coarse dusky specks; vertical fins dusky, the anterior portion of anal black. Coast of southern California. Many specimens, the longest 12 inches long, from *Albatross* Station 3015, in 145 fathoms. (Gilbert.) (σκάφη, spade; ὄψις, face.)

Macrurus (Cœlorhynchus) scaphopsis, GILBERT, Proc. U. S. Nat. Mus. 1890, 115, Albatross Station 3015, coast of southern California, in 145 fathoms.

1008. TRACHONURUS, Günther.

Trachonurus, GÜNTHER, Challenger Report, Deep-Sea Fishes, XXII, 124, 1887 (*villosus*).

Scales not imbricated, separated by furrows, and densely covered with sharp spinules, so that the animal seems villous to the touch; dorsal spine smooth; dorsal much lower than anal; teeth in both jaws in villiform bands; snout obtuse, the mouth subinferior; suborbital ridge little developed. This genus is distinguished from *Cœlorhynchus* by the indistinct squamation. (τραχύς, rough; οὐρά, tail.)

2968. TRACHONURUS SULCATUS (Goode & Bean).

Head 7½ in total length; depth about 9½; eye 3⅔ in head; snout 4 to 4½. D. II, 8 or 9, the second of numerous low rays; A. 120; V. 7; P. 13; scales 7-175 or more—33. Barbel 2½ to 2 in eye. Body elongate, rapidly contracted behind the abdomen; the tail long and whip-like. Scales moderate, strongly armed, each with 8 to 10 spinelets, irregularly placed, less numerous in the young, which feel bristly to the touch, separated by wide deep furrows; armature of head similar to that of body, but the scales upon snout, cheeks, and chin have very feeble spines. Interorbital area nearly flat, its length equaling diameter of eye or about 3 in head; postorbital part of head as long as eye; snout short, obtuse, scarcely overhanging the mouth; nostrils somewhat above level of middle of eye, the anterior one nearly upon the dorsal outline. Upper jaw with 2 series of teeth in villiform bands, the outer series slightly enlarged; teeth of lower jaw in a single series; maxillary reaching to vertical through hind margin of pupil in adult, shorter in younger individuals; length of upper jaw, including maxillary, 3 in head; mandible 2 in depth of body; barbel 2 to 2¼ in eye. Gill rakers very small, tubular, almost rudimentary, about 10 below angle of first arch; attachment of membrane to first arch very extensive, but free from isthmus; no pseudobranchiæ. First dorsal comparatively low, the first spine rudimentary, the second elongate and smooth; insertion of first dorsal immediately over or somewhat behind base of pectoral, its distance from snout 1½ in head, its base equal to snout, its longest spine, when laid down, reaching behind origin of second dorsal, or 1½ to 2 in head; second dorsal very low, its distance from first 3 to 4 in head, 32 rays in a distance equal to length of head; 22 in same distance of anal; anal much higher than second dorsal, yet very low, its longest ray equal to eye; distance of anal origin from snout 4⅜ in total length, or nearly under origin of second dorsal; pectoral inserted under or somewhat in front of origin of first dorsal, its length about 2 in head; ventral inserted behind vertical from end of base of first dorsal, extending to origin of anal, its length about equaling eye; vent about midway between origin of ventrals and anal. Color brown; abdomen and lower parts of head blackish in the young. West Indies and Gulf of Mexico, in deep water; taken both by the *Albatross* and the *Blake*. (*sulcatus*, furrowed.)

Coryphænoides sulcatus, GOODE & BEAN, Proc. U. S. Nat. Mus. 1885, 596, Gulf of Mexico, at Albatross Station 2394, Lat. 28° 38′ 30″ N., Long. 87° 02′ W., in 420 fathoms (Type, No. 37335); GOODE & BEAN, Oceanic Ichthyology, 403, 1896.
Macrurus (Malacocephalus) sulcatus, GÜNTHER, Challenger Report, Deep-Sea Fishes, XXII, 169, 1887.
Trachonurus sulcatus, GOODE & BEAN, Oceanic Ichthyology, 410, fig. 343, 1896.

1009. LIONURUS, Günther.

Lionurus, GÜNTHER, Challenger Report, XXII, Deep-Sea Fishes, 124, 1887 (*filicauda*).

This genus is close to *Macrourus*, differing in the smooth, flaccid scales, and soft, cavernous skull, characters associated with its extreme bathybial degradation. (λεῖος, smooth; οὐρά, tail.)

a. Barbel minute, not ½ pupil; eye small, 5 in head; tail very slender.
FILICAUDA, 2969.

aa. Barbel moderate, 1⅔ in eye; eye 3¾ in head.
LIOLEPIS, 2970.

2969. LIONURUS FILICAUDA (Günther).

D. 11; P. 20; V. 9; cæca 7. Snout considerably projecting beyond the mouth, pointed in the middle, twice as long as eye, which is unusually small, only ½ as wide as interorbital space. Mouth rather wide, extending beyond the center of the eye. Upper teeth villiform, in a very narrow band, those of mandible very small, biserial. Barbel minute. Preoperculum with the angle produced backward, broadly rounded and crenulated on the margin. Terminal portion of the tail prolonged into a long filament, more slender than in any of the other species. Bones of head soft. Scales of moderate size, thin, cycloid, and deciduous, 6 or 7 in a transverse series between the first dorsal spine and the lateral line; snout and inferior half of the infraorbital region naked. Second dorsal spine slender, with the barbs in front very inconspicuous and sometimes entirely absent; distance between dorsal fins less than length of head; outer ventral ray produced into a small filament. Distance between vent and isthmus less than length of head. Head and trunk whitish, tail brownish, lower part of head and gill openings black. (Günther.) This species is clearly one of those in this family which extends to the greatest depths. The decrease in the size of the eye, the very soft bones, the comcomitant want of firmness in the structure of the scales, and the tail, which tapers into a very fine filament, indicate its abyssal abode. The scales are nearly all gone in all the specimens obtained. The species appears to be abundant in individuals, and has, like a true deep-sea fish, a wide distribution. (Günther.) Antarctic Ocean and deep seas off both coasts of South America. (*filum*, thread; *cauda*, tail.)

Coryphænoides (*Lionurus*) *filicauda*, GÜNTHER, Ann. and Mag. Nat. Hist., XX, 1878, 27,
 Deep seas on both sides of South America, in 1,375 to 2,650 fathoms.
Macrurus filicauda, GÜNTHER, Challenger Report, Deep-Sea Fishes, 141, pl. 34, fig. B, 1887.
Lionurus filicauda, GOODE & BEAN, Oceanic Ichthyology, 409, fig. 342, 1896.

2970. LIONURUS LIOLEPIS, Gilbert.

Head 4⅔ in length; depth 6⅓; maxillary nearly reaching vertical from posterior margin of orbit, 2⅘ in head; eye 3¾; interorbital space concave, equaling snout, 4⅓; barbel ¾ eye. D. II, 10. A. 120. Snout short and high, with well-marked lateral ridge, the extreme tip flattened; the median ethmoidal ridge is prominent, and the supraocular ridge is continued forward on the snout, meeting the lateral ridge in a projecting point. Top of snout wholly naked mesially, a narrow band of scales around each margin and in front. Lower side of head, including under side of snout, mandibles, gill membranes, and most of interopercles, naked; a very small patch of scales on posterior part of interopercle. Mouth moderate, the snout overhanging the premaxillaries for a distance equaling ⅓ maxillary. Teeth in rather narrow cardiform bands in each jaw, not, however, forming single series laterally; the outer series in upper jaw only is enlarged; angle of preopercle little produced, not concealing the interopercle; infraorbital ridge not continued on to it. Gill membranes forming posteriorly a rather wide free fold across isthmus; outer gill arch joined to gill cover as usual in this genus. Scales small, everywhere cycloid, very deciduous, lost in most specimens; no spines developed, but occasionally can be seen traces of a median ridge and a pair of lateral ridges; about 6 or 7 series of scales between lateral line and base of first dorsal. Origin of first dorsal over or in advance of base of pectorals; base of first dorsal 1⅕ to 1⅖ in interval between dorsals; second dorsal spine usually smooth, occasionally with from 1 to 3 weak prickles near the middle; length of spine 1⅘ in head. Origin of anal slightly behind first dorsal, the vent midway between base of ventrals and anal; ventrals short, the outer ray slightly produced, with from 10 to 12 rays; pectorals with 20 or 21 rays. Color very dark brown; snout, opercles, lower side of head, and abdominal region black or blue black; mouth and gill cavity black; peritoneum dusky silvery. Coast of southern California. Many specimens taken at *Albatross* Station 2980, in 603 fathoms. (λεῖος, smooth; λεπίς, scale.) (Gilbert.)

Macrurus (Lionurus) liolepis, GILBERT, Proc. U. S. Nat. Mus. 1890, 117, coast of southern California, at Albatross Station 2980, in 603 fathoms.

Remotely related to the Scombriform fishes, and perhaps derived from the same ancestral stock as the *Trichiuridæ*, is the singular

Suborder TÆNIOSOMI.

(THE RIBBON-FISHES.)

This group is thus defined by Dr. Gill:

"Scapular arch subnormal, post-temporal undivided and closely applied to the back of the cranium, between the epiotic and pterotic, or upon the parietal; hypercoracoid perforate at or near the margin; cranium with the epiotics enlarged, encroaching backward and juxtaposed behind, intervening between the exoccipitals and supraoccipital; prootic and opisthotic represented chiefly by the enlarged prootic; suborbital chain imperfect; the copular bones separated by intervening cartilaginous elements; the hypopharyngeals styliform and parallel with the branchial arches; epipharyngeals in full number (4 pairs), and mostly compressed; the dorsal fin composed of inarticulate rays or spines, separable into lateral halves, and the ventrals (when present) subbrachial. A myodome may be present or absent, none being developed in the *Regalecidæ*, but 1 being distinct and supplemented by a dichost in the *Trachyteridæ*." (Gill.)

"The ribbon-fishes," says Günther, "are true deep-sea fishes, met with in all parts of the oceans, generally found when floating dead on the surface or thrown ashore by the waves. Their body is like a band, specimens of from 15 to 20 feet long being from 10 to 12 inches deep and about an inch or two broad at their thickest part. The eye is large and lateral; the mouth small, armed with very feeble teeth; the head deep and short. A high dorsal fin runs along the whole length of the back, and is supported by extremely numerous rays, its foremost portion, on the head, is detached from the rest of the fin, and composed of very elongate flexible spines. The anal fin is absent. The caudal fin (if preserved, which is rarely the case in adults) has an extra-axial position, being directed upward like a fan. The ventrals are thoracic, either compressed of several rays or reduced to a single long filament. The coloration is generally silvery, with rosy fins. When these fishes reach the surface of the water the expansion of the gases within their bodies has so loosened all the parts of their muscular and bony system that they can be lifted out of the water with difficulty only, and nearly always portions of the body and fins are broken and lost. The bones contain very little bony matter, and are very porous, thin, and light. At what depth ribbon fishes live is not known; probably the depths vary for different species; but although none has yet been obtained by means of the deep-sea dredge, they must be abundant at the bottom of all oceans, as dead fishes or fragments of them are frequently obtained.

Some writers have supposed from the great length and narrow shape of these fishes that they have been mistaken for 'sea serpents,' but as these monsters of the sea are always represented by those who have had the good fortune of meeting with them as remarkably active, it is not likely that harmless ribbon-fishes, which are either dying or dead, have been the objects described as 'sea serpents.'" ($\tau\alpha\iota\nu\iota\alpha$, ribbon; $\sigma\tilde{\omega}\mu\alpha$, body.)

FAMILIES OF TÆNIOSOMI.

a. Ventral fins reduced each to a single long filament, thickened at the tip; anterior rays of dorsal produced; mouth small; caudal fin short or wanting.

REGALECIDÆ, CCXVI.

aa. Ventral fins normally developed or else wanting.

b. Caudal fin short, fan-shaped, inserted at an angle with axis of body; the tail not much produced beyond it. TRACHYPTERIDÆ, CCXVII.

bb. Caudal fin short, the tail beyond it ending in a long filament, longer than rest of body. STYLEPHORIDÆ, CCXVIII.

Family CCXVI. REGALECIDÆ.

(OAR-FISHES.)

Body very elongated and compressed, the head oblong, the opercular apparatus well developed (the operculum extended backward, the suboperculum obliquely behind it, and the interoperculum extended upward below the 2), the preorbital chain oblique and widest at the second bone; ventrals represented by single elongate rays, the cranium with the myodome atrophied and the dichost suppressed, the supraoccipital pushed forward by the extensive development of the epiotics which encroach forward on the roof as well as back and sides of the cranium, and with short ribs. (Gill.) Superficial characters are the very long dorsal, extending the whole length of the back and with the rays at the nape much produced; pectorals very short; caudal fin short or wanting; anal very low; head small; mouth very short; no air bladder; pyloric cæca numerous. One genus, with 2 or more species. Very large, surface-swimming fishes of the open seas; the great size, undulating motion and projecting mane causing them frequently to be taken for sea serpents. (*Regalecidæ*, Gill, Standard Nat. Hist., III, 1885; GILL, Amer. Nat. 1890, 482.)

1010. REGALECUS, Brünnich.

(OAR-FISHES.)

Regalecus, BRÜNNICH, Nya Sammlung, III, 414, 1788 (*glesne*).
Gymnetrus, BLOCH & SCHNEIDER, Syst. Ichth., 487, 1801 (*remipes*).
Xypterus, RAFINESQUE, Indice, 59, 1810 (*imperati*).

Characters of the genus included above. "It is not certain that there is more than 1 species of *Regalecus*, although, as the synonymy which follows clearly shows, various names have been suggested in connection with the comparatively few individuals which, during the past century and a half, have been captured in the North Atlantic. There appears to be considerable possibility of individual variation in proportions of height to length, and in the number of rays in the dorsal fin, but it is a fact well known to ichthyologists that constancy is not to be expected in forms in which the number of vertebræ and fin rays has been extended far beyond the normal average. It should also be said that most of the individuals studied have been in very imperfect condition, and also that in many instances the observations have been made by untrained observers, so that it seems doubtful whether there is really more than 1 species to be assigned to the Atlantic fauna. At all events, Günther, Collett, Lütken, and Day agree in the idea that it is impossible to discriminate between the forms already described, and we follow their lead in considering them all, for the present, as a single species. It is not impossible, of course, that, should better material be obtained, it may be desirable to separate the group into more subspecies, but until this shall be done discrimination leads to confusion rather than to definite knowledge. The fishes belonging to the genus *Regalecus* are very remarkable, not only on account of their peculiar appearance and structure, but because of their enormous size. They have been known to attain the length of 20 feet, and it is more than probable that they grow very much longer, and that many of the creatures popularly identified with the "sea serpent" are only large individuals of this type. Indeed, it seems quite safe to assign to this group all the so-called "sea serpents" which have been described as swimming rapidly near the surface, with a horse-like head raised above the water, surmounted by a mane-like crest of red or brown. The individual which came ashore at Hungry Bay, in Bermuda, in 1860, and which was about 17 feet long, was described by the people who saw it before its capture as being very much larger, and as having a head of an immense horse with a flaming red mane." (Goode & Bean.) (*rex*, king; *halec*, herring. The species have long been known as "king of the herrings," as have those of *Trachypterus.*)

2971. REGALECUS GLESNE (Ascanius).

(OAR-FISH; SEA SERPENT.)

Head 16 to 20; depth 12 to 24; eye 4 to 6 in head; snout short, truncated. D. 275 to 400; P. 11 to 14; V. I. Body very elongate. Cleft of mouth vertical, the upper jaw very protractile; jaws minute or absent. Anterior 8 to 15 rays of dorsal forming an elevated crest, sometimes in 2 parts, the posterior rays with membranaceous tips; each ventral ray with a lobate membranaceous tip; skin with numerous bony tubercles; lateral line placed low. Color silvery gray, with a few spots or streaks of darker hue, most numerous anteriorly. Günther (Challenger Report, XXII, 73 to 76) has in the most painstaking manner brought together a list of the specimens taken in the North Atlantic; as far as they are known to science. He mentions 14 known upon the Scandinavian coasts from 1740 to 1852; 19 on the British coasts from 1759 to 1884; 1 in the Mediterranean (he states, however, that about ½ a dozen specimens have been observed in the Mediterranean); 1 in the Bermudas; 3 at the Cape of Good Hope; 1 in the Indian Ocean, and 5 off the coast of New Zealand. He calls attention to the fact that of those observed on the British and Scandinavian coasts 4 were observed in the month of January, 5 in February, 8 in March, 2 in April, 1 in May, 1 in June, 1 in July, 2 in August, 1 in September, and 1 in October. He also calls attention to the fact that by far the greater proportion of their capture, in the Northern Hemisphere at least, is in the stormy season. This agrees with what we know of the capture of *Trachypterus*, which likewise seems to be brought to the surface only by great commotions of the ocean. The popular name of *Regalecus* is oarfish, in allusion to the blade-like expansion of the extremities of the 2 ventral fins. *Regalecus* is also called in the books the "king of the herrings." Strangely enough, no representative of this genus has been found on the coast of North America. Günther is of the opinion that the distribution of this fish in the depths of the sea is the same as that of *Trachypterus.* The similarity in their geographical distribution is quite remarkable. (Goode & Bean.) (Eu.) (*glesne*, from "Glesnæs," a farm at Glesvær, near Bergen, where the type of the species was taken.)

Spada marina, IMPERATO, Hist. Nat., 679, 687, 1599, Naples.
Regalecus glesne, ASCANIUS, Icones Rerum Nat., II, pl. 11, about 1788, Glesvær, Norway.
Ophidium glesne, ASCANIUS, Nya Saml. Vid. Selsk. Skrivt., III, 419, 1788.
Regalecus remipes, BRÜNNICH, Nya Saml. Vid. Selsk. Skrivt., III, 1788, 414, taf. B., figs. 4, 5;
　WALBAUM, Artedi Piscium, III, 647, tab. 3, fig. 4, 1792.
Cepola gladius, WALBAUM, Artedi Piscium, III, 617, 1792.
Gymnetrus hawkenii, BLOCH, Ichthyol., XII, 88, 425, 1792.
Gymnetrus grillii, LINDROTH, Vet. Akad. Handl. 1798, 291, pl. 8.
Gymnetrus ascanii, SHAW, Gen. Zool., IV, 197, 1803; after Ascanius.
Xypterus imperati, RAFINESQUE, Indice, 59, 1810; after Ferrante Imperato.
Gymnetrus longiradiatus, RISSO, Eur. Mérid., III, 296, 1826, Nice.
Gymnetrus telium, CUVIER & VALENCIENNES, Hist. Nat. Poiss., X, 361, pl. 299, 1834, Nice.
Regalecus banksii, CUVIER & VALENCIENNES, Hist. Nat. Poiss., X, 365, 1834, Filey Bay,
　Yorkshire.
Gymnetrus capensis, CUVIER & VALENCIENNES, Hist. Nat. Poiss., X, 376, 1834, Cape of Good
　Hope.
Regalecus glesne, ASCANIUS, Icones Rerum Naturalium, 1806, pl. 11; LACÉPÈDE, Hist.
　Nat. Poiss., II, 214, 215, 1800; GOODE & BEAN, Oceanic Ichthyology, 480, fig. 395, 1896.
Gymnetrus remipes, BLOCH & SCHNEIDER, Syst. Ichth., 482, tab. 88, 1801; YARRELL, Brit.
　Fishes, Ed. 2, I, 223, and Ed. 3, II, 301.
Gymnetrus glesne, CUVIER & VALENCIENNES, Hist. Nat. Poiss., X, 366.
Gymnetrus gladius, CUVIER & VALENCIENNES, Hist. Nat. Poiss., X, 352, pl. 298, 1835.
Regalecus gladius, GÜNTHER, Cat., III, 308, 1861.

Family CCXVII. TRACHYPTERIDÆ.

(THE KING OF THE HERRINGS.)

Body moderately elongate, strongly compressed, naked, the skin smooth or prickly. Lateral line present. Head short; the mouth rather small, terminal, with feeble teeth; premaxillaries protractile; opercles unarmed; opercular apparatus abbreviated (the operculum extended downward, the suboperculum below it, and the interoperculum contracted backward and bounded behind by the operculum and suboperculum); the cranium with a myodome and dichost, the supraoccipital continued behind into a prominence; the epiotics confined to the sides and back of the cranium, and without ribs. Eye large, lateral; branchiostegals 6; gill membranes separate, free from the isthmus; gills 4, a slit behind the fourth. Pseudo-branchiæ well-developed, in a pouch formed by a fold of the mucous membrane. Dorsal fin single, extending from the head to the tail, its rays all technically spinous, being neither articulated nor branched, but all very soft, flexible, and fragile; anal fin wanting; pectorals short; ventrals thoracic, the rays elongate, less than I, 5 in number, usually atrophied in the adult; caudal fin either rudimentary or else divided into 2 parts, the upper and larger fan-shaped, directed obliquely upward from the slender tip of the tail. Bones very soft, the muscles little coherent. Pyloric cæca very numerous. Vertebræ in large number. Deep-sea fishes, often of large size, found in most warm seas. Their extreme fragility renders them rare in collections, and the species are little known. One genus; species about 12. The ribbon-fishes are well known in the Eastern Atlantic and the Mediterranean, and have even been found as far west as Madeira [and Cuba]. Some few representatives have been found on the west coast of South America, and 1 or 2 examples have been taken in New Zealand. They are generally admitted to be true deep-sea fishes, which live at very great depths, and are only found when floating dead on the surface or washed ashore by the waves. Almost nothing is known of their habits except through Nilsson's observations in the Far North. This naturalist, as well as Olafsen, appears to have had the opportunity of observing them in life. They say that they approach the shore at flood tide on sandy shelving bottoms, and are often left by the retreating waves. Nilsson's opinion is that its habits resemble those of the flat fishes, and that they move with one side turned obliquely upward, the other toward the ground; and he says that they have been seen on the bottom in 2 or 3 fathoms of water, where the fisherman hook them up with the implements employed to raise dead seals, and that they are slow swimmers. This is not necessarily the case, however, for the removal of pressure and the rough treatment by which they were probably washed upon the shore would be demoralizing, to say the least. *Trichiurus*, a fish similar in form, is a very strong, swift swimmer, and so is *Regalecus*. Whether or not the habits of *Trachypterus arcticus*, on which these observations were made, are a safe guide in regard to the other forms is a matter of some doubt, but it is certain that they live far from the surface, except near the Arctic Circle, and that they only come ashore accidentally. They have never been taken by the deep-sea dredge or trawlnet, and, indeed, perfect specimens are very rare, the bodies being very soft and brittle, the bones and fin rays exceedingly fragile. A considerable number of species have been described, but in most instances each was based upon 1 or 2 specimens. It is probable that future studies may be as fruitful as that of Emery, who, by means of a series of 23 specimens, succeeded in uniting at least 3 of the Mediterranean species, which for half a century or more had been regarded as distinct. The common species of the Eastern Atlantic, *Trachypterus atlanticus*, is not rare, 1 or more specimens, according to Günther, being secured along the coast of northern Europe after almost every severe gale. We desire to quote the recommendation of Dr. Günther, and to strongly urge upon any one who may be so fortunate as to secure 1 of these fishes, that no attempt should be made to keep it entire, but that it should be cut into short lengths and preserved in the strongest spirits, each piece wrapped separately in muslin. (Goode & Bean.)

1011. TRACHYPTERUS, Gouan.

(KING OF THE HERRINGS.)

Trachypterus, GOUAN, Hist. Poiss., 104, 153, 1770 *(trachypterus).*
Bogmarus, BLOCH & SCHNEIDER, Syst. Ichth., 518, 1801 *(islandicus=arcticus).*

Body elongate, compressed, ribbon-shaped, the dorsal fin extending the entire length of the back. Anal absent; each ventral well developed, if present, but sometimes absent. Caudal present and placed for the most part above the longitudinal axis of the body. No air bladder. Pyloric appendages numerous. Ventrals appearing to be absent in some individuals, but Day calls attention to the fact that most of the specimens of *T. arcticus* taken along the coast of Great Britain had no ventrals. In the very young, as has been shown by Emery, the fin rays commence to grow when it is about 6 mm. long, and continue to lengthen until it is about 24 mm. long, after which a partial shortening takes place. Ventrals very elongate in the young, and the caudal rays much longer than in the grown fish. Young individuals (from 2 to 4 inches) are not rarely met with near the surface; they possess the most extraordinary development of fin rays observed in the whole class of fishes, some of them being several times larger than the body, and provided with lappet-like dilatations. There is no doubt that fishes with such delicate appendages are bred and live in depths where the water is absolutely quiet, as a sojourn in the disturbed water of the surface would deprive them at once of organs which must be of some utility for their preservation. (Goode & Bean.) ($\tau\rho\alpha\chi\dot{\upsilon}\varsigma$, rough; $\pi\tau\epsilon\rho\acute{o}\nu$, fin.)

  a. Color bright metallic silvery, a jet-black blotch at base of dorsal; 3 dark spots on side, 2 smaller ones on belly; anterior profile, snout, and tip of mandible, jet black; caudal and ventral fins carmine red in life.   REX-SALMONORUM, 2972.
  aa. Color shining leaden gray; no black.   TRACHYURUS, 2973.

2972. TRACHYPTERUS REX-SALMONORUM, Jordan & Gilbert.

(KING OF THE SALMON.)

Head 8½; cross depth at nape 8. D. V-170; C. 8; V. 6; P. 11. Body long and slender, closely compressed and ribbon-shaped, as usual in the genus. Head short, deeper than long, the anterior profile steep and nearly straight to the base of the nuchal crest; dorsal fin beginning on the top of nuchal crest, which is directly over the second third of the diameter of eye; height of crest slightly more than diameter of eye, the latter greater than length of snout, and 3 in head. Mouth oblique; maxillary rugose and very broad, its width ½ its length; length of lower jaw greater than length of snout, 2½ in head, its angle under the front of the orbit. Opercular bones rugose, entirely covering the gills. Premaxillary covered with minute and feeble teeth, in addition to which in this specimen are 3 canines, 2 on one side and 1 on the other, directed very obliquely backward. On the side having 2 canines, 1 is placed directly behind the other; lower jaw with 3 strong canines on one side and 2 strong and 1 weak canine on the other, all directed obliquely backward and inward. Dorsal fins slightly connected at base; the filamentous rays of the first dorsal not quite twice the length of head; ventrals inserted just below axil of pectorals, filamentous, about ½ longer than head; pectorals ⅙ longer than eye; caudal rays simple to near tip, where is sometimes a single fork, the longest filamentous rays about 3 times length of head; dorsal fin much lower than body, longest rays of second dorsal nearly ⅞ length of head; a series of spinules along base of dorsal, 1 pair for each ray. Lateral line well developed, with a series of small inconspicuous plates, each of which has a minute central prickle. Lower part of the body thickly beset with small spinous tubercles; rest of the skin naked; rays of all the fins accompanied by a series of small prickles. Coloration everywhere bright metallic silvery, an oblong jet-black blotch a little longer than eye lying close along base of dorsal and beginning 1½ diameters of eye behind eye; 3 larger spots, dusky but not black, lying behind this along side between lateral line and dorsal fin; 2 smaller dusky spots on belly, the one just behind base of ventrals, the other under the second of the 4 spots of back; these spots, except the first one mentioned, are all diffuse and a little less than twice the diameter of eye in length and about twice as long as deep; anterior profile below crest, including front of snout and tip of mandible, jet-black; caudal and ventral fins carmine red in life; other fins unmarked. Length 17 inches. This species bears some resemblance to *Trachypterus altivelis* described by Kner from Valparaiso. The latter species has, however, the nuchal crest much lower and farther back, the first dorsal and the ventrals much lower, the second dorsal fin higher, the skin rougher, the 4 black spots different in size and position from those found in our specimen, and the caudal rays divided near the base. It is probable that the 3 specimens of *Trachypterus* mentioned in the Synopsis of the Fishes of North America, p. 619, and referred with doubt to *Trachypterus altivelis*, really belong to the present species. Four specimens known; 1 from Santa Cruz, California, taken by Dr. C. L. Anderson; 2 from the Straits of Fuca, taken by Mr. J. G. Swan, and the type, obtained by a fisherman (Mr. Knox) in the open sea outside the bay of San Francisco. According to Mr. Swan the species is known by the Makah Indians west of the Straits of Fuca as "king of the salmon," and its destruction is believed to have a baneful influence on the salmon fishing. "When the king of the salmon is killed the salmon will cease to run." (*rex*, king; *salmonorum*, of the salmon.)

? Trachypterus altivelis, JORDAN & GILBERT, Proc. U. S. Nat. Mus. 1881, 52; JORDAN & GIL-
    BERT, Synopsis, 618, 1883; specimen from Santa Cruz; not of KNEE.
Trachypterus rex-salmonorum, JORDAN & GILBERT, Proc. Cal. Ac. Sci. 1894, 145, pl. 9, open
    sea outside Bay of San Francisco. (Type, No. 1382, L. S. Jr. Univ. Coll. Mr. Knox.)

2973. TRACHYPTERUS TRACHYURUS, Poey.

D. 82; P. 15; V. 6. Eye 2¼ in head, high, as long as snout. Mouth almost vertical. Bones of head thin as paper. Lateral line a little concave on the middle of trunk. No scales; pectorals small; ventrals behind pectorals, very long, reaching past vent, which is at second third of length, including caudal; dorsal almost as high as body, without plume in front. Vertebræ 36 + 18. Shining leaden gray, a silvery band produced by the vertebral column showing through. Cuba. (Poey); not seen by us. (τραχύς, rough; οὐρά, tail.)

Trachypterus trachyurus, POEY, Memorias, II, 420, 1861, Cuba.

Family CCXVIII. STYLEPHORIDÆ.

Body elongate, compressed, ribbon-shaped; the dorsal extending from head nearly to end of tail; tail terminating in an exceedingly long, cord-like appendage, about twice as long as head. Anal absent; ventrals absent; caudal erected upward, having its rays connected by a rather firm membrane. Snout produced; mouth small, toothless; maxillary bones small, short, hidden behind premaxillaries; mandible long, extending far behind the eye. Eye large, turned forward; suborbital very large, covering nearly the whole of cheek and extending backward behind eye. Opercles small. Gill openings wide; gills 4. Vent premedian. Branchiostegals 4. (Goode & Bean.) This family is based on a single specimen obtained in the West Indies in 1790 and preserved in the British Museum. The relations of the fish are uncertain, and it may not belong to the *Tæniosomi*. Its nearest relations are, however, apparently with *Trachyterus*.

(*Stylephoridæ*, SWAINSON, Nat. Hist. Class'n Fishes, II, 47, 1839.)

1012. STYLEPHORUS, Shaw.

Stylephorus, SHAW, Trans. Linn. Soc. Lond., I, 1791, 90 (*chordatus*).

Characters of the genus included above. (στῦλος, a style or projecting part; φορέω, to bear.)

2974. STYLEPHORUS CHORDATUS, Shaw.

Head 6; depth 5. D. 110; C. 6; P. 13; B. 4. Snout produced, subcylindrical; mouth small and toothless; maxillary bones small, short, and hidden behind the intermaxillaries; mandible long, extending far behind eye; eyes large, close together, directed forward toward snout; suborbital very large, covering nearly the whole of cheek, and extending backward behind eye; opercles small; gill openings very wide; gills 4. Vent situated before middle of total length; pectorals pointed, directed upward, about ⅓ as long as head; dorsal extending from head nearly to end of tail; caudal directed upward, and having its rays connected by a rather firm membrane, the tail terminating in a narrow band-like appendage about twice as long as body. Color uniform silvery. (Günther.) This remarkable form is known only from a single specimen, 11 inches long, with the caudal appendage 22 inches in length, which was taken in the Atlantic, between Cuba and Martinique, about the year 1790, and is now in the British Museum. It is undoubtedly an inhabitant of great depths. (Goode & Bean.) · (*chordatus*, with a chord; from χόρδη, string.)

Stylephorus chordatus, SHAW, Trans. Linn. Soc. London, I, 1791, 90, pl. 6, between Cuba and Martinique; SHAW, Zool., IV, 87; SHAW, Naturalists' Miscellany, VIII, pl. 274; BLAINVILLE, Journ. Phys., LXXXVII, 60, pl. 1, fig. 1; CUVIER & VALENCIENNES, Hist. Nat. Poiss., X, 381; GÜNTHER, Cat., III, 306, 1861; GOODE & BEAN, Oceanic Ichthyology, 482, pl. 66, figs. 393 and 394, 1896.

Suborder HETEROSOMATA.

(THE FLATFISHES.)

"Cranium posteriorly normal; anteriorly with twisted vertex, to allow 2 orbits on the same side, or 1 vertical and 1 lateral; basis cranii not quite simple. Dorsal fin long, of jointed rays; superior pharyngeals 4, the third longest, much extended forward, the inferior separate." (Cope.) This suborder includes the 2 families, *Pleuronectidæ* and *Soleidæ*. Its nearest relationship is probably with the *Gadidæ*, although the developed pseudobranchiæ and the thoracic ventral fins, indicate an early differentiation from the anacanthine fishes. In the very young fishes the 2 sides of the body are alike and the eyes are 1 on each side, with normal cranium. (ἕτερος, different; σῶμα, body.) (*Anacanthini pleuronectoidei*, Günther, Cat., IV, 399, 504.)

FAMILIES OF HETEROSOMATA.

a. Preopercular margin more or less distinct, not hidden by the skin and scales of the head; eyes large, well separated; mouth moderate or large; teeth present.

PLEURONECTIDÆ, CCXIX.

aa. Preopercular margin adnate, hidden by the skin and scales of the head; eyes small, close together; mouth very small, much twisted; teeth rudimentary or wanting.

SOLEIDÆ, CCXX.

Family CCXIX: PLEURONECTIDÆ.*

(THE FLOUNDERS.)

Body strongly compressed, oval or elliptical in outline; head unsymmetrical, the cranium twisted, both eyes being on the same side of the body, which is horizontal in life, the eyed side being uppermost and colored, the blind side lowermost and usually plain. In the very young fish the bones of the head are symmetrical, 1 eye on each side, and the body is vertical in the water. In most species the cranium becomes twisted, bringing the upper eye over with it. Eyes large, well separated. Mouth small or large, the dentition various, the teeth always present; premaxillaries protractile; no supplemental maxillary bone; pseudobranchiæ present. Gills 4, a slit behind the fourth; lower pharyngeals separate; no air bladder; preopercle with its margin usually distinct, not wholly adnate or hidden by the skin of the head; vent not far behind head, the viscera confined to the anterior part of the body. Scales various, rarely absent, usually small. Lateral line usually present, extending on the caudal fin, sometimes duplicated or wanting. Dorsal fin long, continuous, of soft rays only, beginning on the head; anal similar, shorter; caudal various, sometimes coalescent with dorsal and anal; pectorals inserted rather high, rarely wanting; ventrals under the pectorals, usually of several soft rays, one of them sometimes wanting. Fishes mostly carnivorous, inhabiting sandy bottoms in all seas, some species ascending rivers. Many of them are important food-fishes. Genera about 55; species

---

* For complete synonymy and descriptions of the American species of this family of fishes, see "A review of the flounders and soles (*Pleuronectidæ*) of America and Europe," by David Starr Jordan and David Kop Goss, in Report United States Fish Comm. for 1886, 225-342, pls. 1 to 9, first published in 1889.

nearly 500. The group *"Bibroniidæ"* recently recognized by some of the Italian ichthyologists as a separate family *("Bibronidi"),* is composed entirely of larval forms in the early stages of their development. In this condition the eyes are symmetrical and the body translucent. Several generic names have been given to these peculiar forms (*Peloria, Bibronia, Coccolus, Charybdia, Bascanius, Delothyris*), but, of course, these genera can have no permanent place in the system. *Peloria* has been shown by Dr. Emery to be the young of *Platophrys.* The others seem to belong to the *Cynoglossinæ* or to some allied group, but we are not yet certain as to the correct identification of any of them. We recognize among the *Pleuronectidæ* 6 subfamilies—*Hippoglossinæ, Psettinæ, Samarinæ, Pleuronectinæ, Oncopterinæ,* and *Pelecanichthyinæ.* These subfamilies are natural groups and are in most cases easily distinguished, although some few aberrant genera exist, which serve as links joining one group to another. Thus *Isopsetta* of the *Pleuronectinæ* is certainly a near ally of *Psettichthys,* which is as certainly a genuine member of the *Hippoglossinæ.* The *Hippoglossinæ* and the *Pleuronectinæ* are largely arctic in their distribution, few of the former group and none of the latter extending into the Tropics. The *Oncopterinæ* seem to take the place of the *Pleuronectinæ* in antarctic waters, but the species of this group are few in number. The *Psettinæ* and the soles are, on the other hand, essentially warm-water fishes, their representatives in the north being comparatively few. The *Samarinæ* are few in number and belong to the East Indian fauna, and the single species of *Pelecanichthyinæ* belongs to the bassalian fauna of the Pacific. As the tropical *Hippoglossinæ* and all the *Psettinæ* are sinistral species, the eyes and color being on the left side of the body, it follows that the tropical flounders are nearly all left-sided species, while those of arctic and antarctic waters are chiefly dextral species, the eyes and color on the right. The *Hippoglossinæ* are the most generalized of the flatfishes. From the northern representatives of this group, the allies of *Hippoglossoides,* the *Pleuronectinæ,* are certainly descended. The *Psettinæ* are apparently derived from ancestors of the type of *Paralichthys.* The soles show closest affinities with the *Psettinæ,* from ancestors of which group they have become degraded. Very remarkable is the relation between the number of vertebræ and the geographical distribution of the various species. It has been already noticed by Dr. Gill, Dr. Günther and others that in some groups of fishes northern representatives have the number of vertebræ increased. In no group is this more striking than in the flounders, as the following table, showing the numbers of the vertebræ in various species, will clearly show. The numbers inclosed in brackets are copied from Dr. Günther; the others represent our own count of specimens.

Numbers of vertebræ in flounders.

I.—HIPPOGLOSSINÆ.

Hippoglossus hippoglossus	$16 + 34 = 50$
Atheresthes stomias	$12 + 37 = 49$
Hippoglossoides platessoides	$13 + 32 = 45$
Lyopsetta exilis	$11 + 34 = 45$
Eopsetta jordani	$11 + 32 = 43$

Psettichthys melanostictus	11 + 29 = 40
Paralichthys oblongus	11 + 30 = 41
Paralichthys dentatus	10 + 30 = 40
Paralichthys lethostigmus	10 + 27 = 37
Paralichthys albiguttus	10 + 27 = 37
Paralichthys californicus	10 + 25 = 35
Xystreurys liolepis	12 + 25 = 37
Ancylopsetta quadrocellata	9 + 26 = 35

II.—PLEURONECTINÆ.

Glyptocephalus zachirus	13 + 52 = 65
Glyptocephalus cynoglossus	[58]
Microstomus pacificus	12 + 40 = 52
Microstomus kitt	[13 + 35 = 48]
Parophrys vetulus	11 + 33 = 44
Pleuronectes platessa	[14 + 29 = 43]
Isopsetta isolepis	10 + 32 = 42
Lepidopsetta bilineata	11 + 29 = 40
Limanda limanda	[40]
Liopsetta glacialis	13 + 27 = 40
Pleuronichthys decurrens	14 + 26 = 40
Pleuronichthys verticalis	13 + 25 = 38
Flesus glaber	11 + 26 = 37
Flesus flesus	[12 + 24 = 36]
Pseudopleuronectes americanus	10 + 26 = 36
Hypsopsetta guttulata	11 + 24 = 35
Platichthys stellatus	12 + 23 = 35

III.—PSETTINÆ.

Monolene sessilicauda	[43]
Lepidorhombus whiff-iagonis	[11 + 30 = 41]
Citharichthys sordidus	11 + 29 = 40
Platophrys lunatus	9 + 30 = 39
Arnoglossus laterna	10 + 28 = 38
Arnoglossus grohmanni	10 + 28 = 38
Zeugopterus punctatus	[12 + 25 = 37]
Platophrys ocellatus	10 + 27 = 37
Lophopsetta maculata	11 + 25 = 36
Bothus rhombus	12 + 24 = 36
Syacium papillosum	11 + 25 = 36
Citharichthys arctifrons	10 + 26 = 36
Syacium micrurum	10 + 25 = 35
Phrynorhombus regius	10 + 25 = 35
Citharichthys spilopterus	10 + 24 = 34
Citharichthys macrops	10 + 24 = 34
Etropus microstomus	10 + 24 = 34
Etropus crossotus	10 + 24 = 34
Azevia panamensis	33
Psetta maxima	12 + 19 = 31

The subdivision of the flounders into genera leaves room for considerable variety of opinion. Most of the species are well defined and easily recognized, but they do not fall readily into generic groups unless we regard almost every well-marked species as the type of a distinct genus. A natural result of an attempt at sharply defining the genera is to reach what seems an extreme degree of generic subdivision. On the other hand, attempts to unite these smaller groups to form larger ones often leave these larger ones at once unnatural and ill-defined.

It will probably appear to some that the process of generic subdivision has been in this paper carried too far. It is possible that this is true, but the arrangement which we have adopted seems to bring out the relations of the different forms better than can be done by a more conservative view of the genera. (*Pleuronectidæ*, Günther, Cat., IV, 1862.)

SUBFAMILIES OF PLEURONECTIDÆ.

A. Ventral fins symmetrical, similar in position and in form of base, the ventral of the colored side not extended along the ridge of the abdomen.
  a. Mouth nearly symmetrical, the dentition nearly equally developed on both sides, the gape usually but not always wide. (Halibut tribe.) HIPPOGLOSSINÆ, I.
  aa. Mouth unsymmetrical, the jaws on the eyed side with nearly straight outline, the bones on the blind side strongly curved; teeth chiefly on the blind side.
    b. Eyes and color on the right side (with occasional exceptions). (Flounder tribe.) PLEURONECTINÆ, II.
AA. Ventral fins unsymmetrical, dissimilar in position and usually also in form, the ventral fin of the eyed side being extended along the ridge of the abdomen. Eyes and color on the left side. (Turbot tribe.) PSETTINÆ, III.

ANALYSIS OF GENERA.

I. HIPPOGLOSSINÆ.

(HALIBUT TRIBE.)

Large-mouthed flounders with the ventral fins symmetrical.—Mouth symmetrical, the jaws and the dentition nearly equally developed on both sides; gape usually wide, the maxillary more than $\frac{1}{3}$ length of head. Lower pharyngeals narrow, usually with but 1 or 2 rows of sharp teeth; teeth in jaws usually acute. Eyes large; edge of preopercle free. Pectoral and ventral fins well developed, *the ventral fins similar in position* and *in form of base*, the ventral fin of the eyed side not being attached along the ridge of the abdomen. Septum of gill cavity without foramen.

  a. Vertebræ and fin rays much increased in number (the vertebræ about 50; dorsal rays about 100, anal rays about 85); body comparatively elongate; caudal fin lunate; lateral line simple; anal spine mostly obsolete. Dextral species, arctic in distribution. (Genera allied to *Hippoglossus*.)
    b. Large teeth in both jaws arrow-shaped, biserial, some of them depressible; upper eye with vertical range; gill rakers short; scales deciduous, ciliated; lateral line without arch; flesh soft. Vertebræ (in *A. stomias*) 12 + 37 = 49. ATHERESTHES, 1013.
    bb. Large teeth not arrow-shaped, biserial above, uniserial below; scales very small, cycloid; gill rakers long and slender; eyes strictly lateral.
      c. Lateral line without anterior arch; lower pharyngeal teeth uniserial. REINHARDTIUS, 1014.
      cc. Lateral line with an anterior arch; lower pharyngeal teeth biserial; vertebræ (in *H. hippoglossus*) 16 + 34 = 50. HIPPOGLOSSUS, 1015.
  aa. Vertebræ and fin rays in moderate number (vertebræ less than 46; dorsal rays fewer than 95; anal rays fewer than 75); caudal fin double truncate or rounded, the median rays longest.
    d. Lateral line without distinct anterior arch; vertebræ 40 to 46; body normally dextral;[*] caudal peduncle distinct; scales ciliated; anal spine usually strong. Species of subarctic distribution. (Genera allied to *Hippoglossoides*.)

---

[*] Frequently sinistral in *Hippoglossoides elassodon*.

e. Lateral line simple, without accessory dorsal branch; teeth sharp, those
of lower jaw uniserial; dorsal beginning above eye.

    f. Teeth in the upper jaw biserial.

        g. Scales comparatively large, thin, and deciduous; lateral line 70;
body slender, the flesh soft; vertebræ (in *L. exilis*) 11 + 34 = 45.
                LYOPSETTA, 1016.

        gg. Scales small and adherent; lateral line 96; body robust, the
flesh firm; vertebræ (in *E. jordani*) 11 + 32 = 43.
                EOPSETTA, 1017.

    ff. Teeth in the upper jaw uniserial; scales small and flesh firm; verte-
bræ (in *H. platessoides*) 13 + 32 = 45.   HIPPOGLOSSOIDES, 1018.

ee. Lateral line with an accessory dorsal branch; scales small, firm, ctenoid;
dorsal fin beginning before the eye; teeth sharp, unequal, some of them
canine-like; mouth not large; lower pharyngeal teeth sharp, uniserial;
vertebræ (in *P. melanostictus*) 11 + 29 = 40.   PSETTICHTHYS, 1019.

dd. Lateral line with an arch in front; no accessory branch; vertebræ in smaller
number (35 to 41); anal spine usually obsolete; body normally sinistral.
(Species chiefly of the temperate or subtropical seas, none of them Arctic
and none European.) (Genera allied to *Paralichthys*.)

    h. Dorsal fin beginning above the pupil; teeth rather small; no canines;
body indifferently dextral or sinistral (in some species at least).

        i. Scales ctenoid.

           j. Teeth in upper jaw in 2 series; gill rakers broad.
                VERASPER, 1020.

           jj. Teeth all uniserial; gill rakers slender.   HIPPOGLOSSINA, 1021.

        ii. Scales cycloid; teeth uniserial; gill rakers short and thick.

           k. Teeth small, pointed, equal.   LIOGLOSSINA, 1022.

           kk. Teeth unequal, blunt, conical; caudal fin subsessile, the cau-
dal peduncle extremely short; skin of shoulder girdle with
patches of cup-shaped scales; vertebræ (in *X. liolepis*) 12 +
25 = 37.   XYSTREURYS, 1023.

    hh. Dorsal fin beginning in advance of eye; teeth sharp, uniserial or smooth.

        l. Scales weakly ciliated; caudal fin with a distinct peduncle; mouth
large; teeth unequal, some of the anterior canine like; gill
rakers rather long and slender; no dorsal lobe nor produced
ventral rays; vertebræ 35 to 41.   PARALICHTHYS, 1024.

        ll. Scales very strongly ctenoid on both sides of body; mouth smallish,
with small, sharp teeth; anterior rays of dorsal more or less
exserted, thus forming a more or less distinct lobe; gill membranes considerably united; gill rakers short and broad; caudal peduncle short; left ventral produced; vertebræ (in *A.
quadrocellata*) 9 + 26 = 35.

           m. Lateral line with its tubes much branched, covering parts
of contiguous scales; dorsal lobe low; left ventral much
produced.   RAMULARIA, 1025.

           mm. Lateral line with its tubes simple, not branched.

                n. Body broad, ovate, the depth more than ⅓ length; dorsal
lobe and left ventral moderately produced.
                ANCYLOPSETTA, 1026.

                nn. Body elliptical, the depth not more than ⅓ length; dor-
sal lobe and left ventral greatly produced.
                NOTOSEMA, 1027.

        lll. Scales entirely smooth; caudal peduncle short; mouth small; gill
rakers short and thick; dorsal with an anterior lobe; left
ventral elongate.   GASTROPSETTA, 1028

II.—PLEURONECTINÆ.

(FLOUNDER TRIBE.)

Mouth small, unsymmetrical, the jaws on the eyed side with nearly straight outline, the bones on the blind side strongly curved; dentition chiefly developed on the blind side; eyes large; edge of preopercle not hidden by the scales; pectoral fins well developed; vertical fins well separated; ventral fins nearly or quite symmetrical, that of the eyed side not prolonged along the ridge of the abdomen; anal spine usually strong (obsolete in *Microstomus* and *Embassichthys*). Body dextral (except frequently in *Platichthys stellatus*). Species arctic or subarctic in distribution.

  a. Vertebræ in moderate number, from 10 + 26 = 36 to 11 + 33 = 44; dorsal rays 65 to 80; anal rays 45 to 60.

    b. Teeth small, acute, in several series; lateral line nearly straight, with an accessory dorsal branch; lower pharyngeals narrow, with small biserial teeth; scales cycloid.

      c. Lips thick, each with several longitudinal folds; dorsal fin beginning on the blind side; vertebræ 38 to 40.   PLEURONICHTHYS, 1029.

      cc. Lips simple; dorsal fin beginning on the median line; vertebræ (in *H. guttulata*) 11 + 24 = 35.   HYPSOPSETTA, 1030.

    bb. Teeth chiefly uniserial, all more or less blunt, conical or incisor-like.

      d. Lateral line with an accessory dorsal branch.

        e. Lateral line without distinct arch in front.

          f. Teeth compressed, incisor-like, close set.

            g. Scales closely imbricated, mostly cycloid; upper eye on median line; vertebræ (in *P. vetulus*) 11 + 33 = 44.   PAROPHRYS, 1031.

            gg. Scales scarcely imbricated, all very strongly ctenoid; eyes both lateral.   INOPSETTA, 1032.

          ff. Teeth conical, separated, not incisor-like; scales closely imbricated, all strongly ctenoid; mouth comparatively large (approaching that of *Psettichthys*); vertebræ (in *I. isolepis*) 10 + 32 = 42.   ISOPSETTA, 1033.

        ee. Lateral line with a distinct arch in front; scales imbricated, rough-ctenoid; vertebræ (in *L. bilineata*) 11 + 29 = 40.   LEPIDOPSETTA, 1034.

      dd. Lateral line without accessory dorsal branch.

        h. Lateral line with a distinct arch in front; scales imbricated, rough-ctenoid; vertebræ (in *L. limanda*) about 40.   LIMANDA, 1035.

        hh. Lateral line without distinct arch in front.

          i. Scales regularly imbricate, all (on eyed side) ctenoid in both sexes; no stellate tubercles on head nor on bases of dorsal and anal fins; teeth, incisor-like, close set; lower pharyngeals very narrow, each with 2 rows of separate, conical teeth; fin rays scaly.   PSEUDOPLEURONECTES, 1036.

          ii. Scales imperfectly imbricated, or else not all ctenoid.

            j. Scales chiefly cycloid in both sexes; lower pharyngeals small and narrow, separate, each with about 1 row of small, bluntish teeth; teeth incisor-like, close set, forming a cutting edge; no stellate scales at base of dorsal and anal.   PLEURONECTES, 1037.

            jj. Scales rough-ctenoid in the male, more or less cycloid in the female (fin rays scaly in the male, naked in the female); lower pharyngeals very large, more or less united in the adult, their surface somewhat concave, the teeth in 5 or 6 rows, large, blunt, close set; teeth in jaws incisor-like; fin rays of dorsal and anal without tubercles at base.   LIOPSETTA, 1038.

jjj. Scales all in both sexes and on both sides of the body represented by coarse scattered stellate tubercles; similar tubercles between bases of dorsal and anal rays; lateral line without scales; lower pharyngeals broad, each with 3 rows of blunt, coarse teeth; teeth incisor-like.　　　　PLATICHTHYS, 1039.

aa. Vertebræ in increased number (varying from 13 + 35 = 48 to 13 + 52 = 65); dorsal rays 90 to 120; anal rays 70 to 100; teeth broad, incisor-like; scales small, all cycloid. (Genera allied to *Glyptocephalus*.)

   k. Left side of skull normal; anal spine obsolete; vertebræ 48 to 52.

    l. Body elongate, the depth 2½ to 3 in length; vertebræ 48 to 52.
　　　　　　　　　　　　　　　　　　　MICROSTOMUS, 1040.

    ll. Body stouter, the depth 2 to 2½ in length; vertebræ more numerous, about 63.　　　　　　　　　　　　EMBASSICHTHYS, 1041.

   kk. Left side of skull with large mucous cavities; anal spine strong; vertebræ 58 to 65.　　　　　　　　　　　GLYPTOCEPHALUS, 1042.

III.—PSETTINÆ.

(TURBOT TRIBE.)

Large-mouthed flounders, with the ventral fins unsymmetrical.—Mouth symmetrical, the dentition nearly equally developed on both sides; gape usually wide (narrow in *Platophrys, Etropus,* etc.), the maxillary commonly more than ½ length of head; lower pharyngeals narrow, each with one or more rows or a narrow band of small, sharp teeth; teeth in jaws acute; eyes not minute; pectorals and ventrals usually well developed; edge of preopercle free; ventral fins dissimilar in form or in position, that of the left or eyed side inserted on the ridge of the abdomen, its base extended along this ridge, its rays more or less wide apart; caudal fin rounded or subtruncate; no accessory lateral line; anal spine usually weak or obsolete; a pelvic spine sometimes developed; vertebræ in moderate or small number, 31 to 45. Body sinistral. Species chiefly tropical or subtropical in distribution.

  a. Pectoral fin of both sides present; septum of gill cavity below gill arches without foramen; a deep emargination near the isthmus; ventral fins free from anal.

   b. Vomer with teeth; lateral line with a strong arch in front; teeth subequal, in villiform bands; body broadly ovate; caudal fin subsessile; interorbital area broad; scales small, cycloid; gill rakers long and slender; anterior dorsal rays produced; vertebræ 36.　　　　　　　LOPHOPSETTA, 1043.

   bb. Vomer toothless; ventral fins free from anal; caudal fin subsessile.

    c. Lateral line with a distinct arch in front; teeth small, uniserial, or imperfectly biserial.

     d. Interorbital space more or less broad, deeply concave, at least in the males; form broad ovate; gill rakers short and thick.

      e. Scales small, ctenoid, adherent, 75 to 100 or more; anterior rays of dorsal not elevated; pectoral of left side usually filamentous in the male; vertebræ (in *P. lunatus*) 9 + 30 = 39.
　　　　　　　　　　　　　　　　PLATOPHRYS, 1044.

      ee. Scales moderate, 60 to 70; anterior rays of dorsal greatly produced; no lateral line on blind side.　　　PERISSIAS, 1045.

     dd. Interorbital space a narrow ridge; dorsal not elevated in front.

      f. Gill rakers obsolete; interorbital area armed with a spine; scales rough.　　　　　　　　　　ENGYOPHRYS, 1046.

      ff. Gill rakers slender; right ventral elongate; scales ctenoid.
　　　　　　　　　　　　　　　　TRICHOPSETTA, 1047.

cc. Lateral line without arch in front.

   g. Teeth in upper jaw biserial, in the lower uniserial, the front teeth of upper jaw enlarged; vertebræ 35 or 36; gill rakers short; interorbital space broad in the male.   SYACIUM, 1048.

   gg. Teeth in each jaw uniserial; interorbital space very narrow, the ridges coalescing between the eyes.

      h. Mouth not very small, the maxillary more than ⅓ length of head.

         i. Gill rakers very short and thick, tubercle-like.

            j. Scales cycloid, small, and firm.   CYCLOPSETTA, 1049.

            jj. Scales small, firm, ctenoid.   AZEVIA, 1050.

        ii. Gill rakers slender, of moderate length; scales thin, deciduous, ciliated; vertebræ 34 to 40. CITHARICHTHYS, 1051.

      hh. Mouth very small, the teeth subequal, the maxillary less than ⅓ length of head; scales thin; teeth uniserial; vertebræ 9 + 25 = 34.   ETROPUS, 1052.

aa. Pectoral fin of blind side wanting; eyes very close together; caudal fin subsessile; teeth small, uniserial; mouth moderate; lateral line of eyed side arched, that of right side nearly straight; dorsal fin beginning on snout, its anterior rays not exserted, its rays all simple and very numerous; gill rakers few and feeble; scales small; body thin, very elongate; vertebræ (in *M. sessilicauda*) 43; (deep-sea flounders).   MONOLENE, 1053.

1013. ATHERESTHES, Jordan & Gilbert.

Atheresthes, JORDAN & GILBERT, Proc. U. S. Nat. Mus. 1880, 51 (*stomias*).

Eyes and color on the right side. Body very long and slender, closely compressed, tapering into a long and slender caudal peduncle; head elongate, narrow; mouth extremely large, oblique; the long and narrow maxillary extending beyond the eye; each jaw with 2 irregular series of sharp, unequal, arrow-shaped teeth, some of them long and wide set, and others short and close set, sharp; the long teeth freely depressible. Gill rakers numerous, long, slender, and stiff, strongly dentate within. Scales rather large, thin and readily deciduous, slightly ciliated, those on the blind side similar, smooth; lateral line without arch. Fins low and fragile; dorsal commencing over the eye, its anterior rays low, the posterior rays somewhat forked; no anal spine; pectorals and ventrals small, both of the latter lateral; caudal lunate. The single species which constitutes this genus is one of the most remarkable of the flounders. Of all the group, it approaches in form and general characters most nearly to the Gadoid fishes, from ancestors of which we may presume the flounders to be descended, although Dr. Gill has suggested the possibility of their descent from Trachypteroid fishes. (ἀθήρ, the beard or spike of an ear of corn; ἐσθίω, to eat; from the arrow-shaped teeth.)

2975. ATHERESTHES STOMIAS (Jordan & Gilbert).

(THE ARROW-TOOTHED HALIBUT.)

Head about 3¾ in length; depth 3½; eye large, 4¾ in head. D. 103; A. 86; scales 135; vertebræ 12 + 37 = 49. Head long, the snout protruding, somewhat truncate at tip; mouth excessively large; the maxillary more than ½ length of head, and reaching behind eye; premaxillary in front above the level of the lower eye; teeth in upper jaw anteriorly in a single series, long, slender, and wide set, much smaller and closer set behind; on sides of jaw the teeth are very small and in 2 distinct series, the inner of which corresponds to the single series in front, the teeth thus gradually increasing in size forward; teeth in inner series of lower jaw very sharp and slender, longer than the upper teeth, wide set, alternating with shorter, depressed teeth; outside of these larger teeth is a series of fixed small teeth; all of the long teeth in both jaws depressible and conspicuously arrow-shaped toward their tips; inner series of small teeth in upper jaw also arrow-shaped, depressible; interorbital space scaly, ridged, not a third width of eye. Gill rakers long and strong, about 4 + 13 in number, the longest more than ¼ diameter of eye. Upper eye with its range entirely vertical. Scales extremely thin, irregular in size, not evenly imbricated; lateral line very prominent. Dorsal fin beginning just behind the middle of the eye; caudal peduncle nearly as long as the pectoral fin, about ⅔ length of head. Plain olive brown, the margins of the scales darker; blind side dusted with black points. Length 2 feet. Bering Sea to San Francisco, common northward; not rare in deep water off San Francisco, and is brought in in considerable numbers from the sweep-nets (*parranzelle*) used in Drakes Bay. At Unalaska it occurs commonly in shallow water. In the north the flesh is firmer and the coloration more pronounced. Dr. Gilbert dredged it in abundance on both sides of the peninsula of Alaska and in Bristol Bay, in 32 to 406 fathoms. Mr. Scofield found it abundant in Chignik Bay, and it was taken by us in 1897 at Unga and Karluk. (ὀτομίας, large mouthed.)

Platysomatichthys stomias, JORDAN & GILBERT, Proc. U. S. Nat. Mus. 1880, 51, 301, San Francisco. (Coll. Jordan & Gilbert.)

Atheresthes stomias, JORDAN & GILBERT, Proc. U. S. Nat. Mus. 1880, 57, 454; BEAN, Proc. U. S. Nat. Mus. 1881, 242; JORDAN & GILBERT, Proc. U. S. Nat. Mus. 1881, 66; JORDAN & GILBERT, Synopsis, 820, 1883; BEAN, Proc. U. S. Nat. Mus. 1883, 354; JORDAN, Nat. Hist. Aquat. Anim., 188, pl. 53, 1884; JORDAN & GOSS, Review Flounders and Soles, 236, pl. 1, 1889; GILBERT, Rept. U. S. Fish Comm. 1893 (1896), 459.

1014. REINHARDTIUS, Gill.

Reinhardtius, GILL, Cat. Fishes East Coast N. A., 50, 1861 (*hippoglossoides;* no description). ·

Platysomatichthys, BLEEKER, Comptes Rendus, Ac. Sci. Amsterdam, XIII, 1862, 426 (*pinguis = hippoglossoides*).

Reinhardtius, GILL, Proc. Ac. Nat. Sci. Phila. 1864, 218 (*hippoglossoides*).

Eyes and color on right side. Body more or less elongate, compressed; head long and large; mouth large; maxillary reaching beyond eye; jaws with strong, unequal teeth, the upper with 2 series in front, these converging behind; lower jaw with a single series of strong, distant teeth; no teeth on vomer or palatines. Gill rakers few, short, stout, and rough. Fins rather low; caudal fin lunate. Lower pharyngeal teeth in 1 row. Scales small, cycloid; lateral line without anterior curve. One species known, an arctic fish, in some degree intermediate between the true halibut and *Atheresthes*. (Named for Prof. Johann Reinhardt, of the University of Copenhagen, an able investigator of the fishes of Greenland.)

2976. REINHARDTIUS HIPPOGLOSSOIDES (Walbaum).

(GREENLAND HALIBUT.)

Head 3¼ in length; depth nearly 3. D. 100; A. 75; scales 160; orbit 8 in head; snout about 3½, more than twice as long as orbit; eyes even in front; interorbital space flat, scaly, wider than the orbit; lower jaw prominent; length of maxillary 2¼ in head; teeth conical, pointed; upper jaw with 2 series, convergent posteriorly, those of the outer series gradually smaller posteriorly; a pair of strong canine teeth anteriorly in the inner series, the other teeth of this series being very small; lower jaw with a series of strong, distant teeth. Gill rakers short, thick, and strongly dentate. Fins naked. Longest dorsal rays ⅓ length of head; no anal spine; dorsal and anal rays all simple, the dorsal beginning over posterior third of the eye. Scales very small, not ciliated. Yellowish brown. Reaching a very large size. Arctic parts of the Atlantic, south to Finland and the Grand Banks; not very common. (Eu.) (ἱππόγλωσσος, halibut; εἶδος, resemblance.)

Pleuronectes cynoglossus, FABRICIUS, Fauna Grœnlandica, 163, 1780, Greenland; not of LINNÆUS.
Pleuronectes hippoglossoides, WALBAUM, Artedi Piscium, 115, 1792; based on FABRICIUS.
Pleuronectes pinguis, FABRICIUS, Zoologiske Bidrag., 43, 1824, Greenland.
Hippoglossus grœnlandicus, GÜNTHER, Cat., IV, 404, 1862, Greenland.
Reinhardtius hippoglossoides, GILL, Cat. Fishes East Coast N. A., 50, 1861; GILL, Proc. Ac. Nat. Sci. Phila. 1864, 218.
Platysomatichthys hippoglossoides, GOODE & BEAN, Bull. Essex Inst., II, 7, 1879; COLLETT, Norske Nord-Havs Exped., 142, 1880; JORDAN & GILBERT, Synopsis, 819, 1883; GOODE, Nat. Hist. Aquat. Anim., 197, pl. 56, 1884; JORDAN & GOSS, Review Flounders and Soles, 237, pl. II, 1889; and of late American writers generally.
Hippoglossus pinguis, REINHARDT, Kgl. Dansk. Vidensk. Selsk., 116, 1838.
Platysomatichthys pinguis, BLEEKER, *l. c.*, 426, 1862.

1015. HIPPOGLOSSUS, Cuvier.

(HALIBUT.)

Hippoglossus, CUVIER, Règne Animal, Ed. 1, II, 221, 1817 (*hippoglossus*).

Eyes and color on the right side. Form oblong, not strongly compressed. Mouth wide, oblique; teeth in the upper jaw in 2 series, those below in 1; anterior teeth in upper jaw, and lateral teeth in lower, strong; no teeth on vomer or palatines; lower pharyngeal teeth in 2 rows. Dorsal fin beginning above the eye, its middle rays elevated, the posterior rays of dorsal and anal bifid; caudal fin lunate; ventral fins both lateral. Scales very small, cycloid; lateral line with a strong curve in front. Gill rakers few, short, compressed, wide set. Vertebræ 16+34. Largest of the flounders. This genus contains but 1 species, the well-known halibut; abundant on both coasts of the North Atlantic and of the North Pacific. (*Hippoglossus*, the ancient name of the halibut, from ἵππος, horse; γλῶσσα, tongue.)

2977. HIPPOGLOSSUS HIPPOGLOSSUS (Linnæus).

(HALIBUT.)

Head 3¾; depth about 3. D. 105; A. 78; scales 150 or more. Body comparatively elongate, not strongly compressed, deep mesially, thence rapidly tapering each way; head broad; eyes large, separated by a very broad flattish area; lower eye slightly advanced; mouth large, the maxillary reaching middle of orbit. Nearly uniform dark brown; blind side white. One of our most important food-fishes, reaching a weight sometimes of 400 pounds. Found in all northern seas, southward in deep water to France, Sandy Hook, and occasionally to the Farallones off San Francisco; abundant throughout the North Atlantic as also the North Pacific and Bering Sea, in water of moderate depth; taken with hook and line on all cod banks.

Pleuronectes hippoglossus, LINNÆUS, Systema Naturæ, Ed. x, 269, 1758, European Ocean.

Hippoglossus vulgaris, FLEMING, British Animals, 197, 1828; GÜNTHER, Cat., IV, 403, 1862; DAY, Fishes Great Britain, II, 5, pl. 44; STORER, Fish. Mass., 145, 1839; DE KAY, New York Fauna: Fishes, pl. 49, f. 157, 294, 1842; STORER, Synopsis Fish. N. A., 475, 1847; LOCKINGTON, Rep. Com. Fisheries California, 39, 1878–79; LOCKINGTON, Proc. U. S. Nat. Mus. 1879, 71; BEAN, Proc. U. S. Nat, Mus. 1879, 63; JORDAN & GILBERT, Proc. U. S. Nat. Mus. 1880, 454; GOODE, Proc. U. S. Nat. Mus. 1880, 471; JORDAN & GILBERT, Proc. U. S. Nat. Mus. 1881, 66; BEAN, Proc. U. S. Nat. Mus. 1881, 242; JORDAN & GILBERT, Synopsis, 819, 1883; BEAN, Cat. Col. Fish. U. S. Nat. Mus. 1883, 20; DRESEL, Proc. U. S. Nat. Mus. 1884, 244; GOODE, Nat. Hist. Aquatic Anim., 189, pl. 54, 1884; and of American writers generally.

Hippoglossus maximus, GOTTSCHE, Archiv fur Naturgesch. 1835, 164, no locality.

Hippoglossus gigas, SWAINSON, Nat. Hist. Class'n Anim., II, 302, 1839, no locality.

Hippoglossus ponticus, BONAPARTE, Catalogo Metodico, 47, 1846, Black Sea; after PALLAS.

Hippoglossus americanus, GILL, Proc. Ac. Nat. Sci. Phila. 1864, 220.

Hippoglossus hippoglossus, JORDAN, Cat. Fish. N. A., 133, 1885; JORDAN & GOSS, Review Flounders and Soles, 237, pl. 3, 1889.

1016. LYOPSETTA, Jordan & Goss.

Lyopsetta, JORDAN & GOSS, in JORDAN, Cat. Fish. N. A., 135, 1885 (*exilis*).

Teeth sharp, those of the lower jaw uniserial; the upper jaw biserial; lateral line simple (without accessory dorsal branch) and without distinct anterior arch. Scales comparatively large, thin, ciliated, and deciduous; body dextral; anal spine usually strong; vertebræ about 45; body slender, the flesh soft; dorsal fin beginning above eye. This genus contains but a single species, a small, soft-bodied flounder, of the waters of the North Pacific. In its technical characters *Lyopsetta* is very close to *Hippoglossoides*, but the species has the soft flesh of *Atheresthes*. (λύω, to loosen; ψῆττα, flounder.)

2978. LYOPSETTA EXILIS (Jordan & Gilbert).

Head 4; depth 3¼. D. 78; A. 62; V. 6; scales 16–71–18. Body slender, compressed, the flesh soft; caudal peduncle slender; mouth not large, very oblique, the gape curved; lower jaw scarcely projecting, with a knob at symphysis; maxillary rather narrow, reaching middle of pupil, 2⅔ in length of head; teeth small, slender, close set, nearly uniform; above in 2 series, below in 1. Eyes large, separated by a sharp, scaly ridge; lower eye advanced. Scales comparatively large, thin and deciduous, ctenoid, but not so rough as in the other species, those on blind side similar, less rough. Lateral line prominent, rising anteriorly, without trace of arch. Fins low, fragile; anal preceded by a spine; caudal fin long, rather pointed; pectorals small, the right pectoral little more than ¼ length of head. Dorsal beginning immediately in front of pupil; anal higher than dorsal. Gill rakers short, slender, toothed, 9 below angle, the longest about ⅓ diameter of orbit. Pale olivaceous brown, with dark points, forming edgings on each scale; bronze spots sometimes present; fins mostly dusky; dorsal and anal edged anteriorly with yellowish; ventrals largely yellow. Length 12 inches. North Pacific, in rather deep water; San Francisco to Puget Sound. This small flounder is brought in in large quantities by the sweep nets off San Francisco. It is of little value as a food-fish. (*exilis*, slender.)

Hippoglossoides exilis, JORDAN & GILBERT, Proc. U. S. Nat. Mus. 1880, 154, off San Francisco (Type, No. 27121. Coll. Jordan & Gilbert); JORDAN & GILBERT, Proc. U. S. Nat. Mus. 1880, 454; JORDAN & GILBERT, Proc. U. S. Nat. Mus. 1881, 67; JORDAN & GILBERT, Synopsis, 827, 1883.

Lyopsetta exilis, JORDAN & GOSS, Review Flounders and Soles, 238, 1889.

1017. EOPSETTA, Jordan & Goss.

Eopsetta, JORDAN & GOSS, in JORDAN, Cat. Fish. N. A., 135, 1885 (*jordani*).

Teeth sharp, those of the lower jaw uniserial, the upper biserial; scales small, ciliated, and adherent; lateral line without accessory dorsal branch and without distinct anterior arch; anal spine usually strong; body normally dextral, robust, the flesh firm; dorsal fin beginning above eye; vertebræ about 43. This genus contains but a single species, a large flounder which is abundant on the coast of California. It is very close to the genus *Hippoglossoides*. (ἐως, morning; ψῆττα, flounder.)

2979. EOPSETTA JORDANI (Lockington).

(CALIFORNIA "SOLE.")

Head 3⅛; depth 2¼. D. 94; A. 72; scales 96. Body broadly elliptical. Dorsal and ventral outline equally and regularly curved. Mouth oblique, the jaws about even, the symphyseal knob but little projecting; gape curved; maxillary broad, reaching to behind pupil, 2⅔ in head; teeth in 2 series in the upper jaw, the inner series small and distant from the outer, which is considerably enlarged in front; lower jaw with a single series similar to the outer series in the upper jaw, but larger. Gill rakers roughish, strong, about 15 below angle, the longest about ½ as long as eye. Lower pharyngeals rather narrow, each with a single row of sharp teeth. Eyes large; interorbital space a narrow, blunt, scaly ridge. Dorsal beginning over anterior margin of pupil, the rays all simple; caudal fin with the middle rays slightly produced; anal preceded by a spine; pectoral ½ length of head. Scales of colored side small, firm, strongly ciliated, nearly uniform over head and body; lower jaw and snout scaleless; scales on blind side smooth. Olive brown, nearly uniform; membrane of dorsal and anal fins clouded with darker. Length 20 inches. Pacific Coast of the United States from Puget Sound to Point Concepcion. One of the commonest flatfishes of the California coast, being found in abundance in shallow water from Monterey northward. It is a good food-fish, and large numbers are dried each year by the Chinese. (Named for David Starr Jordan.)

Hippoglossoides jordani, LOCKINGTON, Proc. U. S. Nat. Mus. 1879, 73, San Francisco (Coll. W. N. Lockington); LOCKINGTON, Rep. Com. Fisheries California 1878-79, 40; LOCKINGTON, Scientific Press Supplement, April, 1879, 120; JORDAN & GILBERT, Proc. U. S. Nat. Mus. 1880, 454; JORDAN & GILBERT, Proc. U. S. Nat. Mus. 1881, 67; JORDAN & GILBERT, Synopsis, 826, 1883; JORDAN, Nat. Hist. Aquat. Anim., 187, 1884.

Eopsetta jordani, JORDAN & GOSS, Review Flounders and Soles, 239, 1889.

1018. HIPPOGLOSSOIDES, Gottsche.

Hippoglossoides, GOTTSCHE, Archiv fur Naturgesch. 1835, 164 ("*limanda*"=*platessoides*).

Citharus, REINHARDT, Kong. Dansk. Vid. Selsk. 1838, 116 (*platessoides*); not *Citharus* BLEEKER, 1862.

Drepanopsetta, GILL, Cat. Fish. East Coast N. A., 50, 1861 (*platessoides*).

Pomatopsetta, GILL, Proc. Ac. Nat. Sci. Phila. 1864, 217 ("*dentata*"=*platessoides*).

Eyes and color on the right side (except sometimes in *H. elassodon*). Body oblong, moderately compressed; mouth rather large, with 1 row of sharp teeth on each jaw; no teeth on vomer or palatines; gill rakers rather long and slender; scales ctenoid; lateral line nearly straight, simple; dorsal fin low in front, beginning over or before the eye; ventrals both lateral; caudal double truncate, produced behind. This genus, as here restricted, contains 3 closely related species, 2 of the North Pacific, 1 of the North Atlantic. All are essentially arctic species, inhabiting shallow waters in the regions where they are most abundant. (ίππόγλωσσος, *Hippoglossus*; εῖδος, resemblance.)

a. Dorsal rays about 88; anal about 70; gill rakers $x+10$; interorbital space with an obtuse, prominent, rather broad ridge.   PLATESSOIDES, 2980.

aa. Dorsal rays about 82; anal about 61; gill rakers $x+12$ to 14; interorbital space with a narrow, nearly naked ridge.   ELASSODON, 2981.

aaa. Dorsal rays 72 to 76; anal 56 to 60; gill rakers $x+12$; interorbital space moderate, with 2 rows of scales.

    b. Depth 2¼ in length; D. 76; A. 60; pectoral ½ length of head.   ROBUSTUS, 2982.

    bb. Depth 2⅔ in length; D. 72; A. 56; pectoral ⅔ in length of head.

        HAMILTONI, 2983.

2980. HIPPOGLOSSOIDES PLATESSOIDES (Fabricius).

(SAND-DAB.)

Head 3¾; depth 2½. D. 88 (80 to 93); A. 70 (64 to 75); scales 90 (pores). Body ovate; mouth moderate, oblique; maxillary narrow, reaching to below pupil, 2⅗ in length of head; teeth rather small, conical, larger anteriorly, in 1 row in each jaw, those in the lower largest. Eyes rather large, the upper longer than snout, 4⅓ in head; lower jaw included, but with a projecting knob at the chin; snout thick, scaly; interorbital space narrow, with a raised obtuse ridge entirely covered with rough scales in about 6 series; mandible with a series of scales; gill rakers rather short and robust, not toothed, about 10 below angle, the longest less than ¼ length of eye; fins with small, rough scales; a strong preanal spine; pectoral not quite ½ length of head. Reddish brown, nearly plain. The identity of the American and European representatives of this species (*platessoides* and *limandoides*) is now conceded by all writers. A little difference is recognizable between arctic and subarctic examples, the former having a somewhat greater number of fin rays. Thus Greenland specimens, according to Collett, have D. 88, A. 69; specimens from Finmark have D. 92, A. 72; these representing the var. *platessoides.* Specimens from England (var. *limandoides*) have D. 80, A. 66, while those from intermediate localities present in general fin formulæ likewise intermediate, showing that no sharp division is possible. This is a rather common foodfish of the deep waters northward, on both sides of the ocean. North Atlantic, south to Cape Cod, and the coasts of England and Scandinavia. (Eu.) (*platessa*, the plaice; εἶδος, resemblance.)

Pleuronectes linguatula, MÜLLER, Zool. Dan. Prodromus, 45, 1776; not of LINNÆUS.

Pleuronectes platessoides, FABRICIÙS, Fauna Grœnlandica, 164, 1780, Greenland.

Pleuronectes limandoides, BLOCH, Ausl. Fische, III, 24 tab. 186, 1787, Europe, and of various copyists.

Pleuronectes limandanus, PARNELL, Edinburgh New Phil. Journ. 1835, 210.

Citharus platessoides, REINHARDT, Kongl. Dansk. Vid. Selsk., 116, 1838.

Drepanopsetta platessoides, GILL, Cat. Fish. East Coast N. A., 50, 1861.

Hippoglossoides platessoides, GILL, Proc. Ac. Nat. Sci. Phila. 1864, 217; COLLETT, Norske Nord-Havs. Exped., 144, 1880; GOODE, Proc. U. S. Nat. Mus. 1880, 471; JORDAN & GILBERT, Synopsis, 826, 1883; STEARNS, Proc. U. S. Nat. Mus. 1883, 125; GOODE, Nat. Hist. Aquatic Anim., 197, pl. 55, 1884; JORDAN & GOSS, Review Flounders and Soles, 240, pl. 4, 1889; GOODE & BEAN, Ocean Ichthyology, 438, 1896, and of recent American writers generally.

Hippoglossoides limandoides, GÜNTHER, Cat., IV, 405, 1862; DAY, Fishes Great Britain and Ireland, II, 9, pl. 45, 1884.

Hippoglossoides limanda, GOTTSCHE, Archiv fur Naturgesch. 1835, 168; not *Pl. limanda,* LINNÆUS.

Platessa dentata, STORER, Rept. Fish. Mass., 143, 1839; DE KAY, N. Y. Fauna: Fishes, 298, 1842; STORER, Synopsis, 476, 1846.

Hippoglossoides dentatus, GÜNTHER, Cat., IV, 406, 1862; GÜNTHER, Challenger Report, XXII, Fishes, 3, 1887.

Pomatopsetta dentata, GILL, Proc. Ac. Nat. Sci. Phila. 1864, 217.

2981. HIPPOGLOSSOIDES ELASSODON, Jordan & Gilbert.

Head $3\frac{1}{2}$; depth $2\frac{1}{2}$; eye 4 in head. D. 77 to 87; A. 59 to 67; V. 6; scales 45–100–40. Body oblong-elliptical; caudal peduncle about as long as deep; upper profile of head continuous with the outline of back; depression over eye slight; mouth rather large, the gape curved, considerably wider on the blind side; lower jaw projecting, with a symphyseal knob; maxillary narrow, reaching beyond middle of pupil, $2\frac{1}{4}$ in head; teeth small, close set, nearly uniform, in a single row. Gill rakers slender, smooth, 14 to 16 below arch, the longest nearly $\frac{1}{2}$ diameter of orbit. Eyes large, separated by a narrow, knife-like ridge, which is naked, or with a single series of scales. Scales small, firm, rough, those on tail roughest, those on blind side similar, mostly smooth anteriorly. Lateral line rising anteriorly, but without arch; dorsal beginning immediately in front of pupil; anal preceded by a spine; caudal long; pectoral of eyed side $\frac{1}{4}$ length of head; ventral reaching past front of anal; pectoral and ventral of eyed side with prickle-like scales. Brownish, nearly uniform, sometimes spotted with darker; fins grayish, irregularly blotched with dusky. Body sometimes sinistral. Length 18 inches. Bering Sea south to Cape Fattery; a rather abundant shore fish in Puget Sound, and it seems to be still more common northward, being, in Alaska, a food-fish of some importance. Abundant north and south of the Aleutian Islands and in Bristol Bay. Our specimens from Kamchatka agree in all respects; D. 77 to 84; A. 60 or 61. Pectoral not quite $\frac{1}{2}$ head. Interorbital ridge sharp, with 1 series of scales; gill rakers $x + 14$. ($\dot{\epsilon}\lambda\alpha\acute{6}\acute{6}\acute{o}\omega$, to diminish; $\acute{o}\delta o\acute{v}\varsigma$, tooth.)

Hippoglossoides elassodon, JORDAN & GILBERT, Proc. U. S. Nat. Mus. 1880, 278, Seattle; Tacoma (Type, No. 27263. Coll. D. S. Jordan); JORDAN & GILBERT, Proc. U. S. Nat. Mus. 1880, 454; BEAN, Proc. U. S. Nat. Mus. 1881, 242; JORDAN & GILBERT, Synopsis, 826, 1883; BEAN, Proc. U. S. Nat. Mus. 1883, 20; JORDAN, Nat. Hist. Aquat. Anim., 188. pl. 52, 1884; JORDAN & GOSS, Review Flounders and Soles, 241, pl. 5, 1889; JORDAN & GILBERT, Rept. Fur Seal Invest., 1898.

2982. HIPPOGLOSSOIDES ROBUSTUS, Gill & Townsend.

Head $3\frac{2}{3}$; depth $2\frac{1}{5}$; eye $5\frac{2}{8}$ in head. D. 76; A. 60; scales 95 (pores). Interorbital space a broad, somewhat elevated ridge with 2 rows of scales. Body rather high, its greatest height nearly equaling $\frac{1}{4}$ the length from snout to base of caudal; profile decurved above the eye; body thick; scales on head separate and rarely touch each other. Gill rakers long, $x + 11$. Maxillary $2\frac{1}{2}$ in head, directed upward anteriorly; teeth of the single row mostly separated from each other by intervals equal to width of teeth, curved inward, and uniform on the sides; toward front 4 or 5 enlarged, preceded by 2 smaller, leaving the middle toothless; in the lower jaw of nearly uniform size and inclining backward. Pectoral $\frac{1}{4}$ head; ventrals reaching first or second anal ray. Scales on body ciliated or weakly ctenoid, those on cheek smoother; no ctenoid scales on blind side. Caudal shorter than in *H. hamiltoni*, $1\frac{2}{3}$ in head. No exserted nasal tubes. Color plain brown. Bering Sea. Only the type known, $12\frac{1}{4}$ inches long, from which we have taken the above description. (*robustus*, robust.)

Hippoglossoides robustus, GILL & TOWNSEND, Proc. Biol. Soc. Wash., XI; 1897 (Sept. 17, 1897), 234, Bering Sea, Lat. 56° 14′ N., Long. 164° 08′ W., Albatross Station 3541, in 49 fathoms. (Type, No. 48766, U. S. Nat. Mus. Coll. *Albatross*.)

2983. HIPPOGLOSSOIDES HAMILTONI, Jordan & Gilbert, new species.

Head $3\frac{1}{2}$ in length; depth $2\frac{2}{3}$; longest diameter of upper eye $3\frac{1}{4}$ in head; snout (measured from upper eye) 5 in head; maxillary of colored side $2\frac{1}{4}$, of blind side $2\frac{1}{5}$, in head; depth of caudal peduncle equaling its length, $3\frac{1}{4}$ in head. D. 72; A. 56; P. 11; pores in lateral line 91. Upper profile of head continuing the dorsal curve without interruption, there being a slight depression above the eye and an increased convexity on the snout; mandible very heavy, projecting anteriorly, so that its symphyseal profile completes the curve of the snout; a very short prominence at symphysis directed vertically downward; gape strongly curved and the mouth narrowed anteriorly, so that the maxillary and premaxillary are almost wholly concealed along the middle of their length by the overarching prefrontal; teeth acute, in a single series in each jaw, all except the anterior teeth in each jaw short; at the symphysis of lower jaw the teeth are longer and directed inward, while in the anterior end of each premaxillary the teeth are still more enlarged, and the series on each side describes a strong curve with its convex side toward the median line; maxillary reaching vertical from slightly behind middle of lower eye; nostril tubes conspicuous, the anterior in closest proximity to the upper lip, which it entirely overhangs; posterior nostril tube wider and slightly shorter; eyes of nearly equal size, and opposite, separated by a wider ridge than in *H. elassodon*, the ridge bearing in its narrowest portion 2 well-defined rows of strongly spinous scales; a conspicuous series of pores joining lateral line with upper margin of upper eye, and another encircling the lower eye below and behind; a third series along mandible and preopercle; 1 large pore above posterior nostril; gill rakers slender, unarmed, 2 above the angle, 11 or 12 below it, the longest $2\frac{1}{4}$ in eye; dorsal fin beginning above front of pupil, the longest ray $2\frac{3}{6}$ in head; anal preceded by a strong spine, its height equaling that of dorsal; pectoral very long and slender, $\frac{2}{3}$ length of head, that of blind side shorter, $\frac{1}{4}$ length of head; ventrals reaching to base of fourth or fifth anal ray; caudal long, evenly rounded behind, the middle rays not longer than those adjacent, their length equaling distance from tip of snout to preopercular margin; scales on colored side strongly ctenoid except in a strip along middle of sides anteriorly; elsewhere each scale provided with 2 to 4 long spines; on blind side they are smooth except on nape and caudal peduncle; cheeks, opercles, and interorbital space covered with larger, rougher scales than those on sides; mandible and snout naked; a single series surrounding each eye anteriorly, and 1 on maxillary or colored side; blind side of head with maxillary naked; cheeks covered with minute smooth thin scales, the opercles with a few scattered spinous scales, the preopercle naked. Color nearly uniform brownish, without distinctive markings on body or fins. One specimen, 17 cm. long, from *Albatross* Station 3641, off Dalnoi Point, Kamchatka; depth 16 fathoms. Allied to *Hippoglossoides elassodon*, from which it differs in the fewer fin rays and scales, the wider interorbital space, the longer caudal and pectoral fins and the much smaller symphyseal knob. The nasal tubes are larger, the scales rougher, and the anterior part of lateral line more arched. Its relations with *H. robustus* are much nearer but the species are apparently distinct. (Named for Gerald Edwin H. Barrett-Hamilton, of Dublin, member of the British Commission of Fur Seal Investigation, 1896 and 1897, who made valuable collections of Kamchatkan fishes.)

Hippoglossoides hamiltoni, JORDAN & GILBERT, Rept. Fur Seal Invest., 1898, Dalnoi Point, Kamchatka. (Coll. *Albatross*.)

1019. PSETTICHTHYS, Girard.

Psettichthys, GIRARD, Proc. Ac. Nat. Sci. Phila. 1854, 140 (*melanostictus*).

Body dextral; teeth uniserial, sharp, unequal, some of them canine-like; mouth moderate, the lower pharyngeal teeth sharp, uniserial; scales small, ctenoid, ciliated, and firm; lateral line with an accessory dorsal branch and without distinct anterior arch; anal spine strong; dorsal fin

3030——87

beginning before the eye; vertebræ about 40; flesh firm. This genus contains but 1 species, found on the coast of California. It is nearly related to *Hippoglossoides,* but possesses the peculiar accessory dorsal branch to the lateral line, characteristic of so many of the Pacific coast flounders. (ψῆττα, the turbot; ἰχθύς, fish.)

2984. PSETTICHTHYS MELANOSTICTUS, Girard.

Head 4; depth 2⅓. D. 85; A. 60; scales 112. Body not very deep, elliptical; mouth rather small, the maxillary extending to below pupil, 2⅞ in head; teeth large, in a single series in each jaw, those in lower jaw largest; a few large canines in front of each jaw. Eyes very small, separated by a broad, flat, scaly space, without ridge; lower eye slightly in advance of upper; gill rakers rather stout, weak, hooked at tip, 14 below the angle; scales very small, ctenoid on colored side; lateral line nearly straight, with a long accessory dorsal branch; dorsal commencing in advance of upper eye, the anterior rays elevated, slender and exserted, the longest about ⅓ length of head; first ray of dorsal nearly free from its membrane; pectoral fin short, 2⅓ in head; anal fin preceded by a spine; caudal large, strongly convex; lower pharyngeals very narrow, each with 1 row of sharp, recurved teeth. Grayish brown, finely speckled with darker on body and fins. Pacific coast of North America, from Sitka south to Monterey. This is one of the commoner flounders of the Pacific coast, being everywhere known by the name of "Sole." It lives near the shore, and reaches a length of about 20 inches. In color this species is quite unlike the species of *Hippoglossoides,* but in most other respects the two groups are closely allied. (μέλας, black; στικτός, spotted.)

Psettichthys melanostictus, GIRARD, Proc. Ac. Nat. Sci. Phila. 1854, 140, San Francisco; Astoria, Oregon; GIRARD, U. S. Pac. R. R. Surv., x, Fishes, 154, 1858; GÜNTHER, Cat., IV, 420, 1862; LOCKINGTON, Rep. Com. Fisheries Cal. 1878–79, 40; LOCKINGTON, Proc. U. S. Nat. Mus. 1879, 76; JORDAN & GILBERT, Proc. U. S. Nat. Mus. 1880, 453; JORDAN & GILBERT, Proc. U. S. Nat. Mus. 1881, 67; JORDAN, Nat. Hist. Acquatic Animals, 186, pl. 51, 1884; JORDAN & GOSS, Review Flounders and Soles, 241, pl. 6, 1889.
Hippoglossoides melanostictus, JORDAN & GILBERT, Synopsis, 828, 1883.

1020. VERASPER, Jordan & Gilbert, new genus.

Verasper, JORDAN & GILBERT, Report Fur Seal Invest., 1898 MS. *(moseri).*

This genus is allied to *Xystreurys* and *Hippoglossina,* having few short gill rakers like the former and strongly ctenoid scales like the latter. It differs strongly from all its congeners in having the premaxillary teeth in 2 series, teeth uniformly small, without canines. Body dextral; dorsal inserted above the front of pupil; lateral line strongly arched above the root of the pectoral, without recurrent dorsal branch; scales firm, extremely spinous; gill rakers short, thick, and triangular, few in number; none of the fin rays notably produced or exserted. Japan and Kuril Islands; 2 species known, the following and *V. variegatus* (Schlegel), a common food fish of Japan, the 2 very closely related. (*verus,* true; *asper,* rough, the word being suggested by *Veratrum.*)

2985. VERASPER MOSERI, Jordan & Gilbert, new species.

Head 3⅓ in length to base of caudal; depth 2. D. 82; A. 58; pectoral 12; pores in lateral line 84. Depth of caudal peduncle 4 in greatest depth of body; length of caudal peduncle, measured axially, 1⅔ in its depth. Head much depressed, with rather wide, flat interorbital space, resembling in appearance *Psettichthys melanostictus*, its thickness at interorbital space equaling distance between pupils of upper and lower eyes. Mouth small, very oblique, the gape strongly arched, the broad maxillary reaching a vertical behind middle of pupil, 2⅜ in head; mandible narrowing toward tip, with very rudimentary symphyseal knob. Teeth in upper jaw in 2 distinct series throughout, those of the outer series increasing slightly in size toward front of jaw, but none of them canine-like; mandibular teeth in 1 row, except at symphysis, where a few teeth form a short outer series. Nasal openings of eyed side approximated in front of middle of interorbital space, the anterior with a short tube, the posterior with a raised rim. Eyes small, their anterior margins opposite, the diameter of lower eye equaling distance from tip of snout to posterior nostral, 6⅓ in head. Interorbital space rather broad and flat, not ridge-like, its total width equaling ⅓ diameter of orbit. Gill rakers short, broad, triangular, minutely toothed on inner margin, ⅓ diameter of eye; 7 present on horizontal limb of outer arch. Lateral line with a short high anterior arch, the cord of which is ⅓ the straight portion; height of arch ⅓ its length; behind the arch lateral line descending in a gentle curve to middle of sides, the scales very rough, each possessing several long, sharp spines diverging from median portion of posterior margin; anterior and posterior portions of dorsal and anal fins naked, the rays of the middle portion each with a series of strongly ctenoid scales; caudal densely scaled to tip; pectorals and ventrals naked; head covered with strongly spinous scales, excepting snout, maxillary, and mandible; on blind side of head the snout, jaws, preopercle, subopercle, lower half of opercle, and all but a central strip on interopercle, scaleless; on blind side the scales are rough on head, ventral area, and along bases of ventral fins, largely smooth elsewhere. Dorsal beginning above front of pupil, the rays increasing in length to the forty-fifth, which is 2⅔ in head; longest anal ray (the seventeenth) 2⅓ in head. Caudal broadly rounded, 1⅔ in head; pectoral short and broad, 2⅔ in head; ventrals of nearly equal length, reaching origin of anal, 3⅓ in head; no anal spine. Color in spirits, centers of the scales light gray, the margins dark brown; fins light or dusky, the vertical fins with conspicuous black bars, parallel with the rays, these most evident on under side where the pigment seems principally to occur, and are seen through the fin more faintly on the colored side; lining of cheeks and gill cover of colored side dusky; peritoneum gray. Kuril Islands; 1 male 28 cm. long, from Shana Bay, Iturup Island; also taken at Hakodate. (Named for Jefferson Franklin Moser, U. S. N., Lieutenant-Commander, in charge of the U. S. Fish Commission Steamer *Albatross*, and a member of the United States Fur Seal Commission for 1896.)

Verasper moseri, Jordan & Gilbert, Rept. Fur Seal Invest., 1898 MS., Shana Bay, Iturup Island, Kuril Group. (Type No. 48797. Coll. *Albatross*, Capt. J. F. Moser.)

1021. HIPPOGLOSSINA, Steindachner.

Hippoglossina, STEINDACHNER, Ichth. Beiträge, v, 13, 1876 (*macrops*).

Teeth rather small, uniserial, no canines; lateral line with a strong arch in front, and with no accessory dorsal branch; anal spine obsolete; body indifferently dextral or sinistral (in some species at least). Scales ctenoid; dorsal fin beginning above pupil; gill rakers rather long and slender. This genus is intermediate between *Hippoglossoides* and *Paralichthys*, agreeing with the former in the insertion of the dorsal and in general appearance, and with the latter in the direction of the lateral line. Several species are now known. Some of them are dextral, and perhaps all of them are normally so, or perhaps, as in the case of *Xystreurys liolepis*, all are indifferently dextral or sinistral. (A diminutive of *Hippoglossus*, the halibut.)

   a. Mouth large, the maxillary extending to opposite posterior margin of eye, 2 in head; gill rakers numerous, 4 + 13; dorsal rays about 68; anal 53.   STOMATA, 2986.
  aa. Mouth moderate, the maxillary extending to opposite middle of pupil, about 2⅓ in head.
    b. Dorsal rays about 66; anal 52; depth of body 2⅓ in length.   MACROPS, 2987.
   bb. Dorsal rays about 62; anal 48; depth of body 2⅔ in length; gill rakers 2 + 8 or 9.   BOLLMANI, 2988.

2986. HIPPOGLOSSINA STOMATA, Eigenmann & Eigenmann.

Head 2¾ to 3 in length; depth 2⅓ to 2⅔. D. 67 to 70; A. 52 to 54; scales 80. Sinistral. Eye (not orbit) large, 5 in head; lower orbit slightly in advance of upper; interorbital a narrow ridge. Form, elongate elliptical, the profile depressed over the eye. Mouth large, maxillary extending to posterior margin of eye, as long as or longer than pectoral, 2 in head; lower jaw about 1¾ in head. Teeth small, uniserial; anterior nares of each side with long dermal flaps. Scales of left side all ctenoid, those of right side cycloid on anterior half or two-thirds of body; middle third of interorbital naked, anterior and posterior thirds scaled. Gill rakers 4 + 13 or 14. Dorsal beginning over middle of eye, anterior rays with but 1 or 2 scales, rest scaled to near tip, all but last 8 rays simple; anal similar to dorsal, with a strong procumbent spine; highest dorsal and anal rays about 3⅓ in head; pectoral of colored side about 2 in head, that of blind side shorter; caudal double truncate, 5 to 5⅓ in length. Brown, strongly tinged in life with robin's-egg blue; numerous spots of light blue and light and dark brown; 5 pairs of large, dark-brown ocelli along dorsal and ventral parts of eyed side, the alternate ones longer and more conspicuous; fins colored like body, profusely mottled with light and dark; sinistral pectoral barred; a dark-brown spot above and below on caudal peduncle just in front of caudal, showing conspicuously on blind side. The eggs are probably pelagic; they are transparent, and measure 1.2 mm. in diameter; the single oil globule measures 0.16 mm. Coast of southern California; 2 specimens obtained in deep water off San Diego, November 7, 1889, both females, 1 with ripe eggs. (Eigenmann & Eigenmann.) (ὄτομάτός, large mouthed.)

Hyppoglossina stomata, EIGENMANN & EIGENMANN, Proc. Cal. Ac. Sci. 1890, 22, San Diego.
  (Coll. C. H. Eigenmann.)

2987. HIPPOGLOSSINA MACROPS, Steindachner.

Head $2\frac{3}{4}$; depth $2\frac{1}{3}$. D. 66; A. 52; scales 75 to 80; upper orbit $3\frac{1}{2}$ in head. Body elliptical, deeper than in related species; mouth moderate, the maxillary reaching to middle of eye; teeth small, sharp, uniserial; lower eye slightly in front of upper; eyes separated by a naked narrow ridge; nostrils close together, the anterior ending in a tube; horizontal limb of preopercle somewhat concave, the vertical convex. Dorsal beginning over middle of eye; pectoral of left side $\frac{1}{2}$ head, much longer than maxillary, which is $2\frac{2}{3}$ in head; interorbital space a narrow ridge; scales of left side all strongly ctenoid, those on blind side ciliated only on posterior third of body; no anal spine. Color brownish, with obscure darker blotches. Body sinistral (in the only specimen known). (Steindachner.) Pacific coast of Mexico. One specimen from Mazatlan; not seen by us. (μακρός, large; ὤψ, eye.)

Hippoglossina macrops, STEINDACHNER, Ichth. Beitr., V, 13, pl. 3, 1876, Mazatlan; JORDAN & GOSS, Review Flounders and Soles, 242, 1889.

2988. HIPPOGLOSSINA BOLLMANI, Gilbert.

Head 3 ($3\frac{3}{4}$ to $3\frac{1}{5}$) in length; depth $2\frac{2}{5}$ to $2\frac{3}{4}$ ($3\frac{1}{3}$ to $3\frac{1}{2}$); snout 5 in head. D. 60 to 63; A. 47 to 49; scales along lateral line 70 to 75. Body regularly elongate, elliptical; dorsal and ventral outlines equally curved; orbital rim entering anterior profile, which is equally curved before and behind eyes; greatest depth of body above pectorals. Mouth rather large, the maxillary reaching about to middle of pupil, $2\frac{1}{5}$ to $2\frac{3}{5}$ in head. Teeth equally developed on both sides, small and equal, uniserial. Premaxillary spine prominent. Interorbital space a narrow, sharp, naked ridge; eyes large, the lower slightly in advance of upper, $3\frac{2}{3}$ to 4 in head. Gill rakers moderately long and slender, the longest 3 in length of ventral of eyed side; 2 + 8 or 9 developed, the last 2 much shorter. Scales small, firm, strongly ctenoid, those below pectoral much reduced, about 40 in a cross series; arch of lateral line strongly marked, $2\frac{2}{3}$ to $2\frac{1}{5}$ in straight part. Dorsal beginning above middle of pupil of upper eye, its anterior rays low, its longest rays $2\frac{1}{7}$ in head; a strong antrorse spine before anal; pectoral of eyed side 2 in head, that of blind side $2\frac{1}{2}$ to $2\frac{2}{3}$ in head; ventrals subequal, each 6-rayed, 4 in head, extending more than $\frac{1}{2}$ their length beyond anus; each is lateral, but that of eyed side nearest ridge of abdomen, and a little behind its fellow; last ray of left ventral joined to abdomen alongside of anal spine; caudal acute, its peduncle long. Color grayish brown, a row of 6 round, bluish spots, smaller than pupil, along base of dorsal, 4 similar spots along base of anal, and a few indistinct smaller ones on rest of body and head; body with 6 large black spots somewhat smaller than eye, these regularly 4 below dorsal and 2 above anal, the first of dorsal above arch of lateral line, the second above anterior third of straight part, the third at base of last rays and almost forming a cross bar with the 1 at base of anal rays. Dorsal, anal, and caudal dusky, with small whitish spots; a pale spot at base of last 4 dorsal and anal rays; a small black spot at base of outer caudal rays on peduncle; pectorals and ventrals dusky, but not spotted; right side immaculate. Length

7 inches. Pacific coast of Colombia. Numerous specimens were dredged
at *Albatross* Station 2805, at a depth of 51½ fathoms. This species differs
from *Hippoglossina stomata* in the gill rakers, which are shorter and fewer
in number, and in the larger scales on sides. Scales in 16 rows between
lateral line and back, instead of 21 or 22, as in *H. stomata*. Gill rakers
somewhat shorter, 8 or 9 on anterior limb, 2 on upper limb. In *H. macrops*
the gill rakers are slender, close set, 13 or 14 on anterior limb, 4 on vertical
limb. In other respects of color, fin rays, and squamation agreeing perfectly with *H. stomata*. (Named for Charles Harvey Bollman.)

Hippoglossina macrops, JORDAN & BOLLMAN, Proc. U. S. Nat. Mus. 1889, 175; not of STEIN-
DACHNER.
Hippoglossina bollmani, GILBERT, Proc. U. S. Nat. Mus. 1890, 122, Albatross Station 2805,
southwest of Panama, in 51½ fathoms. (Type, No. 41143.)

1022. LIOGLOSSINA, Gilbert.

Lioglossina, GILBERT, Proc. U. S. Nat. Mus. 1890, 122 (*tetrophthalmus*).

This genus is allied to *Hippoglossina*, but its scales are all cycloid, the
teeth are small, pointed, uniserial, and uniform, and the gill rakers short
and thick. (λειός, smooth; πλῶσσα, tongue; for *Hippoglossina*.)

2989. LIOGLOSSINA TETROPHTHALMA, Gilbert.

Head large, 3¼ in length in a specimen 1 foot long. D. 76 to 83; A. 58 to
62; lateral line (pores) 97. Body of moderate height, the profile distinctly
angulated above upper pupil, the snout projecting; length of caudal
peduncle ½ its depth, its outlines diverging backward; depth of body 2¼
in length; snout projecting beyond profile, bluntly rounded, the lower
jaw included. Mouth large, the maxillary reaching nearly to vertical from
posterior border of lower eye, 2½ in head; a blunt projecting process anteriorly from head of maxillary. Teeth small, pointed, in a single close-set
series in each jaw, none of them enlarged; vomer toothless; lower eye
slightly in advance of upper; vertical from front of upper falling midway
between front of orbit and front of pupil of lower eye; vertical diameter
of upper orbit but little more than ¼ its longitudinal diameter, which is
contained 3½ in head; interorbital space a blunt high ridge, entirely scaleless, its width ⅔ diameter of orbit. Anterior nostril of blind side with a
very long flap, that of eyed side shorter; a well-marked cutaneous flap on
lower eye above pupil. Gill rakers very large, broad, and strong, well
toothed on inner edges, longest equaling diameter of pupil, the number
on outer gill arch 10 or 11. First dorsal ray over anterior margin of pupil
of upper eye, the fin not high, its highest ray 3 in head; anal similar;
caudal sharply double truncate, the median rays produced; ventrals
rounded, equal, barely reaching front of anal; no spine before anal fin;
pectorals moderate, with 9 or 10 developed rays, ¼ length of head; ventral
6. Scales rather small, growing distinctly larger posteriorly, everywhere
smooth; head scaled, except snout, interorbital area, mandible, and part
of maxillary, the latter with a patch of scales on posterior end of its
expanded portion; on blind side an area around nostrils, and the greater
part of exposed portion of preorbital, scaleless; fin rays of vertical fins,

all with bands of fine scales, those on caudal especially broad; lateral line with a broad arch in front, the cord of which is $3\frac{2}{3}$ in straight portion. Color dusky brownish, with 2 conspicuous pairs of round black spots narrowly edged with gray, the anterior pair about $\frac{1}{4}$ size of orbit, the posterior larger than pupil; the anterior pair under beginning of posterior third of dorsal, and about halfway between lateral line and dorsal and anal margins, respectively; the posterior pair nearer outline of body and about under the tenth before the last dorsal ray; vertical fins obscurely blotched with darker; ventral of eyed side with conspicuous black blotch margined with white, occupying the distal portion of its inner 2 rays; pectoral unmarked; membrane of gill cavity and peritoneum white. Two specimens, each about 12 inches long, from the Gulf of California, taken in 29 and 76 fathoms, at *Albatross* Stations 3014 and 3016. (Gilbert.) ($\tau\epsilon\tau\rho\alpha$-, four; $\dot{o}\phi\theta\alpha\lambda\mu\dot{o}\varsigma$, eye, or eye-like spot.)

Lioglossina tetrophthalmus, GILBERT, Proc. U. S. Nat. Mus. 1890, 122, Gulf of California. (Coll. Dr. Gilbert.)

1023. XYSTREURYS, Jordan & Gilbert.

Xystreurys, JORDAN & GILBERT, Proc. U. S. Nat. Mus. 1880, 34 (*liolepis*).

Body broad, covered with small smooth scales. Teeth rather small, uniserial and bluntly conical, unequal; no canines; caudal fin subsessile, the caudal peduncle extremely short; skin of shoulder girdle with patches of cup-shaped scales; lateral line with a strong anterior arch, no accessory branch; vertebræ about 37; gill rakers short and thick. This genus is very close to *Hippoglossina*, differing chiefly in the subsessile caudal fin, the smooth scales, and the peculiar, short, thick gill rakers. The typical species, like some other Pacific coast flounders, is almost indifferently dextral or sinistral. ($\xi\dot{v}\sigma\tau\rho\sigma\nu$, raker; $\epsilon\dot{v}\rho\dot{v}\varsigma$, wide, from the broad gill rakers.)

2990. XYSTREURYS LIOLEPIS, Jordan & Gilbert.

Head $3\frac{1}{3}$; depth $1\frac{3}{5}$. D. 80; A. 62; scales 123. Vertebræ $12 + 25 = 37$. Body elliptical ovato, broad and compressed, its curves regular; the profile continuous with curve of back; mouth small, very oblique, the lower jaw included; maxillary reaching about to pupil, $2\frac{2}{3}$ in head; eyes rather large, $4\frac{1}{2}$ in head, separated by a very narrow, blunt scaly ridge; teeth small, conical, blunt, in a single row; those in lower jaw subequal, close set; those in upper jaw more distant, decreasing in size backward; teeth $\frac{12+13}{14+15}$. Gill rakers $2 + 7$, very short, broad, and strong, minutely serrate on inner margin, about 7 below angle, the longest scarcely $\frac{1}{4}$ as long as the eye. Scales small, oblong, cycloid, the smaller accessory scales extremely numerous; lateral line without dorsal branch, with a broad curve above pectorals; branchial arches and skin of the shoulder girdle with small, cup-shaped, tubercular scales. Dorsal rather high, firm, low in front, beginning just in advance of middle of pupil, highest near the middle of the body; caudal peduncle very short and deep, its depth 4 times its length. Pectoral of eyed side falcate, usually much longer than head, its length varying considerably. Caudal fin somewhat double truncate, with rounded angles, the middle rays being produced. Anterior nostril of blind side with a long flap. Color olive brown, mottled with darker, sometimes with very distinct round black blotches; vertical fins blotched with dark; pectoral of colored side with oblique bars. Length 15 inches. Southern California, rather common from Point Concepcion southward to San Diego. It is a very variable species, the coloration and the length of the pectoral fins having a wide range of variation. The body is indifferently dextral or sinistral. (λεῖος, smooth; λεπίς, scale.)

Xystreurys liolepis, JORDAN & GILBERT, Proc. U. S. Nat. Mus. 1880, 34, Santa Barbara; JORDAN & GILBERT, Proc. U. S. Nat. Mus. 1880, 454; JORDAN & GILBERT, Proc. U. S. Nat. Mus. 1881, 66; JORDÁN & GOSS, Review of Flounders and Soles, 243, 1889.
Paralichthys liolepis, JORDAN & GILBERT, Synopsis, 825, 1883.

1024. PARALICHTHYS, Girard.

(BASTARD HALIBUTS.)

Paralichthys, GIRARD, U. S. Pac. R. R. Surv., x, 146, 1858 (*maculosus = californicus*).
Pseudorhombus, BLEEKER, Comptes Rendus, Acad. Sci. Amsterd., XIII, 1862, 5, Notice sur quelques genres de la famille des Pleuronectidæ (*polyspilos*).
Uropsetta, GILL, Proc. Ac. Nat. Sci. Phila. 1862, 330 (*californicus = maculosus*).
Chænopsetta, GILL, Proc. Ac. Nat. Sci. Phila. 1864, 218 (*ocellaris = dentatus*).

Eyes and color normally on the left side. Body oblong; mouth large, oblique; each jaw with a single row of usually slender and sharp teeth, which are more or less enlarged anteriorly; no teeth on vomer or palatines. Gill rakers slender. Scales small, weakly ctenoid or ciliated; lateral line simple, with a strong curve anteriorly. Dorsal fine beginning before the eye, its anterior rays not produced; both ventrals lateral; caudal fin double truncate, or double concave, its middle rays produced; no anal spine. Species numerous, found in all warm seas. This genus, as now restricted, contains a considerable number of species, inhabiting both coasts of America and the eastern and southern coasts of Asia. As indicated by the reduced number of vertebræ, the species range further southward than do those of the type of *Hippoglossoides*. (παράλληλος, parallel; ἰχθύς, fish.)

a. Gill rakers in large number, about 9 + 20.
    b. Gill rakers as long as eye and very slender. D. 72; A. 55; depth 2¾ in length.
                                           CALIFORNICUS, 2991.
    bb. Gill rakers shorter, about ⅔ length of eye. D. 80; A. 61; depth 2¼ in length.
                                           ÆSTUARIUS, 2992.
aa. Gill rakers in moderate number (5 + 11 to 6 + 21), rather long and slender.
    c. Dorsal rays 70 to 75; anal rays 54 to 60.
        d. Head small, lateral line 4½ in length; depth 2¼; interorbital space rather broad and flatish, ⅔ diameter of eye; eyes small, 5⅔ in head; gill rakers rather short, 4 + 15, the longest about ⅔ eye.
                                       BRASILIENSIS, 2993.
        dd. Head rather large, 3½ in length; depth 2 to 2¼; eyes small.
            e. Gill rakers 5 to 6 + 15 to 18; eyes wide apart.   ADSPERSUS, 2994.
           ee. Gill rakers 5 + 11; eyes close together.   WOOLMANI, 2995.
    cc. Dorsal rays 85 to 93; anal rays 67 to 73; gill rakers 5 + 15 or 16, long and slender, the longest ⅔ length of eye; body ovate, the depth about 2¼ in length; head about 3⅔.
                                         DENTATUS, 2996.

aaa. Gill rakers few, shortish, wide set, the number 2 + 8 to 3 + 10.

    f. Body ovate, more or less compressed and opaque; depth about 2½ in length; no distinct, definitely placed ocelli; scales cycloid.

        g. Dorsal rays in large number (85 to 93, as in *P. dentatus*); anal rays 65 to 73; pores of the lateral line about 100; accessory scales few; gill rakers 2 + 10, lanceolate, dentate, wide set, and much shorter than the eye.   LETHOSTIGMUS, 2997.

        gg. Dorsal rays in moderate number (70 to 80); anal rays 54 to 61.

           h. Scales very small, about 120 in lateral line; depth of body about ⅓ length; head 3⅔ in length; gill rakers roughly toothed, 3 + 9 in number.   SQUAMILENTUS, 2998.

           hh. Scales moderate, 90 to 100 pores in the lateral line; interorbital width about equal to length of eye; dorsal rays 75 to 81; anal rays 59 to 61; gill rakers 2 or 3 + 9 or 10. Coloration, grayish brown with numerous (more or less distinct) whitish blotches, which are rarely obsolete; vertebræ 10 + 27 = 37.

                                               ALBIGUTTUS, 2999.

  ff. Body oblong, strongly compressed, semitranslucent; scales weakly ciliated; about 93 pores in lateral line. Coloration, light grayish, thickly mottled with darker; 4 large horizontally oblong, black ocelli, each surrounded by pinkish area; 1 just behind middle of the body, below the dorsal, 1 opposite this, above anal, and 2 similar smaller spots below last rays of dorsal and above last of anal; vertebræ 11 + 30 = 41.   OBLONGUS, 3000.

2991. PARALICHTHYS CALIFORNICUS (Ayres).

(BASTARD HALIBUT; MONTEREY HALIBUT.)

Head 3¾ to 4¼; depth 2⅔. D. 70; A. 55; scales 100. Vertebræ 10 + 25 = 35. Body rather long and thickish; caudal peduncle long; head small; eye small, little wider than the broad, flattish interorbital space; maxillary as long as pectoral, ½ length of head, reaching beyond eye; teeth slender, sharp, rather long, the canines moderate. Scales small, finely ciliate, each scale surrounded by narrow accessory scales; scales on blind side similar; fins with ctenoid scales. Dorsal low, beginning over front of upper eye just past pectoral, pointed, reaching curve of lateral line, 2¼ in head, that of blind side shorter and rounded behind; arch of lateral line 3½ or 4 in straight part. Gill rakers very long and slender, numerous, as long as eye, about 9 + 20; lower pharyngeals narow, with small slender teeth. Anal spine small, concealed. Grayish brown, uniform, or mottled with blackish and pale, the head sometimes sprinkled with black dots; young brownish, with bluish spots. Coast of California, Tomales Bay to Cerros Island. This large flounder is one of the common food-fishes of the Pacific coast, where it takes the place occupied on the Atlantic side by *Paralichthys dentatus*. It reaches a length of 3 feet and a weight of 60 pounds. From its resemblance to the halibut, it usually goes by the name of bastard halibut. It is readily distinguished from the Atlantic members of the same genus by its fewer fin rays and by its more numerous gill rakers. As was first shown by Mr. Lockington, the small fish called *Paralichthys maculosus*, is simply the young of the larger fish, then called *Uropsetta californica*. Unlike other species of the genus, *Paralichthys californicus* is almost as frequently dextral as sinistral. (*californicus*, Californian.)

Pleuronectes maculosus, GIRARD, Proc. Ac. Nat. Sci. Phila. 1854, 155, young, San Diego.

Paralichthys maculosus, GIRARD, U. S. Pac. R. R. Surv., x, Fishes, 147, 1858, not *Rhombus maculosus*, CUVIER, also a species of *Paralichthys*; GÜNTHER, Cat., IV, 431, 1862; GILL, Proc. Ac. Nat. Sci. Phila. 1864, 197; LOCKINGTON, Rep. Com. Fisheries California 1878–79, 41; LOCKINGTON, Proc. U. S. Nat. Mus. 1879, 79; JORDAN & GILBERT, Proc. U. S. Nat. Mus. 1880, 454; JORDAN & GILBERT, Proc. U. S. Nat. Mus. 1881, 66; JORDAN, Nat. His. Aquat. Anim., 182, 1884.

Hippoglossus californicus, AYRES, Proc. Cal. Ac. Nat. Sci. 1859, 29, and 1860, fig. 10, adult, San Francisco.

Pseudorhombus californicus, GÜNTHER, Cat., IV, 426, 1862.

Uropsetta californica, GILL, Proc. Ac. Nat. Sci. Phila. 1862, 330; GILL, Proc. Ac. Nat. Sci. Phila. 1864, 198.

Paralichthys californicus, JORDAN & GILBERT, Synopsis, 821, 1883; JORDAN & GOSS, Review Flounders and Soles, 245, 1889.

2992. PARALICHTHYS ÆSTUARIUS, Gilbert & Scofield.

Head 3⅜; depth 2¼; eye 5½; interorbital space flat, 12 in head, ½ diameter of eye; maxillary 2 in head, equal to pectoral fin; gill rakers 9 + 20, the longest ⅜ length of eye. D. 72 to 83; A. 58 to 64. (In 7 specimens examined the rays are: Dorsal 72, 79, 81, 81, 82, 83, 83; anal 58, 60, 60, 62, 63, 63, 64.) Vertebræ 10 + 28; scales weakly ciliated, with small accessory scales, 105 in the lateral line; length of the arch contained 4 times in straight part of lateral line, 2 in head; height of arch 4½ in head. Four of the 7 specimens are sinistral. Color pale chocolate brown. Specimens small, 6 to 9 inches in length. Taken at Shoal Point, at mouth of the Colorado River, Mexico, by the United States Fish Commission steamer *Albatross*. This species is distinguished from other members of the genus by its numerous fin rays and many gill rakers. It is nearest related to *Paralichthys californicus*. (*æstuarius*, pertaining to the river mouth.)

Paralichthys æstuarius, GILBERT & SCOFIELD, Proc. U. S. Nat. Mus. 1897, 499, pl. XXIX, Gulf of California, at mouth of Colorado River, Sonora. (Type, No. 48128. Coll. C. H. Gilbert.)

2993. PARALICHTHYS BRASILIENSIS (Ranzani).

Head 4⅕; depth 2⅕. D. 70 to 75; A. 54 to 60; scales not very small, about 100 in course of lateral line; interorbital space rather broad and flattish, ⅜ diameter of eye; eyes small, 5⅜ in head; gill rakers rather short, 4 + 15, the longest about ⅗ eye; pectoral 1⅓ in head; curve of lateral line high and short, 4 in straight part, its height 1¾ in its length; mouth moderate, the maxillary 2⅕ in head; teeth rather few, the anterior canines large. Color dark brown, more or less mottled and spotted with paler. South America; said to range northward to Guatemala. Here described from numerous specimens from Rio Janeiro and from Maldonado, in the Museum of Comparative Zoology. The locality "Guatemala" given by Günther seems to be somewhat doubtful, and the species may not occur in West Indian waters at all. (*brasiliensis*, living in Brazil.)

Hippoglossus brasiliensis, RANZANI, Nov. Spec. Pisc., 10, tab. 3, 1840, Brazil.

Platessa orbignyana, VALENCIENNES, D'Orbigny Voy. S. Amer. Mérid. Poiss., pt. 5, pl. 16, fig. 1, 1847.

Rhombus aramaca, CASTELNAU, Anim. nouv. ou rares, Poiss., 78, pl. 40, fig. 3; not of CUVIER.

Pseudorhombus vorax, GÜNTHER, Cat., IV, 429, 1862, South America.

Pseudorhombus brasiliensis, GÜNTHER, Fishes Centr. Amer., 473, 1869.

Paralichthys brasiliensis, JORDAN & GOSS, Review Flounders and Soles, 246, 1889.

2994. PARALICHTHYS ADSPERSUS (Steindachner).

Head $3\frac{1}{2}$; depth $2\frac{1}{4}$. D. 75; A. 58; scales 106; eye 6 in head; interorbital $\frac{3}{4}$ vertical diameter of eye; maxillary $2\frac{1}{2}$; mandible $1\frac{3}{5}$; pectoral 2; caudal $1\frac{3}{4}$. Body moderately elongate and compressed; mouth large, the maxillary reaching a little past eye; teeth large, sharp, and slightly recurved, larger in front of jaws; snout very slightly produced; interorbital moderately wide, its posterior half with scales; anterior nostril with a flap which reaches to middle of posterior nostril; gill rakers 3 to 6 + 15 to 17, hardly as long as eye. Snout and mandible naked; end of maxillary and rest of head with scales; the rays of all the fins with small scales; the membrane naked; each scale on body with a row of accessory scales around its posterior edge; scales cycloid, the accessory scales giving the fish a rough feeling; curve of lateral line nearly 5 in the straight part, pectoral reaching slightly past curve of lateral line, its tip pointed; pectoral of blind side shorter, not reaching to end of curve, its tip blunt; origin of dorsal over anterior edge of upper eye, bending slightly toward the blind side; caudal double lunate. Color brownish gray, thickly mottled with many larger and smaller spots, points, and rings; side with 3 or 4 larger spots of irregular form and ocellated with paler.

Specimens taken by Dr. Jordan at Mazatlan are described as follows: "Head $3\frac{1}{2}$; depth about 2 in length of body. D. 73 (70 to 76); A. 57 (53 to 60); P. 12; V. 6. Scales on lateral line about 106 + 8 with 35 dorsally and 36 ventrally. Flesh firm. Body oblong, moderately compressed; mouth large, oblique, the mandible very heavy, slightly projecting; 4 canine teeth on each side of lower jaw in adult specimens, 8 in young, the 2 anterior teeth long; anterior teeth of upper jaw strong, but smaller than those in the lower jaw; the lateral teeth very small and close set. Eye small, shorter than snout, about 7 (6 to 8) in length of head; interorbital area smooth, flattish, $\frac{2}{3}$ width of eye. Scales cycloid, small anteriorly and larger posteriorly; lateral line strongly arched anteriorly, arch about $3\frac{1}{2}$ in straight part. Gill rakers of medium length, broad, retrorse-serrate on inner side, longest about $\frac{2}{3}$ length of eye, from 4 + 13 to 5 + 14 in number, counted in 8 specimens; pectoral fin about as long as mandible, slightly more than $\frac{1}{3}$ length of head. Dorsal low, anterior origin opposite anterior margin of eye; caudal barely double concave; caudal peduncle very strong; anal spine obsolete; ventral fins small, inserted symmetrically; fins all scaly. Color: Large specimens are dark brown, with blotches on fins; small specimens are covered with pearly white and very dark brown blotches; the brown blotches almost circular, larger and with less definite outlines near the center of the body, very dark and distinct on caudal. Seven specimens were taken by the Hopkins Expedition in the estuary at Mazatlan, where they reach a length of 44 cm. Several specimens were also taken at La Paz. These specimens seem to be identical with *Paralichthys adspersus*. The original types have on an average more gill rakers than we find in our Mazatlan specimens, but this character is subject to variation, and no other distinction appears. In one of Dr. Steindachner's types from Callao (No. 11,417, Mus. Comp. Zool.) we find the gill rakers longer, 6 + 17; depth $2\frac{1}{4}$ in length; D. 67; A. 51; scales 120;

arch of lateral line barely twice as long as high, nearly 5 in straight part; maxillary $2\frac{1}{6}$ in head. Mr. Garman has kindly examined for us 6 other specimens, with the following results:

"*Paralichthys adspersus* from Callao, has gill rakers—

"'$\frac{1}{17}$ as long as eye;

"'$\frac{6}{15}$ about $\frac{2}{3}$ as long as the eye.

"'$\frac{7}{13}$ nearly as long as the eye.

"'$\frac{3}{14}$ about $\frac{2}{3}$ as long as the eye.

"'$\frac{6}{13}$ about $\frac{2}{3}$ as long as the eye.

"'$\frac{6}{17}$ near $\frac{3}{4}$ as long as eye.'" *

We are now disposed to regard these Mazatlan specimens as identical with *Paralichthys adspersus*, the range of variation in the number of gill rakers in the latter probably including the former. Pacific coast of tropical America, from Gulf of California to the coast of Peru; everywhere abundant and very variable. (*adspersus*, covered with spots.)

Pseudorhombus adspersus, STEINDACHNER, Ichthyol. Notizen, v, 9, pl. 2, 1867, Chinchas Islands.
Paralichthys adspersus, JORDAN & GILBERT, Proc. U. S. Nat. Mus. 1882, 370; JORDAN & GILBERT, Bull. U. S. Fish. Comm. 1882, 108 and 111; JORDAN, Cat. Fish N. A., 133, 1885; JORDAN & GOSS, Review Flounders and Soles, 246, 1889; JORDAN & WILLIAMS, Proc. Cal. Ac. Sci. 1895, 503.

2995. PARALICHTHYS WOOLMANI, Jordan & Williams.

Head $3\frac{1}{2}$; depth about 2; gill rakers $5+11$. D. 74; A. 57; P. 12; V. 6; scales 100. Flesh firm; body oblong; mouth large, mandible heavy, not projecting; about 8 teeth on each side of lower jaw, the anterior ones long and slender; teeth in upper jaw smaller than those in lower jaw, the lateral teeth very small and close set. Eye small, $5\frac{1}{4}$ in length of head; interorbital area moderately prominent, narrow, about $\frac{2}{3}$ length of eye. Scales cycloid, small anteriorly and increasing in size posteriorly, covering head and fins; lateral line greatly arched anteriorly, arch about $3\frac{1}{4}$ times in length of straight portion. Gill rakers slender, the longest about $\frac{1}{4}$ in length of eye. Pectoral and ventral fins small; pectoral about $\frac{1}{4}$ in length of head; origin of dorsal opposite anterior margin of eye; caudal ending in an obtuse angle, not double concave; caudal peduncle strong; anal spine obsolete. Body and fins blotched with deep brown and pearly white and specked with very dark brown, blotches more definite on median fins and especially on caudal where there are 3 indefinite lines of blotches crossing the fin. Galapagos Islands. One specimen taken by the *Albatross* in 1890, which was at first identified as *Paralichthys adspersus*, from which species it differs but little except in the number of gill rakers. (Named for Mr. Albert Jefferson Woolman, of Duluth, Minnesota, in recognition of his work on the fishes of Mexico and Florida.)

Paralichthys woolmani, JORDAN & WILLIAMS, Proc. U. S. Nat. Mus. 1896, 457, Galapagos Islands. (Type, No. 47575. Coll. *Albatross*.)

* Garman, in lit., May 3, 1895.

2996. PARALICHTHYS DENTATUS (Linnæus).

(SUMMER FLOUNDER.)

Head $3\frac{1}{5}$ to 4; depth $2\frac{2}{5}$; eye 6 in head; maxillary 2; pectoral $2\frac{1}{5}$; ventral $3\frac{1}{2}$; caudal peduncle 4; caudal $1\frac{1}{4}$. D. 86 to 91; A. 65 to 71; lateral line 108 (tubes). Curve of lateral line $3\frac{2}{5}$ to $4\frac{1}{5}$ in straight portion; body ovate; maxillary about $\frac{1}{2}$ head, reaching past posterior margin of eye; mouth large, oblique, the gape curved; canines large, conical, wide set; gill rakers comparatively long and slender, longest $\frac{2}{3}$ eye, 5+15 to 6+18 in number; interorbital area a rather flattish ridge, in the adult about equal to vertical diameter of eye, narrower in the young, forming a bony ridge; scales cycloid, each with numerous small accessory scales; vertebræ 11+30=41. Color in life, light olive brown; adults with very numerous small white spots on body and vertical fins; sometimes a series of larger white spots along bases of dorsal and anal fins; about 14 ocellated dark spots on sides, these sometimes little conspicuous, but always present; a series of 4 or 5 along base of dorsal, and 3 or 4 along base of anal, those of the 2 series opposite, and forming pairs; 2 pairs of smaller less distinct spots midway between these basal series and lateral line anteriorly, with a small one on lateral line in the center between them; a large distinct spot on lateral line behind middle of straight portion; fins without the round dark blotches. Atlantic coast of United States, from Cape Cod to Florida; the common flounder of the coasts of the Northern States, its range apparently not extending much south of Charleston. Of the species found in that region it is the most important from a commercial point of view. It reaches a length of about 3 feet and a weight of about 15 pounds. It has been confounded by nearly all writers with the more southern species now called *P. lethostigmus*, from which it is best distinguished by its much greater number of gill rakers and by its mottled coloration. On account of this confusion it is impossible wholly to disentangle its synonymy from that of *P. lethostigmus*. So far as the proper nomenclature of the two is concerned, this confusion makes little difference. There is no doubt that this is the original *Pleuronectes dentatus* of Linnæus, as the original Linnæan type is still preserved in London. This has been examined by Dr. Bean and its identity with the present species fully established. It seems also certain that this is the *Platessa ocellaris* of De Kay, who properly distinguishes his *ocellaris* from his *oblonga*, the latter being *P. lethostigmus*. A little doubt must be attached to the *P. melanogaster* of Mitchill, very scantily described from a doubled (black-bellied) example of this species or of *P. lethostigmus*. As the former species is much more common about New York than the latter it is probable that Mitchill's fish belonged to it. We have also received a doubled example from New York corresponding exactly to Mitchill's description. We may therefore regard the name *melanogaster* as a synonym of *dentatus*. The differences in the gill rakers of these species were first noticed by Jordan & Gilbert in 1883. These authors erroneously referred all these synonyms to the species with the few gill rakers and described the present one as new under the name *Paralichthys ophryas*. The discovery of the Linnæan type of *Pleuronectes dentatus* has rendered a reconsideration of this matter necessary, and it is evident that to the "*P. ophryas*" belong also the prior names *dentatus, melanogaster,* and *ocellaris.* (*dentatus,* toothed.)

Pleuronectes dentatus, LINNÆUS, Syst. Nat., Ed. XII, 1, 458, 1766, and of numerous copyists; MITCHILL, Trans. Lit. and Phil. Soc. N. Y. 1815, 390.

Pleuronectes melanogaster, MITCHILL, Trans. Lit. and Phil. Soc. N. Y. 1815, 390, New York; doubled example.

Platessa ocellaris, DE KAY, N. Y. Fauna: Fishes, 300, pl. 47, fig. 152, 1842, New York.

Paralichthys ophryas, JORDAN & GILBERT, Synopsis, 822, 1883, Charleston.

Platessa dentata, STORER, Rept. Fish. Mass., 143, 1839.

Pseudorhombus dentatus, GOODE & BEAN, Proc. U. S. Nat. Mus. 1879, 123.

Paralichthys dentatus, GOODE, Nat. Hist. Aquat. Anim., 178, 1884, detailed account; includes *P. lethostigma*; JORDAN, Cat. Fish. N. A., 134, 1885; JORDAN & GOSS, Review Flounders and Soles, 246, 1889.

Pseudorhombus ocellaris, GÜNTHER, Cat., IV, 430, 1862; JORDAN & GILBERT, Proc. U. S. Nat. Mus. 1878, 370.

Chœnopsetta ocellaris, GILL, Proc. Ac. Nat. Sci. Phila. 1864, 218.

Paralichthys ocellaris, JORDAN & GILBERT, Proc. U. S. Nat. Mus. 1882, 617.

2997. PARALICHTHYS LETHOSTIGMUS, Jordan & Gilbert.

(SOUTHERN FLOUNDER.)

Head $3\frac{1}{3}$; depth $2\frac{1}{4}$. D. 85 to 92; A. 65 to 73; pores about 100. Body ovate, more or less compressed and opaque; no distinct, definitely placed ocelli; scales cycloid. Mouth wide, oblique, the mandible very heavy and much projecting; 8 to 10 teeth on each side of the lower jaw, the 2 anterior teeth very long; anterior teeth of upper jaw strong, but smaller than those in the lower jaw; the lateral teeth very small and close set; eyes small, shorter than the snout, about 6 in head; interorbital space in adult broad, flattish, and scaly, as wide as length of eye. Accessory scales few; gill rakers $2+10$, lanceolate, dentate, wide set, and much shorter than the eye; caudal peduncle rather long; length of arch of lateral line nearly $\frac{1}{4}$ that of straight part. Color dusky olive, darker than in *P. dentatus,* and with very few darker mottlings or spots. This species is the common large flounder of the South Atlantic and Gulf coasts of the United States, ranging as far north as New York. It very closely resembles *Paralichthys dentatus,* with which it has been repeatedly confounded. It is, however, sharply distinguished by the character of the gill rakers. It is also always darker in color, and almost uniform, while *P. dentatus* is usually profusely spotted. Its only tenable name is the recent one, *Paralichthys lethostigmus.* South Atlantic and Gulf coasts of United States, north to New York. ($\lambda\eta\theta\eta$, forgetfulness; $\sigma\tau i\gamma\mu\alpha$, spot, from the absence of spots.)

Platessa oblonga, DE KAY, New York Fauna: Fishes, 299, pl. 48, fig. 156, 1842, New York, not *Pleuronectes oblongus,* MITCHILL; STORER, Syn. Fish. N. A., 477, 1846.

Paralichthys lethostigma, JORDAN & GILBERT, Proc. U. S. Nat. Mus. 1884, 237, Jacksonville, Florida; JORDAN & GOSS, Review Flounders and Soles, 247, 1889.

Pseudorhombus oblongus, GÜNTHER, Cat., IV, 426, 1862.

Chœnopsetta dentata, GILL, Proc. Ac. Nat. Sci. Phila. 1864, 218.

Pseudorhombus dentatus, GOODE, Proc. U. S. Nat. Mus. 1879, 110; GOODE & BEAN, Proc. U. S. Nat. Mus. 1879, 123.

Paralichthys dentatus, JORDAN & GILBERT, Proc. U. S. Nat. Mus. 1882, 302; JORDAN & GILBERT, Proc. U. S. Nat. Mus. 1882, 617; BEAN, Cat. Coll. Fish. Proc. U. S. Nat. Mus. 1883, 45; JORDAN & GILBERT, Synopsis, 822, 1883.

2998. PARALICHTHYS SQUAMILENTUS, Jordan & Gilbert.

Head $3\frac{3}{5}$; depth 2. D. 78; A. 59; scales 123 (pores). Body deep, strongly compressed; caudal peduncle very short; profile angulated at front of upper eye. Head wide, the eyes large, wide apart. Mouth very large, oblique, the broad maxillary reaching well beyond pupil, its length more than $\frac{1}{2}$ the head. Lower jaw projecting; mandible with a sharp compressed knob at symphysis; teeth few, unequal, in a single row, about 8 in each jaw canine-like, the 2 in front of lower jaw longest; lateral teeth of upper jaw minute. Interorbital space a narrow scaleless bony ridge, slightly concave anteriorly, scarcely $\frac{1}{2}$ diameter of pupil. Scales very small, smooth, adherent; curve of lateral line $4\frac{1}{3}$ in straight part; snout, jaws, and preopercle naked. Gill-rakers short, $3+9$ in number, triangular, roughly toothed, little higher than wide, the longest nearly $\frac{1}{2}$ eye. Dorsal beginning over front of eye, the anterior rays $4\frac{1}{2}$ in head; pectoral short, shorter than maxillary; anal spine weak; caudal double rounded. Brownish; body and fins spotted with darker; caudal mottled with white; pectorals banded, with dark spots. South Atlantic and Gulf coasts of United States. This species is very close to *Paralichthys albiguttus*, from which it differs chiefly in the small scales. It seems to be rather rare. Besides the original types from Pensacola, another referred to the same species is in the National Museum from Charleston. (*squamilentus*, scaly.)

Paralichthys squamilentus, JORDAN & GILBERT, Proc. U. S. Nat. Mus. 1882, 303, Pensacola (Type, No. 30862); JORDAN & GILBERT, Synopsis, 823, 1883; BEAN, Cat. Coll. Fish, U. S. Nat. Mus. 1883, 45; JORDAN & GOSS, Review Flounders and Soles, 248, 1889.

2999. PARALICHTHYS ALBIGUTTUS, Jordan & Gilbert.

(GULF FLOUNDER.)

Head $3\frac{3}{4}$; depth $2\frac{1}{3}$. D. 72 to 80; A. 59 to 61; scales 9 to 100 (pores); eye 6 or 7 in head; maxillary $1\frac{7}{8}$; pectoral $2\frac{1}{5}$; ventral 3; caudal $1\frac{1}{2}$; curve of lateral line 3 in straight part. Body moderately elongate-elliptical; mouth large, the maxillary reaching past eye; jaws subequal; teeth strong, slender, and curved, about 7 on side of lower jaw, 4 or 5 moderate canines in front of upper jaw, the lateral teeth being minute, close set; interorbital space $\frac{2}{3}$ length of eye, the upper ridge rather prominent behind upper eye, scaled posteriorly; mandibles naked; a small patch of scales on maxillary; gill rakers broad and toothed behind, the longest $2\frac{1}{2}$ in eye, $3+10$ in number. Fins low; anterior rays of dorsal not elevated nor exserted, the longest rays behind the middle, $2\frac{3}{4}$ in head; pectoral not reaching to end of curve; caudal double lunate. Scales moderate, cycloid, covered with epidermis which bears small flaps about the borders of many of the scales. Dark olive, mottled with dusky, and marked by numerous more or less distinct pale spots, which are sometimes obsolete; three dark spots, bordered with white, sometimes present, particularly in the young, 1 on lateral line posteriorly and 1 above and below anterior end of straight part of lateral line. Vertebræ $10+27=37$. South Atlantic and Gulf coasts of the United States. This species is common on the South Atlantic and Gulf coasts. It has the few gill rakers of *P. lethostigmus*, the mottled coloration of *P. dentatus*, while from each it is distinguished by its smaller number

of dorsal and anal rays. In the number of its vertebræ it agrees with
P. lethostigmus. It seems to reach a smaller size than either of these
species. Here described from a specimen, 16 inches in length, collected at
Cedar Key, Florida. (*albus*, white; *gutta*, spot.)

Pseudorhombus dentatus, JORDAN & GILBERT, Proc. U. S. Nat. Mus. 1878, 370; not of
LINNÆUS.
Paralichthys albigutta, JORDAN & GILBERT, Proc. U. S. Nat. Mus. 1882, 302, Pensacola
(Type, No. 30818. Coll. Dr. Jordan); JORDAN & GILBERT, Synopsis, 823, 1883; JORDAN
& SWAIN, Proc. U. S. Nat. Mus. 1884, 233; JORDAN & GOSS, Review Flounders and
Soles, 248, 1889.

3000. PARALICHTHYS OBLONGUS (Mitchill).

(FOUR-SPOTTED FLOUNDER.)

Head 4; depth 2¼. D. 72; A. 60; scales 93. Body comparatively elongate, strongly compressed. Eyes large, nearly 4 in head, separated by a
prominent, narrow, sharp ridge. Upper jaw with very numerous small,
close-set teeth laterally, and 4 or 5 canines in front; the lateral teeth
abruptly smaller than the anterior; each side of lower jaw with 7 to 10
teeth. Chin prominent. Maxillary narrow, reaching past middle of
pupil, 2¼ in length of head. Gape curved; gill rakers short and toothed
behind, 2 + 8. Scales weakly ctenoid or cycloid. Dorsal low, beginning
over front of eye, some of the anterior rays exserted, but not elongate, the
longest rays behind middle of fin, not quite ¼ head; caudal 1¼ in head; pectoral 1⅔; anal spine obsolete. Grayish, thickly mottled with darker and
somewhat translucent; 4 large, horizontally oblong, black ocelli, each
surrounded by a pinkish area, 1 just behind middle of the body below the
dorsal, 1 opposite this above anal, 2 similar smaller spots below last rays
of dorsal and above last of anal. Coasts of New England and New York.
This species is rather common on the coast of Cape Cod and the neighboring islands, but it has been rarely noticed elsewhere. The limits of its
range are not yet definitely known. It is a very strongly marked species.
Its translucency of coloration indicates that it lives in deeper water than
the other species of the genus. Here described from specimen from Woods
Hole.

Another specimen in our collection from Woods Hole, Massachusetts,
referred to this species, shows the following characters: Brownish,
somewhat mottled, without traces of ocelli (possibly faded); fins similar.
Body rather elongate, slenderer than in other species and more compressed; mouth rather large, oblique, the lower jaw not projecting, the
maxillary 2½ in head, reaching to opposite posterior border of pupil; about
12 teeth on each side of lower jaw, the anterior rather long, about equal to
anterior teeth of upper jaw; lateral teeth of upper jaw becoming gradually
smaller posteriorly, much larger, less numerous, and more widely set than
in other species of this genus. Eyes large, longer than snout, 4 to 4½ in
head, separated by a narrow, elevated, bony ridge, narrower than pupil,
anteriorly scaleless, and curved behind the upper eye posteriorly. Scales
moderate, cycloid, rather thin; curve of lateral line 4¾ in straight part.
Gill rakers 2 + 8 in number, rather long and slender, about 4½ in maxillary.
Dorsal beginning above middle of eye, its anterior rays not longer than

others, the middle rays a little longer than longest of anal, which are about ¼ head; caudal as long as head; anal spine obsolete; ventrals small; pectoral 1¾ in head. Head 4½; depth 2¾. D. 77; A. 63; scales 90. Length about 14 inches. (*oblongus*, oblong.)

Pleuronectes oblongus, MITCHILL, Trans. Lit. and Phil. Soc., I, 1815, 391, New York.
Platessa quadrocellata, STORER, Proc. Boston Soc. Nat. Hist. 1847, 242; STORER, Hist. Fish. Mass., 397, pl. 31, fig. 3, Provincetown.
Chænopsetta oblonga, GILL, Proc. Ac. Nat. Sci. Phila. 1864, 218.
Paralichthys oblongus, GOODE, Proc. U. S. Nat. Mus. 1880, 472; JORDAN & GILBERT, Synopsis, 824, 1883; JORDAN & GOSS, Review Flounders and Soles, 249, pl. 8, 1889; GOODE & BEAN, Oceanic Ichthyology, 436, 1896.

1025. RAMULARIA, Jordan & Evermann.

Ramularia, JORDAN & EVERMANN, new genus (*dendriticus*).

This genus is close to *Ancylopsetta*, differing mainly in the structure of the lateral line, the tubes of which are borne by series of smaller, concealed cycloid scales, the free edges of which are notched to the opening of the pore; these scales are concealed in the skin, and from the pores proceed backward membranaceous tubes which ramify over the bases of contiguous scales. Dorsal scarcely elevated in front; left ventral much produced. Body broad ovate, sinistral, with very rough scales. Gill rakers few, very broad. (*ramulus*, a branchlet, from the tubes of the lateral line.)

3001. RAMULARIA DENDRITICA (Gilbert).

Head 3⅗ in length; depth 1¾. D. 84; A. 63; scales 100; 36 scales in a series upward and backward from lateral line. Body very broad, its depth 1⅔ in length, the two outlines equally curved; profile not very strongly angulated in front of upper eye. Lower eye slightly in advance of upper; interorbital space a rather broad, convex, scaly ridge, about ¼ upper eye, which is contained about 5 times in head and is equal to snout. A blunt spine on snout on head of maxillary. Nostril openings very broad, without tube, the anterior with a narrow flap. Mouth moderate, very oblique, the gape curved, the maxillary reaching slightly beyond vertical from middle of lower eye, 3 in head. Teeth in a single, rather close-set series in each jaw, strong, conical, directed very obliquely inward, becoming gradually larger toward front of jaw, but not canine-like. Gill rakers very short, barely movable, as broad as long, strongly toothed, 6 on anterior limb. Dorsal beginning over middle of upper eye, the anterior rays partly free toward tips, but little, if any, elevated above those that follow, the first 2¼ in head; dorsal highest in its posterior third, the longest ray 2⅔ in head; anal similar, the rays of posterior third of each fin slightly forked at tip; caudal peduncle deep and short, its depth about ¼ head, its length ¼ its depth; caudal rounded, almost double truncate; ventrals with narrow bases, the left one slightly in advance of the right; fin greatly produced, reaching far beyond front of anal, a trifle shorter than head; left pectoral 1¾ in head. Scales very strongly ctenoid, the edge spinous, the entire exposed portion rough; width of anterior arch of lateral line 3½ in straight portion; tubes of lateral line borne by a series of smaller con-

3030——88

cealed cycloid scales, the free edges of which are notched to the opening
of the pore; these scales entirely covered by the integument, and from
the pores there proceed backward membranaceous tubes, ramifying over
the bases of contiguous scales; this is true also of lateral line of blind side;
eyed side entirely scaled except snout and mandible.   Vertical fins covered with thick skin, each ray accompanied by 1 or 2 series of ctenoid
scales; left ventral also scaled.   Color olive brown, with 3 large black
ocellated spots larger than orbit, the posterior one on lateral line in front
of caudal peduncle, the 2 anterior under middle of dorsal, halfway between
lateral line and dorsal and anal outlines, respectively; each spot with a
light center; distal portion of vertical fins more or less brown on right
side. (Gilbert.) Gulf of California. A single specimen, 13 inches long,
from *Albatross* Station 3022, in 11 fathoms. (*dendriticus*, like a tree,
branched; δένδρον, tree.)

Ancylopsetta dendritica, GILBERT, Proc. U. S. Nat. Mus. 1890, 121, Gulf of California at
   Albatross Station 3022, in 11 fathoms.

1026. ANCYLOPSETTA, Gill.

Ancylossetta, GILL, Proc. Ac. Nat. Sci. Phila. 1864, 224 (*quadrocellata*).

Body sinistral, broadly ovate, the depth more than ½ length; mouth
moderate; teeth uniserial, unequal, some of the anterior enlarged; caudal fin with a very short peduncle; scales very strongly ctenoid on both
sides of the body; anterior rays of dorsal notably exserted, the rays of the
anterior part of the fin elongate, thus forming a distinct lobe; gill membranes considerably united; gill rakers short and broad, with rough
teeth; left ventral produced; vertebræ about 35. This genus is very
close to *Paralichthys*, differing in the subsessile caudal fin, the short gill
rakers, the rough scales, and in the prolongation of the anterior rays of
the dorsal fin. (ἄγκυλος, hook; ψῆττα, turbot.)

3002. ANCYLOPSETTA QUADROCELLATA, Gill.

Head 3¾ to 3⅔; depth 1⅔. D. 70 to 76; A. 57 to 59; pores in lateral line
83 to 90; vertical series of scales 70; fourth or fifth dorsal ray longest,
nearly ⅔ length of head. Caudal 1⅓ in head; ventral of colored side 1⅜.
Body oval, compressed, very deep; an abrupt angle above eye; mouth very
small, the maxillary reaching to below middle of orbit, 2¼ in length of
head; teeth comparatively small, about 14 on each side of lower jaw; no
strongly differentiated canines in either jaw. Eyes moderate, separated
by a very narrow, sharp, scaly ridge; gill rakers very short, thick, few in
number, 2+6 or 7, the longest less than ½ diameter of pupil; scales
rather small, very strongly ctenoid, those on blind side also rough; curve
of lateral line rather low; tubes of lateral line simple; dorsal beginning
in front of pupil, its anterior rays long and filiform, much exserted; caudal short and rounded, 1⅔ in head; ventral fin of colored side rather long,
as long as pectoral, ½ length of head; anal spine wanting. Brownish
olive, with 4 large, oblong, ocellated spots, the first above the arch of the
lateral line; the 3 posterior forming an isosceles triangle, the hindmost

being on the lateral line; the ocellated spots are frequently furnished with a bright white center, and the sides and vertical fins have often a few scattered white spots; a small, indistinct, dark spot on middle of each eighth or tenth ray of dorsal and anal. Vertebræ $9 + 26 = 35$. South Atlantic and Gulf coasts of the United States; not rare; a very handsome species. (*quadrocellatus*, having 4 ocelli.)

Ancylopsetta quadrocellata, GILL, Proc. Ac. Nat. Sci. Phila. 1864, 224; not *Platessa quadrocellata*, STORER; JORDAN & GOSS, Review Flounders and Soles, 250, 1889.

Paralichthys ommatus, JORDAN & GILBERT, Proc. U. S. Nat. Mus. 1882, 616, Charleston; JORDAN & GILBERT, Synopsis, 824, 1883; JORDAN & SWAIN, Proc. U. S. Nat. Mus. 1884. 234; JORDAN & GOSS, Review Flounders and Soles, 250, 1889.

Pseudorhombus quadrocellatus, JORDAN & GILBERT, Proc. U. S. Nat. Mus. 1878, 370.

1027. NOTOSEMA, Goode & Bean.

Notosema, GOODE & BEAN, Bull. Mus. Comp. Zool., x, No. 5, 192, 1883 (*dilecta*).

Body sinistral, elliptical in form, the caudal fin pedunculate. Mouth moderate, beneath the central axis of the body. Eyes large, close together, the upper one nearly encroaching upon the profile, the lower slightly in advance of the upper. Teeth in a single series in the jaws, about equally developed on each side, largest in front, absent on vomer and palatines. Pectoral fins somewhat unequal, that upon the blind side $\frac{3}{4}$ as large as the other; dorsal fin commencing slightly behind anterior margin of upper eye, the first 8 rays separated into a distinct subdivision of the fin, several of them being prolonged; caudal rounded, sinistral; ventral much elongated. Scales small, ctenoid on colored side of body; lateral line prominent, strongly arched, alike on both sides, the tubes simple. Gill rakers moderately numerous, rather stout, subtriangular, pectinate posteriorly. Pseudobranchiæ well developed. Vertebræ 35. This genus is scarcely distinct from *Ancylopsetta*, the body more elongate, the dorsal and ventral rays more produced. ($\nu\tilde{\omega}\tau o\varsigma$, back; $\delta\tilde{\eta}\mu\alpha$, banner.)

3003. NOTOSEMA DILECTUM (Goode & Bean).

Head $3\frac{1}{2}$; depth 2. D. 68; V. 6. A. 54 to 56; scales 48 (pores) on straight part of lateral line; width of interorbital area almost imperceptible; mandible reaching to middle of pupil of lower eye, its length 2 in head; upper jaw $2\frac{1}{4}$ times length of head. Origin of dorsal over anterior margin of eye, second and third rays the longest, which are 2 in greatest depth of body; anal beginning close to vent, its posterior rays longest; caudal pedunculate, double truncate; pectoral of eyed side subtriangular, its length $5\frac{1}{4}$ in length of body; ventral of eyed side much produced, its length more than 3 times that of its mate. Color dark brown, speckled with darker, 3 large subcircular ocellated spots nearly as large as eye, with white center, dark iris, narrow, light margin, and a brown encircling outline, these arranged in an isosceles triangle, the apex on the lateral line, the others before it and distant from the lateral line a distance equal to their own diameter; blind side white; fins blotched with dark brown. (Goode & Bean.) Gulf Stream. Known from the original types obtained in the deep waters (75 fathoms) of the Gulf Stream, off the Carolina coast. (*dilectus*, delightful.)

Notosema dilecta, GOODE & BEAN, Bull. Mus. Comp. Zool., x, No. 5, 193, 1883, Gulf Stream off the coast of South Carolina; GOODE & BEAN, Oceanic Ichthyology, 437, 1896.

Paralichthys stigmatias, GOODE, Nat. Hist. Aquat. Anim., 182, 1884; by inadvertance for *dilectus*.

Ancylopsetta dilecta, JORDAN, Cat. Fish. N. A., 134, 1885; JORDAN & GOSS, Review Flounders and Soles, 250, 1889.

1028. GASTROPSETTA, B. A. Bean.

Gastropsetta, BARTON A. BEAN, Proc. U. S. Nat. Mus. 1894, 633 (*frontalis*).

Body oblong-ovate, highly arched in front, covered with small, cycloid, embedded scales; lateral line arched in front, deflected downward on caudal peduncle. Teeth small, in a single series in each jaw. Dorsal fin beginning in advance of eye, its anterior rays produced, not connected by the irregular and broadly fringed membrane. Gill rakers very short, almost as broad as long, few in number. Ventral of eyed side produced, ending in a long filamentous ray in the young. This genus is closely allied to *Ancylopsetta*, from which it differs in form of body, and especially in the entirely smooth scales, singularly branched and produced anterior dorsal rays, and very short and broad gill rakers. ($\gamma\alpha\delta\tau\eta\rho$, belly; $\psi\eta\tau\tau\alpha$, turbot or flounder.)

3004. GASTROPSETTA FRONTALIS, B. A. Bean.

Head 4⅘; depth 2½; middle caudal rays 2⅘; eye large, 3¾ in head. D. 60; A. 48; V. 6; P. I, 10. Mouth of moderate size, maxillary 2½ in head, the jaws curved; interorbital ridge prominent, very narrow. Dorsal beginning in front of eye on snout, its anterior rays singularly branched, the third and fourth longest, almost equaling length of head; anal fin beginning at vent, which is situated on blind side, its anterior rays scarcely produced; ventral of colored side much produced; middle caudal rays long. Color in spirits, light brown; 3 black spots on body, 2 along back, and 1 near anal base; fins with dusky blotches; several vertical stripes across eyes. A smaller specimen from *Albatross* Station 2317 has D. 62; A. 52; V. 6; P. I, 11. Gill rakers short, broad laminæ, 2+7. Teeth weak, uniserial. Anterior rays of dorsal greatly produced, the third 1¼ times as long as head. Ventral of eyed side very long, ending in a thread-like filament. Color as in the preceding. An example from *Albatross* Station 2373 near Apalachicola, is 224 mm. long; its depth 90 mm. D. 60; A. 49; P. I, 10; V. 6; C. 15. Vent situated in a deep notch, which forms the front margin of abdomen, and not on side, as in other specimens. Color darker than that of the Key West examples, being dark reddish brown; body spotted and fins blotched as in the preceding. Two specimens obtained by the *Albatross*, January 15, 1885, at Station 2317, Lat. 24° 25′ 45″ N., Long. 81° 46′ 45″ W., near Key West, Florida, in 45 fathoms of water, the type 8 inches long, the other one 6 inches. (B. A. Bean.) (*frontalis*, pertaining to the forehead.)

Gastropsetta frontalis, BARTON A. BEAN, Proc. U. S. Nat. Mus. 1894, 633, Key West. (Type, No. 37668, U. S. Nat. Mus. Coll. *Albatross*.)

1029. PLEURONICHTHYS, Girard.

Pleuronichthys, GIRARD, Proc. Ac. Nat. Sci. Phila. 1854, 139 (*cœnosus*).
Heteroprosopon, BLEEKER, Comptes Rendus Acad. Amsterdam, XIII, 1862, 8 (*cornutus*).
Parophrys, GÜNTHER, Cat. Fishes, IV, 454, 1862; not of GIRARD.

Eyes and color on the right side. Body deep; head short, with very short, blunt snout; mouth small, with several series of slender, acute teeth, which are most developed on the blind side, and are often wanting in 1 or both jaws on the colored side; no teeth on vomer or palatines; lips thick, with several lengthwise folds within which is a series of short fringes. Lower pharyngeals narrow, each with a double row of very small teeth. Gill rakers wide set, very short and weak. Lateral line nearly straight, with a dorsal branch in our species. Scales small, cycloid, nonimbricate, embedded. Dorsal fin anteriorly twisted from the dorsal ridge toward the blind side; anal fin preceded by a spine; caudal fin convex behind. Intestinal canal elongate. Herbivorous species, feeding chiefly on algæ. Pacific Ocean. This well-marked genus contains 3 American species, which are very closely related to each other. The Asiatic species, *Platessa cornuta*, Schlegel, of the coasts of China and Japan, is also a member of this group, having an accessory branch to the lateral line as in the American species. This species bears some resemblance to *Pl. verticalis*. The species of *Pleuronichthys* spawn in the spring, and live in comparatively deep water. ($\pi\lambda\epsilon\tilde{\nu}\rho o\nu$, side; $i\chi\theta\acute{\nu}\varsigma$, fish.)

  a. Dorsal fin beginning on the level of the lower lip, its first 9 rays on the blind side.
                                                                                         DECURRENS, 3005.
  aa. Dorsal fin beginning on level of upper lip, its first 5 rays being on the blind side.
    b. Interorbital ridge posteriorly with a strong spine directed backward, some tubercles on interorbital ridge.   VERTICALIS, 3006.
    bb. Interorbital ridge prominent, but without spines and conspicuous tubercles.
                                                                                         CŒNOSUS, 3007.

3005. PLEURONICHTHYS DECURRENS, Jordan & Gilbert.

Head $3\frac{1}{2}$; depth $1\frac{2}{3}$. D. 72; A. 40; scales 80; eye 3 in head; maxillary $4\frac{1}{2}$; pectoral $1\frac{2}{3}$; highest dorsal rays $1\frac{1}{2}$; anal rays $1\frac{2}{3}$; caudal 1. Body short and wide; mouth very small, the maxillary reaching nearly to pupil; teeth villiform, in moderate bands on blind side, a narrow band on eyed side of lower jaw; eyes very large, the upper edge of upper eye even with profile; snout extremely short; a blunt tubercle in front of upper eye, another at each end of the narrow interorbital ridge, the posterior largest, but usually not spine-like; 2 or 3 above the latter behind the upper eye; some prominences above the opercle; gill opening short, not extending above upper edge of pectoral. Dorsal beginning very low, on level of end of maxillary, its first 9 rays on the blind side; anal spine well developed, the origin of anal a little behind vertical from base of pectoral; pectoral of eyed side a little larger than its mate, both rounded behind; ventral of blind side shorter than that of eyed side, and placed slightly before it, caudal well rounded. Scales cycloid, embedded, a space between them anteriorly; lateral line without arch, slightly curved. Color brownish, usually much mottled with chocolate and grayish, often finely spotted with brownish on body and fins; all fins darker than body;

dorsal, anal, caudal, and ventrals narrowly edged with white; pectoral uniformly blackish. Pacific coast of United States, south to Monterey. This species is rather scarce along the California coast, being taken chiefly in deep water. It reaches a larger size than either *P. verticalis* or *P. cœnosus*. Here described from a specimen from San Francisco market, 8 inches in length. (*decurrens*, running down.)

Pleuronichthys cœnosus, LOCKINGTON, Proc. U. S. Nat. Mus. 1879, 97; not *Pleuronichthys cœnosus*, GIRARD.

Pleuronichthys quadrituberculatus, JORDAN & GILBERT, Proc. U. S. Nat. Mus. 1880, 50, not of PALLAS; JORDAN, Nat. Hist. Aquat. Anim., 189, 1884.

Pleuronichythys decurrens, JORDAN & GILBERT, Proc. U. S. Nat. Mus. 1880, 453, San Francisco; Monterey Bay (Coll. Jordan & Gilbert); JORDAN & GILBERT, Proc. U. S. Nat. Mus. 1881, 69; JORDAN & GILBERT, Synopsis, 829, 1883; JORDAN & GOSS, Review Flounders and Soles, 282, 1889.

3006. PLEURONICHTHYS VERTICALIS, Jordan & Gilbert.

Head 4 in body; depth 2. D. 65; A. 45; scales about 80; vetebræ 13 + 25 = 38. Form broad ovate, the outlines regular; head small, somewhat constricted behind the upper eye; eyes large, but smaller than in *P. decurrens*. Interorbital ridge narrow; a small tubercle or prominence in front of upper eye; a large one in front of upper edge of lower eye; another larger and sharper at interior edge of the interocular space; another at the posterior edge of interocular spine ridge; this latter developed into a long, sharp, triangular spine, which is nearly as long as the pupil, and is directed backward; a prominent tubercle at posterior lower angle of upper eye; upper edge of opercle somewhat uneven, but no other tubercles present. Mouth small, as in other species; the lips thick, with lengthwise plicæ. Teeth in a broad band on the left (blind) side of each jaw; no teeth on the right side in either jaw. Gill rakers very small, weak, and flexible, about 10 in number. Scales essentially as in other species, small, cycloid, embedded, scarcely imbricated; lateral line nearly straight, with an accessory branch which extends to the middle of the dorsal fin. Dorsal fin beginning on blind side at level of premaxillary, there being but about 4 of its rays on left side of median line; vertical fins less elevated than in other species, the longest rays of dorsal about ½ length of head; anal fin preceded by a spine; caudal peduncle short and deep; caudal fin elongate, rounded behind; pectorals short, nearly equal; ventrals moderate, reaching anal spine. Color dark olive brown, with round grayish spots, the body and fins mottled with blackish. This species agrees in habits and general characters with *Pleuronichthys decurrens*. Coast of California, in rather deep water. The above description from the original type. (*verticalis*, pertaining to the vertex.)

Pleuronichthys verticalis, JORDAN & GILBERT, Proc. U. S. Nat. Mus. 1880, 49, San Francisco (Coll. Jordan & Gilbert); JORDAN & GILBERT, Proc. U. S. Nat. Mus. 1881, 169; JORDAN & GILBERT, Synopsis, 829, 1883; JORDAN, Nat. Hist. Aquat. Anim., 189, 1884; JORDAN & GOSS, Review Flounders and Soles, 282, 1889.

3007. PLEURONICHTHYS CŒNOSUS, Girard.

Head 3¾; depth 2. D. 68; A. 49; scales 61; eye 3 in head; pectoral 1½; dorsal and anal rays 1½; caudal a little longer than head. Body ovate; snout scarcely produced; mouth small, maxillary reaching past front of lower eye; 3 or 4 rows of teeth on blind side of jaws, 1 on eyed side of lower; eyes very large; interorbital a high, narrow ridge, somewhat angulated behind, but with no conspicuous spine or tubercle; snout very short, about ½ eye; gill opening not extending above upper edge of pectoral. Scales cycloid, embedded, some distance apart anteriorly, their edges not in contact; lateral line nearly straight, with a long dorsal branch which reaches past middle of body. Dorsal and anal high; origin of dorsal on blind side on a level with premaxillary, its first 5 rays on blind side; origin of anal under base of pectoral; pectoral of eyed side a little larger than its mate; caudal well rounded. Color dark brown, usually mottled, the colors variable; our specimens from Puget Sound, very dark, the fins colored like body, with light and dark spots; a conspicuous black spot on lateral line on middle of sides. Pacific coast, from Sitka to San Diego. This species is comparatively common in rather deep water and about rocks, being most abundant about Puget Sound. Its apparent abundance as compared with the other species of the genus is doubtless due to its inhabiting shallower waters than they. It is quite variable in form. The above description from a specimen, 6 inches long, from Seattle. (*cœnosus*, muddy.)

Pleuronichthys cœnosus, GIRARD, Proc. Ac. Nat. Sci. Phila. 1854, 139, San Francisco; GIRARD, U. S. Pac. R. R. Surv., X, Fishes, 151, 1858; LOCKINGTON, Rep. Com. Fisheries California, 1878–79, 45; LOCKINGTON, Proc. U. S. Nat. Mus. 1879, 97; JORDAN & GILBERT, Proc. U. S. Nat. Mus. 1880, 50; JORDAN & GILBERT, Proc. U. S. Nat. Mus. 1880, 453; JORDAN & GILBERT, Proc. U. S. Nat. Mus. 1881, 68; JORDAN & GILBERT, Synopsis, 830, 1883; JORDAN, Nat. Hist. Aquat. Anim., 189, 1884; JORDAN & GOSS, Review Flounders and Soles, 282, 1889; JORDAN, Proc. Cal. Ac. Sci. 1895, 852.

Parophrys cœnosa, GÜNTHER, Cat., IV, 456, 1862.

1030. HYPSOPSETTA, Gill.

(DIAMOND FLOUNDERS.)

Hypsopsetta, GILL, Proc. Ac. Nat. Sci. Phila. 1864, 195 (*guttulatus*).

Eyes and color on the right side; body broad, ovate, rhomboid; mouth very small; teeth slender, equal, acute, in several series; lips thick, not plicate; lateral line nearly straight, with an accessory dorsal branch; scales small, smooth; dorsal fin beginning on the dorsal ridge, not turned to the blind side at its insertion; anal spine present; caudal fin convex; gill rakers little developed. This genus consists of a single species, abundant on the coast of California. It is very close to *Pleuronichthys*, from which it differs only in a few characters of comparatively minor importance. Its range is in shallower and warmer water than that of the species of *Pleuronichthys*, and, in accordance with this fact, its flesh is firmer and its number of vertebræ fewer than in the latter genus. ($\ddot{v}\psi\iota$, deep; $\psi\widetilde{\eta}\tau\tau\alpha$, flounder.)

3008. HYSOPSETTA GUTTULATA (Girard).

(DIAMOND FLOUNDER.)

Head 3¾; depth 1⅔. D. 68; A. 50; scales 95. Body very deep, somewhat angulated near middle of back and belly; eyes moderate, separated by a flattish, raised area; head without spines or tubercles; scales of opercular region little developed; those of blind side reduced; no teeth on right side of either jaw; accessory lateral line long, ¼ length of body; anal spine small; pectorals about ½ length of head; ventrals rather short; caudal peduncle much deeper than long; caudal large, nearly as long as head. Brown, with numerous pale-bluish blotches in life, these disappearing in spirits; blind side white, with a strong tinge of yellow along profile of head; fins plain, sometimes with black specks. Coast of California and southward, Cape Mendocino to Magdalena Bay. This species is one of the most abundant in the shore waters of the California coast. It is a food-fish of fair quality. (*guttulatus*, with small spots.)

Pleuronichthys guttulatus, GIRARD, Proc. Ac. Nat. Soi. Phila. 1856, 137, Tomales Bay, California (Coll. E. Samuels); GIRARD, Journ. Boston Soc. Nat. Hist. 1857, pl. 25, figs. 1–4; GIRARD, U. S. Pac. R. R. Surv., X, Fishes, 152, 1858; LOCKINGTON, Rep. Com. Fisheries California, 1878–79, 44; LOCKINGTON, Proc. U. S. Nat. Mus. 1879, 94.

Parophrys ayresi, GÜNTHER, Cat., IV, 1862, 457, San Francisco. (Coll. Dr. W. O. Ayres.)

Pleuronectes guttulatus, GÜNTHER, Cat., IV, 445, 1862.

Hypsopsetta guttulata, GILL, Proc. Ac. Nat. Sci. Phila. 1864, 195; JORDAN & GILBERT, Proc. U. S. Nat. Mus. 1880, 453; JORDAN & GILBERT, Proc. U. S. Nat. Mus. 1881, 68; JORDAN & GILBERT, Synopsis, 830, 1883; JORDAN, Nat. Hist. Aquat. Anim., 185, 1884; JORDAN & GOSS, Review of Flounders and Soles, 283, 1889.

1031. PAROPHRYS, Girard.

Parophrys, GIRARD, Proc. Ac. Nat. Sci. Phila. 1854, 139 (*vetulus*).

Body rather elongate, covered with small, cycloid scales; scales of the head roughish. Head rather pointed; mouth small, the teeth uniserial, all more or less blunt, compressed, incisor-like, close set. Lateral line with an accessory dorsal branch; upper eye on median line of top of head. A single species, on the Pacific coast of America. The narrow interorbital space and the vertical range of the upper eye give it a peculiar physiognomy, but in most regards it is not very different from some of the species of *Pleuronectes*. (παρά, near; ὀφρύς, eyebrow, from the narrow interbital.)

3009. PAROPHRYS VETULUS, Girard.

Head 3½; depth 2½; eye 4½ in head. D. 74 to 86; A. 54 to 68; scales 105 (tubes). Body elongate-elliptical; snout very prominent, much protruding, forming an abrupt angle with the descending profile; depth of head opposite middle of upper eye about equaling distance from middle of orbit to snout; eyes large, separated by a very narrow, high ridge, the upper with vertical range; mouth very small; maxillary not reaching pupil; teeth trenchant, small, and rather narrow, widened at tip, about 45 teeth on left side of lower jaw; few teeth on right side of lower jaw. Accessory lateral line long. Pectoral about ⅓ length of head; caudal truncate, 1½ in head; fin rays entirely scaleless; scales on body all cycloid, those on cheeks often slightly ciliated. Uniform light olive brown; the young somewhat spotted with blackish. Pacific coast of North America, Sitka to Santa Barbara. This small flounder lives in waters of moderate depth. It is, next to *Platichthys stellatus*, probably the most abundant of the flounders of the California coast. (*vetulus*, an old man.)

Parophrys vetulus, GIRARD, Proc. Ac. Nat. Sci. Phila. 1854, 140, California; GÜNTHER, Cat., IV, 455, 1862; LOCKINGTON, Rep. Com. Fish. Cal. 1878–79, 45; LOCKINGTON, Proc. U. S. Nat. Mus. 1879, 100; JORDAN & GILBERT, Proc. U. S. Nat. Mus. 1880, 453; JORDAN & GILBERT, Proc. U. S. Nat. Mus. 1881, 68; JORDAN, Nat. Hist. Aquat. Anim., 185, 1884; JORDAN & GOSS, Review Flounders and Soles, 284, 1889.

Pleuronectes digrammus, GÜNTHER, Cat., IV, 445, 1862, Victoria. (Coll. Earl Russell.)

Parophrys hubbardi, GILL, Proc. Ac. Nat. Sci. Phila. 1862, 281, San Francisco.

Pleuronectes vetulus, JORDAN & GILBERT, Synopsis, 831, 1883.

1032. INOPSETTA, Jordan & Goss.

Inopsetta, JORDAN & GOSS, in JORDAN, Cat. Fish. N. A., 136, 1885 (*ischyrus*).

This genus resembles *Parophrys*, differing chiefly in having the scales less imbricated, all strongly ctenoid, and having the eyes both lateral, the snout much less acute than in *Parophrys*. A single species, closely allied to *Platichthys stellatus*, but separated from it by the curious character common to many of our Pacific coast flounders, of having an accessory branch to the lateral line. (*ἴς*, strength; *ψῆττα*, flounder.)

3010. INOPSETTA ISCHYRA (Jordan & Gilbert).

Head 3¼; depth 2. D. 70 to 76; A. 52 to 57; V. 6; scales 85. Body oblong, robust; caudal peduncle rather long; snout projecting, forming an angle with the profile; mouth oblique, the chin projecting; teeth $\frac{5+25}{10+22}$, narrowly incisor-like, bluntish, in a single, rather close-set series; maxillary reaching past front of orbit, 5¾ in head; eyes large; interorbital space rather broad, scaly, continuous with a ridge above opercle; head mostly covered with scales like those of the body, but smaller and rougher; gill rakers feeble; lower pharyngeals each with 2 rows of coarse, blunt teeth; scales thick and firm, adherent, not closely imbricated, those in front well apart; all the scales strongly ctenoid; blind side with similar scales, almost as strongly ctenoid; vertical fins mostly scaly; lateral line conspicuous, its scales less rough than the others; a distinct short accessory lateral line on both sides, extending to about the tenth dorsal ray, less than ⅙ head; a series of pores around lower eye behind; dorsal beginning over pupil, its anterior rays low, its highest rays nearly ½ length of head; caudal large, double truncate; pectoral of right side about ⅓ head. Light olive-brown, vaguely clouded with light and dark; fins reddish brown; a few roundish dusky blotches on dorsal and anal; pectoral and caudal tipped with dusky; blind side white, immaculate, or with small, round rusty spots; left side of head sometimes rusty tinged. Puget Sound. This species is known only from 4 specimens taken by Dr. Jordan at Seattle in 1880. It is a large, rough flounder, with firm, white flesh. (*ἰσχυρός*, robust.)

Parophrys ischyrus, JORDAN & GILBERT, Proc. U. S. Nat. Mus. 1880, 276 and 453, Puget Sound (Coll. Dr. Jordan); JORDAN & GILBERT, Proc. U. S. Nat. Mus. 1881, 67; JORDAN, Nat. Hist. Aquat. Anim., 185, 1884.

Pleuronectes ischyrus, JORDAN & GILBERT, Synopsis, 832, 1883.

Inopsetta ischyra, JORDAN, Cat. Fish. N. A., 136, 1885; JORDAN & GOSS, Review Flounders and Soles, 284, 1889.

1033. ISOPSETTA, Lockington.

Isopsetta, LOCKINGTON MS., in JORDAN & GILBERT, Synopsis, 832, 1883 (*isolepis*).

Body much compressed, elliptical in form; mouth rather large; the teeth chiefly uniserial, all more or less blunt, separated, not incisor-like; scales closely imbricated, all strongly ctenoid; lower pharyngeals each with a double row of bluntish teeth. A single species found on the coast of California. *Isopsetta* approaches in many respects very close to the large-mouthed flounders of the type of *Hippoglossoides*, and it may fairly be said to be intermediate between *Psettichthys* and *Lepidopsetta*. Its affinities on the whole are nearest the latter, but the close relation of the *Hippoglossinæ* and *Pleuronectinæ* is clearly shown. (ἴσος, equal; ψῆττα, flounder.)

3011. ISOPSETTA ISOLEPIS (Lockington).

Head 4; depth 2¼. D. 88; A. 65; scales 88; vertebræ 10 + 32 = 42. Body elliptical, much compressed, moderately deep, the curvature very regular; head moderate, strongly compressed, the profile little depressed above the eye; eyes rather large; interorbital space broad, flattish, with several series of scales. Scales on cheeks similar to those on body, rather large, ctenoid, and closely imbricated. Mouth comparatively large, maxillary reaching pupil, 3⅔ in head; teeth not large, about $\frac{11+14}{9+24}$, conical, close set, in 1 somewhat irregular series, or partly in 2 series, those on colored side small; lower pharyngeals each with a double row of bluntish teeth. On the blind side the scales are more or less ctenoid, sometimes smooth; those on the cheeks weakly ctenoid; most of the opercle, the preopercle, interopercle, and subopercle on blind side naked; lateral line with a very slight arch in front, the depth of which is less than ⅓ the length; accessory branch nearly as long as head; fins rather low, mostly covered with ctenoid scales. Color brownish, mottled and blotched with darker. This small flounder is rather common off the coast of California, where it reaches a length of about 15 inches. It much resembles *Psettichthys melanostictus*, but its small mouth and blunt dentition indicate a real affinity with the small-mouthed flounders, among which it is here placed. Its nearest relative among our species is doubtless *Lepidopsetta bilineata*. Puget Sound to Point Concepcion, in rather deep water; not rare. (ἴσος, equal; λεπίς, scale.)

Lepidopsetta umbrosa, LOCKINGTON, Proc. U. S. Nat. Mus. 1879, 106; not of GIRARD.
Lepidopsetta isolepis, LOCKINGTON, Proc. U. S. Nat. Mus. 1880, 325, San Francisco.
Parophrys isolepis, JORDAN & GILBERT, Proc. U. S. Nat. Mus. 1880, 453 and 1881, 67; JORDAN & GILBERT, Synopsis, 832, 1883; JORDAN, Nat. Hist. Aquat. Anim., 186, 1884.
Isopsetta isolepis, JORDAN, Cat. Fish. N. A., 136, 1885; JORDAN & GOSS, Review Flounders and Soles, 285, 1889.

1034. LEPIDOPSETTA, Gill.

Lepidopsetta, GILL, Proc. Ac. Nat. Sci. Phila. 1864, 195 (*umbrosus*).

Body robust; mouth small. Teeth stout, conical, little compressed, bluntish, in 1 series, rather irregularly placed. Lateral line with a distinct arch in front and accessory dorsal branch; scales imbricated, rough ctenoid, smooth in the very young. A single species, abundant on the Pacific coast of North America. It is close to *Inopsetta*, from which it is separated by the arch of the lateral line, and still closer to *Limanda*, from which the accessory branch of the lateral line alone separates it. (λεπίς, scale; ψῆττα, flounder.)

3012. LEPIDOPSETTA BILINEATA (Ayres).

Head $3\frac{2}{3}$; depth $2\frac{1}{5}$. D. 80; A. 60; teeth $\frac{27+7}{25+10}$; scales 85. Vertebræ 11 + 29 = 40. Body broadly ovate, thickish; mouth moderate, turned toward the left side; teeth stout, conical, little compressed, bluntish, in 1 series, rather irregularly placed. Lower pharyngeals broad, with 2 rows of blunt teeth. Gill rakers few, very short, thick and weak, without teeth. Snout projecting; eyes large, separated by a prominent ridge, which, like the cheeks and upper portion of opercle, is covered with rough stellate scales; lower eye advanced; opercle, subopercle, and interopercle of left side scaly; preopercle naked. Scales rather small, mostly ctenoid, not closely imbricated, those on the blind side smooth; scales on cheeks and other parts of head very rough; scales of body smoother and less closely imbricated anteriorly, the degree of roughness variable, northern specimens (var. *umbrosus*) being roughest. Lateral line moderately arched anteriorly, with an accessory dorsal branch, which is less than $\frac{1}{2}$ length of head; height of arch less than $\frac{1}{8}$ its length. Dorsal beginning over eye, its anterior rays low; caudal convex; anal preceded by a spine; a concealed spine behind ventrals; rays of dorsal and anal all simple; dorsal and anal somewhat scaly; caudal $\frac{3}{4}$ length of head; pectoral $\frac{1}{2}$ head. Lower pharyngeals broad, each with 2 rows of blunt teeth. Yellowish brown, with numerous round pale blotches. Pacific coast of North America, Bering Strait to Monterey. This species is one of the commonest of the flounders of the Pacific coast, its abundance apparently increasing toward the northward. In Bering Sea it far outnumbers all other flounders. We have specimens from Bering Island, Medui Island, Unalaska, St. Paul, St. George, and Chignik Bay. It reaches a weight of 5 or 6 pounds and is an inhabitant of shallow waters. Specimens from Puget Sound and northward are rougher than southern specimens and constitute a slight geographical variety, for which the name *Lepidopsetta bilineata umbrosa* may be used. This is the same as *P. perarcuatus* of Cope. (*bilineatus*, two-lined.)

Platessa bilineata, AYRES, Proc. Ac. Nat. Sci. Cal. 1855, 40, San Francisco.
Platichthys umbrosus, GIRARD, Proc. Ac. Nat. Sci. Phila. 1856, 136, Puget Sound.
Pleuronectes perarcuatus, COPE, Proc. Ac. Nat. Sci. Phila. 1873, 30, Unalaska.
Pleuronectes umbrosus, GÜNTHER, Cat., IV, 454, 1862.
Pleuronectes bilineatus, GÜNTHER, Cat., IV, 444, 1862; JORDAN & GILBERT, Synopsis, 833, 1883.
Lepidopsetta bilineata, GILL, Proc. Ac. Nat. Sci. Phila. 1864, 195; LOCKINGTON, Proc. U. S. Nat. Mus. 1879, 103; LOCKINGTON, Rep. Com. Fisheries California, 1878–79, 46; JORDAN & GILBERT, Proc. U. S. Nat. Mus. 1880, 453; JORDAN & GILBERT, Proc. U. S. Nat. Mus. 1881, 68; BEAN, Proc. U. S. Nat. Mus. 1881, 241; BEAN, Cat. Coll. Fish. U. S. Nat. Mus. 1883, 19; BEAN, Proc. U. S. Nat. Mus. 1883, 353; JORDAN, Nat. Hist. Aquat. Anim., pl. 50, 184, 1884; JORDAN & GOSS, Review Flounders and Soles, 286, 1889.

1035. LIMANDA, Gottsche.

(RED DABS.)

Limanda, GOTTSCHE, Archiv fur Naturg. 1835, 160 (*limanda*).
Myzopsetta, GILL, Proc. Ac. Nat. Sci. Phila. 1864, 217 (*ferruginea*).

Teeth chiefly uniserial; lateral line with a distinct arch in front, and without accessory dorsal branch; scales imbricated, rough ctenoid; vertebræ about 40. This genus closely allied to *Pseudopleuronectes*, from which it differs only in the presence of an arch on the anterior part of the lateral line. (*Limanda*, an old name of the European Dab, *Limanda limanda*, from *limus*, mud.)

  a. Head comparatively large, 3½ to 4 in length.
    b. Dorsal rays 85; anal rays
      c. Scales rather small, 90 to 100 in lateral line; scales of right side ctenoid, closely imbricated; those of blind side mostly smooth; teeth conical, close set, forming a continuous series, about 11 + 30 in lower jaw; snout abruptly projecting, forming in front of upper eye a sharp angle with the descending profile.   **FERRUGINEA, 3013.**
      cc. Scales larger, wider part, about 80 in lateral line; scales of blind side more or less rough.   **ASPERA, 3014.**
    bb. Dorsal rays 66 to 70; anal 47 to 53; scales small, 86 to 95; snout long, protruding; scales of blind side smooth.   **PROBOSCIDEA, 3015.**
  aa. Head very short, 5½ in length; snout very short; interorbital space very narrow. D. 64; A. 63; scales 88.   **BEANII, 3016.**

3013. LIMANDA FERRUGINEA (Storer).

(RUSTY DAB.)

Head 4 in length; depth 2½; D. 85; A. 62; scales 100. Body ovate-elliptical, strongly compressed; teeth small, conical, close set, in a single series on each side in each jaw, about 11 + 30 in the lower jaw; snout projecting, forming a strong angle above upper eye, with the descending profile; gill rakers of moderate length, very weak, not toothed; eyes moderate, 4½ in head, the lower slightly in advance of upper, separated by a high, very narrow ridge, which is scaled posteriorly, and is continued backward as an inconspicuous but rough ridge to the beginning of the lateral line; scales imbricate, nearly uniform, those on right side rough ctenoid, those on left side nearly or quite smooth; scales on body rougher than on cheeks; caudal peduncle short, higher than long; dorsal inserted over middle of eye, its middle rays highest; pectoral less than ¾ length of head; caudal fin rounded; anal spine present; lateral line simple, with a rather low arch in front, the depth of which is barely ⅓ the length; a concealed spine behind ventrals; ventral of colored side partly lateral, the other wholly so; anal spine strong. Brownish olive, with numerous, irregular, reddish spots; fins similarly marked; left side with caudal fin, caudal peduncle and margins of dorsal and anal fins lemon yellow. Atlantic coast of North America, Labrador to New York. This species is rather common northward on our Atlantic coast. It is allied to the European Dab, but has smaller scales and a more prominent snout. Our specimens are from the east coast of Massachusetts. (*ferrugineus*, rusty red.)

1035. LIMANDA, Gottsche.

(MUD DABS.)

Limanda, GOTTSCHE, Archiv fur Naturgsch. 1835, 100 (*limanda*).
Myzopsetta, GILL, Proc. Ac. Nat. Sci. Phila. 1864, 217 (*ferruginea*).

Teeth chiefly uniserial; lateral line with a distinct arch in front, and without accessory dorsal branch; scales imbricated, rough ctenoid; vertebræ about 40. This genus is closely allied to *Pseudopleuronectes*, from which it differs only in the presence of an arch on the anterior part of the lateral line. (*Limanda,* an old name of the European Dab, *Limanda limanda*, from *limus*, mud.)

> *a.* Head comparatively large, 3¼ to 4½ in length.
> > *b.* Dorsal rays 85; anal rays 62.
> > > *c.* Scales rather small, 90 to 100 in lateral line; scales of right side ctenoid, closely imbricated, those of blind side mostly smooth; teeth conical, close set, forming a continuous series, about 11 + 30 in lower jaw; snout abruptly projecting, forming in front of upper eye a sharp angle with the descending profile.   FERRUGINEA, 3013.
> > > *cc.* Scales larger, wide spart, about 80 in lateral line; scales of blind side more or less rough.   ASPERA, 3014.
> > *bb.* Dorsal rays 60 to 70; anal 47 to 53; scales small, 86 to 95; snout long, protruding; scales of blind side smooth.   PROBOSCIDEA, 3015.
> *aa.* Head very short, 5¼ in length; snout very short; interorbital space very narrow.
> D. 64; A. 63; scales 88.   BEANII, 3016.

3013. LIMANDA FERRUGINEA (Storer).

(RUSTY DAB.)

Head 4 in length; depth 2⅕. D. 85; A. 62; scales 100. Body ovate-elliptical, strongly compressed; teeth small, conical, close set, in a single series on each side in each jaw, about 11 + 30 in the lower jaw; snout projecting, forming a strong angle above upper eye, with the descending profile; gill rakers of moderate length, very weak, not toothed; eyes moderate, 4½ in head, the lower slightly in advance of upper, separated by a high, very narrow ridge, which is scaled posteriorly, and is continued backward as an inconspicuous but rough ridge to the beginning of the lateral line; scales imbricate, nearly uniform, those on right side rough ctenoid, those on left side nearly or quite smooth; scales on body rougher than on cheeks; caudal peduncle short, higher than long; dorsal inserted over middle of eye, its middle rays highest; pectoral less than ⅔ length of head; caudal fin rounded; anal spine present; lateral line simple, with a rather low arch in front, the depth of which is barely ⅔ the length; a concealed spine behind ventrals; ventral of colored side partly lateral, the other wholly so; anal spine strong. Brownish olive, with numerous, irregular, reddish spots; fins similarly marked; left side with caudal fin, caudal peduncle and margins of dorsal and anal fins lemon yellow. Atlantic coast of North America, Labrador to New York. This species is rather common northward on our Atlantic coast. It is allied to the European Dab, but has smaller scales and a more prominent snout. Our specimens are from the east coast of Massachusetts. (*ferrugineus,* rusty red.)

Platessa ferruginea, D. H. STORER, Rept. Fish. Mass., 141, pl. 2, 1839, Cape Ann; DE KAY
New York Fauna: Fishes, 297, pl. 48, fig. 155, 1842; STORER, Syn. Fish. N. A., 476, 1846.
Platessa rostrata, H. R. STORER, Bost. Journ. Nat. Hist., VI, 1850, 268, Labrador.
Pleuronectes ferrugineus, GÜNTHER, Cat., IV, 447, 1862; JORDAN & GILBERT, Synopsis,
834, 1882.
Myzopsetta ferruginea, GILL, Proc. Ac. Nat. Sci. Phila. 1864, 217.
Limanda ferruginea, GOODE, Proc. U. S. Nat. Mus. 1880, 472; GOODE, Hist. Aquat. Anim.,
pl. 49, 1884; GOODE & BEAN, Oceanic Ichthyology, 427, 1896; JORDAN & GOSS, Review
Flounders and Soles, 287, 1889.
Limanda rostrata, GILL, Proc. Ac. Nat. Sci. Phila. 1864, 217.

3014. LIMANDA ASPERA (Pallas).

(ALASKA DAB.)

Head 3⅓; depth 2. D. 69; A. 53; scales about 80. Form of *Lepidopsetta bilineata*. Teeth small, almost conical, on both sides of the mouth; interorbital space narrow, scaly; opercle and preopercle naked below; gill rakers very feeble; pharyngeals not very broad, their teeth bluntish, not paved; scales small, wide apart, partly embedded, each one with 1 to 4 spinules, which are almost erect; anterior scales with 3 to 4 of these spinules; posterior mostly with 1; scales of blind side smoother; only middle rays of dorsal and anal scaly; no accessory lateral line; anal spine present; twentieth anal ray and thirty-seventh dorsal ray longest; caudal double truncate. Brown, nearly plain, the blind side with tinges of lemon yellow. Bering Sea, generally common, south to Vancouver Island and to the Okhotsk Sea. We have specimens from Petropaulski and Robben Reef, Bristol Bay, and Herendeen Bay. It is especially abundant in Bristol Bay, and, according to Dr. Gilbert, it is an excellent food-fish. Dr. Bean has also collected it in various localities in Alaska. Its scales are larger and rougher than in *L. ferruginea* which, in many respects, it resembles. A specimen from the island of Saghalien is in the museum at Cambridge. The above description is from examples taken by Dr. Bean. (*asper*, rough.)

Pleuronectes asper, PALLAS, Zoogr. Rosso-Asiat., III, 425, 1811, east coast of Siberia; GÜN-
THER, Cat., IV, 454, 1862; STEINDACHNER, Pleuronectiden, etc., aus Decastris Bay, 1870–
1875; JORDAN & GILBERT, Synopsis, 835, 1883.
Limanda aspera, BEAN, Proc. U.S. Nat. Mus. 1881, 242; BEAN, Cat. Coll. Fish, U.S. Nat. Mus.
1883, 20; BEAN, Proc. U. S. Nat. Mus. 1883, 354; BEAN, Hist. Aquat. Anim., 184, pl. 48,
1884; JORDAN & GOSS, Review Flounders and Soles, 288, 1889.

3015. LIMANDA PROBOSCIDEA, Gilbert.

Depth 2¼ to 2½ in length; head large, 3 to 3⅓ in length in a specimen 7 inches long. D. 63 to 67; A. 47 to 49; scales 86 to 95. Resembling *L. ferruginea*, but having fewer rays in dorsal and anal, larger scales and longer snout. Profile sharply angulated above front of upper eye, the snout convexly protruding; form varying from very slender to broadly elliptical, the 2 outlines equally curved; caudal peduncle short, widening backward, its least depth twice its length; mouth oblique, maxillary reaching beyond front of lower eye, 4 in head; teeth narrow, little compressed, in a single series on both sides of the jaw, extending farther back on the blind side; eyes on right side; lower eye well in advance of upper, the diameter of upper eye 5¼ to 6 in head, 1½ in snout; vertical from front of upper eye, falling midway between front of orbit and front of pupil of lower eye; interorbital space a very narrow, sharp ridge, naked in females, with a single series of ctenoid scales in males; gill rakers short, about equal to diameter of pupil, 13 or 14 in number, 9 or 10 on lower limb; scales loosely imbricated, ctenoid in males on colored side, smooth in females; blind side of both sexes smooth; head scaled on eyed side in males; the opercle, subopercle, interopercle, and preopercle mostly naked in females; head on blind side naked; rays of vertical fins with a single series of ctenoid scales; dorsal fin beginning slightly behind front of upper eye, the first 3 rays usually higher and with membranes more deeply incised than in those which follow; highest portions of both dorsal and anal fins behind the middle of the body; these fins about equal, their longest rays equal to the snout and eye; caudal ⅞ head; pectorals short, ½ in head; ventrals reaching beyond front of anal, 3½ in head; the usual small antrorse spine in front of anal fin. Color light grayish or brownish, thickly covered with small whitish spots; entire left side with margins of dorsal, caudal, and anal fins bright lemon yellow (as in *ferruginea*); vertical fins grayish, with an occasional dark-brown ray. Specimens described 7½ inches long. Bering Sea; several specimens from *Albatross* Stations 3239 and 3240, in Bristol Bay, in 11½ to 14¼ fathoms; 1 young individual from Herendeen Bay. (Gilbert.) (*proboscideus*, having a long snout or proboscis.)

Limanda proboscidea, GILBERT, Report U. S. Fish Comm. 1893 (1896), 460, pl. 33, Bristol Bay and Herendeen Bay. (Coll. *Albatross*.)

3016. LIMANDA BEANII, Goode.

Head 5½; depth 2⅔. D. 64; A. 63; scales 88. Body elliptical, with angular outlines, strongly compressed; head very short; snout abbreviated; mouth small, subvertical; teeth small, apparently in two rows, chiefly on the blind side of lower jaw; eyes large, as long as mandible; interorbital space very narrow. Dorsal fin beginning about pupil, its rays long, wide apart, exserted; right ventral near the median line; caudal broad, fan-shaped. Lateral line with an abrupt curve, the length of which is twice its height and about equal to length of head, its scales highly specialized; lateral line on colored side less developed; scales small, strongly ctenoid on the right side; larger and cycloid on the blind side. Grayish, mottled with darker; a conspicuous black blotch on the outer rays of caudal on each side. (Goode.) Deep water off the coasts of New England; not common. (Named for Dr. Tarleton Hoffman Bean.)

Limanda beanii, GOODE, Proc. U. S. Nat. Mus. 1880 (Feb. 16, 1881), 473, southern coast New England, Fish Hawk Stations, 875, 876 ; GOODE & BEAN, Oceanic Ichthyology, 428, pl. 102, figs. 355a and 355b, 1896.
Pleuronectes beani, JORDAN & GILBERT, Synopsis, 835, 1883; JORDAN & GOSS, Review Flounders and Soles, 288, 1889.

1036. PSEUDOPLEURONECTES, Bleeker.
(WINTER FLOUNDERS.)

Pseudopleuronectes, BLEEKER, Comptes Rendus Acad. Amst., Pleuron., 7, 1862 (*planus*).

Body oblong, with firm flesh; the scales firm, regularly imbricated, strongly ctenoid on eyed side in both sexes; fin rays scaly; mouth small;

teeth uniserial, incisor-like, close set, all more or less blunt; lower pharyngeals very narrow, each with 2 rows of separate, conical teeth. This genus is distinguished from *Pleuronectes* chiefly by the well-imbricated ctenoid scales, and from *Limanda*, which it more closely resembles, by the want of arch to the lateral line. Besides the typical species, we refer to this genus a second from the North Pacific. ($\psi\varepsilon\tilde{v}\delta o\varsigma$, false; *Pleuronectes*.)

  a. Dorsal rays 65; anal rays 48; scales 83; vertical fins nearly plain.

AMERICANUS, 3017.

  aa. Dorsal rays 58; anal rays 38; scales 70; vertical fins with black bars.

PINNIFASCIATUS, 3018.

3017. PSEUDOPLEURONECTES AMERICANUS (Walbaum).

(COMMON FLATFISH; WINTER FLOUNDER.)

Head 4 in length; depth 2¼. D. 65; A. 48; scales 83. Body elliptical; an angle above eye. Head covered above with imbricated, strongly ctenoid scales, similar to those on the body; blind side of head nearly naked; interorbital space rather broad, strongly convex, its width ½ eye, entirely scaled; teeth compressed, incisor-like, widened toward tips, close set, forming a continuous cutting edge; some of teeth often emarginate, sometimes movable; right side of each jaw toothless. Highest dorsal rays less than length of pectorals, and more than ⅓ length of head; anal spines present. Dark rusty brown, spotted or nearly plain; young olive brown, more or less spotted and blotched with reddish. Atlantic coast of North America, from Labrador to Chesapeake Bay. This small flounder is one of the most abundant of the group on our Atlantic coast. It reaches a length of about 15 inches and a weight of less than 2 pounds. It is a very good food-fish and sells readily in the markets. Along the south coast of Massachusetts this species is more abundant than any other of the flatfishes. The specimens examined by us are from Labrador, Cape Breton, Anticosti, Grand Menan, Boston, Provincetown, Woods Hole, New Bedford, and Somers Point, New Jersey.

Flounder, SCHÖPF, Schrift. Gesellschaft Naturforscher Freunde, VIII, 1788, 148, New York.
Pleuronectes americanus, WALBAUM, Artedi Piscium, 113, 1792, based on the *Flounder* of SCHÖPF; BLOCH & SCHNEIDER, Syst. Ichth., 150, 1801; GÜNTHER, Cat., IV, 443, 1862; JORDAN & GILBERT, Synopsis, 837, 1883; STEARNS, Proc. U. S. Nat. Mus. 1883, 125.
Pleuronectes planus, MITCHILL, Trans. Lit. & Philos. Soc. N. Y., I, 1815, 387, New York.
Platessa pusilla, DE KAY, New York Fauna: Fishes, 296, pl. 47, fig. 153, 1842, New York; STORER, Synopsis, 477, 1846.
Platessa plana, STORER, Rept. Fishes Mass., 140, 1839; DE KAY, New York Fauna: Fishes, 295, pl. 49, fig. 158, 1842; STORER, Synopsis, 476, 1846.
Pseudopleuronectes planus, BLEEKER, Comptes Rendus Amsterd., XIII, 1862, 7.
Pseudopleuronectes americanus, GILL, Proc. Ac. Nat. Sci. Phila. 1864, 216; GOODE, Nat. Hist. Aquat. Anim., 182, pl. 44, 1884; JORDAN & GOSS, Review Flounders and Soles, 289, 1889.

3018. PSEUDOPLEURONECTES PINNIFASCIATUS (Kner).

Head 3¼ in body; depth 2⅕. D. 58; A. 38; scales 70; eye 5⅔ in head; snout 5; highest anal ray 2; pectoral 2; caudal 4½ in body. Body subelliptical, the snout rather pointed and not forming an angle above eye; mouth rather small, maxillary reaching scarcely to the middle of the lower eye; interorbital space rather broad, ¼ width of eye; a rather prominent rugose ridge above opercle, with a smaller similar ridge behind it; both sides of jaws with teeth, those on blind side stronger; origin of dorsal over middle of upper eye. Color brown, with vague dusky spots; 6 or 7 blackish vertical bars on dorsal and anal; similar lengthwise blotches on caudal. Okhotsk Sea, east to Kamchatka. (Steindachner.) Not seen by us. From the excellent figure we conclude that it belongs to *Pseudopleuronectes*, although its pharyngeals have not been described. It seems to us nearer to *P. americanus* than to *Liopsetta glacialis*. (*pinna*, fin; *fasciatus*, banded.)

Pleuronectes pinnifasciatus, KNER, in STEINDACHNER, Ueber einige Pleuronectiden, etc., aus Decastris Bay, 2, pl. 1, fig. 1, 1870, Decastris Bay, mouth of Amur River; JOR-DAN & GOSS, Review Flounders and Soles, 290, 1889.

1037. PLEURONECTES (Artedi) Linnæus.

(PLAICE.)

Pleuronectes, ARTEDI, Genera, etc., in part, 16, 1738.
Pleuronectes, LINNÆUS, Syst. Nat., Ed. x, 268, 1758 (*platessa*); included all known *Pleuronectidæ*.
Platessa, CUVIER, Règne Animal, Ed. 1, II, 220, 1817 (*platessa*).
Pleuronectes, SWAINSON, Nat. Hist. Class'n Anim., II, 302, 1839 (*platessa*).
Pleuronectes, BLEEKER, Comptes Rendus Acad. Amsterd., XIII, 1862 (*platessa*); and of most recent authors.

Body oblong, with firm flesh. Mouth small, teeth uniserial, incisor-like, compressed, forming a continuous cutting edge. Lateral line straightish, without arch or accessory dorsal branch. Scales imperfectly imbricated, chiefly cycloid in both sexes; lower pharyngeals small and narrow, separate, each with 1 or 2 rows of small bluntish teeth. No stellate scales along bases of dorsal and anal. Species mostly European; valued as food. ($\pi\lambda\epsilon\tilde{\upsilon}\rho o\nu$, side; $\nu\epsilon\kappa\tau\eta\varsigma$, swimmer.)

3019. PLEURONECTES QUADRITUBERCULATUS, Pallas.

Head 3⅔; depth 2. D. 68; A. 50; scales 78. Mouth very small, with small, incisor-like teeth, rounded at tip. Eyes separated by a narrow ridge; about 5 small, prominent, conical, obtuse, bony tubercles in a row above the opercle, continuous with the direction of the lateral line, which is straight, without accessory dorsal branch; tubercle above opercle largest. Scales small, cycloid in all specimens examined. Anal spine present. Grayish, mottled with paler and with round black spots; fins very dark. Bering Sea on both coasts, south to Kadiak; not common. Our specimens from Avatcha Bay, Bristol Bay, Herendeen Bay, Chernofsky Harbor, Grantley Harbor, Chignik Bay, and Robben Island. The above description from a small specimen (No. 28025, U. S. Nat. Mus.) collected by Mr. W. J. Fisher at Kadiak. The species proves, as suspected by Jordan & Goss, to be a true *Pleuronectes*, having the lower pharyngeals narrow, separate, with 2 rows of bluntish teeth. (*quadrituberculatus*, having four tubercles.)

Pleuronectes quadrituberculatus, PALLAS, Zoogr. Rosso-Asiat., III, 423, 1811, sea between Kamchatka and Alaska; BEAN, Proc. U. S. Nat. Mus. 1881, 241; JORDAN & GILBERT, Synopsis, 836, 1883.
Pleuronectus pallasii, STEINDACHNER, Ichth. Beitr., VIII, 45, 1879, Kamchatka.
Parophrys quadrituberculatus, GÜNTHER, Cat., IV, 456, 1862.
Platessa quadrituberculata, JORDAN & GOSS, Review Flounders and Soles, 292, 1889.

1038. LIOPSETTA, Gill.

(EEL-BACK FLOUNDERS.)

Liopsetta, GILL, Proc. Ac. Nat. Sci. Phila. 1864, 217 (*glaber*); females.
Euchalarodus, GILL, Proc. Ac. Nat. Sci. Phila. 1864, 222 (*putnami*); males.

Teeth chiefly uniserial, incisor-like; scales imperfectly imbricated, rough ctenoid in the male, more or less cycloid in the female (fin rays scaly in the male, naked in the female); lower pharyngeals very large, more or less united in the adult, their surface somewhat concave, with teeth in 5 or 6 rows, large, blunt, close set; lateral line without arch or dorsal branch. This genus comprises several species of small flounders of the Arctic seas. The genus is distinguished by the large, half-united pharyngeals, as also by the peculiar squamation, the scales in the males being very rough, in the females smooth. This difference has given rise to the nominal genus *Euchalarodus*, based on the males, while *Liopsetta* was based on the smoother females, which were erroneously supposed to be scaleless. (λεῖος, smooth; ψῆττα, flounder.)

  a. Dorsal rays 55 or 56; anal 40 to 42.

   b. Pectoral fin short, ½ length of head in males, shorter in females. GLACIALIS, 3020.
   bb. Pectoral fin long, 1½ in head in males, nearly 2 in females. PUTNAMI, 3021.
  aa. Dorsal rays 59 to 62; anal 45 or 46; pectoral 1⅔ in head in males. OBSCURA, 3022.

3020. LIOPSETTA GLACIALIS (Pallas).

(ARCTIC FLOUNDER.)

Head 4; depth 2⅓. D. 56; A. 42. Form of *Liopsetta putnami*. A roughened ridge above the cheeks and opercles on the eyed side. Eyes separated by a narrow, smooth, bony ridge. Scales minute, embedded, nonimbricate, ctenoid in the males, smooth in the females; scales on blind side similar, less developed; scales of lateral line a little larger. Teeth colored, incisor-like, forming an even edge, mostly on blind side. An anal spine; pectorals short. Dark brown, the fins spotted. Arctic shores of Alaska and Siberia, south in Bering Sea to Petropaulski, St. Michaels, and Bristol Bay. Our specimens from Port Clarence, Petropaulski, Bristol Bay, mouth of Nushagak River, and Kotzebue Sound; the description from specimens from the last-named locality taken by Dr. Bean. It is said to be abundant in the Arctic Ocean and as far south as Bristol Bay. "Although small, its great abundance and fine flavor make it important as an article of food." The male is the rough fish described by Pallas as *P. cicatricosus*. The smoother female is Dr. Günther's *Pleuronectes franklinii*, the sexual differences being much as in *Liopsetta putnami*. *Liopsetta dvinensis* of the northern coasts of Russia may be the same species. (*glacialis*, icy.)

Pleuronectes glacialis, PALLAS, Itin., III, App., 706, mouth of River Obi; BLOCH & SCHNEIDER, Syst. Ichth., 150, 1801; PALLAS, Zoogr. Rosso-Asiat., III, 424, 1811; RICHARDSON, Fauna Bor.-Amer., Fish., 258, 1836; DE KAY, N.Y. Fauna: Fishes, 302, 1842; STORER, Synopsis Fish. N. A., 479, 1846; BEAN, Proc. U. S. Nat. Mus. 1881, 241; JORDAN & GILBERT, Synopsis, 837, 1883; BEAN, Cat. Coll. Fish. U. S. Nat. Mus., 20, 1883; BEAN, Nat. Hist. Aquat. Anim., 184, pl. 47, 1884.

Pleuronectes cicatricosus, PALLAS, Zoogr. Rosso-Asiat., III, 424, 1811, male, sea between Kamchatka and Alaska.

? *Platassa dvinensis*, LILLJEBORG, Veb. Ah. Handl. 1850, 360, tab.20, mouth of River Dwina.
Pleuronectes franklinii, GÜNTHER, Cat. Fish., IV, 442, 1862, Arctic seas of America, female; BEAN, Proc. U. S. Nat. Mus. 1881, 241.
Liopsetta glacialis, JORDAN & GOSS, Review Flounders and Soles, 295, pl. 17, 1889.

3021. LIOPSETTA PUTNAMI (Gill).

(EEL-BACK FLOUNDER.)

Head 3½; depth 2. D. 55; A. 40; scales 70 (pores). Body oblong, ovate. Eyes rather small, separated by a naked elevated ridge. Jaws sometimes each with 2 distinct rows of teeth, the interrupted outer series of truncate, close set, thickish, incisor-like teeth, which are sometimes movable; the inner row of similar teeth more widely set and rather distant from the outer row (and often or generally wanting); about 20 teeth in outer row in lower jaw; right side of each jaw toothless; interorbital ridge continuous, with a broad, naked, smoothish, tuberculose ridge, which joins the lateral line. Scales small, distant, nonimbricate, smooth in the female, and more or less ctenoid in the male, those on blind side smaller. Fins moderate, somewhat scaly; anterior rays of dorsal low; pectoral a little more than ¼ head; bases of vertical fins not tuberculate; anal spine present; lower pharyngeals separate, broad, with coarse teeth. Grayish brown, mottled with darker brown; fins with blackish spots. Length 10 inches. Atlantic coast of North America, from Cape Cod northward to Labrador and beyond; occasionally found in abundance. This species is rather common along the coast of northern Massachusetts and northward to Labrador. Specimens are frequently found in the markets, mixed with those of *Pseudopleuronectes americanus*. The numerous specimens in our possession were found in the markets of Indianapolis, having been sent thither from Boston. The remarkable sexual differences in the species have been fully discussed by Dr. Bean (Proc. U. S. Nat. Mus. 1878, 345), the form formerly called *Euchalarodus putnami* being the male, and that called *Pleuronectes glaber* being the female of the same species. These conclusions of Dr. Bean are fully corroborated by our series of specimens in which both sexes are fully represented.

Although *Liopsetta putnami* is abundant where found, its ascertained range is somewhat limited. The specimens in the United States National Museum represent localities from Salem, Massachusetts, to Belfast, Maine. In the Museum of Comparative Zoology the localities represented are Providence, Boston, Salem, Grand Manan, and Labrador. (Named for Prof. Frederic Ward Putnam.)

Platessa glabra, STORER, Proc. Boston Soc. Nat. Hist. 1843, 130, female, Massachusetts; STORER, Syn. Fish. N. A., 477, 1846; STORER, Hist. Fish. Mass., 199, pl. 31, fig. 1, 1867; PUTNAM, Bull. Essex Inst., VI, 1874, 12; not *Platessa glabra* of RATHKE, 1837, a species of *Flesus*.
Euchalarodus putnami, GILL, Proc. Ac. Nat. Sci. Phila. 1864, 216-221, Salem, Massachusetts (Coll. F. W. Putnam), male; GILL, Report U. S. Fish Comm. 1873, 794; GOODE & BEAN, Amer. Journ. Sci. and Arts, XIV, 1877.
Liopsetta glabra, GILL, Proc. Ac. Nat. Sci. Phila. 1864, 217.
Pleuronectes glaber, GILL, Report U. S. Fish Comm. 1873, 794; GOODE & BEAN, Amer. Journ. Sci. and Arts, XIV, 1877, 476; XVII, 1879, 40; GOODE & BEAN, Proc. U. S. Nat. Mus. 1878, 347; JORDAN & GILBERT, Synopsis, 836, 1883; GOODE, Nat. Hist. Aquat. Anim., 183, pl. 45, 1884.
Liopsetta putnami, JORDAN & GOSS, Review Flounders and Soles, 294, pl. 16, 1889.

3022. LIOPSETTA OBSCURA (Herzenstein).

To this species we refer 2 males from Shana Bay, Iturup Island. The scales on the colored side are everywhere strongly ctenoid and imbricated, while in Herzenstein's types (supposed to be females) they were cycloid. In our specimens the head is somewhat smaller, $3\frac{9}{10}$ in length instead of $3\frac{7}{10}$ to $3\frac{9}{5}$; the depth is greater, $2\frac{1}{4}$ in length instead of $2\frac{3}{7}$ to $2\frac{3}{8}$; the interorbital space is covered with very fine scales, not naked; the curve of the lateral line seems more marked, its cord contained 5 instead of 6 times in the straight portion. All of the fins are higher than in the female types, the pectoral of colored side being $1\frac{3}{7}$ head, the caudal $1\frac{1}{6}$, the ventral $\frac{1}{4}$ head, and the highest dorsal ray $1\frac{3}{7}$. Some of these differences may well be sexual. The lower pharyngeals are short and broad, the two closely appressed but united in our specimens, 27 and 29 cm. long.. The teeth are large and very blunt, like cobble stones, and are arranged in 1 row along the outer edge; a row of larger teeth along the inner edge, and a short row along the posterior edge of the triangle. The arrangement is very similar is that found in *L. glacialis*, but here a few small teeth, without definite arrangement, are interposed in the middle of the bone, between the third series described. Dorsal 59 and 62; anal 45 and 46; tubes in the lateral line 79. Color on eyed side uniform dark brown on body and fins, the extreme tips of the fin rays white; on blind side yellowish white, with a few irregular scattered dark spots; dorsal and anal yellowish at base, becoming more or less mottled with dusky on distal half, the fins marked with broad dark bars parallel with the rays, about 7 on the anal fin, 10 or 11 on the dorsal; caudal light on basal half more or less blotched with darker, becoming black posteriorly. With this species we identify also a number of young individuals, 9 to 15 cm. long, from the same locality (Iturup Island). They are probably young females, but the viscera are in such condition as to prevent positive determination. The scales are perfectly smooth, but in other respects they agree perfectly with the adult males, except in their more varied coloration; head and body brownish, profusely spotted in coarser or finer pattern with light gray; also with a few scattered black spots edged with gray; markings on the fins as described for adults. In 7 specimens the dorsal contains 60, 62, 62, 62, 64, 65, and 66 rays; anal 45, 45, 45, 46, 47, 47, 48. Sea of Okhotsk. Our specimens from Shana Bay, Iturup Island, one of the Kurils; originally described from Mantchuria. (*obscurus*, dark.)

Pleuronectes obscurus, HERZENSTEIN, Mélanges Biologiques, 127, 1890, Mantchuria.
Liopsetta obscura, JORDAN & GILBERT, Rep. Fur Seal Invest., 1898.

1039. PLATICHTHYS, Girard.

(STARRY FLOUNDERS.)

Platichthys, GIRARD, Proc. Ac. Nat. Sci. Phila. 1854, 136 (*rugosus = stellatus*).

Body very robust, broad, not greatly compressed. Mouth small; teeth chiefly uniserial, incisor-like. Scales all in both sexes and on both sides of body reduced to coarse scattered stellate tubercles, which are not imbricated; similar tubercles between bases of dorsal and anal rays; lateral line without scales, with no anterior arch or accessory lateral line; lower pharyngeals broad, each with 3 rows of blunt coarse teeth. A single species, the largest of the small-mouthed flounders, and distinguished from related forms chiefly by the development of coarse stellate tubercles instead of scales. ($\pi\lambda\alpha\tau\upsilon\varsigma$, flat; $i\chi\theta\upsilon\varsigma$, fish.)

3023. PLATICHTHYS STELLATUS (Pallas).

(GREAT FLOUNDER.)

Head $3\frac{3}{5}$; depth 2. D. 58; A. 42. Vertebræ 34. Body broad and short, the snout forming a slight angle with the profile; lower jaw projecting; interocular space rather broad, with very rough scales; large rough scales at base of dorsal and anal rays and on sides of head; similar but smaller scales scattered over the body; lateral line smooth; fins without scales; a cluster of bony prominences above opercle. Teeth incisor-like, truncate, rather broad, $\frac{10+15}{12+16}$. Lower pharyngeals broad, with coarse paved teeth. Dark brown or nearly black, with lighter markings; fins reddish brown; dorsal and anal with 4 or 5 vertical black bands; caudal with 3 or 4 black longitudinal bands. Pacific coast of America, from Point Concepcion to the Arctic Ocean and south to the Amur River. This is one of the largest of the American flounders, reaching a weight of 15 to 20 pounds. Of the small-mouthed flounders it is much the largest species known. It is an excellent food-fish, and from its size and abundance it is one of the most important of the group in the region where it is found, constituting half the total catch of flounders on our Pacific coast, and it is equally abundant in Bering Sea. It lives in shallow water and sometimes ascends the larger rivers. It is one of the most widely distributed of all the flounders, its range extending from San Luis Obispo, where it was obtained by Jordan & Gilbert, to the mouth of the Anderson and Colville rivers on the Arctic coast, where it was observed by Dr. Bean, and to Port Clarence, where Mr. Scofield obtained specimens. We have also specimens from Petropaulski, Bering, Medni, and Robben islands and from Bristol Bay. A specimen from the island of Saghalien in Asia is in the museum at Cambridge. (*stellatus*, starry.)

Pleuronectes stellatus, PALLAS, Zoographia Rosso-Asiatica, III, 416, 1811, Kamchatka, Aleutian and Kuril Islands; GÜNTHER, Cat., IV, 443, 1862; STEINDACHNER, Pleur. von Decastris Bay, 1870, 1; JORDAN & GILBERT, Proc. U. S. Nat. Mus. 1880, 453; JORDAN & GILBERT, Proc. U. S. Nat. Mus. 1881, 68; BEAN, Proc. U. S. Nat. Mus. 1881, 420; JORDAN & GILBERT, Synopsis 835, 1883; BEAN, Proc. U. S. Nat. Mus. 1883, 353; BEAN. Cat. Coll. Fish. U. S. Nat. Mus. 1883, 20; JORDAN, Nat. Hist. Aquat. Anim., 184, pl. 46, 1884.

Platichthys rugosus, GIRARD, Proc. Ac. Nat. Sci. Phila. 1854, 139, 155, San Francisco; Presidio; Petaluma; GIRARD, U. S. Pac. R. R. Surv., X, Fishes, 148, 1858.

Platessa stellata, DE KAY, N. Y. Fauna: Fishes, 301, 1842; STORER, Synopsis, 478, 1846.

Platichthys stellatus, LOCKINGTON, Rep. Com. Fish. Cal. 1878-79, 43; LOCKINGTON, Proc. U. S. Nat. Mus. 1879, 91; JORDAN & GOSS, Review Flounders and Soles, 296, 1889.

1040. MICROSTOMUS,* Gottsche.

(SMEAR DABS.)

Microstomus, GOTTSCHE, Archiv fur Naturgesch. 1835, 150 (*latidens*); not *Microstoma*, RISSO, 1826.

Cynicoglossus, BONAPARTE, Fauna Italica, 1837, fasc., XIX (*cynoglossus*, NILSSON, not of L.).

Cynoglossa, BONAPARTE, Catalogo Metodico Pesci Europei, 48, 1846 (*microcephalus*); not *Cynoglossus*, HAMILTON, 1822.

Brachyprosopon, BLEEKER, Comptes Rendus Acad. Sci. Amsterd., XIII, Pleuron., 7, 1862 (*microcephalus*).

Cynicoglossus, JORDAN & GILBERT, Synopsis, 460, 1883 (*microcephalus*).

Body elongate, compressed; mouth very small; teeth broad, incisor-like, on blind side only; scales small, all cycloid; vertebræ numerous (48 to 52); dorsal rays 90 to 100; anal rays 70 to 85; anal spine obsolete; left side of skull normal, without mucous cavities; ventral fins with 5 rays each. Arctic seas. This genus is widely separated from *Pleuronectes*

---

* We here retain the generic name *Microstomus*, although in accordance with recent usage of most ornithologists and ichthyologists, it should be suppressed, as identical with *Microstoma*. The two words are from the same root, and differ only in the termination. But is not this difference enough? The code of nomenclature of the American Ornithologists' Union very properly declares that "a name is only a name, and has no necessary meaning," and therefore no necessarily correct spelling, except the spelling selected by the writer from whom it dates its origin. As a result of this, the original spelling of each generic name is (undoubted misprints aside) the orthography to be adopted, regardless of all questions as to the correct etymology of the word. As a necessary sequence, it seems to us that all generic names, not actually preoccupied by names spelled in the same way, should be tenable. There is no other certain boundary line between names tenable and names untenable. We therefore regard all generic names as available unless used in zoology earlier and in exactly the same orthography. Among American genera of fishes we may therefore use the following, notwithstanding their earlier analogies:

Microstomus for	*Cynicoglossus* notwithstanding the prior *Microstoma*.	
Heterodontus	*Cestracion*	*Heterodon.*
Lucania		*Lucanus.*
Thymallus	*Choregon*	*Thymallus.*
Nebris		*Nebria.*
Xiphidion	*Xiphister*	*Xiphidium.*
Amitra	*Monomitra*	*Amitrus.*
Scytalina	*Scytaliscus*	*Scytalinus.*
Lagochila	*Quassilabia*	*Lagocheilus.*
Auchenopterus	*Cremnobates*	*Auchenipterus.*
Lyopsetta		*Liopsetta.*
Leucos	*Myloleucus*	*Leucus.*
Pterophryne	*Pterophrynoides*	*Pterophrynus.*
Scaphirhynchus	*Scaphirhynchops*	*Scaphorhynchus.*
Lepidion	*Haloporphyrus*	*Lipidia.*
Gramma		*Grammia.*
Stenotomus		*Stenotoma.*

If *Microstomus* be discarded, the next name in order of date is *Cynicoglossus*. The following is Bonaparte's definition of *Cynicoglossus* as quoted by Gill (Proc. Ac. Nat. Sci. Phila. 1864, 222):

"Secundo è *Cynicoglossus* nob. che come il *Pl. cynoglossus* L. ha la linea laterale retta, la bocca piccola, i denti come quello di sopra [*Platessa*] ma la mascelle iguale, con labbra turgide, e l' ano senza spina."

Later, in his Catalogo Metodico dei Pesci Europei, Bonaparte changes this name from *Cynicoglossus* to *Cynoglossa*, giving the sole species as *Cynoglossa microcephala*, and quoting as its synonym "*Pleuronectes cynoglossus*, N. Nilss.", showing that his identification of the Linnæan species coincided with that of Nilsson, who at first used the name "*Pleuronectes cynoglossus*" for the present species instead of the species of *Glyptocephalus*. In Bonaparte's Catalogo, *Glyptocephalus*, Gottsche, is regarded by Bonaparte as synonymous with *Platessa*.

It is thus evident, as Dr. Gill has suggested, that Bonaparte meant to refer to the *Pleuronectes microcephalus* instead of *Pl. cynoglossus*, he "having followed Nilsson in his erroneous identification" of the latter with the former. In further evidence of this we have the fact that *Cynicoglossus microcephalus* (*kitt*) has no anal spine, while such a spine is present in the species of *Glyptocephalus*. We would be, therefore, justified in the use of *Cynicoglossus* instead of the later *Brachyprosopon*, if *Microstomus* should be regarded as ineligible on account of the prior name *Microstoma*. (Jordan & Goss.)

and its allies by its greatly increased number of vertebræ, a character accompanied by a similar increase in the number of fin rays. It is close to *Glyptocephalus*, but the lack of the cavernous structure of the bones of the head, a structure peculiar to the species of that genus, sufficiently distinguishes it. (μικρός, small; στόμα, mouth.)

a. Dorsal rays 85 to 93; anal rays 70 to 76; head very small, about 5 in length; eye 4 in head.   KITT, 3024.

aa. Dorsal rays 102; anal ray 85; head 4½ in length; eye 3 in head.   PACIFICUS, 3025.

3024. MICROSTOMUS KITT (Walbaum).

(SMEAR DAB.)

Head 5¼ in length; depth 2¼. D. 85 to 93; A. 70 to 76; scales 130; caudal 1¼ in head; pectoral 1¾. Body moderately elongate; mouth small, the maxillary not reaching to front of lower eye; teeth on blind side conical, rather compressed and blunted, 11 to 13 on either jaw; eyes close together, the lower slightly in advance; gill rakers short, not numerous. Origin of dorsal above middle of upper eye, its rays larger in the posterior half of body; pectorals about equal in size; no spine before anal; caudal rounded; head, except snout, entirely scaled; scales cycloid; lateral line with a small curve; vertebræ 13 + 35 = 48. Color dull yellowish, blotched, and with dark spots, especially over the chest and along the base of anal fin; dark blotches and spots on anal, caudal, and ventral fins; dark base to pectoral, which has also some cloudy markings. (Day.) Seas of the north of Europe in rather deep water, south to Cornwall. Recorded by Steindachner (as *Pleuronectes gilli*), from the sea between Iceland and Greenland. This small flounder is rather common in the waters of northern Europe. It reaches the length of a foot or more, and is said to be excellent as food. Like its congener, *Microstomus pacificus*, this species is very slimy in life. *Pleuronectes gilli*, as described by Dr. Steindachner, seems to differ from *Microstomus kitt* only in the larger head, which is but 4⅔ in the length to base of caudal. It is probably not specifically distinct from the latter. Only a single specimen, 10¼ inches long, is known. (Eu.) (The specific name "*kitt*," given by Walbaum on the authority of Jago's description, should be adopted for this species. According to Day, the species is still called "*kitt*" on the coast of Cornwall.)

Rhombus lævis cornubiensis, JAGO, in Ray, "Syn. Pisc., 162, tab. 1, fig. 1, 1713."

The Smear Dab, PENNANT, British Zoology, III, 230, pl. 41, 1776.

Pleuronectes kitt, WALBAUM, Artedi Piscium, III, 120, 1792, after RAY; the description in part confused with that of *Lepidorhombus*.

Pleuronectes lævis, SHAW, Gen'l Zool., IV, 299, 1803.

Pleuronectes quenseli, HÖLBÖLL, Bohusläns Fiske, IV, 59, 1821, Bohusläns, Sweden.

Pleuronectes quadridens, FABRICIUS, Kongl. Dansk.Vid. Selsk. Afhandl., I, 39, 1824, Iceland.

Microstomus latidens, GOTTSCHE, Arc:.:v fur Naturgsch. 1835, 150, Zealand.

Pleuronectes gilli, STEINDACHNER, Ichth. Notizen, VII, 40, 1868, Polar Sea north of Iceland.

Pleuronectes microcephalus, DONOVAN, British Fishes, II, pl. 42, 1802; GÜNTHER, Cat., IV, 447; STEINDACHNER, Ichth. Beitr., VIII, 47; DAY, Fishes Great Britain, II, 28, pl. 102; COLLETT, Norges Fiske, 145, and of recent European writers generally.

Pleuronectes microstomus, FABER, Isis, 886, 1828.

Platessa microcephala, FLEMING, British Anim., 198, 1828, and of numerous writers.

Cynoglossa microcephala, BONAPARTE, Catalogo Metodico Pesci Eur., 48, 1845.

Microstomus kitt, JORDAN & GOSS, Review of Flounders and Soles, 1886, 298.

3025. MICROSTOMUS PACIFICUS (Lockington).

(SLIPPERY SOLE.)

Head $4\frac{1}{4}$ in body; depth $2\frac{5}{6}$. D. 102; A. 85; scales 140; eye $3\frac{1}{2}$ in head; maxillary 5; pectoral $1\frac{3}{8}$; greatest height of dorsal $2\frac{1}{8}$; anal $2\frac{1}{3}$; caudal $1\frac{2}{5}$; vertebræ 12 + 40 = 52. Body elongate, elliptical; mouth small, the maxillary reaching just past front of lower eye; teeth long and broad, forming a continuous cutting edge, on blind side only, about 10 teeth on lower jaw; eyes very large, nearly twice as long as snout, the upper even with profile above; interorbital a narrow scaly ridge; gill opening adnate to shoulder girdle above pectoral; gill rakers short, 8 below angle, 5 or 6 very small scarcely developed ones above; scales small, cycloid, not closely imbricated, lateral line nearly straight. Origin of dorsal slightly behind middle of upper eye, caudal truncate or slightly rounded. Color olive brown, blotched on body and fins with darker, all fins blackish toward the ends of the rays. Pacific coast of North America, Monterey to Unalaska, in rather deep water, 15 to 50 fathoms; common. Here described from a specimen, about 14 inches in length, from *Albatross* Station 2927, off the coast of California. This small flounder abounds in deep water about San Francisco, but comes near the shore farther north. It is exceedingly slimy when first taken. The large individuals are considered excellent as food; the smaller are thrown away. It rarely reaches the weight of a pound.

Glyptocephalus pacificus, LOCKINGTON, Rep. Cal. Com. Fisheries, 1878–79, 43, off Point Reyes, California; LOCKINGTON, Proc. U. S. Nat. Mus. 1879, 86; JORDAN, Nat. Hist. Aquat. Anim., 188, 1884.

Cynicoglossus pacificus, JORDAN & GILBERT, Proc. U. S. Nat. Mus. 1880, 453: JORDAN & GILBERT, Proc. U. S. Nat. Mus. 1881, 68; JORDAN & GILBERT, Synopsis, 838, 1883.

Microstomus pacificus, JORDAN & GOSS, Review Flounders and Soles, 299, 1889.

1041. EMBASSICHTHYS, Jordan & Evermann.

Embassichthys, JORDAN & EVERMANN, Check-List Fishes, 506, 1896 (*bathybius*).

This genus is a deep sea representative of *Microstomus*, from which it differs in the increased number of vertebræ (63 instead of 48 to 52). Its fin rays are correspondingly increased, the body is deeper than in *Microstomus*, and it has teeth on both sides of the jaws, as in *Glyptocephalus*. (ἐν, in; βάσσος, for βαθύς deep; ἰχθύς, fish; a fish in the depths.)

3026. EMBASSICHTHYS BATHYBIUS (Gilbert).

Head 4 to $4\frac{2}{5}$ in length; depth 2 to $2\frac{1}{3}$. D. 111 to 117; A. 96 to 98; vertebræ 14 + 49 = 63. Body oval, very thin and deep, the greatest depth at anterior third of body; upper profile very abruptly angulated opposite hinder margin of upper pupil, the anterior half of head conspicuously protruding beyond general outline. Caudal nearly sessile, the peduncle very short. Mouth small, maxillary about $\frac{1}{5}$ length of head in specimens 1 foot long. Teeth broad incisors, slightly notched at tip, nearly equally developed on blind and colored sides, 21 on blind side of lower jaw, 16 on colored side. As in other members of this group, the lower jaw is the longer, the upper teeth included. Interorbital space wholly scaled, with a very high, rather sharp ∿-shaped ridge. Eyes very large, the upper entering largely into the upper profile, the lower much in advance; front margin of upper orbit on vertical of front of lower pupil; diameter of upper eye 2¾ to 2⅓ in head. Anterior nostrils of both sides in rather long tubes, the posterior margins produced to form short flaps. Preopercular margins adnate, as usual, concealed by scales. No conspicuous mucous excavations on blind side. Gill rakers weak and rather short, 10 or 11 on anterior of arch. Scales very small, cycloid, in about 165 cross rows, the tubes of lateral line much fewer, not regularly arranged; over 50 longitudinal rows above lateral line. Dorsal beginning over posterior edge of pupil; fins low, the highest dorsal rays behind middle of body, ⅔ length of head; caudal rounded, 1¼ in head; pectorals 2 in head; ventrals small, each with 5 rays, as in *Microstomus pacificus*. (*Glyptocephalus cynoglossus* and *zachirus* have 6 rays in each ventral.) Color of eyed side warm brown, darker toward margins, becoming black on vertical fins; everywhere on body and fins coarsely blotched with light blue, the marks so arranged on upper and lower thirds of sides as to form 5 broad bars of bluish, alternating with those of the ground color, and corresponding above and below; lips and branchiostegal membranes black; blind side dusky brownish. This well-marked species differs from the species of *Microstomus* in its much greater depth and bright coloration, and in having teeth well developed on both sides of jaws, as in the species of *Glyptocephalus*. Two specimens from the Santa Barbara Channel, in deep water. (Gilbert.) (βαθύς, deep; βίος, life.)

Cynicoglossus bathybius, GILBERT, Proc. U. S. Nat. Mus. 1890, 123, Santa Barbara Channel, at Albatross Station 2980, Lat. 33° 49′ 45″ N., Long. 119° 24′ 30″ W., in 603 fathoms. (Type in U. S. N. M. Coll. Gilbert.)

1042. GLYPTOCEPHALUS, Gottsche.

(FLUKES.)

Glyptocephalus, GOTTSCHE, Archiv fur Naturgsch. 1835, 156 (type *saxicola=cynoglossus*, L.).

Eyes and color on the right side. Body extremely elongate, more than twice as long as deep, much compressed. Head very small and short, its blind side with many excavations and mucous cavities in the skull, mandible, and preopercle. Mouth very small; teeth moderate, incisor-like, broad, equal, close set, in a single series; no teeth on vomer or palatines. Gill rakers short, weak. Lower pharyngeals narrow, with 1 or 2 rows of conical teeth. Lateral line nearly straight, simple; scales very small, smooth; dorsal and anal very long, there being more than 90 rays in the dorsal and more than 80 in the anal; caudal fin rounded; anal spine present; ventral rays 6. Vertebræ in increased number, 58 to 65. Northern seas, in deep water. This genus is one of the most strongly marked in the family, being distinguished from most of the genera by the greatly increased number of vertebræ, and from all of them by the remarkable cavernous structure of the bones of the head. Two species known. ($\gamma\lambda\upsilon\pi\tau\delta\varsigma$, sculptured; $\varkappa\varepsilon\phi\alpha\lambda\dot{\eta}$, head.)

> *a.* Pectoral fins very short, not falcate, that of right side about $\frac{1}{4}$ length of head; vertebræ 58.   CYNOGLOSSUS, 3027.
> *aa.* Pectoral fins of colored side falcate, longer than the head; vertebræ 65.
>   ZACHIRUS, 3028.

3027. GLYPTOCEPHALUS CYNOGLOSSUS (Linnæus),

(CRAIG FLUKE; POLE FLOUNDER.)

Head 5 to $5\frac{1}{4}$ in body; depth $2\frac{1}{2}$ to 3. D. 101 to 112; A. 87 to 100; scales 125; V. 6; highest dorsal and anal rays 2 in head; pectoral a little more than 2; vertebræ 58. Body oblong, fusiform; head small, ovate; the profile slightly decurved; mouth very small, with the cleft oblique; teeth on blind side close set, with incisoral edges, $\frac{11}{8}$; on the eyed side, distant, obtusely conic, $\frac{4}{7}$; eyes moderate, the lower advanced, close together, 3 in head; scales regularly imbricated, lateral line straight; pectoral short, falcate; origin of dorsal above middle of upper eye; anal spine present; caudal convex or angulated behind; pectoral fins very short, not falcate, that of right side about $\frac{1}{4}$ length of head; upper jaw with about 30 teeth; opercle adnate to the shoulder girdle for a short distance only. Color grayish brown; fins with dark spots; tip of pectoral dusky above. North Atlantic, on both coasts, chiefly in deep water, south to Cape Cod and France. This species is found in rather deep water on sandy bottoms. It reaches a length of 12 to 18 inches. This flounder has been taken in great numbers with the beam trawl in deep water off our New England coast. It is pronounced by the United States Fish Commission to be not inferior as a food-fish to the European sole. (Eu.) (*cynoglossus*, a sole; $\varkappa\dot{\upsilon}\omega\nu$, dog; $\gamma\lambda\tilde{\omega}\delta\delta\alpha$, tongue.)

Pleuronectes, etc., *Corpore oblongo glabro*, GRONOW, Museum Ichthyol., 1, IV, 39, etc., Belgium.

Pleuronectes cynoglossus, LINNÆUS, Syst. Nat., Ed. X, 269, 1758, after GRONOW; GÜNTHER, Cat., IV, 449, 1862; DAY, Fishes Great Britain, II, 30, pl. 103; LILLJEBORG, Sveriges och Norges Fiske, II, 386, 1891; and of European writers generally.

Platessa pola, CUVIER, Règne Animal, Ed. II, Vol. 2, 339, 1829, after la Pole of Duhamel.

Pleuronectes saxicola, FABER, Tidsskr. f. Naturv., 5 B., 244, 1828, Denmark.

Pleuronectes nigromanus, NILSSON, Prodr. Ichth. Scand., 55, 1832.

Platessa elongata, YARRELL, Hist. Brit. Fish., 619, 1859, young.

Glyptocephalus acadianus, GILL, Proc. Ac. Nat. Sci. Phila. 1873, 360, Nova Scotia. (Type, No. 12685.)

Glyptocephalus cynoglossus, GILL, Proc. Ac. Nat. Sci. Phila. 1873, 360; GOODE & BEAN, Proc. U. S. Nat. Mus. 1878, 21; GOODE, Proc. U. S. Nat. Mus. 1880, 337; GOODE, Proc. U. S. Nat. Mus. 1880, 475; COLLETT, Norske Nord-Havs Exped. 1880, 150; GOODE & BEAN, Bull. Mus. Comp. Zool., X, No. 5, 195, 1883; JORDAN & GILBERT, Synopsis, 838, 1883; GOODE, Nat. Hist. Aquat. Anim., 198, 1884; JORDAN & GOSS, Review Flounders and Soles, 300, pl. 19, 1889.

Solea cynoglossa, RAFINESQUE, Indice di Ittiologia Siciliana, 53, 1810; based on the Sole or *Cynoglossum* of RONDELET.

Glyptocephalus saxicola, GOTTSCHE, Archiv fur Naturgsch. 1835, 156.

Platessa saxicola, KRÖYER, Danmark's Fiske, 338, 1843.

Pleuronectes elongatus, GÜNTHER, Cat., IV, 450, 1862.

Gyptocephalus elongatus, GILL, Proc. Ac. Nat. Sci. Phila. 1873, 362.

3028. GLYPTOCEPHALUS ZACHIRUS, Lockington.

(LONG-FINNED SOLE.)

Head 5¼ to 5⅜; greatest width of body 3¼ to 3½; eye 3⅛ in head; snout 8. D. 94 to 106; A. 79 to 89; P. 11 to 13; V. 6; vertebræ 13 + 52 = 65. Body elongate-ovate, anterior portion of the oval shorter than posterior; snout declivous, almost vertical, its tip level with upper margin of lower eye, its curve uniting without sensible depression with that of nape; dorsal outline rising with a regular gentle curve from snout to about twenty-second dorsal ray, thence declining very gradually and regularly with but slight curvature to caudal peduncle; abdominal outline almost straight from knob of mandible to ventral; from thence to end of anal curved in same manner as dorsal outline; peduncle of tail expanded toward caudal, its least width about ¼ of greatest depth of body; greatest distance from anal to lateral line less than length of head. Eyes large, elliptical, the lower in advance of the upper about ½ length of pupil, and scarcely reaching dorsal profile anteriorly. Interocular space very narrow, about ⅕ of longitudinal diameter of eye, smooth; not raised above the eye in a fresh fish; a slight ridge rising at its posterior part, forming lower posterior margin of upper eye, and dying out on cheek. Nostrils of right side level with upper margin of lower eye; anterior nostril with a short tube, the posterior with a raised margin, and vertical with the front margin of the lower orbit; posterior nostril of blind side in advance of eye; anterior nostril nearly as on colored side; nostrils small and inconspicuous. Gape of mouth very small on colored side, considerably larger on blind side; on the colored side the cleft is nearer vertical than horizontal; posterior end of maxillary reaching very little behind anterior margin of orbit of lower eye, and the symphysis of intermaxillaries about level with upper edge of orbit; mandible projecting in the closed mouth, short, not passing a vertical from front margin of pupil, with a prominent knob below the symphysis, and a smaller one at its posterior extremity. Teeth on both sides of jaws throughout the full length of the gape, in a single row, broad, but thick, forming a blunt, continuous edge, about 34 in lower jaw and rather fewer in the upper, in an individual 11₁³₆ inches long; in an example 14⅔ inches long there were 14 teeth on the colored side and 26 on the blind side of the mandible, the latter the larger; in the intermaxillaries, 13 on the colored side and 23 on the blind side; each lower pharyngeal with a double row of teeth, the inner larger than the outer; the 4 anterior teeth of outer row conspicuously larger than those following; about 12 teeth in each inner row; upper pharyngeals each with a close-set row of 6 or 7 blunt conical teeth. Branchiostegals 7; gill rakers few, flexible, very short. Dorsal commencing between front of orbit and pupil, considerably behind nostrils, long and low, forming a continuous arch of slightly greater curvature than dorsal outline, the longest rays in central portion, and ending opposite anal at about ⅔ of width of caudal peduncle from origin of caudal; anal with a horizontal spine, the first ray rather distant from the visible portion of the spine, and nearly length of ventral behind pectoral base, similar to the dorsal; almost all the rays of dorsal and anal directly backward; caudal convex on posterior margin, rather narrow, the rays once bifurcate, sometimes bifurcate again near the tips; pectoral of colored side exceedingly long and lanceolate, about $\frac{1}{4}$ of total length of fish.; first 5 rays simple, the others once bifurcate; fourth ray longest, fifth nearly equal, sixth a little longer than third, thence diminishing rapidly. Usual proportion of the first 4 rays 3-8-10-12; pectoral of blind side lanceolate, rather more than $\frac{1}{4}$ of length of that of colored side, and formed of the same number of rays, first 4 simple, the others once forked; fourth and fifth rays longest; ventrals inserted so that their hinder axil is vertical with, or a little posterior to, anterior axil of pectoral, their tips reaching to first anal ray; 4 posterior rays once bifurcate. Lateral line almost straight, rising very slightly anteriorly, formed of a double row of tubes, about 138 in number, excluding those upon caudal; a row of similar pores commencing at ridge under upper eye, and continuing around lower eye almost to its front margin; scales small, smooth, uniform over the body, and extending over the head to snout, on which they are smaller; intermaxillaries and mandible scaleless; scales on blind side similar; caudal scaly on both sides; no scales on the other fins. Color uniform brownish or cinereous; fins darker; the color formed by minute dark spots on the scales; membrane between the fin rays closely set with dark points; blind side whitish, the ground tint clouded with numerous black points. Deep waters of the Northern Pacific, from San Francisco northward; found throughout Bering Sea in 35 to 350 fathoms. This species is a thin, dry flounder, reaching a length of something over a foot. It is taken in the sweep nets in deep water about San Francisco. It is readily known by its long pectoral fin. ($\zeta\alpha$-, an intensive particle; $\chi\epsilon\acute{\iota}\rho$, hand, from the long pectoral.)

Glyptocephalus zachirus, LOCKINGTON, Proc. U. S. Nat. Mus. 1879, 88, San Francisco; LOCKINGTON, Rep. Com. Fisheries California 1878-79, 42; JORDAN & GILBERT, Proc. U. S. Nat. Mus. 1880, 453; JORDAN & GILBERT, Proc. U. S. Nat. Mus. 1881, 68; JORDAN & GILBERT, Synopsis, 838, 1883; JORDAN, Nat. Hist. Aquat. Anim., 188, 1884; JORDAN & GOSS, Review Flounders and Soles, 301, 1889; GILBERT, Rept. U. S. Fish. Comm. 1893 (1896), 460.

1043. LOPHOPSETTA, Gill.

(WINDOW PANES.)

Lophopsetta, GILL, Proc. Ac. Nat. Sci. Phila. 1862, 216 (*maculatus*).

Eyes and color on the left side. Body broadly ovate, strongly compressed, pellucid; mouth large, oblique, the maxillary reaching to beyond eye; teeth subequal, in narrow bands, or in single series; a small patch of teeth on the vomer. Scales small, cycloid, imbricate, the skin without bony tubercles. Lateral line strongly arched in front, without accessory branch. Dorsal fin beginning on the snout, its anterior rays exserted; anal fin not preceded by a spine; ventral of left side free from the anal, inserted nearly on the ridge of the abdomen, its base broad, the rays well separated; pectoral and ventral fins moderate. One species. Very close to the European genus *Bothus*, Rafinesque (= *Scophthalmus*, Rafinesque, = *Rhombus*, Cuvier, = *Passer*, Valanciennes), from which it differs in the more numerous gill rakers, pellucid body, and produced dorsal rays, all characters of minor importance. The European Turbot (*Psetta*, Swainson), is also closely related, but the typical species, *Psetta maxima*, is a large robust fish, scaleless and beset with bony tubercles. (λόφος, crest; ψῆττα, turbot.)

3029. LOPHOPSETTA MACULATA (Mitchill).

(WINDOW PANE.)

Head 3½; depth 1⅔. D. 65; A, 52; scales 85; eye 4 in head; pectoral 1¼; highest dorsal rays 1⅞; highest anal rays 1⅘; interorbital space ⅓ eye. Body broadly rhomboid, strongly compressed, translucent in life; mouth large, the maxillary reaching nearly to posterior margin of eye, maxillary of eyed side with a bony tubercle on its anterior end; jaws subequal, the lower with a sharp knob at symphysis; teeth in each jaw in 1 series laterally, in a very narrow band in front; interorbital space rather broad, slightly concave, its posterior third or fourth with scales; gill rakers short and slender, about $8 + 25$; maxillary, mandibles, snout, and the greater part of interorbital naked; scales on head and body cycloid, loosely imbricated, those on the blind side a little smaller. Anterior rays of dorsal produced, their ends branched and free, the first on tip of snout, the rays at the beginning of posterior third of fin the highest; origin of anal directly under angle of preopercle; base of ventrals long, that of the eyed side extending along ridge of body from notch in isthmus to front of anal, base of ventral on blind side shorter; pectoral reaching past curve on eyed side, its mate much smaller; caudal rather long. Color light olive brown, almost translucent, everywhere marbled with paler, and with many small, irregular, sharply defined black spots; dorsal, anal, and caudal with larger, round, blended spots of dark brown; pectoral with brown, interrupted cross lines. This small flounder much resembles the European Brill (*Bothus rhombus*), but is smaller, thinner, and more translucent in body. Its weight rarely exceeds a pound or two, and its value as a food-fish is but slight; nevertheless, it is a near ally of the European Turbot (*Psetta maxima*), and in its technical characters it very closely agrees with the latter species. Atlantic coast of United States, from Casco Bay to South Carolina; common. (*maculatus*, spotted.)

Pleuronectes maculatus, MITCHILL, Rept. in part, Fish. N. Y., 9, 1814, New York; DE KAY, New York Fauna: Fishes, 301, pl. 47, fig. 151, 1842; STORER, Synopsis, 479, 1846; STORER, Hist. Fish. Mass., 204, 1867; JORDAN & GOSS, Review of Flounders and Soles, 258, 1889.

Pleuronectes aquosus, MITCHILL, Trans. Lit. and Phil. Soc. N. Y., I, 1815, 389, pl. 2, fig. 3, New York.

Lophopsetta maculata, GILL, Proc. Ac. Nat. Sci. Phila. 1862, 216; *ibid*, 1864, 220; JORDAN & GILBERT, Proc. U. S. Nat. Mus. 1878, 371.

Bothus maculatus, JORDAN & GILBERT, Synopsis, 815, 1883.

Rhombus aquosus, GÜNTHER, Cat., IV, 411, 1862.

1044. PLATOPHRYS, Swainson.

Solea, RAFINESQUE, Indice di Ittiologia Siciliana, 52, 1810 (*rhomboide*); not of QUENSEL, 1806.

Platophrys, SWAINSON, Nat. Hist. Class'n Fishes, II, 302, 1839 (*ocellatus*).

Peloria, COCCO, Intorno ad Alcuni Pesci del mar di Messina, Giorn. del Gabin., 1844, 21–30, Lettere di Messina (*heckeli*, a larval form of *P. podas*); not *Pelorus* of MONTFORT, 1808.

? *Coccolus,** BONAPARTE, in COCCO, Alcuni Pesci Messina, 21, 1844 (*annectens;* larval form— probably of *P. podas,* with the right eye in transit to the left side).

Bothus, BONAPARTE, Catologo Metodico, 49, 1846 (*podas*); not of RAFINESQUE.

Rhomboidichthys, BLEEKER, Act. Soc. Sci. Indo-Nederl. Manad. and Makassar, 67, 1357-58 (*myriaster*).

Platophrys, BLEEKER, Comptes Rendus Acad. Sci. Amsterd., XIII, 1862, Pleuron., 5 (*ocellatus*).

Eyes and color on the left side. Body ovate, strongly compressed; mouth of the large type, but comparatively small; the maxillary ⅓ or less of the length of the head; teeth small, subequal, in 1 or 2 series; no teeth on vomer or palatines. Interorbital space broad and concave, broadest in adult males. Gill rakers moderate. Dorsal fin beginning in front of eye, all its rays simple; ventral of colored side on ridge of abdomen; caudal convex behind; pectoral of left side usually with 1 or more filamentous rays, longest in the male. Scales very small, ctenoid, adherent; lateral line with a strong arch in front. Coloration usually variegated.

This well-marked genus is widely diffused in the warm seas. The sexual differences are greater than usual among flounders, and the different sexes have often been taken for different species. As a rule, in the males, the pectoral fin on the left side is much prolonged, the interorbital area is much widened and very concave, and there are some tubercles about the snout and lower eye. The young fishes, as is usually the case, resemble the adult females. Lately, Dr. Emery has shown that the larval flounder, known as *Peloria heckeli,* is in all probability the young of *Pleuronectes podas.* The generic name, *Coccolus,* based on forms slightly more mature than those called *Peloria,* probably belongs here also. We have seen no larval forms so young as those which have been described as *Peloria heckeli.* We have, however, examined small transparent flounders, one with the eyes quite symmetrical, taken in the Gulf Stream, and another with the eyes on the left side, taken at Key West. Both these may be larvæ of *Platophrys ocellatus.* The figures published by Emery seem to make it almost certain that the corresponding European forms belong to *P. podas,* although some doubt as to this is expressed by Facciolà. The species of *Platophrys* are widely distributed through the warm seas, no tropical waters being wholly without them. All the species of *Platophrys* are extremely closely related and can be distinguished with difficulty. On the other hand, the variations due to differences of age and sex are greater than in any other of our genera. The following analysis of the species of *Platophrys* is very unsatisfactory. There are certainly 3 species (*podas,* the European species, *maculifer,* and *lunatus*) which are known to be distinct in their adult state. The young forms of *maculifer* and *lunatus* are not well known, nor is it known how they differ from *ocellatus,* *spinosus,* and other species which presumably reach a smaller size. Only a thorough study of the species, in all stages of development in their native waters, can give us the characters by which the species can be really discriminated. (πλατύς, broad; ὀφρύς, eyebrow.)

* "Parvus mole et pleuronectiformis, medius inter Pleuronectidas et Bibroniinos hic piscis videtur! Attamen dum illi oculos unilaterales habeant, iste vero bilaterales; in hoc novo genere oculi, alter a latere, altere in vertice vix ad appositum latus convenus positi sunt." (Bonaparte: quoted by Facciolà, Su di Alcuni Rari Pleuronettidi.)

a. Anal rays, at least anteriorly, each with a spinule at base (these formeᴅ by a slight widening of the tip of the interhæmal spines, each being covered by a little rough scale); front of dorsal with similar projections.

   b. Color brown, with pale rounded spots; fins dotted with brown; a faint dark spot at first ⅓ of lateral line; snout with horny points; mouth small, the maxillary reaching front of eye.   SPINOSUS, 3030

aa. Anal rays without spinules at their base.

   c. Anterior profile of head convex before the interorbital area, the very short snout scarcely forming a reentrant angle at its base; form elliptic-ovate, the outlines more regular than in *P. lunatus.*

      d. Dorsal rays 85 to 95.

         e. Scales not very small, about 75 pores in lateral line; no blue markings, at least in the young.

            f. Mouth small, the maxillary 3 in head; no spines about the snout; eye 3½ in length; interorbital width 3 in head (in the type); pectoral short; curve of lateral line 6 times in straight part. Color dark brown, with numerous stellate white spots, the most distinct of them with darker edgings; these generally scattered over the body; but some of them on sides of body gathered together in little rings; these spots blue rather than white in life.   CONSTELLATUS, 3031.

            ff. Mouth smaller, the maxillary 3¾ in head. Color light grayish, tinged with reddish, with small round spots of darker gray, and with lighter rings inclosing spaces of ground color.
               OCELLATUS, 3032.

         ee. Scales smaller, 90 to 95 pores in lateral line. Color of adult, reddish gray, the body everywhere covered with rings formed of round, sky-blue spots, which are not confluent and not edged with black; besides these, very few detached spots or other blue markings.
            MACULIFER, 3033.

      dd. Dorsal rays 105; anal rays 80; pectoral short; interorbital space 2¾ in head; depth 1¾ in length; scales 91; body deep. Color (specimen 4¾ inches long) grayish, much spotted and mottled with whitish, no blue in young example.   ELLIPTICUS, 3034.

   cc. Anterior profile of head strongly concave before interorbital area, the projecting snout leaving a marked reentrant angle above it.

      g. Mouth not very small; maxillary 3 in head. Color dark olive, with many rings, curved spots, and small round dots of sky blue edged with darker on body, these largest near middle of sides, where some are as large as eye; 3 obscure dark blotches on straight part of lateral line.   LUNATUS, 3035.

      gg. Mouth small; maxillary 3½ in head. Color highly variegated with different shades of gray, the pale blotches rounded, very irregular in size and position; no blue spots.   LEOPARDINUS, 3036.

3030. PLATOPHRYS SPINOSUS (Poey).

Depth 1½. D. about 74; A. about 57; scales about 80. Anal rays, at least anteriorly, each with a spinule at base, these formed by a slight widening of the tips of the interhæmal spines, each being covered by a little rough scale; front of dorsal with similar projections. Snout with horny points; mouth small, the maxillary reaching front of eye. Eyes very wide apart, 2⅔ in head, the interorbital space 1⅔ in head; pectoral fin short; curve of lateral line 5 in straight part. Color brown, covered with pale rounded spots; fins dotted with brown; a faint dark spot at first third of lateral line. Described from specimens from Cuba, probably the types, 4¼ inches long, which have been partly dried before being placed in alcohol. Cuba. The original description of this species is a very scanty one. In all respects, unless it be the color, it agrees with the European species, *Platophrys podas.* We have found 2 small specimens sent by Professor Poey to the Museum of Comparative Zoology, which may be the types of this species. They are 4¼ inches long, and have been partly dried in the sun. A result of this has been to increase the prominence of the interhæmal spines. Whether these be the original types or not, the species is an extremely doubtful one. The eyes are farther apart in these specimens than in any of *Platophrys ocellatus,* which we have examined. They agree in this respect with Agassiz's figure of *Rhombus ocellatus.* (*spinosus,* spinous.)

Rhomboidichthys spinosus, POEY, Synopsis, 409, 1868, Cuba; POEY, Enumeratio, 139, 1875.
Platophrys spinosus, JORDAN & GOSS, Review Flounders and Soles, 266, 1889.

3031. PLATOPHRYS CONSTELLATUS, Jordan.

Head 4; depth 1½; eye 3½ in head; interorbital width 3. D. 89; A, 65; scales 75. Body elliptic-ovate, the outlines more regular than in *P. lunatus;* anterior profile of head convex before the interorbital area, the very short snout scarcely forming a reentrant angle at its base; anal rays without spinules at their base; mouth small, the maxillary 3 in head; no spines about the snout; pectoral short; curve of lateral line 6 times in straight part. Color dark brown, with numerous stellate white spots, the most distinct of them with darker edgings; these generally scattered over the body, but some of them on sides of body are gathered together in little rings (perhaps these spots are blue rather than white in life); fins mottled with dark brown, the pectoral finely barred. Specimens examined 3½ inches long. Galapagos Archipelago. Originally described from 3 specimens, the largest 3½ inches long, numbered 11146 on the register of the Museum of Comparative Zoology. They are from James Island, in the Galapagos. The species is closely related to *P. ocellatus* and others, but in color, at least, it is different, and its habitat is remote; locally common. (*constellatus,* with star-like spots.)

Platophrys constellatus, JORDAN, in JORDAN & GOSS, Review Flounders and Soles, 266, 1889, James Island, Galapagos Archipelago. (Types in M.C. Z.)

3032. PLATOPHRYS OCELLATUS (Agassiz).

Head 4 in length; depth 1½; eye (lower) 3⅔ in head; snout 5. D. 85; A. 64; scales 75 (pores); vertebræ 37. Body ovate, deep anteriorly, the profile descending steeply, rendered abruptly concave in front of interorbital space by the conspicuously projecting short snout. Mouth very small and oblique, the maxillary reaching vertical from front of lower eye, 3¾ in head; tip of lower jaw entering the profile. Teeth fine, conical, in 2 series in the upper jaw, 1 in the lower, those of the outer row in upper jaw larger and more widely separated than those of the inner series. Snout very short, equaling interorbital width. Interorbital space narrow, deeply concave, closely scaled. Eyes large, the lower in advance of upper. Gill rakers obsolete, 7 rudiments on horizontal branch of anterior arch. Scales moderate, not extending on the fins, those on colored side ctenoid, those on blind side smooth; arch of lateral line short and high, its base contained 4½ to 5 times in the straight portion. Dorsal fin beginning opposite anterior nostril, the rays nearly uniform in length, the longest about ⅓ head; pectoral of colored side 4¾ in length; ventral of colored side beginning under middle of lower eye, with 6 rays; the right ventral with 5 rays. Color in life, light grayish with reddish tinge, covered with small round spots of darker gray and with lighter rings inclosing spaces of the ground color; vertical fins similarly colored, with a small black spot near base of each ninth or tenth ray; 2 black spots on median line of body divide the length into nearly equal thirds; some other small black spots scattered over colored side. Western Atlantic, from Long Island to Rio Janeiro, on sandy shores. Here described from Key West specimens, types of *P. nebularis.* This species is very common at Key West in clear, shallow water on sandy bottom. The largest of the numerous specimens taken is 3 inches in length. A specimen similar to these has been taken by Dr. Bean on the south coast of Long Island. This seems to be the same as the Cuban species called *Rhomboidichthys ocellatus* by Poey, and some of the specimens sent by Poey to the Museum of Comparative Zoology are apparently identical with the types of *P. nebularis.* In the Museum of Comparative Zoology we have compared specimens of the real *Platophrys ocellatus* (No. 11423, Rio Janeiro, Agassiz) with a representative specimen of *P. nebularis* (No. 26147, from the Tortugas, Florida), and are unable to find any differences. We adopt, therefore, the name *Platophrys ocellatus* for all, and regard it as one of the widely distributed flounders, like *Etropus crossotus* and *Citharichthys spilopterus.* (*ocellatus,* with eye-like spots.)

Rhombus ocellatus, AGASSIZ, Spix, Pisc. Brasil., 85, pl. 46, 1829, Brazil.
Rhombus bahianus, CASTELNAU, Anim. nouv. rares Amérique du Sud, 1855, Bahia.
Platophrys nebularis, JORDAN & GILBERT, Proc. U. S. Nat. Mus. 1884, 31, 143, Key West (Type, 34972. Coll. Dr. Jordan) ; GOODE & BEAN, Oceanic Ichthyology, 441, 1896.
Platophrys ocellatus, SWAINSON, Nat. Hist. Class'n Fishes, II, 302, 1839; JORDAN & GOSS, Review Flounders and Soles, 266, 1889.
Rhomboidichthys ocellatus, GÜNTHER, Cat., IV, 433, 1862; POEY, Synopsis, 408, 1868.

3033. PLATOPHRYS MACULIFER (Poey).

Head 4; depth 1⅝. D. 90 to 95; A. 70; scales 90 to 95. Body elliptical, ovate. Mouth small, oblique, the maxillary 3⅔ in head; teeth in each jaw in 2 irregular series; filamentous rays of pectorals reaching very nearly to last rays of dorsal; arch of lateral line short and high, its length 1⅘ times its height and 2⅔ in head; snout very short, 4 in head; interorbital area 3¾ in head. Color of adult reddish gray, the body everywhere covered with rings formed of round, sky-blue spots, which are not confluent and are not edged with black; besides these, very few detached spots or other blue markings; head with similar blue spots, but no rings; area inclosed in the blue rings not different from the ground color; caudal with blue spots, other fins with none; dorsal and anal mottled; a large, diffuse, dusky spot at front of straight part of lateral line; 1 better defined on middle of lateral line; a faint one farther back; pectorals grayish, with dark bars. Cuba. We identify specimens taken by Dr.

Jordan at Havana with this species. In the Museum of Comparative Zoology are other specimens similar to these, sent to Cambridge by Poey. In several respects these specimens agree fairly with Poey's *P. ellipticus,* but that species is said to have 104 dorsal rays. (*macula,* spot; *fero,* I bear.)

? Pleuronectes maculiferus, POEY, Memorias, II, 316, 1860, Cienfuegos. (Coll. Poey.)
? Rhomboidichthys maculiferus, POEY, Synopsis, 408, 1868; POEY, Enumeratio, 139, 1875.
Platophrys maculifer, JORDAN & GOSS, Review Flounders and Soles, 267, 1889.
Platophrys ellipticus, JORDAN, Proc. U. S. Nat. Mus. 1886, 51; not of POEY.

3034. PLATOPHRYS ELLIPTICUS (Poey).

Depth 1¾. D. 105; A. 80; scales 91. Body elliptical, ovate; anterior profile of head convex before the interorbital area; pectoral short; interorbital space 2¾ in head; body deep. Color (specimen 4¾ inches long) grayish, much spotted and mottled with whitish; no blue (in young example). Cuba. Poey describes his *P. ellipticus* as having 104 dorsal rays. In none of our other species does the number of these rays reach 100. Among the specimens sent by Poey to the museum at Cambridge is 1, described above, 4¾ inches long, which has 105 dorsal rays. We have therefore assumed that the species to which this specimen belongs is the real *P. ellipticus,* and that the one heretofore called *P. ellipticus* is Poey's *P. maculifer.* Both these assumptions are open to considerable doubt. (*ellipticus,* elliptical.)

? Pleuronectes ellipticus, POEY, Memorias, II, 315, 1860, Cuba. (Coll. Poey.)
? Romboidichthys ellipticus, GÜNTHER, Cat., IV, 434, 1862; POEY, Synopsis, 408, 1868; POEY, Enumeratio, 139, 1875.
Platophrys ellipticus, JORDAN & GOSS, Review Flounders and Soles, 267, 1889.

3035. PLATOPHRYS LUNATUS (Linnæus).

(PEACOCK FLOUNDER.)

Head 3⅔ in length; depth 2. D. 93; A. 70; scales 90; lower eye 6 in head; maxillary 2⅔; interorbital 2⅔; highest dorsal rays 2⅔; highest anal rays 2½; caudal 1½; base of ventral of eyed side 3¼. Vertebræ 9 + 30 = 39. Body elliptical, ovate, strongly compressed; anterior profile concave, the snout projecting, leaving a reentrant angle above it; mouth moderate, the maxillary reaching to middle of pupil of lower eye; jaws subequal, the lower with a well-developed knob at symphysis, teeth small, in an irregular double series in each jaw; anterior end of maxillary with a large blunt spine, pointing outward and forward, a smaller one behind it on upper edge of maxillary, pointing upward and backward; interorbital very wide and deeply concave; orbital rim, below on upper orbit, above on lower, broken up into blunt papillæ; gill rakers short and thick, 9 developed on lower part of arch, none on upper. Anterior part of interorbital, snout, maxillary, and mandible, naked; scales all cycloid; the rays of dorsal and anal with scales, a few on ventral of eyed side; arch of lateral line 5 in straight part. Pectoral of eyed side filamentous, reaching to base of caudal, its mate of opposite side shorter, about 1¾ in head; origin of dorsal over snout; ventral of eyed side with a long base, extending from

3030——90

angle at isthmus, along ridge of body; slightly past front of anal; base of ventral of blind side ½ the length of that of its mate; caudal with the middle rays produced, double convex. Color dark olive, with many rings, curved spots, and small round dots of sky blue edged with darker on body, these largest near middle of sides, where some are as large as the eye; 3 obscure dark blotches on straight part of lateral line; head and vertical fins with sharply defined blue spots, which are mostly round; spots on opercles larger and curved; pectorals with dark bars. West Indies, north to Florida; common. Here described from a specimen from Green Turtle Cay, Florida, 14 inches in length. This handsome and curiously colored species is not rare in the waters of the West Indies. The specimens examined by us are from Cuba, Sombrero, St. Thomas, and other localities in the West Indies. The original figure of this species published by Catesby is a very good one and leaves no room for doubt as to the species intended. The figure of Bloch, called *Pleuronectes argus*, is also fairly accurate, and can refer to no other species. This species reaches a length of some 18 inches, and is the largest in size of the American species of *Platophrys*. We have never seen any young examples which certainly belong to it, and till its development is traced some of the species known from small examples only must be doubtful. (*lunatus*, crescent-shaped, from the spots.)

Solea lunata et punctata (the Sole), CATESBY, Nat. Hist. Carolina, tab. 27, 1725, Bahamas.
Pleuronectes lunatus, LINNÆUS, Syst. Nat., Ed. x, 269, 1758, Bahamas; based on CATESBY; and of the various copyists.
Pleuronectes argus, BLOCH, Ichthyol., tab. 48, 1783, Martinique; after Plumier.
? Pleuronectes surinamensis, BLOCH & SCHNEIDER, Syst. Ichth., 156, 1801, Surinam; "*satis parva et glabra;*" fins scaly; mouth small; lateral line arched in front; D. 96; A. 55.
Rhomboidichthys lunatus, GÜNTHER, Cat., IV, 433, 1862; POEY, Synopsis, 408, 1868.
Rhomboidichthys lunulatus, POEY, Enumeratio, 138, 1875.
Platophrys lunatus, JORDAN, Proc. U. S. Nat. Mus. 1886, 51; JORDAN & GOSS, Review Flounders and Soles, 267, 1889.

3036. PLATOPHRYS LEOPARDINUS (Günther).

Head 3¾ in length; depth 1¾; eye (lower) 3¼ in head. D. 86 to 88; A. 64 (62 to 66); scales about 80. Mouth very small, the maxillary 3¼ in head; teeth very small, biserial above. Interorbital space concave, rather broad, its width 3½ in head. Eyes large, the lower considerably before the upper. Lateral line with a short sharp curve anteriorly. Gill rakers very small. Anterior rays of dorsal not elevated; left pectoral not produced, little longer than right, 1¼ in head. Coloration highly variegated with different shades of gray, the pale blotches rounded, very irregular in size and position; no distinct black spots along the lateral line; a large whitish cloud between the eyes; blind side pale, scaled like the eyed side. Gulf of California. This species is known only from the original type from unknown locality, and from a single specimen, 2⅔ inches long, in the United States National Museum, taken by Mr. H. F. Emeric, at Guaymas, Sinaloa. From this the above description was taken. (*leopardinus*, leopard-like.)

Rhomboidichthys leopardinus, GÜNTHER, Cat. Fish., IV, 434, 1862, locality unknown.
Platophrys leopardinus, JORDAN, Proc. U. S. Nat. Mus. 1884, 260, specimen from Guaymas; JORDAN & GOSS, Review of Flounders and Soles, 268, 1889.

1045. PERISSIAS, Jordan & Evermann.

Perissias, JORDAN & EVERMANN, new genus (*tœneopterus*).

This genus differs from *Platophrys* in the larger scales, narrower interorbital, and especially in the greatly produced ribbon-like lobe at the front of the dorsal. From *Engyprosopon* it differs in the short thick gill rakers and in the produced dorsal rays. The lateral line is wanting on the blind side. Deep sea. (περισσός, strange.)

3037. PERISSIAS TÆNIOPTERUS (Gilbert).

Head 3⅔; depth 2¼. D. 86 to 88; A. 67 to 70; scales 60 to 65, the arch with 15 pores; 20 scales in a series running upward and backward from lateral line. Body elongate; caudal fin subsessile, the last anal and dorsal rays inserted near rudimentary caudal rays; height of caudal peduncle 4 in height of body; upper profile descending very obliquely anteriorly, a slight reentrant angle in front of lower eye; in males the profile slightly angulated in front of upper orbit, below which it ascends more steeply; lower eye much in advance of upper; in females eyes close together, the vertical from middle of lower eye passing through front of upper orbit; diameter of upper orbit 3½ in head; in males the lower eye may be entirely in advance of upper; in females 3 inches long, and in very young males the interorbital space is a narrow, concave, scaleless groove, less than diameter of pupil, running into a deep pit behind lower eye; in males 2 inches long the interorbital space has already widened, and in specimens 3½ inches long is as wide as longitudinal diameter of orbit; it is traversed by an oblique ridge running upward and backward from front of lower eye, separating the anterior scaleless portion from the deep scaly pit behind; supraorbital ridge of lower eye serrated, forming a strong series of spines, less marked in females; anterior rim of upper orbit similarly but less strongly marked; a strong double spine on maxillary in front of nostrils; a spine near end of maxillary in males; mouth small, maxillary not reaching front of pupil, equaling diameter of orbit; teeth small, in a single close-set series in each jaw, equally developed on both sides, with enlarged canines; gill rakers very short and weak, 8 on horizontal limb; anterior nostrils with very short flaps; dorsal beginning above front of lower eye; in all specimens, females as well as males, the first 2 rays detached from the rest of fin, the second ray produced into a flat, ribbon-shaped filament about as long as head; dorsal and anal rays all unbranched; median caudal rays forked; no anal spine; ventral of colored side on ridge of abdomen, the 2 anterior rays in males connected by membrane at base only, produced into flat filaments as long as head, extending far beyond front of anal; pectoral of left side well developed, but small, slightly more than ½ head; that of blind side little developed, about ⅓ diameter of orbit; scales of left side strongly ctenoid, absent on interorbital space, snout, maxillary, and mandible; lateral line with strong curve anteriorly, the cord of which is contained 5 times in straight portion. Along lateral line are occasional broad cutaneous flaps, colored blue in life; scales of blind side cycloid, the tubes of lateral line obsolete, the course of lateral line indistinctly indicated by pits at bases of scales and occasional pore-like markings; median rays of dorsal and anal on left side with series of ctenoid scales, otherwise scaleless; caudal rays with double series on both sides. Color on left side olive brown, with many small irregular spots of light gray, with darker border; 3 or 4 dark blotches along lateral line; along dorsal and ventral outlines about 5 pairs of light spots, broadly ocellated with blackish; males with a bright blue spot on anterior profile at base of each of first 10 or 12 dorsal rays and 1 on end of snout; blind side in males with a broad oblique bar covering about ¼ of sides, bluish black in life, dark brown in spirits; from its upper anterior part a number of narrow parallel streaks run forward toward head, much as in *Engyophrys sancti-laurentii;* filamentous rays of dorsal and ventral white; fins all speckled; a small black spot at base of median caudal rays. Differing from all known species of *Platophrys* in the ribbon-shaped prolongations of second dorsal ray and first and second ventral rays of eyed side, and in the obsolete lateral line of blind side. Several specimens from the Gulf of California and the western coast of Lower California, in 40 fathoms. (Gilbert.) ($\tau \alpha \iota \nu i \alpha$, ribbon; $\pi \tau \epsilon \rho \acute{o} \nu$, fin.)

Platophrys tæniopterus, GILBERT, Proc. U. S. Nat. Mus. 1890, 118, Gulf of California, north of La Paz, at Albatross Station 2998, Lat. 24° 51' N., Long. 110° 39' W., in 40 fathoms. (Type, No. 43095. Coll. Gilbert.)

1046. ENGYOPHRYS, Jordan & Bollman.

Engyophrys, JORDAN & BOLLMAN, Proc. U. S. Nat. Mus. 1889, 176, (*sancti-laurentii*).

This genus is allied to *Platophrys,* Swainson, but differs from it in having the interorbital space very narrow and armed with a spine, and the scales of moderate size and ctenoid. Gill rakers obsolete. No anal spine. Gill membranes entirely separate. It is still nearer the genus *Engyprosopon,* Günther, but in that group the interorbital space is broader and the gill rakers are developed and slender. ($\dot{\epsilon} \gamma \gamma \acute{v} \varsigma$, near together; $\dot{o} \phi \rho \acute{v} \varsigma$, eyebrow.)

3038. ENGYOPHRYS SANCTI-LAURENTII, Jordan & Bollman.

Head 2⅘ to 2⅞ (3 to 3½); depth 1⅞ to 2 (2 to 2½). D. 78 to 85; A. 68 to 72; scales 60 to 68, along lateral line. Body broadly ovate, much compressed, the greatest depth over pectorals; dorsal and ventral outlines equally curved; profile scarcely concave before eyes. Mouth very small, oblique, the maxillary reaching opposite pupil of lower eye, 4 to 4½ in head. Teeth present on blind side, well developed, close set and even, none on vomer. Snout short, 4½ to 5 in head. Interorbital space a very narrow, sharp, scaleless ridge, the ridge forking above pupil, leaving a very narrow concavity anteriorly; lower ridge armed with a strong spine, turned backward, inserted just above pupil of lower eye. Anterior orbital rim of upper eye rather high, entering profile. Eyes large, lower in advance of upper, 3¾ to 4 in head. Gill rakers almost obsolete, represented by 5 or 6 small fleshy papillæ. Scales moderately small, ctenoid, and not very firmly attached; small scales on rays of dorsal and anal fins; arch of lateral line short and small, but abrupt, 4 to 5 times in straight part. Dorsal beginning on blind side just behind posterior nostril and in front of eye; pectoral of colored side 2 in head, that of blind side 2⅓ in head; ventrals of colored side slightly longest, 3 in head; that of colored side with 6 rays, of blind side with 5 or 6 rays. Color of left or eyed side, blackish brown, with scattered white and black spots, the latter most prominent along base of dorsal and anal fin; 3 large, black, nonocellated blotches on straight part of lateral line, the first at beginning, second at middle, and third on peduncle; fins dusky; dorsal and anal with scattered white and black spots; caudal with 5 black spots arranged in a curved series; blind side with 5 or 6 curved parallel dusky bands as wide as eye, the first beginning on interopercle and curving across cheek to along base of dorsal; second beginning at throat and curving along posterior margin of preopercle, and extending on back, parallel with the first from vent; third curving around in front of pectorals, across posterior part of opercle, and extending to base of dorsal fin behind the middle; rest behind pectorals. All of these bands fade out behind middle of body, so that the posterior portion is immaculate. In young examples these bands are very faint or obsolete. Coast of Colombia, southwest of Panama. Numerous specimens, the largest about 4½ inches long, were dredged at *Albatross* Station 2795, at a depth of 33 fathoms, and at *Albatross* Station 2805 at a depth of 51½ fathoms. This peculiar species is distinguished from the species of *Platophrys* and *Engyprosopon* by its very narrow interorbital ridge, from the species of *Arnoglossus* by the form of the body, the short gill rakers, etc., and from all related species by the peculiar coloration of the blind side. (Named for St. Lawrence, in allusion to the gridiron-like markings of the blind side.)

Engyophrys sancti-laurentii, JORDAN & BOLLMAN, Proc. U. S. Nat. Mus. 1889, 176, Pacific Ocean, off coast of Colombia, at Albatross Station 2805, Lat. 7° 56′ N., Long. 79° 41′ 30″ W., and Station 2795, Lat. 7° 57′ N., Long. 78° 55′ W. (Type, No. 41155.)

1047. TRICHOPSETTA, Gill.

Trichopsetta, GILL, Proc. U. S. Nat. Mus. 1888, 603 (*ventralis*).

Body ovate, covered with rather large, ctenoid adherent scales; mouth moderate, the chin prominent; vomer toothless; teeth small, somewhat enlarged and hooked in front, uniserial; maxillaries obliquely truncated behind; interorbital area a narrow ridge, with a median groove in front; none of the dorsal rays produced; ventrals free from the anal; caudal fin subsessile; both pectoral fins present; right ventral much produced, the left on the ridge of the abdomen; lateral line with a strong arch in front. (θρίξ, hair; ψῆττα, turbot, from the prolonged ventral.)

3039. TRICHOPSETTA VENTRALIS (Goode & Bean).

Head 4 in body; depth 2⅓. D. 93; A. 73; pectoral 11 (eyed side), 7 or 8 (blind side); scales 19-66-23; eye 3⅔ in head; maxillary scarcely 2; interorbital very narrow, scaleless, its width 8 in eye; scales strongly ctenoid; dorsal beginning upon snout upon the blind side, in advance of eyes, its highest rays equaling length of mandible; origin of anal under base of pectoral, its longest ray equaling or slightly exceeding ½ the distance of its anterior ray from snout; caudal equal to length of head without snout; pectorals inserted considerably below origin of lateral line, close to gill opening, that of the eyed side 6 in length of body; that of the blind side almost as long as head. Color light brownish gray; a dark blotch as long as eye on the anterior rays of the anal, a few obscure on different parts, of lighter hue at the junction of the curved and straight portion of the lateral line. (Goode & Bean.) Deep waters of the Gulf of Mexico. (*ventralis*, pertaining to the ventrals.)

Citharichthys ventralis, GOODE & BEAN, Proc. U. S. Nat. Mus. 1885, 592, deep waters of Gulf of Mexico. (Coll. *Albatross*.)

Arnoglossus ? ventralis, JORDAN & GOSS, Review Flounders and Soles, 262, 1889.

Trichopsetta ventralis, GOODE & BEAN, Oceanic Ichthyology, 440, pl. 109, fig. 372, 1896.

1048. SYACIUM, Ranzani.

Syacium, RANZANI, Novis Speciebus Piscium, Diss. Sec., 20, 1840 (*micrurum*).

Hemirhombus, BLEEKER, Comptes Rendus Acad. Sci. Amsterd., XIII, Pleuron., 4, 1862 (*guineënsis*).

Aramaca, JORDAN & GOSS, in JORDAN, Cat. Fish. N. A., 133, 1885 (*pœtula*).

Body elliptic-ovate, much compressed; interorbital space broad in the males and more or less concave, narrowed in the female; mouth moderate, the gape curved; teeth in the upper jaw biserial, in the lower uniserial; the front teeth of the upper jaw enlarged; vomer toothless; scales rather large, ciliate; lateral line without arch in front; pectoral fins on both sides present; septum of gill cavity below gill arches without foramen; a deep emargination near the isthmus; gill rakers short and thick; dorsal low, its anterior rays not elevated; pectorals both present; caudal subsessile; no anal spine; pectorals produced in the males; ventral fins short, that of colored side on ridge of abdomen. This genus contains a considerable number of species, mostly American and African, which form a transition from *Platophrys* to *Citharichthys*. They fall readily into 2 groups distinguished by the width of the interorbital space. As this width is dependent on age, and as it is subject to various intergradations, the group *Aramaca* founded on it can not be admitted as a distinct genus. (ὀυάκιον, diminutive of ὀῦαξ, a kind of pulse, the application unexplained.)

a. Snout and orbits without spines or spinous processes.
    b. Scales rather large, 50 to 57 in the lateral line; interorbital space broad. Color nearly plain brown, with darker dots or mottlings, no ring-like spots or ocelli; fins mottled; left pectoral barred; blind side sometimes wholly or partly dusky, especially in northern specimens.   PAPILLOSUM, 3040.
    bb. Scales rather small, 58 to 70 in the lateral line.
        c. Scales 65 to 70. Color dark brown, with many rings and spots of light gray and blackish, some of the dark rings with a black central spot; a diffuse dusky blotch on lateral line above pectoral, and 1 near base of caudal peduncle; fins with numerous inky spots and dark markings; blind side pale.   MICRURUM, 3041.
        cc. Scales 58 to 60.

d. Interorbital space in male broader than eye. Color light brown, with grayish and light bluish dots, some darker areas, and a few round brown spots ocellated with lighter; interorbital space with a . vertical brown bar bordered by lighter; fins mottled and spotted.

LATIFRONS, 3042.

dd. Interorbital space not broader than pupil. Color light olive brown, nearly uniform, the vertical fins with elongate dark spots.

OVALE, 3043.

3040. SYACIUM PAPILLOSUM (Linnæus).

Head $3\frac{2}{3}$ in length; depth $2\frac{1}{4}$. D. 82; A. 63 to 70; scales 53; eye 5 in head; maxillary $2\frac{2}{3}$; pectoral of eyed side $1\frac{1}{2}$; caudal $1\frac{2}{3}$. Body elliptic-ovate, the anterior profile regularly decurved, forming an angle above the snout; mouth rather large, arched; maxillary extending to below middle of eye, its posterior end concave; teeth in upper jaw in 2 series, some of the outer forming small canines; lower teeth in 1 row; eye large, 4 in head; lower eye in advance of upper, especially in the adult; interorbital space broad, concave, greater than the long diameter of the eye in the males, about equal to the vertical diameter in the females; accessory scales very numerous; mandible, maxillary, and interorbital with scales; gill rakers short, scarcely as long as pupil, about $2+8$; dorsal rather low, beginning slightly in front of lower eye, the first 3 or 4 rays on blind side, the anterior rays produced beyond the membrane; ventrals with moderate base, that of eyed side on ridge of body, that of blind side slightly in advance of its mate; anal beginning a little in advance of pectoral; pectoral of eyed side pointed behind, the upper rays filamentous (at least in the male); caudal double truncate. Vertebræ $10+26=36$. Color nearly plain brown, with darker dots or mottlings, no ring-like spots or ocelli; fins mottled; left pectoral barred; blind side sometimes wholly or partly dusky, especially in northern specimens. Charleston to Rio Janeiro, in rather deep water. Here described from an adult specimen from Charleston, a foot in length. Of the species found in the deep waters about Pensacola, and called by Dr. Bean *Hemirhombus pœtulus*, we have numerous specimens. Lately we have received from Mr. Charles C. Leslie, of Charleston, a specimen which shows its presence also in Carolina waters. It has not yet been recorded from Cuba, but in the Museum of Comparative Zoology is a specimen (26104) taken by Mr. Samuel Garman, at Kingston, Saint Vincent. But its range extends much farther to the southward, for among the collections made by Professor Agassiz, at Rio Janeiro, there are many specimens (11375, 4666), the largest about a foot long. These seem to be completely identical with Florida examples, differing only in having the blind side pale, it being usually partly blackish in northern samples. These Brazilian specimens agree very closely with the figure of *Rhombus soleæformis*, except that Agassiz has represented that species as having a dusky blotch at the shoulder. No such marking is apparent in any of our specimens. The coloration and the breadth of the interorbital both render it unlikely that Agassiz's *soleæformis* could have been *micrurum*. The *Aramaca* of Marcgrave, which is the sole basis of *Pleuronectes papillosus. Pleuronectes macrolepidotus*, and *Rhombus aramaca*, can not well be any known species other than the present one. According to Marcgrave's rude figure and his description, this species has the form of a sole, the eyes wide apart, the left pectoral produced, the mouth very large, the body oblong, and the coloration stone-like (sand color) on the left side and white on the eyed side. *Syacium micrurum* is not colored in that way, and its eyes are not noticeably far apart. We therefore adopt for this species the oldest name, *Syacium papillosum.* (*papillosus,* having papillæ.)

Aramaca, MARCGRAVE, Hist. Brasil., 181, 1648, Brazil.
Pleuronectes papillosus, LINNÆUS, Syst. Nat., x, 271, 1758, Brazil; based on MARCGRAVE.
? *Pleuronectes macrolepidotus,* BLOCH, Ausländische Fishe, VI, 25, tab. 190, 1787; apparently based on MARCGRAVE.
Pleuronectes aramaca, DONNDORF, Beiträge zur Ausgabe des Linnæischen Natursystems, XIII, 386, 1798; after MARCGRAVE.
Rhombus aramaca, CUVIER, Règne Animal, Ed. 2, II, 341, 1829; after MARCGRAVE.
Rhombus soleæformis, AGASSIZ, Spix, Pisc. Brasil., 86, tab. 47, 1829, Atlantic Ocean.
Hippoglossus intermedius, RANZANI, Novis Speciebus Piscium Dissertatio Secundo, 1840, 14, pl. 4, Brazil.
Hemirhombus soleæformis, GÜNTHER, Cat. Fish., IV, 423, 1862.
Hemirhombus pœtulus, BEAN MS., JORDAN & GILBERT, Proc. U. S. Nat. Mus. 1882, 304, Pensacola (Coll. Silas Stearns); GOODE & BEAN, Proc. U. S. Nat. Mus. 1882, 414; BEAN, Cat. Coll. Fish U. S. Nat. Mus. 1883, 45.
Citharichthys aramaca, JORDAN & GILBERT, Synopsis, 816, 1883.
Citharichthys pœtulus, JORDAN & GILBERT, Synopsis, 964, 1883; JORDAN, Proc. U. S. Nat. Mus. 1884, 38; GOODE & BEAN, Oceanic Ichthyology, 448, pl. 109, fig. 373, 1896.
Aramaca papillosa, JORDAN, Proc. U. S. Nat. Mus. 1886, 602; synonymy confused with *S. micrurum.*
Aramaca soleæformis, JORDAN, Proc. U. S. Nat. Mus. 1886, 602.
Syacium papillosum, JORDAN & GOSS, Review Flounders and Soles, 268, 1889.

3041. SYACIUM MICRURUM, Ranzani.

Head $3\frac{1}{8}$ in length; depth $2\frac{3}{8}$. D. 87 to 92; A. 54 to 68; scales 65 to 70 (pores); eye 4 in head; maxillary $2\frac{1}{4}$ to 3. Form regularly elliptical, the profile evenly convex to end of snout; eyes large, nearly even in front, the male with the interorbital space deeply concave, its width $\frac{2}{3}$ the vertical depth of the eye (or more in Brazilian specimens); female with interorbital area much narrower, with a more or less perfect median groove, its width about equal to depth of pupil; mouth small, the maxillary reaching to below middle of eye; teeth small, slender, in 2 rows above, in 1 row below, the outer series in upper jaw somewhat enlarged, but hardly canine-like; gill rakers very short and thick, about $1 + 7$ in number. Scales small, firm, moderately ctenoid; pectoral $1\frac{1}{4}$ in head in the female, reaching nearly to base of caudal in the male; vertebræ $9 + 24 = 33$. Color dark brown, with many rings and spots of light gray and blackish, some of the dark rings with a black central spot; a diffuse dusky blotch on lateral line above pectoral, and 1 near base of caudal peduncle; fins with numerous inky spots and dark markings; blind side pale. West Indian fauna, Key West to Rio Janeiro; rather common. We have found in the Museum of Comparative Zoology specimens purporting to be the types of *Hemirhombus ocellatus,* Poey (No. 11144; Poey's number, 88). These are female examples, and they differ from the types of *Hemirhombus athalion,* obtained in Cuba by Dr. Jordan, only in their greater size. Numerous specimens (11373) from Rio Janeiro belong to the same species. Among these are males, which have the interorbital space much broader than in the types of *ocellatus* and *æthalion*. Besides these specimens we have examined others from Hayti, Cuba, and Key West, and there can be no reasonable doubt of their identity, and that all are identical with Günther's *Hemirhombus aramaca.* This fish is described and fairly well figured by Ranzani under the name of *Syacium micrurum.* It is the type of his genus *Syacium*, a generic name which, strangely enough, has received no notice from subsequent authors until lately. (μικρός, small; οὐρά, tail.)

Syacium micrurum, RANZANI, Nov. Spec. Pisc. Diss. Sec., 20, pl. 5, 1840, Brazil; JORDAN & GOSS, Review Flounders and Soles, 269, 1889.
Hippoglossus ocellatus, POEY, Memorias, II, 314, 1860, Cuba.
Hemirhombus aramaca, GÜNTHER, Cat., IV, 42, 1862, Cuba; Jamaica; not *Rhombus aramaca*, CUVIER.
Citharichthys æthalion, JORDAN, Proc. U. S. Nat. Mus. 1886, 52, Havana. (Type, No. 37748. Coll. D. S. Jordan.)
Hemirhombus ocellatus, POEY, Synopsis, 407, 1868; POEY, Enumeratio, 138, 1875.
Citharichthys ocellatus, JORDAN & GILBERT, Synopsis, 964, 1883; JORDAN, Proc. U. S. Nat. Mus. 1884, 143.
Hemirhombus æthalion, JORDAN, Proc. U. S. Nat. Mus. 1886, 602.

3042. SYACIUM LATIFRONS (Jordan & Gilbert).

Head 4; depth 2⅓. D. 92; A. 72; scales 60. Body elliptical, the dorsal and ventral outlines equally arched; mouth placed low, below axis of body; snout with an abrupt constriction in front of upper orbit, the outline then again convex; eyes on left side, distant, the lower in advance of the upper; a vertical line from anterior margin of upper orbit passing through middle of lower; distance of upper eye from dorsal outline equaling ⅔ its vertical diameter; interorbital space concave, very wide, its width 1½ times diameter of orbit in a specimen 8 inches long, much narrower in the young; a ridge from upper angle in lower eye runs upward and backward to join a ridge from upper orbit. Nostrils on a level with upper margin of lower eye, the anterior with a flap, distant from the posterior, which is circular; length of snout to front of lower eye 4½ to 5 in head; mouth very oblique, the gape convex upward and backward; maxillary ⅔ length of head, reaching to middle of lower pupil, very narrow and covered with small scales; teeth small, the upper jaw with 2 series, the front teeth of the outer series somewhat enlarged; lower jaw with a single series; vomer and palatines toothless; gill rakers short and broad, the longest about ½ vertical diameter of pupil; about 7 on anterior limb of arch; pseudobranchiæ present; preopercle with posterior margin nearly vertical, only the lower third free, the upper ⅔ grown fast to opercle and scaled over; the lower margin running very obliquely downward and forward, the angle thus an obtuse one; dorsal fin commencing on the snout in front of upper eye, the first 4 or 5 rays exserted and turned over to the blind side; the highest rays are behind the middle of the fin and are about ⅔ length of head; anal fin similar to dorsal, its origin under base of pectorals; caudal short, about ⅔ length of head, the middle rays the longest, the outer rays slightly prolonged; ventrals unsymmetrical, that of colored side on the ridge of the abdomen, the other inserted in front of it; pectoral of colored side long, the rays very slender, the two upper prolonged and filamentous, the upper (in adults) more than ⅓ total length; pectoral of blind side more than ⅔ of length of head; scales ciliated, somewhat irregular, of moderate size, with small scales intermixed; snout naked, head and body otherwise scaly; scales on interorbital region very small; a series of small scales on basal half of each dorsal and anal ray; base of caudal thickly scaled, a series of small scales running nearly to tip of each ray, lateral line slightly rising anteriorly, but without distinct curve. Color light brown, with grayish and light bluish dots, some darker areas and a few round brown spots ocellated with lighter; interorbital space with a vertical brown bar bordered by lighter; fins mottled and spotted. This species is known only from the original types, taken by Professor Gilbert at Panama. The several variations in this species have not been studied. The species differs from *Syacium ovale* chiefly in the much broader interorbital space. We should regard this as unquestionably the adult male of *S. ovale* were it not that in making large collections of the latter species at Mazatlan we found not one referable to *S. latifrons*. (*latus*, broad; *frons*, forehead.)

Citharichthys latifrons, JORDAN & GILBERT, Bull. U. S. Fish Comm. 1881, 334, Panama. (Coll. C. H. Gilbert.)
Syacium latifrons, JORDAN & GOSS, Review Flounders and Soles, 271, 1889.

3043. SYACIUM OVALE (Günther).

Head 3⅖ in length; depth 2‎⅒. D. 86; A. 69; scales 58; eye 4½ in head; maxillary 2⅔; pectoral 1½; caudal 1⅛. Body elliptic-ovate, body outline from snout to caudal peduncle uniform, the snout not produced; mouth moderate; maxillary concave behind, reaching to middle of pupil of lower eye; lower jaw slightly included; teeth biserial in upper jaw, the inner series small and sharp, the outer much larger, irregular, uniserial in lower jaw; the lower eye slightly in advance of the upper; interorbital space narrow, as broad as pupil, concave; gill rakers as long as pupil, 2 + 8 in number. Scales strongly ctenoid; scales on mandible, maxillary, and a few in front of interorbital, the middle of which is naked; lateral line not curved. Dorsal beginning slightly in front of upper eye on blind side, the anterior rays produced a little beyond membrane; base of ventral of blind side wider than that of eyed side; caudal double lunate. Color light olive brown, nearly uniform, the vertical fins with elongate dark spots; caudal with large, irregular black spots. Pacific coast of tropical America; common at Mazatlan and Panama. Here described from specimens 6 or 7 inches in length, collected at Mazatlan, Mexico, by the Hopkins expedition to Sinaloa. None of these shows the broad interorbital area of *Syacium latifrons*. (*ovalis*, oval.)

Hemirhombus ovalis, GÜNTHER, Proc. Zool. Soc. Lond. 1864, 154, Panama; GÜNTHER, Fish. Centr. Amer., 472, pl. 80, fig. 1, 1869; JORDAN & GILBERT, Bull. U. S. Fish Comm. 1882, 108-111.
Citharichthys ovalis, JORDAN, Proc. U. S. Nat. Mus. 1885, 391.
Syacium ovale, JORDAN & GOSS, Review Flounders and Soles, 271, 1889.

1049. CYCLOPSETTA, Gill.

Cyclopsetta, GILL, Proc. U. S. Nat. Mus., XI, 1888, 601 (*fimbriata*).

Mouth very large; jaws squarely truncated behind; teeth uniserial, those of the upper jaw moderate, of lower jaw enlarged and largest at sides; dorsal and anal almost symmetrical, dorsal commencing in front of eye on snout, scarcely deflected on blind side; caudal slightly pedunculate and convex; pectorals subequal and with a subtruncate free margin; ventrals nearly equal, the left on the preanal ridge, the right lateral, each with the inner ray connected by membrane to the body; interbranchial membrane imperforate; gill rakers tubercular and surmounted by blunt denticles. This genus differs from *Azevia* only in the smooth scales. (κύκλος, circle; ψῆττα, flounder, from the cycloid scales.)

a. Dorsal rays 91 to 95; anal 73 to 75; scales 90 to 95. Color nearly plain, the fins blotched. QUERNA, 3044.

aa. Dorsal rays 80 to 82; anal 62; dorsal and anal with dark ocelli.

b. Scales small, about 90; pectoral fin uncolored; anterior dorsal rays scarcely produced. CHITTENDENI, 3045.

bb. Scales larger, about 70; pectoral fin with black ocellus; anterior rays of dorsal somewhat produced. FIMBRIATA, 3046.

3044. CYCLOPSETTA QUERNA (Jordan & Bollman).

Head $3\frac{1}{8}$ to $3\frac{3}{8}$; depth $2\frac{1}{8}$. D. 91 to 95; A. 73 to 75; scales along lateral line 90 to 95. Body shaped as in *Azevia panamensis*. Mouth large, maxillary $1\frac{4}{5}$ in head. Teeth as in *A. panamensis*, in single series, rather long and slender, the anterior somewhat more enlarged. Snout 5 in head, its tip hooked over the lower jaw so that the outer canines project. Interorbital space rather narrow, slightly concave, with a few small scales, its width a little less than pupil, $\frac{1}{3}$ diameter of eye. Eyes moderate, $5\frac{1}{2}$ in head, the upper somewhat in advance. Gill rakers short and broad, as in *A. panamensis*, each with 3 or 4 strong teeth. Scales small, cycloid on both sides, those below pectorals more reduced than in *A. panamensis*, about 65 in a cross series; anterior part of lateral line bent slightly upward, this portion about $3\frac{1}{2}$ in straight part. Dorsal beginning above and between the nostrils, the anterior rays short, but with free tips; longest ray $2\frac{1}{2}$ in head; pectoral of eye side $1\frac{4}{5}$ to 2 in head, of blind side $2\frac{1}{8}$ to $2\frac{1}{2}$; ventrals subequal, each 6-rayed, $2\frac{4}{5}$ in head, extending $\frac{1}{3}$ their length beyond vent. Color plain brown, unspotted; fins dusky, thickly punctulate; young with 2 large oval indistinct dark spots on dorsal and anal; 3 on caudal, of which the middle is much larger. Distinguished from *A. panamensis* (Steindachner) by having much smaller cycloid scales on eyed side and by its plain coloration. Coast of Colombia. Numerous specimens, the largest about 8 inches in length, were dredged in 7 fathoms at *Albatross* Station 2800 and in 16 fathoms at Station 2802. (*quernus*, oaken, i. e., tanned.)

Azevia querna, JORDAN & BOLLMAN, Proc. U. S. Nat. Mus. 1889, 174, Pacific Ocean off coast of Colombia, at Albatross Station 2802, Lat. 8° 38′ N., Long. 79° 31′ 30″ W. (Type, No. 41159.)

3045. CYCLOPSETTA CHITTENDENI, B. A. Bean.

Head 3½ in body; depth 2½. D. 82; A. 62; scales 90; eye 5 in head. Mouth widely cleft, oblique, the jaws curved; cleft of mouth less than 2 in head. Teeth of each jaw in a single series, those of lower jaw strong and sharp, curved inward and backward, those of upper jaw not so large, and very irregular in size. Ventral fins well developed, that of eyed side being on abdominal ridge, and about ¾ as long as pectoral; pectorals ½ as long as head, their length equaling a little more than ⅕ of body depth, posterior margin oblique; gill rakers very short, tubercular, almost as broad as long, 3 or 4 + 8 in number. Color brown; fins lighter, marked with blackish; 3 small faint blotches of black on first half of dorsal fin, and 3 rather distinct blotches on second half, last blotch extending to caudal peduncle; anal fin with 3 black blotches situated as and similar to those of dorsal fin; ventral of eyed side blackish, that of blind side pale; caudal fin with 3 black spots at its extremity; pectoral fin of colored side blackish; quite a large blotch of black on body under this fin. This species is distinguished from *Cyclopsetta fimbriata* by its shorter head, smaller and closely adhering scales, larger teeth, the little-produced anterior dorsal rays and by the oblique posterior margin of the pectorals. In *C. fimbriata* the scales are rather large and deciduous, the teeth small, the anterior rays of the dorsal considerably produced, and the posterior margin of the pectoral is subtruncate. A single specimen collected by Dr. John F. Chittenden, of the Victoria Institute, Port of Spain, Trinidad Island, and named in his honor. It is 7¾ inches in length. (B. A. Bean.)

Cyclopsetta chittendeni, B. A. BEAN, Proc. U. S. Nat. Mus. 1894, 635, Trinidad. (Type, No. 44100. Coll. Dr. Chittenden.)

3046. CYCLOPSETTA FIMBRIATA (Goode & Bean).

Head 3½ in length; depth nearly 2. D. 80; A. 60 or 61; pectoral 10; ventral 6; scales 25–70–31; maxillary 2 in head; caudal 4⅔ in total length; pectoral 5½. Mouth very large, the upper jaw strongly curved, lower jaw included; teeth uniserial in each jaw, some of the anterior ones in the upper jaw being much larger than those following, while those in the lower jaw are still larger than these, some of the teeth in each jaw depressed; upper eye placed at a distance from profile equal to ½ its own diameter, which is a little less than 5 in head; eyes in the same vertical; interorbital ridge low, 4 in eye; gill rakers very short, tubercular, about 9 on lower part of angle. Scales cycloid; curve of lateral line slight, curve 3½ in straight part. Dorsal beginning on snout in advance of nostrils, first ray higher than second, highest rays behind middle of fin; origin of anal under base of pectoral, its highest rays behind middle of fin, higher than highest dorsal rays; ventral of eyed side on ridge of abdomen; middle caudal rays produced. Color grayish brown; dorsal and anal fins each with 2 round dark blotches upon their posterior halves, which are slightly larger than eye; a similar dark blotch upon middle of caudal, sometimes with smaller blotches irregularly placed near its outer margin; pectoral with a very narrow dark band near its base, whole of outer half marked with a dark blotch, reticulated and mottled with lighter; intervening portion pearly white with dark specks upon the rays; bliud side cream colored. Deep waters of the Gulf of Mexico. (Goode & Bean.) (*fimbriatus*, fringed; from the produced dorsal rays.)

Hemirhombus fimbriatus, GOODE & BEAN, Proc. U. S. Nat. Mus. 1885, 591, deep waters of the Gulf of Mexico, between Mississippi Delta and Cedar Keys. (Type, No. 37330. Coll. *Albatross*.)

Arnoglossus ? fimbriatus, JORDAN & GOSS, Review Flounders and Soles, 262, 1889.

Cyclopsetta fimbriata, GOODE & BEAN, Oceanic Icthyology, 451, fig. 368, 1896.

1050. AZEVIA, Jordan.

Azevia, JORDAN, in JORDAN & GOSS. Review Flounders and Soles, 271, 1889 (*panamensis*).

Body elliptical, compressed, covered with small, firm, ctenoid scales; mouth large; teeth in both jaws uniserial; vomer without teeth; gill rakers very short and thick, tubercle-like; interorbital space very narrow in both sexes, the ridges coalescing between the eyes; lateral line without arch in front; ventrals free from the anal; septum of gill cavity below gill arches, without foramen; a deep emargination near isthmus. None of the fins especially modified or with elongate rays. This genus differs from *Citharichthys* in its tubercular gill rakers, as also in its small, firm scales, and other characters of minor importance. (*Azevia*, a Portuguese name for the sole, used at Lisbon, according to Brito-Capello. It probably corresponds to the Cuban name *Acedía*.)

3047. AZEVIA PANAMENSIS (Steindachner).

Head 3⅔ in length; depth 2¼. D. 95; A. 73 to 78; scales 73 to 78; eye 5 in head; maxillary 2; pectoral 1⅔; caudal ¼. Body rather elongate; anterior profile evenly convex; mouth large, the maxillary reaching to posterior border of eye, the upper jaw somewhat hooked over the lower; about 3 teeth in upper jaw enlarged and hooked, canines in lower jaw long and sharp; eyes about even in head; interorbital space very narrow, less than diameter of pupil, a ridge along its middle; gill rakers divided into many sharp points around its edge, very short, as wide as long, about 4 + 9 in number. Scales on posterior part of interorbital, maxillary, and mandible; tip of snout, the greater part of interorbital, and tip of lower jaw naked; scales all strongly ctenoid; lateral line not curved anteriorly. Origin of dorsal at the vertical between tip of snout and front of eyes, scarcely on blind side, the anterior rays somewhat produced beyond membrane, the fin rather low; origin of anal below angle of opercle; pectorals short, that of eyed side pointed, its mate of the opposite side broadly rounded behind; caudal double lunate. Here described from a specimen collected by the Hopkins Expedition to Sinaloa, at Mazatlan, Mexico, about 11 inches in length. We have also examined specimens from Panama, in the museum at Cambridge, a part of the series of Dr. Steindachner's original types. Pacific coast of Central America; common at Mazatlan and Panama. (*panamensis*, from Panama.)

Citharichthys panamensis, STEINDACHNER, Ichth. Beitr., III, 62, 1875, Panama; JORDAN & GILBERT, Bull. U. S. Fish Comm. 1882, 108 and 111; GILBERT, Bull. U. S. Fish Comm. 1882, 112.

Azevia panamensis. JORDAN & GOSS, Review Flounders and Soles, 272, 1889; JORDAN, Proc. Cal. Ac. Sci. 1895, 503.

1051. CITHARICHTHYS,* Bleeker.

(WHIFFS.)

Citharichthys, BLEEKER, Comptes Rendus Acad. Sci. Amsterd., XIII, Pleuronectoidei, 6, 1862 (*cayennensis* = *spilopterus*).
Orthopsetta, GILL, Proc. Ac. Nat. Sci. Phila. 1862, 330 (*sordidus*).
Metoponops, GILL, Proc. Ac. Nat. Sci. Phila. 1864, 198 (*cooperi* = *sordidus*).

Eyes and color on the left side. Body oblong; mouth of the large type, but comparatively small, with 1 series of small, sharp teeth in each jaw; no teeth on vomer or palatines. Gill rakers moderate, slender. Dorsal fin beginning just in front of eye; all the fin rays simple; ventrals of colored side on the ridge of the abdomen; no anal spine; caudal fin convex or double truncate behind; none of the fins produced. Scales thin, deciduous, slightly ctenoid. Lateral line nearly straight, simple. Lower pharyngeals separate, each with a single row of teeth. Vertebræ 30 to 40. This genus includes small flounders of weak organization, especially characteristic of the sandy shores of tropical America. The subgenus *Orthopsetta* includes species of more northern range and somewhat different form, and especially noteworthy as having an increased number of vertebræ. The two groups intergrade so perfectly that no sharp line of division can be drawn between them. (*Citharus*, an allied genus; ἰχθύς, fish—a fish which lies on its κίθαρος, or ribs; that is, on its side.)

ORTHOPSETTA (ὀρθός, straight; ψῆττα, flounder):
a. Vertebræ 37 to 40; interorbital ridge sharply elevated; the head not closely compressed; eyes large; species of the North Pacific.
  b. Interocular space concave, scaly, at least behind.
    c. Gill rakers x + 16 to 18.
      d. Scales 65 to 70; dorsal rays 95; anal 77; depth 2⅓.   SORDIDUS, 3048.
      dd. Scales 46 to 50; dorsal rays 83 to 87; anal 67 to 70; depth 2⅔ in length.   FRAGILIS, 3049.
    cc. Gill rakers x + 10 or 11; dorsal rays about 84; anal 65; scales 50; depth 2¼ in length.   XANTHOSTIGMUS, 3050.
  bb. Interocular space a sharp, naked ridge; dorsal rays 85 to 90; anal 68 to 72; scales 55 to 60; head 3¾ in length; depth 2¼.   STIGMÆUS, 3051.
CITHARICHTHYS:
aa. Vertebræ 33 to 36; interorbital ridge low and narrow, the head closely compressed. Species of the Atlantic or the Tropics.
  e. Eyes large, 3 to 4½ in head.
    f. Head large, 3 to 3⅓ in length.
      g. Interorbital space very narrow, 5 in eye; snout with a spine; pectoral of eyed side elongate, ⅓ longer than head; maxillary 2¼ in head.   D. 91; A. 73; scales 48.   DINOCEROS, 3052.
      gg. Interorbital space very broad, 2 in eye; snout without spine; pectoral of eyed side shorter than head; maxillary 2⅓ in head. D. 78; A. 62; scales 43.   PLATOPHRYS, 3053.
    ff. Head smaller, about 4 in length.
      h. Body comparatively elongate, the depth about 2½ in length; mouth very small; the maxillary 3½ in head; dorsal rays 83; anal 67; scales 40; eye 4 in head.   ARCTIFRONS, 3054.

---

* "As the name *Citharichthys* was introduced a short time before that of *Orthopsetta*, proposed for the *Psettichthys sordidus*, and was framed for a species related to that type, that name must be adopted if the *O. sordida* is not regarded as generically distinct." (Gill.)

hh. Body comparatively broad, the depth about $\frac{1}{2}$ the length; mouth larger.

    i. Snout with a strong, sharp spine on eyed side, above upper lip; eyes large, 3 in head; greatest depth of body over the pectorals; interorbital space with a wide ridge, about $\frac{1}{2}$ diameter of eye. D. 74; A. 60; scales 40.   UNICORNIS, 3055.

    ii. Snout without distinct spine; eyes moderate, $3\frac{1}{2}$ to $4\frac{1}{2}$ in head; greatest depth of body under middle of dorsal; interorbital space a narrow, scaly ridge with a slight median groove; maxillary $2\frac{1}{2}$ in head; teeth small, those in front slightly enlarged; body not very thin; gill rakers moderate, 6 + 13.

        j. Dorsal rays 68; anal 52; scales smaller, the lateral line with about 53 pores; sides with whitish blotches.
                                                  UHLERI, 3056.

        jj. Dorsal rays 80; anal 56; scales large, 41 in lateral line; sides and fins with dark blotches.   MACROPS, 3057.

ee. Eyes quite small, 5 to 6 in head; snout short, forming an angle with the profile; mouth moderate, oblique, the maxillary $2\frac{1}{2}$ to $2\frac{2}{3}$ in head; teeth small, the anterior somewhat enlarged; dorsal rays about 80; anal rays 60; body and fins speckled.

    k. Scales not very large, 45 to 48 in lateral line; gill rakers long and slender, longer than pupil.   SPILOPTERUS, 3058.

    kk. Scales large, 40 to 46 in lateral line; gill rakers short, not longer than pupil.   GILBERTI, 3059.

Subgenus ORTHOPSETTA, Gill.

3048. CITHARICHTHYS SORDIDUS (Girard).

(SOFT FLOUNDER.)

Head $3\frac{3}{7}$; depth $2\frac{1}{5}$. D. 95; A. 77; scales 65 to 70. Form elliptical; interocular space concave, scaly, a conspicuous sharp ridge above the lower eye; mouth not large, the maxillary about 3 in length of head; teeth anteriorly subequal, growing much smaller behind. Gill rakers about 7 + 16. Lower pharyngeals narrow, each with 1 row of slender teeth. Scales rather large, thin, and membranaceous, readily deciduous, their edges slightly ciliate; accessory scales numerous. Eye large, much longer than snout, $3\frac{1}{2}$ in head; depth of caudal peduncle less than $\frac{1}{5}$ head; pectorals long, nearly $\frac{2}{3}$ length of head. Vertebræ 11 + 29 = 40. Dull olive brownish of varying shade, the males with dull orange spots and blotches; each scale with a darker edge; dorsal and anal fins in the male blackish, with dull orange blotches, and edged anteriorly with yellowish; female paler, the fins nearly plain. Pacific coast of North America, in water of moderate depth; British Columbia to Lower California. This small flounder is one of the commonest species on the Pacific coast, being found in water of 10 fathoms or more depth in all localities from the Mexican boundary to British Columbia. Although much larger in size than any other species of the genus, it rarely exceeds 2 pounds in weight. In its deciduous scales and soft flesh it much resembles *Lyopsetta exilis* and *Atheresthes stomias*, 2 species which are often taken in company with it. Of all the species of *Citharichthys*, this one has the most extended range to the northward. (*sordidus*, sordid, from its dull coloration.)

Psettichthys sordidus, GIRARD, Proc. Ac. Nat. Sci. Phila., VII, 1854, 142, San Francisco; Tomales Bay; GIRARD, U. S. Pac. R. R. Surv., X, Fishes, 155, 1858.

Metoponops cooperi, GILL, Proc. Ac. Nat. Sci. Phila. 1864, 198, Santa Barbara; shrivelled specimen. (Type, No. 9407.)

Orthopsetta sordida, GILL, Proc. Ac. Nat. Sci. Phila. 1862, 330.

Citharichthys sordidus, LOCKINGTON, Rep. Com. Fisheries of California, 1878–79, 42; LOCKINGTON, Proc. U. S. Nat. Mus. 1879, 83; JORDAN & GILBERT, Proc. U. S. Nat. Mus. 1880, 453; JORDAN & GILBERT, Proc. U. S. Nat. Mus. 1881, 67; JORDAN & GILBERT, Synopsis, 817, 1883; BEAN, Proc. U. S. Nat. Mus. 1883, 353; JORDAN & GOSS, Review Flounders and Soles, 274, 1889.

3049. CITHARICHTHYS FRAGILIS, Gilbert.

Head $3\frac{2}{5}$ to $3\frac{3}{5}$ in length; depth $2\frac{3}{8}$ to $2\frac{3}{4}$ (in specimens 5 inches long.) D. 83 to 87; A. 67 to 70; scales 46 to 50. Vertebræ 10 + 27. Body elongate, posteriorly sharply wedge-shaped, tapering to base of caudal; anterior profile very conspicuously angulated above front of upper eye, the snout strongly projecting, its anterior profile nearly vertical; depth of caudal peduncle $2\frac{7}{8}$ in head. Anterior nostril with a short tube and flap, the latter nearly obsolete on blind side. Mouth more oblique than in *C. sordidus;* maxillary reaching vertical from front of pupil, $2\frac{2}{5}$ to $2\frac{3}{5}$ in head. Teeth in a single series, close set, those anteriorily somewhat enlarged, but none of them canine-like. Eyes large, the vertical from front margin of upper eye falling through front of lower pupil; longest diameter of upper orbit $2\frac{3}{8}$ in head; interorbital space narrow, concave, scaled, the lower ridge strongest, its width about $\frac{1}{8}$ diameter of orbit. Symphyseal knob sharp. Gill rakers long, slender, close set (as in *C. sordidus*), 18 on anterior limb of arch, the longest $\frac{1}{3}$ orbit. Scales large, deciduous, somewhat irregularly arranged, 12 or 13 series above lateral line; scales smooth on blind side, minutely spinous on eyed side; lateral line without anterior arch. Dorsal beginning slightly in advance of eye, the longest ray $\frac{1}{2}$ head; pectorals long and narrow, with 11 rays on colored side. Color dusky olivaceous, with occasional slaty-blue spots. This species is closely related to *C. sordidus*, from which it differs in the fewer vertebræ and fin rays and in the larger scales. Many specimens from the Gulf of California in from 18 to 76 fathoms, at *Albatross* Stations 3011, 3016 to 3018, and 3033. (Gilbert.) (*fragilis*, fragile.)

Citharichthys fragilis, GILBERT, Proc. U. S. Nat. Mus. 1890, 120, Gulf of California, east coast of Lower California. (Type, No. 44409. Coll. Dr. Gilbert.)

3050. CITHARICHTHYS XANTHOSTIGMUS, Gilbert.

D. 81 to 86; A. 63 to 67; scales 50. Vertebræ 11 + 26. Body deep, varying from $2\frac{1}{3}$ (in young, 3 inches long) to $2\frac{1}{10}$ (7 inches long) in length of body. Profile angulated above front of upper eye, the snout convexly projecting. Depth of caudal peduncle $\frac{1}{3}$ head. Lower eye in advance, the vertical from front of the upper passing through front of lower pupil. Mouth rather small, the outline somewhat curved, the maxillary reaching the vertical from front of lower pupil, 3 to $3\frac{1}{4}$ in head; mandible with a sharp downward-directed point at symphysis. Teeth in a single close-set series in each jaw, growing slightly larger anteriorly, but without canines.

Anterior nostril with a short tube, and a narrow flap arising from its inner edge. Interorbital width 3¾ to 5 in orbit, slightly concave, the lower ridge much stronger and higher than the upper, scaled posteriorly. Eye large, the orbit 3½ to 3¾ in head. Gill rakers rather long and slender, coarsely dentate on inner margin, distant, 10 or 11 on anterior limb of arch. Scales large, in regular series, appearing cycloid, but the edges very minutely spinous; lateral line gently rising on anterior ½, but without curve; fifty vertical series of scales, with as many pores in lateral line; 13 to 15 horizontal series above lateral line. Dorsal beginning immediately behind posterior nostril of blind side, ending so as to leave caudal peduncle free for a distance equaling ½ diameter of eye; ventrals long, reaching beyond origin of anal; pectoral very long and slender, normally with 9 rays, the longest ray on colored side longer than head, about ¼ length of body. Color light olive brown, irregularly flecked with slaty, and with numerous bright yellow spots broadly ocellated with brownish black, a series of these usually on lateral line, and 2 others halfway between it and the dorsal and ventral outlines, respectively, those of the latter series forming pairs; fins not conspicuously marked, the pectorals sometimes with faint broad dusky cross bars. Both coasts of Lower California. In external appearance the species closely resembles *C. sordidus*, to which, however, it is not closely related, differing in number of scales, fin rays, and vertebræ, and in the size and number of gill rakers. (Gilbert.) Many specimens, from *Albatross* Stations 3039, 3043, and 3044, in 47 to 74 fathoms. (ξανθός, yellow; στίγμα, spot.)

Citharichthys xanthostigma, GILBERT, Proc. U. S. Nat. Mus. 1890, 120, Gulf of California, west coast of Lower California, and Magdalena Bay. (Type, No. 44408. Coll. Dr. Gilbert.)

3051. CITHARICHTHYS STIGMÆUS, Jordan & Gilbert.

Head 3¾ in length without caudal; depth 2⅗; dorsal 87; anal 68; scales 54 (pores). Body moderately deep, the 2 profiles regularly and equally arched; snout short, gibbous, projecting a little beyond the outline; caudal peduncle very short, not high, its length (from end of last vertebra to vertical from last anal ray) about ⅔ its height, which is ⅓ length of head; caudal fin appearing sessile. Mouth moderate, very oblique, the maxillary reaching beyond front of pupil, 2¾ in head; teeth in a single series, subequal in the two jaws, rather long, very slender and numerous, decreasing toward angle of mouth; about 40 teeth in the upper jaw, and 30 in the lower on blind side. Eyes large, close together, separated by a narrow, sharp, scaleless ridge; the upper eye largest, slightly behind the lower, with considerable vertical range; diameter of upper eye 3½ in head. Snout and lower jaw scaleless; end of maxillary and rest of head scaled. Gill rakers moderate, not strong, about 9 on anterior limb. Dorsal fin beginning on the vertical from front of upper eye, the first 3 rays being somewhat turned to blind side; the fin low, the highest at beginning of its posterior third, the longest ray nearly ½ length of head; anal spine present, very small; caudal rounded, about equalling length of head; pectoral of colored side 1⅔ in head, of blind side, 2½. Scales

3030——91

moderate, those forming the lateral line persistent, the others deciduous, those on colored side with ciliated margins, on blind side smooth; lateral line without anterior curve, the scales crowded and smaller anteriorly. Color in spirits uniform olivaceous, the scales dark edged; lips and some of the membrane bones of head margined with blackish; fins dusky, each seventh (to tenth) ray of vertical fins with a very small but conspicuous black spot on its middle. The above description is from the original type from Santa Barbara. Numerous specimens dredged by the *Albatross* in 9 to 41 fathoms off the coast of California show the following characters: Gill rakers $x+9$. Specimens 5 mm. long show white spots each with a black half ring on the outer side symmetrically arranged along bases of dorsal and anal; 4 distinct pairs of these, 2 unpaired ones more anteriorly along dorsal base, and a few fainter ones midway between these rows and the lateral line and alternating with them; there are some other scattered light spots. The abdomen is covered by a broad black streak; this, however, is wanting in specimens larger and smaller. Coast of California; rare; in rather deep water. The original type of this species is a young example, taken near Santa Barbara by Capt. Andrea Larco. In the Museum of Comparative Zoology are other specimens collected by Mr. Cary at San Francisco. These have 72 anal rays, while the original type had but 68. A few other specimens have been since obtained. Some of these are full of spawn at a length of 5 inches. ($\delta \tau \iota \gamma \mu \alpha \tilde{\iota} o \varsigma$, speckled.)

Citharichthys stigmæus, JORDAN & GILBERT, Proc. U. S. Nat. Mus. 1882, 410, 411, Santa Barbara (Coll. A. Larco. Type, 31099 U. S. Nat. Mus.); JORDAN & GILBERT, Synopsis, 965, 1883; JORDAN & GOSS, Review Flounders and Soles, 274, 1889; GILBERT, Rept. U. S. Fish. Comm. 1893 (1896) 473.

Subgenus CITHARICHTHYS.

3052. CITHARICHTHYS DINOCEROS, Goode & Bean.

Head $3\frac{1}{4}$ in length; depth $2\frac{1}{4}$. D. 91; A. 73; scales 14–48–16; eye $3\frac{1}{2}$ in head; maxillary a little less than 2; greatest height of dorsal 2; pectoral $2\frac{1}{2}$ in body; caudal $5\frac{1}{2}$. Teeth uniserial in both jaws, those in the front much the largest; a strong spine upon the snout overhanging the upper lip, above this a second shorter spine; interorbital very narrow, its width less than 5 in eye, ridge rather prominent, narrow, sharp. Scales thin, deciduous, cycloid, large; lateral line slightly curved over the pectoral. Dorsal beginning on snout, in advance of eye, upon the blind side, its highest rays behind the middle; origin of anal under base of pectoral; third and fourth pectoral rays upon the eyed side elongate, the fin $\frac{3}{4}$ longer than its mate of the opposite side; caudal subsessile, pointed. Color grayish brown above, white below. Vertebræ 33 to 36. West Indies, in deep water. The type specimen, 92 mm. long to base of caudal, was taken by the *Blake*, off Guadeloupe; others were taken off St. Lucie and Barbados, from 310 to 955 fathoms. (Goode & Bean.) ($\delta \varepsilon \iota \nu \acute{o} \varsigma$, terrible; $\varkappa \acute{\varepsilon} \rho \alpha \varsigma$, horn.)

Citharichthys dinoceros, GOODE & BEAN, Bull. Mus. Comp. Zool., XII, No. 5, 157, 1886, off Martinique, St. Lucie, and Barbados; JORDAN & GOSS, Review Flounders and Soles, 275, 1889; GOODE & BEAN, Oceanic Ichthyology, 447, 1896.

3053. CITHARICHTHYS PLATOPHRYS, Gilbert.

Head 3; depth 2. D. 78; A. 62; scales 43. Body ovate; caudal fin sub-sessile, the free portion of caudal peduncle about $\frac{1}{4}$ as long as diameter of pupil, its depth $\frac{1}{3}$ length of head. Mouth very oblique; maxillary $2\frac{1}{2}$ in head, reaching vertical from middle of lower eye. Teeth slender, close set, in a single series in each jaw, those in front of upper jaw largest, but not canine-like. Eyes large, the lower much in advance of the upper, their horizontal diameter $3\frac{3}{4}$ in head. Interorbital space very wide for the genus, concave, divided by an oblique ridge running backward from middle of upper orbit; interorbital width $8\frac{3}{4}$ in head, nearly $\frac{1}{2}$ as wide as eye. Distance from tip of snout to front of lower eye $\frac{2}{3}$ diameter of eye, from tip of snout to upper eye $\frac{1}{4}$ head. Gill rakers short and very slender, less than diameter of pupil, 9 present on horizontal limb of outer arch. Scales large, those on blind side very weakly ctenoid. Dorsal beginning behind nostril on blind side of snout, its longest ray $2\frac{1}{3}$ in head; pectoral of eyed side long and narrow, 4 in length, containing 11 rays, that of blind side but $\frac{1}{2}$ its length; ventrals short; caudal rounded, $1\frac{3}{4}$ in head. Color in spirits, uniform light brownish (olivaceous in life), without distinctive marks; fins somewhat dusky; ventral of eyed side jet-black, that of blind side blackish on distal portion of inner rays. (Gilbert.) One specimen known, from *Albatross* Station 2799, southwest of Panama. ($\pi\lambda\alpha\tau\acute{u}\varsigma$, broad; $\grave{o}\phi\rho\acute{u}\varsigma$, eyebrow.)

Citharichthys platophrys, GILBERT, Proc. U. S. Nat. Mus. 1890, 454, Albatross Station 2799, southwest of Panama. (Coll. *Albatross.*)

3054. CITHARICHTHYS ARCTIFRONS, Goode.

Head 4 in body; depth $2\frac{3}{4}$. D. 82; A. 67; pectorals 9 or 10, 7; scales 8-40-8; eye 4 in head; maxillary $3\frac{1}{2}$; caudal 1. Body comparatively elongate; mouth small; teeth small, the anterior scarcely enlarged; interorbital space narrow, sharp, scaleless; scales cycloid, deciduous; small scales on the rays of the ventral fins; lateral line sharply defined, straight. Dorsal beginning above front of upper eye, its highest ray about 3 times the distance from snout to first ray; origin of anal under base of pectoral; caudal subsessile, triangular; rays of vertical fins all exserted; pectoral inserted low, that of eyed side twice the length of the other. Color dirty light brown. Deep waters of the Gulf Stream. (*arctus,* contracted; *frons,* forehead.)

Citharichthys arctifrons, GOODE, Proc. U. S. Nat. Mus. 1880, 341, 472, Gulf Stream off southern coast of New England; GOODE & BEAN, Bull. Mus. Comp. Zool., Vol. x, No. 5, XIX, 194, 1883; JORDAN & GILBERT, Synopsis, 818, 1883; JORDAN & GOSS, Review Flounders and Soles, 275, 1889; GOODE & BEAN, Oceanic Ichthyology, 442, fig. 366, 1896.

3055. CITHARICHTHYS UNICORNIS, Goode.

Head 4 in length; depth a little less than length. D. 74; A. 60; P. 4 (right), 10 (left); scales 12-40-12; eye 3 in head; maxillary scarcely 2; highest dorsal ray 2. Body deep, its greatest height over the pectorals;

scales thin, deciduous; eye equal to snout or interorbital space; interorbital with a strong ridge; teeth minute, close set, in a single series, stronger on the blind side; a strong, sharp spine on the snout at the anterior termination of the ridge at lower margin of upper eye; caudal pointed, triangular, subsessile; pectoral of left side twice as long as the eye, not ⅓ longer than right pectoral. Dorsal beginning at side of preorbital spine, its anterior rays being slightly upon the blind side; anal equal to dorsal in height. Ashy gray, with dark lateral line; eyes black. (Goode.) Deep waters of the Gulf Stream. (*unicornis*, having one horn.)

Citharichthys unicornis, GOODE, Proc. U. S. Nat. Mus. 1880, 342, Gulf Stream off southeast of New England; JORDAN & GILBERT, Synopsis, 818, 1883; JORDAN & GOSS, Review Flounders and Soles, 275, 1889; GOODE & BEAN, Oceanic Ichthyology, 444, fig. 369, A & B, 1896.

3056. CITHARICHTHYS UHLERI, Jordan.

D. 68; A. 52; scales 53 (pores). Body comparatively broad, regularly oval, without angle; greatest depth of body under middle of dorsal; eyes moderate, 4½ in head, close together, the orbital ridges coalescent, the lower larger. Teeth small, uniserial; maxillary 2½ in head; gill rakers short and very slender, $x + 12$. Color dark brown, with whitish blotches, the fins mottled. Hayti. A single specimen in the Museum of Comparative Zoology, 4½ inches in length. The species is close to *Citharichthys macrops*, but its fin rays and scales are considerably more numerous than in the latter. (Named for Mr. Philip Reese Uhler, the well-known entomologist, its discoverer.)

Citharichthys uhleri, JORDAN & GOSS, Review Flounders and Soles, 275, 1889, Hayti. (Coll. P. R. Uhler. Type in Mus. Comp. Zool.)

3057. CITHARICTHYS MACROPS, Dresel.

Head 4 in body; depth scarcely 2. D. 80; A. 56; scales 14–41–16; lower eye 4 in head; maxillary 2½; highest dorsal rays a little over 2; pectoral of eyed side 1⅔; caudal 4 in body; vertebræ 9 + 25 = 34. Body suboval; upper profile very convex, descending in a sharp curve from nape to front of upper eye, and forming an abrupt angle with the short, blunt snout; mouth moderate, very oblique and curved; maxillary reaching to below middle of eye; teeth minute, uniserial, slightly larger on blind side; interorbital narrow, with a scaleless ridge, which curves upward and backward to upper angle of gill opening; upper eye very close to profile, its anterior margin on the same vertical line with lower; snout shorter than eye; gill rakers about ⅓ the length of eye, 6 + 13 in number. Scales large, not ciliated, no accessory scales; origin of dorsal on blind side near tip of snout, anterior rays exserted, the first ray as long as eye, the fin highest at its middle portion; origin of anal under base of pectoral, its highest part a little higher than dorsal; caudal pointed; pectoral of blind side somewhat shorter than that of eyed side. Color in spirits, light-olive brown; body with some 20 dark-brown spots, the largest as large as eye, 4 of these arranged at equal intervals along the lateral line, the second near the middle the most prominent; dorsal and anal fins with a series of round, brown spots, 1 at the middle of every sixth or seventh ray, besides small irregular spots and mottlings;

caudal spotted and mottled with dark brown, and with 2 round, brown spots, 1 above the other at the base of the fin. (Dresel.) South Atlantic and Gulf coasts of the United States; rather common; a well-marked species. We have examined several specimens dredged in the harbor of Beaufort, N. C., by Prof. Oliver P. Jenkins. (μακρός, large; ὤψ, eye.)

Citharichthys macrops, DRESEL, Proc. U. S. Nat. Mus. 1884, 539, Pensacola (Type No. 21500); JORDAN, Proc. U. S. Nat. Mus. 1886, 29; JORDAN & GOSS, Review Flounders and Soles, 275, 1889.

3058. CITHARICHTHYS SPILOPTERUS, Günther.

Head 3½ in body; depth 2⅙. D. 75 to 80; A. 58 to 61; scales 45 to 48; eye 6 in head; maxillary 2½; pectoral 2⅓; highest dorsal and anal rays 2; caudal 1⅓. Body moderately elongate, much compressed; snout short, forming an angle with the profile; jaws strongly curved, the upper somewhat hooked over the lower; lower jaw slightly included; maxillary reaching to posterior margin of lower orbit; teeth small, in a single row, the anterior a little enlarged; interorbital area a low, narrow ridge, which is divided only anteriorly; gill rakers short and rather slender, about 3 in eye, 4 + 12 in number; scales cycloid. Origin of dorsal above anterior edge of upper eye, very slightly on blind side, its highest rays in its posterior half; origin of anal slightly behind base of pectoral; pectoral of eyed side very slightly shorter than that of eyed side; vertebræ 34. Color olive brownish, somewhat translucent, with darker dots and blotches; a series of distant obscure blotches along bases of dorsal and anal. Atlantic coast of tropical America north to New Jersey; very common on sandy shores; not found in the Pacific, all west coast references belonging to *C. gilberti.* Here described from a specimen from Havana, 6 inches in length. This little flounder is almost everywhere abundant on the sandy shores of the warmer parts of the Western Atlantic, in shallow water. Careful comparison of specimens from South Carolina, Cuba, and Brazil shows no tangible difference, and we are compelled to regard all as forming a single species. It rarely exceeds 5 or 6 inches in length. It usually comes into the markets mixed with other shore fishes, and it nowhere receives any notice as a food-fish. This species is common in the markets of Havana, and it is evidently the original of Poey's *Hemirhombus fuscus,* although in Poey's description there seems to be some confusion, because the teeth are said to be biserial above, and 60 scales are counted in the lateral line. A specimen from Poey in the museum at Cambridge is labeled "*Hemirhombus fuscus,* type. Collector's number, 87." This belongs to *C. spilopterus,* and it has 48 scales in the lateral line. Bleeker's *C. guatemalensis* agrees in all respects with *C. spilopterus.* We are unable to find any description of *C. cayennensis,* if, indeed, the species has ever been described. Specimens of *C. spilopterus* are in the museum at Cambridge from Cuba, Pará, Sambaia, Pernambuco, Camaru, Rio das Velhas, Rio Janeiro, and San Matheo. (σπίλος, spot; πτερόν, fin.)

Citharichthys spilopterus, GÜNTHER, Cat., IV, 421, 1862, New Orleans, San Domingo, Jamaica; JORDAN & GILBERT, Proc. U. S. Nat. Mus. 1882, 618; JORDAN & GILBERT, Synopsis, 817, 1883; JORDAN, Proc. U. S. Nat. Mus. 1886, 53; JORDAN & GOSS, Review Flounders and Soles, 276, 1889.

Citharichthys cayennensis, BLEEKER, Comptes Rendus Acad. Sci. Amsterd., XIII, 1862, 6,
Cayenne; name only.
Citharichthys guatemalensis, BLEEKER, Neder. Tydschr. Dierk. 1864, 73, Guatemala;
GÜNTHER, Fish. Centr. Amer., 472, 1869.
Hemirhombus fuscus, POEY, Synopsis, 406, 1868, Havana; POEY, Enumeratio, 138, 1875.

3059. CITHARICHTHYS GILBERTI, Jenkins & Evermann.

Head $3\frac{1}{4}$ to $3\frac{2}{5}$; depth of head 4; depth of body $1\frac{9}{10}$ to $2\frac{1}{5}$. D. 77 to 82;
A. 57 to 61; scales 18–40 to 46–19. Body comparatively broad, formed as in
C. spilopterus, the two profiles about equally arched; snout slightly longer
than longest diameter of eye, and without a distinct spine. Eyes on left
side, equal in size, small, 5 to $5\frac{2}{3}$ in head; interorbital space narrow, $1\frac{2}{3}$ in
eye, low, slightly grooved, and scaled on posterior portion only. Maxillary
$2\frac{2}{3}$ in head, reaching barely to posterior border of eye; upper jaw projecting. Teeth small, in a single series; gill rakers 4 + 13, short and slender,
not longer than pupil, with a rather broad base, narrowing to a slender
stalk. Dorsal fin beginning in front of upper eye, the first 3 rays growing from the blind side, the distance of origin from snout 7 in head; fin
rays all simple, $2\frac{3}{10}$ in head; pectorals nearly equal, the one on colored
side being slightly longer, $1\frac{9}{10}$ in head; rays on colored side 9; on blind
side 8; ventrals $2\frac{3}{10}$ in head; caudal rounded, caudal peduncle short,
its depth 8 in the body, equaling height of anal; scales large, ciliated,
pretty uniform, those toward head and margins of disk becoming smaller;
lateral line gradually descending along the course of about 16 scales, from
which point it is straight. Color light brown, with about 15 irregular
dark blotches of various sizes, the largest being a pair on the latter third
of the disk, 1 on each side of lateral line, as great in diameter as length
of ventral fin. Specimens from fresh waters (*C. sumichrasti*) are much
darker in color; gray, everywhere closely peppered with dark specks;
pectoral and caudal mottled. Pacific coast of tropical America; very
abundant in sandy bays from Guaymas to Panama, ascending all the
streams. This species very closely resembles *C. spilopterus*, representing
the latter on the Pacific coast, and it has been frequently recorded under
the name *C. spilopterus*. *C. gilberti* differs mainly in the shorter gill rakers
and in the slightly larger scales. Fresh-water specimens (as the type of
C. sumichrasti from Rio Zanatenco, Chiapas, and numerous examples collected by us in Rio Presidio, near Mazatlan) differ considerably in color,
being much darker, but there is no other difference. ("This species is
dedicated to Prof. Charles H. Gilbert, whose collection and notes on fishes
from Mazatlan, containing undescribed species, this among them, were
destroyed by fire in 1883.")

Citharichthys gilberti, JENKINS & EVERMANN, Proc. U. S. Nat. Mus. 1888, 157, Guaymas,
Mexico (Type, No. 39627. Coll. Jenkins & Evermann); JORDAN, Proc. Cal. Ac. Sci.
1895, 503.
Citharichthys sumichrasti, JORDAN & GOSS, Review Flounders and Soles, 276, 1889, Rio
Zanatenco, Chiapas. (Coll. Prof. Francis E. Sumichrast. Type, 25299, M. C. Z.)
Citharichthys spilopterus, GÜNTHER, Fish. Centr. Amer., 471, pl. 80, fig. 2, 1869; JORDAN
& GILBERT, Proc. U. S. Nat. Mus. 1882, 382; JORDAN & GILBERT, Proc. U. S. Nat. Mus.
1882, 630; JORDAN & GILBERT, Bull. U. S. Fish Comm. 1882, 108–111; not of GÜN-
THER, 1862.

1052. ETROPUS, Jordan & Gilbert.

Etropus, JORDAN & GILBERT, Proc. U. S. Nat. Mus. 1881, 364 (*crossotus*).

Eyes and color on left side. Body regularly oval, deep, and compressed. Head small; mouth very small, the teeth close set, slender, and pointed, somewhat incurved, mostly on the blind side; no teeth on vomer. Eyes small, separated by a narrow, scaleless ridge; margin of preopercle free. Ventrals free from anal, that of colored side inserted on ridge of abdomen, its base rather long. Dorsal fin beginning above eye; caudal double truncate; anal without spine. Scales thin, deciduous, ctenoid on left side, cycloid on blind side. Lateral line simple, nearly straight. Size small. This genus is very close to *Citharichthys*, from which it differs only in the very small size of the mouth, and in the correspondingly weak dentition. The 3 or 4 known species are similar in appearance to the species of *Citharichthys*, and they inhabit the same waters. Another genus extremely close to *Etropus* and *Citharichthys* is *Thysanopsetta*, a South American genus. The teeth in *Thysanopsetta* are, however, arranged in a band. The larval forms are translucent and symmetrical, as in *Platophrys*, *Monolene*, *Arnoglossus*, etc. (ἦτρον, abdomen; πούς, foot; in allusion to the insertion of the ventrals, common to all the *Psettinæ*, but not found in other small-mouthed species.)

- *a.* Snout not acute; dorsal rays 75 to 85.
  - *b.* Body comparatively elongate, the depth rather less than ½ length.
    - *c.* Dorsal rays 81; anal 58; head 4¼ in length; eye 3⅜ in head; maxillary 4. MICROSTOMUS, 3060.
    - *cc.* Dorsal rays 77; anal 61; head 4 in length; eye 3½ in head; maxillary 4⅓. RIMOSUS, 3061.
  - *bb.* Body very deep, the depth more than ½ length; eye 3¾ in head; maxillary 4; head 4⅘; depth 1⅚ to 2; D. 76 to 85; A. 56 to 67; scales 42 to 48; cirri on subopercle of blind side very numerous, white; olive ground, with darker blotches; fins sanded. CROSSOTUS, 3062.

3060. ETROPUS MICROSTOMUS (Gill).

"D. 81; A. 58; caudal 4, 6, 5, 3; pectoral 10; ventral 6. The height of the body enters about 2⅗ times (0.36–0.37) in the extreme length; that of the caudal peduncle about 11 times. The head forms a fifth of the length, is rather abbreviated, scarcely sinuous above the eyes, blunt at the snout, which scarcely exceeds ⅐ of the head's length and the rostral area is rhombic, and not higher than long. The eyes are even; the longitudinal diameter contained about 3⅗ times (0.05½) in the head's length. The mouth is rather small, the length of the upper only equaling ¼ of the length, and that of the lower ⅗ of the head's length. The teeth are very small, and close together; larger in front. The dorsal commences above the front of the orbit, and is highest and convergent near the fortieth ray, which equals about $\frac{1}{10}$ of the total length; the anal is highest at about the twenty-fifth ray, and is as high or even higher than the dorsal. The caudal is rounded behind, and forms about ⅙ of the length. The pectoral fins are unequally developed, that of the dark side being prolonged, and contained only 6⅗ times in the total length, while that of the white side only equals $\frac{1}{10}$ of the same; the rays are all simple.

The ventral fins are also unequally developed, the right being on the abdominal ridge at its origin, rather in advance of the opercular margin and with its longest rays contained about 14 times in the total length; stretched backward, it extends to the second anal ray; the fin on the white side is more advanced, wider, and its rays longer, contained less than 12 times in the length and extends backward to nearly the third anal ray. The scales are large, angular behind, covered with smaller ones, especially near the point of junction of contiguous ones, where alone they are developed on the blind side; the scales of the eyed side are mostly minutely ciliated behind, unarmed, however, near the lateral line, the scales of which last are quadrate and mostly covered; the scales of the blind side are less angular behind and unarmed. The lateral line runs through about 42 scales, while of longitudinal rows there are 10 above and 14 below the lateral line. The color is uniform reddish brown. A single specimen, little more than 3 inches long, was first obtained by Professor Baird, at Beesleys Point." (Gill.)

This species has not been certainly recognized by recent writers, unless, as supposed by Jordan & Goss, it is identical with *Etropus rimosus*. It is in any event certainly an *Etropus*. In the Museum of Comparative Zoology are numerous young specimens collected at Somers Point, New Jersey, by Dr. Stimpson. These seem to belong to the genus *Etropus*. The teeth are equal; the scales are 44, and the depth of the body is $2\frac{1}{3}$ in its length. The eye is 4 in head, the dorsal rays 75 to 80, and the anal rays 56 or 57. The color is light brown, mottled and spotted with darker. These probably represent the *Citharichthys microstomus* of Gill, collected in the same neighborhood by the same naturalist. We are unable to distinguish them from *Etropus rimosus*. (μικρός, small; στόμα, mouth.)

Citharichthys microstomus, GILL, Proc. Ac. Nat. Sci. Phila. 1864, 223, Beesleys Point, New Jersey (Coll. Prof. S. F. Baird); GOODE & BEAN, Oceanic Ichthyology, 446, 1896; JORDAN, Proc. U. S. Nat. Mus. 1890, 332.
Etropus microstomus, JORDAN & GOSS, Review Flounders and Soles, 278, 1889.

3061. ETROPUS RIMOSUS, Goode & Bean.

Head 4 in body; depth 2 to $2\frac{1}{4}$. D. 77; A. 61; scales 12–41–14; eye $3\frac{1}{4}$ in head; maxillary $4\frac{1}{3}$; snout 8; caudal 1. Body somewhat elongate, pear-shaped; mouth very small, its cleft less than $\frac{1}{3}$ the orbit, its angle below anterior margin of lower eye; teeth well developed on blind side on each jaw, also on eyed side of lower jaw in front; eyes placed in the same vertical; upper eye close to dorsal profile, and separated from its mate by a space less than $\frac{1}{3}$ its diameter; interorbital ridge low; nostrils in a line with the interorbital ridge, each in a short tube, the posterior the larger, the anterior midway between tip of snout and front of lower orbit; head entirely scaled; accessory scales numerous. Dorsal commencing on blind side at anterior margin of eye, the highest rays somewhat behind middle of fin, its length 7 times in total length; origin of anal under base of pectoral, its highest rays equal to those of dorsal; pectoral of eyed side longest, equal to head without snout; caudal fin rounded. Color gray, hoary above, with a few irregularly placed indistinct brownish blotches, none of them larger than eye; white below. West coast of Florida;

type, 100 mm. long, collected by the *Albatross* at Station 2408, depth. 21 fathoms, between Pensacola and Cedar Keys, Florida. (Goode & Bean.) On reexamining our specimens of *Etropus*, we find that those obtained by Jordan & Evermann from Pensacola differ from the others in the greater elongation of the body and in the somewhat grayer coloration. These correspond fairly to the description and figure of *Etropus rimosus*. All other specimens from the United States coast collected by Dr. Jordan and his associates are, in our opinion, referable to *Etropus crossotus*.. The original description of *Citharichthys microstomus*, Gill, fits this species better than any other known. The fish in question is much too elongate for *Etropus crossotus* (depth 2⅔ in total length), and the mouth is too small for any of the known species of *Citharichthys* (maxillary 4 in head; mandible 2½). We have little doubt of the identity of *Etropus rimosus* and *microstomus*, but leave the matter for further investigation. The separation of *E. rimosus* from *E. crossotus* is not beyond question. (*rimosus*, frosted.)

Etropus rimosus, GOODE & BEAN, Proc. U. S. Nat. Mus. 1885, 593, coast of Florida, between Pensacola and Cedar Keys, dredged at the depth of 21 fathoms; GOODE & BEAN, Oceanic Ichthyology, 455, pl. 104. figs. 360, 361, 1896. (The latter figure an excellent representation of the symmetrical, translucent larval form, before the right eye has crossed the forehead.)

Etropus crossotus, JORDAN & EVERMANN, Proc. U. S. Nat. Mus. 1886, 476; not of JORDAN & GILBERT.

3062. ETROPUS CROSSOTUS, Jordan & Gilbert.

Head 4⅖ in length; depth 1¾ to 2. D. 76 to 85; A. 56 to 67; V. 6; scales 42 to 48; vertebræ $9+25=34$. Body oval, strongly compressed, with the dorsal and ventral curves nearly equal; both outlines strongly arched anteriorly, the body much deeper in adult specimens. Head very small; snout short; mouth very small, its cleft not so long as diameter of orbit. Teeth conical, pointed, close set, strongly incurved, in a single series, those in upper jaw on blind side only, those in lower jaw on both sides. Eyes large, the lower in advance of the upper, the two separated by a very narrow scaleless ridge, which extends backward above preopercle; edge of opercle on blind side, with a row of conspicuous white cilia. Upper nostril turned somewhat to blind side; anterior nostril on left side, with a very slender cirrus. Dorsal fin commencing over front of upper eye, its middle rays highest, the anterior not elevated; anal fin not preceded by a spine, its middle rays highest; caudal fin very sharply double-truncate, as long as head; pectorals short, that of left side the longer, about ¾ length of head; ventral of colored side on ridge of abdomen, the membrane of its last rays nearly reaching base of first ray of anal; ventral of blind side longer than the other, ½ length of head, inserted farther forward than that of colored side. Vent lateral, with a well-developed anal papilla. Scales thin, large, ctenoid on colored side, smooth on blind side, those on the middle part of the body larger; head entirely scaly, except snout and interorbital ridge; rays of vertical fins with scales on the basal half, on colored side; lateral line developed equally on both sides, nearly straight. Color olive brown, with some darker blotches most distinct in the larger specimens; vertical fins finely mottled and streaked with black and gray; pectoral and ventral on left side spotted. Tropical America on both coasts, north to Cerros Island and North Carolina, south to Panama and Rio Janeiro; the type a single specimen, about 5 inches long, taken with a seine in the Astillero at Mazatlan. This little fish seems to be abundant in all warm and sandy shores of tropical America. It is the smallest and feeblest of all our flounders, and has therefore been generally overlooked by collectors. In the Museum of Comparative Zoology are specimens of this species from Rio Janeiro, Santos, Victoria, Para, and Sambaia, in Brazil. The largest of these is 6 inches in length. Head 5 in length; depth $1\frac{9}{10}$; scales 44; D. 85; A. 67. We have specimens from Charleston, Cedar Keys, New Orleans, Galveston, Beaufort, North Carolina, Mazatlan, Panama, and from several localities along both sides of the coast of Lower California. These vary in form, color, and squamation, but we are unable to point out specific distinctions among them. (κροσσωτός, fringed, from the cirri of the subopercle.)

Etropus crossotus, JORDAN & GILBERT, Proc. U. S. Nat. Mus. 1881, 364, Mazatlan; JORDAN & GILBERT, Proc. U. S. Nat. Mus. 1882, 305; JORDAN & GILBERT, Proc. U. S. Nat. Mus. 1882, 618; JORDAN & GILBERT, Bull. U. S. Fish Comm. 1882, 108–111; JORDAN & GILBERT, Synopsis, 839, 1883; BEAN, Cat. Fish. Intern. Exh. 1883, 44; JORDAN & SWAIN, Proc. U. S. Nat. Mus. 1884, 234; JORDAN & GOSS, Review Flounders and Soles, 278, 1889.
Etropus microstomus, JORDAN, Proc. U. S. Nat. Mus. 1886, 29; not *Citharichthys micros-tomus*, GILL.

1053. MONOLENE, Goode.

Monolene, GOODE, Proc. U. S. Nat. Mus. 1880, 338 *(sessilicauda).*
Thyris, GOODE, Proc. U. S. Nat. Mus. 1880, 344 *(pellucidus ;* larval form); name preoccupied.
Delothyris, GOODE, Proc. U. S. Nat. Mus. 1883, 110 *(pellucidus)*; substitute for *Thyris.*

Body thin, elongate; eyes on the left side, very close together, near the profile; mouth moderate, the length of the maxillary less than $\frac{1}{3}$ that of the head; teeth minute, in a single series, nearly equal on both sides; no teeth on vomer or palatines. Scales rather large, ctenoid on colored side, cycloid on blind side. Lateral line well marked, that of colored side strongly and angularly curved anteriorly, that of blind side nearly straight. Pectoral of blind side wholly absent; dorsal beginning on the snout, its rays all simple. Caudal fin sessile, almost confluent with dorsal and anal. Ventral fins normal, that of the left side on the ridge of the abdomen; gill rakers few, feeble. Vertebræ 43. Deep-sea fishes, closely allied to *Trichopsetta* and *Arnoglossus*, but with the right pectoral obsolete. The translucent larva of *Monolene* is similar to the larva of *Platophrys*. It was at first described as a distinct genus (*Thyris = Delothyris*) by Dr. Goode before its true character was recognized. The following are the characters ascribed to the larval genus *Delothyris :*[*]

---

[*] The following are the characters of the species, *Delothyris pellucidus*, Goode: Colorless, translucent; 3 conspicuous, dusky, longitudinal lines on left side, the middle one faintest. Two streaks on right side; eyes black. Body thin, pellucid, divided into 3 longitudinal tracks by depressions at the bases of the rows of interspinous processes. Scales small, thin, caducous. Head very small; eyes small, protruding, their diameter equal to the interorbital space and $\frac{1}{3}$ the length of the snout; mouth small, formed as in the soles, the upper jaw somewhat hook-shaped. Dorsal fin beginning in advance of the eye, of long, flexible, simple rays, the tips of which are much exserted. Pectorals inserted far below lateral line, that on blind side as long as orbit, the other as long as snout; ventrals reaching past front of anal. Head 5; depth 3. D. 100; A. 80; P. 12 (left), 4 (right). Length 3 inches. (Goode.) Gulf Stream, off the coast of Rhode Island.

Body elongate, soft, and translucent. Head very short; mouth small, toothless. Eyes sinistral, close together, the. lower slightly advanced. Pectoral of blind side smallest; ventrals crowded together on median keel of body, their bases prolonged on this keel. Rays simple; dorsal beginning on the snout; caudal subsessile, almost confluent with dorsal and anal. Scales very thin, easily detached, probably cycloid. Lateral line well marked, straight. (μόνος, one; ὠλένη, arm.)

a. Dorsal rays 99 to 103; anal rays 79 to 81; scales 92.   SESSILICAUDA, 3063.
aa. Dorsal rays 124; anal rays 100; scales 105.   ATRIMANA, 3064.

3063. MONOLENE SESSILICAUDA, Goode.

Head 5 in body; depth 2⅔. D. 99 to 103; A. 79 to 84; scales 23–92–25; eye 4 in head; highest dorsal ray 2; highest anal rays slightly more than 3; pectoral 1⅓; caudal nearly 1. Body moderately elongate; maxillary extending slightly past front of lower eye, with uniserial, subequal teeth; lower eye in advance of upper; interorbital space very small, less than ⅙ the diameter of eye; head everywhere closely scaled; scales ctenoid; lateral line strongly curved over anterior ⅔ of pectoral, the curve with 2 angles, 72 scales along straight portion; lateral line of blind side nearly straight. Origin of dorsal over anterior edge of lower eye, longest rays in the posterior fourth of the fin; origin of anal under base of pectoral; pectoral present only on eyed side. Color on left side ashy brown, with numerous more or less distinct darker brown spots; on blind side white, pectoral blackish with traces of lighter transverse bands. Specimens from shallow water near Key West (Coll. Prof. C. C. Nutting), according to Mr. Garman, are much more brightly colored. These are "grayish brown, with numerous spots of darker to blackish over head and body, the spots being ½ as large as the eye or smaller, arranged for the greater part in broad transverse bands as wide as the interspaces, of which bands the first and foremost passes from the nape to the opercle, the second lies immediately behind the pectoral, the third just in front, and the fourth just behind the middle of total length, and the fifth, more indistinct, crosses near the ends of the dorsal and anal; the caudal is crossed by 2 rather indefinite narrow streaks; the pectoral is white at its base and bears 3 or 4 narrow curved transverse bands of white, separating 3 or 4 similar bands of black, which with the white are more distinct in the lower half of the fin." Specimens from 150 fathoms or more have markings similar but less distinct. D. 104; A. 84; V. 6; P. 11; scales 22–93–24. Deep waters of the Gulf Stream, Cape Cod to Key West. (*sessilis*, sessile; *cauda*, tail.)

Monolene sessilicauda, GOODE, Proc. U. S. Nat. Mus. 1880, 337, 338, deep sea south of New England; GOODE, Proc. U. S. Nat. Mus. 1880, 472; JORDAN & GILBERT, Synopsis, 841, 1883; JORDAN & GOSS, Review Flounders and Soles, 280, 1889; GOODE & BEAN, Oceanic Ichthyology, 452, figs. 357 A & B, 1896; GARMAN, Bull. Iowa Lab. Nat. Hist. 1896, 91.

Thyris pellucidus, GOODE, Proc. U. S. Nat. Mus. 1880, 344, Gulf Stream off the coast of Rhode Island.

Delothyris pellucidus, GOODE, Proc. U. S. Nat. Mus. 1883, 109.

3064. MONOLENE ATRIMANA, Goode & Bean.

Head 4¼ in length of body; depth about 3. D. 124; A. 100; scales 30–105–32; eye 2⅔ in head; maxillary 3; highest dorsal ray 2; left ventral 3½; pectoral 4¼ in body; caudal 6. Body rather elongate; snout slightly produced. Mouth oblique, the maxillary extending to a little behind front of lower eye, teeth uniserial, well developed on both sides; lower eye in advance of upper; interorbital a very narrow ridge, about 9 in eye; nostrils in very short tubes in the same line with the interorbital ridge, the posterior one is slightly less distant from lower eye than the anterior one is from the snout; head everywhere scaly; lateral line strongly arched over anterior third of pectoral. Origin of dorsal on blind side above front of lower eye, longest rays in posterior fourth of fin; highest rays of anal a little higher than dorsal rays; pectoral on eyed side only; caudal sessile, rounded. Color light brownish gray, right ventral pale, other fins dusky; pectoral and eyelids black. West Indies. The type was taken by the *Blake* in 288 fathoms, off Barbados; its length is 114 mm. (Goode & Bean.) (*ater*, black; *manus*, hand.)

Monolene atrimana, GOODE & BEAN, Bull. Mus. Comp. Zool., XII, 155, 1886, deep waters off Barbados; JORDAN & GOSS, Review Flounders and Soles, 280, 1889; GOODE & BEAN, Oceanic Ichthyology, 455, fig. 358, 1896.

Family CCXX. SOLEIDÆ.

(THE SOLES.)

Body oblong or elongate, usually scaly; mouth very small, much twisted toward the eyed side; the teeth in villiform bands, very small or obsolete; eyes small, close together, with or without a bony ridge between them; edge of preopercle adnate, concealed by the skin and scales; gill openings narrow, the gill membranes adnate to the shoulder girdle above; pectoral fins small or wanting; ventral fins small, 1 or both sometimes wanting. Small fishes living on sandy bottoms, similar to the *Pleuronectidæ* in structure, but much degraded, the fins and teeth having lost many of their distinctive qualities. The vertebræ are usually in increased numbers.*

---

* The following are the numbers of vertebræ in several species of *Soleidæ*:

I.—ACHIRINÆ.

Achirus fasciatus	8 + 20 = 28
Achirus inscriptus	9 + 19 = 28

II.—SOLEINÆ.

Synaptura zebra	[8 + 41 = 49]
Solea solea	9 + 40 = 49
Solea kleini	10 + 37 = 47
Solea aurantiaca	[46]
Quenselia ocellata	9 + 28 = 37
Microchirus luteus	8 + 29 = 37
Monochirus hispidus	9 + 25 = 34

III.—CYNOGLOSSINÆ.

Symphurus atricaudus	10 + 42 = 52
Symphurus nigrescens	9 + 40 = 49
Symphurus plagiusa	9 + 38 = 47

They are numerous in the warm seas, and those of sufficient size are valued as food. Genera about 12; species 150. The North American species belong to 2 subfamilies very different one from the other. The soles are naturally divisible into 3 subfamilies, each quite distinct from the others, and possibly independently descended or degraded from normal *Pleuronectidæ*. The *Achirinæ*, or American soles, are apparently allied to the *Psettinæ*, and as in the latter, the ventral fin of the eyed side extends along the ridge of the abdomen. The *Soleinæ*, or European soles, show in the insertion of ventral and in other respects a strong resemblance to the *Pleuronectinæ*. The more aberrant *Cynoglossinæ*, or tongue fishes, are perhaps degraded *Soleinæ*, but the eyes are sinistral, as in the *Psettinæ*. In the *Soleinæ* and *Achirinæ* the eyes are dextral, as in the *Pleuronectinæ*.

ACHIRINÆ:

I. Soles with the eyes on the right side and separated by a distinct bony ridge; the ventral with long base confluent with the anal. Body oblong or ovate, with the color on the right side; eyes moderate or small, the upper eye usually more or less in advance of the lower; mouth small, more or less twisted toward the blind side; teeth little developed, in villiform bands; edge of opercle adnate, usually concealed by the scales; gill openings more or less narrowed, the gill membranes adnate to the shoulder girdle above; blind side of head usually with fringes; pectoral fins small, sometimes wanting; ventral fins developed, one or both of them sometimes obsolete; scales usually ctenoid, rarely wanting; lateral line straight, usually single; right ventral with extended base, confluent with the anal fin.

   a. Gill openings of moderate extent, confluent below; vertical fins well separated; body ovate in outline, the depth nearly ½ the length; pectoral fins rudimentary or wanting; lateral line straight; scales well developed, ctenoid, those on the head more or less enlarged, those of the blind side of the head with fringes; vertebræ about 28.   ACHIRUS, 1054.

   aa. Gill openings very small, separate, each reduced to a small slit below angle of opercle; right ventral beginning at the chin; pectoral fins minute or wanting; lateral line straight; snout dilated, the dorsal beginning upon it.

      b. Scales present, ctenoid; caudal somewhat confluent with dorsal.

        c. Left ventral rudimentary, with 2 rays.   APIONICHTHYS, 1055.

      bb. Scales none; caudal free from dorsal and anal.   GYMNACHIRUS, 1056.

CYNOGLOSSINÆ:

II. Soles with the eyes on the left side, not separated by a bony ridge. Body elongate, more or less lanceolate in outline, with the color on the left side; eyes small, very close together, with no distinct interorbital ridge between them; mouth small, twisted toward the blind side; teeth little developed, in villiform bands; gill openings narrow, the gill membranes adnate to the shoulder girdle above, joined together and free from the isthmus below; pectoral fins wanting (in the adult); ventral fins small, that of the blind side often wanting; vertical fins more or less confluent; scales ctenoid; lateral lines sometimes wanting, sometimes duplicated.

   d. Ventral fin of eyed side only present, free from the anal; no pectoral fins; no lateral line; head without fringes.   SYMPHURUS, 1057.

1054. ACHIRUS, Lacépède.

(AMERICAN SOLES.)

Achirus, LACÉPÈDE, Hist. Nat. Poiss., IV, 659, 1803 (*fasciatus,* etc.).
Achirus, CUVIER, Règne Animal, Ed. 2, II, 343, 1829 (restriction to *fasciatus,* etc.).
Trinectes, RAFINESQUE, Atlantic Journal and Friend of Knowledge, I, 1832 (*scabra*).
Grammichthys, KAUP, Archiv fur Naturgsch. 1858, 94 (*lineatus; fasciatus*); *Achirus* being restricted to *Pardachirus barbatus,* etc.

Monochirus, KAUP, Archiv fur Naturgsch. 1858, 94 (*maculipinnis*); not of RAFINESQUE, 1814, a genus of *Soleinæ*.
? *Aseraggodes*, KAUP, Archiv fur Naturgsch. 1858, 103 (*guttulata*).
Baiostoma, BEAN, Proc. U. S. Nat. Mus. 1882, 413 (*brachiale*).
Bœostoma, JORDAN & GILBERT, Synopsis, 965, 1883; amended orthography.

Eyes and color on the right side. Body oblong, bluntly rounded ante-riorly. Head small; eyes small, close together, the upper eye in advance of the lower, the two separated by a bony ridge; mouth small, somewhat turned toward the colored side; nasal flaps present, the nostril of the blind side fringed; lip of the colored side fringed; teeth very small, on blind side only; gill openings rather narrow, but confluent below, not reduced to a slit; the branchiostegal region scaled. Head closely scaled every-where, the scales on the colored side similar to those on the body, those of the nape and chin much enlarged; scales on the blind side anteriorly with their pectinations more or less produced, forming cirri; scales of both sides extremely rough, extending on the fins. Lateral line straight, sim-ple; edge of preopercle covered by the scales. Dorsal beginning on the snout, low in front and thickly scaled, its rays divided; anal fin similar, without spine; caudal fin free, convex; caudal peduncle very short and deep; pectoral fin of left side wanting, that of right side small or obso-lete; ventral rays 3 or 4, the ventral fin of the colored side long, connected with the anal by a membrane. This strongly marked genus contains numerous species, all very closely related, and nearly all American. It has been united by Dr. Günther with *Solea*, but for no good reason, as the number of vertebræ is very much fewer than in the European soles, and the right ventral fin is decurrent along the abdomen and united with the anal in the American soles, while it is short and wholly free in all the European forms. The 2 groups belong in fact to distinct subfamilies. It is also worth noticing that the name *Achirus* is prior in date to that of *Solea*. The species with rudimentary pectoral fins have been set apart by Dr. Bean to form the genus *Baiostoma*, but the very slight development of these organs in some of the species and the evidently very close relation-ship of them all lead us to regard *Baiostoma* as a subgenus only. If we follow Kaup in restricting the name *Achirus* to the Asiatic group called *Pardachirus*, the present genus would receive the name of *Trinectes*. It seems to us, however, that both Lacépède and Cuvier regarded the species called by us *fasciatus* as the type of their genus *Achirus*. (ἄχειρ, without hands; without pectoral fins.)

BAIOSTOMA (βαιός, small; στόμα, mouth):
  a. Pectoral fins small, present at least on the right side.
    b. Pectoral fin present on both sides, that of the left side rudimentary, of a single ray; that of the eyed side with about 3 rays.
      c. Dorsal rays 60 to 67; anal rays about 48; scales 80; depth 1⅜ in length. Color brownish, irregularly spotted with darker, and with about 10 black vertical lines crossing the lateral line.   ACHIRUS, 3065.
      cc. Dorsal rays 53 to 57; scales 75 to 80; depth 1⅔ in length; scales not very rough, those of colored side with scattered, hair-like appendages, some black, others pale. Color olivaceous; head, body, dorsal, and anal fins covered with a network of dark lines; traces of about 8 dark cross streaks sometimes present.   INSCRIPTUS, 3066.

bb. Pectoral of right side only present.

  d. Dorsal rays 65 or 66; anal rays 48 to 51.

    e. Pectoral well developed, with about 6 rays; scales of eyed side without hair-like filaments; scales of lateral line 77 to 80; chin little prominent. KLUNZINGERI, 3067.

  dd. Dorsal rays 50 to 58; anal rays 38 to 48.

    f. Pectoral fin of 4 to 6 rays, considerably longer than eye; body with 8 to 10 narrow, vertical dark bars, these sometimes obsolete with age.

      g. Vertical fins all with round, dark spots, these usually especially distinct on the caudal fin; some of the scales of eyed side with black, hair-like appendages; pectoral fin with 5 or 6 rays, about 3 in head, its length equal to that from outer edge of 1 eye to outer edge of another. Head 3½; depth 1½. Body with 8 narrow, vertical cross streaks. LINEATUS, 3068.

      gg. Vertical fins dark, without distinct markings. Body broad, ovate, the depth about 1½ in length; pectoral fin with 4 rays; scales of right side with numerous black, hair-like appendages. D. 56; A. 42; scales 70. MAZATLANUS, 3069.

    ff. Pectoral fins of 1 to 3 rays, about as long as eye.

      h. Body with 6 narrow, dark bands, these sometimes obsolete. Body rather narrowly ovate, its depth 1⅝ in length. D. 58; A. 44; scales 85. FONSECENSIS, 3070.

      hh. Body with about 10 black cross lines; depth 1⅝ in length. D. 61; A. 44; scales 60; pectoral of a single ray. FISCHERI, 3071.

      hhh. Body with very numerous (20 to 40) black cross bands, which are as broad as the interspaces.

        j. Blind side of snout with few fringes. Depth 1½ in length. D. 55; A. 48; scales 80. Body covered by many blackish, wavy bands; caudal with black spots. SCUTUM, 3072.

ACHIRUS:

aa. Pectoral fins wholly wanting.

  k. Dorsal rays 46; anal rays 33; right lower lip with serrated fringes; nostril in a fringed tube; depth 1½ in length; head 3. Color brown, head and body with numerous large, rounded or kidney-shaped white spots, edged with dark brown; scales 70. FIMBRIATUS, 3073.

  kk. Dorsal rays 50 to 55; anal rays 37 to 46; right lower lip fringed; left nostril with some fringes; depth 1⅜ in length; head 4; none of the scales of eyed side with hair-like appendages. Color dusky olive, more or less mottled with about 8 dark vertical stripes, these varying very much in width and number; caudal spotted. FASCIATUS, 3074.

  kkk. Dorsal rays 59 or 60; anal rays 41 to 45; snout and chin without evident fringe or barbel; right lower lip fringed; head 4 in length; depth 1¾; scales 64. Body with 12 black cross bands with narrower ones between; caudal spotted. PANAMENSIS, 3075.

Subgenus BAIOSTOMA, Bean.

3065. ACHIRUS ACHIRUS (Linnæus).

D. 60 to 67; A. 48; P. right 3, left 1; scales 80. Pectorals rudimentary on both sides; right ventral fin composed of 5 rays, which are continuous with the anal. Scales on the nape and on chin twice as large as those on body; snout with a few fringes on blind side; right lower lip fringed. Height of body 1¾ in total length (without caudal); width of interorbital space nearly equal to, or rather more than diameter of eye; upper eye slightly in advance of lower; longest dorsal rays are in posterior fifth of the fin, ⅔ of length of head; caudal rounded, rather longer than head. Brownish, irregularly spotted with darker, and with about 10 black vertical lines crossing the lateral line. Coasts of Surinam. (Günther.) Not seen by us.

We know this species only from Dr. Günther's description. *Pleuronectes achirus*, Linnæus, is based on a description by Gronow of some *Achirus* from Surinam. Gronow's fish agrees with the present species in having 60 dorsal rays and 48 anal rays, in being brown, with transverse black bands, with dark spots on the fins, as well as in coming from Surinam. But Gronow explicitly denies the presence of pectorals, and the present species has rudimentary pectoral fins on both sides. Probably these were overlooked by Gronow, and as no other species found in the same region has so large a number of rays, we feel justified in the use of the name *Achirus achirus* for this species. (ἁ-, without; χείρ, hand.)

Pleuronectes oculis dextris, corpore glabro, pinnis pectoralibus nullis, GRONOW, Museum, I, No. 42, Surinam.
Pleuronectes achirus, LINNÆUS, Syst. Nat., Ed. X, 268, 1758, Surinam; based on GRONOW.
Solea gronovii, GÜNTHER, Cat., IV, 472, 1862, Surinam.
Achirus gronovii, JORDAN, Proc. U. S. Nat. Mus. 1886, 602.
Achirus achirus, JORDAN & GOSS, Review Flounders and Soles, 311, 1889.

3066. ACHIRUS INSCRIPTUS, Gosse.

Head 3¾ in body; depth 1¾. D. 53 to 57; A. 40; scales 75 to 80; interorbital width less than eye; upper eye in advance of lower. Pectoral fin present on each side, that of the left side rudimentary, of a single ray; that of the eyed side with about 3; left ventral with 1 or 2 small rays, in some specimens entirely absent; right ventral joined to anal. Scales smaller and less rough than usual in this genus, those of nape scarcely enlarged on eyed side, those of blind side much fringed; scales of colored side with scattered hair-like appendages, some black, others pale. Color olivaceous; head, body, dorsal, and anal fins covered with a network of dark lines; traces of about 8 dark cross streaks sometimes present; caudal fin yellowish, nearly plain, or with a few dark dots or reticulations, its base dusky. Vertebræ 8 + 20 = 28. West Indies north to Key West. Known to us from numerous specimens taken by Dr. Jordan at Key West, and from specimens from Hayti, in the museum at Cambridge. These specimens belong undoubtedly to the species called *reticulatus* by Poey, and this is apparently not different from the *inscriptus* of Gosse, as the agreement with the latter is even closer than with the former description. (*inscriptus*, written on.)

Achirus inscriptus, GOSSE, Nat. Sojourn Jamaica, 52, pl. 1, fig. 4, 1851, Jamaica; JORDAN, Proc. U. S. Nat. Mus. 1884, 143; JORDAN & GOSS, Review Flounders and Soles, 311, 1889.
Monochir reticulatus, POEY, Memorias, II, 317, 1861, Cuba; POEY, Synopsis, 409; POEY, Enumeratio, 139.
Solea reticulata, GÜNTHER, Cat., IV, 472, 1862.
Solea inscripta, GÜNTHER, Cat., IV, 473, 1862.
Bæostoma reticulatum, BEAN & DRESEL, Proc. U. S. Nat. Mus. 1884, 152.

3067. ACHIRUS KLUNZINGERI (Steindachner).

Head 3⅔ in body; depth 1⅔. D. 65; A. 51; scales 37–80–42; eye 3⅔ in head; height of dorsal and anal 1¼; caudal 1. Body moderately broad; eyes in the same vertical line; interorbital as wide as length of eye; angle of mouth reaching a little past front of lower eye; right under lip fringed; scales near upper profile of head enlarged, all scales strongly ctenoid; scales of eyed side without hair-like filaments. Pectoral of right side only present, with about 6 rays; caudal round behind. Color brownish, with 9 or 10 narrow blackish cross lines; small rounded blackish spots on the membranes of each of the vertical fins, much as in *A. lineatus*. (Steindachner.) Pacific coast of tropical America; Panama to Guayaquil. (Named for Dr. C. B. Klunzinger, Professor of Zoology at Stuttgart, author of Memoirs on the Fishes of the Red Sea.)

Solea klunzingeri, STEINDACHNER, Zur Fische des Cauca und der Flüsse bei Guayaquil, 44, 1879, Guayaquil.
Achirus klunzingeri, JORDAN, Proc. U. S. Nat. Mus. 1885, 391; JORDAN & GOSS, Review Flounders and Soles, 312, 1889.

3068. ACHIRUS LINEATUS (Linnæus).

Head 3¼; depth about 1¼. D. 49 to 58; A. 38 to 44; scales 75 to 85. Pectoral fin of right side only developed, of 4 to 6 rays, considerably longer than eye. Body with 8 to 10 narrow vertical dark bars, these sometimes obsolete with age; vertical fins all with round dark spots, these usually especially distinct on the caudal fin; some of the scales of eyed side with black, hair-like appendages; pectoral fin with 5 or 6 rays, about 3 in head, its length equal to that from outer edge of one eye to outer edge of another. Color brown, the young spotted with whitish, the adult sometimes with darker; body with about 8 narrow vertical cross streaks of blackish. West Indies and Brazil, Florida Keys to Uruguay; common and variable.

We have placed the Florida species, *comifer* and *brachialis*, in the synonymy of *lineatus*. They differ from the latter only in the slightly smaller number of the scales and fin rays. The following table shows our count of a number of specimens from different localities:

Locality.	Species.	D.	A.	Scales.
Key West	comifer	50	35	55 to 67
Pensacola	brachialis	51	37	75 to 77
Cienfuegos	lineatus	54	43	85
Rio Janeiro	maculipinnis	57	42	85
Do	maculipinnis	54	44	72
Rio Grande do Sul	maculipinnis	49	38	70
Coary	maculipinnis	53	40	68
Manacapuru	maculipinnis	55	42	75

It is evident from this table that neither the fin rays nor the scales form characters by which the subspecies can be absolutely distinguished. It is evident also, from the examination of large series of specimens, that the

3030——92

coloration is subject to very great variations—as great as in *Achirus fasciatus*. In some of these the caudal is dark and immaculate, in others pale and usually profusely spotted. In some the ground color is nearly plain blackish, in others it is pale, usually with narrow dark cross bands, but sometimes closely spotted everywhere. The specimens examined by us are from Pensacola and Egmont Key (*brachialis*), Key West (*comifer*), Cienfuegos (Cuba, Poey), Coary, Teffy, Tapajos, Porto Alegre, Pernambuco, Cannarivieras, Manacapuru, Porto do Moz, Rio Grande do Sul, Rio Janeiro, San Matheo, Rosario, Itabapuana, Obidos, Xingu, Gurupa, Jutaby, Curaçao, Para, Bahia, Santarem, Iça, Fonteboa, San Paolo, Rio Trompetas, Sambaia, Manes, Javary, and Tabatinga. The species would appear to be one of the commonest in Brazil. (*lineatus*, striped.)

a. Var. lineatus.

Passer lineis transversis notatus, SLOANE, Jamaica, 2, 77, pl. 246, fig. 2, 1725, Jamaica.
Pleuronectes fuscus subrotundus glaber, BROWNE, Jamaica, 445, 1756, Jamaica.
Pleuronectes lineatus, LINNÆUS, Syst. Nat., Ed. X, 268, 1758, Jamaica; based on BROWNE and
      SLOANE; not of Ed. XII, which is *Achirus fasciatus*. ·
Monochir maculipinnis, AGASSIZ, Spix, Pisc. Brasil., 88, pl. 49, 1829, Brazil; POEY, Synopsis,
      409, 1868.
Monochir lineatus, QUOY & GAIMARD, Voy. Uranie, Zool., 238, 1824.
Achirus lineatus, D'ORBIGNY, Voyage Amér. Mérid., Poiss., pl. 16, fig. 2, 1847; JORDAN &
      GOSS, Review Flounders and Soles, 312, 1889.
Solea maculipinnis, GÜNTHER, Cat., IV, 473, 1862; KNER, Novara Fische, III, 289, 1866.
Achirus maculipinnis, JORDAN, Proc. U. S. Nat. Mus. 1886, 602.

b. Var. brachialis.

Baiostoma brachialis, BEAN, Proc. U. S. Nat. Mus. 1882, 413, Appalachicola Bay and South
      Florida. (Types, Nos. 26605 and 30463. Coll. Silas Stearns.)
Bæostoma brachiale, JORDAN & GILBERT, Synopsis, 965, 1883.
Achirus brachialis, JORDAN, Proc. U. S. Nat. Mus. 1884, 149.

c. Var. comifer.

Achirus comifer, JORDAN & GILBERT, Proc. U. S. Nat. Mus. 1884, 31, Key West. (Coll.
      Dr. Jordan.)

3069. ACHÍRUS MAZATLANUS (Steindachner).

(MEXICAN SOLE; LENGUADO DE RIO; TEIPALCATE.)

Head 3$\frac{1}{2}$ in body; depth 1$\frac{1}{2}$. D. 56; A. 42; scales 70; eye 7$\frac{1}{2}$ in head; dorsal and anal rays 3$\frac{1}{2}$ in depth of body; caudal 3. Body broad, oval; eyes small, the upper in advance of the lower; interorbital about $\frac{1}{2}$ the diameter of eye; nostril in a tube, placed just above middle of mouth; pectoral developed on eyed side only, with about 4 rays; origin of dorsal on tip of snout; greatest height of dorsal and anal behind their middle; scales of right side with numerous black hair-like appendages. Color brownish, with 8 or 9 narrow vertical black bars; fins dark, without distinct markings. West coast of Mexico, entering all streams; common and variable. Many specimens from Mazatlan and Rio Presidio examined by us, as also a specimen from Chiapas. (Name from Mazatlan,* the river of the deer.)

---

* "With eternal sun above thee,
    'T is not strange the tall deer loved thee,
    That he gave his name, Mazatl,
    To thy river, Mazatlan!"

Solea mazatlana, STEINDACHNER, Ichth. Notizen, IX, 23, 1869, Mazatlan; JORDAN & GIL-
BERT, Bull. U. S. Fish Comm. 1882, 108.
Solea pilosa, PETERS, Berliner Monatsber. 1869, 709, Mazatlan.
Achirus mazatlanus, JORDAN, Proc. U. S. Nat. Mus. 1885, 391; JORDAN & GOSS, Review
Flounders and Soles, 313, 1889; JORDAN, Proc. Cal. Ac. Sci. 1895, 505.

3070. ACHIRUS FONSECENSIS (Günther).

Head $3\frac{1}{2}$; depth $1\frac{3}{5}$. D. 58; A. 44; P. 2; scales about 85. No trace of a
pectoral on the left side, that on the right not much longer than the eye;
right ventral fin composed of 5 rays, which are continuous with the
anal. Scales on the nape twice or thrice as large as those on the body.
The upper part of the snout slightly overlaps the lower jaw. The left
anterior part of the head with numerous tentacles; the right lower lip
with very distinct slender fringes; nostril on the right side in a wide and
short tube. The width of the interorbital space is less than the diameter
of the eye; the upper eye is in advance of the lower. The rays of the
vertical fins are branched; the longest dorsal rays are $\frac{2}{3}$ of the length of
the head. Caudal rounded, as long as the head. Brownish olive, with 6
pairs of deep brown vertical lines extending on the dorsal and anal fins.
Pacific Coast of Tropical America. (Günther.) Described from 1 specimen $4\frac{1}{4}$ inches long, from Gulf of Fonseca; 2 others since taken by us in
Rio Presidio, near Mazatlan. (Name from Fonseca, the type locality.)

Solea fonsecensis, GÜNTHER, Cat., IV, 475, 1862, Gulf of Fonseca. (Coll. Sir John Richard-
son.)
Achirus fonsecensis, JORDAN & GOSS, Review Flounders and Soles, 314, 1889; JORDAN,
Proc. Cal. Ac. Sci. 1895, 505.

3071. ACHIRUS FISCHERI (Steindachner).

(PEGE OJA.)

Head $3\frac{1}{4}$ in body; depth $1\frac{3}{5}$; caudal $3\frac{1}{4}$. D. 61; A. 44; P. right. 1; V. 5;
scales 60 to 62. Pectoral wanting on left side, rudimentary on right, of a
single ray scarcely longer than eye; right ventral connected with the anal.
Scales on neck and lower portion of head $1\frac{1}{4}$ to 2 times as large as those on
body; right side of lower lip fringed; upper jaw not projecting forward
over lower jaw; left side of head with dermal flaps only around corner
of mouth and on lower jaw. Eyes small, the upper a little further forward than the other, and 2 in snout; breadth of forehead equaling diameter of eye. Dorsal rays increasing gradually in length to the forty-eighth,
which is about $\frac{4}{9}$ length of head. Scales strongly ctenoid, the teeth considerably largest at the middle; rays of all the fins, except of pectorals,
scaled to their tips, the membranes less fully scaled; only the anterior
third of the caudal membranes scaled, and between the last dorsal and
anal rays the scales extend slightly upon the fins. A few black thread-
like appendages on right side of body between scales. Color of right
side dark gray; 2 or 3 blackish cross lines on head, about 10 on body,
between them numberless spots of similar color; spots on fins, especially
those on caudal, a little larger; a few large dark spots on body, irregular

and poorly defined; blind side reddish yellow. Total length about 10 cm. Rio Mamone, near Panama; known to us only from Steindachner's description and figure. (Steindachner.) Not seen by us. (Named for W. Fischer.)

Solea fischeri, STEINDACHNER, Beiträge Kenntniss Fluss-Fische Sudamer., I, 13, 1879, Rio Mamone, near Panama.
Achirus fischeri, JORDAN, Proc. Ac. Nat. Sci. Phila. 1887, 391.

3072. ACHIRUS SCUTUM (Günther).

Head 3⅔; depth 1½. D. 55; A. 48; P. 3; scales 80. No trace of a pectoral on left side; right pectoral quite rudimentary, scarcely longer than the eye; right ventral composed of 5 rays, which are continuous with anal. Scales on nape nearly twice as large as those on the body; snout with scarcely any fringes on the blind side, right lower lip fringed. Width of interorbital space less than horizontal diameter of orbit; upper eye slightly in advance of lower. Longest dorsal rays in posterior third of fin, ⅔ length of head; caudal rounded, longer than head. Grayish; head, body, and fins with numerous blackish, irregular, waving, sometimes bifurcate, transverse bands, which are broader than the interspaces; caudal with rounded deep black spots; the left side uniform white. Pacific coast of Central America. (Günther.) Not seen by us. (*scutum*, a shield.)

Solea scutum, GÜNTHER, Cat., IV, 474, 1862, Gulf of Fonseca, Panama.
Achirus scutum, JORDAN & GOSS, Review Flounders and Soles, 314, 1889.

Subgenus ACHIRUS.

3073. ACHIRUS FIMBRIATUS (Günther).

Head 3; depth 1½. D. 46; A. 33; scales 70. Pectorals none; right ventral of 5 rays, which are continuous with the anal. Scales on nape 4 times, those on the chin twice, as large as on the body. Upper part of the snout slightly bent downward over the mandible and forming a short hook; right lower lip broadly fringed, each fringe being serrated; nostril in a short, wide, fringed tube. No tentacles on left side of head. Width of interorbital space equaling diameter of circular small orbit; upper eye slightly in advance of lower. Longest dorsal rays ⅔ of length of head; rays of vertical fins branched; caudal rounded, its length being ¼ of the total. Brown; head and body with numerous large, rounded, or kidney-shaped white spots, edged with dark brown. Gulf of Fonseca, Central America. (Günther.) Known from 1 specimen, 3⅙ inches long. (*fimbriatus*, fringed.)

Solea fimbriata, GÜNTHER, Cat., IV, 477, 1862, Gulf of Fonseca. (Coll. Sir John Richardson.)
Achirus fimbriatus, JORDAN & GOSS, Review Flounders and Soles, 315, 1889.

3074. ACHIRUS FASCIATUS, Lacépède.

(AMERICAN SOLE; HOG CHOKER.)

Head 4 in body; depth 1½. D. 50 to 55; A. 37 to 46; scales 66 to 75; eye 7 in head; height of dorsal and anal nearly 2; caudal 1½. Body broad, regularly elliptical; mouth moderate, reaching just past front of lower eye; right lower lip fringed; eyes very small, the upper one in advance of the lower; nostril ending in a wide tube, nearer lower eye than tip of snout; interorbital space with scales, more than ¼ eye; head and body scaled with strongly ctenoid scales, none of them with hair-like appendages; lateral line nearly straight; gill opening short, about twice as long as maxillary. Origin of dorsal on tip of snout; last few rays of dorsal and anal rapidly decreasing, giving the fins a truncate appearance posteriorly; pectorals wholly wanting; caudal rounded. Color dusky olive, more or less mottled, and with about 8 dark, vertical stripes, these varying very much in width and in number; vertical fins with the membrane of every second or third pair of rays blackish, besides dark cloudings at base of fin; caudal with numerous longitudinally oblong spots; blind side often with round, dark spots, especially in northern specimens, usually immaculate in southern ones (var. *browni*). Vertebræ $8 + 20 = 28$. South Atlantic and Gulf coast, from Cape Ann to Brazos Santiago, ascending sandy streams in shallow water. The species is the best known of the American soles, and it is common along our coast, ascending the rivers for a considerable distance above tide water. It seldom exceeds 5 or 6 inches in length, and is of but little value as food on account of its small size. Here described from a specimen, 4 inches long, from Beaufort, North Carolina. This species has not yet been recorded from the West Indies. The form found along the Gulf coast has been described as a distinct species under the name *Solea browni*. The differences are not very evident. We have compared a number of specimens from Boston (*fasciatus*) with others from Pensacola, and find the following differences, none of which is constant: In the Gulf variety (*browni*) the blind side is always immaculate, while in almost all Atlantic examples (*fasciatus*) the blind side is profusely covered with round, dark spots. In 1 specimen, however (11360, Boston), the blind side is immaculate. The darker cross streaks on the eyed side are usually broader and more numerous in southern specimens, and the scales on the blind side of the head rougher. There are no constant differences either in the fin rays or in the scales. We have examined specimens of this species from Boston, Chestertown, Tarrytown, New York, Port Monmouth, Havre de Grace, Potomac River, Neuse River, Beaufort, Charleston, Pensacola, Mobile, and Galveston. In 1 large specimen from Pensacola (11482, M. C. Z.) there is a rudiment of a pectoral fin on the eyed side. It consists of a single ray ⅔ as long as the eye. (*fasciatus*, banded.)

Pleuronectes lineatus, LINNÆUS, Syst. Nat., Ed. XII, 458, 1766, on a specimen from Charleston, received from Dr. Garden; not *Pleuronectes lineatus* of Ed. X.

Achirus fasciatus, LACÉPÈDE, Hist. Nat., Poiss., IV, 659, 662, 1803, Charleston; excl. syn., description based entirely on the Linnæan account of the fish sent by Garden; JORDAN & GOSS, Review Flounders and Soles, 315, 1889.

Pleuronectes mollis, MITCHILL, Trans. Lit. and Phil. Soc. N. Y., I, 1815, 388, pl. 2, fig. 4, New York.

Pleuronectes apoda, MITCHILL, Amer. Monthly Mag. and Crit. Rev., Feb., 1818, 244, Straits of Bahama; perhaps *A. lineatus*.

Trinectes scabra, RAFINESQUE, Atlantic Journal and Friend of Knowledge, I, 1832, Pennsylvania, in fresh water.

Solea browni, GÜNTHER, Cat., IV, 477, 1862, New Orleans; Texas.

Achirus lineatus, CUVIER, Règne Animal, Ed. 2, II, 343, 1829; GILL, Cat. Fishes East Coast
    N. Am., in Rept. U. S. Fish Comm., 1871-72, 794; JORDAN & GILBERT, Proc. U. S. Nat.
    Mus. 1878, 368; GOODE, l. c., 1879, 110; GOODE & BEAN, l. c., 1879, 123; BEAN, l. c., 1880,
    77; JORDAN & GILBERT, l. c, 1882, 618; BEAN, l. c., 1883, 365.
Grammichthys lineatus, KAUP, Archiv fur Naturgesch. 1858, 101.
Achirus mollis, STORER, Synopsis, 228, 1846; STORER, Hist. Fish. Mass., 206, pl. 32, 1867;
    DE KAY, New York Fauna: Fishes, 303, pl. 49, fig. 159, 1842.
Achirus achirus mollis, JORDAN, Cat. Fish. N. A., 137, 1885.
Solea achirus, GÜNTHER, Cat., IV, 476, 1862; not *Pleuronectes achirus* L.
Achirus achirus, JORDAN, Proc. U. S. Nat. Mus. 1885, 19; JORDAN, Cat. Fish. N. A., 137, 1885.
Achirus lineatus, var. *browni,* JORDAN & GILBERT, Proc. U. S. Nat. Mus. 1882, 305.

3075. ACHIRUS PANAMENSIS (Steindachner).

Head 4 in body; depth 1¾. D. 59; A. 45; scales 63 to 65; highest dorsal
and anal spines 2 in head; caudal 1. Body broad, elliptical; angle of
mouth below middle of lower eye; edge of lower lip, on the eyed side,
fringed; eyes small, the upper in advance of the lower; interorbital
scaled, scarcely as wide as diameter of eye; scales ctenoid; pectorals
wholly wanting; origin of dorsal on end of snout; highest rays of anal
and dorsal behind their middle; ventral rays short; middle rays of caudal
the longest, fin sharply rounded behind. Color brown; about 12 dark
cross bands on head and body; between these faint, paler cross bands,
which form spots on dorsal and anal; caudal similarly spotted, the spots
forming obscure cross bands. (Steindachner.) Pacific coast of tropical
America, Panama.

Solea panamensis, STEINDACHNER, Ichthyol. Beiträge, V, 10, taf. II, 1876, Panama.
Achirus panamensis, JORDAN & GOSS, Review Flounders and Soles, 316, 1889.

1055. APIONICHTHYS, Kaup.

Apionichthys, KAUP, Archiv fur Naturgesch. 1858, 104 (*dumerili*).
Soleotalpa, GÜNTHER, Cat., IV, 489, 1862 (*unicolor*).

Gill openings very small, separate, each reduced to a slight slit below
angle of opercle; right ventral beginning at the chin, confluent with the
anal; pectoral fins wanting or very small; lateral line present, straight;
eyes small; snout dilated, the dorsal beginning upon it. Scales present,
ctenoid; caudal fin somewhat confluent with dorsal. Left ventral rudimentary, with 2 rays. West Indies and Brazil. This genus is closely
related to *Achiropsis,* Steindachner, of the rivers of Brazil, but in the
latter genus the ventrals are both well developed. (ἀ-, not; πίων, fat;
ἰχθύς, fish.)

3076. APIONICHTHYS UNICOLOR (Günther).

D. 76; A. 57; V. right 5, left 2; scales 92. Body very flat and thin, its
height being contained 2¾ times in the total length (without caudal), the
length of the head 4¼ times. The upper part of the snout is dilated, bent
downward like an aquiline nose, the end covering the symphysis of the
mandibles; the cleft of the mouth is curved, the lower eye being immediately above its angle. The eyes are mere points, rather distant from

each other. The gill opening is reduced to a very small slit, the gill membrane being attached to the sides of the throat. The dorsal fin commences on the extremity of the snout and terminates at the root of the caudal, its rays are simple, and each is accompanied by a series of very small ctenoid scales; the longest rays are not quite ⅓ as long as the head, and occupy the middle and the third quarter of the fin. Caudal quite free, as long as the head, somewhat pointed. The right ventral appears as a mere continuation of the anal; the left is reduced to 2 minute rays near the vent. The scales on both sides are ctenoid, those on the neck and on the chin being twice the size of those on the body. Color uniform brownish gray. Coast of Surinam and Brazil. The above description from Günther, taken from the type of *Soleotalpa unicolor*. A specimen (No. 4677, M. C. Z.), from Obydos, Brazil, examined by us, differs in coloration, being pale brown, the body and fins profusely covered with round, dark spots of varying sizes, the largest as wide as from eye to eye. Head 4⅓; depth 2⅔. D. 78; A. 56; scales 100; V. 2. Eyes reduced to points, the upper in advance of lower, near middle of length of head; gill openings small, subequal; right ventral beginning at the chin, continuous with anal; dorsal and anal slightly connected with caudal. Steindachner gives D. 72; A. 53; scales 95. Color brownish, mottled with darker spots. Probably Günther's specimen is faded. (*unicolor*, one-colored.)

Apionichthys dumerili, KAUP, Archiv fur Naturgesch. 1858, 104, no locality; no description.

Soleotalpa unicolor, GÜNTHER, Cat., IV, 489, 1862, West Indies. (Coll. Scrivener.)

Apionichthys nebulosus, PETERS, Berliner Monatsberichte 1869, 709, Surinam.

? Apionichthys bleekeri, HORST, Nederl. Tydschr. Dierk., IV, 30, 1878, locality unknown; specimen in Mus. Utrecht. (Description not seen by us.)

Apionichthys unicolor, JORDAN, Proc. U. S. Nat. Mus. 1886, 603.

Apionichthys dumerili, BLEEKER, Nederl. Tydschr. voor Dierkunde, II, 1865, 305; STEINDACHNER, Ichth. Beitr., VIII, 48, 1878.

1056. GYMNACHIRUS, Kaup.

Gymnachirus, KAUP, Uebersicht der Soleinæ, Archiv fur Naturgesch. 1858, 101 (*nudus*).

This genus differs from *Achirus* in the absence of scales; the dorsal and anal are free from the caudal. Brazil. ($\gamma\nu\mu\nu\acute{o}\varsigma$, naked; *Achirus*.)

3077. GYMNACHIRUS FASCIATUS, Günther.

Head 4½; depth 1½. D. 68; A. 50; pectoral of right side present, very small, of 2 rays, ⅓ length of eye; jaws hidden in thick skin; lips and left side of head covered with fringes. Gill opening not extending upward as far as pectoral; vertical fins in thick skin. Olive, with 14 brown cross bands as broad as the interspaces, all extending on dorsal and anal, the first across snout, the second and third across eye; caudal with 3 brown bands. (Günther.) Locality unknown, probably Surinam or Brazil; a related species (*G. nudus*, Kaup; no pectoral fins. D. 51; A. 42), being described from Bahia. (*fasciatus*, banded.)

Gymnachirus fasciatus, GÜNTHER, Cat., IV, 488, 1862, locality unknown; JORDAN & GOSS. Review Flounders and Soles, 317, 1889.

1057. SYMPHURUS,* Rafinesque.

(TONGUE-FISHES.)

Symphurus, RAFINESQUE, Indice d'Ittiologia Siciliana, 52, 1810 (*nigrescens*).
Bibronia, Cocco, Alcuni Pesci del mare di Messina, 15, 1844 (*ligulata;* larval form).
Plagusia, CUVIER, Règne Animal, Ed. 2, II, 344, 1829 (based on *Plagusia* of BROWN);
    name preoccupied in Crustaceans, Latreille, 1806.
Plagiusa, BONAPARTE, Catalogo Metodico, 51, 1846 (*lactea*); substitute for *Plagusia* pre-
    occupied.
Aphoristia, KAUP, Archiv fur Naturgesch. 1858, 106 (*ornata*).
Glossichthys, GILL, Cat. Fish. E. Coast N. A., 51, 1861 (*plagiusa*).
Ammopleurops, GÜNTHER, Cat., IV, 490, 1862 (*lacteus = nigrescens*).
? Bascanius, SCHIÖDTE, Naturhist. Tydsskr., V, 269, 1867 (*tœdifer;* larval form).
Acedia, JORDAN, in JORDAN & GOSS, Review Flounders and Soles, 321, 1889 (*nebulosus*).

Body elongate, more or less lanceolate in outline, with the eyes and color on the left side; eyes small, very close together, with no distinct interorbital ridge between them; mouth small, twisted toward the blind side; teeth little developed, in villiform bands; edge of preopercle covered by the scales; gill openings narrow, the gill membranes adnate to the shoulder girdle above, joined together and free from the isthmus below; pectoral fins wanting (in the adult); vertical fins more or less confluent; scales ctenoid; lateral line wanting. Ventral fin of eyed side only present, free from the anal; head without fringes. (σύν, together; φύω, to grow; οὐρά, tail; from the united vertical fins.)

SYMPHURUS:

a. Scales not minute, ctenoid, 65 to 105 in number; dorsal rays 86 to 100; anal rays 70
    to 87.
    b. Scales rather large, about 65; head 4⅓; depth 4¼; color, clouded brown.
                                                                    PIGER, 3078.
    bb. Scales small, moderately ctenoid, 75 to 105 in a longitudinal series.
        c. Dorsal and anal pale anteriorly, becoming more or less abruptly black
            posteriorly.
            d. Caudal fin abruptly pale, at least at tip.
                e. Body elongate, depth 4⅓ in length; head 5⅓. D. 96 to 100; A. 86 or
                    87; scales 88 to 90. Color, grayish, speckled with brown; dor-
                    sal and anal fins black on last tenth, the caudal abruptly pale;
                    tips of fin rays vermilion.   MARGINATUS, 3079.
            ee. Body deeper, the depth 3⅓ to 3¾ in length.
                f. Color, light brown, irregularly barred and marbled with
                    darker; dorsal and anal with 3 to 6 inky blotches poste-
                    riorly. D. 92 to 95; A. 75 to 78.   ATRAMENTATUS, 3080.

---

* We follow Jordan & Goss in using the name *Symphurus* instead of *Aphoristia*, as the so-called *Ammopleurops lacteus* is a genuine member of the latter genus, and as it seems to be evident that the latter species is the original of *Symphurus nigrescens* of Rafinesque. The following is Rafinesque's description: "III. Gen. *Symphurus*. Ala caudale acuta, e riunita all' ale dorsali, ed anali, occhj alla sinistra. *Osserv.* Si dovranno ragguagliare in questo genere due specie del genere *Achirus* di Lacepede, cioè gli *A. bilineatus*, e *A. ornatus*. Sp. no. 44. *Symphurus nigrescens*. Nerastro senza fascie, allungato, una sola linea laterale da ogni lato."
This single lateral line assumed to distinguish *Ammopleurops* from *Aphoristia* is not a real lateral line, but a depression along the median line produced by the junction of the muscles. The species of *Symphurus* are somewhat numerous and very closely allied. With the exception of the European *Symphurus nigrescens*, all of them are American. The development of the species is imperfectly known. According to Giglioli, the larvæ called *Bibronia*, may belong to this genus, and so possibly may *Charybdia*. The name *Plagusia* belongs properly to the present genus rather than to the type of *Plagusia bilineata*, to which it has been restricted by Kaup and Günther. It is, however, preoccupied in Crustaceans, and in any case, both *Plagusia* and the substitute name *Plagiusa* are antedated by the name *Symphurus*.

ff. Color, light olive, with numerous roundish brownish black spots much larger than eyes, dorsal and anal black, with narrow white margin. D. 94; A. 77; scales 95. FASCIOLARIS, 3081.

dd. Caudal fin black, as is a large part of dorsal and anal, the black either continuous or in the form of large spots. Color brownish, often mottled, usually with more or less distinct darker cross bands and with longitudinal streaks along the rows of scales, sometimes nearly plain brown.

g. Scales quite small, 98 to 105.

h. Body decidedly elongate, the depth about 4⅖ in length. D. 97; A. 82; scales 98. ELONGATUS, 3082.

hh. Body less elongate, the depth 3⅓ in length; head 5⅓; longitudinal streaks very distinct. D. 100; A. 80; scales 105. ATRICAUDUS, 3083.

gg. Scales rather larger, 75 to 90.

i. Body rather elongate, the depth 3⅘ to 4⅓; dorsal rays 80 to 90; anal 80 to 85; opercular flap large; body with 3 or 4 dark cross bands. LEEI, 3084.

ii. Body less elongate, the depth $3\frac{1}{10}$ to 3⅗ in length; the head 5⅓ to 5⅔. D. 90 to 95; A. 75 to 80. PLAGUSIA, 3085.

cc. Dorsal and anal pale throughout, or more or less mottled or spotted with darker, the caudal similarly colored, not distinctly black. Body not very elongate, the depth 3 to 3⅓ in length. (Probably all varieties of *S. plagiusa*.)

j. Body with dark cross bands more or less distinct; the fins mottled or speckled; upper eye slightly in advance of lower.

k. Dorsal rays 86 to 95; anal rays 75 to 80; head 5 in length; depth 3¼; scales 85 to 93; cross bands more distinct than in related species. PLAGIUSA, 3086.

kk. Dorsal rays 78 to 85; anal rays 70 to 72; head 5 in length; depth 3½; scales 80 to 90. Color, light brown, with darker cross bars, which become obsolete with age. PUSILLUS, 3087.

jj. Body uniform grayish, without cross bands; last part of dorsal and anal with 3 or 4 oblong black blotches, each somewhat larger than the eye; upper eye directly above lower; head 5⅖ in length. Scales 85; D. 92; A. 75. DIOMEDEANUS, 3088.

ddd. Caudal and posterior part of dorsal and anal not black, scarcely darker than anterior part; scales 92; D. 93; A. 73. WILLIAMSI, 3089.

ACEDIA (Spanish name of *Symphurus plagusia* at Havana):

aa. Scales very small, ctenoid, each with a median dark streak, which simulates a keel, but is not a ridge; snout and jaws naked; fin rays in increased number.

l. Head 5⅔; depth 4⅗. D. 119; A. 107; scales 120. Grayish, everywhere mottled with brown. NEBULOSUS, 3090.

Subgenus SYMPHURUS.

3078. SYMPHURUS PIGER (Goode & Bean).

Head 4⅓ in total length; depth 3⅐. D. 90; A. 69 to 75; ventral 4; scales 65–34 (transverse); eye 6 in head; snout 4½; mouth oblique, curved, its angle below middle of lower eye; teeth feeble, closely placed, a little stronger on colored side; nostril tubular, a little nearer eye than tip of snout; eyes moderate in size, very close together, the upper very slightly in advance, its distance from the dorsal outline equal to its diameter; scales large, ctenoid, deciduous. Dorsal begining over middle of upper eye; longest dorsal and anal rays 3 in depth of body; pectorals obsolete.

Color grayish and brownish, with a submetallic luster upon the scales when examined separately; the denticulations of the scales dark and prominent, giving a clouded general aspect; some of the smaller specimens with large, irregular, brownish blotches above, and a dark subcircular blotch near the root of the tail, its diameter twice eye; colorless below. (Goode & Bean.) West Indies and Gulf of Mexico, in deep water; a well-defined species. (*piger*, sluggish.)

Aphoristia pigra, GOODE & BEAN, Bull. Mus. Comp. Zool., XII, 5, 154, 1886, St. Kitts, in about 250 fathoms (Coll. *Blake*); GOODE & BEAN, Oceanic Ichthyology, 460, 1896.

Symphurus piger, JORDAN & GOSS, Review Flounders and Soles, 326, 1889.

3079. SYMPHURUS MARGINATUS (Goode & Bean).

Head 5¼ in total length; depth 4½. D. 96 to 100; A. 86 or 87; ventral 4; scales 88 to 90; eye 4½ in head; snout 4½. Body slender, lanceolate; mouth moderate, oblique, curved, its angle below front of pupil of upper eye; dentition feeble; eyes moderate, close together, the upper very slightly in advance; nostril in a long slender tube, midway between lower eye and tip of snout; scales moderate, strongly and sharply denticulate, not keeled; origin of dorsal above posterior margin of upper eye; anal scarcely so high as dorsal; median caudal rays short. Color in life, reddish gray, much speckled with brown; belly bluish gray; bases and membranes covering fin rays dark brown; caudal abruptly pale; tips of dorsal and anal rays and some of the membrane covering caudal rays vermilion. West Indies, in deep water. Described from a specimen, 102 mm. in length, collected by the *Blake* at Station CLXXXI, in 321 fathoms. (Goode & Bean.) (*marginatus*, edged.)

Aphoristia marginata, GOODE & BEAN, Bull. Mus. Comp. Zool., XII, No. 5, 153, 1886, off St. Vincent (Coll. *Blake*); GOODE & BEAN, Oceanic Ichthyology, 459, fig. 376, 1896.

Symphurus marginatus, JORDAN & GOSS, Review Flounders and Soles, 323, 1889.

3080. SYMPHURUS ATRAMENTATUS, Jordan & Bollman.

Head 4¾ to 5; depth 3⅙ to 3⅝. D. 92 to 95; A. 75 to 78; scales 95 to 100, 38 in a cross series. Body more elongate than in *S. atricaudus*. Eyes larger, the upper in advance of lower, vertical diameter of each 3½ to 4 in head. Cleft of mouth somewhat more curved than in *S. atricaudus*, otherwise similar. Scales larger than in *S. atricaudus*; spines on posterior margin not so strong. Ventral fins (measured from angle of gill opening) 2⅔ to 3 in head. Color light brown, irregularly barred and marbled with darker; several irregular grayish bars most distinct on posterior parts, a distinct narrow, dark bar behind gill opening; anterior part of dorsal and anal fin pale, posterior dark; anterior part with 4 to 7 dusky oblique areas, posterior part with 3 to 6 roundish inky-black spots; caudal black, narrowly tipped with white; each scale with a narrow dark edge. Length about 4½ inches. Pacific Ocean off Colombia, in water of moderate depth; common. Related to *Symphurus atricaudus* (Jordan & Gilbert), but distinguished by having 3 to 6 black oblong blotches on posterior part of dorsal and anal; the general coloration darker; the scales and eyes larger. (*atramentatus,* inked.)

Symphurus atramentatus, JORDAN & BOLLMAN, Proc. U. S. Nat. Mus. 1889, 177, off coast of Colombia, at Albatross Station 2795, Lat. 7° 57′ N., Long. 78° 55′ W., in 33 fathoms. (Type, 41157, U. S. Nat. Mus. Coll. *Albatross.*)

3081. SYMPHURUS FASCIOLARIS, Gilbert.

. Depth 3⅝ in length; head 5¼. D. 94; A. 77; scales 95. Eye small, 7 in head; cleft of mouth reaching to below middle of lower eye. Color light olive, with numerous roundish brownish-black spots much larger than eye, the largest arranged in 5 vertical dusky cross bars, the spots being connected by a darker ground color; a vertical dusky streak through eye; a wide dusky cross bar, bounded by darker lines on cheeks; dorsal and anal posteriorly black, with narrow white margin; caudal jet-black, with white edge; ventral white. Gulf of California, where several specimens were dredged by the *Albatross,* in shallow water. (Gilbert.) (*fasciolaris,* with narrow bands.)

Symphurus fasciolaris, GILBERT, Proc. U. S. Nat. Mus. 1891, 566, Gulf of California. (Coll. Dr. Gilbert.)

3082. SYMPHURUS ELONGATUS (Günther).

Head 5⅛ in body; depth 4⅔. D. 97; A. 82; scales 98 to 105; eye 10 or 11 in head; gape of mouth 3½; caudal 2½. Body extremely elongate; mouth strongly curved, reaching past lower eye; eyes in contact, the upper in advance; opercle vertical behind, devided into 2 convex flaps by a concave portion, its upper end hardly reaching axis of body; scales not keeled, ctenoid. Pectorals obsolete; dorsal beginning above eye; rays of dorsal and anal short, subequal, the fins confluent with the caudal, which ends in a sharp point; ventral of blind side obsolete, that of eyed side on the body ridge, separated from the anal. Color brownish, often mottled, usually with more or less distinct darker cross bands, and with longitudinal streaks along the rows of scales, sometimes nearly plain brown; caudal fin black, as is a large part of the dorsal and anal, the black either continuous or in the form of large spots. Pacific coast of Central America; not rare. Here described from a specimen, 6 inches long, from *Albatross* Station 2804, in Panama Bay, in 47 fathoms. (*elongatus,* elongate.)

Aphoristia ornata, var. *elongata,* GÜNTHER, Fishes Centr. Amer., 473, 1869, Panama.
Aphoristia elongata, JORDAN & GILBERT, Bull. U. S. Fish. Comm. 1882, 111.
Symphurus elongatus, JORDAN & GOSS, Review Flounders and Soles, 323, 1885.

3083. SYMPHURUS ATRICAUDUS (Jordan & Gilbert).

(SAN DIEGO SOLE.)

Head 5⅓; depth 3¼. D. 100; A. 80; scales 105. Body oblong-lanceolate, anteriorly somewhat blunt, regularly narrowed behind and ending in a point; the snout rather abruptly truncate; eyes and color on the left side. Eyes very small, nearly even behind, the upper eye the larger and extending farther forward. A single nostril in front of interorbital space, and apparently a single smaller one below it. Mouth moderate, extending to opposite eye, somewhat turned toward eyed side; lips large, not fringed, the upper with a small black papilla in advance of lower eye, this apparently normal, but it may be a detached piece of skin, hardened by the alcohol; upper jaw scarcely produced, not forming a hook. Teeth small, on blind side only, the edge of the jaw on eyed side forming a smooth ridge. Gill openings narrow, not extending up to level of mouth. Scales very small, ctenoid, pretty regular over the body, much smaller on the head, the rows of scales rendered very distinct by black dots, the stripes converging toward the snout; scales on the 2 sides of the body similar; no lateral line on either side; about 105 scales (100 to 110) in a longitudinal series from the head to the tail, 45 to 50 in cross series. Dorsal fin beginning on head, continuous with anal around the tail; ventral fin of colored side only present, nearly on ridge of abdomen, and separated from the anal by an interval ⅓ longer than cleft of mouth; rays of middle parts of dorsal and anal fins with a fleshy border at base on blind side. Coloration brownish olive, with vertical dark half bars, irregular in size and position, some of them coming down from the back and others up from the belly, these posteriorly nearly meeting, but anteriorly alternating; streaks of dark points along the rows of scales, these forming very distinct longitudinal streaks; posterior part of dorsal and anal broadly edged with black; right side plain white. San Diego to Cape San Lucas, in sandy bays; common in the bay of San Diego, in which locality the numerous specimens before us were taken. A small specimen, 1½ inches long, with light spots on the colored side and a pale ocellation on the black of the tail, taken by Mr. Lyman Belding near Cape San Lucas, probably belongs to the same species. (*ater*, black; *cauda*, tail.)

Aphoristia atricauda, JORDAN & GILBERT, Proc. U. S. Nat. Mus. 1880, 23, San Diego; JORDAN & GILBERT, Synopsis, 842, 1883; JORDAN & GILBERT, Proc. U. S. Nat. Mus. 1882, 380; JORDAN, Proc. U. S. Nat. Mus. 1886, 54.
Symphurus atricauda, JORDAN & GOSS, Review Flounders and Soles, 324, 1889.

3084. SYMPHURUS LEEI, Jordan & Bollman.

Head 4 to 4½ (4⅛ to 4⅔); depth 3⅔ to 4 (4⅛ to 4½). D. 95 to 100; A. 80 to 85; scales 80 to 90, 35 to 38 in a cross series; ventrals 3⅛ to 3½ in head. Body more elongate than in *S. atricaudus* or *S. atramentatus*, approaching that of *S. elongatus;* outline of under part of head more oblique than in the other Pacific coast species. Eyes larger than in the preceding species, the upper in advance of lower, their vertical diameter 5 to 5¼ in head; cleft of mouth extending slightly farther back than in *S. atricaudus* or *atramentatus*, but not beyond eye as in *S. elongatus;* maxillary reaching posterior border of eye, 3⅔ to 4 in head; snout 5⅓ to 5⅔ in head. Scales comparatively large, not so firmly embedded as in *S. atricaudus* or *atramentatus*, those on opercles rather large. Opercular flap larger than in other Pacific species. Color light brown, speckled with darker, and with 3 or 4 broad black cross bands, width of median bands 2⅓ to 3 in head, the posterior band widest; caudal and the posterior ⅔ of the dorsal and anal black; no black spots on dorsal; scales thickly punctulate, but with no distinct darker edgings. Related to *Symphurus atricaudus* (Jordan & Gilbert), but the body with 4 wide black cross bands, and the form more elongate. Bay of Panama. Many specimens of this species were obtained at *Albatross* Station 2804, at a depth of 47 fathoms. It is evidently very different from *S. atramentatus*, and needs comparison only with *S. elongatus*, from which it seems to be sufficiently distinct. Length of type 4½ inches. (Named for Prof. Leslie A. Lee and Mr. Thomas Lee, naturalists on board the *Albatross* when the species was discovered.)

Symphurus leei, JORDAN & BOLLMAN, Proc. U. S. Nat. Mus. 1889, 178, Lat. 8° 16′ 30″ N. Long. 79° 37′ 45″ W. (Type, No. 41134. Coll. Prof. L. A. Lee and Mr. Thomas Lee.)

3085. SYMPHURUS PLAGUSIA* (Bloch & Schneider).

(ACEDIA.)

Head 5⅓ to 5⅔; depth 3⅟₁₀ to 3⅔ in length. D. 90 to 95; A. 75 to 80; scales 75 to 85. Body rather elongate. Color brownish, often mottled, usually with more or less distinct darker cross bands, and with longitudinal streaks along the rows of scales, sometimes nearly plain brown; caudal black, including a large part of dorsal and anal, the black continuous as in the form of spots. West Indies to Brazil; Cuba to Rio Janeiro; common. The numerous specimens of this species examined by us are from Havana, Pernambuco, Santos, Rio Janeiro, Curuça, and Victoria. (*plagusia*, an old name, from πλάγιος, oblique.)

Plagusia, BROWNE, Jamaica, 445, No. 1, 1756, Jamaica.
Pleuronectes plagusia, BLOCH & SCHNEIDER, Syst. Ichth., 162, 1801, Jamaica; after BROWNE.
Achirus ornatus, LACÉPÈDE, Hist. Nat. Poiss., IV, 659, 1803, on a specimen "presented by Holland to France."
Plagusia tessellata, QUOY & GAIMARD, Voyage Uranie, Zoologie, 240, 1824, Rio Janeiro.
Plagusia brasiliensis, AGASSIZ, Spix, Pisc. Brasil., 89, tab. 50, 1827, Brazil.

---

* The synonymy of this species is somewhat doubtful. The original type of *Pleuronectes plagiusa* was sent to Linnæus by Dr. Garden, of Charleston. It would therefore appear probable that this specimen represented the species of this genus which is found on the Carolina coast. But this typical specimen is still preserved in the rooms of the Linnæan Society in London, where it has been examined by Goode and Bean. From their notes (Proc. U. S. Nat. Mus. 1885, 196) we quote:

"The type of this species may have come from Africa or India. There is considerable doubt as to its origin. (See Garden's Correspondence with Linné, p. 314.) D. ca 92, A. ca 80; scales 77. The species is more elongate than our specimens of *Aphoristia plagiusa*, so called, the depth being contained in the total length without caudal 4½ times and the head 6 times."

As, however, no species of this genus are yet known from Africa or India, it is rather probable that Garden's fish actually came from Charleston. The greater slenderness of the original type is perhaps due to distortion, and the smaller number than usual of the scales does not afford a marked distinction. The name *Achirus ornatus* is also doubtful in its proper application. The only thing distinctive in the description of Lacépède is that the typical specimen was "given by Holland to France." Many of the species in this Dutch collection seem to have come from Surinam, and this is probably no exception. But Lacépède's description might apply as well to any other species of *Symphurus* as to this. The name *Pleuronectes plagusia*, given by Schneider to the species described by Browne, seems to admit of no doubt, as this is the only one of the group yet known from Jamaica. If, therefore, the name *Symphurus plagusia* be used for the northern species, or dropped altogether as not identified, the present species will stand as *Symphurus plagusia*. We have compared numerous specimens from Rio Janeiro (representing the nominal species *tessellatus* or *brasiliensis*) with others (*plagusia=ornata*) from Havana. There is certainly no permanent difference. The Brazilian specimens are a little more slender on an average, but there are numerous exceptions, and all variations in color are found in oth.

Aphoristia ornata, KAUP, Archiv fur Naturgesch. 1858, 106; GÜNTHER, Cat., IV, 490; 1862;
    POEY, Synopsis, 409, 1868; POEY, Enumeratio, 140, 1875; KNER, Novara Fische, III, 292;
    D. 90; A. 75; depth 3¼ in length.
Aphoristia plagiusa, JORDAN, Proc. U. S. Nat. Mus. 1886, 53; not *S. plagiusa* of this paper.
Symphurus plagusia, JORDAN & GOSS, Review Flounders and Soles, 324, 1889.

3086. SYMPHURUS PLAGIUSA* (Linnæus).

(TONGUE FISH.)

Head 5; depth 3 to 3⅜. D. 86 to 95; A. 75 to 80; scales 85 to 93. Body
not very elongate. Body grayish, with dark cross bands more distinct
than in related species; dorsal and anal more or less mottled or spotted
with darker; caudal similarly colored, not distinctly black. South Atlantic and Gulf coasts of the United States, from Cape Hatteras to Pensacola
and Key West, replacing *S. plagusia* northward, the species as similar as
the two names; very common on the sandy shores of our South Atlantic
and Gulf States. Our numerous specimens are from Beaufort, Charleston, Pensacola, and Key West. Those from Key West nearly plain gray,
as would be expected in fishes taken from the coral sands. ($\pi\lambda\acute{\alpha}\gamma\iota\sigma\varsigma$,
oblique.)

Pleuronectes plagiusa, LINNÆUS, Syst. Nat., Ed. XII, 455, 1766, on a specimen from Dr. Gar-
    den, probably from Charleston, but the locality not quite certain; and of various
    copyists.
Plagusia fasciata, HOLBROOK MS., DE KAY, New York Fauna: Fishes, 304, 1842, Charles-
    ton.
Glossichthys plagiusa, GILL, Cat. Fish East Coast N. Am., 51, 1861.
Plagusia plagiusa, GILL, Cat. Fish. East Coast N. Am., 794, 1873.
Aphoristia plagiusa, JORDAN & GILBERT, Proc. U. S. Nat. Mus. 1878, 368; JORDAN, *l. c.*,
    1880, 22; JORDAN & GILBERT, *l. c.*, 1882, 305 and 618; JORDAN & GILBERT, Synopsis, 842,
    1883; JORDAN, Proc. U. S. Nat. Mus. 1884, 144.
Aphoristia fasciata, JORDAN, Proc. U. S. Nat. Mus. 1886, 53.
Symphurus plagiusa, JORDAN & GOSS, Review Flounders and Soles, 325, 1889.

3087. SYMPHURUS PUSILLUS (Goode & Bean).

Head 5 in total length; depth 3½. D. 78; A. 70; scales 85 to 90–35 (transverse); eye 5½ in head; snout 5½; length of gape of mouth 4⅖. Body slender,
lanceolate; mouth small, oblique, curved, its angle under anterior margin
of pupil of lower eye; dentition feeble; eyes small, close together, in the
same vertical line; tubular nostril midway between lower eye and tip of
snout; scales small, strongly ctenoid; jaws and snout scaled. Dorsal
beginning above middle of eye, its highest rays 2⅜ in depth of body;
greatest height of anal 3; median caudal rays short; ventrals well separated from anal. Color light brown, with 6 or 7 cross bars of slightly

---

    * A specimen of *Symphurus*, nearly 6 inches long, collected at Beaufort, North Carolina,
by Prof. O. P. Jenkins, seems referable to *Symphurus pusillus* rather than to the typical
plagiusa. It is highly mottled in coloration, the body and fins being profusely speckled
and blotched with blackish, besides 9 or 10 rather distinct cross bands. D. 85; A. 72;
scales about 80. Depth 3¼ in length. Another large specimen, 7 inches long, from the
Florida Keys, is in the museum at Cambridge. This has: D. 82; A. 72; scales 76. Depth
3 in length. Color brown, almost plain, except that the fins are mottled, especially posteriorly; caudal fin not black. If these two specimens are really typical of *Symphurus
pusillus*, it probably can not be separated as a species from *S. plagiusa*.

darker hue; fins pale, with dusky blotches; blind side white. (Goode & Bean.) Gulf Stream, in deep water. Very close to *Symphurus plagiusa.* (*pusillus,* weak.)

Aphoristia pusilla, GOODE & BEAN, Proc. U. S. Nat. Mus. 1885, 590, Gulf Stream, Lat. 40° N., in deep water; GOODE & BEAN, Oceanic Ichthyology, 461, fig. 379, 1896.
Symphurus pusillus, JORDAN, Proc. U. S. Nat. Mus. 1889, 651.

3088. SYMPHURUS DIOMEDEANUS (Goode & Bean).

Head 5⅔ in body; depth 3¼. D. 96 (including ½ of caudal); A. 79; scales 85; eye 6 in head; snout 5; caudal 10 in total length. Mouth oblique, curved, its angle below front of eye, teeth very feeble; nostril tubular, nearer eye than tip of snout; eyes moderate, equal, very close together, upper eye directly over the lower; scales moderate, somewhat deciduous, ctenoid; jaws and snout with small thin scales. Origin of dorsal above middle of upper eye, highest rays 3½ times depth of body; ventrals well separated from the anal. Color uniform gray, lighter below, the scales above somewhat metallic in luster; the last fourth of dorsal with 3 oblong black blotches somewhat larger than eye, the anal with 4, similar in position; in the young there is a slight brownish marginal line upon each scale, and an appearance of indistinct cloudings of brown upon the colored side. Off Trinidad and Dominica and in the Gulf of Mexico. The specimen here described was collected by the *Albatross* at Station 2414, in the Gulf of Mexico, north of the Tortugas, at a depth of 26 fathoms; its length is 140 mm. Other specimens were dredged by the *Albatross* at Station 2362, in Lat. 22° 08′ 30″ N., Long. 86° 53′ 30″ W., in 25 fathoms, and at Stations 2121 and 2122, between Lat. 10° 37′ 40″ N., Long. 61° 42′ 40″ W., and Lat. 10° 37′ N., Long. 61° 44′ 22″ W., in 31 to 34 fathoms. Specimens were also secured by the *Blake* at Stations XXIV and XXV, off Dominica. (Goode & Bean.) Evidently very close to *Symphurus plagiusa.* (*Diomedea,* the Albatross; from the name of the steamer by which most of the deep-sea explorations of the United States Fish Commission have been accomplished.)

Aphoristia diomedeana, GOODE & BEAN, Proc. U. S. Nat. Mus. 1885, 589, Gulf of Mexico, Lat. 25° 04′ 30″ N., Long. 82° 59′ 15″ W. (Type, No. 37347. Coll. *Albatross*); GOODE & BEAN, Oceanic Ichthyology, 460, fig. 378, 1896.

3089. SYMPHURUS WILLIAMSI, Jordan & Culver.

Head 4⅓; depth 3⅔. D. 93; A. 73; scales 92. Body more slender than in *S. plagiusa,* which it much resembles, but less slender than *S. elongatus;* upper eye slightly in advance of lower. Sand color in life; light gray, everywhere finely mottled with light and dark; traces of a few very narrow dark cross bands; fins all mottled; caudal and posterior part of dorsal and anal not black, scarcely darker than anterior part. Known only from Mazatlan, where 2 specimens, the larger about 1¼ inches long, were obtained by Mr. T. M. Williams, in tide pools with sandy bottom in very shallow water near the estuary. (Named for Thomas Marion Williams, a student in biology in Stanford University, discoverer of the species.)

Symphurus williamsi, JORDAN, Proc. Cal. Ac. Sci. 1895, 506, pl. 55, Mazatlan. (Coll. Hopkins Exped. to Mazatlan.)

Subgenus ACEDIA, Jordan.

3090. SYMPHURUS NEBULOSUS (Goode & Bean).

Head 5⅔ in total length; depth 4⅔. D. 119 (to middle of base of caudal); A. 107; V. 5; scales 120; eye 7¼ in head; snout 5. Body slender; angle of mouth below front of lower pupil; teeth feeble, very slender, and rather closely placed, apparently equally developed on both sides; eyes small, close together, separated by a single row of scales, the upper one very slightly in advance; tubular nostril nearer eye than tip of snout; scales small, ctenoid, each with a median dark streak (but not keeled, as erroneously stated in the original description);* jaws and snout naked. Origin of dorsal a little behind eyes, highest rays 3 in depth of body; longest anal rays twice length of snout; median caudal rays longest, twice length of snout; pectorals obsolete; ventrals well separated. Color grayish, everywhere mottled with brown; a dark median line on scales. (Goode & Bean.) Gulf stream. A well-marked species. The increased number of fin rays indicates a probability that the number of vertebræ will also be found similarly increased. (*nebulosus*, clouded.)

Aphoristia nebulosa, GOODE & BEAN, Bull. Mus. Comp. Zool., x, No. 5, 192, 1883, Gulf Stream, off the coast of Carolina; GOODE & BEAN, Oceanic Ichthyology, 458, fig. 375, 1896.

Symphurus nebulosus, JORDAN & GOSS, Review Flounders and Soles, 326, 1889.

Order CC. PEDICULATI.

(THE PEDICULATE FISHES.)

Carpal bones notably elongate, forming a kind of arm (pseudobrachium) which supports the broad pectoral. Gill opening reduced to a large or small foramen situated in or near the axil, more or less posterior to the pectorals. Ventral fins jugular if present; anterior dorsal reduced to a few tentacle-like, mostly isolated spines; soft dorsal and anal short; no scales. First vertebra united to cranium by a suture; epiotics united behind supraoccipital; elongate basal pectoral radii (actinosts) reduced in number; no interclavicles; post-temporal broad, flat, simple; upper pharyngeals 2, similar, spatulate, with anterior stem and transverse blade; basis of cranium simple; no air duct to the swim bladder. Marine fishes, chiefly of the tropics and the oceanic abysses. The group is an offshoot from the *Acanthopteri*, its chief modifications being in the elongation of the actinosts and in the position of the gill opening. Its nearest relatives among the spiny-rayed fishes are, perhaps, the *Batrachoididæ*. (*pediculatus*, having a footstalk.)

ANALYSIS OF FAMILIES OF PEDICULATI.

a. Gill openings in or behind the lower axil of the pectoral; mouth large, terminal.
  b. Pseudobranchiæ present; pseudobrachia with 2 actinosts; head broad, depressed, the enormous mouth with very strong teeth; ventrals present.

LOPHIIDÆ, CCXXI.

---

* The appearance of "keeled scales," described by Goode & Bean, is due to a black line on the skin under the center of each row of scales. There seems to be no real keel, and the species is congeneric with the other species of *Symphurus*.

bb. Pseudobranchiæ none; pseudobrachia with 3 actinosts.
 c. Ventrals present; arm angulate, the pseudobrachia elongate.
                  ANTENNARIIDÆ, CCXXII.
 cc. Ventrals wanting, arm not angulate, the pseudobrachia moderate.
                  CERATIIDÆ, CCXXIII,
aa. Gill openings in or behind the upper axil of the pectoral; mouth small, usually
 inferior.   OGCOCEPHALIDÆ, CCXXIV.

Family CCXXI. LOPHIIDÆ.

(THE ANGLERS.)

Head wide, depressed, very large. Body contracted, conical, tapering rapidly backward from the shoulders. Mouth exceedingly large, terminal, opening into an enormous stomach; upper jaw protractile; maxillary without supplementary bone; lower jaw projecting; both jaws with very strong, unequal, cardiform teeth, some of the teeth canine-like, most of them depressible; vomer and palatines usually with strong teeth. Gill openings comparatively large, in the lower axil of the pectorals. Pseudobranchiæ present. Gill rakers none. Gills 3. Skin mostly smooth, naked, with many dermal flaps about the head. Spinous dorsal of 3 isolated, tentacle-like spines on the head, and 3 smaller ones behind, which form a continuous fin; second dorsal moderate, similar to the anal; pectoral members scarcely geniculated, each with 2 actinosts and with elongate pseudobrachia; ventrals jugular, I, 5, widely separated, large, much enlarged in the young. Young with the head spinous. Pyloric cæca present. Two genera, with 4 or 5 species, living on sea bottoms, at moderate or great depths; remarkable for their great voracity. (*Pediculati*, part, genus *Lophius*, Günther, Cat., III, 178–182, 1861.)

 a. Vertebræ 27 to 31.   LOPHIUS, 1058.
 aa. Vertebræ 18 or 19 only.   LOPHIOMUS, 1059.

1058. LOPHIUS (Artedi) Linnæus.

(FISHING-FROGS.)

Lophius (ARTEDI) LINNÆUS, Syst. Nat., Ed. X, 1, 236, 1758 (*piscatorius*).

 Characters of the genus included above. Vertebræ numerous, about 30 in number. (*Lophius*, the ancient name of *L. piscatorius*, from λόφος, a crest.)

3091. LOPHIUS PISCATORIUS, Linnæus.

(COMMON ANGLER; FISHING-FROG; MONKFISH; GOOSEFISH; ALL-MOUTH; BELLOWS-FISH.)

 D. I-I-I, III–10; A. 9. Body depressed, tapering, scarcely longer than head. Humeral spine with 3 points, of which the posterior is the longest. Head surrounded with a fringe of barbels; top of head, in young, with many strong spines. Anterior dorsal spine elongate, fleshy at tip. Brownish, mottled, below white; mouth behind the hyoid bone immaculate; pectorals and caudal black at tip; peritoneum black. Length 3 feet. North

3030——93

Atlantic, on both coasts; generally common, ranging southward along the shore to Cape Hatteras; found in deep water as far south as Barbados, in 209 fathoms, and to the Cape of Good Hope; northward to Norway and Nova Scotia. A well-known fish of singular ugliness of appearance, and of enormous voracity. (Eu.) (*piscatorius*, pertaining to an angler, in allusion to the baited dorsal spines which overhang the cavernous mouth.)

*Lophius piscatorius,** LINNÆUS, Syst. Nat., Ed. x, 1, 236, 1758, seas of Europe; after ARTEDI, *Lophius ore cirrhoso*, etc.; GÜNTHER, Cat., III, 179, 1861; GILL, Proc. U. S. Nat. Mus. 1878, 219; JORDAN & GILBERT, Synopsis, 844, 1883.

Lophius americanus, CUVIER & VALENCIENNES, Hist. Nat. Poiss., XII, 380, 1837, Philadelphia (Coll. Le Sueur); STORER, Hist. Fish. Mass., pl. 18, fig. 2, 101, 1867.

1059. LOPHIOMUS, Gill.

Lophiomus, GILL, Proc. U. S. Nat. Mus. 1882, 552 (*setigerus*).

This genus is closely allied to *Lophius* in external characters, but it is strikingly distinguished by the reduced number of its vertebræ, which are only 18 or 19, a fact which is associated with its tropical habitat. One species from the Pacific. (*Lophius;* ὦμός, shoulder, in apparent allusion to the trifid humeral spine.)

3092. LOPHIOMUS SETIGERUS (Vahl).

Dorsal III–III–9; A. 5. Head above orbits and laterally with numerous spines and prickles; humeral bone ending in 3 blunt points; numerous cirri scattered along sides of head and body. Vertebræ 18. Color dusky; floor of mouth black posteriorly, but without white spots; pectorals and ventrals pale on basal half, black distally; caudal and anal black, with some white spots; soft dorsal translucent, with black specks; first dorsal spine with its membranaceous tip white, the latter provided with 2 black eye-like spots.` Pacific Ocean; not uncommon in rather deep water off coasts of China and Japan. Known on the American coast from 1 speci-

---

* According to Professor Horace A. Hoffman this fish is called in Athens Πεσκανδρίτζα or Πεσκαντρίτζα. These names, "probably of Italian origin, meaning fisher; χλάσχα, at Chalcis, σκλεμποῦ, and βατραχόψαρο at Patras. The βάτραχος ὁ ἁλιεύς (the fisher frog) of Aristotle. (See Aristotle 505a 6b 4, 506b 16, 564b 18, 565b 29, 570b 30, 620b 11 ff, 695b 14, 696a 27, 749a 23, 754a 23 ff, 755a 9, 835b 13, 1527b 41–43, 540b 18.) Aristotle says with regard to the βάτραχος: 'Inasmuch as the flat front part is not fleshy, nature has compensated for this by adding to the rear and the tail as much fleshy substance as has been subtracted in front.' The βάτραχος is called the angler. He fishes with the hair-like filaments hung before his eyes. On the end of each filament is a little knob just as if it had been placed there for a bait. He makes a disturbance in sandy or muddy places, hides himself and raises these filaments. When the little fishes strike at them he leads them down with the filaments until he brings them to his mouth. The βάτραχος is one of the σελάχη. All the σελάχη are viviparous or ovoviviparous except the βάτραχος. The other flat σελάχη have their gills uncovered and underneath them, but the βάτραχος has its gills on the side and covered with skinny opercula, not with horny opercula like the fish which are not σελαχώδη. Some fishes have the gall bladder upon the liver, others have it upon the intestine, more or less remote from the liver and attached to it by a duct. Such are βάτραχος, ἔλλοψ, συναγρίς, σμύραινα, and ξιφίας. (This has been proved true of *Lophius piscatorius* by a dissection by Dr. C. H. Gilbert.) The βάτραχος is the only one of the σελάχη which is oviparous. This is on account of the nature of its body, for it has a head many times as large as the rest of its body, and spiny and very rough. For this same reason it does not afterwards admit its young into itself. The size and roughness of the head prevents them both from coming out (i. e., being born alive) and from going in (being taken into the mouth of the parent). The βάτραχος is most prolific of the σελάχη, but they are scarce because the eggs are easily destroyed, for it lays them in a bunch near the shore." (Hoffman & Jordan, Proc. Ac. Nat. Sci. Phila., 1892, 278.)

men, 2¼ inches long, dredged at *Albatross* Station 2805, southwest of Pan-ama. From this specimen, the above description is taken. Comparing this with a larger specimen taken at Tokio by Prof. K. Otaki, we find no differences likely to prove permanent. (*seta*, bristle; *gero*, I bear.)

Lophius setigerus, VAHL,* Skrivt. Naturh., IV, 214, tab. 3, figs. 5 and 6, 1797, China Sea; CUVIER & VALENCIENNES, Hist. Nat. Poiss., XII, 383, 1837 ; GÜNTHER, Cat., III, 180, 1861.
Lophius viviparus, BLOCH & SCHNEIDER, Syst. Ichth., 142, 1801, tab. 32, China Sea; after Vahl.
Lophiomus setigerus, GILBERT, Proc. U. S. Nat. Mus. 1890, 454.

Family CCXXII. ANTENNARIIDÆ.

(THE FROG-FISHES.)

Head and body more or less compressed. Mouth vertical or very oblique, opening upward; lower jaw projecting; jaws with cardiform teeth; premaxillaries protractile. Gill openings small, pore-like, in or behind the lower axils of the pectorals. No pseudobranchiæ. Gills 2¼ or 3; skin naked, smooth, or prickly. Pectoral members forming an elbow-like angle. Pseudobrachia long, with 3 actinosts. Ventral fins present, jugular, near together. Spinous dorsal of 1 to 3 separated, tentacle-like spines; soft dorsal long, larger than anal. Pyloric cæca none. Genera about 5; species 50. Inhabitants of tropical seas, "living on floating seaweed, and enabled, by filling the capacious stomach with air, to sustain themselves on the surface of the water;" therefore widely dispersed by currents in the sea. (*Pediculati*, pt., Günther, Cat., III, 182 to 200, 1861.)

    a. Head compressed; a rostral spine or tentacle, followed by 2 larger spines; palatine teeth developed; dorsal spines disconnected.
        b. Skin naked and smooth; ventral fins elongate.   PTEROPHRYNE, 1060.
        bb. Skin covered with prickles; ventral fins short.   ANTENNARIUS, 1061.
    aa. Head cuboid; a single rostral spine or tentacle, received in a groove; soft dorsal low.   CHAUNAX, 1062.

1060. PTEROPHRYNE, Gill.

(MOUSE-FISH.)

Pterophryne, GILL, Proc. Ac. Nat. Sci. Phila. 1863, 90 (*bougainvillei*).
Pterophrynoides, GILL, Proc. U. S. Nat. Mus., I, 1878, 216 (*histrio*) ; name a substitute for *Pterophryne*, if the latter be regarded as preoccupied by the earlier *Pterophrynus*.

Body smooth or scarcely granular, short, somewhat compressed, with tumid abdomen; mouth small, oblique; palate with teeth; wrist and pec-toral fin slender; ventrals elongated; soft dorsal and anal vertically

---

*Lophiomus setigerus, is thus described by Dr. Günther:
"Dorsal III–III, 8 or 9; A. 6 or 7. Teeth arranged in 2 alternate series in the upper jaw, in 3 in the lower; 2 or 3 teeth on each side of the vomer; humeral spine terminating in 3 points; the mouth behind the hyoid bone purplish black, with white spots. Vertebræ 19, the anterior ones very short, the middle and posterior ones nearly equal in length. Coasts of China and Japan." (Günther.)

expanded. Small fishes of fantastic shape in the West Indies and Gulf Stream. (πτερόν, wing; φρύνη, toad.)

a. "Bait" on first dorsal spine bifurcate at tip.　　　　　　　HISTRIO, 3093.

aa. "Bait" on first dorsal spine bulbous. covered with fleshy filaments.　　GIBBA, 3094.

3093. PTEROPHRYNE HISTRIO (Linnæus).

(MOUSE-FISH; SARGASSUM-FISH.)

Head 2⅓; depth 1⅘. D. III-14; **A. 7;** V. 5. Skin of head and body, as well as dorsal fins, with fleshy tags, which are most numerous on the dorsal spines and abdomen. Wrist slender; ventrals large, nearly ¼ as long as head. Dorsal and anal with the posterior rays not adnate to caudal peduncle; first dorsal spine bifurcate at tip. Yellowish, marbled with brown; 3 dark bands radiating from eye; vertical fins barred with brown; belly and sides with small white spots. Tropical parts of the Atlantic; abundant on our Gulf coast and occasional northward to Cape Hatteras or beyond, especially in floating masses of *Sargassum.* Once taken in Europe (Vadsö, Norway) in floating seaweed from the Gulf Stream. Recorded from the coast of Senegambia; its history and synonymy confused with that of the following species. A remarkable fish, excessively variable in coloration. (*histrio,* a harlequin.)

Lophius tumidus, OSBECK, Iter Chinensis, 400, 1757, Open Sea; pre-Linnæan.

Lophius histrio,[*] LINNÆUS, Syst. Nat., Ed. x, 237, 1758, after various authors, especially *Balistes guaperva* seu *chinensis,* LINNÆUS, Mus. Ad. Fr., 56.

Pterophryne histrio, GILL, Proc. U. S. Nat. Mus. 1878, 216; GOODE & BEAN, Oceanic Ichthyology, 486, 1896.

Antennarius histrio, JORDAN & GILBERT, Synopsis, 846, 1883; COLLETT, Campagnes Hirondelle, 38, 1896.

---

[*] Concerning the use of the name *histrio* for this species, Dr. Gill remarks:
"In 1794 (as appears from the dates on the plates), Shaw published a number of his 'Naturalists' Miscellany,' in which he described 3 fishes under the generic name of *Lophius.* These were described as (1) *Lophius striatus* (the Striated Lophius), pl. 175; (2) *Lophius pictus* (the Variegated Lophius), pl. 176, upper figure, and (3) *Lophius marmoratus* (the Marbled Lophius), pl. 176, lower figure. The originals of these are evidently the varieties (*a, b,* and *c*) of *Lophius histrio* admitted by Bloch & Schneider. It is quite clear that the first two were based on species of typical *Antennarius* (not *Pterophryne*), while the third is incomprehensible, and, if the figure is at all correct, must represent a factitious fish; it most certainly has nothing to do with *Pterophryne.* The other species, however, notwithstanding the bad figures, are readily identifiable. The *Lophius striatus* (as has recently been recognized by Günther) is the first name of an *Antennarius* peculiar to the Pacific, and quite distinct from the Caribbean *Antennarius scaber* (=*A. histrio* Günther), with which it was at first confounded by Günther. The *Lophius pictus* was evidently based on the species or variety of *Antennarius* which was afterwards named *Antennarius phymatodes* by Bleeker, and it agrees very closely, in the distribution of colors, with a specimen figured by that ichthyologist, and would probably be considered by Günther as a variety of his *Antennarius commersonii.* But whatever may be the value of the forms embraced under the name *Antennarius commersonii* by Günther—whether species or varieties—the name *Antennarius pictus* must be revived from Shaw, either especially for the *Antennarius phymatodes* of Bleeker or for the collection designated as *Antennarius commersonii.* It has thus been demonstrated (1) that the Linnæan name, *Lophius histrio,* was originally created for the common *Pterophryne,* and (2) that the names generally employed for the *Pterophryne* were originally applied to very different forms, and members of even a different genus. Hence if the laws of priority, as formulated by the British and American Associations for the Advancement of Science, are to guide us, there can be no question that the species of *Pterophryne* must hereafter be designated as *Pterophryne histrio;* if, however, it is allowable to go behind even the tenth edition of the Systema Naturæ and to take the oldest binomial name, without other considerations, the designation *tumidus* must be revived. It seems best, however, to follow general usage." (Gill, Proc. U. S. Nat. Mus., I, 1878, 226.)

Chironectes pictus, CUVIER & VALENCIENNES, Hist. Nat. Poiss., XII, 393, 1837, Surinam.
Chironectes tumidus, CUVIER & VALENCIENNES, Hist. Nat. Poiss., XII, 397, 1837, "Cabinet du Roi," Sargasso Sea. (Coll. Péron.)
Chironectes arcticus, DÜBEN & KOREN, Kong. Vet. Akad. Abh. Stockholm 1844, 72, Vadsö, Norway, from a specimen carried northward in *Sargassum;* the only European record; fide COLLETT.
Antennarius marmoratus, GÜNTHER, Cat., III, 185, 1861; in part; not of Cuvier.
Chironectes lævigatus, DE KAY, N. Y. Fauna: Fishes, 165, pl. 27, fig. 83, 1842; not of CUVIER.

3094. PTEROPHRYNE GIBBA (Mitchill).

Garman refers to this species certain specimens obtained in Gulf weed about Key West and the Tortugas. These resemble *P. histrio,* but "differ markedly in certain respects. The bait on the first dorsal spine, for instance, is bulbous and covered with slender fleshy filaments in our individuals, but in *P. histrio* it is bifurcate. *P. gibbus* is fairly represented by Cuvier, 1817, in his *Chironectes lævigatus.* The formula for the individuals in hand is D. III, 12; A. 7; V. 5; P. 10; C. 9." (Garman.) West Indies, north to Key West and the Tortugas; not examined by us; probably common, but hitherto confounded with *P. histrio.* (*gibbus,* gibbous.)

Lophius gibbus, MITCHILL, Trans. Lit. and Phil. Soc. N. Y. 1815, I, pl. 4, f. 9, off St. Croix, Lat. 22° N., Long. 64° W. (Coll. Dr. John D. Jaques.)
Chironectes lævigatus, CUVIER, Mém. du Mus., III, 423, pl. 16, fig. 1, 1817, South Carolina (Coll. Bosc); CUVIER & VALENCIENNES, Hist. Nat. Poiss., XII, 399, 1837.
Pterophryne lævigata, GILL, Proc. Ac. Nat. Sci. Phila. 1863, 90.
? *Chironectes sonntagii*,* Baron J. W. VON MÜLLER, Reisen in den Vereinigten Staaten, Canada und Mexico, Band I, 180. 1864, in floating seaweed; no exact locality stated.
Pterophrynoides gibbus, GARMAN, Bull. Iowa Lab. Nat. Hist. 1896, 81.

1061. ANTENNARIUS, Lacépède.

Antennarius (COMMERSON) LACÉPÈDE, Hist. Nat. Poiss., I, 421, 1798 (*chironectes*).
Histrio, FISCHER, Zoognosia, 78, 1813 (*histrio*, etc.). (No type; includes all known *Antennariidæ;* description transposed with that of *Lophius* by error.)
Chironectes, CUVIER, Règne Animal, Ed. 2, vol. II, 252, 1829 (*chironectes*); preoccupied in mammals, Illiger, 1811.

Body oblong, compressed, very deep through the occipital region, tapering behind; breast tumid; mouth rather large, more or less oblique, or even vertical; cardiform teeth on jaws, vomer, and palatines; eye small; skin with small granules or spinules, these usually forked, and numerous fleshy slips. First dorsal spine developed as a small rostral tentacle;

---

* The following is the substance of the long account of *P. sonntagii* (Von Müller):
"D. II, 10 to 12; P. 10 or 11; C. 6 to 7 (' *Strahlenpaare* '); B.6. Head and body slightly compressed; dorsal spines like little horns, covered over and over with spinous growths. Mouth wide, with numerous rows of small teeth; throat and belly with many fleshy slips. Pectorals produced on a long peduncle like the flippers of a tortoise; ventrals similar, but formed more like feet; anal fin like a rudder. Color clear yellowish green, with greenish brown stripes; a broad dark stripe across breast to root of pectoral; another on the back; another on the side, running backward in the form of a hammer, paler at last on lower part of back; several stripes and spots, more or less dusky, on the tail and other extremities; on the soft underside to the anal intense reddish golden yellow spots; between the dark streaks and the yellow ground color of the body are often white shades and markings; eye fiery orange. Atlantic Ocean or Gulf of Mexico; living in floating seaweed."
This species must be a *Pterophryne*, and it is not evidently different from *Pterophryne gibba*.

second and third dorsal spines strong, covered with skin, with numerous fleshy filaments; soft dorsal high and long; anal short and deep; caudal fin rounded, the peduncle free; pectoral fins wide, with a rather wide wrist, at the lower posterior angle of which are the very small gill openings; ventral fins short. Fantastic-looking fishes, often gaily colored; very numerous in warm seas. (*antenna*, a feeler or tentacle.)

> *a.* Bulbous tip or "bait" of first dorsal spine simple, undivided at tip.
>> *b.* Skin smoothish except about eyes; first dorsal spine short, second rough. Body brown, with whitish spots; no ocelli.   INOPS, 3095.
>> *bb.* Skin with prickles, velvety or shagreen-like.
>>> *c.* Prickles simple, none of them bifid.
>>>> *d.* Color black; tips of pectorals and ventrals and one or two spots on side white (prickles undescribed).   PRINCIPIS, 3096.
>>>> *dd.* Color dusky; dorsal with 3 ocelli; caudal with many spots; first dorsal longer than second; no dermal flaps.   TENEBROSUS, 3097.
>>>> *ddd.* Color reddish or grayish, reticulate with heavy black lines; first dorsal spine short.   RETICULARIS, 3098.
>>> *cc.* Prickles or spinules on body mostly bifid.
>>>> *e.* Body without ocelli; first dorsal spine filiform.
>>>>> *f.* Mouth immaculate within; body with numerous rosy and dusky tracts, the latter forming bars and concentric streaks below; fins barred.   STRIGATUS, 3099.
>>>>> *ff.* Mouth largely black within; body blood red, with black spots on sides and below dorsal.   .   .   SANGUINEUS, 3100.
>>>> *ee.* Body with 3 large ocelli, 1 on dorsal, 1 on caudal, and 1 on middle of side, besides many black spots and streaks; tip of first dorsal spine fringed; mouth largely black within.   OCELLATUS, 3101.
> *aa.* Bulbous tip or "bait" on first dorsal spine bifid at tip; skin shagreen-like.
>> *g.* Color reddish, with brown spots, those about the eye radiating.
>>> *h.* Dermal flaps numerous on body; spinules on skin short and stiff, rendering the surface shagreen-like.   SCABER, 3102.
>>> *hh.* Dermal flaps few; spinules on skin longer and slender, rendering the surface velvety.   TIGRIS, 3103.
>> *gg.* Color uniform black; surface of body rough, shagreen-like; inside of mouth white; first dorsal spine short, little longer than second.   NUTTINGII, 3104.
> *aaa.* Bulbous tip or "bait" of first dorsal spine trifid.
>> *i.* First dorsal ray twice as long as second and as long as caudal; sides with numerous black ocelli, besides other streaks and dark spots; skin smoothish.   MULTIOCELLATUS, 3105.
>> *ii.* First dorsal spine barely ⅓ longer than second; shorter than caudal; sides with dark streaks and reticulations; a large ocellus under middle of soft dorsal; body rough, with shagreen.   RADIOSUS, 3106.

3095. ANTENNARIUS INOPS, Poey.

Depth 2¾ with caudal. Skin lustrous, smooth, except for some points behind and below eye; third of the first 3 dorsal rays largest, its membrane not reaching to vent; second ray also large, but shorter, placed between eyes; first spine developed as a fishing rod, filiform, ending in a small, membranaceous lobe, its base close to that of second, and, therefore, distant from end of snout, its spine short, the tip not reaching middle of second spine; short tentacles, like horns, on anterior part of third spine, over the nostrils, and under the mouth; caudal rounded; pectoral so joined that it can not be turned forward as usual in this group, but rising obliquely backward and upward. Eye slightly longer than snout; mouth brown within. Color brown, with white spots on the body and median fins, 6 of the largest of these each with the center yellowish, the largest from once to twice diameter of eye; spots on dorsal fins small; eye golden. Porto Rico. (Poey.) Not seen by us. The type 70 mm. long. (*inops,* helpless. "I call this species '*inops*' on account of the miserable fishing rod which has fallen to its lot." Poey.)

Antennarius inops, POEY, Anal. Soc. Esp. Hist. Nat., x, 1881, 340, Porto Rico. (Coll. Don Juan Gündlach.)

3096. ANTENNARIUS PRINCIPIS (Cuvier & Valenciennes).

D. III-11; A. 7; P. 10. Anterior dorsal spine twice as long as second, ending in a small, slender lobe; membrane behind third spine extending to root of soft dorsal; last ray of dorsal not reaching caudal. Skin rough, covered with small spines; no cutaneous fringes. Black; tips of pectorals and ventrals white; a small white spot above pectoral. (Günther.) West Indies to Brazil; not seen by us. Günther's specimen, above described, from Para. (*principis,* of the prince. Named for its discoverer, Prince Maurice of Nassau.)

Chironectes principis,* CUVIER & VALENCIENNES, Hist. Nat. Poiss., XII, 416, 1837, Brazil; on 2 drawings by Prince MAURICE, the second representing the present species to which Günther restricts the name *principis.* The first figure is more like *tigris.*
Antennarius principis, Günther, Cat., III, 193, 1861.

3097. ANTENNARIUS TENEBROSUS (Poey).

D. III-12; A. 7; P. 11. Anterior dorsal spine longer than second, terminating in a simple and slender tentacle; soft dorsal fin terminating at some distance from caudal. Skin rough, covered with small spines, without cutaneous fringes. Blackish brown, marbled with darker and lighter; a series of 3 black, blue-edged ocelli on upper posterior part of dorsal fin; many similar ocelli on caudal fin, irregularly disposed. (Poey.) Cuba. Not seen by us. (*tenebrosus,* dusky.)

Chironectes tenebrosus, POEY, Memorias, I, 219, pl. 17, fig. 1, 1851, Cuba.
Antennarius tenebrosus, GÜNTHER, Cat., III, 197, 1861.

3098. ANTENNARIUS RETICULARIS, Gilbert.

D. III-12; A. 7. First dorsal spine short, very slender and filiform, not reaching tip of second, terminating in a short, fleshy flap; second spine moderately robust, flexible, not curved backward, wholly free and with-

---

* Concerning this nominal species and *Chironectes mentzelii,* both of which were based on drawings by Prince Maurice, Cuvier & Valenciennes remark:
"We here cite these figures, and we give them specific names only to fix the attention of travelers and to get them to find the species which have served as models for these figures."
The following is the substance of the original description of *Chironectes principis:*
In the first figure, color very deep brown, speckled with black spots on body and fins; dorsal fin with only 1 series of spots. Filament of first spine twice as long as that of the second, and terminating in a little knob or bait, the second spine free from the first and similar in shape.
Second figure of the same form, the second dorsal longer, the first ray ending in a spiral, and the whole body white, with 2 white round spots on each side, one above the other.

out membrane; third spine nearly erect, not free, depressible with difficulty; not curved as in *A. sanguineus;* spines on the body rather coarse
and shagreen-like, with expanded, undivided tips. Color in spirits, top
of head, including dorsal spines and front of soft dorsal, coral red, the
body otherwise light gray, broadly reticulated on sides and below with
heavy black lines, which inclose 5 or 6 large pale spots; pectorals, ventrals, and anal with narrow terminal and wide median black bars. Soft
dorsal uniformly light. This species closely resembles *A. sanguineus,* but
differs from it in the straight, erect spinules, the color, and the character
of the plates on the body. Length 1¼ inches. Gulf of California. (Gilbert.) Only the type known. (*reticularis,* netted.)

Antennarius reticularis, GILBERT, Proc. U. S. Nat. Mus. 1891, 566, Gulf of California, at
   Albatross Station 2825. (Coll. Gilbert.)

3099. ANTENNARIUS STRIGATUS, Gill.

D. III–12; A. 7. First dorsal spine elongate, filiform, twice length of
second, with very slender, dermal tip; third spine more robust than
second, wholly concealed in the skin, its length equal to that of first
spine. Lips, maxillary, and a large transverse area behind second dorsal
spine naked, each side of this area with a few spinous tubercles; skin
elsewhere covered with fine shagreen-like armature. Color in spirits, olivaceous everywhere on body and on inside of mouth, finely mottled with
light olive brown; many irregular blackish areas on head and body, those
on lower side of head showing a tendency to form concentric bars; some
on sides forming irregular bars downward from back; posterior portion
of body not darker than the anterior; terminal parts of all the fins largely
blackish, but with distinct black bars; some scattered round blotches on
sides, each consisting of a number of smaller black spots on an olive
ground; head and body with numerous pinkish and rose-red spots and
bars, the latter sinuous, irregular, with wavy margins; a pinkish bar
behind maxillary; a. broad, saddle-like pinkish blotch across interval
between second and third dorsal spines; a third bar from in front of origin of second dorsal downward toward base of pectorals; a fourth across
top of caudal peduncle; first dorsal spine narrowly barred with brown.
Pacific coast of tropical America, from Cape San Lucas to Panama. Here
described from an adult, 10 inches in length, from Panama. This differs
considerably from the descriptions of the young (*strigatus, tenuifilis*) given
by Gill and Günther.

Two young individuals, types of *A. strigatus,* are thus characterized by
Dr. Gill:

"The anterior dorsal spine is very slender and filiform, without
appendages; the second is straight and moderate; the third concealed
and developed as a hump, obtuse behind. The spines which cover the
body are small and mostly bifid. The back and front of the dorsal fin are
reddish; the rest light brown, with black stripes which diverge downward above the pectorals, those in front being parallel with the profile
and at right angles with those behind; around the pectoral fins and on
the flanks, the streaks are generally blended to form a continuous black

area; a black dorsal saddle is in front of the dorsal fin, and a black band covers the posterior half of the caudal fin; the abdomen is broadly reticulated with black, and the brown intervals themselves are frequently striated with the same color; the interior of the mouth is immaculate." (*strigatus*, striped.)

Antennarius strigatus, GILL, Proc. Ac. Nat. Sci. Phila. 1863, 92, Cape San Lucas (Coll. J. Xantus), young; JORDAN & GILBERT, Proc. U. S. Nat. Mus., 1882, 650, adult; JORDAN, Cat. Fishes, 138, 1885.
Antennarius tenuifilis, GÜNTHER, Fishes Centr. Amer., 440, 1869, Panama; young.

3100. ANTENNARIUS SANGUINEUS, Gill.

Anterior dorsal spine very slender, $2\frac{1}{4}$ in length of caudal fin, terminating in a flap extended on each side, laciniated outward; second spine rough, robust, and curved strongly backward at its end; third not free, but apparent as a hump pointed backward, and extending $\frac{2}{3}$ of the distance from its insertion to that of dorsal fin; skin covered with small bifid spines, whose prongs diverge considerably and are acute. Color blood red, except on abdomen, but with several more or less distinct black spots under origin of dorsal fin and on sides; abdomen light or yellowish brown, spotted with black; intervals between caudal and anal rays also marked with black; floor of mouth behind tongue with 2 lateral black bands converging toward the front, while the posterior margin of the tongue itself is also sometimes lined with black. (Gill.) Pacific coast of tropical America, Cape San Lucas to Panama; scarce. (*sanguineus*, bloody.)

Antennarius sanguineus, GILL, Proc. Ac. Nat. Sci. Phila. 1863, 91, Cape San Lucas (Coll. Xantus); JORDAN, Cat. Fishes, 138, 1885.
Antennarius leopardinus, GÜNTHER, Proc. Zool. Soc. London, 1864, 151, Panama.

3101. ANTENNARIUS OCELLATUS (Bloch & Schneider).

Depth $1\frac{3}{5}$ in length. D. II–I–14; A. 8; P. 11; orbit equaling snout, eye much smaller; maxillary $3\frac{1}{2}$ in body; pectoral rays $5\frac{1}{2}$; caudal 4. Body short, oblong, compressed, very deep through occipital region; mouth large, subvertical; teeth small, sharp, cardiform, in wide bands on jaws, vomer, and palatines; maxillary extending downward to below axis of body; a very large knob at symphysis; lower part of head with many large, thick tentacles. First 2 dorsal spines on interorbital space, the first slender, terminating in a fringed lobe, the second shorter and much thicker, behind it a smooth depression; the third spine exceedingly rough and thick, blunt at tip and adnate to body; soft dorsal long and low, its origin in front of middle of body, tips of last rays reaching base of caudal; anal posterior, tips of its rays coterminous with dorsal rays, its height equal to its length; pectorals near middle of body, placed far below axis; ventrals short, the rays thickened, their position under posterior edge of eye. Skin covered with minute bifurcate spines, running upon dorsal, anal, and caudal rays; gill opening in front and below pectoral, its length about equal to snout. Color brown, marbled with lighter, and with scattered black dots, especially on belly and outer portions of dorsal and caudal; each side with 3 large black spots ocellated with brownish, 1 on dorsal near its base, a second immediately below it on the sides, and a third in the middle of the caudal fin; mouth behind tongue black, with yellow lines. West Indies, north to Florida. Common in the West Indies; the most abundant of the American species. Here described from a specimen from off Pensacola, Florida, about 15 inches in length.

Mr. Garman gives the following note on *Antennarius ocellatus:*

"The species was tolerably figured by Parra, but has not been recognized by some of the subsequent writers. On 5 specimens before me the amount of variations in markings is comparatively small. The 3 large ocelli, on dorsal, caudal, and middle of side, are present on each, as is also the case with the numerous small spots of black on the ventral portions of the body and on the outer portions of dorsal and caudal. The dorsal ocellus lies between the sixth and seventh rays, on the middle of the fin; that on the flank is situated on the vent, and that on the caudal between the fourth and fifth rays, from the top, near the middle of the fin. The black portion of either of these spots is larger than the orbit, which latter is rather small when contrasted with that of other species. The white circle around the black, again, is surrounded by a narrow one of brown. On the caudal, at each side of the ocellus, there are transverse streaks. The first ray of the dorsal is as long as the second, and is covered by scales. The bulb apparently is simple, and bears numerous laciniæ. The second dorsal spine is shorter than the third; both are club shaped. The space behind the second dorsal spine is covered by scales." (*ocellatus*, with eye-like spots.)

Pescador, Parra, Dif. Piezas, Hist. Nat., 1, tab. 1, 1780, Cuba.
Lophius histrio, var. *ocellatus,* Bloch & Schneider, Syst. Ichth., 142, 1801; after Parra.
Antennarius pleurophthalmus, Gill. Proc. Ac. Nat. Sci. Phila. 1863, 92, Key West; Jordan
    & Gilbert, Synopsis, 846, 1883; Jordan, Cat. Fishes, 138, 1885; Goode & Bean, Oceanic
    Ichthyology, 487, 1893.
Antennarius ocellatus, Poey, Synopsis, 105, 1868; Garman, Bull. Iowa Lab. Nat. Hist.
    1896, 82.

3102. ANTENNARIUS SCABER (Cuvier).

D. III–12; A. 7; P. 9 or 10. Anterior dorsal spine as long as second, and provided with 2 long and thick cutaneous flaps at its tip; third dorsal spine not continuous with the soft dorsal; soft dorsal fin terminating at some distance from the caudal, its last ray not extending to root of caudal, if laid backward; dorsal spines, head, back, and sides of the body with more or less numerous cutaneous fringes, those of dorsal spines sometimes forming a dense cluster; skin very rough, covered with small spines. Ground color yellowish or reddish, with numerous brown spots, those around the eye forming radiating streaks; dorsal and anal fins with 3 series of round brown spots, the middle of which is formed by the largest and most constant spots; sometimes uniform brown. Caribbean Sea. (Günther.) A small specimen from Port Castries, St. Lucia, has the body light brown, clouded with darker, fins all with round black spots, those of the base of the dorsal somewhat larger than others; ventrals tipped with black. (*scaber*, rough.)

Chironectes scaber, CUVIER, Mém. Mus., III, 425, pl. 16, fig. 2, 1817, Martinique (Coll. Plée); CUVIER & VALENCIENNES, Hist. Nat. Poiss., XII, 412, 1837.

Lophius spectrum, GRONOW, Cat. Fish., Ed. Gray, 49, 1854, Antilles; after *Lophius acute scabra*, GRONOW, Zoophyl., 210, 1781.

Antennarius scaber, JORDAN, Proc. U. S. Nat. Mus. 1889, 652, specimen from Port Castries, St. Lucia.

Antennarius histrio, GÜNTHER, Cat., IV, 188, 1861; not *Lophius histrio*, LINNÆUS.

3103. ANTENNARIUS TIGRIS; Poey.

D. III–12; A. 7; P. 11. Anterior dorsal spine longer than second, terminating in 2 long cutaneous flaps; third dorsal spine connected with soft dorsal by a broad membrane, the latter terminating at some distance from the caudal, and its last ray not extending to root of caudal if laid backward. Skin rough, covered with small spines, without cutaneous fringes. Ground color yellow, with numerous brown spots and streaks, the latter radiating from the eye; dorsal fin irregularly spotted, with a series of large round brown spots. (Poey.) Cuba. Not seen by us. According to Mr. Garman, *Antennarius scaber* and *A. tigris* "are closely allied, but if placed side by side the squamation and filaments suffice to distinguish them, great similarity in color notwithstanding. *A. scaber* has coarser scales, with shorter, rougher spines, the scales are farther apart, and the cutaneous flaps appear on the body much as figured by Cuvier. On *A. tigris* there are few of the cutaneous appendages, the scales are closer together, the spines are longer and more slender, giving rise to an appearance more like velvet, and the head and body are more compressed." (*tigris*, tiger.)

Chironectes tigris, POEY, Memorias, I, 217, pl. 17, fig. 2, 1851, Cuba.
Antennarius tigris, GARMAN, Bull. Iowa Lab. Nat. Hist. 1896, 83.

3104. ANTENNARIUS NUTTINGII, Garman.

D. 3+12; A. 7; V. 5; P. 11; C. 9. In form this species is shorter, more massive anteriorly, and less compressed than either *A. ocellatus* or *A. radiosus*. A transerve section across the middle of the body is a nearly equilateral triangle. Caudal region short. Head nearly as wide as high; cheeks swollen; forehead rather broad, converging forward on the edges. Occipital concavity wide and deep, free from scales in a wide space below the ends of the first and second dorsal rays, this bare space being apparently for the reception of the fleshy bait bulb, which latter has 2 elongate lobes. Snout as long as the orbit, broad, truncate; chin vertical; symphyseal knob prominent. Mouth wide, subvertical. Eye small; orbit twice as long, hardly more than ½ the interorbital space. First and second dorsal rays equal in length, not inclusive of the 2 elongate fleshy fringed lobes surmounting the first. The base of the first ray stands forward prominently over the mouth, being free for some distance. The greater portion of the second ray is free, while the third is connected with the dorsum, by the skin, from base nearly to tip. This last ray is larger than either of its fellows. Soft dorsal large; middle rays longest, as long as the distance from the maxillary to the hind edge of the operculum, or as long as the rays of the caudal fin; fin not reaching back to the bases of the caudal rays, fringed. Hind margin of caudal convex, fringed. Anal moderate, rays prominent in the margin, fin with a blunt angle on the outer edge, subtending, when laid up against the tail, $\frac{1}{4}$ or more of the length of the caudal rays. The rays on the pectoral fins extend out beyond the margins more noticeably than those of the other fins. Ventrals small, in most instances with 6 points on the outer margin, in one case having but 5. Greatest length of the caudal nearly $\frac{1}{4}$ of the total length. Length of each maxillary $\frac{2}{3}$ of the caudal. Scales short, small, close set, harsh to the touch, having none of the velvety appearance. Uniform black; inside of the mouth black; "bait" white. Great Bahama Banks. Besides the specimens in Nutting's collection there are several others in that of the Mus. Com. Zool. "This species is readily separated from *A. principis* of authors by the short first dorsal spine." ("The specific name is given in honor of Prof. C. C. Nutting, to whom science is so much indebted for the origination and successful accomplishment of the expedition.")

Antennarius nuttingii, GARMAN, Bull. Iowa Lab. Nat. Hist. 1896, 83, pl. II, Great Bahama Banks. (Coll. C. C. Nutting.)

?? *Chironectes mentzelii*,[*] CUVIER & VALENCIENNES, Hist. Nat. Poiss., XII, 417, Brazil; on a drawing by Prince MAURICE.

3105. ANTENNARIUS MULTIOCELLATUS (Cuvier & Valenciennes).

(MARTIN PESCADOR.)

Mouth large, vertical. First dorsal spine slender and straight, nearly equal to length of caudal, terminating in 3 simple tentacles; second dorsal spine curved at the middle and extending to the base of the third; third dorsal spine partly embedded in the skin, reaching halfway to dorsal; wrists and pectorals widened; ventrals short. Skin covered with bifid spines. Fawn color, lighter below; many black spots ocellated with white, both on the body and fins; body with several pink areas, 1 of which forms a triangular saddle in front of the dorsal and another a broad ring around the base of the caudal fin; angles of mouth with a pink spot. West Indies, north to Florida Keys; common. This description (after Gill) from the type of *Antennarius annulatus* from Garden Key. According to Mr. Garman, "this species is distinguished by the trifid bulb and the long first dorsal ray, near twice as long as the second and quite as long as the caudal, by the high nape, by the large third dorsal ray, much larger and more swollen than the second, and by the coloration. The eye is very small. The black centers of the largest of the ocelli are smaller than the eye. Besides the ocellus on the soft dorsal that on the anal and the 3

---

[*] The following is the substance of the very brief description of this nominal species: "First dorsal filament not longer than the second, and ending in a small bait or knob. Body black, with large marblings."

forming a triangle on the caudal, there are others scattered over the caudal and other fins, and over the sides of the body. Below the eye on the cheek and under the chin and the chest the spots are little more than black dots. Over the sides, a specimen in hand, the type of *A. corallinus*, Poey, is freckled with lighter rounded spots. Behind the pectoral, on the side, there is a small ocellus with a black center. On each side in the same position, a short distance above the pectoral, there is a brown ocellus larger than the orbit, in the center of which there is a white dot. A brown streak passes back from the upper part of the orbit and curves down toward the anal ocellus, another passes back from the middle of the eye and curves down toward the pectoral, and a third below the third dorsal spine runs down and then forward toward the lower end of the maxillary. The forehead is comparatively narrow; behind the second dorsal ray the bare space is hardly large enough to receive the bait."

Poey thus describes *Antennarius corallinus*, which according to Garman, is the same as *A. multiocellatus*:

"D. II-I, 12; A. 7; V. I, 5; C. 17. Two dorsal spines in front of eye, formed like horns, another higher on the nape; gill opening spiral, at the lower base of the pectoral; general form of the fish almost globular; the mouth vertical; tongue marbled with black and white; caudal rounded; eyes very small; pectoral low, reaching middle of body without caudal; ventrals short; vent near anal; first dorsal spine ending in a single short filament. Color reddish with black spots; 2 of these spots eye-like, with a larger black center and the iris of the color of the body, surrounded with a black circle; 1 spot at the base of the soft dorsal at the second third of its length, a very weak one at base of anal; between first spot and pectoral fin 3 small inconspicuous ocellate spots, of which the pupil is a small point; middle of the body with dusky spots; dorsal, anal, and caudal with black points; tubercles about eye and on the cheek, but not spinous. Type 95 mm. long. Cuba." (Poey.) (*multus*, many; *ocellus*, an eye-like spot.)

Chironectes multiocellatus, CUVIER & VALENCIENNES, Hist. Nat. Poiss., XI, 422, 1837, Martinique. (Coll. M. Garnot.)
Antennarius annulatus, GILL, Proc. Ac. Nat. Sci. Phila. 1863, 91, Garden Key, Florida (Coll. Lieut. Wright); JORDAN & GILBERT, Synopsis, 846, 1883.
Antennarius corallinus, POEY, Repertorio, I, 188, 1865, Cuba. (Coll. Poey.)
Antennarius multiocellatus, GÜNTHER, Cat., III, 194, 1861; GARMAN, Bull. Iowa Lab. Nat. Hist. 1896, 82.

3106. ANTENNARIUS RADIOSUS, Garman.

D. 3+13; A. 8; V. 5; P. 11; C. 9. Resembling *A. tigris*, Poey, in shape, squamation, etc., but differing in coloration and in possession of a much longer dorsal ray. The staff in this ray is very slender, much longer than the second ray, and bears a small, trifid "bait." Second and third dorsal rays shorter than the first, the third well tied down by the skin. Scales uniform, sharp. No cutaneous fringes on large specimens. Grayish or brownish white, darker on nape and dorsal fin, with numerous spots of light color, as large as the orbit, surrounded by more or less complete edgings of brown, producing a semblance to reticulation, or to spottings by drops of liquid; 7 streaks of brown radiate from the eye; as in *A. tigris*, they are continued upon the head and down toward the ventrals; a large spot of black, white-edged, a little larger than the orbit, ⅓ on the fin and ⅓ on the muscles of the body, occupies the space between the eighth and the tenth rays of the soft dorsal fin; the light areas vary in intensity, and lie close together over nearly the whole of body and fins; belly lighter, with faint indications of lines of brownish, radiating from the head; caudal with oblique transverse cloudings of brownish; hindmost ¼ light. The color in life was probably reddish or yellowish. Secured off Key West, in about 50 fathoms. A young individual, of less than an inch, taken opposite Havana, is of lighter gray, and has a large ocellus, of light color in the center, between the black one at the base of the dorsal and the upper end of the humerus. There are small cutaneous fringes on the flanks. (Garman.) (*radiosus*, rayed.)

Antennarius radiosus, GARMAN, Bull. Lab. Nat. Hist. Iowa Univ. 1896, 85, pl. 1, off Key West, in 50 fathoms. (Coll. C. C. Nutting.)

1062. CHAUNAX, Lowe.

Chaunax, LOWE, Trans. Zool. Soc. Lond., III, 1846, 339 (*pictus*).

Head very large, depressed, cuboid. Mouth large, subvertical; jaws and palate with bands of small teeth. Skin with small, sharp spines. Spinous dorsal reduced to a small tentacle above the snout, retractile into a groove; soft dorsal moderate, low; anal short; ventrals small. Gills 2½; no pseudobranchiæ. Muciferous channels very conspicuous, the lateral line prominent, undulate; another series of mucous tubes extending from lower jaws to axil; still another extends backward from snout and maxillary to a point behind eye, when it ceases, uniting with a vertical line which extends from the lateral line to the lower line; these lines thus inclose a quadrate area on the cheek. Gill opening small, well behind pectoral under front of soft dorsal. Deep seas. (χαύναξ, one who gapes.)

a. Dorsal rays 11; anal 5; depth 2½ in length.   PICTUS, 3107.
aa. Dorsal rays 13; anal 7; depth 2⅔ in length.   NUTTINGII, 3108.

3107. CHAUNAX PÍCTUS, Lowe.

Head 1⅓; depth 2¼. D. I, 11; A. 5; P. 11; V. 4; C. 7. Rostral tentacle short, pedicellate; muciferous channels appearing as chain-like rows of pits. Bright orange above; sides rosy; fins vermilion. Deep waters of the Atlantic; recorded from Madeiro, Soudan, Cape Verdes, Barbados, off Rhode Island, and elsewhere in the Gulf Stream, in 130 to 428 fathoms. A similar species (*Chaunax fimbriatus*), regarded by Günther as the same, occurs in the Japan Seas, Bay of Bengal and the Fiji Islands. (*pictus*, painted.)

Chaunax pictus, LOWE, Trans. Zoöl. Soc. Lond. 1846, 339, Camera de Lobos, Madeira; GÜNTHER, Cat., III, 200, 1861; GOODE, Proc. U. S. Nat. Mus. 1880, 470; JORDAN & GILBERT, Synopsis, 847, 1883; GOODE & BEAN, Oceanic Ichthyology, 487, fig. 398, 1896.

? *Chaunax fimbriatus*, HILGENDORF, Sitzber. Ges. Naturf. Freunde 1879, 80, Sea of Japan.

? *Chaunax nuttingii*, GARMAN, Bull. Lab. Nat. Hist. Iowa Univ. 1896, 85, pl. III, fig. 2, near Sand Key Light, Florida, in 120 fathoms. (Coll. C. C. Nutting.)

3108. CHAUNAX NUTTINGII, Garman.

B. 6. D. II, 13; A. 7; V. 4; P. 14; C. 9. Form resembles that of *Chaunax pictus*, but is shorter, broader, and possessed of more fin rays. Anteriorly it is broad and depressed, posteriorly compressed. From head to soft dorsal on the nape it is arched very little. Head broader than high, flattened or slightly concave on the occiput, nearly vertical on the chin. Snout short, broad, truncate. Eye medium, the length of the scaleless area covering it equals the width of that between the canals on the interorbital space, or about ⅔ of the space itself; the distance from the maxillary is about the ocular width. The niche in which the first dorsal spine is received is subelliptical and about ¾ as long as the eye; the tentacle is little more than ⅓ as long as the niche, is broad near the base, tapers rapidly and bears a 2-lobed "bait" with slender fringes. Mouth wide, oblique, maxillary about 3 times as long as the eye, widened and rounded at the outer end; intermaxillaries alone forming upper border of mouth. Teeth small, slender, sharp, in villiform bands. Origin of soft dorsal in the middle of the distance from the rostral tentacle to the base of the caudal fin, fourth ray above the gill opening, anterior rays shorter. Vent below the seventh ray of the second dorsal. Pectorals short, broad, rounded. The canals of the lateral system are in the main like those of *C. pictus*, but have stronger curves; they begin to curve outward immediately behind the niche, not remaining parallel or converging as in Lowe's species. Scales very fine, sharp and close together. In life this fish was probably red or yellowish with transverse cloudings or blotches of brownish, it is now dingy brownish white; one of the blotches lies just behind the eye, another lies below the orbit, and apparently 3 transverse bands cross the back through the soft dorsal; orbit blackish; tentacular niche black. The coloration of the individual described indicates a habitat within reach of the effects of sunlight. Florida Keys; the type dredged nearly 8 miles south of Sand Key Light, Florida, in about 120 fathoms. (Garman.) This species is evidently not very different from *Chaunax pictus* and may be the same. (Named for C. C. Nutting, professor of zoology in the University of Iowa, director of the Bahama Expedition of 1893.)

Chaunax nuttingii, GARMAN, Bull. Lab. Nat. Hist. Iowa Univ. 1896. 86, pl. III, f. 2, Sand
   Key Light, Florida. (Coll. C. C. Nutting.)

Family CCXXIII. CERATIIDÆ.

(THE SEA DEVILS.)

Head and body compressed. Mouth terminal, more or less oblique. Gill openings small, in the lower part of the axils. No pseudobranchiæ. Spinous dorsal represented by 1 or more tentacles. Pectoral members not geniculated, with short pseudobrachia and 3 actinosts. No ventral fins. Fishes of the open seas, usually inhabiting considerable depths; 13 genera and 15 species known. All are uniform blackish in color.
"The bathybial sea devils," writes Günther, "are degraded forms of *Lophius;* they descend to the greatest depths of the ocean. Their bones

are of an extremely light and thin texture, and frequently other parts of their organization, their integuments, muscles, and intestines are equally loose in texture when the specimens are brought to the surface. In their habits they probably do not differ in any degree from their surface representative, *Lophius.* The number of the dorsal spines is always reduced, and at.the end of the series of these species only 1 spine remains, with a simple, very small lamella at the extremity (*Melanocetus johnsonii, Melanocetus murrayi*). In other forms sometimes a second cephalic spine, sometimes a spine on the back of the trunk, is preserved. The first cephalic spine always retains the original function of a lure for other marine creatures, but to render it more effective a special luminous organ is sometimes developed in connection with the filaments with which its extremity is provided (*Ceratias bispinosus, Oneirodes eschrichtii*). So far as it is known at present these complicated tentacles attain to the highest degree of development in *Himantolophus* and *Ægæonichthys.* In other species very peculiar dermal appendages are developed, either accompanying the spine on the back or replacing it. They may be paired or form a group of 3, are pear-shaped, covered with common skin, and perforated at the top, a delicate tentacle sometimes issuing from the foramen." (*Pediculati,* genus *Ceratias,* Günther, Cat., III, 205, 1861; *Ceratiidæ,* Gill, Proc. U. S. Nat. Mus. 1878, 216.)

a. Mouth moderate.
    b. Gills in 2½ pairs.
       CERATIINÆ:
        c. Cleft of mouth nearly vertical; skin prickly.
           d. Cephalic spine single.
               e. Dorsal spine present; lateral caruncles present; no teeth on vomer.
                                     CERATIAS, 1063.
               ee. Dorsal spine wanting; caruncles present.
                   f. Caruncles remote from soft dorsal.   MANCALIAS, 1064.
                   ff. Caruncles close to soft dorsal.   CRYPTOPSARAS, 1065.
       ONEIRODINÆ:
        cc. Cleft of mouth nearly horizontal; skin smooth; 1 cephalic spine and 1 postcephalic spine.   ONEIRODES, 1066.
     HIMANTOLOPHINÆ:
    bb. Gills in ½ 2½ pairs; body with scattered tubercular scutella; no second dorsal spine.
        g. Body and head compressed; mouth oblique; joint of mandible below or behind eye; eye rudimentary.
           h. Body oblong, oval; dorsal rays 9; pectoral 12. HIMANTOLOPHUS, 1067.
           hh. Body short and deep; dorsal rays 4; pectoral about 17.
                                     CORYNOLOPHUS, 1068.
aa. Mouth with enormous gape.
    MELANOCETINÆ:
        i. Cleft of mouth nearly vertical; pectoral small, in advance of dorsal and of gill opening; second dorsal spine wanting; gills in 2½] airs.
           j. Gular tentacle wanting; no teeth on vomer.   LIOCETUS, 1069.
           jj. Gular tentacle present; 1 tooth on the vomer.   LINOPHRYNE, 1070.
    CAULOPHRYNINÆ:
        ii. Cleft of mouth nearly horizontal; pectorals below dorsal and behind gill opening; gills in ½ 2½ pairs.
           k. Dorsal and anal greatly produced; skin naked; head and body with many luminous filaments.   CAULOPHRYNE, 1071.

1063. CERATIAS, Kröyer.

Ceratias, KRÖYER, Naturhist. Tidsskrift. 2 Række, I, 639, 1844 *(holbolli)*.

Head and body much compressed and elevated, oblong, covered with prickly skin. Mouth wide, its cleft nearly vertical; teeth in jaws conic, movable, of moderate size; no teeth on vomer or palatines. Gills $2\frac{1}{2}$; gill arches unarmed. Spinous dorsal reduced to 2 spines, 1 on the head, the other on the back, the basal element of the second spine exserted; the cephalic spine much elongate; soft dorsal and anal short; pectorals very short, broad, of about 20 rays. Caudal fin much produced, fan-shaped, with exserted rays. Pyloric cæca 2, small. Skeleton soft, fibrous. Greenland. ($\varkappa\varepsilon\rho\alpha\tau\iota\alpha\varsigma$, one that has horns.)

3109. CERATIAS HOLBOLLI, Kröyer.

D. I–I, 4; A. 4; P. 19; C. 10. Head $2\frac{2}{3}$; depth nearly 2; head deeper than long; eyes small, not more than $\frac{1}{10}$ the length of the head; free rays of the head a little shorter than to the base of the caudal fin; the forked part of the caudal fin shorter than the length of the fish; the length of the pectoral fin equals almost $\frac{1}{10}$ the entire length of the fish, the membranes from the dorsal and anal fins posteriorly extend almost to the base of the caudal fin. Color entirely black. (Kröyer.) North Atlantic; 4 specimens known; 3 from Greenland, 1 from Nova Scotia. (Named for C. Holböll, a Danish naturalist.)

Ceratias holbolli, KRÖYER, Naturh. Tidsskr. 1844, 639, Greenland; GÜNTHER, Cat., III, 205, 1861; JORDAN & GILBERT, Synopsis, 847, 1883; GOODE & BEAN, Oceanic Ichthyology, 489, pl. 117, fig. 394, 1896.

1064. MANCALIAS, Gill.

Mancalias, GILL, Proc. U. S. Nat. Mus., I, 1878, 227 *(uranoscopus)*.
Typhlopsaras, GILL, Forest and Stream, New York, 1883, Nov. 8, 284 *(shufeldti)*.

General characters of *Ceratias*, but with the spinous dorsal reduced to a single rostral spine, and 2 fleshy claviform tubercles or caruncles behind it. Pectoral fins narrow, with 10 to 15 slender rays. (*mancus*, defective, "with a quasi-diminutive termination to correspond with *Ceratias*.")

a. Dorsal caruncles placed before dorsal fin a distance 6 times in length of trunk from gill opening to base of caudal.   URANOSCOPUS, 3110.
aa. Dorsal caruncles placed nearer dorsal, the distance from dorsal $4\frac{1}{2}$ times in trunk, as above.   SHUFELDTI, 3111.

3110. MANCALIAS URANOSCOPUS (Murray).

D. I, 3 or 4; A. 4; C. 8; P. 10. Anterior spine of first dorsal produced in a long filament, ending in a pear-shaped bulb, terminating in a semitransparent whitish spot, this spine originating on posterior part of head, and reaching, when depressed, nearly to the tip of tail; far behind this are 2 short, fleshy tubercles, lying in a depression in front of second dorsal. Teeth moderate, depressible. Skin everywhere with minute embedded conical spines. Eye very small, placed high on the middle of the head.

3030——94 .

Color uniform black. (Murray.) To this description Goode & Bean add from the same specimen: "Anal opposite second dorsal, the 4 median caudal rays being much larger than the others and bifid; pectorals small, above the gill opening; the upper jaw is formed by the intermaxillaries, and is armed, together with the lower jaw, with a series of teeth of moderate size, which can be depressed as in *Lophius.* The skin is thickly covered with minute embedded conical spines; the eyes are very small and are placed high upon the middle of the head." Mid-Atlantic, in very deep water; 2 specimens known,. the type in 2,400 fathoms, taken between the Canary and Cape Verde islands, the second (26159) in 372 fathoms off the coast of Rhode Island in the Gulf Stream. (οὐρανοσκόπος, star-gazing, from the upturned eyes.)

Ceratias uranoscopus, MURRAY, in Wyville Thompson, The Atlantic, II, 67, fig. 20, 1878, mid-Atlantic, between Canary and Cape Verde Islands in 2,400 fathoms; GÜNTHER, Challenger Report, XXII, 54, pl. 11, fig. C, 1887.
Mancalias uranoscopus, GILL, Proc. U. S. Nat. Mus. 1878, 228; GOODE, Proc. U. S. Nat. Mus. 1880, 469; JORDAN & GILBERT, Synopsis, 848, 1883; GOODE & BEAN, Oceanic Ichthyology, 490, 1896.

3111. MANCALIAS SHUFELDTI (Gill).

Maxillary $\frac{1}{4}$ the length from gill opening to caudal base; intermaxillary $3\frac{1}{2}$ times in this length. Form more slender than that figured by Günther, with 4 rays in the dorsal, and apparently 15 in pectoral. There are no vomerines; intermaxillary and mandible armed with a narrow band of depressible teeth of various lengths. The skin with a fine granular appearance and everywhere covered with minute prickles. The caruncles only 2 in number and situated as in *Mancalias uranoscopus,* as figured in the *Challenger* fishes. Length of dorsal spine, without the joint bearing the pear-shaped appendage, equaling distance from gill opening to root of tail; the joint bearing the appendage is $\frac{2}{5}$ of this distance; in *Mancalias uranoscopus* (No. 26159) the first dorsal, without the joint bearing the appendage, contains the distance from the gill opening to the root of the tail $1\frac{1}{4}$ times. The joint containing the appendage is $\frac{1}{3}$ as long as the distance from the gill opening to root of tail. Dermal caruncles distant from the dorsal a space equal to $\frac{1}{6}$ of distance from the gill opening to root of tail. In *M. shufeldti* the caruncles are placed at a distance from the dorsal a space contained $4\frac{1}{4}$ times in the distance from the gill opening to the root of the tail. In the specimens described by Goode & Bean as *Mancalias uranoscopus* (No. 26159), the length $3\frac{1}{4}$ inches, the length of the maxillary is $\frac{1}{4}$ of length from gill opening to root of tail, and the intermaxillary $3\frac{1}{4}$ times in same distance. Teeth in jaws depressible, in narrow bands, and of unequal size; vomer toothless. Two small caruncles not far from front of dorsal fin, and instead of being placed opposite each other, according to the usual arrangement, one is placed behind the other. Skin covered with minute granules or papillæ, each one surmounted by a slender prickle, as in *Typhlopsaras.* The pectoral of the individual described contains 15 rays. The pectorals of *T. shufeldti* are imperfect. (Goode & Bean.) Gulf Stream, off the coast of southern New England;

1 specimen known. (Named for Dr. Robert W. Shufeldt, United States Army, the well-known ornithotomist.)

Typhlopsaras shufeldti,[*] GILL, Forest and Stream, Nov. 8, 1883, Western Atlantic (Type, No. 33552); JORDAN, Cat. Fishes, 138, 1885.

Ceratius shufeldti, GÜNTHER, Challenger Report, XXII, Deep-Sea Fishes, 54, 1887.

Mancalias shufeldti, GOODE & BEAN, Oceanic Ichthyology, 490, fig. 401, 1896.

1065. CRYPTOPSARAS, Gill.

Cryptopsaras, GILL, Forest and Stream, Nov. 8, 1883, 284 (*couesii*).

Body shortened; back longitudinally convex, eyes small but conspicuous; anterior spine with concealed basal joint and elongated terminal joint; a large intermediate globular and a pair of subpedunculated lateral dorsal appendages or caruncles close to the front of the dorsal fin; pectorals well developed, of about 15 rays. Deep seas. (κρυπτός, concealed; modern Greek ψαρᾶς, fisherman, in reference to the concealed rod bearing the dorsal spine or fishing apparatus.)

3112. CRYPTOPSARAS COUESII, Gill.

The basal joint of the rod-like spine is almost entirely concealed and procumbent, and the distal joint alone free, reaching backward to the dorsal tubercle; the bulb is pyriform, and surmounted by a long whitish filament; dorsal and anal each with 4 spines, the caudal 8 (the 4 middle dichotomous) and the pectorals each about 15 rays. (Gill.) A specimen of *Cryptopsaras* (No. 33558, U. S. Nat. Mus.) was obtained, by the *Albatross*, from Station 2101, in Lat. 38° 18′ 30″ N., Long. 68° 24′ W., at a depth of 1,686 fathoms. The type of *Cryptopsaras couesii* is only 35 mm. long. The caudal is imperfect. The length without caudal is 30 mm. and contains the greatest height 2¼ times. The bulb on the dorsal spine when laid backward can be made to reach to the dermal caruncles on the back. The length of the upper jaw is about ¼ of the length without caudal; gill opening nearly midway between front of head and root of tail; mouth placed vertically; intermaxillary teeth occupying about entire length of bone; mandibulary teeth unequal in size; at symphysis of mandible a pair of minute spines closely connected at base and slightly separated at the extremity. Specimen No. 39483 is 58 mm. long; 47 mm. to base of caudal. Greatest height 2⅔ in length without caudal. Gill opening a little nearer end of caudal than to front of head; distal portion of dorsal spine about ¼ length without caudal; median dermal caruncle very much

---

[*] The following is the original account of *Typhlopsaras:*

"*Typhlopsaras.*—Ceratiines with an elongated trunk, rectilinear back, obsolete or no eyes, far exserted basal joint of the anterior spine and shortened terminal joint, a small intermediate and a pair of pedunculated dorsal appendages some distance in advance of the dorsal fin, and reduced pectoral fin with about 5 or 6 rays.

"*Typhlopsaras shufeldti.*—The first joint of the rod-like spine reaches to the axil of the dorsal fin, and the bulb to the base of the caudal fin, when the spine is bent backward; the bulb is pear-shaped and without any appendages; the dorsal has 4 rays, the anal 4, the caudal 8 (the median 4 of which are forked), and there are 4 or 5 pectoral rays. A single specimen was found. I have dedicated the species to my esteemed friend, Dr. R. W. Shufeldt, U. S. A., the well-known ornithotomist.

"The name, *Typhlopsaras,* is a compound from the Greek τυφλος (blind) and ψαρας (angler), meaning 'blind angler.'"

larger than the two lateral ones; skin covered with minute granules of uniform size; pectoral with 16 rays, its length about ⅓ that of head; length of upper jaw about ⅓ of total without caudal; pair of spines at symphysis of mandible replaced by a very small knob; teeth in intermaxillary very small, diminishing in number toward the symphysis, apparently uniserial. On each side of head of vomer 2 or 3 depressible teeth; palatines apparently wanting. We have seen something like traces of similar teeth on the vomer of *Mancalias uranoscopus*, but owing to the condition of the specimen can not be certain about this character. (Goode & Bean.) Gulf Stream, off the coast of New England. (Named for the eminent ornithologist, Dr. Elliott Coues.)

Cryptopsaras couesii, GILL, Forest and Stream, Nov. 8, 1883, 284, Gulf Stream off. New England (Coll. Albatross); JORDAN, Cat Fishes, 138, 1885; GOODE & BEAN, Oceanic Ichthyology, 491, fig. 402, 1896.
Ceratias couesii, GÜNTHER, Challenger Report, XXII, 55, 1887.
?*Ceratias carunculatus*, GÜNTHER, Challenger Report, XXII, 55, pl. 11, fig. *d*, 1887, south of Yezo, Japan, in 345 fathoms; 1½ inches long. (Coll. *Challenger*.)

1066. ONEIRODES, Lütken.

Oneirodes, LÜTKEN, Overs. Kong. Dansk. Vidensk. Selsk. Forhandl. 1871, 56 (*eschrichtii*).

Body compressed, oval, short, covered with smooth skin. Head compressed, very large. Mouth moderate, almost horizontal, the joint of mandible behind eyes. Teeth unequal, depressible; vomer with teeth. Gill arches unarmed; gills in 2¼ pairs. Spinous dorsal represented by a cephalic spine, the basal element of which is procumbent and subcutaneous, the tip bulbous, and a second spine about midway between the rostral spine and the soft dorsal; soft dorsal and anal short; no ventrals; no pyloric cæca. Greenland. (ὀνειρώδης, dream-like, in illusion to the small, almost covered, eyes.)

3113. ONEIRODES ESCHRICHTII, Lütken.

D. I–I, 6; A. 8; C. 8. Terminal half of the bulb of the cephalic spine whitish. Cephalic spine with a bulbous termination, surmounted by slender filaments in several transverse rows. Caudal fin shorter than trunk without head. Color black. Deep sea, off Greenland. Known from a single specimen 8 inches long. (Gill.) (Named for D. F. Eschricht, a Danish naturalist, a student of the Cetacea.)

Oneirodes eschrichtii, LÜTKEN, Overs. Dansk. Vidensk. Selsk. Forhandl. 1871, 56, 9–18, pl. 2, deep sea off Greenland; GILL, Proc. U. S. Nat. Mus. 1878, 218; JORDAN & GILBERT, Synopsis, 848, 1883; GOODE & BEAN, Oceanic Ichthyology, 492, 1896.

1067. HIMANTOLOPHUS, Reinhardt.

Himantolophus, REINHARDT, Dansk. Vidensk. Selsk. Nat. 1837, 74 (*grœnlandicus*).

Head large, compressed. Skin thick, with scattered, round, prickly scales. Body oval, compressed. Mouth moderate, the cleft oblique, the joint of the mandible below or behind the eyes. Gills in ¼ 2½ pairs; gill arches armed with dentigerous tubercles. Spinous dorsal represented only by a single long rostral spine, the basal element of which is procumbent and subcutaneous; the extremity with numerous long filaments. Soft dorsal short, with 9 rays; anal short; pectoral rather broad, with 12 rays. Greenland. (*ἱμάς*, a thong; *λόφος*, crest.)

3114. HIMANTOLOPHUS GRŒNLANDICUS, Reinhardt.

Depth 2⅓ in total length. D. I-9; P. 12. Body oblong oval. Cephalic ray provided with about 11 tentacles. (Gill.) Greenland. "This species has never been fully described, the only existing example being an imperfect one, 23 inches long, obtained off the coast of Greenland, about 1837."

Himantolophus grœnlandicus, REINHARDT, Dansk. Vid. Selsk. Nat. Math. Afh. 1837, 74, Greenland; GILL, Proc. U. S. Nat. Mus. 1878, 218; JORDAN & GILBERT, Synopsis, 849, 1883.

1068. CORYNOLOPHUS, Gill.

Corynolophus, GILL, Proc. U. S. Nat. Mus. 1878, 219 *(reinhardti).*

This genus is scarcely distinct * from *Himantolophus,* differing in the short oval form, the short dorsal of about 5 rays, and the broader pectoral with about 17 rays each. (*κορύνη*, club; *λόφος*, crest.)

3115. CORYNOLOPHUS REINHARDTI (Lütken).

Depth 1¼. D. I-5; P. 17. Body short, oval; cephalic ray ⅔ length of head, with about 8 tentacles, which branch out forming a brush at tip; skin sparsely covered with thorn-like prickles. Greenland. One specimen known, 14 inches long. (Named for Prof. Johann Reinhardt, naturalist at the University of Copenhagen.)

Himantolophus reinhardti, LÜTKEN, Kong. Dansk. Vidensk. Selsk. 1878, 321, Greenland.
Corynolophus reinhardti, GILL, Proc. U. S. Nat. Mus. 1878, 219.

1069. LIOCETUS, Günther.

Liocetus, GÜNTHER, Challenger Report, xx.., 57, 1887 *(murrayi).*

Mouth enormous, the cleft nearly vertical; pectorals small, in advance of dorsal and of gill opening; second dorsal spine wanting; gills in 2½ pairs; no gular tentacle. This genus is similar to *Melanocetus,* differing in having no teeth on the vomer, a greater projection of the mandible, and a smaller mouth. Deep sea. (*λεῖος*, smooth; *κῆτος*, whale.)

3116. LIOCETUS MURRAYI (Günther).

D. I-13; A. 4; C. 9; P. 14. (Günther.) Extremely similar to *Melanocetus johnsonii,* but without trace of vomerine teeth, while there is no distinction between the two species as regards dentition of jaws; posterior angle

*Dr. Gill, replying to certain strictures as to the validity of this genus, made by Lütken [who calls it a "wanton" subdivision], states that the "differences alleged to exist between *Himantolophus* and *Corynolophus* are very marked. If they do exist as stated there can be no doubt that the two should be kept apart. I know of no reason except the singularity and greatness of the difference specified for doubting the correctness of Reinhardt's observations."

of mandible projecting more and forming a salient point; mouth comparatively less wide, and the maxillary considerably shorter, being about $\frac{2}{3}$ of total length, without caudal, while it is rather more than $\frac{1}{4}$ in the Madeiran species. Eye rudimentary. One cephalic spine, shorter than maxillary; last dorsal ray connected by a short and delicate membrane with caudal fin; most of the caudal rays bifid, the longest shorter than maxillary;. pectoral fin as much developed as in *Melanocetus johnsonii*. Entirely black. Total length 44 lines; length of mandible 14 lines; length of maxillary 12 lines; length of caudal fin 10$\frac{1}{2}$ lines. A young individual, 44 lines in length, was taken by H. M. S. *Challenger* in the mid-Atlantic, at a depth of 1,850 fathoms (Station 106); another of 13 lines at the depth of 2,450 fathoms (Station 348). (Goode & Bean.) (Named for Dr. John Murray, second director of the civilian staff on board H. M. S. *Challenger*.)

Melanocetus bispinosus, GÜNTHER, Study of Fishes, 473, 1880; name only.
Melanocetus (Liocetus) murrayi, GÜNTHER, Challenger Report, XXII, 57, pl. 11, fig. A, 1887, mid-Atlantic.
Liocetus murrayi, GOODE & BEAN, Oceanic Ichthyology, 495, fig. 407, 1896.

1070. LINOPHRYNE, Collett.

Linophryne, COLLETT, Proc. Zool. Soc. London 1886, 138 (*lucifer*).

Head enormous; the body slender, compressed, mouth oblique. Spinous dorsal reduced to a single cephalic tentacle, the basal part of which is erect, not procumbent. Teeth in the jaws on the vomer and the upper pharyngeals. Gill openings exceedingly narrow, situated a little below the root of the pectoral. Soft dorsal and anal very short; ventrals none. Abdominal cavity forming a sac, suspended from the trunk. Skin smooth; a long tentacle on the throat. This genus differs from *Melanocetus* in the presence of the gular tentacle. ($\lambda i\nu o\varsigma$, linen, net; $\varphi\rho\acute{\upsilon}\nu\eta$, a toad.)

3117. LINOPHRYNE LUCIFER, Collett.

D. I–3; A. 2; C. 9; P. 14 or 15. A spinous projection or horn above each orbit. Cephalic tentacle black, with a large ovate bulb, the upper half of which is white; gular tentacle much larger, terminating in 2 tongue-like appendages, which are furnished on the upper edge with a row of round, white papillæ. (Goode & Bean.) Mid-Atlantic, northwest of Madeira, Lat. 36° N., Long. 20° W. One specimen known. (*Lucifer*, an evil spirit; *lux*, light; *fero*, I bear.)

Linophryne lucifer, COLLETT, Proc. Zool. Soc. London 1886, 138, pl. 15, mid-Atlantic, between Madeira and the West Indies (Coll. Capt. P. Andresen. Mus. Univ. Christiania); GÜNTHER, Challenger Report, XXII, 57, 1887; GOODE & BEAN, Oceanic Ichthyology, 496, fig. 408, 1896.

1071. CAULOPHRYNE, Goode & Bean.

Caulophryne, GOODE & BEAN, Oceanic Ichthyology, 496, 1896 (*jordani*).

Head large, compressed; mouth with the cleft nearly horizontal; body short, much compressed. Spinous dorsal reduced to a single cephalic tentacle, which is supported on a short procumbent base. Teeth of unequal size in the intermaxillary and the mandible; vomer, palatines, and upper pharyngeals toothed. . Gill openings narrow, horizontal slits placed below and in front of root of pectorals. Branchiæ in ¼ 2½ pairs. Branchial arches armed with dentigerous tubercles. Skin naked. Numerous luminous filaments on head and body. Soft dorsal and anal many-rayed, the rays greatly produced; caudal long, tapering; ventrals none; pectorals very broad, sessile, postmedian, under dorsal fin, with numerous rays. Pyloric appendage reduced to 1 small rudiment. Air bladder absent. (*καυλός*, stem; *φρύνη*, toad, from the many stems or fin rays.)

3118. CAULOPHRYNE JORDANI, Goode & Bean.

' Depth about 2 in length without caudal, the greatest height occurring behind the head.· Cephalic appendage with a pale tuft at its tip, the length of the distal portion 3 in length of body. The tuft somewhat mutilated, but showing no evidence of a laminated structure; basal portion of cephalic appendage about twice as long as the very small eye; maxillary very slender, narrow, extending about as far backward as intermaxillary; intermaxillary slightly protractile and with about 10 teeth on each side, several of which are nearly twice as large as the rest, its length 2¼ in body; mandible as long as head without snout, with 8 teeth on each side, the anterior pair and several other pairs along shaft of bone being greatly enlarged; a pair of enlarged teeth on head of vomer; several similar teeth on palatines; upper pharyngeals armed with several strong teeth; eye very small, inconspicuous, its distance from tip of snout equaling nearly ¼ its distance from soft dorsal origin. Intestine shorter than length without caudal. Soft dorsal with 16 rays, all of which, except the last 4, are greatly produced; the second, third, and fourth rays longest, nearly twice as long as body. Anal with 14 rays, all of which, except last 3, are much produced, the fin not quite perfect, yet its anterior rays are longer than body; caudal with 8 rays, the 4 inner ones divided, the rest simple; middle rays of caudal as long as distance from tip of lower jaw to base of pectoral; pectoral comparatively short, with 16 simple articulated rays, the longest about ¼ as long as head. About 9 luminous filaments on each side of head, 7 more between nape and dorsal, and about 12 on sides; the filaments nearly twice as long as eye. Head and body black; caudal, cephalic tuft, and most of the rays pale. Gulf Stream. The .type of the species (No. 39265) was taken by the steamer *Albatross*, September 19, 1887, in Lat. 39° 27′ N., Long. 71° 15′ W., 1,276 fathoms. (Named for David Starr Jordan.)

Caulophryne jordani, GOODE & BEAN, Oceanic Ichthyology, 496, pl. 21, fig. 409, 1896, Gulf Stream, off Carolina, in 1,276 fathoms. (Coll. *Albatross*; the plate named *Caulophryne setosus*, by slip in proof reading.)

Family CCXXIV. OGCOCEPHALIDÆ.

(THE BAT-FISHES.)

Head very broad and depressed, the snout more or less elevated, the trunk short and slender. Mouth not large, subterminal or inferior, the lower jaw included; teeth villiform or cardiform. Gill openings very small, above and behind the axils of the pectoral fins. Body and head covered with bony tubercles or spines. Spinous dorsal reduced to a small rostral tentacle, which is retractile into a cavity under a prominent process on the forehead; in 1 genus the rostral tentacle is obsolete; soft dorsal and anal fins small and short; ventrals well developed; pectoral fin well developed, its base strongly angled, with long pseudobrachia and 3 actinosts. Branchiostegals 5; no pseudobranchiæ. Genera 8; species about 30, chiefly American, some of them in the deep sea. (*Pediculati*, part; genera *Malthe* and *Halieutæa*, Günthér, III, 200-205, 1861.)

OGCOCEPHALINÆ:

  a. Disk with the frontal region elevated and the snout more or less produced forward, the tail stout; orbits lateral; teeth on vomer and palatines; rostral tentacle present.
    b. Gills 2½; disk longer than broad.   OGCOCEPHALUS, 1072.
    bb. Gills 2½; disk broader than long.   ZALIEUTES, 1073.
HALIEUTINÆ:
  aa. Disk with the frontal region depressed, not elevated above the rest; eyes partly superior; snout rounded, obtuse in front; tail slender.
    c. Dorsal fin present.
      d. Vomer and palatines with teeth.   HALIEUTICHTHYS, 1074.
      dd. Vomer and palatines toothless.
        e. Disk subcircular; gills 2¼.
        f. Mouth rather large, subvertical; prickles rather strong.
                                                    HALIEUTÆA, 1075.
        ff. Mouth rather small, terminal; prickles feeble.
                                                    HALIEUTELLA, 1076.
      ee. Disk subtriangular; gills 2; prickles very strong.
                                                    DIBRANCHUS, 1077.

1072. OGCOCEPHALUS, Fischer.

(SEA-BATS.)

Ogcocephalus, FISCHER, Zoognosia, 78, 1813 (*vespertilio*).
Oncocephalus, GILL, modified spelling.
Malthe, CUVIER, Règne Animal, Ed. I, II, 311, 1817 (*vespertilio*).
Malthœa, corrected spelling.

  Body stoutish, tapering backward; head very broad and depressed, triangular in form, the forehead elevated and produced. Eyes large, lateral. Mouth rather small, subinferior under the snout; villiform teeth in bands, on jaws, vomer, and palatines. Skin covered with rough, bony tubercles. Dorsal and anal fins very small; rostral tentacle present, retractile into a cavity under a bony prominence on the forehead; ventrals present, I, 5, well separated; pectorals large, placed horizontally. Gills 2¼. No air bladder; no pyloric cæca. Tropical America, in shallow water; small fishes of singular form, often regarded by the ignorant as venomous. (ὄγκος, hook; κεφαλή, head; properly written *Oncocephalus*, but Fischer chose the above monstrous spelling.)

  a. Snout produced, the rostral process pointed, 6 to 10 in length of body.
                                                    VESPERTILIO, 3119.
  aa. Snout short, the rostral process 12 to 15 times in length of body.   NASUTUS, 3120.
  aaa. Snout short, the rostral tubercle reduced to a button-like tubercle, which is about 25 times in length of body.   RADIATUS, 3121.

3119. OGCOCEPHALUS VESPERTILIO (Linnæus).

(BAT FISH; DIABLO.)

Head, from tip of upper jaw to gill opening, nearly $\frac{1}{4}$ the length; depth 5 in length from upper jaw to base of caudal; width 1⅖. D. 4; A. 4; rostral process from 6 to 10 (9 in our specimens from Havana); P. 4½; V. 6; C. 4¼. Body stoutish, much depressed, rostral process longer than in the other species, variable in length; mouth small, the maxillary reaching nearly to posterior margin of eye; villiform teeth in bands, on jaws, vomer, and palatines; interorbital flattish, its width less between anterior edge of eyes than posterior edge; rostral groove longer than broad; body covered with bony protuberances, variable in size, and not very definite in position, lower parts with a shagreen-like covering; posterior edge of pectorals much behind middle of body; ventrals long, reaching outward to edge of the disk-like, anterior part of body; origin of dorsal over posterior edge of pectoral; anal under the vertical of tips of dorsal rays, anal reaching nearly to base of caudal. Pale grayish brown above, reddish below; back with round black spots, conspicuous in life, but growing fainter and sometimes disappearing in spirits; belly in life a coppery red; pectorals nearly plain dusky. Length 12 inches. West Indies, north to the Florida Keys; common· in shallow water. Here described from a specimen from Havana, Cuba, about 10 inches in length. The length of the snout is subject to great variation, but it is never short and button-like, as in *O. radiatus*. (*vespertilio*, a bat.)

Lophius vespertilio, LINNÆUS, Syst. Nat., Ed. x, I, 236, 1758, American Seas; after *Lophius fronti unicorni* of ARTEDI.

Malthœa vespertilio, CUVIER & VALENCIENNES, Hist. Nat. Poiss., XII, 440, 1837; prominence on snout 10 in length; DE KAY, N. Y. Fauna: Fishes, 167, 1842; GÜNTHER, Cat., III, 200, 1861; JORDAN & GILBERT, Syopsis, 850, 1883; JORDAN & SWAIN, Proc. U. S. Nat. Mus. 1884, 234.

Lophius rostratus, SHAW, Zool., IV, 383, pl. 163, 1803.

Malthœa longirostris, CUVIER & VALENCIENNES, Hist. Nat. Poiss., XII, 452, 1837, Bahia (Coll. Blanchet); snout 6 in length.

3120. OGCOCEPHALUS NASUTUS (Cuvier & Valenciennes).

Head 2. D. 4; A. 4. Rostral process short, about 12 to 15 times in length of body. Cavity of nostril tentacle higher than broad; width of body 2 in length; vent behind middle of body. Dusky above, with round black spots, edged with whitish. West Indies. (Lütken); not seen by us; perhaps a variation of *O. vespertilio*. (*nasutus*, long-nosed.)

Malthœa nasuta, CUVIER & VALENCIENNES, Hist. Nat. Poiss., XII, 452, 1837, Martinique.

Malthe nasuta, LÜTKEN, Nat. For. Vid. Medd. 1865, 4; JORDAN & GILBERT, Synopsis, 850, 1883.

Malthœa notata, CUVIER & VALENCIENNES, Hist. Nat. Poiss., XII, 453, 1837, Surinam; snout 15 in length; body spotted.

3121. OGCOCEPHALUS RADIATUS (Mitchill).

(SHORT-NOSED BAT-FISH.)

Head 2. Dorsal 4; A. 4. Rostral cavity somewhat broader than high, or width equal to height; distance between anterior angles of orbits about equal to that between the posterior angles; eye a little wider than interorbital width; snout, exclusive of rostral tubercle, not produced beyond the rostral cavity, but with a cylindrical button-like tubercle, slightly contracted at base, pointing obliquely upward and forward, its length 25 times in body; posterior edge of pectoral slightly nearer base of caudal than upper jaw; caudal peduncle very thick and heavy; vent about midway between tip of jaw and base of caudal fin. Color brownish, with dark round spots sometimes edged with white; pectorals with a network of white lines dividing the dark color into dark brown spots; tip of caudal blackish, belly coppery red. Length 8 to 12 inches. Coast of Florida and neighboring waters; very common in shallow bays among weeds, especially about the Florida Keys. Here described from a specimen from Cedar Key, Florida, 7 to 8 inches in length. (*radiatus*, rayed.)

Lophius radiatus, MITCHILL, Amer. Monthly Mag., March, 1818, 326, Straits of the Bahamas.

Malthe cubifrons, RICHARDSON, Fauna Bor.-Amer., III, 103, 1836, said to be from Labrador (Coll. J. J. Audubon), but this is certainly an error; Audubon collected also in Carolina and Florida; GÜNTHER, Cat., III, 203, 1861; JORDAN & GILBERT, Synopsis, 850, 1883; JORDAN, Cat. Fishes, 139, 1885.

? *Malthæa truncata*, CUVIER & VALENCIENNES, Hist. Nat. Poiss., XII, 454, 1837, America; snout wholly obsolete; perhaps a species of *Zalieutes*.

Malthæa angusta, CUVIER & VALENCIENNES, Hist. Nat. Poiss., XII, 454, 1837, Dutch Guiana; snout more than 20 in length.

1073. ZALIEUTES, Jordan & Evermann.

Zalieutes, JORDAN & EVERMANN, Check-List Fish. N. and M. A., 511, 1896 (*elater*).

Disk wider than long, about as long as rest of body (including caudal fin); middle line of head elevated, but the forehead not projecting beyond mouth; rostral tentacle present, the cavity about as wide as high; mouth small; minute teeth on vomer and palatines. Gills 2¼. Eastern Pacific. The genus is very close to *Malthopsis*, Alcock,[*] but the latter, like *Ogcocephalus*, has the disk longer than broad, but the gills are reduced to 2. (ζάλη, surge of the sea; ἁλιευτής, fisher.)

3122. ZALIEUTES ELATER (Jordan & Gilbert).

Body very broad and depressed, the disk considerably broader than long, its width 1⅔ times in length of body; back and snout considerably raised above rest of body; greatest depth of body scarcely more than width of mouth. Mouth small, its width ⅓ greater than diameter of orbit; snout very short, scarcely projecting beyond mouth, its length

---

[*] *Malthopsis*, ALCOCK, Ann. and Mag. Nat. Hist. 1891, 26; *Malthopsis luteus*, from the Andaman Sea; *Malthopsis mitriger*, GILBERT & CRAMER, Proc. U. S. Nat. Mus. 1896, 434, with plate, off Hawaiian Islands.

about equal to interorbital width, shorter than its own width in front. Eye rather large, much longer than snout, wider than interorbital area. Process representing first dorsal spine present, small. Skin covered with spines, which are comparatively slender and sharp, their stellate bases inconspicuous, those on snout and middle of back and tail largest, much slenderer and sharper than in *Ogcocephalus vespertilio;* no spines on ocelli of back; belly rough; under side of tail with tubercular plates; tail depressed toward base of fin. Pectorals $\frac{1}{4}$ longer than ventrals, their length $1\frac{2}{3}$ width of mouth; caudal a little longer than pectoral, $4\frac{1}{5}$ in body. Color light olive, above everywhere thickly and uniformly covered with small round spots of dark brown, these about as large as the pupil and about as wide as the lighter interspaces; a conspicuous ocellus, larger than eye, on each side of back, this ocellus with a bright yellow spot in the center, surrounded by a black ring, around which is a pale ring, and finally a fainter dark one; under parts plain white; pectorals spotted; caudal yellowish at base, with a terminal blackish band. Length 4 inches. Pacific coast of Mexico, south to Panama, in water of moderate depth; very rare near the shore, but obtained in abundance by the *Albatross* at Stations 2794 and 2795, near Panama. (*Elater*, the spring beetle, from the resemblance of the ocelli to the eye-like spots on the back of *Elater*.)

Malthe elater, JORDAN & GILBERT, Proc. U. S. Nat. Mus. 1881, 365, Mazatlan. (Type, No. 28127. Coll. Dr. J. U. Bastow.)
Ogcocephalus elater, JORDAN, Proc. Cal. Ac. Sci. 1895, 506.

1074. HALIEUTICHTHYS, Poey.

Halieutichthys, POEY, Proc. Ac. Nat. Sci. Phila. 1863, 83 (*reticulatus*).

Disk subcircular, anteriorly cordiform, the head merging into the body, very large and much depressed; cranial portion not elevated; interorbital space low and narrow; eyes partly superior; mouth terminal, horizontal, the jaws subequal, the lower jaw nearly semicircular; teeth fine, on jaws and palate. Gills $2\frac{1}{2}$; no gill rakers; gill openings anterior to pectoral; rostral tentacle very small, retractile; dorsal and anal few-rayed; pectorals large, the carpus slender; caudal rounded; skin above sparsely armed with stellate tubercles; lower surface smooth. ($\dot{\alpha}\lambda\iota\epsilon\upsilon\tau\dot{\eta}\varsigma$, fisher; $\dot{\iota}\chi\theta\dot{\upsilon}\varsigma$, fish.)

a. Surface of body covered with brownish reticulations; fins not barred with black.
ACULEATUS, 3123.
aa. Surface of body blackish, not reticulate; pectorals with a broad black bar mesially, the tip pale; caudal blackish toward the tip. CARIBBÆUS, 3124.

3123. HALIEUTICHTHYS ACULEATUS (Mitchill).

D. I, 4 or 5; A. 4; V. I, 5; P. 16 to 18; C. 9; gills $2\frac{1}{2}$. Disk cordiform, about as wide as long, its length more than $\frac{2}{3}$ that of body. Body covered above with stout conical spines with stellular bases, largest upon the trunk, upon which they are arranged in about 2 irregular longitudinal rows on each side of the dorsal; upon the disk they are placed above the principal bones of the skeleton, most abundant upon its cranial portion; a single row of stout spines, usually 3-pointed, on the outer margin of disk, a particularly large one at each outer angle; body entirely smooth below; snout very short, obtuse; bridge over the rostral cavity covered in front with a 3-pointed spine, having on each side a simple spine; short, stout, simple spines upon each supraorbital margin, the front of which is immediately above and behind the cavity containing the nostrils; vertex with several similar spines; many spines closely placed upon the humeral area; numerous short tentacles upon margin of disk and on sides of trunk; supraoral cavity elliptical, small (horizontal diameter $\frac{2}{3}$ diameter of orbit), containing a well-developed, club-shaped, very perceptible tentacle; width of opening of anterior nostril, which is in a short tube, $\frac{1}{2}$ that of posterior nostril, which is not tubular; width of mouth much less than distance between pupils and equaling diameter of orbit. Diameter of orbit $8\frac{1}{4}$ times in distance from snout to base of caudal, 6 times in distance from snout to origin of soft dorsal, $6\frac{1}{4}$ times in distance to origin of anal, 3 times in distance to base of ventrals, and 6 times in distance to angle between pectorals and trunk, $4\frac{2}{3}$ times in distance from snout to gill opening, 6 in greatest width of disk, and nearly 2 in that of trunk; width of interorbital area $\frac{2}{3}$ diameter of orbit. First dorsal ray longest, equal to diameter of orbit; anal fin inserted under third ray of dorsal, with 4 rays, the third or longest very slightly longer than the longest dorsal ray; ventral fins inserted nearly under the middle of the disk, with 1 rudimentary and 5 dorsal rays, increasing in length posteriorly, the last and longest contained 5 times in total length; distances between origins of ventrals $6\frac{1}{4}$ in total length; pectorals with peduncles entirely included in common membrane, with blades far back, horizontal, lying close to trunk, composed of 16 rays, the middle or longest $3\frac{2}{3}$ in total length; caudal fin rounded, composed of 9 rays, the external rays, 1 above and 2 below, simple, the others bifid; length of middle ray equal to that of trunk (measured from junction of pectorals to base of caudal rays) and slightly exceeding the longest pectoral ray. Length of intestine contained $1\frac{2}{3}$ times in total length. Color, body covered above with reticulations of brown, the general hue varying from light yellowish gray to grayish brown, the markings being darker upon darker specimens; pectoral and caudal fins with about 3 dark bars; the terminal bars in young very black; body beneath milky white (Goode & Bean.) West Indies, Gulf of Mexico, and Gulf Stream, in water of moderate depth; taken by the *Blake* and the *Albatross* at numerous stations in depths ranging from 10 to 95 fathoms. "As in *Halieutœa*, *Dibranchus*, and allies, a rostral tentacle is present in this genus. Among specimens belonging to the Museum of Comparative Zoology there is evidence of the existence of a couple of distinct forms in the West Indian waters. The true *H. aculeatus* is much the lighter in the ground colors and has brownish reticulations across the back, 2 or 3 narrowish transverse bands of the same color across the pectorals, and 2 or 3 similar bands appear on the caudal, the posterior being darkest. The margins of the fins are light in color. The rostrum is acute; it ends in a spine which turns úpward, and, seen from above, it is hardly long enough to cover the tentacular niche. Evidently this type belongs to the shallower waters. The localities noted carry its distribution from the Bahamas to the Yucatan Banks, in depths of 40 fathoms and less." (Garman.) (*aculeatus*, with needle-like spines.)

Lophius aculeatus, MITCHILL, Amer. Mon. Mag., II, 1818, 325, Straits of Bahama.
Halieutichthys reticulatus (POEY MS.) GILL, Proc. Ac. Nat. Sci. Phila. 1863, 91, Cuba (Coll. Prof. Felipe Poey); JORDAN & GILBERT, Synopsis, 851, 1883.
Halieutichthys aculeatus, GOODE, Proc. U. S. Nat. Mus. 1879, 109; GOODE & BEAN, Oceanic Ichthyology, 504, pl. 122, fig. 414a and b, 1896; GARMAN, Bull. Lab. Nat. Hist. Iowa Univ. 1896, 87, pl. 4, fig. 1.

3124. HALIEUTICHTHYS CARIBBÆUS, Garman.

D. I-5; A. 4; V. 5; P. 17; C. 9. Color darker than *H. aculeatus;* the reticulations are not present; the outer half of the pectoral, except at the margin, is black; and, excepting the narrow posterior margin, the hinder fifth of the caudal fin is black; the upper surface is clouded brownish without traces of the network pattern common to *H. aculeatus*. On the specimens described, the rostrum is acute, and the spine extends forward to cover the cavity receiving the tentacle so that it is not visible when viewed from above. West Indies. As now known, this species ranges from Jamaica to Barbados in depths of 70 to 150 fathoms or more. (Garman.) (*caribbæus*, from the Caribbean Sea.)

Halieutichthys caribbæus, GARMAN, Bull. Lab. Nat. Hist. Iowa Univ. 1896, 87, pl. 4, fig. 2, Jamaica to Barbados.

1075. HALIEUTÆA, Cuvier & Valenciennes.

Halieutœa, CUVIER & VALENCIENNES, Hist. Nat. Poiss., XII, 455, 1837 (*stellatus*).

Head very large, broad, depressed, its outline nearly circular; cleft of the mouth wide, horizontal; jaws with small cardiform teeth; no teeth on vomer or palatines. Skin everywhere covered with small, stellate spines. Forehead with a transverse bony ridge, beneath which is a tentacle, retractile into a cavity, the only rudiment of the spinous dorsal fin; soft dorsal and anal very short, far back. Gills 2½, the anterior gill arch without laminæ. Branchiostegals 5; vertebræ 17. Pacific Ocean. (ἁλιευτής, one who fishes.)

3125. HALIEUTÆA SPONGIOSA, Gilbert.

D. 6; A. 4; C. 9; V. 4; P. 12 or 13. This species is remarkable for the soft, spongy texture of the body, and the membranaceous or cartilaginous character of its bones. Width of head 1½ in its length; tail long and slender, the vent midway between base of caudal and articulation of mandible; width of base of tail 4½ in its length; mouth little or not at all overpassed by the snout, its width 2¾ to 3 in that of head, lower jaw usually not included; gape of mouth oblique, almost wholly anterior. Teeth in wide cardiform bands in the jaws, none of them enlarged; palate toothless. Interorbital width slightly greater than length of snout, 5 in width of head; eye 1¼ in interorbital width. Rostral tentacle short, with an expanded 3-lobed tip; front of dorsal midway between base of caudal and occiput; caudal long, rounded, the lower rays more shortened than the upper, the longest nearly ¼ width of head; anal rays high, closely bound together, the fin slender, shaped like the intromittent organ of *Gambusia,* the length of its base equaling ⅔ diameter of orbit, its longest ray reaching base of caudal; pectorals long, the posterior rays rapidly shortened, the longest ¼ width of head; head and body everywhere with broadly conical, tubercular plates, varying in size, marked with strong lines, radiating from the center; the apex sometimes blunt, more often provided with a slender spine, sometimes bifid or trifid; on the tail these spines become longer and are directed backward; plates along edge of disk not compressed nor specially modified. A deep groove-like channel just behind mandible and following curve of latter, becoming continuous with another deeper channel running just below edge of disk to near base of pectorals; a third groove runs backward from nostrils, uniting with the others, these grooves spanned at intervals by pairs of fleshy tentacles with fringed tips, which spring from the edges of the grooves and meet across them; at the bottom of the grooves under each pair of tentacles is a small fleshy tubercle. Fin rays, at least at base, with series of small curved prickles. Color uniform dusky, the tail sometimes lighter; fins blackish, more or less edged with white. One specimen with the body and tail uniformly light. Pacific coast of Mexico in deep water. Numerous specimens, the largest 4½ inches long, from *Albatross* Station 2992, in 460 fathoms. (Gilbert.) (*spongiosus,* spongy.)

Halieutæa spongiosa, GILBERT, Proc. U. S. Nat. Mus. 1890, 124, west of Revillagigedo Islands, at Albatross Station 2992, Lat. 18° 17' 30" N., Long. 114° 43' 15" W., in 460 fathoms. (Coll. Gilbert.)

1076. HALIEUTELLA, Goode & Bean.

Halieutella, GOODE & BEAN, Proc. Biol. Soc. Washington, 1882, 88 (*lappa*).

Body subcircular, depressed, its width equal to its length, covered with flaccid, inflatable skin. Spines feebler and less numerous than in *Halieutæa.* Head merged in body; forehead with a transverse bony ridge; no perceptible supraoral cavity; no tentacle. Mouth small, terminal; lower jaw slightly curved forward. Teeth in the jaws minute, cardiform, not discernible on palate, though possibly present. Carpus broad, slightly exserted; pectoral fins remote from tail, obliquely placed, with membranes subvertical. Branchial aperture posterior to carpus, upon the disk, and not remote from its margin. Gills 2½. Dorsal fin 5-rayed, inserted at junction of disk with caudal peduncle; anal fin 4-rayed, originating at root of caudal peduncle. (ἁλιευτήρ, a fisherman.)

3126. HALIEUTELLA LAPPA, Goode & Bean.

D. 5; A. 4; C. 9; P. 15; V. 5. Disk subcircular, more than ⅔ as long as the body. Body covered with a loose, flaccid, inflatable skin, which so obscure its proportion, that it is impossible to determine its exact height, but it is not nearly so much depressed as in the related genera. When the body is inflated the height and length of the disk is nearly equal. Spines rather feeble; about 10 between snout and dorsal fin; about 6 strong spines, with conical bases and stellular tips, on outer margin of disk on each side, the anterior of them being opposite the eye; in front of these spines on the discal margin, and between them and the snout, are several small, simple spines, pointing backward; belly armed with spines similar to those on the back, but weaker; a stellate spine upon tip of snout, with 2 weaker, simple spines on each side; nasal openings midway between eye and tip of snout; mouth small, upon the margin of the disk; upper jaw shorter than diameter of eye. Teeth as described in the generic diagnosis. Dorsal fin inserted at posterior limit of disk, with 5 simple, articulated rays, its longest ray $\frac{1}{5}$ as long as disk; anal fin with 4 simple, articulated rays, inserted directly beneath fourth ray of dorsal, its second and longest ray $\frac{1}{4}$ as long as disk; caudal twice as long as anal, and slightly longer than caudal peduncle, with 9 simple, articulated rays. Carpus inserted at a distance from snout equal to twice length of longest pectoral ray, which is slightly greater than distance of posterior margin of carpus, at its junction with disk, from vent; number of pectoral rays 15; ventral inserted at a point equidistant from the snout and origin of anal, its longest ray (the fourth) equal to $\frac{1}{5}$ distance of anal fin from snout. Color yellowish white. Gulf Stream. A single specimen, 1¼ inches long, known. (*lappa*, the burdock, from its prickles.)

Halieutella lappa, GOODE & BEAN, Proc. Biol. Soc. Washington, II, 1882, 88, Gulf Stream, at Fish Hawk Station 1151, Lat. 39° 58′ 30″ N., Long. 70° 37′ W., in 125 fathoms; GOODE & BEAN, Oceanic Ichthyology, 500, pl. 122, figs. 512a and 512b, 1896.

1077. DIBRANCHUS, Peters.

Dibranchus, PETERS, Monatsber. Kon. Akad. Wiss. Berlin 1876, 736 (*atlanticus*).

Head merged in body, very large, much depressed, forming a broadly ovate disk, with margin laterally prolonged; cranial portion not elevated; the interorbital area low, narrow, with orbits partly superior; supraoval cavity large, protected above by a transverse bony ridge. Mouth terminal, horizontal, wide; lower jaw convex; teeth in cardiform bands, none on vomer or palatines. Gills 2; no gill rakers; gill openings small, anterior to pectorals. Rostral tentacle retractile, trilobate at tip. Skin with numerous strong stellate spines above and below, those at margins of disk especially strong, 3-pointed. Atlantic; distinguished from related genera by the reduction of the gills to 2 pairs. (δίς, two; βράγχος, gill.)

3127. DIBRANCHUS ATLANTICUS, Peters.

D. 6 or 7; A. 4; C. 9; P. 13 to 15; V. I, 5; Br. 6; gills 2. Disk orbicular, nearly as wide as long, its length about $\frac{1}{3}$ that of body, its lateral outline prolonged on each side, and terminating in a strong spine armed at the tip with a group of irregularly arranged acicular spinelets. Body covered above with numerous stout conical spines with stellular bases, these

largest upon the trunk, where they are arranged approximately in about
4 irregular longitudinal rows upon each side of the dorsal fin; closely set
rows of these stout spines mark the outer margin of the disk, and there is
also a cluster of 5 to 7 upon each carpal peduncle; outside of these marginal spines, upon each side, is an irregular marginal row of 5 depressed
knife-like spines, each tipped with a crown of 3 acicular spinelets; on
the anterior margin of the disk the 2 rows coalesce to a greater or less
extent and form a bristling array of closely set spines, some pointing dorsally, some laterally, some ventrally; 2 kinds of spines upon the dorsal
surface, in addition to the large ones already described, some large, somewhat remote from each other, conical, stellular; others, much more numerous and filling the interspaces, pickle-like, stellular; belly armed with
numerous closely-set spines of a similar kind; snout somewhat projecting,
armed with 3 many-tipped spines; a spine-armed ridge in front of the
eyes, over the top of the snout; in this 4 spines are conspicuous, 1 in front
of each eye, and between these a larger pair in front of the supraorbital
ridges; from these last mentioned spines extend spine-armed ridges along
the upper margins of each orbit; under the snout is a cavity (horizontal
diameter $\frac{1}{2}$ that of the orbit) containing a barbel, pediceled, with thick,
club-shaped, trilobate tip; on each side of this cavity are the nasal openings, which are as in *Halieutichthys.* Width of mouth equal to distance
between centers of pupils of eyes. Diameter of orbit contained as follows
in other dimensions of the body: In total length $9\frac{1}{2}$; in distance from
snout to dorsal 6; same to anal 7; the base of ventrals 3; to angle between pectorals and trunk $5\frac{1}{2}$; to gill openings 5; in greatest width of
disk $5\frac{1}{2}$; of trunk 4. Width of interorbital area in diameter of orbit $\frac{2}{3}$.
Dorsal fin with 6 or 7 rays, the longest (third) $1\frac{1}{4}$ times diameter of orbit
and 6 times in total length; anal fin inserted entirely behind dorsal, with
4 rays, the longest (third) about as long as longest in dorsal fin; ventral
fins inserted nearly under middle of disk, a little nearer vent than to
mandibular symphysis, with 1 rudimentary and 5 well-developed rays,
increasing in length posteriorly, the last and longest $6\frac{1}{2}$ times in total; distance between ventral organs $7\frac{1}{4}$ in total length. Pectorals with peduncles slightly exserted, bases included in common membrane, composed of
13 to 15 rays, the longest third or fourth $4\frac{3}{4}$ in total. Caudal fin rounded,
consisting of 9 rays, all bifid except the 2 external ones; length of middle ray
about $\frac{1}{4}$ that of trunk and exceeding that of pectoral, being contained $4\frac{1}{4}$
times in total length. Stomach egg-shaped, intestine somewhat longer
than body; liver very wide and large. Color uniform reddish, gray above,
slightly lighter below. Deep waters of the Atlantic; very abundant, in
about 300 fathoms. Known from the west coast of Africa, off the Cape
Verdes, off Barbados, and north in the Gulf Stream to Newport. (*atlanticus,* of the Atlantic.)

Dibranchus atlanticus, PETERS, Monatsber. Kon. Akad. Wiss. Berlin 1876, 736, with plate,
   West Africa, Lat. 10° N., Long. 17° W., in 360 fathoms (Coll. H. M. S. *Gazelle*); GÜN-
   THER, Challenger Report, XXII, 59, 1887; VAILLANT, Travailleur, etc., 343, 1888; GOODE &
   BEAN, Oceanic Ichthyology, 501, pl. 122, fig. 413, 1896.

ADDENDA.

Page 12. After *Entosphenus tridentatus* add:

11(a). ENTOSPHENUS CAMTSCHATICUS (Tilesius).

A lamprey taken by Steller in the Bolschaya River, Kamchatka, has not been recorded by subsequent writers. It is reported by Steller as 13½ inches in length; the head ⅓ of an inch; mouth long, with 2 teeth above, 6 below; dorsals 2. Color shining brassy, dark above; sides with dusky serpentine lines. A figure published by Tilesius shows the upper teeth as bifid, and 9 teeth below.

Pallas describes specimens from the sea at Petropaulski as 7 inches long, not marbled nor variegated. The species of Steller is probably an *Entosphenus*. That of Pallas may be the same, or it may be a *Lampetra*, allied to or identical with *L. aurea*.

Petromyzon marinus camtschaticus, TILESIUS, Mém. Acad. St. Petersburg 1809, 240, with plate, Kamchatka.
Petromyzon camtschaticus, TILESIUS, *l. c.*, 241.
Lampetra variegata (STELLER MS.) TILESIUS, *l. c.*, 247, Bolschaya River, Kamchatka.
?*Petromyzon marinus camtschaticus*, PALLAS, Zoogr. Ross. Asiat., III, 1810, 67, Petropaulski.

Page 14. From the synonymy of *Lampetra wilderi* omit "*Petromyzon branchialis*, Günther, Cat., VIII, 504, 1870," and after the last synonym add: Not *P. branchialis*, Linnæus, which is the larva of some European species, perhaps of *P. marinus*.

Page 25. In the description of *Catulus uter* the teeth should read $\frac{9}{9}$ instead of $\frac{60}{64}$.

Page 27. In the key, under *dd*, read: Root of tail with a conspicuous notch above.

Page 28. The following key to West Coast species of *Galeus* and *Mustelus* will prove helpful.

a. Eye large, spiracle small, the latter not more than ¼ major diameter of orbit.
    b. Mouth broad, snout broadly rounded, mandibular angle of 90° or more. Fins less deeply incised; the lower caudal lobe rounded; pectoral and ventral margins nearly straight.   GALEUS CALIFORNICUS, 33.
    bb. Mouth narrow, snout long, acute, mandibular angle 60° to 65°. Fins deeply incised; lower caudal lobe acute; pectoral and ventral margins concave.   MUSTILUS LUNULATUS, 30.
aa. Eye small, spiracle large, the latter ⅔ to ½ the major diameter of orbit. Snout sharp, mouth narrow, the mandibular angle about 70°. Terminal lobe of caudal broad, obliquely truncate posteriorly. Nostrils very large, their width nearly equaling width of interspace. Fins less incised.   GALEUS DORSALIS, 32.

Page 37. After *Carcharhinus henlei* add:

45(a). **CARCHARHINUS CERDALE**, Gilbert, new species.

Body moderately compressed, not elevated, the depth at front of dorsal not more than $\frac{1}{4}$ greater than the oblique anterior margin of dorsal fin, less than distance from the nostril to first gill slit. Head depressed, the snout flattened, long and narrow, acute; length of snout beyond mouth $\frac{1}{6}$ to $\frac{1}{10}$ greater than distance between angles of mouth in all but one (the largest) of our specimens, in which it is slightly less than width of mouth; $\frac{3}{5}$ to $\frac{4}{5}$ greater than distance from tip of lower jaw to a line connecting angles of mouth; $\frac{1}{7}$ to $\frac{1}{12}$ greater than width of snout opposite outer angle of nostrils. Interorbital width equaling distance from tip of snout to front of eye in young, to middle or posterior border of eye in older individuals; less than $\frac{1}{2}$ distance to first gill opening. Middle of eye nearer nostril than angle of mouth by $\frac{1}{2}$ to $\frac{4}{5}$ its diameter; distance from eye to nostril $\frac{1}{2}$ or slightly more than $\frac{1}{2}$ distance from nostril to tip of snout; middle of nostrils much nearer front of mouth than tip of snout; nasal flap with a very narrow, short, acute lobe, placed at end of inner third of flap; outer angle of nostrils nearly at margin of snout, the inner angles separated by a distance equaling or slightly exceeding that between inner angle of nostril and back of eye. Lips very little developed, the lower entirely concealed in closed mouth, the upper visible as a very short fold. Teeth in lower jaw narrow, erect, serrulate on both margins, more coarsely so toward base; the serration more conspicuous in our smallest specimens (450 mm.), and is obsolescent on some of the teeth in adults; teeth in upper jaw broadly triangular, in front of jaw narrower and erect, those in sides of jaw growing at once broader and more and more oblique; the lateral teeth with a strong notch on the outer side; both margins strongly serrate, the serrations increasing toward base, one or more of those below notch sometimes enlarged and cusp-like in adults; teeth about $\frac{28}{28}$. Conspicuous areas of large and of small pores on underside of head. Gill openings of moderate width, the longest equaling distance between eye and nostril, the fifth much shortened, about $\frac{3}{4}$ length of first. Eye small, equaling length of nasal opening, $1\frac{3}{4}$ to 2 in middle gill slit. Pectoral short and broad, the posterior margin not strongly incurved; tip of fin extending to a vertical intersecting dorsal base at origin of its posterior third or fourth; anterior margin of pectoral 3 times length of inner or posterior margin, the latter less than width of base; first dorsal beginning behind a vertical from axil of pectoral a distance about equaling that which separates eye from nostril; free margin of fin gently concave, the anterior angle extending to a point midway between base and tip of posterior lobe, when the fin is depressed; base of first dorsal $2\frac{1}{2}$ to $2\frac{3}{5}$ in interspace between dorsals; base of second dorsal 7 in interspace between dorsals, $2\frac{1}{6}$ in its distance from anterior margin of pit; origin of second dorsal falling over or behind middle of anal base, the fin but slightly concave, with rounded anterior angle, its posterior angle much produced, the posterior margin exceeding base of fin, which about equals length of anterior margin; anal inserted more anteriorly than second dorsal, its base longer, its margin much more deeply concave, the length of base contained about 1⅔ times in its distance from lower caudal lobe; lower caudal pit in advance of the upper; caudal broad throughout, the lower lobe not falcate, slightly less ($\frac{1}{10}$ to ¼) than ½ length of upper lobe, which is about 4¼ in total length. Shagreen coarse. Color varying from light to dark gray above, the belly and lower part of sides whitish; fins all dusky or grayish, the caudal often with a blackish border; pectoral with or without a black tip, the latter when present not as conspicuous as in *C. æthalorus*, usually not extended into inner face of fin. A specimen 730 mm. long has the claspers undeveloped, extending slightly beyond margin of ventrals. Another specimen, 850 mm. long, has the claspers fully developed, extending beyond the margin of the ventrals for a distance of 50 mm. Strongly resembling *C. æthalorus*, with which it is associated in the Bay of Panama. It is distinguishable at sight by the narrower gill slits, broader and less falcate fins, and by the much less conspicuous black tips to the pectorals. The dentition is very dissimilar in the two, and makes it necessary to arrange them in different subgenera. Abundant at Panama, where numerous specimens were secured. (Gilbert.) (*κερδάλη*, wary, fox-like.)

Carcharhinus, sp. indescr., JORDAN, Proc. U. S. Nat. Mus. 1885, 363.
Carcharhinus cerdale, GILBERT, Fishes of Panama, MS. 1898, Panama. (Type, No. 11884, L. S. Jr. Univ. Mus. Coll. Dr. Gilbert.)

Page 39. *Carcharhinus nicaraguensis* or its marine original form was found in abundance in the Bay of Panama by Dr. Gilbert.

Page 41. After *Carcharhinus oxyrhynchus* add:

54(b). CARCHARHINUS VELOX, Gilbert, new species.

Preoral portion of snout slightly more than 1⅔ times width of mouth, 5 times distance between nostrils, 1⅔ times width of snout opposite outer angles of nostrils, 1¼ times interorbital width, 2⅓ times distance from chin to line joining angles of mouth. Nostrils transverse in position, the inner angle nearer mouth than tip of snout by a distance slightly less than length of nostril. Front of eye equidistant from nostril and front of mouth, the middle of eye nearer angle of mouth than nostril; diameter of eye less than nostril, slightly more than ½ longest gill slit. Snout very porous. Folds at angle of mouth slightly longer than usual. Gill slits rather wide, the middle one 1¾ times diameter of orbit. Teeth of lower jaw very narrow, erect, very minutely serrulate, appearing entire except under a lens. Teeth in upper jaw very oblique, wide at base, with a deep notch on outer margin, the terminal cusp rather narrowly triangular. Pectoral broadly falcate, the anterior margin convex, the distal edge concave, both angles rounded; tip of pectoral reaching a short distance beyond base of first dorsal; anterior margin of pectoral 2⅔ times the posterior (inner) margin, about 1⅛ times the distal edge; first dorsal inserted about the diameter of orbit behind a vertical from axil of pectoral, nearer pectoral, therefore, than ventral; anterior margin concave basally, convex on distal half, the anterior angle rounded; free margin concave, largely owing to the much produced acute posterior lobe; vertical height of fin exceeding length of base, the anterior lobe very high, extending beyond tip of posterior when the fin is declined, equaling $\frac{7}{9}$ length of anterior margin of pectoral; posterior margin of dorsal $3\frac{1}{4}$ in the anterior margin; base of first dorsal contained $2\frac{1}{2}$ times in interspace between dorsals; base of second dorsal $6\frac{3}{7}$ times; margin of second dorsal gently concave; front margin low, the angle broadly rounded, barely reaching posterior end of base when fin is declined; posterior lobe much produced and acute, slightly longer than base of fin, the latter $1\frac{5}{7}$ in the distance from its base to front of caudal pit; upper lobe of caudal $3\frac{3}{8}$ in total length, the lower lobe $2\frac{1}{2}$ in the upper; terminal lobe of caudal $3\frac{3}{8}$ in the upper lobe; anal larger than second dorsal, higher, with deeply incurved margin, its base a little longer, its origin slightly in advance of that of second dorsal, the posterior insertions of the two fins nearly opposite; length of anal base $1\frac{3}{8}$ in its distance from anterior edge of caudal pit. Color bluish above, whitish or grayish below; free margin of pectoral narrowly white, the anterior edge narrowly bordered with black, most evident when seen from the outer surface, the inner surface being dusky; first dorsal unmarked, the second dorsal with dusky anterior lobe; upper edge of caudal black, the lower margin faintly dusky; fins otherwise unmarked. A single specimen, a female 4 feet long, was procured from the Panama market. As preserved, it is partially skinned. The following measurements were taken when the specimen was intact, before preservation. Where not agreeing with dimensions given above, the latter will be found more reliable:

	Inches.
Tip of snout to insertion of dorsal	$16\frac{1}{4}$
Base of first dorsal	$4\frac{3}{8}$
Distance between dorsals	11
Base of second dorsal	$1\frac{3}{4}$
From second dorsal to front of caudal pit	$2\frac{7}{8}$
Front of caudal pit to tip of caudal	$13\frac{3}{4}$
Tip of snout to axil of pectoral	15
Axil of pectoral to front of base of ventrals	$11\frac{1}{8}$
Front of ventrals to front of anal	$6\frac{1}{2}$
Front of anal to front of caudal pit	$4\frac{9}{16}$
Girth at front of first dorsal	$17\frac{3}{4}$

Distinguished from other known sharks of the Pacific coast of America by the excessively long, slender, acute snout, the slender body, and the very long caudal fin. Panama; only the type known. (*velox*, swift.)

Carcharhinus velox, GILBERT, Fishes of Panama, MS. 1898, Panama. (Type, No. 11893, L. S. Jr. Univ. Mus. Coll. Dr. Gilbert.)

Page 42. *Scoliodon longurio* has the teeth serrulate at base only. The base of the first dorsal is $2\frac{3}{4}$ in the interspace between the dorsals.

Page 44. *Sphyrna tiburo* occur also in the Pacific. We have recently secured specimens at Mazatlan, where *S. tudes* and *S. zygæna* are also found.

Page 47. *Carcharias littoralis* reaches a length of 8 feet 7 inches. (Specimens from Beaufort, North Carolina. Coll. H. H. Brimley.)

Page 49. *Lamna cornubica,* the salmon shark, is abundant and destructive to salmon on the coasts of southern Alaska, especially about Kadiak, where it was seen by us.

Page 53. Under *Squalidæ,* read ovoviviparous for "oviparous."

Page 54. *Squalus sucklii* has been but once recorded from Bering Sea. (Bering Island. Coll. Dr. L. Stejneger.) It is very abundant at Kadiak.

Page 60. *Pristis perrotteti* is not authentically known except from the rivers of Africa. Our west coast species is doubtless distinct and should stand as *Pristis zephyreus. Pristis pectinatus* occurs northward at least to Beaufort, North Carolina. (Specimen 12¼ feet long. Coll. H. H. Brimley.)

Page 61. After the synonymy of *Pristis pectinatus* insert:

80(a). PRISTIS ZEPHYREUS, Jordan & Starks.

(Pez de Espada.)

Snout to nostrils, 3 in length to base of caudal; breadth of saw at anterior end between first 2 pairs of teeth ⅓ breadth of its base behind the last pairs; teeth on saw trenchant behind, arranged in 22 pairs; hinder teeth wide apart, the interspaces 5 times their base; posterior teeth turned slightly backward, a groove on their posterior edge; front teeth not quite ⅓ as long as the saw is broad at their base; distance between first and second tooth 3 times base of first. (Other specimens examined for us by Dr. G. W. Rogers show 18 to 21 pairs of teeth.) Eye equal to spiracle, contained 3 times in base of saw just behind last pair of teeth; width of mouth a little greater than base of saw; mouth with about 65 series of blunt teeth; slant height of pectoral in front a little more than half distance from tip of snout to mouth. Dorsals subequal; first dorsal inserted in advance of ventrals, about ⅓ its base over ventrals; caudal with a lower lobe, which is equal to slant height of pectoral; tail with a keel on side. Color plain olive gray above, light below. Measurements: Length 50 inches; caudal 7 inches; pectoral 7 inches; dorsal front 5¼ inches; snout without nostril 11 inches. Type: A skin in L. S. Jr. Univ. Museum. Common in brackish waters at the mouth of the Rio Presidio, where 1 fine specimen was obtained. The species is also recorded (as *Pristis perrotteti*) by Dr. Gilbert from Mazatlan, and by Dr. Günther from Chiapas. Dr. Günther identifies this species with *Pristis perrotteti* described by Müller & Henle, from the Senegal River. In view of the great difference in the fauna of the Gulf of California from that of equatorial Africa this identification may be questioned, especially as there are several details in which the description of *P. perrotteti* differs from our specimen.

Pristis zephyreus, Jordan & Starks, Fishes of Sinaloa, 383, 1895, Mazatlan, Mexico. (Coll. Hopkins Exped. to Sinaloa.)

Page 62. After *Rhinobatus lentiginosus*, add:

81(a). RHĪNOBATUS STELLIO, Jordan & Rutter.

Disk triangular, its greatest width a little less than ⅓ the distance from snout to dorsal, and equal to distance from snout to a line connecting points of greatest width. Sides of disk straight, tip of snout rounded, posterior point of pectoral more broadly rounded than snout. Length of snout equal to, or a little less than, ½ greatest width of disk, equal to distance between outer points of anterior gill openings; interorbital width 4 to 4⅔ in snout, a little less than length of eye and spiracle, but about equal to length of nostril; internasal width equal to orbit; spiracle ½ length of eye, a prominent curved papilla and a slight ridge in its posterior side. Anterior nasal valve with a long slender flap extending across the nostril; 3 broad flaps on posterior side. Rostral ridges separate for their entire length, width between them at base equal to width of spiracle. Mouth nearly straight, its width 2¼ in its distance from snout and equal to distance between inner folds on posterior side of spiracle. Eye 4½ to 5¼ in snout. Width of body at axil of pectorals 1⅛ in snout. Dorsal fins about equal in size and shape, the distance between them 2¼ times base of first, the distance between the origins of the two fins equal to snout and about equal to distance from axil of pectoral to origin of first dorsal. Sides of tail with a conspicuous fold. Skin above with a fine uniform shagreen, nearly smooth below except near margins of the disk. A series of very small spines above eye and spiracle, 1 or 2 minute spines on shoulder girdle; the largest spines of body situated along median line of back, extending beyond first dorsal; no spine on snout, but in 2 of the 3 specimens there is a pair of minute spineless plates near its tip. Color dusky brown above, about 7 faint dusky bars on the side of the tail behind first dorsal; uniform pale below; large translucent areas on each side of the snout; back with numerous small light spots, much smaller than pupil, arranged symmetrically but not in the same pattern on the 3 type specimens; 2 or 3 pairs between eyes, a few pairs behind eyes near median line, some below eye, where they approach nearest the margin of disk, usually 1 or 2 on median line, sometimes 2 are confluent, about 40 or 45 pairs in all; axil of pectoral in 1 specimen with a dusky blotch on upper side. This species is most nearly related to *Rhinobatus glaucostigma* of the Pacific coast, differing in having a narrow interorbital, narrower body behind disk, and in the very different color. The description is based on 3 specimens, each about 20 inches. Jamaica. (*stellio*, the starry one.)

Rhinobatus stellio, JORDAN & RUTTER, Proc. Ac. Nat. Sci. 1897, 91, Kingston, Jamaica.
(Type, No. 11851, L. S. Jr. Univ. Mus. Coll. Joseph Seed Roberts.)

Page 66. To the synonymy of *Raja* add: *Cephaleutherus*, Rafinesque, Indice, 61, 1810 (*maculatus*).

The genus *Cephaleutherus*, Rafinesque, was, as Dr. Gill has shown (in lit.), probably based on a monstrous example of the genus *Raja*, in which the pectoral fins were not developed on the snout. It should be transferred to the synonymy of *Raja*, leaving *Myliobatis* as the generic name of the Eagle Rays.

Page 74. After *Raja equatorialis* add:

104(a). RAJA ROSISPINIS, Gill & Townsend.

Snout moderately produced, with a soft, moderately narrow, rostral cartilage and a bluntish tip; interorbital space nearly plane; snout with a number of plates having stellate bases about middle, and many smaller asperities, leaving only the borders of the pectorals and ventrals naked; larger spines with stellate bases are interspersed between the disk and the pectoral rays; back with sparse, coarse prickles; a row of about 26 thornlike spines, with radiating ridges, extends from the interhumeral area to the dorsal fins; 2 spines on each shoulder, 1 spine above antocular region, another above postocular region, and another behind it about ½ the distance; skeleton soft. Bering Sea; only the type known. (*roseus*, rosy; *spinus*, spine.)

Raia rosispinis, GILL & TOWNSEND, Proc. Biol. Soc. Wash., XI, 1897 (Sept. 17, 1897), 231, Bering Sea. (Type, No. 48762, U. S. Nat. Mus. Coll. *Albatross*.)
*Raia obtusa,** GILL & TOWNSEND, Proc. Biol. Soc. Wash., XI, 1897 (Sept. 17, 1897), 231, Bering Sea. (Type, No. 48763, U. S. Nat. Mus. Coll. *Albatross*.)

104(b). RAJA INTERRUPTA, Gill & Townsend.

Snout moderately produced, with a very soft attenuated rostral cartilage and a blunt tip; interorbital space concave; mouth small; the width equal to ½ preoral area; entire back covered with very small embedded spines, extending nearly uniformly over the disk and snout, leaving only the tip of the latter naked; a row of compressed, acutely curved, smooth spines along middle of back, extending from the interhumeral region to dorsal, but interrupted along the posterior half of disk, where the spines are absent or obsolete; about 4 spines are in the anterior portion and the series recommences on a line with the emargination of the disk; a single spine on each shoulder and occasionally a rudimentary second; no specialized supraorbital spines. Bering Sea; only the type known. (*interruptus*, interrupted.)

Raia interrupta, GILL & TOWNSEND, Proc. Biol. Soc. Wash., XI, 1897 (Sept. 17, 1897), 232, Bering Sea. (Type, No. 48760, U. S. Nat. Mus. Coll. *Albatross*.)

Page 75. *Raja aleutica* and *Raja abyssicola* were described by Gilbert (not Gilbert & Thoburn) in Rept. U. S. Fish. Comm. 1893 (Dec. 9, 1896), 396 and 397, pls. 20 and 21.

---

* The following is the original description of this nominal species:
Snout not at all produced, but very bluntly rounded; interorbital space narrow; mouth small, rectilinear; minute distant prickles on the snout, the anterior portion of disk and interorbital area, as well as in a broad median band extending on tail to dorsal and commencing at the interhumeral area; a row of scarcely enlarged acute spines above the eye; an uninterrupted row of unguiform spines with smooth bases extending from the interhumeral area to dorsal fin; 2 similar spines arm each shoulder. Bering Sea; only the type known. (Gill and Townsend.) To which we add: Spines in longitudinal series 23 to 25; width of mouth 1⅔ in preoral area; width of disk 1½ times its length; tail a little longer than disk; interorbital width 3 in snout; snout from eye 3⅔ in disk to end of base of ventrals. Color plain brown, rather pale. One specimen 11 inches long, a very young male in very bad condition. Evidently the young of *R. rosispinis*.

Page 78. After *Narcine brasiliensis* insert:

112(a). NARCINE ENTEMEDOR, Jordan & Starks.

(ENTEMEDOR.)

Snout 3¾ in length of disk; preocular part of snout equals preoral; interocular space in snout 1½; width of mouth 2¼. Eye much smaller than spiracle; spiracles edged with small tubercles. Length of disk equal to its width; disk equal to length of tail, without caudal fin; tail with a loose fold of skin on each side. First and second dorsals equal, rounded behind; ventrals large, ending midway between posterior edge of disk and caudal fin. Color pale olive brown, a little clouded with darker; second dorsal edged with pale; dots on head dusky. Two specimens taken in the estuary at Mazatlan, and a third procured by Mr. James A. Richardson in the harbor of La Paz. Specimens had also been obtained by Dr. Gilbert, at Panama, in 1883, but having been destroyed by fire, the species has remained undescribed until recently. Length of largest specimen 20 inches. (The Spanish name *Entemedor* seems to be equivalent to *Intimidator.*)

Narcine entemedor, JORDAN & STARKS, Fishes of Sinaloa, 386, 1895, Mazatlan, Mexico.
(Type, No. 1699, L. S. Jr. Univ. Mus. Coll. Hopkins Exped. to Sinaloa.)

Page 81. After *Urolophus nebulosus* add:

115(a). UROLOPHUS UMBRIFER, Jordan & Starks.

Disk round, not wider than long, its length greater than tail; snout pointed, not exserted. Snout from eye 4½ in disk; eyes equal to spiracles; mouth 2 in distance to tip of snout; caudal spine inserted in front of middle of tail; skin perfectly smooth. Color brown above, with blackish cross shades or bars, radiating from the shoulder; a dark band behind eyes, and 1 from eyes; caudal fin dark. Mazatlan. One adult female specimen, the uterus containing 4 young. Occasionally taken with *Urolophus mundus*, but much less common. This is probably not identical with Garman's *Urolophus nebulosus*, being perfectly smooth and different in color. (*umbra*, shade; *fero*, I bear.)

Urolophus umbrifer, JORDAN & STARKS, Fishes of Sinaloa, 389, 1895, Mazatlan, Mexico.
(Coll. Hopkins Exped. to Sinaloa.)

Page 82. *Urolophus asterias*, Jordan & Gilbert, is identical with *Urolophus mundus* (Gill), as is shown by specimens recently collected by D.. Gilbert at Panama.

120(a). UROLOPHUS ROGERSI, Jordan & Starks.

Disk broader than long by a distance 2½ times the interorbital width; anterior margins of disk nearly straight, the tip of snout projecting; snout from eye 3¾ in length of disk; eye little smaller than spiracles; width of mouth 2¼ times in preoral part of snout; caudal spine inserted in front of middle of tail; skin with minute prickles on margin of pectorals and on middle of back, leaving smooth areas near middle of pectorals and over branchial arches; 16 to 20 large spinules along median line of back and tail. Color plain brown; caudal fin darker, edged with white. This species differs from *Urolophus asterias* in having a wider disk, more acute snout, much smaller prickles, and fewer spinules on back and tail. Mazatlan. Three specimens obtained in the Astillero, the longest 18 inches in entire length. (This species is named for Dr. George Warren Rogers, a scholarly physician, native of Vermont, but long resident in Mazatlan.)

Urolophus rogersi, JORDAN & STARKS, Fishes of Sinaloa, 388, 1895, Mazatlan, Mexico. (Type, No. 1700, L. S. Jr. Univ. Mus. Coll. Hopkins Exped. to Sinaloa.)

Page 88. In the key at top of page for *Rhinopterinæ* read *Myliobatinæ*.

132. AETOBATUS LATICEPS (Gill).

This species is probably not different from *A. narinari* and may be omitted. We find no differences between specimens from Mazatlan and the West Indies. The following description is based upon Mazatlan specimens:

Length of disc $1\frac{3}{4}$ in width; proximal $\frac{1}{2}$ of anterior margin of pectoral fins straight, distal $\frac{1}{2}$ convex; posterior margin concave, the end of each ray forming a small scallop; lateral angle sharp. Snout forming an angle, from its tip to division of nasal lobes, $1\frac{1}{4}$ times breadth of head; width of snout $1\frac{1}{3}$ times distance from its tip to the division of nasal lobes; nasal lobes projecting back over the mouth; width of mouth $1\frac{1}{4}$ its distance to tip of snout; numerous blunt buccal papillæ around upper dental plate and on ridge between nostrils; interorbital $4\frac{3}{4}$ in disk; eyes smaller than spiracles, which are as long as base of dorsal. Ventrals well rounded, $3\frac{3}{4}$ in length of disk; tail $3\frac{1}{4}$ times disk. First caudal spine equals base of dorsal, which is $\frac{1}{2}$ second spine. Color bluish black with many round yellowish spots scattered equally over the back and ventral fins; spots about as large as eye on back, smaller on head, sometimes two spots run together forming an elliptical spot, about 16 spots from eye along anterior margin of pectoral to lateral angle; posterior margin of pectoral very narrowly margined with white; ventral side pearly white. From the description of *Aetobatus laticeps* this species differs in the following respects: Disk not so broad; tail not so long; width of head and snout less; ventrals not truncated behind; pectorals not margined with blackish; spots on ventrals not assuming the form of ocelli. (Jordan.)

Page 87. For the description of *Pteroplatea crebripunctata* in text substitute the following:

Width of disk twice length to posterior end of anal slit; snout forming a regular curve from a little in front of middle of pectorals, a very small blunt projection at tip; anterior margin of disk convex near snout and lateral angles, pectorals concave medially; posterior margin weakly convex; posterior angle broadly rounded; lateral angle sharply rounded; distance from snout to a line drawn through lateral angles, $2\frac{1}{4}$ times in distance to tip of tail. Interorbital a little wider than its distance to tip of snout; eyes twice spiracles; mouth equals snout, $6\frac{1}{2}$ in disk. Tail ratliké, with a scarcely perceptible fold of skin on its dorsal side. Ground color olive brown, everywhere with small dark points, not so close set as in *Pteroplatea rava*, indistinct grayish spots, ½ as large as iris, scattered over the body among the dark points, these spots more distinct on anterior edge of disk; tail mottled with darker; lower parts light. Markings nowhere so distinct as in *P. rava*. Very common on sandy shores everywhere about Mazatlan, from which locality it was originally described; also taken by Dr. Gilbert.

Page 87. After *Pteroplatea marmorata* add:

130(a). PTEROPLATEA RAVA, Jordan & Starks.

(MANTARAIA COLORADA.)

Length of disk 1⅔ width; snout forming an angle which is almost a right angle; pectorals slightly concave medially; posterior margin of disk weakly convex; posterior angle not broadly rounded, but curved in somewhat suddenly; lateral angles acute. A line drawn through lateral angles would bisect a line from snout to tip of tail. Interorbital 1¼ in snout; eye 1¼ in spiracles; mouth 7 in disk, 1¼ in snout; tail straight and slender, with a very slight fold on dorsal side. Ground color light olive brown, thickly set with sharp-cut black points; conspicuous gray or white spots, ½ as large as iris, scattered over the body, around which the black spots form rings; brighter yellowish spots and half-spots around anterior edge of disk; tail mottled above with darker; lower parts chiefly light orange red or rust-colored in life. All the markings are very distinct and clear cut, the reddish of the belly conspicuous. Mazatlan. One specimen 12 inches long. (*ravus*, reddish.)

Pteroplatea rava, JORDAN & STARKS, Fishes of Sinaloa, 390, 1895, Mazatlan, Mexico. (Type, No. 1587, L. S. Jr. Univ. Mus. Coll. Hopkins Exped. to Sinaloa.)

Page 90. After *Myliobatis californicus* add:

134(a). MYLIOBATIS ASPERRIMUS, Gilbert, new species.

Upper surface of head and body, excepting the snout, an area on outer side of spiracle, the pectoral margin and its posterior angle, and the ventral fins thickly covered with minute, usually stellate prickles, of uniform size, most numerous on median portions of head and back; those on basal ½ or ⅔ of pectoral least crowded and arranged in definite longitudinal series, corresponding with the muscle bands; tail very rough throughout, covered with similar stellate prickles and also crossed by numerous narrow grooves, or indented lines, mostly convex forward, somewhat irregular in position and direction, and not corresponding on the two sides. In the type they follow at an average interval of about 10 mm. Lower side of disk mostly smooth, with some prickles on the basal part of pectorals anteriorly, arranged in lengthwise series, and other patches on lower side of head, belly, and base of ventrals. Color dusky brown above, the anterior portion of pectorals with 8 to 10 narrow transverse bars of bluish white, most of which break up into series of spots toward outer margin of disk, the posterior ones also breaking up toward middle line; the bars and spots fainter anteriorly, becoming whiter and more intense posteriorly; toward outer angles of disk the bars sometimes separated by intermediate series of light round spots, the bars usually failing to meet across the back; posterior portion of disk including base of tail and upper surface of ventrals covered with round white. spots not much larger than pupil, some of those immediately succeeding the bars showing a transverse serial arrangement; top of head with one or more pairs of indistinct light spots; margin of snout and of pectorals blackish; spiracular border black; dorsal with a black blotch posteriorly; underside of head and disk bright white; proximal portion of tail blackish above,'lighter below, the entire tail becoming black more posteriorly.

Dimensions of type specimen.

	Millimeters.
Length of disk to front of anus	272
Length of disk to posterior edge of pectorals	338
Width of disk	345
Length of tail (not perfect)	1,215
Greater width of head at origin of pectorals	79
Width of cranium between orbits	45
Width of snout opposite front of eye	55
Tip of snout to middle of nasal flap	60
Length of nasal flap	26
Greatest width of nasal flap	35
Diameter of iris	10½
Width of mouth	33
Distance between anterior gill openings	75
Distance between posterior gill openings	45
Distance from anterior to posterior gill openings	45
Length of spiracle	26
Length of fontanelle	60
Greatest width of fontanelle (at anterior end)	23

Rostrofrontal fontanel scarcely constricted anteriorly, the bounding ridges diverging abruptly at their anterior ends. Nasal flap with a shallow median notch, covering the mouth except the median portion of lower dental plate; posterior margin coarsely fringed. Teeth in each jaw in 1 broad median row, and 3 lateral rows, those of median row about 5 times as broad as long anteroposteriorly.

One specimen, a male, with undeveloped claspers which do not nearly reach edge of ventrals, from Panama. (Gilbert.) (*asperrimus*, very rough.)

Myliobatis asperrimus, GILBERT, Fishes of Panama, MS. 1898. (Type, No. 11895, L. S. Jr. Univ. Mus. Coll. Dr. Gilbert.)

134(b). MYLIOBATIS GOODEI, Garman.

Disk about ⅔ as long as broad; lateral angles acute, bluntly rounded at the apices; posterior angles of pectorals nearly right; snout very broad, short, with a slight prominence in front; the fin, or flange, beneath the eye at the side of the head is very wide, much wider than in either *M. freminvillei* or *M. californicus;* eye very small, without a prominence above in either male or female (immature specimens); tail less than 2 and more than 1⅔ times length of the disk; dorsal fin smaller than that of *freminvillei;* teeth in 7 series, much shorter and narrower than those of *freminvillei,* third row about 2 and middle row about 4 times as wide as long. Body smooth. Entire length 29 inches; snout to end of ventrals 11.5, vent to end of tail 18.5, and width of disk 17.5 inches. Olivaceous, darker on the center; white below. The Museum of Comparative Zoölogy has a large specimen which agrees well with this description. Compared with *M. freminvillei,* this species has very small eyes, the pectoral below the orbit is wider than the eyeball, and the fin in front of the skull is but little wider than at its sides. In *freminvillei* the eyeball is twice as wide as the fin beneath it, and the fin in front of the skull is much wider than below the eye. Comparing specimens of about the same size, of the same sex, of *freminvillei, californicus,* and *goodei,* the latter is readily distinguished from the former two by the broad flange at the side of the head, the small eyes, the small teeth, and the broader lateral angles of the pectorals. Central America. (Garman); probably on the Atlantic Coast. (Named for George Brown Goode.)

Myliobatis goodei, GARMAN, Proc. U. S. Nat. Mus. 1885, 39. Central America. (Types, Nos. 9524 male, and 9529 female.)

Page 91. Family XXVIII should stand as *Aodontidæ,* the name *Mantidæ* being used for a family of *Orthoptera.*

Page 92. After *Aodon hypostomus* insert:

58 (a). CERATOBATIS, Boulenger.

Ceratobatis, BOULENGER, Ann. Mag. Nat. Hist., ser. 6, vol. XX, August, 1897, 227 *(robertsii).*

Characters of *Dicerobatis,* Blainville, but the teeth restricted to the upper jaw. ($K\varepsilon\rho\dot{a}\varsigma$, horn; $Ba\tau\iota\varsigma$, ray.)

138 (a). CERATOBATIS ROBERTSII, Boulenger.

Band of teeth occupying only ⅛ width of mouth, its width 10 times in its length; teeth tessellated, hexagonal, 2 to 3 times as broad as long, rugose with numerous obtuse ridges; mouth inferior, wide. Pupil vertically elliptic. Body smooth; pectoral fins with nearly straight, slightly convex anterior and slightly concave posterior border; cephalic fins measuring a little less than width of mouth; spiracles behind the eyes; space between last branchial clefts ⅓ that between first; dorsal fin between the ventrals; tail slender, without spine, nearly twice length of body.

	Millimeters.
Length of disk, without cephalic appendages	350
Width of disk	780
Cephalic fin	90
Width of mouth	105
Diameter of eye	12
Ventral fin	70
Tail	620

Black above, white beneath. Jamaica. One specimen known. This ray grows to a very great size, but specimens are almost impossible to obtain, owing to the superstitious fear of the fishermen. (Named for Rev. Joseph Seed Roberts.)

Ceratobatis robertsii, BOULENGER, Ann. Mag. Nat. Hist,. ser. 6, vol. XX, August, 1897, 227, Jamaica. (Type in British Mus. Coll. J. S. Roberts.)

Page 105. There is no truth in the statement that *Acipenser medirostris* is poisonous. It is a good food-fish, and on the coast of Washington it is somewhat abundant.

Page 116.

ADDITIONAL NOTES ON TACHYSURINÆ.

In the text of this work, pages 116 to 133, the descriptions of the species of *Tachysurinæ* are for the most part too brief to render certain the discrimination of species. The following additional descriptions of these species will be found useful as supplementary to those given in the text. A slight change in the arrangement of the genera has been found desirable, and 3 new species are added.

70. SCIADEICHTHYS, Bleeker.

Dorsal shield much enlarged, formed like an armorial shield; teeth on palate villiform; posterior nasal openings not connected by membrane; band of palatine teeth extended backward.

165. SCIADEICHTHYS TROSCHELI (Gill).

Head $3\frac{1}{4}$ ($3\frac{5}{8}$ in total with caudal); width of head $4\frac{1}{3}$ ($4\frac{5}{8}$ in total); depth 5 ($6\frac{1}{3}$). D. I, 7; P. I, 12; A. 18. Body comparatively robust, broad anteriorly; head not much depressed, broader than high; eye moderate, 7 to 8 times in length of head; width of interorbital space $1\frac{3}{4}$; breadth of mouth $1\frac{3}{4}$; length of snout $3\frac{1}{4}$. Teeth all villiform; band of vomerine teeth simple, trapezoidal, quadrangular, longer than broad, without division on median line; band of palatine teeth very large, each separated in young specimens from the vomerine band by a narrow toothless line; in old specimens the vomerine and palatine bands are wholly confluent; each palatine band with a narrow backward prolongation on the median line; band of premaxillary teeth broad, about six times as long as wide; lower jaw included. Maxillary barbel nearly or quite reaching gill opening; outer mental barbels about $\frac{2}{3}$ head, the inner nearly $\frac{1}{2}$. Dorsal shield much larger than in most species, shaped like an armorial shield, its posterior margin concave, its anterior end acute, wedged into a deep emargination of the occipital process, the two becoming coossified with age; length of antedorsal plate on the median line 5 to 6 in head, a little more than its width; occipital process short and broad, much broader than long, its median line with a broad keel, its edges nearly straight. Shields all coarsely granular, the granulations anteriorly forming radiating striæ. Fontanel large, claviform, broadest posteriorly, its posterior end about midway between tip of snout and front of dorsal, its greatest breadth about equal to the diameter of the eye, and ⅙ its length, a short groove extending backward from its obtuse tip; sides of fontanel bony and granulated for its whole length, the granules extending forward to opposite nostrils. Dorsal spine strong, 1¾ in head, moderately compressed; pectoral spine 1⅘ in head. Axillary pore obsolete. Humeral process coarse, granular, broad, nearly ½ length of pectoral spine; base of adipose fin scarcely ⅔ length of anal, its posterior margin little free; caudal deeply lunate, small, its upper lobe slightly the longer and narrower, 1⅖ in head; ventrals not quite reaching anal; vent much nearer base of ventrals than anal. Dark brown, with strong bronze luster above, white below; dorsal dusky, especially above; pectorals blackish; anal dark; caudal rather pale; ventrals usually dark toward the tip, their inner side pale; maxillary barbel dusky; mental barbels pale. This species is not rare along the Pacific coast of tropical America, specimens having been observed at Mazatlan, Punta Arenas, and Panama.

Sciades troscheli, GILL, Proc. Ac. Nat. Sci. Phila. 1863, 171, Panama. (Coll. Capt. Dow.)
Arius brandtii, JORDAN & GILBERT, Bull. U. S. Fish Comm., II, 1882, 39; description from 28230, U. S. Nat. Mus., 24 inches in length.

The following is the original description of *Sciades troscheli*, Gill: "Dorsal I, 7; anal 16; caudal 11, I, 6; 7, I, 11. The greatest height is contained about 4½ times in the length to the base of the caudal fin, and 5½ times in the total. The caudal peduncle, behind the anal, equals the interval between the snout and the eye, and its least height that between the center of the anterior nostril and the eye. The head in front and on the sides is smooth, and a smooth, oblong, triangular area extends nearly to the vertical from the upper angle of the preoperculum; a triangular area on each side is incurved externally to the narrow anterior extremity, and covered with white pisiform granulations. The dorsal buckler is a pentagon, with a semicircular excavation behind and with its surface rugose. The head enters 3 times in the length before the end of the anal fin and more than 4 times in the total; its width equals the interval between the snout and upper angle of preoperculum, and the interocular area equals ½ the head's length. The eye is elliptical, and its diameter is contained 6½ times in the head's length. The distance of the posterior nostril from it equals a diameter. The maxillary barbels extend to about the middle of the pectoral; the outer mental to its base, and the inner mental are ⅔ as long as the outer. There are 3 villiform patches on the palate which are almost contiguous, and together describe an arch in front; the median patch is small, rather transverse, and widest toward the front; the outer are oblong, subtriangular. The band of the upper jaw is nearly uniform and quite wide; the lower, interrupted at the symphysis, is nearly ½ as wide as the upper, and is narrowed toward its ends. The dorsal spine enters 1½ times in the head's length; has in front, first, minute teeth pointed downward, and then a row of small pisiform tubercles; teeth pointed downward on its hinder border. The first ray is little higher than the spine. The anal commences at a distance from the snout 3⅔ times as great as that from the base of the caudal fin; its length enters 6⅔ in the length, exclusive of the caudal, and when bent back it reaches to the supernumerary caudal rays; the greatest height nearly equals the length. The pectoral fins extend rather beyond the base of the dorsal and exceed ⅕ of the length, exclusive of the caudal; the spine equals that of the dorsal. The ventrals are inserted midway between the base of the pectoral spines and the axil of the anal, and extend to the origin of the anal. The fins are almost blackish." A single specimen is in the collection of Captain Dow from Panama. The type of *Sciades troscheli* is now lost. At our request, Dr. Gill has again considered this description, in connection with the species now known from the coast. He is positive that his type of *troscheli* had the large dorsal shield characteristic of *brandtii.* Apparently Dr. Eigenmann is right in regarding *troscheli* and *brandtii* as identical.

166. SCIADEICHTHYS EMPHYSETUS (Müller & Troschel).

Head 3⅖; depth 6. D. I, 7; A. 18. Closely related to *Sciadeichthys troscheli.* Depth little greater than the width. Profile straight, less steep than in *S. troscheli.* Depth of the head 1¾ in its length, its width 1¼. Top of the head sparsely and coarsely granular, the granulation extending forward only to middle of cheek; fontanel bordered anteriorly by smooth ridges; occipital process coarsely and closely granular, without a prominent keel, its margins convex, its tips emarginate, not coossified with the dorsal plate; dorsal plate shield-shaped, not keeled, its surface irregularly pitted, its margin more finely graven, its length about 3⅓ in the head. In the specimen examined the dorsal plate seems to have been at some time slightly broken in front, a small, narrow, sharp process of the occipital process fitting into the split. Eye small, 3 in the snout, 11 in the head, 5 in the interocular. Maxillary barbels flattened, reaching to below middle or end of dorsal fin, postmentals not quite to base of pectorals. Upper jaw slightly projecting; all the teeth minute, villiform, the vomerine patch emarginate in front and behind, joined to the subtriangular palatine patches; pterygoid patches long-elliptical. Gill membrane with a narrower free margin than in *troscheli.* Distance of dorsal fin from tip of snout 2¼ in the length, the dorsal spine 1⅔ in the head, granular in front, recurved teeth on its inner margin. Distance between the dorsal and adipose fins 3⅓ in the length; adipose fin about as long as the dorsal fin. Caudal deeply forked, the upper lobe longer, 3⅔ in the length; anal little longer than high; ventrals reaching nearly to the anal, about 2 in the head; pectoral spine 1¼ in the head, its outer margin granular, the inner rather finely toothed. The skin on the dorsal surface of the head and humeral region finely reticulate with mucous canals. Yellowish brown, lighter below, the fins yellowish, finely punctulate. One specimen 0.51 m. Surinam. (Eigenmann & Eigenmann, Nematognathi, 53.)

167. SCIADEICHTHYS TEMMINCKIANUS (Cuvier & Valenciennes), text, p. 122.

168. SCIADEICHTHYS FLAVESCENS (Cuvier & Valenciennes), text, p. 123.

169. SCIADICHTHYS MESOPS (Cuvier & Valenciennes).

170. SCIADEICHTHYS PROOPS (Cuvier & Valenciennes).

Head 4 to 4⅓; depth 7. D. I, 7; A. 18; eye 1¼ to 1½ in snout, 5½ to 8 in head, 1¾ to 2⅔ in the interorbital, 2¼ to 3¾ in the interocular. Slender and elongate, broader than deep. Head depressed, its width 1⅓ in its length, its depth 2, width at mouth 2; anterior portion of the head flat above; top of the head, humeral process, and dorsal plate coarsely granular, the granules arranged in series along the fontanel. Occipital process mucronate, broader than long; dorsal plate large, butterfly-shaped. Opercle striate; fontanel 1⅓ times as long as the eye, its center in front of the middle of the eye, continued as a shallow groove. Jaws subequal; teeth all villiform, the intermaxillary band very wide and shallow; teeth on the roof of the mouth in 6 contiguous patches. Gill membranes meeting in an angle, forming a broad fold across the isthmus; gill rakers 5 + 10. Pectoral pores large; vertical series of pores. Distance of dorsal spine from the snout 2⅕ in the length; the dorsal spine granular in front, striate on the sides, weakly serrate behind, its length 1¼ to 1½ in head; space between dorsal and adipose fins 2⅘ to 3 in length, the adipose fin little shorter than the dorsal, the posterior margin free. Caudal deeply forked, its upper lobe longer, 4 to 4½ in the length; anal emarginate, as high as long, 2 to 2⅛ in head; ventrals 2 in head; pectoral spine roughened or granular in front, serrated behind, 1⅕ to 1⅛ in head. Plumbeous above, with blue luster, white below; maxillary barbels dark, the mental barbels white; fins all more or less dotted with brown. Five specimens 0.25 to 0.46 m. long. Pernambuco. (Hartt & Fletcher.) Northern coast of South America to Pernambuco. (Eigenmann & Eigenmann, Nematognathi, 57.)

171. SCIADEICHTYS PASSANY (Cuvier & Valenciennes), text, p. 124.

172. SCIADEICHTHYS ALBICANS (Cuvier & Valenciennes), text, p. 124.

71. SELENASPIS, Bleeker.

Dorsal shield much enlarged, truncate before, in the adult; palatine teeth villiform, the patch extended backward in the adult; posterior nasal openings connected by membrane.

173. SELENASPIS HERZBERGII (Bleeker).

Head 3⅗ to 3¾; depth 5 to 6. D. I, 7; A. 18; eye 1¾ to 2½ in snout, 5½ to 8 in head, 2⅓ to 4 in interocular. Elongate, the width as great or greater than the depth. Width of the head 1¼ to 1⅓ in its length, at the angle of the mouth about 2; depth 1⅜ to 1¾ in its length. Humeral process, dorsal plate, top of head to between the eyes, granular. Occipital process wider than long, scarcely keeled. Fontanel not continued behind the eyes, and withou backward projecting groove; posterior nostrils connected by a membrane. Barbels flattish, those of the maxillary reaching to near the ventrals, to middle of pectorals in older individuals; postmental to or beyond base of pectoral, mental to gill opening. Teeth villiform; vomerine and palatine patches of about equal size and shape in the young; a separate patch behind the palatines is developed later. Gill membranes meeting in an angle, forming a fold across the isthmus; gill rakers 6 + 10. Distance of dorsal spine from snout 2¼ to 2¾ in the length; dorsal and pectoral spines subterete, the outer margins roughened, the sides striate; the dorsal spine slightly serrate behind, a little shorter than the pectoral spine, 1⅓ to 1⅔ in the head; pectoral spine strongly serrate behind; space between dorsal and adipose fins 3⅖ to 4 in the length; adipose fin as long as the dorsal; upper caudal lobe longer, about 4 in the length; anal as high as long, 2 in head; ventrals 1⅔ to 2 in head; pectoral pore minute; sides with vertical series of pores. Color plumbeous above, silvery on sides; fins dusky. The specimens examined measure from 0.14 to 0.38 meters. Para; Curuca; Bahia. (Eigenmann & Eigenmann, Nematognathi, 59.)

174. SELENASPIS DOWII (Gill).

Head 3½; depth 6. D. I, 7; A. 19; eye small, elliptical, 3 in snout, 12 in head, 6 in interocular. A narrow flap of skin across the snout connecting the posterior nasal openings. Width below the dorsal spine a little greater than the depth, less than the width at the humeral process, which equals the greatest width of the head measured at the opercles. Head depressed, its depth at base of occipital process 1¼ in the greatest width, becoming gradually more depressed forward; width at angle of mouth 1⅔ in length of head, its greatest width about 1¼ in its length; snout short, 4 in head. Top of head coarsely granular, the granules forming striæ in front, vermiculations posteriorly or, in places, more or less regular striæ. Occipital process truncate, its width at tip greater than its length, the dorsal plate large, saddle-shaped, its bony tubercles forming striæ which are parallel with the strongly convex margin of the "saddle"; opercular bones granular striate, the humeral process with bony tubercles. Fontanel nearly obsolete, the granular bony surface being separated in front by thick skin, which covers an elongate area about 7 times longer than wide. Maxillary barbels reaching beyond humeral process; postmental barbels beyond gill opening, the mentals shorter. Upper jaw produced, equal to the short diameter of the eye. Teeth of the intermaxillaries in a villiform band which is narrowed in front, not produced backward to the angle of the mouth; vomer with a rather broad band confluent with the much wider subquadrate palatine patches which are produced backward in an angle; ovate patches on the pterygoids separate from the palatine teeth; teeth of the lower jaw in a comparatively shallow band, tapering very gradually to the angle of the mouth; the teeth of the jaws minute villiform, those of the palate and

3030——96

vomer bluntly conical. Gill membranes broadly united, meeting in an angle, joined to the isthmus, but with a free margin; gill rakers 9 + 15. Distance of dorsal from end of snout 2¾ in the length. Dorsal spine granular on sides and in front, about ½ the length of the head in height; distance of the adipose fin from the dorsal 3¼ in the length, the height of the adipose fin about 2½ in its length, which is contained 2⅗ in the length of the head. Caudal deeply forked, 5½ in the length; anterior ⅔ of the anal strongly convex, the posterior ⅓ slightly emarginate, the highest ray about 2½ in head. Ventrals reaching to anal, about 2 in head, their distance behind the dorsal equal to the length of dorsal and ½ the dorsal plate; pectoral spine granulose on sides, the outer margin with a series of larger granules which become recurved notches toward the tip, the inner edge with recurved hooks, its height 1⅜ in the length of the head; a small pectoral pore; no evident series of vertical pores. Bluish gray above, becoming white below; the fins brownish with dots. Description from the type of *Arius alatus*, .68 m. long, from Panama; collected by Dr. Steindachner. (Eigenmann & Eigenmann, Nematognathi, 61.)

Selenaspis dowii is thus characterized by Jordan & Gilbert:* Head 4 (4⅗ with caudal); depth 6½ (7½); width of head 5⅕. D. I, 8; A. 4, 12. Length (29529, U. S. Nat. Mus.) 10 inches. Body elongate, narrow, and slender, the caudal peduncle 1⅔ in head. Head low and narrow, tapering anteriorly, the snout subtruncate. Eye small, 7 in head, placed rather high; interorbital space little arched, with ridges and depressions, 2⅘ in head; snout 3⅘ in head; breadth of mouth 2⅕ in head. Mouth moderate, with thinnish lips; teeth villiform, bluntish; vomerine teeth forming 2 smallish, rounded patches, separated by a moderate interspace; each patch confluent with the neighboring palatine patch, which is rounded and rather large; the suture marked by a constriction; palatine bands without backward prolongation; premaxillary band of teeth broad. Barbels very long; maxillary barbel extending well beyond tip of pectoral fin; outer mental barbel reaching well past front of pectoral; inner 2⅗ in head. Dorsal shield comparatively large, not distinctly crescent-shaped, its divisions produced backward, their length about twice the length of the shield on the median line; anterior margin with 2 emarginations, the point fitting into an emargination of the occipital process; dorsal shield without keel. Occipital process very broad and short, its edges nearly straight, its breadth at base considerably greater than its length; its median line with a rather low keel. Fontanel broad and very short, ending obtusely at a point not far behind eye, the distance from this point to tip of snout 1⅔ in its distance from base of dorsal; each side of fontanel with a conspicuous smooth ridge, the 2 ridges converging anteriorly; shields of head rather finely granulated, few of the granulations forming lines, none of them extending farther forward than posterior margin of eye. Opercle striate. Gill membranes meeting below in a sharp angle, forming a rather broad fold across isthmus. Dorsal spine very short, its length a trifle less than pectoral spine, 2¼ in head. Axillary pore obsolete. Humeral process granulated, rather narrowly triangular, a

*Bull. U. S. Fish Comm., II, 1882, 50.

little less than $\frac{1}{4}$ length of pectoral spine, which extends barely $\frac{2}{3}$ the distance to the ventral fins; adipose fin long and low, very nearly or quite coterminous with the anal; caudal narrow, rather short, the upper lobe the longer, $1\frac{3}{4}$ in head; anal rather low and short; ventrals short, the vent not far behind them. Color dusky above, pale below, the fins all more or less dusky; maxillary barbels dusky, others pale. A single young male was obtained at Panama.

71(a). ASPISTOR, Jordan & Evermann, new genus.

Aspistor, JORDAN & EVERMANN, new genus (*luniscutis*).

This genus differs from *Selenaspis* in the presence of granular teeth on the palate and in the absence of a membranaceous flap connecting the posterior nostrils. (ἀσπιστήρ, a shielded warrior.)

175. ASPISTOR LUNISCUTIS (Cuvier & Valenciennes).

Head $3\frac{2}{3}$; depth $5\frac{1}{4}$ to 6. D. I, 7; A. 16 to 19; eye 2 to 3 in snout, 6 to 9 in head, 3 to $4\frac{1}{2}$ in the interocular. Body comparatively stout, the greatest width equaling the greatest depth. Head large, flattish above; profile descending; width of head $1\frac{1}{5}$ in its length, width at the mouth 2 to $2\frac{2}{3}$, its depth at the base of the occipital process scarcely less than its greatest width; top of head coarsely granular in young, the granules becoming finer and more regularly arranged in the adult; opercles smooth; humeral process with radiating lines of granules. Occipital process variable in shape, broader than long, the posterior margin convex; dorsal plate variable in outline, rounded anteriorly, saddle-shaped, either broader than long or longer than broad; middle of the fontanel above the posterior margin of the eye; the fontanel divided into 3 by 2 bony ridges, the middle portion being more than $\frac{1}{4}$ of its whole length. Sides of head with reticulating mucous canals. No skinny flap connecting the posterior nostrils. Maxillary barbels extending little beyond the base of the pectoral, or shorter; mental barbels short. Upper jaw little produced; teeth in the jaws rather large, conical; teeth of vomer and palatines finely granular, the vomerine patches separated from each other and from the palatine patches in the young, united and covering almost the entire roof of the mouth in the adult; the inner margins of the palatine patches approximated, sometimes a small elliptical patch of teeth between. Gill membranes forming a broad marginal flap across the isthmus. Gill rakers 3 to $4 + 7$ to 9. Axillary pore minute or wanting; vertical series of pores present. Distance of dorsal from tip of snout $2\frac{1}{4}$ in the length; the spine $1\frac{1}{4}$ to $1\frac{3}{4}$ in the head, granular in front, scarcely serrate behind; distance of adipose fin from the dorsal $3\frac{3}{4}$ to 4 in the length, the adipose fin twice as long as high, adnate, as long as the dorsal fin; caudal forked, the upper lobe longer, $4\frac{1}{4}$ to $4\frac{1}{3}$ in the length; anal fin about as long as high, $2\frac{1}{4}$ to $2\frac{1}{4}$ in head; ventrals $1\frac{3}{4}$ to 2 in head; pectoral spine stout, $1\frac{1}{4}$ to $1\frac{1}{2}$ in head, granular in front (serrate in the very young), striate on sides, serrate along inner margin. Color purplish brown above, sprinkled with brown dots below; fins about the color of the back. Numerous specimens examined 0.11 to 0.44 m. long.   Porto Alegre; Bahia; Nazareth, near Bahia; Rio Janeiro; Pará; Porto Seguro; São Matheos; Cannavierias. (Eigenmann & Eigenmann, Nematognathi, 63.)

176. SELENASPIS PARKERI (Traill), text, p. 125.

72. NETUMA, Bleeker.

Dorsal shield small, lunate; teeth on palate villiform, the patch on each side with a backward-extending process or angle.

Subgenus NOTARIUS, Gill.

Occipital process constricted at base.

177. NETUMA GRANDICASSIS (Cuvier & Valenciennes).

Head $3\frac{2}{3}$ to $3\frac{3}{4}$; depth $5\frac{2}{3}$ to 6.   D. I, 7; A. 18; eye 3 to $3\frac{1}{4}$ in snout, $8\frac{1}{4}$ to 10 in head, 4 to $4\frac{1}{4}$ in interocular.   Body cylindrical in front, tapering to a slender caudal peduncle.   Head greatly depressed, profile almost straight, descending, the width of the head $1\frac{1}{4}$ to $1\frac{2}{3}$ in its length, its depth $1\frac{1}{4}$ to 2 in its length.   Occipital process with a deep constriction where it joins the occiput, shaped like a clover leaflet, much as *Felichthys panamensis*, sometimes broader than long, sometimes much longer than broad, sometimes keeled.   Center of the fontanel over the middle of eye, the fontanel not continued backward as a groove; occipital process, top of head, and humeral process granular; interorbital region with 4 ridges, the inner ones bounding the fontanel, the outer ones running obliquely backward from near the posterior nasal opening.   Maxillary barbels reaching to the base of the pectoral, mentals to gill opening, postmentals a little longer.   Upper jaw projecting a diameter of the eye or more, the lip very wide, especially in front, making the nose pointed; teeth of both jaws rather large, those on the palate somewhat smaller; the depth of the intermaxillary band 7 to 9 in its width; the mandibular band very shallow; vomerine teeth none in 3 of the examples, a small patch on one side in another specimen, and a small patch on each side in another; palatine patches triangular, produced backward.   Gill membranes meeting in an angle, forming a fold across the isthmus; gill rakers 6 + 10.   Distance of dorsal spine from snout $2\frac{1}{5}$ to $2\frac{1}{4}$ in the length, the spine broken in the specimen studied.   Distance of adipose fin from the dorsal $3\frac{2}{3}$ to 4 in the length; adipose fin at least as long as the dorsal fin, adnate.   Caudal fin forked, the upper lobe longer, about 5 in the length, the tips broken; anal fin apparently longer than high, but the rays are somewhat worn off; ventrals small; pectoral pore large, slit-like.   Color light brown above, somewhat smutty below from the occurrence of minute scattered dots. We have examined 4 specimens from 0.23 m. to 0.33 m. long, collected by Agassiz & Bourget, Thayer Expedition, at Maranhao, and a fifth, 0.21 m. long, collected by Professor Agassiz at Bahia. (Eigenmann & Eigenmann, Nematognathi, 65.)

178. NETUMA STRICTICASSIS (Cuvier & Valenciennes), text, p. 126.

Subgenus **NETUMA**.

Occipital process not constricted.

179. NETUMA DUBIA (Bleeker), text, p. 126.

180. NETUMA KESSLERI (Steindachner).

Head $3\frac{1}{3}$ (4 in total with caudal); depth 6 (7 in total); width of head 4. D. I, 7; A. IV, 13. Length (29252, U. S. Nat. Mus.) 14 inches. Body rather long and low; the head long, broad and much depressed, much broader than deep. Eye very small, about 10 in head, placed well above the mouth. Interorbital space 2 in head; snout $3\frac{1}{3}$; breadth of mouth 2. Mouth large, with thickish lips, the upper jaw considerably projecting. Teeth all villiform, rather pointed. Vomerine patches rather large, roundish, usually fully confluent into a trapezoidal band, without division on the median line, and separated by a very narrow groove from the palatine bands. Palatine bands very large, broadly triangular, with a backward prolongation from the inner margin. (Teeth on vomer and palatines all forming one continuous band in old specimens, according to Steindachner.) Bands of teeth in jaws broad, the jaws strong. Barbels rather short and very slender, the maxillary barbels reaching little past base of pectoral; outer mental barbels about reaching gill opening; inner about as long as snout. Antedorsal shield short, crescent-shaped, rough, but without median keel. Occipital process long, narrowly triangular, its edges straight, its length $\frac{1}{4}$ to $\frac{1}{2}$ more than its width at base, its median line sharply keeled. Fontanel broad and shallow, its posterior end obtuse or almost truncate, its tip not prolonged in a groove, its edge bounded by a bony ridge, which is not granulated in front of middle of eye; end of fontanel about midway between tip of snout and front of dorsal, its greatest width about equal to length of eye. Shields of head all very coarsely granular, the roughnesses extending forward about to tho eye. Gill membranes forming a broad free fold across isthmus. Dorsal spine moderate, a little more than $\frac{1}{2}$ head, about equal to pectoral spine; humeral process triangular, granular, not quite $\frac{2}{3}$ length of pectoral spine; axillary pore obsolete; adipose fin long and low, its posterior margin little free; caudal short and broad, the upper lobe the longer, $1\frac{2}{3}$ in head; anal and ventrals rather small, the vent close behind the latter. Color dark brown, with bronze reflections; belly white; fins all dusky in 1 specimen, in the other mostly pale; maxillary barbels dusky, others pale. Two large specimens obtained at Panama by Gilbert. (Jordan & Gilbert, Bull. U. S. Fish Comm., II, 1882, 40.)

181. NETUMA INSCULPTA (Jordan & Gilbert).

Head 4 ($4\frac{2}{3}$ in total); depth $5\frac{2}{3}$ ($6\frac{3}{4}$); width of head $4\frac{3}{5}$. D. I, 6; A. IV, 14. Length (29415) $13\frac{1}{2}$ inches. Body moderately elongate, little compressed, the caudal peduncle slender and short. Head shortish, low and

broad, anteriorly depressed. Eye rather large, 6¼ in head, placed rather
high. Interorbital space flat and nearly smooth, 2 in head; snout 3;
breadth of mouth 1⁹⁄₁₆; snout very bluntly rounded, almost truncate in
front. Mouth large; teeth all villiform; vomerine bands of teeth large
(fully confluent with each other in the type, partly separated in smaller
examples), and with the large, club-shaped band on the palatines, from
which they are separated by a slight furrow and constriction; palatine
band of teeth with a backward prolongation; premaxillary band of teeth
large; maxillary barbel long, somewhat compressed, extending to middle
of pectoral spine; outer mental barbel reaching base of pectoral spine,
inner 2 in head. Dorsal shield short, crescent-shaped, without median
keel, its tips produced, its length on the median line about ⅔ the length
of 1 of its halves. Occipital process about as broad at base as long,
with a moderate median keel, its lateral margins somewhat concave;
fontanel becoming gradually contracted at a point a little nearer base
of dorsal than tip of snout, thence forming a narrow groove, which
extends to within a diameter of the pupil of the base of the occipital
process; this groove sometimes nearly obsolete; greatest width of fontanel about ⅔ diameter of eye. Granulated striæ extending along the
sides of the fontanel to a point opposite or in front of middle of eye.
Shield of head finely and evenly granulated, the roughnesses more uniform
than usual, and many of them arranged in lines, especially anteriorly;
opercle not striate, the skin marked with fine vermiculations; gill membranes forming a broad fold across the isthmus. Dorsal and pectoral
spines long, about equal, 1⅛ in head. No axiliary pores; humeral process
very large, triangular, finely granular, about ½ as long as pectoral
spine; adipose fin large, without free tip; upper lobe of caudal the longer,
1⅛ in head; anal and ventrals moderate, the vent close behind the latter.
Color rather pale; belly pale; fins and barbels all pale, or but slightly
tinged with dusky. A single adult male was obtained by Dr. Gilbert
at Panama. Two smaller ones are in the Museum collection, also from
Panama. (Jordan & Gilbert, Bull. U. S. Fish Comm., II, 1882, 41.)

182. NETUMA PLANICEPS (Steindachner).

Head 4 (4⅓ in total); depth 5⅓ (5⅔); width of head 5. D..I, 7; A. IV, 13.
Length (29417) 11 inches. Body comparatively elongate; the head small,
rather narrow, depressed anteriorly; the snout rather narrow and moderately rounded. Eye moderate, placed well above mouth, its length 5¼ in
head. Interorbital space flat and smooth, 2¼ in head; snout 3½; breadth
of mouth 2. Mouth rather large, with thickish lips; teeth villiform;
vomerine bands moderate, confluent with each other and with the much
larger ovate palatine bands, a slight constriction or furrow making the
divisions; palatine bands each with a backward prolongation; premaxillary band moderate; barbels very short; maxillary barbel scarcely or
not reaching to base of pectoral; outer mental barbel scarcely past gill
opening below; inner shorter than snout. Dorsal shield short, anteriorly truncate, not keeled, the length on the median line about ¼ of 1 of
its halves. Occipital process subtriangular, rather narrow, truncate

behind, its margins straight, becoming concave forward, its width at base about equal to its length; fontanel an almost obsolete groove, its posterior end not reaching base of occipital process by about the diameter of the eye, the groove extending forward to a point about midway between tip of snout and base of dorsal spine; anterior to this point is an equilateral triangle, flat, covered with smooth skin, the base of the triangle formed by the smooth, flattish interorbital area. Shields of head rather coarsely granular-striate, the granulations beginning anteriorly about opposite posterior margin of eye; opercle scarcely striate; gill membranes forming a moderate fold across the isthmus. Dorsal spine high, about equal to pectoral spine, and but little shorter than head; no axillary pore; humeral process triangular, granulated, a little more than $\frac{1}{3}$ length of pectoral spine; adipose fin rather long; upper lobe of caudal the longer, a little shorter than head; ventrals and anal moderate. Color brownish, not very dark; belly pale, thickly speckled with brown; fins more or less dusky; maxillary barbels black; mental barbels pale. Two specimens were obtained by Dr. Gilbert at Panama. They disagree in several details from Steindachner's description, and it is barely possible that they belong to a different species. The head in Steindachner's types is $3\frac{3}{4}$ to $3\frac{4}{5}$ in length, and the occipital process is narrower and less widened anteriorly. (Jordan & Gilbert, Bull. U. S. Fish Comm., II, 1882, 42.)

183. NETUMA PLATYPOGON (Günther).

Head $3\frac{3}{5}$ ($4\frac{3}{5}$); width of head $4\frac{3}{5}$; depth $5\frac{1}{5}$ ($6\frac{3}{5}$). D. I, 7; A. IV, 14. Length (28286) $15\frac{1}{2}$ inches. Body rather elongate, the head not very broad nor much depressed, a little broader than deep. Eye rather large, 5 to 6 in head. Interorbital space slightly more than $\frac{1}{2}$ head, a trifle less than width of mouth; length of snout $3\frac{1}{5}$ in head. Teeth all pointed; bands of vomerine teeth small, roundish, their boundaries traceable by a slight depression in the young, in the adult fully confluent with each other and with the palatine bands; palatine bands broad, ovate, several times as large as the patches on vomer, continued backward over the pterygoid region; premaxillary band rather broad, 5 to 6 times as broad as long; maxillary barbel reaching past base of pectoral in the young, not to gill opening in the adult, its base a little broader and more compressed than usual; outer mental barbels 2 in head; inner $2\frac{1}{2}$. Dorsal shield very short, lunate, subtruncate in front, its breadth more than 3 times its length on the median line; occipital process long, triangular, with straight margins, its length about $1\frac{3}{5}$ times its width in front, its broad median line rather sharply keeled. In the young it is proportionally shorter, little longer than broad. At the beginning of this keel is the end of the long, narrow, groove-like fontanel, which extends forward to a point just behind the eye, where it merges into the flattish and smooth anterior part of the head. Shields of the head all finely granular, the granules rarely forming distinct lines. Dorsal spine long, $1\frac{1}{3}$ to $1\frac{1}{2}$ in head, the soft rays projecting beyond the spine; pectoral spine about as long as dorsal, sharply serrate behind, the anterior serræ not very sharp; axillary pore small or absent; humeral process nearly smooth, rather narrow and short, $\frac{1}{4}$ length of pectoral spine; adipose fin short and rather high, its base barely $\frac{2}{3}$ length of base of anal; caudal deeply forked, its upper lobe the longer and slightly falcate, about as long as head; ventrals very short, reaching anal in females, shorter in the males; vent nearer base of ventrals than anal. Color in life very pale olive brown, with bronze and blue reflections, white below; fins all pale, the tip of anal and edges of caudal somewhat dusky; female with fins rather darker, the upper edge of the pectorals and ventrals largely black; in the males these fins are pale, or somewhat brown above; maxillary barbels blackish; lower pale. Generally abundant along the Pacific coast of tropical America. Specimens were observed by Dr. Gilbert at Mazatlan, Libertad, Punta Arenas, and Panama. It reaches a length of about 18 inches, and is seldom eaten. It resembles *Galeichthys gilberti*, but is readily distinguished by the small, pale ventrals, as also by the generic character of the dentition. The males of this species, according to Dr. Steindachner, carry the eggs in their mouths until after hatching. (Jordan & Gilbert, Bull. U. S. Fish Comm., II, 1882, 44.)

184. NETUMA OSCULA (Jordan & Gilbert).

Head $3\frac{3}{4}$ ($4\frac{3}{4}$ in total); depth $6\frac{1}{4}$ ($7\frac{2}{3}$); width of head $4\frac{3}{4}$. D. I, 7; A. IV, 14. Body moderately elongate, the head short, rather narrow, tapering forward, considerably broader than deep. Eye small, $7\frac{1}{2}$ in head, placed well above the mouth. Interorbital space $1\frac{9}{10}$ in head; snout 3; breadth of mouth $2\frac{3}{5}$. Mouth very small for the genus, with thick lips. Teeth on vomer and palatines villiform, but rather coarse and bluntish. Vomerine patches small, rather longer than broad, separated on the median line, and each also separated by a narrow groove from the large and roundish palatine bands, which have a distinct backward prolongation. Premaxillary band of teeth very broad, barely 3 times as long as wide. Barbels short, the maxillary barbels reaching slightly beyond base of pectorals, the outer mental barbels scarcely past gill opening below; inner mental barbels about as long as snout. Dorsal shield short, crescent-shaped, granulated, but without median keel, its length about $\frac{1}{4}$ its breadth. Occipital process narrow, its edges almost parallel until abruptly widened at base; the narrow part considerably longer than broad, with curved edges; a well-developed median keel. Fontanel broad and shallow, abruptly contracted at a point midway between tip of snout and end of occipital process, thence continued backward as a narrow groove to a point less than an eye's diameter in front of the base of the occipital process. Greatest width of fontanel about $\frac{3}{5}$ eye. Shields of top of head all coarsely and rather sparsely granular, and anteriorly striate. Interorbital space nearly plane, with a few low, smooth ridges. Opercles scarcely rugose. Gill membranes forming a narrow fold across isthmus posteriorly. Dorsal spine very high, $1\frac{1}{5}$ in head, a little longer than pectoral spine; humeral process granular, not quite $\frac{2}{3}$ length of pectoral spine; no axillary pore; adipose fin adnate posteriorly; caudal long, its upper lobe the longer, somewhat falcate, $1\frac{7}{10}$ in head; anal rather high. Color brown, with bluish reflections; lower parts dusky, with dark punctulations; fins all blackish; maxillary and outer mental barbels dusky. A single male example 11 inches long was obtained at Panama by Dr. Gilbert. (Jordan & Gilbert, Bull. U. S. Fish Comm., II, 1882, 46.)

185. NETUMA ELATTURA (Jordan & Gilbert).

Head $3\frac{2}{3}$ ($4\frac{1}{3}$ in total); depth $5\frac{3}{4}$ ($6\frac{3}{4}$); width of head $4\frac{2}{3}$. D. I, 6; A. IV, 14. Length (29408, U. S. Nat. Mus.) $12\frac{1}{2}$ inches. Body low, not very elongate, the head rather short and very broad, much broader than deep, the snout depressed and very broadly rounded, almost truncate. Eye moderate, placed rather high, its diameter 7 in head. Interorbital space $2\frac{1}{2}$ in head; snout $3\frac{1}{3}$; breadth of mouth $1\frac{2}{3}$. Mouth large, with thickish lips, the upper jaw considerably projecting. Teeth on vomer and palatines villiform, but bluntly conical, less acute than in most of the species. Vomerine patches oblong, small, separated by a narrow interspace from each other and from the palatine bands, which are roundish and comparatively small, with a backward prolongation. Teeth in jaws in broad bands. Barbels rather short, the maxillary barbels reaching a little past base of pectorals, the outer mental barbels a little past gill opening, the inner a little more than $\frac{1}{3}$ head. Dorsal shield not very short, crescent-shaped, with a distinct median keel, its length on the median line about $\frac{1}{4}$ its breadth. Occipital process short, broadly triangular, with concave sides which spread out abruptly near the base, forming a sort of shoulder, its length scarcely equal to its width at base. Median keel well developed. Fontanel broad and shallow, abruptly narrowed posteriorly at a point a little nearer base of dorsal than tip of snout, but extending as a groove to a point distant less than a diameter of the eye from the base of the occipital process, this groove indistinct in the smaller specimen. Greatest width of fontanel scarcely more than $\frac{1}{2}$ the eye. Shields of head granular-striate, the roughness less coarse than in *A. kessleri*. Interorbital space with 2 prominent ridges and numerous striæ, none of them granular, the granulations chiefly confined to the region behind widest part of fontanel. Opercle striate. Gill membranes forming a moderate fold across isthmus. Dorsal spine low, shorter than pectoral spine, which is $1\frac{2}{3}$ in. head, the anterior edges of both bluntly serrate; humeral process broadly triangular, granulated, not $\frac{2}{3}$ length of pectoral spine, much smaller than in *A. insculptus*; no axillary pore; adipose fin long and low, without free posterior margin; lower fins of moderate length; vent much nearer ventrals than anal. Caudal short, the upper lobe longest, $1\frac{2}{3}$ in head (a little more than $\frac{1}{2}$ head in the smaller specimens). Color dusky above, the lower parts soiled with dark points; fins all more or less dusky with dark points; maxillary barbels dusky, others pale. One male individual (29408) was obtained at Panama by Dr. Gilbert; another (30995) at Panama by Mr. Rowell. (Jordan & Gilbert, Bull. U. S. Fish Comm., II, 1882, 45.)

185 (a). NETUMA INSULARUM, Flora Hartley Greene.

Head 3⅔ in length; width of head 4⅔ in length; interorbital space in length 7; interorbital space in head scarcely 2; snout in head 3; breadth of mouth in head 2; eye in head 6⅔. D. I, 7; A. 17. Head much broader than deep; snout depressed and broadly rounded; eye above the level of the mouth. Upper jaw projecting. Teeth on vomer and palatines villiform and bluntly conical, the 2 vomerine patches forming together a band almost as long and slightly broader than the premaxillary band, the 2 sides separated by a narrow interspace; palatine teeth well separated from the vomerine teeth and in 2 large triangular patches which extend backward over the pterygoid region; each triangle has a sharp notch in its anterior side; its antero-posterior length is twice its lateral width; teeth of lower jaw in a narrower band than the upper jaw. Maxillary barbel extending to end of first third of the length of the pectoral spine; outer mental barbel to base of pectoral; inner mental barbel past gill opening, 2¼ in head. Dorsal shield crescent-shaped, without median keel, length on median line 2⅔ in distance between the horns of the crescent; 2 notches on its anterior side to meet the corresponding points from the occipital process. Occipital process broadly triangular, with the outer sides concave and 2 small projections at its posterior end. Median keel evident, rather short. Occipital process much broader at base than long; its length 3½ in head; posterior breadth 2 in length of process. Fontanel broad and shallow, narrowed gradually posteriorly to a point halfway between snout and base of dorsal spine. A narrow line runs back from it the distance of a long diameter of the eye. Greatest width of the fontanel equals the short diameter of the eye. Shields of the head granular-striate, the striæ evident and extending to the middle of the interorbital space, and on the side to meet the humeral process at the top of the gill opening. Opercles nearly smooth. Gill membranes forming a fold across the isthmus. Dorsal and pectoral spines crenulate in front and sharply decurved serrate behind. Dorsal shorter than pectoral, which is 1¼ in head. No axillary pore evident. Adipose fin long and low with posterior margin attached. Vent much nearer ventrals than anal. Color in alcohol, dark blue above, light blue on side, and white below; maxillary barbel dusky; fins all dusky. The type of this species (No. 47577, U. S. Nat. Mus.) was collected by the *Albatross* in the Galapagos Archipelago, being part of the collection studied by Jordan & Bollman in 1889. It was recorded by them (Proc. U. S. Nat. Mus. 1889, 179) as "*Tachysurus elatturus* (var?)." Its relations to *Netuma elattura* are close, but its fins are better developed and there are several differences in details of structure.

Netuma insularum, FLORA HARTLEY GREENE in GILBERT, Proc. U. S. Nat. Mus. 1896 (Feb. 5, 1897), 439, Galapagos Archipelago.

69. GALEICHTHYS, Cuvier & Valenciennes.

Dorsal shield small, lunate; teeth on palate villiform, the patches on each side not extending backward over the pterygoid region.

Subgenus GALEICHTHYS.

Shields of head mostly covered by soft skin, hiding the granulations.

163. GALEICHTHYS LENTIGINOSUS (Eigenmann & Eigenmann).

Head 4 to $4\frac{1}{5}$; depth 5 to 6. D. I, 6; A. 22. Eye $2\frac{1}{2}$ in snout, $8\frac{1}{2}$ in head, $4\frac{1}{2}$ in interocular, $2\frac{1}{4}$ in interorbital. Body nearly terete anteriorly, becoming compressed backward; the width, above the pectorals, a little greater than the depth. Head flat, depressed, its depth at base of occipital process $1\frac{1}{6}$ in its greatest width, which is about $1\frac{1}{4}$ in its length. Occipital process somewhat roughened, about twice as long as its greatest width, its margin straight and oblique; the middle of the fontanel above the posterior part of the eye. Head everywhere covered with skin; sides of the head and opercle with vermiculating canals. Snout somewhat pointed; upper jaw very little projecting; lips thick; teeth all villiform; the intermaxillary band strongly curved; vomerine teeth in 2 oval patches joined to the larger patches of the palatines; mandibulary band of teeth separated in front, the outer margins, if continued forward, forming an angle at the symphysis. Maxillary barbels reaching beyond base of pectorals; mental barbels reaching about $\frac{3}{5}$ toward the gill opening; the postmentals to the gill opening in 1 specimen, a little before in the other. Gill membranes forming a broad, free margin across the isthmus. Gill rakers 3 + 4. Pectoral pore minute; humeral process pointed behind. Distance of dorsal from snout $2\frac{1}{5}$ in the length, the dorsal spine covered with a membrane, its outer margin granular, its height $1\frac{2}{5}$ in the head, the first soft ray $1\frac{3}{5}$ the length of the fin; distance of adipose fin from the dorsal $3\frac{3}{5}$ to 4 in the length, the fin adnate, longer than the dorsal; caudal lunate, the upper lobe longer, somewhat falcate, $4\frac{1}{2}$ to 5 in the length; anal fin twice as long as high, the highest ray 2 to $2\frac{1}{4}$ in the head; ventrals short and broad, $1\frac{1}{3}$ in the head; pectoral spine covered with a membrane, $1\frac{1}{2}$ to $1\frac{2}{5}$ in head. Light brown, becoming nearly white below, the sides freckled; fins reddish. Panama. (Eigenmann & Eigenmann, Nematognathi, 50.)

164. GALEICHTHYS PERUVIANUS, Lütken.

Head $3\frac{1}{2}$; depth $4\frac{1}{3}$ to $5\frac{1}{2}$. D I, 7; A. 14 to 16. Eye 2 in snout, 7 in head, 4 in the interocular, 2 in the interorbital. Subterete, tapering to a long, slender caudal peduncle; the greatest width about equal to the greatest depth. Head not much depressed; interorbital area flattish, the greatest depth of the head $1\frac{3}{4}$ in its length, its greatest width $1\frac{4}{5}$ to $1\frac{3}{5}$; the width at angle of mouth 2 in its length; the surface of the cranial bones longitudinally furrowed, covered with muscle and skin. Occipital process more than 3 times as long as wide; anterior fontanel elongate, its center over the middle of the eye, continued as a deep groove to the base of occipital process; a small opening a pupil's distance behind the anterior fontanel, and a larger one in the occipital bone at the end of the groove. Snout, upper part of the neck, and the opercle sometimes with conspicuous reticulating mucous canals. Snout blunt, decurved. Maxillary barbels extending beyond base of pectoral, mentals about to gill openings, the postmentals about 1 diameter of the eye farther. Jaws sub-equal; the upper longer; teeth all fine, villiform; intermaxillary band of teeth very wide, its depth about 8 in its width; 2 small patches on the vomer; palatine patches very wide and shallow, tapering to a point. Gill membranes meeting at an acute angle, forming a fold across the isthmus; gill rakers 3 + 10. Humeral process very thin, covered with skin, more than ¼ as long as the pectoral spine, broadly expanded and rounded behind; pectoral pore present. Distance of dorsal spine from snout 2⅔ to 2¾ in the length; the dorsal spine broken off in all the specimens; distance of adipose from the dorsal 3⅝ to 3¾ in the length; adipose fin as long as the dorsal fin, adnate; caudal fin broadly lunate, the upper lobe longer, falcate, 3¾ in the length; anal fin higher than long, the highest ray 2 to 2⅓ in head; ventrals reaching to the anal, 1½ to 2 in head; pectoral spines broken in all the specimens. Back, top of head, and a band from humeral process to the lower caudal lobe, blue black; a broad conspicuous, bluish-silvery band along the lateral line; lower parts white; fins blackish; ventrals and anal sometimes with light areas. Eleven specimens, 0.25 to 0.35 m. long. Callao, Peru; Haslar Expedition. (Eigenmann & Eigenmann, Nematognathi, 5.)

Subgenus HEXANEMATICHTHYS, Bleeker.

Shields not entirely covered by soft skin, the granulations evident, especially in the male.

187. GALEICHTHYS SEEMANNI (Günther).

D. I, 7; A. 19; P. I, 10. The height of the body is contained 4½ times in the total length (without caudal); the length of the head 2⅓; head much broader than high, its greatest width being equal to its length without snout. Eyes of moderate size, much nearer to the end of the snout than to the operculum; the length of the snout is ⅔ of the width of the interorbital space. The median longitudinal fonticulus on the upper side of the head extends to the base of the occipital process. Teeth on the vomer separated in the middle by a short interspace, forming a pair of small subquadrangular patches which are confluent with those of the palatines. The latter are much longer than broad, elliptical. The band of intermaxillary teeth is 5¼ times as broad as long. The maxillary barbels extend nearly to the end of the head, and are about twice as long as the outer ones of the mandible. Crown of the head, and nape finely granular; occipital process broader than long, with a prominent ridge along its middle. The basal bone of the dorsal spine is small, with a few fine granules. Dorsal spine of moderate strength, more than ½ as long as the head, serrated along both edges; the first soft ray is as high as the body. Adipose fin rather shorter than dorsal. The upper caudal lobe is the longer, ⅔ of total length. Porus axillaris present. Ventral fin shorter than pectoral. Sides of the body silvery; basal half of the inner side of the paired fins black. Central America. A fine specimen 12 inches long, from the Haslar collection, collected by Dr. Seemann. (Günther.)

Arius seemanni, GÜNTHER, Cat., v, 147, 1864, Central America.

Jordan (Proc. Ac. Nat. Sci. Phila. 1883, 282) adds the following note on the type of this species: "Fontanel extending backward in a deep and narrow groove which reaches the occipital process. Middle of top of head smooth, much as in *A. platypogon.*"

The following account is given by Eigenmann & Eigenmann (Nematognathi, 78):

Head $3\frac{3}{4}$; depth 5. D. I, 7; A. 18. Body about as deep as wide, tapering to a slender peduncle. Head flat, depressed in front, top of the head coarsely granular; opercles smooth or with faint striations; humeral process slightly granular, covered with skin; the greatest depth of the head $1\frac{3}{5}$ in its length, greatest width $1\frac{1}{3}$ to $1\frac{2}{5}$; the width at angles of the mouth $2\frac{1}{3}$. Occipital process wider than long; fontanel open to above the posterior margin of the eye, with a deep backward-extending groove. Interorbital area smooth, without ridges. Eye 2 in snout, 7 in head, $3\frac{3}{5}$ to 4 in the interocular, $2\frac{1}{3}$ in the interorbital. Maxillary barbels reaching slightly beyond base of pectorals, mental barbels $\frac{2}{3}$ toward the gill opening, the postmentals $\frac{1}{2}$ a diameter of the eye behind the gill opening or farther. Upper jaw longer; teeth all villiform; vomerine teeth in 2 small ovate patches, which are separated from each other but joined to the much larger palatine patches. Gill membranes forming a moderate fold across the isthmus. Gill rakers 5 + 12. Pectoral pore large; vertical series of pores present. Distance of dorsal fin from tip of snout $2\frac{1}{3}$ to $2\frac{3}{5}$ in the length, the spine rather stout, $1\frac{2}{3}$ in head, its outer edge granular toothed, its inner edge with short, recurved teeth; distance of the adipose fin from the dorsal $3\frac{1}{2}$ to $3\frac{1}{6}$ in the length; adipose fin slightly longer than high, shorter than the dorsal fin. Caudal 4 in the length; anal emarginate, little longer than high; ventrals $1\frac{3}{4}$ to 2 in the length of the head; pectoral spine $1\frac{1}{3}$ in head, its anterior margin granular toothed, its inner edge with long, straight teeth. Plumbeous, silvery below; fins dusky, inner surface of ventrals and pectorals dark. One female 0.28 m., Panama. One male 0.21 m., Panama.

187(a). GALEICHTHYS GILBERTI, Jordan & Williams.

Head $3\frac{4}{7}$ to 4; width of head $5\frac{1}{5}$; depth 5. D. I, 7; A. IV, 14. Body comparatively elongate, the head depressed but not very broad, somewhat broader than high; eye rather large, 5 to 6 in length of head; width of interorbital space $2\frac{1}{4}$ in head; breadth of mouth $1\frac{3}{4}$; length of snout $3\frac{1}{3}$. Teeth all villiform; bands of vomerine teeth separated by a rather wide interval, each small, roundish, confluent with the neighboring palatine band, the junction marked by a slight constriction; palatine band ovate, broad behind, varying considerably in size and somewhat in form, the width ranging from $\frac{1}{3}$ diameter of eye to $\frac{2}{3}$, being generally largest in adults; band of palatine teeth without backward prolongation; band of maxillary teeth rather broad and short, its length about 5 times its breadth. Maxillary barbel flattened at base, reaching a little past base of pectoral in young, scarcely to gill opening in adult; outer mental barbels 2 in head, inner 3. Gill rakers 4 + 12. Dorsal shield very short, narrowly crescent-shaped, its length on median line not more than $\frac{1}{2}$ that of one of its sides. Occipital process subtriangular, not quite as long as broad at base, with a strong median keel, its sides slightly curved. A short distance in front of the beginning of the keel is the end of the very narrow groove-like fontanel, which is somewhat widened anteriorly, finally merging into the broad, flat, smooth interorbital area, the boundaries of which are not well defined; shields of head unusually smooth, all finely and very sparsely granular, the granules not forming distinct lines. Some specimens (probably females) about as smooth as in the subgenus *Galeichthys*. Gill membranes forming a rather broad fold across isthmus. Dorsal spine long, usually, but not always, shorter than the pectoral spine, about $1\frac{3}{4}$ in head; axillary pore absent; humeral process rather broadly triangular, not much produced backward, less than $\frac{1}{2}$ length of pectoral spine, its surface not granular, covered with skin; adipose fin $\frac{1}{2}$ length of anal, its posterior margin little free; upper lobe of caudal the longer and somewhat falcate, about as long as head; ventrals long, about reaching anal in females, rather shorter in males; vent much nearer base of ventrals than anal. Color olive green, with bluish luster, white below; upper fins dusky olivaceous; caudal yellowish dusky at tip; anal yellowish with a median dusky shade; ventrals yellowish, the basal half of upper side abruptly black; pectorals similarly colored, the black area rather smaller; maxillary barbel blackish; other barbels pale. Length 12 to 18 inches. Coast of Sinaloa; very common; by far the most abundant species at Mazatlan; not recorded from localities farther south.

Arius assimilis, JORDAN & GILBERT, Bull. U. S. Fish Comm., II, 1882, 47; not of GÜNTHER.
Galeichthys gilberti, JORDAN & WILLIAMS, Rept. Fishes Sinaloa, in Proc. Cal. Ac. Sci. 1895, 395, pl. 26, Mazatlan. (Type, No. 29213. Coll. Chas. H. Gilbert.)

188. GALEICHTHYS JORDANI (Eigenmann & Eigenmann).

Head $3\frac{2}{5}$ to $3\frac{3}{5}$; depth $5\frac{1}{2}$ to $5\frac{5}{6}$. D. I, 7; A. 18; eye large, $1\frac{3}{4}$ in snout, $5\frac{1}{2}$ in head, 2 in the interorbital, $2\frac{3}{4}$ to 3 in the interocular. The specimens agree very closely with the description of *assimilis* by Jordan & Gilbert (*gilberti* of the present paper). They differ in the width of the mouth and in having a pectoral pore. Rather robust, the width little less than the depth; caudal peduncle compressed. Head heavy, little broader than high, its height $1\frac{1}{2}$ in its length, its width $1\frac{2}{5}$ to $1\frac{1}{3}$, width at the angle of the mouth 2 to $2\frac{1}{4}$; interorbital area flat and smooth; posterior portion of the head finely and sparsely granular; opercle and humeral process smooth; occipital process about as long as broad, unusually sharply keeled; fontanel extending to above the posterior part of the orbit, continuing as a deep groove to the base of the occipital process; maxillary barbels extending to the pectoral pore, postmentals at least to the gill opening, mental about $\frac{3}{5}$ as long as the postmental barbels; snout blunt, decurved; upper jaw a little produced; teeth all villiform, those on the vomer forming 2 small, separate, ovate patches, which are contiguous to the twice or thrice as large palatine patches; gill membranes forming a fold across the isthmus; gillrakers 6 + 9; pectoral pore large; vertical series of pores present; distance of dorsal spine from tip of snout $2\frac{3}{5}$ to $2\frac{4}{5}$ in the length; the spine of the dorsal and pectoral fins granular on the basal half of their outer margin; almost the entire inner margins serrate, the spines of equal length, $1\frac{1}{2}$ in head; distance of adipose fin from the dorsal $3\frac{3}{5}$ in the length; the adipose more than $\frac{1}{2}$ as long as the dorsal fin, its posterior margin free; caudal deeply forked, the upper lobe longer, somewhat falcate, $3\frac{3}{4}$ to 4 in length; anal fin about as long as high, deeply emarginate, its highest ray $2\frac{2}{5}$ in head; ventral fins not reaching to the anal, 2 in head. Dorsal surface dark blue, with metallic luster, becoming silvery below; lower caudal lobe dusky; basal half of the inner surface of the paired fins black. (Eigenmann & Eigenmann, Nematognathi, 79.) Panama; known only from specimens in the Museum of Comparative Zoology. The specimens from Mazatlan referred to *jordani*, *seemanni*, and *assimilis* by authors belong to *Galeichthys gilberti*.

188(a). GALEICHTHYS AZUREUS, Jordan & Williams.

Head $3\frac{1}{4}$; width of head $4\frac{5}{9}$; depth 9. D. I, 7; P. I, 10; A. IV, 14. Gill rakers 6 + 13. Body robust, its width anteriorly greater than its depth; caudal peduncle short, stout; distance from end of anal fin to base of median caudal rays about $\frac{1}{2}$ length of head. Head flat, very broad, its depth at posterior angle of jaw about $\frac{1}{2}$ its width; interorbital region flat, smooth anteriorly and granulated posteriorly; fontanel almost obsolete, wide anteriorly and ending in a short groove posteriorly at a point $\frac{1}{2}$ distance from tip of snout to posterior end of occipital process; top of head, occipital process, and antedorsal shield finely granular, granulations mostly arranged in radiating striæ and extending forward to a line with the pupils; nostrils very large and close together, posterior one with a broad valve; occipital process pentagonal, its length $4\frac{1}{2}$ in head, about as long as wide, with a very low ridge; dorsal shield crescent-shaped, with points extending back on each side of fin, its median length about $\frac{1}{2}$ the length of the side; eye small, about 9 in head; interorbital width almost 2 in head; snout almost 4 in head; breadth of mouth $2\frac{3}{10}$ in head; maxillary barbel slender, thick at base, $1\frac{4}{5}$ in head; outer mental barbel reaching to posterior angle of jaw, about $2\frac{2}{3}$ in head; inner mental barbel about 4 in head; teeth all villiform; premaxillary band narrow, about $\frac{1}{8}$ as wide as long: vomerine and palatine bands of teeth fully confluent on each side, forming together a crescent-shaped patch, narrowly divided on the median line of the vomer; form of vomerine bands similar to that of the palatine bands but smaller; palatine band of teeth without backward prolongations; opercle with radiating ridges; humeral process granular, triangular, lower posterior corner prominent; axillary pore very small; gill membranes forming a broad fold across isthmus; dorsal fin short, base not including spine equal to base of adipose dorsal; dorsal spine robust, but little shorter than pectoral spine, about 2 in head, its anterior serræ small and tubercle-like, its posterior edge, as well as that of pectoral, retrorsely serrate; soft rays of dorsal extending but little beyond spine, the longest about $\frac{3}{5}$ length of head; adipose dorsal about $\frac{1}{2}$ as high as long; caudal lobes unequal, the upper lobe about $\frac{1}{8}$ longer than lower lobe; anal short, of medium height; distance from vent to base of ventrals $\frac{1}{2}$ distance from origin of anal; pectoral spine very strong, its anterior margin with serræ toward the tip, becoming small tubercles toward base; soft rays but little longer than spine, which reaches slightly beyond ¼ distance from origin to base of ventrals.   Color dark blue, with silvery reflections on sides; belly pale; mental barbels dusky; maxillary barbels light below and black above; paired fins darkest on inner side; other fins almost uniformly dusky.   One specimen, 19¼ inches long, taken by the Hopkins Expedition at Mazatlan; probably not distinct from *G. guatemalensis.*

Galeichthys azureus, JORDAN & WILLIAMS, Rept. Fishes Sinaloa, in Proc. Cal. Ac. Sci. 1895, 398, pl. 27, Mazatlan. (Type, No. 1575, L. S. Jr. Univ. Mus. Coll. Hopkins Exped. to Sinaloa.)

189. GALEICHTHYS CÆRULESCENS (Günther).

D. I, 7; A. 17; P. I, 10.   The height of the body is contained about 5 times in the total length (without caudal), the length of the head 3¼ or 3¾ times; head much broader than high, its greatest width being ¾ of its length.   Eyes rather small, their diameter being ⅓ of the extent of the snout, ⅔ of their distance from the gill opening, and ¼ of the width of the interorbital space.   The teeth on the palate form a slightly curved band, composed of 2 vomerine patches which are much broader than long, and of a pair of palatine patches which are subcontinuous with, scarcely broader and longer than, those of the vomer.   The barbels of the maxillaries extend to the middle, the outer ones of the mandible to the base of the pectoral.   Crown of the head granular; occipital process broader than long, subtriangular, subtruncated behind, and slightly raised along the median line; the basal bone of the dorsal spine is subtriangular, small.   Dorsal spine of moderate strength, more than ½ as long as the head, granulated in front and slightly serrated behind; the first soft ray is as high as the body; adipose fin shorter than the dorsal; caudal deeply forked, with the upper lobe the longer, its length being nearly equal to that of the head; pectoral spine serrated along its inner edge and on the extremity of its outer edge; it is as long as the head without snout. Ventral fin shorter than pectoral.   Sides steel-blue iridescent, blackish toward the back, and silvery below; vertical fins black; inner side of the paired fins blackish.   Guatemala.   a–b. Fine specimens, 12 inches long. Huamuchal.   From the collection of Messrs. Godman and Salvin.   (Günther, Cat., v, 149.)

The following note on the types of this species is given by Jordan (Proc. Ac. Nat. Sci. Phila. 1883, 282): "Head more depressed than in *A. assimilis.* Fontanel very short, ending abruptly behind and not produced in a groove behind the smooth area of the top of the head, the boundary of the smooth area being rather abruptly convex.   Occipital process broader than long, its edges nearly straight.   Bands of palatine teeth small, not produced backward on the inner margin.   Paired fins black at base above."   No recent collector has found this species.

189(a). **GALEICHTHYS XENAUCHEN**, Gilbert, new species.

Head $3\frac{7}{8}$ in length; depth at front of dorsal $5\frac{1}{2}$; anal 23. Width of head at opercle $1\frac{7}{9}$ in its length; width at front of eyes 2 in head; width of mouth at inner angles $2\frac{2}{3}$ in head; interorbital width $2\frac{1}{10}$; eye very small, 9 in head, $3\frac{1}{2}$ in its distance from tip of snout, $4\frac{5}{8}$ in postocular part of head, $4\frac{1}{4}$ in interorbital width. Teeth all villiform; mandibular bands well separated on middle line, very broad mesially, rapidly tapering to a point laterally, the band produced beyond angle of mouth, its greatest width $2\frac{1}{3}$ times in its length; premaxillary band very convexly curved, following the outline of the snout, its width $5\frac{3}{4}$ in its length; vomerine patches roundish, separated by an evident medial groove, marked off from the palatine patches by a narrower groove and a constriction; the palatine patches are equal in width to the vomerine patches, and less than twice as long, of nearly equal width throughout. Maxillary barbels very slender, reaching slightly beyond the base of the pectoral spine; the mental barbels do not reach edge of gill membrane, the outer pair equaling length of snout and $\frac{1}{2}$ of eye. Nostrils very large, the anterior broadly oval, with widely reflexed rim; the posterior widely elliptical, not concealed by the valve; distance from anterior nostril to tip of snout equaling that from posterior nostril to front of eye. Fontanel wide, with nearly parallel edges on frontal region, abruptly narrowing at front of occiput, where it is continuous with a narrow and shallow groove; the latter fails to reach base of occipital process by a distance equaling $\frac{1}{2}$ diameter of eye. The raised margins of the fontanel continuous with a pair of sharp ridges bounding the groove, these accompanied by a pair of lower ridges on their outer sides and parallel with them; posteriorly these ridges roughened with granules and merging into the granulated area on posterior part of occiput; occipital process granulated, the granules arranged in more or less definite lines radiating backward and downward on each side from median point of base; lateral portions of occiput with an area of radiating striæ separated from the central ridges by a smooth groove-like depression; a narrow granulated area extending forward on each side of fontanel to above back of orbits; occipital process very long and narrow, its width opposite its middle being but $\frac{2}{9}$ of its length, abruptly expanding near base, the basal width being $\frac{1}{4}$ its length plus that of dorsal plate on median line; opercles and humeral plate weakly striate. Gill membranes with a wide, free fold posteriorly; gill rakers weak and short, $1+4$ movable ones; no evident axial pore. Dorsal spine slender, with a series of sharp granulations on anterior edge, minutely roughened, not serrate, behind, broken in the type, but its length was about $\frac{2}{3}$ that of head; pectoral spines rather slender, rough granular on outer margins, with short, fine serræ within, both mutilated in the type, but their length was about equal to that of dorsal spine; pectoral extending nearly $\frac{2}{3}$ distance to ventrals, the ventrals nearly to origin of anal; distance from anus to base of ventrals $\frac{2}{3}$ its distance from front of anal; anal fin very long, its base $1\frac{1}{4}$ in head, its longest ray $\frac{1}{4}$ head; distance between dorsals $3\frac{1}{3}$ in length; adipose fin long, highest about opposite the middle, a short, almost verti-

3030——97

cal, free posterior margin, its vertical height $3\frac{5}{6}$ in its length; the latter over twice its distance from rudimentary caudal rays, greater than base of first dorsal, equal to $\frac{1}{2}$ length of head; caudal with broad lobes, the lower rounded; the upper mutilated in the type, but evidently acute and longer than the lower. Color purplish above, more bluish anteriorly, the lower parts silvery, coarsely punctate with brown; fins all blackish except the lower surface of the paired fins. In appearance most closely allied to species of *Netuma*, having the low, depressed head with the lateral outlines converging forward to the narrow pointed snout, and a long, largely adherent adipose dorsal. The palatine patches are, however, narrow and without backwardly projecting lobes. The species is distinguished from all those known from the Pacific coast of America by the long and extraordinarily narrow occipital process. Type, a female 380 mm. long, from Panama. (ξένός, strange; αὐχήν, nape.)

Galeichthys xenauchen, GILBERT, Fishes of Panama, MS. 1898, Panama.

190. GALEICHTHYS GUATEMALENSIS (Günther).

Head $3\frac{4}{7}$ ($4\frac{3}{8}$ in total); width of head 5 ($6\frac{1}{5}$); depth $6\frac{1}{8}$ (7). D. I, 6; P. I, 10; A. III, 15. Length (28140, U. S. Nat. Mus.) $12\frac{1}{2}$ inches. Body slender, its width anteriorly greater than depth; caudal peduncle compressed, short; distance from end of anal to base of median caudal rays about $\frac{1}{2}$ length of head. Head depressed, not very broad, its depth at posterior margin of branchiostegal membranes less than $\frac{2}{3}$ its width; interorbital region flat, smooth, the smooth area forming a broad equilateral triangle, its base at the interorbital space, the apex at a point $\frac{4}{5}$ the distance from snout to dorsal, the triangle forming the termination of the almost obsolete fontanel; top of head, occipital process, and antedorsal shield finely granular, some of the anterior granulations only arranged in lines, none of them in radiating striæ. Occipital process broadly trapezoidal, its width slightly greater than the length of its side, with a slight or obsolete median carina; its posterior margin truncated; its sides slightly convex posteriorly, concave toward the front. Dorsal shield small, narrow, crescent-shaped, its median length about $\frac{1}{2}$ the length of its side. Eye small, 6 in head; interorbital width $2\frac{2}{7}$ in head; snout 4 in head; breadth of mouth 2. Maxillary barbel very slender, reaching base of pectoral spine; outer mental barbel to well beyond margin of branchiostegal membranes, its length about $\frac{1}{2}$ head; inner mental barbel 3 in head. Teeth all villiform; width of premaxillary band about $\frac{1}{6}$ its length; vomerine and palatine bands of teeth fully confluent on each side, forming together a crescent-shaped patch, narrowly divided on the median line of the vomer; form of vomerine band similar to that of the palatine band; palatine band of teeth without backward prolongation; opercle with radiating ridges; humeral process granular, narrow, produced backward, not quite $\frac{1}{2}$ length of pectoral spines; no axillary pore. Gill membranes forming a narrow fold across isthmus. Dorsal short, its base about equal to that of the adipose dorsal; dorsal spine robust, but little shorter than the pectoral spine, about $\frac{3}{5}$ length of head, its anterior serræ small and tubercle-like; its posterior edge, as well as that of the pectoral, retrorsely serrate; soft rays of dorsal extending much beyond the spine, the longest about ¾ length of head; adipose dorsal about ½ as high as long, its posterior margin largely free; caudal very widely forked, the upper lobe falcate, nearly ¼ longer than the lower, as long as head; anal short and low; distance from vent to base of ventrals slightly more than ½ its distance from origin of anal; pectoral spine very strong, much stronger than dorsal spine, its anterior margin with serræ toward the tip, becoming small tubercles toward base; inner edge with strong retrorse serræ, the soft rays longer than spines, reaching ¾ distance to base of ventrals. Color very dark bluish or greenish above; sides with bronze luster; belly silvery; mental barbels white, with black edge; maxillary barbels blackish; fins all blackish, the caudal nearly uniform; paired fins darkest on inner side; sides with vertical series of mucous pores, conspicuous in life. This species is not uncommon at Mazatlan, where several specimens were obtained by Dr. Gilbert. Four specimens from Colima are also in the National Museum. It has not been observed at Panama. The original description of this species is brief and not entirely correct. That it was intended to refer to the species here described we have ascertained by the examination of Dr. Günther's original types in the British Museum. (Jordan & Gilbert, Bull. U. S. Fish Comm., II, 1882, 48.)

191. GALEICHTHYS ASSIMILIS (Günther).

D. I, 7; A. 19; P. I, 10. The height of the body is contained 4⅔ times in the total length (without caudal), the length of head 3¾; head much broader than high, its greatest width being ¾ of its length. Eyes rather small, situated nearer to the end of snout than to that of operculum; the length of snout is ⅔ of the width of interorbital space. The median longitudinal fonticulus on the upper side of the head does not extend to the base of occipital process. Teeth on vomer but slightly separated in the middle, forming a pair of oblong transverse patches which are confluent with those on the palatine bones; the latter are short, club-shaped. The band of intermaxillary teeth is 5 times as broad as long. All the teeth villiform. The maxillary barbels extend nearly to the end of head; the length of the outer ones of the mandible is ½ or ⅔ that of the head. Crown of the head granular, the granulations being arranged in radiating streaks. Occipital process broader than long, triangular, with its hinder end concave. The basal bone of the dorsal spine of moderate size, crescent-shaped. Dorsal spine of moderate strength, more than ½ as long as head, granulated in front and slightly serrated behind; the first soft ray longer than spine and as high as body; adipose fin shorter than dorsal; caudal deeply forked, with the upper lobe the longer, its length being contained 5¼ times in the total; pectoral spine serrated along its inner edge and on the extremity of the outer edge; ventral fin shorter than pectoral. Sides of the body silvery; vertical fins grayish; basal half of the inner side of the paired fins black. Guatemala. A fine specimen, 13 inches long. Lake of Yzabal. From the collection of Messrs. Godman and Salvin. (Günther, Cat., V, 146.)

Jordan (Proc. Ac. Nat. Sci. Phila. 1883, 281) has the following note on the type of this species: "Area between the eyes smooth, extending backward in the form of a rather narrow triangle, which is moderately obtuse behind. Fontanel narrow and short, ending far in front of occipital process, not extending backward as a groove behind the smooth area of the top of the head; posterior end of fontanel midway between tip of snout and middle of dorsal shield. Occipital process broad, its edges not straight. Band of palatine teeth large, but not produced backward on the inner margin. * * * There is no evidence of the occurrence of the true *A. assimilis* in Pacific waters."

192. GALEICHTHYS SURINAMENSIS (Bleeker), text, p. 129.

193. GALEICHTHYS DASYCEPHALUS (Günther).

Head $4\frac{1}{4}$ ($5\frac{3}{4}$ in total); depth 6 ($7\frac{1}{4}$ in total); width of head $5\frac{1}{4}$. D. I, 7; A. IV, 17. Length (29400) 11 inches. Body elongate, compressed behind, the head small, narrow, and moderately depressed anteriorly, the snout not very blunt. Eye rather large, placed somewhat above level of angle of mouth, its length 5 in head; width of interorbital space $2\frac{1}{6}$ in head; breadth of mouth $2\frac{1}{2}$; length of snout $3\frac{1}{4}$. Teeth villiform, those of vomer and palatines rather coarse, bluntly conic; bands of vomerine teeth separated by a rather broad area, each confluent with the neighboring palatine band, the two forming a small oblong patch much smaller than the eye, the division between the palatine and vomer scarcely appreciable. Palatine bands without backward prolongation. Bands of teeth in jaws short and broad. Maxillary barbel reaching about to middle of pectoral spine; outer mental barbel to base of pectoral; inner slightly more than $\frac{1}{2}$ head. Dorsal shield short, crescent-shaped, a little more than 3 times as broad as long on the median line. Occipital process subtriangular, its sides straight, slightly longer than broad, its median line rather sharply keeled. Close in front of its base begins the deep fontanel, which is narrow and groove-like posteriorly, becoming rather abruptly broader above the opercle, then gradually narrowed anteriorly. Ridges bounding fontanel prominent anteriorly to a point just behind vertical from nostrils, coarsely granular for their whole length, the granules mostly arranged in 1 series. Between these ridges and the eye on each side is another ridge extending obliquely backward and inward from above front of eye, likewise very coarsely granular, the granules mostly in 2 series. Shields of head all rough granular, the granules forming irregular lines. Gill membranes forming a narrow fold across isthmus. Dorsal spine moderate, about equal to pectoral spine, $1\frac{1}{4}$ in head; axillary pore present, small; humeral process broad, scarcely granular, about $\frac{2}{5}$ pectoral spine; adipose fin rather long and low; caudal long, the upper lobe the longer, somewhat longer than head; anal long and high, its outline emarginate, its longest rays a little more than $\frac{1}{2}$ head; ventrals long, the vent nearer their base than that of anal. Color dusky, the entire ventral surface soiled with dark points; fins all largely blackish; barbels black. Two specimens were obtained at Panama by Dr. Gilbert. This species may be known at once by the 4 granulated ridges, which extend the length of the interorbital space. In the female, later taken, the granulations on the head are largely covered by soft skin.

194. GALEICHTHYS LONGICEPHALUS (Eigenmann & Eigenmann).

Head $3\frac{2}{3}$; depth $6\frac{1}{4}$. D. I, 7; A. 20. Elongate, slender, greatest width little greater than the depth. Head long and depressed, its greatest width $1\frac{1}{4}$ in its length, its greatest depth little more than $\frac{1}{2}$ its length. Top of head with faint granules almost entirely concealed by the skin; interorbital area flat and with 4 ridges which are obscurely granular, the inner two bordering the fontanel, the outer ones curved in front extending obliquely backward from near the posterior nasal opening; occipital process as long as broad, its margins concave; fontanel produced as a deep groove to the base of the occipital process; opercle faintly striate; humeral process entirely covered with thick skin, not granular. Eye lateral, well above the angle of the mouth, its diameter $1\frac{1}{4}$ in snout, 6 in head, 3 in interocular; snout depressed and rounded in front. Maxillary barbels extending scarcely beyond base of pectoral, mentals not to gill opening. Upper jaw little projecting; width of the mouth $2\frac{1}{2}$ in the head; intermaxillary teeth long and slender, the depth of the band $4\frac{1}{2}$ in its width; vomerine and palatine teeth obtusely conical, the vomerine patches separate, contiguous to, but not confluent with, the palatine patches. Gill membranes not forming an angle where they meet, with a rather broad, free margin. Gill rakers short and thick, 4 + 5. Pectoral pore small; vertical series of pores present; distance of dorsal fin from tip of snout $2\frac{2}{3}$ in the length, the spine $1\frac{3}{4}$ in the head, its outer margin granular-toothed near its base, its inner margin with short teeth; distance of adipose fin from the dorsal $3\frac{1}{6}$ in the length; adipose fin much longer than high, as long as the dorsal fin; caudal forked, the upper lobe $\frac{1}{4}$ longer than the lower, very nearly as long as head, $3\frac{4}{5}$ in the length; anal fin emarginate, scarcely longer than high, its height $2\frac{1}{4}$ in the head; ventrals reaching almost to the anal, about 2 in head; pectoral spine a little longer than the dorsal spine $1\frac{2}{3}$ in the head; its outer edge roughened, inner edge with rather sharp teeth. Brown above, the sides silvery, entire ventral surface sprinkled with brown dots; a black median line on the back; fins dusky; barbels blackish. One specimen, a male, .29 m. long (No. 4972, M. C. Z.). Panama. Steindachner. (Eigenmann & Eigenmann, Nematognathi, 82.)

195. GALEICHTHYS RUGISPINIS (Cuvier & Valenciennes)

Head $3\frac{1}{2}$ to 4; depth $5\frac{1}{2}$ to 6. D. I, 7; A. 19 to 21. Slender, compressed on the tail. Head broad and depressed, tapering forward; width of the head $1\frac{1}{2}$ to $1\frac{2}{3}$ in its length, at the angle of the mouth $2\frac{2}{3}$ to $2\frac{1}{2}$; depth of head $1\frac{3}{4}$ to 2; profile rather steep. Top of head, humeral process, front and sides of spines, and dorsal plate granular, the granulation not extending forward to above middle of cheeks. Occipital process triangular, about as long as broad, the median ridge not very prominent. Middle of fontanel behind the eye, the posterior portion separated by a bridge, not continued backward as a groove; interorbital region with 4 ridges. Eye small, 3 in snout, 10 in head, $3\frac{1}{2}$ in the interocular. Barbels villiform. Maxillary barbel reaching to or beyond base of pectoral; postmental to gill opening, mental barbels much shorter. Mouth inferior, lower jaw included, lips thick; teeth villiform, the anterior ones in the jaws longer; depth of the intermaxillary band 4 in its width; palatine patches 1 diameter of eye apart, the width of the patches less than 1 diameter of eye. Gill membranes meeting in an angle, forming a fold across the isthmus. Gill-rakers 6 + 11. Pectoral pore none; vertical series of pores present. Distance of dorsal spine from the snout 2⅓ to 2½ in the length, the spine broken in the specimens examined. Space between dorsal and adipose fins 4 to 4⅔ in the length. Adipose fin adnate, as long as the anal fin; ventrals 2⅓ in the head; pectoral spine serrated behind (broken). Two specimens 0.22 m. and 0.26 m. long. Para. Agassiz and Bourget. (Eigenmann & Eigenmann, Nematognathi, 83.)

196. GALEICHTHYS PHRYGIATUS (Cuvier & Valenciennes), text, p. 130.

74. TACHYSURUS, Lacépède.

Teeth on palate granular; dorsal shield small; palatine bands of teeth without backward projecting angle.

197. TACHYSURUS NUCHALIS (Günther).

D. I, 7; A. 21; P. I, 10. The height of the body ⅙ of the total length (without caudal), the length of head ¼. Head as broad as high, its greatest width being ⅔ its length; its upper surface granulated; occipital process triangular, as long as broad, with the lateral margins slightly concave; it is elevated into an obtuse ridge running along the middle; the longitudinal groove in the middle of the forehead is rather wide, narrow behind, and does not extend to the base of occipital process. Teeth on palate are coarsely granular, and form 2 subtriangular patches of moderate extent, which, sometimes, are subcontinuous with their anterior angles. The maxillary barbels extend nearly to end of pectoral. Dorsal spine of moderate strength, slightly serrated along both edges, ⅚ length of head; adipose fin small, shorter than dorsal; pectoral spine as long as, but stronger than, that of dorsal; pectoral fin shorter than head. British Guiana. a–c. Six inches long. Purchased of Mr. Scrivener. d–f. Young. Presented by Sir R. Schomburgk. (Günther, Cat., v, 171.)

198. TACHYSURUS FISSUS (Cuvier & Valenciennes).

D. I, 7; A. 20 or 21. Length of head ⅛ of the total (without caudal). The distance between the end of snout and that of occipital process ¼ of the total length (with caudal); basal bone of dorsal spine small. The teeth on the palate form 2 separate subovate patches. The maxillary barbel extends to, or nearly to, the middle of pectoral fin. Adipose fin small. Cayenne. a–b. Presented by Prof. R. Owen. These specimens having had the cavity of the mouth and of the gills extended in an extraordinary manner, I was induced to examine the cause of it, when, to my great surprise, I found them filled with about 20 eggs, rather larger than an ordinary pea, perfectly uninjured, and with the embryos in a forward state of development. The specimens are males, from 6 to 7 inches long, and in each the stomach was almost empty. Although the eggs might have been put into the mouth of the fish by their captor, this does not appear probable. On the other hand, it is a well-known fact that the American Siluroids take care of their progeny in various ways; and I have no doubt that in this species and in its allies the males carry the eggs in their mouth, depositing them in places of safety, and removing them when they fear the approach of danger or disturbance. (Günther, Cat., v, 172.)

199. TACHYSURUS SPIXII (Agassiz).

Head $3\frac{3}{5}$ to 4; depth 5 to $5\frac{1}{2}$. D. I, 7; A. 21. Body compressed, especially toward the caudal fin, the depth greater than the width. Head narrowed forward, its greatest width $1\frac{1}{8}$ in its length, its greatest depth $1\frac{1}{2}$; width at the mouth $2\frac{1}{5}$ in the length of the head. Top of the head granular in the young, the granules becoming more or less united in the adult, forming fine reticulating ridges, especially on the occipital process, longer than broad, with a blunt median ridge, the margins concave. Fontanel narrow, without interruptions, continued as a deep tapering groove to near the base of the occipital process; interorbital area with 4 ridges; opercles and humeral process roughened, covered with skin; sides of the head, and snout with reticulating mucous canals. Eye $1\frac{1}{2}$ to 2 in the snout, 5 to $6\frac{1}{2}$ in the head, $2\frac{3}{4}$ to 3 in the interocular. Maxillary barbels varying in extent. from about the middle of the pectoral to the base of the ventrals; postmental barbels extending to the base of pectoral or to near its tip; mentals to edge of gill membrane or to beyond base of pectoral. Upper jaw projecting; lips more or less papillose; teeth on the intermaxillary and the outer ones of the mandible, villiform; the inner series of the mandible and the palate with granular teeth; the palatine patches of teeth small, subovate, sometimes contiguous in front. Gill membranes united, joined to the isthmus, not forming a free margin across it; gill rakers 6 + 11 to 13. Pectoral pore moderate; distance of dorsal spine from snout $2\frac{1}{8}$ to $2\frac{1}{7}$ in the length; the spine $1\frac{1}{3}$ to $1\frac{1}{2}$ in head, serrated on its inner margin, granular or almost smooth on its outer margin. Distance of adipose from the dorsal fin $3\frac{1}{5}$ to $3\frac{3}{5}$ in the length, the adipose fin shorter than the dorsal fin, free posteriorly; caudal forked, the upper lobe slightly the longer, 4 to 5 in the length; anal fin scarcely longer than high, its highest ray about 2 in head; ventral fin $1\frac{2}{3}$ to 2 in head; pectoral spine strong, about as long as the dorsal spine, serrated on its inner margin, granular or scarcely roughened on the outer margin. Color brownish above, sides and ventral surface silvery, sometimes with brown dots. We have examined over 70 specimens measuring from 0.07 to 0.24 m. from Maranhao, Bahia, Rio Janeiro, Para, Santos in São Paulo, Abrolhos, Brazil. The specimens from Para are much darker in color, the lips more papillose, the barbels longer than those of other specimens. The Santos specimens are ashy above, white below the lateral line, with rather large brown dots on sides, becoming fewer below. (Eigenmann & Eigenmann, Nematognathi, 89.)

200. TACHYSURUS MELANOPUS (Günther).

D. I, 7; A. 21; P. I, 10. The height of the body is contained 5 times in the total length (without caudal), the length of the head $4\frac{1}{3}$ times; head somewhat broader than high, its greatest width being $\frac{3}{4}$ of its length; the occiput and nape are finely granulated; occipital process subtriangular, as long as broad, with the lateral margins somewhat concave, and with the median ridge a little elevated. The longitudinal groove in the middle of the crown of the head is indistinct, narrow, linear behind, scarcely extending to the base of the occipital process. The teeth on the palatines are obtusely conical, and form 2 rather small subovate patches, apart from each other, and situated on the front part of the palate. The maxillary barbels do not quite extend to the middle of the pectoral fin. Dorsal spine of moderate strength, scarcely serrated anteriorly, equal in length to the distance of the gill opening from the anterior margin of the orbit, or even somewhat shorter; adipose fin small, the length of its base being less than that of the dorsal; pectoral spine nearly as long and strong as that of the dorsal fin, very strongly serrated anteriorly. Porous axillaries nearly as wide as a nasal opening. The upper (inner) surface of the ventral fins deep black, the lower (outer) white; the inner surface of the pectorals blackish. Rio Motagua (east slope). a–b. From 8 to 9 inches long. From Mr. Salvin's collection. (Günther, Cat., v, 172.) The specimens from the Pacific Coast mentioned in the text (page 132) belong to the following species.

200(a). TACHYSURUS LIROPUS, Susan B. Bristol.

Head $3\frac{3}{7}$ to $3\frac{3}{4}$; depth $4\frac{3}{5}$ to $5\frac{1}{10}$. D. I, 6; A. II, 19; P. I, 9 or 10. Body elongate, its width anteriorally a little less than depth, the posterior portion much compressed; back elevated at front of dorsal; anterior profile from front of dorsal to tip of snout oblique; head flat, very broad, its width $1\frac{1}{2}$ in its length; snout broad, rounded, $1\frac{1}{10}$ to $1\frac{1}{6}$ in interorbital width; eye rather large, laterally placed, its width about $1\frac{3}{5}$ in its length, $4\frac{1}{6}$ to $4\frac{9}{10}$ in head; mouth small, upper jaw considerably projecting, its breadth $2\frac{3}{5}$ to 3 in head; jaws thin; wide bands of minute pointed teeth present on both jaws; vomerine bands widely separated, and indistinguishable from the palatine band, which is small, oblong-ovate, and scarcely prolonged backward; interval separating vomerine bands about $2\frac{1}{2}$ or 3 in eye; the teeth on these bands larger than those on jaws, and very bluntly conical. Interorbital space broad, $2\frac{1}{2}$ to 3 in head. Barbels long and slender, the maxillary barbel extending to, nearly to, or, in some cases, past base of pectoral, $1\frac{1}{14}$ to $1\frac{3}{5}$ in head; outer mental barbel $1\frac{1}{2}$ to $1\frac{3}{4}$ in head; inner mental barbel $2\frac{1}{8}$ to $2\frac{5}{6}$ in head. Antedorsal shield very short, narrowly crescent-shaped, its length on the median line about 2 or 3 in its width; occipital process subtriangular, a little longer than broad at base, its edge slightly concave, its median keel strong. The long, narrow groove of the fontanel beginning abruptly a short distance in front of occipital keel, the distance from its end to base of dorsal $1\frac{3}{5}$ to 2 in the distance to tip of snout. Shields of head rather smooth, finely granular, the granules forming distinct lines anteriorly. The flat area between eyes triangular, with a median groove extending from fontanel forward to tip of snout, its posterior end a little behind eye, the granulations on each side of it extending forward as far as posterior border of pupil; opercles with no radiating striæ. Gill membranes forming a very narrow fold across the isthmus. Gill rakers 5 + 12. Nostrils 2 on either side, large, placed close together and near tip of snout, the posterior with a large flap; axillary pore well developed; humeral process smooth, very short, 4 to $5\frac{1}{2}$ in pectoral spine. Base of dorsal $2\frac{3}{5}$ to $2\frac{9}{10}$ in head; dorsal spine long and very strong, $1\frac{1}{7}$ to $1\frac{1}{8}$ in head, its upper anterior serræ small and tubercle-like, its upper posterior and its lower edges retrosely serrate; the soft rays extending considerably beyond the spine, $1\frac{1}{6}$ to $1\frac{1}{4}$ in head; adipose fin small, its base $3\frac{1}{2}$ to $4\frac{1}{2}$ in head, its height $1\frac{3}{5}$ to 2 in its base; caudal widely forked, the upper lobe, measured from base of caudal to its tip, the longer, about $1\frac{1}{4}$ in head; base of anal $1\frac{3}{5}$ to $1\frac{6}{7}$ in head, its longest ray $2\frac{1}{8}$ to $2\frac{1}{8}$ in head; ventrals reaching $\frac{5}{6}$ to $\frac{7}{8}$ the distance to origin of anal; vent about midway between origin of ventrals and origin of anal; pectoral spine $1\frac{1}{2}$ in head, serrate, the serræ on inner edge larger and sharper than those on outer, the upper anterior serræ tubercle-like, the rays a little longer than spine. Bluish silvery, light yellowish below; top of head and back brown; fins dusky olive, lighter at base, all margined with darker; ventrals pale; adipose fin covered with minute black dots; maxillary barbels dark brown, with bluish silvery luster; other barbels lighter; eye yellowish. Here described from 6 specimens from San Juan Lagoon, mouth of Rio Ahome, Sonora, Mexico (No. 47584, U. S. Nat. Museum). Length $7\frac{1}{4}$ to 9 inches.

Tachysurus liropus, BRISTOL, in GILBERT, Proc. U. S. Nat. Mus. 1896 (Feb. 5, 1897), 438, San Juan Lagoon, near mouth of Rio Ahome, Sonora, Mexico.

200(b). TACHYSURUS EMMELANE, Gilbert, new species.

Head $3\frac{2}{5}$ in length ($4\frac{1}{10}$ in total); depth 5 (6 in total). A. 27 (3 + 24). Eye 7 in head, $2\frac{1}{2}$ in its distance from tip of snout, 4 in postorbital part of head, $3\frac{3}{8}$ in interorbital width, $2\frac{1}{5}$ in frontal width opposite middle of eyes. Mouth of moderate width, gently convex, the distance between its angles (measured internally) $2\frac{4}{5}$ in head. Teeth in premaxillary and front of mandible finely villiform; posterior mandibular teeth stronger than those in front, bluntly conic, not, however, granular or flat and pavement-like, as are the posterior mandibular teeth in *T. furthii*, *T. melanopus*, and *T. liropus*. Mandibular bands with a wide interspace mesially, each widest near symphysis, rapidly tapering laterally, and extending beyond angle of mouth. The width of the bands is less than in related species, $\frac{1}{4}$ eye at their widest point. The length of 1 of the mandibular bands is slightly greater ($1\frac{1}{10}$) than length of eye. Premaxillary band very short, its length but $\frac{1}{6}$ greater than that of 1 of the mandibular bands, extending on each side less than $\frac{1}{2}$ distance from median line to angle of mouth; width of band $\frac{1}{5}$ its length. Palatine teeth granular, in small oblanceolate patches, which taper to a point laterally, and are widely separated on medial line, the patches agreeing in size and shape with those in *T. liropus*. Head depressed, tapering and at the same time narrowing anteriorly, as in other species of *Tachysurus;* profile rising in a uniform, gently convex curve to occiput, where it becomes concave, owing to the more rapidly ascending outline of the occipital process. Eye low, but little above angle of mouth, the interorbital space decidedly convex. Barbels slender, the maxillary barbels reaching edge of gill membrane in front of pectoral spine, the outer mental barbels extending beyond gill·· membrane, 1⅖ in head, the inner not to edge of membrane. Gill membrane widely attached to isthmus, without free edge. Occipital region with very fine granulations, those on middle of occiput forming parallel series along the fontanel groove, those on median portion of occipital process in series which diverge backward˙ from the median line. The sculptured area extends forward to a vertical which traverses the cheek at a distance of its own diameter behind the eye; anterior edge of granulated area equidistant between tip of snout and front of dorsal plate; fontanel produced backward as a deep, narrow groove,˙ which fails to reach base of occipital process by a distance equaling ⅓ the length of the process on the median line; the groove widening but little anteriorly; an area behind and on each side of the groove with parallel series of granulations, and marked off from the rest of the head by a shallow trench; base of occipital process similarly indicated by a transverse indented line; occipital process not keeled, very wide at base, becoming abruptly very narrow behind, its posterior ⅓ having parallel margins and being as wide as long, the lateral margins therefore deeply concave; width of process at base equaling its length on median line, plus that of dorsal plate, its hinder edge deeply incised to receive the anterior rounded wedge process of the dorsal plate, the latter finely granulated anteriorly, the lateral wings concealed under the smooth skin; a narrow groove as long as eye occupies the anterior end of the fontanel; no similar groove found in *T. furthii,* a short roundish one present in the type of *T. liropus,* and a continuous one the entire length of fontanel in the specimen which we identify with *T. melanopus;* opercle without radiating ridges; a short, slit-like axillary pore present; humeral process short, the exposed portion not broadly triangular, the surface smooth, or indistinctly rough. Gill rakers 6+13, of moderate length and thickness, the longest below the angle, ⅔ diameter of eye. Dorsal spine with a series of obtuse granulations in front and very weak retrorse serræ behind, its length to tip of calcified portion 1⅔ in head; longest soft ray 1⅜ in head; adipose dorsal not adnate, its anterior insertion about over middle of anal; distance between dorsals equal to length of head; base of adipose dorsal much greater than its height, less than base of first dorsal; pectoral spine strong, ridged and granulated in front, the hinder edge with very strong serræ; length of spine 1¾ in head, the fin projecting beyond tip of spine and reaching ⅔ distance from axil to base of ventrals; ventrals reaching to or nearly to origin of anal; vent midway between base of ventrals and front of anal; base of anal equaling length of pectoral spine; margin of anal gently concave, the longest ray 2¼ in head; caudal with pointed lobes, the lower longest in the type, 1⅘ in head. Color dark steel blue or brownish above, becoming bright silvery below; posterior ⅔ of anal white the anterior portion black with a narrow white edge; pectorals and ventrals with anterior (outer) face white or slightly dusky; pectorals with inner face of upper rays black; a black blotch covers all of inner face of ventrals, except terminal half of inner rays; barbels blackish. Closely related to *T. melanopus* and *T. multiradiatus*, differing from the former in the longer anal fin, from the latter in the black markings on lower fins. The description of the type of *T. multiradiatus* (*Bagrus? arioides*) Kner & Steindachner, Abhandl der K. bayer Akad. der Wissen, X, I, 1864, indicates a species with much rougher sculpturing of the head, a longer fontanel groove, narrower occipital process, and more anteriorly inserted adipose dorsal. The type is a single specimen, 280 mm. long, from Panama. (Gilbert.) (ἐν, in μελανη, ink.)

Tachysurus emmelane, Gilbert, Fishes Panama, MS. 1898, Panama.

201. TACHYSURUS FURTHII (Steindachner).

Head $3\frac{1}{4}$ to $3\frac{3}{4}$; depth 5 to $5\frac{1}{4}$. D. I, 7; A. 20. Body compressed posteriorly; profile slightly convex. Head broad, tapering forward, its greatest width $1\frac{2}{3}$ to $1\frac{1}{4}$ in its length; width, at the angle of the mouth, $2\frac{3}{5}$ to $2\frac{1}{4}$ in the head. Top of head densely covered with fine granules. Occipital process about as long as broad, with a median ridge, emarginate on its sides and at tip; interorbital region with 4 smooth ridges, the inner bordering the fontanel, the other extending obliquely backward from near the posterior nasal opening; sides of head and snout with vermiculating mucous pores. Middle of the fontanel over the pupil. Eye strictly lateral, not entirely above the angle of the mouth, its center in front of the posterior end of the mandible, 2 in snout, 6 to 7 in head, $3\frac{1}{4}$ to 4 in the distance between the eyes. Maxillary barbels thin, reaching to the middle of the pectoral or shorter, postmentals beyond base of pectorals, or sometimes not beyond edge of gill membrane; mentals to edge of gill membrane or shorter. Jaws about equal, the upper rather thin; teeth on the intermaxillaries villiform; the mandible with villiform teeth except the inner 2 or 3 series, which are granular; like the palatine patches irregular, suboval, sometimes the anterior end, sometimes the posterior, and sometimes both ends pointed. Gill membranes united, joined to the isthmus without a free margin. Gill rakers long and slender, $4 + 11$. Axillary pore small; vertical series of pores present. Distance of dorsal from snout $2\frac{2}{3}$ to $2\frac{2}{3}$ in the length; the dorsal spine $1\frac{2}{3}$ to $1\frac{3}{4}$ in the head, on sides and front granular, with small, sharp teeth on its inner margin; the first soft ray little, if any, higher than the spine. Distance of adipose fin from the dorsal 3 to $3\frac{1}{4}$ in the length, the fin longer than high, shorter than the dorsal fin. Caudal fin forked, the lobes rounded, $4\frac{1}{2}$ in the length. Anal little longer than high, the highest ray 2 in the length of the head. Ventrals short, $1\frac{3}{4}$ to $2\frac{2}{3}$ in head. Pectoral spine long and slender, $1\frac{1}{4}$ to $1\frac{3}{4}$ in the head, outer margin granular, inner margin with short teeth. Ashy above, white below. We have examined 15 specimens, the largest measuring 0.29 m. The sexes do not differ externally. Panama. (Eigenmann & Eigenmann, Nematognathi, 90.)

202. **TACHYSURUS VARIOLOSUS** (Cuvier & Valenciennes), text, p. 132.

203. **TACHYSURUS MULTIRADIATUS** (Günther), text, p. 132.

75. CATHOROPS, Jordan & Gilbert, text, p. 133.

204. **CATHOROPS HYPOPHTHALMUS** (Steindachner), text, p. 133.

205. **CATHOROPS GULOSUS** (Eigenmann & Eigenmann), text, p. 133.

Page 134. After *Ictalurus furcatus* add:

206(a). **ICTALURUS ANGUILLA,** Evermann & Kendall.

(EEL CAT; WILLOW CAT.)

Head 4; depth 4½; eye 7 in head; snout 2½; interorbital 1½; maxillary (without barbel) 3; free portion of maxillary barbel longer than head; dorsal spine 2 in head; pectoral spine 2; width of mouth 2. D. I, 6; A. 24; vertebræ 42. Head large, broad, and heavy; the mouth unusually broad; cheeks and postocular portion of top of head very prominent; interorbital space flat, a broad, deep groove extending backward to origin of dorsal fin; body stout, compressed posteriorly; back scarcely elevated. Eye small; maxillary barbel long, reaching considerably past gill opening; other barbels short. Origin of dorsal fin equidistant between snout and origin of adipose fin, its distance from snout 2⅔ in length of body; base of dorsal fin 3½ in head; longest dorsal ray 1¾ in head; dorsal spine strong, entire both before and behind; pectoral spine strong, entire in front, a series of strong, retrorse serræ behind; humeral process 2⅓ in pectoral spine; ventrals barely reaching origin of anal, their length 2 in head; anal fin long and low, the longest rays about 2⅓ in head; base of fin greater than head, 3⅓ in body; caudal moderately forked, the middle rays about 2⅓ in outer rays, which are about 1⅔ in head. Color uniform pale yellowish or olivaceous; no spots anywhere.

An examination of the 6 cotypes shows that there is not much variation, all the important characters remaining quite constant. The maxillary barbel varies somewhat in length, in some individuals scarcely reaching gill opening, and the number of anal rays varies from 24 to 26.

A comparison of the skull of this species with that of *I. furcatus* and of *I. punctatus* of the same size shows a number of very marked differences. Nearly all the bones in *I. anguilla* are heavier than in the other species; the supraoccipital is broadly triangular, and its upper surface finely grooved, while in each of the other species it is much longer and narrower and the upper surface nearly smooth.

From the blue cat (*Ictalurus furcatus*) this species differs chiefly in the fewer rays in the anal fin, the wider mouth, the shorter, heavier head, the much longer maxillary barbel, and in the cranial characters already given. From the spotted cat (*I. punctatus*) it may be distinguished by its wider mouth, more blunt snout, heavier head, the color, and the cranial characters already mentioned.

The eel cat rarely attains a greater weight than 5 pounds, and usually does not exceed 3 pounds. Its flesh is firm and of excellent flavor. The spawning season appears to be during the spring, as several of the individuals examined were in mature spawning condition.* Lower Mississippi Valley; thus far known only from the Atchafalaya River, Louisiana and the Ohio River at Louisville. (*anguilla*, the generic name of the eel.)

Ictalurus anguilla, EVERMANN & KENDALL, Bull. U. S. Fish Comm. 1897 (Feb. 9, 1898), 125, pl. 6, fig. 1, Atchafalaya River, Louisiana. (Type, No. 48788. Coll. Evermann & Chamberlain.)

Recent studies of the catfishes of the Lower Mississippi Valley by Dr. Evermann have shown that the most abundant and most important species of catfish in that region is *Ictalurus furcatus* (Le Sueur), and not *Ameiurus lacustris* (Walbaum), as has hitherto been supposed. The large specimen described by Dr. Bean as *Amiurus ponderosus* is an *Ictalurus* (as shown by the skeleton now in the United States National Museum) and apparently *I. furcatus*. The common names "Great Fork-tailed Cat," "Mississippi Cat," and "Blue Cat" all belong to *I. furcatus*.

Page 138. The species called *Ameiurus dugesii* belongs to the genus *Villarius*, Rutter.

Page 142. After *Ameiurus nigrilabris* add:

77(a). VILLARIUS, Rutter.

Villarius, RUTTER, Proc. Cal. Ac. Sci., ser. 2, vol. VI, 1896, 256 (*pricei*).

Allied to *Ameiurus*, differing in the presence of scattered cilia on the sides. Backward process from occipital short, broad, emarginate, connected by ligament with the first interspinal buckler; in adults the distance between this process and the buckler is equal to the length of the former; in young examples the process overlaps the keel on the underside of the buckler. Head narrow, width of intermaxillary band of teeth ¼ of head; caudal deeply forked, the upper lobe the longer; barbels long, those of the maxillary extending past the gill opening. Sides with scattered hair-like cirri; these are very noticeable under a lens, but not readily distinguished by the naked eye. This genus differs from all others of the family in having hair-like cirri on the sides. It differs from *Ictalurus* in having the occipital process and the interspinal buckler widely separated and connected by ligament; from *Ameiurus* in having a narrow head and a deeply forked caudal. Two species known, the following and *Villarius dugesii* (Bean). (*villus*, a hair.)

---

* This species is well known to the fishermen of the Atchafalaya River, by whom it is usually called the "eel cat," though the name "willow cat" is sometimes applied to it. It was explained by the fishermen that the name "eel cat" was given on account of the long feelers (i. e., barbels) and the name "willow cat" because it is most frequently found about the roots of willow trees. The eel cat is not an abundant species in the Atchafalaya River. During six days (April 19–24) spent at Morgan City several hundred catfish were examined at the three fish houses, and the total number of eel cats seen was fewer than twenty-five. The fishermen report that this proportion is about as great as at any time of the year. Of the four commercial species of catfishes handled on this river the most abundant one is the blue cat (*Ictalurus furcatus*), and the next is the yellow cat or goujon (*Leptops olivaris*); the eel cat comes next and the spotted cat (*Ictalurus punctatus*) last. The blue cat and the yellow cat probably constitute 98 per cent of the entire catch.

220(a). VILLARIUS PRICEI, Rutter.

B. 8; D. I, 6; A. 22 or 23; C. 17; P. I, 9; V. 8.   Head $3\frac{1}{4}$ to $3\frac{3}{4}$ in body; eye 5 to 7 in head; snout $2\frac{3}{4}$; maxillary $5\frac{1}{4}$ to 6. Maxillary barbel very long, reaching beyond the pectoral spine, in the adult about to its tip when depressed, 3 to 4 times as long as the barbel at nostril. Origin of dorsal midway between snout and middle of base of adipose fin; pectorals inserted halfway between snout and ventrals; longest dorsal ray 6 to 7 times in length of body; spine of dorsal longer than its base, equal to base of adipose fin; longest pectoral ray about half of head, pectoral spine $2\frac{1}{2}$ to 3 in head, with about 12 distinct hooked serræ behind, these fewer and somewhat smaller in the young; base of anal 3 times in its distance from snout, its longest ray equal to length of ventral; caudal deeply forked. Lateral line faint. This species differs from *V. dugesii* (Bean) in having very prominent serrations on the pectoral spines, the types of *dugesii* having the pectoral spines without serræ. We have examined a specimen of *dugesii*, 4 inches long, from Salamanca, Mexico, which is in the type basin; it has the cirri minute and light in color, a row of papillæ along the lateral line, and the pectoral spines with 4 or 5 degenerate serræ. (Named for William Wightman Price, who collected the type specimen.)

Villarius pricei, RUTTER, Proc. Cal. Ac. Sci., ser. 2, vol. VI, 1896, 257, San Bernardino Creek, a tributary of the Yaqui River, southern Arizona. (Type, No. 4826, L. S. Jr. Univ. Mus.)

Page 143.   *Leptops olivaris* is known as the *Goujon* in Louisiana, where it is an important food-fish.

Page 146.   In *Schilbeodes gyrinus* the anal rays are 14 to 16; not 13.

Page 152.   Under *Rhamdia salvini* read "Osbert Salvin" for "Oscar Salvin."

Page 170.   *Pantosteus arizonæ*, Gilbert, is described and figured in Proc. U. S. Nat. Mus. 1898, 488, pl. 36.

Page 174.   *Catostomus discobolus* is distinct from the true *C. latipinnis*. The two species are confused in the description of *C. latipinnis* given by us. They may each be described as follows:

279. CATOSTOMUS LATIPINNIS, Baird & Girard.

Head 4; depth about $5\frac{1}{2}$; eye high up and small, 5 to 7 in head, 3 to $3\frac{1}{2}$ in snout, $2\frac{1}{4}$ to $2\frac{3}{4}$ in interorbital space; interorbital width $2\frac{3}{4}$ in head. D. 14 or 15; A. 7; scales 19 or 20–89 to 102–16 to 18, 46 to 50 transverse rows in front of dorsal fin. Head depressed and flat above, its greatest depth $1\frac{1}{2}$ in its length, the depth below lower edge of orbit 3 in its length. Least depth of caudal peduncle $4\frac{1}{4}$ in head, or $3\frac{1}{4}$ in its own length. Fins very large, the dorsal with its upper margin concave; ventrals and pectorals rounded; dorsal as long as its longest ray, $1\frac{1}{10}$ in head, its last ray a little less than $\frac{1}{2}$ the length of the first ray; origin of dorsal fin nearer tip of snout than base of caudal; ventrals not reaching quite to vent, $1\frac{3}{4}$ in head. Muzzle not projecting; about 6 rows of short, thick papillæ on upper lip, the smallest above; lower lip large, incised to its base, with about 12 rows of short, thick papillæ, posteriorly quite small; distance from front of upper lip to back of lower $1\frac{1}{4}$ in snout; jaws with a slight cartilaginous sheath; width of preorbital a little less than $\frac{1}{4}$ its length. Reaches a length of about 2 feet. Lower Colorado River basin. This description by Gilbert & Scofield, based upon specimens from the Gila River at Tempe, Arizona.

279(a). CATOSTOMUS DISCOBOLUS, Cope.

Head $3\frac{4}{5}$ to $4\frac{1}{2}$; depth about $5\frac{1}{4}$; eye small, high up, $5\frac{1}{4}$ to 6 in head, $2\frac{3}{4}$ in snout, $2\frac{2}{3}$ in interorbital width, which is $8\frac{1}{2}$ in head; width of preorbital less than $\frac{1}{4}$ its length; least depth of caudal peduncle $2\frac{1}{2}$ in its length, or 2 in head; greatest depth of head $1\frac{3}{4}$ in its length; depth from lower edge of orbit $3\frac{1}{4}$ in head. D. 12 or 13; A. 7; scales 19 to 21–101 to 109–17 to 21, 52 to 63 in front of dorsal. Muzzle projecting slightly beyond upper lip. Upper margin of dorsal very slightly concave, the length of its base $1\frac{1}{2}$ in its longest ray, or $1\frac{1}{2}$ in head; last dorsal ray $\frac{1}{2}$ length of first; origin of dorsal midway between tip of snout and base of caudal; ventral rounded, $1\frac{3}{4}$ in head, not quite reaching vent. Mouth as in *C. latipinnis* except that the posterior tubercles on lower lip are long and not nearly so closely set, there being 9 or 10 rows; jaws with a slight cartilaginous pellicle. Upper portion of the Colorado River basin. Attains the length of a foot or more. The above description by Gilbert & Scofield from specimens from Green River at Green River Station, Wyoming.

Catostomus discobolus, COPE, Hayden's Geol. Surv. Wyo., 435, 1870, Green River, Wyoming; GILBERT & SCOFIELD, Proc. U. S. Nat. Mus. 1898, 490.

Page 175. After *Catostomus griseus* add:

280(a). CATOSTOMUS RETROPINNIS, Jordan.

A doubtful species which is, however, not yet shown to be invalid. Head $4\frac{1}{2}$; depth $5\frac{1}{2}$; eye $6\frac{1}{4}$ in head; snout 2; interorbital $2\frac{1}{2}$. D. 11; A. 7; scales 17–108–14. Body slender, head slender, snout very long, caudal peduncle long, its least depth less than snout, $2\frac{1}{2}$ in head; dorsal profile very little elevated; mouth large, wholly inferior, overhung by the piglike projecting snout; lips thin but very broad, lower lip incised nearly to base, with about 6 rows of moderate papillæ; lobes of lower lip very long, about $\frac{1}{3}$ of snout; gill rakers short and weak. Origin of dorsal a little nearer base of caudal than tip of snout; base of dorsal equal to snout; longest dorsal ray a little greater than base of fin; anal fin long and pointed, the fourth ray longest, $1\frac{1}{10}$ in head; caudal lunate, the middle ray $1\frac{1}{2}$ in outer rays; pectoral somewhat falcate, the longest $1\frac{1}{4}$ in head; ventrals rather short, not reaching vent by more than an eye's diameter. Length 14 inches.

Catostomus retropinnis, JORDAN, Bull. U. S. Nat. Mus., XII, 178, 1878, Milk River, Montana. (Type, No. 21197. Coll. Dr. Elliott Coues.)

Page 176. After *Catostomus catostomus* add:

282(a). **CATOSTOMUS RIMICULUS**, Gilbert & Snyder.

Head $4\frac{1}{6}$ in body; depth 5; depth of caudal peduncle $2\frac{3}{4}$ in head; eye $7\frac{1}{2}$; dorsal rays 11; scales 18–91–13, before dorsal 42. D. 11; A. 7; pectoral 17. Head as deep as wide. Both lips full, the lobe of lower lip broadly rounded behind, the cleft not nearly reaching base of lip, the portion between mandible and apex of cleft with 4 series of tubercles; tubercles coarse and blunt, becoming reduced in size toward margins of lips, but less so than in related species; upper lip with 5 rows of tubercles. Eyes very small, the front of the eye nearly midway of head; interorbital space convex, $2\frac{3}{4}$ in head. Scales comparatively smooth, gradually growing smaller posteriorly. Dorsal fin inserted midway between end of snout and base of caudal, first ray preceded by 2 short, simple ones; last ray divided to base; length of base of fin equal to the height, which is contained $6\frac{1}{2}$ times in the body; height of anal twice the length of the base, 5 in body; length of pectorals $4\frac{2}{3}$ in body; ventrals $6\frac{1}{2}$ in body; caudal $4\frac{1}{4}$. Color above dusky, the central parts of scales lighter; under parts white; dorsal and caudal fins dusky, others white. This species belongs to the *C. catostomus* type, with very small scales, and is most nearly related to *C. tahoensis.* From the latter it differs in the smaller eye, less deeply cleft lower lip, blunter labial tubercles, larger scales, and the much smaller fontanel, which is reduced in adults to a very narrow linear slit, or more commonly entirely obsolete. Lower portion of the Klamath River basin, northern California. (Diminutive of *rimus*, crevice, from the small fontanel.)

Catostomus rimiculus, GILBERT & SNYDER, Bull. U. S. Fish Comm. 1897 (Jan. 6, 1898), 3, Trinity River, Humboldt County, California. (Type, No. 5654, L. S. Jr. Univ, Mus. Coll. Capt. W. E. Dougherty.)

Page 177. *Catostomus rex* is identical with *Deltistes luxatus* and should be added to the synonymy of that species, p. 183.

The type of *Catostomus labiatus* did not come from Klamath Lake, but from the Sacramento River, at Stockton, California. It is identical with *C. occidentalis.* The species from Klamath Lake has been recently described as

285. **CATOSTOMUS SNYDERI**, Gilbert.

Head $4\frac{1}{3}$ in length; snout $2\frac{3}{10}$ in head, equaling interorbital width; eye $5\frac{3}{4}$. D. 11; A. 7; scales 13 or 14–69 to 77–10 or 11. Mouth very small, the width between angles but $\frac{1}{2}$ length of snout in our largest specimen; greatest width of lobe of lower lip $\frac{2}{3}$ diameter of eye; lower lip deeply incised, with 1 or 2 papillæ between symphysis and base of cleft; upper lip narrow, with 5 or 6 papillæ in a cross series, the uppermost becoming very small; basal portion of the lower lip with coarse tubercles, those toward posterior margin becoming very fine and arranged in evident series separated by grooves. Mucous canals on head forming conspicuous raised ridges with prominent pores, the system much more conspicuously developed than in any related species. Origin of dorsal fin constantly nearer snout than base of caudal; the dorsal fin short, its base not exceeding the height of the longest ray, usually less. In our specimens the pectorals reach scarcely $\frac{3}{4}$ distance to ventrals and the ventrals scarcely $\frac{4}{5}$ distance to vent; the anal may extend beyond base of rudimentary caudal rays. Scales strongly ridged, their margins crenate; the anterior scales are smaller, but do not appear greatly crowded; the average number of tubes in the lateral line is about 73, the number varying from 69 to 77. Dusky, the lower part of sides with coarse black specks, the under parts white; fins all dusky. (Gilbert.)

A larger specimen has been described as follows: Head $4\frac{1}{2}$; depth 4; eye $6\frac{1}{4}$ in head; snout $2\frac{1}{2}$; maxillary $3\frac{1}{2}$; mandible $2\frac{1}{4}$; interorbital $2\frac{4}{5}$; width of mouth $3\frac{1}{2}$ in head, more than $\frac{1}{2}$ length of snout; greatest width of lower lip $\frac{3}{4}$ diameter of eye. D. II, 11; A. 7; scales 13–70–11. Body rather slender; head long, mouth moderate, horizontal; lips thick papillose, the upper with about 4 or 5 rows of papillæ, lower with about 7; lower lip divided nearly to base, leaving only 1 row of papillæ crossing the symphysis; premaxillary not much projecting and not forming a prominent hump; maxillary rather short, not reaching vertical at front of anterior nostril; eye equally distant between snout and posterior edge of opercle; mucous canals on head forming raised ridges, the pores conspicuous. Fins moderate; origin of dorsal a little nearer snout than base of caudal, sixth spine over insertion of ventrals; pectoral $1\frac{1}{4}$ in head, reaching slightly more than $\frac{2}{3}$ distance to ventrals; ventrals not quite reaching vent, the seventh ray longest, $1\frac{2}{3}$ in head; anal long, pointed, reaching to base of caudal, $1\frac{1}{5}$ in head. Scales crowded anteriorly, about 32 transverse rows in front of dorsal, strongly ridged, the margins crenate. (Evermann & Meek.) Length 1 to 2 feet. Klamath Lakes, Oregon; specimens examined from Upper Klamath Lake, Lost River, and Williamson River. (Named for Mr. John O. Snyder, instructor in Zoology in Stanford University.)

Catostomus snyderi, GILBERT, Bull. U. S. Fish Comm. 1897 (Jan. 6, 1898) 3, Upper Klamath Lake, Oregon (Type, No. 48222.· Coll. Gilbert, Cramer & Otaki); EVERMANN & MEEK, Bull. U. S. Fish Comm. 1887, 69.

Page 178. After *Catostomus occidentalis* add:

286(a). CASTOSTOMUS TSILTCOOSENSIS, Evermann & Meek.

Head $4\frac{1}{4}$; depth 5; eye $6\frac{1}{2}$ in head; snout 2. D. 13; A. 7; scales 13–65–8, 34 before the dorsal. Pectoral $1\frac{1}{5}$ in head; longest dorsal ray $1\frac{2}{3}$; base of dorsal $1\frac{2}{5}$; longest anal ray $1\frac{1}{4}$; ventral $1\frac{2}{3}$. Body rather slender, subterete; head small, snout long and pointed; mouth inferior, overhung by the projecting snout; lips rather thin, 1 row of large papillæ on upper lip, and about 2 irregular rows of smaller ones behind or inside of it; lower lip incised nearly to base, 1 or 2 rows of small papillæ across the isthmus; lobes of lower lip moderately long and thin, the bases with papillæ merging into plications toward the tips. Eye quite small, the anterior edge of orbit at middle of head. Top of head flat or very slightly convex between the eyes. Fins small; pectorals short and rounded; ventrals short, rounded,

3030——98

the middle rays but little longer than the others; anal small, somewhat pointed; margin of dorsal somewhat concave; caudal lunate, not deeply forked. Muciferous canals on head not strongly developed. Scales moderately large; lateral line nearly straight, not running upward toward nape. This species differs from *C. occidentalis*, to which it is related, in the smaller head, longer, more pointed snout, smaller eye, larger scales, and its much smaller fins. In *C. occidentalis* the pectoral fins are falcate, while in this species they are more rounded; the ventrals also are less pointed. Length a foot or less. Coastal streams of middle western Oregon; known from Tsiltcoos Lake and the Siuslaw River. (*tsiltcoosensis*, from the type locality.)

Catostomus tsiltcoosensis, EVERMANN & MEEK, Bull. U. S. Fish Comm. 1897 (Jan. 6, 1898), 68, fig. 1, Tsiltcoos Lake, Lane County, Oregon. (Type, No. 48479. Coll. Dr. Seth E. Meek.)

Page 180. The species called *Catostomus fecundus* in the text belongs in the genus *Chasmistes*, to which it should be transferred as *Chasmistes fecundus* (Cope & Yarrow).

Page 182. The species of *Chasmistes* are not confined to the Great Basin. One species (*C. brevirostris*) occurs in the Klamath Lakes basin.

Page 183. The species called *Chasmistes luxatus* in the text belongs to a genus distinct from *Chasmistes*, which may be characterized as follows:

93(a). DELTISTES, Seale.

Deltistes, SEALE, Proc. Cal. Ac. Sci., ser. 2, vol. VI, 1896, 269 (*luxatus*).

This genus is close to *Chasmistes*, agreeing with it in every respect except in the peculiar structure of the gill rakers. In *Chasmistes* they are as in *Catostomus*, while in *Deltistes* they are broad, shaped like the Greek letter $\varDelta$ (delta), and their edges are unarmed and entire. Lower pharyngeals weak, with numerous small teeth. *Deltistes luxatus* (Cope) is the single known species. ($\delta\acute{\epsilon}\lambda\tau\alpha$, the Greek letter corresponding to D.)

After *Chasmistes cujus* add:

297(a). CHASMISTES STOMIAS, Gilbert.

Head $4\frac{1}{3}$; depth $4\frac{1}{3}$; eye 7; snout $2\frac{3}{5}$; maxillary (measured from free end to tip of snout) $3\frac{1}{3}$; mandible $2\frac{1}{2}$. D. II, 11; A. I, 7; scales 13–85–10; interorbital width $2\frac{1}{4}$; vertical depth of head at mandibular articulation $2\frac{1}{6}$. Head small, body heavy forward, the back strongly and regularly arched from snout to origin of dorsal fin, thence declined in a nearly straight line to base of caudal; ventral surface nearly straight. Premaxillary spines strongly protruding, forming a prominently projecting snout; mouth rather small, inclined upward at an angle of about 40°, maxillary scarcely reaching vertical from front of anterior nostril; width of mouth $1\frac{3}{5}$ in snout or $4\frac{3}{5}$ in head; upper lip thin, without papillæ; lower lip thin, interrupted at symphysis, forming narrow lateral lobes, the width of which is about $2\frac{3}{5}$ times in their length; faint indications of a few papillæ; mucous canals forming ridges, the pores conspicuous; gill rakers long, narrowly triangular at the tip when viewed from behind, densely tufted on the anterior edge; fontanel narrow, its length $2\frac{2}{5}$ in the snout, its width about $\frac{4}{5}$ its length. Fins all large; the origin of the dorsal a little nearer tip of snout than base of caudal, the sixth ray over base of ventral, its base $1\frac{3}{4}$ in head, the free edge nearly straight, the last ray $1\frac{3}{4}$ in the first, which is $1\frac{1}{4}$ in head; pectorals scarcely falcate, reaching a little more than $\frac{2}{3}$ distance to base of ventrals, their length $1\frac{1}{4}$ in head; ventrals long, reaching vent, the rays gradually increasing in length from the outer to the seventh and eighth, which are longest, the ninth and tenth being but slightly shorter, the length of the longest ray $1\frac{2}{3}$ in head or about $\frac{1}{5}$ longer than the first; anal long and pointed, the fourth ray longest, reaching base of caudal, $1\frac{1}{5}$ in head; each ray of anal fin with 8 to 12 strong tubercles; caudal lobes about equal, their length $1\frac{2}{5}$ times the middle ray. Length a foot or more. Upper Klamath Lake, Oregon, where it is abundant and of some importance as a food-fish. The Klamath Indian name is K-ahp-tu. ($\sigma\tau o\mu\acute{\iota}\alpha\varsigma$, large-mouthed.)

Chasmistes stomias, GILBERT, Bull. U. S. Fish Comm. 1897 (Jan. 6, 1898), 5, with plate, Upper Klamath Lake, Oregon (Type, No. 48223. Coll. Gilbert, Cramer & Otaki); EVERMANN & MEEK, Bull. U. S. Fish Comm. 1897, 70.

297(b). CHASMISTES COPEI, Evermann & Meek.

Head $3\frac{2}{5}$; depth 4; eye $6\frac{1}{2}$; snout $2\frac{1}{2}$; maxillary (measured from free end to tip of snout) 3; mandible $2\frac{2}{3}$. D. II, 10; A. I, 7; scales 13–80–12; interorbital width $2\frac{1}{4}$; vertical depth of head at mandibular articulation $2\frac{1}{5}$. Head large, cheek very deep, the depth equal to distance from tip of snout to nostril; body stout, back scarcely elevated, caudal peduncle rather short and stout; ventral surface somewhat convex. Premaxillary spines less protruding than in *C. stomias*, not forming a prominent hump; mouth large, inclined upward at an angle of 45°, maxillary not nearly reaching vertical at front of anterior nostril; width of mouth $1\frac{1}{4}$ in snout, or 4 in head; upper lip thin, without papillæ; lower lip thin, entirely without papillæ, interrupted at symphysis, forming rather broad lateral lobes; pores on head very conspicuous; gill rakers larger than in *C. stomias*, broadly triangular at tip when viewed from behind, densely tufted on anterior edge, each appendage more or less bifid and club-shaped, closely resembling those of *C. liorus;* fontanel narrow, its length $2\frac{1}{3}$ in snout, width $\frac{1}{5}$ its length. Fins all small; origin of dorsal a little nearer snout than base of caudal, its sixth ray over base of ventrals, free edge straight, base $2\frac{1}{4}$ in head, last ray a little less than 2 in first, which is 2 in head; pectorals somewhat falcate, reaching slightly more than $\frac{1}{2}$ distance to ventrals, their length $1\frac{2}{3}$ in head; ventrals very short, reaching only $\frac{2}{3}$ distance to vent, free end nearly straight; outer ray longest, $2\frac{2}{7}$ in head; inner shortest, $3\frac{1}{5}$ in head; anal fin short, bluntly pointed, not reaching base of caudal, third and fourth rays longest, $1\frac{1}{4}$ in head; no tubercles on anal rays; caudal lobes equal, length about $1\frac{2}{7}$ times the middle ray. Scales small and crowded anteriorly, about 14 rows downward and backward from front of dorsal to lateral line, 11 vertically upward from base of ventral to lateral line, about 38 oblique series before dorsal; lateral line nearly straight, with about 80 scales. Entire upper parts of head and body, and sides nearly to level of base of pectorals, dark olivaceous; under parts abruptly whitish or yellowish in alcohol; a dark spot in upper part of axil; dorsal and caudal dark; pectorals dark on inner surface; ventrals and anal plain. From *Chasmistes stomias* this species is readily distinguished by its larger head, larger, more oblique mouth, less prominent snout, and very small fins. The differences in the fins are very great, particularly in the ventrals. It differs from *C. brevirostris*, as characterized by Dr. Gilbert, in its much larger, more oblique mouth, the absence of papillæ on the lips, and shorter fins. Length 2 feet. Upper Klamath Lake, Oregon. Klamath Indian name "Tswam." (Named for the late Prof. Edward Drinker Cope, who wrote the first paper on the fishes of Upper Klamath Lake.)

Chasmistes copei, EVERMANN & MEEK, Bull. U. S. Fish Comm. 1897 (Jan. 6, 1898), 70, fig. 3, Pelican Bay, Upper Klamath Lake, Oregon. (Type, No. 48224. Coll. Meek & Alexander.)

Page 205. *Campostoma pricei* can not be distinguished by us from *C. ornatum*. See Rutter, Proc. Cal. Ac. Sci., ser. 2, vol. VI, 1896, 259.

Page 211. After *Algansea tincella* add:

337(a). ALGANSEA TARASCORUM, Steindachner.

Head 3⅔; depth 4⅔; eye less than 5; snout about 4; interorbital 3. D. III, 7; A. III, 6; P. 17; V. 9; scales 84 or 85, 18 or 19. Body stouter than in *A. lacustris*, head shorter, lateral line more decurved and nearer ventral line at middle of body, and scales more numerous. Mouth very oblique, lower jaw not projecting, maxillary not quite reaching vertical at anterior edge of eye. Teeth 4-4, hooked, and with narrow grinding surface. Origin of dorsal in advance of ventrals, equally distant between base of caudal and middle of eye; height of dorsal twice its base. Ventrals not reaching anal fin by an eye's diameter; caudal deeply notched. A dark gray longitudinal band with metallic luster extending from opercle to caudal fin, lying chiefly above lateral line; color otherwise plain. Length 5¼ inches. Lake Pátzcuaro, Mexico. (Steindachner.)

Algansea tarascorum, STEINDACHNER, Einige Fischarten Mex., II, pl. 3, figs. 2-2c, 1895, Lake Pátzcuaro, Mexico. (Coll. Princess Theresa von Bayern.)

Page 218. To the synonymy of *Pimephales notatus* add:

Spinicephalus fibulatus, LE SUEUR, in VAILLANT, Bull. Soc. Philom., VIII, 1896, 29, pl. 26.

Page 225. After *Ptychocheilus oregonensis* add:

358. PTYCHOCHEILUS GRANDIS (Ayres).

(SACRAMENTO PIKE.)

This species differs from *P. oregonensis* principally in the larger size of the scales above the lateral line, the smaller number of rays in the dorsal fin, and the lighter and slenderer pharyngeal bones. Head 3¼ to 3⅓ in length; depth 5 to 5⅒; eye 3⅔ to 4 in head; scales 13 to 16 above lateral line, 70 to 80 transverse rows along lateral line (16 to 18 above lateral line, 69 to 72 transverse rows in *P. oregonensis*). D. 8; A. 8. In other respects similar to *P. oregonensis*. *Ptychocheilus harfordi* is apparently not distinct from *P. grandis*, being based on a specimen with very small scales. *P. grandis* is confined to waters of California, *P. oregonensis* to Washington, Idaho, and Oregon.

Gila grandis, AYRES, Proc. Cal. Ac. Nat. Sci. 1854, 18, San Francisco.

Ptychocheilus major, AGASSIZ, Am. Jour. Sci. Arts 1855, 229, San Francisco.

Ptychocheilus harfordi, JORDAN & GILBERT, Proc. U. S. Nat. Mus. 1881, 72, Sacramento River (Type, No. 27246. Coll. Jordan & Gilbert); JORDAN & GILBERT, Synopsis, 226 1883.

Page 239. After *Leuciscus balteatus* add:

376(a). LEUCISCUS SIUSLAWI, Evermann & Meek.

Head $4\frac{1}{3}$; depth $4\frac{1}{3}$; eye 4; snout $3\frac{1}{2}$; maxillary $3\frac{2}{3}$. D. II, 9; A. II, 12 or 13; scales 11-58-8; teeth 2, 4-5, 2, somewhat hooked. Body rather slender, slightly elevated and somewhat compressed; head small and pointed, cheek not deep; snout pointed, somewhat longer than eye; mouth moderate, somewhat oblique, maxillary just reaching vertical at front of orbit; jaws subequal, the lower sometimes slightly projecting; eye large, not as great as snout. Origin of dorsal fin behind base of ventrals and much nearer base of caudal than tip of snout, the longest ray $1\frac{1}{4}$ in head, greater than base of fin; origin of anal fin under last dorsal ray but 2, its height equal to that of dorsal, its base equal to its longest ray; free edges of dorsal and anal nearly straight; pectoral $1\frac{1}{4}$ to $1\frac{1}{3}$ in head, not reaching insertion of ventrals; ventrals short, $1\frac{1}{2}$ in pectoral, reaching anus; caudal deeply forked; lateral line complete, decurved. Color in spirits, brownish or olivaceous above, middle of side with a broad dark band involving the lateral line anteriorly and posteriorly, but lying chiefly above it mesially; middle of side from gill opening to beneath dorsal fin with a broad rosy band, following closely beneath the lateral line; lower part of sides and under parts silvery, dusted over with fine dark specks; a light yellowish band extending backward from upper posterior border of eye nearly halfway to origin of dorsal fin; cheek with a silvery or golden crescent; top of head dark; opercles dusky silvery; snout dusky; fins plain, dorsal and caudal somewhat dusky. This species is close to *L. balteatus*, but has smaller anal and dorsal fins, a more slender body, smaller and more slender head, and longer, more pointed snout. The extent of variation in proportional measurements and in the number of anal fin rays appears to be much less than in *L. balteatus*. It also resembles *L. cooperi*, but has a much shorter lower jaw and a more pointed snout. Known only from the Siuslaw River and Tsiltcoos Lake, western Oregon, where it is common. (*siuslawi*, of the Siuslaw River.)

Leuciscus siuslawi, EVERMANN & MEEK, Bull. U. S. Fish Comm. 1897 (Jan. 6, 1898), 72, fig. 4, Siuslaw River, Mapleton, Oregon. (Type, No. 48480. Coll. Dr. Meek.)

Page 240. After *Leuciscus elongatus* add:

378(a). LEUCISCUS NACHTRIEBI, Cox.

Head $4\frac{1}{4}$ to $4\frac{1}{2}$; depth 5 ($4\frac{1}{8}$ to $5\frac{1}{2}$); eye 4; snout $4\frac{2}{3}$. D. 8; A. 8. Body rather heavy, not greatly compressed; back slightly elevated, its curve a little greater than that of the belly; caudal peduncle rather stout, its depth $\frac{1}{2}$ the length of the head. Head rather short, not any more compressed than the body, upper surface slightly flattened; snout quite blunt in mature individuals, its length $1\frac{1}{6}$ times width of eye; mouth not very large, but little oblique, lower jaw included; maxillary scarcely reaching to front of orbit; pharyngeal teeth 2, 4–5, 2. Dorsal fin inserted nearer base of caudal than tip of snout, also slightly back of ventrals; caudal fin forked; anal slightly smaller than dorsal; ventrals small, not reaching vent by $\frac{1}{3}$ their length; pectorals inserted rather high, not reaching the ventrals by $\frac{3}{4}$ their length; scales small, 12–72–9, lateral line complete on mature individuals, decurved, the pores extending on head in several lines, 1 passing back of eye, another down to nostril. General color dusky, darkest on back; sides above lateral line dull silvery, below lateral line light silvery; a faint dark dorsal band in some specimens, in others absent; no black lateral band, but some specimens have a very faint dusky shade along lateral line; no light stripe above lateral line; upper portion of opercles with a dusky shade, lower part bright silvery; upper part of head dark-colored; all the above colors typical in the young as well as adults. Length 4 inches. *L. nachtriebi* differs from *L. neogæus* in having a well-developed lateral line, a smaller eye, fewer scales, less oblique mouth, a shorter maxillary, and in being a larger fish and differently colored. It differs from *L. elongatus* in having a smaller mouth, the lower jaw never projecting, head less pointed, a shorter maxillary, finer scales, and the absence of the black lateral band. Lakes of northern Minnesota; at present known from Mille Lacs, Man Trap, Mud and Elbow lakes. (Named for Prof. Henry F. Nachtrieb, State zoologist of Minnesota.)

Leuciscus nachtriebi, Cox, Rept. U. S. Fish Comm. 1894 (Dec. 14, 1896), 615, Mille Lacs Lake, Aikin County, Minnesota. (Type, No. 47688. Coll. Minn. Nat. Hist. Surv.)

Page 241. To the synonymy of *Leuciscus neogæus* add:

Cyprinus burtonianus, LE SUEUR in VAILLANT, Bull. Soc. Philom., VIII, 1896, 28, with plate, Burton Mine, Missouri.

Page 244. *Leucos* and *Myloleucus* can not be maintained as subgenera, the characters of the teeth not being constant.

The following notes on *Rutilus olivaceus* as seen at Emerald Bay, Lake Tahoe, may prove useful.

385. RUTILUS OLIVACEUS (Cope).

(TAHOE CHUB.)

This species is very different from *Rutilus symmetricus*, looking like *Leuciscus lineatus*. Very common; reaches 2 to 3 pounds weight; devours eggs of trout. No doubt the records of *Leuciscus lineatus* (*atrarius*) from Lake Tahoe belong to this species. Head 4; depth $4\frac{1}{5}$ to $4\frac{1}{3}$. D. 8; A. 8; scales 11–56–6; teeth always 5–5, with broad grinding surface. Body oblong, moderately compressed, the back somewhat elevated anteriorly in old examples. Head conical, rounded above; eye moderate, $1\frac{1}{2}$ in snout (6 inches long), 5 in head; about as long as maxillary. Mouth terminal, very oblique, the lower jaw included; the snout not prominent; the short maxillary not reaching eye. Dorsal high and pointed; anal short, rather high; pectoral long, reaching $\frac{3}{4}$ distance to ventrals, which reach vent; ventrals below front of dorsal, which is behind middle of body. Scales with edges largely exposed; lateral line running low, complete. Dusky olive above and on sides to level of ventrals, with brassy luster everywhere; middle of belly only white, a pale yellowish area between pectorals and ventrals; head brassy, dusky above, closely dotted above and on sides; body everywhere closely dotted with black, except on middle line below; fins all dusky, with dark points. This species is well separated from all the *R. symmetricus* tribe.

Page 247. *Luxilinus occidentalis* is the young of *Lavinia exilicauda*, Baird & Girard (p. 209), and must be placed in the synonymy of that species. *Luxilinus* is a pure synonym of *Lavinia*.

Page 249. Under *Opsopœodus bollmani*, for "Buckland Creek" read "Buckhead Creek."

Page 254. For *Azteca*, line 22, substitute *Aztecula*, Jordan & Evermann, new subgenus. The former name is preoccupied by *Azteca*, Forel, 1878, a genus of ants. The same substitution to be made in the key on page 255 and on page 258.

Page 260. Before *Notropis cayuga* insert:

404(a). NOTROPIS WELAKA, Evermann & Kendall.

Head $4\frac{1}{3}$; depth 5; eye 3 in head; snout $3\frac{1}{4}$. D. 8; A. 8 or 9; scales 6-35-3; teeth 4-4, hooked. Body rather slender, moderately compressed; head short, snout bluntly pointed; mouth moderate, somewhat oblique, lower jaw slightly included, maxillary scarcely reaching front of eye; premaxillaries protractile. Eye large; posterior edge of pupil at middle of longitudinal length of head; interorbital width greater than eye; caudal peduncle long and slender. Dorsal fin inserted well behind base of ventrals, a little nearer base of caudal than tip of snout, its longest rays shorter than head, but slightly longer than longest anal rays; anterior dorsal and anal rays longest; pectoral $1\frac{1}{3}$ in head; ventrals reaching origin of anal; caudal deeply notched, the lobes long and pointed. Scales large, lateral line incomplete, developed only on 6 to 10 scales. Back olivaceous; side with a broad black band extending from snout through eye, and ending in a rather distinct black spot on base of caudal, the black spot in some specimens (probably mature males) surrounded by orange; the black line bordered above by a narrow orange or reddish line, less distinct, or even whitish, in females and immature individuals; under parts plain; fins all plain; dorsal and caudal somewhat dusky; dusky specks on body along base of anal and under side of caudal peduncle; lower jaw tipped with dusky. This species resembles *Notropis anogenus*, but differs in having the mouth somewhat larger and less oblique, the lower jaw more included, the body more slender, the lateral line less developed, the dorsal fin more posterior, and the anal rays more numerous. It was found in considerable abundance in the St. Johns River, near Welaka, Florida. (*welaka*, from the type locality.)

Notropis welaka, EVERMANN & KENDALL, Bull. U. S. Fish Comm. 1897 (Feb. 9, 1898), 126, pl. 6, fig. 2, St. Johns River, near Welaka, Florida. (Type, No. 48786. Coll. Dr. W. C Kendall.)

Page 262. After *Notropis blennius* add:

408(a). NOTROPIS BUCHANANI, Meek.

Head 4; depth 4. D. 8; A. 8; scales 6-31-2; teeth 4-4. Body rather robust, back considerably elevated, snout blunt, mouth small and nearly horizontal. Snout short, about $\frac{2}{3}$ diameter of eye. Preorbital bone slightly longer than broad. Eye moderate, 3 in head. Lateral line compiete, or nearly so; about 12 scales in a series before dorsal fin. Dorsal fin slightly nearer tip of snout than base of caudal; pectorals reaching ventrals; ventrals reaching anal. Color light olivaceous, a faint silvery lateral band; no dark lateral band or black caudal spot. This species belongs to the *N. blennius* type. It is a smaller species, lighter in color, and has fewer scales in the lateral line. Poteau River, Arkansas. (Named for Dr. John L. Buchanan, president of the Arkansas Industrial University.)

Notropis buchanani, MEEK, Bull. U. S. Fish Comm. 1895 (April 13, 1896), 342, small creek near Poteau, Indian Territory. (Type, No. 47532. Coll. Dr. Meek.)

Page 267. Under *Notropis nux; nuece*, not *neche*, is nut in Spanish.

Page 274, line 11, for *luxoides*, read *luxiloides*.

Page 287. After *Notropis lutipinnis* insert:

466(a). NOTROPIS CHAMBERLAINI, Evermann, new species.

Head $4\frac{1}{3}$; depth $4\frac{1}{5}$; eye 4; snout 4. D. 7; A. 9; scales 7-39-3, about 15 before the dorsal. General form much like that of *Hybognathus;* body only moderately compressed, dorsal and ventral outlines slightly arched; head rather small, pointed; mouth small, a little oblique, the maxillary scarcely reaching anterior border of orbit, lower jaw slightly included; snout equal to eye; eye in axis of body. Fins all rather small; origin of dorsal slightly behind vertical at insertion of ventrals; free edge of dorsal fin somewhat concave, the anterior rays about equal to length of head; pectoral short, slightly falcate, the longest rays about $1\frac{3}{4}$ in head; ventrals shorter than pectoral, barely reaching vent; anal similar to dorsal, the rays shorter; caudal widely forked, the middle rays $2\frac{1}{2}$ in the outer, the lobes as long as head, the lower lobe slightly longer than the upper. Scales moderately imbricated, the exposed portions not deeper than long; lateral line complete, somewhat decurved. Teeth 2, 4-4, 2 or 1, rather weak, hooked, and with small grinding surface. Intestine short; peritoneum silvery. General color light straw; middle of side with a broad, well-defined silvery band from upper end of gill opening to middle of base of caudal fin, the anterior half lying wholly above the lateral line, the posterior portion lying partly below it; this silvery band bounded above by a narrow dark border; cheeks and opercles silvery; a darkish band along median line of back; fins all plain straw color or pale lemon. Fourteen examples of this species, 2 to 3 inches in length, were obtained from the Atchafalaya River at Melville, Louisiana, by Mr. Fred M. Chamberlain, for whom the species is named.

Notropis chamberlaini, EVERMANN MS., Atchafalaya River, Melville, Louisiana. (Type, No. 48901.)

Page 291. *Notropis scopifer*, Eigenmann & Eigenmann, is identical with *Notropis hudsonius selene* (Jordan), (p. 269), and should be omitted.

Page 294. After *Notropis dilectus* insert:

487(a). NOTROPIS LOUISIANÆ, Evermann, new species.

Head 4⅔; depth 5¼; eye 3; snout 3. D. 7; A. 11; scales 7-37-3, 19 or 20 before the dorsal. Teeth 1, 4-4, 2, little hooked. Body long and slender, back not arched; head short, but pointed; mouth rather large, oblique, maxillary scarcely reaching orbit, lower jaw somewhat included; eye large, equal to or greater than snout. Fins rather small; origin of dorsal far behind insertion of ventrals, its longest rays 1⅔ in head; pectorals short, their length equal to height of anal; ventrals very short, 2 in head; caudal deeply forked. Scales firm, moderately imbricated; lateral line complete, gently decurved. Color pale; side with a faint plumbeous band; back and upper part of sides with numerous dark specks chiefly on the margins of the scales, thus forming cross-hatchings; a narrow dark vertebral band on caudal peduncle; peritoneum silvery, with numerous minute round black specks. Length 2⅓ inches. This species resembles *Notropis dilectus*, but has a much smaller mouth, blunter snout, and in being less silvery along the side. Known only from the Atchafalaya River, Louisiana.

Notropis louisianæ, EVERMANN MS., Atchafalaya River, Melville, Louisiana. (Type, No. 48902. Coll. Fred M. Chamberlain.)

Page 348. *Anguilla chrysypa* is abundant in the Gulf of St. Lawrence, according to Dr. Wm. Wakeham.

Page 355. The original type of *Congermuræna* is *C. habenata*, Kaup, a species with blunt or granular teeth. The American species all belong to a distinct genus, *Congrellus*, Ogilby (type *balearica*), distinguished by the villiform teeth. These genera are charactered by Mr. Ogilby in a paper as yet unpublished.

Pages 356 and 357. In *Congermuræna flava* the upper jaw projects far beyond the lower. By a slip in the original description the reverse is said to be the case.

Page 359. *Murænesox coniceps* is called Culevra Blanca at Mazatlan, and reaches a length of 7 feet.

Page 368. *Avocettina gilli*, Bean, should probably stand as a species distinct from *Avocettina infans*. The description in the footnote on page

368 is sufficiently full. See Jordan, Proc. Cal. Ac. Sci., ser. 2, vol. VI, 1896, 206, pl. 21.

Page 369. No. 604, *Labichthys elongatus*, is a true *Avocettina*, having the vent far behind the head. It should stand as—

602(a). **AVOCETTINA ELONGATA** (Gill & Ryder).

Page 376. After *Myrichthys tigrinus*, Girard, add:

615(a). **MYRICHTHYS XYSTURUS**, Jordan & Gilbert.

Teeth all more or less blunt and granular; a band of 3 or 4 series on each side of lower jaw; a band of 2 rows on each side of upper jaw; vomer with a long series divided into 2 for about ⅓ its length. Anterior nasal tubes conspicuous, turned downward. Eye 2½ in snout; front of eye above middle of gape, the length of which is a little more than ⅓ of head; the angle of mouth well behind eye. Interorbital width about ⅔ length of the rather long and slender snout, which projects much beyond lower jaw, the tip of the latter about reaching middle of snout. Length of head contained 4¼ times in that of trunk; head and trunk together shorter than tail, and contained 2⅛ to 2¼ times in total length. Pectoral very small, its length about equal to depth of gill opening. Dorsal beginning close behind nape, much in front of gill opening; fins low; tail pointed, the tip sharp. Color light olive; sides each with 3 series of large round brown spots, those of the 2 upper series of equal size, those of lower scarcely ½ as large, faint, and often obsolete anteriorly; the spots irregular in their arrangement, those of the upper series usually twice as numerous as those of the next; those of the upper series along base of dorsal fin extending partly on the base of the fin; lower series of spots along base of anal, some of them extending on the fin or even entirely upon it; on the belly are sometimes small dark spots, scarcely arranged in series; dorsal fin with a terminal series of dark spots, which are partly confluent, the fin narrowly margined with white; anal reddish, with a lighter margin; pectoral with a blackish blotch; head covered with round black spots, which become smaller and more numerous toward the snout; lower jaw with dark spots; iris light yellow. Pacific coast of Mexico; common among the rocks about Mazatlan. (ζύστον, a spike; οὐρά, tail.)

A species distinct from *M. tigrinus*, which is known only from the original type figured by Jordan & Davis, and described in the text of Part I of this work, page 376. This specimen, said to be from "Adair Bay, Oregon," may not be American, as there is no such bay in Oregon, and no second specimen of the true *Myrichthys tigrinus* has been found anywhere.

Ophichthys xysturus, JORDAN & GILBERT, Proc. U. S. Nat. Mus. 1881, 346, Mazatlan, Mexico. (Type No. 28142. Coll. Dr. Gilbert.)

615(b). **MYRICHTHYS PANTOSTIGMIUS**, Jordan & McGregor.

Head 3⅔ in trunk; head and trunk 1⅓ in tail; cleft of mouth 3 in head; eye 2⅔ in snout, which is 5 in head; pectorals 2 in snout; anterior nasal tube equal to the eye. Color olivaceous, with distinct rows of roundish blackish spots, some oblong, smaller on head and covering the whole belly; 39 spots in the dorsal row, these spots usually alternating each with its fellow on the other side of dorsal, sometimes opposite; spots of second row usually opposite; spots of third row smaller and more numerous, extending from the cheeks to opposite the vent, thence running along base of anal, not running on fin, most of the spots of this row little more than ⅓ length of snout; 2 rows of smaller spots along belly from gill opening to front of anal; spots on nape rather large, on head larger and more numerous than in *M. xysturus;* pale color of head reduced to reticulations; chin and throat spotted as much as head; no pale centers to any of the spots; dorsal without spots or with only a few, which come up from back; from beginning to end the dorsal has a broad black margin about ⅓ height of fin; anal mostly pale, but toward tip having some black markings; pectoral with upper half jet-black, a white margin posteriorly, a small black spot in lower corner. This species is distinguishable from all others by the great number of spots of small size and without pale centers; the black edge of dorsal; the black spot on the rather large pectoral, and especially by having the belly spotted as much as the other parts. Clarion Island. One specimen, about a foot long, known. (πᾶς, whole, entire, all; στίγμα, spot.)

Myrichthys pantostigmius, JORDAN & McGREGOR, Rept. U. S. Fish Comm. 1898, pl. 4, Clarion Island. (Type, No. 5710, L. S. Jr. Univ. Mus. Coll. R. C. McGregor.)

Page 377. After *Pisoodonophis cruentifer* add:

618(a). PISOODONOPHIS DASPILOTUS, Gilbert, new species.

Brownish above, gray below, the head and body usually thickly covered with black spots smaller than the eye; these are smaller and more numerous on the head, fewer and fainter on the lighter interior surface, and become indistinct or entirely disappear on the terminal portion of tail. In 1 specimen the head and trunk are spotted and the entire tail unicolor. In another no spots are present, the upper parts being a uniform dark brown, the under parts lighter brown, a few dark freckles only being present on sides of head. In all specimens the snout and lower jaw are blackish. The anus is near the middle of the total length, sometimes nearer the tip of snout, sometimes nearer tip of tail. The cheeks are not greatly swollen. The gape extends behind the eye, its length, measured from tip of lower jaw to angle of mouth, being contained 4⅔ to 4¼ in head. The snout projects beyond the lower jaw for a distance about equaling diameter of orbit. Eye 2 to 2⅛ in snout, 1⅔ to 2⅛ in interorbital width. Tubes of anterior nostrils about ⅓ diameter of eye, directed downward near tip of snout. Posterior nostrils under front of eye, concealed in the upper lip as usual. Teeth all bluntly conic, in rather wide bands on jaws and vomer; they are usually not disposed in regular series within the bands, but each band has about the width of 4 series, and these are sometimes distinguishable. The mandibular teeth become larger on approaching the symphysis, those at point of mandible and those on head of vomer being much the largest teeth present. The patch on shaft of vomer tapers backward to a point considerably behind angle of mouth. Origin of dorsal entirely behind tip of pectorals, its distance from snout $\frac{1}{8}$ to $\frac{1}{5}$ greater than length of the head. The tip of the tail is compressed, acute, horny, used for defense. Pectoral very short, from a wide base which slightly exceeds length of gill slit. The fin rapidly narrows downward, the longest portion containèd 12 to 14 times in length of head. The width of gill slit is about $\frac{1}{8}$ head.

The following table gives measurements of 4 specimens in millimeters:

Total length.	Head and trunk.	Tail.	Head.	Gape.	Eye.	Interorbital width.	Width at cheeks.	Length of snout.	Projection of snout beyond mandible.	Length of pectoral.	Basal width of pectoral.	Distance from snout to dorsal.	Vertical height of dorsal.	Depth of body.
362	177	185	38	8	3	5	$9\frac{1}{2}$	$6\frac{1}{4}$	$2\frac{3}{4}$	$3\frac{1}{2}$	5	48	3	12
401	203	198	48	$10\frac{1}{2}$	3	7	11	$7\frac{1}{2}$	$3\frac{1}{2}$	4	6	53	4	$14\frac{1}{2}$
492	248	244	52	11	$3\frac{1}{2}$	$7\frac{1}{2}$	$16\frac{1}{2}$	$8\frac{1}{4}$	$3\frac{1}{2}$	$3\frac{3}{4}$	$6\frac{1}{4}$	68	$5\frac{1}{2}$	18
494	255	239	56	12	$4\frac{1}{4}$	7	16	$8\frac{1}{2}$	$4\frac{1}{2}$	4	6	$68\frac{1}{2}$	$5\frac{1}{2}$	16

Four specimens were secured, 3 obtained in brackish water at the mouth of a small stream which empties into Panama Bay, the fourth in a freshwater pond at Miraflores. There is some reason to suppose that they burrow in the mud.

Pisoodonophis daspilotus, GILBERT, Fishes of Panama, MS. 1898, Panama.

Page 382. *Murœna ophis*, Linnæus, is without much doubt the original *Ophichthus havannensis*. The species would therefore stand as—

626. OPHICHTHUS OPHIS (Linnæus).

Page 396. *Sidera castanęa*, Jordan & Gilbert, should be removed from the synonomy of *Lycodontis funebris*. It is apparently a valid species and should be inserted as—

650(a). LYCODONTIS CASTANEUS (Jordan & Gilbert).

(MORENA PRIETA.)

Tail about as long as rest of body, or slightly longer. Head $2\frac{1}{2}$ in length of trunk; cleft of mouth wide, $2\frac{1}{8}$ to $2\frac{1}{2}$ in head. Teeth everywhere uniserial or nearly so, those on sides of mandible small, compressed, close set, subtriangular, directed backward, about 18 in number on each side; mandible with about 4 large canines anteriorly; upper jaw with the teeth partly in 2 series, some of the teeth being movable, the other mostly stronger, caninelike, especially anteriorly; front of vomer with 2 very long, slender canines, behind them a single series of small teeth; teeth all entire. Eye large, slightly nearer tip of snout than angle of mouth, its diameter 2 to $2\frac{1}{2}$ in snout; gill opening $\frac{1}{8}$ wider than orbit; tube of anterior nostril short, less than $\frac{1}{2}$ diameter of orbit; posterior nostril without tube; occiput not especially elevated, the anterior profile scarcely concave (perfectly straight in young 2 feet long). Dorsal fin commencing much in advance of gill opening, becoming unusually high posteriorly, where its vertical height is more than ⅓ greatest depth of body; the length of the longest ray more than greatest depth of body. Color light brownish chestnut, slightly paler on abdomen; no spots or bands anywhere; fins without dark margins; no dark spot on gill opening or at angle of mouth; no black about eye; head without conspicuous pores. The specimen here described is 44 inches in length; others about 2 feet in length agree very closely. This enormous eel is very common among the rocks about Mazatlan, where it reaches a length of 6 feet. It is close to the West Indian species, *L. funebris*, but the colors are not the same, *funebris* being a greenish black, while *castaneus* is a purplish chestnut, without shades of olive or green. (*castaneus*, chestnut.)

Sidera castanea, JORDAN & GILBERT, Proc. U. S. Nat. Mus. 1882, 647, Mazatlan. (Type, Nos. 28246, 29535 and 29591. Coll. C. H. Gilbert.)

After *Lycodontis mordax* add:

649(a). LYCODONTIS PICTUS (Ahl).

Head 4 in trunk; tail about as long as body; eye 2½ in snout, situated midway between snout and angle of mouth; cleft of mouth 2⅔ in head; snout 5⅔ in head; anterior nasal tube 5 in snout; gill opening 11 in head. Teeth in each jaw in a single series; palatine series either parallel with these or divergent; no distinct canines; teeth comparatively small; anterior vomerine 1 or 2 in number, bluntish and conical; posterior vomerine teeth rather blunt. Anterior nasal tubes moderate. Dorsal low anteriorly and beginning in front of gill opening. Color brownish gray or purplish, everywhere covered with small purplish black spots, which are not confluent; in the adult the spots are arranged in roundish or ring-like blotches on the sides; fins colored like body, without dark edges; young pale with black ring-shaped markings; variation in color and form of markings numerous. East Indies; everywhere common. East to offshore islands of Mexico. Two specimens, about 3 feet in length, taken at Clarion Island by Mr. R. C. McGregor. (*pictus*, painted.)

Muræna picta, AHL, le Muræna et ophichtho, VI, 6, tab. 2, f. 2; GÜNTHER, Cat. Fish, VIII, 116.
Gymnothorax pictus, BLEEKER, All. Ichth., Muræna, 87, tab. 26, 28, 29, 45.
Murænophis pantherina, LACÉPÈDE, Hist. Nat. Poiss., V, 628, 1803.
Muræna variegata, QUOY & GAIMARD, Voy. Uranie, Zool., 246, pl. 52, f. 1.
Muræna lita, RICHARDSON, Voy. Erebus and Terror, 84, Moluccas.
Muræna siderea, RICHARDSON, Voy. Erebus and Terror, 85, pl. 48, f. 1–5, Australia.
Muræna pfeifferi, BLEEKER, Nat. Tyds. Ned. Ind., V, 173, Celebes.
Sidera pfeifferi, KAUP, Apodes, 70.

Page 401. After *Muræna argus* insert:

660(a). MURÆNA CLEPSYDRA, Gilbert, new species.

Closely related to *M. insularum* and *M. argus*, from the tropical Pacific, differing from both in color. Nostrils tubular, of almost equal length.

Mouth closing completely, the teeth entirely concealed by the lips. Gape straight, horizontal, extending to well behind the eyes, $2\frac{1}{3}$ to $2\frac{3}{4}$ in head. Teeth in jaws large, compressed, and wide at base, tapering uniformly to an acute point, directed backward, close set, everywhere uniserial; those in sides of mandible noticeably smaller than those of upper jaw, the teeth in both jaws increasing in size anteriorly; as many as 18 or 20 teeth may be present in the half of either jaw, but many of them are usually wanting, leaving gaps in the series; a single row of small teeth on shaft of vomer, beginning opposite front of eye; head of vomer with 2 long canines, larger than any of the other teeth, one or both of these usually wanting in larger specimens, having apparently fallen out. Head 2 ($1\frac{11}{12}$ to $2\frac{1}{12}$) in trunk; head and trunk $1\frac{1}{3}$ to $1\frac{1}{7}$ in tail; depth at anus approximately $\frac{1}{2}$ length of head; eye small, its diameter contained 12 to 16 times in head; snout 5 to $5\frac{1}{3}$. Dorsal beginning on the head, its distance from snout $1\frac{1}{3}$ to $1\frac{1}{2}$ in head. Color dark brown, lighter on belly, dull whitish on under side of head; head, body, and fins closely covered with white spots, those on posterior parts larger, with some smaller ones intermingled, the larger spots with a more or less evident central constriction which makes them hourglass-shaped; toward the head the spots become very small and crowded, not more than $\frac{1}{2}$ as large as pupil; fins indistinctly light margined; a large elliptical jet-black blotch surrounds the gill slit, distinctly margined by a series of confluent white spots; the longitudinal diameter of the blotch is contained 5 to $5\frac{1}{2}$ times in the length of the head; angle of mouth with a small black blotch, often obscure, preceded by a pale spot on mandible; the throat is marked with a number of parallel lengthwise folds, the bottom of each fold with a dark line.

The following table gives measurements in millimeters of 5 specimens:

Total length.	Head and trunk.	Tail.	Head.	Gape.	Snout.	Eye.	Depth at anus.	Distance from snout to origin of dorsal.
675	311	364	106	45	$20\frac{1}{2}$	7	59	72
630	289	341	96	38	$18\frac{1}{2}$	$6\frac{1}{4}$	$52\frac{1}{2}$	70
612	287	325	98	39	19	$6\frac{1}{4}$	47	64
473	203	270	66	28	13	$5\frac{3}{4}$	40	50
397	177	220	58	21	11	$4\frac{1}{2}$	27	39

This species is abundant at Panama, where it is frequently brought to market. About 25 specimens were seen during our visit, all essentially alike in coloration. The type is 397 millimeters long (see table of measurements), and has the spots on body less numerous than in larger specimens. (*clepsydra*, *κλεψύδρα*, an hourglass, from its markings.)

Muræna clepsydra, GILBERT, Fishes of Panama, MS. 1898, Panama.

Page 410. It is probable that several species are confounded under the name *Elops saurus*. According to Ogilby the Australian species has only 63 vertebræ.

Page 411. 199. Genus ALBULA, Bloch & Schneider.

The proper binomial authority for this generic name, as well as for the names *Synodus, Umbra,* and *Anableps,* is Scopoli, as Dr. Gill informs us. These pre-Linnæan names, with others, were first used in binomial nomenclature as names of genera by Scopoli, Introd. His. Nat. 1777, pp. 449 (*Synodus*) and 450 (*Albula, Umbra, Anableps*). The genera should then stand as follows:

Page 411. 199. ALBULA (Gronow) Scopoli.

Page 533. 248. SYNODUS (Gronow) Scopoli.

Page 623. 298. UMBRA (Krämer) Scopoli.

Page 684. 312. ANABLEPS (Artedi) Scopoli.

Page 414. To the description of *Chanos chanos* the following may be added:

The skeletal peculiarities of *Chanos* are numerous and remarkable, many archaic characters persisting. The following account of the skeleton has been prepared by Mr. Starks:

SKELETON OF CHANOS CHANOS.

a. Cranium:

The frontals are very large, covering nearly the whole top of the head, and extending over the dorsoanterior part of the parietals, supraoccipital and the parotic process. On the side of the skull there is an area bounded by the supraoccipital, the opisthotic and the sphenotic, which is not ossified, but is composed of cartilage. Between the frontals, at about their middle, there is a place in which the bone is fibrous and largely cartilaginous; it is easily broken through. The basal cavity under the brain cavity is large. On the upper part of the operculum is a large scale-like bone. The suborbitals are well developed and plate-like, extending back nearly to the posterior edge of the preopercle.

b. Vertebral column:

There are 42 vertebræ in the spinal column. The first vertebra is coossified to the skull, and apparently bears no ribs; the second vertebra supports a pair of very small, slender ribs, which articulate directly with the sides of the vertebra; the third vertebra supports the first pair of large ribs; they are articulated with the transverse processes. The first 14 or 15 neural spines and pairs of transverse processes are articulated with the vertebræ by sutures; they are easily separated from the vertebræ by boiling or maceration. The vertebræ gradually increase in size and reach their largest size about ⅔ of the distance from the anterior to the posterior end of the spinal column, where they are 3 or 4 times the size of the anterior ones. This character is more marked in the adult than in the young.

c. Shoulder girdle:

The shoulder girdle is exceedingly well braced, the post-temporal is widely forked, and strongly articulated to the epiotic processes of the skull. The supraclavicle is long and slender, its posterior face is hollowed out and attached some distance from the upper end of the clavicle, which projects upward. This projecting upper end of the clavicle is braced to the skull by two long bones. The first bone is very slender, at its anterior end it is connected to the exoccipital; near its middle it is connected with the posterior end of the post-temporal, at which point it turns at a sharp angle and runs to the clavicle. The second bone is much larger; it is articulated to the basioccipital. Its posterior edge is nearly straight for its whole length, but its anterior edge is produced and much swollen near its middle, and joins the post-temporal over the first bone, then runs to the upper end of the clavicle. The inner part of the clavicle and the coracoid are thin and pierced by many holes, so that the bone in places is little more than network. The hypercoracoid has a very large foramen; at its posterior edge is a projection which supports a thin bone, probably a dermal bone. The mesocoracoid is well developed. There are 4 actinosts; the first is long, but they rapidly decrease in size to the fourth, which is short and triangular. The first ray of the pectoral is large at the basal end, and hollowed out; it works directly on the hypercoracoid.

d. Branchial apparatus:

The branchial apparatus is peculiar in the adult, in having gill rakers somewhat resembling the filaments of a feather, on both sides of each arch and on the basibranchial. They meet in a middle line between the arches and unite, forming a continuous lattice-work screen, through which nothing but the very smallest bodies can pass. The pharyngeals have no teeth, but have gill rakers similar to those on the arches; they are inclosed in sac-like projections on each side. This description is taken from the skeleton of a large specimen 4 feet long. The gill rakers are not united in young individuals. ·

e. Other parts:

The septæ between the myotomes are ossified about ½ an inch under the skin, forming long, slender rays of bone. There is an upper series running from the middle of the sides up on the back, and a lower series from the sides down on the belly, and form a sort of a basket around the the body. Those below have a single branch near the middle of each, the ones above have 2 branches each; these branches are lost toward the posterior end. These bones are not present in the young. The large caudal fin is attached very firmly to the hypural, the long rays of each lobe join the hypural at about the same oblique angle, the base of each ray is deeply divided and articulated immovably with the hypural. The middle short rays are all nearly horizontal and are much less firmly fastened. The first interspinal ray of the anal is hollow and cone-shaped, the posterior end of the air bladder runs into it as in the genera *Eucinostomus* and *Calamus*. The scales are very thick and closely imbricated; the skin anteriorly is ¼ inch thick. (Jordan, Fishes of Sinaloa, 404-409.)

Page 417. After *Dorosoma petenense* add:

202(a). SIGNALOSA, Evermann & Kendall.

Signalosa, EVERMANN & KENDALL, Bull. U. S. Fish Comm. 1897 (Feb. 9, 1898), 127 (*atcha-falayæ*).

Body short, deep, and compressed, the form somewhat elliptical; ventral outline more strongly curved than the dorsal; head rather large, snout sharp and pointed, not tumid; mouth small, oblique, the lower jaw scarcely included; maxillary of 3 pieces, broad and curved, but without notch in the outer margin as in *Dorosoma;* caudal peduncle short and deep. Branchiostegals 5; pseudobranchiæ large; gill rakers short and very numerous, about 340 in number. No teeth; adipose eyelid present; stomach gizzard-like; scutes about 17 + 10. Last ray of dorsal very long and filamentous. This genus is allied to *Dorosoma*, from which it is plainly distinguished by the absence of the notch in the maxillary, the more pointed snout, the less-included lower jaw, the shorter anal fin, larger scales, and the fewer scutes. It differs from *Alosa* in the very numerous gill rakers, the character of the dorsal fin, and in other respects. (*signum*, a flagstaff or pole; *Alosa*, the shad; a reference to the long dorsal ray.)

679(a). SIGNALOSA ATCHAFALAYÆ, Evermann & Kendall.

Head $3\frac{2}{3}$; depth $3\frac{1}{4}$; eye $3\frac{1}{4}$ in head; snout $5\frac{1}{4}$; maxillary $3\frac{1}{3}$. D. I, 12; A. I, 24; scales 42-15; scutes 17 + 10. Body oblong-elliptical, compressed, the back in front of dorsal narrow; ventral edge sharp, serrate; head small, mouth terminal, oblique, lower jaw slightly included; snout rather pointed, not blunt, as in *Dorosoma cepedianum;* maxillary in 3 pieces, long and curved, reaching vertical at front of pupil, the outer edge not notched; no teeth. Caudal peduncle short, compressed, and deep. Origin of dorsal fin over base of ventrals, much nearer tip of snout than base of caudal, the last ray filamentous, about $\frac{1}{4}$ longer than head and nearly reaching base of caudal; the first dorsal ray about 2 in the last one; pectoral $1\frac{1}{4}$ in head, reaching base of ventrals; ventrals short, reaching only halfway to vent, their length $1\frac{2}{3}$ in pectorals; anal rays short, base of fin $1\frac{1}{4}$ in head; scutes moderate; caudal widely forked, the lower lobe the longer; scales large, thin, deciduous, somewhat crowded anteriorly; accessory scales at bases of pectorals and ventrals; base of caudal with small scales. Color bluish black or dark olivaceous on back and sides to level of the jet-black humeral spot; rest of sides and under parts bright silvery; dorsal and caudal dusky; other fins plain. The cotypes from Grand Plains Bayou are 2 females with ripe roe. They are $4\frac{1}{4}$ and $5\frac{1}{2}$ inches long, respectively, and differ from the types only in the deeper body and the much darker coloration of the upper parts.

3030——99

The amount of variation in this species, shown by the material at hand, is exhibited in the following table:

No.	Head.	Depth.	Eye.	Snout.	Max.	Dorsal.	Anal.	Scutes.	Scales.	Locality.
1	$4\frac{1}{6}$	3	$3\frac{3}{4}$	$5\frac{1}{4}$	$3\frac{1}{5}$	I, 12	I, 24	16 + 11	40–15	Grand Plains
2	4	$2\frac{5}{6}$	$3\frac{1}{4}$	$5\frac{1}{3}$	$3\frac{1}{5}$	I, 12	I, 24	16 + 11	42–15	Bayou, Miss.
3	4	3	$3\frac{4}{5}$	$5\frac{1}{3}$	$3\frac{3}{4}$	I, 12	I, 24	16 + 10	42–14	
4	$3\frac{3}{5}$	$3\frac{1}{7}$	$3\frac{1}{2}$	$5\frac{1}{2}$	$3\frac{1}{5}$	I, 12	I, 24	17 + 10	42–15	Melville, La.
5	$3\frac{3}{5}$	3	4	5	$3\frac{2}{5}$	I, 12	I, 24	16 + 11	43–15	
6	$3\frac{3}{4}$	3	$3\frac{3}{4}$	$5\frac{1}{2}$	3	I, 12	I, 24	16 + 10	41–15	Grand Plains
7	$3\frac{1}{2}$	3	$3\frac{1}{4}$	5	$3\frac{1}{5}$	I, 12	I, 24	17 + 9	41–15	Bayou, Miss.
8	$3\frac{1}{2}$	3	$3\frac{1}{2}$	5	3	I, 12	I, 24	17 + 9	41–15	Black Bayou,
9	$3\frac{1}{4}$	$2\frac{5}{6}$	$3\frac{1}{2}$	$5\frac{1}{4}$	3	I, 12	I, 24	17 + 9	40–15	Miss.

This species appears to be rather common in the larger lowland streams and bayous of Louisiana and Mississippi. It probably does not reach a large size, adult examples being less than 6 inches long. It is not used as food, but is of considerable value as bait in the catfish fishery of the Atchafalaya River and its connecting lakes and bayous. Length 4 to 6 inches. (*atchafalayæ*, from the type locality.)

Signalosa atchafalayæ, EVERMANN & KENDALL, Bull. U. S. Fish Comm. 1897 (Feb. 9, 1898), 127, pl. 7, fig. 4, Atchafalaya River, Melville, Louisiana. (Type, No. 48790. Coll. Fred M. Chamberlain.)

Page 425. The statement that *Pomolobus mediocris* does not ascend rivers to spawn is not correct. This species is known to ascend the St. Johns River, Florida, at least as far as Lake Monroe, during the winter. They usually run somewhat earlier than the shad.

Page 427. After *Alosa*, Cuvier, add:

a. Gill rakers numerous, 93 to 120; upper jaw with sharp, deep notch at tip; lower jaw not projecting. SAPIDISSIMA, 693.
aa. Gill rakers fewer than 76; notch in upper jaw smaller; lower jaw more strongly projecting. ALABAMÆ, 693(a).

Page 428. After synonymy of *Alosa sapidissima* add:

693(a). ALOSA ALABAMÆ, Jordan & Evermann.

(ALABAMA SHAD; GULF SHAD.)

Head $4\frac{3}{5}$; depth 3; snout $4\frac{1}{2}$; eye $4\frac{1}{8}$; maxillary $2\frac{1}{8}$. D. 15; A. 20; scales 55, —16 in a crosswise series; scutes 21 + 15; vertebræ 54; gill rakers 56 to 68. Body deep; back gently and evenly arched from tip of snout to origin of dorsal fin, thence descending in a regular curve to base of caudal fin; ventral outline nearly straight from tip of mandible to ventrals, and also from there to base of caudal. Head small, snout pointed; upper lip with a small notch, into which fits the tip of the slightly projecting lower jaw; maxillary narrow; cheek much deeper than long; teeth on tongue and maxillary scarcely perceptible. Origin of dorsal nearer snout than base of caudal, the fin low, the longest ray shorter than the base, or about equal to snout and eye; base of anal somewhat greater than that of dorsal, or equal to length of pectoral. Gill rakers 68, the longest about equal to length of snout. Peritoneum pale. Color as in *Alosa sapidissima;* the caudal, dorsal, and pectoral fins rather darker tipped. The male differs from the female only in being somewhat more slender. This species differs from *Alosa sapidissima* chiefly in the fewer gill rakers, its sharper, more pointed snout, smaller notch in upper jaw, more projecting mandible, and more slender maxillary. It seems to reach maturity at a much smaller size than the common shad. Streams tributary to the Gulf of Mexico; known from Tuscaloosa, Alabama, and Pensacola, Florida.

Alosa alabamæ, JORDAN & EVERMANN, in EVERMANN, Rept. U. S. Fish Comm. 1895 (Dec. 28, 1896), 203, Black Warrior River, Tuscaloosa, Alabama (Type, Nos. 47689 and 47690. Coll. J. H. Fitts); EVERMANN & KENDALL, Bull. U. S. Fish Comm. 1897, 127, pl. 7, figs. 5 and 6.

According to Ogilby, *Kowala* is a genus distinct from *Sardinella.*

Page 436. *Ilisha panamensis* is not separable from *I. furthi.* The latter name has priority.

Page 437. *Opisthopterus lutipinnis* is very abundant on the outer sand beaches about Mazatlan.

Page 445. Species 728, *Stolephorus poeyi,* is a species of *Lycengraulis,* and should stand as —

743(a). LYCENGRAULIS POEYI (Kner & Steindachner).

Numerous specimens lately taken by Dr. Gilbert at Panama. A large species used as food. The teeth are unequal in *Lycengraulis,* but none of them can be properly described as canine-like.

Page 447. After *Stolephorus lucidus* add:

732(a). STOLEPHORUS RASTRALIS, Gilbert & Pierson, new species.

Head 3.16 (3.1 to 3.3); depth 3.8 (3.5 to 4.2); eye 3.4 in head (3.33 to 4). D. 14 (12 to 15); A. 26 to 32. Body much compressed and deep; belly sharply keeled in front of ventrals; dorsal outline much less curved than ventral, the lower profile rising very rapidly from a point opposite middle of pectorals to tip of snout, in shape of head thus closely resembling the species of *Cetengraulis.* Maxillary reaching almost but not quite to gill opening; snout high, compressed, its length $\frac{1}{2}$ to $\frac{3}{4}$ diameter of eye. Gill rakers averaging in larger examples $51 + 64$, in smaller specimens $44 + 50$; the largest about as long as eye. Insertion of dorsal fin variable, but never posterior to a point midway between base of caudal and middle of eye; pectoral fins reaching to or nearly to insertion of ventrals, the latter not to vent. Color olivaceous, the lower part of side with violet reflections; sides of head silvery; a conspicuous silvery lateral band varying in width from about $1\frac{1}{4}$ times length of orbit in the largest examples to less than $\frac{1}{2}$ orbit in the smaller specimens; the band is widest before dorsal, tapering to $\frac{1}{2}$ or less than $\frac{1}{3}$ its greatest width on caudal peduncle, where it frequently disappears in the young. In larger specimens the ventral edge of this band is frequently ill-defined anteriorly; top of head with widely spaced black specks; a dark vertebral streak, more or less of it often consisting of 2 narrow lines; tips of caudal lobes often blackish; fins otherwise unmarked. This species differs from closely allied species in the following characters: From *Stolephorus lucidus*, in the much longer head, more compressed body, well-defined lateral stripe, and smaller eye; from *S. compressus*, in the longer head and wider lateral band; from *S. panamensis* and *S. mundeolus*, in the much more numerous gill rakers, and the more anterior position of the dorsal relative to the anal, the origin of the anal being under the middle of the dorsal, while in *S. panamensis* the origins of the two fins lie in the same vertical. Length 2 to 3 inches. Panama. Many specimens. (Gilbert & Pierson.) (*rastrum*, a rake, from the long gill rakers.)

Stolephorus rastralis, GILBERT & PIERSON, Fishes of Panama, MS. 1898, Panama.

732(b). STOLEPHORUS MUNDEOLUS, Gilbert & Pierson, new species.

Head 4.15 (4 to 4.25); depth 3.77 (3.40 to 4.25); eye 3.44 in head (3.12 to 3.70). D. 13 or 14; A. 33 (33 to 35); scales 36 (35 to 39). Dorsal and ventral contours about equally and gradually rounded from the middle region of body to the tip of snout and base of caudal fin. Snout short, high, compressed, blunt at tip, its length 1.8 in eye. Eye very large. Maxillary broad, tapering to a sharp point, which reaches margin of gill opening. Gill rakers 17 to 22 + 21 to 24; the longest 1.5 to 2 in eye. Anterior insertion of dorsal fin varying from a point midway between base of caudal and middle of eye, to a point midway between the caudal and tip of snout. In 10 examples its insertion is before that of the anal. Anal fin long, averaging 33 rays, its origin beneath the anterior third of the dorsal; length of base shorter than in *S. panamensis*, being 3.04 in length, while in the latter its length is contained 2.5 in length. Pectoral long, reaching well beyond the insertion of the ventrals, equaling length of head behind front of pupil; a large axillary scale; ventrals scarcely reaching vent. Uniform light olive, with silvery reflections; a faint, narrow, silvery-gray lateral stripe, sometimes scarcely distinguishable; sides of head plain silvery; upper margin of orbital rim black; dorsal region blackish; a faint, narrow dark line on each side of the light middorsal streak; caudal slightly dusky; fins otherwise unmarked. This species is closely allied to *Stolephorus panamensis* and *S. compressus*, but may be distinguished from the former by its longer head, larger eye, greater depth, fewer scales along the lateral line, and its much shorter anal base; also by the much fainter lateral silvery stripe. The eye is contained 14 to 16 times in length, excluding the caudal, while in *S. panamensis* the length contains the eye 16 to 20 times. From *S. compressus* it differs in the relative length of the head and maxillary. In *S. mundeolus* the maxillary is contained in the head 1.27 times (1.19 to 1.37); in *compressus* 1.48 times (1.30 to 1.81). In *mundeolus* the head is contained 4.15 times in the length; in *compressus* 4.44 times. Length 4 to 6½ inches. Panama; many specimens. (Gilbert.) (*mundeolus*, somewhat shining, from *mundus*, neat or clean.)

Stolephorus mundeolus, GILBERT & PIERSON, Fishes of Panama, MS. 1898, Panama.

732(c). STOLEPHORUS NASO, Gilbert & Pierson, new species.

Head 3.3 to 3.5; depth 4.7 to 5.8; eye 4½ to 5 in head. D. 14 or 15; A. 22 to 24; lateral line about 35 (?). Dorsal and ventral outlines weakly arched; body slender, its greatest depth 1.5 in head, compressed; belly carinated in front of ventrals, and sometimes behind them in larger specimens. Head long and slender, its greatest width 1.5 to 1.7 in its length, the lower profile much more oblique than the upper. Snout long, compressed, bluntly rounded, its length exceeding the small eye. Cheek with a very acute posterior angle. Opercle narrow, oblique. Maxillary rather bluntly pointed, failing to reach gill opening by about ¼ diameter of pupil. Teeth on the maxillary quite prominent and directed forward. Gill rakers short, 17 + 20 in number; the longest 1.5 in eye. Scales large, thin, deciduous, only a few scattering ones remaining in our specimens. Dorsal fin inserted midway between front or middle of orbit and base of median caudal rays. Origin of anal under or slightly behind middle of dorsal; length of anal base about equal to the distance from front of orbit to base of ventral fin; pectorals not reaching ventrals, their length about ½ length of head. Length of ventrals equaling or slightly exceeding distance from tip of snout to middle of pupil. Color light olive, with the usual bright reflections; a large dark patch of brown dots on occiput; a double series of dots along median line posterior to dorsal, this absent in some specimens; large specimens with a bright, well-defined silvery streak, slightly narrowing anteriorly and on caudal peduncle, its greatest width about equaling diameter of eye; in the young, this band is fainter and narrower; a conspicuous series of black dots at base of anal. Characterized by the slender form, well-defined silvery streak, sharply carinated breast, the small eye, and the very long, compressed, deep, and rather bluntly rounded snout. Most closely resembling *S. starksi*, from which it differs in the smaller eye, longer snout, and slightly longer anal. Length 2 to 2½ inches. Panama; common. (Gilbert & Pierson.) (*naso*, long-nosed.)

Stolephorus naso, Gilbert & Pierson, Fishes of Panama, MS. 1898, Panama.

732(d). STOLEPHORUS STARKSI, Gilbert & Pierson, new species.

Head 3.3 to 3.6; depth 4.8 to 5.5 in length, 1⅓ in head; eye 3 to 3.5 in head. D. 15 or 16; A. 17 to 22; scales about 41. Body long and slender, slightly deeper and more compressed than in *S. ischanus*, which much resembles this species. Dorsal outline very little arched; ventral outline nearly straight from gill opening to insertion of anal fin, the lower profile of head oblique, nearly straight. Belly compressed, keeled for anterior ⅔ of distance anterior to base of ventrals. Head long and pointed, its width 1¼ times in its length; maxillary abruptly widened at the mandibular joint, tapering posteriorly to a blunt point, which reaches almost to the gill opening, its length equal to length of base of anal; snout long, sharp, and projecting, abruptly compressed in its terminal portion as seen from above, its length ¾ diameter of orbit, or slightly more. Branchiostegal membranes united at base for a very short distance. In 4 examples examined as to this point, the gill·rakers are as follows: 20+25, 23+24, 21+23, 19+30, the longest contained 1¼ to 1¾ in eye. Scales large, thin, and deciduous, a few only remaining on the specimens at hand. Origin of the dorsal fin equally distant from the base of the caudal fin and tip of snout or front of eye. Anal inserted under beginning of posterior third of base of dorsal; pectorals not reaching ventrals, the latter ⅔ distance to front of anal. Color light olive, with broad, well-defined lateral silvery streak of nearly uniform width, usually narrowing anteriorly and on middle of caudal peduncle, its width in our largest specimen ⅝ diameter of eye; the silvery streak has a slight golden tinge; a narrow dark vertebral line, which widens on the nape; occiput blackish. Vertebræ 40, counted in 1 example only. This species differs from *Stolephorus cultratus* in its slenderer body, shorter snout, wider opercle and smaller teeth; the belly is also not sharply carinate, the dorsal is more anteriorly placed, the ventrals are farther back, and the silvery streak is wider anteriorly. It differs from *S. delicatissimus* in its longer, slenderer head and body, smaller eye, longer, sharper snout, and much wider, better-defined silvery streak. Length 1¼ to 2¼ inches. Panama; common. (Gilbert & Pierson.) (Named for Edwin Chapin Starks.)

Stolephorus starksi, GILBERT & PIERSON, Fishes of Panama, MS. 1898, Panama.

Page 448. After *Stolephorus spinifer* add:

737(a). STOLEPHORUS SCOFIELDI, Jordan & Culver.

Head 3¾ to 3¹⁹⁄₂₀ in length to base of caudal; depth 4½ to 5; eye 3¾ to 4 in head. D. 12; A. 25 or 26; scales 41 or 42. Body somewhat compressed and elevated, the belly not carinated nor serrated. Teeth in both jaws, and on palatines, a few on vomer; maxillary covered with teeth its entire length and reaching beyond base of mandible, but not to opercular margin. Gill rakers 10+12, the longest a little more than ½ the eye. Origin of dorsal midway between base of median caudal rays and center of eye; anal not quite as long as head, its origin below the middle of dorsal; lower caudal lobe longer than upper; longest ray equaling length of the head; shortest caudal ray 2½ in longest. Pectorals not reaching ventrals, 1¾ in head. Both anal and dorsal fins preceded by a rudimentary spine, not ½ length of first true ray. Color translucent, with a distinct broad silvery stripe as wide as the eye, growing more diffuse at lower anterior edge, narrowing on caudal peduncle, and becoming fan-shaped on the base of caudal; tip of snout black; a distinct median band of black specks extending from tip of snout to base of caudal; no distinct black markings on fins. Close to *Stolephorus delicatissimus*, but with larger head, wider lateral band, and greater number of dorsal and anal rays. Length 3 inches. Found in the Astillero at Mazatlan; not very abundant. (Named for Mr. Norman Bishop Scofield, a member of the Hopkins expedition to Sinaloa.)

Stolephorus scofieldi, JORDAN & CULVER, Fishes of Sinaloa, 410, 1895, Mazatlan, Mexico. (Type, No. 2941, L. S. Jr. Univ. Mus. Coll. Hopkins Exped. to Sinaloa.)

737(b). **STOLEPHORUS ASTILBE**, Jordan & Rutter.

Head $4\frac{1}{3}$ in length; depth $4\frac{1}{2}$ to 5. D. 12; A. 19 to 22; eye $3\frac{1}{2}$ in head; pectoral $1\frac{2}{3}$; base of anal $1\frac{1}{6}$. Body rather elongate, not greatly compressed; edge of belly moderately sharp; head sharp; snout projecting beyond lower jaw, shorter than diameter of eye; tip of lower jaw reaching a little past anterior edge of orbit; maxillary reaching gill opening, its end tapering to a sharp point; eye longer than snout, nearly 2 in postorbital part of head; gill rakers $\frac{2}{3}$ eye; a slight keel on top of head. Origin of dorsal midway between base of caudal and eye; scales caducous. Translucent, head silvery; sides without lateral band; a dark spot on top of head; back with black points. This species is similar to *Stolephorus brownii*, but more slender, head shorter, and lateral silvery stripe wanting. Length 3 inches. Jamaica. Numerous specimens obtained. (ἀ, not; στίλβη, shining.)

Stolephorus astilbe, JORDAN & RUTTER, Proc. Ac. Nat. Sci. Phila. 1897, 95, Jamaica. (Type, No. 4854, L. S. Jr. Univ. Mus. Coll. Joseph Seed Roberts.)

737(c). **STOLEPHORUS ROBERTSI**, Jordan & Rutter.

Head 3 in length; depth 4. D. 14; A. 23; scales about 35; eye 4 in head; pectoral $2\frac{1}{2}$; base of anal $1\frac{2}{3}$; caudal $1\frac{1}{2}$. Body deep, strongly compressed, abdomen compressed to an edge, head large, compressed, the snout rather sharp, projecting beyond lower jaw, a little shorter than eye; cheek triangular; opercle large; distance from lower angle of cheek to edge of opercle equal to distance from same point to posterior edge of eye; maxillary short, not reaching root of mandible, its end rounded; lower jaw not reaching beyond anterior edge of orbit; gill rakers longer than eye, as long as orbit. Origin of dorsal midway between base of caudal and front of eye; scales caducous. Color translucent; head silvery, punctulate above; a silvery lateral band nearly as broad as eye; caudal with dark points, other fins colorless. This species seems to be related to *Stolephorus opercularis*, but the lateral band is distinct and the opercle is shorter. Jamaica; only the type, 2 inches long, known. (Named for Rev. Joseph Seed Roberts, of Kingston, Jamaica, who collected the type specimen.)

Stolephorus robertsi, JORDAN & RUTTER, Proc. Ac. Nat. Sci. Phila. 1897, 95, Jamaica. (Type, No. 4853, L. S. Jr. Univ. Mus.)

Page 449. *Anchovia* can not be maintained as a distinct genus. The name must be placed as a synonym of *Stolephorus*.

Page 450. Add:

741(a). **CETENGRAULIS ENGYMEN**, Gilbert & Pierson, new species.

Head 3 to 3.3; depth 4 to 4.9; eye 4 in head. D. 14 or 15; A. 20 to 23; B. 7 (9); vertebræ 41. Body compressed, fusiform, not so deep as in *C. mysticetus* or *C. edentulus*. The dorsal and ventral outlines being about equally and regularly curved in the larger specimens; in the smaller specimens the ventral contour is more nearly straight. Belly trenchant, but not carinate nor serrate; caudal peduncle moderate, its depth being contained 1.5 times in its length. Head similar to *C. mysticetus;* the snout longer, contained 5.5 to 7 times in head, 1¼ times in eye (the snout contained 8 to 9 times in head in *C. mysticetus*). Both jaws bear minute teeth, those on the maxillary largest. Branchiostegal membranes united for only ⅛ to ¼ of the distance between tip of mandible and mandibular articulation, wholly free from the isthmus. Tip of mandible directly beneath the anterior border of orbit. Gill rakers long, $\frac{9}{10}$ diameter of eye, 20 to 30 on the upper limb, 25 to 30 on the lower limb; in 5 examples as follows: 25 + 30, 27 + 25, 30 + 26, (23 + 29 to 20 + 25), 25 + 30. The origin of the dorsal is midway between base of median caudal rays and a point varying between front and middle of the eye. Insertion of anal below the posterior fourth or third of the dorsal, its length equaling the distance from the posterior border of the eye to insertion of pectoral. The pectoral is short, 2¼ to 2⅓ in head, failing to reach the insertion of the ventrals by ½ or nearly ⅓ its length. Caudal deeply forked, its median rays 2½ to 3 times in head. Color uniformly silvery with a distinct, well-defined lateral silvery band extending from upper angle of gill opening to base of caudal, its greatest width equaling the diameter of orbit, becoming narrower on caudal peduncle. This species differs from *C. mysticetus* in the much narrower union of the gill membranes, the less numerous gill rakers, and in the longer snout. Length 1¼ to 2⅓ inches. Panama Bay. Not rare. (Gilbert & Pierson.) (ἐγγύς, near; δμήν, membrane.)

Page 451. *Lycengraulis* has the teeth large and somewhat unequal, but none of them is properly described as "canine-like."

Page 459. Add:

229(a). ERICARA, Gill & Townsend.

Ericara, GILL and TOWNSEND, Proc. Biol. Soc. Wash., XI, 1897 (Sept. 17, 1897), 232 (*salmonea*).

Alepocephalids with small, perfectly smooth, imbricated cycloid scales, wide cranium, projecting snout, deeply cleft mouth, uniserial and acrodont teeth on vomer and anterior portion of palatines, and dorsal and anal of normal extent and opposite each other. Bering Sea. (ἐρί, an intensive particle; καρά, head.)

753(a). ERICARA SALMONEA, Gill & Townsend.

D. 17; A. 24. Maxillary extending to vertical of posterior border of orbit; head large; length 8½; depth 5; width 4½. Bering Sea; only the type known, a large example in good condition.

Ericara salmonea, GILL & TOWNSEND, Proc. Biol. Soc. Wash., XI, 1897 (Sept. 17, 1897), 232, Bering Sea, southwest of Pribilof Islands, at Albatross Station 3603, in 1,771 fathoms. (Type, No. 48769, U. S. Nat. Mus. Coll. *Albatross*.)

Page 465. Dr. G. A. Boulenger has kindly sent us the following note regarding the types of *Coregonus richardsonii* which are in the British Museum:

I have examined the types (dry) of *Coregonus richardsonii*. There are about 20 gill rakers on the lower part of the anterior arch, the longest ⅓ the diameter of the eye. The maxillary extends to below anterior border of eye, and its length is 4 times in length of head as stated by Günther, therefore a little shorter than in *C. clupeiformis*. Tongue with 4 series of teeth, as in *C. labradoricus*. It seems to agree best with *C. nelsoni* (description), but has fewer scales in lateral line. In short, I can not identify *C. richardsonii* with any of the forms known to me.

Page 471. After *Argyrosomus laurettæ* add:

768(a). **ARGYROSOMUS ALASCANUS**, Scofield, new species.

Head 4¼; depth about 4. D. 12; A. 12; scales 10–85–9. Eye a little shorter than snout, 5 in head, 1⅓ in interorbital space. Head wedge-shaped, the upper and lower profiles straight and meeting with a sharp angle at the snout. Viewed from above the snout is blunt, almost square, with the narrow, pale rounded tip of the lower jaw slightly projecting. Mouth oblique; the distance from tip of snout to tip of maxillary is equal to the distance from tip of snout to center of pupil; the maxillary from its anterior articulation is contained 3½ in the head, its width 3 in its length, its upper anterior edge closing under the preorbital; mandible 2⅓ in head, its articulation with the quadrate bone beneath the posterior edge of the eye; width of supplemental bone a little more than ½ width of maxillary. Preorbital broad, its greatest width equaling ⅜ its length or diameter of pupil; width of supraorbital equals ²⁄₇ its length. Gill rakers 12 to 14 + 21 to 23, long and slender, the longest ⅔ diameter of the eye. The tongue, vomer, and palatines without teeth. Distance from tip of snout to nape equaling ½ the distance from nape to the front of the dorsal or ⅔ length of head. Adipose fin large; ventral scale ½ length of fin; longest dorsal ray 1½ in head; longest anal ray 2 in head; the pectorals reach more than ½ to the ventrals; the ventrals reach ⅔ to vent; the caudal is forked for a little more than ½ its length. Color dusky above, silvery beneath; the dorsal, adipose fin, tips of caudal rays and upper side of anterior pectoral rays dusky; the rest pale. This species appears nearest related to *Argyrosomus artedi*, from which it differs chiefly in the number of gill rakers.* Length about a foot. Northern Alaska near Bering Straits; 3 specimens known, 1 from salt water at Point Hope, the others from fresh water at Grantley Harbor.

Argyrosomus alascanus, SCOFIELD, in JORDAN, Rept. U. S. Fur Seal Investigations, 1898, Point Hope and Grantley Harbor, Alaska. (Coll. Scofield & Seale.)

Page 482. Beginning with line 10 from the bottom, the statement that the small form of the redfish has been traced from the mouth of the Columbia to Wallowa Lake is not true. The remark was meant to apply to the large form. The question as to whether the small form descends to the sea is still unsettled.

* The fin formulæ, etc., of these 3 specimens are as follows:

Locality.	Length.	Gill rakers.		Dorsal.	Anal.	Scales.
Grantley Harbor	8½ in.	14 + 23	14 + 22	12	14	88
Do	9 in.	14 + 22	12 + 22	12	12	87
Point Hope, Alaska	11 in.	13 + 21	12 + 21	12	12	85

Page 492:

779. SALMO MYKISS, Walbaum.

(MYKISS; SOMKA; KAMCHATKA SALMON TROUT.)

By an unfortunate error the writers have heretofore used the name *Salmo mykiss* for the Cutthroat Trout of the Northwest. It was known that the Cutthroat is the only true or black-spotted trout in Alaska, and it was assumed that its range extended along the coast to all streams in Bering Sea. But our recent explorations have shown that it probably does not occur in Bering Sea, nor is there any undoubted record to the north of Wrangell. If it reaches Kadiak or Sitka or Prince William Sound, it is only rarely, and the streams of the Aleutian Islands and the east coast of Bering Sea contain no species of *Salmo*. The name *Salmo mykiss* must, therefore, be restricted to the Kamchatkan species, while the species of the American rivers heretofore called *Salmo mykiss* must stand as *Salmo clarkii*. We have, therefore, studied with great interest a specimen of the genuine *Salmo mykiss*, the first on record since the times of Pallas, Krasheninnikof, and Steller. The specimen, an adult male, 960 mm. long, was taken by Dr. Leonhard Stejneger, September 15, 1897, in the Kalakhtyrka River, near Petropaulski, Kamchatka. It was called "Sonka" or "Somka" by the natives. It is said to occur rarely and to be found in but few rivers, the Kalakhtyrka among them. It is considered to be superior as food to other *Salmonidæ*, except the King Salmon (*O. tschawytscha*). Head 4 in length; depth 4¼. D. 11. A. 10 (developed rays); scales 125-24. Mouth large, the maxillary 1⅜ in head, being somewhat produced at tip; vomerine teeth few, evidently deciduous, only 3 being present. Eye 8¼ in head; snout 2⅜. Pectoral 2 in head, longest anal ray 2⅔. Anal fin high and somewhat falcate; ventrals inserted under anterior third of dorsal, reaching about halfway to vent; adipose fin over posterior end of anal; caudal lunate. Color dark grayish above, sides silvery; a few small, faint, round black spots on back and on top of head, these sparse and obscure; a few faint spots on base of dorsal, and some on adipose dorsal; spots on caudal small, but distinct, especially in middle of fin; no trace of red at throat, in example preserved in formalin, and doubtless none in life. The specimen is now a half skin, in good condition.

The following measurements were taken from the fresh specimen by Dr. Stejneger:

	Millimeters.
Total length	960
Total length without caudal	853
Head	215
Tip of nose to anterior end of dorsal	400
Length of base of dorsal	100
Posterior end of dorsal to anterior end of adipose fin	167
Length of base of adipose fin	17
Posterior end of adipose to caudal	81
Posterior end of anal to caudal	81
Length of base of anal	71
Anterior end of anal to posterior end of ventral	165
Height of body in front of dorsal	195
Height of body at posterior end of adipose and anal fins	105

Millimeters.

Height of body at beginning of caudal 77

Ventrals under anterior third of dorsal; adipose fin over posterior end of anal; ventrals reach about ½ distance to vent; 24 scales in transverse series from origin of dorsal to lateral line; 125 scales in lateral line. Color silvery gray on back, black spots obsolete.

This species is evidently a close ally of the Atlantic salmon, belonging to the restricted subgenus *Salmo*. From *Salmo salar* it differs in the slightly larger mouth and rather different coloration and in very little else.

Mykiss, PENNANT, Arctic Zool., Intro., 126, 1792, Kamchatka; after KRASHENINNIKOF, etc.

Salmo mykiss, WALBAUM, Artedi Piscium, 59, 1792, Kamchatka; based on *Mykiss* of PENNANT.

Salmo penshinensis, PALLAS, Zool. Rosso–Asiat., III, 1811, Gulf of Penshin.

Salmo purpuratus, PALLAS, Zool. Rosso–Asiat., III, 374, 1811, Bering Sea.

The correct names of the American Cutthroat Trout and its numerous known varieties are the following:

780. **SALMO CLARKII** (Richardson).

780(a). **SALMO CLARKII LEWISI** (Girard).

780(b). **SALMO CLARKII GIBBSII** (Suckley).

780 (c). **SALMO CLARKII HENSHAWI** (Gill & Jordan).

780(d). **SALMO CLARKII VIRGINALIS** (Girard).

780 (e). **SALMO CLARKII SPILURUS** (Cope).

780(f). **SALMO CLARKII PLEURITICUS** (Cope).

780(g). **SALMO CLARKII BOUVIERI** (Bendire).

780(h). **SALMO CLARKII STOMIAS** (Cope).

780 (i). **SALMO CLARKII MACDONALDI**, Jordan & Evermann.

Page 500. Before *Salmo irideus* insert the following:

781(b). **SALMO GAIRDNERI BEARDSLEEI**, Jordan & Seale.

(BLUEBACK TROUT OF LAKE CRESCENT.)

Head 3¼ in length to base of caudal; depth about 4; eye 4⅚ in head, 1⅔ in snout; scales 24–130–20, 130 cross series, those in front of dorsal numerous, about 70 if counted along median line, 60 if the rows along upper side are counted; dorsal with 10 branched rays; anal with 11 branched rays; branchiostegals 11; gill rakers 8 + 13, rather long and slender, the longest nearly $\frac{6}{16}$ in length, 7 to 9 in maxillary. Head pointed; mouth rather large; maxillary extending to hinder margin of eye, 1¼ in head, with about 20 teeth; snout 3⅔ in head; preorbital very narrow, the maxillary almost touching the orbit; posterior suborbitals shorter than eye, about 6 in head; opercle not very broad, equal to eye, its free part 4⅔ in head; interorbital width 3⅔ in head, equal to snout; several large teeth along margin of tongue; no hyoid teeth; teeth on vomer in zigzag series. Origin of dorsal in middle of the length, margin slightly concave, the first ray 1⅝ times the last, the last ray being pointed, slightly greater than base, 2$\frac{1}{10}$ in head. Origin of anal midway between origin of dorsal and base of caudal, margin straight, the tip of the last ray slightly exserted; anterior rays 3¼ times posterior, and equal to base of fin, 2¼ in head. Adipose fin high and slender, situated above or anterior to end of anal. Pectorals 1½ in head; ventrals under middle of dorsal, 2⅛ in head. Caudal broad, nearly truncate, the middle portions abruptly lunate when spread open, with pointed angles, each lobe being somewhat convex on its edge; longest rays 1⅓ in head. Least depth of caudal peduncle 2¾ in head. Pyloric cæca 50 to 60, short and thick, the longest about 3 in head. Color in spirits very dark blue above, sides abruptly brighter, with many scales abruptly silvery; below white, lower jaw white, its margin dusky; cheeks below suborbitals very dark; sides, top of head, dorsal, and caudal fins spotted, the spots all very small; pectorals and ventrals nearly colorless, without spots, and slightly dusky; adipose fin with 2 spots; tips of lower fins faintly tinged with yellowish. Two specimens, each 16 inches long, Nos. 1861 and 1862, L. S. Jr. Univ. Mus. They were taken on March 12 and 16, 1896, in Lake Crescent, by Mrs. George E. Mitchell, of Fairholme, and sent to us by Mr. M. J. Carrigan, of Port Angeles.

A third specimen shows the following characters: Head 3⅚; depth 3⅔. D. 12; A. 12 branched rays; branchiostegals 11 or 12; scales 23–123–26, 64 before dorsal; snout 2⅚; eye 7⅔; maxillary 1⅔ in head, its depth 8 in its length. Body robust, little compressed; head large; maxillary moderate, extending beyond eye; opercle moderate, its width 5⅔ in head. Last ray of dorsal pointed. Caudal subtruncate, lunate mesially, each lobe somewhat convex, pointed at tip. Caudal peduncle short and thick. Series of vomerine teeth long, in double row. Color above dark green, with black spots, which are small and sparse on body, extending to below lateral line; many small spots on head, dorsal, and caudal; spots not more numerous behind than before; sides and belly bright silvery; no red on lower jaw; a faint pink shade along lateral line; pectorals colorless except the upper ray; ventrals and anal colorless; flesh pale; gill rakers removed. This specimen, male, was taken in Lake Crescent. Length 26½ inches; weight in life 14 pounds. This specimen differs from a large *gairdneri* most in the large scales. In addition the head is much larger, and the body deeper.

A fourth, still larger, specimen (No. 1865, L. S. Jr. Univ.), an old spent male, 27 inches long, has been still later received. It shows the following characters: D. 11; A. 12. Head 3⅗ in length; gill rakers 8 + 12, of medium size, rather broad but sharp pointed; opercle 3½ in head; eye 7 in head; branchiostegals 11; maxillary long, reaching beyond eye, 1⅔ in head, its width 9½ in length. A double row of sharp teeth extending to within a short distance of end, where they are replaced by a single row of slightly larger teeth; teeth on tongue rather large; no teeth on hyoid; teeth on vomer in zigzag series. Scales 137–26. This specimen, a spent male, has the flabby muscles and slimy, half-concealed scales of the spent male salmon. The dark dots are very numerous and small and show very distinctly on back and sides, as also on head and fins; there is a dull red lateral band on head and body—this is about an inch broad, its outlines diffuse; a black blotch on cheek; maxillary dusky with a red blotch toward its tip; lower jaw and branchiostegals dusky; pectoral, ventral, and anal dark; back dark green, belly dusky.

The following account of the life coloration of *Salmo beardsleei* is given by Mr. George E. Mitchell:

The Blueback Trout caught in Lake Crescent are on the back a deep dark-blue ultramarine color of a peculiar transparency, dotted with small round black spots from the size of a pin's head to a little larger. The 2 fins on the top of the back are a dark smoky color, also dotted as on back end, and are transparent. The tail is the color and transparency with dots also—same as the top fins. The side fins and the bottom fins are dead white and sometimes faintly tinged with a pinkish hue at the edges; the belly is white. Looking at the fish sideways the sides of the fish show the scales to be iridescent, the red flash predominating. The head has very much the polish of mother-of-pearl around the lower jaws and jowls, red and pale-blue colors predominating; under the eyes a few black spots; on top of head the blue much darker than on top of back—so dark, in fact, that the black spots on it look blacker than the rest. The nearer the shore these fish are caught the lighter the blue on back, the fish often having an impression of the surroundings distinctly marked on them.

The following notes are added by Admiral Beardslee:

HABITS.

The Blueback is a deep-water dweller; those taken by me in late October were caught at depths varying from 30 to 50 feet, on large spoons. They fought hard until brought near the surface, then gave up, and when landed were found puffed up with air. Specimens taken in spring and put in pools in mountain streams with other trout died very soon, while the others lived. The trout caught by Mr. Mitchell, in March, was taken near bottom, by a large spoon, and it is not on record that at so early a date one has previously been caught.

FLESH.

Light lemon color before cooking; devoid of the oily salmon flavor, and very excellent; whitening by cooking.

OVA.

October 28. The eggs in the large fish were in *individual* size, and in size of cluster much smaller than those of a salmon of the same size.

The following extracts from a letter from Mr. Carrigan, dated Port Angeles, April 30, are of much interest:

* * * Answering your direct inquiries: The Beardslees and Crescents are readily distinguishable, and can always be told apart. There are no red spots at the points indicated on the Crescent trout—no markings to suggest the Cutthroat trout.

(Named for Admiral L. A. Beardslee, U. S. N., in recognition of his active and intelligent interest in American game fishes.)

Salmo gairdneri beardsleei, JORDAN & SEALE, Proc. Cal. Ac. Sci., ser. 2, vol. VI, 1896, 209, pl. 23, Crescent Lake, Clallam County, Washington. (Coll. Mrs. George E. Mitchell. Type, No. 1864, L. S. Jr. Univ. Mus.)

780(c). SALMO GAIRDNERI CRESCENTIS, Jordan & Beardslee.

(SPECKLED TROUT OR LAKE CRESCENT.)

Head $3\frac{4}{5}$ in length to base of caudal; depth 5; exposed portion of eye 6 in head, $1\frac{3}{4}$ in snout; scales 32–151–34, 151 cross series, 83 in front of dorsal; dorsal with 10 branched rays, anal with 11; branchiostegals 10; gill rakers 6+11, counting rudiments, these very short and thick, the longest but $\frac{3}{16}$ inch in length, $18\frac{1}{4}$ in maxillary; mouth large, maxillary extending much beyond eye, $1\frac{2}{3}$ in head, with about 20 teeth; tongue with the usual teeth; teeth on vomer in zigzag series; hyoid region of tongue without teeth. Snout 3½ in head; preorbital very narrow, not so wide as maxillary adjacent to it; the posterior suborbitals longer than eye, 5⅓ in head; opercle and subopercle very narrow, scarcely as wide as eye, the free part of opercle 6½ in head; interorbital width 4¼ in head. Origin of dorsal in middle of length of body, its margin straight, anterior 2½ times posterior, and slightly longer than base, 2¼ in head; last ray of dorsal pointed. Origin of anal midway between origin of dorsal and base of caudal, margin irregular, anterior rays 3 times length of posterior and equal to base of fin, 2⅝ in head. Adipose fin high and slender, situated immediately behind anal; pectoral 1¾ in head; ventrals under middle of dorsal, 2⅝ in head; caudal broad, slightly emarginate, nearly truncate when spread, its corners not rounded, its longest rays 1½ in head; least depth of caudal peduncle 3⅗ in head. Pyloric cæca about 51, the longest about 1⅚ in head, and very slender. Color in alcohol, very dark steel blue above, becoming paler below, nearly white anteriorly on belly, where only the margins of the scales are punctate; no silvery anywhere; lower jaw dusky, a large black blotch on cheek between suborbital and premaxillary; sides, back, top of head, dorsal and caudal fins with few small dark spots; pectorals dusky, slightly spotted at base; anal slightly dusky, without spots; ventrals dusky with a few spots in middle; adipose fin with a few spots; lower fins all tipped with pale, probably yellowish red in life; spots all very small and faint, not confined to posterior part of body. The specimen before us, No. 1863, L. S. Jr. Univ., is a male, 18¼ inches long. It was taken at Fairholme on Lake Crescent, Clallam County, Washington, March 12, 1896, by Mrs. G. E. Mitchell, of Fairholme. (Named for Crescent Lake, Washington, the type locality.)

Salmo gairdneri crescentis, JORDAN & BEARDSLEE, Proc. Cal. Ac. Sci., ser. 2, vol. VI, 1896, 207, pl. 22, Crescent Lake, Clallam County, Washington. (Coll. Mrs. George E. Mitchell. Type, No. 1863, L. S. Jr. Univ. Mus.)

Page 504. Under *Cristivomer* for "Eastern North America" read "Northern North America." The genus occurs also in the lakes of Alaska and British Columbia.

Page 508. Before *Salvelinus alpinus* insert:

784(a). SALVELINUS KUNDSCHA, Pallas.

This seems to be a species very distinct from *S. malma*. A specimen in the United States National Museum (No. 33814) from Petropaulski has been described by Bean & Bean as follows:

Similar in form to *S. malma*, but the body stouter and less elongate. Head 4½ to 4⅝ (4⅐ in the Tareinsky Bay specimen); depth 4½ to 4⅝; eye 5⅓ in head, 2 in interorbital, or 1½ in snout; maxillary reaching to or beyond vertical through posterior edge of orbit; upper jaw nearly ½ length of head; lower jaw slightly shorter than upper. Hyoid teeth feebly developed. Scales small, 36–195, 122 pores. Fins all short; origin of dorsal about midway between tip of snout and base of upper caudal lobe, the base of the fin nearly as long as the longest ray, or ½ as long as head, its upper margin very slightly concave, the last ray 2 in the longest; adipose fin over end of anal, its width about $\frac{1}{4}$ its length, which is about equal to eye; pectoral 7 to 7$\frac{1}{2}$ in body length; ventral under middle of dorsal, not nearly reaching vent, its length 2 in head; caudal emarginate, its middle rays $\frac{1}{2}$ the outer; anal scarcely concave when expanded. Pyloric cæca 22; branchiostegals 12; gill rakers 6+10, the longest less than $\frac{1}{2}$ eye. Color bluish gray above, whitish below; the sides with numerous large white spots, some of which are $\frac{2}{3}$ as large as eye. (Bean & Bean.)

This species is said to be common from Kamchatka northward, but only 6 specimens are actually extant, 4 obtained at Petropaulski by Dr. Leonhard Stejneger and 1 by Col. N. Grebnitski, and now in the United States National Museum, and 1 obtained from Tareinsky Bay by Mr. Gerald E. H. Barrett-Hamilton and now in the museum of Stanford University.

Salmo kundscha, PALLAS, Zoogr. Rosso-Asiat., III, 250, 1811, Kamchatka.
Salmo leucomœnis, PALLAS, Zoogr. Rosso-Asiat., III, 250, 1811, Kamchatka.
Salmo curilus, PALLAS, Zoogr. Rosso-Asiat., III, 251, 1811, Kuril Islands.

The true *Salvelinus malma* is very common at Unalaska, Kadiak, Komandorski Islands, and Petropaulski. Specimens from these various places are all alike. Head 4$\frac{1}{4}$ to 4$\frac{1}{3}$; depth 4$\frac{1}{4}$ to 4$\frac{3}{4}$. Spots grayish, tinged with red, much smaller than eye. Caudal well forked; lower fins short; pectoral reaching halfway to vent. Hyoid teeth present. The head seems much shorter than in examples from the United States. The dwarf form from the little brook (Pyramid Creek) at the head of Captains Harbor agrees fully in form with large examples taken in the sea about Unalaska. The small ones are brighter in color and mature at 4 to 6 inches. The form occurring throughout the northwestern United States, and described on page 508 as *Salvelinus malma*, should apparently be regarded as a species distinct from *S. malma*, and would stand as—

784(a). SALVELINUS PARKEI (Suckley).

Page 515. Add this footnote to *Salvelinus oquassa marstoni:*

A specimen of *Salmo marstoni* sent me some days ago indicates a more distinct species than was at first supposed. This is the most slender of our charrs, apparently the swiftest. The male is gorgeous; brilliant red extends upon the back and onto the dorsal and caudal fins as well as upon the other fins. Though quite distinct, the species is nearer to *S. oquassa* than any other. (Garman, in lit., March 24, 1895.)

Page 524. After *Osmerus dentex* add:

794(a). OSMERUS ALBATROSSIS, Jordan & Gilbert, new species.

(KADIAK SMELT.)

Head 4$\frac{2}{3}$; depth 5$\frac{1}{4}$. D. 2, 10; A. 1, 20; scales 75; maxillary 2$\frac{1}{10}$; eye 5$\frac{1}{4}$; snout 3$\frac{3}{4}$; mandible 2; pectorals 1$\frac{1}{2}$; ventrals 1$\frac{1}{2}$; dorsal 1$\frac{2}{3}$; caudal 1$\frac{2}{3}$. Body elongate, moderately compressed; back elevated at nape so that anterior profile is somewhat depressed between and behind eyes; interorbital space 3$\frac{3}{4}$ in head. Mouth large, lower jaw heavy, strongly projecting; opercle with concentric striæ; pectorals moderate; ventrals long; dorsal high; anal fin low, very long, its longest ray 2$\frac{3}{4}$ in head; caudal moderate, well forked; ventrals inserted before dorsal.

Scales small, deciduous, those on back still smaller; lateral line distinct. Gill rakers long and slender, about 12 below angle of arch, longest about as long as eye. Tongue with moderate teeth, the anterior 2 to 4 small hooked canines; upper jaw with small sharp teeth similar to those in lower jaw, none of them canine-like; small teeth on palatines and pterygoids; vomer with 2 very small canines scarcely fang-like. Color bluish above with bright reflections; scales margined with dark points; sides silvery with golden and coppery luster; inside of gill openings dusky; fins white, somewhat dotted. About Kadiak Island, Alaska. Two specimens caught in the upward haul of a dredge in Shelikof Straits, north of Karluk, Kadiak Island, Alaska, at *Albatross* Station No. 3675. The depth of the dredge haul was 109 fathoms, but these fishes were no doubt taken from near the surface. One specimen is 8, the other about 7 inches in length. The species is allied to *Osmerus dentex*, the Rainbow Smelt, but differs in the extremely long anal and in the very weak vomerine and lingual canines. The flesh is firm, as in *O. dentex*. (Named for the U. S. Fish Commission steamer *Albatross*.)

Osmerus albatrossis, JORDAN & GILBERT, Rept. Fur. Seal Invest., MS. 1898, Shelikof Straits, north of Karluk, Alaska.

Mesopus should replace *Hypomesus*. It is originally characterized on page 14 (not 168) Proc. Ac. Nat. Sci. Phila. 1862, *Hypomesus* on page 15. The ventrals are inserted below front of dorsal in *Mesopus* as in allied genera, and there are 8 branchiostegals as in allied groups. The feeble teeth distinguish *Mesopus* from *Osmerus*. The statement that the stomach is cæcal in *Argentinidæ* is true of a few genera only, and the character has no high systematic value. In *Mesopus pretiosus* and *Osmerus dentex*, the stomach is siphonal, as in *Salmonidæ*. In *Thaleichthys pacificus*, however, the stomach forms a blind sac. The small number of pyloric cæca and the peculiar structure of the ovaries remain to define *Argentinidæ* as a family distinct from *Salmonidæ*.

Page 525. To the synonymy of *Hypomesus olidus* add:

Osmerus oligodon, KNER.

The species ranges south to Amur River.

Page 530. After *Bathylagus pacificus* add:

804(a). BATHYLAGUS BOREALIS, Gilbert.

Head $4\frac{1}{12}$ to base of caudal; depth $5\frac{2}{3}$; eye $2\frac{1}{4}$ in head; snout $2\frac{3}{4}$ in eye. D. 8; A. 19; ventral 8; pectoral 8. Scales in about 40 rows, judging from the scars; head scaleless. Interorbital width grooved, the groove widening posteriorly, opening onto the flat occipital region, which is not swollen. Width of cartilaginous portion of interorbital space $\frac{1}{3}$ orbit; including the thin membranaceous plates which overarch the orbits, the interorbital width is $\frac{2}{3}$ orbit. The anterior profile of snout declines gently, bringing the mesial portion of premaxillaries on a level with lower margin of pupil. Distance from tip of snout to end of maxillary slightly exceeding length of snout, $2\frac{1}{3}$ in orbit. Opercle with 2 strong ridges diverging downward and backward from behind the eye. Front of dorsal midway between front of snout and adipose fin; base of dorsal contained 3½ times in length of head. Ventrals inserted under posterior portion of dorsal. Free portion of adipose fin very long and narrow, rising above the base of the second and third anal rays before the last, its tip reaching rudimentary caudal rays when depressed; anal fin rather long, the base 1⅔ in head, the vent immediately before it. Length of tail much exceeding head, 3⅔ in total length without caudal. Uniform blackish brown on sides, the head and ventral region blue black. Differing from *B. pacificus* in its much greater depth, longer tail, longer anal fin, and flat occiput. Length 132 mm. Bering Sea, in deep water north of Unalaska; 2 specimens known. (*borealis*, northern.)

Bathylagus borealis, GILBERT, Rept. U. S. Fish Comm. 1893 (Dec. 9, 1896), 402, Bering Sea at Albatross Station 3327, north of Unalaska, in 322 fathoms.

804(b). BATHYLAGUS MILLERI, Jordan & Gilbert, new species.

Distinguished by the posterior insertion of the dorsal fin and the greatly swollen occipital region provided with a median keel. The type is in very poor condition, the skin being largely denuded from head and body. No traces remain of the scales, the pectoral and ventral fins are lost and the others greatly mutilated. Enough remains, however, to demonstrate that it is distinct from all known species and to furnish characters by which the species may be recognized. The interorbital space is converted into a very deep channel by 2 vertical thin lamellæ which arise on either side, and mark off the narrow interorbital space from the contiguous supraocular areas. From the base of these vertical lamellæ arise externally the thin supraocular plates, which extend outward and upward and roof over the orbit. A deep narrow channel is included between the lamellæ and the plates. The floor of the interorbital groove is raised mesially into a sharp ridge, which is continuous anteriorly with the ethmoidal ridge and posteriorly with a ridge running along middle of occiput. On anterior half of occiput this ridge is a high strong keel; posteriorly, it becomes lower and rounded. The occipital region is swollen and prominent, much higher than the interorbital space. It is bounded laterly by 2 strong rounded ridges which originate at the upper posterior margin of the orbit and converge rapidly backward. The occipital cartilage is heavy and strong, not yielding readily to pressure. The width of interorbital space is ⅙ orbit; the distance between outer margins of orbital plates above middle of eyes is ⅔ diameter of eye. The opercle is marked with delicate striæ diverging downward and backward, but is without strong ridges. The front of dorsal is midway between adipose fin and gill opening, slightly nearer base of caudal than tip of snout. The fin contains 8 rays. Anal badly mutilated, containing at least 24 rays. The mutilated condition of the type will not permit further description. Length 155 mm. Cortez Banks off San Diego, California, in deep water; known only from the type taken by the *Albatross* at Station 3627, in 776 fathoms. (Named for Walter Miller, professor of classical philology in Leland Stanford Jr. University, in recognition of his intelligent interest in zoological nomenclature.)

3030——100

Page 531. In key under *h*, for "incomplete" read "complete."

Page 537. We can not separate *Synodus jenkinsi* from *Synodus scituliceps*, and the former name should probably be abandoned.

Page 555. *Macrostoma angustidens* and related species need further study. The synonymy and application of the names *angustidens, elongatus,* and *resplendens* are uncertain. *Macrostoma brachychir* is probably a good species.

In *M. caudispinosum* the dorsal has 20, not 36, rays.

Page 580. The generic name *Bonapartia*, Goode & Bean, is preoccupied in ornithology. For its use in fishes the name *Zaphotias* is proposed, taking the same species (*pedaliota*) as type. The genus and its species would then stand as follows:

274. ZAPHOTIAS, Goode & Bean, new generic name.

(*Zaphotias*, having organs which emit light; ζα, intensive particle; φῶς, light.)

872. ZAPHOTIAS PEDALIOTUS (Goode & Bean).

Page 582. *Cyclothone microdon* occurs also in Bering Sea in very deep water.

Page 586. *Astronesthes* is from ἄστρον, star; ἐσθής, vestment.

Page 594. *Plagyodus* (Steller) should probably supersede *Alepisaurus,* in which case the family becomes *Plagyodontidæ.*

Page 603. *Sternoptyx diaphana* is common off both the Japanese and Hawaiian islands.

Page 608. For *Aldrorandia*, Goode & Bean, substitute the earlier name *Halosauropsis*, Collett.

Halosauropsis, COLLETT, Camp. Sci. Hirondelle, June, 1896, 143 (*macrochir*).

Page 618. Add:

916(a). MACDONALDIA ALTA, Gill & Townsend.

D. 32; A. 31 to end of dorsal, 52 spines, 125 rays. Body comparatively high, greatest height equal to 3⅔ the distance between vent and tip of snout; pectoral fin with its root twice as far from upper cleft of branchial aperture as from the lateral line, and much nearer to the posterior end of operculum than to lateral line. Bering Sea; only the type known. (*altus*, deep.)

Macdonaldia alta, GILL & TOWNSEND, Proc. Biol. Soc. Wash., XI, 1897 (Sept. 17, 1897), 232, Bering Sea, Lat. N. 54° 54', Long. W. 168° 59', Albatross Station 3604, Aug. 13, 1895, in 1,401 fathoms. (Type, No. 48774, U. S. Nat. Mus. Coll. *Albatross.*)

916(b). MACDONALDIA LONGA, Gill & Townsend.

D. 33; A. 26 to opposite end of dorsal, 55 spines, 111 rays. Body comparatively slender, with the greatest height about ⅓ distance between vent and tip of snout; pectoral fin with its root 3 times as far from upper cleft of branchial aperture as from lateral line, and very much nearer lateral line than end of operculum. Bering Sea; only the type known. (*longus*, long.)

Macdonaldia longa, GILL & TOWNSEND, Proc. Biol. Soc. Wash., XI, 1897 (Sept. 17, 1897), 232, Bering Sea, Albatross Station 3607, 1895, in 900 fathoms. (Type, No. 48775, U. S. Nat. Mus. Coll. *Albatross*.)

Page 627. *Lucius vermiculatus* occurs also in Texas, specimens having been obtained in both the Trinity and Neches rivers near Palestine, by Evermann & Scovell.

Page 632. *Aplocheilus* = *Apocheilichthys* = *Haplocheilus* = *Panchax*, is a genus distinct from *Fundulus*, and should be erased from the synonymy of the latter. The genus is defined by the flat, much produced snout, and the long anal fin.

To the synonymy of *Fundulus* add:

Plancterus, GARMAN, Monogr. Cyprinodonts, in Mem. M. C. Z., XIX, No. 1, 96, 1895 (*kansœ* = *zebrinus*.)

Page 635. In the key, under *aa*, the phrase "inhabiting mountain springs and brooks" applies only to Nos. 943 and 944. It should be transferred and made a part of *s*.

Page 637. *Fundulus punctatus* and *F. vinctus* are wrongly placed by Garman in the synonymy of *F. parvipinnis*.

Page 638. *Fundulus pallidus* is placed by Garman in the synonymy of *F. grandis*, to which it bears but little resemblance.

Page 639. To the synonymy of *Fundulus majalis* add:

Hydrargyra formosa, STORER, Proc. Bost. Soc. Nat. Hist. 1837, 76.

Page 641. To the synonymy of *Fundulus heteroclitus macrolepidotus* add:

Hydrargyra ornata, LE SUEUR, Journ. Ac. Nat. Sci. Phila., I, 1817, 131, Delaware River, near Philadelphia. (Coll. G. Ord.)

Garman regards *Fundulus grandis* as a good species. We have recently compared specimens from Cape Cod with others from Tampa, and reach the same conclusion.

Page 642. Before *Fundulus ocellaris* insert:

932(c). FUNDULUS HETEROCLITUS BADIUS, Garman.

(This is the form found about Grand Manan, named but not character-ized by Garman.)

Garman refers *Fundulus ocellaris* to the synonymy of *Fundulus grandis*, which is very doubtful.

Page 643. Garman refers *Fundulus fonticola* also to the synonymy of *F. grandis*, which is not correct. He also wrongly regards *Fundulus ber-mudæ* as a variety of *heteroclitus*.

Page 644. *Fundulus robustus* is referred, probably by error, by Garman to the synonymy of *F. labialis*, which is certainly incorrectly made a variety of *F. parvipinnis*.

Page 645. Garman refers *Fundulus zebra, zebrinus*, and *extensus* to the synonymy of *Fundulus adinia*, all of which is certainly wrong. Such refcrences defy all our knowledge of the geographic distribution of these fishes. For example, *F. extensus* is a brackish-water fish of Cape San Lucas; *F. zebra,* which is the basis of *F. zebrinus,* is a fish of the mountain streams of New Mexico, Colorado, and northeastward, while *F. adinia* is found near the mouth of the Rio Grande. There is no doubt that the original *Fundulus zebra* is the species called *zebrinus* by us and *kansæ* by Garman. It came from some point between "Fort Union and Fort Defiance." In other words, it came from the head waters of the Canadian River or the Rio Grande. No species of this type has been recorded from the upper Rio Grande, but the species called *zebrinus* and *kansæ* is in all the upper waters of the Arkansas basin, to which the Canadian River belongs, and doubtless in the streams above Fort Union.

Page 646. To the synonymy of *Fundulus zebrinus* add:

Fundulus kansæ, GARMAN, Monogr. Cyprinodonts, 103, pl. 2, fig. 10, 1895, Kansas.

This species (*F. zebrinus*) is rightly made the type of a new subgenus, or possibly genus, *Plancterus,* by Garman. It has long, convoluted intestines and very small pharyngeals. *Fundulus seminolis* (subgenus *Fontinus*) has short intestines and coarse pharyngeals.

Page 648. *Fundulus stellifer* is wrongly referred by Garman to the synonymy of *F. catenatus.*

Page 649. *Fundulus lineatus* is referred by Garman to the synonymy of *F. sciadicus,* which reference seems to be correct.

Fundulus albolineatus, which Garman also refers to *F. sciadicus,* seems to be a perfectly good species. It is certainly not *F. sciadicus.*

Garman refers *Fundulus confluentus* to the synonymy of *F. grandis,* which is probably not correct.

Page 650. Garman's reference of *Fundulus funduloides* to the synonymy of *F. grandis* may be correct.

The species called *Fundulus dovii* in the text is an *Aplocheilus* and should stand as:

968(a). APLOCHEILUS DOVII (Günther).

Garman recognizes *Zygonectes* as a distinct genus, but its boundaries are not easily defined.

The description of *Fundulus confluentus* should be modified to include the following, taken from the type: Head $3\frac{3}{4}$; depth $4\frac{1}{2}$. D. 11; scales 44 or fewer. A black spot on middle of membrane of last 3 dorsal rays. This species resembles *F. diaphanus* rather than *F. majalis. Fundulus ocellaris* seems to be identical with *F. confluentus.*

Page 651. Garman refers *Fundulus macdonaldi* to the synonymy of *F. sciadicus,* which is probably correct, but the reference of *F. floripinnis* to the same synonymy is certainly wrong.

Page 652. Garman refers *F. pulvereus* to the synonymy of *F. grandis,* which is without warrant.

Page 655. To the synonymy of *Fundulus chrysotus* add:

Gambusia arlingtonia, GOODE & BEAN, Proc. U. S. Nat. Mus. 1879, 118, Arlington River, Florida. (Type, No. 21308. Coll. Dr. Goode.)

Zygonectes henshalli, JORDAN, Proc. U. S. Nat. Mus. 1879, 237, San Sebastian River, Florida. (Type, No. 23449. Coll. Dr. James A. Henshall.)

To the synonymy of *Fundulus cingulatus* add:

Zygonectes rubrifrons, JORDAN, Proc. U. S. Nat. Mus. 1879, 237, San Sebastian River, Florida. (Type, No. 23450. Coll. Dr. James A. Henshall.)

Zygonectes auroguttatus, HAY, Proc. U. S. Nat. Mus. 1885, 556, Westville, Florida. (Type, No. 37362. Coll. Mann & Davison.)

An examination of a large amount of material recently collected in Florida by Drs. Evermann and Kendall shows that the synonymy of these species should stand as indicated above.

Examination of the type of *Gambusia arlingtonia* shows it to be the young of the form hitherto known as *Z. henshalli*, which, from an examination of the type and other specimens, proves to be the female of *Fundulus chrysotus*. The dorsal in *Gambusia arlingtonia* is not inserted so far back as the sixth anal ray, but is rather over the third or fourth. Both *G. arlingtonia* and *Z. henshalli*, agree with descriptions of *F. chrysotus* except in coloration. Both are females, as shown by form of anal fin. All specimens examined of the *henshalli* form are females, as shown in part by dissection and by the form of the anal fin. All specimens examined of the form agreeing with descriptions of *F. chrysotus* prove to be males, as shown partly by dissection and by the form of the anal fin. Front series of teeth much enlarged in all; anal fin usually with 11 rays.

The type of *Z. rubrifrons* differs from that of *Z. henshalli* in having a heavier head, really longer snout, mandible more oblique, giving the muzzle a truncated appearance, and the slope of the back to the snout beginning farther forward.

The type of *Z. rubrifrons* agrees with the description of *F. cingulatus*, except in the number of anal rays, there being 10 instead of 8, as given in the description, which is a redescription of the type of *F. cingulatus*. Cuvier & Valenciennes, however, give 10 anal rays in the original description. Specimens in the United States National Museum labeled *Zygonectes cingulatus*, from Pensacola, Florida, agree with the type of *Z. rubrifrons*, with the exception of 1 specimen, which has 11 anal rays.

National Museum specimens collected by Dr. Shufeldt at New Orleans, labelled *Zygonectes chrysotus*, contain both the *Z. henshalli* and *Z. chrysotus* forms, i. e., those with pearly spots and no cross bars (females) and those with dark cross bars (males)—that is, male and female of *Fundulus chrysotus*. Comparison of specimens collected at Tampa and Welaka, Florida, reveal 2 color forms. Most of those from Tampa have the heavier head, truncated muzzle, and outlines of *F. cingulatus*. The 2 color forms are those with dark cross bars, all males as shown by dissection and form of anal fin, and those with no cross bars and no pearly spots, which are all females. The majority of individuals have 10 anal rays each.

Most of the Welaka specimens have more slender and pointed head, preorbital less deep, really shorter snout, and the curve of the body toward the snout beginning farther back than in the preceding, and the majority have each 11 anal rays. The 2 color forms represent the 2 sexes—females with pearly spots and no cross bars, and males with dark cross bars and many with small brown spots.

While a few of the *chysotus* form are found in the Tampa collection, and a few of the *cingulatus* form with the Welaka lot, they can be easily distinguished. A very few of the *cingulatus* form have 11 anal rays and a very few of *chrysotus* 10, but they can be otherwise distinguished. Whereas the females of *F. cingulatus* have no trace of pearly spots the females of *F. chrysotus* almost invariably have them. As a rule, the cross bars in the male, *F. cingulatus,* are narrower and more numerous than in the male of *F. chrysotus,* though young individuals of the latter do not differ in this respect. In *Fundulus cingulatus* there are often faint spots on the scales of the back forming longitudinal lines which seem to be absent in *F. chrysotus.* The teeth in the front row of *F. chrysotus* are larger than in *F. cingulatus.*

Page 658. Garman refers *Fundulus guttatus* to *F. nottii,* which is very doubtful, but he is right in so referring *F. hieroglyphicus.* He also refers *F. dispar* to *F. nottii,* which is probably wrong.

Page 658. *Fundulus guttatus* (Agassiz) can not be separated from *Fundulus nottii* (Agassiz).

. **Page 659.** *Fundulus melapleurus* is, as Garman observes, a *Gambusia,* and should stand as *Gambusia melapleura.*

Adinia guatemalensis and *A. pachycephala* are recklessly referred by Garman to the synonymy of *Fundulus parvipinnis.* They might just as well have been placed at random under any other species of a totally different fauna.

Page 660. Before *Adinia* insert:

300(a). APLOCHEILUS, McClelland.

Snout flat, both jaws much depressed. Bones of mandible firmly united; upper jaw protractile; each jaw with a narrow band of villiform teeth. Body oblong, depressed anteriorly, compressed posteriorly. Dorsal fin short, commencing behind the origin of the anal, which is more or less elongate. Intestinal tract but slightly convoluted; air bladder present. (Günther.)

Aplocheilus, McCLELLAND, Ind. Cypr. As. Res., XIX, 301, 1839 (*chrysostigmus* = *panchax*).
Panchax, CUVIER & VALENCIENNES, Hist. Nat. Poiss., XVIII, 380, 1846 (*panchax*).
Haplochilus, GÜNTHER, Cat., VI, 310, 1866, corrected spelling.

968(a). APLOCHEILUS DOVII (Günther).

For description and synonymy see p. 650.

Page 662. According to Garman the air bladder is present in *Rivulus.* He refers *R. marmoratus* to the synonymy of *R. cylindraceus,* which is probably correct.

Add the following species:

973(a). RIVULUS ISTHMENSIS, Garman.

Head 3⅓ in body; eye 3 in head; snout 6. D. 9; A. 11; V. 6; P. 15; scales 32–8. Elongate, compressed posteriorly, depressed forward; head broad, much depressed, flattened on the crown; snout medium, blunt;

interorbital width greater than eye. Origin of dorsal fin over middle of base of anal, $\frac{3}{4}$ distance from snout to base of caudal; origin of anal fin midway between head and caudal, the last ray nearly as far back as that of dorsal; caudal elongate, póinted, as long as head. Light olivaceous, with a dark blotch at base of dorsal and another on back above or in front of first anal ray; apparently a light, transverse streak at base of caudal. Rio San Jose, Costa Rica.

Rivulus isthmensis, GARMAN, The Cyprinodonts, Mem. Mus. Comp. Zool., XIX, No. 1, July, 1895, 140, Rio San Jose, Costa Rica. (Type in M. C. Z.)

Page 663. *Lucania ommata* is wrongly referred to the synonymy of *Heterandria formosa* by Garman.

Page 664. The species called *Lucania goodei* in the text has 2 rows of teeth and is a true *Fundulus*, or rather *Zygonectes*, as Garman has shown. It may stand as *Fundulus goodei*.

Page 665. *Lucania venusta* is wrongly referred by Garman to *L. parva*, to which, however, it is closely related.

Page 668. Garman wrongly refers *Characodon bilineatus* and *C. variatus* to the synonymy of *C. lateralis*.

Page 669. Add:

883(a). CHARACODON EISENI, Rutter.

Head $3\frac{1}{2}$; depth $3\frac{1}{4}$; eye 3. D. 11 to 13; A. 13; scales 30 to 32-12. Snout shorter than eye, lower jaw projecting. About 9 teeth in upper jaw and about 14 in lower; teeth strongly bicuspid, the villiform teeth not developed. Mouth almost vertical when closed, mandible about $\frac{1}{2}$ length of eye; interorbital space flat, the anterior part equal to orbit, wider posteriorly. Insertion of dorsal in middle of total length; anal inserted under fourth ray of dorsal; pectoral reaching past insertion of ventral; tips of depressed dorsal and anal in vertical through middle of caudal peduncle; caudal broad, truncate, length of middle rays equal to length of top of caudal peduncle. Head about $\frac{1}{4}$ of total; greatest depth of body above ventrals; depth of caudal peduncle $\frac{1}{2}$ its length. Color in alcohol, male with a broad indefinite lateral band; female with dark blotches on sides which in 1 of 3 specimens form distinct cross bands. This species is most closely related to *Characodon variatus*, Bean. It differs from that species in having fewer rays and scales, much fewer teeth, larger eye, much more posterior position of dorsal, and in color. Length $1\frac{1}{4}$ inches. Rio Grande de Santiago, Tepic, Mexico.

Characodon eiseni, RUTTER, Proc. Cal. Ac. Sci. 1896, 266, Rio Grande de Santiago, Tepic, Mexico. (Type, No. 4999, L. S. Jr. Univ. Mus. Coll. Dr. Gustav Eisen.)

Page 670. Add the following:

The specimens from Parras, Mexico, referred by Garman to *C. lateralis*, appear to be new. They may be described as follows:

984(a). CHARACODON GARMANI, Jordan & Evermann, new species.

B. 4; D. 12; A. 12; V. 6; P. 17; scales 32-11 or 12; vertebræ 15+18. Body compressed, moderately stout, caudal pedicel deep, back gently arched. Head about ⅓ of length to base of caudal; very little arched transversely. Snout short, not as long as the eye; chin steep. Mouth medium; upper jaw protractile. Teeth in outer series bicuspid. Eye large, nearly equal to interorbital space, ¼ longer than snout, ⅔ of head. The specimen examined had 4 branchiostegal rays on each side; whether this is normal must be decided from others. Fins small; dorsal origin about ⁵⁄₉ of the distance from snout to caudal; anal opposed to dorsal; posterior margin of caudal subtruncate. Olive to reddish brown, with scattered small spots of darker on the back, a darker band with or without spots of dark along the flank, more distinct posteriorly. Fins with fine dots of dark color. Parras, Coahuila, Mexico. (Named for Prof. Samuel Garman of the Museum of Comparative Zoology, in recognition of his valuable studies of the *Cyprinodonts*.)

Caracodon lateralis, GARMAN, The Cyprinodonts, Mem. Mus. Comp. Zool., XIX, No. 1, pl. 1, fig. 9, 1895, Parras, Coahuila, Mexico; not of GÜNTHER.

984(b). CHARACODON LUITPOLDII, Steindachner.

Head 4⅛ to 4⅔; depth 3 to 3¹⁵⁄₁₆; eye 4 to 4⅔ in head; snout 3 to 3⅔; interorbital 1¾ to 2. D. 14; A. 15 or 16; P. 15 or 16; V. 6; scales 40–17. Body moderately slender; caudal peduncle strongly compressed; head short; upper profile slightly arched, somewhat depressed at occiput; ventral outline more convex; bases of anal and dorsal quite oblique. Dorsal rounded, longest ray 1⅔ in head; anal somewhat smaller. Outer teeth slender, movable, broadened toward front of jaw which is notched; behind these a band of minute teeth, scarcely distinguishable. Two rows of scales below eye; preorbital, jaws, and narrow border of preopercle scaleless. Pectoral shorter than head, not reaching ventrals, which are nearer snout than base of caudal; origin of dorsal nearer base of caudal than gill opening; anal slightly behind dorsal. Color in alcohol, upper half of body light brown or brownish gray, lighter gray or silvery gray below, fading to yellowish white toward ventral line; a silvery gray band along middle of side, not well defined, its width that of 1 or 2 scales. Lake Pátzcuaro, Mexico. (Steindachner.)

Characodon luitpoldii, STEINDACHNER, Einige Fischarten Mexico, 12, pl. 2, figs. 3–3b, 1895, Lake Pátzcuaro, Mexico. (Coll. Princess Theresa von Bayern.)

Page 675. Garman refers *Cyprinodon elegans* to the synonymy of *C. eximius* and *C. felicianus* to that of *C. riverendi*, both of which seem to be correct.

Page 680. *Gambusia infans* is probably identical with *G. gracilis*, as indicated by Garman.

Page 681. Garman calls our *Gambusia affinis G. patruelis* and makes *G. holbrooki*, the northern form, a distinct species, neither of which views seems to be justifiable.

Page 682. *Gambusia nobilis* and *G. nicaraguensis* are referred by Garman to the synonymy of *G. gracilis*, which is questionable; but his reference to *G. puncticulata* of *G. picturata* is probably correct.

Page 682. After *Gambusia affinis* add:

1000(a). GAMBUSIA TRIDENTIGER, Garman.

Head $4\frac{1}{2}$; depth at anal $4\frac{1}{2}$; snout short, not as long as eye, narrow, rounded forward, and blunt. D. 7 or 8; A. 10; V. 6; P. 12; scales 28 to 30–8; vertebræ $14 + 17$. Mouth medium, directed obliquely upward; lower jaw longer than the upper, which is short, narrow, and protractile. Teeth in the outer series larger, strongly hooked, pointed, broadened somewhat toward the apex; inner series very small, in bands, tricuspid as in *Pœcilia;* pharyngeal with a shoulder. Eye large, longer than snout, 3 in head. Fins small, excepting the caudal; dorsal smaller than anal and farther back, its origin about midway from occiput to end of caudal, nearly above the hindmost anal ray, 17 or 18 scales from the head; anal origin midway between snout and end of caudal; farther forward on the male, between the ventrals, and the fin is modified to form an intromittent organ about $\frac{1}{3}$ length of entire fish; caudal deep, as long as head, rounded on hinder margin. Scales large, median series on flank as wide as eye. Intestine short. Light olivaceous, yellowish or brownish, with 7 or 8 vertical bars of brownish, separated by light or silvery spaces of equal width, on the sides of the caudal portion, edges of scales darker, the centers or median series more or less silvery; belly and lower surface of head silvery or golden; peritoneum black, showing through abdnominal wall; occiput dark; top of snout light; a dark line between ånal and caudal; dorsal with a faint spot or group of puncticulations behind the middle near the base; other fins plain to dark tipped. (Garman.) Isthmus of Panama, in fresh water (*tridentiger,* bearing trifid teeth).

Gambusia tridentiger, GARMAN, Cyprinodonts, Mem. Mus. Comp. Zool., XIX, No. 1, 89, pl. 4, fig. 10, 1895, Isthmus of Panama.

Pages 688 and 689. *Heterandria versicolor* and *H. occidentalis* are correctly referred by Garman to the genus *Pœcilia.* It is not improbable that *H. versicolor* is the same as *Pœcilia vivipara,* Bloch & Schneider.

Lebistes is doubtless identical with *Pœcilia,* as is also *Acropœcilia. Acropœcilia tridens* is probably identical with *Pœcilia dominicensis,* as stated by Garman.

Page 691. Garman wrongly refers *Pœcilia butleri* to the synonymy of *P. sphenops.*

Most of the Mexican and Central American species are imperfectly known and imperfectly described. Of these Garman refers the following to the synonymy of *P. sphenops,* whether correctly or not only a study of adequate material can determine: *Pœcilia mexicana, P. thermalis, P. petenensis, P. dovii, P. couchiana, P. plumbeus, P. fasciatus,* and *P. spilurus.*

Pœcilia pavonina is referred, perhaps correctly, to the synonymy of *P. vittata.*

Page 696. Garman thinks that *Pœcilia vandepolli* is identical with *P. reticulata,* Peters, which may be described as follows:

1032. PŒCILIA RETICULATA, Peters.

D. 7 or 8; A. 8 or 9; V. 5; scales 26 to 28–8. Depth of body $\frac{2}{7}$ and length of head nearly $\frac{1}{4}$ of the length to the base of the caudal. Males rather more slender. Eye longer than snout, not quite ¼ of head, ⅔ of interorbital space. Forehead flat. Dorsal origin somewhat nearer to end of snout than to end of caudal, opposite first ray of anal on females. Anal of male advanced, between the ventrals, which are elongate; anal process as long as the head, without hooks. Caudal large, rather longer than head, obtusely rounded; free portion of tail somewhat elongate, base of anal being ¼ of its distance from the caudal; ventrals reaching anal; pectorals as long as the head, not reaching ventrals. Female yellowish olive, scales with a narrow blackish edge, belly silvery, trunk above the belly blackish. Male with 2 brown streaks along the trunk, sometimes conflu·ent into a band, a brown streak along the middle of the side of the tail, a round black spot behind the shoulder, another at the commencement of the caudal streak, and a third at the root of the caudal; 1 or 2 of these spots may be absent. Trinidad; Venezuela (*reticulatus*, netted).

The male from Venezuela differs in color from those from Trinidad. It has large silvery patches between the brown streaks, and a large ovate black spot in the middle of the side of the tail. (Günther.)

NOTE.—The following is the original description: "Grüngelblich mit einem schwarzen Netzwerk, dessen Maschen den Rändern der Schuppen parallel liegen, am Bauche silbrig. Schuppen in 7 Längs- und in 27 Querreihen; obwohl einige derselben durchbohrt erscheinen, ist doch keine deutliche Seitenlinie zu sehen. Ganze Länge 39, Höhe 9, Länge des Kopfes 7 Millimeter. D. 8; A. 10. Caracas; in dem Guayre-Flusse von Gollner gesammelt."

Pœcilia reticulata, PETERS, Monatsb. Berl. Ak. 1859, 412, Caracas; GARMAN, Cyprinodonts, 63, 1895.

Girardinus guppii, GÜNTHER, Cat., VI, 353, 1866, Trinidad; Venezuela; EIGENMANN, Proc. U. S. Nat. Mus. 1891, 65.

Girardinus vandepolli, VAN LINDTH DE JEUDE, Notes from Leyden Museum, IX, 137, 1887, Curaçao, one of the Leeward Islands.

Pœcilia vandepolli arubensis, VAN LINDTH DE JEUDE, Notes from Leyden Museum, IX, 137, 1887, Aruba, one of the Leeward Islands.

Pœcilia branneri, EIGENMANN, Ann. N. Y. Ac. Sci. 1894, 629.

Page 697. Garman refers *Pœcilia elongata*, one of the best marked species of large size, and marine in its habitat, to the synonymy of *P. gillii*. This is certainly wrong, as is also the reference to *P. gillii* of *P. chisoyensis* and *P. boucardi*.

P. melanogaster is probably correctly referred to *P. dominicensis*.

Page 698. Add:

1037(a). PŒCILIA CUNEATA, Garman.

B. 5; D. 8 to 10; A. 10 or 9; V. 6; P. 15 or 16; scales 28 or 29–9. Short and deep; caudal pedicel deep. Head depressed, broad, flat on the crown, equaling depth between dorsal and anal, or ¼ of the length to the base of the caudal; snout as long as the eye, broad, truncate; chin short, steep; mouth wide, directed upward; jaws weak, loosely joined, lower short, upper shorter, protractile; outer series of teeth slender, oar-shaped, hooked, movable; inner in bands, small, pointed; eye large, as long as snout, ¼ of interorbital space, ⅔ of head. Dorsal larger than anal, origin midway from head to base of caudal, over third ray of anal, 13 scales behind the occiput. Anal small, acute angled, third ray longest; on the male the base of the anal is forward of that of the dorsal, the fin is modified to form a sharp-pointed organ in which the rays are less changed than in most species; its length is less than that of the head. Ventrals small, not reaching the anal. Pectorals reaching back over 7 scales. Caudal deep, as long as the head, hind margin rounded. Scales large. Intestine long. Brownish, olive tinted, bases of scales dark, back darker, and top of head darkest; more or less of the hind margin, or $\frac{1}{2}$ of the scale, is whitish to silvery on the scales of the flank; lighter to silvery under head and abdomen; dorsal with 1 to several transverse series of small spots of black; fin sometimes black tipped; a brownish streak extending back and upward on the opercle behind the eye; caudal with small spots of black on the basal half, or with a couple of clouded transverse bands; other fins uniform or puncticulate; very small ones are lighter with a faint silvery band along the middle of the flank, but without vertical bars; a large one has numerous small white spots, somewhat like *Fundulus heteroclitus*. Females $2\frac{1}{4}$ and males $1\frac{9}{10}$ inches. Turbo, Gulf of Darien.

Pœcilia cuneata, GARMAN, Cyprinodonts, 62, pl. 5, fig. 3, 1895, Turbo, Gulf of Darien.

Page 704. After *Typhlichihys*, Girard, add:

a. No scleral cartilages; no pigment in or about the eye; retinal elements readily separable into ganglionic, inner reticular, and nuclear layers, the nuclear and outer reticular layers rarely distiguishable; diameter of eye about .150 mm.

SUBTERRANEUS, 1047.

aa. Scleral cartilages large, forming a hood over front of eye; a mass of pigment in front of eye; pigment layer of retina with more or less pigment; eye a mere vestige, about .040 mm. in diameter. ROSÆ, 1047(a).

Page 706. After *Typhlichthys subterraneus* add:

1047(a). TYPHLICHTHYS ROSÆ, Eigenmann.

Extremely close to *T. subterraneus*, from which it seems to differ only in the less development of the eye. Scleral cartilages large, forming a hood over the front of the eye; a mass of pigment in front of eye; pigment layer of retina with more or less pigment; eye a mere vestige, $\frac{1}{5}$ the size of that of *T. subterraneus*, about .040 mm. in diameter. The types of this species are 2 small, thoroughly dissected specimens, in the Museum of Indiana University, collected from a cave in Jasper County, Missouri, by Miss Ruth Hoppin. (Named for Mrs. Rosa Smith Eigenmann.)

Typhlichthys rosæ, EIGENMANN, Science, N. S., vol. VII, No. 164, 227, February 18, 1898, cave near Sarcoxie, Jasper County, Missouri.

Page 723. *Hemiramphus balao* is a valid species as defined.

Page 729:

Exocœtus volitans, Linnæus, as Lönnberg has shown, is identical with *E. evolans* L. As the genus *Exocœtus*, Syst. Nat., Ed. x, 316, is based solely on *Exocœtus volitans*, the name *Exocœtus* must go with this species, taking the place of *Halocypselus*. The ordinary flying fishes must therefore be called *Cypsilurus*. The species with long anal fin may, however, be held as generically distinct from the type of *Cypsilurus*, and for them (*exsiliens, rondeletii*, etc.) the name *Exonautes* has been proposed by Jordan & Evermann, Check List, 322. (Type, *exsiliens*.) (ἔξο, out of; ναυτής, swimmer.)

Our species of *Exonautes* are the following:

1080. EXONAUTES EXSILIENS (Müller).
1081. EXONAUTES RONDELETII (Cuvier & Valenciennes).
1082. EXONAUTES VINCIGUERRÆ (Jordan & Meek).
1083. EXONAUTES SPECULIGER (Cuvier & Valenciennes).
1084. EXONAUTES RUFIPINNIS (Cuvier & Valenciennes).

To these should be added the following:

1084(a). EXONAUTES AFFINIS (Günther).

Head 4; depth 6; eye 3½; snout 3¼. D. 11 to 13; A. 11 to 13; scales 6–50 to 52, 35 before dorsal. Interorbital space flat, slightly greater than eye. Pectoral fin extending scarcely beyond dorsal and anal; base of ventral midway between eye and base of caudal, its rays reaching beyond middle of base of anal; dorsal opposite anal, its anterior rays 2¼ in head. Pectoral with an oblique white blotch across its lower half, and with a narrow whitish margin; ventrals grayish. Cuba? Atlantic; West Africa. (Günther.) Probably distinct from *E. speculiger*.

Exocœtus affinis, GÜNTHER, Cat., VI, 288, 1866, Cuba?

The species of *Cypsilurus* are the following:

1085. CYPSILURUS HETERURUS (Rafinesque).
1086. CYPSILURUS LUTKENI (Jordan & Evermann).
1087. CYPSILURUS FURCATUS (Mitchill).
1088. CYPSILURUS NIGRICANS (Bennett).
1089. CYPSILURUS XENOPTERUS (Gilbert).
1090. CYPSILURUS LINEATUS (Cuvier & Valenciennes).

Under this species (p. 739) for Corea (in 3 places) read Gorea.

1091. CYPSILURUS CYANOPTERUS (Cuvier & Valenciennes).

This is a good species. The specimens recorded from James Island belong to *C. bahiensis*.

1092. CYPSILURUS BAHIENSIS (Ranzani).
1093. CYPSILURUS CALIFORNICUS (Cooper).
1094. CYPSILURUS CALLOPTERUS (Günther).
1095. CYPSILURUS GIBBIFRONS (Cuvier & Valenciennes).

Page 732. In the key, for "*jj*" read "*hh*," for "*jjj*" read "*hhh*," for "*kk*" read "*ii*," for "*ii*" read "*gg*," and for "*hh*" read "*ff*."

Page 746. According to the studies of Mr. Rutter and Dr. Gilbert all the forms of *Gasterosteus* should probably be reduced to a single species (*Gasterosteus aculeatus*), having 3 or 4 geographic varieties, each running into a number of forms which differ in the degree of armature of the body.

Page 749. After *Gasterosteus bispinosus cuvieri*, insert:

1100(a). GASTEROSTEUS GLADIUNCULUS, Kendall.

Head 3½; depth 3⅘; D. II–I, 10; A. I, 8. Head rather long; eye about 3 times in head; opercle not striate; body deep, compressed, with 5 lateral dermal plates anteriorly counting from pectoral fin, none posteriorly; caudal peduncle short, naked, not keeled; innominate bone lanceolate, its width about 3 times in length; ventral spines rather long, about $1\frac{3}{5}$ times in head, serrated above and below, a strong cusp at base on both upper and lower edge. Color in life, grass green, mottled and finely punctated with black on top of head and back; sides of head and body golden, with dark blotches; breast silvery, ventrals scarlet. In alcohol the back becomes smoky black, the mottling and black dots more distinct, the golden hue of the sides fades, becoming more or less silvery, the dark blotches more pronounced. Coast of Maine and Woods Hole, Massachusetts. (*gladiunculus*, little sword; sticklebacks being called by the boys about Portland, Maine, "Little swordfish.")

Gasterosteus gladiunculus, KENDALL, Proc. U. S. Nat. Mus. 1895, 623, off Seguin Island, Maine. (Type, No. 47589. Coll. *Grampus*.)

Page 754. *Aulostomus maculatus* is pinkish-red in life.

Page 757. *Fistularia tabacaria* has been recorded by Storer from Holmes Hole, Massachusetts, and H. M. Smith records it from Buzzards Bay, near Quisset, and from about Woods Hole.

Page 762. In the key to species of *Siphostoma* read:

> *eee.* Dorsal covering 4 or 5 caudal (not body) rings.
> > *o.* Rings 16 to 18 + 29 to 33.
> > > *q.* Rings 16 + 30 to 33; dorsal 30 to 34, on 3 + 5 rings.

Page 767. It is doubtful if *Siphostoma pelagicum* occurs in America. *S. rousseaui* has probably been sometimes mistaken for it.

Page 768. After *Siphostoma jonesi* add:

1124(a). SIPHOSTOMA ROBERTSI, Jordan & Rutter.

Head $7\frac{1}{2}$ in length; depth $2\frac{4}{5}$ in head; eye $5\frac{2}{3}$ in head. Dorsal 20, on 0 + 4 rings; segments 17 + 32. Snout $2\frac{1}{5}$ in head, with a slight keel; a slight keel on top of head, another above opercle, and 1 on anterior side of opercle, but not reaching posterior edge; shields without spines; lateral keel ending on last body segment; ventral keel on next to last; upper body keel extending nearly to end of dorsal fin, upper caudal beginning below it on first caudal segment; all ridges of body very prominent, the tail with 4 plain ridges; caudal pouch 3 in total length. Color mottled brown, paler below, the membrane connecting the segments pale bluish, forming cross stripes which are especially marked on the egg pouch; prominent pale cross bars on lower side of head; dorsal colorless, except that the base is finely dusted with brown; caudal thickly dusted with brown, except near base. This species is most closely related to *Siphostoma jonesi*, differing in having a shorter dorsal with more rays, and in the lateral keel ending distinct from lower caudal keel. Jamaica; 1 specimen, $4\frac{1}{2}$ inches long, known. (Named for Rev. Joseph Seed Roberts, who collected the type.)

Siphostoma robertsi, JORDAN & RUTTER, Proc. Ac. Nat. Sci. Phila. 1897, 97, Kingston, Jamaica. (Type, No. 4988, L. S. Jr. Univ. Mus.)

1124(b). SIPHOSTOMA STARKSII, Jordan & Culver.

Head 10½; depth 21. Dorsal 38, on 0 + 10 or 11 rings. Rings 13 or 14 + 37 or 38. Head and body in tail 2. Snout 2¾ in head. Dorsal ⅓ longer than head. Body rather stout. Head scarcely carinate above. Snout with a slight smooth carina. Two lateral keels, confluent into 1 behind. Belly slightly keeled; no keel on opercle. Color dark olive, much mottled with darker but without distinct markings; yellow below. Male and female common in the fresh waters of Rio Presidio at Mazatlan, among algæ; not seen in salt or brackish water. The pouch of the male teeming with eggs in January. Length 4 to 6 inches. Mazatlan, Mexico. Common in the Rio Presidio in sluggish water, on the bottom, about a mile below the village of Presidio. The species is probably found in brackish and fresh waters rather than in the sea.

Siphostoma starksii, JORDAN & CULVER, Fishes Sinaloa, in Proc. Cal. Ac. Sci. 1895, 416, pl. 30, Rio Presidio, Mazatlan. (Type, No. 2686, L. S. Jr. Univ. Mus. Coll. Hopkins Exped. to Mazatlan.)

1124(c). SIPHOSTOMA SINALOÆ, Jordan & Starks, new species.

Allied to *Siphostoma arctum* Jenkins & Evermann.

Head 8½ in length to base of caudal; depth 3½ in head. Dorsal 26, on 1½ + 5 rings, 14 + 35. Snout 1¼ in head, a strong median ridge above running to between middle of eyes, a ridge on each side from angle of mouth to below eye, occipital and nuchal plates keeled, a slight keel on anterior part of opercle; dorsal keels ceasing in front of the last 4 or 5 rays of dorsal, the lateral ridge running up and continuing as dorsal ridges; belly with a keel on each side. Preanal part of belly 1¾ in postanal part; pectoral shorter than eye, caudal 3 in head. Color olive brown above, abruptly lighter below lateral ridges anteriorly, the edges of the plates dark, forming reticulations on lower parts of body; between every 4 rings is a narrow white cross bar; from each eye is a narrow light bar running upward and backward to occiput; caudal dark. The 2 type specimens, 1 of which was sent to the British Museum, collected by the Hopkins Expedition at Mazatlan. They were erroneously referred to *Siphostoma arctum* in our paper on the Fishes of Sinaloa. Type, No. 2945, L. S. Jr. Univ. Museum.

Page 772. *Corythroichthys,* Kaup, should apparently be recognized as a genus distinct from *Siphostoma.* The species belonging in it are the following:

1134. CORYTHROICHTHYS ALBIROSTRIS, Heckel.

1135. CORYTHROICHTHYS CAYANNENSIS (Sauvage).

1135(a). CORYTHROICHTHYS CAYORUM, Evermann & Kendall.

Head 8⅔; depth 12⅔; snout 3½ in head; eye 4½. D. 21 rays, on 1½ + 3½ rings; A. 3, on first caudal ring; C. 10; P. 10. Rings 17 + 26 = 43. Body short and stout; head short, snout very short; tail but little longer than head and trunk. Cranial ridges strong; a high, sharp keel on snout, the occipital keel very high, its edge convex, notched near the middle, not continuous with keel on snout; a strong supraocular ridge, beginning opposite posterior end of nasal keel and continuing backward with 1 hiatus upon upper edge of opercle; just below this on the opercle another longer but scarcely stronger ridge; another short ridge on anterior part of opercle at level of lower part of eye; opercles very convex, as if swollen outward; keels on body and tail all strong; the 2 lateral keels on body terminating on third caudal ring; the 2 lateral keels on tail beginning on the last body ring, thus overlapping the body keels; median keel on side well developed, terminating on sixteenth body ring; ventral keels strong; abdominal keel very strong. Egg sac on first 18 caudal rings. Color yellowish brown, with darker punctulations; tip of snout white; cheek, throat, and under parts of snout white, crossed by about 7 or 8 irregular brownish bars extending downward and backward; opercles brown; fins pale. This species is related to *C. albirostris* of Heckel, differing from it chiefly in the shorter snout, smaller dorsal, and fewer rings. Key West, Florida. (*cayorum*, of the Keys; from Cayo Hueso, Bone Key, the original Spanish name of the island of Key West.)

Corythoichthys cayorum, EVERMANN & KENDALL, Bull. U. S. Fish Comm. 1897 (Feb. 9, 1898), 128, pl. 7, fig. 7, near Crawfish Bar, Key West, Florida. (Type, a male 3½ inches long, No. 48784. Coll. Drs. Evermann & Kendall.)

Page 774. *Syngnathus æquoreus* is doubtfully American. Until a comparison of specimens can be made our species may stand as—

1138. SYNGNATHUS HECKELI (Kaup).

Page 792. *Lethostole*, Jordan & Evermann, is identical with *Chirostoma*, and the definition assigned is that of *Chirostoma*.

To the synonymy of *Chirostoma estor* add:

Atherinichthys albus, STEINDACHNER, Anzeiger der Kais. Akad. Wiss. Wien, 1894, 148, Lake Pátzcuaro, Mexico. (Coll. Princess Therese von Bayern.)

Page 793. In *Chirostoma humboldtianum* the scales are serrulate. After this species insert the following:

1155(a). CHIROSTOMA GRANDOCULE (Steindachner).

Head 4; depth 5⅗; eye 3⅖ in head; interorbital width 4⅓; pectoral fin 1½; ventral 2¼; caudal 1⅙; anal base 1¼, its greatest height 1⅞. D. V-I, 10; A. I, 20; P. 15 or 16; scales 60 to 62-15 or 16. Upper profile of head merging gradually into that of back, rising slightly toward beginning of second dorsal. Lower jaw slightly projecting; posterior end of upper jaw reaching eye. Teeth on maxillary sharp, brush-like, in 3 or 4 rows, the inner teeth of the maxillary and the outer teeth of lower jaw somewhat enlarged and close set. Cheek narrower than in *C. humboldtianum* and *C. estor*, and with 4 rows of scales. Origin of first dorsal midway between anterior border of eye and base of caudal, the second dorsal ½ diameter of eye nearer base of caudal than hinder border of eye; greatest height of second dorsal scarcely greater than base of fin. Longest anal ray about 1⅕ in base of fin; dorsal and anal concave on free border; origin of anal nearly an eye's diameter in front of that of second dorsal; caudal deeply incised, the middle rays about 2 in the longest; caudal peduncle more than 4⅔ in body, its least depth somewhat more than 2 in greatest depth of body. Scales slightly ctenoid. Side with a broad, sharply defined silvery-gray band. Body much more slender, snout shorter, and eye larger than in *C. humboldtianum* or *C. estor.* Length 5 inches. Lake Pátzcuaro, Mexico.

Atherinichthys grandoculis, STEINDACHNER, Anzeiger der Kais. Akad. d. Wissensch. Wien. 1894, 149, Lake Pátzcuaro, Mexico. (Coll. Princess Therese von Bayern.)

354(a). ESLOPSARUM, Jordan & Evermann.

Eslopsarum, JORDAN & EVERMANN, Check-List Fishes, 330, 1896 (*jordani*).

This genus is close to *Chirostoma*, from which it differs in the large entire scales. To it belong the 2 following species:

1156. ESLOPSARUM BARTONI (Jordan & Evermann).

1157. ESLOPSARUM JORDANI (Woolman).

To the synonymy of this species should be added

Atherinichthys brevis, STEINDACHNER, Anzeiger der Kais. Akad. d. Wissensch. Wien. 1894, 149, Lake Cuitzeo, Mexico. (Coll. Prinzessin Therese von Bayern.)

Page 793. In *Eslopsarum jordani* the anal is I, 16, not I, 6.

Page 795. *Kirtlandia laciniata* has been found to intergrade with *K. vagrans* and should stand as—

1158a. KIRTLANDIA VAGRANS LACINIATA (Swain).

Page 796. Under *d* in the key read:

d. Snout about equal to eye, which is 3 to 3½ in head.

Page 800. An examination of numerous specimens of *Menidia* from various places between Florida and Halifax shows that *M. notata* and *M. menidia* intergrade perfectly. The first will therefore stand as—

1167a. MENIDIA MENIDIA NOTATA (Mitchill).

Page 801. *Menidia guatemalensis* and *Menidia pachylepis* belong in the genus *Thyrina*, Jordan & Culver.

Page 819. *Agonostomus nasutus* has the anal usually II, 10, sometimes II, 9.

Page 821. Add the following:

In the Transactions of the Jamaica Society of Arts for 1855, Mr. Richard Hill gives a paper on "Fishes of the Jamaica Shores and Rivers" which has been overlooked by subsequent writers. The list is chiefly a nominal one, but it contains a number of vernacular names not elsewhere given. The only new species are given under the head of *Labrax* (page 142) and *Mugil* (page 143), and these are named rather than described. They are the following:

There is another *Labrax*, common enough in the Kingston market when the rains send strong freshets from the river into the harbor. The fishermen call it the river chub, and confound it with the *mucronatus*. It is a different species; it is marked with bauds like the *Perca fluviatilis* of Europe, and the *Perca granulata* of America. We will call it the *Labrax pluvialis*, rainy weather chub.

> *Mugil petrosus*—rock mullet;
> *lineatus*—short mullet, 1;
> *albula*—short mullet, 2;
> *curema*—long mullet;
> *equinoculus*—horse-eye mullet;
> *capitulinus*—drab mullet, long ears;
> *plumieri*—pond mullet;
> *liza*—callipeva;
> *Dajaus monticola*—mountain mullet;
> *choirorynchus*—hog-nose mullet.

The *Dajaus monticola* inhabits only the mountain streams; the *choirorynchus* or hog-nose mullet is a fish of double the size of the *monticola*, and found in the same waters. The *mugil liza* is the largest of the mullets, from 20 inches to 3 feet long; the callipeva is the name by which it is exclusively known. This is, no doubt, its Indian name. The *equinoculus* and *capitulinus*, known in the market as long mullets, are readily distinguishable from each other by the size of the head, and especially by the size of the eye: the horse-eye mullet has the large eye, the *capitulinus* unusually small. The *plumieri*, Plumier's mullet of Cuvier & Valenciennes, is a long mullet; and the *lineatus* and *albula* what the market people distinguish as short mullets. The callipeva is a river mullet seldom extending further than the embouchure of streams, or into the ponds and marshes. The *curema* is a large mullet found on the sea banks; it is the most highly colored of all the mullets, the back is a golden green and it has scales on the second dorsal fin.

Dajaus choirorynchus is identical with *Agonostomus nasutus*, but the scanty description hardly justifies the substitution of this name for the later one. The other new species we fail to identify. *Labrax pluvialis* we do not recognize.

Page 823. The great *Barracuda* should stand as—

1199. SPHYRÆNA BARRACUDA (Walbaum).

To its synonymy add:

Esox barracuda, WALBAUM, Artedi Piscium, III, 94, 1792; after CATESBY.

Page 827. The ventrals in the *Polynemidæ* are truly thoracic, the long pubic bone being attached to the shoulder girdle. This family is probably nearest allied to the *Sciænidæ*.

Page 833. After *Ammodytes personatus* add:

372(a). RHYNCHIAS, Gill, new genus.

Rhynchias, GILL, MS., new genus (*septipinnis*).

This generic name is provisionally given to a species known only from a description of Pallas, and supposed to differ from *Ammodytes* in the presence of ventral fins. It may prove to belong to some different family. (ῥύγχος, snout.)

1214(a). RHYNCHIAS SEPTIPINNIS (Pallas).

This species has not been recognized by any recent collector, and it is not certain to what family it belongs. The following is the substance of Pallas's description:

D. 43; A. 24; V. 8; P. 16; C. 24. Form of *Ammodytes tobianus*. Head

3030——101

compressed; snout long, slender, depressed. Maxillary with fine teeth; rictus long. Branchiostegals 4. Body compressed, slender, with transverse streaks. Scales inconspicuous; 1 lateral line. Pectoral large, unarmed. Dorsal short, well backward, lower posteriorly; caudal sub-bifurcate. Color white, the dorsal edged with darker. Kamchatka. (Pallas.)

If we can trust the description, this fish would seem to represent a distinct genus of *Ammodytidæ*, characterized by the presence of ventral fins, but it may be that the account is erroneous in this regard and that Pallas had in mind *Ammodytes personatus*. (*septem*, seven; *pinna*, fin.)

Ammodytes septipinnis, PALLAS, Rosso-Asiat., III, 1811, Kamchatka.

Page 833.   *Ammodytes alascanus* is not separable from *A. personatus*.

Page 839.   *Caulolepis longidens* occurs also in the Pacific, specimens having been collected by the *Albatross* at Cortez Banks, off San Diego, California, in 1896.

Page 847. Add:

1230(a). MYRIPRISTIS CLARIONENSIS, Gilbert.

Head $3\frac{1}{6}$ in length; depth $2\frac{4}{7}$. D. X-I, 14; A. IV, 12; scales $3\frac{1}{2}$-41-7. Least depth of caudal peduncle $\frac{1}{4}$ length of snout and eye. Greatest (oblique) diameter of eye $2\frac{1}{4}$ in head. Least interorbital width equaling length of snout, $4\frac{1}{2}$ in head. Mouth less oblique than in related species, the line of upper jaw with a more pronounced double curve. Lower jaw the longer, with well-developed symphyseal knob. Teeth finely villiform, very slightly enlarged toward middle of both jaws; wide patches of similar teeth on head of vomer and on palatine bones. Length of maxillary (measured from front of upper jaw) very slightly (about $\frac{1}{10}$) less than length of snout and eye. Color before immersion in spirits, reddish, the upper parts dusky, especially on top of head and on the margins of the scales; evident horizontal dusky streaks between the rows of scales; opercular membrane blackish; fins all light, without dark markings. Differing from all known American species of *Myripristis* in having $3\frac{1}{2}$ series of scales between the lateral line and the base of the spinous dorsal, instead of $2\frac{1}{2}$. Length $6\frac{1}{4}$ inches. Revillagigedo Islands; only the type known.

Myripristis clarionensis, GILBERT, Proc. U. S. Nat. Mus. 1896, 441, pl. 69, Clarion Island, Revillagigedo Archipelago. (Type, No. 47746. Coll. Dr. Gilbert.)

Page 852. Insert the following description by Jordan & Rutter of *Holocentrus marianus*, based upon a specimen 6 inches long from Jamaica:

Head $2\frac{2}{3}$; depth 3 in length; eye $2\frac{1}{4}$ in head. D. XI, 13; A. IV, 9; scales 4-45-7. Dorsal outline much more curved than ventral; mouth low, but little oblique, the lower jaw projecting and entering upper profile; maxillary to below middle of eye; eye large, lower margin of orbit cut by a line connecting tip of snout and upper base of pectoral; angle of opercle high, higher than top of pupil, with 3 sharp teeth, small teeth along the margin next the subopercle; subopercle long and narrow, dentate near upper end; preopercle very finely serrate, with a strong spine at angle; a single row of scales on opercle along margin of preopercle; suborbital bones very narrow, finely serrate; premaxillary groove on top of head as long as eye; length of pectoral equals head behind middle of eye; spinous dorsal depressible into a groove, highest (anterior) rays of soft dorsal equal to ventrals, longer than soft rays of anal; third anal spine very long and heavy, as long as pectorals; caudal forked almost to base, the lobes equal, as long as pectorals. Each row of scales with a red band, yellow lines between the rows; fins all yellowish. This is a strongly marked species, very different from *Holocentrus ascensionis*, perhaps the type of a distinct genus, characterized by the large mouth and projecting chin.

Page 856. Dr. Bean reports the Red Mullet or Goat Fish (*Mullus auratus*) as being plentiful at Sandy Hook in September and October.

Page 857. The nominal genus *Mulloides* can not be separated from *Upeneus.*

Page 866. In *Scomber colias* read: Head about 3; depth $4\frac{2}{3}$; first dorsal longer than high.

Page 873. To the synonymy of *Scomberomorus* add:

Polipturus, RAFINESQUE, Anal. de la Nature 1815, 84; substitute for *Scomberomorus.*

Page 874. In line 12 of description of *Scomberomorus maculatus,* for "side" read "part."

Page 878. *Bipinnula,* Jordan & Evermann, is a synonym of *Escolar,* Jordan and Evermann, in Goode and Bean, Oceanic Ichthyology, 519, 1896. The error resulted from Goode & Bean taking our original MS. name *Escolar,* for which we afterwards substituted *Bipinnula.*

This genus and its species will therefore stand as follows:

396. ESCOLAR, Jordan & Evermann.

Escolar, JORDAN & EVERMANN, in GOODE & BEAN, Oceanic Ichthyology, 519, Aug. 23, 1896 (*violaceus*).

Bipinnula, JORDAN & EVERMANN, Fishes North and Middle Amer., 878, Oct. 3, 1896 (*violaceus*).

1267. ESCOLAR VIOLACEUS (Bean).

Page 886. Instead of *Lepidopus caudatus,* which is not yet known to occur in American waters, insert:

1276. LEPIDOPUS XANTUSI, Goode & Bean.

Head $4\frac{2}{3}$ in body; depth 3 in head; eye $5\frac{1}{4}$; interorbital space $8\frac{1}{4}$; snout 3; maxillary $3\frac{1}{4}$. D. 82; A. II, 45. Jaws with long, sharp teeth in front, followed by single rows of weaker ones, arranged in groups of twos and threes. Height of dorsal, near middle of body, 3 in head. Anal preceded by 2 scutes, the first minute, the second wide, strongly keeled, its length $\frac{3}{4}$ the diameter of eye. Pectorals of 12 rays, length 2 in head. Each ventral consists of a flat keeled spine followed by a minute ray. This species is known from 2 small mutilated specimens, both found on the beach near San Jose del Cabo, Cape San Lucas. The type was taken by John Xantus, about 1860, and recorded by Jordan & Gilbert as *Lepidopus caudatus.* The second, of about the same size ($5\frac{1}{2}$ inches), was taken by

Richard C. McGregor, in 1897. From the latter the above account was taken. The species differs from *Lepidopus caudatus* in the much shorter dorsal and longer anal. D. 103; A. 24. (Named for John Xantus de Vesey.)

Lepidopus caudatus, Jordan & Gilbert, Proc. U. S. Nat. Mus. 1882, 358; not of Euphrasen.
Lepidopus xantusi, Goode & Bean, Ocean. Ichth., 519, 1896; same type; no description.

Page 889. *Trichiurus lepturus* is recorded by Storer from Buzzards Bay (1840) and Wellfleet, Massachusetts (1845), and H. M. Smith records it from Woods Hole (1897).

Page 892. The synonymy at top of page under *Tetrapturus imperator* belongs to the footnote on same page.

Page 899. Add:

1286(a). OLIGOPLITES MUNDUS, Jordan & Starks, new species.

Head 4; depth $2\frac{4}{5}$; eye $4\frac{1}{4}$. D. V–I, 19; A. II–I, 20. Body deep and compressed. Length of head about $\frac{1}{6}$ greater than its depth at nape; eye equal to snout and to interorbital; maxillary extending considerably beyond vertical from hinder margin of eye, its length $1\frac{3}{4}$ in head; second suborbital not over $\frac{1}{3}$ as wide as lowest, and much shorter, thus forming a prominent notch in posterior margin of suborbital bones; a slight emargination in opercle in front of pectoral. Teeth small, sharp, in a band in each jaw, narrow in upper. Origin of soft dorsal midway between snout and base of caudal, the anal opposite; the anterior rays of both somewhat produced; second soft ray of each equal to head behind pupil, and equal to pectoral; ventrals equal to $\frac{3}{4}$ of pectorals, their inner margins fastened to body; caudal deeply forked, the middle rays $3\frac{1}{2}$ in longest, which are longer than head. Lateral line nearly straight, but forming a broad angle above pectoral. Color silvery on sides, becoming darker above; fins colorless. This species differs from *Oligoplites altus* in the much larger mouth and in having the suborbital bones notched posteriorly. *Oligoplites saliens* of the West Indies seems to be more elongate in body and with the suborbitals even behind as in *O. altus*. Pacific coast of tropical America.

This description is based on a specimen 11 inches long from San Juan Lagoon, Mexico, at the mouth of Ahome River, collected by the *Albatross*. Three other specimens from Algodones Lagoon, Mexico (*Albatross* Coll.), agree in every respect, except that 1 of them has but 4 free spines in front of dorsal.

Numerous other specimens have been since brought by Dr. Gilbert from Panama.

Oligophites mundus, Jordan & Starks, in Jordan & Evermann, Check-List Fishes, 344, 1896, Mazatlan, Mexico; name only.

Page 909. The Californian species *Trachurus symmetricus* is probably a species distinct from *T. picturatus*, described from Madeira. The two forms have never been properly compared.

Page 912. The identity of *Hemicaranx amblyrhynchus* with *Caranx falcatus*, Holbrook, needs proof. The latter species, if distinct, may be described as follows:

1305(a). HEMICARANX FALCATUS (Holbrook).

Head 6 in total length; depth about 3. D. VII–I, 28; A. II–I, 25; C. 19; V. 5; P. 16; lateral line with 50 plates. Body oval, compressed; the head short, the facial outline descending in a gentle curve to snout, which is rounded though narrow. Eye large, in the middle third of the head, the posterior margin rather nearer snout than posterior margin of opercle; nostrils close together, nearly midway between eye and snout, and on a line within the orbit, the posterior larger, subround, the anterior ovoidal. Mouth small; each jaw with a single row of slender, conical teeth; a small patch of minute teeth on the vomer, and a small, narrow group of similar teeth on the palatines; tongue small, narrow, a few minute teeth near its base; pharyngeal bones armed with numerous card-like teeth, longer than those of the jaws. Soft dorsal long and low, the first 3 or 4 rays moderately elevated, the fin scaled at base; pectoral falcate, very long, extending to anterior third of soft dorsal; ventral small, very short, reaching beyond vent; anal shaped like the soft dorsal; caudal very long and widely forked, the upper lobe more than $\frac{1}{3}$ longer than the lower. Lateral line at first almost semicircular; at origin of soft dorsal descending to median plane, then straight; plates beginning with the soft dorsal increasing in size to the thirty-fifth, whence they decrease rapidly; scales minute, those of lateral line elongated quadrilateral, with 1 angle prolonged and rounded. Color, upper part of head and body above lateral line pale brown with slight bluish tint; lower jaw, opercle, and side yellowish; belly silvery, with a slight golden tint; anterior dorsal transparent; posterior transparent but with a yellowish tint; caudal yellowish. Known certainly only from Charleston, South Carolina.

Caranx falcatus, HOLBROOK, Ichth. South Carolina, 92, pl. 13, fig. 2, Charleston, South Carolina.

Page 914. Add:

1306(a). HEMICARANX ZELOTES, Gilbert, new species.

Head 4 to $4\frac{1}{5}$; depth $2\frac{2}{5}$ to $2\frac{3}{5}$. D. VII–I, 26 to 29; A. II–I, 23 to 25; P. 20 to 22; scutes about 52. Body regularly elliptical, its greatest depth about in middle of its length, exclusive of caudal peduncle. Head small; anterior profile more decurved, and hence the snout is blunter than in *H. atrimanus;* depth of head just behind eye about $\frac{5}{6}$ its length. Jaws subequal, tip of lower slightly projecting; maxillary narrow, not quite reaching anterior margin of pupil, about $3\frac{1}{6}$ in head ($3\frac{1}{3}$ to $3\frac{2}{5}$ in *atrimanus*). A single series of small, close-set, subequal teeth in each jaw; no teeth on vomer, palatines, or tongue. Orbit considerably greater than snout, $3\frac{1}{8}$ to $3\frac{4}{7}$ in head. Interorbital width (taken at anterior margin of orbit) slightly less than orbit. Occiput with an evident carina. Distance from snout to first dorsal spine greater than length of pectoral. Spinous dorsal very low, the highest spine considerably less than orbit (greater than orbit in *atrimanus*); a well-developed antrorse spine before the dorsal; soft dorsal and anal similar, not falcate, the rays decreasing in size from the first; highest ray of soft dorsal 2 to $2\frac{1}{2}$ in head; highest ray of anal about $2\frac{1}{3}$ in head; dorsal and anal depressible into a high sheath of scales, the last 3 or 4 rays uncovered; caudal fin wide, well forked, the upper lobe the longer, the longest ray not quite $\frac{1}{4}$ total length of body; pectoral fin long, $3\frac{1}{5}$ to $3\frac{1}{2}$ in body ($2\frac{2}{5}$ to $2\frac{4}{7}$ in *atrimanus*); ventrals $2\frac{2}{5}$ to $2\frac{3}{5}$ in head. Scales as in *atrimanus;* lateral line with a very strong curve anteriorly, the height of the curve $2\frac{3}{8}$ to $3\frac{1}{2}$ in its length; its length $2\frac{1}{8}$ to $2\frac{1}{4}$ in the straight portion; entire length of straight portion with scutes, which are very small in front and behind; scutes considerably wider and lower than in *atrimanus*, the widest about $\frac{1}{2}$ diameter of orbit (about $\frac{1}{8}$ diameter of orbit in *atrimanus*). Coloration much as in *H. atrimanus*, but darker, and the fins without yellow; blackish olive above, dusky silvery below; top of head and snout black; spinous dorsal and the broad margins of soft dorsal and anal black; caudal dark, margined with black; pectorals very dark, black inside, the extreme lower rays light; a large jet-black blotch at base, on each side of pectorals, extending for about $\frac{1}{5}$ the whole length of the fin; axil black. Closely related to *Hemicaranx atrimanus.* Like it, it has a large jet-black area at axil and base of pectoral, and differs from it in the following characters: In having a shorter pectoral, shorter ventrals, profile of snout more rounded, a lower spinous dorsal, a shorter maxillary, a higher, shorter curve in lateral line, wider scutes, which are less sharply carinated, and darker fins. Panama. (ζελοτής, an imitator.)

Page 921. *Caranx crysos* and *Caranx pisquetus* are probably distinct species, the former ranging from New York to Florida, the latter from the West Indies to Brazil.

In *Caranx pisquetus* the pectoral fins are very long, as in the Pacific species *Caranx caballus*, from which we can not separate it. The species need further study.

Page 934. After *Vomer setipinnis* add:

1329(a). VOMER SPIXII (Swainson).

Head $2\frac{7}{8}$; depth $1\frac{3}{5}$; D. VI–I, 22; A. I, 18; eye $3\frac{2}{3}$ in head; maxillary $2\frac{1}{3}$; snout $1\frac{1}{3}$; caudal 1. Body very deep, in form much like *Selene œrstedii;* profile very steep, almost vertical; snout slightly protruding. Mouth oblique, maxillary reaching to the vertical from front of eye; gill rakers 7 to 27, the longest a little more than $\frac{1}{2}$ eye. Lateral line strongly arched in front, the arch $1\frac{1}{4}$ the straight part; plates of lateral line little differentiated; pectoral falcate, as long or slightly longer than head; ventrals small, under base of pectorals. Color bluish above, sides silvery, fins except ventrals and anal dusky. Here described from specimens from Jamaica about 10 inches in length. These specimens are evidently different from the Northern *Vomer setipinnis* (=*Vomer browni*), the body in specimens of the same length being much deeper. It corresponds to the figure given by Agassiz of *Vomer browni*, this figure being the basis of *Vomer spixii* of Swainson. Probably all West Indian records of *Vomer setipinnis* belong to *Vomer spixii.* (Named for Jean Baptiste Spix, of Munich, naturalist and explorer.)

Platysomus spixii, SWAINSON, Class. Fishes, III, 250 and 406, 1839, Brazil; after AGASSIZ & SPIX.

Vomer gabonensis, GUICHENOT, Ann. Soc. Maine et Loire, 1865, 42, Gaboon.

Page 938. To the synonymy of *Chloroscombrus chrysurus* add:

Seriola cosmopolita, CUVIER, Règne Ánimal, Ed. 2, vol. II, 1829, Gorea; after *Scomber chloris*, BLOCH.

Add the following species:

1334(a). CHLOROSCOMBRUS ECTENURUS, Jordan & Osgood.

Head $3\frac{7}{4}$; depth $2\frac{3}{8}$. D. VIII–I, 27; A. II–I, 26. Snout slightly shorter than eye, which is $3\frac{1}{4}$ in head. Chord of curved part of lateral line $1\frac{3}{4}$ in straight part. Depth of caudal peduncle 2 in its length, measuring from the base of the last dorsal ray to the base of the first caudal ray. Pectorals long and falcate, 3 in length; ventrals short, $2\frac{1}{8}$ in head, extending beyond the vent, which is situated in a groove in which these fins fit. Depth of head equal to or slightly less than its length; maxillary reaching anterior edge of eye, $2\frac{3}{4}$ in head. Lateral line unarmed; curve of ventral outline very slightly more pronounced than that of the dorsal; dorsal and anal fin sheaths well developed. Tips of upper spines and rays dusky; a black blotch at base of upper rays of caudal, and a black axillary and opercular spot. The species is closely related to *Chloroscombrus chrysurus*, the common species of the South Atlantic and Gulf States, which it evidently represents in the West Indies. The species *chrysurus* is deeper in every way, having a deeper body, a deeper head, and a deeper caudal peduncle. In *chrysurus* also the eye is larger, the mouth more nearly vertical, and the arch of the lateral line higher. When specimens of equal size from Florida and Havana are compared the characters are very evident. In 2 specimens, each $7\frac{1}{2}$ inches in length, from Havana and Florida, respectively, the depth of the body of the one is contained $1\frac{1}{4}$ times in that of the other, the depth of the head $1\frac{1}{8}$, the depth of the caudal peduncle $1\frac{1}{8}$, and the length of the eye $1\frac{1}{4}$. The names *chrysurus* (South Carolina), *latus* (Carolina), and *caribæus* (Texas) evidently all belong to the species of the United States coast. The type of *chloris* came from Acará, in Guinea, and *cosmopolita* of Cuvier was originally as a mere substitute for *chloris*. Until the African species can be examined, it is better not to use the name for either of the American forms. Probably *Chloroscombrus chloris*, when studied, will be found distinct from either. If not, that name would take the place of *ectenurus*. West Indies; known from Jamaica and Cuba. (ἐκτενής, extended; οὐρά, tail.)

Chloroscombrus ectenurus, JORDAN & OSGOOD, Proc. Ac. Nat. Sci. Phila. 1897, 101, Jamaica. (Coll. J. S. Roberts.)

The validity of *Chloroscombrus ectenurus* is still doubtful.

Page 942. After *Trachinotus falcatus* add:

1337(a). TRACHINOTUS RHOMBOIDES (Bloch).

Head $3\frac{1}{5}$; depth $1\frac{3}{5}$ in length; eye $3\frac{1}{5}$ in head. D. VI–I, 20; A. II–I 18. Back much elevated, but not angulated at origin of soft dorsal; end of snout not vertical, curved; head slightly concave at occiput. Maxillary to below anterior margin of pupil; eye on level of lower edge of premaxillary and axil of pectoral. Origin of soft dorsal behind tip of pectoral, its lobe much elongated, extending to middle of caudal; lobe of anal reaching to below base of caudal; caudal lobes equal, $2\frac{1}{2}$ in body; pectoral rounded, $1\frac{1}{2}$ in head; ventrals $2\frac{3}{4}$ in head. Scales minute, large posteriorly near lateral line. Pale olive above, becoming silvery on belly; lobes of vertical fins dusky. This West Indian species is apparently different from the northern *Trachinotus falcatus* with which it has been confounded. *Trachinotus falcatus* seems to be confined to the coasts of the United States. In specimens of the same size the vertical fins are much higher in the West Indian species. ($\rho\acute{o}\mu\beta\acute{o}\varsigma$, rhomb; $\varepsilon\check{\iota}\delta\omega\varsigma$ resemblance.)

Chætodon rhomboides, BLOCH, Ichth., 1787, pl. 209, Martinique.

Page 945. After *Trachinotus paloma* insert:

428(a). ZALOCYS, Jordan & McGregor.

Zalocys, JORDAN & McGREGOR, Rept. U. S. Fish Comm. 1898 (*stilbe*).

This genus is closely allied to *Hypodis*, Rafinesque ($=Lichia$, Cuvier), differing in the absence of a procurrent spine before the dorsal, and in the cultrate thoracic region. From *Trachinotus* it is distinguished by the same characters and also by the lower forehead and nonfalcate dorsal and anal fins. *Hypodis* is scarcely different from *Trachinotus*, the only tangible characters being the larger teeth, the low dorsal, and the less elevated forehead. *Porthmeus*, Cuvier ($= Lichia$ *amia* and *L. vadigo*) is a well-defined genus, distinguished by the large mouth and projecting lower jaw. ($\zeta\acute{a}\lambda\eta$, surge of the sea; $\grave{a}\varkappa\acute{v}\varsigma$, swift.)

1344(a). ZALOCYS STILBE, Jordan & McGregor.

Head $4\frac{1}{2}$; depth $2\frac{1}{2}$. D. VI-I, 26; A. II-I, 23. Body elliptical, deeper than in *Hypodis glaucus;* belly sharply compressed; ventral outline similar to that of dorsal; anterior profile of the head elevated and sharp, the eye being rather below than above its middle; eye 5 in head, with conspicuous adipose eyelid before and behind; posterior nostril much larger than anterior; vertically oblong maxillary broad, without supplemental bone, extending to pupil, $2\frac{5}{8}$ in head. Mouth moderate, oblique; each jaw with bands of villiform teeth; similar teeth on vomer, palatines, and tongue. Preopercle very broad; cheek moderate; suborbital narrow; preorbital very narrow, 4 in eye. No pseudobranchiæ. Gill rakers very long and slender, numerous. No procumbent spine before dorsal; spines low and separate, progressively higher; soft dorsal and anal each with a sheath of scales; first rays of dorsal very slightly elevated, $2\frac{1}{4}$ in head; anal without distinct anterior lobe, longest ray $2\frac{3}{4}$ in head; caudal peduncle long and slender; depth $3\frac{3}{4}$ in head; length below $2\frac{1}{4}$ in head; caudal fin widely forked; lobes long and slender, upper a little the longer, more than $\frac{1}{4}$ longer than the head and $2\frac{3}{4}$ in body; pectoral moderate, $1\frac{1}{2}$ in head; ventrals very small, $6\frac{1}{2}$ in head; snout $3\frac{3}{4}$ in head; premaxillary protractile. Color dark steel blue or blackish above; lower parts soiled white; axil and base of pectoral within jet-black; dorsal and anal each with a narrow whitish edging; caudal black, each lobe with a narrow whitish edging within. Body covered with small smooth scales, much as in *Trachinotus;*

lateral line undulate, very slightly arched anteriorly. Clarion Island; 1 specimen, 16 inches in length, known. (στίλβη, shining.)

Zalocys stilbe, JORDAN & McGREGOR, Rept. U. S. Fish Comm. 1898, pl. 5, Clarion Island, Revillagigedo Archipelago. (Type, No. 11996, L. S. Jr. U. M. Coll. R. C. McGregor.)

Page 965. *Rhombus, Palometa,* and *Poronotus* should probably stand as distinct genera. The species placed in *Rhombus* in the text would then stand as follows:

> **1363. RHOMBUS PARU** (Linnæus).
>
> **1364. RHOMBUS XANTHURUS** (Quoy & Gaimard).
>
> **1365. PALOMETA PALOMETA** (Jordan & Bollman).
>
> **1366. PALOMETA MEDIA** (Peters).
>
> **1367. PALOMETA SIMILLIMA** (Ayres).
>
> **1368. PORONOTUS TRIACANTHUS** (Peck).

The identity of the South Atlantic Coast *Rhombus alepidotus* with the West Indian *Rhombus paru* is very doubtful.

Page 973. The genus *Acrotus,* Bean, represents a family distinct from *Icosteidæ.*

<h3 style="text-align:center">Family CXXXVI(a). ACROTIDÆ.</h3>

Two additional specimens of *Acrotus willoughbyi* have lately come to light—the one from Port Townsend, the other from Monterey.
After *Acrotus willoughbyi* insert:

<h3 style="text-align:center">Family CXXXVI(b). ZAPRORIDÆ.</h3>

Body robust, moderately compressed, the back not elevated, the belly not carinate. Body covered with small adherent cycloid scales, which cover the membranes of all the fins except the distal third, as also the gill membranes, lower jaw, cheeks, opercles, and nuchal region. No lateral line; no spinules. Head short, the nape not elevated, the forehead broad and abruptly convex in profile; eye moderate, placed high; preopercle, parietal region, and region about eye with very large open mucous pores. No spines on head; edges of membrane bones of head covered with thick scaly skin. Mouth moderate, terminal, oblique, its cleft mainly anterior; upper jaw protractile, but not movable; maxillary rather narrow, simple; lower jaw very heavy, its thick lip projecting beyond upper jaw. Teeth alike in both jaws, rather strong, blunt, even, close set, forming a uniform cutting edge; no teeth on vomer, palatines, or tongue, the tongue very thick. Lower pharyngeals narrow, with bluntish teeth, those on the edge larger; upper pharyngeals rather large, with small, blunt, velvety teeth; no distinct tooth-like processes in the œsophagus; pseudobranchiæ present; gill rakers very slender and flexible, rather short; gills 4, a large slit behind the fourth; gill membranes separate, free from the isthmus; opercle adnate to shoulder girdle above its angle; coracoids not largely developed. Pectoral fin long, rounded, attached a little nearer ventral than dorsal outline; ventrals wholly wanting. Dorsal fin beginning above gill opening, composed entirely of simple inarticulate rays or spines, these moderately flexible, attached to the membrane to their tips, and all except the first and last of about equal length. Caudal peduncle short and stout, not contracted, the large caudal subtruncate or rounded at tip, and without procurrent rays; vent nearly median. Anal much shorter than dorsal, somewhat higher, and composed of soft rays, subequal in length. Skeleton rather limp and flexible, but much less so than in *Icosteus.*

445(a). ZAPRORA, Jordan.

Zaprora, JORDAN, Proc. Cal. Ac. Sci. 1896, 202 (*silenus*).

Characters of the genus included above.

This genus bears some resemblance to *Icichthys,* but differs in the stout caudal peduncle, absence of ventrals and lateral line, and in the form and structure of the head. Among the genera known to us it seems to come nearest to *Icichthys,* and it might be placed among the *Icosteidæ* were it not for the presence of pharyngeal teeth. (ζά, an intensive particle; πρώρα, prow.)

1372(a). ZAPRORA SILENUS, Jordan.

Head $5\frac{2}{5}$ in length to base of caudal; depth $4\frac{1}{5}$. D. LVI; A. 27; P. 20 to 22; C. 22; scales about 200–85. Greatest thickness of body about $\frac{2}{5}$ its depth; length of caudal peduncle $1\frac{2}{5}$ in its least depth, which is $1\frac{9}{10}$ in head. Eye $5\frac{1}{3}$ in head; snout $5\frac{1}{3}$; interorbital space 3; maxillary $2\frac{3}{4}$, ending under front of pupil; mandible $2\frac{1}{2}$, its depth $4\frac{3}{8}$; teeth about $\frac{15}{30}$ on each side; lips, snout, and bones about eye naked; rest of head covered with small scales. Lower jaw with a thick lip, slightly fringed on its edge, and with a mesial frenum; the rounded tip entering the profile when the mouth is closed. Three large pores on each ramus of mandible; behind these 3 others in a line on horizontal limb of preopercle; 3 on vertical limb; 2 close together in front of eye; 1 near the nostrils, so similar to them that there seem to be 3 nasal openings; 7 on suborbitals; 4 in 2 rows behind eye; 1 above eye, and before upper edge of preopercle; a horizontal row of 5 along temporal region, the last and largest of all in opercular flap above gill opening; 1 at vertex; 1 between vertex and eye, and 2 on each side of nape. Gill rakers 8+20, the longest $\frac{1}{2}$ eye. No trace of lateral line. Scales small, resembling those of a salmon, covering the membranes of all the fins on the basal two-thirds. Pectoral as long as head, its base $2\frac{1}{3}$ in head; longest dorsal spine $1\frac{5}{6}$; caudal $1\frac{1}{10}$; longest anal ray $1\frac{2}{3}$. Color in spirits uniform dusky, without markings on the body, the belly pale, and the side of the head irregularly blotched with lemon yellow, apparently bright in life, and brightest about the pores of the head. Coast of British Columbia; only the type, 29 inches long, known. (σειλένος, a drunken demigod, covered with slime, in allusion to the open mucous pores.)

Zaprora silenus, JORDAN, Proc. Cal. Ac. Sci. 1896, 203, pl. 20, Nanaimo, Vancouver Island. (Type in Provincial Museum at Victoria. Coll. H T. Stainton.)

Page 982. Prof. Harrison Garman records *Elassoma zonatum* from Waccamaw River, Whitesville, North Carolina, and Little Pedee River, South Carolina. Vertebræ 29; scales 34 to 36; D. IV, 9; A. III, 5.

Page 1019. Under *kk* read: "gill membranes narrowly or broadly connected."

Page 1047. Before *Ulocentra* insert:

1436(a). COTTOGASTER CHENEYI, Evermann & Kendall.

Head 4; depth 6; eye 4 in head; snout 4; maxillary 3½; interorbital width 5½. D. XI–12; A. II, 8; scales 7–56–6. Body rather stout, heavy forward, compressed behind; head heavy; mouth moderate, slightly oblique, lower jaw included, maxillary reaching front of pupil; premaxillaries protractile. Cheeks, opercles, breast, and nape entirely naked; scales of body large and strongly ctenoid; lateral line complete, straight; median line of belly naked anteriorly, with ordinary scales posteriorly. Fins large; dorsals separated by a space equal to ½ diameter of eye; origin of spinous dorsal a little nearer origin of soft dorsal than tip of snout, its base about equal to length of head; longest dorsal spine 2½ in head, the outline of the fin gently and regularly rounded; soft dorsal higher than spinous portion, the second to tenth rays about equal in length, scarcely 2 in head, the first, eleventh, and twelfth rays but slightly shorter than the others; anal moderate, its origin under base of third dorsal ray, the spines slender, the second a little longer than the first, whose length is 3¾ in head; longest anal rays about 2⅓ in head; caudal lunate, the lobes more produced and pointed than usual among darters; pectorals long and pointed, the middle rays longest, about 1⅙ in head, reaching tips of ventrals; ventrals well separated, not nearly reaching vent, the longest rays 1¼ in head. Color in alcohol, back dark brownish, covered with irregular spots and blotches of darker; side with about 8 or 9 large dark spots lying on the lateral line; belly pale; top of head dark; snout black; lower jaw and throat dark; a broad black line downward from eye to throat; cheek and opercles rusty; spinous dorsal crossed by a median dark line; ventrals blue black; other fins pale, but dusted with rusty specks. An examination of the 14 cotypes shows some variation in the species. In 2 examples there is a well-developed frenum, rendering the premaxillaries nonprotractile, and in a third specimen the frenum is partially developed; in some individuals the origin of the spinous dorsal is exactly midway between the tip of snout and origin of soft dorsal. The females and immature males are less highly colored than the adult male described above. Length 1¾ to 2¼ inches. This species is most closely related to *Cottogaster shumardi*, from which it may be readily distinguished by the shorter snout, the naked cheeks and opercles, the smaller soft dorsal, the smaller anal, and the different coloration. Fifteen examples of this interesting darter were obtained July 18, 1894, by Messrs. Evermann & Bean in the Racket River near Norfolk, St. Lawrence County, New York. It did not seem to be very common, as only 15 examples resulted from numerous hauls of the collecting seine. (Named for Mr. A. Nelson Cheney, State fish-culturist of New York, in recognition of his valuable contributions to our knowledge of the food and game fishes of that State.)

Cottogaster cheneyi, EVERMANN & KENDALL, Bull. U. S. Fish Comm. 1897 (Feb. 9, 1898), 129, pl. 8, fig. 8, Racket River near Norfolk, New York. (Type, No. 48781. Coll. Evermann & Bean.)

Page 1049. After *Ulocentra gilberti* add:

1438(a). ULOCENTRA MEADIÆ, Jordan & Evermann, new species.

Head 3⅔; depth 4¾; eye 3⅓ in head; snout 3¾; interorbital 5. D. XII–12; A. II, 7; scales 7–48–6. Body rather heavy, somewhat fusiform; head large; snout blunt, decurved, profile rising abruptly to interorbital, thence nearly horizontal to origin of dorsal, from which it descends gently in a straight line to caudal peduncle; opercular spine small but sharp; mouth low, horizontal, rather large, the maxillary reaching vertical at front of orbit; premaxillaries protractile; branchiostegal membranes not connected, free from the isthmus; ventral fins close together, the space separating their bases about ⅓ diameter of orbit; fins all moderate; distance from tip of snout to origin of spinous dorsal 3 in body; spinous and soft dorsals close together, the space separating them about 2 in orbit; longest soft dorsal rays 1⅓ in head, about equaling those of anal; the two anal spines of about equal length, the first the stouter; pectorals long, longer than head, their tips passing those of ventrals but not reaching vent; ventrals short, 1⅕ in head; caudal slightly lunate when expanded. Scales rather large, strongly ctenoid; cheeks and breast naked; opercle scaled above, naked below; nape scaled; lateral line complete, straight; ventral line of body covered with ordinary adherent scales. Color in alcohol, yellowish or olivaceous above and on sides, the back with 6 dark saddle-like blotches, the first just anterior to origin of spinous dorsal, the second under the fifth and sixth spines, the third under the last two spines, the fourth under the sixth and seventh soft rays, the fifth just posterior to the last dorsal ray, and the sixth, which is quite small, upon the caudal peduncle at the base of the caudal fin; sides blotched with dark, 6 to 8 larger dark blotches along side just below lateral line, sometimes more or less continuous with the dark dorsal blotches; a dark blotch at base of middle caudal rays; belly pale; top of head dark; a dark spot at lower posterior angle of eye and a smaller one back of it on upper edge of opercle; a dark band downward from eye; opercle dark; upper lip dark, interrupted by a light line at the symphysis; spinous dorsal pale, with a broad dark band through its lower third; soft dorsal crossed by 3 or 4 irregular lines of dark specks; caudal with about 4 broad dark cross bars; other fins pale. Length 2 inches. This species somewhat resembles *U. gilberti*, but differs from it in the larger head, stouter body, larger scales, naked cheeks, larger mouth, and in other respects. Known only from Indian Creek, basin of Powell River, east Tennessee, where 3 examples were collected October 17, 1893. (Named for Mrs. Meadie Hawkins Evermann.)

Ulocentra meadiæ, JORDAN & EVERMANN, new species, Indian Creek, Cumberland Gap, Tennessee. (Type, No. 48903. Coll. Dr. R. R. Gurley.)

Page 1051. To the synonymy of *Ulocentra simotera* add:

Etheostoma duryi, HENSHALL, Journ. Cin'ti Soc. Nat. Hist., April, 1889, 32, small tributary of Tennessee River at Whiteside, Tennessee. (Type in Mus. Cin'ti Soc. Nat. Hist. Coll. Charles Dury.)

Page 1089. To the synonymy of *Etheostoma cœruleum* add:

Etheostoma formosa, HENSHALL, Journ. Cin'ti Soc. Nat. Hist., April, 1889, 32, small tributary of Tennessee River at Whiteside, Tennessee. (Type in Mus. Cin'ti Soc. Nat. Hist. Coll. Charles Dury.)

Page 1109. Add:

1501(a). APOGON ATRICAUDUS, Jordan & McGregor.

Head $2\frac{1}{2}$; depth 3. D. VI–I, 9; A. II, 8; scales largely ctenoid; eye $3\frac{1}{2}$ in head; second dorsal spine stoutest, about 2 in head; gill rakers 17, moderate. Body similar in shape to *A. retrosellus*. Jaws reaching to posterior border of eye, $1\frac{5}{6}$ in head. Pectoral reaching to opposite front of anal, $1\frac{3}{8}$ in head. Color rosy, darkened with dusky points; more or less olivaceous above; head and throat verging on orange; first dorsal black; second dorsal rosy; caudal dusky, more or less flushed with rosy, other fins paler; no black spot on head or on base of caudal, there being no definite markings anywhere except the dusky red of the tail. West coast of Mexico. Numerous specimens collected at San Benedicto, Socorro, and Clarion islands. Usual length 3 to 4 inches. (*ater*, black; *cauda*, tail.)

Apogon atricaudus, JORDAN & MCGREGOR, Rept. U. S. Fish Comm. 1898, Socorro, Clarion and San Benedicto islands. (Coll. R. C. McGregor.)

Page 1125. *Centropomus affinis* can not be separated from *C. ensiferus*.

Page 1148. To the synonymy of *Epinephelus* add:

Phrynotitan, GILL, Stand. Nat. Hist., III, 255, 1885 (*Batrachus gigas*).

Page 1150. In the key under *dd*, read: Lower jaw strongly projecting.

Page 1156. Add:

1551(a). EPINEPHELUS NIPHOBLES, Gilbert & Starks.

Head $2\frac{3}{5}$ in body; depth $2\frac{1}{2}$. D. XI, 14; A. III, 9; scales 16–116–40; eye 5 in head; maxillary 2; third dorsal spine $2\frac{1}{3}$; middle dorsal rays $2\frac{1}{3}$; highest anal rays 2; third anal spine $3\frac{1}{10}$; pectoral $1\frac{5}{6}$; ventrals $1\frac{3}{4}$; caudal $1\frac{3}{4}$. Form rather robust, moderately compressed; dorsal outline uniformly curved from tip of snout to caudal peduncle; mouth large, the maxillary reaching to below posterior orbital rim; lower jaw strongly projecting; teeth conical and sharp, in 1 or 2 bands at sides of jaws, 3 or 4 in front; upper jaw with a rather strong canine on each side of front; snout longer than eye; nostrils close together, the posterior one the larger, a little in front of the vertical from front of eye, the anterior in a short, wide tube with a flap behind; vertical and horizontal limbs of preopercle meeting at right angles, its edge with blunt serræ, those at angle enlarged; opercle with 3 flat spines before the flap; gill rakers moderate, nearly $\frac{1}{2}$ eye, 8+16 in number. Top of head, orbitals, maxillary, and mandible, naked; fine scales on cheeks and opercles; scales on body ctenoid;

fins without scales. Dorsal beginning a little in front of the vertical from pectoral base, the third spine a little the highest, but the ones behind it not much shortened; soft dorsal higher than spinous, its outline rounded; pectoral rounded behind, reaching to below the base of eighth dorsal spine; third anal spine the longest, not nearly so long as the soft rays, the anal fin similar in shape to the soft dorsal; ventrals reaching past vent, scarcely to front of anal, their ends rounded, as are all the fins; caudal broadly rounded. Color in spirits brownish red, sides with clear-cut, distinct, white spots about as large as pupil, about 6 at base of dorsal, 6 or 7 along lateral line, following its arch, a horizontal series of 4 extending back from opercular flap, about 3· from base of pectoral following curve of ventral outline, 2 at base of anal, 1 behind lower edge of caudal peduncle and 1 above anus; a well-marked streak above maxillary following its outline; lips colored like rest of head; dorsal dusky, with vague white spots; ventrals and anal nearly black, with a reddish tinge; anal with a narrow white border below; pectoral and caudal uniform yellowish. Magdalena Bay, Lower California; only the type, 6 inches long, known. ($\nu\iota\phi o\beta\acute{\eta}\varsigma$, snowed over, from the white spots.)

Epinephelus niphobles, GILBERT & STARKS, Proc. U. S. Nat. Mus. 1896, 442, Magdelena Bay, Lower California. (Type, No. 47582. Coll. *Albatross.*)

Page 1164. Species 1558 should probably be called *Alphestes chloropterus* (Cuvier & Valenciennes). The name *afer*, given to a specimen from Guinea, may belong to some other species.   .

Page 1168. Add:

1560(a). DERMATOLEPIS ZANCLUS, Evermann & Kendall.

Head 2⅜; depth 2₁⁹₆; eye 8 in head; snout 3½; maxillary 3; mandible 2. D. XI, 19; A. III, 10; scales difficult to count, but about 30–130-35, those above lateral line counted obliquely backward and downward from origin of dorsal, those below from origin of anal upward and forward to lateral line. Branchiostegals 8; gill rakers 8 + 12, short and stout, the longest 1⅔ in orbit. Body stout, compressed, oblong-elliptical, the dorsal and ventral outlines about equally curved; head moderate, the profile rising from tip of snout to origin of dorsal fin, thence descending in a regular, gentle curve to caudal peduncle; a depression above nostrils and a slight one on nape; interorbital very narrow, equal to orbit; mouth moderate, somewhat oblique; premaxillaries protractile; maxillary broad at tip, reaching vertical at posterior edge of the pupil; supplemental bone well developed; lower anterior edge of maxillary covered by the broad dermal flap of the premaxillary; eye small, high up; nostrils close together and close to eye, the anterior small and round, the posterior oblong-oval, much larger than the other. Small cardiform teeth on each jaw, those in front movable, scarcely canine-like; similar teeth on vomer and a long, narrow band on each palatine. Preopercle coarsely serrate, the serræ short and blunt, more or less obscured by the skin; opercle with a broad dermal border, somewhat produced at lower angle. Fins all large; origin of dorsal slightly in advance of base of pectoral, its distance from tip of snout equal to length of head; third dorsal spine longest, its length about 2⅓ in head or 2½ times length of first ray; interspinal membranes of the spinous dorsal deeply incised, the anterior portion of each somewhat produced beyond its spine; soft dorsal high, the middle rays longest, 1¾ in head, the anterior portion of the fin gently convex, the posterior slightly concave; pectoral short, broad, and rounded, barely reaching origin of anal, the length 1⅓ in head; ventral pointed, the second and third rays longest, 1⅓ in pectoral, the fin somewhat falcate; anal fin strongly falcate, the fourth and fifth rays longest, longer than pectoral, 1⅓ in head, 2½ times length of last anal ray; second anal spine short, 5¼ in head; caudal shallowly lunate, the lobes 1¼ in head. Scales small, smooth, and thin, closely but irregularly imbricated; nape, opercles, and cheeks scaled, snout and lower jaw naked; bases of all the fins except the ventrals densely scaled; lateral line beginning at upper angle of opercle, gently arched above pectoral fin, following approximately the curvature of the back and on median line of caudal peduncle. General color of body in life brown, with large, irregular blotches of dirty white on back and upper part of sides, these blotches with small rusty spots; lower part of sides, belly, and caudal peduncle with irregular whitish spots; belly brassy brown; snout and nape with numerous small, round dark spots; cheek with large blotches of whitish overlaid with black and brassy spots; lips whitish, with dark spots; spinous dorsal blotched with white, olivaceous and black; soft dorsal brown, with numerous white spots and a few black ones, the posterior rays tipped with white and orange; anal olivaceous, with irregular white spots, greenish at edge, the produced rays black toward distal ends; pectoral dark olivaceous, with greenish white splotches, the edge yellowish; ventral rays greenish white, the membranes black; inside of mouth white; eye brown. Related to *D. inermis* (Cuvier & Valenciennes), but differing notably from that species in the shorter, stouter gill rakers, the emarginate caudal, the shorter anal spines, and the strongly falcate anal fin. Length 20 inches. Key West; only the type known. (ζάγκλον, a scythe or sickle, from the falcate anal fin.)

Dermatolepis zanclus, EVERMANN & KENDALL, Bull. U. S. Fish Commu. 1897 (Feb. 9, 1898), 129, pl. 8, fig. 9, Key West, Florida. (Type, No. 48843. Coll. Drs. Evermann & Kendall.)

Page 1186. Add:

1576(a). MYCTEROPERCA HOPKINSI, Jordan & Rutter.

Head 2⅔; depth 4½. D. XI, 15; A. III, 11; scales about 125; eye 6 in head, 1½ in snout. Body long, not much compressed; angle of preopercle sharply serrate; gill rakers 6 + 9, counting rudiments; nostrils close together, the posterior larger, with a horizontal septum across base; profile concave above nostrils; maxillary nearly to posterior margin of eye, 2⅕ in head; lower jaw projecting; 2 anterior canines of upper jaw very strong; third and fourth dorsal spines longest; posterior portion of anal truncate; caudal concave. Pectorals 2, ventrals 2½, and caudal 1½ in head. Color of alcoholic specimen nearly uniform brownish, side of jaws paler; soft dorsal, anal, ventrals, and caudal with a narrow pale edging, these fins otherwise brownish olive, with a subterminal band of black; pectorals pale, darker in middle. Allied to *Mycteroperca calliura,* differing in having fewer gill rakers, more slender body, smaller scales, and a less lunate caudal. Jamaica; only 1 specimen, 6 inches long, known. (Named for Timothy Hopkins.)

Mycteroperca hopkinsi, JORDAN & RUTTER, Proc. Ac. Nat. Sci. Phila. 1897, 105, Jamaica. (Type, No. 5073, L. S. Jr. Univ. Mus. Coll. J. S. Roberts.)

Page 1187. Insert:

1576(a). MYCTEROPERCA BOULENGERI, Jordan & Starks.

Head $2\frac{1}{3}$ in length; depth $2\frac{3}{5}$. D. XI, 14 or 15; A. III, 9 or 10; scales about 90, 20 above and 42 below; snout $3\frac{1}{2}$ in head; maxillary $2\frac{1}{3}$; eye $5\frac{1}{2}$; pectoral $1\frac{3}{4}$; ventral $1\frac{5}{8}$; longest anal ray $1\frac{2}{3}$; caudal $1\frac{2}{3}$; longest dorsal spine $2\frac{1}{2}$; gill rakers short, about 6+17, the longest about $\frac{2}{3}$ eye; longest dorsal ray 2 in head. Body short and deep, compressed; head moderate, compressed, its profile not steep, nearly straight, a depression before eye. The supraoccipital and temporal crests are high, the supraoccipital crest extending to the posterior margin of orbit; the temporal crests are parallel to each other, and extending to pupil; interorbital space concave. Upper canines moderate, the lower quite small. Nostrils small, well separated, the anterior slightly larger. Lower jaw very strongly projecting; maxillary reaching opposite posterior edge of pupil. Preopercle slightly notched, the angle slightly salient, with enlarged teeth. Dorsal not deeply notched, the fourth spine not much elevated; second dorsal high, not long, its angle not rounded; caudal scarcely lunate, the upper lobe long, the lower truncate; anal very high, strongly elevated, its posterior border incised, the anterior rounded; pectoral and ventral moderate. Scales smoothish, not very small. Color olive gray, covered everywhere with oblong irregular markings of black, between which the ground color forms rivulations; gray lines radiating from the eye; a black blotch below maxillary; pectoral olive yellow; other fins blackish, clouded with pale; first dorsal with faint small black spots. Mazatlan, Mexico. (Named for George Albert Boulenger, ichthyologist of the British Museum, in recognition of his epoch marking work on the Percoid fishes.)

Mycteroperca boulengeri, JORDAN & STARKS, Fishes Sinaloa, 445, pl. 38, 1895, Mazatlan, Sinaloa, Mexico. (Type, No. 1621, L. S. Jr. Univ. Mus. Coll. Hopkins Exped. to Sinaloa)

Page 1235. The original type of *Lobotes* is *surinamensis,* not *erate.*

Lobotes erate is a species distinct from *L. surinamensis,* inhabiting the coasts of India and China. *Lobotes farkhari* and *L. incurvus* are probably identical with *L. erate,* and all 3 should be erased from the synonymy of *L. surinamensis.*

Page 1236. After *Lobotes surinamensis* add:

The *Lobotes* of the Pacific coast of Central America is distinguished from the other known species, *L. surinamensis* and *L. erate,* by the small size of the preopercular serrations, those at the angle not elongated and spine-like, even in the young. The following description is furnished by Dr. Gilbert:

1623(a). LOBOTES PACIFICUS, Gilbert, new species.

(BERRUGATE.)

Head $2\frac{4}{4}$ in length; depth $2\frac{1}{5}$ to $2\frac{1}{10}$ (to base of caudal rays); depth of caudal peduncle $2\frac{1}{2}$ in head. D. XII, 15; A. III, 11; pectoral 15. Scales 11–46 (+6 on base of caudal)–18; vertebræ 12+12; Br. 6. Body more elongated than *L. surinamensis*, agreeing in this respect with *L. erate*, the depth less than $\frac{1}{4}$ the length. Upper profile deeply concave at occiput, thence strongly convex to front of dorsal; head shorter and narrower than in *L. surinamensis*, the interorbital width but slightly longer than snout, $3\frac{0}{10}$ to 4 in head ($3\frac{1}{2}$ to $3\frac{2}{3}$ in head in *L. surinamensis*). Eye small, $6\frac{2}{3}$ to $7\frac{1}{4}$ in head, 2 or $2\frac{1}{10}$ in interorbital width. Mandible strongly protruding, but without symphyseal knob; maxillary narrow, not concealed in closed mouth, its tip reaching vertical from middle of pupil, $2\frac{4}{5}$ to $2\frac{3}{10}$ in head. Upper jaw with a moderate villiform band of teeth, in front of which is a single series of conical, close-set canines; lower jaw with a single series, similar to outer series of upper jaw, and behind them a very narrow band of villiform teeth, which grow slightly larger toward symphysis; palate toothless. Posterior margin of preopercle vertical, the angle protruding but little in the young. In 5 young examples, 7 to 11 inches long, the preopercular teeth are fine, acute, short, and inconspicuous, about as in species of *Pomadasis*. They increase but little in size toward the angle, where they are never spine-like; on lower limb they are perceptible only in the immediate vicinity of the angle, the remainder of the horizontal limb being entire. In the adult the vertical limb is finely and evenly toothed, the angle and lower limb slightly roughened or entire; opercle with 2 short spinous points, behind the lower of which a narrow tongue-shaped process of the subopercle extends to near the edge of opercular membrane; humeral process very weakly toothed, contrasting with the strong serrate condition in *L. surinamensis*. Gill rakers short, $2\frac{1}{2}$ in eye in young, comparatively shorter in adults, 6 on vertical limb, all but one of which are broad, firmly fixed tubercles, 14 on horizontal limb, the anterior 2 or 3 tubercular. Spinous dorsal low, with gently rounded outline; notch between dorsals shallow, the eleventh spine $\frac{2}{3}$ the length of the longest, which is contained 2 to $2\frac{1}{4}$ times in head in the young, 3 times in adults; when declined the spines are partially received within a scaly grove; soft dorsal, anal, and caudal with dorsal portions densely scaled and with series of scales running up on membrane to beyond middle of fin; soft dorsal and anal of equal height, forming bluntly rounded lobes, the longest rays of which are about $\frac{1}{2}$ head in adults, $1\frac{1}{2}$ to $1\frac{3}{4}$ in young; third anal spine about $\frac{1}{2}$ length of longest ray; pectorals shorter than ventrals, 2 to $2\frac{1}{4}$ in head; ventrals $1\frac{1}{2}$ in head in young, shorter in adults. Scales less strongly ctenoid than in *L. surinamensis;* tubes of lateral line mostly simple, occasionally with 1 to 3 branches. Color grayish or brownish, with plumbeous or silvery reflections. The youngest examples show faintly the dark streaks so conspicuous in young of *L. surinamensis*, viz, a pair running backward from interorbital space; a pair from upper posterior border of eye converging toward front of dorsal, and a broader band from eye downward and backward across cheek; soft dorsal, anal, and caudal uniform blackish, or the caudal with an ill-defined lighter edge; pectorals translucent; ventrals blackish. Abundant at Panama, where it is known as *Berrugate*.

Lobotes auctorum, STEINDACHNER, Ichth. Beitr., IV, 6, 1875; not of GÜNTHER.

Lobotes surinamensis, JORDAN & GILBERT, Bull. U. S. Fish Comm., II, 1882, 110; GILBERT, l. c., 112; JORDAN, Proc. U. S. Nat. Mus. 1885, 378; not *L. surinamensis* of BLOCH.

Lobotes pacificus, GILBERT, Fishes of Panama, 1898 MS., Panama. (Type, No. 5883, L. S. Jr. Univ. Mus. Coll. Gilbert.)

Page 1238. After *Priacanthus cruentatus* add the following:

1625(a). PRIACANTHUS CAROLINUS, Lesson.

This species is very close to *Priacanthus cruentatus*, distinguished by the larger spine on preopercle, which reaches the edges of the opercle and is $2\frac{1}{2}$ in eye; that of *P. cruentatus* not reaching opercle and measuring 4 in eye, its edge less rough. Body a little deeper than that of *P. cruentatus*; depth of the latter 3 in the length. In *P. carolinus* the depth is $2\frac{3}{4}$ in the length; caudal truncate. In color and general appearance the 2 species are similar. The distinctness of this species from *P. cruentatus* is very doubtful. Abundant at Clarion Island, where it was taken by Mr. R. C. McGregor. (*carolinus*, from Caroline Islands.)

Priacanthus carolinus, LESSON, Voyage Coquille, Poiss., 204, 1826, Caroline Islands.

Priacanthus schlegeli, HILGENDORF, Sitzgber. Ges. Naturf. 1879, 79, Japan.

Page 1262. In the first line of the description of *Neomænis vivanus* read: Head $2\frac{3}{5}$ to $2\frac{3}{4}$; depth $2\frac{1}{2}$ to 3. D. X, 14; A. III, 8 or 9; eye $4\frac{3}{4}$ in head; scales (7) 8–72–17, 50 pores.

Page 1264. After *Neomænis vivanus* add:

1639(a). NEOMÆNIS HASTINGSI, Bean.

(Bermuda Silk Snapper.)

Head 3; depth 3; least depth of caudal peduncle 9 in length of type to caudal base. D. X, 14; A. III, 8; V. I, 5; P. 16; scales 8 or 9–65–17. Maxillary reaching scarcely past front of eye, 3 in head. Vomerine teeth in an arrow-shaped patch, with a backward extension which is fully $\frac{1}{4}$ as long as the eye; canines in upper jaw very feeble; 2 or 3 posterior teeth of mandible are weak canines; 7 rows of scales on cheeks, 9 rows on gill cover. Least interorbital width equal to eye, which is $1\frac{1}{2}$ in snout and 4 in head. Gill rakers 7 + 9, the one in the angle conspicuously longest, about 2 in eye. First dorsal spine 7 in head; fifth and longest spine about 3 in head; last dorsal spine equal to eye in length; longest ray of soft dorsal equal to maxillary, or 3 in head; first anal spine 8, the second and third about 4 in head, the second slightly longer than third; anal base nearly $2\frac{1}{2}$ in head; third and longest anal ray about equal to anal base; pectoral extending to vent; ventral not reaching vent by a space $\frac{1}{4}$ as long as the eye. Colors in life, ground color vermilion, the upper parts overlaid with coppery brown, lower parts vermilion; 4 or 5 narrow golden stripes below lateral line; caudal dark brown with a narrow black margin; anal dusky, the spines and the membranes of last 2 rays pale; a narrow black blotch at pectoral base; ventral pale, somewhat mingled with dusky; membranes of spinous and soft dorsal uniformly dark; snout copper color; eye lemon yellow; pupil blue black; many scales, especially on front of body, with a minute brown dot at base; brownish spots on scales forming many oblique streaks above lateral line. Some living examples show a faint dark lateral blotch much like that of *N. synagris*, and similarly placed. In spirits the body is pink with the upper parts brownish; the dusky color remains on the anal and the black blotch at base of pectoral; black margin of caudal becoming merged with the general dark color of the fin. (Bean.) Most closely related to *N. vivanus*. Length of type 11½ inches. Bermuda, where numerous specimens were obtained in 1897. (Named for General Russell Hastings of Soncy, Bermuda.)

Neomænis hastingsi, BEAN, Bull. Amer. Mus. Nat. Hist., x, Article iii, 45, 1898, Bermuda.

Page 1290. Under *g* read: Anal fin short, its rays III, 7 to III, 13.

Page 1413. In first line of description of *Cynoscion phoxocephalus* for "A. III, 10" read "A. II, 10."

Page 1416. In last line of description of *Sagenichthys ancylodon* for "companion" read "comparison."

Page 1605. Instead of *Chlorichthys* read:

640. THALASSOMA, Swainson.

Thalassoma, SWAINSON, Nat. Hist. Class. Fishes, II, 224, 1839 (*purpureus*).
Julis, GÜNTHER, Cat., IV, 179, 1862; not of CUVIER & VALENCIENNES.

The species of *Thalassoma* (*pavo, unimaculatus, bifasciatus*) examined have 3 anal spines, as is the case with the American species referred to *Chlorichthys*. The first spine, small and hidden in the skin, is easily overlooked. There is therefore no distinction between *Thalassoma* and *Chlorichthys*, and all the American species must be referred to the former genus.

The species will stand as follows:

2014. THALASSOMA LUCASANUM (Gill).
2015. THALASSOMA SOCORROENSE, Gilbert.
2016. THALASSOMA NITIDUM (Günther).
2017. THALASSOMA NITIDISSIMUM (Goode).
2018. THALASSOMA STEINDACHNERI (Jordan).
2019. THALASSOMA BIFASCIATUM (Bloch).
2020. THALASSOMA GRAMMATICUM, Gilbert.
2021. THALASSOMA VIRENS, Gilbert.

Page 1670. In *Pomacanthus* (*P. paru*, species examined by Mr. E. C. Starks) and *Chætodon* the air bladder is wholly contained in the body cavity, while in *Holacanthus* and *Angelichthys* (*A. ciliaris* species examined) it is posteriorly separated from the body cavity. The 2 latter genera constitute the subfamily *Holacanthinæ*, distinct alike from *Chætodontinæ* and *Pomacanthinæ*.

Page 1717. *Ceratacanthus*, including *Osbeckia*, should stand as a valid genus distinguished from *Alutera* by the convex or lanceolate caudal. The species will then stand as follows:

> **2135. CERATACANTHUS SCHŒPFII** (Walbaum).
> **2136. CERATACANTHUS PUNCTATUS** (Agassiz).
> **2137. CERATACANTHUS SCRIPTUS** (Osbeck).
> **2138. ALUTERA MONOCEROS** (Osbeck).

Page 1741, line 17, read: Swainson takes "*κάνθα*" to mean spine, not "*ἄκανθα*," which is the correct word for spine. There is no classical warrant for *Cantherines* and *Canthigaster*, unless derived from *κάνθος*, the ass.

Page 1776. In sixth line from bottom read "increased" for "self."

Page 1786. Note on *Sebastodes rufus*:

This species is ovate in form, like *S. ovalis*, from which it differs in color and form of mouth and head. Its depth is $2\frac{3}{4}$ in length, not $3\frac{3}{4}$, as stated (through misprint) by Dr. Eigenmann. A fine specimen before us was taken by Dr. Gilbert off San Diego.

Page 1790. The type number of *Sebastodes hopkinsi* is 2282, not 2286.

Page 1795. The subgenus *Zalopyr* does not include *Sebastodes atrorubens* nor *S. atrovirens*. These 2 species belong in the subgenus *Rosicola*.

Page 1799. In first line under *Sebastodes crameri* for "P. $19\frac{9}{10}$" read "P. 19."

Page 1805. After line 4 add: (*intro*, within; *niger*, black.)

Page 1815. The type of *Sebastodes zacentrus* came from *Albatross* Station 2946, not 2996.

Page 1829. After line 2 insert Subgenus *Sebastosomus*.

Page 1831. Specimens of *Sebastodes taczanowskii* were obtained in 1896 by the *Albatross* at the Kuril Islands, and this species should therefore go in the regular text.

Page 1832. After line 15 insert Subgenus *Sebastomus*.

Page 1833. Before *Sebastodes matsubaræ* (Hilgendorf) insert Subgenus *Zalopyr*.

Page 1833. In fifth line from bottom insert Subgenus *Pteropodus* after "*nebulosus*."

Page 1836. In key at bottom of page, to *a* add: mouth plumbeous within. To *aa* add: mouth black within.

Page 1837, line 2, for "Cardonniera" read "Cardouniera." For "Scorfanudi Funal" read "Scorfana di Funal."

Page 1840. Above "*a.* Breast scaly," insert Parascorpæna (*παρά*, near; to *Scorpæna*). Before 2236. *Scorpæna agassizii*, Goode & Bean, insert Subgenus *Parascorpæna*, Bleeker.

Page 1850. In last line of synonymy of *Scorpæna mystes* for 1501 read 1601.

Page 1854. After *Scorpœna inermis* add:

2247(a). SCORPÆNA NEMATOPHTHALMUS (Günther).

Head 3 in total length; depth $3\frac{1}{2}$; eye 4 in head; snout rather less than 4. D. XII, 10; A. III, 5; scales 40 or 41. Dorsal outline much arched at greatest depth of body. Eye placed high, entering upper outline of head. Intermaxillaries styliform, armed, like the dentaries, with a rather narrow band of villiform teeth; band of vomerine teeth angularly bent, produced forward at the angle; maxillaries styliform at superior extremity, moderately dilated at the lower. Head scaled to posterior angle of orbit above and to the preorbital and angle of mouth laterally. Spines on head very prominent and acute in the young, more obtuse in older examples; 2 turbinal spines; on each side of the occiput a series of 5 spines between orbit and nape; 2 between eye and scapula; preorbital armed with 2 strong, recurved spines at the inferior margin; 3 spines on interorbital ridge; preopercular margin rounded, with 4 spines, the uppermost and strongest opposite end of interorbital ridge; opercle with 2 flat spines; a pair of spines at throat. The only skinny appendage is a long, slender, tapering filament above posterior angle of orbit. Origin of dorsal immediately behind vertical from suprascapula, its distance from occiput equaling length of first spine, which is about $\frac{1}{4}$ length of second; third and fourth spines longest, $2\frac{3}{4}$ in head; the following spines gradually decreasing to the eleventh, which equals the first; twelfth spine much longer, apparently belonging to the soft portion, which is supported by it; margin of soft portion rounded, very little higher than the spinous, posteriorly fixed to the back of the tail by a membrane; caudal subtruncated; origin of anal somewhat behind that of soft dorsal, its second spine strong, rather longer than the third dorsal, and with a longitudinal groove; pectoral reaching anal; ventral reaching vent. Scales of moderate size, rather irregularly arranged. Color probably uniform red. Supposed to be from the West Indies. (Günther.) Only the type known. ($\nu\tilde{\eta}\mu\alpha$, thread. $\dot{o}\phi\alpha\lambda\mu\dot{o}\varsigma$, eye.)

Sebastes nematophthalmus, GÜNTHER. Cat., III, 99, 1860, West Indies; the exact locality unknown. (Coll. Sir R. Schomburgk.)

Page 1862. *Anoplopoma fimbria* is occasionally taken off Santa Catalina in deep water. A specimen was seen by us at Redondo Beach.

Page 1866, line 7 from bottom, read "always," not "usually."

Page 1867. In the key, *b* should read as follows:

b. Fourth line of pores forking in advance of base of ventrals, the lower branch running to base of ventral fin, the upper to middle of ventral.　　OCTOGRAMMUS, 2259.

In the footnote for "Keinosuke Otaki" read "Keinoske Otaki."

Page 1879, line 27, for "jointed" read "joined." Line 37, after "hypercoracoid" add "and hypocoracoid."

Page 1880. In the key read:

m. Lateral line armed with a series of bony plates; preopercular antler-like processes usually numerous.

Page 1881. The key should be modified to read:

q. Interorbital space deeply concave or grooved.   Head with cirri (in *lateralis*), or
·   without (in *asperulus*).   ARTEDIUS, 712.

A better distinction between *Artedius* and *Axyrias* is found in the presence of patches of ctenoid scales on the head in the latter.

Page 1884. In line 19, for *Hexagrammidæ* read *Zaniolepidinæ.*

Page 1898. The type number of *Icelinus strabo* is 5045, not 5451.

Page 1902. *Artedius asperulus* is better separated from *A. lateralis* by the coalescence of the bands of scales behind the dorsal and their continuance upon the caudal peduncle. In the description of the genus *Artedius* the bands of scales are said not to meet behind the dorsal. This applies to *A. lateralis* only.

Page 1902. To the description of *Artedius* add: No patches of ctenoid scales on the head.

Page 1903. To the description of *Artedius asperulus* add: Head without cirri.

Page 1906. The type of *Artediellus atlanticus* is No. 418 L. S. Jr. Univ. Mus.

Page 1940. In the key, under *a*, add: Preopercular spine with 3 hooks above. Under *aa*, add: Preopercular spine with 6 or 7 hooks above.

Page 1958. *Cottus aleuticus* extends southward in the Coast Range to Monterey.

Page 1964. In key under *h* for "anal" read "axil."

Page 2000. After *Porocottus tentaculatus* add:

2371(a). POROCOTTTUS BRADFORDI, Rutter, new species.

Head 3; depth 3¾ to 4¼; eye 4. D. IX, 15* or 16; A. 11 to 13; P. 13 or 14; B. 6. Head broad, somewhat depressed; bones of head cavernous; lower jaw included, maxillary to below middle or hinder edge of pupil, 2¼ in head; teeth in jaws and in a narrow crescent on vomer; eye equal to snout; nasal spines blunt, covered by the skin; no ocular, opercular, nor suprascapular spines; preopercular spines 3, upper slender, curved inward, lower straight, pointing downward, middle 1 short and ₕₗᵤₙₜ, a mere tubercle; a very slight tubercle represents the fourth spine belonging to the genus; no slit behind last gill; 3 pairs of cirri on top of head, 1 above eye, multifid, another at occiput, single or bifid, the other between them, trifid to multifid; a minute barbel on tip of maxillary; whole top and side of head, lower jaw, and edge of preopercle thickly covered with pores; a double series of pores, 34 to 36 each, along lateral line with many accessory pores, these arranged in groups of 1 to 5 between the pairs of the

*		D.				A.	
Fin rays	IX, 15	IX, 16	IX, 17	VIII, 17	11	12	13
Number specimens	11	12	1	1	1	22	2

lateral line; nostrils with short tubes; dorsals united at base, the spines with short filaments, middle spines 3 in head, middle rays of soft dorsal 2¼ in head; caudal and ventrals 1⅔ in head, ventrals usually reaching vent or anal, but sometimes falling short of each; pectoral 1¼ in head, reaching to or beyond anal. Color dusky, below colorless, a pale bar across occiput (often absent), another between dorsals, 2 across body under soft dorsal and another behind soft dorsal; sometimes the pale color predominates and the dusky portion is left as 4 bars; sometimes plain dusky without cross bars; spinous dorsal dusky with 3 or 4 colorless spaces on the web; other fins barred with series of dusky blotches, ventrals sometimes colorless; 5 to 8 oval white spots behind pectoral, sometimes obscure; males with inner ray or rays of ventral tuberculate or serrate. This species differs from *Porocottus sellaris* in the presence of cirri on top of head; it has more numerous fin rays and more cirri on head than *Porocottus quadrifilis*. This species is the most common fish in the rock pools at Karluk, where many specimens were taken. These are in the U. S. National Museum, in the collection of the U. S. Fish Commission, and in that of Leland Stanford Junior University. (Named for Mr. William B. Bradford, secretary of the Alaska Packers' Association, from whom the collector received many favors.)

Page 2015. Before *Oxycottus* insert the following:

745(a). SIGMISTES, Rutter, new genus.

Sigmistes, RUTTER, MS., new genus (*caulias*).

This genus differs from *Oxycottus*, to which it is most closely related, in the deep compressed body, strongly arched lateral line, long dorsal fin, and large mouth. Body deep and compressed; skin smooth; lateral line complete, strongly arched anteriorly; gill membranes united, free from isthmus; no slit behind last gill; preopercular spine simple, short, strongly curved upward, anal papilla large; vent immediately behind ventral fins, about ⅔ of distance from gill membrane to anal; ventral rays I, 3. ($\sigma i \gamma \mu \alpha$, the letter S, from the form of the lateral line.)

2382(a). SIGMISTES CAULIAS, Rutter, new species.

Head 3⅔; depth 3½. D. IX, 20 (IX, 21 in 1 specimen); A. 15 (14 in each of 2 specimens); P. 13. Back elevated, body compressed; eyes lateral, 4⅓ in head; snout 3½; cleft of mouth lateral; maxillary 2 in head, reaching to below pupil (only a little past front of eye in 1 specimen). Teeth coarse, cardiform, the inner row of upper jaw enlarged, almost canine-like; a similar pair in inner series of mandible, near symphysis; a small patch on vomer, and 1 on front of palatines; preopercular spine small, sharp, appressed, strongly curved upward, the preopercular margin without spines or tubercules below it; nostrils in short tubes, 1 pair directly behind nasal spines, 1 pair lateral, directly in front of eyes; nasal spines strong, sharp; a pair of tufted cirri above eyes, a pair simple or branched at occiput, and a pair of simple ones halfway between these; a filament on nasal spine, a series of 3 or 4 short ones on margin of preopercle and 1 at opercular angle; a series of pores around under side of jaw and along edge of preopercle, 2 concentric series under eye and across cheek, and others scattered on head behind eyes; skin smooth, lateral line strongly arched.   Dorsal fins connected at base, third spine longest, 2¼ in head, margin of fin even from third to sixth spines, origin of spinous dorsal over upper edge of gill opening; soft dorsal higher, longest rays 2 in head, base of soft dorsal ⅔ length; tips of anal rays all free, longest 2⅓ in head; origin of anal under third ray of soft dorsal; longest pectoral ray a little longer than head; caudal truncate, 1¼ in head; ventral about reaching anal, about same length as anal papilla; tail slender, least depth slightly less than eye, length from anal 1⅓ in head, its length from dorsal about equal to its depth.   Color in life, pale pinkish; spinous dorsal dusky, nearly black along margin; soft dorsal plain or with dusky cross bars; anal with about 7 dusky cross bars, extending downward and forward almost at right angles to the rays; 3 or 4 pale blotches surrounded by a black ring along base of dorsal, 1 between dorsals, 1 at end of soft dorsal, and others at base of soft dorsal (some or all sometimes absent); a curved dark line from snout through eye to preopercular spine.   Six specimens, 1 each 2, 2½, and 3 inches long, and 3 1¼ inches long.   From the rock pools at Karluk, on the Island of Kadiak.   Coll. Cloudsley Rutter. The type is in Leland Stanford Junior University Museum.   Cotypes are in the U. S. Fish Commission and U. S. National Museum.   (καυλος, stem, from the many dorsal rays.)

Page 2015.   In third line under *Oxycottus acuticeps* instead of "region" read "reaching."

Oxycottus is much nearer *Blennicottus* than *Oligocottus*, and perhaps is best placed as a subgenus of *Blennicottus*.   There is no slit behind the last gill in any of the species.   In the subgenus *Oxycottus* should be placed:

2383. BLENNICOTTUS ACUTICEPS (Gilbert).

2384. BLENNICOTTUS EMBRYUM (Jordan & Gilbert).

Page 2042.   In line 9 for *Phalangistes* substitute *Brachyopsis*.

Page 2051.   In first line of footnote, for "Dr. Gilbert" read "Scofield & Seale."

Page 2071.   In line 14 of the description of *Averruncus sterletus* read "upward," not "downward."

Page 2108.   The synonymy on this page and the last synonym on page 2107 all belong with the footnote.

Page 2113.   The type number of *Neoliparis greeni* is 3019, not 3010.

Page 2128.   Before *Bathyphasma* insert the following:

785(a). CRYSTALLICHTHYS, Jordan & Gilbert, new genus.

Crystallichthys, JORDAN & GILBERT, new genus (*mirabilis*).

Closely allied to *Liparis*, but with nostril single.   A single dorsal fin; a well-developed sucking disk; wide bands of teeth, many of which are trilobate near tip; an inferior mouth, much overhung by the produced conical snout; a single nostril, corresponding to the anterior nostril of other Liparids, the posterior opening being wholly wanting. The typical species, *C. mirabilis*, differs from all known species of *Liparis* except *L. cyclostigma* in its large size, compressed form, and translucent gelatinous texture. (κρυϭτάλλος, crystal; ἰχθύς, fish.)

2458(a). CRYSTALLICHTHYS MIRABILIS, Jordan & Gilbert, new species.

Head 4 in length; depth $2\frac{1}{2}$; snout $2\frac{1}{4}$ in head; eye $3\frac{1}{4}$ in snout; width of mouth $\frac{1}{4}$ length of head; length of gill slit $\frac{1}{4}$ snout, equaling distance from front of eye to front of nostril tube; P. 33. Head and body compressed, especially along upper profile, which descends in a gentle, nearly even curve to tip of snout; lower profile less curved, nearly straight and horizontal on anterior third of body; snout conical, tapering to a sharp tip, its lower profile nearly horizontal, protruding beyond the mouth for a distance (measured axially) equaling $\frac{2}{3}$ its length; mandibular symphysis vertically below nostril tube; upper jaw strongly arched anteriorly, the mandible much shorter, nearly transverse in position. When the mouth is closed, there is exposed the entire width of the thick upper lip and the anterior portion of the band of fringes which precedes the premaxillary teeth. Teeth slender, shorter than in *Liparis cyclostigma*, arranged in about 25 oblique series in the $\frac{1}{4}$ of each jaw; the posterior longer teeth more or less distinctly 3-lobed in both jaws, the anterior teeth shorter, simple. A deep cleft on lower side of snout running from its tip to front of premaxillaries, deepening backward, opening into the deep groove above premaxillaries; from base of cleft arises a high free fold, the sharp edge of which nearly reaches the margins of the cleft; a series of 3 large pores along each side of this cleft, with 3 more equally spaced on each side and parallel with front of mouth; belonging to this series but distant from them and much smaller, is another on middle of cheek below eye, and 1 halfway between eye and middle of gill slit; a pore behind eye, and a series of 4 on each side of nape complete the pores of the head; no pore in the position of the posterior nasal opening; a second series of 6 on each side of mandible and preopercle; no other pores on head. Nostril single, in a distinct wide tube, as long as the diameter of pupil; distance from eye to angle of mouth $3\frac{1}{4}$ in head; vertical from angle of mouth, passing through front of orbit. Gill cleft narrow, reaching base of first pectoral ray, its length $4\frac{3}{4}$ in head. Lateral line rising in an abrupt curve from upper end of gill opening, decurved again behind pectorals, to reach middle of sides, on the posterior half of which it becomes obsolete; anteriorly the lateral line is accompanied above by a second series of pores which is not curved, but runs straight forward from just above the summit of the curve. Dorsal and anal fins enveloped anteriorly in thick gelatinous tissue, so that their points of origin and number of fin rays can not be determined, the fins high, the longest anal ray equaling length of snout and eye; 32 dorsal and 33 anal rays can be distinguished in the posterior transparent portions of the fins, the total number of rays being greater; last anal ray joining outer caudal ray at middle of length of the latter; dorsal joined narrowly to base of caudal at end of basal seventh of outer caudal ray; longest caudal ray 2¼ in head; lower 7 pectoral rays thickened, forming a lobe, the distal third of each ray free from the membrane; longest pectoral ray 1¼ in head; disk of moderate size, anteriorly placed, its posterior margin under the gill slit, its length ⅓ that of head. Color translucent, apparently light grayish or purplish in life, the dorsal region, including dorsal fin, marked with many large round spots, probably reddish in life, each spot surrounded with a faint darker ring. A large species, soft and gelatinous in texture, the color translucent grayish or purplish, marked on back with many large light circles which were probably reddish in life. Type, a specimen 330 mm. long, from *Albatross* Station 3643, off Provostmaya, Kamchatka, at a depth of 100 fathoms. Bering Sea; 1 specimen from Kamchatka, a smaller one dredged off St. Paul Island, Pribilof group. (*mirabilis*, wonderful.)

Crystallichthys mirabilis, JORDAN & GILBERT, Rept. Fur Seal Invest., MS., 1898, Provostmaya, Kamchatka. (Coll. *Albatross*.)

Page 2129. *Allurus* is preoccupied by *Allurus*, Forster, 1862, a genus of *Hymenoptera*; also by *Allurus*, Eisen, 1874, a genus of worms. We substitute for our use of it the name *Allinectes*.

Allinectes, JORDAN & EVERMANN, new subgenus (*ectenes*).

Page 2131, line 7, for "*Alldurus*" read "*Allurus.*" In lines 4 and 7, for "ἀλλός" read "ἄλλος."

Page 2137. After *Careproctus ectenes* add:

787(a). **PROGNURUS**, Jordan & Evermann, new genus.

Prognurus, JORDAN & EVERMANN, new genus (*cypselurus*).

This genus is distinguished from *Careproctus* by the very elongate caudal which is forked at the tip. (πρόγνη, a swallow or martin; οὐρά, tail.)

2469(a). **PROGNURUS CYPSELURUS**, Jordan & Gilbert, new species.

Head 4⅔ in length; depth 4¹⁰⁄₆₀; cleft of mouth 1¼ in head, ⅔ distance from symphysis of lower jaw to angle of mouth; total interorbital width 2¼ in head; eye large, equaling length of snout, 3¾ in head; gill opening entirely above base of pectoral, not reaching base of upper ray, its length 3 in head; opercular lobe broadly rounded. Snout blunt, broadly rounded, the mouth horizontal along its lower margin, scarcely overlapped by it; upper lip wide. Teeth acute, without cusps, in about 27 oblique rows in 1 side of each jaw; maxillary reaching the vertical from posterior edge of the pupil; nostril opening in a wide, low tube. Front margin of ventral disk very slightly behind angle of mouth, its diameter ⅔ that of eye, about ¼ length of head. Pectorals broadly rounded, regularly shortened below, not deeply notched, the lower 7 rays thickened and exserted; the longest free ray about ⅓ length of head; upper portion of fin with 26 rays, the tips only protruding, the longest equaling length of head; dorsal beginning shortly behind vertical from gill slit, its distance from tip of snout 3⅔ in length; dorsal with about 58 rays; caudal very long and narrow, only its basal third connate with last rays of dorsal and anal. Unlike all other Liparids, the caudal is forked at tip, the terminal notch involving about ⅓ of fin. Translucent dusky, darker around snout, gill openings, and on the fins, the vertical fins largely jet-black; mouth and gill cavity dusky, not black. This species is most nearly related to *Careproctus melanurus*, from which it differs in darker coloration and shorter gill slit. From all known species of *Careproctus* it differs in the very elongate caudal fin which is forked at the tip. Bering Sea and North Pacific. The type, a single specimen, 21 cm. long, dredged at *Albatross* Station 3644, off Bogoslof Island, at a depth of 664 fathoms. A second specimen was obtained by the *Albatross* in 1889, at Station 3074, off the coast of Washington, in 877 fathoms, but it was too seriously mutilated to admit of description. (κυψελός, a swift; οὐρά, tail.)

Page 2175. The genus *Chelidonichthys* should be compared with *Trigla* rather than with *Prionotus*.

Chelidonichthys pictipinnis is probably not American, and should be omitted or, at most, admitted only in a footnote.

The genus *Chelidonichthys* differs from *Trigla* in the absence of lateral plates.

Page 2183. To the synonymy of *Cephalacanthus* add:

Cephalacandia, RAFINESQUE, Anal. de la Nature 1815, 85; substitute for *Cephalacanthus*.

Page 2196, line 5, for "Pañeca" read "Puñeca."

Page 2207. *Sicya* being preoccupied in *Lepidoptera* we substitute for our use of it the name *Sicyosus*.

Sicyosus, JORDAN & EVERMANN, new subgenus *(gymnogaster)*.

Page 2207. Add the following species:

2531(a). SICYDIUM PUNCTATUM, Perugia.

D. VI–I, 11; A. I, 10; scales 56. Head 5¼ in total length without caudal, its width equaling its height or ⅔ that of body under first dorsal. Scales of body larger than those of head or nape; maxillary reaching posterior border of eye. Eye 4 in head, or 1¼ in interorbital space. Snout 4; pectoral equaling head in length; spines of first dorsal somewhat elongated, the longest (third) twice height of body; second dorsal as high as body and like the anal. Teeth of upper jaw fine, very slender, and ending behind in an obtuse angle; lower jaw with conic robust teeth and minute horizontal ones. Color grayish, the ventral gall color (giallognolo); under part of head with numerous small black spots; scales strongly ciliated and each with a brown spot in the center; dorsals brown; anal transparent, with a narrow black line; ventral disk yellowish. Length 8 cm. This species is not *S. plumieri* of Cuvier & Valenciennes, nor is it *S. antillarum*, Ogilvie-Grant, because of the difference in the number of scales, the different proportions and a different coloration. The type was collected by Captain Guiseppe Capurro at St. Pierre, Martinique. (Perugia.)

Sicydium punctatum, PERUGIA, Annali del Museo Civico di Storia Naturali di Genova, ser. 2, vol. XVI, 1896, 18, Martinique. (Coll. Capt. Guiseppe Capurro.)

Page 2226. *Gobius zebra* has 26 scales. Many fine specimens of this species, 3 to 4 inches long, from Clarion Island, are in the museum of Stanford University.

Page 2227. The type locality of *Gobius bosci* is Martinique.

Page 2230, second line, read "ÉMÉRAUDE" for "EMERANDE."

Page 2241. Insert the following synonymy after No. 2572:

Gobius lucretiœ, EIGENMANN & EIGENMANN, Proc. Cal. Ac. Sci., ser. 2, vol. I, Jan. 25, 1888, 57, Pearl Island, Gulf of Panama.

Page 2263. To the synonymy of *Gobioides* add:

Plecopodus, RAFINESQUE, Anal. de la Nature 1815, 87; substitute for *Gobioides.*

Page 2269. Omit the last reference but one.

Page 2314. To the synonymy of *Batrachoides* add:

Batrictius, RAFINESQUE, Anal. de la Nature 1815, 82; substitute for *Batrachoides.*

Page 2350. After *Enneanectes carminalis* insert:

868(a). DIALOMMUS, Gilbert.

Dialommus, GILBERT, Proc. U. S. Nat. Mus. 1890, 452 *(fuscus).*

Teeth conic, strong, in a narrow band in the front of both jaws, this narrowing to a single series laterally; outer teeth enlarged in both jaws. Teeth on vomer in a single series; palatines smooth. A single slender tentacle above orbits, and 1 on each side of nape. Body with moderate cycloid scales; lateral line high in front, declining behind pectoral fins, not strongly developed, evident on a few scales near head; the remainder of its course traceable by occasional pores on bases of scales or by their notched margins. Dorsal beginning on the nape, its anterior $\frac{4}{5}$ composed of slender flexible spines, the remainder of soft rays, unbranched; anal without spines; caudal distinct, rounded; ventrals well developed, I, 3. Eyes as in *Anableps,* the cornea divided by an oblique pigmented band into an anterior lower and a posterior upper half. One species known. ($\delta\iota\acute{\alpha}\ \gamma\acute{\upsilon}\omega$, to loose one from another, to part asunder; $\H{o}\mu\mu\alpha$, eye.)

2687(a). DIALOMMUS FUSCUS, Gilbert.

Head 5 in length; depth 6 to 7. D. XXV, 13 or 14; A. I, 28; lateral line, 52. Elongate, slender, scarcely tapering. Head short, transversely evenly rounded, with very short, blunt, decurved snout; width of head greater than its depth, and more than $\frac{2}{3}$ its length. Mouth horizontal at lower outline of snout, the maxillary nearly reaching vertical from posterior margin of orbit, $2\frac{2}{3}$ in head. Teeth strong, conical, the outer series enlarged in both jaws, a narrow band of villiform teeth behind the outer series; vomer with a single series; palatines toothless. Eyes large, round, closely approximated, their diameter greater than length of snout, twice the width of interorbital space, $3\frac{1}{3}$ in head. Gill membranes very widely joined, free from isthmus. No hook on inner edge of shoulder girdle. Dorsal fin beginning on the nape, over front of opercle, its spines slender and flexible, much lower than soft rays; height of anterior and middle spines about equal, $\frac{1}{4}$ length of head; the posterior spines shortened, about $\frac{1}{4}$ that length; height of soft rays $\frac{1}{2}$ head; first anal ray short and spinous, the succeeding rays articulated, but not branched (like those of dorsal). Interradial membranes of anal fin very deeply incised; caudal fin wholly free, rounded, its length nearly equaling that of head; pectorals slightly shorter than head, posteriorly pointed, the longest rays below the middle of fin; ventrals comparatively broad, inserted but little in front of pectorals, their bases separated by a space equal to ⅓ diameter of orbit. Color in spirits, brownish above and on sides, becoming blackish on head; under side of head, belly, and a line along each side of anal fin light; back with traces of about 10 black cross bars, which invade base of dorsal fin and extend onto middle of sides; in 1 specimen the scales of the interspaces are marked each with a light spot (probably blue in life); fins all dusky, the caudal variegated with lighter in fine pattern; ventrals light at base. Two specimens from the Galapagos Islands—1 from Duncan Island, 72 mm. long, the other from Albemarle Island, 75 mm. long.

Dialommus fuscus, GILBERT, Proc. U. S. Nat. Mus. 1890, 452, Galapagos Islands. (Coll. *Albatross*.)

Page 2352. *Gibbonsia evides* intergrades with *Gibbonsia elegans* and must apparently be regarded as a subspecies of the latter.

Page 2356. Under *Malacoctenus ocellatus* for "A. II, 8" read "A. II, 18."

Page 2413. Genus 904 should read *Ulvicola*, Gilbert, not "Gilbert & Starks."

Page 2421. *Anoplarchus alectrolophus* is the type of a new genus:

908(a). ALECTRIAS, Jordan & Evermann.

Alectrias, JORDAN & EVERMANN, new genus (*alectrolophus*).

This genus is distinguished from *Anoplarchus* by having the gill membranes free from the isthmus, as in *Cebedichthys*. (ἀλέκτωρ, a cock; from the crest.)

Page 2422, under *Anoplarchus alectrolophus:* The 2 specimens obtained in Monterey Bay by Arthur W. Greeley were erroneously referred to this species. They are the young of *Cebedichthys*.

Page 2470. To synonymy of *Lycenchelys* add:

Lycodophis, VAILLANT, Exp. Sci. Trav. et Talisman, 311, 1888 (*albus*).

Page 2472. *Furcella* is preoccupied by *Furcella*, Lamarck, 1801, a genus of Mollusca. We substitute for our use of it the name *Furcimanus*.

Furcimanus, JORDAN & EVERMANN, new genus (*diapterus*).

Page 2473. In lines 33, 36, and 37 read "in 282 to 376 fathoms" instead of the depths given.

Page 2475. In lines 26 and 28, for "coast of Alaska" read "coast of California."

Page 2480. The locality for *Melanostigma pammelas* is coast of California, not Alaska.

Page 2567. In first line of footnote read "fin," not "spine."

Page 2601, line 20, for "*Trachyterus*" read "*Trachypterus*."

Page 494. The large "Silver Trout" of Lake Tahoe, a specimen of which is described on page 494, should probably be separated subspecifically from its parent form, *Salmo clarkii henshawi*. It may be described as follows:

780(d). SALMO CLARKII TAHOENSIS, Jordan & Evermann, new subspecies.

(SILVER TROUT OF LAKE TAHOE.)

Head $4\frac{1}{15}$; depth $3\frac{2}{3}$; eye $7\frac{2}{3}$ in head. D. 9; A. 12; Br. 10; scales 33-205-40; 140 pores. Pectoral $1\frac{2}{3}$ in head; maxillary $1\frac{2}{3}$. Body very robust, compressed, unusually deep for a trout, the outline elliptical. Head large, rather more compressed than in typical *Salmo clarkii henshawi*; eye small, silvery. Vomerine teeth in 2·long series, those of the 2 series alternating in position; hyoid teeth distinct, in a rather long series; gill rakers short, thickish, 5 + 13. Mouth large, the maxillary extending well beyond the eye. Preopercle moderate, its lower posterior edge not evenly rounded, but with a slightly projecting, rounded lobe and a slight concavity above and below, this character not strongly marked; opercle evenly, but not strongly, rounded. Scales small, reduced above and below, those in or near lateral line largest. Fins moderate, the anal rather high, with 1 more ray than usual; caudal slightly lunate, almost truncate when spread open. Color dark green above, belly silvery; sides with a broad coppery shade covering cheeks and opercles; sides of lower jaw yellowish; fins olivaceous, a little reddish below; orange dashes between rami of lower jaw moderately conspicuous; back, from tip of snout to tail, closely covered with large, unequal black spots; spots on top of head and nape round; posteriorly the spots run together, forming variously shaped markings, usually vertically oblong; these may be regarded as formed of 3 or 4 spots placed in a series, or with 1 or 2 at the side of the other; the longest of the oblong markings not quite as long as eye; along side of head and body the spots are very sparse, those on head round, those behind vertically oblong; belly profusely covered with small black spots which are nearly round; still smaller round spots numerous on lower jaw; all the spots on caudal peduncle vertically oblong or curved; dorsal and caudal densely covered with oblong spots, smaller than those on the body; anal with rather numerous round spots; pectorals and ventrals with a few small spots, the first ray in each case with a series of faint small spots; adipose fin spotted. The above description from a specimen 2 feet 4 inches long and weighing $7\frac{1}{2}$ pounds. This form attains a weight of 10 to 30 pounds and spawns only in the depths of the lake. *Salmo clarkii henshawi* reaches a much smaller size, is much darker in color, and spawns in the streams. Thus far known only from the deep waters of Lake Tahoe. (*tahoensis*, from Lake Tahoe.)

Salmo clarkii tahoensis, JORDAN & EVERMANN, new subspecies, Lake Tahoe. (Coll. A. J. Bayley.)

Page 518. It is wholly uncertain where Valenciennes got the specimen which he called *Thymallus ontariensis*. It is probably the ordinary Grayling (*Thymallus thymallus*) of Europe, erroneously attributed to Milbert's New York collection. In any case, its identity with the Michigan Grayling is more than doubtful, as the rivers in which the latter occurs were then unexplored. The American Graylings would therefore stand as follows:

787. THYMALLUS SIGNIFER (Richardson).

(ARCTIC GRAYLING; POISSON BLEU.)

788. THYMALLUS TRICOLOR, Cope.

(MICHIGAN GRAYLING.)

788(a). THYMALLUS TRICOLOR MONTANUS (Milner).

(MONTANA GRAYLING.)

To the synonymy of the Montana Grayling add the following wholly unnecessary synonym:

Thymallus lewisi, HENSHALL, Forest and Stream, July 23, 1898, 70, headwaters Jefferson River, Red Rock Lake, Montana; after notes of Lewis & Clark.

Dr. J. C. Merrill, U. S. A., informs us that this Grayling is found also in Sun River at Fort Shaw, Montana.

Page 852. The species called *Holocentrus marianus*, Cuvier & Valenciennes, is the type of a distinct genus:

382 (a). FLAMMEO, Jordan & Evermann, new genus.

Flammeo, JORDAN & EVERMANN, new genus (*marianus*).

This genus is distinguished by the very large mouth and projecting chin. The lower jaw is considerably more than ½ the length of the head, and the chin projects beyond the upper jaw. In the species properly referable to *Holocentrus*, the lower jaw is slightly included and its length is less than ½ the head. The single known species of this genus is

1238. FLAMMEO MARIANUS (Cuvier & Valenciennes).

Page 858. In line 13 the interorbital space in *Upeneus* is said to be "concave;" this should, of course, read "convex."

Page 2013. *Oligocottus maculosus*, Girard, was evidently based on a specimen from the Farallones of the species called by us *Oligocottus borealis*, Jordan & Snyder. The name *maculosus* must therefore be transferred to the latter species. The species called by us *Oligocottus maculosus* must therefore be renamed and may stand as

2381. OLIGOCOTTUS SNYDERI, Greeley, new species.

Oligocottus snyderi, GREELEY, MS. 1898.

Page 2126, lines 13 and 14. The specimens from St. Paul Island and Petropaulski, referred by us to *Liparis cyclostigma*, Gilbert, belong to *Crystallichthys mirabilis*, Jordan & Gilbert, described on page 2865.

Page 2626. Before *Paralichthys æstuarius*, Gilbert & Scofield, insert:

2991(a). PARALICHTHYS MAGDALENÆ, Abbott, new species.

Head 3⅔; depth 2⅓. D. 80; A. 64; scales 120. Body oval-elliptical, the dorsal outline evenly bowed, the greatest depth in the middle of the body.

Ventral outline straighter anteriorly. Mouth large; mandible somewhat projecting, about 1¼ in head; maxillary large, extending considerably beyond orbit; snout (measuring from upper orbit) about 4 in head, ⅔ length of eye; interorbital about 9½ in head; eye 6 in head. Teeth moderately strong, the anterior ones in the lower jaw somewhat larger than the others; gill rakers 7 + 17, slender, weakly serrate, the longest a trifle less than eye. Scales cycloid, on cheeks, opercles and maxillary; snout, interorbital, and mandible naked; accessory scales present, especially prominent among the small crowded scales in the region below the pectoral; arch of lateral line 3¼ in straight part; pores about 38 + 82. Ventral 3⅓ in head; pectoral 2; dorsal beginning above anterior rim of orbit; middle rays of anal and dorsal longest, 2¼ in head, equaling width of caudal peduncle; caudal double lunate, the middle rays the longest; pectoral of blind side rounded, 2¼ in head. Color dark reddish brown, closely peppered with darker dots; a series of indistinct white spots, 4 or 5 in number, following margins of the body, as in *P. californicus;* traces of darker mottling along sides of body. Length 17 inches.

This species, represented by a specimen from Magdelena Bay, Lower California, is closely related to *P. californicus,* resembling it in the large number and close arrangement of the gill rakers, but differing from it in having cycloid scales, a greater number of fin rays, somewhat narrower interorbital, and greater depth in proportion to the length.

Paralichthys magdalenæ, ABBOTT MS., new species, **Magdalena Bay, Lower California.** (Type, No. 10196, L. S. Jr. Univ. Mus. Coll. Charles H. Gilbert.)

Page 2627. Instead of *Paralichthys adspersus* (Steindachner), read :

2904. PARALICHTHYS SINALOÆ, Jordan & Abbott, new species.

The specimens described in the text (from Mazatlan and La Paz), under the name *Paralichthys adspersus,* belong to a distinct species, thus far known only from the west coast of Mexico and Central America and which may be called *Paralichthys sinaloæ,* one of the many specimens taken by the Hopkins Expedition at Mazatlan and La Paz being taken as type, and the following as cotypes: Nos. 11726, 11727, 11728, 11729, 11730, 11731, L. S. Jr. Univ. Mus., all from Mazatlan. *Paralichthys adspersus* is known only from the coast of Peru, the specimens before us being from Callao. (Coll. Admiral L. A. Beardslee.)

Paralichthys sinaloæ is distinguished from *P. adspersus* by its cycloid scales and broader interorbital space. The gill rakers in *P. adspersus* are close-set, rather long and slender, about ¾ to ⅘ of eye, and with rather slender spinules on the inner margin. In *P. sinaloæ* they are set farther apart on the limb of the gill arch, are shorter and thicker—about 2 in eye—and have the inner edge armed with coarser teeth. All specimens of *P. sinaloæ* have each 14 or 13 gill rakers on the lower limb of the arch, while in 4 examples of *P. adspersus,* from Callao, there are 16 or 17 gill rakers on the lower limb (19 in 1 specimen), and 7 or 6 above (5 in 1 specimen), showing that while there may be variation in the number yet it is confined within limits which do not intergrade and the average number in the 2 species is quite different. In the single known specimen of *P. woolmani* the number of gill rakers given is 5 + 11, which makes it probable that the present species is not the same. The more striking difference between *P. adspersus* and *P. sinaloæ* lies in the scales, which in the latter are cycloid, while in the true *P. adspersus* they are strongly toothed as stated by Steindachner. The specimens from Callao referred by us in the text to *adspersus* belong to that species, but they are not original types of *adspersus*, belonging to the later collections of Agassiz and Steindachner.

Paralichthys sinaloæ, JORDAN & ABBOTT MS., new species, Mazatlan, Sinaloa, Mexico.
    (Type, No. 2930, L. S. Jr. Univ. Mus. Coll. Hopkins Exped. to Mazatlan.)

Page 686. After *Platypœcilus maculatus* add:

1009(a). PLATYPŒCILUS QUITZEOENSIS, B. A. Bean.

Head 3½; depth 2⅘; eye 3¼ in head; snout 4; interorbital width 2⅔. D. 13; A. 13; scales 30, 10. Body compressed, back elevated, head small and depressed, flat on top; snout short. Mouth small, cleft oblique, the lower jaw heavy, projecting; teeth conic, those in upper jaw in an irregular series, those below very small, apparently irregularly arranged and close-set. Origin of dorsal fin in advance of that of anal, midway between tip of upper jaw and end of caudal rays, the first ray of anal being under sixth dorsal ray. Color in alcohol light brown, with traces of darker on back; interorbital space and edge of scales dark brown; 3 dark bars on posterior part of body, the first extending from median line to origin of anal, the second from median line to end of anal base, the third midway between end of anal and origin of caudal; 2 dark spots on end of caudal peduncle; fins all pale. Lake Quitzeo, Michoacan, Mexico. Only the type known.

Platypœcilus quitzeoensis, B. A. BEAN, Proc. U. S. Nat. Mus. 1898, 540, with text figure,
    Lake Quitzeo, Michoacan, Mexico. (Type, No. 48209. Coll. E. W. Nelson.)
    3030——103

ARTIFICIAL KEY TO THE FAMILIES OF THE TRUE FISHES OR TELEOSTEI.

The following key is intended simply to facilitate the identification of species of the true fishes. No attempt is made to indicate the natural characters or relations of the families, and only those species of any group which are included in the present work are taken into consideration. Most of the ordinary fishes can be readily placed by its means, but it should not be trusted in the study of ichthyological rarities, or of fishes from the deep seas.

I.—VENTRAL FINS PRESENT, ABDOMINAL.

A. Back with an adipose fin behind the single rayed dorsal fin.
B. Adipose fin composed of a single spine with a thin membrane; body mailed...................XXXV, LORICARIIDÆ, 155.
BB. Adipose fin without spine.
C. Head with 4 to 8 long barbels about the mouth and nostrils; body scaleless; a single spine in each pectoral and in dorsal fin.........................XXXIV, SILURIDÆ, 115.
CC. Head without barbels as described above.
D. Sides of body without photophores or luminous glands; no barbel at throat.
E. Body scaleless; teeth very strong, some of them fang-like.
F. Dorsal fin very long and high, occupying nearly whole length of back......................LXXXI, ALEPISAURIDÆ, 593.
FF. Dorsal fin short, median or posterior.
LXXXII, ODONTOSTOMATIDÆ, 597.
EE. Body scaly.
G. Pseudobranchiæ present.
H. Dorsal, anal, and ventrals each with a small but distinct spine; scales ctenoid....................CIV, PERCOPSIDÆ, 783.
HH. Dorsal, anal and ventral without distinct spine.
I. Head naked.
J. Branchiostegals 6 to 20.
K. Dorsal fin long and high, of about 24 rays.
LXV, THYMALLIDÆ, 517.
KK. Dorsal fin moderate, of fewer than 20 rays.
L. Stomach with many pyloric cœca.
LXIV, SALMONIDÆ, 460.
LL. Stomach with few pyloric cæca; size small.
LXVI, ARGENTINIDÆ, 519.
JJ. Branchiostegals 3 or 4; mouth very small.
LXVII, MICROSTOMATIDÆ, 527.
II. Head scaly on sides.
M. Maxillary very narrow, rudimentary, or obsolete; hypocoracoids not divergent...............LXVIII, SYNODONTIDÆ, 532.
MM. Maxillary well developed, dilated behind; pectorals normal; hypocoracoids mostly divergent.LXIX, AULOPODIDÆ; 541.
GG. Pseudobranchiæ absent.
N. Pectorals normally formed, teeth incisor-like or else rudimentary; pseudobranchiæ absent......XXXIX, CHARACINIDÆ, 331.

2875

NN. Pectorals not normally formed.
  O. Pectorals undivided, subhumeral; pseudobranchiæ absent.
                                 LXX, BENTHOSAURIDÆ, 543.
  OO. Pectoral rays elongate, arranged in two groups.
                            LXXI, BATHYPTEROIDIDÆ, 544.
DD. Sides of body with photophores more or less developed.
  P. Barbel at throat present, very long; body naked.
                         LXXVIII, ASTRONESTHIDÆ, 586.
  PP. Barbels none.
  Q. Vertebral spines projecting through skin of back before dorsal
      fin; body short and deep, greatly compressed.
                     LXXXIV, STERNOPTYCHIDÆ, 603.
  QQ. Vertebral spines not exserted in front of dorsal.
  R. Pseudobranchiæ present.
  S. Premaxillaries forming entire margin of upper jaw; body scaly;
      opercles complete.
  T. Form elongate, the snout pointed, barracuda-like; photophores
      very small................LXXXIII, PARALEPIDIDÆ, 599.
  TT. Form oblong, the snout not much produced; photophores con-
      spicuous.....................LXXV, MYCTOPHIDÆ, 550.
  SS. Premaxillaries not forming the whole margin of upper jaw, the
      maxillary entering into it; body naked; opercular appa-
      ratus incomplete.............LXXVI, MAUROLOCIDÆ, 576.
RR. Pseudobranchiæ absent; mouth large, with canine teeth; scales
      deciduous or wanting...LXXVII, CHAULIODONTIDÆ, 578.
AA. Back without adipose fin.
  B. Back with a single dorsal fin made up of rays and not preceded
      by a series of free spines or followed by finlets.
  C. Tail evidently strongly heterocercal.
  D. Body naked; snout with a spatulate blade; mouth wide, without
      barbels ....XXX, POLYODONTIDÆ, 101.
  DD. Body with 5 series of body shields; mouth, inferior, toothless,
      preceded by 4 barbels ....XXXI, ACIPENSERIDÆ, 102.
DDD. Body scaly.
  E. Scales cycloid; a broad bony gular plate; dorsal fin many rayed.
                        XXXIII, AMIIDÆ, 112.
  EE. Scales ganoid; no gular plate; dorsal fin short.
                        XXXII, LEPISOSTEIDÆ, 108.
  CC. Tail not evidently heterocercal.
  F. Tail tapering to a point, without caudal fin; anal fin very long,
      of about 200 rays; body scaly.
                     LXXXVI, HALOSAURIDÆ, 606.
  FF. Tail not tapering to a point; caudal fin developed.
  G. Body naked.
  H. Throat with a long barbel; no caudal filament; mouth large.
  I. Barbel free at tip ....LXXIX, STOMIIDÆ, 587.
  II. Barbel connecting throat with symphysis of lower jaw.
                      LXXX, MALACOSTEIDÆ, 592.
  HH. Throat without barbel.
  J. Caudal fin with a long filament; body elongate; mouth very
      small ....CI, FISTULARIIDÆ, 755.
  JJ. Caudal fin without filament.
  K. Pectorals present.
  L. Gill membranes joined to the isthmus; opercles complete.
                     XXXVII, CYPRINIDÆ, 199.
  LL. Gill membranes free from isthmus; opercles incomplete.
                    LXXIII, RONDELETIIDÆ, 547.
  KK. Pectorals wanting; body snake-like; dorsal long and low.
                    LXXXV, IDIACANTHIDÆ, 604.
  GG. Body scaly.

M. Head with a large divided luminous plate in place of eyes.
LXXII, IPNOPIDÆ, 546.
MM. Head with eyes concealed beneath the skin; vent at the throat.
XCIII, AMBLYOPSIDÆ, 702.
MMM. Head with normally developed eyes.
N. Body with a coat of mail; maxillary with barbels.
XXXV, LORICARIIDÆ, 155.
NN. Body with ordinary scales.
O. Anal fin with many spines; mouth toothless, sucker-like.
LXXXVIII, LIPOGENYIDÆ, 619.
OO. Anal fin without distinct spines.
P. Pectoral fins inserted high, near axis of body; lower pharyn-
geals united; lateral line along sides of belly.
Q. Jaws each with long sharp teeth mixed with smaller ones.
XCIV, ESOCIDÆ, 708.
QQ. Jaws with small equal teeth, conic or tricuspid.
R. Lower jaw more or less produced; teeth tricuspid.
XCV, HEMIRAMPHIDÆ, 718.
RR. Lower jaw a little produced; teeth conic; pectorals elongate,
forming an organ of flight......XCVII, EXOCŒTIDÆ, 726.
PP. Pectoral fins inserted below axis of body; lower pharyngeals
separate.
S. Gill membranes broadly joined to the isthmus; head naked; no
teeth in jaws.
T. Lower pharyngeal teeth very numerous, in 1 row like the teeth
of a comb. (Suckers.).......XXXVI, CATOSTOMIDÆ, 161.
TT. Lower pharyngeal teeth few, fewer than 8, in 1 to 3 rows. (Carp;
Chubs; Minnows.)...........XXXVII, CYPRINIDÆ, 199.
SS. Gill membranes free from the isthmus.
U. Throat with a long barbel; sides with phosphorescent spots.
LXXIX, STOMIATIDÆ, 587.
UU. Throat without barbels.
V. Phosphorescent spots present; teeth unequal.
LXXVII, CHAULIODONTIDÆ, 578.
VV. Phosphorescent spots none.
W. Head scaly, more or less.
X. Maxillaries connate with premaxillaries; jaws long.
LXVII, SYNODONTIDÆ, 532.
XX. Maxillaries distinct.
Y. Upper jaw not protractile, its lateral margins formed by the
maxillaries; lateral line more or less developed.
Z. Teeth cardiform; jaws depressed, prolonged.XCI, LUCIIDÆ, 624.
ZZ. Teeth villiform; jaws short; no lateral line.
a. Pectoral very broad, of about 35 rays..LXXXIX, DALLIIDÆ, 620.
aa. Pectoral narrow, of about 13 rays...........XC, UMBRIDÆ, 622.
YY. Upper jaw protractile, its margin formed by premaxillaries
alone; no lateral line.............XCII, PŒCILIIDÆ, 630.
WW. Head naked.
a. Anterior vertebræ coalesced and modified; no pseudobranchiæ;
jaws with strong canines..XXXVIII, ERYTHRINIDÆ, 330.
aa. Anterior vertebræ normal, not modified.
b. Dorsal fin inserted more or less before anal (rarely slightly
behind it); shore fishes or river fishes, usually silvery
in coloration and with skeleton firm; air bladder well
developed.
c. Gular plate present, between branches of lower jaw; mouth
large; teeth present, all pointed; axillary scales and
sheaths large....................LVI, ELOPIDÆ, 408.
cc. Gular plate none.
d. Lateral line well developed.
e. Teeth present, no accessory branchial organ.

f. Mouth small, horizontal; posterior part of tongue and roof of mouth covered with coarse-paved teeth.

LVII, ALBULIDÆ, 410.

ff. Mouth large, the teeth all pointed, some of them canine, none paved or molar................LVIII, HIODONTIDÆ, 412.

ee. Teeth none; an accessory branchial organ behind gill cavity.

LIX, CHANIDÆ, 414.

dd. Lateral line wanting; no gular plate.

g. Mouth small, inferior, toothless, the maxillary simple or nearly so; stomach gizzard-like......LX, DOROSOMATIDÆ, 415.

gg. Mouth moderate, terminal, the maxillary of about 3 pieces; stomach not gizzard-like............LXI, CLUPEIDÆ, 417.

ggg. Mouth subinferior, very large, below a tapering, pig-like snout; maxillary very long...........LXII, ENGRAULIDIDÆ, 439.

bb. Dorsal fin posterior, opposite anal; deep-sea fishes, of loose organization; mostly blackish in color; mouth small, with small pointed teeth; air bladder wanting.

LXIII, ALEPOCEPHALIDÆ, 451.

BB. Dorsal fin single, preceded by free spines.

h. Body scaleless, naked or with bony plates.

i. Ventral fins I, I, the spine strong; snout moderate.

XCVIII, GASTEROSTEIDÆ, 742.

ii. Ventral fins, I, 5, the spine slender; snout prolonged.

XCIX, AULORHYNCHIDÆ, 752.

hh. Body scaly; snout tubular ....C, AULOSTOMIDÆ, 754.

BBB. Dorsal fin composed of free spines; ventrals with 1 or 2 spines each; body elongate.....LXXXVII, NOTACANTHIDÆ, 613.

BBBB. Dorsal fins 2, the anterior of spines only, the posterior chiefly of soft rays.

j. Pectoral fin with 5 to 8 lowermost rays detached and filamentous ....CIX, POLYNEMIDÆ, 827.

jj. Pectoral fin entire.

k. Snout tubular, bearing the short jaws at the end; body compressed ....CII, MACRORHAMPHOSIDÆ, 758.

kk. Snout not tubular.

l. Teeth strong, unequal; lateral line present.

CVIII, SPHYRÆNIDÆ, 822.

ll. Teeth small or wanting; lateral line obsolete.

m. Dorsal spines 4, stout; anal spines 3......CVII, MUGILIDÆ, 808.

mm. Dorsal spines 4 to 8, slender; anal spine single.

CVI, ATHERINIDÆ, 788.

BBBBB. Dorsal fin soft-rayed, followed by a series of detached finlets.

XCVI, SCOMBRESOCIDÆ, 724.

II.—VENTRAL FINS PRESENT, THORACIC OR SUBJUGULAR, THE NUMBER OF RAYS DEFINITELY I, 5.

A. Gill openings in front of the pectoral fins.

B. Body more or less scaly or armed with bony plates.

C. Ventral fins completely united; gill membranes joined to the isthmus; no lateral line .....CLXXXVIII, GOBIIDÆ, 2188.

CC. Ventral fins separate.

D. Suborbital with a bony stay, which extends across the cheek to or toward the preopercle; cheeks sometimes entirely mailed.

E. Pectoral fin with 3 lower rays detached and free; head bony.

CLXXXIV, TRIGLIDÆ, 2147.

EE. Pectoral fin with 2 lower rays detached and free; body mailed.

CLXXXV, PERISTEDIIDÆ, 2177.

EEE. Pectoral fin entire.

F. Slit behind fourth gill small or wanting.

G. Dorsal spines 4; lips fringed; eyes superior.

CXCVII, URANOSCOPIDÆ, 2305.

GG. Dorsal spines 8 to 17.

H. Anal spines 3; body scaly..........CLXXVI, SCORPÆNIDÆ, 1758.

HH. Anal spines obsolete; body partly or wholly naked.

CLXXIX, COTTIDÆ, 1879.

FF. Slit behind fourth gill large; body scaled.

I. Nostril single on each side, a small pore above it; dorsal fin continuous............CLXXVIII, HEXAGRAMMIDÆ, 1863.

II. Nostrils 2 on each side; dorsal fins 2, separate, except in the genus *Erilepis* ....CLXXVII, ANOPLOPOMATIDÆ, 1861.

DD. Suborbital stay wanting; cheeks not mailed.

J. Spinous dorsal transformed into a sucking disk on top of head, composed of 8 to 30 transverse plates.

CLXXXIX, ECHENEIDIDÆ, 2265.

JJ. Spinous dorsal (if present) not transformed into a sucking disk.

K. Dorsal spines all or nearly all disconnected from each other.

L. Body elongate, spindle-shaped..CXXVII, RACHYCENTRIDÆ, 947.

LL. Body oblong or ovate, compressed.

M. Caudal peduncle very slender, the fin widely forked; preopercle entire............................CXXV, CARANGIDÆ, 895.

MM. Caudal peduncle stoutish, the fin little forked.

N. Gill membranes free from the isthmus; preopercle serrulate.

CXXXIV, CENTROLOPHIDÆ, 962.

NN. Gill membranes broadly united to the isthmus; preopercle entire

CLXIV, EPHIPPIDÆ, 1666.

KK. Dorsal spines (if present) all, or most of them, connected by membrane.

O. Pectoral fin with 4 to 9 lowermost rays detached and filiform.

CIX, POLYNEMIDÆ, 827.

OO. Pectoral fin entire.

P. Dorsal and anal each with 1 or more detached finlets.

Q. Anal preceded by 2 free spines........CXXV, CARANGIDÆ, 895.

QQ. Anal not preceded by 2 free spines.

R. Caudal peduncle keeled..............CXVIII, SCOMBRIDÆ, 863.

RR. Caudal peduncle not keeled............CXIX, GEMPYLIDÆ, 877.

PP. Dorsal and anal without finlets.

S. Lateral line armed posteriorly with a series of keeled plates; 2 free anal spines; gill membranes free from isthmus.

CXXV, CARANGIDÆ, 895.

SS. Lateral line armed posteriorly with a sharp, movable, lancet-like spine, or with a few bony tubercles; scales small, rough; gill membranes adherent to isthmus.

CLXVII, TEUTHIDIDÆ, 1688.

SSS. Lateral line unarmed.

T. Throat with 2 long barbels (placed just behind chin); dorsal fins 2 ....CXVII, MULLIDÆ, 855.

TT. Throat without long barbels.

U. Head with a short bony horn before each eye; gill membranes united to isthmus; scales very small, rough.

CLXVI, ZANCLIDÆ, 1687.

UU. Head without bony prominence or horns.

V. Anal fin preceded by 2 free spines (these obsolete in the very old, joined by membrane in the very young).

W. Preopercle entire; teeth moderate if present.

CXXV, CARANGIDÆ, 895.

WW. Preopercle serrate; teeth unequal, some of them very strong.

CXXVI, POMATOMIDÆ, 945.

VV. Anal fin not preceded by free spines.

X. Nostril single on each side; lateral line interrupted; lower pharyngeals united.

Y. Anal spines 2......................CLIX, POMACENTRIDÆ, 1543.
YY. Anal spines 3 to 11. Fresh-water fishes.CLVIII, CICHLIDÆ, 1512.
XX. Nostril double on each side.
  Z. Lateral line extending to tip of middle rays of caudal.
    a. Anal spines 3, the second strong.
    b. Dorsal fins 2, separate; body elongate.
          CXLV, CENTROPOMIDÆ, 1116.
   bb. Dorsal fin continuous....................CL, HÆMULIDÆ, 1289.
   aa. Anal spines 1 or 2, the second large or small.
          CLV, SCIÆNIDÆ, 1392.
  ZZ. Lateral line not extending beyond base of caudal fin.
    c. Gills 3½, the slit behind the last very small or wanting.
    d. Mouth not vertical, the lips not fringed; dorsal fin continu-
        ous, the spines 8 to 18; scales cycloid; lower pharyngeals
        united.
      e. Teeth in each side of each jaw united, forming a sort of beak.
          CLXI, SCARIDÆ, 1620.
     ee. Teeth distinct or nearly so, the anterior usually more or less
        canine ....CLX, LABRIDÆ, 1571.
   dd. Mouth nearly vertical, the lips with fleshy fringes; dorsal
        divided, the spinous part short, of about 4 spines; lower
        pharyngeals separate ....CXCVII, URANOSCOPIDÆ, 2305.
   cc. Gills 4, a long slit behind the fourth.
    f. Teeth setiform, like the teeth of a brush; body elevated, lon-
        ger than deep, the soft fins completely scaled; gill mem-
        branes attached to the isthmus.
      g. Dorsal fin continuous............CLXV, CHÆTODONTIDÆ, 1669.
     gg. Dorsal fin divided ....CLXIV, EPHIPPIDÆ, 1666.
    ff. Teeth not setiform.
    h. Body deeper than long, covered with rough scales; dorsal
        spines 8; anal spines 3; soft fins very long.
          CLXIII, CAPROIDÆ, 1663.
   hh. Body longer than deep.
    i. Gill membranes broadly joined to isthmus; body long and low;
        no lateral line ....CLXXXVII, GOBIIDÆ, 2188.
    ii. Gill membranes free from isthmus or very nearly so.
      j. Premaxillaries excessively protractile, their basal process very
        long, in a groove at top of cranium.
      k. Teeth small; scales large, silvery; spines strong.
          CLIII, GERRIDÆ, 1366.
     kk. Teeth none; spines slender ....CLII, MÆNIDÆ, 1364.
    jj. Premaxillaries moderately protractile or not protractile.
      l. Lower pharyngeals united; scales large; anal fin with 3 spines
        and more than 15 soft rays; preopercle entire. (Vivip-
        arous fishes of the Californian fauna.)
          CLVII, EMBIOTOCIDÆ, 1493.
     ll. Lower pharyngeals separate.
     m. Body elongate, not compressed, covered with hard grooved
        scales; jaws box-like..CXXXVIII, TETRAGONURIDÆ, 975.
    mm. Body not as above.
      n. Lateral line incomplete or interrupted, running close to dorsal
        fin; dorsal spines very slender, continuous with the soft
        rays; body low, covered with small scales; anal fin very
        long.
       o. Anal rays fewer than 30; maxillary produced behind.
          CXCI, OPISTHOGNATHIDÆ, 2279.
      oo. Anal rays more than 30; maxillary not produced behind.
          CXCII, BATHYMASTERIDÆ, 2287.
     nn. Lateral line, if present, not as above.
     p. Scales circular, cycloid, nonimbricate, each with 1 or 2 erect
        spines; dorsal spines obsolete.
          CXII, STEPHANOBERYCIDÆ, 835.

pp. Scales not as above.
   q. Anal fin much longer than dorsal; body much compressed, the belly prominent.
   r. Dorsal spines none; scales cycloid..CXI, BATHYCLUPEIDÆ, 834.
   rr. Dorsal spines few, graduated; anal spines 3.
                              CXXXIX, PEMPHERIDIDÆ, 977.
   qq. Anal fin not much, if any, longer than dorsal.
      s. Pseudobranchiæ wanting or covered by skin.
         t. Dorsal fin of soft rays, only beginning as a crest on the head; caudal widely forked. Pelagic fishes.
                              CXXIX, CORYPHÆNIDÆ, 951.
         tt. Dorsal fin with spines anteriorly, not beginning on the head. Fresh-water fishes.
            u. Anal spines 3 to 10.
               v. Dorsal spines 6 to 12; lateral line well developed.
                              CXLI, CENTRARCHIDÆ, 984.
               vv. Dorsal spines about 4; no lateral line; length less than 2 inches.
                              CXL, ELASSOMATIDÆ, 981.
            uu. Anal spines 1 or 2; body oblong or elongate; length less than 8 inches.......................CXLIII, PERCIDÆ, 1015.
      ss. Pseudobranchiæ developed.
         w. Spinous dorsal of 2 or 3 short spines only; anal without spines; scales small, smooth........... CXLVI, SERRANIDÆ, 1126.
      ww. Spinous dorsal, if present, not as above.
         x. Opercle ending in a long scaly flap; snout depressed, spatulate; mouth very large, the lower jaw projecting.
                              CXCIV, CHÆNICHTHYIDÆ, 2293.
         xx. Opercle not ending in a long scaly flap; snout not greatly depressed.
            y. Pectoral fin broad, its lower rays thickened and not branched.
                              CLVI, CIRRHITIDÆ, 1490.
            yy. Pectoral rather narrow at base, its lower rays branched, like the others.
               z. Dorsal fin continuous, the spines few, slender; maxillary usually with an enlarged tooth behind; nape sometimes with an adipose appendage; anal fin long, even.
                              CXC, MALACANTHIDÆ, 2274.
               zz. Dorsal fin continuous or divided, not as above.
                  a. Perch-like fishes, the caudal peduncle not very slender, the scales well developed, ctenoid or cycloid; the dorsal with distinct spines; the anal with at least 1 spine, its soft rays usually few.
                     b. Maxillary not sheathed by the preorbital, or only partially covered by the edge of the latter; ventral with its accessory scale very small or wanting; pectoral without accessory scale; sheath at base of spinous dorsal little developed; vomer usually with teeth; opercle usually ending in a spine.
                        c. Precaudal vertebræ with transverse processes from the third or fourth to the last; ribs all but the last 1 to 4 sessile, inserted on the centra behind the transverse processes; anal spines 3; species silvery in color, the dorsal deeply notched, with 10 spines; vertebræ 10 + 15 = 25.
                              CXLII, KUHLIIDÆ, 1013.
                        cc. Precaudal vertebræ normal, anteriorly without transverse processes; all or most of the ribs inserted on the transverse processes when these are developed.
                           d. Anal spines 2 or 1; pseudobranchiæ small; preopercle with a hook-like spine below; vertebræ increased in number (30 to 46). Fresh-water fishes.......CXLIII, PERCIDÆ, 1015.
                           dd. Anal spines 2, rarely 3; vertebræ 24 or 25; dorsal fin divided. Marine fishes ....CXLIV, CHEILODIPTERIDÆ, 1105.

ddd. Anal spines 3, never 2 nor 1; dorsal fin continuous or divided; vertebræ 24 to 35.

e. Vomer, and usually palatines also, with teeth.

f. Anal fin shorter than dorsal; head not everywhere covered with rough scales; postocular part of head not shortened.

CXLVI, Serranidæ, 1126.

ff. Anal fin scarcely shorter than dorsal and similar to it; head and body everywhere covered with rough scales; body deep, compressed, the posterior part of head shortened.

CXLVIII, Priacanthidæ, 1236.

ee. Vomer without teeth; dorsal fin continuous; body deep, compressed .... CXLVII, Lobotidæ, 1235.

bb. Maxillary slipping for most of its length under the edge of the preorbital, which forms a more or less distinct sheath; ventrals with an accessory scale; opercle without spines; maxillary without supplemental bone; anal spines 3, rarely 2.

g. Fishes carnivorous; intestines of moderate length; teeth in jaws not all incisor-like; vertebræ usually 24 or 25.

h. Vomer with teeth, these sometimes very small; maxillary long.

CXLIX, Lutianidæ, 1241.

hh. Vomer without teeth; palatines and tongue toothless.

i. Teeth on sides of jaws not molar; maxillaries formed essentially as in the *Serranidæ;* preopercle mostly serrate.

CL, Hæmulidæ, 1289.

ii Teeth on sides of jaws molar; maxillaries peculiar in form and in articulation; anterior teeth conical or else more or less incisor-like; preopercle entire...CLI, Sparidæ, 1343.

gg. Fishes herbivorous; intestinal canal elongate; anterior teeth in jaws incisor-like; no molars or canines; premaxillaries moderately protractile.......... CLIV, Kyphosidæ, 1380.

aa. Mackerel-like fishes, with the caudal peduncle usually very slender, the fin widely forked, the scales various, usually not ctenoid; the dorsal spines various, anal fin long.

j. Scales firm, linear, parchment-like; body compressed; bones of head rough; dorsal spines few; mouth small.

CXXXVII, Grammicolepididæ, 973.

jj. Scales not linear, mostly cycloid.

k. Dorsal spines numerous, most of them produced in long filaments; pectorals very long.. CXXIV, Nematistiidæ, 894.

kk. Dorsal spines mostly low, not more than 2 of them filamentous.

l. Dorsal fin very long, all the rays soft; skeleton soft.

CXXXVI, Icosteidæ. 968.

ll. Dorsal fin with 3 or more spines.

m. Dorsal fin divided, the spines 6 to 12 in number.

n. Scales weak, cycloid; jaws without canines.

CXXVIII, Nomeidæ, 948.

nn. Scales ciliate; jaws with canines...CXXVI, Pomatomidæ, 945.

nnn. Scales firm, each with a median ridge; no canines.

CXXXIII, Steinegeriidæ, 960.

mm. Dorsal spines 3 or 4, the fin not divided.

o. Scales minute, body oblong, the shoulder girdle moderate.

CXXXIV, Centrolophidæ, 962.

oo. Scales rather large, firm; body broad, ovate, the shoulder girdle very strong...................... CXXXII, Bramidæ, 956.

BB. Body scaleless, smooth or armed with tubercles, prickles, or scattered bony plates.

C. Breast with a sucking disk.

D. Gill membrane free from the isthmus; no spinous dorsal; a large sucking disk between the ventral fins.

CXCIX, Gobiesocidæ, 2326.

DD. Gill membranes joined to the isthmus; a sucking disk formed of the ventral fins.
  E. Skin perfectly smooth; spinous dorsal not distinct.
                      CLXXXIII, LIPARIDIDÆ, 2105.
  EE. Skin with tubercles or spines, or else with a distinct spinous dorsal................CLXXXII, CYCLOPTERIDÆ, 2094.
CC. Breast without sucking disk.
  F. Gill membranes broadly attached to the isthmus.
    G. Ventrals completely united.......CLXXXVIII, GOBIIDÆ, 2188.
    GG. Ventrals widely separated; body depressed; preopercle with a strong spine ....CLXXXVII, CALLIONYMIDÆ, 2184.
  FF. Gill membranes nearly or quite free from the isthmus.
    H. Anal preceded by 2 free spines (these lost with age; connected by membranes in the very young)..CXXV, CARANGIDÆ, 895.
    HH. Anal without free spines.
      I. Dorsal and anal fins followed by finlets.CXVIII, SCOMBRIDÆ, 863.
      II. Dorsal and anal without finlets.
      J. Suborbital with a bony stay; no free anal spines.
                      CLXXIX, COTTIDÆ, 1879.
      JJ. Suborbital without bony stay.
      K. Mouth very large, nearly horizontal, the teeth sharp; no pseudobranchiæ...............CXCIII, CHIASMODONTIDÆ, 2291.
      KK. Mouth large, nearly vertical; body compressed; preopercle armed with spines........CXCV, TRICHODONTIDÆ, 2295.
AA. Gill openings small, behind, above, or below the pectoral fins, which are more or less pediculate.
  L. Gill openings in or behind upper axil of pectorals; mouth small.
                  CCXXIV, OGCOCEPHALIDÆ, 2735.
  LL. Gill openings in or behind lower axil of pectoral; mouth large.
    M. Head compressed; no pseudobranchiæ.
                  CCXXII, ANTENNARIIDÆ, 2715.
    MM. Head depressed; pseudobranchiæ present.
                  CCXXI, LOPHIIDÆ, 2713.

III.—VENTRAL FINS PRESENT, THORACIC OR JUGULAR, THE NUMBER OF RAYS NOT DEFINITELY I, 5.

  A. Eyes unsymmetrical, both on the same side of head.
  B. Eyes large, well separated; edge of preopercle usually evident.
                  CCXIX, PLEURONECTIDÆ, 2602.
  BB. Eyes small, very close together; edge of preopercle hidden by skin; mouth very small..........CCXX, SOLEIDÆ, 2692.
AA. Eyes symmetrical, one on each side of the head.
  C. Ventral rays with or without spine, the number of soft rays more than 5.
  D. Caudal fin wanting; scales spinous..CCXV, MACROURIDÆ, 2561.
  DD. Caudal fin well developed.
    E. Tail isocercal, the vertebræ progressively smaller to base of caudal; ventrals jugular; no spines in any of the fins.
    F. Jaws and vomer with strong canines; second dorsal and anal deeply notched; no barbel..CCXIII, MERLUCCIIDÆ, 2529.
    FF. Jaws and vomer without distinct canines; chin usually with a barbel ....CCXIV, GADIDÆ, 2531.
  EE. Tail not isocercal, the last vertebræ not reduced in size.
    G. Ventral rays about 15; dorsal fin single, elevated.
                  CXXX, LAMPRIDIDÆ, 953.
    GG. Ventral rays I, 3 or I, 5; dorsal very high.
                  CXXXI, PTERACLIDIDÆ, 955.
    GGG. Ventral rays I, 6 to I, 10; dorsal with spines.

H. Vent anterior; dorsal spines 3 or 4; scales ctenoid.
CV, APHREDODERIDÆ, 785.

HH. Vent normal.

 I. Chin with two long barbels, behind symphysis; dorsal continuous, with 5 spines .... CXVI, POLYMIXIIDÆ, 854.

 II. Chin without barbels.

 J. Dorsal fin divided, the anterior part of a single slender spine; ventrals elongate......... CCXII, BREGMACEROTIDÆ, 2525.

 JJ. Dorsal fin divided, the anterior part of many spines.

 K. Body covered with firm serrated scales; anal spines 4; dorsal spines not elevated............ CXV, HOLOCENTRIDÆ, 845.

 KK. Body naked or covered with small scales, besides bony plates or warts....... : .... : ....CLXII, ZEIDÆ, 1659.

KKK. Body uniformly covered with cycloid scales; dorsal spines mostly very high and filamentous.
CXXIV, NEMATISTIIDÆ, 894.

JJJ. Dorsal fin continuous, its spines 2 to 8.

 L. Suborbitals narrow, not covering the cheeks.
CXIV, BERYCIDÆ, 837.

 LL. Suborbitals very broad, covering the cheeks.
CXIII, TRACHICHTHYIDÆ, 836.

CC. Ventral fins with or without spine, the number of soft rays fewer than 5.

 M. Gill opening before the pectoral fin.

 N. Anal fin present; caudal fin not directed upward.

 O. Upper jaw not prolonged into a sword.

 P. Dorsal fin with some spines or simple rays.

 Q. Dorsal fin without soft rays, composed of spines only.
CC, BLENNIIDÆ, 2344.

 QQ. Dorsal fin with soft rays anteriorly, with spines posteriorly; gill membranes joined to isthmus.. CCVI, ZOARCIDÆ, 2455.

 QQQ. Dorsal fin of spines anteriorly, with soft rays posteriorly.

 R. Dorsal spines all separate and unconnected; body scaleless, naked, or with bony plates; ventral with a sharp spine.
XCVIII, GASTEROSTEIDÆ, 742.

 RR. Dorsal spines connected by membrane.

 S. Suborbital with a bony stay, extending across the cheek, to or toward the preopercle, the cheek sometimes entirely covered with a coat of mail.

 T. Pectoral fin divided into 2 parts, 1 of them very long; head bony................ CLXXXVI, CEPHALACANTHIDÆ, 2182.

 TT. Pectoral fin not divided.

 U. Body entirely covered with an armor of bony plates; head bony.
CLXXXI, AGONIDÆ, 2031.

 UU. Body naked, or more or less rough or scaly, not entirely covered by bony plates.

 V. Gill opening very small, not extending below upper edge of pectoral; skin everywhere prickly; head very large, bony above................... CLXXX, RHAMPHOCOTTIDÆ, 2029.

 VV. Gill opening large, extending downward nearly or quite to lowest pectoral ray............ CLXXIX, COTTIDÆ, 1879.

 SS. Suborbital without bony stay.

 W. Dorsal spines 2 to 4 only; head very broad, depressed; gills 3; gill membranes broadly united to the isthmus.

 x. Ventrals each a strong spine; teeth incisor-like; scales shagreen-like....... : .... CLXVIII, TRIACANTHIDÆ, 1697.

 xx. Ventrals not reduced each to a single spine.
CXCVIII, BATRACHOIDIDÆ, 2313.

WW. Dorsal spines numerous; gills 4.

 X. Gill membranes separate, free from the isthmus.

 Y. Body greatly elongate; lower jaw with a slit at base to permit free motion; lips not fringed.

Z. Soft dorsal and anal with a distinct lobe anteriorly, distinct from spinous part............CXIX, GEMPYLIDÆ, 877.

ZZ. Soft dorsal and anal without anterior lobe, continuous with spinous part....................CXX, LEPIDOPODIDÆ, 884.

YY. Body moderately elongate; opercles and lips fringed; eyes superior......................CXCVI, DACTYLOSCOPIDÆ, 2297.

XX. Gill membranes broadly united, attached to the isthmus or not.

  x. Gill opening moderate or large............CC, BLENNIIDÆ, 2344.

  xx. Gill openings very small, reduced to oblique slits before the pectoral fins.....................CCIII, CERDALIDÆ, 2448.

PP. Dorsal fins of soft rays only.

  a. Breast with a large sucking dish between ventral fins.
      CXCIX, GOBIESOCIDÆ, 2326.

  aa. Breast without sucking disk.

  b. Body covered with a coat of mail; dorsal very short.
      CLXXXI, AGONIDÆ, 2031.

  bb. Body not mailed; dorsal many-rayed.

  c. Lateral line and base of dorsal beset with prickles; skeleton very soft; body compressed ...CXXXVI, ICOSTEIDÆ, 968.

  cc. Lateral line unarmed.

  d. Tail isocercal, the vertebral column pointed behind, the last vertebræ very small; hypercoracoid not perforate; no pseudobranchiæ.

  e. Caudal fin present.....................CCXIV, GADIDÆ, 2531.

  ee. Caudal fin wanting.................CCXV, MACROURIDÆ, 2561.

  dd. Tail not isocercal, truncate at base of. caudal; hypercoracoid perforate.

  f. Gill membranes joined to the isthmus; pseudobranchiæ present.

  g. Ventral fins under shoulder girdle......CCVI, ZOARCIDÆ, 2455.

  gg. Ventral fins inserted below the eyes.
      CCVII, DEREPODICHTHYIDÆ, 2480.

  ff. Gill membranes free from the isthmus.

  h. Ventral fins inserted below or before the eyes; pseudobranchiæ generally well developed......CCVIII, OPHIDIIDÆ, 2481.

  hh. Ventral fins inserted below shoulder girdle; no pseudobranchiæ.
      CCXI, BROTULIDÆ, 2498.

OO. Upper jaw prolonged into a bony sword; dorsal fin long and high; size large.............CXXII, ISTIOPHORIDÆ, 890.

NN. Anal fin wanting; caudal fin distorted or directed upward; body ribbon-like.

  i. Ventral fins each of a few slender rays.
      CCXVII, TRACHYPTERIDÆ, 2597.

  ii. Ventral fins each reduced to a long slender filament.
      CCXVI, REGALECIDÆ, 2595.

MM. Gill openings behind the pectoral fins.

  j. Gill openings above and behind pectorals; mouth small, low.
      CCXXIV, OGCOCEPHALIDÆ, 2735.

  jj. Gill openings below and behind pectorals; mouth large, nearly vertical.................CCXXII, ANTENNARIIDÆ, 2715.

IV.—VENTRAL FINS WHOLLY WANTING.

A. Premaxillary and maxillary wanting or grown fast to the palatines; body greatly elongate, eel-shaped; gill openings restricted to the sides; scales minute or wanting; scapular arch not attached to the skull. Eels.

B. Gill openings not very far behind cranium; gape not inordinately distensible; gill arches 4 pairs.

C. Gill openings. well developed, leading to large interbranchial slits; tongue present; opercles and branchial bones well developed; scapular arch present.

D. Skin covered with rudimentary embedded scales, usually linear in form, arranged in small groups, and placed obliquely at right angles to those of the neighboring groups; pectorals and vertical fins well developed, the latter confluent about the tail; lateral line present; posterior nostril in front of eyes; tongue with its margins free.

E. Gill openings well separated; branchiostegals long, bent upward behind.

F. Gill openings lateral and vertical; snout conic, the jaws not very heavy; gape longitudinal; lips thick; lower jaw projecting; teeth in cardiform bands on jaws and vomer; eggs minute.....................XLIII, Anguillidæ, 346.

FF. Gill openings horizontal, inferior.

G. Snout very blunt, with very strong jaws; gape transverse; lips obsolete; teeth blunt, in 1 series, on jaws only.
XLIV, Simenchelyidæ, 348.

GG. Snout conical and slender, the jaws of moderate strength; gape lateral; lips obsolete; tongue but little developed; teeth acute, in bands on jaws and vomer.
XLV, Ilyophidæ, 349.

EE. Gill openings inferior, very close together, apparently confluent; branchiostegal rays abbreviated behind; head conical; tongue small; posterior nostrils in front of eye.
XLVI, Synaphobranchidæ, 350.

DD. Scales wholly wanting; eggs (so far as known) of moderate size, much as in ordinary fishes.

H. Tip of tail with a more or less distinct fin, the dorsal and anal fins confluent around it; the tail sometimes ending in a long filament. Coloration almost always plain, brownish, blackish, or silvery, the fins often black-margined.

I. Posterior nostril without tube, situated entirely above the upper lip.

J. Tongue broad, largely free anteriorly and on sides; vomerine teeth moderate.

K. Pectoral fins well developed; body not excessively elongate; lower jaw not projecting; anterior nostril remote from eye.....................XLVII, Leptocephalidæ, 352.

JJ. Tongue narrow, adnate to the floor of the mouth or only the tip slightly free; vomerine teeth well developed, sometimes enlarged.

L. Jaws not attenuate and recurved at tip; gill openings well separated; anterior nostril remote from eye.

M. Pectoral fins well developed; skin thick; skeleton firm; snout moderate; tail not ending in a filiform tip.
XLVIII, Murænesocidæ, 358.

MM. Pectoral fins wholly wanting; snout and jaws much produced, the upper longer; jaws straight; skin thin and skeleton weak; tail ending in a filiform tip; gill openings small, subinferior; teeth sharp, subequal, recurved, a long series on the vomer. Deep-sea eels, soft in body, black in color.....................XLIX, Nettastomatidæ, 364.

LL. Jaws long and slender, tapering to a point, recurved at tip; nostrils large, both pairs close in front of eye; gill openings convergent forward, separate or confluent; pectorals and vertical fins well developed; membranes of fins thin, not enveloping the rays; skeleton well developed. Deep-sea eels.....................L, Nemichthyidæ, 366.

II. Posterior nostril close to the edge of the upper lip; tongue more or less fully adnate to the floor of the mouth; teeth subequal.....................LI, Myridæ, 370.

IIII. Tip of tail without rays, projecting beyond the dorsal and anal fins (not filiform); posterior nostril on the edge of the upper lip; anterior nostril near tip of snout, usually in a small tube; tongue usually adnate to the floor of the mouth. Coloration frequently variegated.

LII, Ophichthyidæ, 372.

CC. Gill openings small, roundish, leading to restricted interbranchial slits; tongue wanting; pectoral fins (typically) wanting; opercles feebly developed; fourth gill arch modified, strengthened, and supporting pharyngeal jaws.

N. Scapular arch obsolete or represented by cartilage; heart not far back; pectorals wanting; (skin thick; coloration often variegated) .... LIII, Murænidæ, 388.

BB. Gill openings far behind cranium; gape of mouth inordinately distensible; gill arches 5 or 6 pairs; tail excessively long, tapering to a point.

O. Distance from gill opening to vent much greater than that from tip of snout to gill opening.

LIV, Saccopharyngidæ, 405.

OO. Distance from gill opening to vent much less than from tip of snout to gill opening.......... LV, Eurypharyngidæ, 406.

AA. Premaxillary and maxillary present, often immovably united to rest of cranium.

P. Gill openings united in a single slit below throat; no pectoral fins; body eel-shaped......... XLI, Symbranchidæ, 342.

PP. Gill openings not united in a longitudinal slit.

Q. Dorsal fin wanting; anal fin very long; vent near the head; caudal obsolete; body band-like...... XL, Gymnotidæ, 340.

QQ. Dorsal fin present.

R. Body eel-shaped, contracted at the neck; the vertical fins confluent around the tail; premaxillary and maxillary immovably united to the skull..XLII, Derichthyidæ, 343.

RR. Body eel-shaped, ending in a long filament, longer than rest of body.

X. No anal or caudal fin...........CCXXVIII, Stylephoridæ, 2601.

XX. No caudal fin; anal present.......CCIV, Ptilichthyidæ, 2451.

RRR. Body not truly eel-shaped.

S. Gill openings far behind pectoral fins; mouth oblique, very large; spinous dorsal represented by fleshy tentacles.

CCXXIII, Ceratiidæ, 2727.

SS. Gill openings before pectoral fins.

T. Gill membranes broadly united to the isthmus, restricting the gill openings to the sides.

U. Snout tubular, bearing the short, toothless mouth at the end; body mailed................... CIII, Syngnathidæ, 760.

UU. Snout not tubular.

V. Breast without sucking disk.

W. Dorsal fin single, of spines or undivided rays only.

X. Jaws and vomer with coarse molar teeth.

CCII, Anarhichadidæ, 2445.

XX. Jaws and vomer without molars.

Y. Mouth nearly vertical; dorsal spines slender, rather high.

CCI, Cryptacanthodidæ, 2442.

YY. Mouth not nearly vertical; dorsal spines moderate or low, some or all of them usually pungent.....CC, Blenniidæ, 2344.

WW. Dorsal fins 2, the anterior of spines, the posterior of soft rays; body short and deep.

Z. Spinous dorsal of 2 or 3 spines; scales rather large, rough or bony .... CLXIX, Balistidæ, 1698.

ZZ. Spinous dorsal of 1 or 2 spines; scales minute, rough, forming a velvety covering......... CLXX, Monacanthidæ, 1712.

WWW. Dorsal fin continuous, of soft rays only.

a. Body oblong or elongate, the back not elevated; dorsal and anal joined to caudal.

   b. Pectoral rather narrow, the lower rays similar to the others.
                        CCVI, ZOARCIDÆ, 2455.
 bb. Pectorals very broad, the lower rays procurrent and produced
      at tip ....CLXXXIII, LIPARIDIDÆ, 2105.
aa. Body short, not elongate; dorsal and anal free from caudal.
   c. Teeth in each jaw confluen into 1.
   d. Body compressed, rough ....CLXXV, MOLIDÆ, 1752.
 dd. Body not compressed, spinous ....CLXXIV, DIODONTIDÆ, 1742.
 cc. Teeth in each jaw confluent into 2.
   x. Back broadly rounded.........CLXXII, TETRAODONTIDÆ, 1726.
 xx. Back with a sharp median ridge.
            CLXXIII, CANTHIGASTERIDÆ, 1740.
ccc. Teeth separate; body enveloped in a bony box.
            CLXXI, OSTRACIIDÆ, 1721.
VV. Breast with a sucking disk.
   e. Skin perfectly smooth; dorsal continuous or slightly notched.
            CLXXXIII, LIPARIDIDÆ, 2105.
 ee. Skin more or less tubercular; dorsal usually divided.
            CLXXXII, CYCLOPTERIDÆ, 2094.
TT. Gill membranes free from the isthmus.
   f. Vent at the throat.
   g. Vertical fins confluent; body elongate, almost eel-shaped.
            CCX, FIERASFERIDÆ, 2494.
 gg. Vertical fins separate; body oblong, scaly.
            XCIII, AMBLYOPSIDÆ, 702.
 ff. Vent posterior, not at the throat.
   h. Caudal fin wanting; body naked, greatly elongate.
            CXXI, TRICHIURIDÆ, 888.
 hh. Caudal fin present.
   i. Upper jaw prolonged into a sword; size very large.
            CXXIII, XIPHIIDÆ, 893.
 ii. Upper jaw not prolonged into a sword.
   j. Belly with a series of bony scutes along its edge; body much
      compressed.......................LXI, CLUPEIDÆ, 417.
 jj. Belly not armed with scutes.
   k. Mouth inordinately large, formed like the mouth of a whale,
      with sharp teeth; no scales...LXXIV, CETOMIMIDÆ, 548.
 kk. Mouth not inordinately large, not peculiar in form.
   l. Body ovate, much compressed.
   m. Scales small, cycloid, silvery......CXXXV, STROMATEIDÆ, 964.
 mm. Scales wanting; caudal peduncle very slender.
            CXXXVI, ICOSTEIDÆ, 968.
   l. Body oblong or elongate, much longer than deep.
   n. Gill membranes broadly united; teeth present.
   o. Dorsel fin of spines only.................. CC, BLENNIIDÆ, 2344.
 oo. Dorsal fin of soft rays only; body eel-shaped.
            CCV, SCYTALINIDÆ, 2453.
ooo. Dorsal fin single, the posterior half of soft rays, the anterior of
      spines; body elongate, covered with small scales.
            CC, BLENNIIDÆ, 2344.
oooo. Dorsal fins 2, the anterior of slender spines, posterior soft, body
      naked............................CLXXIX, COTTIDÆ, 1879.
 nn. Gill membranes separate.
   p. Jaws toothless, the lower jaw projecting; body scaly, with cross
      folds of skin.....................CX, AMMODYTIDÆ, 831.
 pp. Jaws with teeth.
   q. Body naked, without folds of skin; no pseudobranchiæ.
            CCIX, LYCODAPODIDÆ, 2491.
 qq. Body with small scales; pseudobranchiæ present; head with
      very large mucous pores; lower jaw very strong.
            CXXXVIb, ZAPRORIDÆ, 2849.

GLOSSARY OF TECHNICAL TERMS.*

Abdomen. Belly.

Abdominal. Pertaining to the belly; said of the ventral fins of fishes when inserted considerably behind the pectorals, the pelvic bones to which the ventral fins are attached having no connection with the shoulder girdle.

Abortive. Remaining or becoming imperfect.

Actinosts. A series of bones at the base of the pectoral rays.

Acuminate. Tapering gradually to a point.

Acute. Sharp-pointed.

Adipose fin. A peculiar, fleshy, fin-like projection behind the dorsal fin, on the backs of salmons, catfishes, etc.

Adult. A mature animal.

Airbladder. A sac filled with air, lying beneath the backbone of fishes, corresponding to the lungs of higher vertebrates.

Alisphenoid. A small bone on the anterior lateral wall of the brain case.

Amphicœlian. Double concave; said of vertebræ.

Anadromous. Running up; said of marine fishes which run up rivers to s awn.

Anal.p Pertaining to the anus or vent.

Anal fin. The fin on the median line behind the vent, in fishes.

Anchylosed. Grown firmly together.

Angular. A small bone on the posterior end of the mandible.

Antrorse. Turned forward.

Anus. The external opening of the intestine; the vent.

Arterial bulb. The muscular swelling, at the base of the great artery, in fishes.

Articular. The bone of the mandible supporting the dentary.

Articulate. Jointed.

Atlas. The first vertebra.

Atrophy. Nondevelopment.

Attenuate. Long and slender, as if drawn out.

Auditory capsule. The ventrolateral swelling of the skull.

Barbel. An elongated fleshy projection, usually about the head, in fishes.

Basal. Pertaining to the base; at or near the base.

Basibranchials. A lower median series of bones of the branchial arches.

Basioccipital. A median posterior ventral bone of the skull to which the atlas is attached.

Basis cranii. Formed by shelves of bone developed from the inner sides of the prootics which meet and form a roof to the myodome and a floor to the brain cavity.

Bicolor. Two-colored.

Bicuspid. Having 2 points.

Brachial ossicles. Synonymous with actinosts, q. v.

Branchiæ. Gills; respiratory organs of fishes.

Branchial. Pertaining to the gills.

Branchihyals. Small bones at base of gill arches.

Branchiostegals. The bony rays supporting the branchiostegal membranes, under the head of a fish, below the opercular bones, and behind the lower jaw.

---

* In the preparation of this Glossary the authors are indebted to Mr. Edwin Chapin Starks for valuable assistance.

Bristle. A stiff hair, or hair-like feather.

Buccal. Pertaining to the mouth.

Caducous. Falling off early.

Cæcal. Of the form of a blind sac.

Cæcum. An appendage of the form of a blind sac, connected with the alimentary canal at the posterior end of the stomach, or pylorus.

Canines. The teeth behind the incisors—the "eye-teeth;" in fishes, any conical teeth in the front part of the jaws, longer than the others.

Cardiform (teeth). Teeth coarse and sharp, like wool cards.

Carinate. Keeled; having a ridge along the middle line.

Carotid. The great artery running to the head.

Carpus. The wrist.

Catadromous. Running down; said of fresh-water species which run down to the sea to spawn.

Caudal. Pertaining to the tail.

Caudal fin. The fin on the tail of fishes and whales.

Caudal peduncle. The region between the anal and caudal fins in fishes.

Cavernous. Containing cavities, either empty or filled with a mucous secretion.

Centrum. The body of a vertebra.

Cephalic fins. Fins on the head of certain rays; a detached portion of the pectoral.

Ceratobranchials. Bones of the branchial arches just below their angle.

Ceratohyal. One of the hyoid bones.

Chiasma. Crossing of the fibers of the optic nerve.

Chin. The space between the rami of the lower jaw.

Ciliated. Fringed with eyelash-like projections.

Cirri. Fringes.

Claspers. Organs attached to the ventral fins in the male of sharks, skates, etc.

Clavicle. The collar bone, or lower anterior part of shoulder girdle, not entering into socket of arm.

Compressed. Flattened laterally.

Condyle. Articulating surface of a bone.

Coracoid. The principal bone of the shoulder girdle in fishes; otherwise a bone or cartilage on the ventral side, helping to form the arm socket. Synonymous with hypercoracoid, q. v.

Cranial. Pertaining to the cranium or skull.

Ctenoid. Rough-edged; said of scales when the posterior margin is minutely spinous or pectinated.

Cycloid. Smooth-edged; said of scales not ctenoid, but concentrically striate.

Deciduous. Temporary; falling off.

Decurved. Curved downward.

Dentary. The principal or anterior bone of the lower jaw or mandible, usually bearing the teeth.

Dentate. With tooth-like notches.

Denticle. A little tooth.

Depressed. Flattened vertically.

Depth. Vertical diameter (usually of the body of fishes).

Dermal. Pertaining to the skin.

Diaphanous. Translucent.

Distal. Remote from point of attachment.

Dorsal. Pertaining to the back.

Dorsal fin. The fin on the back of fishes.

Emarginate. Slightly forked or notched at the tip.

Endoskeleton. The skeleton proper; the inner bony framework of the body.

Enteron. The alimentary canal.

Epibranchials. The bones directly above the angle of the branchial arches.

Epihyal. One of the hyoid bones.

Epipleurals. Rays of bone attached to the ribs and anterior vertebræ usually touching the skin in the vicinity of the lateral line.

Erectile. Susceptible of being raised or erected.

Ethmoid. A median anterior bone of the skull.

Exoccipitals. Two bones of the skull, 1 on each side of the foramen magnum.

Exoskeleton. Hard parts (scales, scutes) on the surface of the body.

Exserted. Projecting beyond the general level.

Extralimital. Beyond the limits (of this book).

Facial. Pertaining to the face.

Falcate. Scythe-shaped; long, narrow, and curved.

Falciform. Curved, like a scythe.

Fauna. The animals inhabiting any region, taken collectively.

Femoral. Pertaining to the femur, or proximal bone of the hinder leg.

Filament. Any slender or thread-like structure.

Filiform. Thread form.

Fontanel. An unossified space on top of head covered with membrane.

Foramen. A hole or opening.

Foramen magnum. The aperture in the posterior part of the skull for the passage of the spinal cord.

Forehead. Frontal curve of head.

Forficate. Deeply forked; scissors-like.

Fossæ (nasal). Groves in which the nostrils open.

Frontal bone. Anterior bone of top of head, usually paired.

Fulcra. Rudimentary spine-like projections extending on the anterior rays of the fins of ganoid fishes.

Furcate. Forked.

Fusiform. Spindle-shaped; tapering toward both ends, but rather more abruptly forward.

Ganglion. A nerve center.

Ganoid. Scales or plates of bone covered by enamel.

Gape. Opening of the mouth.

Gill arches. The bony arches to which the gills are attached.

Gill openings. Openings leading to or from the branchiæ.

Gill rakers. A series of bony appendages, variously formed, along the inner edge of the anterior gill arch.

Gills. Organs for breathing the air contained in water.

Glabrous. Smooth.

Glossohyal. The tongue bone.

Graduated (spines). Progressively longer backward, the third being as much longer than the second as second is longer than first.

Granulate. Rough with small prominences.

Gular. Pertaining to the *gula*, or upper foreneck.

Hæmal arch. An arch under a hæmal spine for the passage of a blood vessel.

Hæmal canal. The series of hæmal arches as a whole.

Hæmal spine. The lowermost spine of a caudal vertebra, in fishes.

Hæmopophyses. Appendages on the lower side of abdominal vertebræ, in fishes.

Height. Vertical diameter.

Heterocercal. Said of the tail of a fish when unequal; the backbone evidently running into the upper lobe.

Homocercal. Said of the tail of a fish when not evidently unequal; the backbone apparently stopping at the middle of the base of the caudal fin.

Humerus. Bone of the upper arm.

Hyoid. Pertaining to the tongue.

Hyoid apparatus. Formed by a series of bones extending along the inner side of the mandible and supporting the tongue.

Hyomandibular. A bone by which the posterior end of the suspensorium is articulated with the skull; the supporting element of the suspensorium, the mandible, the hyoid apparatus, and the opercular apparatus.

Hypercoracoid. The upper of the 2 bones attached to the clavicle, indirectly bearing the pectoral fin.

Hypleural. The modified last vertebra supporting the caudal fin.

Hypobranchials. Bones of the branchial arches below the ceratobranchials.

Hypocoracoid. The lower of the 2 bones attached to the clavicle behind.

Hypohyals. Small bones, usually 4, by which the respective sides of the hyoid apparatus are joined.

Imbricate. Overlapping, like shingles on a roof.

Imperforate. Not pierced through.

Inarticulate. Not jointed.

Incisors. The front or cutting teeth.

Inferior pharyngeals. Synonymous with pharyngeals, q. v.

Infraoral. Below the mouth.

Interhæmal spines. Elements supporting the anal fin.

Interhæmals. Bones to which anal rays are attached, in fishes.

Interhyal. Upper hyoid bone attached to hyomandibular.

Intermusculars. Synonym of epipleurals, q. v.

Interneural spines. Elements supporting the dorsal fins.

Interspinous bones. The interneurals and the interhæmals.

Intermaxillaries. The premaxillaries; the bones forming the middle of the front part of the upper jaw, in fishes.

Interneurals. Bones to which dorsal rays are attached, in fishes.

Interopercle. Membrane bone between the preopercle and the branchiostegals.

Interorbital. Space between the eyes.

Interspinals. Bones to which fin rays are attached (in fishes); inserted between neural spines above and hæmal spines below.

Isocercal (tail). Last vertebræ progressively smaller and ending in median line of caudal fin, as in the codfish.

Jugular. Pertaining to the lower throat; said of the ventral fins, when placed in advance of the attachment of the pectorals.

Keeled. Having a ridge along the middle line.

Lacustrine. Living in lakes.

Lamellæ. Plate-like processes like those inside the bill of a duck.

Larva. An immature form, which must undergo change of appearance before becoming adult.

Lateral. To or toward the side.

Lateral line. A series of muciferous tubes forming a *raised line* along the sides of a fish.

Lateral processes. Synonym of parapophyses, q. v.

Laterally. Sidewise.

Lunate. Form of the new moon; having a broad and rather shallow fork.

Mandible. Under jaw.

Maxilla, or *maxillary.* Upper jaw.

Maxillaries. Outermost or hindmost bones of the upper jaw, in fishes; they are joined to the premaxillaries in front, and usually extend farther back than the latter.

Mesethmoid. Synonym of ethmoid, q. v.

Mesopterygoid. A bone of the suspensorium.

Metapterygoid. A bone of the suspensorium, or chain supporting the lower jaw.

Molars. The grinding teeth; posterior teeth in the jaw.

Muciferous. Producing or containing mucus.

Myocomma. A muscular band.

Myodome. Cavity under the brain cavity for the reception of the rectus muscles of the eye.

Nape. Upper part of neck, next to the occiput.

Nares. Nostrils, anterior and posterior.

Nasal. Pertaining to the nostrils.

Nasal plate. Plate in which the nostrils are inserted.

Neural arch. An opening through the base of the neural spine for the passage of the spinal cord.

Neural canal. The neural arches as a whole.

Neural processes. Two plates rising vertically, 1 on each side of the centrum of the vertebra, which unite toward their ends and form a spine.

Neural spine. The uppermost spine of a vertebra.

Nictitating membrane. The third or inner eyelid of birds, sharks, etc.

Notochord. A cellular chord which in the embryo precedes the vertebral column.

Nuchal. Pertaining to the nape or *nucha.*

Obsolete. Faintly marked; scarcely evident.

Obtuse. Blunt.

Occipital. Pertaining to the occiput.

Occipital condyle. That part of the occipital bone modified to articulate with the atlas.

Occiput. Back of the head.

Ocellate. With eye-like spots, generally roundish and with a lighter border.

Oid (suffix). Like; as *Percoid,* perch-like.

Opercle, or *operculum.* Gill cover; the posterior membrane bone of the side of the head, in fishes.

Opercular bones. Membrane bones of the side of the head, in fishes.

Opercular flap. Prolongation of the upper posterior angle of the opercle, in sunfishes.

Opisthocœlian. Concave behind only; said of vertebræ which connect by ball-and-socket joints.

Opisthotic. A bone of the skull to which the lower limb of the post-temporal usually articulates.

Orbicular. Nearly circular.

Orbit. Eye socket.

Osseous. Bony.

Ossicula auditus. Bones of the ear, in fishes.

Osteology. Study of bones.

Oviparous. Producing eggs which are developed after exclusion from the body, as in all birds and most fishes.

Ovoviviparous. Producing eggs which are hatched before exclusion, as in the dogfish and garter snake.

Ovum. Egg.

Palate. The roof of the mouth.

Palatines. Membrane bones of the roof of the mouth, 1 on each side extending outward and backward from the vomer.

Palustrine. Living in swamps.

Papilla. A small fleshy projection.

Papillose. Covered with papillæ.

Parapophyses. The lateral projections on some of the abdominal vertebræ to support ribs.

Parasphenoid. Bone of roof of mouth behind the vomer. Synonym of prefrontal.

Parietal. Bone of the side of head above.

Parotic process. A posterior lateral process of the skull formed by the pterotic and opisthotic.

Pectinate. Having teeth like a comb.

Pectoral. Pertaining to the breast.

Pectoral fins. The anterior or uppermost of the paired fins, in fishes, corresponding to the anterior limbs of the higher vertebrates.

Pelagic. Living on or in the high seas.

Pelvic girdle. The bones supporting the ventral fins or pelvics.

Pelvis. The bones to which the hinder limbs (ventral fins in fishes) are attached.

Perforate. Pierced through.

Peritoneum. The membrane lining the abdominal cavity.

Pharyngeal bones. Bones behind the gills and at the beginning of the œsophagus of fishes, of various forms, almost always provided with teeth; usually 1 pair below and 2 pairs above. They represent a fifth gill arch.

Pharyngobranchials. Upper elements of the branchial arches, usually bearing teeth.

Pharyngognathous. Having the lower pharyngeal bones united.

Physoclistous. Having the air bladder closed.

Physostomous. Having the air bladder connected by a tube with the alimentary canal.

Pigment. Coloring matter.

Pineal body. A small ganglion in the brain; a rudiment of an optic lobe, which in certain lizards (and in extinct forms) is connected with a third or median eye.

Pituitary body. A small ganglion in the brain.

Plicate. Folded; showing transverse folds or wrinkles.

Plumbeous. Lead colored; dull bluish gray.

Polygamous. Mating with more than 1 female.

Postclavicle. A ray composed of 1 or 2 bones attached to the inner upper surface of the clavicle and extending downward.

Postorbital. Behind the eye.

Post-temporal. The bone, in fishes, by which the shoulder girdle is suspended to the cranium.

Præcoracoid. A portion of coracoid more or less separated from the rest.

Præcoracoid arch. An arch in front of the coracoid in most soft-rayed fishes.

Prefrontals. Bones forming lateral projections at the anterior end of the skull.

Premaxillaries. The bones, 1 on either side, forming the front of the upper jaw in fishes. They are usually larger than the maxillaries and commonly bear most of the upper teeth.

Premolars. The small grinders; the teeth between the canines and the true molars.

Preocular. Before the eye.

Preopercle. The membrane bone lying in front of the opercle and more or less nearly parallel with it.

Preorbital. The large membrane bone before the eye, in fishes.

Procælian. Concave in front only.

Procurrent (fin). With the lower rays inserted progressively farther forward.

Projectile. Capable of being thrust forward.

Prootic. A bone forming an anterolateral ossification of the brain case.

Protractile. Capable of being drawn forward.

Proximal. Nearest.

Pseudobranchiæ. Small gills developed on the inner side of the opercle, near its junction with the preopercle.

Pterotic. A bone at the posterior lateral process of the skull.

Pterygoids. Bones of roof of mouth in fishes, behind the palatines.

Pubic bones. Same as pelvic bones, q. v.

Pubis. Anterior lower part of pelvis.

Pulmonary. Pertaining to the lungs.

Punctate. Dotted with points.

Pyloric cœca. Glandular appendages in the form of blind sacs opening into the alimentary canal of most fishes at the *pylorus,* or passage from the stomach to the intestine.

Quadrate. A bone of the suspensorium on which the mandible is hinged.

Quincunx. Set of 5 arranged alternately, thus *

Radius. Outer bone of forearm.

Ray. One of the cartilaginous rods which support the membrane of the fin of a fish.

Recurved. Curved upward.

Reticulate. Marked with a network of lines.

Retrorse. Turned backward.

Rudimentary. Undeveloped.

Rugose. Rough with wrinkles.

Sacral. Pertaining to the *sacrum*, or vertebræ of the pelvic region.

Scapula. Shoulder blade; in fishes, the bone of the shoulder girdle below the post-temporal.

Scapular arch. Shoulder girdle.

Scute. Any external bony or horny plate.

Second dorsal. The posterior or soft part of the dorsal fin, when the two parts are separated.

Septum. A thin partition.

Serrate. Notched, like a saw.

Sessile. · Without a stem or peduncle.

Setaceous. Bristly.

Setiform. Bristle-like.

Shaft. Stiff axis of a quill.

Shoulder girdle. The bony girdle posterior to the head, to which the anterior limbs are attached (post-temporal, scapula, and coracoid or clavicle).

Soft dorsal. The posterior part of the dorsal fin in fishes, when composed of soft rays.

Soft rays. Fin rays which are articulate and usually branched.

Spatulate. Shaped like a spatula.

Sphenoid. Basal bone of skull.

Sphenotic. A lateral bone of the skull.

Spine. Any sharp projecting point; in fishes those fin rays which are unbranched, inarticulate, and usually, but not always, more or less stiffened.

Spinous. Stiff or composed of spines.

Spinous dorsal. The anterior part of the dorsal fin when composed of spinous rays.

Spiracles. Openings in the head and neck of some fishes and batrachians.

Stellate. Star-like; with radiating ridges.

Striate. Striped or streaked.

Sub (in composition). Less than; somewhat; not quite; under, etc.

Subcaudal. Under the tail.

Subopercle. The bone immediately below the opercle (the suture connecting the two often hidden by scales).

Suborbital. Below the eye.

Suborbital stay. A bone extending from one of the suborbital bones in certain fishes, across the cheek, to or toward the preopercle.

Subulate. Awl-shaped.

Superciliary. Pertaining to the region of the eyebrow.

Superior pharyngeals. Synonym of pharyngobranchials, q. v.

Supplemental maxillary. A small bone lying along upper edge of the maxillary in some fishes.

Supraclavicle. A bone interposed between the clavicle and the post-temporal.

Supraoccipital. The bone at posterior part of skull in fishes, usually with a raised crest above.

Supraoral. Above the mouth.

Supraorbital. Above the eye.

Suprascapular. The post-temporal or bone by which the shoulder girdle in fishes is joined to the skull.

Suspensorium. The chain of bones from the hyomandibular to the palatine.

Suspensory bones. Bones by which the lower jaw, in fishes, is fastened to the skull.

Suture. The line of union of 2 bones, as in the skull.
Symphysis. Point of junction of the 2 parts of lower jaw; tip of chin.
Symplectic. The bone in fishes that keys together the hyomandibular and quadrate posteriorly.
Synonym. A different word having the same or a similar meaning.
Synonymy. A collection of different names for the same group, species, or thing; "A burden and a disgrace to science." (*Coues.*)
Tail. In fishes (usually), the part of the body posterior to the anal fin. (Often used more or less vaguely.)
Temporal. Pertaining to the region of the temples.
Terete. Cylindrical and tapering.
Terminal. At the end.
Tessellated. Marked with little checks or squares, like mosaic work.
Thoracic. Pertaining to the chest; ventral fins are thoracic when attached immediately below the pectorals, as in the perch, the pelvic bones being fastened to the shoulder girdle.
Transverse. Crosswise.
Trenchant. Compressed to a sharp edge.
Truncate. Abrupt, as if cut squarely off.
Tubercle. A small excrescence, like a pimple.
Type (of a genus). The species upon which was based the genus to which it belongs.
Type (of a species). The particular specimen upon which the original specific description was based.
Type locality. The particular place or locality at which the type specimen was collected.
Typical. Of a structure the most usual in a given group.
Ultimate. Last or farthest.
Unicolor. Of a single color.
Vent. The external opening of the alimentary canal.
Ventral. Pertaining to the abdomen.
Ventral fins. The paired fins behind or below the pectoral fins in fishes, corresponding to the posterior limbs in the higher vertebrates.
Ventral plates. In serpents or fishes, the row of plates along the belly between throat and vent.
Ventricle. One of the thick-walled chambers of the heart.
Versatile. Capable of being turned either way.
Vertebra. One of the bones of the spinal column.
Vertical. Up and down.
Vertical fins. The fins on the median line of the body; the dorsal, anal, and caudal fins.
Villiform. Said of the teeth of fishes when slender and crowded into velvety bands.
Viscous. Slimy.
Viviparous. Bringing forth living young.
Vomer. In fishes, the front part of the roof of the mouth; a bone lying immediately behind the premaxillaries.
Zygapophyses. Points of bone affording to the vertebræ more or less definite articulation with each other.

INDEX.

Page.

abacurus, Eleotris.................... 2200
Abadejo............................ 1184
abboti, Syngnathus 764
abbotti, Osmerus mordax 524
abbreviata, Chimæra 95
abbreviatus, Nauclerus.............. 900
Abeona 1496
   aurora...................... 1497
   minima...................... 1497
   trowbridgii 1497
aberrans, Hypoplectrus unicolor ... 1193
  Liopropoma 1136
  Perca..................... 1136
abildgaardi, Scarus................. 1635
  Sparisoma 1635
Aboma............................. 2240
   chiquita 2241
   de Mar..................... 2195
   de Rio..................... 2236
   etheostoma 2240
   lucretiæ................... 2241
Abramidopsis...................... 249
Abramis 249
   americanus 250
   balteatus 239
   crysoleucas 250
     bosci.......... 251
   gardoneus 251
   lateralis 239
   leptosomus 250
   occidentalis............... 247
   versicolor 250
Abudefduf 1560
   analogus................. 1563
   declivifrons............. 1562
   rudis 1563
   saxatilis 1561
   taurus 1563
abyssicola, Raja..................... 76, 2751
acadian, Cottus 2023
acadianus, Glyptocephalus......... 2657
  Hemitripterus 2023
Acantharchus...................... 989
   pomotis............... 989
Acanthias......................... 53
   americanus 54
   blainvillei 57
   sucklii................. 54
   vulgaris 54

Page.

acanthias, Squalus................. 54
Acanthidium........................ 55
   pusillum 55
Acanthinion 939
   rhomboides 942
acanthistius, Bodianus 1147
Acanthochætodon.................. 1682
Acanthoclinus 2374
   chaperi 2375
Acanthocottus 1970, 1971
   labradoricus 2001
   laticeps 1989
   mucosus 1975
   ocellatus 1976
   patris 2009
   profundorum........ 1991
   sellaris............... 1998
   variabilis .:......... 1975
   virginianus......... 1976
Acanthocybiinæ 865
Acanthocybium 876
   petus 877
   solandri 876
Acanthoderma 879, 1713
   temminkii 880
acanthoderma, Thyrsites 880
Acantholabrus exoletus............ 1576
Acantholebius...................... 1866
   nebulosus 1872
Acantholepis...................... 525
Acanthonotus...................... 614
   nasus 615
acanthophorus, Serranus 1196
Acanthopteri 779, 1241
Acanthorhinus..................... 53
Acanthosoma 1753
   carinatum 1754
Acanthostracion 1721, 1722, 1724
   polygonius 1725
Acanthurus........................ 1689
   aliala................ 1694
   bahianus............. 1693
   brevis............... 1691
   broussonetii.......... 1691
   cæruleus............. 1691
   chirurg.............. 1692
   glaucoparcius......... 1694
   hepatus 1692
   hirundo.............. 1691

	Page.
Acanthurus matoides	1693
nigricans	1692
phlebotomus	1692
subarmatus	1691
tractus	1693
triostegus	1691
zebra	1691
acanthurus, Gasteropelecus	579
Gonostomus	579
Acara aya	1264
bartoni	1516
cæruleopunctatus	1514
fusco-maculata	1540
pinima	1323
pitamba	1276
rectangularis	1515
tetracanthus	1540
acara, Pristipoma pinima	1323
accensum, Plectropoma	1193
accensus, Hypoplectrus	1193
unicolor	1193
accipiter, Podothecus	2055
acclivis, Larimus	1422
Acedia	2704, 2705, 2709, 2712
Acentrogobius	2210
Acentrolophus	962
maculosus	963
acervum, Cybium	875
achigan, Bodianus	1011
Achirinæ	2693
Achirophichthys	387
typus	388
Achirus	2693, 2695, 2700
achirus	2695
brachialis	2698
comifer	2698
fasciatus	2700
fimbriatus	2700
fischeri	2690, 2700
fonsecensis	2699
gronovii	2696
inscriptus	2696
klunzingeri	2697
lineatus	2697, 2698, 2702
maculipinnis	2698
mazattanus	2698
ornatus	2709
panamensis	2702
scutum	2700
achirus Achirus	2695
mollis	2702
Pleuronectes	2696
Solea	2702
Acipenser	103
acutirostris	104, 105
agassizii	105

	Page.
Acipenser aleutensis	104
alexandri	105
anasimos	106
anthracinus	106
telaspis	106
attilus	105
ayresi	104
bairdi	105
brachyrhinchus	104
brevirostrum	106
buffalo	106
carbonarius	106
caryi	104
cataphractus	107
cincinnati	106
copei	106
dekayi	106
girardi	105
holbrooki	105
honneymani	106
hopeltis	106
hospitus	105
kennicotti	105
kirtlandi	106
lævis	106
lagenarius	102
lamarii	106
latirostris	105
lecontei	105
lesueuri	106
lichtensteini	105
macrorhinus	105
macrostomus	106
maculosus	106
medirostris	104, 2757
megalaspis	105
microrhynchus	106
milberti	105
mitchilli	105
nertinianus	106
obtusirostris	106
ohiensis	100
oligopeltis	105
oxyrhynchus	105
paranasimos	106
platorynchus	107
platyrhinus	106
putnami	104
rafinesqui	106
rauchi	106
rhynchæus	106
richardsoni	106
rosarium	106
rostellum	106
rubicundus	106
rupertianus	106

	Page.
Acipenser serotinus	106
storeri	105
sturio	105
thompsoni	105
transmontanus	104
yarrelli	105
Acipenseridæ	102
acipenserinus, Agonus	2062
Aspidophorus	2062
Paragonus	2062
Phalangistes	2062
Podothecus	2061, 2062
ackleyi, Raja	70
Acomus	173
griscus	175
Acoupa	1403
acoupa, Cestreus	1404
Cheilodipterus	1404
Cynoscion	1403
Cynoscium	1404
Acrocheilus	207
aleutaceus	208
acrolepis, Macrourus	2585
Macrurus	2585
acronotus, Carcharhinus	36
Squalus	36
Acronurus carneus	1692
cœruleatus	1691
fuscus	1692
nigriculus	1693
Acropœcilia	690
tridens	690, 2833
Acrotidæ	2849
Acrotinæ	969
Acrotus	973, 2849
willoughbyi	973, 2849
Actinochir	2114, 2116, 2127
major	2128
aculeata, Mola	1754
aculeatus, Chelmon	1671
Chrysops	1347
Clinus	2433
Doryichthys	773
Gasterosteus	747, 2836
cataphrac-	
tus	750
Halieutichthys	2739
Lophius	2741
Lumpenus	2433
Prognathodes	1671
Stenotomus	1346
Stichæus	2433
acuminata, Gobius longissima	2230
Jenkinsia	419
Muræna	377
Sciæna	1488

	Page.
acuminatus, Eques	1487
umbrosus	1487
Etrumeus	419
Myrichthys	376
Ophicthys	377
Ophisurus	377
Pareques	1487
acuta, Loricaria	158
Myliobatis	89
Ophisoma	356
Perca	1024
acuticeps, Blennicottus	2864
Oligocottus	2015, 2016
acutirostris, Acipenser	104, 105
Cerna	1181
Corvina	1437
Lutjanus	1259
Pristis	61
Serranus	1181
Acutomentum	1765, 1774, 1785
macdonaldi	1787
acutum, Hæmulon	1299
acutus, Exocœtus	728
Fodiator	728
Pseudoscarus	1652
Scarus	1652
Acus	774
acus, Sphyræna	717
Tylosurus	716
Adinia	660; 2830
dugèsii	661
guatemalensis	660, 2830
multifasciata	661
pachycephala	660, 2830
adinia, Fundulus	645
adirondacus, Salmo	505
adobe, Agosia	310
Adonis	2377
cristatus	2383
adscenionis, Epinephelus	1152, 1154
Scomber	927
Trachinus	1153
adspersus, Ctenolabrus	1577
Labrus	1577
Paralichthys	2627
Pseudorhombus	2627
Tautogolabrus	1577
adusta, Corvina	1448
Eupomacentrus	1551
Gobiesox	2334
Julidio	1602
Ophioscion	1447
Pomacentrus	1552
Pseudojulis	1603
adustus, Couesius	325
Æglefinus	2542

	Page.
Æglefinus linnæi	2543
æglefinus, Gadus	2543
Melanogrammus	2542, 2543
Morrhua	2543
Ælurichthys	116
ælurus, Amiurus	140
ænea, Cichla	990
æneolus, Notropis	266
æneus, Centrarchus	990
Cottus	1973
Myoxocephalus	1972
Pimelodus	143
Tetragonopterus	333
ænigmaticus, Icosteus	972
æpyterus, Ammocœtes	11
æquatoris, Talismania	456
Æquidens	1513
cærulopunctatus	1514
æquidens, Culius	2202
Eleotris	2202
Prionodes	1210
Serranus	1211
æquoreus, Syngnathus	774
æreus, Sebastodes	1807
æsculapius, Alepisaurus	595
æsopus, Boleosoma	1057
æstivalis, Clupea	427
Gobio	316
Hybopsis	316
marconis	316
Pomolobus	426
æstuarius, Paralichthys	2626
æthalion, Citharichthys	2673
Hemirhombus	2673
æthalorus, Carcharhinus	40
Æthoprora	565
effulgens	566
lucida	565
Aetobatinæ	88
Aetobatis	88
Aetobatus	88
laticeps	88, 2753
narinari	88
afer, Alphestes	1164
Epinephelus	1165
Gymnothorax	395
affine, Myctophum	570
Plectropoma	1193
Siphostoma	769, 770
affinis, Atherinops	807
Auchenopterus	2371
Carapus	2497
Caulolatilus	2277
Centropomus	1124
Cheilodipterus	1113
Chimæra	95
affinis, Clinostomus	239
Cremnobates	2372
Esox	628
Exocœtus	735, 2836
Exonautes	2836
Fierasfer	2495
Gambusia	680
Gila	228
Heros	1529
Heterandria	681
Hypoplectrus	1193
unicolor	1193
Isopisthus	1399
Leuciscus	240
Lucania	665
Pimelodus	134
Scopelus	571
Stomias	588
Synaphobranchus	351
Syngnathus	769
Thynnus	869
afra, Muræna	396
Perca	1833
africana, Scorpæna	1833
afrum, Plectropoma	1166
agassizii, Acipenser	105
Alepocephalus	453
Amphysticus	1502
Aulopus	541
Bathysaurus	540
Brama	959
Chlorophthalmus	541
Chologaster	704
Cratinus	1188
Cylindrosteus	111
Dicromita	2506
Holconotus	1506
Hyperprosopon	1502
Liparis	2121
Pimephales	217
Salmo	507
Salvelinus fontinalis	507
Scorpæna	1840
Serranus	1189
Xenichthys	1287
aggregatum, Ditrema	1499
aggregatus, Cymatogaster	1498
Micrometrus	1499
agilis, Gadus	2534
Agnus	2306
anoplus	2308
Agonidæ	2031
Agoninæ	2033
Agonomalus	2036
proboscidalis	2037
Agonopsis	2068

	Page.
Agonopsis chiloensis	2069
Agonostoma globiceps	821
nasutum	820
Agonostominæ	809
Agonostomus	818
microps	820
monticola	819
nasutus	819, 2840, 2841
percoides	819
Agonus	2064
acipenserinus	2062
annæ	2043
barkani	2044
cataphractus	2065, 2067
chiloensis	2069
curilicus	2036
decagonus	2053
dodecaedron	2046
gilberti	2060
japonicus	2036
lævigatus	2048
niger	2069
quadricornis	2041
rostratus	2048
spinosissimus	2054
stegophthalmus	2036
vulsus	2068
Agosia	308, 309, 313
adobe	310
chrysogaster	313
couesii	310
falcata	313
shuswap	313
metallica	314
nevadensis	310
novemradiata	312
nubila	311, 312
carringtonii	311
oscula	309
shuswap	313
umatilla	313
velifera	212
yarrowi	309
agua-bonita, Salmo irideus	503
mykiss	504
Aguaji	1174, 1177
aguaji, Trisotropis	1175
Aguja Blanca	892
de Casta	715, 892
de Paladar	892
Prieta	891
Voladora	891
Agujon	714, 715, 716
Agulha, Peixe	711
Ahlia	370
egmontis	370

	Page.
aigula, Lachnolaimus	1580
Ailurichthys	116
bagre	117
eudouxii	118
filamentosus	118
gronovii	117
longispinis	119
marinus	118
nuchalis	117
panamensis	117
pinnimaculatus	117
ailurus, Pimelodus	140
Aimaras	330
aix, Pallasina	2050
Aka nevo	1833
Aka soi	1830
Alabama Shad	2810
alabamæ, Alosa	2810
Etheostoma	1095
whipplei	1095
Notropis	298
Alalonga, Albacora	871
alalonga, Orcynus	871
Ala-lunga	871
alalunga, Germo	871
alascanus, Ammodytes	832
Argyrosomus	2817
Sebastolobus	1761
Xenochirus	2081
Alaska Blackfishes	620, 621
Codfish	2541
Dab	2645
Dog salmon	478
Greenfish	1869
Stickleback	749
Alaskan Pollacks	2535
alatunga, Scomber	871
alatus, Arius	125
Lampanyctus	559
Mugil	733
Prionotus	2159
Alausa	427
californica	423
striata	431
Alausella	424
alba, Rogenia	422
Albacora	869
alalonga	871
thynnus	870
albacora, Thynnus	871
Albacore, Great	870
Long-finned	871
Albacores	870
albacores, Scomber	870
Albatrossia	2573
pectoralis	2573

	Page.
albatrossis, Osmerus	2823
albeolus, Notropis megalops....	259, 283, 284
albescens, Echeneis	2272
Remora	2272
albicans, Bagrus	124
Sciadeichthys	124, 2760
Tachisurus	124
albicauda, Echeneis	2269
albidactylus, Exocœtus	739
albidum, Hæmulon	1299
Moxostoma	192
albidus, Amiurus	138
Gadus	2531
Ictalurus	138
Labrax	1132
Osmerus	538
Pimelodus	132, 138
Ptychostomus	192
Tetrapturus	892
albigutta, Cathetostoma	2313
Kathetostoma	2312
albiguttus, Paralichthys	2631
albirostre, Siphostoma	772
albirostris, Corythroichthys	772, 2838
Prionotus	2163
Syngnathus	772
albolineatus, Fundulus	649
albomaculatus, Paralabrax	1197
Serranus	1197
Albramis oligaspis	294
Albula	411, 2807
conorynchus	411
erythrocheilos	412
fosteri	412
goreensis	412
neoguinaica	412
parræ	411
rostrata	412
seminuda	412
vulpes	411
albula, Mugil	812, 2841
Albulidæ	410
albulus, Bryttus	1007
Lepomis	1007
album, Hæmulon	1295, 1296
Moxostoma	191
Alburnellus	254
altipinnis	287
amabilis	291
amœnus	296
arge	294
jaculus	293
jemezanus	294
matutinus	301
megalops	291
micropteryx	297

	Page.
Alburnellus percobromus	295
rubrifrons	295
simus	267
umbratilis	299
Alburnops	254, 256, 261
blennius	262
heterodon	261
illecebrosus	269
longirostris	267
nubilus	215
plumbeolus	283
saludanus	270
shumardi	268
taurocephalus	253
Alburnus amabilis	291
americanus	1475
dilectus	294
formosus	280
lepidulus	294
lineolatus	263
megalops	291
nitidus	293
oligaspis	294
rubellus	293
rubrifrons	295
socius	292
umbratilis	299
zonatus	285
alburnus, Centropomus	1475
Menticirrhus	1475
Perca	1475
Umbrina	1475
albus, Atherinichthys	2839
Centropomus	1135
Cestreus	1411
Córegonus	466
Cynoscion	1411
Gymnotus	340
Lepisosteus	110
Otolithus	1411
Ptychostomus	191
Alcidea	1886
thoburni	1887
Aldrovandia	608, 2826
goodei	610
gracilis	610
macrochir	609
pallida	611
rostrata	609
Alecrin	32
Alectis	931
ciliaris	931
crinitus	932
Alectrias	2869
alectrolophus, Anoplarchus	2421, 2422, 2869
Blennius	2422
Centronotus	2422

	Page.
alectrolophus, Gunnellus	2422
Aledon	1753
capensis	1754
storeri	1754
Alepos	915
Alepidosaurus	594
(Caulopus) borealis	597
(Caulopus) poeyi	596
(Caulopus) serra	597
alepidotum, Gobiosoma	2259
alepidotus, Chætodon	966
Derepodichthys	2480
Gobius	2259
Lucioblennius	2404
Rhombus	966
Stromateus	966
Alepisauridæ	593
Alepisaurus	594, 2826
æsculapius	595
altivelis	596
azureus	595
borealis	596
fexox	595
serra	597
Alepocephalidæ	451
Alepocephalus	452
agassizii	453
bairdii	454
macropterus	458
productus	452
tenebrosus	453
Aleposomus	459
copei	459
aleutensis, Acipenser	104
Lyconectes	2444
Aleutera	2860
aleutianus, Sebastodes	1795
aleutica, Raja	75; 2751
aleuticus, Cottus	1957
Alewife	426
Alewives	424
alexandri, Acipenser	105
alexandrini, Orthragoriscus	1754
Alexurus	2202
armiger	2203
Alfione	1507
Alfoncino	1107
Alfonsin a Casta Cumprida	844
Larga	844
Alfonsines	844
Algansea	211
antica	245
bicolor	245
dugesi	211
formosa	246
obesa	246
	Page.
Algansea sallæi	212
tarascorum	2796
tincella	211, 2796
algeriensis, Gasterosteus	748
Algoma	212
amara	215
fluviatilis	215
alia, Labrus tautoga	1579
aliala, Acanthurus	1694
Teuthis	1693
aliciæ, Leuciscus	236
Squalius	236
aliciolus, Trachurus	904
Alilonghi	871
alipes, Salmo	509
Salvelinus alpinus	509
alleghaniensis, Salmo	507
alleterata, Gymnosarda	869
alleteratus, Scomber	869
allidus, Merlucius	2531
Alligator Gar	111
Allinectes	2866
alliteratus, Euthynnus	869
Orcynus	869
All-Mouth	2713
Allochir	2129, 2131, 2135
Allosomus	467, 473
Allurus	2129, 2131, 2136, 2866
almeida, Belone	715
Tylosurus	715
Almejero, Mojarra	1294
Alopecias vulpes	45, 46
alopecias, Squalus	46
Alopias	45
macrourus	46
vulpes	45
Alopiidæ	45
Alosa	427, 2810
alabamæ	2810
apicalis	429
bishopi	430
cyanonoton	427
lineata	426
menhaden	432
præstabilis	428
sapidissima	427, 428, 2810
teres	420
alosoides, Amphiodon	413
Hiodon	413
Alphestes	1164
afer	1164
chloropterus	2854
multiguttatus	1165
alpinus, Salmo	509, 514
Salmo, nivalis	509
Salvelinus	508

	Page.
alpinus, Salvelinus alipes	509
arcturus	510
aureolus	511
stagnalis	510
Alpismaris	533
risso	537
alta, Cliola	322
Macdonaldia	2826
alter, Atinga minor orbicularis	1749
alternans, Scarus	1651
alternata, Perca mitchilli	1133
alticolus, Catostomus	179
Alticus	2396
altifrous, Heros	1538
altipinna, Belone	717
altipinnis, Alburnellus	287
Micropogon	1464
Minnilus	287
Notropis	287
altivelis, Alepisaurus	596
Auchenopterus	2370
Cremnobates	2371
Sebastolobus	1763
altus, Bubalichthys	165
Chorinemus	899
Hudsonius	322
Hybopsis	321
Oligoplites	899
Priacanthus	1240
Pseudopriacanthus	1239
alusis, Muræna	403
aluta, Bairdiella	1437
Sciæna	1438
alutaceus, Acrocheilus	208
Alutarius amphacanthus	1720
macracanthus	1720
obliteratus	1720
Alutera	1717, 1718, 1720
cinerea	1720
cuspicauda	1718
guntheriana	1720
monoceros	1720, 2860
picturata	1719
punctata	1718, 1719
schœpfii	1718
scripta	1719
Aluteres	1717
berardi	1720
pareva	1719
Aluterus anginosus	1720
cultifrons	1718
holbrooki	1718
venosus	1719
alutus, Apogon	1110
Apogonichthys	1110
Sebastodes	1790

	Page.
Alvarius	1099
fonticola	1105
lateralis	1099
alveata, Trygonorhina	65
alvordii, Cottus	1952
Alvordius	1028, 1029, 1030
aspro	1033
crassus	1034
evides	1037
macrocephalus	1031
maculatus	1032, 1034
nevisensis	1034
phoxocephalus	1031
spillmani	1039
variatus	1034
Alysia	568
loricata	569
amabalis, Alburnellus	291
Alburnus	291
Minnilus	291
Notropis	291
amara, Algoma	215
Hybognathus	215
Amarilla, Chopa	1386
Guativere	1144, 1145
Salmonete	859
Amarillas, Mojarra de las Aletas	1376
Amarillo, Cibi	919
Pargo	1260
Ronco	1303
amarus, Hudsonius	270
Hybognathus	215
Notropis hudsonius	270
amazonica, Sciæna	1419
amazonicus, Johnius	1419
ambassis, Sargus	1346
Amber-fish, Great	903
Amber-fishes	901
Amber Jack	903
ambiguus, Lutjanus	1272
Merluccius	2530
Mesoprion	1272
Neomænis	1271
Amblodon	1483
bubalus	165
concinnus	1484
grunniens	1484
lineatus	1484
neglectus	1484
niger	169
saturnus	1456
Ambloplites	989
interruptus	991
pomotis	989
rupestris	990
cavifrons	990

	Page.
amblops, Ceratichthys	321
Hybopsis	320
Nocomis	321
Rutilus	321
Amblygobius	2210
Amblyopsidæ	702
Amblyopsis	706
spelæus	706
amblyopsis, Culius	2200
Eleotris	2199, 2200
Amblyopus brasiliensis	2264
brevis	2263
mexicanus	2264
peruanus	2265
sagitta	2263
Amblypomacentrus	1549
amblyrhynchus, Caranx	913
Hemicaranx	912
Amblyscion	1420, 1421
argenteus	1421
amboinensis, Balistes	1704
Ameiurus	135, 136, 139
bolli	140
catus	138
dugesi	138
dugesii	2789
erebennus	139
lacustris	137
lupus	137
marmoratus	141
melas	141
mispilliensis	141
natalis	139
nebulosus	140
catulus	141
mamoratus	141
nigrilabris	142, 2789
okeechobeensis	138
platycephalus	142
vulgaris	140
xanthocephalus	141
American Eel	348
Perch	1023
Pike Perches	1020
Shad	427
Smelt	523
Sole	2700
Soles	2693
americana, Cherna	1160
Lucioperca	1022
Manta	93
Morone	1134, 1135
Morrhua	2540, 2541
Perca	1024, 1135
Raia	69
Scorpæna	2023

	Page.
americana, Stilbe	230
Tautoga	1579
americanus, Abramis	250
Acanthias	54
Alburnus	1475
Ammodytes	833
Amphiprion	1139
Apogonichthys	1107
Balistes	1707
Blennius	2457
Carcharias	47
Cyprinus	250, 251, 1475
Enchelyopus	2457, 2555
Eques	1490
Esox	626
lucius	626
Hemitripterus	2023
Hippoglossus	2612
Histiophorus	891
Labrax	1135
Labrus	1579
Leucosomus	250
Lophius	2714
Lucius	626
Menticirrhus	1474
Notemigonus	251
Odontaspis	47
Petromyzon	10
Phycis	2555
Platycephalus	2029
Pleuronectes	2647
Polynemus	830
Polyprion	1139
Pseudopleuronectes	2647
Squalus	47
amethystinus, Salmo	505
amethystinus-punctatus, Mauro-licus	577
Amia	112, 1106
calva	113
canina	113
cinerea	113
immaculata	411
lentiginosa	113
marmorata	113
occellicauda	113
occidentalis	113
ornata	113
piquottii	113
reticulata	113
retrosella	1109
subcærulea	113
thompsoni	113
Amia viridis	113
Amiatus	113
Amiichthys	1113
diapterus	1113

	Page.
Amiidæ	112
Amitra	2138
liparina	2138, 2139
Amitrichthys	2139, 2141
Amitrinæ	2106
Amiurus	135
ælurus	140
albidus	138
bolli	140
borealis	137
brachyacanthus	141
brunneus	142
catus	141
caudifurcatus	135
cragini	141
furcatus	134
lophius	138
meridionalis	135
natalis analis	140
nigrilabris	142
niveiventris	138
obesus	141
ponderosus	137
prosthistius	139
pullus	141
vulgaris	140
Ammocœtes	9
æpyterus	11
aureus	13
bicolor	10
borealis	11
branchialis	14
cibarius	13
concolor	11
niger	14
tridentatus	12
unicolor	10
Ammocœtus	9
Ammocrypta	1061
asprella	1061
beanii	1064
gelida	1064
pellucida	1062
pellucida clara	1063
vivax	1063
vitrea	1065
ammocryptus, Tetrodon	1735
Ammodytes	832
alascanus	832, 2842
americanus	833
aureus	13
dubius	832
personatus	833, 2841, 2842
septipinnis	2842
tobianus	2841

	Page.
Ammodytes vittatus	833
Ammodytidæ	831
Ammodytoidei	781, 831
Ammopleurops	2704
amœnus, Alburnellus	296
Notropis	296
Amore guaco	2236
pixuma	2201
amorea, Gobius	2201
Amorphocephalus	1617
granulatus	1619
amphacanthus, alutarius	1720
Amphiodon	412, 413
alosoides	413
amphiodon, Hyodon	413
Amphioxi	2
Amphioxus	3
lanceolatus	3
amphioxys, Monacanthus	1717
Pseudomonacanthus	1717
Amphiprion americanus	1139
matejuelo	849
Amphistichus	1503
agassizi	1502
argenteus	1503, 1504
heermanni	1504
similis	1504
amplexicollis, Sarothrodus	1674
amplus, Scarus	1635
Tetrapterus	892
ampullaceus, Ophiognathus	406
Saccopharynx	406
Anablepinæ	632
Anableps	684, 2807
dovii	685
Anacanthini	782, 2528
Anacanthus	82
Anacyrtus guatemalensis	338
anagallinus, Lepomis	1004
anale, Ditrema	1501
analigutta, Pomacentrus	1554
analis, Amiurus natalis	140
Caranx	927
Centridermichthys	2013
Clinocottus	2012
Conger	356
Eupomacentrus	1554
Holconotus	1501
Hyperprosopon	1501
Hypocritichthys	1500, 1501
Lutjanus	1267
Mesoprion	1266
Neomænis	1265
Notacanthus	615
Oligocottus	2013

	Page.
analis, Ophisoma	356
Orthragoriscus	1754
Pomacentrus	1555
Scyris	932
Umbrina	1468
analogus, Abudefduf	1563
Epinephelus	1152
Euschistodus	1563
Kyphosus	1385
Pimelepterus	1386
analostana, Cliola	279
Cyprinella	279
analostanus, Leuciscus	279
Notropis	279
Anarhichadidæ	2445
Anarhichas	2445
latifrons	2446
lupus	2447
Anarmostus	1291
Anarrhichas denticulatus	2446
karrak	2446
leopardus	2446
maculatus	2446
minor	2446
orientalis	2447
pantherinus	2446
strigosus	2447
vomerinus	2447
Anarrhichthyinæ	2445
Anarrhichthys	2447
felis	2448
ocellatus	2448
anasimos, Acipenser	106
anceps, Cottus (Acantbocottus)	1973
Plesioperca	1039
Anchisomus	1729
augusticeps	1731
caudicinctus	1742
geometricus	1736
reticularis	1735
Anchoa pelada	436
Anchovia	449, 2815
macrólepidota	449
Sardinella	429
anchovia, Clupea	429
Anchovies	439, 448
Silvery	439
Anchovy, California	448
Striped	443
Anchybopsis	243
Ancistrus	160
chagresi	160
Anclyopsetta	2634
dendritica	2633
Anclyopsetta dilecta	2636

	Page.
Anclyopsetta quadrocellata	2635
Ancylodon	1416
ancylodon	1416
jaculidens	1416
parvipinnis	1399
ancylodon, Lonchurus	1416
Sagenichthys	1416
andreæ, Rhinoscopelus	569
Scopelus	569
Stenobrachius	569
andrei, Gobius	2218
Pomadasis	1332
Pristipoma	1332
Anged	414
Angel, Black	1679
Angel fish	58, 1668, 1684, 1685
Angel Sharks	58
Angelichthys	1684, 2859
ciliaris	1684, 1685
iodocus	1686
isabelita	1685
angelus, Squatina	59
anginosus, Aluterus	1720
Angler, Common	2713
Anglers	2713
anglorum, Lumpus	2097
anguiformis, Ophichthys (Sphage-branchus)	374
anguiformis, Sphagebranchus	374
Anguilla	347
aterrima	348
blephura	348
chrysypa	348, 2801
cubana	348
laticauda	348
lutea	348
novæterræ	348
novæorleanensis	348
oceanica	355
punctatissima	348
rostrata	348
tenuirostris	348
texana	348
tyrannus	348
wabashensis	348
xanthomelas	348
anguilla, Anguilla rostrata	348
Ictalurus	2788
anguillaris, Blennius	2436, 2457
Gunnellus	2436
Lumpenus	2436
Stichæus	2436
Zoarces	2457
Anguillidæc	346
anguilliformis, Pholidichthys	2405

	Page.
anguina, Muræna	390
anguineus, Chlamydoselachus	16
Nerophis	774
angulifer, Heros	1517
anguliferum, Cichlasoma	1517
angusta, Malthæa	2738
angusticeps, Belone	712
Coregonus	466
Sphæroides	1731
Tetrodon	1731
Tylosurus	712
angustidens, Macrostoma	555, 2826
angustifrons, Dermatolepis	1159
Serranus	1159
angustus, Platycephalus	2029
Añil	1193
Anisarchus	2435
medius	2436
Anisochætodon	1672
Anisotremus	1314, 1315, 1318
bicolor	1319
bilineatus	1319
cæsius	1316
davidsonii	1321
dovii	1317
interruptus	1319
pacifici	1316
scapularis	1320
serrula	1323
spleniatus	1321
surinamensis	1318, 1319
interruptus	1319
tæniatus	1322
trilineatus	1320
virginicus	1322, 1323
anisurum, Moxostoma	190, 196
anisurus, Catostomus	190
anna-carolina, Mugilomorus	410
annæ, Agonus (Brachyopsis)	2043
Cottus	1960
annularis, Centropristes	1214
Nauclerus	900
Pomoxis	987
Serranus	1214
annulata, Melanura	624
annulatum, Exoglossum	327
annulatus, Antennarius	2725
Spheroides	1735
Sphæroides politus	1736
testudineus	1736
Tetrodon	1736
anogenus, Notropis	259, 260
anolis, Saurus	535
anomala, Dekaya	2277
anomalum, Campostoma	205

	Page.
anomalus, Caulolatilus	2277
Rutilus	206
Anoplagonus	2088, 2089, 2093
inermis	2094
Anoplarchus	2421
alectrolophus	2421, 2422, 2869
atropurpureus	2422, 2423
cristagalli	2423
purpurescens	2423
Anoplogaster	839
cornutus	840
Anoplogastrinæ	838
Anoplopoma	1861
fimbria	2861, 1862
merlangus	1862
Anoplopomatid.æ	1861
Anoplopomatinæ	1861
anoplos, Astroscopus	2308
Uranoscopus	2308
anoplus, Agnus	2308
Astroscopus	2308
Anopsus	7
Anosmius	1741
Antaceus	103
antecessor, Gasterosteus	900
Antennariidæ	2715
Antennarius	2717
annulatus	2725
corallinus	2725
histrio	2716, 2723
inops	2718
leopardinus	2721
marmoratus	2717
multiocellatus	2724
nuttingii	2723
ocellatus	2721
pleurophtalmus	2722
principis	2719
radiosus	2725
reticularis	2719
sanguineus	2721
scaber	2722
strigatus	2720
tenebrosus	2719
tenuifilis	2721
tigris	2723
antennatus, Chilomycterus	1750
Diodon	1750
Anthias	1226
aquilonaris	1283
asperilinguis	1227
caballerote	1257
chema	1157
formosus	1304
furcifer	1222
jocu	1258

	Page.
Anthias multifasciatus	1226
oculatus	1283
peruanus	1223
(Hermianthias) peruanus	1223
quartus rondeleti	1266
rabirubia	1276
saponaceus	1232
striatus	1157
trifurcus	1202
vivanus	1224
Anthiinæ	1131
anthracinus, Acipenser	106
antica, Algansea	245
anticus, Leucus	245
Antigonia	1664
capros	1665
mulleri	1665
Antigoniinæ	1663
antillanus, Conodon	1324
antillarum, Caranx	921
Chilomycterus	1749
Sicydium	2206
Talismania	455
Antimora	2544
microlepis	2545
viola	2544
antiquorum, Hippocampus	776
antiquus, Hippocampus	776
antistius, Chænobryttus	992
antoniensis, Pimelodus	140
Antonino	909
antrostomus, Idiacanthus	605
Aodon	91
hypostomus	92, 2756
Aodontidæ	2756
Apeltes	752
quadracus	752
apeltes, Gasterosteus	752
Apeltinæ	743
aper, Labrus	1586
Aphanopinæ	885
Aphanopus	885
minor	885
Aphododerus cookianus	787
Aphoristia	2704
atricauda	2708
diomedeana	2711
elongata	2707
fasciata	2710
marginata	2706
nebulosa	2712
ornata	2707, 2710
pigra	2706
plagiusa	2710
pusilla	2711
Aphredoderidæ	785

	Page.
Aphredoderus	786
gibbosus	787
sayanus	786
Aphyoninæ	2499
Aphyonus	2525
mollis	2525
apia, Pirati	1174
apiarius, Petrometopon	1142
Saranus	1142
apiatus, Lepomis	998
apicalis, Alosa	429
Clupea	429
Echeneis	2268
Sardinella	429
Apionichthys	2702
bleekeri	2703
dumerili	2703
nebulosus	2703
unicolor	2702
Aplesion	1010
pottsii	1083
Aplites	1010
Aplocheilus	632; 2827; 2828; 2830
dovii	2828, 2830
Aplodinotinæ	1397
Aplodinotus	1483
grunniens	1484
Aplurus	879
simplex	880
Apocheilichthys	633, 2527
Apocope	308, 309
carringtonii	312
couesii	310
henshavii	312
nubila	311
oscula	309
ventricosa	309
vulnerata	312
apoda, Perca	1259
Pleuronectes	2701
Apodes	344
Apodichthys	2411
flavidus	2411
fucorum	2413
inornatus	2412
sanguineus	2412
univittatus	2412
violaceus	2427
virescens	2412
Apodontis	873
apodus, Neomænis	1258
Apogon	1106
alutus	1110
atricaudus	2853
binotatus	1109
dovii	1108
imberbis	1107

	Page.
Apogon maculatus	1109
pigmentarius	1109
retrosella	1108
rex-mullorum	1107
ruber	1107
Apogonichthys	1110
alutus	1110
americanus	1107
puncticulatus	1111
stellatus	1110
Apomotis	995
chætodon	995
cyanellus	996
ischyrus	997
murinus	996
obesus	993
phenax	997
punctatus	997
symmetricus	998
apos, Gunnellus	2430
appendiculatus, Centropomus	1119
Exocœtus	736
appendix, Lepomis	1005
Petromyzon	10
approximans, Polydactylus	829
Polynemus	829
Pomadasys	1333
Trichidion	829
Aprion	42, 1279
ariommus	1278
macrophthalmus	1280, 1281
aprion, Gerres	1373
Aprionodon	42
isodon	42
Aprodon	2460
cortezianus	2461
Apsicephalus	1729
Apsilus	1278
dentatus	1278
Apterichthys	373
selachops	374
Apterurus	91
apua, Bodianus	1174
Epinephelus	1159
Mycteroperca venenosa	1173, 1174
Serranus	1158
apus, Centronotus	2430
Platytroctes	458
aquæ-dulcis, Gymnothorax	391
Muræna	391
Rabula	390
Aquavina	1204
aquilonaris, Anthias	1283
Etelis	1283
aquosus, Pleuronectes	2660
Rhombus	2660

	Page.
arabatsch, Salmo	483
arabicus, Chanos	415
aracanga, Pseudoscarus	1648
Scarûs	1642, 1647
Sparisoma	1642
aræa, Atherina	790
aræopus, Catostomus	172
Pantosteus	172
Aramaca	2670
papillora	2672
soleœformis	2672
aramaca, Citharichthys	2672
Hemirhombus	2673
Pleuronectes	2672
Rhombus	2626, 2672
arangoi, Chærojulis	1597
Arará, Bonaci	1174
arara, Hæmulon	1306
Ronco	1304
Serranus	1159, 1175
aratus, Lutjanus	1274
Mesoprion	1274
Neomænis	1273
Arbaciosa	2340
eos	2343
humeralis	2341
rhessodon	2340
rupestris	2341
zebra	2341
arcansanum, Etheostoma zonale	1075
arcansanus, Notropis telescopus	292
archidium, Bairdiella	1432
Elattarchus	1431
Odontoscium	1422
Archistes	1900
plumarius	1900, 1901
Archocentrus	1514, 1515, 1525
Archoperca	1169, 1171
Archoplites	990
interruptus	991
Archosargus	1358, 1359, 1361
aries	1361
pourtalesii	1360
probatocephalus	1361, 1362
tridens	1360
unimaculatus	1359, 1360
Archosion parvipinnis	1399
remifer	1399
Arctic Flounder	2649
Grayling	517
Sculpin	1973
arctica, Liparis	2121
arcticum, Benthosema	574
arcticus Chironectes	2717
Salmo	521
arcticus Scopelus	574

	Page.
arctifrons, Calamus	1355
Citharichthys	2683
Arctoscopus	2297
japonicus	2297
Arctozenus	601
borealis	601
coruscans	601
arctum, Siphostoma	771
arcturus, Salmo	510
Salvelinus alpinus	510
arcuata, Harengula	431
arcuatum, Ditrema	1502
Hæmulon	1305
Hyperprospon	1502
arcuatus, Bathygadus	2564
Chætodon	1680
Hyperprosopon	1502
Pomacanthus	1679, 1681
ardens, Catostomus	179
Hypsilepis	301
Leuciscus	301
Minnilus	301
Notropis umbratilis	301
ardeola, Belone	713
Tylosurus	713
ardesiaca, Gila	237
ardesiacus, Lepomis	1006
Squalius	237
arenata, Umbrina	1474
arcnatus, Arius	132
Priacanthus	1237, 1238
Rhinichthys	308
Rypticus	1232
arenicola, Fierasfer	2496
Gillellus	2299
arenosus, Gadus	2541
argalus, Belone	713
arge, Alburnellus	294
Kuhlia	1014
Notropis	294
argentata, Ciliata	2559
Couchia	2559
Motella	2559
argentatus, Astyanax	336
Gaidropsarus	2559
Merluccius	2530
Plargyrus	283
Tetragonopterus	336
argentea, Bathyclupea	835
Chimæra	95
Muræna	348
Selene	936
Sphyræna	826
Steindachneria	2568
argenteum, Ditrema	1504
argenteum, Hyperprosopon	1502
punctatum	1502
argenteus, Amblyscion	1421
Amphistichus	1503, 1504
Centronotus	899
Diplodus	1363
Eucinostomus	1371
Gerres	1371
Hyperprosopon	1501, 1502
Ichthyomyzon	11
Larimus	1421
Leuciscus	221
Micropogon	1463
Pagrus	1357
Petromyzon	11
Pimelodus	125
Sarchirus	110
Sargus	1363
Sparus	1357
Synodus	411
Trachinotus	944
Trachynotus	944
Trichiurus	889
Argentina	525
carolina	410
glossodonta	411
machuata	410
menidia	443
pennanti	577
pretiosa	525
sialis	526
silus	526
striata	526
syrtensium	526
Argentines	525
Argentinidæ	519
argentinus, Pimelodus	135
argentipinnis, Rhombus	966
argentissimus, Gasterosteus	747
Plagopterus	329
argentiventris, Lutianus	1261
Lutjanus	1261
Mesoprion	1261
Neomænis	1260
argenti-vittatus, Thynnus	871
argentosa, Dionda	215
Argo	957
argus, Muræna	401
Pleuronectes	2666
Squalus	26
Argyrea	796
Argyreiosus	935
gabonensis	935
pacificus	936

	Page.
Argyreiosus setipinnis	934
unimaculatus	934
vomer	936
argyreiosus, Leucosomus	224
Pogonichthys	224
Symmetrurus	224
Argyreus	305
dulcis	307
nasutus	306
notabilis	309
nubilus	311
osculus	309
rubripinnis	282
argyreus, Fario	480
Lepidopus	887
Salmo	480
Argyriosis capillaris	936
spixii	936
triacanthus	936
Argyriosus brevoorti	936
filamentosus	936
mauricei	936
mitchilli	936
setifer	936
argyritis, Hybognathus	214
Argyrlepes	915
Argyrocottus	1995
zanderi	1995
argyroleuca, Bairdiella	1434
Corvina	1434
argyroleucus, Bodianus	1433
argyromelas, Seriola	950
Argyropelecus	603
durvillii	604
hemigymnus	604
olfersi	604
argyrophanus, Engraulis	445
Stolephorus	444
argyropomus, Gasterosteus	748
Argyrops caprinus	1345
argyrops, Sparus	1346
argyrosoma, Damalichthys	1510
Embiotoca	1510
Argyrosomus	467
alascanus	2817
artedi	468
sisco	469
hoyi	469, 472
laurettæ	471, 2817
lucidus	470
nigripinnis	472
osmeriformis	468
prognathus	471
pusillus	470
tullibee	473
bisselli	473

	Page.
argyrosomus, Damalichthys	1509
Argyrotænia	832
vittata	833
argyrurus, Coryphæna	953
argyrus, Pimelodus	135
aries, Archosargus	1361
Sargus	1362
arioides, Bagrus	133
ariomnus, Aprion	1278
Minnilus	290
Notropis	290
Photogenis	290
Ariopsis	119
Ariosoma	353
Arius	119
alatus	125
arenatus	132
assimilis	129, 2774
brandtii	122, 2758
cærulescens	129
dasycephalus	130
dowi	125
dubius	127
elatturus	128
emphysetus	122
equestris	128
felis	128
fissus	131
flavescens	123
furthii	132
grandicassis	126
guatemalensis	129
herzbergii	125
hypophthalmus	133
insculptus	127
kessleri	127
laticeps	132
luniscutis	125
melanopus	132
mesops	123
milberti	128
multiradiatus	133
nuchalis	131
oscula	127
parkeri	126
passany	124
phrygiatus	131
planiceps	127
platypogon	127
puncticulatus	131
quadriscutis	126
rugispinis	130
seemani	128
seemanni	2772
stricticassist	126
surinamensis	130

	Page.
Arius temminckii	123
valenciennesi	124
variolosus	132
arizonæ, Pantosteus	170
Arlina	1054
atripinnis	1051
effulgens	1058
arlingtonia, Gambusia	652, 2828, 2829
arlingtonius, Funduluse	652
armata, Bairdiella	1436
Corvina	1437
armatus, Aspidophorus	2067
Centridermichthys	2012
Centropomus	1123
Serranus	1165
armiger, Alexurus	2203
Arnillo	1278
arnillo, Mesoprion	1279
Tropidinius	1279
Arnillos	1278
arnillus, Lutjanus	1279
Arnoglossus fimbriatus	2677
ventralis	2670
Aroides	119
Arothron	1738
erethizon	1739
Arrow-toothed Halibut	2609
artedi, Argyrosomus	468
sisco	469
Polynemus	828
Artediellus	1905
atlanticus	1906, 2862
pacificus	1906
uncinatus	1905, 1906
Artedius	1902, 2862
asperulus	1903, 2862
fenestralis	1900
lateralis	1902
pugetensis	1890
quadriseriatus	1897
artesiæ, Etheostoma	1094
Pœcilichthys	1094
arubensis, Pœcilia vanderpolli	696, 2834
arundinaceus, Syngnathus	765
ascanii, Salmo	509
Silus	526
Ascelichthyinæ	1883
Ascelichthys	2024
rhodorus	2025
ascendens, Siphostoma	768
Syngnathus	768
ascensionis, Caranx	925
Epinephelus	1154
Holocentrus	848
rufus	849
Perca	849
	Page.
ascensionis, Scomber	925
ascita, Mystus	155
aselus, Cheilichthys	1740
Aseraggodes	2694
asper, Centridermichthys	1944
Cottus|	1944
Diodon	1744, 1752
Exerpes	2367
Hexagrammos	1872
Macrurus	2572
Pleuronectes	2645
aspera, Limanda	2645
Uranidea	1944
asperilinguis, Anthias	1227
Odontanthias	1227
asperrimus, Balistes	1706
Myliobatis	2754
aspersus, Epinephelus	1154
Serranus	1153
asperulus, Artedius	1903
Asphidorus quadricornis	2041
Aspicottus	1937, 1938
bison	1938
aspidolepis, Chætostomus	159
Hemiancistrus	159
Aspidophoroides	2088
bartoni	2092
grœnlandicus	2092
guntheri	2090
inermis	2093
monopterygius	2091, 2092
olriki	2089
tranquebar	2092
Aspidophoroidinæ	2033
Aspidophorus	2064
acipenserinus	2062
armatus	2067
cataphractus	2067
chiloensis	2069
decagonus	2054
dodecaedrus	2046
europæus	2067
lisiza	2036
malarmoides	2054
niger	2069
proboscidalis	2038
rostratus	2048
spinosissimus	2054
superciliosus	2036
aspidurus, Urolophus	81
Aspistor	2763
luniscutis	2763
Aspisurus	1689
asprella, Ammocrypta	1061
Crystallaria	1061
asprellus, Etheostoma	1061

	Page.
asprellus, Pleurolepis	1061
Radulinus	1920
asprigenis, Pœcilichthys	1085
aspro, Alvordius	1033
Hadropterus	1032
Percina	1833
Asproperca	1024
zebra	1027
assimilis, Arius	129, 2774
Galeichthys	2779
Hexanematichthys	129
Astatichthys	1066
cœruleus	1089
zonalis	1075
asterias, Blennius	2383
Mustelus	29
Urolophus	82, 2752
Asternopteryx	2420
gunelliformis	2420
Asternotremia	786
mesotrema	787
Asterospondyli	19
Asthenurus	2526
atripinnis	2527
astilbe, Stolephorus	2815
astori, Ichthyomyzon	12
Lampetra	12
Petromyzon	12
Astracion tricornis	1725
Astrolytes	1898
notospilotus	1899
Astronesthes	586
barbatus	586
gemmifer	586
niger	586
richardsoni	587
Astronesthidæ	586
Astroscopus	2306
anoplos	2308
anoplus	2303
guttatus	2310
y-græcum	2307
zephyreus	2309
Astyanax	333
argentatus	336
Asymmetron	4
lucayanum	4
atæniatus, Chætodon	1676
Sarothrodus	1676
atchafalayæ, Signalosa	2809
atelaspis, Acipenser	106
aterrima, Anguilla	348
Muræna	396
Thyrsoidea	396
Atheresthes	2609
stomias	2609

	Page.
Atherina	789
aræa	790
bosci	801
brownii	443
carolina	791
harringtonensis	791
humboldtiana	793
insularum	807
laticeps	790
martinica	795
menidia	801
microps	791
mordax	523
notata	800
stipes	790
storeri	807
veliana	790
viridescens	800
vomerina	793
Atherinella	805
crystallina	805
eriarcha	803
evermanni	804
panamensis	805
Atherinichthys	792
albus	2839
brevis	2840
californiensis	807
gracilis	797
grandoculis	2840
guatemalensis	801
humboldti	793
menidia	800
notata	800
pachylepis	801
Atherinidæ	788
Atherinoides	792
atherinoides, Chriodorus	719
Clupea	451
Engraulis	451
Notropis	254, 293
Pterengraulis	450
Atherinops	807
affinis	807
insularum	807
regis	808
Atherinopsis	806
californiensis	806
tenuis	802
Athlennes	717
hians	718
Atimostoma	950
Atinga	1750
Atinga alter minor orbicularis	1749
Atinga Chilomycterus	1750
Diodon	1746
atinga, Guamaiacu	1749

	Page.
Atka-fish	1864
atkinsi, Gasterosteus bispinosus	748
atlantica, Elacate	948
Emblemaria	2402
atlanticum, Oreosoma	1663
atlanticus, Artediellus	1906
Benthodesmus	887
Bregmaceros	2527
Callorhynchus	95
Dibranchus	2743
Epinephelus	1154
Megalops	409
Neoliparis	2107
Prometheus	883
Promethichthys	883
Rupiscartes	2397
Salarias	2397
Sparus	1153
Tarpon	409
Tetragonurus	976
Thynnus	871
atomarium, Sparisoma	1631
atomarius, Scarus	1631
Atopoclinus	2376
ringens	2376
Atractoperca	1194
clathrata	1198
Atractosion	1402, 1413
nobilis	1413
Atractosteus	109, 111
lucius	111
tropicus	111
atramentatus, Symphurus	2706
atraria, Perca	1200
Siboma	233
atrarius, Centropristis	1200
Pimelodus	140
Serranus	1200
Squalius	233
Xenomystax	361
atricauda, Aphoristia	2708
Hydrargira	624
atricaudus, Apogon	2853
Symphurus	2707
atrilatus, Zygonectes	682
atrilobatus, Chromis	1546
atrimana, Monolene	2692
atrimanus, Caranx	914
Hemicaranx	913
atripes, Ditrema	1507
Lythrurus	300
Minnilus	300
Notropis umbratilis	300
Phanerodon	1507
atripinnis, Arlina	1051
Asthenurus	2527

	Page.
atripinnis, Bregmaceros	2527
Goodea	685
atrocaudalis, Notropis cayuga	260
atrocyaneus, Pomacentrus	1552
atromaculata, Esox	629
Etheostoma	1057
atromaculatus, Cyprinus	222
Semotilus	222
thoreauianus	223
atronasus, Cyprinus	307
Rhinichthys	307
croceus	308
lunatus	308
meleagris	308
atropurpureum, Ophidium	2423
atropurpureus, Anoplarchus	2422, 2423
Atropus	929
atrorubens, Sebastodes	1796
atrovirens, Sebastichthys	1798
Sebastodes	1797
attenuata, Vincigurria	577
attenuatus, Maurolicus	577
Merlucius	2546
Osmerus	523
attilus, Acipenser	105
atwoodi, Carcharias	50
aubrieti, Lutjanus	1271
Auchenionchus	2360
Auchenopterus	2369, 2371
affinis	2371
altivelis	2370
asper	2368
fasciatus	2373
integripinnis	2372
marmoratus	2371
monophthalmus	2372
nigripinnis	2369
nox	2373
auctorum, Lobotes	1236, 2858
Auctospina	1765, 1776, 1817
audens, Menidia	798
augusticeps, Anchisomus	1731
Aulastome marcgravii	757
Auliscops	753
spinescens	754
auliscus, Siphostoma	767
Aulopidæ	541
Aulopus agassizii	541
Aulorhynchidæ	752
Aulorhynchus	753
flavidus	754
Aulostoma	754
cinereum	755
coloratum	755
Aulostomidæ	754

	Page.
Aulostomus	754
cinereus	755
maculatus	754, 2837
aurantiacum, Etheostoma	1041
aurantiacus, Balistes	1718
Ceratacanthus	1718
Cottogaster	1041
Hadropterus	1041
Hypohomus	1040
aurata, Cliola	272
Coryphæna	953
Moniana	272
Suleima	1386
auratum, Pristipoma	1324, 1343
auratus, Carassius	201
Centropomus	1107
Cyprinus	201
Gadus	2542
Holocentrus	1145
Mullus	856
barbatus	856
Notemigonus	250
Psenes	951
Serranus	1145
aurea, Brevoortia tyrannus	434
Clupea	434
Lampetra	13
aureolum, Moxostoma	192
aureolus, Catostomus	192, 196
Gerres	1375
Salvelinus alpinus	511
Xenotis	1003
aureoruber, Scarus	1635
aureoviridis, Sphyræna	1119
aureus, Ammocœtes	13
Caranx	923
Chætodon	1680
Clupanodon	434
Eupomotis	1010
Fundulus	659
Haplochilus	659
Heros	1533
Pomacanthus	1680
Sparus	1010
auriculatus, Sebastodes	1817, 1818
dallii	1818
auriga, Dules	1220
Monacanthus	1716
Serranus	1221
auritus, Labrus	1001
Lepomis	1001, 1009
solis	1001
aurofrenatum, Sparisoma	1634
aurofrenatus, Scarus	1634
auroguttatus, Zygonectes	654, 2829
aurolineatum, Bathystoma	1310

	Page.
aurolineatum, Hæmulon	1310
aurolineatus, Diabasis	1309
auropunctatus, Callyodon	1624
Cryptotomus	1624
aurora, Abeona	1497
Caproponus	1665
Catostomus	176
Fario	499
Salmo	493
Sebastichthys	1803
Sebastodes	1802
aurorubens, Centropristis	1278
Lutjanus	1278
Mesoprion	1278
Rhomboplites	1277, 1278
aurovittatus, Mesoprion	1276
Ocyrus	1276
australe, Etheostoma	1081
Zophendum	212
australis, Echeneis	2269, 2271
Esox	628
Icelus	1918
Teuthis	1691
Remilegia	2270
austrina, Myxostoma	192
austrinum, Minytrema	192
Moxostoma	192
Auxis	867
rochei	868
tapeinosoma	868
thazard	867
thynnoides	868
vulgaris	868
Averruncus	2069, 2864
emmelane	2069
sterleus	2071
averruncus, Kathetostoma	2311
avitus, Chologaster	704
avocetta, Nemichthys	369
Avocettina	367
elongata	2802
gilli	2801
infans	367, 2801
Awa	414
Awaous	2234
flavus	2235
mexicanus	2237
nelsoni	2235
taiasica	2236
axillare, Pristipoma	1328
axillaris, Boreocottus	1981
Brachydeuterus	1328
Cottus	1981
Gerres	1378
Myoxocephalus	1980, 1981
Pomadasis	1328

	Page.
axinophrys, Xystes	2076
Axyrias	1903, 2862
harringtoni	1904
aya, Acara	1264
Bodianus	1265
Chætodon	1675
Lutjanus	1265
Mesoprion	1264
Neomænis	1264
Aylopon	1226
martinicensis	1228
ayresi, Acipenser	104
Centropristes	1205
Parophrys	2640
Petromyzon	13
Ayresia	1545, 1548
punctipinnis	1548
ayresii, Sebastodes	1808
azalea, Runula	2377
Azevia	2677
panamensis	2677
querna	2675
azorica, Coryphæna	953
Azteca	254, 255, 258, 2799
Aztecula	2799
aztecus, Notropis	258
Azul, Pescado	1553
Azules, Pescados	1549
azurea, Hermosilla	1383, 1384
azureus, Alepisaurus	595
Galeichthys	2775
Azurina	1544
hirundo	1544
azurissimus, Microspathodon	1570
dorsalis	1570
Bacalao	1184, 1185
bacalaus, Gobius	2230
Bacalhao sabara	2230
Bachelor	987
badius, Euctenogobius	2227
Fundulus heteroclitus	2827
Gobius	2227
Rhinichthys	308
bagre, Ailurichthys	117
Felichthys	117
Silurus	117
Bagre Colorado	122
Sapo	2319
Bagres de Rio	149
Bagrus albicans	124
arioides	133
cœlestinus	125
emphysetus	122
flavescens	123

	Page.
Bagrus macronemus	117
mesops	123
passany	124
pemecus	125
proops	124
temminckianus	123
valenciennesi	124
Bahama Lancelet	4
bahamensis, Piscis viridis	1638
Vulpes	411
bahianus, Acanthurus	1693
Teuthis	1693
bahiensis, Cypsilurus	2836
Exocœtus	739
Felichthys	118
Galeichthys	119
Rhombus	2664
baileyi, Cyprinodon	675
bailloni, Gasterosteus	747
Baione	506
Baiostoma	2694, 2695
brachialis	2698
bairdi, Acipenser	105
Callionymus	2185
Cottus punctulatus	1950
bairdianum, Siphostoma	765, 770
bairdianus, Sphyrænops	1114
Syngnathus	770
Bairdiella	1432, 1433
aluta	1437
archidium	1432
argyroleuca	1434
armata	1436
chrysoleuca	1438
chrysura	1433
ensifera	1434
icistia	1435
punctata	1434
ronchus	1436
bairdii, Alepocephalus	454
Bathymyzon	9
Cottus	1951
Gastrostomus	406
Macrourus	2583
Microspathodon	1566, 1567
Mitchillina	454
Petromyzon	9
Pomacentrus	1567
Salmo	508
Salvelinus	508
Bajonado	1352
bajonado, Calamus	1352, 1353
Pagellus	1352
Sparus	1352
balantiophthalmus, Scomber	911
balao, Hemiramphus	723, 2835

	Page.
Balaos	718, 722, 723
balearica, Conger-muræna	356
Congermuræna	356
Muræna	356
balearicum, Ophisoma	356
balias, Chirus	1873
Baliste, Le Bridé	1704
Balistes	1699, 1700, 1703
amboinensis	1704
americanus	1707
asperrimus	1706
aurantiacus	1718
barbatus	1720
bellus	1703
broccus	1716
buniva	1701, 1711
capistratus	1704
caprinus	1702
capriscus	1701
carolinensis	1701
cicatricosus	1709
ciliaris	1702
ciliatus	1715
curassavicus	1709
equestris	1702
forcipatus	1702
frenatus	1705
fuliginosus	1702
guttatus	1702
heckeli	1709
hippe	1705
hispidus	1716
kleinei	1720
lævis	1719
liberiensis	1702
lineo-punctatus	1709
linguatula	1720
longus	1707
macrops	1706
macropterus	1707
maculatus	1707, 1708
melanopterus	1707
mento	1710
mitis	1705
monoceros	1719, 1720
moribundus	1702
naufragium	1700, 1701
niger	1711
nigra	1711
nitidus	1709
notatus	1709
oblongiusculus	1720
oculatus	1707
ornatus	1719
piceus	1711
polylepis	1700
powelli	1702
Balistes punctatus	1702
ringens	1709, 1711
rufus	1707
schmittii	1705
schœpfii	1718
scolopax	759
scriptus	1719
serraticornis	1720
sobaco	1706
spilopterygius	1702
sufflamen	1706
tæniopterus	1702
unicornus	1720
urantiacus	1718
vetula	1703
Balistidæ	1698
Ballerus	249
balteatum, Chichlasoma	1521
balteatus, Abramis	239
Cyprinus (Abramis)	239
Eques	1490
Heros	1522
Leuciscus	238
Pomacanthus	1680
Richardsonius	239
Thynnus	871
Upeneus	860
banana, Butyrinus	411
Gobius	2236
Banana-fish	411
bancrofti, Torpedo	78
Banded Pickerel	626
Sunfishes	994
Bang	423
banksi, Citula	927
Barathrodemus	2517
manatinus	2517
Barathronus	2524
bicolor	2524
barbaræ, Siphostoma	765
barbata, Brotula	2500
Loricaria	158
Pallasina	2049
barbatula, Læmonema	2557
barbatulum, Læmonema	2556
barbatum, Echistoma	587
Lyconema	2474
barbatus, Astronesthes	586
Balistes	1720
Enchelyopus	2500
Gadus	2541
Liparis	2118
Lonchurus	1482
Mullus auratus	856
Siphagonus	2050
Barber	1691
Barbeiro	1693, 1226

	Page.
Barbero	1691
Negro	1692
Barboso, Congro	155
Barbu	829
Barbuda, Lija	1720
Barbudo	150, 829
Barbudos	828, 854
Barbulifer	2260
ceuthœcus	2260
papillosus	2261
barbulifer, Rhinoliparis	2145
bardus, Minomus	171
Barfish	987
barkani, Agonus (Brachyopsis)	2044
Barndoor Skate	71
baronis-mulleri, Pimelodus	151
Rhamdia	151
Barracouta	826
Barracuda	2841
California	826
European	826
Great	823
Northern	825
barracuda, Esox	823
Sphyræna	2841
Barracudas	822
barrattii, Boleosoma	1102
Hololepis	1102
Pœcilichthys	1102
barreto, Gobioides	2264
Barretos	2263
bartholomæi, Caranx	919
bartoni, Acara	1516
Aspidophoroides	2092
Chirostoma	793
Cichlasoma	1515
Cylindrosteus	111
Eslopsarum	2840
Bascanichthys	378
bascanium	379
peninsulæ	379
scuticaris	378
bascanium, Bascanichthys	379
Cœcula	380
Bascanius	2704
Bashaw	143
basilaris, Heros	1532
Basking Sharks	50
Bass, Bayou	1012
Black	1010
Sea	1198
Calico	987
Channel	1453
Common Rock	990
Grass	987
Green	1012

	Page.
Bass, Oswego	1012
Rock	989, 1197
Rock Sea	1201
Round	988
Sea	1126
Stone	1139
Strawberry	987
Striped	1131, 1132
White	1132
White Lake	1132
"White Sea" of California	1413
Yellow	1134
Bassogigas	2515
gillii	2515
stelliferoides	2516
Bassozetus	2507
catena	2509
compressus	2508
normalis	2507
tænia	2510
Bastard Halibuts	2624
Margaret	1257
Weakfish	1406
batabana, Corvula	1430
batabanus, Johnius	1431
Larimus	1431
Batfish	89, 2183, 2737
Short-nosed	2738
Bat-Fishes	2735
Bathyagonus	2077
nigripinnis	2078
bathybius, Cynicoglossus	2656
Embassichthys	2655
Histiobranchus	352
Synaphobranchus	352
Bathyclupea	834
argentea	835
Bathyclupeidæ	834
Bathygadinæ	2562
Bathygadus	2563
arcuatus	2564
cavernosus	2581
favosus	2565
longifilis	2566
macrops	2566
Bathylaco	540
nigricans	540
Bathylaginæ	527
Bathylagus	528
benedicti	529
borealis	2824, 2825
euryops	529
milleri	2825
pacificus	530, 2824
Bathymaster	2288
hypoplectus	2290

	Page.
Bathymasterjordani	2289
signatus	2288
bathymaster, Bregmaceros	2527
Bathymasteridæ	2287
Bathymyzon	9
bairdii	9
Bathynectes	2507
compressus	2508
laticeps	2523
Bathyonus	2507
compressus	2509
tænia	2510
Bathyophis	605
ferox	605
Bathyphasma	2128, 2864
ovigerum	2128
bathyphila, Cyclothone	582
bathyphilum, Neostoma	583
Bathypteroidæ	544
Bathypterois	544
longipes	546
quadrifilis	545
Bathysaurus	539
agassizii	540
ferox	539
Bathysebastes	1860
Bathystoma	1308
aurolineatum	1310
rimator	1308
striatum	1310
Bathytroctes	454
stomias	454
Batis	66
Batoidei	59
Batrachoides	2314, 2868
pacifici	2314
surinamensis	2314
tau	2314
vernullas	2316
variegatus	2316
Batrachoididæ	2313
Batrachops	1740
Batrachus	2314, 2315
celatus	2316
guavina	2195
magaritatus	2323
pacifici	2315
porosissimus	2321
surinamensis	2314
tau	2316
beta	2316
pardus	2317
variegatus	2316
Batrictius	2868
battaræ, Orthragoriscus	1754
Bay Shark	37

	Page.
Baya	1176
bayanus, Pomadasis	1331
Bayou Bass	1012
bdellium, Petromyzon	11
Bdellostoma dombey	6
polytrema	6
stouti	6
beadlei, Synechoglanis	135
beani, Caranx	920
Heros	1538
Ophidion	2487
Pleuronectes	2646
Pœcilichthys	1057
Triglops	1924
beanii, Ammocrypta	1064
Limanda	2646
Melamphaes	843
Plectromus	842
Prionotus	2170
Serrivomer	367
beardsleei, Salmo gairdneri	2819
Bear Lake Bullhead	1954
Beau Gregory	1555
Beauty, Rock	1684
beckwithi, Cyprinella	273
Becuna	823
becuna, Sphyræna	823
beldingii, Cottus	1958
belengeri, Caranx	923
belisanus, Belonesox	684
belizianus, Eleotris (Culius)	2201
bella, Hypoclydonia	1115
Bellator	2173
egretta	2174
militaris	2173
bellicus, Nocomis	3213
Bellows-fish	759, 2713
bellus, Balistes	1703
Minnilus	297
Notropis	297
Belly, Yellow	1001
Belone almeida	715
altipinna	717
angusticeps	712
ardeola	713
argalus	713
caribbæa	717
cigonella	713
crassa	716
depressa	711, 713
diplotænia	712
exilis	714
galeata	716
gerania	716
guianensis	715
hians	718

	Page.		Page.
Belone jonesi	717	beryllina, Menidia	797
latimana	717	gracilis	797
maculata	718	beryllinum, Chirostoma	798
melanochira	716	beryllinus, Cryptotomus	1624
microps	712	Beryx	844
notata	711	decadactylus	844
pacifica	716	splendens	844
raphidoma	716	Besan	1687
scrutator	714	Beshow	1862
stolzmanni	713	Besugo	1356
subtruncata	711	beta, Batrachus tau	2316
timucu	715	betaurus, Cirrhites	1492
truncata	714, 715	biaculeatus, Gasterosteus	748
belone Esox	714	Biajaiba	1270
Histiophorus	892	de lo Alto	1140
Tetrapturus	892	Bibronia	2704
Belonesox	884	bicaudalis, Lactophrys	1723
belizanus	684	Ostracion	1723
Belonichthys	773	bicolor, Algansea	245
Belonopsis	369	Ammocœtes	10
leuchtenbergii	369	Anisotremus	1319
bendirei, Potamocottus	1965	Barathronus	2524
Uranidea	1964	Exocœtus	738
benedicti, Bathylagus	529	Grammiconotus	726
benoiti, Myctophum	573	Leuciscus	232, 245
Scopelus	573	Leucos	245
Benthocometes	2514	Pristipoma	1320
robustus	2514	Rhypticus	1232
Benthodesmus	887	Rondeletia	548
atlanticus	887	Rutilus	244
elongatus	888	Rypticus	1231
Benthosauridæ	543	Smecticus	1232
Benthosaurus	543	Squalius	232
grallator	543	Tigoma	232
Benthosema	573	bicornis, Centridermichthys	1913
arcticum	574	Cottus	1913
mulleri	574	Icelus	1911
berardi, Aluteres	1720	Bielaya Ryba	480
Bergall	1577	bifasciatum, Cichlasoma	1521
Berg-gylt	1577	Thalassoma	1610, 2559
berglax Macrourus	2581	bifasciatus, Chlorichthys	1609
berlandieri, Lepidosteus	111	Heros	1521
Mugil	812	Julis	1610
Bermuda Catfish	882	Labrus	1609
Chub	1387	Thalassoma	1610
bermudæ, Fundulus	643	bifrenata Hemitremia	259
bermudensis, Fierasfer	2497	bifrenatus, Hybopsis	259
Lefroyia	2497	Notropis	258
bernardini, Catostomus	178	bifurca, Chætodon cauda	1562
Berrugate	2857	Big Eye	1238
bertheloti, Crius	971	Big-eyed Herring	410, 426
Berycidæ	837	Scad	911
Berycinæ	838	Big-headed Gurnard	2171
Berycoid Fishes	833	Big Skate	68
Berycoidei	781, 833, 834	of California	72
Berycoids	837	biguttata, Cochlognathus	252

	Page.
biguttatus, Ceratichthys	323
Labrisomus	2360
Malacoctenus	2360
Semotilus	322
bilinearis, Merluccius	2530
Merlucius	2531
Stomodon	2531
bilineata, Lepidopsetta	2643
Platessa	2643
bilineatum, Pristipoma	1319
bilineatus, Anisotremus	1319
Characodon	668
Pleuronectes	2643
Billfish	109, 714, 725, 892
billingsiana, Cliola	272
Cyprinella	272
biloba, Corvina	1460
Pachypops	1460
bilobus, Blepsias	2018
Histiocottus	2018
Peropus	2018
bimaculata, Percina	1027
Pileoma	1027
bimaculatus, Chætodon	1674
Clinus	2358
Malacoctenus	2358
Pœcilioides	678
Pseudoxiphophorus	678
Sayris	725
Xiphophorus	678
binoculata, Raja	72
Uraptera	73
binotatus, Apogon	1109
Bipinnula	878, 2843
violacea	878
bipinnulata, Seriola	907
bipinnulatus Elagatis	906
birostratus, Prionotus	2152, 2156
birostris, Raia	93
Manta	92
bishopi, Alosa	430
Sardinella	430
bison, Aspicottus	1938
Carpiodes	166
Enophrys	1938
Lepisosteus	110
bispinosus, Gasterosteus	748
atkinsi	748
cuvieri	749
Melanocetus	2734
Melichthys	1711
Myliobatis	89
bisselli, Argyrosomus tullibee	473
Coregonus tullibee	473
bistrispinus, Bodianus	1234
Rypticus	1233
bisus, Scomber	867
bivittata, Elacate	948
Haliperca	1205
bivittatus, Centropristis	1205
Chærojulis	1597
Halichæres	1597
Hybopsis	233
Iridio	1595
Labrus	1596
Minnilus	233
Platyglossus	1597
Serranus	1205
bixanthopterus, Caranx	926
Black and yellow Rockfish	1825
Angel	1679
Black-banded Rockfish	1827
Sunfish	995
Bass	1010
Large-mouthed	1012
Small-mouthed	1011
Black-belly	426
Bullhead	141
Croaker	1456
Drums	1454
Blackfin	472
Snapper	1261
Black-fish	207, 963, 1199, 1200, 1578
blackfish, Labrus	1578
Blackfishes, Alaska	620, 261
Grouper	1161, 1174
Grunt	1297
Guativere	1146
Harry	1199
Black-head Minnow	217
Blackhorse	168
Black Jewfish	1161
Moray	396
nosed Dace	305, 307
Oldwife	1711
Perch	1504
Pilot	1555
Rockfish	1784
Rudder Fishes	963
Ruffe	963
Ruffs	962
Sculpin	1985
Sea Bass	1198
Black-sided Darter	1028, 1032
Black-spotted Trout	487
Swallowers	2291
Will	1199
blackfordi, Lutjanus	1265
Yarrella	584
blainvillei, Acanthias	51
Blakea	2351
elegans	2353

	Page.
Blanca, Aguja	892
Chopa	1388
Lisa	813
Mojarra	1372
Pesca	321
Sardina	332
Blancas, Mojarras	1372
blanchardi, Gasterostea	746
Neoclinus	2354
Blanco, Burro	1328
Matajuelo	2275, 2276
Pescado de Chapala	792
Ronco	1297
Blancos, Pescados	792
Blanquillo	2276, 2278
Blanquillos	2274, 2276
bleekeri, Aptonichthys	2703
bleekeriana, Ilisha	436
Pellona	436
Blenitrachus	2391
Blennicottus	2016, 2864
acuticeps	2864
embryum	2016, 2864
globiceps	2017
bryosus	2017
Blennies	2344, 2377
Snake	2435
Blenniidæ	2344
Blenniinæ	2346
Blennioclinus	2360
Blennioid Fishes	2343
Blennioidea	2343
Blennioidei	782
blennioides, Diplesion	1053
Etheostoma	1033
Blenniolus	2386, 2390
blennioperca, Hyostoma	1053
Blenniophidium	2428
petropauli	2430
Blennius	2377, 2378, 2553
alectrolophus	2422
americanus	2457
anguillaris	2436, 2457
asterias	2383
bosquianus	2394
brevipinnis	2391
capiti lævi	2438
carolinus	2378
chuss	2555
ciliatus	2457
(Clinus) lumpenus	2438
crinitus	2383
cristatus	2382, 2383
dolichogaster	2417
europæus	2419
favosus	2380

	Page.
Blennius filicornis	2381
fimbriatus	2457
fucorum	2379
geminatus	2385
gentilis	2388
gracilis	2438
gunnellus	2419
hentz	2390
herminier	2362
labrosus	2457
lampetræformis	2438
lumpenus	2438
marmoreus	2381
microstomus	2385
multifilis	2385
murænoides	2419
nuchifilis	2383
oceanicus	2379
pilicornis	2380
polaris	2469
polyactocephalus	2409
punctatus	2390, 2440
regius	2553
roscus	2420
serpentinus	2439
stearnsi	2379
striatus	2388
tænia	2418
torsk	2561
truncatus	2381
vinctus	2382
(Zoarches?) polaris	2469
blennius, Alburnops	262
Etheostoma	1072, 1073
Minnilus	262
Notropis	261
Blennophis	2400
webbii	2401
Blenny, Snake	2438
Blepharichthys	931
crinitus	932
Blepharis	931
crinitus	932
major	932
sutor	932
blepharis, Carangoides	932
blephura, Anguilla	348
Blepsias	2018
bilobus	2018
cirrhosus	2018
oculofasciatus	2021
trilobus	2019
ventricosus	1936
Blepsiinæ	1883
Bleu, Poisson	517
Blindfish, Cuban	2501

	Page.
Blindfish, of the Mammoth Cave...	706
Small	704
Blind Fishes	702
Gobies	2261
Goby of Point Loma	2262
Bloater	471
Blob	1950
blochii, Bodianus	1583
Caranx	919
Galeichthys	118
Orthrogoriscus	1754
Pimelodus	155
Piramutana	155
Blower	1733
Blow-fish, Spring-back	1734
Blue-back	426
Mullet	813
Salmon	481
Trout	514
Trout of Lake Crescent	2819
Blue Bream	1005
Cat	134
Cod	1875
Darter	1088
Herring	425
Mullet	191
Parrot-fish	1636, 1652
Perch	1505, 1577
Pike	1021
Sharks	33
Sunfish	1005
Surgeon	1691
Tang	1691
Blue-breasted Darter	1076
Bluefin	472
Bluefish, California	1410
Bluefishes	945, 946
Blue-gill	1005
Blueheaded Sucker	171
Blue-spotted Sunfish	996
Blunt-nosed Minnow	218
Shiner	934
Boar-fishes	1663
Boar Grunt	1303
Bobo	821
Boca Dulce	29
Negra	1837
Bocaccio	1780
Bocon	442, 450
bocona, Sardina	449
Bodianus	1143, 1581
acanthistius	1147
achigan	1011
apua	1174
argyroleucus	1433
aya	1265

	Page.
Bodianus bistrispinus	1234
blochii	1583
bodianus	1583
costatus	1462
cruentatus	1142
diplotænia	1582
dubius	1146
exiguus	1433
flavescens	1024
fulvus	1144
punctatus	1146
ruber	1145, 1146
guativere	1145
jaguar	849
marginatus	1174
pallidus	1433
panamensis	1141
pectoralis	1582
pentacanthus	849
pulchellus	1584
punctatus	1146
punctiferus	1147
ruber	1265
rufus	1135, 1583
rupestris	990
stellifer	1443
striatus	1259
tæniops	1144
triurus	1236
vivanet	1257
bodianus, Bodianus	1583
Cossyphus	1583
Bodieron	1867
Bœostoma brachiale	2698
reticulatum	2696
Boga	1365
Bogoslovius	2574
clarki	2575
firmisquamis	2575
Bola	1455
Boleichthys	1101
eos	1102
exilis	1103
fusiformis	1101
warreni	1103
whipplii	1096
boleoides, Cottus	1968
Radulinus	1919
Uranidea	1968
Boleosoma	1054
æsopus	1057
barrattii	1102
camurum	1060
chlorosoma	1060
copelandi	1046
fusiformis	1102

Page.

Boleosoma gracile.................... 1102
 lepida 1089
 lepidum 1089
 longimanus.............. 1054
 maculatum 1057, 1077
 mutatum................. 1057
 nigrum 1056
  effulgens........ 1058
  maculaticeps.... 1058
  mesæum 1059
  olmstedi 1057
  vexillare........ 1058
 olmstedi brevipinnis.... 1057
 phlox.................... 1052
 podostemone............ 1055
 pottsii.................... 1083
 punctulatum 1091
 shumardi................ 1047
 stigmæum 1048
 susanæ 1059
 tessellatum 1046, 1057
 variatum 1070
 whipplei................. 1096
boleosoma Gobius.................... 1102
bolli, Ameiurus 140
bollmani, Hippoglossina 2621
 Opsopœodus.............. 249
 Scarus.................... 1646
Bollmannia 2237
 chlamydes.............. 2238
 macropoma............. 2239
 ocellata................. 2238
 stigmatura 2239
bombifrons, Lepomis................ 1003
 Pomotis................ 1003
Bonaci Arará 1174
 Cardenal 1173, 1174
 de Piedra................... 1172
 Gato...................... 1187
bonaci, Epinephelus 1175
 Mycteroperca 1174
  xanthosticta. 1176
 Serranus.................... 1175
 Trisotropis................. 1175
Bonapartia................... 580, 2826
 pedaliota............. 580, 2826
bonapartii, Scopelus 557
bonariense, Hæmulon 1297
bonariensis, Halatractus 905
 Seriola 905
bonasus, Bubalichthys.............. 164
 Raja..................... 90
 Rhinoptera............... 90
Bone-fish........................... 411
Bone Shark 51
Bonito 869, 948

Page.

Bonito California.................... 872
 Oceanic................... 868
Bonitos 871
Bonnet-head 44
Bony-fish 410, 433
Bony Fishes 113
 Ganoids 107
Bony-tail........................... 226
Boohoo............................. 891
boops, Caranx...................... 922
 Centaurus.................... 1755
 Myctophum................... 572
 Notropis 268
 Ostracion 1755
 Scopelus 572
 Trachurus 922
Borborys........................... 633
borea, Lucioperca.................. 1022
boreale, Etheostoma............... 1082
borealis, Alepidosaurus (Caulopus). 597
 Alepisaurus................. 596
 Amiurus.................... 137
 Ammocœtes................. 11
 Arctozenus 601
 Bathylagus 2824, 2825
 Chimæra.................... 95
 Fierasfer.................... 2443
 Icelinus.................... 1896
 Læmargus 57
 Maurolicus.................. 577
 Notorhynchus 18
 Oligocottus 2014
 Paralepis.................... 601
 Petromyzon................. 13
 Pimelodus 137
 Pœcilichthys............... 1082
 Scopelus 577
 Sphyræna................... 825
 Squalus..................... 57
 Sudis....................... 601
Boregat........................... 1867
Boreocottus........................ 1970
 axillaris 1981
Boreogadus........................ 2533
 polaris................. 2534
 saida 2533
Boreogaleus........................ 32
Borer 7
Borers............................. 5
boreum, Stizostedion canadense.... 1022
boreus, Esox 628
Borlase............................. 963
bosci, Abramis crysoleucas......... 251
 Atherina 801
 Gobiosoma.................. 2259
 Gobius............... 2227, 2259

	Page.
bosci, Halatractus	905
Leuciscus	251
Pimelepterus	1388
Zonichthys	905
boscianus, Chasmodes	2394
boscii, Seriola	905
bosqui, Cyphosus	1388
bosquianus, Blennius	2394
Chasmodes	2394
bosquii, Pimelepterus	1388
bostonensis, Catostomus	179
bostoniensis, Muræna	348
Botete	1731
Bothragonus	2086
swanii	2086, 2088
Bothrocara	2475
mollis	2476
pusilla	2476
Bothrolæmus	939
Bothus	2661
maculatus	2660
Boucanelle	1261
boucardi, Leuciscus	247
Leucus	247
Pœcilia	695
Pristipoma	1334
Rutilus	247
boulengeri, Mycteroperca	1171, 2856
Bout de Tabac	1215
bouvieri, Salmo clarkii	2819
mykiss	496
purpuratus	496
bovinum, Plectropoma	1193
bovinus, Cyprinodon	673
Hypoplectrus	1193
unicolor	1193
Bowfin	113
Bowfins	111, 112
bowmani, Plargyrus	283
Box, Tobacco	68
Boxaodon	1365
brachiale, Bœostoma	2698
Sparisoma	1641
brachialis, Achirus	2698
Ammocœtes	14
Baiostoma	2698
Petromyzon	14, 2745
Scarus	1641
brachiatus, Diodon	1746
Brachioptilon	92
hamiltoni	93
brachiurus, Gymnotus	340
brachiusculus, Grammicolepis	974
brachyacanthus, Amiurus	141
brachycentrus, Gasterosteus	748
Nauclerus	900
brachycephalum, Siphostoma	769
brachycephalus, Exocœtus	733
Syngnathus	769
Uranichthys	382
brachychir, Macrostoma	2826
Myctophum	556
brachychirus, Trachurops	911
Brachyconger	359
savanna	360
Brachydeuterus	1325
axillaris	1328
corvinæformis	1326
leuciscus	1327
nitidus	1326
Brachygenys	1307
chrysargyreus	1307
tæniata	1308
Brachyistius	1499
frenatus	1499
Brachymullus	858
Brachyopsis	2046, 2864
annæ	2043
barkani	2044
decayonis	2054
dodecaedrus	2046
rostratus	2046, 2048
segaliensis	2048
verrucosus	2044
xyosternus	2043
Brachyospinæ	2032
brachypoda, Gasterosteus pungitius	746
Pygosteus pungitius	746
Brachypomacentrus	1549
Brachyprosopon	2653
brachyptera, Echeneis	2272
Remora	2272
Rhamdia	151
brachypterus, Holocentrus	852
Lutjanus	1268
Neomænis	1268
Pimelodus	152
Remoropsis	2272
Thynnus	870
Zygonectes	682
brachyrhinchus, Acipenser	104
Brachyrhinus	1221
creolus	1222
furcifer	1222
Brachysomophis	387
crocodilinus	388
horridus	388
brachysomus, Calamus	1353
Epinephelus	1154
Sparus	1353
brachyurus, Oxydontichthys	385
bradfordi, Porocottus	2862

	Page.
Brama	958
agassizii	959
brevoortii	959
chilensis	960
dussumieri	960
orcini	960
parræ	1586
raii	958, 959
raji	960
saussurii	958
brama, Cynædus	1360
Bramble Sharks	57
Bramidæ	956
Bramocharax	338
bransfordi	339
Bramopsis	1501
mento	1502
Branch Herring	426
branchialis, Ammocœtes	14
Petromyzon	14
Branchiostegus	2578
Branchiostoma	3, 4
californieuse	4
californiensis	4
caribæum	3
lanceolatum	3
lubricum	3
Branchiostomatidæ	2
Branchiostomidæ	2
Branderius	373
brandti, Arius	122
Cottus	1984
Myoxocephalus	1984
Tachisurus	122
brandtii, Arius	2758
branicki, Pomadasis	1333
Pristipoma	1334
branneri, Pœcilia	2834
bransfordi, Bramocharax	339
Loricaria	158
Rhamdia	151
brasilianus, Gerres	1378
brasiliense, Pristipoma	1320
brasiliensibus, Capeuna	1311
Guabi coara	1305
brasiliensis, Amblyopus	2264
Centropristis	1221
Chirostoma	794
Chlorichthys	1591
Clupea	411
Conger	360
Esox	723
Hemiramphus	722
Hippoglossus	2626
Labrus	1591
Menidia	801
brasiliensis, Mugil	810, 814, 816
Muræna seu conger	403
Narcine	78, 2752
Paralichthys	2626
Plagusia	2709
Pseudorhombus	2626
Scorpæna	1842
Serranus	1221
Thynnus	869
Torpedo	78
Vomer	934
braytoni, Notropis	264
Bream	250, 1009, 1358, 1360
Bream, Blue	1005
Copper-nosed	1005
Redbreast	1001
Bregmaceros	2526
atlanticus	2527
atripinnis	2527
bathymaster	2527
macclellandii	2526
Bregmacerotidæ	2525
Bresson	125
brevibarbe, Lepophidium	2485
Ophidium	2485
brevicauda, Pomoxys	987
Salmo	493
brevicaudata, Brevoortia tyrannus	434
Breviceps	116
breviceps, Evorthodus	2208
Gasterosteus	746
Larimus	1423
Moxostoma	196
Pomotis	1003
Ptychostomus	196
brevidens, Gonostoma	579
brevimanus, Gerres	1377
Tetragonopterus	335
brevipes, Lycodes	2467
brevipinna, Scymnus	57
Somniosus	57
brevipinne, Ditrema	1499
Pristipoma	1341
brevipinnis, Blennius	2391
Boleosoma olmstedi	1057
Hypsoblennius	2390
Isaciella	1341
Orthopristis	1341
Thynnichthys	869
Thynnus	869
brevirostris, Chasmistes	183, 199
Cololabis	726
Gerres	1376
Hippocampus	776
Histiophorus	892
Hypoprion	41

	Page.
brevirostris, Macrognathus.........	723
Saurus	533
Scombresox	726
Syngnathus	765
brevirostrum, Acipenser	106
Hæmulon.............	1300
brevis, Acanthurus.................	1691
Amblyopus	2263
Atherinichthys	2840
Centropomus	1125
Cephalus	1754
Cetengraulis...............	450
Engraulis...................	450
Tyntlastes..................	2262
brevispinis, Sebastichthys	1788
Sebastodes	1787
brevoorti, Argyriosus	936
Euleptorhamphus	724
Brevoortia	433
tyrannus	433
aurea.........	434
brevicaudata .	434
patronus	434
brevoortii, Brama	959
bricei, Chætodon	1678
bristolæ, Emmnion..................	2375
Broad Shad	1372
Whitefish	464
Broad-head, Grubber	447
broccus, Balistes	1716
Monacanthus...............	1716
Brochet de Mer	1118
brodamus, Cottus....................	2066
Brook Lampreys	12
Silverside	805
Stickleback	744
Sucker	178
Trout	506
Trout of western Oregon....	501
Brosme...............	2561
brosme	2561
brosme, Brosme	2561
Brosmius	2561
Enchelyopus...............	2561
Gadus	2561
brosmiana, Lota	2551
Brosminæ	2532
Brosmius......................	2561
brosme	2561
flavescens	2561
flavesny	2561
marginatus...............	2502
vulgaris	2561
Brosmophycinæ.....................	2498
Brosmophycis	2502
marginatus	3502

	Page.
Brosmophycis ventralis.............	2503
Brotula	2500
barbata......................	2500
Brotulidæ	2498
Brotulinæ...........................	2498
Brotuloid Fishes	2498
broussoneti, Gobioides.............	2264
broussonetii, Acanthurus	1691
Umbrina	1466
broussonnetii, Gobiodes	2263
Brown Cat	142
Hind..............	1142
Rockfish	1817
browni, Atherina...................	443
Engraulis	443
Hemirhamphus	723
Solea......................	2701
Stolephorus	443
Vomer	934
Brown-winged Sea-robin	2167
brucus, Squalus	58
brunnea, Maynea..................	2476
brunneus, Amiurus	142
Catulus	24
Gobius	2218
Ilyophis	350
Serranus	1175
Trisotropis	1175
Brycon...........................	337
dentex	337
striatulus	337
bryoporus, Spratelloides............	422
Bryostemma	2408
nugator.................	2410
polyactocephalum .. 2408, 2409
bryosus, Blennicottus globiceps	2017
Bryssetæres........................	2328
pinniger.................	2328
Bryssophilus..................... 2329, 2330
Bryttus	995
albulus	1007
fasciatus	993
gloriosus	994
humilis....................	1004
mineopas	996
oculatus	1004
punctatus	998
reticulatus	998
signifer....................	996
unicolor	1001
Bubalichthys	163
altus..................	165
bonasus	164
bubalinus.............	165
bubalus	516
niger	164

	Page.
Bubalichtha urus	164
bubalina, Oliola	273
Cyprinella	273
bubalinus, Bubalichthys	165
. Leuciscus	273
Notropis	273
bubalis, Cottus	1972
Myoxocephalus	1971
bubalus, Amblodon	165
Bubalichthys	165
Catostomus	165
Ichthyobus	164
Ictiobus	164
Bubbler	1484
buccanella, Lutjanus	1262
Mesoprion	1262
Neomænis	1261
buccata, Ericymba	302
bucciferus, Labrisomus	2363
bucco, Moxostoma	190
Ptychostomus	191
bucculentus, Chonophorus	2236
buchanani, Notropis	2800
Buffalo Cod	1875
Common	163
fishes	163
Mongrel	164
Razor-backed	164
Red mouth	163
Small-mouthed	164
Sucker-mouthed	164
buffalo, Acipenser	106
bufo, Lophius	2316
Scorpæna	1849
Bugara	1508
Bugfish	433
Bull Red-fish	1453
Bull Trout	507
bullaris, Cyprinus	221
Semotilus	222
bulleri, Prionodes	1213
Serranus	1214
Bullhead	1950
Bullhead, Bear Lake	1954
Black	141
Common	140
Prickly	1944
Rocky Mountain	1949
Bullhead Shark	20
Sharks	19
Bullon	1650
Bumper	938
buniva, Balistes	1701, 1711
Burbots	2550
Burgall	1577
burgall, Ctenolabrus	1577

	Page.
Burnstickle	747
Burr-fish, Common	1748
Burr-fishes	1747
Burrito	1333, 1327
Burritos	1325
Burro	1332
Blanco	1328
Burros	1329
burtonianus, Cyprinus	2798
busculus, Prionodes	1211
Butirinus maderaspatensis	415
butleri, Pœcilia	691
butlerianus, Pœcilichthys	1102
Butter-fish	967, 1144, 2416, 2419
Butter-fishes	965
Butterfly	1677
Butterfly-fishes	1669, 1672
Butterfly Ray	86
Butyrinus	411
banana	411
Bythites	2504
fuscus	2504
Bythitinæ	2498
caballa, Cybium	876
Scomberomorus	876
Caballerote	1255, 1257
caballerote, Anthias	1257
Lutjanus	1257
Mesoprion	1257
caballus, Caranx	921
Cabezon	1423, 1889, 2321
Smooth	2012
Cabezones	1889
Cabezote	790
Cabezuda, Lisa	811
caboverdianus, Ginglymostoma	26
Cabra Mora	1152
Cabrilla	1158, 1197, 1832
Calamaria	1184
de Ralzero	1171
Piritita	1181
Cabrilla, Spotted	1196
Cabrillas Verdes	1194
Cabrillo de Astillero	1176
Cachucho	1282
Caçonetta	40
Cæcilia	373
cælolepis, Centroscymnus	55
cænicola, Pœcilia	641
cænosus, Pimelodus	140
cærulea, Cliola	277
Clupea	421
Codoma	277
Coryphæna	1653
Maletta	423

	Page.
cærulea, Novacula	1653
Tautoga	1577
cæruleatus, Acronurus	1691
cæruleo·aureus, Harpe	1583
cæruleopunctatus, Acara	1514
cærulescens, Arius	129
Galeichthys	2776
Hexanematichthys	129
Pimelodus	135
cæruleus, Acanthurus	1691
Carcharhinus	37
Carcharias	37
Cheonda	232
Clupanodon	423
Ctenolabrus	1577
Cyclopterus	2097
Notropis	277
Photogenis	277
Pseudoscarus	1654
Scarus	1652, 1654
Squalius	232
Sqnalus	33
Teuthis	1691
cærulopunctatus, Æquidens	1514
Cæsar	1308
Cæsiomorus	939
Cæsiosoma californiense	1391
cæsius, Anisotremus	1316
Pomadasys	1317
Cagon de lo Alto	1277
Caillen	429
Cailleu-tassart	432
Caiman	2216
Caji	1258
Calafate	1711
Calamaria, Cabrilla	1184
Calamus	1347, 1349
arctifrons	1355
bajonado	1352, 1353
brachysomus	1353
calamus	1349
leucosteus	1353
macrops	1350, 1354
medius	1356
megacephalus	1350, 1351
penna	1354, 1355
pennatula	1351
plumatula	1352
proridens	1350
taurinus	1354
calamus, Calamus	1349
Chrysophrys	1350
Pagellus	1350
calcarata, Scorpæna	1854
Calico Bass	987
California Anchovy	448

	Page.
California Barracuda	826
Big Skate	72
"Bluefish"	1410
Bonito	872
Conger Eel	395
Dogfish	54
Hagfish	6
Herring	422
Jewfish	1137
Lancelet	4
Pompano	967
Redfish	1585
Sardine	423
Smelt	806
Sole	2613
Stickleback	751
Stingray	89
Tomcod	2539
Torpedo	77
Whiting	1476
californica, Alausa	423
Morrhua	2539
Squatina	59
Tetronarce	77
Torpedo	78
Uropsetta	2626
californicus, Cypsilurus	2836
Exocœtus	730, 740
Gadus	2539
Galeus	30
Halichæres	1601
Hippoglossus	2626
Mustelus	30
Myliobatis	89
Oxyjulis	1601
Paralichthys	2625, 2626
Pseudojulis	1601
Pseudorhombus	2626
Stereolepis	1138
californiense, Branchiostoma	4
Cæsiosoma	1391
Myctophum	572
Siphostoma	764
californiensis, Atherinichthys	807
Atherinopsis	806
Branchiostoma	4
Chilomycterus	1751
Cyprinodon	674
Diapterus	1370
Doryichthys	774
Doryrhamphus	773
Eucinostomus	1369
Gerres	1370
Medialuna	1391
Ophisurus	384
Otolithus	1413

	Page.
californiensis, Polynemus	829
Scorpis	1391
Syngnathus	764
Typhlogobius	2262
Xenichthys	1286
Xenistius	1286
callarias, Gadus	2541
callaris, Salmo	508
Callaus	1455
Callechelys	378
muræna	378
peninsulæ	379
Calliclinus	2360
Calliodon	1621, 1642, 1644, 1650
dentiens	1623
gibbosus	1296
lineatus	1651
retractus	1623
calliodon, Liparis	2120
Callionymidæ	2184
Callionymus	2185
bairdi	2185
calliurus	2187
himantophorus	2186
pauciradiatus	2188
pelagicus	2184
callipteryx, Campostoma	206
callisema, Oliola	273
Codoma	273
Episema	273
Notropis	272
callisoma, Herpetoichthys	384
callistia, Cliola	276
Codoma	276
callistius, Notropis	276
Photogenis	276
calliura, Oliola	275
Cyprinella	275
Etheostoma	1011
Myteroperca	1186
Calliurus	995, 1010
diaphanus	996
floridensis	992
formosus	996
longulus	996
melanops	992
microps	996
murinus	996
punctulatus	992, 1011
calliurus, Callionymus	2187
Ioglossus	2193
Trisotropis	1186
Callogobius	2210
callolepis, Harengula	430
callopterus, Cypsilurus	2836
Exocœtus	740

	Page.
Calloptilum	2526
mirum	2527
Callorhynchus atlanticus	95
centrina	95
Callyodon	1621, 1642, 1651
auropunctatus	1624
flavescens	1640
psittacus	1638
ustus	1624
callyodon, Cyclopterus	2110
Liparis	2111
Neoliparis	2110
calopteryx, Serranus	1213
Calotomus	1626
xenodon	1626
calva, Amia	113
Calycilepidotus	1936
lateralis	1899
spinosus	1937
Camarina	1381
nigricans	1382
camelopardalis, Mycteroperca tigris	1187
Serranus	1187
campbelli, Moxostoma	186
Salmo	508
Campbellite	987
campechianus, Lutjanus	1265
Mesoprion	1265
camperi, Scombresox	725
Campostoma	204
anomalum	205
callipteryx	206
dubium	206
formosulum	206
gobionium	206
hippops	206
mormyrus	206
nasutum	206
ornatum	205
pricei	205, 2796
prolixum	206
Campostominæ	202
Campylodon	614
fabricii	615
camtschatica, Lampetra	13
camtschaticus, Entosphenus	2745
Pteromyzon	2745
marinus	2745
camura, Oliola	280
Vaillantia	1060
camurum, Boleosoma	1060
Etheostoma	1076
camurus, Nanostoma	1076
Notropis	279
camurus, Pœcilichthys	1076
Caña-bota	19

	Page.
canada, Elacate	948
canadense, Stizostedion	1022
boreum	1022
griseum	1022
canadensis, Lucioperca	1022
Salmo	507
canadus, Gasterosteus	948
Rachycentron	948
canaliculatus, Icelus	1917
canariensis, Clinus	2362
candidissimus, Leptocephalus	354
Candil	846
Candlefish	521
Cane di Mare	48
canescens, Chætodon	1688
Zanclus	1688
canina, Amia	113
caninianus, Scopelus	570
caninus, Caranx	921
Lachnolaimus	1580
Pagellus	1352
Canis carcharias	38
canis, Mustelus	29
Salmo	479
Squalus	29
canna, Hæmulon	1297, 1299
Cannorhynchus	756
cantharinum, Pristipoma	1340
cantharinus, Orthopristis	1339, 1340
Cantharus nigromaculatus	987
Cantherines	1713
carolæ	1713
pullus	1713
Canthidermis	1705
maculatus	1706, 1707
sobaco	1705
sufflamen	1706
willughbeii	1707
Canthigaster	1741
caudicinctus	1742
lobatus	1732
punctatissimus	1741
rostratus	1741
Canthigasteridæ	1740
Canthirhyncus	2088
monopterygius	2092
Canthorhinus	1713
Cautileña, Mojarra	1369
Capelin	520
capensis, Aledon	1754
Carcharodon	50
Elops	410
Scorpæna	1833
Sebastes	1833
Sebastodes	1833
Capeuna brasiliensibus	1311

	Page.
capeuna, Hæmulon	1311
Hæmylum	1311
Serranus	1311
capillaris, Argyriosis	936
Zeus	936
capillatus, Clinus	2362
Labrisomus	2362
capistratus, Balistes	1704
Chætodon	1677
Pachynathus	1704
Sarothrodus	1678
Tetrodon	1742
Capitaine	1579
capite, Eleotris plagioplateo	2201
Labrus obtuso	1609
capiti lævi, Blennius	2438
capito, Poromitra	840
capitulinus, Mugil	2841
caprinus, Argyrops	1345
Balistes	1702
Catostomus	168
Otrynter	1345
Stenotomus	1345
Capriscus	1699, 1700
capriscus, Balistes	1701
murium dentibus minutis	1720
caprodes, Etheostoma	1027
Percina	1026
manitou	1028
zebra	1027
Sciæna	1027
Caproidæ	1663
Caproidea	1663
Caprophonus	1664
aurora	1665
capros, Antigonia	1665
caramura, Murenophis	395
Carangichthys	916, 917, 922
Carangidæ	895
Carangoides	928
blepharis	932
cibi	920
dorsalis	930
gallichthys	932
iridinus	919
orthogrammus	928
carangoides, Seriolophus	895
Carangops	912
heteropygus	913
secundus	914
carangua, Caranx	920
Carangus	915, 916
chrysos	921
esculentus	921
fallax	923
hippos	921

	Page.
Carangus lugubris	925
carangus, Scomber	920
Caranx	915, 916, 919, 927
amblyrhynchus	913
analis	927
antillarum	921
ascensionis	925
atrimanus	914
aureus	923
bartholomæi	919
beani	920
belengeri	923
bixanthopterus	926
blochii	919
boops	922
caballus	921, 2846
caninus	921
carangua	920
chilensis	927
chrysus	921
cibi	920
crinitus	932
crumenophthalmus	911
crysos	921, 2846
cuvierri	910
daubentonii	920
defensor	921
dentex	927
dorsalis	930
dumerili	904
ekala	921
erythrurus	920
falcatus	913, 2844, 2845
fallax	923
fasciatus	914
fosteri	923
frontalis	925
furthii	914
georgianus	927
girardi	922
guara	926
heteropygus	913
hippos	920, 923
iridinus	919
latus	922
lepturus	923
lessoni	923
leucurus	915
lugubris	924
luna	927
macarellus	909
macrophthalmus	911
marginatus	922
medusicola	924
melampygus	925
muroadsi	908
Caranx orthogrammus	929
otrynter	930
panamensis	928
paraspistes	923
personi	923
picturatus	910
pisquetus	921, 2846
platessa	927
plumieri	912
poloosoo	928
punctatus	908
richardi	923
ruber	919
sanctæ-helenæ	908
scombrinus	908
secundus	914
sem	923
semispinosus	911
setipinnis dorsalis	934
gambonensis	934
solea	927
speciosus	928
stellatus	926
suareus	908
sutor	932
symmetricus	910
trachurus	910
cuvieri	910
vinctus	918
xanthopygus	921
Caranxomorus	952
plumierianus	911
carapinus, Coryphænoides	2579
Carapo	340
carapo, Gymnotus	341
Carapus	340
affinis	2497
fasciatus	341
inæquilabiatus	341
Carassius	201
auratus	201
Carauna	1145
carbonaria, Pileoma	1027
carbonarium, Hæmulon	1300
carbonarius, Acipenser	106
Gadus	2534
Pollachius	2535
Salmo	509
Carbonero	919
Carbonero, Ronco	1300
carbunculus, Etelis	1283
Carcharhinæ	28
Carcharhinus	33, 35, 37, 2747
acronotus	36
æthalorus	40
cæruleus	37

	Page.
Carcharhinus cerdale	2746, 2747
commersoni	38
falciformis	36
fronto	39
glaucus	33
henlei	37, 2746
lamia	38
lamiella	37
leucos	38
limbatus	40
milberti	37
nicaragnensis	39, 2747
obscurus	35
oxyrhynchus	40, 2747
perezi	36
platyodon	39
platyrhynchus	36
remotus	37
velox	2747, 2748
Carcharias	33, 46
americanus	47
atwoodi	50
cæruleus	37
falcipinnis	35
fronto	39
glaucus	33
griseus	47
henlei	37
isodon	42
lalandi	43
lamia	38
leucos	38
limbatus	40
littoralis	46, 2748
longurio	42
microps	40
milberti	37
mulleri	40
obscurus	35
oxyrhynchus	41
porosus	37
punctatus	42
terræ-novæ	43
trigris	49
verus	50
carcharias, Canis	38
Carcharodon	50
Squalus	38, 50
Carchariidæ	46
Carcharodon	50
capensis	50
carcharias	50
rondeleti	50
smithi	50
Carcharodontinæ	47
Cardenal	1108
Bonaci	1173, 1174
Mojarra	850
Cardinal fishes	1105
cardinalis, Serranus	1174
Trisotropis	1174
Cardonniera	1837
Careliparis	2114, 2115
Caremitra	2129, 2130, 2131
Carenchelyi	343
Careproctus	2129, 2130, 2131
colletti	2131
ectenes	2136, 2866
gelatinosus	2134, 2135
melanurus	2135, 2867
ostentum	2134
phasma	2132
ranula	2134
reinhardi	2133, 2134
simus	2131
spectrum	2133
caribæum, Branchiostoma	3
caribæus, Diplodus	1360
Sargus	1360
caribbæa, Belone	717
caribbæus, Chloroscombrus	938
Halieutichthys	2741
Tylosurus	717
cariuatum, Acanthosoma	1754
Siphostoma	763
carinatus, Diodon	1754
Labichthys	368
Placopharynx	198
Salmo	493
carminale, Tripterygium	2350
carminalis, Euncanectes	2350
carminatus, Cœlorhynchus	2588
Macrurus	2589
Macrurus (Cœlorhynchus)	2589
carnatus, Sebastichthys	1825
Sebastodes	1824
carneus, Acronurus	1602
Gobiesox	2337
Sicyases	2337
carolæ, Cantherines	1713
carolina, Argentina	410
Atherina	791
Trigla	2156, 2172
Carolina Whiting	1474
carolinæ, Potamocottus	1952
carolinensis, Balistes	1701
Cestreus	1409
Clupea	434
Cynoscion	1409

	Page.
čarolinensis Gobius	2218
Hyperistius	988
Mystus	117
Otolithus	1409
Seriola zonata	902
carolinus, Blennius	2378
Gasterosteus	944
Labrus	1578
Pholis	2379
Priacanthus	2858
Prionotus	2156, 2157
Pteraclis	956
Trachinotus	944
Trachynotus	944
Carp, Lake	167
Sucker	166
Suckers	165
carpio, Carpiodes	166
Catostomus	166, 190
Cyprinodon	675
Cyprinus	201
Moxostoma	190
Carpiodes	165
bison	166
carpio	166
cutisanserinus	167
cyprinus	167, 168
damalis	167
difformis	166
grayi	167
nummifer	166
selene	167
taurus	165
thompsoni	167
tumidus	167
urus	164
vacca	168
velifer	167
vitulus	165
Carp-like Fishes	160
Carps	161, 199
Crucian	201
carribæus, Cœlorhynchus	2589
carribbæus, Macrurus	2590
Carrilla Pinta	1152
carringtonii, Agosia nubila	311
Apocope	312
Cartilaginous Ganoids	100
carunculatus, Ceratias	2732
caryi, Acipenser	104
Ditrema	1509
Embiotoca	1509
Hypsurus	1508, 1509
Casabe	938
Casabes	937

	Page.
cassidyi, Embiotoca	1505
cassini, Muræna	356
Casta Cumprida, Alfonsin a	844
Casta Larga, Alfonsin	844
Castagnole	959
castanea, Sidera	396
castaneola, Sparus	960
castaneum, Macrostoma	556
castaneus, Ichthyomyzon	11
Lycodontis	2804
Notoscopelus	556
Petromyzon	11
castelnæana, Pellona	436
castelnaui, Cylindrosteus	111
castor, Pontinus	1856
Scorpæna	1856
Cat Shark	31
Sharks	22
Cat, Blue	134
Brown	142
Channel	134
Channel of the Potomac	138
Chuckle-headed	134
Duck-bill	101
Eel	2788
Flannel-mouth	137
Florida	137
Great Fork-tailed	137
Mississippi	137
Mud	142, 143
Russian	143
Sacramento	140
Schuylkill	140
Spoon-bill	101
Stone	144
White	134, 138
Willow	2788
Yellow	139, 143
Catablemella	554
Catætyx	2504
rubrirostris	2505
Catalina	1322
Catalineta	1322, 1684
Catalinetas	1682
Catalufa	1237, 1238
de lo Alto	978
catalufa, Priacanthus	1238
Catalufas	1236
Deep-water	977
Cataphracti	781
Cataphractus schoneveldii	2067
cataphractus, Acipenser	107
Agonus	2065, 2067
Aspidophorus	2067
Cottus	2053, 2066

	Page.
cataphractus, Gasterosteus.........	749
aculeatus	750
Scaphirhynchus.....	107
cataractæ, Ceratichthys.............	306
Gobio...................	306
Rhinichthys.............	306
dulcis.....	306
cataractus, Leucosomus.............	221
catena, Bassozetus.................	2509
catenata, Echidna...................	403
Muræna...................	403
Poecilia...................	648
Xenisma.................	648
catenatus, Fundulus...............	648
Gymnothorax...........	403
catenula, Murænophis..............	403
catesbæi, Scarus.....................	1638
Sparisoma................	1638
catesbei, Pomotis...................	1010
catesby, Scarus.....................	1638
catesbyi, Sparisoma.................	1638
Catfish of the Lakes.................	137
Catfish, Bermuda...................	882
Sea..................... 118, 119, 128
Small...................... 140, 141
Catfishes.............................	114, 115
Gaff-topsail..............	116
catharinæ, Pristipoma.............	1323
cathetoplateo, Ostracion oblongus..	1728
Cathetostoma albigutta.............	2313
Cathorops......................... 133, 2788
gulosus............... 133, 2788
hypophthalmus........ 133, 2788
Catochænum.......................	1373
Catonotus........................	1066
fasciatus.................	1098
flabellatus...............	1098
kennicotti...............	1098
lineatus.................	1099
Catostomidæ.......................	161
Catostominæ.......................	162
Catostomus........................ 173, 174
alticolus..............	179
anisurus..............	190
aræopus...............	172
ardens...............	179
aureolus............... 193, 196
aurora...............	176
bernardini.............	178
bostonensis............	179
bubalus...............	165
caprinus...............	168
carpio............... 166, 190
catostomus........... 176, 2792
chloropteron..........	179
clarki.................	172
	Page.
Catostomus commersonii..........	178
communis..............	179
congestus..............	192
cypho...................	184
discobolus........ 172, 175, 2791
duquesni...............	193
duquesnii..............	198
elongatus..............	169
erythrurus.............	193
fasciatus..............	187
fasciolaris.............	186
fecundus............. 180, 2794
flexuosus..............	179
forsterianus...........	176
generosus..............	170
gibbosus..............	186
gila.....................	180
gracilis.................	179
griseus............... 175, 2791
guzmaniensis.........	171
hudsonius..............	176
insignis.................	180
labiatus.............. 177, 2792
lactarius..............	175
latipinnis............. 174, 2790
lesueurii................	195
longirostris............	176
longirostrum..........	176
macrocheilus..........	178
macrolepidotus........	194
maculosus.............	181
megastomus..........	181
melanops..............	187
melanotus........ 206, 218, 322
nanomyzon.............	177
nebulifer..............	171
nebuliferus............	171
nigricans..............	181
etowanus....	181
occidentalis........... 178, 2793
oneida.................	193
pallidus...............	179
planiceps.............	181
plebeius...............	171
pocatello.............	175
reticulatus.............	179
retropinnis........... 175, 2791
rex.................... 177, 2792
rhothœcus.............	181
rimiculus..............	2792
snyderi...............	2792
sucklii...............	179
taboensis..............	177
teres...................	179
texanus...............	192
tsiltcoosensis..........	2793

	Page.
Catostomus tuberculatus	186
utawana	179
velifer	167
xanthopus	181
catostomus, Catostomus	176
Cyprinus	176
Phenacobius	304
Cats, Channel	133
Mud	142
Stone	143
Catulus	23, 24
brunneus	24
cephalus	25
retifer	25
uter	25, 2745
xaniurus	24
catulus, Ameiurus nebulosus	141
Evorthodus	2218
Gobius	2218
Pimelodus	141
catus, Ameiurus	138
Amiurus	141
Epinephelus	1159
Pimelodus	140
Serranus	1159
Silurus	138
cauda bifurca, Chætodon	1562
Gobius longissima acuminata	2230
Perca nigra	1303
Turdus convexa	1145
caudacuta, Motella	2560
Rhinonemus	2560
caudafurcatus, Amiurus	135
Pimelodus	135
caudalis, Halichæres	1600
Iridio	1599
Julis	1599
Platyglossus	1599, 1600
Pomacentrus	1556
caudanotatus, Mesoprion	1262
caudata, Lamna	37
caudatus, Lepidopus	887, 2844
Trichiurus	887
caudicinctus, Prilonotus (Ancbisomus)	1742
Tetrodon	1742
caudicula, Conger	355
Leptocephalus	355
caudilimbatus, Conger	355
Echelus	355
Leptocephalus	355
caudimacula, Diplodus	1363
Hæmulon	1299, 1302, 1309
Sargus	1363
caudispinosum, Macrostoma	556, 2826
caudispinosus, Notoscopelus	556

	Page.
caudispinosus, Scopelus	556
Caularchus	2327
mæandricus	2328
caulias, Sigmistes	2863
Caulolatilinæ	2275
Caulolatilus	2276
affinis	2277
anomalus	2277
chrysops	2277
cyanops	2278
microps	2277
princeps	2276, 2277
Caulolepis	838
longidens	839, 2842
Caulophryne	2734
jordani	2735
Caulophryninæ	2728
Caulopus	594, 596
caulopus, Homesthes	2394
caurinus, Cyprinus (Leuciscus)	220
Leucosomus	220
Mylocheilus	219, 220
Sebastes	1821
Sebastodes	1821
Cavalla	875
cavalla, Cybium	876
Scomberomorus	875
Cavally	920
cavernosus, Bathygadus	2581
Hymenocephalus	2580
cavifrons, Ambloplites rupestris	990
Diagramma	1343
Hemitripterus	2023
Icelinus	1892
Tarandichthys	1891
Caxis	1259
caxis, Lutjanus	1260
Mesoprion	1260
Sparus	1259
cayanus, Pristigaster	438
cayennense, Siphostoma	772
cayennensis, Citharichthys	2686
Corythroichthys	2838
Lutjanus	1404
Otolithus	1404, 1411
Syngnathus	773
Trachinotus	945
Cayennia	2265
guichenoti	2265
cayennsis, Vomer	934
cayorum, Corythroichthys	2838
Ogilbia	2503
cayuga, Eucalia inconstans	744
Notropis	260
atrocaudalis	260
Cazon de Playa	36

	Page.
Cebedichthyinæ	2349
Cebedichthys	2426, 2869
cristagalli	2427
violaceus	2427
Céfalo	811
celatus, Batrachus	2316
Cenisophius	243
Centaurus	1753
boops	1755
Centrarchidæ	984
Centrarchinæ	985
Centrarchus	988
æneus	990
fasciatus	1012
hexacanthus	987
interruptus	991
macropterus	988
maculosus	991
pentacanthus	990
pomotis	989
tetracanthus	1540
viridis	992
centrarchus, Cichlasoma	1526
Heros	1526
Centridermichthys analis	2013
asper	1944
bicornis	1913
globiceps	2017
gulosus	1945
maculosus	2014
uncinatus	1906
centrina, Callorhynchus	95
Centriscus	759
scolopax	759
Centroblennius	2435
nubilus	2438
centrognathus, Zanclus	1688
Centrolabrus	1575
exoletus	1576
Centrolophidæ	962
Centrolophinæ	962
Centrolophus	962
liparis	963
morio	963
niger	963
pompilus	963
Centronotus	900, 2414
alectrolophus	2422
apus	2430
argenteus	899
conductor	900
cristagalli	2423
dolichogaster	2417
dybowskii	2431
fasciatus	2418
gardenii	948

	Page.
Centronotus gunelliformis	2421
gunnellus	2419
islandicus	2438
lætus	2420
nebulosus	2414
(Opisthocentrus)	
quinquemaculatus	2430
pictus	2416
roseus	2420
spinosus	948
taczanowskii	2416
Centrophorus cœlolepis	55
centropleura, Cottus semiscabra	1945
Centropomidæ	1116
Centropomus	1117
affinis	1124, 2853
alburnus	1475
albus	1135
appendiculatus	1119
armatus	1123
auratus	1107
brevis	1125
cuvieri	1121
ensiferus	1125, 2853
grandoculatus	1120
luteus	1024
medius	1120
mexicanus	1121
nigrescens	1119
parallelus	1122
pectinatus	1122
pedimacula	1119
robalito	1123
rubens	1107
scaber	1124
undecimalis	1118
undecimradiatus	1119
unionensis	1122
viridis	1118
Centropristes	1198
annularis	1214
atrarius	1200
aurorubens	1278
ayresi	1205
bivittatus	1205
brasiliensis	1221
dispilurus	1220
fascicularis	1208
fusculus	1211
luciopercanus	1216
macrophthalmus	1281
macropoma	1206
merus	1162
nigricans	1200
oculatus	1283
ocyurus	1200

	Page.		Page.
Centropristes philadelphicus	1201	cephalus, Semotilus	222
phœbe	1212	Cepolophis	2477
præstigiator	1214	Cepphus	2540
psittacinus	1213	cerapalus, Opsanus	2316
radialis	1205	cerasinus, Gobiesox	2336
radians	1208	Hypsilepis cornutus	283
rufus	1199	Notropis	283
striatus	1199	Ceratacanthus	1717, 1718, 2860
subligarius	1219	aurantiacus	1718
tabacarius	1215	punctatus	2860
tridens	1202	schœpfii	2860
trifurca	1202	scriptus	2860
trifurcus	1202	Ceratias	2729
Centropyge	1682	carunculatus	2732
Centroscyllium	56	couesii	2732
fabricii	56	holbolli	2729
Centroscymnus	54	uranoscopus	2730
cœlolepis	55	Ceratichthys	252, 314
centrura, Dasibatis	83	amblops	321
Dasyatis	83	biguttatus	323
Raja	83	cataractæ	306
Centrurophis	381	cumingii	318
Centridermichthys armatus	2012	cyclotis	323
cepediana, Megalops	416	dissimilis	319
cepedianum, Dorosoma	416	gelidus	317
exile	416	hyalinus	321
cepedianus, Chatoessus	416	hypsinotus	320
Priacanthus	1238	labrosus	319
Cephalacandia	2867	leptocephalus	323
Cephalacanthidæ	2182	lucens	321
Cephalacanthus	2183, 2367	micropogon	323
volitans	2183	monacus	318
Cephaleutherus	89, 2750	nubilus	312
Cephalocassis	119	physignathus	326
Cephalogobius	2210	plumbeus	324
cephaloides, Cottus	2008	prosthemius	324
Cephalopholis	1143	rubrifrons	320
Cephaloptera	92	sallæi	212
johni	93	squamilentus	323
manta	93	sterletus	316
vampyrus	93	stigmaticus	323
Cephalopterus	92	symmetricus	246
giorna	93	ventricosus	309
hypostomus	92	vigilax	253
Cephaloscyllium	23, 25	zanemus	319
Cephalus	228, 1753	Ceratiidæ	2727
brevis	1754	Ceratiinæ	2728
cocherani	1756	Ceratius shufeldti	2731
elongatus	1756	Ceratobatis	2756
orthogoriscus	1754	robertsii	2756
pallasianus	1754	Ceratocottus	1939
varius	1756	diceraus	1940, 1941
cephalus, Catulus	25	lucasi	1940
Gobiesox	2332	Ceratoptera	92
Mugil	811	vampyrus	93
Paraliparis	2141	Ceratoscopelus	557

	Page.
Ceratoscopelus madeirensis	557
cercostigma, Cyprinella	275
Minnilus	275
Notropis	274
Cerdale	2448
ionthas	2449
cerdale, Carcharhinus	2746, 2747
Scytalina	2454
Cerdalidæ	2448
Cerna	1148
acutirostris	1181
gigas	1154
macrogenis	1181
nebulosa	1181
sicana	1162
Cernier	1139
cernium, Polyprion	1139
Cero	875
cervinum, Lepophidium	2484, 2485
Moxostoma	197
cervinus, Ptychostomus	197
Teretulus	197
cervus, Synanceia	1941
Cestracion	43
francisci	21
pantherinus	21
quoyi	21
Cestraciont Sharks	19
Cestreus acoupa	1404
albus	1411
carolinensis	1409
leiarchus	1415
microlepidotus	1415
nebulosus	1409
nobilis	1413
nothus	1407
obliquatus	1405
othonopterus	1405
parvipinnis	1410
phoxocephalus	1414
regalis	1407
thalassinus	1408
reticulatus	1409
squamipinnis	1404
stolzmanni	1412
xanthulum	1411
Cestrorhinus	43
cetaceus, Squalus	51
Cetengraulis	450
brevis	450
edentulus	450
engymen	2815
mysticetus	450
Cetomimidæ	549
Cetomimus	549
gillii	549

	Page.
Cetominus storeri	550
Cetorhinidæ	50
Cetorhinus	51
maximus	51
shavianus	51
ceuthœcum, Gobisoma	2261
ceuthœcus, Barbulifer	2260
Chænichthyidæ	2293
Chænobryttus	991
antistius	992
gulosus	992
Chænomugil	816
proboscideus	816
Chænopsinæ	2347
Chænopsis	2403
ocellatus	2403
Chærojulis	1587
arangoi	1597
bivattatus	1597
cinctus	1593
crotaphus	1598
cyanosigma	1591
grandisquamis	1597
humeralis	1597
internasalis	1594
maculipinna	1595
radiatus	1591
ruptus	1593
Chætodipterus	1667
faber	1668
zonatus	1668
Chætodon	1672, 1673, 1677, 1679, 2859
alepidotus	966
arcuatus	1680
atæniatus	1676
aureus	1680
aya	1675
bimaculatus	1674
bricei	1678
canescens	1688
capistratus	1677
cauda bifurca	1562
chirurgus	1692
ciliaris	1685
cornutus	1688
couaga	1691
cyprinaceus	1388
faber	1668
glaucus	941
gracilis	1675
humeralis	1674
lanceolatus	1490
littoricola	1680
lutescens	1680
macrolepidotus	1677
maculocinctus	1674

	Page.
Chætodon marginatus	1562
mauritii	1562
nigrirostris	1673, 1674
ocellatus	1674
oviformis	1668
parræ	1685
paru	1681
plumieri	1668
rhomboides	942, 2848
sargoides	1562
saxatilis	1562
sedentarius	1675
squamulosus	1685
striatus	1677
tricolor	1684
triostegus	1691
unicolor	1676
zebra	1691
chætodon, Apomotis	995
Cottus	2316
Mesogonistius	995
Pomotis	995
Chætodontidæ	1669, 1670
Chætodontinæ	2860
Chætodontops	1672, 1673
Chætostomus	160
aspidolepis	159
fischeri	160
gaucharote	159
chagresensis, Chalcinopsis	337
chagresi, Ancistrus	160
Pimelodella	154
Pimelodus	154
chalceum, Pristopoma	1338
chalceus, Orthopristis	1337
Chalcinopsis	337
chagresensis	337
dentex	337
striatulus	337
chalcogramma, Theragra	2535
chalcogrammus, Gadus	2536
Pollachius	2536, 2537
chalinius, Epinephelus	1181
Chalinura	2576
filifera	2577
serrula	2576
simula	2578
Chalinurus	2576
Chalisoma	1699
velata	1703
challengeri, Macdonaldia	617
Notacanthus	618
chalybæus, Hybopsis	288
Minnilus	288
Notropis	288
chalybeius, Chlorophthalmus	542

	Page.
chalybeius, Hyphalonedrus	542
chamæleonticeps, Lopholatilus	2278
chamberlaini, Notropis	2800
Chani	414
Chanidæ	414
Channel Bass	1453
Cat	134
Cat of the Potomac	138
Cats	133
Channomuræna	404
cubensis	404
vittata	404
Chanos	414
arabicus	415
chanos	414, 2807
chloropterus	415
cyprinella	415
indicus	415
mento	415
nuchalis	415
orientalis	415
salmoneus	415
chanos, Chanos	414
Mugil	415
chantenay, Raia	71
chaperi, Acanthoclinus	2375
Paraclinus	2374
Chapin	1722, 1723
Chapinus	1721, 1722, 1723
Chappaul	224
Characinidæ	331
Characininæ	332
Characius	331
Characodon	667
bilineatus	668, 2831
eiseni	2831
ferrugineus	669
furcidens	669
garmani	2832
lateralis	668, 2831
luitpoldii	2832
variatus	669, 2831
Charr, European	508
Greenland	508, 510
Long-finned	509
Oregon	507
Charrs	506
charybdis, Lepomis	992
Chasmistes	182
brevirostris	183, 199
copei	2795
cujus	183, 2794
fecundus	2794
liorus	183
luxatus	183, 2794
stomias	2794

	Page.
Chasmodes	2391
boscianus	2394
bosquianus	2394
jenkinsi	2391, 2392
novemlineatus	2393
quadrifasciatus	2392
saburræ	2392
Chatoessus	415
cepedianus	416
ellipticus	416
eumorphus	433
mexicanus	416
petenensis	417
signifer	433
Chauffe-soleil	1548
Chauffe-soleils	1545
Chauliodontidæ	578
Chauliodontinæ	578
Chauliodus	584
macouni	585
richardsoni	587
schneideri	585
setinotus	585
sloanei	585
Chaunax	2726
fimbriatus	2726
nuttingii	2726, 2727
pictus	2726
Cheilichthys	1729; 1730; 1734
asellus	1740
psittacus	1740
pachygaster	1738
Cheilodipteridæ	1105
Cheilodipterinæ	1105
Cheilodipterus	946, 1112
acoupa	1404
affinis	1113
chrysopterus	1324
heptacanthus	947
Cheilonemus	220
pulchellus	222
Cheilotrema	1455, 1456
Cheiragonus	2038
gradiens	2041
Chelidonichthys	2175, 2867
pictipinnis	2175, 2867
Chelmo pelta	1671
Chelmon aculeatus	1671
chemnitzii, Notacanthus	614
cheneyi, Cottogaster	2851
Cheonda	228, 230, 236
cæruleus	232
cooperi	236
modesta	234
Cherna	1157
americana	1160

	Page.
cherna, Anthias	1157
Cherna Criolla	1157
de Vivero	1160
Cherno de lo Alto	1151
chesteri, Phycis	2556
Urophycis	2556
Chevalier, Ombre	508
chevola, Gallichthys	932
Chi	209
Chiasmodon	2291
niger	2291, 2292
Chiasmodontidæ	2291
Chiasmodus	2291
Chicarro	911
chickasavensis, Luxilus	275
Chicolar	879
Chigh	209
chihuahua, Notropis	265
Chilara	2488
taylori	2489
chilensis, Brama	960
Caranx	927
Exocœtus	730
Pelamys	873
Sarda	872
chiliticus, Hybopsis	287
Notropis	287
Chilodipterus	1112
chiloensis, Agonopsis	2069
Agonus	2069
Aspidorphorus	2069
Chilomycterus	1747, 1748, 1750
antennatus	1750
antillarum	1749
atinga	1750
californiensis	1751
fuliginosus	1749
geometricus	1749
puncticulatus	1750
reticulatus	1751
spinosus	1749
Chilorhinus	372
suensonii	372
Chimæra	94
abbreviata	95
affinis	95
argentea	95
borealis	95
cristata	95
mediterranea	95
monstrosa	94
plumbea	95
Chimæras	93
Chimæridæ	93
Chimærinæ	94
Chimæroidei	93

	Page.
Chimæroids	93
Chino, Escolar	1114, 1284
Mojarra	1377
Chinook Salmon	479
chiostictus, Entomacrodus	2398
Salarias	2398
chiquita, Aboma	2241
Gobius	2241
Chirivita	1679
Chirivitas	1679
Chiro	410
Chirocentrodon	435
tæniatus	435
Chirolophinæ	2347
Chirolophis polyactocephalus	2409
Chirolophus japonicus	2409
Chironectes	2717
arcticus	2717
lævigatus	2717
mentzelii	2724
multiocellatus	2725
pictus	2717
principis	2719
scaber	2723
sonntagii	2717
tenebrosus	2719
tigris	2723
tumidus	2717
Chiropsis	1866
constellatus	1868
guttatus	1869
nebulosus	1872
Chirostoma	792, 2839, 2840
bartoni	793
brasiliensis	794
beryllinum	798
estor	792, 2839
grandocule	2839
humboldtianum	793, 2839
jordani	793
peninsulæ	797
sicculum	806
vagrans	795
chirurgus, Acanthurus	1692
Chætodon	1692
Chirus	1866
balias	1873
constellatus	1869
decagrammus	1869
guttatus	1868
hexagrammus	1872
maculoseriatus	1868
monopterygius	1866
nebulosus	1872
octogrammus	1870
ordinatus	1870

	Page.
Chirus pictus	1873
trigrammus	1872
chirus, Xiphister	2424
Xiphistes	2424
Chisel-mouths	207, 208
chisoyensis, Pœcilia	693
Chitonotus	1889
Chitonotus megacephalus	1891
pugetensis	1890, 1891
chittendeni, Cyclopsetta	2676
Chivo	860
chlamydes, Bollmannia	2238
Chlamydoselachidæ	16
Chlamydoselachus	16
anguineus	16
chlevastes, Gymnothorax	399
Lycodontis	398
Sidera	399
Chlopsis	364
equatorialis	364
chlora, Oliola	263
Chlorichthys	1605; 2859
bifasciatus	1609, 1610
brasiliensis	1591
grammaticus	1610
lucasanus	1607
nitidissimus	1608
nitidus	1608
socorroensis	1607
steindachneri	1609
virens	1610
chloris, Pseudoscaris	1648
Pseudoscarus	1654
Scarus	1637, 1640
Scomber	938
chloristia, Oliola	278
Codoma	278
chloristius, Notropis niveus	278
chlorocephalus, Hybopsis	286
Minnilus	286
Notropis	286
chloropteron, Catostomus	179
chloropterum, Plectropoma	1165
chloropterus, Chanos	415
Prospinus	1165
Chlorophthalmus	541
agassizii	541
chalybeius	542
Chloroscombrinæ	897
Chloroscombrus	937
caribbæus	938
chloris	2847
chrysurus	938, 2847
ectenurus	2847
orqueta	937
stirurus	938

	Page.
chlorosoma, Boleosoma	1060
Vaillantia	1060
chlorostictus, Sebastichthys	1812
Sebastodes	1811
chlorostomus, Trisotropis	1179
chlorurum, Plectropoma	1193
Chlorurus	1642
chlorurus, Hypoplectrus	1193
unicolor..	1193
Serranus	1193
Chœnopsetta	2624
dentata	2630
oblonga	2633
ocellaris	2630
Chœroichthys	773
chœrostomus, Engraulis	444
Stolphorus	444
Chœtopterus	1279
Chogset	1577
chogset, Ctenolabrus	1577
Labrus	1577
fulve	1577
Choice, Sailor's	1297, 1338
choirorynchus, Dajaus	2841
Choker, Hog	2700
Chologaster	703
agassizii	704
avitus	704
cornutus	703
papilliferus	704
Chondroganoidea	98, 100
Chondrostei	102
Chondrostoma gardoneum	251
pullum	206
Chondrostominæ	202
Chonophorus	2234
bucculentus	2236
Chonophorus flavus	2235
taiasica	2237
Chopa	1387
Amarilla	1386
Blanca	1387
Chopa Spina	1357, 1358
Chopas	1384
chordatus, Saccopharynx	406
Stylephorus	2601
Choregon	517
Chorinemus altus	899
inornatus	899
occidentalis	898
palometa	899
quiebra	899
saliens	899
saltans	899
Chorististium	1136
rubrum	1136

	Page.
Chornia Ryba	621
Chorophthalmus agassizii	542
truculentus	542
chouicha, Oncorhynchus	480
Chriodorus	719
atherinoides	719
Chriolax	2148
Chriolepis	2205
minutillus	2205
Chriomitra	873
concolor	874
Chriope	254, 255, 258
Chromides	781, 1511
Chromis	946, 1545
atrilobatus	1546
cyaneus	1547
enchrysurus	1548
epicurorum	947
fenestrata	1518
fusco-maculatus	1540
insolatus	1548
marginatus	1546
multilineatus	1547
nebulifer	1524
punctipinnis	1548
chromis, Diabasis	1299
Hæmulon	1299
Labrus	1483
Chronophorus mexicanus	2237
Chrosomus	209
dakotensis	210
eos	210
erythrogaster	209
eos	210
oreas	211
pyrrhogaster	210
chrosomus, Hybopsis	288
Hydrophlox	288
Minnilus	288
Notropis	288
chrysargyreum, Hæmulon	1308
chrysargyreus, Brachygenys	1307
chryseus, Rhinoberyx	847
chrysitis, Dionda	214
chrysocephalus, Luxilus	282
chrysochloris, Pomolobus	425
Clupea	425
chrysogaster, Agosia	313
chrysoleuca, Bairdiella	1438
Corvina	1439
Sciæna	1439
chrysoleucus, Notemigonus	250
chrysomelanurus, Sparus	1157
chrysomelas, Sebastichthys	1826
Sebastodes	1825, 1826
Chrysophrys calamus	1350

	Page.
Chrysophrys cyanoptera	1354
taurina	1354
Chrysops aculeatus	1347
chrysops, Caulolatilus	2277
Latilus	2278
Ophichthys	385
Ophisurus	385
Perca	1132
Roccus	1132
Sparus	1346
Stenotomus	1346
chrysopsis, Hyodon	413
chrysoptera, Perca	1339
chrysopteron, Hæmulon	1309
chysópterum, Hæmulon	1309
Sparisoma	1636, 1637
chrysopterus, Cheilodipterus	1324
Diabasis	1309
Leuciscus	221
Orthopristis	1338
Scarus	1637
chrysos, Carangus	921
Chrysotosus	954
chrysotus, Fundulus	655
Haplochilus	656
Zygoneetes	656
chrysura, Bairdiella	1433
Sciæna	1434
chrysurus, Chloroscombrus	938
Coryphæna	952
Dipteron	1433
Glyphidodon	1567
Lutjanus	1276
Mesoprion	1276
Micropteryx	938
Microspathodon	1567
Ocyurus	1275
Scomber	938
Sparus	1276
chrysus, Caranx	921
chrysypa, Anguilla	348
Chub	1387
Bermuda	1387
Columbia	219
Flat-headed	326
Great	232
Indian	322
Nigger	327
River	322
Sacramento	231
Silver	221, 320
Steelbacked	205
Tahoe	2798
Chub Mackerel	866
Chub of the Rio Grande	233
Suckers	184

	Page.
Chub of Utah Lake	232
Chuckle-headed Cat	134
Chuss	2555
chuss, Phycis	2555
Urophycis	2555
Chylomycterus schœpfi	1748
cibaria, Lampetra	13
cibarius, Ammocœtes	13
Cibi Amarillo	919
Mancho	919
cibi, Carangoides	920
Caranx	920
cicatricosus, Balistes	1709
Pleuronectes	2649
Xanthichthys	1709
Cichla ænea	990
fasciata	1012
floridana	1012
minima	1012
ohioensis	1012
storeria	987
Cichlasoma	1514, 1515
anguliferum	1517
balteatum	1521
bartoni	1515
bifasciatum	1521
centrarchus	1526
deppii	1524
fenestratum	1518
goodmanni	1516
helleri	1521
intermedium	1517
lentiginosum	1524
longimanus	1520
macracanthum	1518
margaritiferum	1519
melanopogon	1523
melanurum	1523
montezuma	1518
multispinosum	1525
nebuliferum	1524
nigrofasciatum	1525
parma	1519
rectangulare	1515
rostratum	1522
sieboldii	1516
spilurum	1520
Cichlidæ	1512
Cichlids	1512
Ciego, Pez	2501
Cigar-fish	907
cigonella, Belone	713
ciliaris, Alectis	931
Angelichthys	1684, 1685
Balistes	1702
Chætodon	1685

	Page.
ciliaris, Holacanthus	1685
Pomacanthus	1685, 1686
Zeus	932
Ciliata argentata	2559
ciliatus, Balistes	1715
Blennius	2457
Epinephelus	1784
Monacanthus	1714, 1715
Petromyzon	12
Sebastodes	1783
cimbria, Motella	2560
cimbricus, Enchelyopus	2561
cimbrius, Enchelyopus	2560
Gadus	2560
Onos	2561
Rhinonemus	2561
cincinnati, Acipenser	106
cinctus, Chærojulis	1593
Julis	1593
cinerea, Alutera	1720
Amia	113
Etheostoma	1078
cinereum, Aulostoma	755
Etheostoma	1078
Xystæma	1372
cinereus, Aulostomus	755
Gerres	1370
Marcrourus	2585
Microspathodon	1570
dorsalis	1570
Mugil	1373
Nothonotus	1078
Turdus peltatus	1373
cingulatus, Fundulus	656
Pomacanthus	1680
Zygonectes	655, 656
circumnotatus, Scarus	1641
cirratum, Ginglymostoma	26
cirratus, Milvus	2183
Phycis	2554
Squalus	26
Urophycis	2553
Cirrhisomus	1729
Cirrhites	1491
betaurus	1492
rivulatus	1491
Cirrhitichthys	1491
rivulatus	1492
Cirrhitidæ	1490
Cirrhitoid Fishes	1490
Cirrhitoide	781
Cirrhitoidei	1490
Cirrhitoids	1490
cirrhosum, Lepisoma	2362
cirrhosus, Blepsias	2018
Trachinus	2019
	Page.
Cirrimens	1469
cirris, Cottus plurimis	2066
Cirrostomes	2
Cisco	468
Moon-eye	469
Cisco of Lake Michigan	469
Tippecanoe	469
Ciscoes	467
cismontanus, Coregonus williamsoni	463
Citharædus	1672
Citharichthys	2678, 2682
æthalion	2673
aramaca	2672
arctifrons	2683
cayennensis	2686
dinoceros	2682
fragilis	2680
gilberti	2686
guatemalensis	2686
latifrons	2674
macrops	2684
microstomus	2688
ocellatus	2673
ovalis	2674
panamensis	2677
platophrys	2683
pœtulus	2672
sordidus	2679
spilopterus	2685, 2686
stigmæus	2681
sumichrasti	2686
uhleri	2684
unicornis	2683
ventralis	2670
xanthostigmus	2680
Citharus	2614
platessoides	2615
citrinellus, Heros	1534
Citula	929
banksi	927
dorsalis	930
ciuciara, Echelus	356
civilis, Hybognathus	215
Clam Cracker	83
Clamagore	1652
clara, Ammocrypta pellucida	1063
Menidia	801
clarias, Pimelodus	155
Silurus	155
Claricola	1066, 1069, 1093
clarionensis, Holacanthus	1683
Myripristis	2842
clarionis, Xesurus	1695
clarki, Bogoslovius	2575
Catostomus	173
Fario	501

	Page.
clarki, Pantosteus	172
clarkii, Salmo	492, 2819
bouvireri	2819
gibbsii	2819
henshawi	2819
lewisi	2819
macdonaldi	2819
pleuriticus	2819
spilurus	2819
stomias	2819
tahoensis	2870
virginalis	2819
clarum, Etheostoma pellucidum ...	1063
clathrata, Atratoperca	1198
clathratus, Labrax	1198
Paralabrax	1197, 1198
Serranus	1198
claudalus, Hyodon	413
claviformis, Moxostoma	186
claviger, Cottus	1939
Enophrys	1938
clepsydra, Muræna	2805
Clepticinæ	1574
Clepticus	1586
genizara	1587
parræ	1586
clevelandi, Phoxinus	237
Glevelandia	2254
ios	2254
longipinnis	2255
rosæ	2255
Cling-fishes	2326, 2329
Clininæ	2344
Clinocottus	2012
analis	2012
Clinostomus	228, 230, 239
affinis	239
elongatus	240
funduloides	239
hydrophlox	238
margarita	241
montanus	238
pandora	234
phlegethontis	243
proriger	240
tænia	238
Clinus aculeatus	2433
bimaculatus	2358
canariensis	2362
capillatus	2362
delalandii	2359
evides	2353
gillii	2358
gobio	2365
hermineri	2362
lumpenus	2438

	Page.
Clinus macrocephalus	2364
maculatus	2433
medius	2435
mohri	2438
nebulosus	2438
nigripinnis	2370
nuchipinnis	2362
ocellatus	2357
ocellifer	2353
pectinifer	2362
philipii	2359
præcisus	2441
punctatus	2440
unimaculatus	2441
zonifer	2359
Oliola	252
alta	322
analostana	279
aurata	272
billingsiana	272
bubalina	273
cærulea	277
callisema	273
callistia	276
calliura	275
camura	280
chlora	263
chloristia	278
cobitis	305
deliciosa	272
euryopa	270
eurystoma	277
forbesi	272
formosa	271
fretensis	261
galactura	279
gibbosa	272
gunnisoni	273
hæmatura	218
hudsonia	269
hypseloptera	280
iris	272
jugalis	272
leonina	271
lepida	273
lineolata	263
longirostris	267
ludibunda	273
lutrensis	272
microstoma	264
missuriensis	262
montiregis	272
nigrotæniata	264
nivea	278
notata	274
nubila	215

	Page.
Oliola ornata	271
procne	264
pyrrhomelas	281
rubripinna	281
sallæi	212
saludana	270
sima	267
smithii	253
spectruncula	265
stigmatura	275
storeriana	270
straminea	262
suavis	272
taurocephala	253
trichroistia	276
tuditana	253
umbrosa	273
urostigma	275
velox	253
venusta	274
vigilax	253
vittata	258
vivax	253
whipplei	279
xænura	280
Clodalus	412
clodalus, Hiodon	413
Clupanodon	422
aureus	434
cæruleus	423
pseudohispanicus	423
Clupea	421
æstivalis	427
anchovia	429
apicalis	429
atherinoides	451
aurea	434
brasiliensis	411
cærulea	421
carolinensis	434
chrysochloris	425
elongata	421
esca	421
fasciata	426
halec	421
harengus	421, 422
heterura	416
hudsonia	269
humeralis	431
indigena	428
lamprotænia	419
latulus	422
leachi	422
libertatis	433
lineolata	422
macrocephala	411

	Page.
Clupea macropthalma	430
mattowacca	426
mediocri	426
megalops	426
membras	421
menhaden	434
minima	422
mirabilis	422
neglecta	434
pallasii	422
parvula	426
pseudoharengus	426
pseudohispanica	424
pusilla	426
sadina	420
sagax	423
stolifera	432
thrissa	432
thrissina	431
tyrannus	434
vernalis	426
villosa	521
virescens	426
vittata	421
Clupeidæ	417
clupeiformis, Coregonus	465, 469
Salmo	466
Clupeinæ	418
Clupeoidea	407
clupeoides, Engraulis	447
Stolephorus	447
clupeola, Harengula	429, 430
Sardinella	429
Clupeonia	428
Clypeocottus	1937
robustus	1938
Coal-fish	1862, 2534
coara, Guabi brasiliensibus	1305
Coast Range Trout	500
Coballito del Mar	776, 777
Cobbler	640, 641, 931
Cobessicontic Smelt	524
Cohia	948
Cobitis heteroclita	641
killifish	641
majalis	639
cobitis, Oliola	305
Leuciscus	305
Tiaroga	305
coccineus, Lycodes	2469
Scarus	1635
coccogenis, Hypsilepis	285
Leuciscus	285
Minnilus	285
Notropis	284
coccoi, Rhinoscopelus	568

	Page.		Page.
cocooi, Scopelus	569	cœruleum, Etheostoma spectabile..	1089
Stenobrachius	569	cœruleus, Astatichthys	1089
cocherani, Cephalus	1756	Pœcilichthys	1089
Cochinito	1694	cognata, Uranidea	1955
Cochino	1703	cognatus, Cottus	1954
Cochlognathus	251	Coho Salmon	480
biguttata	252	colias, Scomber	866
ornata	252	colii, Salmo	509
ornatus	252	colinus, Gadus	2535
Cocinera	918	Coliscus	217
Cocinero Dorado	921	parietalis	217
Cock and hen Paddle	2096	collapsum, Moxostoma	190
Cockeye Pilot	1555, 1561	collapsus, Ptychostomus	190
Cocuyo	1709	colletti, Careproctus	2131
Cod, Blue	1875	Collettia	567
Buffalo	1875	nocturna	567
Cultus	1875	rafinesquei	567
Green	2534	colliei, Hydrolagus	95
Wachna	2537	Colocephali	346, 388
Codfish, Alaska	2541	Cololabis	726
Common	2541	brevirostris	726
Greenland	2542	Colomesinæ	1727
Codfishes	2531, 2540	Colomesus	1740
Codling	2552, 2555	psittacus	1740
Codoma	254, 256, 270	colonus, Serranus	1222
cærulea	277	Colorada, Mautararia	2754
callisema	273	colorada, Lija	1713
callistia	276	Vieja	1639
chloristia	278	Colorado, Bagre	122
eurystoma	285	Pargo	1264, 1267, 1356
ornata	271	Perro	1583
pyrrhomelas	281	Pescado	1453
stigmatura	275	colorado, Lutianus	1268
trichroistia	276	Lutjanus	1268
vittata	258	Neomænis	1267
xænura	280	Colorado River Trout	496
Codorniz	1467	coloratum, Aulostoma	755
Cods, Cultus	1875	coluber, Gempylus	884
Cœcilophis	381	Columbia	784
Cœcula bascanium	280	Chub	219
scuticaris	379	River Sucker	178
teres	379	Trout	492
cœcus, Gastrobranchus	8	Salmon	479
cœlestinus, Bagrus	125	Columbia transmontana	784
Pseudoscarus	1655, 1656	columbianus, Pantosteus	172
Scarus	1656	Vomer	934
Coelho	882	comatus, Cypselurus	736
cœlolepis, Centrophorus	55	Exocœtus	736
Cœlorhynchus	2587	comes, Roccus	1407
carminatus	2588, 2589	commersoni, Carcharhinus	38
carribæus	2589	commersonii, Catostomus	178
occa	2587	Cyprinus	179
scaphopsis	2590, 2591	Fistularia	758
cœnosa, Parophrys	2639	Common Alligator Fish	2061
cœnosus, Pleuronichthys	2638, 2639	American Sea-Horse	777
cœruleum, Etheostoma	1088	Angler	2713

	Page.
Common Atlantic Salmon..........	486
Buffalo Fish...............	163
Bullhead	140
Burr-fish	1748
Cobbler...................	641
Codfish	2541
Dolphin	952
Eastern Pickerel	627
Stickleback......	748
Flatfish ....,....	2647
Gar Pike	109
Grunt	1304
Gurnard..................	2156
Half-beak	721
Herring	421
Killifish	640
Mackerel.................	865
Mullet....................	811
Pámpano	944
Pike......................	628
Pipefish	770
Rat-tail	2583
Red Horse...............	192
Rock Bass...............	990
Sawfish...................	60
Scup	1346
Shad	427
Skate.....................	68
of California........	73
Spotted Moray	395
Sting Ray	83
Sturgeon	105
Sucker	178
Sunfish	1009
Surf-fish.................	1504
Surgeon	1691
Swordfish	894
Trunk-fish	1723
Weakfish.................	1407
Whitefish	465
communis, Catostomus.............	179
Leucosomus............	326
Liparis	2118
Platygobio	326
Pogonichthys	326
complanata, Cyprinella.............	272
Moniana	272
compressa, Lota....................	2551
compressus, Bassozetus	2508
Bathynectes...........	2509
Bathyonus	2509
Engraulis	447
Gadus................	2551
Nauclerus	900
Rutilus................	282, 296
Stolephorus	447

	Page.
concatenatus, Ostracion	1723
Conchognathus	349
grimaldii	349
concinnus, Amblodon	1484
Gasterosteus.............	745
coucolor, Ammocœtes	11
Chriomitra...............	874
Euschistodus	1559
Ichthyomyzon	11
Lycodes..................	2463
Nexilarius	1559
Petromyzon...............	11
Scomberomorus	873
Thyrsoidea..............	396
Condenado, Ronco...................	1306
conductor, Centronotus	900
Conejo	596, 882
Coney	1141
confertus, Hyborhynchus	217, 218
Pimephales promelas....	217
confinis, Pimelodus.................	141
Salmo	505
confluentus, Fundulus	650
Salmo..................	480
conformis, Lavinia	231
Leuciscus..............	231
Squalius.................	231
Tigoma	231
cougener, Paru brasiliense.........	966
Conger	353
analis	356
brasiliensis..............	360
caudicaula	355
caudilimbatus..............	355
esculentus	355
impressus..................	356
limbatus	360
macrops....................	355
microstomus	356
mordax	387
niger.......................	355
occidentalis	355
opisthophthalmus..........	356
orbignyanus	355
rubescens	355
verreauxi	355
verus	355
vulgaris....................	355
Conger Eel of California	395
Eels.........·...........	352, 354
conger, Leptocephalus	354
Muræna	354
Congermuræna	355
balearica...........	356
flava..............	357, 2801
macrura	356

Page.

Congermuræna mellissii 356
    nitens.............. 357
    prorigera 357
congestum, Moxostoma 192
congestus, Catostomus 192
Congresox 359
Cougro Barhoso 155
Congros Barbosos 154
Congrus............................ 353, 381
    curvidens................. 360
    leucophæus 355
coniceps, Murænesox 359
conico, Ostracion oblongus......... 1745
coniferum, Oreosoma 1663
Conocara........................... 456
    macdonaldi............... 457
    macroptera.............. 457, 458
conocephala, Gila 219
conocephalus, Mylopharodon 219
Conodon 1324
    antillanus 1324
    nobilis..................... 1324
    pacifici 1316
    plumieri 1324
    serrifer 1324
Conorhychus........................ 411
conorynchus, Albula 411
conspersa, Muræna................. 397
    Tigoma................. 234
conspersus, Gymnothorax.......... 397
    Lycodontis 397
    Serranus............... 1156
    Squalins 234
Constantino de las Aletas Prietas.. 1119
constellatus, Chiropsis 1868
    Platophrys 2663
    Sebastichthys 1807
    Sebastodes............ 1806
consuetus, Salmo 479
continuum, Hæmulon 1297
contractus, Rhinogobius 2236
contrainii, Tylosurus.............. 717
conus, Moxostoma.................. 196
    Ptychostomus 196
convexa Turdus cauda 1145
convexifrons, Pomotis............. 1003
Cony, Horny 1715
Cook, Rock 1575, 1576
cookianus, Aphododerus 787
cooperi, Cheonda 236
    Leuciscus.................. 236
    Metoponopus 2680
    Raia 73
    Salmo 483
    Squalius 236
copei, Acipenser 106

Page.

copei, Aleposomus 459
    Chasmistes 2795
    Cottus 1968
    Paraliparis 2143
    Squalius..................... 236
Copelandellus 1100
    quiescens 1100
copelandi, Boleosoma............... 1046
    Cottogaster............. 1045
    Rheocrypta.............. 1046
Copelandia........................ 992
    eriarcha 994
copii, Leuciscus 293
Copper-nosed Bream 1005
Corallicola 2369
corallina, Narcine brasiliensis...... 78
corallinum, Cryptotrema 2366
corallinus, Antennarius 2725
Corbineta 1435
Cordylus 865
Coregoni 461
Coregoninæ 461
coregonoides, Paralepis............. 602
Coregonus...................... 461, 462, 465
    albus 466
    angusticeps.............. 466
    clupeiformis 465, 469
    couesii................... 463
    coulterii 462
    harengus 469
    hoyi 468, 470
    kennicotti 464
    labradoricus 466
    latior 466
    lucidus 471
    merckii................... 470
    nelsonii 466
    neohantoniensis 466
    nigripinnis 472
    novæ-angliæ 465
    osmeriformis 468
    otsego 466
    prognathus 472
    quadrilateralis.......... 465
    richardsonii 465, 2816
    ruber 538
    sapidissimus 466
    signifer.................. 518
    thymalloides............. 518
    tullibee 473
    bisselli 473
    williamsoni.............. 463
    cismontanus 463
coregonus, Moxostoma.............. 191
    Ptychostomus.......... 191
coretta, Thynnus 870

	Page.
coriaceus, Eleutheractis	1233
Rypticus	1233
corinus, Hexanchus	18
cormura, Thyrsoidea	394
Cornet-fishes	755
Corneta	757
cornifer, Achirus	2698
cornubica, Lamna	49, 2749
cornubicus, Squalus	49
cornubiensis, Lepadogaster	2108
Pimelepterus	964
Rhombus lævis	2654
cornutus, Anoplogaster	840
Chætodon	1688
Cbologaster	703
Cyclichthys	1749
Cyprinus	282
Holæanthus	1685
Hypsilepis	283
cerasinus	283
cyaneus	283
gibbus	283
Leuciscus	283
Minnilus	283
Notropis	281
cyaneus	283
frontalis	283
Silurus	759
Zanclus	1687, 1688
coro, Pristipoma	1324
Sciæna	1324
coroides, Umbrina	1466
Coronado	903
coronata, Seriola	905
coronatus, Cyclopterus	2097
Enneacentrus guttatus	1142
Halatractus	905
Petrometopon cruentatus	1142
Serranus	1142
Zonichthys	905
Corporal	221
corporalis, Cyprinus	221, 222
Leucosomus	222
Semotilus	221, 222
Corpore oblongo glabro	2657
Corsair	1808
cortezianus, Aprodon	2461
coruscans, Arctozenus	601
Paralepis	602
Sudis	602
coruscus, Holocentrus	851
Corvalos	1477
Corvina	1408, 1425, 1455, 1461
acutirostris	1437
adusta	1448
argyroleuca	1434

	Page.
Corvina armata	1437
biloba	1460
chrysoleuca	1439
deliciosa	1456
dentex	1426
fulgens	1435
furcræa	1460
furthi	1441
macrops	1428
microps	1445
monacantha	1419
neglecta	1484
ocellata	1454
ophioscion	1448
oscula	1484
oxyptera	1222
richardsoni	1484
ronchus	1436
saturna	1457
stearnsi	1458
stellifera	1445
subæqualis	1429
trispinosa	1443
vermicularis	1453
Corvina de las Aletas Amarillas	1410
corvinæforme, Brachydeuterus	1326
Hæmulon	1327
Pomadasis	1327
Corvinus (Johnius) jacobi	1457
Corvula	1427
batabana	1430
macrops	1427, 1428
sanctæ-luciæ	1429
sialis	1428
subæqualis	1429
Corynolophus	2733
reinhardti	2733
Coryphæna	952
argyrurus	953
aurata	953
azorica	953
cærulea	1653
chrysurus	952
dolfyn	953
dorado	953
equisetis	953
fasciolatus	952
hippurus	952
immaculata	953
imperialis	952
lessonii	953
lineata	1619
marcgravii	953
novacula	1619
perciformis	964
plumieri	2276

	Page.
Coryphæna punctulata	953
scomberoides	953
sueuri	953
virgata	953
vlamingii	953
Coryphænidæ	951
Coryphænoides	2578
carapinus	2579
norvegicus	2579
rupestris	2578
Coryphœna lineolata	1619
nigrescens	1200
psittacus	1619
Coryphopterus	2210
glaucofrænum	2220
Corythroichthys	761, 763, 772, 2838
albirostris	772, 2838
cayannensis	2838
cayorum	2838
cosmopolita, Micropteryx	938
Seriola	938, 2847
Cossyphus	1581
bodianus	1583
darwinii	1586
diplotænia	1582
eclancheri	1583
pectoralis	1582
puellaris	1584
pulchellus	1584
rufus	1583
verres	1583
costatesi, Smaragdus	2225
costatus, Bodianus	1462
Micropogon	1462
costellatus, Chirus	1869
Cottidæ	1879
Cottinæ	1882
Cottogaster	1044
aurantiacus	1041
cheneyi	2851
copelandi	1045
putnami	1046
shumardi	1046, 2851
uranidea	1044, 1045
Cottopsis	1942
asper	1944
gulosus	1945
parvus	1945
semiscaber	1950
Cottunculus	1992
microps	1992
thomsonii	1993
torvus	1994
Cottus	1493, 1941, 1953, 1970
acadian	2023
æneus	1973

	Page.
Cottus aleuticus	1957, 2862
alvordii	1952
anceps	1973
annæ	1960
asper	1944
axillaris	1981
bairdi	1950
beldingii	1958
bicornis	1913
boleoides	1968
brandti	1984
brodamus	2066
bubalis	1972
cataphractus	2053, 2066
cephaloides	2008
chætodon	2316
cirris plurimis	2066
claviger	1939
cognatus	1954, 1955
copei	1968
criniger	2013
decastrensis	1983
diceraus	1941
elegans	1939
evermanni	1945
fabricii	2009
formosus	1969
franklini	1967
glaber	2316
glacialis	1976
gobio	1941, 1968, 2009
gobioides	1968
gracilis	1968
grœnlandicus	1975
gulosus	1944
hemilepidotus	1936
hexacornis	2003
hirundo	2011
hispidus	2023
humilis	1979
ictalops	1950
indicus	2092
jaok	1978
japonicus	2036
klamathensis	1955
labradoricus	2004
leiopomus	1962
maculatus	1972
marginatus	1966
marmoratus	1983
meridionalis	1951
mertensii	1986
minutus	1958
mitchilli	1973
monopterygius	2092
niger	1983, 1986

	Page.
Cottus nigricans	1973
nivosus	1985
octodecimspinosus	1976
onychus	1953
pachypus	1973
perplexus	1955
philonips	1959
pistilliger	2008
platycephalus	1983, 1988
polaris	1999
pollicaris	1941, 1953
polyacanthocephalus	1977
porosus	1975
princeps	1962
punctulatus	1948, 1951
quadricornis	2001
quadrifilis	1998, 2000
rhotheus	1946
ricei	1952
richardsoni	1951
scorpio	1973
scorpioides	1973
scorpius	1974
grœnlandicus	1975
semiscaber	1949
semiscabra centropleura	1945
shasta	1947
spilotus	1961
stelleri	1941
tæniopterus	1979, 1988
tentaculatus	2000
thomsonii	1994
trachurus	1936
tricuspis	2009
tripterygius	2023
uncinatus	1906
ventralis	2008, 2009
verrucosus	1980
villosus	2022
virginianus	1976
viscosus	1968
wilsoni	1952
Cotylis nigripinnis	2332
nuda	2331
stannii	2332
stelleri	2104
ventricosus	2104
Cotylopus	2207
gymnogaster	2207
salvini	2208
couchi, Dionda	216
Moniana	272
Couchia argentata	2559
couchiana, Limia	695
Pœcilia	695
couchii, Pœcilia	695

	Page.
Couchii, Serranus	1139
Couchu	160
couesii, Agosia	310
Apocope	310
Ceratias	2732
Coregonus	463
Cryptopsaras	2731
Prosopium	463
Couesius	323
adustus	325
dissimilis	324
greeni	324
physignathus	326
plumbeus	323
prosthemius	324
squamilentus	323
Couia	183
coulterii, Coregonus	462
courbina, Pogonathus	1483
Pogonias	1483
cromis	1483
Courpata	976
courtadei, Serranus	1152
courvina, Johnius	1419
Sciæna	1419
Cow-fish	1724
Cow-nose Ray	90
Cow-pilot	1561
Cow Shark	19
Sharks	17
Crab Eater	948
Crabra	1837
Cracker, Clam	83
cragini Amiurus	141
Etheostoma	1091
Craig Fluke	2656
crameri, Leuresthes	802
Sebastodes	1799
Crampfish	77
Craniomi	781, 2146
Crapet	987
Crappies	986, 987
crassa, Belone	716
Tigoma	231
crassicauda, Lavinia	231
Leuciscus	231
Siboma	231
crassiceps, Melamphaes	843
Plectromus	843
Scopelus	843
crassilabre, Moxostoma	194, 196
crassilabris, Embryx	2458
Geophagus	1543
Lycodopsis	2458
Lycolia	2869
Ptychostomus	194

	Page.
crassilabris, Satanoperca	1542
crassus, Alvordius	1034
Esox	627
Lepidosteus	110
Squalius	231
Tylosurus	716
craticula, Zygonectes	657
Cratinus	1188
agassizi	1188
Cravo	954
Crawl-a-bottom	181, 1038
Crayracion	1746
crebripunctata, Pteroplatea	87
Creek Chuh	222
Creekfish	185
Cremnobates	2369
affinis	2372
altivelis	2371
fasciatus	2373
integripinnis	2373
marmoratus	2371
monophthalmus	2372
nox	2374
cremnobates, Labrosomus	2366
Starksia	2365, 2366
Crenilabrus	1581
microstoma	1576
crenulare, Myctophum	575
crenularis, Tarletonbeania	575
crenulatus, Rhombus	966
Creole	1586
Fish	1221
creolus, Brachyrhinus	1222
Paranthias	1222
Serranus	1222
crescentale, Gobiosoma	2259
crescentalis, Gobiosoma	2260
Pomacanthus	1682
crescentis, Salmo gairdneri	2821
Crested Gobies	2209
crestonis, Teuthis	1692
Crevallé, Horse	920
Crevallés	915, 920, 921
crinigerum, Siphostoma	771
crinitus, Alectis	932
Blennius	2383
Blepharichthys	932
Blepharis	932
Caranx	932
Gallichthys	932
Zeus	932
Criolla, Cherna	1157
Criollo, Pargo	1265
cristagalli, Anoplarchus	2423
Cebedichthys	2427
Centronotus	2423

	Page.
cristagalli, Gobius	2209
cristata, Chimæra	95
cristatus, Adonis	2383
Blennius	2382, 2383
cristiceps, Melamphaes	844
Plectromus	843
Cristivomer	504
namaycush	504
siscowet	505
cristulata, Scorpæna	1841
Crius	970
bertheloti	971
Croaker	1460, 1484
Black	1456
White	1397
Yellow-tailed	1467
croaker, Sciæna	1462
Croakers	1392, 1461
croceus, Leuciscus	308
Rhinichthys atronasus	308
crocodilinus, Branchysomophis	388
Ophichthys	388
Ophisurus	388
crocodilus, Gasteropelecus	558
Lampanyctus	558
Scopelus	558
crocota, Plectropoma	1192
crocotus, Hypoplectrus	1192
unicolor	1192
crocro, Pomadasis	1333
Pristipoma	1333
croicensis, Erychthys	1651
Scarus	1650
Cromileptes	1148
cromis, Labrus	1483
Pogonias	1482
courbina	1483
crossotus, Etropus	2689
crotalina, Lycolia	2869
crotalinus, Embryx	2458
Lucodopsis	2459
Crotalopsis	386
mordax	387
punctifer	387
crotaphus, Chærojulis	1598
Julis	1591, 1598
Platyglossus	1598
Crucian Carps	201
cruentatus, Bodianus	1142
Labrus	1238
Petrometopon	1141
coronatus	1142
Priacanthus	1238, 2858
Sparus	1142
cruentifer, Pisoodonophis	377
crumenophthalmus, Caranx	911

	Page.		Page.
crumenophthalmus, Scomber	911	Cubera	1254
Trachurops	911	cubera, Lutjanus	1255
cruoreum Xiphidion	2425	Cubiceps	950
cruoreus, Squalius	233	indicus	951
Cryptacanthodes	2443	multiradiatus	951
inornatus	2443	pauciradiatus	951
maculatus	2443	cubifrons, Malthe	2738
Cryptacanthodidæ	2442	Cub-shark	38
Cryptops	341	Cuckold	1724
Cryptopsaras	2731	cuculus, Trigla	2177
couesii	2731	cucuri, Prionodon	40
Cryptopterus	381, 382	Cucuyo	1701, 1709
puncticeps	382	Cugupuguacu	1158, 1163
cryptosus, Stromateus	968	cnjus, Chasmistes	183
Cryptotomus	1621	Culius	2199
auropunctatus	1624	æquidens	2202
beryllinus	1624	amblyopsis	2200
dentiens	1623	belizianus	2201
retractus	1623	perniger	2201
roseus	1626	cultrata, Novacula	1619
ustus	1624	cultratus, Stolephorus	443
Cryptotrema	2366	Xyrichthys	1619
corallinum	2366	cultriferum, Pristipoma	1333
crysoleucas, Abramis	250	cultrifrons, Alutera	1718
bosci	251	Cultus Cod	1875
Cyprinus	250	culveri, Trachinotus	942
crysos, Caranx	921	cumberlandicum, Etheostoma flabellare	1098
Scomber	921		
Crystallaria	1060	cumingii, Ceratichthys	318
asprella	1061	Hybopsis	318
Crystallichthys	2864	cuneata, Pœcilia	2834
mirabilis	2865	Cunner	1576, 1577
crystallina, Atherinella	805	cupreoides, Pimelodus	140
Thyrina	804	cupreus, Pimelodus	140
Crystallogobiinæ	2192	Silurus	140
Ctenodax	975	Trachinotus	944
Ctenodon	2432	Trachynotus	944
maculatus	2433	curassavicus, Balistes	1709
Ctenogobius	2210, 2211, 2218	curema, Mugil	813, 2841
fasciatus	2223	curilicus, Agonus	2036
Ctenolabrus adspersus	1577	curilus, Salmo	508, 2823
burgall	1577	Curimata	332
cæruleus	1577	magdalenæ	332
chogset	1577	Curimatella	332
uninotatus	1577	Curimatinæ	331
cubæ, Vomer	934	Curimatus	332
Cuban Blindfish	2501	magdalenæ	332
cubana, Anguilla	348	curtus, Stolephorus	445
Muræna	348	Vomer	934
cubanus, Engraulis	442	curvidens, Congrus	360
Epinephelus	1159	curvilineata, Murenophis	395
Stolephorus	442	curvus, Tetrodon	1728
cubensis, Channomuræna	404	Cusk Eels	2481, 2487
Hynnis	932	Cusks	2561
Limia	692	cuspicauda, Alutera	1718
Pœcilia	692	cutisanserinus, Carpiodes	167

	Page.
Cutlass Fishes	888
Cut-lips	199, 327
Cut-throat Trout	487, 492, 493
cuvieri, Caranx	910
Centropomus	1121
Gasterosteus bispinosus	749
Tetragonurus	976
Trachurus	910
cuzamilæ, Scarus	1648
cyanea, Furcaria	1547
cyanellus, Apomotis	996
Ichthyobus	164
Lepomis	996
cyaneus, Chromis	1547
Heliastes	1547
Hypsilepis cornutus	283
Notropis cornutus	283
Cyanichthys	1747
cyanocephalus, Iridio	1594
Labrus	1594
Lythrurus	300
Minnilus	300
Notropis umbratilis	300
cyanoguttatus, Herichthys	1538
Heros	1537
cyanolene, Sparisoma	1633
cyanonoton, Alosa	427
Cyanoperca	1022
cyanophrys, Naucrates	900
Psenes	950
cyanops, Caulolatilus	2278
cyanoptera, Chrysophrys	1354
cyanopterus, Cypsilurus	2836
Exocœtus	739
Lutjanus	1255
Mesoprion	1255
Neomænis	1254
cyanostigma, Chærojulis	1591
Julis	1591
Platyglossus	1591
Cybium	873
acervum	875
caballa	876
cavalla	876
immaculatum	876
maculatum	874
petus	877
regale	875
sara	877
solandri	877
verany	877
Cycleptinæ	162
Cycleptus	168
elongatus	168
nigrescens	169
Cyclichthys	1747, 1748

	Page.
Cyclichthys cornutus	1749
Cycloganoidea	111
Cyclogaster	2114
pulchellus	2127
cyclogaster, Liparis	2118
Cyclogobius	2249
cyclolepis, Microgobius	2247
Moseleya	2570
Nematonurus	2571
Zalypnus	2246
Cyclonarce	78
cyclopomatus, Serranus	1175
Cyclopsétta	2675
chittendeni	2676
fimbriata	2676
querna	2675
Cyclopteichthys	2103
glaber	2104
stelleri	2104
ventricosus	2104
Cyclopteridæ	2094
Cyclopterinæ	2095
Cyclopteroides	2102
gyrinops	2102
Cyclopterus	2096
cæruleus	2097
callyodon	2110
coronatus	2097
gelatinosus	2135
lineatus	2118
liparis	2118, 2123
major	2118
minor	2121
liparoides	2108
lumpus	2096, 2097
minutus	2097
montacuti	2108
musculus	2118
nudus	2336
orbis	2100
pavoninus	2097
pyramidatus	2097
spinosus	2099, 2100
stelleri	2104
ventricosus	2104
cyclopus, Liparis	2112, 2118
Cyclospondyli	52, 53
Cyclospondylous Sharks	52
cyclostigma, Liparis	2125
Cyclothone	581, 582
bathyphila	582
elongata	583
lusca	582
microdon	582, 2826
cyclotis, Ceratichthys	323
cylindraceus, Rivulus	662

	Page.
Cylindrosteus	109, 110
agassizii	111
bartoni	111
castelnaui	111
productus	111
rafinesquei	111
zadocki	111
Cymatogaster	1498, 1502
aggregatus	1498
ellipticus	1503
larkinsii	1503
minimus	1497
pulchellus	1503
rosaceus	1500
cymatogramma, Pileoma	1053
cymatotænia, Etheostoma (Hadropterus)	1042
Hypohomus	1041
Cynædus brama	1360
Cynichthys	1148
Cynicoglossus	2653
bathybius	2656
pacificus	2655
Cynocephalus	33
cynodon, Lutjanus	1255
Mesoprion	1255, 1260
Cynoglossa	2653
microcephala	2655
cynoglossa, Solea	2657
Cynoglossinæ	2693
cynoglossus, Glyptocephalus	2656
Pleuronectes	2611
Cynoperca	1020, 1021
Cynoponticus	359
ferox	360
Cynoscion	1401, 1403
acoupa	1403
albus	1411
carolinensis	1409
jamaicensis	1406
leiarchus	1414
macdonaldi	1411
maculatum	1409
microlepidotus	1415
nebulosus	1409
nobilis	1413
nothus	1406
obliquatus	1405
othonopterus	1404
parvipinnis	1410
phoxocephalus	1413, 2859
regale	1407
regalis	1407
reticulatus	1408
squamipinnis	1404, 1405
stolzmanni	1412

	Page.
Cynoscion thalassinus	1407
virescens	1415
xanthulus	1410
Cynoscium acoupa	1404
cypho, Catostomus	184
Esox	627
Cyphosus bosqui	1388
cyprinaceus, Chætodon	1388
Cyprinella	254, 256, 273
analostana	279
beckwithi	273
billingsiana	272
bubalina	273
calliura	275
cercostigma	275
complanata	272
forbesi	272
gunnisoni	273
lepida	273
ludibunda	273
lugubris	274
luxoides	274
macrostoma	274
notata	274
rubripinna	281
suavis	272
texana	274
umbrosa	273
venusta	274
whipplii	279
cyprinella, Chanos	415
Ictiobus	163
Sclerognathus	164
Cyprinidæ	199, 200
Cyprininæ	201
Cyprinodon	670
baileyi	675
bovinus	673
californiensis	674
carpio	675
elegans	675, 2832
eximius	673, 2832
felicianus	676, 2832
gibbosus	672
latifasciatus	676
macularius	674
baileyi	675
martæ	675
mydrus	676
nevadensis	674
parvus	666
riverendi	673, 2832
variegatus	671
Cyprinodontinæ	631
cyprinoides, Gobius	2209
Lophogobius	2209

	Page.
Cyprinus	201
americanus	250, 251, 1475
atromaculatus	222
atronasus	307
auratus	291
balteatus	239
bullaris	221
burtonianus	2798
carpio	291
catostomus	176
caurinus	220
commersonii	179
cornutus	282
corporalis	221, 222
crysoleucas	250
gracilis	326
hemiplus	250
maxillingua	327
megalops	282
melanurus	282
oblongus	186
oregonensis	225
pala	415
smithii	413
sucetta	186
sueurii	195
teres	179
tolo	415
vittatus	307
cyprinus, Carpiodes	167, 168
Cypselurus	730, 731, 735, 2835, 2836
bahiensis	2836
californicus	2836
callopterus	2836
comatus	736
cyanopterus	2836
furcatus	737, 2836
gibbifrons	2836
heterurus	2836
lineatus	2836
nigricans	2836
xenopterus	2836
cypselurus, Prognurus	2866
Cyttinæ	1660
Cyttus hololepis	1662
Dab, Alaska	2645
Rusty	2644
Smear	2654
Dabbler, Mud	640
Dabs, Mud	2644
Smear	2653
Dace	228, 281
Black-nosed	305, 307
Long-nosed	306
Dace, Red-bellied	209

	Page.
Dacentrus	1495
lucens	1496
Dactylagnus	2304
mundus	2304
dactyloptera, Scorpæna	1837
Dactylopterus	2183
communis	2184
pirapeda	2183
volitans	2183
dactylopterus, Helicolenus	1837
Sebastes	1837, 1838
Sebastoplus	1837
Dactyloscopidæ	2297
Dactyloscopus	2300, 2301
lunaticus	2302
pectoralis	2301
poeyi	2302
tridigitatus	2301
zelotes	2303
dactylosus, Paraliparis	2144
Daddy Sculpin	1974
Dæctor	2325
dowi	2325
Dajaus	818, 819
choirorynchus	2841
microps	820
monticola	2841
dakotensis, Chrosomus	210
Dalatiidæ	56, 620
dalli, Gobius	2230
Dallia	621
delicatissima	621
pectoralis	621
dallii, Pteropodus	1819
Sebastodes auriculatus	1818
Damalichthys	1509
argyrosoma	1510
argyrosomus	1509
vacca	1510
damalis, Carpiodes	167
d'Amplora, Maire	555
Dark-green Parrot-fish	1638
Darter, Black-sided	1028, 1032
Blue	1088
Blue-breasted	1076
Fan-tailed	1079
Green-sided	1053
Johnny	1056
Least	1104
Manitou	1027
Rainbow	1086
Tessellated	1057
Sand	1061, 1062
darwini, Cossyphus	1586
Pimelometopon	1586
Sebastes	1832

	Page.
darwini, Sebastodes	1832
Trochocopus	1586
Dasibatis	82
centrura	83
dipterura	85
hastata	84
longa	85
sabina	85
sayi	86
tuberculata	84
daspilotus, Pisoodonophis	2803
Dasyatidæ	79
Dasyatinæ	79
Dasyatis	82, 83
centrura	83
dipterura	85
gymnura	84
hastata	83
longa	85
sabina	84
say	86
Dasybatus	66, 82
dipterurus	85
dasycephalus, Arius	130
Galeichthys	2780
Hexanematichthys	130
Dasyscopelus	574
spinosus	575
Dasycottus	1991
setiger	1991
daubentonii, Caranx	920
davidsonii, Anisotremus	1321
Monacanthus	1715
Pomadasys	1321
Pristipoma	1321
davisoni, Etheostoma	1049
Ulocentra	1049
Decactylus	173, 174, 177
decadactylus, Beryx	844
decagonus, Agonus	2053
Aspidophorus	2054
Brachyopsis	2054
Leptagonus	2052
decagrammus, Chirus	1869
Hexagrammos	1867
Hexagrammus	1875
Labrax	1868
Decapterus	907
hypodus	908
macarellus	909
punctatus	907
sanctæ-helenæ	908
scombrinus	908
Decaptus	906
decastrensis, Cottus	1983
decimalis, Serranus	1175

	Page.
declivifrons, Abudefduf	1562
Euschistodus	1562
Glyphidodon	1562
declivis, Seriola	905
Zonichthys	905
Decodon	1584
puellaris	1584
decoratus, Entomacrodus	2399
Promicropterus	1234
Rhypticus	1234
decurrens, Pleuronichthys	2637, 2638
de Casta, Aguja	892
Mojarra	1372
de Chapala, Pescado Blanco	792
de dos Colores, Pescado Azul	1557
Vaqueta	1684
de España, Sardina	423
de Gallo, Pez	895
de la Alto, Isabelito	1674
de la Piedras, Mojarra	1681
de las Piedras, Pez	1700
de Ley, Mojarra	1370
de lo Alto, Cagon	1277
Pargo	1262
Sesi	1261
de Mar, Aboma	2195
Esmeralda	2204
Esmeraldas	2203
de Marais, Poisson	113
de Paladar, Aguja	892
de Perdriz, Liza Ojo	814
de Playa, Cazon	36
de Pluma, Pez	1347
de Raizero, Pargo	1273
de Rio, Aboma	2236
Bagres	149
Lenguado	2698
de Vivero, Cherna	1160
Deep-water Catalufas	977
Gurnards	2177
Porgies	1344
defensor, Caranx	921
de Gato Pai	1837
Dekaya	2276
anomala	2277
dekayi, Acipenser	106
Gasterosteus	746
Isuropsis	48
Isurus	48
Phycis	2555
Pimelodus	140
Scomber	867
Syngnathus	771
delalandi, Labrisomus	2359
Malacoctenus	2358
delalandii, Clinus	2359

	Page.
de Ley, Sardina	430
delicatissima, Dallia	621
Umbra	621
delicatissimus, Engraulis	444
Stolephorus	444
deliciosa, Cliola	272
Corvina	1456
Moniana	262
Sciæna	1455
Delolepis	2442
virgatus	2442
Delothyris	2690
pellucidus	2691
delphinus, Minomus	171
Pantosteus	171
del Rey, Pescadillos	807
Pescado	806
Pez	799
Deltentosteus	2210
Deltistes	2794
luxatus	2792
de Mer, Brochet	1118
Demoiselle	1543, 1561
dendritica Anclyopsetta	2634
Ramularia	2633
denegatus, Pomacentrus	1567
dennyi, Liparis	2124
dentata, Chœnopsetta	2630
Platessa	2615, 2630
Pomatopsetia	2615
Stygicola	2500
dentatus Apsilus	1278
Grammatostomias	590
Hippoglossoides	2615
Lucifuga	2500
Lutjanus	1255
Mesoprion	1279
Paralichthys	2629, 2630
Pleuronectes	2630
Pseudorhombus	2630, 2632
Stygicola	2500
Tropidinius	1279
Upeneus	859
Dentex	1288
filamentosus	1289
dentex, Brycon	337
Caranx	927
Chalcinopsis	337
Corvina	1426
Engraulis	451
Larimus	1426
Menidia	801
Odontoscion	1425
Osmerus	524
Scomber	927
Denticinæ	1244

	Page.
denticulatus, Anarrhichas	2446
dentiens, Calliodon	1623
Cryptotonus	1623
denudatum, Gonostoma	579
deppii, Cichlasoma	1524
Heros	1524
deprandus, Esox	628
depressa, Belone	211, 713
Fistularia	757
depressus, Lonchurus	1482
de Raizero, Cabrilla	1171
Derepodichthyidæ	2480
Derepodichthys	2480
alepidotus	2480
Derichthyidæ	343
Derichthys	343
serpentinus	343
dermatinus, Lycodapus	2492
Salmo	479
Dermatolepis	1166, 1168
angustifrons	1169
inermis	1167, 2855
punctatus	1163
zanclus	2854
dermatolepis, Epinephelus	1169
Dermatostethus	761, 763
punctipinnis	763
desmarestia, Raia	71
detersor, Julis	1610
detrusus, Gillichthys	2251
Devilfish	92
Devil, Sea	91, 92, 2727
Diabasis	1291
aurolineatus	1309
chromis	1299
chrysopterus	1309
elegans	1304
flavolineatus	1306
fremebundus	1297
maculicauda	1314
obliquatus	1304
parra	1299
plumieri	1306
scudderi	1300
steindachneri	1302
trivittatus	1311
Diablo	2737
Diabolichthys	92
elliotti	93
diabolus, Raja marinus	93
Diacope	1247
viridis	1246
diadema, Pseudoscarus	1646
Scarus	1646
Diagramma cavifrons	1343
melanospilum	1321

	Page.
Dialommus	2868
fuscus	2868
diaphana, Hydrargyra	645
Raia	71
Sternoptyx	603, 2826
Diaphanichthys	353
diaphanus, Calliurus	996
Fundulus	645
menona	645
Diaphasia	2495
Diaphus	564
theta	564, 565
diaptera, Furcella	2472
Diapterus	1373, 1375
californiensis	1370
dowi	1368
gracilis	1370
homonymus	1371
lefroyi	1372
diapterus, Amiichthys	1113
Lycodes	2473
Dibranchus	2743
atlanticus	2743
diceraus, Ceratocottus	1940, 1941
Cottus	1941
Enophrys	1941
Dicerobatis	2756
Dicerobatus	92
Dick, Nigger	327
Slippery	1595
Dicrolene	2522
intronigra	2522
Dicromita	2506
agassizii	2506
Dicrotus	882
parvipinnis	883
Dictyosomatinæ	2349
diego, Scomber	867
diencæus, Eupomacentrus	1552
difformis, Carpiodes	166
digitatus, Lycodes	2466
digitis, Trigla vicensis	2183
digrammus, Pleuronectes	2641
dilecta, Anclyopsetta	2636
Notosema	2636
dilectum, Notosema	2635
dilectus, Alburnus	294
Notropis	294
Dimalacocentrus	1613
di Mare, Cane	48
dimidiata, Mycteroperca	1179
dimidiatus, Epinephelus	1179
Gasterosteus	749
Halichæres	1594
Icthycallus	1594
Julis	1594

	Page.
dimidiatus, Leucus	244
Platyglossus	1594
Serranus	1179
Syngnathus	765
Trisotropis	1179
Dinectus	103
truncatus	106
Dinematichthys marginatus	2502
ventralis	2503
Dinemus	854
venustus	854
dinemus, Miunilus	293
dinoceros, Citharichthys	2682
Diodon	1744, 1747
antennatus	1750
asper	1744, 1752
atinga	1746, 1750
brachiatus	1746
carinatus	1754
echinus	1746
fuliginosus	1749
geometricus	1748, 1749
holocanthus	1746
hystrix	1744, 1746
liturosis	1746
maculatus	1746
maculifer	1747
maculostriatus	1748
melanopis	1746
meulini	1748
multimaculatus	1746
nigrolineatus	1749
novemmaculatus	1746
pilosus	1744, 1752
punctatus	1746
quadrimaculatus	1746
reticulatus	1751
rivulatus	1748
schœpfi	1748
sexmaculatus	1746
spinosissimus	1746
spinosus	1749
verrucosus	1749
Diodontidæ	1742
diomedeana, Aphoristia	2711
diomedeanus, Hoplunnis	361
Symphurus	2711
Dionda	212, 213, 214
argentosa	215
chrysitis	214
couchi	216
episcopa	215
grisea	216
melanops	216
papalis	214
plumbea	216

	Page.
Dionda punctifer	215
serena	214
spadicea	216
texensis	215
Dioplites	1010
nuecensis treculii	1012
variabilis	1012
diplæmia, Hypsilepis	300
diplæmius, Minnilus	300
Diplanchias	1753
nasus	1754
Diplectrum	1203, 1204, 1207
euryplectrum	1206
fasciculare	1208
formosum	1207
macropoma	1205
radiale	1204
radialis	1205
radians	1208
sciurus	1204
diplemius, Semotilus	222
Diplesion	1052
blennioides	1053
fasciatus	1081
Diplesium blennioides	1053
Diplodus	1362
argenteus	1363
caribæus	1360
caudimacula	1363
flavolineatus	1360
holbrookii	1362
probatocephalus	1361
rhomboides	1358
sargus	1363
Diplolepis	1418
squamosissimus	1419
diploproa, Sebastichthys	1802
Sebastodes	1801
Diplospondyli	16
diplotænia, Belone	712
Bodianus	1582
Cossyphus	1582
Harpe	1582
diplotænia, Tylosurus	712
Dipterodon	1106
hexacanthus	1107
ruber	1107
Dipteron chrysurus	1433
dipterura, Dasibatis	85
Dasyatis	85
dipterurus, Dasybatis	85
Dipterygonotus	1365
Dipturus	66
dipus, Microdesmus	2450
Discoboli	1758
Discoboli liparidina	2105

	Page.
discobolus, Catostomus	172, 175, 2791
Discocephali	781, 2265
Discopyge	78
ommata	78
dispar, Fundulus	658
Zygonectes	659
dispilurus, Centropristis	1220
Dules	1219
dispilus, Halichæres	1598
Iridio	1597
Platyglossus	1598
dissimilis, Ceratichthys	319
Couesius	324
Hybopsis	318
Leucosomus	324
Luxilus	319
distichus, Salmo	509
distinctum, Sparisoma	1635, 1636
distinctus, Scarus	1636
Ditrema	1510
aggregatum	1499
anale	1501
arcuatum	1502
argenteum	1504
atripes	1507
brevipinne	1499
caryi	1509
furcatum	1506
jacksoni	1505
læve	1511
laterale	1506
megalops	1502
orthonotus	1507
rhodoterum	1503
temminckii	1510, 1511
toxotes	1508
vacca	1510
Diver, Sand	535
Dobula	228
Doctor-fish	1689, 1691
dodecaedron, Agonus	2046
Occa	2044
dodecaedrus, Aspidophorus	2046
Brachyopsis	2046
Dodecagrammos	1866
Dogfish	113, 623
California	53, 54
Dog Salmon	478
of Alaska	478
Shark	28, 29
Snapper	1257
dolfyn, Coryphæna	953
dolichocephalus, Gobius	2237
dolichogaster, Blennius	2417
Centronotus	2417
Gunnellus	2417

	Page.
dolichogaster, Murænoides	2417
Pholis	2416
Doliodon	939
Dollardee	1005
Dollar-fish	967
Dollfish	1674
Dolly Varden Trout	507
dolomieu, Micropterus	1011
Dolphin, Common	952
Small	953
Dolphins	951, 952
dombey, Bdellostoma	6
Gastrobranchus	6
Le Gastrobranche	6
Polistotrema	6
Dómine	880
dominicensis, Pœcilia	696
Vomer	934
domninus, Protoporus	233
Doncella	1590, 1595
Doncellas	1587
Dorada, Mojarra	928
Dorado	952
Cocinero	921
dorado, Coryphæua	953
Doratonotus	1611
megalepis	1611
thalassinus	1612
Dories, John	1659
Dormeur	1235
Dormitator	2195
gundlacbi	2198
latifrons	2197
lineatus	2198
maculatus	2196, 2198
microphthalmus	2198
dormitator, Philypnus	2195
Platycephalus	2195
dormitatrix, Electris	2195
dormitor, Gobiomorus	2198
Philypnus	2194
Dorosoma	415
cepedianum	416
exile	416
insociabilis	416
mexicanum	416
notata	416
petenense	417, 2809
Dorosomidæ	415
dorsale, Hæmulon	1303
dorsalis azurissimus, Microspathodon	1570
Carangoides	930
Caranx	930
setipinnis	934
cinereus, Microspathodon	1570

	Page.
dorsalis Citula	930
Galeus	30
Halatractus	902
Hybopsis	262
Hypsypops	1570
Macrurus	2585
Microspathodon	1568, 1570
Mustelus	30
Pomatoprion	1570
Semotilus	222
Seriola	902
Umbrina	1469
Vomer	934
dorsatus, Petromyzon narinus	10
dorso, Perca monapterygia	1833
dorsomacula, Girella	1382
dorsopunicans, Pomacentrus	1557
Dorsuarius	1384
Dory	1021
Doryichthys	773
aculeatus	773
californiensis	774
lineatus	773
Doryrhamphus	773
californiensis	773
lineatus	773
Dough-belly	205
Dourade	952
Dovetail Fish	1563
dovii, Anableps	685
Anisotremus	1318
Aplocheilus	2828, 2830
Apogon	1108
Fundulus	650
Gymnothorax	397
Haplochilus	650
Heros	1535
Lycodontis	397
Muræna	397
Opisthopterus	437
Pœcilia	695
Pomadasis	1318
Pristigaster	437
Pristipoma	1318
Sidera	397
dowi, Arius	125
Dæctor	2325
Diapterus	1368
Eucinostomus	1367
Exocœtus	735
Gerres	1368
Leptarius	125
Selenaspis	125, 2761
Tachisurus	125
Thalassophryne	2326
Doydixodon	1382

	Page.
Doydixodon fasciatum	1383, 1384
freminvillei	1382, 1384
Drachinus trichodon	2297
Dragonets	2184
Drepanopsetta	2614
pl* atessoides	2615
Drum	1482
Fresh-water	1484
Drummer, Ground	1436
Jewsharp	1473
Mongolar	1406
White-mouth	1462
drummond-hayi, Epinephelus	1159
drummondi, Otolithus	1409
Drums, Black	1454
Red	1453
River	1483
Sea	1482
Drunken-fish	1722
dubia, Netuma	126, 2765
Seriola	905
dubium, Campostoma	206
Exoglossum	206
dubius, Ammodytes	832
Arius	127
Bodianus	1146
Fierasfer	2496
Menephorus	1147
Serranus	1147
Tachisurus	127
Dublin Pound Trout	507
Duck-bill Cat	101
ductor, Gasterosteus	900
Naucrates	900
dugesi, Adinia	661
Algansea	211
Ameiurus	138
Fundulus	661
Dulce, Boca	29
dulcis, Argyreus	307
Rhynichthys	307
cataractæ	306
Dules	1217
auriga	1220
dispilurus	1219
flaviventris	1221
subligarius	1218
dumerili, Apiouichthys	2703
Caranx	904
Paraloncburus	1478
Polycirrhus	1479
Seriola	903, 904
Squatina	59
duodecim, Engraulis	446
duplex, Orthopristis	1339
duquesnei, Ptychostomus	193
duquesni, Catostomus	193
Placopharynx	198
durvillii, Argyropelecus	604
duryi, Etheostoma	2853
Dusky Shark	35
dussumieri, Brama	960
Seriola	900
Dussumieria stolifera	419
Dussumieriinæ	417
dux, Lachnolaimus	1580
dvinensis, Platassa	2650
dybowskii, Centronotus	2431
Pholidapus	2430
Eagle Rays	87, 89
earili, Phycis	2555
Urophysis	2554
Easter Mackerel	866
Eastern Carp Sucker	168
Mud Minnow	624
Eater, Crab	948
Écaille, Grande	409
Echelus	353
caudilimbatus	355
cinciara	356
Echeneididæ	2265
Echeneis	2268, 2271
albescens	2272
albicauda	2269
apicalis	2268
australis	2269, 2271
brachyptera	2272
fusca	2270
guaiacan	2270
holbrooki	2270
jacobæa	2272
lineata	2268, 2270
lunata	2269
metallica	2270
naucrateoides	2270
naucrates	2269
neucrates	2269
niewhofi	2272
osteochir	2273
pallida	2272
postica	2272
quatuordecimlaminatus	2272
remora	2272
remoroides	2272
scutata	2271
sexdecemlamellata	2272
sphyrænarum	2268
squalipeta	2272
tetrapturorum	2273
trepioa	2268
verticalis	2270

	Page.
Echeneis vittata	2269
Echidna	402
catenata	403
flavoscripta	403
fuscomaculata	403
nocturna	402
echinatum, Leiodon	57
echinatus, Orbis	1745
Echinorhinidæ	57
Echinorhinus	57
obesus	58
spinosus	58
echinus, Diodon	1746
Echiodon	2495
Echiopsis	386
Echiostoma	589
barbatum	589
margarita	589
eclancheri, Cossyphus	1583
Harpe	1583
ectenes, Careproctus	2136
Micropogon	1463
ectenurus, Chloroscombrus	2847
edentula, Platirostra	102
edentulus, Cetengraulis	450
Engraulis	450
edwardi, Sciæna	1490
Stilbiscus	363
Eel, American	348
Conger of California	395
Fresh-water	348
Lamprey	10
Sand	833
Snipe	369
Eel Cat	2788
Eel-back Flounder	2649, 2650
Eel-pout	2453, 2455, 2456, 2457
Eels	344, 346, 347
Conger	352, 354
Cusk	2481, 2487
Long-necked	343
Ooze	349
Snake	372
Snipe	366
Snub-nosed	348
Spiny	612
Symbranchoid	342
True	346
Worm	370
eeltenkee, Myliobatis	88
effuigens, Æthoprora	566
Arlina	1058
Boleosoma nigrum	1058
Larimus	1421
eglanteria, Raia	68, 71
Raja	71

	Page.
egmontis, Ahlia	370
Myrophis	371
egregia, Tigoma	237
egregius, Leuciscus	237
Squalius	237
egretta, Bellator	2174
Prionotus	2175
eigenmanni, Evara	304
Gobiesox	2339
Gobius	2218
Rimicola	2339
Sebastodes	1789
Eigenmannia	341
humboldti	341
eiseni, Characodon	2831
ekala, Caranx	921
ekstromi, Liparis	2108
elaborata, Muræna	389
elaboratus, Gymnothorax	389
Lycodontis	389
Elacate	948
atlantica	948
bivittata	948
canada	948
falcipinnis	948
malabarica	948
motta	948
nigra	948
pondiceriana	948
Elagatis	906
bipinnulatus	906
pinnulatus	907
Elanura	1930
forficata	1930
Elaphocottus	2006
pistilliger	2008
Elapsopsis	381
elassochir, Noturus	147
elassodon, Hippoglossoides	2615
Elassoma	982
evergladei	982
zonatum	982, 2851
Elassomidæ	981
Elastoma	1281
macrophthalmus	1281
elater, Malthe	2739
Ogcocephalus	2739
Zalieutes	2738
Elattarchus	1431
archidium	1431
Elattonistius	412
elattura, Netuma	128, 2769
elatturus, Arius	128
Electric Rays	76
Star-gazers	2306
electricus, Rhinobatus	63

	Page.
Electris dormitatrix	2195
elegans, Blakea	2353
Cottus	1939
Cyprinodon	675
Diabasis	1304
Etheostoma	1074
Gasterosteus	748
Gibbonsia	2353, 2869
Gila	226
Hæmulon	1304
Kyphosus	1387
Labeo	186
Leuciscus	227
Mesoprion	1278
Myxodes	2353
Nanostoma	1075
Orthragoriscus	1754
Pimelepterus	1387
Rhomboplites	1278
Sebastes	1830
Sebastodes	1830
Eleginus	2537
navaga	2537
Eleotridinæ	2188
Eleotris	2199
abacurus	2200
æquidens	2202
amblyopsis	2199, 2200
belizianus	2201
capite plagioplateo	2201
grandisquama	2198
guavina	2199
gyrinus	2201
lateralis	2195
latifrons	2198
longiceps	2195
mauritii	950
mugiloides	2198
omocyaneus	2198
perniger	2201
pictus	2201
pisonis	2200, 2201
seminuda	2204
sima	2198
smaragdus	2204
somnolentus	2198
Elephant Fish	95
Fishes	94
Shark	51
elephas, Squalus	51
Eleutheractis	1229
coriaceus	1233
eleutherus, Noturus	148, 149
Schilbeodes	148
Elliops	133
elliotti, Diabolichthys	93

	Page.
ellipsoidea, Lebias	672
ellipticus, Chatoessus	416
Cymatogaster	1503
Platophrys	2665
Pleuronectes	2665
Rhomboidichthys	2665
Ellwife	426
elongata, Aphoristia	2707
Clupea	421
Cyclothone	583
Platessa	2657
Pœcilia	697
Umbrina	1476
elongatum, Gonostoma	583
elongatus, Avocettina	2802
Benthodesmus	888
Catostomus	169
Cephalus	1756
Clinostomus	240
Cycleptus	168
Labichthys	369
Leuciscus	240
Luxilus	240
Megalops	409
Menticirrhus	1476
Ophiodon	1875
Osmerus	525
Pleuronectes	2657
Pomadasis	1328
Pomotis	1001
Sclerognathus	169
Scopelus	555
Sebastes	1816
Sebastodes	1815
Squalius	240
Symphurus	2707
Zoarces	2457
Elopidæ	408
Elopinæ	408
Elops	409
capensis	410
inermis	410
purpurascens	410
saurus	410, 2806
elucens, Siphostoma	768
Syngnathus	768
El Verde	817
emarginatum, Scarus	1641
Sparisoma	1641
emarginatus, Lobotes	1257
Serranus	1181
Embassichthys	2655
bathybius	2655
Embiotoca	1504
argyrosoma	1510
caryi	1509

	Page.		Page.
Embiotoca cassidyi	1505	Enchelycore	389
jacksoni	1504, 1505	euryrhina	390
lateralis	1506	nigricans	389
lineata	1506	Enchelyopus	889, 2456, 2540, 2560
ornata	1506	americanus	2457, 2555
perspicabilis	1506	barbatus	2500
webbi	1505	brosmæ	2561
Embiotocidæ	1493	cimbricus	2561
Embiotocinæ	1494	cimbrius	2560
Emblemaria	2401	regalis	2553
atlantica	2402	Enchrasicholus	448
nivipes	2402	enchrysurus, Chromis	1548
oculocirris	2403	Endormi Emerande	2230
Emblemariinæ	2347	Enedrias	2414
emblematicus, Gobius	2247	nebulosus	2414
Lepidogobius	2247	Engraulididæ	439
Scarus	1654	engraulinus, Photogenis leucops	296
Zalypnus	2247	Engraulis	448
embryum, Blennicot-		argyrophanus	445
tus	2016, 2864	atherinoides	451
Oligocot-		brevis	450
tus	2017	brownii	443
Embryx	2458	chœrostomus	444
crassilabris	2458	clupeoides	447
crotalinus	2458	compressus	447
embryx, Gerres	1379	cubanus	442
Emerald Fish	2229	delicatissimus	444
Emerande, Endormi	2230	dentex	451
Emichthys megalops	1502	duodecim	446
emiliæ, Opsopœodus	248	edentulus	450
Emmeekia	1601	grossidens	451
venusta	1602	janeiro	451
emmelane, Averruncus	2069	lemniscatus	443
Tachysurus	2785	louisiana	446
Emmelas	1765, 1773, 1777	macrolepidotus	449
Lepophidium	2483	mitchilli	446
Emmelichthyinæ	1364	mordax	448
Emmelichthys	1365	mysticetus	450
vittatus	1365, 1366	nanus	449
Emmnion	2375	panamensis	448
bristolæ	2375	perfasciatus	442
Emniinæ	2345	piquitinga	443
emoryi, Gila	227	poeyi	445
Leuciscus	226	productus	447
Emperador	894	spinifer	448
Empetrichthys	666	surinamensis	447
merriami	667	tricolor	443
Emphycus	2552, 2554	engymen, Cetengraulis	2815
emphysetus, Arius	122	Engyophrys	2668
Bagrus	122	sancti-laurentii	2668
Sciadeichthys	122, 2759	enigmaticus, Schedophilus	972
Tachysurus	122	Enjambre	1141
Enantioliparis	2114	Enjambres	1140
encæomus, Gobius	2223	Enneacanthus	992
Encheliopus	2457	eriarchus	994
Enchelycephali	345, 346	gloriosus	993

	Page.
Enneacanthus margarotis	994
obesus	993
pinniger	994
simulans	994
Enneacentrus	1143
fulvus	1145
ontalibi	1146
guttatus	1142
coronatus	1142
panamensis	1141
punctatus	1146
tæniops	1144
enneagrammus, Ernogrammus	2441
Stichæus	2441
Enneanectes	2349
carminalis	2350, 2868
Enneistus	1143, 1147
Ennichthys	1501
Enophrys	1937, 1938
bison	1938
claviger	1938
dicerans	1941
Enseigne, Porte	1687
ensenadæ, Rhinoptera	91
ensifera, Bairdiella	1434
Sciæna	1435
ensiferus, Centropomus	1125
ensiformis, Trichiurus	887
ensis, Gaidropsarus	2558
Motella	2559
Onos	2559
Sphyræna	824
Entemedor	2752
entemedor, Narcine	2752
Entomacrodus	2397
chiostictus	2398
decoratus	2399
margaritaceus	2398
nigricans	2399
entomelas, Sebastichthys	1786
Sebastodes	1785
Entosphenus	11
camtschaticus	2745
epihexodon	12
tridentatus	12
Entoxychirus	53
Enxaréo	926
Enypnias	2231, 2233
Eopsetta	2613
jordani	2613
eos, Arbaciosa	2343
Boleichthys	1102
Chrosomus	210
erythrogaster	210
Gobiesox	2343
Orthonops	2262

	Page.
eos, Pœcilichthys	1102
Pronotogrammus	1225
Sebastodes	1810
Eosebastes	1765, 1775, 1798
Eperlanus	522
Ephippidæ	1666
Ephippinæ	1667
ephippium, Plectropoma	1192
Ephippus faber	1668
gigas	1668
zonatus	1669
Epicopus	2529
gayi	2530
epicurorum, Chromis	947
Epigonichthys	4
Epigonus	1111
occidentalis	1112
epihexodon, Entosphenus	12
Lampetra	12
Epinephelinæ	1128
Epinephelus	1148, 1152, 2853
adscensionis	1152, 1154
afer	1165
analogus	1152
apua	1159
ascensionis	1154
aspersus	1154
atlanticus	1154
bonaci	1175
brachysomus	1154
calliurus	1186
catus	1159
chalinius	1181
ciliatus	1784
cubanus	1158
dermatolepis	1169
dimidiatus	1179
drummond-hayi	1159
falcatus	1185
flavolimbatus	1155
galeus	1164
gigas	1154
guaza	1154
guttatus	1142, 1159
inermis	1168
interstitialis	1179
jordani	1177
labriformis	1155
lunulatus	1159
maculosus	1158
merus	1162
microlepis	1178
morio	1160
multiguttatus	1166
mystacinus	1151
nigritus	1162

	Page.
Epinephelus niphobles	2853
niveatus	1156
olfax	1183
ordinatus	1155
panamensis	1141
pardalis	1183
punctatus	1146, 1154
quinquefasciatus	1164
rosaceus	1184
ruber	1181
sellicauda	1155
striatus	1157, 1208
tæniops	1144
tigris	1187
venenosus	1172
xenarchus	1180
Epinnula	880
magistralis	880
episcopa, Diondá	215
Hybognathus	215
episcopi, Gambusia	683
episcopus, Hybognathus	215
Episema	254
callisema	273
jejuna	290
Epitrachys	1023
epsetus, Esox	443
equatorialis, Chlopsis	364
Raja	74
Eques	1485, 1489
acuminatus	1487
umbrosus	1487
americanus	1490
balteatus	1490
lanceolatus	1489, 1490
lineatus	1487
pulcher	1489
punctatus	1488, 1489
viola	1486
equestris, Arius	128
Balistes	1703
Equietus	1485
equinoculus, Mugil	2841
equirostrum, Scombresox	726
equisetis, Coryphæna	953
Equitinæ	1397
crate, Lobotes	1236
erebennus, Ameiurus	139
erebus, Muræna	396
erethizon, Arothron	1739
Ovoides	1739
Tetradon	1739
eriarcha, Atherinella	803
Copelandia	994
Eurystole	803
eriarchus, Enneacanthus	994

	Page.
Ericaria	2816
salmonea	2816
Erichæta	999
Ericosma	1028, 1030, 1036
Ericymba	302
buccata	302
ericymba, Sciæna	1445
Erilepidinæ	1861
Erilepis	1862
zonifer	1863
Erimystax	314, 315
Erimyzon	184
goodei	186
sucetta	185, 186
oblongus	186
erinacea, Raia	68
erinaceus, Trichocyclus	1744
Erinemus	314
Eritrema	308
Erizo	1745
Guanabana	1746
Ernogrammus enneagrammus	2441
erochrous, Hololepis	1102
Pœcilichthys	1102
Erogala	254
Erotelis	2203
smaragdus	2204
valenciennesi	2204
Erychthys	1642
croicensis	1651
erythræus	1531
Heros	1531
Erythrichthys	1365
vittatus	1366
Erythrinidæ	330
erythrinoides, Scarus	1635
erythrocheilos Albula	412
erythrogaster, Chrosomus	209
eos	210
Leuciscus	210
Serranus	1160
erythrogastrum, Pœcilosoma	1089
erythrops, Gobiesox	2336
Ichthelis	990
erythroptera, Pimelodus	135
erythrorhynchos, Salmo	508
erythrorrhynchos, Salmo	508
erythrurus, Caranx	920
Catostomus	193
Ptychostomus	193
esca, Clupea	421
escambiæ, Zygonectes	658
escamuda, Sardina	431
eschrichtii, Oneirodes	2732
Escolar	879, 2843
Chino	1114, 1284
de Natura	976

	Page.
Escolor violaceus	4843
Escolares	879
Escolars	877
Escribano	720, 722
esculentus, Carangus	921
Conger	355
Merluccius	2530
Eslopsarum	2840
bartoni	2840
jordani	2840
Esloscopus	2300, 2303
esmarkii, Lycodes	2463
Esmeralda	2227, 2230
de Mar	2203, 2204
Negra	2204
Esocidæ	708
esopus, Labeo	186
Esox	625
affinis	628
americanus	626
atromaculata	629
australis	628
barracuda	823, 2841
belone	714
boreus	628
brasiliensis	723
crassus	627
cypho	627
deprandus	628
epsetus	443
estor	628
fasciatus	626
flavulus	639
immaculatus	630
imperialis	717
lineatus	627
longirostris	714
lucioides	628
lucius	628
lugubrosus	628
marinus	714
masquinongy	629
immaculatus	630
niger	626
nobilior	629
ohiensis	630
ornatus	626
osseus	110
ovinus	672
phaleratus	628
pisciculus	641
pisculentus	641
porosus	627
raveneli	626
reticulatus	628
salmoneus	538, 627, 629

	Page.
Esox saurus	725
scomberius	626
spet	826
sphyræna	826
stomias	585
synodus	536
timucu	711
tridecemlineatus	628
tristœchus	111
umbrosus	627
vermiculatus	627
viridis	110
vittatus	628
vulpes	411
zonatus	639
Espada	894
Pez de	2749
Espadon	894
Espagnol, Quatilibi	1140
Espino, Puerco	1745
estor, Chirostoma	792
Esox	628
Gila	240
Lethostole	792
Leuciscus	240
Squalius	240
Estrella	1054
Etelinæ	1243
Etelis	1281
aquilonaris	1283
carbonculus	1283
oculatus	1282
Etheostoma	1028, 1066, 1069, 1097
alabamæ	1095
artesiæ	1094
asprellus	1061
atromaculata	1057
aurantiacum	1041
australe	1081
blennioides	1033, 1053
blennius	1072, 1073
boreale	1082
cæruleum	1089, 2853
spectabile	1089
calliura	1011
camurum	1076
caprodes	1027
cinerea	1078
cinereum	1078
cragini	1091
cymatotænia	1042
davisoni	1049
duryi	2853
elegans	1074
evides	1037
exile	1103

	Page.		Page.
Etheostoma flabellare	1097	Etheostoma schumardi	1047
cumberland-		scierum	1038
icum	1098	scovellii	1082
lineolatum	1098	squamatus	1040
flabellaris	1097	squamiceps	1096
flabellata	1097	stigmæum	1048
fonticola	1105	swannanoa	1070
fontinalis	1097	tessellatum	1078
formosa	2853	thalassinum	1071
fusiforme	1103	tippecanoe	1090
guntheri	1034	tuscumbia	1100
histrio	1051	uranidea	1045
inscriptum	1072	variatum	1069
ioæ	1084	verecundum	1050
iowæ	1083	vexillare	1058
jessiæ	1084	virgatum	1093
jordani	1079, 1080	vulneratum	1077
juliæ	1093	whipplei alabamæ	1095
laterale	1099	whipplii	1095
lepidogenys	1087	wrighti	1047
lepidum	1089	zonale	1075
linsleyi	1097	arcansanum	1075
longimana	1054	etheostoma, Aboma	2240
luteovinctum	1086	Etheostominæ	1018
lynceum	1075	ethon, Syngnathus	767
macrocephalum	1031	Etmopterus	55
maculatum	1077	pusillus	55
microperca	1104	etowanus, Catostomus nigricans	181
micropterus	1083	Etropus	2687
nevisense	1034	crossotus	2689
nianguæ	1043	microstomus	2687, 2690
spilotum	1044	rimosus	2688
nigrofasciatum	1039	Etrumeus	419
nigrum	1057	acuminatus	419
notatum	1070	sadina	420
obeyense	1092	teres	420
olmstedi	1057	Eucalla	743
ouachitæ	1035	inconstans	744
pagei	1092	cayuga	744
parvipinne	1096	pygmæa	744
pellucidum clarum	1063	Eucentrarchus	988
peltatum	1034	Euchalarodus	2649
phoxocephalum	1031	putnami	2650
podostemone	1055	Eucinostomus	1367
pottsii	1082	argenteus	1371
prœliare	1104	californiensis	1369
prœliaris	1104	dowi	1367
punctulatum	1090	gula	1370
quappella	1084	gulula	1371
quiescens	1101	harengulus	1368
rex	1026	lefroyi	1372
roanoka	1036	productus	1372
rufilineatum	1079	pseudogula	1368
rufolineatum	1079	Euctenogobius	2210, 2215, 2226
rupestre	1073	badius	2227
sagitta	1080	latus	2237

	Page.
Euctenogobius lyricus	2225
sagittula	2229
Eucyclogobius	2248
newberryi	2248
eudouxii, Ailurichthys	118
Felichthys	118
Galeichthys	118
Eugaleus	31
Eugomphodus	46
littoralis	47
Eulachon	521
Eulamia	33
lamia	38
longimana	38
milberti	37
nicaraguensis	39
platyrhynchus	36
eulepis, Microgobius	2244
Euleptorhamphus	723
brevoorti	724
longirostris	724
velox	724
Eumesogrammus	2441
præcisus	2441
subbifurcatus	2440
Eumicrotremus	2097
orbis	2099, 2100
spinosus	2098, 2099
eumorphus, Chatoessus	433
Eupomacentrus	1549, 1550, 1551
adustus	1551
analis	1554
diencæus	1552
flavilatus	1557
flaviventer	1557
fuscus	1552
leucorus	1551
leucostictus	1555
otophorus	1555
partitus	1558
planifrons	1559
rectifrænum	1553
Eupomotis	1006, 1007
aureus	1010
euryorus	1008
gibbosus	1009
heros	1007
holbrooki	1008
humilis	1004
macrochirus	1005
pallidus	1006
europæus, Aspidophorus	2067
Blennius	2419
Trachurus	911
European Barracuda	826
Charr	508

	Page.
European Hake	2530
Lancelets	3
Porgies	1356
Sculpin	1974
Stickleback	747
Eurymyctera	392
euryopa, Cliola	270
Hudsonius	270
euryops, Bathylagus	529
Icelus	1915, 1916
Myxostoma	193
Tylosurus	711
euryorus, Eupomotis	1008
Lepomis	1009
Eurypharyngidæ	406
euryplectrum, Diplectrum	1206
euryrhina, Euchelycore	390
Eurystole	802
eriarcha	803
eurystole, Stolephorus	445
eurystoma, Oliola	277
Codoma	285
Eurystomus	173
eurystomus, Notropis	277
Photogenis	277
Euscarus	1627, 1629, 1639
Euschistodus	1560, 1562
analogus	1563
concolor	1559
declivifrons	1562
Eusebastes	1760
Eusphyra	43
Eustomatodus	907
Euthynnus	868
alliteratus	869
pelamys	869
Eutychelithus	1483
evansi, Hybognathus	213
Evapristis	1334, 1336, 1340
Evarra	304
eigenmanni	304
Eventognathi	161
Evepigymnus	907
evergladei, Elassoma	982
evermanni, Atherinella	804
Cottus	1945
Scarus	1651
Synodus	535
Thyrina	804
Evermannia	2256
longipinnis	2256
zosterura	2256
Evertzens, Jacob	1143
evides, Alvordius	1037
Clinus	2353
Etheostoma	1037

	Page.
evides, Gibbonsia	2352, 2869
Hadropterus	1036
Plectobranchus	2432
evionthas, Ophichthus	381
Quassiremus	380
evolans, Exocœtus	730
Halocypselus	729
Prionotus	2167, 2168, 2169
Trigla	2169
Evoplites	1245
viridis	1246
Evorthodus	2208
breviceps	2208
catulus	2218
Evoxymetopon	885
tæniatus	886
exasperata, Platyrhina	65
Syrrhina	65
exasperatus, Rhinobatus	65
Zapteryx	64
Exerpes	2367
asper	2367
exiguus, Bodianus	1433
Stolephorus	442
exile, Dorosoma cepedianum	416
Etheostoma	1103
exilicauda, Lavinia	208
Leuciscus	209
exsiliens, Exocœtus	732, 734
exilis, Belone	714
Boleichthys	1103
Hippoglossoides	2613
Lyopsetta	2613
Noturus	147
Pœcilichthys	1103
Schilbeodes	147
Tylosurus	714
eximius, Cyprinodon	673
Exocœtidæ	726
Exocœtus	730, 731, 732, 734
acutus	728
affinis	735, 2836
albidactylus	739
appendiculatus	736
bahiensis	739
bicolor	738
brachycephalus	733
californicus	730, 740
callopterus	740
chilensis	730
comatus	736
cyanopterus	739
dowi	735
evolans	730, 2835
exsiliens	732, 734

	Page.
Exocœtus fasciatus	733
furcatus	737
georgianus	730
gibbifrons	741
gryllus	729
heterurus	735
hillianus	729
lamellifer	733
lineatus	739
lutkeni	736
maculipinnis	737
melanurus	735, 736
mesogaster	729
monocirrus	730
nigricans	737
noveboracensis	735, 736
nuttalli	737
obtusirostris	730
orbignianus	729
parræ	740
procne	737
quadriremis	735
roberti	735
robustus	736
rondeletii	733, 734
rubescens	734
rufipinnis	735
scylla	735
speculiger	734
spilonotopterus	740
spilopus	738
splendens	730
vermiculatus	740
vinciguerræ	734
volador	733
volitans	734, 736, 2835
zenopterus	738
Exoglossinæ	204
Exoglossum	327
annulatum	327
dubium	206
lesueurianum	327
maxillingua	327
mirabile	303
nigrescens	327
spinicephalum	206
vittatum	327
exoletus, Acantholabrus	1576
Centrolabrus	1576
Labrus	1576
Exonautes	2835
affinis	2836
exsiliens	2836
rondelettii	2836
rufipinnis	2836

	Page.
Exonautes speculiger	2836
vinciguerræ	2836
expansum, Ostraciou	1724
exsiliens, Exonautes	2836
extensus, Fundulus	646
Lycodapus	2494
faber, Chætodipterus	1668
Chætodon	1668
Ephippus	1668
Faber marinus	1668
fabricii, Campylodon	615
Centroscyllium	56
Cottus	2009
Gadus	2534
Gunnellus	2438
Liparis	2121, 2128
Lumpenus	2437
Macrorus	2582
Spinax	56
falcata, Agosia	313
shuswap	313
Mycteroperca	1184
phenax	1185
Seriola	905
falcatus, Caranx	913, 2845
Epinephelus	1185
Hemicaranx	2845
Labrus	942
Lachnolaimus	1580
Serranus	1185
Sparus	1583
Trachinotus	941
Trisotropis	1185
falciformis, Carcharhinus	36
Carcharias	35
Platypodon	36
falcipinnis, Elacate	948
fallax, Carangus	923
Caranx	923
Pomotis	1003
Trachurus	910
Fallfish, Red	286
Fall-fishes	220, 221
Fall Herring	425
Fanegal	1837
fanfarus, Naucrates	900
Fanguito	692
Fan-tailed Darter	1097
Mullet	816
Fario	483
argyreus	480
aurora	499
clarkii	501
gairdneri	499
Fario newberryi	499

	Page.
Fario stellatus	492
tsuppitch	493
farkharii, Lobotes	1236
fasciata, Aphoristia	2710
Cichla	1012
Clupea	426
Molinesia	695
Plagusia	2710
Pœcilia	641
Seriola	904
Trigla	2183
fasciatum, Doydixodon	1383, 1384
Pristipoma	1339
fasciatus, Achirus	2700
Auchenopterus	2373
Bryttus	993
Caranx	914
Carapus	341
Catonotus	1098
Catostomus	187
Centrarchus	1012
Centronotus	2418
Cremnobates	2373
Ctenogobius	2223
Diplesion	1081
Esox	626
Exocœtus	733
Genyonemus	1479
Giton	340
Gobiesox	2338
Gobius	2222
Gunnellus	2418
Gymnachirus	2703
Gymnotus	340
Halatractus	904
Harpurus	1691
Hemirhamphus	720
Larimus	1424
Murænoides	2418
Mytilophagus	1504
Orthagoriscus	1754
Pholis	2417
Pimephales	217
Pogonias	1483
Prionodes	1212
Scomber	904
Sebastes	1761, 1827
Sicyases	2338
Syngnathus	771
Synodus	536
Tetragonopterus	334
Trachurus	904
Trachynotus	941
fasciculare, Diplectrum	1208
fascicularis, Centropristis	1208

	Page.
fascicularis, Hippocampus ·········	778
Serranus ···············	1208
fasciolaris. Catostomus ·············	186
Notropis umbratilis····	301
Sebastichthys···········	1827
Symphurus ·············	2707
fasciolatus, Coryphæna ·············	952
Fatback ···························	433, 946
Fat-head···························	217, 1585
Father-lasher ····················	1971
favosus, Bathygadus ····· ··········	2565
Blennius ···················	2380
fecundus, Catostomus···············	180
Felichthys ·························	116
bagre ····················	117
bahiensis ················	118
eudouxii················	118
filamentosus·············	118
marinus ·················	118
panamensis ··············	117
pinnimaculatus·········	117
felicianus, Cyprinodon ·············	676
Trifarcius················	676
felinus, Pimelodus ··················	140
Serranus····················	1187
felis, Anarrhichthys ················	2448
Arius ··························	128
Hexanematichthys············	128
Mustelis ······················	31
Pimelodus·····················	141
Silurus·······················	128
fenestralis, Artedius···············	1900
fenestrata, Chromis ················	1518
fenestratum, Cichlasoma···········	1518
fenestratus, Heros ·················	1518
ferox, Alepisaurus··················	595
Bathyophis··················	605
Bathysaurus ················	539
Cynoponticus ···············	360
Idiacanthus ·················	605
Lepisosteus ·················	111
Stomias ····················	588
ferruginea, Limanda ················	2644
Myzopsetta ····'········	2645
Platessa ················	2645
ferrugineus, Characodon···········	669
Pleuronectes ·········	2645
feuille, Polyodon····················	102
Fiatolas···························	964
fibulatus, Spinicephalus·············	2796
Fiddler Fish ····················	63
fieldii Stomias·····················	586
Fierasfer·························	2495
affinis··················	2495
arenicola ···············	2496
bermudensis ··············	2497

	Page.
Fierasfer borealis ····················	2443
dubius·····················	2496
fierasfer, Lycodapus ···············	2493
Fierasferidæ ·························	2494
filamentosus, Ailurichthys·········	118
Argyriosus ··········	936
Dentex···············	1289
Felichthys ············	118
Hemirhamphus ······	723
Icelinus··············	1893
Monacanthus ········	1716
Scomber···············	932
Tarandichthys·······	1892
Filefish ·············· 1712, 1715, 1717, 1718
Orange ·····················	1718
filicornis, Blennius·················	2381
filifera, Chalinura·················	2577
fimbria, Anoplopoma···············	1862
Gadus ·····················	1862
fimbriata, Cyclopsetta ·············	2676
Raia······················	93
Solea ·····················	2700
Squatina ················	59
fimbriatus, Achirus···············	2700
Arnoglossus ···········	2677
Blennius···············	2457
Chaunax ··············	2726
Hemirhombus··········	2677
Icelinus···············	1894
Serranus··············	1154
Zoarces ···············	2457
Fimbriotorpedo·····················	77
Fine-scaled Sucker················ 173, 178
firmisquamis, Bogoslovius·········	2575
Macrurus ···········	2576
fischeri, Achiris···················	2700
Achirus ···················	2699
Chætostomus··············	160
Solea ·····················	2700
Tetragonopterus ···········	334
Fish, Angel·························	58
Bat ···························	2737
Butter ························	2419
Cobbler ·····················	931
Common Alligator ···········	2061
Common Buffalo ··············	163
Creek ·······················	185
Creole ······················	1221
Devil ·························	92
Dismal Swamp···············	703
Doll··························	1674
Dovetail ·····················	1563
Elephant······················	95
Emerald ·····················	2229
File·················· 1715, 1718
Fiddler······················	63

	Page.
Fish, Fool	1715, 1718
Glance	954
Globe	1734
Good	487
Guitar	63
Hand-saw	596
Harvest	965
Indian	1680
Leather	1714, 1715
Lion	1850
Lizard	538
Log	964
Mutton	1376
Oil	879
Portuguese Man-of-War	949
Prick	555
Priest	1784
Rabbit	882
Rainwater	665
Red-mouth Buffalo	163
Red Parrot	1635
Ribbon	1489, 1490
San Pedro	954
Scabbard	887, 889
Scour	879
Sergeant	948
Singing	2321
Soldier	1089
Tongue	2710
Tyrant	886
Ugly	137
Unicorn	1719
Yellow	1144
Fishes	14
Angel	58
Atka	1864
Black Rudder	963
Blennioid	2343
Blind	702
Bony	113
Brotuloid	2498
Buffalo	163
Cardinal	1105
Carp-like	160
Cirrhitoid	1490
Cutlass	888
Elephant	94
File	1712, 1717
Four-eyed	684
Ganoid	100
Guitar	61
Isospondylous	407
Jugular	2528
Lancet	593, 594. 595
Lantern	530, 550
Lizard	533

	Page.
Fishes, Mackerel-like	860
Mail-cheeked	1756
Milk	414
Parrott	1620
Pediculate	2712
Perch-like	979
Pike-like	622
Plectognathous	1696
Porcupine	1742, 1744
Rag	968
Rudder	1380
Scorpion	1839
Sergeant	947
Spiny-rayed	779
Synentognathous	707
Trachinoid	2273
True	97
Trunk	1720
Fishing-Frogs	2713
fissuratus, Neoliparis	2113
fissus, Arius	131
Tachisurus	131
Tachysurus	131, 2782
Fistularia	756
commersonii	758
depressa	757
immaculata	758
neoboracensis	757
petimba	758
serrata	758
tabacaria	757, 758, 2837
Fistularlidæ	755
fistularis, Flagellaria	757
fistulatum, Siphostoma	765
fistulatus, Syngnathus	765
flabellare, Etheostoma	1097
cumberlandicum	1098
lineolatum	1098
flabellaris, Etheostoma	1097
flabellata, Etheostoma	1097
flabbellatus, Catonotus	1098
Flag, Spanish	1817
Flagellaria	756
fistularis	757
flagellum, Raia	88
Saccopharynx	406
Flags, Spanish	1139
Flamenco	1269
Flammeo	2871
marianus	2871
flammeus, Leuciscus	242
Phoxinus	242
Flannel-mouth Cat	137
Sucker	174
Flasher	1235

	Page.
Flatfish	1680, 2602
Common	2647
Flat-headed Chub	326
flava, Congermuræna	357
flavescens, Arius	123
Bagrus	123
Bodianus	1024
Brosmius	2561
Callyodon	1640
Mesoprion	1260
Morone	1024
Perca	1023
fluviatilis	1024
Prionodes	1215
Scarus	1640
Sciadeichthys	123, 2760
Sparisoma	1639, 1640
Tachisurus	123
flavesny, Brosmius	2561
flavicauda, Hyporthodus	1156
flavidus, Apodichthys	2411
Aulorhynchus	754
Sebastichthys	1782
Sebastodes	1781
flaviguttatum, Hæmulon	1312
Lythrulon	1312
flaviguttatus, Hæmulon	1312
flavilatus, Eupomacentrus	1557
Pomacentrus	1558
flavipinnis, Hybognathus	215
Ilisha	435
Pellona	436
Pristigaster	436
flavirostre, Siphostoma	768
flavirostris, Syngnathus	768
flavissimus, Forcipiger	1671
flaviventer, Eupomacentrus	1557
flaviventris, Dules	1221
Serranus	1221
flavolimbatus, Epinephelus	1155
flavolineatum, Hæmulon	1306
flavolineatus, Diabasis	1306
Diplodus	1360
Pimelepterus	1386
Sargus	1360
flavomarginatus, Pseudoscarus	1652
Scarus	1652
flavoscripta, Echidna	403
flavoscriptus, Gymnothorax	395
flavovittatus, Mulloides	860
Upeneus	860
flavulus, Esox	639
flavus, Awaous	2235
Chonophorus	2235
Gobius	2235
Noturus	144

	Page.
flavus, Turdus	1583
Flesh-colored Rockfish	1824
fleurieu, Osterhinchus	1107
flexuolaris, Lepomis	1011
flexuosus, Catostomus	179
Flier	988
Flioma	1793
floræ, Neoliparis	2111
florealis, Platyglossus	1597
Florida Cat	137
floridæ, Jordanella	677
Siphostoma	766
floridana, Cichla	1012
floridanus, Phycis	2554
Urophycis	2554
floridensis, Calliurus	992
Fundulus	642
floripinnis, Fundulus	651
Haplochilus	651
Zygonectes	651
Flounder, Arctic	2649
Eel-back	2650
Four-spotted	2632
Great	2652
Gulf	2631
Peacock	2665
Pole	2657
Soft	2679
Southern	2630
Starry	2651
Summer	2629
Winter	2646
Fluke, Craig	2656, 2657
fluviatilis, Algoma	215
Hudsonius	269
Hybognathus	215
Perca flavescens	1024
Sargosomus	1496
Fly-fish	1809
Flying-fish, Great	740
Sharp-nosed	728
Flying-fishes	726, 730
Flying Gurnard	2182, 2183
Robin	2183
Fodiator	727
acutus	728
fodiator, Tylosurus	715
fœtens, Salmo	538
Saurus	538
Synodus	538
folium, Polyodon	102
fonsecensis, Achirus	2699
Solea	2699
fonticola, Alvarius	1105
Etheostoma	1105
Fundulus	643
Microperca	1104

	Page.		Page.
fontinalis, Etheostoma	1097	fraterculus, Mylocheilus	220
Salmo	507	fremebundum, Hæmulon	1297
Salvelinus	506	fremebundus, Diabasis	1297
agassizii	507	freminvillei, Doydixodon	1382, 1384
Fontinus	633, 634, 645	Myliobatis	89
Foolfish	1715, 1718	frenatus, Balistes	1705
forbesi, Cliola	272	Brachyistius	1499
Cyprinella	272	Micrometrus	1499
Orthopristis	1336	Odontopyxis	2075
forcipatus, Balistes	1702	Sarritor	2073
Forcipiger	1671	Zaniolepis	1877
flavissimus	1671	French Grunt	1306
forficata, Elanura	1930	Mullet	813
Guaperva lata	1702	Frère Jacques.	846
formosa, Algansea	246	Fresh-water Drum	1484
Oliola	271	Eel	348
Etheostoma	2853	fretensis, Oliola	261
Heterandria	687	Hybopsis	261
Hydrargyra	2827	Notropis	261
Leucos	246	Friars	789
Mollienisia	699	friedrichsthali, Heros	1528
Moniana	271	Frigate Mackerels	867
Perca	1208	frigida, Moniana	271
Uranidea	1969	frigidus, Notropis	271
formosulum, Campostoma	206	Lycodes	2465
formosum, Diplectrum	1207	Frilled Sharks	16
Hæmulon	1305	Frog Fishes	2715
formosus, Alburnus	280	frondosum, Sparisoma	1641, 1642
Anthias	1304	frondosus, Scarus	1636, 1642
Calliurus	996	frontalis, Caranx	925
Cottus	1969	Gastropsetta	2636
Girardinus	688	Leuciscus	283
Holacanthus	1685	Notropis cornutus	283
Leuciscus	246	fronto, Carcharhinus	39
Leucus	246	Carcharias	39
Notropis	271	Frostfish	2540
Spheroides	1736	Frost Fishes	886
Tetrodon	1737	Fry, Hog-mouth	444
forskali, Glossodus	411	fucensis, Liparis	2119
forsteri, Sphyræna	824	Theragra	2536
forsterianus, Catostomus	176	fucorum, Apodichthys	2413
fosteri, Albula	412	Blennius	2379
Caranx	923	Xererpes	2413
Scombresox	726	fulgens, Corvina	1435
Four-Bearded Rocklings	2560	Myriopristis	846
Four-eyed Fishes	684	Priacanthus	1238
Four-spotted Flounder	2632	fulgida, Meda	329
Fox Shark	45	fuliginosus, Balistes	1702
fragilis, Citharichthys	2680	Chilomycterus	1749
Francesa, Lisa	410	Diodon	1749
francisci, Crestacion	21	Holconotus	1505
Gyropleurodus	20	Symbranchus	342
franklini, Cottus	1967	fulva, Labrus chogset	1577
Pleuronectes	2650	fulvomaculatum, Pristipoma	1339
Uranidea	1967	fulvomaculatus, Labrus	1339
Fraser River Salmon	481	fulvum, Ginglymostoma	26

	Page.
fulvus, Bodianus	1144
punctatus	1146
ruber	1145, 1146
Enneacentrus	1145
outalibi	1146
Labrus	1145
Physiculus	2547
fumeus, Notropis	294
Funal, Scorfanudi	1837
Fuucinita	1107
Fundulinæ	631
funduloides, Clinostomus	239
Fundulus	650
Leuciscus	240
Squalius	240
Zygonectes	650
Fundulus	632, 633, 637, 2827
adinia	645, 2827
albolineatus	649, 2828
arlingtonius	652
aureus	659
bermudæ	643, 2827
catenatus	648, 2828
chrysotus	655, 2828
cingulatus	656, 2829
confluentus	650, 2828
diaphnus	645, 2828
menona	645
dispar	658, 2830
dovii	650, 2828
dugesii	661
extensus	646, 2827
floridensis	642, 651, 2828
fonticola	643, 2827
funduloides	650, 2828
fuscus	624
goodei	2831
grandis	2827, 2828
guatemalensis	660
guttatus	658, 2830
henshalli	653
heteroclitus	640
badius	2827
grandis	641
macrole pido-tus	641, 2827
hieroglyphicus	658, 2830
jenkinsi	651
kansæ	2828
labialis	644, 2727
limbatus	643, 649, 2828
luciæ	654
macdonaldi	650, 2828
majalis	639, 2827, 2828
melapleurus	659, 2830
mudfish	641

	Page.
Fundulus multifasciatus	645
nigrofasciatu	641
notatus	659
nottii	656, 2830
ocellaris	642, 2827, 2828
pachycephalus	661
pallidus	638, 2827
parvipinnis	640, 2827, 2830
pisculentus	641
pulvereus	652, 2828
punctatus	637, 2827
rathbuni	649
rhizophoræ	644
robustus	644, 2827
rubrifrons	653
scartes	654
sciadicus	654, 2828
seminolis	647, 2828
similis	638
stellifer	648, 2828
swampina	645
tenellus	659
vinctus	637, 2827
viridescens	641
xenicus	662
zebra	641, 647, 2827
zebrinus	646, 2827, 2828
zonatus	657
funebris, Gobiesox	2334
Gymnothorax	396
Lycodontis	396
Noturus	147
Schilbeodes	147
Sidera	396
furca, Perca	1200
Furcaria	1545, 1546
cyanea	1547
puncta	1547
furcatum, Ditrema	1506
furcatus, Amiurus	134
Cypselurus	737, 2836
Exocœtus	737
Ictalurus	134
Phanerodon	1506
Pimelodus	134
Furcella	2472, 2869
diaptera	2472
furcidens, Characodon	669
furcifer, Anthias	1222
Brachyrhinus	1222
Paranthias	1221
Pimelodus	135
Serranus	1222
furciger, Icelus	1913
Furcimanus	2869
furcræa, Corvina	1460

	Page.
furcræa, Perca	1460
furcræus, Pachypops.	1459
Pachyurus	1460
furiosus, Noturus	149
Schilbeodes	149
furnieri, Micropogon	1462
Umbrina	1463
furthi, Arius	132
Caranx	914
Corvina (Homoprion)	1441
Hemicaranx	914
Ilisha	436
Pellona	436
Pristipoma	1319
Sciæna	1441
Sphæroides	1737
Spheroides	1737
Stellifer	1441
Tachisurus	132
Tachysurus	132, 2787
furvus, Serranus	1200
fusca, Echeneis	2270
Hydrargyra	624
Labrus tautoga	1579
Sciæna	1483
fuscatus, Silurus	140
fuscoauratus, Tetragonopterus	334
fusco-maculata, Acara	1540
Echidna	403
fusco-maculatus, Chromis	1540
fuscula, Haliperca	1211
fusculus, Centropristes	1211
fuscum, Siphostoma	770
fuscus, Acronurus	1692
Bythites	2504
Dialommus	2868
Eupomacentrus	1552
Fundulus	624
Gadus tomcodus	2540
Hemirhombus	2686
Pomacentrus	1552
Psenes	951
Serranus	1181
Syngnathus	770
Trachinotus	942
fusiforme, Etheostoma	1103
fusiformis, Boleichthys	1101
Boleosoma	1102
Hololepis	1102
Phalangistes	2048
Pœcilichthys	1102
fyllæ, Raja	69
Gabilan	91
gabonensis, Argyreiosus	935

	Page.
gabonensis, Caranx setipinnis	935
Vomer	934, 2846
Gadella	2545
Gadidæ	2531
Gadinæ	2531
Gadus	2540
æglefinus	2543
agilis	2534
albidus	2531
arenosus	2541
auratus	2542
barbatus	2541
brosme	2561
californicus	2539
callarias	2541
carbonarius	2534
chalcogrammus	2536
cimbrius	2560
colinus	2535
compressus	2551
fabricii	2534
fimbria	1862
glacialis	2534
gracilis	2538
heteroglossus	2541
lacustris	137, 2551
longipes	2555
lubb	2561
macrocephalus	2541
maculosus	2551
manus	2541
maraldi	2546
merluccius	2530
merlus	2530
molva	2552
morrhua	2541
ogac	2542
ogat	2542
periscopus	2536
polymorphus	2540
productus	2531
proximus	2539
pruinosus	2540
punctatus	2553
pygmæus	2542
raptor	2552
ruber	2530
rupestris	2541
saida	2534
tau	2316
tenuis	2555
tomcod	2540
tomcodus	2540
fuscus	2540
luteus	2540
mixtus	2540

	Page.
Gadus torsk	2561
vertagus	2541
virens	2534
Gaff-topsail	118
Catfishes	116
Pámpano	940
Gag	1177
Gaidropsarinæ	2532
Gaidropsarus	2557
argentatus	2559
ensis	2558
septentrionalis	2559
gaimardianus, Mugil	814
gairdneri, Fario	499
Salmo	497
beardsleei	2819
crescentis	2821
kamloops	499
shasta	502
stonei	503
galactura, Oliola	279
galacturus, Hysilepis	279
Notropis	279
Galafate	1711
galapagorum, Umbrina	1468
galeata, Belone	716
galeatus, Gymnocanthus	2010
Tylosurus	716
Galei	21
Galeichthys	119, 122, 2770, 2771
assimilis	2779
azureus	2775
bahiensis	119
blochii	118
cærulescens	2776
chevola	932
crinitus	932
dasycephalus	2780
eydouxii	118
gilberti	2773
gronovii	117
guatemalensis	2778
jordani	2774
lentiginosus	122, 2771
longicephalus	2781
peruvianus	122, 2771
phrygiatus	2782
rugispinis	2781
seemanni	2772
surinamensis	2780
xenauchen	2777
galeichthys, Carangoides	932
Galeidæ	27
Galeinæ	27
Galeocerdo	32
maculatus	32

	Page.
Galeocerdo tigrinus	32
galeoides, Otophidium	2491
Galeorhininæ	27
Galeorhinus	31
zyopterus	32
Galeus	29, 31, 2745
californicus	30
dorsalis	32
maculatus	30
galeus, Epinephelus	1164
Serranus	1164
Gallichthys	931
gallinula, Monacanthus	1716
Galliwasp	538
Gallus	931
virescens	932
gallus, Zeus	936
galtiæ, Squalius	237
Gambusia	678
affinis	680, 2832, 2833
arlingtonia	652, 2828, 2829
episcopi	683
gracilis	682, 683, 2832
holbrooki	681, 2832
humilis	682
infans	680, 2832
melapleura	2830
modesta	693
nicaraguensis	682, 2832
nobilis	682, 2832
patruelis	682, 2832
picturata	683, 2832
plumbea	695
punctata	679
puncticulata	680, 2832
senilis	682
speciosa	681
tridentiger	2833
Gambusiinæ	632
Gambusinus	633, 635, 648
gamphodon, Oxyrhina	49
Ganoid Fishes	100
Ganoidei	100
Ganoids, Bony	107
Cartilaginous	100
Gar, Alligator	111
Great	111
Long-nosed	109
Short-nosed	110
Gar Pikes	108
garabata, Mojarra	1353
gardeniana, Hiatula	1578
gardenii, Centronotus	948
Sternoptyx	966
Stromateus	966
gardoneum, Chondrostoma	251

Page.

gardoneus, Abramis................. 251
      Leuciscus 251
      Notemigonus............ 251
Gardonus 243
Garfish............................. 714
Garibaldi 1564
Garibaldis 1564
Garlopa............................ 1186
garmani, Characodon................ 2832
      Gobius................. 2225
      Lepomia................ 1002
      Notropis............... 281
Garmannia...................... 2231, 2232
      hemigymna 2233
      paradoxa............... 2232
      seminuda 2233
garnoti, Halichæres 1593
      Iridio................. 1593
      Julis.................. 1593
      Platyglossus 1593
Garrupa 1161, 1797
      nigrita................ 1161
Gascon............................. 910
Gaspereau 426
Gaspergou 1484
Gasteracanthus..................... 746
Gasteropelecus..................... 337
      acanthurus......... 579
      crocodilus......... 558
      humboldti.......... 572
      maculatus 338
      Trachinus 2297
Gasterostea........................ 745
      blanchardi 746
Gasterosteidæ 742
Gasterosteinæ 743
Gasterosteus....................... 746
    aculeatus............. 747, 2836
      cataphractus. 750
    algeriensis 748
    antecessor............. 900
    apeltes 752
    argentissimus 747
    argyropomus........ 748
    bailloni 747
    biaculeatus 748
    bispinosus 748
      atkinsi ... 748
      cuvieri .. 749, 2836
    brachycentrus 748
    breviceps 746
    canadus.............. 948
    carolinus............ 944
    cataphractus........ 749
    concinnus............ 745
    dekayi 746

Page.

Gasterosteus, dimidiatus........... 749
      ductor 900
      elegans 748
      gladiunculus......... 2836
      globiceps 744
      gymnurus 748
      inconstans........... 744
      insculptus 750
      intermedius........... 750
      islandicus............ 748
      lævis 745
      leiurus.............. 747
      loricatus............. 747
      lotharingus.......... 746
      mainensis............ 745
      micropus............. 744
      millipunctatus....... 752
      nebulosus............ 746
      neustrianus.......... 747
      obolarius............ 750
      occidentalis.......... 745
      plebeius 751
      ponticus 747
      pugetti 751
      pungitius 745
      brachypoda. 746
      pygmæus 744
      quadracus 752
      saltatrix 947
      semiarmatus 747
      semiloricatus 747
      serratus.............. 750
      spinulosus 748
      tetracanthus 748
      trachurus............ 747
      wheatlandi........... 749
      williamsoni.......... 750
      microceph-
      alus.... 750
Gastrobranchus.................... 7
      cæcus 8
      dombey............. 6
Gastrophysus 1727
Gastropsetta...... 2636
      frontalis 2636
Gastrostomus 406
      bairdii 406
Gata 26
Gato 28
    Bonaci 1187
gaucharote, Chætostomus 159
    Hemiancistrus......... 159
    Hypostomus........... 159
gavailis, Lepisosteus 110
gayi, Epicopus 2530
gelatinosum, Melanostigma 2479

	Page.
gelatinosus, Careproctus	2134, 2135
Cyclopterus	2135
Liparis	2134, 2135
gelida, Ammocrypta	1064
gelidus, Ceratichthys	317
Gobio	317
Hybopsis	316, 317
geminatus, Blennius	2385
Hypleurochilus	2385
gemma, Hypoplectrus	1193
gemmifer, Astronesthes	586
Lampanyctus	559
Gempylidæ	877
Gempylinæ	878
Gempylus	883
coluber	884
ophidianus	884
prometheus	883
serpens	884
solandri	883
generosus, Catostomus	170
Pantosteus	170
Genicanthus	1682
tricolor	1684
Genizara	1586
genizara, Clepticus	1587
Rabirubia	1586
gentilis, Blennius	2388
Hypsoblennius	2387
Isesthes	2388
Genyatremus	1342
luteus	1342
Genyonemus	1460
fasciatus	1479
lineatus	1460
Genyoroge	1247
viridis	1246
Genypterus omostigma	2490
Genytremus	1314
interruptus	1319
geometricus, Anchisomus	1736
Chilomycterus	1749
Diodon	1748, 1749
Tetrodon	1735, 1736
Zeus	936
Geophagus	1542
crassilabris	1543
georgianus, Caranx	927
Exocœtus	730
georgii, Tetrapturus	892
gerania, Belone	716
germanus, Notropis	261
Germo	870
alalunga	871
germo, Scomber	871
Germon	871

	Page.
Gerres	1373, 1377
aprion	1373
argenteus	1371
aureolus	1375
axillaris	1378
brasilianus	1378
brevimanus	1377
brevirostris	1376
californiensis	1370
cinereus	1370
dowi	1368
embryx	1379
gracilis	1370
gula	1371
harengulus	1369
jonesi	1368
lineatus	1377
mexicanus	1380
olisthostoma	1377
olisthostomus	1376
patao	1378
peruvianus	1376
plumieri	1379
pseudogula	1368
rhombeus	1374
squamipinnis	1373
zebra	1373
Gerridæ	1366
ghini, Orthragoriscus	1754
Ghost-fish	2443
gibba, Pterophryne	2717
gibber, Salmo	478
gibbiceps, Heros	1536
gibbifrons, Cypselurus	2836
Exocœtus	741
Gibbonsia	2351
elegans	2353, 2869
evides	2352, 2869
gibbonsii, Holconotus	1509
gibbosa, Cliola	272
Gila	235
Moniana	272
Perca	1009, 1296
marina	1295
Tigoma	235
gibbosum, Hæmulon	1296
gibbosus, Aphredoderus	787
Calliodon	1296
Catostomus	186
Cyprinodon	672
Eupomotis	1009
Holocentrus	1319
Leuciscus	231
Pomotis	1005
Squalius	231
gibbsii, Hemilepidotus	1936

	Page.
gibbsii, Salmo	493
clarkii	2819
mykiss	493
gibbus, Hypsilepis cornutus	283
Liparis	2123
Lophius	2717
gigas, Cerna	1154
Ephippus	1668
Epinephelus	1154
Hippoglossus	2612
Mugil	1483
Perca	1154
Sciæna	1483
Seriola	903
Serranus	1154
Stereolepis	1137
Zonichthys	903
Gila	226
affinis	228
ardesiaca	237
conocephala	219
elegans	226
emoryi	227
estor	240
gibbosa	235
gracilus	227
grahami	227
grandis	225, 2797
gula	234
microlepidota	207
montana	238
nacrea	228
nigra	235
phlegethontis	243
pulchella	234
robusta	227
seminuda	228
vandoisula	240
Gila Trout	226
gila, Catostomus	180
gilberti, Agonus	2060
Citharichthys	2686
Galeichthys	2773
Hypsoblennius	2386
Ilypnus	2253
Isesthes	2387
Lepidogobius	2254
Menidia	798
Notropis	266
Noturus	148
Podothecus	2058
Salmo irideus	502
Schilbeodes	148
Sebastodes	1823
Ulocentra	1049
Gilbertina	2027

	Page.
Gilbertina sigolutes	2028
Gillellus	2298
arenicola	2299
ornatus	2299
semicinctus	2298
gilli, Labichthys	368
Leuciscus	239
Pleuronectes	2654
Sebastodes	1811
Synchirus	2024
Gillia	2249
gillianus, Julis	1610
Gillichthys	2249
detrusus	2251
guaymasiæ	2252
mirabilis	2250
y-cauda	2252
gillii, Bassogigas	2515
Cetomimus	549
Clinus	2358
Lepomis	992
Lipogenys	619
Malacoctenus	2358
Neobythites	2513
Pœcilia	692
Stephanoberyx	836
Xiphophorus	692
Xystroplites	1007
Ginglymostoma	26
caboverdianus	26
cirratum	26
fulvum	26
Ginglymostomidæ	25
giorna, Cephalopterus	93
girardi, Acipenser	106
Caranx	922
Girardinichthys	666
innominatus	666
Girardinus	686
formosus	688
guppii	2834
metallicus	687
occidentalis	689
pleurospilus	688
sonoriensis	689
uninotatus	687
vandepolli	2834
versicolor	689
Girella	1381
dorsomacula	1382
nigricans	1382
Girellinæ	1381
Giton	340
fasciatus	340
Gizzard Shads	415
glaber, Cottus	2316

	Page.
glaber, Cyclopterichthys	2104
Ostracion oblongus	1735
Pleuronectes	2650
glabra, Liopsetta	2650
Platessa	2650
glabro, Corpore oblongo	2657
Ostracion subrotundus ventre	1749
glaciale, Myctophum	574
glacialis, Cottus	1976
Gadus	2534
Liopsetta	2649, 2650
Pleuronectes	2649
Scopelus	574
Squalus	57
gladiunculus, Gasterosteus	2836
gladius, Trichiurus	887
Tylosurus	716
Xiphias	894
Glance Fish	954
Glasseye	1021
glauca, Prionace	33
glaucofrænum, Coryphopterus	2220
Gobius	2219
glaucoides, Trachynotus	941
glaucopareius, Acanthurus	1694
glaucos, Sebastodes	1777
glaucostictus, Rhinobatus	63
glaucostigma, Rhinobatus	62
glaucus, Carcharhinus	33
Carcharias	33
Chætodon	941
Isuropsis	48
Squalus	33
Trachinotus	940
Trachynotus	941
Glaustegus	61
Globefish	1734
globiceps, Agonostoma	821
Blennicottus	2017
bryosus	2017
Centridermichthys	2017
Gasterosteus	744
Oligocottus	2017
globosa, Lyosphæra	1752
gloriosus, Bryttus	994
Enneacanthus	993
Glossamia	1111
pandionis	1111
Glossichthys	2704
plagiusa	2710
Glossodon	412
harengoides	413
heterurus	413
glossodonta, Argentina	411
Glossodus	411

	Page.
Glossodus forskali	411
Glossoplites	991
gloveri, Salmo	487
Glut Herring	426
glutinosa, Myxine	7
Glyphidodon	1560
chrysurus	1567
declivifrons	1562
rudis	1563
saxatilis	1562
taurus	1563
troschelli	1562
Glyphisodon	1560, 1561
moucharra	1562
rubicundus	1565
Glyptocephalus	2656
acadianus	2657
cynoglossus	2656
pacificus	2655
saxicola	2657
zachirus	2658
Gnathanodon	927
Gnathobolus	437
mucronatus	438
Gnathocentrum	1687
Gnathodon speciosus	928
gnathodus, Pseudoscarus	1650
Scarus	1650
Gnatholepis	2210
Gnathophis	355
Gnathypops	2283
macrops	2284
maxillosa	2284
maxillosus	2284
mystacina	2286
mystacinus	2286
rhomalea	2285
scops	2283
snyderi	2285
Goatfish, Red	858
Yellow	859
Goatfishes	857
Gobies	2184, 2188, 2210
Blind	2261
Crested	2209
Half-naked	2231
Naked	2257
Gobiesocidæ	2326
Gobiesocinæ	2327
Gobiesox	2329, 2330, 2331
adustus	2334
carneus	2337
cephalus	2332
cerasinus	2336
eigenmanni	2339
eos	2343

	Page.
Gobiesox erythrops	2336
fasciatus	2336
funebris	2334
gyrinus	2331
hæres	2337
humeralis	2341
macrophthalmus	2335
nigripinnis	2332
nudus	2331
papillifer	2330
pinniger	2329
pœcilophthalmus	2335
punctulatus	2338
reticulatus	2328
rhessodon	2340
rhodospilus	2335
rubiginosus	2337
rupestris	2341
strumosus	2333
tudes	2333
virgatulus	2333
zebra	2342
Gobiichthys	2210
Gobiidæ	2188
Gobiinæ	2190
Gobio æstivalis	316
cataractæ	306
gelidus	317
plumbeus	324
vernalis	321
gobio, Clinus	2365
Cottus	1941, 1968, 2009
Gobioclinus	2365
Gobioclinus	2364
gobio	2365
Gobioidea	2184
Gobioidei	781
Gobioides	2263, 2868
barreto	2264
broussoneti	2264
broussonnetii	2263
peruanus	2264
gobioides, Cottus	1968
Uranidea	1968
Gobioidinæ	2192
Gobiomorus	2194
dormitator	2195
dormitor	2195
gronovianus	950
Gobionellus	2210, 2215, 2227
hastatus	2229
oceanicus	2230
smaragdus	2228
stigmaticus	2224
gobionium, Campostoma	206
Gobiosoma	2257

	Page.
Gobiosoma alepidotum	2259
bosci	2259
ceuthœcum	2261
crescentale	2259
crescentalis	2260
histrio	2258
ios	2255
longipinne	2256
molestum	2258
multifasciatum	2260
zosterurum	2257
Gobious oblongus	2264
Gobius	2209, 2211, 2216
alepidotus	2259
amorea	2201
andrei	2218
bacalaus	2230
badius	2227
banana	2236
boleosoma	2221
bosci	2227, 2259, 2868
brunneus	2218
cauda longissima acuminata	2230
carolinensis	2218
catulus	2218
chiquita	2241
cristagalli	2209
cyprinoides	2209
dalli	2230
dolichocephalus	2237
eigenmanni	2218
emblematicus	2247
eucæomus	2223
fasciatus	2222
flavus	2235
garmani	2225
glaucofrænum	2219
gracilis	2249
gronovii	949
gulosus	2244
hastatus	2229
hemigymnus	2233
kraussi	2228
lacertus	2218
lanceolatus	2229, 2230
lepidus	2249
lineatus	2218, 2260
longicauda	2229
lucretiæ	2868
lyricus	2224
manglicola	2220
mapo	2218
martinicus	2236
mexicanus	2237
microdon	2227
minutus	2097

	Page.		Page.
Gobius nelsoni	2235	Gonocephalus macrocephalus	2184
newberryi	2248	Gonochætodon	1672
nicholsii	2218	Gonopterus	1687
oceanicus	2230	mœrens	1688
paradoxus	2232	Gonostoma	578
pisonis	2201	brevidens	579
plumieri	2206	denudatum	579
poeyi	2226	elongatum	583
quadriporus	2221	microdon	582
sagittula	2228	Gonostominæ	578
seminudus	2234	Gonostomus acanthurus	579
shufeldti	2221	Goodea	685
smaragdus	2227	atripinnis	685
smyrnensis	2118	goodei, Aldrovandia	610
soporator	2216	Erimyzon	186
stigmaticus	2224	Halosaurus	610
stigmaturus	2220	Hymenocephalus	2572
strigatus	2228	Lucania	664
taiasica	2236	Macrurus	2572
thalassinus	2245	Myliobatis	2755
townsendi	2250	Nematonurus	2571
viridipallidus	2259	Paralonchurus	1480
wurdemanni	2225	Ptilichthys	2452
zebra	2226	Sebastichthys	1780
gobius, Liparis	2108	Sebastodes	1779
goboides, Hypsicometes	2294	Spinivomer	367
Goby, Long-jawed	2250	Trachinotus	943
Naked	2259	Urolophus	81
Sharp-tailed	2229	Goodeinæ	632
goddeffroyi, Percichthys	1197	Goodfish	487
godmanni, Cichlasoma	1516	Goodies	1458
Heros	1516	Goody	1458
Pimelodus	152	Goosefish	2713
Rhamdia	152	Gorbuscha	478
Goggle-eye	990, 992	gorbuscha, Oncorhynchus	478
Jack	911	Salmo	478
Goggler	911	Gordiichthys	363
Golden Shiner	250	irretitus	363
Trout of Mount Whitney	503	goreensis, Albula	412
Goldfish	201	Trachynotus	943
Golet	507	Vomer	934
Goltra	508	gouani, Lepidopus	887
Goma soi	1833	Goujon	2790
gomesii, Ophichthus	384	Gourd-seed Sucker	168
Ophichthys	385	gracile, Boleosoma	1102
Ophisurus	385	Myctophum	572
Gonenion	946	Peristedion	2179
serra	947	gracilis, Aldrovandia	610
Goniobatis	88	Atherinichthys	797
macroptera	88	Blennius	2438
Goniodus	57	Catostomus	179
Gonionarce	78	Chætodon	1675
Gonioperca	1194	Cottus	1968
Gonioplectrus	1139	Cyprinus	326
hispanus	1140	Diapterus	1370
Gonocephalus	2183	Gadus	2538

	Page.
gracilis, Gambusia	682, 683
Gerres	1370
Gila	227
Gobius	2249
Hippocampus	777
Hybopsis	221
Lepidogobius	2249
Lepisosteus	110
Leptocephalus	354
Leuciscus	283, 326
Lycodes	2465
Menidia	797
beryllina	797
Moniana	272
Perca	1024
Photonectes	591
Pimelodus	135
Platygobio	326
Pleurogadus	2538
Pœcilichthys	1103
Ptychocheilus	225
Scomber	867
Scopelus	572
Septogunnellus	2436
Tigoma	236
Tilesia	2538
Umbrina	1474
Uranidea	1968
Xiphophorus	683
graciosus, Pimelodus	135
gradiens, Hypsagonus (Cheiragonus)	2041
graellsi, Ophidion	2488
grahami, Gila	227
Leuciscus	228
Oligocephalus	1089
grallator benthosaurus	543
Gramma	1228
loreto	1229
Grammateus	1347, 1348, 1353
humilis	1355
medius	1356
grammaticum, Thalassoma	1610, 2859
grammaticus, Chlorichthys	1610
Grammatopleurus	1866
lagocephalus	1875
Grammatostomias	590
dentatus	590
Grammichthys	2693
lineatus	2702
Grammicolepididæ	973
Grammicolepis	974
brachiusculus	974
Grammiconotus	725
hicolor	726
Gramminæ	1131
	Page.
Grammistes acuminatus	1487
hepatus	1343
mauritii	1323
trivittatus	1311
unimaculatus	1360
Grand Oranchee	1057
Grande Écaille	409
grandicassis, Arius	126
Netuma	126, 2764
Tachisurus	126
stricticassis	126
grandicornis, Scorpæna	1850
grandipinnis, Photogenis	280
grandis, Fundulus	2827
heteroclitus	641
Gila	225, 2797
Leuciscus	225
Ptychocheilus	225, 2796
grandisquama, Eleotris	2198
grandisquamis, Chærojulis	1597
Platyglossus	1597
Upeneus	860
grandoculatus, Centropomus	1120
grandocule, Chirostoma	2839
grandoculis, Atherinichthys	2840
granulata, Perca	1024
Raia	72
granulatus, Amorphocephalus	1619
granulosa, Pristis	61
Graodus	254
nigrotæniatus	264
Grass Bass	987
Porgy	1355
Rockfish	1819
Gray Grunt	1296
Pike	1022
Snapper	1255
grayi, Carpiodes	167
Lepidosteus	111
Salmo	509
Grayling, Arctic	517, 2871
European	2870
Michigan	518, 2871
Montana	519, 2871
Graylings	517
Great Albacore	870
Amber-fish	903
Barracuda	823
Bear Lake Herring	470
Blue Shark	33
Chub	232
Flounder	2652
Flying-fish	740
Fork-tailed Cat	137
Gar	111

	Page.
Great Lake Trout	504
Northern Pike	630
Pámpano	943
Pike	629
Pipefish	764
Sculpin	1976
Sculpins	1970
Sea Lamprey	10
Tunnies	869
White Shark	50
grebnitskii, Pholidapus	2431
Green Bass	1012
Cod	2534
Parrot-fish	1638, 1657
Pike	627
Sturgeon	104
Sunfish	996
Green-back Trout	497
Green-fish	1382
Alaska	1869
greenei, Uranidea	1965
greeni, Couesius	324
Neoliparis	2112
Greenland Charr	508, 510
Codfish	2542
Halibut	2611
Greenling	1871
Greenlings	1863, 1866
Green-sided Darter	1053
Gregory, Beau	1555
Grenadiers	2561
grex, Scomber	867
grimaldii, Conchognathus	349
Grindle	113
grisea, Dionda	216
Lucioperca	1022
Sciæna	1484
Unibranchapertura	342
griseolineatum, Siphostoma	764
griseolineatus, Syngnathus	764
griseum, Stizostedion canadense	1022
griseus, Acomus	175
Carcharias	47
Catostomus	175
Hexanchus	19
Labrus	1257
Lutjanus	1257
Mesoprion	1257
Neomænis	1255
Notidanus	19
Saurus	537
Squalus	19
grœllsi, Ophidium	2487
grœnlandica, Nansenia	528
grœnlandicus, Aspidophoroides	2092
Cottus	1975

	Page.
grœnlandicus, Cottus scorpius	1975
Gunnellus	2418
Himantolophus	2733
Hippoglossus	2611
Microstomus	528
Myoxocephalus	1974
Salmo	521
Gronias	135, 136, 142
nigrilabris	142
gronovianus, Gobiomorus	950
gronovii, Achirus	2696
Ailurichthys	117
Galeichthys	117
Gobius	949
Nomeus	949
Ostracion	1725
Solea	2696
Zoarces	2457
grossidens, Engraulis	451
Lycengraulis	451
Ground Drummer	1436
Spearing	533
Grouper, Black	1161, 1174
Mangrove	1171
Nassàu	1157
Red	1160
Yellow	1183
Yellow-finned	1155, 1172
Groupers	1148
Grubber Broad-head	447
Grubby	1973
grunniens, Amblodon	1484
Aplodinotus	1484
Haploidonotus	1484
Labrus	1483
Mugil	1483
Grunt, Black	1297
Boar	1303
Common	1304
French	1306
Gray	1296
Margaret	1295
Open-mouthed	1306
Red-mouthed	1308
Striped	1296
White	1310
Yellow	1303
Grunters	1289
Grunts	1291
Striped	1313
gryllus, Exocœtus	729
Grystes	1010
lineatus	1868
megastoma	1012
nobilis	1012
nuecensis	1012

	Page.
Guabi coara brasiliensibus	1305
Guacamaia	1657, 1658
guacamaia, Hemistoma	1659
Pseudoscarus	1656, 1659
Scarus	1656, 1658
Guacamaias	1655
guachancho, Sphyræna	824
Guachinango, Pargo	1264
guaco, Amore	2236
Guaguanche	824
Pelon	824
guaiacan, Echeneis	2270
Guajacon	679
Guajacones	678
Guajica	692
Guamaiacu atinga	1749
Guamajacu guara	1745
guanabana, Erizo	1746
Guapena	1489
Guaperva	1703
lataforcipata	1702
guara, Caranx	926
Guamajacu	1745
Scomber	927
Guarapucu	876
Guardfish	715
Guasa	1154, 1162
guasa, Promicrops	1164
Serranus	1164
Guaseta	1164
guatemalensis, Arius	129
Adinia	660
Anacyrtus	338
Atherinichthys	801
Citharichthys	2686
Fundulus	660
Galeichthys	2778
Hexanematichthys .	129
Menidia	801
Pimelodus	152
Rhamdia	152
Rœboides	338
Guativere	1144, 1145
Amarilla	1144, 1145
Black	1146
Red	1145
guativere, Bodianus	1145
Serranus	1145
Guatucupa juba	1323
Guavina	2194, 2198, 2201
guavina	2198
Hoyera	2236
Mapo	2196
Tétard	2200
guavina. Batrachus	2195
Eleotris	2199

	Page.
guavina, Guavina	2198
Guavinas	2194
guaymasiæ, Gillichthys	2252
guaza, Epinephelus	1154
Labrus	1154
Gudlax	954
Guebucu	891
guebucu, Skeponopodus	891
guentheri, Percina	1034
Guerubaco	2198
Gueule, Petite	1370
guianensis, Belone	715
guichenoti, Cayennia	2265
guineensis, Ostracion	1725
Guiritinga	119
Guitarfish	63
Guitarfishes	61
Guitarro	62
gula, Eucinostomus	1370
Gerres	1371
Gila	234
Squalius	234
Gulf Flounder	2631
Menhaden	434
Shad	2810
gulo, Holocentrus	1139
gulonellus, Leucosomus	326
Pogonichthys (Platygobio)	236
gulosa, Uranidea	1945
gulosus, Cathorops	133, 2788
Centridermichthys	1945
Chænobryttus	992
Cottopsis	1945
Cottus	1944
Gobius	2244
Lepidogobius	2244
Pomotis	992
Tachisurus	133
Gulpers	404
gulula, Eucinostomus	1371
gummigutta, Hypoplectrus	1192
unicolor	1192
Plectropoma	1192
gundlachi, Dormitator	2198
gunelliformis, Asternopteryx	2420
Centronotus	2421
Murænoides	2421
Gunellus ingens	2419
macrocephalus	2419
Gunnell	2419
Gunnellops	2420
roseus	2420
Gunnellus alectrolophus	2422
anguillaris	2436
apos	2430

	Page.
Gunnellus dolichogaster............	2417
fabricii	2438
fasciatus	2418
grænlandicus...........	2418
islandicus	2439
murænoides	2418
nebulosum..............	2414
ornatus..................	2420
punctatus	2440
ruberrimus..............	2417
vulgaris	2419
gunnellus, Blennius.................	2419
Centronotus.............	2419
Murænoides.............	2419
Pholis	2419
Gunnels.......................	2414
gunneri, Scomber	955
gunnerianus, Squalus..............	51
gunnisoni, Oliola...................	273
Cyprinella..............	273
guntheri, Aspidophoroides.........	2090
Etheostoma	1034
Hadropterus	1033
Halosaurus	608
Hoplopagrus.............	1244
Lampanyctus.............	559
Mugil....................	812
Sirembo	2523
Sphyræna.................	824
Xiphophorus	702
guntheriana, Alutera	1720
guppii, Girardinus ,................	2834
Gurnard, Big-headed	2171
Common	2156
Flying....................	2183
Northern Striped...... 2167, 2168	
Red......................	2177
Gurnards 2147, 2148, 2152	
Deep-water	2177
Flying...................	2182
Mailed...................	2176
Small-scaled	2175
Gurnardus	2148
Gusas	1162
guttata, Mycteroperca venenosa...	1174
Perca 1142,1164	
Scorpæna	1847
guttatus, Astroscopus..............	2310
Balistes	1702
Chiropsis	1869
Chirus	1868
Enneacentrus	1142
coronatus .	1142
Epinephelus 1142,1159	
Fundulus	658
Hippocampus	776

	Page.
guttatus, Johnius...................	1174
Lampris...................	955
Lutianus..................	1269
Lutjanus..................	1269
Mesoprion	1269
Murænoides..............	2419
Neomænis.................	1269
Ophisurus...............	382
Percopsis	784
Petrometopon	1142
Pomotis	993
Promicrops...............	1162
Sebastapistes	1848
Upsilonphorus	2310
Zeus	955
Zygonectes...............	653
guttavarium, Plectropoma.........	1192
guttavarius, Hypoplectrus.........	1192
unicolor	1192
guttifer, Ophichthus...............	383
Ophichthys	383
guttulatus, Hippocampus	778
Pisodonophis...........	377
Pleuronectes...........	2640
Pleuronichthys	2640
guzmaniensis, Catostomus	171
Pantosteus..........	171
Gymnachirus	2703
fasciatus	2703
Gymneleotris	2204
seminuda.............	2204
seminudus	2204
Gymnelinæ	2456
Gymnelis	2477
pictus	2477
stigma	2477
viridis...................	2477
Gymnepignathus	907
Gymnocanthus	2006
galeatus..............	2010
pistilliger ... 2006, 2008, 2009	
tricuspis	2008
Gymnocephalus	962
ruber..............	1146
Gymnodontes 781, 1726	
Gymnogaster	889
gymnogaster, Cotylopus	2207
Sicydium.............	2208
Sicyopterus	2208
Gymnomuræna 402, 403	
nectura.............	404
vittata	404
Gymnonoti........................	339
Gymnopsis	402
Gymnosarda	868
alleterata...............	869

	Page.
Gymnosarda pelamis	868
gymnostethus, Prionotus	2153
Gymnothorax	392, 400, 401
afer	395
aquæ-dulcis	391
catenatus	403
chlevastes	399
conspersus	397
dovii	397
elaboratus	389
flavoscriptus	395
funebris	396
longicauda	392
marmoreus	391
miliaris	398
mordax	396
moringua	395
nigrocastaneus	390
obscuratus	389
ocellatus	399
nigromarginatus	400
saxicola	399
panamensis	391
picturatus	395
pictus	2805
polygonius	394
rostratus	395
sanctæ-helenæ	397
scriptus	398
umbrosus	390
verrilli	394
versipunctatus	394
vicinus	394
virescens	394
Gymnotidæ	340
Gymnotorpedo	77
Gymnotus albus	340
bachiurus	340
carapo	341
fasciatus	340
putaol	341
gymnura, Dasyatis	84
Trygon	84
gyrans, Querimana	818
Gyrinichthys	2137
minytremus	2137
gyrinops, Cyclopteroides	2102
gyrinus, Eleotris	2201
Gobiesox	2331
Noturus	146
Schilbeodes	146
Silurus	146
Gyropleurodus	20
Gyropleurodus francisci	20
quoyi	21

	Page.
Haddo	478
Haddock, Jerusalem	954
Haddocks	2542
Hadropterus	1028, 1030, 1038
aspro	1032
aurantiacus	1041
cymatotænia	1042
evides	1036
guntheri	1033
macrocephalus	1031
maculatus	1031, 1034
nianguæ	1043
nigrofasciatus	1038
ouachitæ	1035
peltatus	1034
phoxocephalus	1030
roanoka	1036
scierus	1037
serrula	1038
shumardi	1047
squamatus	1040
tessellatus	1070
variatus	1070
hæmatura, Oliola	218
hæmaturus, Hybopsis	218
Leuciscus	218
Hæmulidæ	1289
Hæmulon	1291
acutum	1299
albidum	1299
album	1295, 1296
arara	1306
arcuatum	1305
aurolineatum	1310
bonariense	1297
brevirostrum	1300
cana	1299
canna	1297
capeuna	1311
carbonarium	1300
caudimacula	1299, 1302, 1309
chromis	1299
chrysargyreum	1308
chrysopteron	1309
chrysopterum	1309
continuum	1297
corvinæforme	1327
dorsale	1303
elegans	1304
flaviguttatum	1312
flaviguttatus	1312
flavolineatum	1306
formosum	1305
fremebundum	1297
gibbosum	1296
heterodon	1306
hians	1304

	Page.
Hæmulon jeniguano	1310
labridum	1319
luteum	1304
macrostoma	1297
macrostomum	1296
maculicauda	1314
maculosum	1295
margaritiferum	1312
mazatlanum	1314
melanurum	1302, 1303
microphthalmum	1296
modestum	1340
multilineatum	1304
notatum	1297
obtusum	1319
parra	1287
parræ	1297, 1309
plumieri	1304
quadrilineatum	1309, 1311
quinquelineatum	1311
retrocurrens	1297
rimator	1309
schranki	1302, 1303
sciurus	1303
scudderi	1299, 1360
serratum	1299
sexfasciatum	1294
sexfasciatus	1295
similis	1304
steindachneri	1301
striatum	1311
subarcuatum	1306
tæniatum	1308
undecimale	1300
xanthopteron	1307
xanthopterum	1307
Hæmulopsis	1325
Hæmylum capeuna	1311
hæres, Gobiesox	2337
Hagfish	7
California	6
Hagfishes	5, 7
Hairtails	889
Hake, European	2530
New England	2530
Silver	2530
White	2555
Hakes	2529
Halælurus	23
Halatractus	901
bonariensis	905
bosci	905
coronatus	905
dorsalis	902, 906
fasciatus	904
halec, Clupéa	421

	Page.
Half-beak, Common	721
Halfbeaks	719
Half-moon	1391
Half-naked Gobies	2231
Halias	2502
marginatus	2502
Halibut	2611
Arrow-toothed	2609
Bastard	2625
Greenland	2611
Monterey	2625
Halibuts, Bastard	2624
Halicampus	761
Halichæres	1587
bivittatus	1597
californicus	1601
caudalis	1600
dimidiatus	1594
dispilus	1598
garnoti	1593
maculipinna	1595
nicholsi	1592
poeyi	1598
radiatus	1591
sellifer	1592
semicinctus	1593
Halieutæa	2741
spongiosa	2742
Halieutella	2742
lappa	2742
Halieutichthys	2739
aculeatus	2739
caribbæus	2741
reticulatus	2741
Halieutinæ	2736
Haliperca	1203, 1204
bivittata	1205
fuscula	1211
jacome	1215
phœbe	1212
præstigiator	1214
tabacaria	1215
halleri, Urolophus	80
Halocypselus	729, 2835
evolans	729
obtusirostris	730
Haloporphyrus viola	2544
Halosauridæ	606
Halosauropsis	2826
Halosaurus	607
goodei	610
guntheri	608
macrochir	610
oweni	607
rostratus	609
hamatus, Icelus	1913

	Page.
hamiltoni, Brachioptilon	93
Hippoglossoides	2616
Hamlet	395, 1157
hamlini, Podothecus	2056
Hammer-head	181
Shark	45
hammondi, Percopsis	784
Pimelodus	135
Semotilus	222
Hand-saw Fish	596
Hannabill	1199
Haplocheilus	633, 2827, 2830
Haplochilus aureus	659
chrysotus	656
dovii	650
floripinnis	651
luciæ	655
melanopleurus	660
melanops	682
pulchellus	659
sciadicus	654
Haplodoci	782, 2313
haplognathus, Lepomis	1004
Haploidonotus	1483
grunniens	1484
Haplomi	622
Harder	949
Hardhead	497
Hardheads	719
Hardmouth	208
Hardtail	921
Hare-lip Sucker	199
harengoides, Glossodon	413
Harengula	428, 430
arcuata	431
callolepis	430
clupeola	429, 430
humeralis	431
jaguana	430
macropthalma	430
maculosa	430
pensacolæ	431
sardina	430
harengulus, Eucinostomus	1368
Gerres	1369
harengus, Clupea	421, 422
Coregonus	469
Lavinia	209
Myxus	818
Queriuana	817
Salmo	469
harfordi, Ptychocheilus	225, 2797
Harpe	1581
cæruleo-aureus	1583
diplotænia	1582
eclancheri	1583

	Page.
Harpe pectoralis	1582
pulchella	1584
pulchra	1585
rufa	1583
Harpinæ	1572, 1574
Harpurus	1689
fasciatus	1691
harringtonensis, Atherina	791
harringtoni, Axyrias	1904
Harriotta	96
raleighana	96
Harriottinæ	94
Harry, Black	1199
Harvest Fish	965, 967
hasselti, Paraserranus	1205
hastata, Dasibatis	84
Dasyatis	83
Trygon	84
hastatus, Gobionellus	2229
Gobius	2229
hastingsi, Neomænis	2858
Haustor	135, 136, 137
havannensis, Muræna	382
Ophichthus	382
Uranichthys	382
haydeni, Ptychostomus	187
hayi, Hybognathus	214
Hay-ko	478
Head-fish	1753
Head-fishes	1752, 1753
hearnei, Salmo	510
beheri, Scomber	923
Hechudo	447
heckeli, Balistes	1709
Nerophis	774
Syngnathus	2839
Hectoria	1138
heermanni, Amphistichus	1504
Heliases	1545, 1546, 1548
insolatus	1548
multilineatus	1547
Heliastes	1545
cyaneus	1547
Helicolenus	1836
dactylopterus	1837
maderensis	1837
Helioperca	999, 1004
helleri, Cichlasoma	1521
Heros	1521
Xiphophorus	701, 702
Helmichthys	353
Helmictis	353
helolepis, Trachyrincus	2568
Helops	103
helvomaculatus, Sebastodes	1808
Hemdurgan	1760

	Page.
Hemiancistrus	159
aspidolepis	159
gaucharote	159
Hemianthias, peruanus	1222
vivanus	1223
Hemiarius	119
Hemibranchii	741
Hemibranchs	741
Hemibrycon	333
Hemicaranx	912
amblyrhynchus	912, 2844
atrimanus	913. 2846
falcatus	2845
furthii	914
leucurus	914
secundus	914
zelotes	2845
Hemichætodon	1672
Hemigobius	2210
Hemigrammus	333
hemigymna, Garmannia	2233
hemigymnus, Argyropelecus	604
Gobius	2233
Hemilepidotinæ	1880
Hemilepidotus	1934
gibbsii	1936
hemilepidotus	1935
jordani	1934
spinosus	1937
tilesii	1936
hemilepidotus, Cottus	1936
Hemilepidotus	1935
Hemiodon	156, 157
Hemioloricaria	156
Hemioplites simulans	994
Hemioplitus	992
Hemiplus	249
lacustris	250
hemiplus, Cyprinus	250
Hemiramphidæ	718
Hemiramphus	722
balao	723, 2835
brasiliensis	722
browni	723
Hemirhamphus fasciatus	720
filamentosus	723
longirostris	724
macrochirus	723
macrorhynchus	724
marginatus	723
picarti	720
pleii	723
poeyi	720
richardi	720
roberti	721
rosæ	722

	Page.
Hemirhamphus unifasciatus	720, 721
Hemirhombus	2670
æthalion	2673
aramaca	2673
fimbriatus	2677
fuscus	2686
ocellatus	2673
ovalis	2674
pætulus	2672
soleæformis	2672
Hemirrhamphus	722
Hemistoma	1642
guacamaia	1659
Hemitremia	228, 230, 242
bifrenata	259
heterodon	261
maculata	259
vittata	242
Hemitripterinæ	1883
Hemitripterus	2022
acadianus	2023
americanus	2023
cavifrons	2023
marmoratus	1889, 2022
Hemitrygon	82, 83
hemphillii, Stathmonotus	2408
henlei, Carcharhinus	37, 2746
Carcharias	37
Rhinotriacis	31
Triacis	31
henshalli, Fundulus	653
Zygonectes	653, 2829
henshavii, Apocope	312
Rhinichthys	312
henshawi, Salmo clarkii	2819
mykiss	493
hentz, Blennius	2390
Hypsoblennius	2390
hentzi, Hypsoblennius	2390
Isesthes	2390
Hepatus	1689
hepatus, Acanthurus	1692
Grammistes	1343
Teuthis	1693
heptacanthus, Cheilodipterus	947
heptagonus, Hippocampus	775, 777
Heptatremidæ	5
Heptranchias maculatus	18
heraldi, Tetrodon	1736
Herichthys	1526
cyanoguttatus	1538
hermanni, Sternoptyx	603
hermineri, Clinus	2362
herminier, Blennius	2362
herminiger, Labrisomus	2361
Hermosilla	1383, 1384

	Page.
Hermosilla azurea	1383, 1384
Heros	1526
affinis	1529
altifrons	1538
angulifer	1517
aureus	1533
balteatus	1522
basilaris	1532
beani	1538
bifasciatus	1521
centrarchus	1526
citrinellus	1534
cyanoguttatus	1537
deppii	1524
dovii	1535
erythræus	1531
erythreus	1531
fenestratus	1518
friedrichsthali	1528
gibbiceps	1536
godmanni	1516
helleri	1521
intermedius	1517
irregularis	1541
labiatus	1530
lentiginosus	1524
lobochilus	1531
longimanus	1521
macracanthus	1519
maculipinnis	1529
managuensis	1533
margaritifer	1520
melanopogon	1523
melanurus	1524
microphthalmus	1536
montezuma	1518
motaguensis	1534
multispinosus	1526
nicaraguensis	1532
nigrofasciatus	1525
oblongus	1535
parma	1519
pavonaceus	1538
rostratus	1523
salvini	1528
sieboldii	1517
spilurus	1520
tetracanthus	1539
triagramma	1529
trimaculatus	1529
troscheli	1537
urophthalmus	1537
heros, Eupomotis	1007
heros, Lepomis	1008
Pomotis	1007
Herpetoichthys	381

	Page.
Herpetoichthys, callisoma	384
ocellatus	384
sulcatus	382
Herring, Big-eyed	410, 426
Blue	425
Branch	426
California	422
Common	421
Fall	425
Glut	426
Great Bear Lake	470
Lake	468
Michigan	468
Mountain	463
Rainbow	524
Round	420
Summer	426
Tailor	425
Thread	432
Toothed	413
Wall-eyed	426
Herrings	417, 421
herschelii, Histiophorus	892
Tetrapturns	892
herschelinus, Liparis	2123
herzbergii, Arius	125
Selenaspis	124, 2760
Silurus	125
Tachysurus	125
Hesperanthias	1281
oculatus	1283
Heterandria	686
affinis	681
formosa	687, 2831
holbrooki	681
metallica	687
nobilis	682
occidentalis	689, 2833
ommata	664
patruelis	681
plenrospilus	688
uninotata	687
versicolor	688, 2833
heteroclita, Cobitis	641
heteroclitus, Fundulus	640, 2827
badius	2827
grandis	641
macrolepidotus	641
heterodon, Alburnops	261
Hæmulon	1306
Hemitremia	261
Hybopsis	261
heterodon, Leuciscus	261
Notropis	261
Heterodontidæ	19

	Page.
heteroglossus, Gadus	2541
Heterognathi	329
Heterognathus	792
heterolepis, Johnius	1419
Notropis	260
Plagioscion	1419
Sciæna	1419
Heteromi	612
Heteroprosopon	2637
heteropygus, Carangops	913
Caranx	913
Heterosomata	782, 2602
Heterostichus	2350
rostratus	2351
heterura, Clupea	416
heterurus, Cypselurus	2836
Exocœtus	735
Glossodon	413
hexacanthus, Centrarchus	987
Dipterodon	1107
hexacornis, Cottus	2003
Oncocottus	2002
Hexagrammidæ	1863, 2862
Hexagramminæ	1864
Hexagrammos	1866
asper	1872
decagrammus	1867
hexagrammus	1872
lagocephalus	1873
stelleri	1871
superciliosus	1872
Hexagrammus, decagrammus	1875
lagocephalus	1875
monopterygius	1866
octogrammus	1869
ordinatus	1870
otakii	1867
scaber	1873
hexagrammus, Chirus	1872
Hexagrammos	1872
Labrax	1872
Ozorthus	2441
Stichæus	2441
Hexanchidæ	17
Hexanchus	18
corinus	18
griseus	19
Hexanematichthys	119, 121, 128, 2772
assimilis	129
cærulescens	129
dasycephalus	130
felis	128
guatemalensis	129
hymenorhinus	125
jordani	129
longicephalus	130

	Page.
Hexanematichthys phrygiatus	130
rugispinis	130
seemani	128
surinamensis	129
hians, Athlennes	718
Belone	718
Hæmulon	1304
Myctophum	572
Sayris	725
Tylosurus	718
Hiatula	1577
gardeniana	1578
hiatula	1579
onitis	1579
hiatula, Hiatula	1579
Labrus	1578
Hickory Shad	416, 425
hieroglyphicus, Fundulus	658
Zygonectes	658
Hilgendorfia	2139, 2140, 2144
hillianus, Exocœtus	729
Spinax	55
Himantolophinæ	2728
Himantolophus	2732
grœnlandicus	2733
reinhardti	2733
himantophorus, Callionymus	2186
Himantura	82
Hind, Brown	1142
Red	1141, 1158
Rock	1152
Speckled	1159
hinnulus, Squalus	29
Hiodon	412, 413
alosoides	413
clodalus	413
selenops	414
tergisus	413
Hiodontidæ	412
Hipohomus spilotus	1043
hippe, Balistes	1705
Hippocampinæ	761
Hippocampus	775
antiquorum	776
antiquus	776
brevirostris	776
fascicularis	778
gracilis	777
guttatus	776
guttulatus	778
heptagonus	775, 777
hippocampus	775
hudsonius	777
ingens	776
kuda	778
lævicaudatus	777

	Page.
Hippocampus longirostris	778
marginalis	778
punctulatus	777
stylifer	778
zosteræ	778
hippocampus, Hippocampus	775
Syngnathus	775
Hippocephalus	2033
japonicus	2036
superciliosus	2036
Hippoglossina	2620
bollmani	2621
macrops	2621
stomata	2620
Hippoglossinæ	2605
Hippoglossoides	2614
dentatus	2615
elassodon	2615
exilis	2613
hamiltoni	2616
jordani	2614
limanda	2615
limandoides	2615
melanostictus	2618
platessoides	2614
robustus	2616
hippoglossoides, Platysomatichthys	2611
Pleuronectes	2611
Reinhardtius	2611
Hippoglossus	2611
americanus	2612
brasiliensis	2626
californicus	2626
gigas	2612
grœnlandicus	2611
hippoglossus	2611
intermedius	2672
maximus	2612
ocellatus	2673
pinguis	2611
ponticus	2612
vulgaris	2612
hippoglossus, Hippoglossus	2611
Pleuronectes	2612
hippops, Campostoma	206
hippos, Carangus	921
Caranx	920, 923
Scomber	908, 920
hippuroides, Lepimphis	952
hippurus, Coryphæna	952
hirudo, Ichthyomyzon	11
hirundinaceus, Squalus	33
Hirundo	2183
Acanthurus	1691
Azurina	1544
Cottus	2011

	Page.
hirundo, Leiocottus	2011
Hispaña, Serrana	1488
hispanis, Serrana	1489
Hispaniscus	1765, 1776, 1813
hispanum, Plectropoma	1140
hispanus, Gonioplectrus	1140
hispidus, Balistes	1716
Cottus	2023
Monacanthus	1715
Tetrodon	1733
Histiobranchus	351
bathybius	352
infernalis	352
Histiocottus	2018
bilobus	2018
Histiophorus	890
americanus	891
belone	892
brevirostris	892
herschelii	892
pulchellus	891
Histrio	2717
histrio, Antennarius	2716, 2723
Etheostoma (Ulocentra)	1051
Gobiosoma	2258
Lophius	2716, 2722
Petrophryne	2716
Scorpæna	1843, 1846
Ulocentra	1050, 1051
Hitch	209
hiulcus, Stolephorus	443
Hog Choker	2700
Molly	181, 1026
Sucker	181
Hogfish	1026, 1338, 1579
Spanish	1583
Hog-mouth Fry	444
Holacanthinæ	2860
Holacanthus	1682, 1729, 2859
ciliaris	1685
clarionensis	1683
cornutus	1685
formosus	1685
iodocus	1687
leionothos	1735
melanothus	1728
passer	1682
strigatus	1683
tricolor	1684
holacanthus, Diodon	1746
Ostracion oblongus	1746
Holanthias martinicensis	1228
holbolli, Ceratias	2729
holbrooki, Acipenser	105
Aluterus	1718
Echeneis	2270

	Page.
holbrooki, Eupomotis..............	1008
Gambusia	681
Heterandria	681
Ophidion	2487
Ophidium	2488
Pomotis	1008
holbrookii, Diplodus	1362
Lepomis	1008
Sargus..................	1363
Holconoti 781, 1493
Holconotus 1502, 1505
agassizii...............	1506
analis	1501
fuliginosus	1505
gibbonsii	1509
megalops	1502
rhodoterus	1502
trowbridgii............	1497
Holia..................................	478
hollardi, Hollardia	1698
Hollardia	1697
hollardi....................	1698
holocanthus, Diodon................	1746
Holocenthrus	847
Holocentridæ	845
Holocentrum.......................	847
longipinne.............	849
perlatum..............	853
productum............	852
prospinosum......,....	853
retrospinis............	853
riparium	852
rostratum.............	852
sicciferum	850
vexillarium	852
Holocentrus........	847
ascensionis............ 848, 2843
rufus......	849
auratus...............	1145
brachypterus..........	852
coruscus	851
gibbosus	1319
gulo..................	1139
marianus............. 852, 2842
merou	1154
osculus...............	853
pentacanthus	849
punctatus.............	1153
rostratus.............	849
sancti-pauli............	853
sanguineus............	1761
siccifer...............	849
sogo..................	849
striatus	849
suborbitalis	850
surinamensis..........	1236

	Page.
Holocentrus tigrinus	1214
unicolor...............	1192
vexillarius	852
Holocephali	93
holocyaneos, Scarus...............	1654
Hololepis barratti	1102
erochrous	1102
fusiformis	1102
hololepis, Cyttus...................	1662
Zenion	1661
holomelas, Paraliparis..............	2140
Holoporphyrus......................	2543
Holorhinus.........................	89
vespertilio.............	90
Holostei.......................... 98, 107
holotrachys, Macrourus.............	2582
Homalogrystes......................	1148
Homalopomus........................	2529
trowbridgii..........	2531
Homesthes..........................	2394
Homesthes caulopus	2394
homianus, Squalus.................	51
homonymus, Diapterus	1371
Homoprion..........................	1439
acutirostris	1437
furthi	1441
lanceolatus...........	1444
subtruncatus	1434
xanthurus........... 1434, 1459
honneymani, Acipenser..............	106
hoodi, Salmo	505
hoodii, Salmo 507, 510
Hoopid Salmon.....................	480
hopkinsi, Hynnis	933
Mycteroperca	2855
Plagiogrammus...........	2428
Sebastodes..............	1789
Hopladelus.........................	142
olivaris..............	143
Hoplarchus	1526
hopliticus, Paricelinus..............	1886
hoplomystax, Sparisoma............	1632
Hoplopagrinæ......................	1242
Hoplopagrus	1244
guntheri	1244
Hoplostethus	837
japonicus	837
mediterraneus	837
Hoplunnis	361
diomedianus	361
schmidtii	361
Horned Dace.......................	222
Pout.................... 135, 140
Horny Cony........................	1715
Hornyhead.........................	322
Hornyheads........................	314

	Page.
horrens, Prionotus	2172
horridus, Brachysomophis	388
Horse Crevallé	920
Mackerel	870, 909
Horse-eye Jack	923
Horsefish	934
Horsehead	936
hospes, Mugil	814
hospitus, Acipenser	105
houghi, Pimelodus	135
Hound, Smooth	29
Houndfish	715, 716
Hoyera, Guavina	2236
hoyi, Argyrosomus	469, 472
Coregonus	468, 470
Pimelodus	141
Uranidea	1969
hubbardi, Parophrys	2641
hudsonia, Cliola	269
Clupea	269
hudsonicus, Salmo	507
Hudsonius	254, 256, 266
altus	322
amarus	270
euryopa	270
fluviatilis	269
sallæi	212
hudsonius, Catostomus	176
Hippocampus	777
Hybopsis	269
Leuciscus	269
Notropis	269
amarus	270
saludanus	270
selene	269
humboldti, Atherinichthys	793
Eigenmannia	341
Gasteropelecus	572
Leuciscus	236
Myctophum	571
Scopelus	572, 577
Squalius	237
Sternopygus	341
Tigoma	237
humboldtiana, Atherina	793
humboltianum, Chirostoma	793, 2839
humeralis, Arbaciosa	2341
Chærojulis	1597
Chætodon	1674
Clupea	431
Gobiesox	2341
Harengula	431
Julis	1596
Leiostomus	1459
Oligocephalus	1097
Paralabrax	1196, 1197

	Page.
humeralis, Platyglossus	1597
Sardinella	431
Scarus	1641
Serranus	1197
humeri-maculatus, Sargus	1360
humile, Pristipoma	1331
humilis, Bryttus	1004
Cottus	1979
Eupomotis	1004
Gambusia	682
Grammateus	1355
Lepomis	1004
Pagellus	1355
Pomadasis	1331
Tetragonopterus	335
Humpback Salmon	478
Sucker	184
Whitefish	466
huntia, Molva	2551
Huro	1010
nigricans	1012
huronensis, Lepisosteus	110
Huso	103
hyalinus, Ceratichthys	321
hyalope, Squalius	222
Hybognathus	212, 213
amara	215
amarus	215
argyritis	214
civilis	215
episcopa	214
episcopus	215
evansi	213
flavipinnis	215
fluvialitis	215
hayi	214
melanops	216
nigrotæniata	214
nubila	215
nuchalis	213
osmerinus	213
perspicuus	218
placitus	213
plumbea	216
procne	264
punctifer	215
regius	213
serena	214
stramineus	262
volucellus	263
Hybopsis	314, 315, 319
æstivalis	316
marconis	316
altus	321
amblops	320
bifrenatus	259

	Page.
Hybopsis bivittatus	233
chalybæus	288
chiliticus	287
chlorocephalus	286
chrosomus	288
cumingii	318
dissimilis	318
dorsalis	262
fretensis	261
gelidus	316, 317
gracilis	321
hæmaturus	218
heterodon	261
hudsonius	269
hyostomus	316
hypsinotus	320
kentuckiensis	322
labrosus	319
lacertosus	284
longiceps	264
meeki	317
missuriensis	262
monacus	318
montanus	317
niveus	278
phaenna	270
procne	264
rubricroceus	286
rubrifrons	320
scylla	263
spectrunculus	265
storerianus	270, 321
stramineus	262
tetranemus	315
timpanogensis	233
tuditanus	253
volucellus	263
watauga	319
winchelli	321
xænocephalus	289
Hyborhynchus	217
confertus	217, 218
nigellus	217
notatus	218
puniceus	218
siderius	314
superciliosus	218
tenellus	218
Hydrargira	632
atricauda	624
Hydrargyra diaphana	645
formosa	2827
fusca	624
limi	624
luciæ	655
majalis	639

	Page.
Hydrargyra multifasciata	645
nigrofasciata	641
ornata	2827
similis	639
swampina	641, 645
trifasciata	639
vernalis	639
zebra	647
Hydrolagus	95
colliei	95
Hydrophlox	254, 257, 284
chrosomus	288
lutipinnis	287
rubricroceus	286
hydrophlox, Clinostomus	238
Leuciscus	238
Squalius	238
hygomii, Myctophum	573
Scopelus	573
Hylomyzon	173
nigricans	181
Hymenocephalus	2580
cavernosus	2580
goodei	2572
longifilis	2567
Hymenoptera	2866
hymenorhinus, Hexanematichthys	125
Hynnis	932
cubensis	932
hopkinsi	933
Hyodon amphiodon	413
chrysopsis	413
claudalus	413
vernalis	413
Hyoganoidea	98
Hyostoma	1052
blennioperca	1053
newmani	1053
simoterum	1051
hyostomus, Hybopsis	316
Nocomis	316
hypacanthus, Psednoblennius	2406
Hypargyrus	252
tuditanus	253
Hypeneus	858
Hyperistius	986
carolinensis	988
Hyperoartii	8
Hyperotreti	5
Hyperprosopon	1501
agassizi	1502
analis	1501
arcuatum	1502
arcuatus	1502
argenteum	1502
punctatum	1502

	Page.
Hyperprosopon argenteus	1501, 1502
Hypilepis cornutus cerasinus	283
Hypentelium	173, 174, 181
macropterum	181
Hyperchoristus	589
tanneri	589
Hyphalonedrus	541
chalybeius	542
Hypleurochilus	2385
geminatus	2385
multifilis	2385
punctatus	2390
Hypocaranx	927
Hypoclydonia	1115
bella	1115
Hypocritichthys	1500
analis	1500, 1501
Hypodis	915, 2848
glaucus	2848
hypodus, Decapterus	908
Hypogymnogobius	2210
Hypohomus	1039, 1040
aurantiacus	1040
cymatotænia	1041
nianguæ	1042
squamatus	1040
Hypomesus	524
olidus	525, 2824
pretiosus	525
hypophthalmus, Arius	133
Cathorops	133, 2798
hypoplecta, Rathbunella	2290
Hypoplectrus	1187
accensus	1193
affinis	1193
bovinus	1193
chlorurus	1193
crocotus	1192
gemma	1193
gummigutta	1192
guttavarius	1192
indigo	1193
lamprurus	1190
maculiferus	1192
puella	1192
unicolor	1190, 1192
aberrans	1193
accensus	1193
affinis	1193
bovinus	1193
chlorurus	1192
crocotus	1192
gummigutta	1192
guttavarius	1192
indigo	1193
nigricans	1193

	Page.
Hypoplectrus unicolor primivarius	1192
puella	1192
vitulinus	1192
hypoplectus, Bathymaster	2290
Hypoplites	1247
Hypoprion	41
brevirostris	41
longirostris	41
signatus	41
Hypoprionodon	41
Hyporhamphus	719
roberti	721
rosæ	721
tricuspidatus	720
unifasciatus	720
Hyporthodus	1148
flavicauda	1156
Hyposerranus	1148
Hypostominæ	156
Hypostomus gaucharote	159
hypostomus, Aodon	92
Cephalopterus	92
Hypsagonus	2038
gradiens	2041
quadricornis	2038, 2041
swanii	2088
hypseloptera, Cliola	280
hypselopterus, Leuciscus	280
Notropis	280
hypselurus, Pimelodus	152
Rhamdia	152
Hypsicometes	2293
goboides	2294
Hypsifario	474, 477, 481
kennerlyi	483
Hypsilepis ardens	301
coccogenis	285
cornutus	283
cyaneus	283
gibbus	283
diplæmia	300
galacturus	279
iris	272
kentuckiensis	279
Hypsinotus	1664
rubescens	1665
hypsinotus, Ceratichthys	320
Hybopsis	320
Hypsoblennius	2386
brevipinnis	2390
gentilis	2387
gilberti	2386
hentz	2390
hentzi	2390
ionthas	2388
striatus	2388, 2392

	Page.
Hypsolepis	254
Hypsopsetta	2639
Hypsurus	1508
caryi	1508, 1509
Hypsypops	1564
dorsalis	1570
rubicundus	1564, 1565
Hysterocarpinæ	1494
Hysterocarpus	1495
traski	1496
hystrix, Diodon	1744, 1746
Icelinus	1894
borealis	1896
cavifrons	1892
filamentosus	1893
fimbriatus	1894
oculatus	1895
quadriseriatus	1807
strabo	1897, 2862
tenuis	1894
Icelus	1911
australis	1918
bicornis	1911
canaliculatus	1917
euryops	1915, 1916
furciger	1913
hamatus	1913
megacephalus	1891
pugetensis	1891
quadriseriatus	1897
scutiger	1910
spiniger	1914
uncinatus	1906
vicinalis	1916
Ichthælurus punctatus	135
Ichthelis	999
erythrops	990
megalotis	1003
melanops	996
ichtheloides, Lepomis	990
Ichthyapus	374
selachops	374
Ichthycallus	1587
dimidiatus	1594
Ichthyobus	163
bubalus	164
cyanellus	164
Ichthyomyzon	10
argenteus	11
astori	12
castaneus	11
Ichthyomyzon concolor	11
hirudo	11
tridentatus	12
Icichthys	969

	Page.
Icichthys lockingtonii	969
icistia, Bairdiella	1435
Sciæna	1436
Icosteidæ	968, 2849
Icosteinæ	969
Icosteus	972
ænigmaticus	972
ictalops, Cottus	1950
Pegedictis	1951
Ictalurinæ	115
Ictalurus	133
albidus	138
anguilla	2788
furcatus	134, 2788
kevinskii	138
lacustris	137
lophius	138
lupus	137
macaskeyi	138
meridionalis	135
nigricans	137
niveiventris	138
okeechobeensis	139
ponderosus	137
punctatus	134
robustus	135
simpsoni	135
Icthyophis	403
vittatus	404
Ictiobinæ	162
Ictiobus	163, 164
bubalus	164
cyprinella	163
meridionalis	164
urus	164
velifer	167
Idiacanthidæ	604
Idiacanthus	605
antrostomus	605
ferox	605
Idol, Moorish	1687
Ilictis	142
Ilisha	435
bleekeriana	436
flavipinnis	435
furthi	436
panamensis	436, 2811
illecebrosus, Alburnops	269
Notropis	268
Stellifer	1442
Ilyophidæ	349
Ilyophis	349
brunneus	350
Ilypnus	2253
gilberti	2253
imberbe, Peristedion	2182

Page.

imberbi, Ophidium 2443
imberbis, Apogon 1107
    Mullus..................... 1107
    Sciæna 1454
    Vulsiculus 2181
imiceps, Ophioscion 1451
    Sciæna 1451
immaculata, Amia.................. 411
    Coryphæna 953
    Fistularia.............. 758
    Perca 1135
    Unibranchapertura ... 342
immaculatum, Cybium 876
immaculatus, Esox 630
    masquinongy ... 630
    Lucius masquinongy. 630
    Salmo 507
    Symbranchus 342
Imostoma......................... 1044, 1046
    shumardi................. 1047
imperator, Tetrapterus.............. 892
    Xiphias.................. 892
Imperial, Serran 1837
imperialis, Coryphæna 952
    Esox..................... 717
    Sebastes 1837, 1838
    Trachurus 927
    Zeus 955
impetiginosus, Serranus............. 1153
impressus, Conger 356
inæquilabiatus, Carapus 341
inæquilobus, Leucosomus 224
    Pogonichthys 224
iucilis, Mugil 813
incisor, Pimelepterus.............. 1386
    Pomotis................. 1005
Inconnu........................... 473, 474
inconstans, Eucalia 744
    cayuga 744
    pygmæa 744
    Gasterosteus........... 744
iucrassatus, Leucosomus........... 222
incurvus, Lobotes................. 1236
indefatigabile, Otophidium......... 2490
Indian Chub 322
    Fish 1680
indicus, Chanos 415
    Cottus.................. 2092
    Cubiceps 951
    Naucrates 900
    Tetrapterus 892
iudigena, Clupea.................... 428
indigo, Hypoplectrus 1193
    unicolor 1193
    Plectropoma............. 1193
Inermia 1365

Page.

Inermia vittata .....:............... 1366
inermis, Anoplagonus.............. 2094
    Aspidophoroides.......... 2093
    Dermatolepis.............. 1167
    Elops 410
    Epinephelus 1168
    Lioperca 1168
    Lutjanus................. 1275
    Mesoprion ....:........ 1275
    Ostracion 1723
    Rabirubia................. 1274
    Raia inornata 73
    . Scorpæna 1853
    Sebastes 1829
    Sebastodes 1829
    Serranus................. 1168
infans, Avocettina 367
    Gambusia 680
    Nemichthys 368
infernalis, Histiobranchus 352
    Muræna................. 396
    Synaphobranchus 352
infirmus, Novaculichthys.......... 1616
    Xyrichthys 1616
iugens, Gunellus.................. 2419
    Hippocampus 776
Iniistius 1619
    mundicorpus............. 1620
Iniomi 530
Innominado...................... 382
innominatus, Girardinichthys 666
inops, Antennarius 2718
Inopsetta 2641
    ischyra................. 2641
inornata, Lota.................... 2551
    Raia 73
    Raja..................... 73
inornatus, Apodichthys............ 2412
    Chorinemus 899
    Cryptacanthodes........ 2443
    Microlepidotus.......... 1341
    Oligoplites 899
    Orthopristis 1342
    Pseudojulis 1604
inscripta, Solea.................... 2696
inscriptum, Etheostoma 1072
    Nanostoma 1072
inscriptus, Achirus 2696
    Nothonotus 1072
    Pomotis 1003
insculpata, Netuma 127, 2765
insculptus, Arius 127
insculptus, Gasterosteus........... 750
    Luciocharax 339
insigne, Pimelodus................ 147
insignis, Catostomus............... 180

	Page.
Isobyrus, Apomotis	997
Lepiopomus	997
Lepomis	997
Parophrys	2641
Pleuronectes	2641
Isothmus	2389
gentilis	2388
gilberti	2387
hentzi	2390
isothmus	2389
punctatus	2390
scrutator	2389
striatus	2388
Islandicus, Centronotus	2438
Gasterosteus	748
Gunnellus	2439
Stichaeus	2439
Isodon, Aprionodon	42
Carcharias	42
Mesoprion	1287
Isodus, Squalus	51
Isogomphodon	33, 35, 40
limbatus	40
maculipinnis	40
Isolepis, Isopsetta	2642
Lepidopsetta	2642
Parophrys	2642
Sternotremia	787
Isoplsthus	1399
affinis	1399
parvipinnis	1399
remifer	1399
Isopsetta	2642
isolepis	2642
Isospondyli	407
Isospondylous Fishes	407
Isthmeusis, Rivulus	2830
Istiophoridae	890
Istiophorus	890
nigricans	891
Isuropsis	47, 48
dekayi	48
glaucus	48
Isurus	47, 48
dekayi	48
oxyrhinchus	48, 49
spallanzani	49
Itaiara	1142, 1162
Itaiara, Promicrops	1164
Serranus	1164
Jabou	1232
Jaboncillo	1232
Jack	897, 920, 1790
Amber	903

	Page.
Jack Goggle-eye	911
Horse-eye	939
Yellow	919
Jack Salmon	1021
Jacket, Leather	1701
Jackets, Leather	898
Jackson, Ditrema	1508
Embiotoca	1504, 1505
Jacob Evertzen	1143
Jacobaea, Echeneis	2272
Remora	2272
Jacobi, Corvinus (Johnius)	1457
Sciaena	1457
Jacobus, Myripristis	846
Jacome	1815
Jacome, Haliperca	1815
Serranus	1815
Jaculidens, Ancylodon	1416
Jaculus, Alburnellus	303
Jaguanguare	1562
Jaguana, Harengula	430
jaguar, Scallanus	849
Jallao	1296
jamaicensis, Cynoscion	1406
Otolithus	1406
Raja	81
Urolophus	81
Janeiro, Engraulis	451
Janissary	1886
Januaria, Umbrina	1474
Jack, Cottus	1978
Myoxocephalus	1977
Japonensis, Salmo	479
Japonica, Squatina	59
Japonicus, Agonus	2036
Arctoscopus	2297
Chirolophus	2409
Cottus	2036
Hippocephalus	2036
Hoplostethus	837
Perca	2034
Phalangistes	2036
Physiculus	2549
Trichodon	2297
Jaqueta	1561
Jaquette, Petite	1559
Jarrovii, Lepidomeda	328
Minomus	170
Pantosteus	171
Javanicus, Psenes	961
Jaw-fishes	2279
jejuna, Episema	290
jejunus, Minnilus	290
Notropis	290
jemezanus, Alburnellus	284

	Page.
insignis, Noturus	147
Schilbeodes	147
insociabilis, Dorosoma	416
insolatus, Chromis	1548
Heliases	1548
insulæ-sanctæ-crucis, Scarus	1651
insularum, Atherina	807
Atherinops	807
Muræna	400
Netuma	2770
integripinnis, Auchenopterus	2372
Cremnobates	2373
intermedia, Tigoma	235
intermedium, Cichlasoma	1517
intermedius, Gasterosteus	750
Heros	1517
Hippoglossus	2672
Leuciscus	235
Paralepis	600
Pomoxys	987
Saurus	535
Squalius	235
Sudis	600
Synodus	535, 536
internasalis, Chærojulis	1594
Julis	1594
Platyglossus	1594
interrupta, Morone	1134
Perca mitchilli	1133
Raja	2751
interruptus, Ambloplites	991
Anisotremus	1319
surinam- ensis	1319
Archoplites	991
Centrarchus	991
Genytremus	1319
Luxilus	282
interstitialis, Epinephelus	1179
Mycteroperca	1178
Serranus	1179
Trisotropis	1179
intertinctus, Mystriophis	386
Ophichthys	387
Ophisura	387
introniger, Sebastichthys	1805
Sebastodes	1805
intronigra, Dicrolene	2522
inurus, Zygonectes	682
Ioa	1064
vigil	1065
vitrea	1064
ioæ, Etheostoma	1084
iodocus, Angelichthys	1686
Holacanthus	1687

	Page.
Ioglossus	2192
callinrus	2193
ionthas, Cerdale	2449
Hypsoblennius	2388
Isesthes	2389
ios, Clevelandia	2254
Gobiosoma	2255
Iotichthys	228, 231, 243
iowæ, Etheostoma	1083
Ipnopidæ	546
Ipnops	546
murrayi	547
irideus, Labrus	988
Salmo	500
agua-bonita	503
gilberti	502
masoni	501
shasta	502
stonei	503
iridinus, Carangoides	919
Caranx	919
Iridio	1587
bivattatus	1595
caudalis	1599
cyanocephalus	1594
dispilus	1597
garnoti	1593
kirschii	1598
maculipinna	1594
nicholsi	1591
pictus	1599
poeyi	1599
radiatus	1590
sellifer	1592
semicinctus	1592
iris, Oliola	272
Hypsilepsis	272
Leuciscus	222
Irish Lord	1934
Pompano	1376
irradians, Serranus	1208
irregularis, Heros	1541
Theraps	1540
irretitus, Gordiichthys	363
Mugil	819
irroratus, Monacanthus	1713
Isabelita	1684
isabelita, Angelichthys	1685
Isabelito de la Alto	1674
Isaciella	1340
brevipimis	1341
ischànus, Notemigonus	251
Stolephorus	442
ischinagi, Megaperca	1138
ischyra, Inopsetta	2641

	Page.		Page.
ischyrus, Apomotis	997	Jack Goggle-eye	911
Lepiopomus	997	Horse-eye	923
Lepomis	997	Yellow	919
Parophrys	2641	Jack Salmon	1021
Pleuronectes	2641	Jacket, Leather	1701
Isesthes	2386	Jackets, Leather	898
gentilis	2388	jacksoni, Ditrema	1505
gilberti	2387	Embiotoca	1504, 1505
hentzi	2390	Jacob Evertzens	1143
ionthas	2389	jacobæa, Echeneis	2272
punctatus	2390	Remora	2272
scrutator	2389	jacobi, Corvinus (Johnius)	1457
striatus	2388	Sciæna	1457
islandicus, Centronotus	2438	jacobus, Myripristis	846
Gasterosteus	748	Jacome	1215
Gunnellus	2439	jacome, Haliperca	1215
Stichæus	2439	Serranus	1215
isodon, Aprionodon	42	jaculidens, Ancylodon	1416
Carcharias	42	jaculus, Alburnellus	293
Mesoprion	1267	Jaguacaguare	1562
isodus, Squalus	51	jaguana, Harengula	430
Isogomphodon	33, 35, 40	jaguar, Bodianus	849
limbatus	40	Jallao	1295
maculipinnis	40	jamaicensis, Cynoscion	1406
isolepis, Isopsetta	2642	Otolithus	1406
Lepidopsetta	2642	Raja	81
Parophrys	2642	Urolophus	81
Sternotremia	787	janeiro, Engraulis	451
Isopisthus	1399	Janissary	1586
affinis	1399	januaria, Umbrina	1474
parvipinnis	1399	jaok, Cottus	1978
remifer	1399	Myoxocephalus	1977
Isopsetta	2642	japonensis, Salmo	479
isolepis	2642	japonica, Squatina	59
Isospondyli	407	japonicus, Agonus	2036
Isospondylous Fishes	407	Arctoscopus	2297
isthmensis, Rivulus	2830	Chirolophus	2409
Istiophoridæ	890	Cottus	2036
Istiophorus	890	Hippocephalus	2036
nigricans	891	Hoplostethus	837
Isuropsis	47, 48	Pereis	2034
dekayi	48	Phalangistes	2036
glaucus	48	Physiculus	2549
Isurus	47, 48	Trichodon	2297
dekayi	48	Jaqueta	1561
oxyrhinchus	48, 49	Jaquette, Petitie	1559
spallanzani	49	jarrovii, Lepidomeda	328
Itaiara	1142, 1162	Minomus	170
itaiara, Promicrops	1164	Pantosteus	171
Serranus	1164	javanicus, Psenes	951
		Jaw-fishes	2279
Jabon	1232	jejuna, Episema	290
Jaboncillo	1232	jejunus, Minnilus	290
Jack	627, 920, 1780	Notropis	290
Amber	903	jemezanus, Alburnellus	294

	Page.
jemezanus, Minnilus	294
Jeniguana	1302
Jeniguano	1310
jeniguano, Hæmulon	1310
jenkinsi, Chasmodes	2391, 2392
Fundulus	651
Synodus	537, 2826
Zygonectes	652
Jenkinsia	418
acuminata	419
lamprotænia	419
stolifera	419
Jenny, Silver	1370
Jerker	322
Jerusalem Haddock	954
jessiæ, Etheostoma	1084
Pœcilichthys	1085
Xenocys	1285
Xyrichthys	1613
Xyrula	1612, 1613
Jewfish, Black	1161
California	1137
Jewfishes	1137
Jewsharp Drummer	1473
Jiguagua	920
Jocu	1257
jocu, Anthias	1258
Lutjanus	1258
Mesoprion	1258
Neomænus	1257
John A. Grindle	113
Dories	1659
Mariggle	410
Paw	1159
johni, Cephaloptera	93
Johnius	1455
amazonicus	1419
batabanus	1431
crouvina	1419
guttatus	1174
heterolepis	1419
jacobi	1457
nobilis	1413
ocellatus	1454
regalis	1407
saxatilis	1475
Johnny	2013
Darter	1056
Verde	1195
Jolt-head Porgy	1352
jonesi, Belone	717
Gerres	1368
Mollienisia	699
Siphostoma	768
Syngnathus	768
Jopaton	1341

	Page.
Jordanella	677
floridæ	677
jordani, Bathymaster	2289
Caulophryne	2735
Chirostoma	793
Eopsetta	2613
Epinephelus	1177
Eslopsarum	2840
Etheostoma	1079, 1080
Galeichthys	2774
Hemilepidotus	1934
Hexanematichthys	129
Hippoglossoides	2614
Mycteroperca	1176
Neomænis	1251
Notropis	259
Pantosteus	171
Raia	73
Ronquilus	2289
Sebastodes	1778
Tachisurus	129
Jordania	1884
zonope	1884
Jordaniinæ	1880
Jorobado	934, 936
josephi, Ophidion	2488
Joturo	821
Joturus	820
pichardi	821
stipes	821
joyneri, Sebastodes	1829
juba, Guatucupa	1323
Perca	1323
Jug-fish	1728
jugalis, Oliola	272
Moniana	272
Jugular Fishes	2528
juliæ, Etheostoma	1093
Julidinæ	1572, 1574
Julidio	1602
adustus	1602
notospilus	1603
Julis	2859
bifasciata	1610
bifasciatus	1610
caudalis	1599
cinctus	1593
crotaphus	1591, 1598
cyanostigma	1591
detersor	1610
dimidiatus	1594
gamoti	1593
gillianus	1610
humeralis	1596
internasalis	1594
lucasanus	1607, 1608

	Page.
Julis maculipinna	1595
melanochir	1609
modestus	1601
nitida	1608
nitidissima	1608
opalina	1591
patatus	1591
pictus	1600
principis	1591
psittaculus	1597
semicinctus	1593
Jumping Mullet	197
Jump-rocks	197
June Sucker of Utah Lake	183
Jurel	899, 921, 923
Jurvucapeba	1142
Kalog	1976
Kamchatka Salmon Trout	2818
kamloops, Salmo gairdneri	499
Kamloops Trout	499
kanawha, Notropis	264
kansæ, Fundulus	2828
karrak, Anarrhichas	2446
Kathetostoma	2311
albigutta	2312
averruncus	2311
Kathetostomatinæ	2306
kaupi, Physiculus	2548
kaupii, Synaphobranchus	351
Kelpfish	1592, 2351, 2352
Spotted	2353
kendalli, Sphagebranchus	375
Verma	375
kennedyi, Trachinotus	942
kennerlyi, Hypsifario	483
Moxostoma	186
Oncorhynchus nerka	483
Salmo	483
kennicotti, Acipenser	105
Catonotus	1098
Coregonus	464
Kenoza	625, 626
kentuckiensis, Hybopsis	322
Hypsilepis	279
Leuciscus	279
Luxilus	279, 322
Kern River Trout	502
Keshimugo	1833
kessleri, Arius	127
Netuma	127, 2765
Tachisurus	127
keta, Oncorhynchus	478
keta vel kayko, Salmo	479
kevinskii, Ictalurus	138
Kieye of Lake Michigan	469

	Page.
Killer, Salmon	749
Killifish	639, 641
Common	640
killifish, Cobitis	641
Killifishes, Salmo	630, 632
killinensis, Salmo	509
King of the Mackerels	1755
Mullets	1106
Salmon	479
Kingfish	875, 1460, 1469, 1475
kirschii, Iridio	1598
kirtlandi, Acipenser	106
Kirtlandia	794
laciniata	795, 2840
martinica	795
vagrans	794, 2840
laciniata	2840
Kisutch	480
kisutch, Oncorhynchus	480
Salmo	481
kitt, Microstomus	2654
Pleuronectes	2654
klamathensis, Cottus	1955
kleinii, Balistes	1720
klunzingeri, Achirus	2697
Solea	2697
kneri, Pristopoma	1338
Kodiak Smelt	2823
kœlreuteri, Scomber	900
Kogumeso	1833
Kowala	428, 2811
Krasnaya Ryba	481
kraussi, Gobius	2228
Krohnius	2587
kroyeri, Scopelus	556
kuda, Hippocampus	778
Kuhlia	1013
arge	1014
xenura	1015
Kuhliidæ	1013
kumlieni, Uranidea	1967
kundscha, Salmo	2823
Salvelinus	2822
Kuro Soi	1834
Kyach	426
Kyphosidæ	1380
Kyphosinæ	1381
Kyphosus	1384
analogus	1385
elegans	1387
incisor	1386
lutescens	1388
ocynrus	1390
sectatrix	1387
Labeo elegans	186

	Page.
Labeo esopus	186
longatus	186
labialis, Fundulus	644
labiatus, Catostomus	177
Heros	1530
Labichthys	368
carinatus	368
elongatus	369, 2802
gilli	368
Labidesthes	805
sicculus	805
Labracopsis	1135
labradoricus, Acanthocottus	2001
Coregonus	466
Cottus	2004
Oncocottus	2004
Labrax	1866, 2840
albidus	1132
americanus	1135
clathratus	1198
decagrammus	1868
hexagrammus	1872
lagocephalus	1875
lineatus	1113
monopterygius	1866
mucronatus	1135
multilineatus	1132
nebulifer	1195
nigricans	1135
notatus	1132
octogrammus	1870
osculatii	1132
pallidus	1135
pluvialis	2841
rufus	1135
stelleri	1872
superciliosus	1873
Labridæ	1571
labridum, Hæmulon	1319
labriformis, Epinephelus	1155
Serranus	1155
Labrinæ	1572–1573
Labrisomus	2360
biguttatus	2360
bucciferus	2363
capillatus	2362
delalandi	2350
herminiger	2361
microlepidotus	2363
nuchipinnis	2362
xanti	2362
Labroid Fishes	1571
Labroperca	1148
Labrosomus	2360
cremnobates	2366
macrocephalus	2364

	Page.
Labrosomus microlepidotus	2364
pectinifer	2362
xanti	2363
labrosus, Blennius	2457
Ceratichthys	319
Hybopsis	319
Zoarces	2457
Labrus adspersus	1577
americanus	1579
aper	1586
auritus	1001
bifasciatus	1609
bivittatus	1596
blackfish	1578
brasiliensis	1591
capite obtuso	1609
carolinus	1578
chogset	1577
fulva	1577
chromis	1483
cromis	1483
cruentatus	1238
cyanocephalus	1594
exoletus	1576
falcatus	942
fulvomaculatus	1339
fulvus	1145
griseus	1257
grunniens	1483
guaza	1154
hiatula	1578
irideus	988
macropterus	988
maximus	1580
onitis	1578
ornatus	1610
pallidus	1005
pentacanthus	1576
plumieri	1305
psittaculus	1596
pulcher	1585
radians	1633
radiatus	1591
rostro reflexo	1677
rufus	1583
salmoides	1012
semiruber	1583
sparoides	987
squeteague	1407, 1409
striatus	1200
subfuscus	1578
tautoga	1579
alia	1579
fusca	1579
rubens	1579
tessellatus	1578

	Page.
Labrus torquatus	1609
versicolor	1346
Lac de Marbre Trout	515
lacera, Lagochila	199
Quassilabia	199
lacerta, Lampanyctus	560
Myctophum	560
Synodus	537
lacertinus, Synodus	536
Lacerto	537, 867
lacertosus, Hybopsis	284
Minnilus	284
Notropis	284
lacertus, Gobius	2218
Scomber	867
Lachnolæmus	1579
maximus	1580
Lachnolaimus	1579
aigula	1580
caninus	1580
dux	1580
falcatus	1580
maximus	1579
psittacus	1580
suillus	1580
lachrymalis, Ptychostomus	194
laciniata, Kirtlandia	795, 2840
vagrans	2840
Menidia vagrans	795
lacrimosum, Sparisoma	1632
lacrimosus, Scarus	1632
lactarias, Catostomus	175
Lactophrys	1721, 1722, 1723
bicaudalis	1723
oviceps	1724
tricornis	1724
trigonus	1723, 1724
triqueter	1722
lacustris, Ameiurus	137
Gadus	137, 2551
Hemiplus	250
Ictalurus	137
Pomolobus pseudoharengus	426
Lady-fish	411, 1583
Spanish	1583
Lady-fishes	410, 1581
Læmargus	56
borealis	57
Læmonema	2556
barbatula	2557
barbatulum	2556
melanurum	2557
lætabilis, Moniana	272
lætus, Centronotus	2420
lævicaudatus, Hippocampus	777

	Page.
lævigata, Pterophryne	2717
lævigatus, Agonus	2048
Chironectes	2717
Lagocephalus	1728
Phalangistes	2048
Salmo	508
Tetrodon	1728
Læviraja	66
lævis, Acipenser	106
Balistes	1719
Gasterosteus	745
Orbis variegatus	1735
Pleuronectes	2654
Raja	71
Rhombus cornubiensis	2654
Squatina	59
La Fayette	967, 1458
Lagarto	533, 538
lagenarius, Acipenser	102
Lagocephalus	1727
lævigatus	1728
pachycephalus	1729
lagocephalus, Grammatopleurus	1875
Hexagrammos	1873
Hexagrammus	1875
Labrax	1875
Oncorhynchus	479
Salmo	479
Lagochila	198
lacera	199
Lagodon	1357
rhomboides	1358
Lake Carp	167
Crescent Speckled Trout	2821
Herring	468
Lawyer	2550
Sheepshead	1484
Sturgeon	106
Tahoe Trout	493, 2870
lalandi, Carcharias	43
Seriola	902, 903
lamarii, Acipenser	106
lamellifer, Exocœtus	733
Lamia	38, 49, 50
lamia, Carcharhinus	38
Carcharias	38
Eulamia	38
lamiella, Carcharhinus	37
Lamiopsis	33
Lamna	49
caudata	37
cornubica	49, 2749
punctata	48
spallanzani	49
Lamnidæ	47
Lamninæ	47
lamotteni, Petromyzon	10

	Page.
Lampadena	560
speculigera	561
Lampanyctus	557
alatus	559
crocodilus	558
gemmifer	559
guntheri	559
lacerta	560
resplendens	555
townsendi	558
* Lamperina	6
Lampetra	12
astori	12
aurea	13
camtschatica	13
cibaria	13
epihexodon	12
plumbea	13
spadicea	13
tridentata	12
variegata	2745
wilderi	13, 2745
lampetræformis, Blennius	2438
Lumpenus	2438
Lamprey Eel	10
Lamprey, Great Sea	10
Silvery	11
Small Black	13
Lampreys	4, 8, 9
Brook	12
River	10
Lampridæ	953
Lampris	954
guttatus	955
lauta	955
luna	954
regius	955
lamprotænia, Clupea	419
Jenkinsia	419
Spatelloides	419
lamprurus, Hypoplectrus	1190
Serranus	1190
Lampugus	952
neapolitanus	953
punctulatus	953
siculus	953
lanatus. Merlucius	2530
Lancelet, Bahama	4
California	4
West Indian	3
Lancelets	2, 3
European	3
lanceolata, Perca	1482
Sciæna	1444
lanceolatum, Branchiostoma	3
lanceolatus, Amphioxus	3

	Page.
lanceolatus, Chætodon	1490
Eques	1489, 1490
Gobius	2229, 2230
Homoprion	1444
Limax	3
Lonchiurus	1482
Lonchurus	1482
Stellifer	1443
Lancet-fish	1691
Lancet Fishes	593, 594, 595
Landlocked Salmon	487
Lane Snapper	1270
Langbarn	2433
Lant	833
Lantern Fishes	530, 550
Lapon	1849
lappa, Halieutella	2742
La Quesche	413
Large-mouthed Black Bass	1012
Large-scaled Sucker	192
Larimus	1420, 1421
acclivis	1422
argenteus	1421
batabanus	1431
breviceps	1423
dentex	1426
effulgens	1421
fasciatus	1424
pacificus	1424
stahli	1423
larkinsii, Cymatogaster	1503
lata, Guaperva forcipata	1702
Latebrus	1114
oculatus	1115
latepictus, Serranus	1175
laterale, Ditrema	1506
Etheostoma	1099
lateralis, Abramis	239
Alvarius	1099
Artedius	1902
Calycilepidotus	1900
Caracodon	2832
Characodon	668
Eleotris	2195
Embiotoca	1506
Leuciscus balteatus	239
Mylocheilus	220
Notropis	263
Phanerodon	1506
Philypnus	2195
Pimelodus	135
Pœcilichthys	1099
Richardsonius	239
Scarus	1637
Scorpænichthys	1902
Tæniotoca	1505, 1506

	Page.
lateralis, Zygonectes	659
laticauda, Anguilla	348
Rhamdia	1512
laticaudus, Pimelodus	1512
laticeps, Acanthocottus	1989
Aetobatus	88, 2753
Arius	132
Atherina	790
Bathynectes	2523
Megalocottus	1988
Mixonus	2523
laticlavius, Prionurus	1696
Xesurus	1695
latidens, Microstomus	2654
latifasciatus, Cyprinodon	676
latifrons, Anarhichas	2446
Citharichthys	2674
Dormitator	2197
Eleotris	2198
Syacium	2673
Xenochirus	2082
Latilinæ	2275
Latilus chrysops	2278
princeps	2277
latimaculatus, Ophisurus	376
latimana, Belone	717
latior, Coregonus	466
latipinna, Mollienisia	700
latipinnis, Catostomus	174, 2790
Zaniolepis	1876
latirostris, Acipenser	105
Lepidosteus	111
latulus, Clupea	422
latus, Caranx	922
Euctenogobius	2237
Scomber	938
Launces, Sand	831, 832, 833
laurettæ, Argyrosomus	471
Laurida	533
mediterranea	537
laurito, Sparisoma	1637
lauta, Lampris	955
lavaretus, Salmo	464
Lavinia	208
conformis	231
crassicauda	231
exilicauda	208, 2799
harengus	209
Lawyer	113, 1255
Lawyer, Lake	2550
leachi, Clupea	422
leachianus, Thynnus	869
Least Darter	1104
Leather Fish	1714, 1715
Leather Jacket	1701
Jackets	898

	Page.
Leather-sided Minnow	236
Le Baliste Bridé	1704
Lebias ellipsoidea	672
ovinus	672
rhomboidalis	672
Lebistes	689, 2833
pœciliodes	689
Lebius	1866
Lebrancho	810
lebranchus, Mugil	811
lecontei, Acipenser	105
Le Diodon	1746
Orbe	1749
Tacheté	1746
leei, Symphurus	2708
lefroyi, Diapterus	1372
Eucinostomus	1372
Ulæma	1371
Lefroyia	2495
bermudensis	2497
Le Gastrobranche Dombey	6
leiarchus, Cestreus	1415
Cynoscion	1414
Otolithus	1415
Leiobatus	61, 70
sloani	81
Leiocottus	2010
hirundo	2011
Leiodon	56
echinatum	57
Leioglossus	916
leionothos, Holacanthus	1735
leiopomus, Cottus	1962
Leiostomus	1558
humeralis	1459
lineatus	1460
obliquus	1459
xanthurus	1458
Leiurus	746
leiurus, Gasterosteus	747
Le Kai Salmon	478
Lembus	2194
lemmoni, Squalius	235
lemniscatus, Engraulis	443
Osmerus	533
Pimelodus	147
Lemnisoma	883
thyrsitoides	884
Lenguado de Rio	2698
lenibus, Ostracion triangulatus	1724
lentiginosa, Amia	113
Muræna	402
lentiginosum, Cichlasoma	1524
lentiginosus, Galeichthys	122, 2771
Heros	1524
Rhinobatus	62, 2750

	Page.			Page.
lentiginosus, Tachysurus	122	Lepidosteus grayi		111
leonensis, Oligocephalus	1089	latirostris		111
leonina, Cliola	271	leptorhynchus		110
Moniana	272	manjuari		111
Leopard Shark	31	oculatus		111
leopardinus, Antennarius	2721	otarius		110
Platophrys	2666	viridis		111
Rhomboidichthys	2666	lepidulus, Alburnus		294
leopardus, Anarrhichas	2446	lepidum, Boleosoma		1089
Lepadogaster cornubiensis	2108	Etheostoma		1089
nudus	2331	lepidus, Gobius		2249
reticulatus	2328	Lepidogobius		2249
testar	2332	Poecilichthys		1089
Lepibema	1131	Lepimphis		952
lineatum	1133	hippuroides		952
mitchilli	1133	Lepiopomus		999
lepida, Boleosoma	1089	ischyrus		997
Oliola	273	Lepisoma		2360
Cyprinella	273	cirrhosum		2362
Lepidamia	1106	Lepisosteidæ		108
Lepidion	2543	Lepisosteus		109
verecundum	2543	albus		110
Lepidochætodon	1672	bison		110
Lepidocybium	873	ferox		111
Lepidogaster mæandricus	2328	gavailis		110
lepidogenys, Etheostoma	1087	gracilis		110
Lepidogobius	2249	huronensis		110
emblematicus	2247	lineatus		110
gilberti	2254	longirostris		110
gracilis	2249	osseus		109
gulosus	2244	oxyurus		110
lepidus	2249	platostomus		110
newberryi	2248	platyrhincus		111
thalassinus	2245	platystomus		110
Lepidolepus	2568	semiradiatus		110
norvegicus	2579	spatula		111
Lepidomeda	328	tristœchus		111
jarrovii	328	tropicus		111
vittata	328	Lepodus		958
Lepidomegas	901	saragus		960
Lepidopidæ	884	Lepominæ		985
Lepidopinæ	885	Lepomis		999, 1010
Lepidopsetta	2642	albulus		1007
bilineata	2643	annagallinus		1004
isolepis	2642	apiatus		998
umbrosa	2642	appendix		1005
Lepidopus	886	ardesiacus		1006
argyreus	887	auritus		1001, 1009
caudatus	887, 2844	solis		1001
gouani	887	bombifrons		1003
peronii	887	charybdis		992
xantusi	2843, 2844	cyanellus		996
Lepidosoma	2568	euryorus		1009
Lepidosteus	109	flexuolaris		1011
berlandieri	111	garmani		1002
crassus	110	gillii		992

	Page.
Lepomis haplognathus	1004
heros	1008
holbrookii	1008
humilis	1004
ichtheloides	990
ischyrus	997
lirus	1007
longispinis	1006
macrochirus	1005
marginatus	1003
megalotis	1002
miniatus	1002
mystacalis	1001
notata	1011
notatus	1008
ophthalmicus	1001
pallida	1012
pallidus	1005
peltastes	1003
phenax	997
punctatus	998
purpurescens	1006
salmonea	1011
symmetricus	999
trifasciata	1011
Lepomotis nephelus	1005
Lepophidium	2482
brevibarbe	2485
cervinum	2484, 2485
emmelas	2483
marmoratum	2482
microlepis	2486
pardale	2486
profundorum	2484
stigmatistium	2483
leptacanthus, Noturus	146
Schilbeodes	146
Leptagonus	2052
decagonus	2052
spinosissinus	2054
Leptarius	119
dowi	125
Leptaspis	916
Leptecheneis	2268
naucrateoides	2270
naucrates	2269
Leptoblennius	2435
nubilus	2438
serpentinus	2439
Leptocardii	2
Leptocephalichthys	353
Leptocephalidæ	352
Leptocephalus	353
candidissimus	354
caudicula	355
caudilimbatus	355

	Page.
Leptocephalus conger ....,	354
gracilis	354
morrissi	354
spallanzanii	354
leptocephalus, Ceratichthys	323
Merlangus	2535
Leptoclinus	2432
maculatus	2433
Leptoconger	362
prolongus	363
Leptocottus	2011
armatus	2012
Leptodes	584
Leptogunnellus	2435
Leptophidium	2482
marmoratum	2483
microlepis	2486
prorates	2485
Leptops	142
olivaris	143, 2790
Leptorhinophis	381
leptorhynchum, Siphostoma	764
Leptorhynchus	369
leuchtenbergii	369
leptorhynchus, Lepidosteus	110
Odontopyxis	2076
Sarritor	2075
Syngnathus	765
leptosomus, Abramis	250
Luxilus	250
Notemigonus	250
Lepturus	889
lepturus	889
lepturus, Caranx	923
Lepturus	889
Macrourus	2584
Trichiurus	889
Le Sphéroide Tuberculé	1733
Tetrodon Plumier	1733
Les Alutères	1717
Batrachopes	1740
Brosmes	2561
Curimates	332
Dichotomyctères	1738
Dilobomyctères	1738
Elacates	948
Lottes	2550
Mustèles	2557
Ovoides	1738
Pristipomes	1329
Promecocepales	1727
Sphéroides	1729
Stellifères	1439
Stenometopes	1729
lessoni, Caranx	923
Tetrapturus	892

	Page.
lessonii, Coryphæna	953
lesueuri, Acipenser	106
Moxostoma	194
lesueurianum, Exoglossum	327
lesueurii, Catostomus	195
Letharcus	375
velifer	375
lethopristis, Orthopristis	1340
lethostigma,. Paralichthys	2630
lethostigmus, Paralichthys	2630
Lethostole	792, 2839
estor	792
Lethotremus	2100
muticus	2101
vinolentus	2101
leuchtenbergii, Belonopsis	369
Leptorhynchus	369
leuciodus, Minnilus	291
Notropis	291
Photogenis	291
Leuciscinæ	202
Leuciscus	228, 252
affinis	240
aliciæ	236
analostanus	279
ardens	301
argenteus	221
balteatus	238, 2797
lateralis	239
bicolor	232, 245
bosci	251
boucardi	247
bubalinus	273
chrysopterus	221
cobitis	305
coccogenis	285
conformis	231
cooperi	236
copii	293
cornutus	283
crassicauda	231
croceus	308
egregius	237
elegans	227
elongatus	240, 2797
emorii	227
erythrogaster	210
estor	240
exilicauda	209
flammeus	242
formosus	246
frontalis	283
funduloides	240
gardoneus	251
gibbosus	231
gilli	239

	Page.
Leuciscus gracilis	283, 326
grahami	228
grandis	225
hæmaturus	218
heterodon	261
hudsonius	269
humboldti	236
hydrophlox	238
hypselopterus	280
intermedius	235
iris	222
kentuckiensis	279
lineatus	232
lutrensis	272
macrolepidotus	224
margarita	241
milnerianus	242
montanus	238
nachtriebi	2798
nasutus	306
neogæus	240, 2798
niger	235
nigrescens	233
nitidus	221
obesus	246, 282
orcutti	241
oregonensis	225
phlegethontis	243
photogenis	296
procne	264
productus	240
prolixus	206
proriger	240
pulchelloides	222
pulchellus	221
purpureus	234
pygmæus	624
robustus	228
rotengulus	221
rubellus	293
rubrifrons	295
siuslawi	2797
spilopterus	279
spirlingulus	282
storeri	222
storerianus	270
telescopus	292
tincella	211
tuditanus	253
vandoisulus	239
vittatus	282
volucellus	263
zeylonicus	415
zunnensis	227
leuciscus, Brachydeuterus	1327
Pomadasis	1328

	Page.
leuciscus, Pomadasys	1328
Pristipoma	1328
leucomænis, Salmo	2823
leucophæus, Congrus	355
leucopsarum, Myctophum (Steno-brachius)	562
Nannobrachium	562
leucops, Photogenis	296
engraulinus	296
leucopus, Photogenis	277
Rhamphoberyx	847
leucorhynchus, Rhinobatus	62
leucorus, Eupomacentrus	1551
Leucos	243, 244, 2798
hicolor	245
formosa	246
obesus	246
leucos, Carcharhinus	38
Carcharias	38
Leucosomus	220, 221, 250
americanus	250
argyreiosus	224
cataractus	221
caurinus	220
communis	326
corporalis	222
dissimilis	324
gulonellus	326
inæquilobus	224
incrassatus	222
occidentalis	247
pallidus	222
pulchellus	222
rhotheus	222
symmetricus	246
leucosteus, Calamus	1353
leucostictus, Eupomacentrus	1555
Pomacentrus	1556
leucurus, Caranx	915
Hemicaranx	914
Nauclerus	900
Leucus anticus	245
boucardi	247
dimidiatus	244
formosus	246
olivaceus	244
tincella	211
Leuresthes	801
crameri	802
tenuis	802
Leuroglossus	527
stilbius	527
Leurynnis	2460
paucidens	2460
levis, Sebastichthys	1816
Sebastodes	1816

	Page.
lewis, Squatina	59
lewisi, Salar	493
Salmo clarkii	2819
mykiss	493
Zygæna	45
liberiensis, Balistes	1702
libertate, Opisthonema	433
libertatis, Clupea	433
Meletta	433
Opisthonema	433
Lichia quiebra	899
lichtensteini, Acipenser	105
ligulata, Seriola	905
Lija	1714, 1715, 1718
Barbuda	1720
Colorada	1713
Trompa	1719
Lile	428, 429, 431
lima, Loricaria	158
Limamuræna	400
melanotis	402
Limanda	2644
aspera	2645
beanii	2646
ferruginea	2644
proboscidea	2645
rostrata	2645
limanda, Hippoglossoides	2615
limandoides, Hippoglossoides	2615
Pleuronectes	2615
Limax lanceolatus	3
limbatus, Carcharhinus	40
Carcharias	40
Conger	360
Fundulus	643
Isogomphodon	40
Oxydontichthys	385
Saurus	533
limi, Hydrargyra	624
Umbra	623
pygmæa	624
Limia	690
couchiana	695
cubensis	692
matamorensis	700
pavonia	692
Pœciloides	700
venusta	665
Limnurgus	666
variegatus	666
limosa, Myxine	8
limosus, Pilodictis	142
Pylodictis	143
Silurus	143
linea, Mesoprion	1260
Siphostoma	768

	Page.
linea, Syngnathus	768
lineata, Alosa	426
Coryphœna	1619
Echeneis	2268, 2270
Embiotoca	1506
Morone	1133
Novacula	1619
Sciæna	1133, 1460
Tigoma	233
Trigla	2167
Unibranchapertura	342
lineatum, Lepibema	1133
lineatus, Achirus	2697, 2698, 2702
Amblodon	1484
Calliodon	1651
Cyclopterus	2118
Cypselurus	2836
Dormitator	2108
Doryichthys	773
Doryrhamphus	773
Eques	1487
Esox	627
Exocœtus	739
Fundulus	649
Genyonemus	1460
Gerres	1377
Gobius	2218, 2260
Grammichthys	2702
Grystes	1868
Labrax	1113
Leiostomus	1460
Lepisosteus	110
Leuciscus	232
Liparis	2118
multistriatus	2118
Micropogon	1461
Monochir	2698
Mugil	812, 2841
Phtheirichthys	2268
Pleuronectes	2698, 2701
Prionotus	2167
Roccus	1113, 1132
Smaris	1378
Squalius	233
Trichodon	2297
Xyrichthys	1619
Zygonectes	649, 657
lineolata, Cliola	263
Clupea	422
Coryphœna	1619
Mollienisia	700
Pelamys	873
Pœcilia	700
lineolatum, Etheostoma flabellare	1098
lineolatus, Alburnus	263
Catonotus	1099

	Page.
lineolatus, Metrogaster	1499
Pseudoscarus	1651
Tetrodon	1728
lineopinnis, Murœna	396
lineo-punctatus, Balistes	1709
Ling	2550
Lings	2551
linguatula, Balistes	1720
Pleuronectes	2615
linnæi, Æglefinus	2543
Merluccius	2530
Molva	2552
Trachurus	911
Linophora	1672
Linophryne	2734
lucifer	2734
linsleyi, Etheostoma	1097
Liocetus	2733
murrayi	2733
Lioglossina	2622
tetrophthalma	2622
liolepis, Paralichthys	2624
Xystreurys	2623
Liomonacanthus	1713
Lionfish	1850
Lioniscus	103
liopeltis, Acipenser	106
Lioperca	1166
inermis	1168
Liopropoma	1135
aberrans	1136
rubra	1137
Liopropominæ	1127
Liopsetta	2649
glabra	2650
glacialis	2649, 2650
obscura	2651
putnami	2650
liorus, Chasmistes	183
liosternus, Phenacobius teretulus	303
Liparididæ	2105
liparidina, Discoboli	2105
Liparidinæ	2105
liparina, Amitra	2138, 2139
Monomitra	2139
Liparis	2114, 2115, 2116, 2118
agassizii	2121
arctica	2121
barbatus	2118
calliodon	2120
callyodon	2111
(Careproctus) reinhardi	2134
communis	2118
cyclogaster	2118
cyclopus	2112, 2118
cyclostigma	2125, 2865

	Page.
Liparis dennyi	2124
ekstromi	2108
fabricii	2121, 2128
fucensis	2119
gelatinosus	2134, 2135
gibbus	2123
gobius	2108
herschelinus	2123
lineatus	2118
multistriatus	2118
liparis	2116, 2118
maculatus	2108
major	2127
montagui	2107, 2108
mucosus	2111
nostras	2118
ophidoides	2118
pulchellus	2126
ranula	2134
reticulata	2108
stellatus	2118
tunicata	2121, 2128
tunicatus	2120
vulgaris	2118
liparis, Centrolophus	963
Cyclopterus	2123
major	2128
minor	2121
Liparis	2116, 2118
liparoides, Cyclopterus	2108
Liparops	2104
stelleri	2104
Liparopsinæ	2095
Lipogenis	619
Lipogenyidæ	619
Lipogenys gillii	619
Lipophrys	2377, 2378
liropus, Tachysurus	2784
Lirus perciformis	964
lirus, Lepomis	1007
Minnilus	298
Notropis	297
Lisa Blanca	813
Cabezuda	811
Francesa	410
lisiga, Aspidophorus	2036
Lisita	814
listeri, Ostracion	1725
lita, Muræna	2805
Litholepis tristœchus	111
Little Pickerel	627
Red-eye	996
Roncador	1460
Skate	68
Smelt	807
Tunnies	868

	Page.
Little Tunny	869
Little-head Porgy	1350
Little-mouth Porgy	1354
littoralis, Carcharias	46, 2748
Eugomphodus	47
Menticirrhus	1477
Squalus	47
Umbrina	1477
littoricola, Chætodon	1680
litura, Mesoprion	1258
liturosus, Diodon	1746
lividus, Petromyzon	12
Silurus	140
Liza	810
liza, Mugil	811, 2841
Liza Ojo de Perdriz	814
Lizardfish	538
Lizard Fishes	532, 533
lobatus, Canthogaster	1732
Spheroides	1731, 1732
lobochilus, Heros	1531
Lobotes	1235
auctorum	1236, 2858
emarginatus	1257
erate	1236, 2856
farkharii	1236, 2856
incurvus	1236, 2856
pacificus	2857, 2858
somnolentus	1236
surinamensis	1235, 2856, 2858
Lobotidæ	1235
lockingtonii, Icichthys	969
Lodde	520
lœve, Ditrema	1511
Log Fish	964
Perches	1024, 1026
Lonchiurus	1481
lanceolatus	1482
Lonchopisthus	2286
micrognathus	2287
lonchurum, Opisthognathus	2281
Lonchurus ancylodon	1416
barbatus	1482
depressus	1482
lanceolatus	1482
lonchurus, Opisthognathus	2281
Longjaw	471
Long-jawed Goby	2250
Long-jaws	710, 711
Long Mingo	1718
longa, Dasibatis	85
Dasyatis	85
Macdonaldia	2826
longatus, Labeo	186
Longe	504
Long-eared Sunfish	1002

	Page.
Long-finned Albacore	871
Charr	509
Sole	2658
longicauda, Gobius	2229
Gymnothorax	392
Muræna	392
Rabula	391
longicephalus, Galeichthys	2781
Hexanematichthys	130
Tachisurus	130
longiceps, Eleotris	2195
Hybopsis	264
Siboma	233
longicollis, Myrophis	371
longidens, Caulolepis	839
longifilis, Bathygadus	2566
Hymenocephalus	2567
longimana, Etheostoma	1054
Eulamia	38
longimanus, Boleosoma	1054
Cichlasoma	1520
Heros	1521
Squalus	38
Xystroplites	1008
longipes, Bathypterois	546
Gadus	2555
longipinne, Gobiosoma	2256
Holocentrum	849
longipinnis, Clevelandia	2255
Evermannia	2256
Rhombus	966
Stromateus	966
longirostris, Alburnops	267
Catostomus	176
Oliola	267
Esox	714
Euleptorhamphus	724
Hemirhamphus	724
Hippocampus	778
Hypoprion	41
Lepisosteus	110
Malthæa	2737
Notropis	267
Saurus	538
Tylosurus	714
longirostrum, Catostomus	176
longispathum, Peristedion	2178
longispinis, Ailurichthys	119
Lepomis	1006
Pontinus	1858
Long necked Eels	343
Long-nosed Dace	306
Gar	109
Sucker	176
Long-spined Sculpin	1976
Long-tail Shark	45

	Page.
longulus, Calliurus	996
Pomotis	996
longurio, Carcharias	42
Scoliodon	42, 2748
longus, Balistes	1707
Ophisurus	377
Pisodonophis	377
Look-Down	936
lophar, Perca	947
Lopharis	946
mediterraneus	947
Lophiidæ	2713
Lophiomus	2714
setigerus	2714
Lophius	2713
aculeatus	2741
americanus	2714
bufo	2316
gibbus	2717
histrio	2716, 2722
ocellatus	2722
piscatorius	2713
radiatus	2738
rostratus	2737
setigerus	2715
spectrum	2723
tumidus	2716
viviparus	2715
lophius, Amiurus	138
Ictalurus	138
Lophobranchii	759
Lophobranchs	750
Lophogobius	2209
cyprinoides	2209
Lopholatilus	2278
chamæleonticeps	2278
Lophopsetta	2659
maculata	2660
Lord, Irish	1934
lordii, Salmo	508
loreto, Gramma	1229
Loricaria	156, 157, 159
acuta	158
barbata	158
bransfordi	158
lima	158
panamensis	157
rostrata	157
strigilata	158
uracantha	158
variegata	159
Loricariichthys	156
Loricariidæ	155
Loricariinæ	156
loricata, Alysia	569
Loricati	1756

	Page.
loricatus, Gasterosteus	747
Macrognathus	110
Phalangistes	2046
Loro 1652, 1653, 1655, 1657	
loro, Scarus	1654
Loros	1642
Lota	2550
brosmiana	2551
compressa	2551
inornata	2551
maculosa	2550
Lotella	2546
maxillaris	2546
lotharingus, Gasterosteus	746
Lotinæ	2532
loubina, Perca	1119
louisiana, Engraulis	446
louisianæ, Notropis	2801
Siphostoma	770
Syngnathus	770
lowei, Polymixia	854
lowii, Omosudis	598
loxias, Prionotus	2156
lubb, Gadus	2561
lubricum, Branchiostoma	3
Lucania	663, 666
affinis	665
goodei	664, 2831
ommata	663, 2831
parva	665, 2831
venusta	665, 2831
lucasana, Sphyræna	826
lucasanum, Thalassoma	1607, 2859
lucasanus, Chlorichthys	1607
Julis	1607, 1608
lucasi, Ceratocottus	1940
lucayanum, Asymmetron	4
Luccius vorax	628
lucens, Ceratichthys	321
Dacentrus	1496
luciæ, Fundulus	654
Haplochilus	655
Hydrargyra	655
Zygonectes	655
lucida, Æthoprora	565
lucidus, Argyrosomus	470
Coregonus	471
Luxilus	299
Notemigonus	299
Salmo (Coregonus)	471
Stolephorus	447
Lucifer	591
lucifer, Linophryne	2734
Lucifuga	2501
dentatus	2500
subterraneus	2501
	Page.
Lucifuginæ	2498
Luciidæ	624
Lucioblennius	2404
alepidotus	2404
lucioceps, Saurus	539
Synodus	539
Luciocharax	339
insculptus	339
lucioides, Esox	628
Lucioperca americana	1022
borea	1022
canadensis	1022
grisea	1022
pépinus	1022
vitrea	1022
luciopercana, Mentiperca	1216
luciopercanus, Centropristis	1216
Prionodes	1216
Serranus	1216
Luciopercinæ	1018
Luciotrutta	473
mackenzii	474
Lucius	625, 626, 628
americanus	626
lucius	628
masquinongy	629
immaculatus	630
ohiensis	629
reticulatus	627
vermiculatus	627, 2827
lucius, Atractosteus	111
Esox	628
americanus	626
Lucius	628
Ptychocheilus	225
Lucky Proach	1971
lucretiæ, Aboma	2241
Gobius	2868
ludibunda, Cliola	273
Cyprinella	373
ludibundus, Notropis	273
Lugger, Stone	181
lugubris, Caranx	924
Cyprinella	274
Malacoctenus	2357
Melamphaes	842
Myxodes	2357
Plectromus	842
lugubrosus, Esox	628
luitpoldii, Characodon	2832
lumbricus, Muræna	342
Myrophis	371
Lumpeninæ	2349
Lumpenus	2435, 2436
aculeatus	2433
anguillaris	2436

	Page.
Lumpenus fabricii	2437
Lampetræformis	2438
mackayi	2436
maculatus	2433
medius	2435
nubilus	2438
oculeatus	2433
lumpenus, Blennius	3438
Clinus	2438
Stichæus	2438
Lumpfish	2096
Lump Sucker	2094, 2096
Lumpus	2096
anglorum	2097
spinosus	2099
vulgaris	2097
lumpus, Cyclopterus	2096, 2097
Luna, piscis	1754
Pez	1753
luna, Caranx	927
Lampris	954
Pomotis	1006
Zeus	955
lunaris, Orthragoriscus	1754
lunata, Echeneis	2269
lunaticus, Dactyloscopus	2302
lunatus, Platophrys	2665
Pleuronectes	2666
Rhinichthys	308
atronasus	308
Rhomboidichthys	2666
Lune, Poisson	954
Tetrodon	1754
luniscutis, Arius	125
Aspistor	2763
Selenaspis	125
Tachisurus	125
lunulatus, Epinephelus	1159
Lutjanus	1158
Mustelus	28
Rhomboidichthys	2666
Serranus	1159
lupus, Ameiurus	137
Anarhichas	2447
Ictalurus	137
Pimelodus	137
lusca, Cyclothone	582
lusitanicus, Vandellius	887
lutea, Anguilla	348
luteovinctum, Etheostoma	1086
lutescens, Chætodon	1680
Kyphosus	1388
Pimelepterus	1389
luteum, Hæmulon	1304
luteus, Centropomus	1024
Gadus tomcodus	2540

	Page.
luteus, Genyatremus	1342
Lutianus	1343
Noturus	144
Rhinichthys	307
Lutianidæ	1241
Lutianinæ	1242
Lutianus	1247
argentiventris	1261
colorado	1268
guttatus	1269
luteus	1343
novemfasciatus	1253
stearnsi	1256
lutipinnis, Hydrophlox	287
Minnilus	287
Notropis	286
Opisthopterus	437
Pristigaster	437
lutjanoides, Lutjanus	1261
Neomænis	1261
Ocyurus	1261
Lutjanus acutirostris	1259
ambiguus	1272
analis	1267
aratus	1274
argentiventris	1261
arnillus	1279
aubrieti	1271
aurorubens	1278
aya	1265
blackfordi	1265
brachypterus	1268
buccanella	1262
caballerote	1257
campechianus	1265
caxis	1260
cayennensis	1404
chrysurus	1276
colorado	1268
cyanopterus	1255
cynodon	1255
cubera	1255
dentatus	1255
griseus	1257
guttatus	1269
inermis	1275
jocu	1258
lunulatus	1158
lutjanoides	1261
mahogoni	1273
melanurus	1276
novemfasciatus	1253
ojanco	1273
pacificus	1253
prieto	1253
profundus	1264

	Page.
Lutjanus purpureus	1264
rosaceus	1267
stearnsi	1257
surinamensis	1319
synagris	1271
torridus	1264
triangulum	1454
tridens	1202
trilobus	1200
uninotatus	1271
verres	1583
viridis	1246
vivanus	1264, 1265
lutkeni, Exocœtus	736
Lutodeira	414
lutrensis, Cliola	272
Leuciscus	272
Notropis	271
luxatus, Chasmistes	183
Luxilinus	247
occidentalis	247, 2799
Luxilus	250, 254, 257, 281
chickasaveusis	275
chrysocephalus	282
dissimilis	319
elongatus	240
erythrogaster	210
interruptus	282
kentuckiensis	279, 322
leptosomus	250
lucidus	299
occidentalis	247
roseus	288
seco	250
selene	269
zonistius	285
luxoides, Cyprinella	274
lycaodon, Oncorhynchus	481, 483
Salmo	483
Lycenchelys	2469, 2470
paxillus	2471
porifer	2471
verrillii	2470, 2471
Lycengraulis	451, 2811, 2816
grossidens	451
poeyi	2811
lychnus, Myripristis	847
Lycias	2461, 2463, 2468
Lycocara	2478
parii	2478
Lycodalepis	2468
mucosus	2470
polaris	2468
turneri	2469
Lycodapodidæ	2491
Lycodapus	2492
Lycodapus dermatinus	2492
extensus	2479
fierasfer	2493
parviceps	2493
Lycodes	2461, 2462
brevipes	2467
coccineus	2469
concolor	2463
diapterus	2473
digitatus	2466
esmarkii	2463
frigidus	2465
gracilis	2465
mucosus	2470
nebulosus	2468
pacificus	2460
palearis	2466
paxilloides	2471
paxillus	2471
perspicillum	2465
polaris	2469
porifer	2472
reticulatus	2465
rossi	2465
seminudus	2468
terræ-novæ	2466
turneri	2469
vahlii	2463
verrillii	2471
zoarchus	2464
Lycodidæ	2455
Lycodinæ	2456
Lycodontis	392, 393
castaneus	2804
chlevastes	398
conspersus	397
dovii	397
elaboratus	398
funebris	396, 2804
miliaris	397
mordax	395, 2805
moringa	395
obscuratus	389
ocellatus	399
nigromarginatus	399
saxicola	399
pictus	2805
polygonius	394
sanctæ-helenæ	397
verrilli	393
vicinus	394
virescens	394
Lycodonus	2473
mirabilis	2474
Lycodophis	2869

	Page.
Lycodopsis	2460
crassilabris	2458
crotalinus	2459
pacificus	2460
paucidens	2460
Lycolia	2869
crassilabris	2869
crotalina	2869
Lyconectes	2444
aleutensis	2444
Lyconema	2474
barbatum	2474
lynceum, Etheostoma	1075
lynx, Pimelodus	138
Lyoliparis	2114, 2116, 2126
Lyomeri	404
Lyopomi	606
Lyopsetta	2612
exilis	2612
Lyosphæra	1751
globosa	1751
lyricus, Euctenogobius	2225
Gobius	2224
lythrochloris, Xenotis	1003
Lythrulon	1311
flaviguttatum	1312
opalescens	1312
Lythrurus	254, 258, 297
atripes	300
cyanocephalus	300
lythrurus, Notropis	300
umbratilis	300
Lythrypnus	2210, 2216, 2230
Macabi	411
Macana	341
macarellus, Caranx	909
Decapterus	909
Macaria	890
macaskeyi, Ictalurus	318
macclellandii, Bregmaceros	2526
macdonaldi, Acutomentum	1787
Conocara	457
Cynoscion	1411
Fundulus	651
Nannobrachium	563
Notropis	284
Penopus	2521
Salmo clarkii	2819
mykiss	497
Sebastodes	1786
Zygonectes	651
Macdonaldia	616
alta	2826
challengeri	617
longa	2826

	Page.
Macdonaldia rostrata	617
macellus, Prionistius	1928
macer, Polyprosopus	51
Machæra	890
machete, Sardina	433
machnata, Argentina	410
Macho	811
Machuelo	432
Machuto	811
mackayi, Lumpenus	2436
Siphostoma	766
mackenzii, Luciotrutta	474
Salmo	474
Stenodus	474
Mackerel, Chub	866
Common	865
Easter	866
Horse	870
Monterey Spanish	873
Snap	946
Spanish, of England	866
Spanish	874
Thimble-eyed	866
Tinker	866
Yellow	921
Mackerel-like Fishes	860
Mackerel Scads	907
Shad	909
Shark	48, 49
Sharks	47
Mackerels	863, 865
Frigate	867
King of	1755
Snake	883
Mackinaw Trout	504
maclura, Pteroplatea	86
Raia	87
Macolor	1247
macouni, Chauliodus	585
macracanthum, Cichlasoma	1518
Pristipoma	1332
macracanthus, Alutarius	1720
Heros	1519
Pomadasis	1332
macrocephala, Clupea	411
Muræna	348
macrocephalum, Etheostoma	1031
Macrocephalus	1117
macrocephalus, Alvordius	1031
Clinus	2364
Gadus	2541
Gonocephalus	2184
Gunellus	2419
Hadropterus	1031
Labrosomus	2364
Mnierpes	2364

	Page.
macrocephalus, Percina	1031
Semotilus	222
macrocerus, Monacanthus	1713
macrocheilus, Catostomus	178
macrochir, Aldrovandia	609
Halosaurus	610
Sebastolobus	1763
macrochirus, Eupomotis	1005
Hemirhamphus	723
Lepomis	1005
Macrodon	330
malabaricus	330
microlepis	330
Macrodonophis	386
mordax	387
macrodus, Squalus	47
macrogenis, Cerna	1181
macrognathum, Opisthognathus	2281
Macrognathus	759
brevirostris	723
loricatus	110
macrognathus, Opisthognathus	2282
macrolepidota, Anchovia	449
Pœcilia	641
macrolepidotum, Moxostoma	193
macrolepidotus, Catostomus	194
Chætodon	1677
Engraulis	449
Fundulus hetero-	
clitus	641
Leuciscus	224
Notropis	299
Pleuronectes	2672
Pogonichthys	223
Stolephorus	449
macrolepis, Pontinus	1855
macronema, Pimelodus	155
macronemus, Bagrus	117
Nemipterus	1289
Polynemus	828
Synagris	1289
macrophthalma, Clupea	430
Harengula	430
macrophthalmus, Aprion	1280, 1281
Caranx	911
Centropristis	1281
Elastoma	1281
Gobiesox	2335
Priacanthus	1238
Sardinella	430
Scomber	867
macropoma, Bollmannia	2239
Centropristis	1206
Diplectrum	1205
Macrops	1281
oculatus	1283

	Page.
macrops, Balistes	1706
Bathygadus	2566
Calamus	1350, 1534
Citharichthys	2684
Conger	355
Corvina	1428
Corvula	1427, 1428
Gnathypops	2284
Hippoglossina	2621
Opisthognathus	2284
Opisthopterus	437
Pristigaster	437
Sciæna	1428
macroptera, Cónocara	457, 458
Goniobatis	88
macropterum, Hypentelium	181
macropterus, Alepocephalus	458
Balistes	1707
Centrarchus	988
Labrus	988
Thynnus	871
macropus, Malacoctenus	2357
Myxodes	2357
Macrorhamphosidæ	758
Macrorhamphosus	759
scolopax	759
macrorhinus, Acipenser	105
macrorhynchus, Hemirhamphus	724
macrospila, Piramutana	155
Macrostoma	554
angustidens	555, 2826
brachychir	2826
castaneum	556
caudispinosum	559, 2826
margaritiferum	555
quercinum	554
macrostoma, Cyprinella	274
Hæmulon	1297
Salmo	481
macrostomum, Hæmulon	1296
macrostomus, Acipenser	106
Notropis	274
Macrouridæ	2561
Macrourinæ	2562
Macrourus	2581
acrolepis	2585
bairdii	2583
berglax	2581
cinereus	2586
fabricii	2582
holotrachys	2582
lepturus	2584
rupestris	2582
stelgidolepis	2585
stromii	2579
macrourus, Alopias	46

	Page.
Macrozoarces	2456
macrura, Congermuræna	356
Macruroplus	2581
macrurum, Ophisoma	357
Macrurus	2581
acrolepis	2585
asper	2572
carminatus	2589
carribbæus	2590
dorsalis	2585
firmisquamis	2576
goodei	2572
(Nematonurus) magnus	2574
occa	2588
pectoralis	2574
rupestris	2579
scaphopsis	2590
simulus	2578
suborbitalis	2573
macrurus, Oxydontichthys	385
macularius, Cyprinodon	674
baileyi	675
maculata, Apogon	1109
Belone	718
Hemitremia	259
Lophopsetta	2660
Morone	1010
Muræna nigra	395
Nerophis	774
Perca	1153
Sciæna	2198
maculaticeps, Boleosoma nigrum	1058
maculatofasciatus, Paralabrax	1196
Serranus	1196
maculatum, Boleosoma	1057, 1077
Cybium	874
Cynoscion	1409
Etheostoma	1077
maculatus, Alvordius	1032, 1034
Anarrhichas	2446
Apogon	1109
Aulostomus	754
Balistes	1707, 1708
Bothus	2660
Canthidermis	1706, 1707
Clinus	2433
Cottus	1972
Cryptacanthodes	2443
Ctenodon	2433
Diodon	1746
Dormitator	2196, 219
Galeocerdo	32
Galeus	32
Gasteropelecus	338
Hadropterus	1031, 1034
Heptranchias	18

	Page.
maculatus, Leptoclinus	2433
Liparis	2108
Lumpenus	2433
Monoprion	1109
Mullhypeneus	859
Mullus	859
Nomeus	950
Nothonotus	1077
Notorhynchus	17
Notropis	259
Ostracion	1725
Pimelodus	135, 155
Platypœcilus	686
Pleuronectes	2660
Procerus	102
Psenes	951
Rhypticus	1234
Scomber	867, 874
Scomberomorus	874, 875
Serranus	1153
Spherides	1733
Spheroides	1733
Stichæus	2433
Upeneus	858
maculicauda, Diabasis	1314
Hæmulon	1314
Orthostœchus	1313
maculifer, Diodon	1747
Platophrys	2664
Pleuronectes	2665
maculiferus, Hypoplectrus	1192
Rhomboidichthys	2665
maculipinna, Chærojulis	1595
Halichæres	1595
Iridio	1594
Julis	1595
Platyglossus	1595
maculipinnis, Achirus	2698
Exocœtus	737
Heros	1529
Isogomphodon	40
Monochir	2698
Muræna	394
Solea	2698
Thyrsoidea	394
maculocinctus, Chætodon	1674
Sarothrodus	1674
maculosa, Harengula	436
Lota	2550
Molva	2551
Muræna	382
Thalassophryne	2324
maculoseriatus, Chirus	1868
maculostriatus, Diodon	1748
maculosum, Hæmulon	1295
maculosus, Acentrolophus	963

	Page.
maculosus, Acipenser	106
Catostomus	181
Centrarchus	991
Centridermichthys	2014
Epinephelus	1158
Gadus	2551
Nomeus	950
Oligocottus	2013
Paralichthys	2626
Pimephales	217
promelas	217
Pleuronectes	2626
Serpens marinus	382
Serranus	1159
madeirensis, Ceratoscopelus	557
Scopelus	557
Mademoiselle	1433
maderaspatensis, Butirinus	415
maderensis, Helicolenus	1837
Madregal	905
Mad Tom	147
Toms	144
mæandricus, Caularchus	2328
Lepidogaster	2328
Mænidæ	1364
Mæninæ	1364
magdalenæ, Curimata	332
Otolithus	1410
Paralichthys	2872
Sciæna	1420
magistralis, Epinnula	880
magnioculis, Ophichthus	385
Ophichtnys	385
Scytalophis	385
magnus, Macrurus (Nematonurus)	2574
Mahogany Snapper	1272
mahogoni, Lutjanus	1273
Mesoprion	1273
Neomænis	1272
Mail-cheeked Fishes	1756
Mailed Gurnards	2176
mainensis, Gasterosteus	745
Maire d'Amplora	555
majalis, Cobitis	639
Fundulus	639, 2827
Hydrargyra	639
Majarra, Raiada	1561
major, Actinochir	2128
Blepharis	932
Cyclopterus liparis	2128
Liparis	2127
Ptychocheilus	225, 2797
Major, Sergeant	1561
Makaira	890
nigricans	891
makaira, Xiphias	891

	Page.
makua, Ranzania	1755
malabarica, Elacate	948
malabaricus, Macrodon	330
Malacanthidæ	2274
Malacanthinæ	2275
Malacanthus	2275
plumieri	2275
trachinus	2276
Malachorhinus	66
Malacocephalus	2569
occidentalis	2570
Malacocottus	1994
zonurus	1994
Malacoctenus	3356
biguttatus	2360
bimaculatus	2358
delalandi	2358
gillii	2358
lugubris	2357
macropus	2357
ocellatus	2356, 2869
varius	2357
versicolor	2359
Malacosteidæ	592
Malacosteus	592
niger	593
Malapterinæ	1572
malarmoides, Aspidophorus	2054
Maletta cærulea	423
maliger, Sebastichthys	1823
Sebastodes	1822
malleus, Squalus	45
Zygæna	45
Mallotus	520
villosus	520
Malma	507, 508
malma, Salmo	508
Salvelinus	507, 2823
Malthæa	2736
angusta	2738
longirostris	2737
nasuta	2737
notata	2737
truncata	2738
vespertilio	2737
Malthe	2736
cubifrons	2738
elater	2739
Mammoth Cave Blindfish	706
Mammy	205
managuensis, Heros	1533
Pimelodus	153
Rhamdia	153
manatia, Raia	93
manatinus, Barathrodemus	2517
Mancalias	2729

	Page.
Mancalias shufeldti	2730
uranoscopus	2729
Mancho, Cibi	919
Man-eater Sharks	50
manglicola, Gobius	2220
mango, Polynemus	830
Mangrove Grouper	1171
Minnow	643
Snapper	1255
Manitou Darter	1027
manitou, Percina caprodes	1028
Manjua	443
Manjuari	111
manjuari, Lepidosteus	111
mannii, Zygonectes	664
Manta	92
americana	93
birostris	92
manta, Cephaloptera	93
Mantararia Colorado	2754
Mantidæ	91, 2756
manus, Gadus	2154
Mapo	2216
Guavina	2196
mapo Gobius	2218
maraldi, Gadus	2546
Uraleptus	2545
marcgravii, Aulastome	757
Coryphæna	953
Rhinobatus	63
marconis, Hybopsis æstivalis	316
Mareño, Pargo	1252
Margaret, Bastard	1297
Margaret Grunt	1295
margarita, Clinostomus	241
Echiostoma	589
Leuciscus	241
margaritaceus, Entomacrodus	2398
Salarias	2399
margaritatus, Batrachus	2323
Porichthys	2322
margaritifer, Heros	1520
Notoscopelus	555
Serranus	1156
margaritiferum, Cichlasoma	1519
Hæmulon	1312
Macrostoma	555
margaritus, Phoxinus	241
Squalius	241
margarotis, Enneacanthus	994
Margate-fish	1295
marginalis, Hippocampus	778
marginata, Aphoristia	2706
Rissola	2489
Thyrsoidea	394
Uranidea	1965

	Page.
marginatum, Ophidium	2489
marginatus, Bodianus	1174
Brosmius	2502
Brosmophycis	2502
Caranx	922
Chætodon	1562
Chromis	1546
Cottus	1966
Dinematichthys	2502
Halias	2502
Hemirhamphus	723
Lepomis	1003
Neobythites	2513
Noturus	147
Phycis	2555
Pimelodus	135
Pomotis	1003
Serranus	1154
Symphurus	2706
Maria Molle	1552
Marian	852
marianus, Flammeo	2871
Holocentrus	852, 2871
Maria-prieta	1319
Marina, Perca	1761
cauda nigra	1303
gibbosa	1295
pinnis	1259
sectatrix	1388
venenosa	1172
Umbla minor	823
marinus, Ailurichthys	118
Esox	714
Faber	1668
Felichthys	118
Petromyzon	10
camtschaticus	2745
dorsatus	10
unicolor	10
Raja diabolus	93
Sebastes	1760, 1761
viviparus	1761
Serpens maculosus	382
Silurus	118
Tylosurus	714
Mariposas	953, 954
marmorata, Amia	113
Pteroplatea	87
Unibranchapertura	342
marmoratum, Lepophidium	2482, 2483
marmoratus, Ameiurus	141
nebulosus	141
Antennarius	2717
Auchenopterus	2371
Cottus	1983

	Page.
marmoratus, Cremnobates	2371
Hemitripterus	1889, 2022
Pimelodus	141
Rhinichthys	306
Rivulus	663, 2830
Scorpænichthys	1889
Spheroides	1734
Symbranchus	342
Tirus	537
marmorea, Rabula	391
marmoreum, Siphostoma	768
marmoreus, Blennius	2381
Gymnothorax	391
Murænophis	391
Syngnathus	768
Marsipobranchii	4
marstoni, Salmo	516
Salvelinus oquassa	515
martæ, Cyprinodon	675
martii, Pristigaster	438
Martin Pescador	2724
martinica, Atherina	795
Kirtlandia	795
Menidia	795
Spicara	1364, 1365
martinicensis, Aylopon	1228
Holanthias	1228
Menticirrhus	1473
Nerophis	774
Novacula	1617
Novaculichthys	1616
Ocyanthias	1228
Odontanthias	1228
Umbrina	1474
Vomer	934
Xyrichthys	1617
martinicus, Gobius	2236
Smaris	1365
Upeneus	859
Masamacush	504
Mascalongus	625, 626, 629
maschalespilos, Scarus	1642
Sparisoma	1641
Maskinongy	629
masoni, Salmo	501
masquinongy, Esox	629
immaculatus	630
Lucius	629
immaculatus	630
ohiensis	629
massachusettensis, Monacanthus	1716
massiliensis, Scorpæna	1139
Masticura	59, 79
Matejuelo	848
Blanco	2275, 2276
Real	410
matejuelo, Amphiprion	849
mathematicus, Tetrodon	1728
Mathemeg	137
matoides, Acanthurus	1693
matsubaræ, Sebastodes	1833
Mattowacca	425
mattowacca, Clupea	426
matutinus, Alburnellus	301
Minnilus	301
Notropis	301
umbratilis	301
matzubaræ, Sebastodes	1796
mauricei, Argyriosus	936
mauritii, Chætodon	1562
Eleotris	950
Grammistes	1323
maurolici, Scopelus	577
Maurolicidæ	576
Maurolicus	576
amethystino-punctatus	577
attenuatus	577
borealis	577
mulleri	577
pennanti	577
tripunctulatus	578
maxillaris, Lotella	2546
Murænoides	2418
Maxillingua	327
maxillingua Cyprinus	327
Exoglossum	327
maxillosa, Gnathypops	2284
maxillosus, Gnathypops	2284
Opisthognathus	2284
Rhinichthys	307
maxima, Selache	51
maximus, Cetorhinus	51
Hippoglossus	2612
Labrus	1580
Lachnolæmus	1580
Lachnolaimus	1579
Selachus	51
Squalus	51
mayous, Salmo omisco	487
Mayfish	639
Maynea brunnea	2476
pusilla	2476
May Sucker	199
mazatlana, Seriola	904
Solea	2699
mazatlanum, Hæmulon	1314
mazatlanus, Achirus	2698
McCloud River Rainbow Trout	502
meadiæ, Ulocentra	2852
meanyi, Ruscarius	1908
Meda	328
fulgida	329

	Page.
media, Palometa	2849
Medialuna	1390, 1391
californiensis	1391
Medialunas	1390
mediocri, Clupea	426
mediocris, Pomolobus	425
medirostris, Acipenser	104
mediterranea, Chimæra	95
Laurida	537
Sarda	872
mediterraneus, Hoplostethus	837
Lopharis	947
Scomber	872
Sternoptyx	604
Thynnus	870
medius, Anisarchus	2436
Calamus	1355
Centropomus	1120
Clinus	2436
Grammateus	1356
Lumpenus	2435
Myoxocephalus	1983
Rhombus	967
Stichæus	2436
Medregal	904
medusicola, Caranx	924
medusophagus, Schedopholus	970
mecki, Hybopsis	317
megacephalus, Calamus	1350, 1351
Chitonotus	1891
Icelus	1891
Megaderus	402
megalaspis, Acipenser	105
megalepis, Doratonotus	1612
Megalobrycon	337
Megalocottus	1987
laticeps	1988
platycephalus	1987
megalodon, Pristis	61
Megalopinæ	408
Megalops atlanticus	409
cepediana	416
elongatus	409
notata	432
oglina	432
thrissoides	409
megalops, Alburnellus	291
Alburnus	291
Clupea	426
Cyprinus	282
Ditrema	1502
Ennichthys	1502
Holconotus	1502
Micropogon	1463
Minnilus	291
Notropis albeolus	284

	Page.
megalops, Opsopœodus	248
Pimelodus	135
Trycherodon	249
megalotis, Ichthelis	1003
Lepomis	1002
Megaperca	1137
ischinagi	1138
Megaphalus	2320
megastoma, Grystes	1012
Opisthognathus	2282
megastomus, Catostomus	181
Melamphaes beanii	843
crassiceps	843
cristiceps	844
lugubris	842
Melamphainæ	838
melampterus, Salmo	483
melampygus, Caranx	925
Melanichthys	1381, 1711
melanocephalus, Plargyrus	217
Melanocetinæ	2728
Melanocetus bispinosus	2734
(Liocetus) murrayi	2734
melanochir, Julis	1609
melanochira, Belone	716
melanogaster, Pleuronectes	2630
Pœcilia	696
Melanogrammus	2542
æglefinus	2542, 2543
melanopis, Diodon	1746
melanopleurus, Haplochilus	660
melanopogon, Cichlasoma	1523
Heros	1523
melanopoma, Polynemus	831
melanops, Calliurus	992
Catostomus	187
Dionda	216
Haplochilus	682
Hybognathus	216
Ichthelis	996
Minytrema	187
Sebastes	1783
Sebastodes	1782, 1783
Zygonectes	682
melanopterum, Pristipoma	1319
melanopterus, Balistes	1707
melanopus, Arius	132
Tachysurus	132, 2784
melanorhina, Plectropoma	1192
melanospilum, Diagramma	1321
melanostictus, Hippoglossoides	2618
Psettichthys	2618
Melanostigma	2478
gelatinosum	2479
pammelas	2479, 2869
Melanostigmatinæ	2456

	Page.
melanostomus, Sebastodes	1803
melanothus, Holacanthus	1728
melanotis, Limamuræna	402
Muræna	401
Pseudojulis	1605
Scarus	1638
melanotus, Catostomus	206, 218, 322
Melanura	623
annulata	624
melanurum, Cichlasoma	1523
Hæmulon	1302. 1303
Læmonema	2557
Perca	1303
melanurus, Careproctus	2135
Cyprinus	282
Exocœtus	735, 736
Heros	1524
Lutjanus	1276
Rutilus	193
melapleura, Pœcilia	660
melapleurus, Fundulus	659
melas, Ameiurus	141
Silurus	141
meleagris, Muræna	399
Priodonophis	399
Rhinichthys	308
atronasus	308
Meletta	424
libertatis	433
sucœrii	425
venosa	426
Melichthys	1711
bispinosus	1711
piceus	1711
ringens	1711
Melletes	1932
papilio	1932
mellissii, Congromuræna	356
Membras	789
membras, Clupea	421
Menephorus	1143, 1146
dubius	1147
punctiferus	1147
menhaden, Alosa	434
Clupea	434
Gulf	434
Menhadens	433
Menidia	443, 796, 2840
audens	798
beryllina	797
brasiliensis	801
clara	801
dentex	801
gilberti	798
gracilis	797
beryllina	797

	Page.
Menidia guatemalensis	801, 2840
martinica	795
menidia	800
notata	2840
notata	800, 2840
pachylepis	801, 2840
peninsulæ	797
sardina	799
vagrans	795
laciniata	795
menidia, Argentina	443
Atherina	801
Atherinichthys	800
Menidia	800
notata	2840
Menominee Whitefish	465
menona, Fundulus diaphanus	645
mentalis, Platypœcilus	686
Menticirrhus	1469, 1470
alburnus	1475
americanus	1474
elongatus	1476
littoralis	1477
martinicensis	1473
nasus	1472, 1473
nebulosus	1475
panamensis	1473
saxatilis	1475
simus	1472
undulatus	1476
Mentiperca	1208, 1209, 1214
luciopercana	1216
mento, Balistes	1710
Bramopsis	1502
Chanos	415
Paraliparis	2142
Xanthichthys	1710
mentzelii, Chironectes	7224
Serranus	1154
merckii, Coregonus	470
meridionalis, Amiurus	135
Cottus	1951
Ictalurus	135
Ictiobus	164
Scleroguathus	164
Merlangus	2529
leptocephalus	2535
polaris	2534
productus	2531
purpureus	2535
merlangus, Anoplopoma	1862
Merlucciidæ	2529
Merluccius	2529
ambiguus	2530
argentatus	2530
bilinearis	2530, 2531

3032   *Index.*

	Page.		Page.
Merluccius esculentus	2530	Mesoprion linea	1260
linnæi	2530	litura	1258
merluccius	2530	mahogoni	1273
productus	2531	ojanco	1273
sinuatus	2530	pacificus	1253
smiridus	2530	pargus	1255
vulgaris	2530	profundus	1263
merluccius, Gadus	2530	ricardi	1273
Merluccius	2530	rosaceus	1267
Merlucius albidus	2531	sobra	1266
attenuatus	2546	uninotatus	1271
lanatus	2530	vivanus	1263
Merlus	2529	vorax	1281
merlus, Gadus	2530	mesops, Arius	123
Mero	1154, 1162	Bagrus	123
de lo Alto	1161	Sciadeichthys	123, 2760
Merou	1154, 1780	Tachisurus	123
merriami, Empetrichthys	667	Mesopus	524
mertensii, Cottus	1986	olidus	525
meru, Holocentrus	1154	mesotrema, Asternotremia	787
Merulinus	2148, 2149, 2156	metallica, Agosia	314
Merus	1148	Echeneis	2270
merus, Centropristis	1162	Heterandria	687
Epinephelus	1162	metallicus, Girardinus	687
mesænm, Boleosoma nigrum	1059	Notropis	297
mesæus, Pœcilichthys	1059	metamorensis, Limia	700
mesogaster, Exocœtus	729	Metoponops	2678
Parexocœtus	728	cooperi	2680
Mesogobius	2210	Metrogaster	1498
Mesogonistius	994	lineolatus	1499
chætodon	995	meulini, Diodon	1748
Mesoprion	1247	Me Waru	1829
ambiguus	1272	Tokenoko	1829
analis	1266	Mexican Sole	2698
aratus	1274	mexicana, Pœcilia	692
argentiventris	1261	mexicanum, Dorosoma	416
arnillo	1279	Myctophum	563
aurorubens	1278	Nannobrachium	563
aurovittatus	1276	mexicanus, Amblyopus	2264
aya	1264	Awaous	2237
buccanella	1262	Centropomus	1121
caballerote	1257	Chatoessus	416
campechanus	1265	Chronophorus	2237
caudanotatus	1262	Gerres	1380
caxis	1260	Gobius	2237
chrysurus	1276	Mugil	813
cyanopterus	1255	Pempheris	978
cynodon	1255, 1260	Saurus	538
dentatus	1279	Tetragonopterus	335
elegans	1278	miarchus, Stolephorus	441
flavescens	1260	Michigan Cisco	469
griseus	1257	Grayling	518
guttatus	1269	Herring	468
inermis	1275	Kieye	469
isodon	1267	Micristius	633
jocu	1258	Micristodus	52

	Page.
Micristodus punctatus	52
microcephala, Cynoglossa	2655
Platessa	2654
microcephalus, Gasterosteus williamsoni	751
Pleuronectes	2654
Somniosus	57
Squalus	57
Microdesmus	2450
dipus	2450
retropinnis	2450
microdon, Cyclothone	582, 2826
Gobius	2227
Gonostoma	582
Osmerus	521
Pseudotriakis	27
Microdonophis	381
Microgadus	2538
proximus	2539
tomcod	2540
micrognathus, Lonchopisthus	2287
Opisthognathus	2287
Microgobius	2242
cyclolepis	2247
eulepis	2244
gulosus	2243
signatus	2246
thalassinus	2245
microlepidota, Gila	207
Microlepidotus	1341
inornatus	1341
microlepidotus, Cestreus	1415
Cynoscion	1415
Labrisomus	2363
Labrosomus	2364
Orthodon	207
Otolithus	1415
Microlepis	228
microlepis, Antimora	2545
Epinephelus	1178
Lepophidium	2486
Macrodon	330
Mycteroperca	1177
Trisotropis	1178
microlophus, Pomotis	1008
Micromesus	90, 91
Micrometrus	1496, 1498
aggregatus	1499
frenatus	1499
rosaceus	1500
micronema, Peristethus	2182
micronemus, Peristedion	2182
Microperca	1103
fonticola	1104
prœliaris	1103
punctulata	1104

	Page.
microperca, Etheostoma	1104
Microphis	773
microphthalmum, Hæmulon	1296
microphthalmus, Dormitator	2198
Heros	1536
Tetragonopterus	334
Micropogon	1461
altipinnis	1464
argenteus	1463
costatus	1462
ectenes	1463
furnieri	1462
lineatus	1461
megalops	1463
opercularis	1461
undulatus	1461
micropogon, Ceratichthys	323
microps, Agonostomus	820
Atherina	791
Belone	712
Calliurus	996
Carcharias	40
Caulolatilus	2277
Corvina	1445
Cottunculus	1992
Dajaus	820
Nebris	1417
Otolithus	1415
Pagellus	1355
Rhypticus	1232
Stellifer	1445
Tylosurus	712
microptera, Rhamdia	153
Micropterinæ	986
Micropterus	1010
dolomieu	1011
salmoides	1012
micropterus, Etheostoma	1083
Pimelodus	153
Micropteryx	901, 937
chrysurus	938
cosmopolita	938
micropteryx, Alburnellus	297
Minnilus	297
Notropis	296
Platysomus	934
micropus, Gasterosteus	744
microrhynchus, Acipenser	106
microrrhinos, Pseudoscarus	1655
Microspathodon	1565
azurissimus	1570
bairdii	1566, 1567
chrysurus	1567
cinereus	1570
dorsalis	1568, 1570
azurissimus	1570

	Page.
Microspathodon dorsalis cinereus..	1570
niveatus	1567
Microspathodontinæ	1544
microstigmius, Myrophis	371
microstoma, Oliola	264
Crenilabrus	1576
Scartella	2384
Tetragonopterus	334
Uranidea	1958
Microstomidæ	527
Microstominæ	527
Microstomus	2653
grœnlandicus	528
kitt	2654
latidens	2654
pacificus	2655
microstomus, Blennius	2385
Citharichthys	2688
Conger	356
Etropus	2687, 2690
Minnilus	262
Pleuronectes	2654
micrurum, Syacium	2672
Midshipman	2317, 2321
Mikiss	2819
milberti, Acipenser	105
Arius	128
Carcharhinus	37
Carcharias	37
Eulamia	37
milbertianus, Syngnathus	771
miles, Porogadus	2520
Prionotus	2160
milesi, Pimephales	217
miliaris, Bellator	2173
Gymnothorax	398
Lycodontis	397
Muræna	398
Thrysoidea	398
Milk Fishes	414
milktschitch, Salmo	481
milleri, Bathylagus	2825
Miller's Thumb	1941, 1950
millipunctatus, Gasterosteus	752
milneri, Nocomis	324
Pagellus	1355
Sparus	1355
milnerianus, Leuciscus	242
Phoxinus	242
Milvus cirratus	2183
mineopas, Bryttus	996
Mingo, Long	1718
miniatum, Peristedion	2178
miniatus, Lepomis	1002
Sebastichthys	1795
Sebastodes	1794

	Page.
Miniellus	254
minima, Abeona	1497
Cichla	1012
Clupea	422
Perca	1057
minimus, Cymatogaster	1497
miniofrenatus, Scarus	1634
Mink, Sea	1475
Minnilus, altipinnis	287
amabalis	291
ardens	301
ariommus	290
atripes	300
bellus	297
bivittatus	233
blennius	262
cercostigma	275
chalybæus	288
chlorocephalus	286
chrosomus	288
coccogenis	285
cornutus	283
cyanocephalus	300
dinemus	293
diplæmius	300
jejunus	290
jemezanus	294
lacertosus	284
leuciodus	291
lirus	298
lutipinnis	287
matutinus	301
megalops	291
micropteryx	297
microstomus	262
nigripinnis	299
notatus	218
oligaspis	294
percobromus	295
plumbeolus	283
punctulatus	302
roseus	288
rubellus	293
rubricroceus	286
rubrifrons	295
rubripinnis	298
scabriceps	268, 290
scepticus	296
selene	269
shumardi	268, 269
stilbius	293
telescopus	292
timpanogensis	233
umbratilis	299
xænocephalus	289
xænurus	280

	Page.
Minnilus, zonistius	285
Minnow, Black-head	217
Blunt-nosed	218
Eastern Mud	624
Leather-sided	236
Mangrove	643
Sheepshead	671
Silver-sided	238
Silvery	213
Spot-tailed	269
Spotted-tail	275
Star-headed	656
Straw-colored	261
Top	659, 680
Minnows, Mud	623
.Pursy	670, 671
Minomus	173
bardus	171
delphinus	171
jarrovii	170
platyrhynchus	170
minor, Anarhichas	2446
Anarrhichas	2446
Aphauopus	885
Atinga alter orbicularis	1749
Cyclopterus liparis	2121
Stellifer	1442
Umbla marina	823
minutillus, Chriolepis	2205
minutus, Cottus	1958
Cyclopterus	2097
Gobius	2097
Minytrema	186
austrinum	192
melanops	187
minytremus, Gyrinichthys	2137
Mionurus	1106
mirabile, Exoglossum	303
mirabilis, Clupea	422
Crystallichthys	2865
Gillichthys	2250
Lycodonus	2474
Phenacobius	303
mirum, Calloptilum	2527
mispilliensis, Ameiurus	141
Mississippi Cat	137
mississippiensis, Morone	1134
Pristis	61
Missouri Sucker	168
missuriensis, Oliola	262
Hybopsis	262
mitchilli, Argyriosus	936
Acipenser	105
Cottus	1973
Engraulis	446
Lepibema	1133

	Page.
mitchilli, Perca	1133
alternata	1133
interrupta	1133
Stolephorus	446
Mitchillina	453
bairdii	454
mitis, Balistes	1705
mitzukurii, Sebastodes	1831
miurus, Mystriophis	387
Noturus	148
Ophichthys	387
Schilbeodes	148
Scytalichthys	387
Mixonus	2523
laticeps	2523
mixtus Gadus tomcodus	2540
Mnierpes	2364
macrocephalus	2364
Mobula	91
modesta, Cheonda	234
Gambusia	693
Pimelodella	154
modestum, Hæmulon	1340
modestus, Julis	1601
Oxyjulis	1601
Pimelodus	154
Pomadasys	1321
Pseudojulis	1601
Squalius	234
Xyrichthys	1619
Mœbia	2510
promelas	2511
mœrens, Gonopterus	1688
Moharra	1373, 1374
mohri, Clinus	2438
Mojarra	1379
Almejero	1294
Blanca	1372
Cantileña	1369
Cardenal	850
China	1377
de Casta	1372
de la Piedras	1681
de las Aletas Amarillas	1376
de Ley	1370
Dorada	928
Garabata	1353
Prieta	1299
Verde	1538
Mojarras	1366, 1373
Mojarritas	1367
Mojarron	1319
Mola	1753
aculeata	1754
mola	1753
nasus	1754

	Page.
Mola planci	1756
retzii	1754
rotunda	1754
mola, Mola	1753
Orthagoriscus	1754
Tetrodon	1754
Molacanthus	1753
Molarii	402
molestum, Gobiosoma	2258
Molidæ	1752
Molinæ	1752
molle, Maria	1552
Mollienisia	698
fasciata	695
formosa	699
jonesi	699
latipinna	700
lineolata	700
petenensis	700
mollis, Achirus mollis	2702
Aphyonus	2525
Bothrocara	2476
Pleuronectes	2701
Molly, Hog	181, 1026
Molva	2551
buntia	2551
linnæi	2552
maculosa	2551
molva	2551
vulgaris	2552
molva, Gadus	2532
Molva	2551
monacantha, Corrina	1419
Monacanthidæ	1712
Monacanthus	1714
amphioxys	1717
auriga	1716
broccus	1716
ciliatus	1714, 1715
davidsoni	1715
filamentosus	1716
gallinula	1716
hispidus	1715
irroratus	1713
macrocerus	1713
massachusettensis	1716
monoceros	1720
occidentalis	1715
oppositus	1716
pardalis	1713
parraianus	1713
piraaca	1715
proboscideus	1719
pullus	1713
punctatus	1713, 1719
scriptus	1719

	Page.
Monacanthus setifer	1716
signifer	1716
spilonotus	1716
stratus	1713
ruppelii	1713
varius	1716
monacanthus, Plectropoma	1165
monacus, Ceratichtbys	318
Hybopsis	318
monæ, Stephanoberyx	836
monapterygia, Pesca dorso	1833
Monda	899
monensis, Squalus	49
monestichus, Salmo	509
Mongolar Drummer	1406
Mongrel Buffalo	164
Whitefish	473
Moniana	254, 256, 271
aurata	272
complanata	272
couchi	272
deliciosa	262
formosa	271
frigida	271
gibbosa	272
gracilis	272
jugalis	272
lætabilis	272
leonina	272
nitida	265
proserpina	272
pulchella	272
rutila	272
tristis	272
Monkfish	58, 2713
monoceros, Alutera	1720, 2860
Balistes	1719, 1720
Monacanthus	1720
Monochir lineatus	2698
maculipinnis	2698
reticulatus	2696
Monochirus	2694
monocirrus, Exocœtus	730
Monodactylinæ	1667
Monolene	2690
atrimana	2692
sessilicauda	2691
Monomitra	2138
liparina	2139
monophthalmus, Auchenopterus	2372
Crennobates	2372
Monoprion	1106
maculatus	1109
pigmentarius	1109
Monopterhinus	18
monopterygius, Aspidophoroides	2091, 2092

	Page.
monopterygius, Canthirhynchus...	2092
Chirus	1869
Cottus	2092
Hexagrammus	1866
Labrax	1866
Pleurogrammus...	1864
Monosira stahli	1423
monstrosa, Chimæra	91
montacuti, Cylopterus	2108
montagui, Liparis	2107, 2108
Montana Grayling	519, 2871
montana, Gila	238
montanus, Clinostomus	238
Hybopsis	317
Leuciscus	238
Squalius	238
Thymallus ontariensis	519
signifer	519
tricolor	2871
Monterey Halibut	2625
Spanish Mackerel	837
montezuma, Cichlasoma	1518
Heros	1518
monticola, Agonostomus	819
Dajaus	2841
Mugil	819
montiregis, Cliola	272
Moon-eye	413
Cisco	469
Moon-eyes	412
Moonfish	934, 954
Moonfishes	935
Moorish Idol	1687
Mora, Cabra	1152
Moray, Black	396
Common Spotted	395
Spotted	399
Morays	388, 400
mordax, Atherina	523
Conger	387
Crotalopsis	387
Engraulis	448
Gymnothorax	396
Lycodontis	395
Macrodonophis	387
Muræna	396
Osmerus	523
abbotti	524
spectrum	523
Sidera	396
Morena Pinta	402
Pintita	397
Prieta	2804
Verde	396
moribundus, Balistes	1702
Morinæ	2532

	Page.
moringa, Lycodontis	395
Muræna	395
Sidera	395
moringua, Gymnothorax	395
morio, Centrolophus	963
Epinephelus	1160
Serranus	1160
mormyrus, Campostoma	206
Morone	1133
americana	1134, 1135
flavescens	1024
interrupta	1134
lineata	1133
maculata	1010
mississippiensis	1134
multilineata	1132
pallida	1135
rufa	1135
Moroninæ	1127
Moronopsis	1013
Morrhua	2540
æglefinus	2543
americana	2540, 2541
californica	2539
punctatus	2543
morrhua, Gadus	2541
morrissi, Leptocephalus	354
Moseleya	2570
cyclolepis	2570
moseri, Verasper	2619
Mossbunker	433
motaguensis, Heros	1534
Pimelodus	151
Rhamdia	151
Motella	2558
argentata	2559
caudacuta	2560
cimbria	2560
ensis	2559
reinhardti	2559
septentrionalis	2560
Mother of Eels	2457
motta, Elacate	948
moucharra, Glyphisodon	1562
Mountain Herring	463
Suckers	169, 170
Mouse-Fish	2715, 2716
Moxostoma	185, 187
albidum	192
album	191
anisurum	190, 196
aureolum	192
austrinum	192
breviceps	196
bucco	190
campbelli	186

	Page.
Moxostoma carpio	190
cervinum	197
claviformis	186
collapsum	190
congestum	192
conus	196
coregonus	191
crassilabre	194, 196
kennerlyi	186
lesueuri	194
macrolepidotum	193
oblongum	186
oneida	193
papillosum	189
pidiense	191
pœcilurum	196
robustum	193
rupiscartes	196
tenue	186
thalassinum	191
trisignatum	179
valenciennesi	190
velatum	190
victoriæ	187
mucosum, Xiphidion	2425
Xiphister	2425
mucosus, Acanthocottus	1975
Liparis	2111
Lycodalepis	2470
Lycodes	2461
Neoliparis	2111
mucronata, Odontognathus	438
Perca	1135
mucronatum, Ophidium	2419
mucronatus, Gnathobolus	438
Labrax	1135
Neoconger	362
Odontognathus	438
Pristigaster	438
Mud Cat	142, 143
Dabbler	640
Dabs	2644
Minnows	622
Parrot	1639
Sunfish	989
Mudfish	113, 640, 1649
mudfish, Fundulus	641
Mudfishes	623
Muffle-jaw	1950
Muger, Vieja	1639
Mugil	809
alatus	733
albula	812, 2841
berlandieri	812
brasiliensis	810, 814, 816
capitulinus	2841

	Page.
Mugil cephalus	811
chanos	415
cinereus	1373
curema	813, 2841
equinoculus	2841
gaimardianus	814
gigas	1483
grunniens	1483
guntheri	812
hospes	814
incilis	813
irretitus	819
lebranchus	811
lineatus	812, 2841
liza	811, 2841
mexicanus	813
monticola	819
nigro-strigatus	817
obliquus	1459
petrosus	814, 2841
plumieri	812, 2841
proboscideus	816
rammelsbergii	812
salmoneus	415
setosus	815
tang	812
thoburni	813
trichodon	816
Mugilidæ	808
Mugilinæ	809
mugiloides, Eleotris	2198
Mugilomorus	409
anna-carolina	410
muikiss, Salmo	492
Muksun of the Russians	464
muksun, Salmo	464
mulleri, Antigonia	1665
Benthosema	574
Carcharias	40
Maurolicus	577
Pempheris	978
Salmo	574
Scopelus	570, 574
Mullet	192
Blue	191
Blue-back	813
Common	811
Fan-tail	816
French	813
Jumping	197
Red-eye	814
Snip-nose	964
Striped	811
Whirligig	818
White	189, 813
Mullet of Utah Lake	179

	Page.
Mullets	808
Mullhypeneus	858
maculatus	859
Mullidæ	855
Mulloides	857, 2843
flavovittatus	860
rathbuni	857
Mullus	856
auratus	856, 2843
barbatus auratus	856
imberbis	1107
maculatus	859
multifasciata, Adinia	661
Hydrargyra	645
Sciæna	1459
multifasciatus, Anthias	1226
Fundulus	645
Pronotogrammus	1226
multifilis, Blennius	2385
Hypleurochilus	2385
multiguttatus, Alphestes	1165
Epinephelus	1166
Plectropoma	1166
multilineata, Morone	1132
Pœcilia	700
multilineatum, Hæmulon	1304
multimaculatus, Chromis	1547
Diodon	1746
Heliases	1547
Labrax	1132
multiocellata, Muræna	398
multiocellatus, Antennarius	2724
Chironectes	2725
multiradiatus, Arius	133
Cubiceps	951
Tachysurus	132, 2788
multispinosum, Cichlasoma	1525
multispinosus, Heros	1526
multistriatus, Liparis lineatus	2118
Mummichog	640
mundeolus, Stolephorus	2812
mundiceps, Novacula	1618
Xyrichthys	1618
mundicorpus, Iniistius	1620
Novacula	1620
mundus, Dactylagnus	2304
Oligoplites	2844
Urolophus	81
Urotrygon	81
Muñeca	1674
Mupinæ	962
Muræna	347, 400
acuminata	377
afra	396
alusis	403
anguina	390

	Page.
Muræna aquæ-dulcis	391
argentea	348
argus	401, 2805
aterrima	396
balearica	356
bostoniensis	348
cassini	356
catenata	403
clepsydra	2805
conger	354
coniceps	359
conspersa	397
cubana	348
dovii	397
elaborata	389
erebus	396
havennensis	382
infernalis	396
insularum	400
lentiginosa	402
lineopinnis	396
lita	2805
longicauda	392
lumbricus	342
macrocephala	348
maculata nigra	395
maculipinnis	394
maculosa	382
melanatis	402
melanotis	401, 402
meleagris	399
miliaris	398
mordax	396
moringa	395
multiocellata	398
nigra	355
nigricans	390
ocellata	399
ophis	382, 2804
panamensis	391
pfeifferi	2805
picta	2805
pinnata	351
pinta	402
pintiti	397
punctata	395
retifera	401
rostrata	348
sanctæ-helenæ	397
sanguinea	390
savanna	360
serpentina	348
seu conger brasiliensis	403
sidera	2805
sordida	403
variegata	2805

	Page.
Muræna vicina	394
muræna, Callechelys	378
Murænesocidæ	358
Murænesocinæ	358
Murænesox	359
coniceps	2801
savanna	360
Muræiidæ	345, 388
Murænoblenna	7, 403
nectura	404
Murænoides	2414
dolichogaster	2417
fasciatus	2418
gunelliformis	2421
gunnellus	2419
guttatus	2419
maxillaris	2418
ornatus	2420
sujef	2419
tænia	2418
muræioides, Blennius	2419
Gunnellus	2418
Muræiopbis	381, 383, 400
caramura	395
catenula	403
curvilineata	395
marmoreus	391
ocelletus	384
pantherina	2805
punctata	397
triserialis	384
undulata	403
vicina	394
Murciélago	2183
muricatus, Orbis	1749, 1750
murinus, Apomotis	996
Calliurus	996
muroadsi, Caranx	908
murrayi, Ipnops	547
Liocetus	2733
Melanocetus	2734
Muscalonge	629
muscarum, Rimicola	2338
musculus, Cyclopterus	2118
Muskallunge	629
Musquaw River Whitefish	466
Mustela	2558
Mustellus stellatus	29
Mustelus	28, 2745
asterias	29
californicus	30
canis	29
dorsalis	30
felis	31
lunulatus	28
plebejus	29
vulgaris	29

	Page.
mutatum, Boleosoma	1057
muticus, Lethotremus	2101
Mutton-fish	1376, 1265, 2457
Mycteroperca	1169, 1171, 1183
bonaci	1174
xanthosticta	1176
boulengeri	1171, 2856
calliura	1786, 2856
dimidiata	1179
falcata	1184
phenax	1185
hopkinsi	2855
interstitialis	1178
jordani	1176
microlepis	1177
olfax	1183
ruberrima	1183
pardalis	1181
reticulata	1187
rosacea	1184
ruber	1180
scirenga	1181
tigris	1187
camelopardalis	1187
venadorum	1186
venenosa	1172
apua	1173, 1174
guttata	1174
xenarcha	1180
Myctophidæ	550
Myctophum	569
affine	570
benoiti	573
hoops	572
brachychir	556
californiense	572
crenulare	575
glaciale	574
gracile	572
hians	572
humboldti	571
hygomii	573
lacerta	560
(Stenobrachius) leu-copsarum	562
mexicanum	563
nannochir	562
nocturnum	568
opalinum	571
procellarum	575
protoculus	565
punctatum	570
regale	563
remiger	573
townsendi	558
Myctophus rafinesquei	567

	Page.
mydrus, Cyprinodon	676
Mykiss	492, 2818
mykiss, Salmo	487, 492, 2818
agua-bonita	504
bouvieri	496
gibbsii	493
henshawi	493
lewisi	493
macdonaldi	497
pleuriticus	496
spilurus	495
stomias	497
virginalis	493
Myliobatidæ	87
Myliobatinæ	2753
Myliobatis	89, 2750
acuta	89
asperrimus	2754
bispinosus	89
californicus	89, 2754
eeltenkee	88
freminvillei	89
goodei	2755
sayi	86
Mylocheilus	219
caurinus	219, 220
fraterculus	220
lateralis	220
Mylolencus	243, 244, 2798
parovanus	246
pulverulentus	246
thalassinus	245
Mylopharodon	218
conocephalus	219
robustus	219
Mylopharodontinæ	202
Mylorhina	90
myops, Salmo	533
Saurus	533
Synodus	533
Trachinocephalus	533
Myoxocephalus	1970, 1971, 1976
æneus	1972
axillaris	1980, 1981
brandti	1984
bubalis	1971
grœnlandicus	1974
jaok	1977
mednius	1983
niger	1985
nivosus	1984
octodecimspinosus	1976
polyacanthocepha-	
lus	1976
scorpioides	1973
scorpius	1974

	Page.
Myoxocephalus stelleri	1981
verrucosus	1979
Myrichthys	375
acuminatus	376
oculatus	376
pantostigmius	2802
tigrinus	376, 2802
xysturus	2802
Myridæ	370
Myriolepis	1862
zonifer	1863
Myriosteon	60
Myripristis	846
clarionensis	2842
fulgens	846
jacobus	846
lychnus	847
occidentalis	847
pœcilopus	847
trachypoma	846
Myrophis	371
egmontis	371
longicollis	371
lumbricus	371
microstigmius	371
punctatus	371
vafer	372
mystacalis, Lepomis	1001
mystacina, Gnathypops	2286
mystacinus, Epinephelus	1151
Gnathypops	2286
Schistorus	1151
Serranus	1151
Mystaconurus	2580
mystes, Scorpæna	1849
mysticetus, Cetengraulis	450
Engraulis	450
mystinus, Sebastichthys	1785
Sebastodes	1784, 1785
Mystriophis	386
intertinctus	386
miurus	387
Mystus	116
ascita	155
carolinensis	117
Mytilophagus	1503
fasciatus	1504
Myxine	7
glutinosa	7
limosa	8
Myxinidæ	7
Myxodagnus	2305
opercularis	2305
Myxodes elegans	2353
lugubris	2357
macropus	2357

	Page.
Myxodes varius	2357
versicolor	2359
Myxostoma austrina	192
euryops	193
pœcilura	196
Myxus harengus	818
Myzopsetta	2644
ferruginea	2645
Naccaysh	413
nachtriebi, Leuciscus	2798
nacrea, Gila	228
Naked Gobies	2257
Goby	2259
nalnal, Sparactodon	947
Namaycush	504
Salmon	505
namaycush, Cristivomer	504
siscowet	505
Salvelinus	505
siscowet	505
Nannobrachium	561
leucopsarum	562
macdonaldi	563
mexicanum	563
nannochir	562
regale	563
nannochir, Myctophum	562
Nannobrachium	562
nanomyzon, Catostomus	177
Nanostoma	1066, 1067, 1070
camurus	1076
elegans	1075
inscriptum	1072
vinctipes	1075
zonale	1075
Nansenia	528
grœnlandica	528
nanus, Engraulis	449
Narcacion	77
Narcine	78
brasiliensis	78, 2752
corallina	78
entemedor	2752
nigra	78
umbrosa	78
Narcobatidæ	76
Narcobatus	77
naresi, Salmo	515
Salvelinus	515
oquassa	515
narinari, Aetobatus	88
Raia	88
Stoasadon	88
naso, Stolephorus	2813
Nassau Grouper	1157

	Page.
nasus, Acanthonotus	615
Diplanchias	1754
Menticirrhus	1472, 1473
Mola	1754
Notacanthus	615
Squalus	49
Umbrina	1473
nasuta, Malthæa	2737
nasutum, Agonostoma	820
Campostoma	206
nasutus, Agonostomus	819, 2841
Argyreus	306
Leuciscus	306
Ogcocephalus	2737
Rhinichthys	306
Trachynotus	941
natalis, Ameiurus	139
analis	140
Pimelodus	140
Nauclerus abbreviatus	900
annularis	900
brachycentrus	900
compressus	900
leucurus	900
triacanthus	900
naucrateoides, Echeneis	2270
Leptecheneis	2270
Naucrates	900
cyanophrys	900
ductor	900
fanfarus	900
indicus	900
noveboracensis	900
serratus	900
naucrates, Echeneis	2269
Leptecheneis	2269
naufragium, Balistes	1700, 1701
Nautichthys	2020
oculofasciatus	2020, 2021
pribilovius	2020
Nautiscus	2019
pribilovius	2019
nautopædium, Porichthys	2323
navaga, Eleginus	2537
Navarchus	950
Nealotus	881
tripes	881
neapolitanus, Lampugus	953
nebrascensis, Nocomis	323
Nebris	1416
microps	1417
zestus	1417
nebularis, Platophrys	2664
nebulifer, Catostomus	171
Chromis	1524
Labrax	1195

	Page.
nebulifer, Paralabrax	1195, 1196
Serranus	1196
nebuliferum, Cichlasoma...........	1524
nebuliferus, Catostomus	171
nebulosa, Aphoristia...............	2712
Cerna....................	1181
Percina	1027
Pileoma..................	1027
Umbrina	1475
nebulosum, Gunnellus	2414
nebulosus, Acantholebius..........	1872
Ameiurus	140
catulus	141
marmoratus .	141
Apionichthys	2703
Centronotus............	2414
Cestreus................	1409
Chiropsis...............	1872
Chirus	1872
Clinus.....	2438
Cynoscion	1409
Enedrias	2414
Gasterosteus	746
Lycodes	2468
Menticirrhus...........	1475
Otolithus	1409
Pimelodus..............	140
Sebastodes	1826
Silurus	143
Symphurus...............	2712
Urolophus	80, 2752
Nector	1436
nectura, Gymnomuræna	404
Murænoblenna.............	404
necturus, Uropterygius	404
Needlefish.........................	714
Needlefishes......................	708
Neetroplus	1541
nematopus	1541
nicaraguensis	1542
nefastus, Pomotis	1003
neglecta, Clupea....................	434
Corvina (Amblodon)......	1484
neglectus, Amblodon................	1484
Negra, Boca.....	1837
Esmeralda.................	2204
Nègre..............................:	1160
Petite	1142
Negro, Barbero....................	1692
Pargo......................	1252
Negro-fish........................	1146
negromaculatus, Rhypticus	1233
nelsoni, Awaous ....:.............	2235
Gobius	2235
nelsonii, Coregonus	466
Nematistiidæ	894

	Page.
Nematistius......................	895
pectoralis	895
Nematognathi.....................	114
Nematonurus	2571
cyclolepis	2571
goodei	2571
Macrurus magnus ...	2574
suborbit-	
ális ...	2572, 2573
Nematonus..........................	2518
pectoralis	2518
nematophthalmus, Scorpæna.......	2861
Sebastes........	2861
nematopus, Neetroplus.............	1541
Physiculus	2548
Nematostoma	774
Nemichthyidæ	366
Nemichthys.......................	369
avocetta...............	369
infans..............	368
scolopacea	369
scolopaceus	369
Nemipterus.......................	1288
macronemus	1289
Nemobrama	854
Nemocampsis.	1010
neoboracensis, Fistularia...........	757
Neobythites......................	2512
gillii	2513
maginatus	2513
ocellatus..............	2513
robustus	2515
stelliferoides.........	2516
Neoclinus	2354
blanchardi..............	2354
satiricus................	2355
Neoconger	362
mucronatus	362
perlongus	363
vermiformis	362
Neoditrema	1511
ransonnetii............	1511
neogæus, Leuciscus	240
Phoxinus.............	241
neoguinaica, Albula................	412
neohantoniensis, Coregonus ....:	466
Neoliparis.......................	2106
atlanticus...............	2107
callyodon	2110
fissuratus..............	2113
floræ....................	2111
greeni..................	2112
mucosus	2111
rutteri	2108
Neomænis	1247, 1248, 1251
ambiguus..............	1271

3044 *Index.*

	Page.
Neomænis analis	1265
apodus	1258
aratus	1273
argentiventris	1260
aya	1264
brachypterus	1268
buccanella	1261
colorado	1267
cyanopterus	1254
griseus	1255
guttatus	1269
hastingsi	2858
jocu	1257
jordani	1251
lutjanoides	1261
mahogoni	1272
novemfasciatus	1252
synagris	1270
vivanus	1262, 2858
Neomuræna	392
nigromarginata	400
Neosebastes	1839
Neostoma	581
bathyphilum	583
Neozoarces	2426
pulcher	2426
nephelus, Lepomotis	1005
Tetrodon	1733
turgidus	1733
nerka, Oncorhynchus	481
kennerlyi	483
Salmo	483
Nerophis	774
anguineus	774
heckeli	774
maculata	774
martinicensis	774
nertinianus, Acipenser	106
Nestis	818
Nettastoma procerum	366
Nettastomidæ	364
Netuma	119, 120, 126, 2764, 2765
dubia	126, 2765
elattura	128, 2769
grandicassis	126, 2764
insculpta	127. 2765
insularum	2770
kessleri	127, 2765
oscula	127, 2768
planiceps	127, 2766
platypogon	127, 2767
proops	124
quadriscutis	126
stricticassis	126, 2765
neucrates, Echeneis	2269
neustrianus, Gasterosteus	747

	Page.
nevadensis, Agosia	310
Cyprinodon	674
Rhinichthys (Apocope)	311
nevisense, Etheostoma	1034
nevisensis, Alvordius	1034
newberryi, Eucyclogobius	2248
Fario	499
Gobius	2248
Lepidogobius	2248
New England Hake	2530
New Light	987
newmani, Hyostoma	1053
New York Smelt	468
Nexilarius	1559
concolor	1559
nianguæ, Etheostoma (Hadropterus)	1043
Etheostoma spilotum	1044
Hypohomus	1042
nicaraguensis, Carcharhinus	39, 2747
Eulamia	39
Gambusia	682
Heros	1532
Neetroplus	1542
Pimelodus	152
Rhamdia	152
nicholsi, Halichæres	1592
Iridio	1591
Platyglossus	1592
nicholsii, Gobius	2218
nieuhofii, Echeneis	2272
nigellus, Hyborhynchus	217
niger, Agonus	2069
Amblodon	169
Ammocœtes	14
Aspidorphorus	2069
Astronesthes	586
Balistes	1711
Bubalichthys	164
Centrolophus	963
Chiasmodon	2291
Chiasmodus	2292
Conger	355
Cottus	1983, 1986
Esox	626
Leuciscus	235
Malacosteus	593
Myoxocephalus	1985
Perca	963
Scomber	948
Sparus	960
Squalius	235
Tautoga	1577
Thyrsites	879
Zeus	936
Nigger Chub	327

	Page.
Nigger Dick	327
Nigger-fish	1144, 1146
nigra, Balistes	1711
Elacate	948
Gila	235
Muræna	355
maculata	395
Narcine	78
nigrescens, Centropomus	1119
Coryphœna	1200
Cycleptus	169
Exoglossum	327
Leuciscus	233
Pimelodus	137
Salmo	507
Serranus	1200
Squalius	234
Tigoma	234
nigricans, Acanthurus	1692
Bathylaco	540
Camarina	1382
Catostomus	181
etowanus	181
Centropristes	1200
Cottus	1973
Cypselurus	2836
Enchelycore	389
Entomacrodus	2399
Exocœtus	737
Girella	1382
Huro	1012
Hylomyzon	181
Hypoplectrus unicolor	1193
Ictalurus	137
Istiophorus	891
Labrax	1135
Makaira	891
Muræna	390
Petromyzon	10
Pimelodus	137
Plectropoma	1193
nigriculus, Serranus	1153
Acronurus	1693
nigrilabris, Ameiurus	142
Amiurus	142
Gronias	142
nigripinnis, Argyrosomus	472
Auchenopterus	2369
Bathyagonus	2078
Clinus	2370
Coregonus	472
Cotylis	2332
Gobiesox	2332
Minnilus	299
Rhypticus	1234
Rypticus	1234

	Page.
nigrirostris, Chætodon	1673, 1674
Sarothrodus	1674
nigrita, Garrupa	1161
nigritus, Epinephelus	1162
Serranus	1161
nigrocastaneus, Gymnothorax	390
nigrocinctus, Sebastes	1828
Sebastichthys	1828
Sebastodes	1827
nigrofasciata, Hydrargyra	641
nigrofasciatum, Cichlasoma	1525
Etheostoma	1039
nigrofasciatus, Fundulus	641
Hadropterus	1038
Heros	1525
nigrolineatus, Diodon	1749
nigromaculatus, Cantharus	987
nigromanus, Pleuronectes	2657
nigromarginata, Neomuræna	400
Sidera	400
nigromarginatus, Gymnothorax ocellatus	400
Lycodontis ocellatus	399
nigropunctata, Perca (Pomacampsis)	1021
nigro-strigatus, Mugil	817
nigrotæniata, Cliola	264
Hybognathus	214
nigrotæniatus, Graodus	264
Notropis	264
nigrum, Boleosoma	1056
effulgens	1058
maculaticeps	1058
mesæum	1059
olmstedi	1057
Etheostoma	1057
vexillare	1058
Petromyzon	14
Nine-spined Stickleback	745
niphobles, Epinephelus	2853
Sparisoma	1633
Nissuee Trout	503
nitens, Congermuræna	357
Ophisoma	357
nitida, Julis	1608
Moniana	265
Pomotis	1003
nitidissima, Julis	1608
nitidissimum, Thalassoma	2859
nitidissimus, Chlorichthys	1608
nitidum, Pristipoma (Hæmulopsis)	1326
Thalassoma	1608, 2859
nitidus, Alburnus	293
Balistes	1709
Brachydeuterus	1326
Chlorichthys	1608

	Page.
nitidus, Leuciscus	221
Pomadasis	1326
Promoxis	987
Salmo	509
Salvelinus	509
nivalis, Salmo alpinus	509
nivea, Oliola	278
niveatus, Epinephelus	1156
Microspathodon	1567
Pomacentrus	1568
Serranus	1156
niveiventris, Amiurus	138
Ictalurus	138
niveus, Hybopsis	278
Notropis	277
chloristius	278
·Photogenis	278
Nivicola	1066, 1082
nivipes, Emblemaria	2402
nivosus, Cottus	1985
Myoxocephalus	1984
Sebastes	1834
Sebastodes	1833
nobilior, Esox	629
nobilis, Atractoscion	1413
Cestreus	1413
Conodon	1324
Cynoscion	1413
Esox	629
Gambusia	682
Grystes	1012
Heterandria	682
Johnius	1413
Perca	1324
Nocomis	314, 315, 322
amblops	321
bellicus	323
hyostomus	316
milneri	324
nebrascensis	323
rubrifrons	320
nocomis, Notropis	268
nocturna, Collettia	567
Echidna	402
nocturnus, Noturnus	146
Pœcilophis	403
Schilbeodes	146
Noire, Oreille	1261
no Mai, Yanagi	1830
Nomeidæ	948
Nomeinæ	949
Nomeus	949
gronovii	949
maculatus	950
maculosus	950
oxyurus	950

	Page.
normalis, Bassozetus	2507
Northern Barracuda	825
Striped Gurnard	2167
Sucker	176
Whiting	1475
North-River Shad	427
norvegica, Perca	1761
norvegicus, Coryphænoides	2579
Lepidoleprus	2579
Sebastes	1761
norwegianus, Squalus	57
norwegica, Perca	1761
No-shee Trout	503
nostras, Liparis	2118
notabilis, Argyreus	309
Notacanthinæ	613
Notacanthus	614
analis	615
challengeri	618
chemnitzii	614
nasus	615
phasganorus	616
rissoanus	618
rostratus	617
Notarius	119, 2764
notata, Atherina	800
Atherinichthys	800
Belone	711
Oliola	274
Cyprinella	274
Dorosoma	416
Lepomis	1011
Malthæa	2737
Megalops	432
Menidia	800
menidia	2840
Perca	1024
notatum, Etheostoma	1070
Hæmulon	1297
Pristipoma	1321
notatus, Balistes	1709
Fundulus	659
Hyborhynchus	218
Labrax	1132
Lepomis	1008
Minnilus	218
Notropis	274
Pimelodus	135
Pimephales	218
Pomotis	1008
Porichthys	2321
Semotilus	659
Tylosurus	710, 711
Zygonectes	659
notemigonoides, Notropis	292
Notemigonus	249, 250

	Page.
Notemigonus americanus	251
auratus	250
chrysoleucus	250
gardoneus	251
ischanus	251
leptosomus	250
lucidus	299
occidentalis	247
nothochir, Ophichthys	380
Quassiremus	380
Nothonotus	1066, 1067, 1076
cinereus	1078
inscriptus	1072
jordani	1080
maculatus	1077
rufilineatus	1079
sanguifluus	1077
tessellatus	1078
thalassinum	1072
vulneratus	1077
nothus, Cestreus	1407
Cynoscion	1406
Otolithus	1407
Notidanoid Sharks	16
Notidanus	18
griseus	19
Notistium	890
Notocanthidæ	613
Notoglanis	149
Notogrammus	2439
rothrocki	2440
Notorhynchus	17
borealis	18
maculatus	17
Notoscopelus	554
castaneus	556
caudispinosus	556
margaritifer	555
quercinus	555
Notosema	2635
dilecta	2636
dilectum	2635
notospilotus, Astrolytes	1899
Julidio	1603
Pseudojulis	1603
Notropis	254, 257-258, 290
æneolus	266
alabamæ	298
albeolus	259, 283
altipinnis	287
amabalis	291
amœnus	296
analostanus	279
anogenus	259, 260
arge	294
ariommus	290

	Page.
Notropis atherinoides	254, 293
aztecus	258
bellus	297
bifrenatus	258
blennius	261, 2800
boops	268
braytoni	264
bubalinus	273
buchanani	2800
cærnleus	277
callisema	272
callistius	276
camurus	279
caynga	260, 2799
atrocaudalis	260
cerasinus	283
cercostigma	274
chalybæus	288
chamberlaini	2800
chihuahua	265
chiliticus	287
chlorocephalus	286
chrosomus	288
coccogenis	284
cornutus	281
cyaneus	283
frontalis	283
dilectus	294, 2801
eurystomus	277
formosus	271
fretensis	261
frigidus	271
fumeus	294
galacturus	279
garmani	281
germanus	261
gilberti	266
heterodon	261
heterolepis	260
hudsonius	269
amarus	270
saludanus	270
selene	269, 2801
hypselopterus	280
illecebrosus	268
jejunus	290
jordani	259
kanawha	264
lacertosus	284
lateralis	263
leuciodus	291
lirus	297
longirostris	267
louisianæ	2801
ludibundus	273
lutipinnis	286, 2800

	Page.
Notropis lutrensis	271
lythrurus	300
macdonaldi	284
macrolepidotus	299
macrostomus	274
maculatus	259
matutinus	301
megalops albeolus	284
metallicus	297
micropteryx	296
nigrotæniatus	264
niveus	277
chloristius	278
nocomis	268
notatus	274
notemigonoides	292
nux	267
orca	289
ornatus	270
ozarcanus	265
phenacobius	263
photogenis	295, 296
piptolepis	266
procne	264
proserpina	272
pyrrhomelas	280
reticulatus	262
roseipinnis	298
roseus	237
rubricroceus	286
rubrifrons	295
sabinæ	262
scabriceps	290
scepticus	296
scopifer	291
scylla	263
shumardi	268
simus	267
socius	292
spectrunculus	265
stigmaturus	275
stilbius	293
swaini	290
telescopus	292
arcansanus	292
texanus	274
topeka	266
trichroistius	275
umbratilis	298
ardens	301
atripes	300
cyanocephalus	300
fasciolaris	301
lythrurus	300
matutinus	301
punctulatus	301

	Page.
Notropis umbratilis umbratilis	299
umbrifer	274
venustus	274, 275
volucellus	263
welaka	2799
whipplii	278
xænocephalus	289
xænurus	280
zonatus	285
zonistius	277, 285
nottii, Fundulus	656
Zygonectes	657
Noturus	143
elassochir	147
eleutherus	148, 149
exilis	147
flavus	144
funebris	147
furiosus	149
gilberti	148
gyrinus	146
insignis	147
leptacanthus	146
luteus	144
marginatus	147
miurus	148
nocturnus	146
occidentalis	144
platycephalus	144
sialis	146
Novacula	1617
cærulea	1653
cultrata	1619
lineata	1619
martinicensis	1617
mundiceps	1618
mundicorpus	1620
novacula, Coryphœna	1619
Xyrichthys	1619
Novaculichthys	1613
infirmus	1616
martinicensis	1616
rosipes	1614
ventralis	1615
novæ-angliæ, Coregonus	465
novæorleanensis, Anguilla	348
novæterræ, Anguilla	348
noveboracensis, Exocœtus	735, 736
Naucrates	900
Vomer	934
novemfasciatus, Lutianus	1253
Neomænis	1252
novemlineatus, Chaemodes	2393
Pholis	2393
novemmaculatus, Diodon	1746
novemradiata, Agosia	312

	Page.
nox, Auchenopterus	2373
Cremnobates	2374
nubila, Agosia	212, 311
carringtonii	311
Apocope	311
Oliola	215
Hybognathus	215
nubilus, Alburnops	215
Argyreus	311
Centroblennius	2438
Ceratichthys	312
Leptoblennius	2438
Lumpenus	2438
Stichæus	2438
nuchalis, Ailurichthys	117
Arius	131
Chanos	415
Hybognathus	213
Pseudoscarus	1654
Scarus	1654
Tachisurus	131
Tachysurus	131, 2782
nuchifilis, Blennius	2383
nuchipinnis, Clinus	2362
Labrisomus	2362
nuda, Cotylis	2331
nudus, Cyclopterus	2336
Gobiesox	2331
Lepadogaster	2331
nuecensis, Dioplites treculii	1012
Grystes	1012
nugator, Bryostemma	2410
Numbfish	77
nummifer, Carpiodes	166
Salmo	508
Nurse	57
Nurse Shark	25, 26
nuttalli, Exocœtus	737
nuttingii, Antennarius	2723
Chaunax	1726, 2727
nux, Notropis	267
Nyctophus	569
obesa, Algansea	246
Tigoma	233
obesus, Amiurus	141
Apomotis	993
Echinorhinus	85
Enneacanthus	993
Leuciscus	246, 282
Leucos	246
Pomotis	993
Squalius	233
obeyense, Etheostoma	1092
oblarius, Gasterosteus	750
obliqua, Sciæna	1459
obliteratus, Alutarius	1720
obliquatus, Cestreus	1405
Cynoscion	1405
Diabasis	1304
Otolithus	1405
obliquus, Mugil	1459
oblonga, Chœnopsetta	2633
Platessa	2630
oblongior, Pimelepterus	1388
oblongiusculus, Balistes	1720
oblongo, Corpore glabro	2657
oblongum, Moxostoma	186
oblongus, Cyprinus	186
Erimyzon sucetta	186
Gobius	2264
Heros	1535
Orthagoriscus	1756
Ostracion catheloplateo..	1728
conico	1745
glaber	1735
holacanthus	1746
Paralichthys	2632
Pleuronectes	2633
Pseudorhombus	2630
Sebastes	1830
Sebastodes	1830
Sparus	2276
obscura, Liopsetta	2651
obscuratus, Gymnothorax	389
Lycodontis	389
Pomacentrus	1552, 1555
obscurus, Carcharhinus	35
Carcharias	35
Centrarchus	1012
Pleuronectes	2651
Pomotis	1006
Squalus	35
obtusa, Labrus capite	1609
Ophisoma	355
Raia	2751
obtusirostris, Acipenser	106
Exocœtus	730
Halocypselus	730
obtusum, Hæmulon	1319
obtusus, Pseudoscarus	1654
Rhinichthys	308
Scarus	1654
Squalus	39
obvelatus, Prionodon	35
Occa	2043
dodecaedron	2044
verru cosa	2043
occa, Cœlorhynchus	2587
Macrurus	2588
Pristis	61
occidentalis, Abramis	247
Amia	113

	Page.
occidentalis, Catostomus	178
Chorinemis	898
Conger	355
Epigonus	1112
Gasterosteus	745
Girardinus	689
Heterandria	689
Leucosomus	247
Luxilinus	247
Luxilus	247
Malacocephalus	2570
Monacanthus	1715
Myripristis	847
Notemigonus	247
Noturus	144
Oligoplites	898
Scombroides	899
Tetronarce	77
Torpedo	77
occipitalis, Scorpæna	1854
Ocean Pipefish	774
Tang	1693
Turbot	1706
Oceanic Bonito	868
oceanica, Anguilla	355
oceanicus, Blennius	2379
Gobionellus	2230
Gobius	2230
ocella, Rhinichthys	307
ocellaris, Chœnopsetta	2630
Fundulus	642, 2827
Platessa	2630
Pseudorhombus	2630
ocellata, Bollmannia	2238
Corvina	1454
Muræna	399
Perca	1454
Raia	69
Raja	68
Sciæna	1454
Sidera	399
ocellatum, Ophidium	2430
ocellatus, Acanthocottus	1976
Anarrhichthys	2448
Antennarius	2721
Chænopsis	2403
Chætodon	1674
Citharichthys	2673
Clinus	2357
Gymnothorax	399
nigromarginatus	400
saxicola	399
Hemirhombus	2673
Herpetoichthys	384
Hippoglossus	2673
ocellatus, Johnius	1454
Lophius	2722
Lycodontis	399
nigromarginatus	399
saxicola	399
Malacoctenus	2356, 2869
Murænophis	384
Neobythites	2513
Ophichthus	383
Ophichthys	384
Opisthocentrus	2429
Platophrys	2663
Priodonophis	399
Rhomboidichthys	2664
Rhombus	2664
Sciænops	1453
Zenopsis	1660
Zeus	1661
ocellicauda, Amia	113
ocellifer, Clinus	2353
octodecimspinosus, Cottus	1976
Myoxocephalus	1976
octofilis, Polynemus	830
Trichidion	830
Octogrammus	1866
pallasi	1870
octogrammus, Chirus	1870
Hexagrammos	1869
Labrax	1870
octonemus, Polydactylus	830
Polynemus	830
Trichidion	830
oculata, Sebastes	1832
Squatina	59
oculatus, Anthias	1283
Balistes	1707
Bryttus	1004
Centropristis	1283
Etelis	1282
Hesperanthias	1283
Icelinus	1895
Latebrus	1115
Lepidosteus	111
Lumpenus	2433
Macrops	1283
Myrichthys	376
Pisoodonophis	376
Orthragoriscus	1754
Scombrops	1114
Sebastodes	1832
Serranus	1283
oculocirris, Emblemaria	2403
oculofasciatus, Blepsias	2021
Nautichthys	2020, 2021
oculo-radiato, Turdus	1591

	Page.
oculo, Turdus radiato	1703
Ocyanthias	1227
martinicensis	1228
Ocyurus	1275
aurovittatus	1276
chrysurus	1275
lutjanoides	1261
rijgersmœi	1276
ocyurus, Centropristes	1200
Kyphosus	1390
Pimelepterus	1390
Sectator	1389
Serranus	1201
Odontanthias asperilinguis	1227
martinicensis	1228
Odontapsis	46
americanus	47
Odontognathus	437
mucronata	438
mucronatus	438
panamensis	438
Odontogobius	2210
Odontopyxis	2085
frenatus	2075
leptorhynchus	2076
trispinosus	2085
trispinous	2086
Odontoscion	1425
dentex	1425
xanthops	1427
Odontoscium archidium	1432
Odontostomidæ	597
œratedii, Selene	935
Tetragonopterus	334
ogac, Gadus	2542
ogat, Gadus	2542
Ogcocephalidæ	2735, 2730
Ogcocephalus	2736
elater	2739
nasutus	2737
radiatus	2738
vespertilio	2737
Ogilbia	2502
cayorum	2503
ventralis	2503
oglina, Megalops	432
oglinum, Opisthonema	432
Ognichodes	2263
ohiensis, Acipenser	106
Esox	630
Lucius masquinongy	629
Ohio Sturgeon	106
ohioensis, Cichla	1012
Oil Fish	879
Shark	32
Oja, Pege	2699

	Page.
Ojanco	1272
ojanco, Lutjanus	1273
Mesoprion	1273
okeechobeensis, Ameiurus	139
Ictalurus	139
Old Wench	1703
Wife	940, 1458, 1649, 1703
Black	1711
olfax, Epinephelus	1183
Mycteroperca	1183
ruberrima	1183
Serranus	1183
olfersi, Argyropelecus	604
Pleurothyris	604
Sternoptyx	604
olidus, Hypomesus	525
Mesopus	525
Salmo (Osmerus)	525
oligaspis, Abramis	294
Alburnus	294
Minnilus	294
Oligocephalus	1066, 1068, 1083
grahami	1089
humeralis	1097
leonensis	1089
pulchellus	1089
Oligocottus	2013, 2864
acuticeps	2016
analis	2013
borealis	2014
embryum	2017
globiceps	2017
maculosus	2013
snyderi	2871
oligodon, Osmerus	2824
Polynemus	830
Oligolepis	2210
oligopeltis, Acipenser	105
Oligoplites	898
altus	899, 2844
inornatus	899
mundus	2844
occidentalis	898
saliens	899, 2844
palometa	899
saurus	898
Oligopodus	955
olisthostoma, Gerres	1377
olisthostomus, Gerres	1376
olivacea, Pœcilia	659
olivaceus, Leucus	244
Rutilus	244
olivaris, Hopladelus	143
Leptops	143
Pelodichthys	143
Pilodictis	143

	Page.
olivaris, Silurus	143
olmstedi, Boleosoma brevipinnis	1057
nigrum	1057
Etheostoma	1057
olriki, Aspidophoroides	2089
Ombre Chevalier	508
ommata, Discopyge	78
Heterandria	664
Lucania	663
Opisthognathus	2283
ommatum, Opisthognathus	2282
ommatus, Paralichthys	2635
omnisco, Salmo maycus	487
omocyaneus, Eleotris	2198
omostigma, Genypterus	2490
Otophidium	2490
omosudis	598
lowii	598
Oncocephalus	2736
Oncocottus	2000
hexacornis	2002
labradoricus	2004
quadricornis	2001
Oncorhynchus	474, 477, 478
chouicha	480
gorbuscha	478
keta	478
kisutch	480
lagocephalus	479
lycaodon	481, 483
nerka	481
kennerlyi	483
orientalis	480
paucidens	483
proteus	478
quinnat	480
sangninolentus	481
scouleri	478
tschawytscha	479
tsuppitch	481
oneida, Catostomus	193
Moxostoma	193
Ptychostomus	193
Oneirodes	2732
eschrichtii	2732
Oneirodinæ	2728
ongus, Serranus	1154
onitis, Hiatula	1579
Labrus	1578
Tautoga	1578, 1579
Onos	2558
cimbrius	2561
ensis	2559
reinhardti	2559
rufus	2559
septentrionalis	2560

	Page.
ontariensis, Thymallus	518
montanus	519
signifer	519
Onus	2529, 2558
riali	2530
onychus, Cottus	1953
Oolachan	521
Oonidus	1738
Ooze Eels	349
Opah	954
opah, Zeus	955
opalescens, Lythrulon	1312
opalina, Julis	1591
opalinum, Myctophum	571
opalinus, Platyglossus	1591
Open-mouthed Grunt	1306
opercularis, Micropogon	1461
Myxodagnus	2305
Polydactylus	831
Polynemus	831
Sciæna	1461
Stolephorus	445
Trichidion	831
Ophichthus	381
evionthas	381
gomesii	384
guttifer	383
havannensis	382, 2804
magnioculis	385
ocellatus	383
ophis	2804
parilis	386
puncticeps	382
retropinnis	383
rugifer	384
triserialis	384
zophochir	385
Ophichthyidæ	372
Ophichthys	381, 382
acuminatus	377
(Sphagebranchus) anguiformis	374
chrysops	385
crocodilinus	388
gomesii	385
guttifer	383
intertinctus	387
magnioculis	385
miurus	387
nothochir	380
ocellatus	384
pardalis	376
parilis	386
pauciporis	386
pisavarius	377
puncticeps	382

	Page.			Page.
Ophichthys punctifer	387	Ophisoma analis		356
retropinnis	383	balearicum		356
schneideri	387	macrurum		357
triserialis	384	nitens		357
xysturus	2802	obtusa		355
zophochir	385	prorigerum		357
ophidianus, Gempylus	884	Ophisomus		2414
Ophididæ	2481	Ophisternon		342
Ophidioidea	2453	Ophisura intertinctus		387
Ophidioidei	782	sugillatus		387
Ophidion	2487	Ophisuraphis		374
beani	2487	Ophisurus		375, 381
grællsi	2488	acuminatus		377
holbrooki	2487	californiensis		384
josephi	2488	chrysops		385
Ophidium	2487	crocodilinus		388
atropurpureum	2423	gomesii		385
brevibarbe	2485	guttatus		382
grœllsi	2487	latimaculatus		376
holbrooki	2488	longus		377
imberbi	2443	parilis		386
josephi	2489	remiger		384
marginatum	2489	xysturus		376
mucronatum	2419	ophryas, Paralichthys		2630
ocellatum	2430	Prionotus		2164
parii	2478	ophthalmicus, Lepomis		1001
pellucidum	354	Opisthistius		1384
profundorum	2484	Opisthocentrinæ		2349
taylori	2489	Opisthocentrus		2428
unernak	2477	ocellatus		2428
viride	2477	quinquemaculatus		2430
ophidoides, Liparis	2118	tenuis		2430
Ophioblenniinæ	2347	Opisthognathidæ		2279
Ophioblennius	2400	Opisthognathus		2280
steindachneri	2401	lonchurum		2281
webbii	2401	lonchurus		2281
Ophiodon	1875	macrognathum		2281
elongatus	1875	macrognathus		2282
pantherinus	1876	macrops		2284
Ophiodontinæ	1864	maxillosus		2284
Ophiognathus	405	megastoma		2282
ampullaceus	406	micrognathus		2287
Ophioscion	1446, 1447	ommata		2283
adustus	1447	ommatum		2282
imiceps	1451	punctatum		2281
scierus	1452	punctatus		2281
simulus	1449	rhomaleus		2285
strabo	1448	scaphiurus		2282
typicus	1448	Opisthonema		432
vermicularis	1452	libertate		433
ophioscion, Corvina	1448	libertatis		433
Sciæna	1448	oglinum		432
ophis, Muræna	382	Opisthopterus		436
Ophichthus	2804	dovii		437
Ophisoma	353, 355	lutipinnis		437, 2811
acuta	356	macrops		437

	Page.
opisthophthalmus, Conger	356
Opladelus	142
Oplopoma	1875
pantherina	1876
oppositus, Monocanthus	1716
Opsanus	2315
cerapalus	2316
pardus	2316
tau	2315
Opsopœa	247, 248, 249
Opsopœodus	247, 248
bollmani	249
emiliæ	248
megalops	248
osculus	248
Opthalmolophus	2360
Oquassa Trout	514
oquassa, Salmo	515
Salvelinus	514, 515
marstoni	515
naresi	515
Oranchee, Grand	1057
Orange Filefish	1718
Rockfish	1793
Orbe, Le Diodon	1749
orbicularis, Atinga alter minor	1749
Rhombus	966
Orbidus	1729
orbignianus, Exocœtus	729
orbiguyana, Pellona	436
Platessa	2626
orbignyanus, Conger	355
Orbis echinatus	1745
lævis variegatus	1735
muricatus	1749
reticulatus	1750
orbis, Cyclopterus	2100
Eumicrotremus	2099, 2100
orbitarius, Pagellus	1350
Sparus	1350
orca, Notropis	289
Orcella	254, 257, 289
orcini, Brama	960
orcutti, Leuciscus	241
Phoxinus	242
Orcynus	869, 870
alalonga	871
alliteratus	869
pelamys	869
schlegelii	870
subulatus	871
thunnia	869
thynnus	870
ordinatus, Chirus	1870
Epinephelus	1155
Hexagrammus	1870

	Page.
oreas, Chrosomus	211
Oregon Brook Trout	501
Charr	507
Sturgeon	104
oregonensis, Cyprinus (Leuciscus)	225
Leuciscus	225
Ptychocheilus	224, 2796
Oreille Noire	1261
Oreosoma	1662
atlanticum	1662
coniferum	1663
Orestiinæ	631
orientalis, Anarrhichas	2447
Chanos	415
Oncorhynchus	480
Pelamys	873
Salmo	480
ornata, Amia	113
Aphoristia	2707, 2710
Cliola	271
Cochlognathus	252
Codoma	271
Embiotoca	1506
Hydrargyra	2827
Raja	70
ornatum, Campostoma	205
ornatus, Achirus	2709
Balistes	1719
Cochlognathus	252
Esox	626
Gillellus	2299
Gunellus	2420
Labrus	1610
Murænoides	2420
Notropis	270
Pholis	2419
Tetradon	1742
Ornichthys	2148
Orqueta	937
orqueta, Chloroscombrus	937
orsini, Ozodura	1754
orthagoriscus, Cephalus	1754
Orthagoriscus	1754
Orthichthys	759
Orthodon	206
microlepidotus	207
orthogrammus, Carangoides	928
Caranx	929
Orthonops eos	2262
orthonotus, Ditrema	1507
Orthopristis	1334, 1335, 1336
brevipinnis	1341
cantharinus	1339, 1340
chalceus	1337
chrysopterus	1338
duplex	1339

	Page.
Orthopristis forbesi	1336
inornatus	1342
lethopristis	1340
poeyi	1339
reddingi	1336
Orthopsetta	2678, 2679
sordida	2680
Orthostœchus	1313
maculicauda	1313
Orthragoriscus	1753
alexandrini	1754
analis	1754
battaræ	1754
blochii	1754
elegans	1754
fasciatus	1754
ghini	1754
hispidus	1754
lunaris	1754
mola	1754
oblongus	1756
oculeatus	1754
ozodura	1755
redi	1754
retzii	1754
rondeletii	1754
solaris	1754
spinosus	1754
truncatus	1756
Orthragus	1753
osbeck, Trachinus	1153
Osbeckia	1717, 1718, 1719, 2860
oscitans Sciæna	1441
Stellifer	1440
oscula, Agosia	309
Apocope	309
Arius	127
Corvina	1484
Netuma	127, 2768
Sciæna	1484
Tachisurus	127
osculatii, Labrax	1132
osculus, Argyreus	309
Holocentrus	853
Opsopœodus	248
osmeriformis, Coregonus	468
osmerinus, Hybognathus	213
Osmerus	522, 523
albatrossis	2823
albidus	538
attenuatus	523
dentex	524, 2823
elongatus	525
lemniscatus	533
microdon	521
mordax	523

	Page.
Osmerus mordax abbotti	524
spectrum	523
oligodon	2824
pretiosus	525
thaleichthys	522
viridescens	523
Osphyolax	775
pellucidus	775
osseus, Esox	110
Lepisosteus	109
Ostariophysi	114
ostentum, Careproctus	2134
osteochir, Echeneis	2273
Rhombochirus	2273
osteosticta, Trygon	84
Ostichthys	846
Ostorhinchus	1106
fleurieu	1107
Ostraciidae	1721
Ostracion	1721
bicaudalis	1723
boops	1755
cathetoplateus oblongus	1728
concatenatus	1723
conico oblongus	1745
expansum	1724
gronovii	1725
guineensis	1725
listeri	1725
maculatus	1725
oblongus bolacanthus	1746
glaber	1735
polydon inermis triqueter	1723
quadricornis	1725
sexcornutus	1725
subrotundus ventre glabro	1749
tetraodon	1740
triqueter	1723
undulatus	1724
yalci	1724
Ostracium quadricorne	1725
trigonum	1724
trigonus	1724
Ostracodermi	781, 1720
Oswego Bass	1012
otakii, Hexagrammus	1867
otarius, Lepidosteus	110
Othonops	2261
othonops, Perkinsia	420
othonopterus, Cestreus	1405
Cynoscion	1404
Otolithinæ	1393
Otolithus albus	1411
californiensis	1413
carolinensis	1409
cayennensis	1404, 1411

	Page.		Page.
Otolithus drummondi	1409	Oxycottus acuticeps	2015, 2864
jamaicensis	1406	Oxygeneum	207
leiarchus	1415	pulverulentum	207
magdalenæ	1410	oxygenius, Polyprion	1139
microlepidotus	1415	Oxyjulis	1601
microps	1415	californicus	1601
nebulosus	1409	modestus	1601
nothus	1407	Oxylabrax	1117
obliquatus	1405	Oxylebiinæ	1864
regalis	1407	Oxylebius	1878
reticulatus	1409	pictus	1878
rhomboidalis	1404	Oxyloricaria	156
squamipinnis	1404	Oxymacrurus	2587
stolzmanni	1412	Oxymetopontinæ	2188
thalassinus	1408	Oxyodontichthys	381
toeroe	1404	brachyurus	385
virescens	1415	limbatus	385
Otophidium	2490	macrurus	385
galeoides	2491	oxyptera, Corvina	1222
indefatigabile	2490	Oxyrhina	47
omostigma	2490	gamphodon	49
otophorus, Eupomacentrus	1555	spallanzani	49
Pomacentrus	1555	oxyrhynchus, Acipenser	105
Otrynter	1344	Carcharhinus	40
caprinus	1345	Carcharias	41
otrynter, Caranx	930	Isurus	48, 49
otsego, Coregonus	466	Tetrodon	1741
ouachitæ, Hadropterus	1035	Oxyurichthys	2210
ouananiche, Salmo salar	487	Oxyurus	353
Ouatilibi	1145	oxyurus, Lepisosteus	110
Espagñol	1140	Nomeus	950
outalibi, Enneacentrus fulvus	1146	Oyster-fish	1578, 2315
Serranus	1146	ozarcanus, Notropis	265
ovale, Syacium	2674	Ozodura	1753
ovalis, Citharichthys	2674	orsini	1754
Hemirhombus	2674	ozodura, Orthragoriscus	1755
Sebastichthys	1789	Ozorthus hexagrammus	2441
Sebastodes	1788		
ovatus, Trachynotus	942	pachycephala, Adinia	660
ovicephalus, Sparus	1361	pachycephalus, Fundulus	661
oviceps, Lactophrys	1724	Lagocephalus	1729
oviformis, Chætodon	1668	Tetrodon	1729
ovigerum, Bathyphasma	2128	pachygaster, Spheroides	1738
ovinus, Esox	672	Tetrodon	1738
Lebias	672	Pachylabrus	1507
ovis, Sargus	1361	variegatus	1508
Ovoides	1738	pachylepis, Atherinichthys	801
erethizon	1739	Menidia	801
setosus	1739	Pachynathus	1703
Ovum	1738	capistratus	1704
oweni, Halosaurus	607	triangularis	1705
Oxybeles	2495	Pachypops	1459
oxybrachium, Sparisoma	1634	biloba	1460
oxybrachius, Scarus	1635	furcræus	1459
Oxycephas	2568	pachypus, Cottus	1973
Oxycottus	2015, 2863, 2864	Pachyurus furcræus	1460

	Page.
Pachyurus squamosissimus	1419
pacifica, Belone	716
pacifici, Anisotremus	1316
Batrachoides	2314
Batrachus	2315
Conodon	1316
Pomadasis	1316
pacificus, Argyreiosus	936
Artediellus	1906
Bathylagus	530
Cynicoglossus	2655
Glyptocephalus	2655
Larimus	1424
Lobotes	2857, 2858
Lutjanus	1253
Lycodes	2460
Lycodopsis	2460
Mesoprion	1253
Microstomus	2655
(Mallotus) Salmo	521
Thaleichthys	521
Thynnus	871
Tylosurus	716
Paddle, Cock and Hen	2096
Paddle-fish	101
Paddle-fishes	100
pætulus, Hemirhombus	2672
pagei, Etheostoma	1092
Pagellus bajonado	1352
calamus	1350
caninus	1352
humilis	1355
microps	1355
milneri	1355
orbitarius	1350
penna	1355
Pagrus	1356
argenteus	1357
pagrus	1356
vulgaris	1357
pagrus, Pagrus	1356
Sparus	1357
Pai de Gato	1837
Pajarito	721
pala, Cyprinus	415
palearis, Lycodes	2466
Palinurichthys	963
perciformis	964
Palinurus	963
perciformis	964
pallasi, Octogrammus	1870
Pallasia	1754
Pallasia	1753
pallasi	1754
pallasianus, Cephalus	1754
pallasii, Clupea	422

	Page.
pallasii, Pleuronectes	2648
Pallasina	2048
aix	2050
barbata	2049
pallida, Aldrovandia	611
Echeneis	2272
Lepomis	1012
Morone	1135
pallidus, Bodianus	1433
Catostomus	179
Eupomotis	1006
Fundulus	638, 2827
Labrax	1135
Labrus	1005
Lepomis	1005
Leucosomus	222
Pimelodus	135
Platygobio	326
Pomotis	1007
Salmo	505
palmipes, Prionotus	2157
Trigla	2156
paloma, Trachinotus	945
Palometa	940, 941, 942, 943, 965, 966, 967, 2849
media	2849
palometa	2849
simillima	2849
palometa, Chorinemus	899
Oligoplites saliens	899
Palometa	2849
Rhombus	966
Stomateus	967
Palu Brasiliense congener	966
palustris, Pœcilichthys	1102
Pammelas	963
perciformis	964
pammelas, Melanostigma	2479, 2869
Pampanito	941
Pámpano	930, 933
Common	944
Gall-topsail	940
Great	943
Round	941
Pámpanos	895, 939
pampanus, Trachynotus	944
panamense, Pristipoma	1331
panamensis, Achirus	2702
Ailurichthys	117
Atherinella	805
Azevia	2677
Bodianus	1141
Caranx	928
Citharichthys	2677
Engraulis	448
Enneacentrus	1141
Epinephelus	1141

	Page.
panamensis, Felichthys	117
Gymnothorax	391
Ilisha	436
Loricaria	157
Menticirrhus	1473
Muræna	391
Odontognathus	438
Parapsettus	1669
Pellona	436
Petrometopon	1141
Piabucina	332
Pomadasis	1331
Pristigaster (Odontognathus)	438
Rabula	391
Serranus	1141
Sidera	391
Solea	2702
Stolephorus	448
Tetragonopterus	334
Umbrina	1473
Panchax	633, 2827, 2830
pauciradiatus, Cubiceps	957
pandionis, Glossamia	1111
pandora, Clinostomus	234
Squalius	234
Pañeca	2196
pannosa, Scorpæna	1845
pantherina, Muænophis	2805
Oplopoma	1876
pantherinus, Anarrhichas	2446
Cestracion	21
Ophiodon	1876
Pseudariodes	155
Pantosteus	169
aræopus	172
arizonæ	170, 2790
clarki	172
columbianus	172
delphinus	171
generosus	170
guzmaniensis	171
jarrovii	171
jordani	171
platyrhynchus	170
plebeius	171
virescens	171, 172
pantostigmius, Myrichthys	2802
Papagallo	895
Papagallos	894
papalis, Diouda	214
papilio, Melletes	1932
papillifer, Gobiesox	2330
papilliferus, Chologaster	704
papillosa, Aramaca	2672
papillosum, Moxostoma	189

	Page.
papillosum, Syacium	2671
papillosus, Barbulifer	2261
Pleuronectes	2672
Ptychostomus	189
Paraclinus	2374
chaperi	2374
Paraconodon	1314, 1315, 1316
Paradiodon	1744
quadrimaculatus	1746
paradoxa, Garmannia	2232
paradoxus, Gobius	2232
Psychrolutes	2026
Paradules	1013
Paragonus	2054
acipenserinus	2062
sturioides	2063
Parahemiodon	156, 157, 158
Paralabrax	1194
albomaculatus	1197
clathratus	1197, 1198
humeralis	1196, 1197
maculatofasciatus	1196
nebulifer	1195, 1196
Paralepididæ	599
Paralepinæ	599
Paralepis	602
borealis	601
coregonoides	602
coruscans	602
intermedius	600
Paralichthys	2624
adspersus	2627, 2872
æstuarius	2626, 2872
albiguttus	2631
brasiliensis	2626
californicus	2625, 2626
dentatus	2629, 2630
lethostigma	2630
lethostigmus	2630
liolepis	2624
maculosus	2626
magdalenæ	2872
oblongus	2632
ommatus	2635
ophryas	2630
sinaloæ	2872
squamilentus	2631
stigmatias	2636
woolmani	2628
Paraliparis	2139, 2140
cephalus	2141
copei	2143
dactylosus	2144
holomelas	2140
liparinus	2139
mento	2142

	Page.
Paraliparis rosaceus	2142
ulochir	2144
parallelus, Centropomus	1122
Paralonchurus	1477, 1478
dumerili	1478
goodei	1480
petersi	1481
rathbuni	1479
Paramacrurus	2587
Paramia	1112
paranasimos, Acipenser	106
Paranthias	1221
creolus	1222
furcifer	1221
Parapomacentrus	1549, 1550
Parapsettus	1669
panamensis	1669
Paraques	1485
Parascorpæna	1839, 2860
Paraserranus hasselti	1205
parasiticus, Simenchelys	349
paraspistes, Caranx	923
Paratractus	916, 917, 921
pisquetus	921
Parché	1674, 1677
pardale, Leptophidium	2486
pardalis, Epinephelus	1183
Monacanthus	1713
Mycteroperca	1181
Ophichthys	376
pardus, Batrachus tau	2317
Opsanus	2316
Parepinephelus	1169, 1170, 1180
Pareques	1485, 1486
acuminatus	1487
pareva, Aluteres	1719
Parexocœtus	728
mesogaster	728
Pargo	1244, 1265
Amarillo	1260
Colorado	1264, 1267, 1356
Criollo	1265
de lo Alto	1262
de Raizero	1273
Guachinango	1264
Mareño	1252
Negro	1252
Prieto	1252
pargus, Mesoprion	1255
Paricelinus	1885
hopliticus	1886
thoburni	1888
parietalis, Coliscus	217
parii, Lycodara	2478
Ophidium	2478
Uronectes	2478

	Page.
parilis, Ophichthus	386
Ophichthys	386
Ophisurus	386
parkei, Salmo	508
Salvelinus	2823
parkeri, Arius	126
Selenaspis	125, 2764
Silurus	126
Trachisurus	126
Parma rubicunda	1565
parma, Cichlasoma	1519
Heros	1519
parmifera, Raia	75
Raja	74
parnatus, Setarches	1860
Parophrys	2637, 2640
ayresi	2640
cœnosa	2639
hubbardi	2641
ischyrus	2641
isolepis	2642
quadrituberculatus	2648
vetulus	2640
parovanus, Cyprinodon	666
Myloleucus	246
Upeneus	859
parra, Diabasis	1299
Hæmulon	1297
parræ, Albula	411
Brama	1586
Chætodon	1685
Clepticus	1586
Exocœtus	740
Hæmulon	1297, 1309
Parraserranus	1203
parrianus, Monacanthus	1713
Parrot-fish, Blue	1636, 1652
Dark-green	1638
Green	1657
Parrot Fishes	1620, 1642
Parrot, Mud	1639
Rose-back	1635
parryi, Rhamdella	153
Rhamdia	153
partitus, Eupomacentrus	1558
Pomacentrus	1558
Paru	1680
paru, Chætodon	1680, 1681
Pomacanthus	1680
Rhombus	965, 2849
Stromateus	966
Parupeneus	858
parva, Lucania	665
parviceps, Lycodapus	2493
parvipinne, Etheostoma	1096
parvipinnis, Archosion	1399

		Page.
parvipinnis, Cestreus		1410
	Cynoscion	1410
	Dicrotus	883
	Fundulus	640, 2827
	Isopisthus	1399
	Promethichthys	883
parvula, Clupea		426
parvus, Cottopsis		1945
passany, Arius		124
	Bagrus	124
	Sciadeichthys	124, 2760
	Tachysurus	124
passer, Holacanthus		1682
	Pomacanthus	1683
Pastinaca		82
Pastor		949
Patao		1378
patao, Gerres		1378
patatus, Julis		1591
patris, Acanthocottus		2009
patronus, Brevoortia tyrannus		434
patruelis, Gambusia		682
	Heterandria	681
paucidens, Leurynnis		2460
	Lycodopsis	2460
	Oncorhynchus	483
	Salmo	483
pauciporis, Ophichthys		386
pauciradiatus, Callionymus		2158
paucispinis, Ancylodon		1399
	Sebastes	1781
	Sebastodes	1780
pavonaceus, Heros		1538
pavonia, Limia		632
	Pœcilia	692
pavoninus, Cyclopterus		2097
Paw, John		1159
paxilloides, Lycodes		2471
paxillus, Lycenchelys		2471
	Lycodes	2471
Peacock Flounder		2665
Pea-lip Sucker		199
Pearl-fish		2495
Pearl-fishes		2494
Pêche-pêche		338
Pêche-Prêtre		1784
peckianus, Syngnathus		771
peckii, Syngnathus		770
pectinatus, Centropomus		1122
	Pristis	60, 61, 2749
pectinifer, Clinus		2362
	Labrosomus	2362
pectoralis, Albatrossia		2573
	Bodianus	1582
	Cossyphus	1582
	Dactyloscopus	2301

		Page.
pectoralis, Dallia		621
	Harpe	1582
	Macrurus (Malacocephalus)	2574
	Nematistius	895
	Nematonus	2518
Pedalion		1753
pedaliota, Bonapartia		580
pedaliotus, Zaphotias		2826
Pediculate Fishes		2712
Pediculati		2712
pedimacula, Centropomus		1119
Pega		2269
Pegador		2269
Pege Oja		2699
Pegedictis		1941, 1942, 1944
	ictalops	1951
Peixe Agulha		711
Peixe-fonda		1312
Peixe Rey		806
pelada, Anchoa		436
pelagicum, Siphostoma		767
pelagicus, Callionymus		2184
	Scomber	952
	Syngnathus	770
pelamides, Scomber		869
pelamis, Gymnosarda		868
	Scomber	869
pelamitus, Scomber		872
Pelamys		871
	chilensis	873
	lineolata	873
	orientalis	873
	sarda	872
pelamys, Euthynnus		869
	Orcynus	869
	Scomber	872
	Thynnus	869
pelegrinus, Squalus		51
Pélérin		51
Pellona		435
	bleekeriana	436
	castelnæana	436
	flavipinnis	436
	furthi	436
	orbignyana	436
	panamensis	436
pellucida, Ammocrypta		1062
	clara	1063
	vivax	1063
	Salmoperca	784
	Thyris	2691
pellucidum, Etheostoma clarum		1063
	Ophidium	354
pellucidus, Delothyris		2691
	Osphyolax	775

	Page.
pellucidus, Pleurolepis	1063
Psenes	950
Pelodichthys	142
olivaris	143
Pelon, Guaguanche	824
Peloria	2660
pelta, Chelmo	1671
peltastes, Lepomis	1003
peltata, Percina	1034
peltatum, Etheostoma	1034
peltatus, Hadropterus	1034
Turdus cinereus	1373
pemecus, Bagrus	125
Pempheridæ	976
Pempheris	977
mexicanus	978
mulleri	978
poeyi	979
schomburgki	978
peninsulæ, Bascanichthys	379
Callechelys	379
Chirostoma	797
Menidia	797
penna, Calamus	1354, 1355
Pagellus	1355
pennanti, Argentina	577
Maurolicus	577
Squalus	49
pennatula, Calamus	1351
Penopus	2520
macdonaldi	2521
pensacolæ, Harengula	431
penshinensis, Salmo	508, 2819
pentacanthus, Bodianus	849
Centrarchus	990
Holocentrus	849
Labrus	1576
Xenochirus	2081
Pentanemus	828
quinquarius	828
pepinus, Lucioperca	1022
Peprilus	965
perarcuatus, Pleuronectes	2643
Perca	1023
aberrans	1136
acuta	1024
afra	1833
alburnus	1475
americana	1024, 1135
apoda	1259
ascensionis	849
atraria	1200
chryseops	1132
chrysoptera	1339
dorso monapterygia	1833
flavescens	1023

	Page.
Perca fluviatilis	2841
flavescens	1024
formosa	1208
furca	1200
furcræa	1460
gibbosa	1009, 1296
gigas	1154
gracilis	1024
granulata	1024, 2841
guttata	1142, 1164
immaculata	1135
juba	1323
lanceolata	1482
lophar	947
loubina	1119
maculata	1153
marina	1761
cauda nigra	1303
gibbosa	1295
pinnis	1259
puncticulata	1146
sectatrix	1388
venenosa	1172
melanurum	1303
minima	1057
mitchilli	1133
alternata	1133
interrupta	1133
mucronata	1135
niger	963
(Pomacampsis) nigropunctata	1021
nobilis	1324
norwegica	1761
notata	1024
ocellata	1454
philadelphica	1202
punctata	1145, 1146, 1433
pusilla	1107
robusta	1154
rock-fish	1133
rufa	849
salmonea	1021
saltatrix	947, 1388
saxatilis	1133
sectatrix	1388
septentrionalis	1133
serratogranulata	1024
stellio	1153
striata	1311
tota maculis	1153
trifurca	1202
undulata	1462
unicolor	1192
unimaculata	1360
varia	1200
variabilis	1784, 1796

	Page.
Perca vitrea	1021
percellens, Raja	63
Rhinobatus	63
Percesoces	781, 787
Perch, American	1023
Black	1504
Blue	1505, 1577
Pike	1021
Pirate	785, 786
Raccoon	1023
Ringed	1023
River	1023
Sacramento	991
Trout	782, 784
Viviparous	1498
White	1133, 1134, 1484, 1501, 1509
Yellow	1023
Perch-like fishes	979
Perches	1015
Perches, American Pike	1020
Log	1024
Percidæ	1015
Percidinæ	2032
perciformis, Coryphena	964
Lirus	964
Palinurichthys	964
Palinurus	964
Pammelas	964
Percina	1024, 1026
aspro	1033
bimaculata	1027
caprodes	1026
manitou	1028
zebra	1027
guentheri	1034
macrocephalus	1031
nebulosa	1027
peltata	1034
phoxocephala	1031
rex	1025
roanoka	1036
Percinæ	1018
Percis	2033
japonicus	2034
percobromus, Alburnellus	295
Minnilus	295
Percoidea	979, 1241
Percoidei	781
percoides, Agonostomus	819
Percoids, Spariform	1241
Percopsidæ	783
Percopsis	783
guttatus	784
hammondi	784
perezi, Carcharhinus	36
Platypodon	36

	Page.
perfasciatus, Engraulis	442
Stolephorus	441, 445
Perichthys godeffroyi	1197
periscopus, Gadus	2536
perisii, Salmo	509
Perissias	2667
tæniopterus	2667
Peristediidæ	2177
Peristedion	2178
gracile	2179
imberbe	2182
longispathum	2178
micronemus	2182
miniatum	2178
platycephalum	2180
Peristethus	2178
micronema	2182
peristethus, Podothecus	2062
Perkinsia	420
othonops	420
perlatum, Holocentrum	853
perlongus, Neoconger	363
Permit	943
perniger, Culius	2201
Eleotris	2201
peroni, Caranx	923
peronii, Lepidopus	887
Peropus	2018
bilobus	2018
perplexus, Cottus	1955
Perrico	1659
perrico, Pseudoscarus	1659
Scarus	1659
Perro Colorado	1583
Perro	1579
perrotteti, Pristis	60, 2749
personatus, Ammodytes	833
perspicabilis, Embiotoca	1506
perspicillum, Lycodes	2465
perspicuus, Hybognathus	218
perthecatus, Stolephorus	442
peruanus, Amblyopus	2265
Anthias	1223
Gobioides	2264
Hemianthias	1222
Promotogrammus	1223
peruvianus, Galeichthys	122, 2771
Gerres	1376
Tachysurus	122
Pesca Blanca	321
Vermiglia	1811
Pescadillo del Red	1416
Pescadillos del Rey	807
Pescadito	233
Pescado Azul	1553
Azul de dos Colores	1557

	Page.
Pescado Blanco de Chapala	792
Colorado	1453
del Rey	806
Pescador	2722
Martin	2724
Pescados Azules	1549
Blancos	792
Pesce Re	806
Tondo	48
petenense, Dorosoma	417
petenensis, Chatoessus	417
Mollienisia	700
Pimelodus	153
Pœcilia	694
Rhamdia	153
Tetragonopterus	335
Petenia	1513
splendida	1513
petersi, Paralonchurus	1481
petimba, Fistularia	758
Petimbuaba	757
Petite Gueule	1370
Jaquette	1559
Nègre	1142
Scie	1323
Petos	876
Petrometopon	1140
apiarius	1142
cruentatus	1141
coronatus	1142
guttatus	1142
panamensis	1141
Petromyzon	9
americanus	10
appendix	10
argenteus	11
astori	12
ayresi	13
bairdii	9
bdellium	11
borealis	13
branchialis	14, 2745
camtschaticus	2745
castaneus	11
ciliatus	12
concolor	11
lamotteni	10
lividus	12
marinus	10
chamtschaticus	2745
dorsatus	10
unicolor	10
nigricans	10
nigrum	14
plumbeus	13

	Page.
Petromyzon tridentatus	12
Petromyzonidæ	8
Petronason	1642
petropauli, Blenniophidium	2430
petrosus, Mugil	284, 814
Trisotropis	1172
petus, Acanthocybium	877
Cybium	877
Pez Ciego	2501
Pez de Espada	2749
Gallo	805
Pluma	1347, 1349, 1350
del Rey	799, 808
Luna	1753
Puerco	1700, 1704
Sierra	60
pfeifferi, Muræna	2805
Sidera	2805
phaenna, Hybopsis	270
Phænodon	586
ringens	586
phaeton, Pristigaster	438
phalæna, Umbrina	1475
Phalangistes	2064, 2864
acipenserinus	2062
fusiformis	2048
japonicus	2036
lævigatus	2048
loricatus	2046
phaleratus, Esox	628
Phanerodon	1506
atripes	1507
furcatus	1506
lateralis	1506
Pharyngognathi	781, 1571
phasganorus, Notacanthus	616
phasma, Careproctus	2132
Phenacobius	302
catostomus	304
mirabilis	303
scopifer	303
scopiferus	303
teretulus	303
liosternus	303
uranops	304
phenacobius, Notropis	263
phenax, Apomotis	997
Lepomis	997
Mycteroperca falcata	1185
philadelphica, Perca	1202
philadelphicus, Centropistes	1201
Serranus	1202
philipii, Clinus	2359
philonips, Cottus	1960
Philosophe	1693
Philypnus	2194

	Page.
Philypnus dormitator	2195
dormitor	2194
lateralis	2195
phlebotomus, Acanthurus	1692
phlegethontis, Clinostomus	243
Gila	243
Leuciscus	243
Phoxinus	243
phlox, Boleosoma	1052
Ulocentra	1052
Phobetor	2006
tricuspis	2009
Phœbe	1211
phœbe, Centropristis	1212
Prionodes	1211
Serranus	1212
Pholidapus	2430
dybowskii	2430
grebnitskii	2431
Pholidichthyinæ	2347
Pholidichthys	2405
anguilliformis	2405
Pholidinæ	2348
Pholis	2377, 2414, 2415, 2417
carolinus	2379
dolicogaster	2416
fasciatus	2417
gunnellus	2419
novemlineatus	2393
ornatus	2419
pictus	2415, 2416
quadrifasciatus	2392, 2394
ruberrimus	2417
subbifurcata	2440
taczanowskii	2416
Photogenis	254
ariommus	290
cæruleus	277
callistius	276
engraulinus	296
eurystomus	277
grandipinnis	280
leuciodus	291
leucops	296
leucopus	277
niveus	278
pyrrhomelas	281
scabriceps	290
spilopterus	279
stigmaturus	275
telescopus	292
photogenis, Leuciscus	296
Notropis	295, 926
Squalius	296
Photonectes	591
gracilis	591

	Page.
Photonectinæ	587
Phoxinus	228, 230, 240
clevelandi	237
flammeus	242
margaritus	241
milnerianus	242
neogæus	241
orcutti	242
phlegethontis	243
phoxocephala, Percina	1031
phoxocephalum, Etheostoma	1031
phoxocephalus, Alvordius	1031
Cestreus	1414
Cynoscion	1413
Hadropterus	1030
phrygiatus, Arius	131
Hexanematichthys	130
Tachisurus rugispinis	131
Phrynotitan	2853
Phtheirichthys	2268
lineatus	2268
Phycinæ	2532
Phycis	2552
americanus	2555
chesteri	2556
chuss	2555
cirratus	2554
dekayi	2555
earlli	2555
floridanus	2554
marginatus	2555
punctatus	2553
regalis	2553
regius	2553
rostratus	2555
tenuis	2555
Physiculus	2547
fulvus	2547
japonicus	2549
kaupi	2547
nematopus	2548
rastrelliger	2549
physignathus, Ceratichthys	326
Couesius	326
Platygobio	324
Physogaster	1727
Piabucina	332
panamensis	332
Picarels	1364
picarti, Hemirhamphus	720
Picconou	194
piceus, Balistes	1711
Melichthys	1711
pichardi, Joturus	821
Pickerel	628
Banded	626

	Page.		Page.
Pickerel Common Eastern	627	Pike Perch	1021
Little	627	Pikea	1135
Pickering	1022	Pikes	624
Picorellus	625	Gar	108
picta, Muræna	2805	pilatus, Prionotus	2156
pictipinnis, Chelidonichthys	2175	Pileoma	1024
Trigla	2176	bimaculata	1027
picturata, Alutera	1719	carbonari	1027
Gambusia	683	cymatogramma	1053
Seriola	910	nebulosa	1027
picturatum, Siphostoma	768	semifasciatum	1027
picturatus, Caranx	910	zebra	1028
Gymnothorax	395	pilicornis, Blennius	2380
Syngnathus	768	Pilodictis limosus	142
Trachurus	909	olivaris	143
pictus, Centronotus	2416	pilosa, Solea	2699
Chaunax	2726	pilosus, Diodon	1744, 1752
Chironectes	2717	Trichodiodon	1743, 1744
Chirus	1873	Pilot, Black	1555
Eleotris	2201	Cockeye	1555, 1561
Gymnelis	2477	Pilot-fish	465
Gymnothorax	2805	Pilot-fishes	900
Iridis	1599	Pilot, Shark's	902
Julis	1600	Pimelepterus	1384
Lycodontis	2805	analogus	1386
Oxylebius	1878	bosci	1388
Pholis	2415, 2416	bosquii	1388
Platyglossus	1600	cornubiensis	964
Torpedo	78	elegans	1387
Urocentrus	2416	flavolineatus	1386
Picuda	823	incisor	1386
picuda, Sphyræna	823	lutescens	1389
Picudilla	824	oblongior	1388
picudilla, Sphyræna	824	ocyurus	1390
pidiense, Moxostoma	191	Pimelodella	153
pidiensis, Ptychostomus	191	chagresi	154
piger, Symphurus	2705	modesta	154
Pigfish	1338	Pimelodinæ	116
Pigfishes	1334	Pimelodus	116, 154
pigmentarius, Apogon	1109	æneus	143
Monoprion	1109	affinis	134
Pigmy Sunfishes	981	ailurus	140
pigra, Aphoristia	2706	albidus	132, 138
Pigus	243	antoniensis	140
Pike, Blue	1021	argenteus	125
Common	628	argentinus	135
Gar	109	argyrus	135
Gray	1022	atrarius	140
Great	629	baronis-mulleri	151
Northern	630	blochii	155
Green	627	borealis	137
Sacramento	224, 2796	brachypterus	152
Sand	1022	cænosus	140
Wall-eyed	1021	cærulescens	135
Yellow	1021	catulus	141
Pike-like Fishes	622	catus	140

	Page.		Page.
Pimelodus caudafurcatus	135	Pimelometopon pulcher	1585
chagresi	154	Pimelonotus	149
clarias	155	Pimephales	216
confluis	141	agassizii	217
cupreoides	140	fasciatus	217
cupreus	140	maculosus	217
dekayi	140	milesi	217
erythroptera	135	notatus	218, 2796
felinus	140	promelas	217
felis	141	confertus	217
furcatus	134	maculosus	217
furcifer	135	Pincers	431
godmani	152	Pinfish	1358
gracilis	135	pingell, Triglops	1923, 1925
graciosus	135	pinguis, Hippoglossus	2611
guatemalensis	152	Platysomatichthys	2611
hammondi	135	Pleuronectes	2611
houghi	135	pini, Trigla	2177
hoyi	141	pinima, Acara	1323
hypselurus	152	Pristipoma acara	1323
insigne	147	Pink-fish	2262
lateralis	135	pinnata, Muraena	351
laticaudus	152	pinnatus, Synaphobranchus	351
lemniscatus	147	pinnifasciatus, Pseudopleuronectes	2647
lupus	137	pinniger, Bryssetæres	2328
lynx	138	Enneacanthus	994
macronema	155	Gobiesox	2329
maculatus	135, 155	Sebastichthys	1794
managuensis	153	Sebastodes	1793, 1794
marginatus	135	Sebastosomus	1794
marmoratus	141	pinnimaculatus, Ailurichthys	117
megalops	135	Felichthys	117
micropterus	153	pinnis, Perca marina	1259
modestus	154	Turdus branchialibus	1257
motaguensis	151	pinnivarius, Hypoplectrus unicolor	1192
natalis	140	pinnulata, Seriola	907
nebulosus	140	pinnulatus, Elegatis	907
nicaraguensis	152	Pinta, Carilla	1152
nigrescens	137	Morena	402
nigricans	137	Pintado	875
notatus	135	Pintano	1561
pallidus	135	Pintanos	1560
petenensis	153	pintita, Morena	397
platycephalus	142	pintiti, Muraena	397
polycaulus	153	Pipe	758
pullus	141	Pipefish, Common	770
punctulatus	143	Great	764
rigidus	155	Ocean	774
salvini	152	Pipefishes	760
spixii	132	Piper	723
vulgaris	140	piptolepis, Notropis	266
vulpeculus	141	Piquier	1687
vulpes	135	Piquitinga	443
wagneri	151	piquitinga, Engraulis	443
Pimelometopon	1585	piquottii, Amia	113
darwinii	1586	piraaca, Monacanthus	1715

	Page.
Pirabebé	2183
Piracoaba	830
Piramutana blochii	155
macrospila	155
pirapeḍa, Dactylopterus	2183
Pira-pixanga or Gat-visch	1153
Pirate Perches	785, 786
Pirati apia	1174
piritita, Cabrilla	1181
pisavarius, Ophichthys	377
piscatorius, Lophius	2713
piscatrix, Pseudorhamdia	155
Pisces	14, 1241
pisces, Unicornu bahamensis	1719
pisciculus, Esox	641
piscis, Luna	1754
Piscis viridis bahamensis	1638
pisculentus, Esox	641
Fundulus	641
pisonis, Eleotris	2201, 2200
Gobius	2201
Pisoodonophis	375, 377
cruentifer	377, 2803
daspilotus	2803
guttulatus	377
longus	377
oculatus	376
xysturus	376
pisquetus, Caranx	921
Paratractus	921
pistilliger, Cottus	2008
Elaphocottus	2008
Gymnocanthus	2006, 2008, 2009
Pitamba, Acara	1276
pituitosus, Rhypticus	1234
pixanga, Serranus	1153
pixuma, Amore	2201
placitus, Hybognathus	213
Placopharynx	197
carinatus	198
duquesnii	198
Plagiogrammus	2427
hopkinsi	2428
plagioplàteo, Eleotris capite	2201
Plagioscion	1418
heterolepis	1419
squamosissimus	1418
surinamensis	1419
Plagiusa	2704
plagiusa, Aphoristia	2710
Glossichthys	2710
Plagusia	2710
Pleuronectes	2710
Symphurus	2710
Plagopterinæ	204
Plagopterus	329

	Page.
Plagopterus argentissimus	329
Plagusia	2704, 2709
brasiliensis	2709
fasciata	2710
plagiusa	2710
tessellata	2709
plagusia, Pleuronectes	2709
Symphurus	2709
Plagyodontidæ	2826
Plagyodus	594, 596, 2826
Plaice	2648
Plain-tail	879
plana, Platessa	2647
planci, Mola	1756
Tympanomium	1754
Plancterus	2827, 2828
planiceps, Arius	127
Catostomus	181
Netuma	127, 2766
Rhinobatus	64
planifrons, Eupomacentrus	1559
Pomacentrus	1559
Planirostra	101
spatula	102
planus, Pleuronectes	2647
Pseudopleuronectes	2647
Plargyrus	250, 254
argentatus	283
bowmani	283
melanocephalus	217
typicus	283
plargyrus, Rutilus	282
Plate-fish	1722
Platessa	2648
bilineata	2643
dentata	2615, 2630
dvinensis	2650
elongata	2657
ferruginea	2645
glabra	2650
microcephala	2654
oblonga	2630
ocellaris	2630
orbignyana	2626
plana	2647
pola	2657
pusilla	2647
quadrituberculata	2648
quadrocellata	2633
rostrata	2645
stellata	2652
platessa, Caranx	927
platessoides, Citharus	2615
Drepanopsetta	2615
Hippoglossoides	2614
Pleuronectes	2615

	Page.
Platichthys	2651
rugosus	2652
stellatus	2652
umbrosus	2643
Platirostra	101
edentula	102
Platophrys	2660, 2661
constellatus	2663
ellipticus	2665
leopardinus	2606
lunatus	2665
maculifer	2664
nebularis	2664
ocellatus	2663
spinosus	2662
tæniopterus	2668
platophrys, Citharichthys	2683
Platopterus	66
platorynchus, Acipenser	107
Scaphirhynchus	107
platostomus, Lepisosteus	110
platycephalum, Peristedion	2180
Platycephalus	2028
americanus	2029
angustus	2029
dormitator	2195
platycephalus, Ameiurus	142
Cottus	1983, 1988
Megalocottus	1987
Noturus	144
Pimelodus	142
Platygaster	435
Platyglossus bivittatus	1597
caudalis	1599, 1600
crotaphus	1598
cyanostigma	1591
dimidiatus	1594
dispilus	1598
florealis	1597
garnoti	1593
grandisquamis	1597
humeralis	1597
internasalis	1594
maculipinna	1595
nicholsi	1592
opalinus	1591
pictus	1600
poeyi	1599
principis	1591
radiatus	1591, 1597
ruptus	1593
semicinctus	1593
Platygobio	325
communis	326
gracilis	326
pallidus	326

	Page.
Platygobio physignathus	325
Platyinius	1279, 1280
vorax	1281
platyodon, Carcharhinus	39
Squalus	39
Platypodon	33, 34, 35
falciformis	36
perezi	36
Platypœcilus	685
maculatus	686
mentalis	686
quitzeoensis	2873
platypogon, Arius	127
Netuma	127, 2767
Tachisurus	127
Platyrhina exasperata	65
triseriata	66
platyrhincus, Lepisosteus	111
Platyrhinoidis	65
triseriatus	65, 66
platyrhinus, Acipenser	106
platyrhynchus, Carcharhinus	36
Eulamia	36
Minomus	170
Pantosteus	170
Scaphirhynchus	107
platyrrhynchus, Scaphirhynchops	107
Platysomatichthys	2610
hippoglossoides	2611
pinguis	2611
stomias	2610
Platysomus	933
micropteryx	934
spixii	934, 2846
Platysqualus	43, 44
platystomus, Lepisosteus	110
Platytroctes	458
apus	458
plebeius, Catostomus	171
Gasterosteus	751
Pantosteus	171
plebejus, Mustelus	29
Plecopodus	2263, 2868
Plectobranchinæ	2349
Plectobranchus	2431
evidles	2432
Plectognathi	1696
Plectognathous Fishes	1696
Plectospondyli	160
plectrodon, Porichthys	2321
Plectromus	840
beanii	842
crassiceps	843
cristiceps	843
lugubris	842
suborbitalis	841

	Page.
Plectropoma accensum	1193
affine	1193
afrum	1166
bovinum	1193
chloropterum	1165
chlorurum	1193
crocota	1192
ephippium	1192
gummigutta	1192
guttavarium	1192
hispanum	1140
indigo	1193
melanorhina	1192
monacanthus	1165
multiguttatus	1166
nigricans	1193
puella	1192
vitulinum	1192
Plectrypops	853
retrospinis	853
pleianus, Pseudoscarus	1656
Scarus	1656
pleii, Hemirhamphus	723
Plesioperca	1028
anceps	1039
Pleuracromylon	29
pleuriticus, Salmo clarkii	2819
mykiss	496
Pleurogadus	2537
gracilis	2538
Pleurogrammus	1864
monopterygius	1864, 1866
Pleurolepis	1061
asprellus	1061
pellucidus	1063
Pleuronectes	2648
achirus	2696
americanus	2647
apoda	2701
aquosus	2660
aramaca	2672
argus	2666
asper	2645
beanii	2646
bilineatus	2643
cicatricosus	2649
cynoglossus	2611, 2657
dentatus	2630
digrammus	2641
ellipticus	2665
elongatus	2657
ferrugineus	2645
franklinii	2650
gilli	2654
glaber	2650
glacialis	2649

	Page.
Pleuronectes guttulatus	2640
hippoglossoides	2611
hippoglossus	2612
ischyrus	2641
kitt	2654
lævis	2654
limandoides	2615
lineatus	2698, 2701
linguatula	2615
lunatus	2666
macrolepidotus	2672
maculatus	2660
maculifer	2665
maculosus	2626
melanogaster	2630
microcephalus	2654
microstomus	2654
mollis	2701
nigromanus	2657
oblongus	2633
obscurus	2651
pallasii	2648
papillosus	2672
perarcuatus	2643
pinguis	2611
plagiusa	2710
plagusia	2709
planus	2647
platessoides	2615
quadridens	2654
quadrituberculatus	2648
quenseli	2654
saxicola	2657
stellatus	2652
surinamensis	2666
umbrosus	2643
vetulus	2641
Pleuronectidæ	2602
Pleuronectinæ	2607
Pleuronichthys	2637
cœnosus	2638, 2639
decurrens	2637, 2683
guttulatus	2640
quadrituberculatus	2638
verticalis	2638
pleurophthalmus, Antennarius	2722
pleurospilus, Girardinus	688
Heterandria	688
pleurostictus, Triglops	1923
Pleurothyris	603
olfersi	604
plumarius, Archistes	1900, 1901
plumatula, Calamus	1352
plumbea, Chimæra	95
Dionda	216
Gambusia	695

	Page.
plumbea, Hybognathus	216
Lampetra	13
plumbeolus, Alburnops	283
Minnilus	283
plumbeum, Zophendum	216
plumbeus, Ceratichthys	324
Couesius	323
Gobio	324
Petromyzon	13
plumier, Le Tetrodon	1733
plumieri, Caranx	912
Chaetodon	1668
Conodon	1324
Coryphœna	2276
Diabasis	1306
Gerres	1379
Gobius	2206
Hæmulon	1304
Labrus	1305
Malacanthus	2275
Mugil	812, 2841
Sciæna	1324
Scomber	911
Scomberomorus	875
Scorpæna	1848
Sicydium	2206
Tetrodon	1733
Trachurops	912
Trichidion	830
plumierianus, Caranxomorus	911
plumierii, Polydactylus	830
Polynemus	830
plurimis, Cottus cirris	2066
plutonia, Raja	69, 70
pluvialis, Labrax	2841
Poacher, Sea	2091
Poachers, Sea	2031
pocatello, Catostomus	175
podostemone, Boleosoma	1055
Etheostoma	1055
Podothecus	2054
accipiter	2055
acipenserinus	2061, 2062
gilberti	2058
hamlini	2056
peristethus	2062
sturioides	2063
thompsoni	2060
veternus	2063, 2064
vulsus	2068
Pœcilia	690, 2833
boucardi	695, 2834
branneri	2834
butleri	691, 2833
cænicola	641
catenata	648

	Page.
Pœcilia chisoyensis	693, 2834
couchiana	695, 2833
couchii	695
cubensis	692
cuneata	2834
dominicensis	696, 2833, 2834
dovii	695, 2833
elongata	697, 2834
fasciata	641
fasciatus	2833
gillii	692, 2834
lineolata	700
macrolepidota	641
melanogaster	696, 2834
melapleura	660
mexicana	692, 2833
multilineata	700
olivacea	659
pavonia	692, 2833
petenensis	694, 2833
plumbeus	2833
presidionis	697
reticulata	2833
schneideri	691
sphenops	694, 2833
spilurus	697, 2833
surinamensis	691
thermalis	693, 2833
(Acropœcilia) tridens	690
vandepolli	696, 2833
arubensis	696, 2834
vittata	692, 2833
vivipara	691, 2833
Pœcilichthys	1066, 1067, 1069
artesiæ	1094
asprigenis	1085
barratti	1102
beani	1057
borealis	1082
butlerianus	1102
camurus	1076
cœruleus	1089
eos	1102
erochrous	1102
exilis	1103
fusiformis	1102
gracilis	1103
jessiæ	1085
lateralis	1090
lepidus	1089
mesæus	1059
palustris	1102
punctulatus	1091
quiescens	1101
rufilineatus	1079
sagitta	1081

Page.

Pœcilichthys sanguifluus 1077
saxatilis 1048
spectabilis 1089
swaini 1086
versicolor 1089
virgatus 1093
vitreus............... 1065
vulneratus 1077
warreni 1103
zonalis 1075
Pœciliidæ.................... 630
Pœciliinæ.................... 632
Pœcilioides 678
bimaculatus 678
Pœcilocephalus 381
pœciloides, Lebistes................ 689
Limia 700
Pœcilophis.................... 402
nocturnus.............. 403
pœcilophthalmus, Gobiesox........ 2335
pœcilopus, Myripristis............. 847
Rhamphoberyx......... 847
Pœcilosoma..................... 1066
erythrogastrum....... 1089
transversum 1089
pœcilura, Myxostoma.............. 196
Pœcilurichthys 333
pœcilurum, Moxostoma............. 196
pœtulus, Citharichthys 2672
poeyi, Alepidosaurus (Caulopus)... 596
Dactylóscopus 2302
Engraulis 445
Gobius 2226
Halichæres 1598
Hemirhamphus 720
Iridio 1599
Lycengraulis.............. 2811
Orthopristis................. 1339
Pempheris 979
Platyglossus 1599
Siphostoma 766
Stolephorus 445
Synodus 536
Pogge...................... 2065
Pogonathus 1482
courbina.............. 1483
Pogonias 1482
courbina............. 1483
cromis 1482
courbina........... 1483
fasciatus 1483
Pogonichthys 223
argyreiosus 224
communis 326
(Platygobio) gulonel-
lus 326

Page.

Pogonichthys inæquilobus......... 224
macrolepidotus 223
symmetricus 246
Pogy..................... 433
Po-he-wa.................. 238
Point Loma, Blind Goby of......... 2262
Poison Toad-fishes 2323, 2325
Poisson Bleu.................. 517
de Marais.................. 113
Lune 954
pola, Platessa 2657
polaris, Blennius.................. 2469
Boreogadus 2534
Cottus ..:.................. 1999
Lycodalepis 2468
Lycodes 2469
Merlangus 2534
Pollachius 2534
Porocottus................... 1998
Pole Flounder.................. 2657
Polistotrema 6
dombey 6
stouti 6
politus, Seriphus 1397
Sphæroides 1736
annulatus 1736
Tetrodon 1736
Pollachius 2534
carbonarius............. 2535
chalcogrammus 2536, 2537
polaris 2534
vireus 2534
Pollack..................... 2534
Puget Sound 2536
Wall-eyed 2536
Pollacks 2534
Alaskan 2535
pollicaris, Cottus................. 1941, 1953
Uranidea 1954
pollux, Póntinus 1857
poloosoo, Caranx ....:.... 928
polyacanthocephalus, Cottus....... 1977
Myoxocepha-
lus 1976
Polyacanthonotinæ................... 613
polyactocephalum, Bryostemma.. 2408, 2409
polyactocephalus, Blennius 2409
Chirolophis 2409
polycaulus, Pimelodus.............. 153
Rhamdia............... 153
Polycirrhus 1477
dumerili 1479
rathbuni 1479
Polyclemus 1477, 1478
Polydactylus.................... 828
approximans 829

	Page.
Polydactylus octonemus	830
opercularis	830
plumierii	830
virginicus	829
polygonius, Acanthostracion	1725
Gymnothorax	394
Lycodontis	394
polylepis, Balistes	1700
Polymixia	854
lowei	854
Polymixiidæ	854
polymorphus, Gadus	2540
Polynemidæ	827, 2841
Polynemus	828
americanus	830
approximans	829
artedi	828
californiensis	829
macronemus	828
mango	830
melanopoma	831
octofilis	830
octonemus	830
oligodon	830
opercularis	831
plumierii	830
quinquarius	828
sexradiatus	2183
tridigitatus	2177
virginicus	830
Polyodon	101
feuille	102
folium	102
spathula	101, 102
Polyodontidæ	101
Polyprion	1138
americanus	1139
cernium	1139
oxygenius	1139
Polyprioninæ	1128
Polyprosopus	51
macer	51
Polypterichthys	754
polytrema, Bdellostoma	6
Polyuranodon	392
Pomacampsis	1020
Pomacanthinæ	1670, 2860
Pomacanthodes	1679, 1681
zonipectus	1682
Pomacanthus	1679, 2859
arcuatus	1679, 1680, 1681
aureus	1680
balteatus	1680
ciliaris	1685, 1686
cingulatus	1680
crescentalis	1682

	Page.
Pomacanthus paru	1680, 1681, 2859
passer	1683
quinquecinctus	1680
tricolor	1684
zonipectus	1681
Pomacentridæ	1543
Pomacentrinæ	1544
Pomacentrus adustus	1552
analigutta	1554
analis	1555
atrocyaneus	1552
bairdii	1567
caudalis	1556, 1557
denegatus	1567
dorsopunicans	1557
flavilatus	1558
fuscus	1552
leucostictus	1556
niveatus	1568
obscuratus	1552, 1555
otophorus	1555
paritus	1558
planifrons	1559
quadrigutta	1570
rectifrænum	1554
rubicundus	1565
variabilis	1552
xanthurus	1557
Pomadasis	1329
andrei	1332
axillaris	1328
bayanus	1331
branicki	1333
corvinæformis	1327
crocro	1333
dovii	1318
elongatus	1328
humilis	1331
leuciscus	1328
macracanthus	1332
nitidus	1326
pacifici	1316
panamensis	1331
productus	1332
ramosus	1334
Pomadasys approximans	1333
cæsius	1317
davidsoni	1321
leuciscus	1328
modestus	1321
virginicus	1323
Pomataprion	1565
dorsalis	1570
Pomatomichthys	1111
Pomatomidæ	945
Pomatomus	946, 1111

	Page.
Pomatomus saltator	947
saltatrix	946
skib	947
Pomatopsetta	2614
dentata	2615
Pomatoschistus	2210
Pomfrets	956, 958, 959
Pomolobus	424
æstivalis	426
chrysochloris	425
mediocris	425, 2810
pseudoharengus	426
lacustris	426
vernalis	426
Pomotis	999, 1006
bombifrons	1003
breviceps	1003
catesbei	1010
chætodon	995
convexifrons	1003
elongatus	1001
fallax	1003
gibbosus	1005
gulosus	992
guttatus	993
heros	1007
holbrooki	1008
incisor	1005
inscriptus	1003
longulus	996
luna	1006
marginatus	1003
microlophus	1008
nefastus	1003
nitida	1003
notatus	1008
obesus	993
obscurus	1006
pallidus	1007
popeii	1003
ravenelii	1010
rubicauda	1001
sanguinolentus	1003
solis	1001
speciosus	1006, 1008
vulgaris	1010
pomotis, Acantharchus	989
Ambloplites	989
Centrarchus	989
Pomoxis	986
annularis	987
sparoides	987
Pomoxys	986
brevicauda	987
intermedius	987

	Page.
Pomoxys protacanthus	987
sparoides	988
Pompano, California	967
Common	944
Irish	1376
Pomphilus	900
Pompilus	962
pompilus, Centrolophus	963
Thynnus	900
Pompon	1318
Ponco Prieto	1297
Pond Smelt	525
ponderosus, Amiurus	137
Ictalúrus	137
pondiceriana, Elacate	948
ponticus, Gasterosteus	747
Hippoglossus	2612
Pontinus	1854
castor	1856
longispinis	1858
macrolepis	1855
pollux	1857
rathbuni	1857
sierra	1859
popeii, Pomotis	1003
Pop-eye	2586
Porbeagles	49
porca, Scorpæna	1839
Porcupine-fish	1742, 1744
Porgee	1509
Porgies	1343
Deep-water	1344
European	1356
Porgy	1346
Grass	1355
Jolt-head	1352
Little-head	1350
Little-mouth	1354
Red	1356
Saucer-eye	1349
Shad	1355
Sheepshead	1354
Southern	1348
White-bone	1353
Porichthys	2317
margaritatus	2322
nautopædium	2323
notatus	2321
plectrodon	2321
porosissimus	2319, 2321
porifer, Lycenchelys	2471
Lycodes	2472
Porkfish	1322
Porobronchus	2495
Poroclinus	2433
rothrocki	2434

3074 *Index.*

	Page.		Page.
Porocottus	1996	Priacanthus	1237
bradfordi	2862	altus	1240
polaris	1998	arenatus	1237, 1238
quadratus	1998	carolinus	2858
quadrifilis	1999, 2000, 2863	catalufa	1238
sellaris	1996, 2863	cepedianus	1238
tentaculatus	2000, 2862	cruentatus	1238, 2858
Poroderma	23	fulgens	1238
Porogadus	2519	macrophthalmus	1238
miles	2520	schlegeli	2858
promelas	2512	serrula	1239
Porogobius	2210	pribilovius, Nautichthys	2020
Poromitra	840	Nautiscus	2019
capito	840	pricei, Campostoma	205
Poronotus	965, 967, 2849	Villarius	2790
simillimus	967	Prick Fish	555
triacanthus	2849	Prickly Bullhead	1944
porosissimus, Batrachus	2321	Priest Fish	1784
Porichthys	2319, 2321	Prieta, Aguja	891
porosus, Carcharias	37	Mojarra	1299
Cottus	1975	Morena	2804
Esox	627	Prieto Pargo	1252
Porte Euseigne	1687	Robalo	1119
Porthmeus	2848	Ronco	1297
Portugais	1679	prieto, Lutjanus	1253
Portuguese Man-of-War Fish	949	Prilonotus	1741, 1742
Post Croaker	1458	(Anchisomus) caudicinctus	1742
postica, Echeneis	2272	Primospina	1765, 1774, 1783
Potamocottus	1942	princeps, Caulolatilus	2276, 2277
bendirei	1965	Cottus	1962
carolinæ	1952	Latilus	2277
punctutatus	1949	principis, Antennarius	2719
zopherus	1952	Chironectes	2719
Potomac Shad	427	Julis	1591
pottsii, Aplesion	1083	Platyglossus	1591
Boleosoma	1083	Prinodon	670
Etheostoma	1082	Priodonophis	392, 393, 399
pourtalesii, Archosargus	1360	meleagris	399
Sargus	1360	ocellatus	399
Pout, Horned	135, 140	Prionace	33
powelli, Balistes	1702	glauca	33
præcisus, Clinus	2441	Prionistius	1927
Eumesogrammus	2441	macellus	1928
præstabilis, Alosa	428	Prionodes	1208, 1209, 1210
præstigiator, Centropristis	1214	æquidens	1210
Serranus	1214	bulleri	1213
presidionis, Pœcilia	697	fasciatus	1212
pretiosa, Argentina	525	flavescens	1215
pretiosus, Hypomesus	525	fusculus	1211
Osmerus	525	luciopercanus	1216
Ruvettus	879	phœbe	1211
Thyrsites	880	stilbostigma	1216
Trachichthys	837	tabacarius	1215
Prêtre, Pêche	1784	tigrinus	1214
Priacanthichthys	1148	Prionodon cucuri	40
Priacanthidæ	1236		

	Page.
Prionodon obvelatus	35
Prionotus 2148, 2150, 2160, 2867	
alatus	2159
albirostris	2163
beanii	2170
birostratus 2152, 2156	
carolinus 2156, 2157	
egretta	2175
evolans 2167, 2168, 2169	
gymnostethus	2153
horrens..................	2172
lineatus	2167
loxias 2155, 2156	
miles	2160
ophryas	2164
palmipes·	2157
pilatus ....:	2156
punctatus 2158, 2164, 2169	
quiescens.................	2161
roseus	2158
rubio	2164
sarritor..................	2169
scitulus	2157
stearnsi	2166
stephanophrys...........	2161
strigatus	2167
tribulus 2171, 2172	
xenisma	2154
Prionurus laticlavius	1696
punctatus...............	1695
Pristidæ	60
Pristigaster........	438
cayanus	438
dovii	437
flavipinnis.............	436
lutipinnis	437
macrops	437
martii	438
mucronatus............	438
(Odontognathus) pana-meusis	438
phaeton...............	438
Pristigasterinæ	418
Pristipoma 1329, 1331	
acara pinima	1323
andrei......	1332
auratum 1324, 1343	
axillare	1328
bicolor	1320
bilineatum	1319
boucardi	1334
branicki................	1334
brasiliense	1320
brevipinne	1341
cantharinum	1340
catharinæ	1323

	Page.
Pristipoma chalceum...............	1338
coro	1324
crocro...................	1333
cultriferum	1333
davidsonii	1321
dovii	1318
fasciatum	1339
fulvomaculatum	1339
furthi ....:...	1319
humile	1331
kneri....................	1338
leuciscus	1328
macracanthum	1332
melanopterum..........	1319
(Hæmulopsis) nitidum..	1326
notatum................	1321
panamense	1331
productum	1332
ramosum	1334
rodo	1323
scapulare	1321
serrula............... 1324, 1343	
spleniatum	1322
surinamense	1319
trilineatum	1320
virginicum.............	1323
Pristipomoides	1279
Pristis	60
acutirostris.................	61
granulosa.................	61
megalodon	61
mississippiensis	61
occa	61
pectinatus 60, 61, 2749	
perrotteti.................. 60, 2749	
zephyreus	2749
Pristobatus	60
Pristocantharus	1334
Proach, Lucky	1971
Proamblys	1247
Proarthri	19
probatocephalus, Archosargus ... 1361, 1362	
Diplodus	1361
Sparus	1361
proboscidalis, Agonomalus	2037
Aspidophorus	2038
proboscidea, Limanda	2645
proboscideus, Chænomugil.........	816
Monacanthus........	1719
Mugil	816
procellarum, Myctophum............	575
procera, Venefica....................	365
Proceros	101
vittatus	102
procerum, Nettastoma	366
Procerus maculatus...............	102

	Page.
Prochilus	2195
procne, Cliola	264
Exocœtus (Cypselurus)	737
Hybognathus	264
Hybopsis	264
Leuciscus	264
Notropis	264
productum, Holocentrum	852
Pristipoma	1332
productus, Alepocephalus	452
Cylindrosteus	111
Engraulis	447
Eucinostomus	1372
Gadus	2531
Leuciscus	240
Merlangus	2531
Merluccius	2531
Pomadasis	1332
Rhinobatus	63
Stolephorus	447
prœliare, Etheostoma	1104
prœliaris, Etheostoma	1104
Microperca	1103
profundorum, Acanthocottus	1991
Lepophidium	2484
Ophidium	2484
Scylliorhinus	22
Zesticelus	1990
profundus, Lutjanus	1264
Mesoprion	1263
Prognathodes	1671
aculeatus	1671
prognathus, Argyrosomus	471
Coregonus	472
Prognurus	2866
cypselurus	2866
prolixum, Campostoma	206
prolixus, Leuciscus	206
prolongus, Leptoconger	363
promelas, Mœbia	2511
Pimephales	217
confertus	217
maculosus	217
Porogadus	2512
Prometheus	882
Prometheus atlanticus	883
prometheus, Gempylus	883
Promethichthys	882
Promethichthys	882
atlanticus	883
parvipinnis	883
prometheus	882
Promicrops	1162
guasa	1164
guttatus	1162
itaiara	1164

	Page.
Promicropterus	1229, 1231, 1233
decoratus	1234
Promoxis nitidus	987
Pronotogrammus	1224
eos	1224
multifasciatus	1226
peruanus	1223
vivanus	1224
proops, Bagrus	124
Netuma	124
Sciadeichthys	123, 2760
Tachisurus	124
Propterygia	66
prorates, Leptophidium	2485
proridens, Calamus	1350
proriger, Clinostomus	240
Leuciscus	240
Sebastichthys	1788, 1793
Sebastodes	1787, 1792
Squalius	240
prorigera, Congermuræna	357
prorigerum, Ophisoma	357
proserpina, Moniana	272
Notropis	272
Prosopium	461, 462
couesii	463
prospinosum, Holocentrum	853
Prospinus	1164
chloropterus	1165
prosthemius, Ceratichthys	324
Couesius	324
prosthistius, Amiurus	139
protacanthus, Pomoxys	987
proteus, Oncorhynchus	478
Salmo	478
protoclus, Myctophum	565
Protoporus	228
domninus	233
proxima, Seriola	904
proximus, Gadus	2539
Microgadus	2539
pruinosus, Gadus	2540
Psednoblennius	2406
hypacanthus	2406
Psenes	950
auratus	951
cyanophrys	950
fuscus	951
javanicus	951
maculatus	951
pellucidus	950
regulus	951
Psettichthys	2617
melanostictus	2618
sordidus	2680
Psettinæ	2608

	Page.
Pseudariodes	154
pantherinus	155
Pseudarius	119
Pseudobastes	1839
Pseudocanthicus	159
pseudocrocodilus, Scopelus	556
pseudogula, Eucinostomus	1368
Gerres	1368
pseudoharengus, Clupea	426
Pomolobus	426
lacustris	426
Pseudohemiodon	156
pseudohispanica, Clupea	424
Sardinia	424
pseudohispanicus, Clupanodon	423
Pseudojulis	1604
adustus	1603
californicus	1601
inornatus	1604
melanotis	1605
modestus	1601
notospilus	1603
venustus	1602
Pseudoloricaria	156
Pseudomonacanthus	1717
amphioxys	1717
Pseudomuræna	392
Pseudophoxinus	243
Pseudopleuronectes	2646
americanus	2647
pinnifasciatus	2647
planus	2647
Pseudopriacanthus	1239
altus	1239
serrula	1239
Pseudorhamdia	153, 154
piscatrix	155
Pseudorhombus	2624
adspersus	2627
brasiliensis	2626
californicus	2626
dentatus	2630, 2632
oblongus	2630
ocellaris	2630
quadrocellatus	2635
vorax	2626
Pseudoscarus acutus	1652
aracanga	1648
cæruleus	1654
chloris	1648, 1654
cœlestinus	1655, 1656
diadema	1646
flavomarginatus	1652
gnathodus	1650
guacamaia	1656, 1657, 1659

	Page.
Pseudoscarus lineolatus	1651
microrrhinos	1655
nuchalis	1654
obtusus	1654
perrico	1659
pleianus	1656
punctulatus	1646
psittacus	1647
quadrispinosus	1648
rostratus	1658
sanctæ-crucis	1651
simplex	1656
superbus	1650
tænniopterus	1646, 1647
trispinosus	1648
turchesius	1659
vetula	1650
Pseudosciæna surinamensis	1420
Pseudoscopelus	2292
scriptus	2292
Pseudotriakidæ	26
Pseudotriakis	27
microdon	27
Pseudoxiphophorus	678
bimaculatus	678
reticulatus	678
Pseudupeneus	858
Psilonotus	1741
punctatissimus	1741
psittacinus, Centropristis	1213
Serranus	1213
psittaculus, Julis	1597
Labrus	1596
psittacus, Callyodon	1638
Cheilichthys	1740
Colomesus	1740
Coryphæna	1619
Lachnolaimus	1580
Pseudoscarus	1647
Scarus	1647
Tetrodon	1740
Xyrichthys	1618, 1619
Psychrolutes	2025
paradoxus	2026
zebra	2027
Psychrolutinæ	1883
Psychromaster	1099
tuscumbia	1100
Pteraclidæ	955
Pteraclis	955
carolinus	956
trichopterus	956
Pterengraulis	450
atherinoides	450
Pterocephala	92
Pterognathus	2354, 2355

	Page.
Pteronotus	149
Pterophryne	2715
gibba	2717
histrio	2716
lævigata	2717
Pterophrynoides	2715
Pteroplatea	86
crebripunctata	87, 2753
maclura	86
marmorata	87, 2754
rava	2754
Pteropodus	1765, 1776, 1819, 2860
dallii	1819
Ptilichthyidæ	2451
Ptilichthys	2452
goodei	2452
Ptychocheilus	224
gracilis	225
grandis	225, 2796
harfordi	225, 2797
lucius	225
major	225, 2797
oregonensis	224, 2796
rapax	225
vorax	227
Ptycholepis	414
Ptychostomus	187
albidus	192
albus	191
breviceps	196
bucco	191
cervinus	197
collapsus	190
conus	196
coregonus	191
crassilabris	194
duquesnei	193
erythrurus	193
haydeni	187
lachrymalis	194
oneida	193
papillosus	189
pidiensis	191
robustus	193
thalassinus	192
velatus	190
Ptyonotus	2005
thompsonii	2005
Pudding-wife	1590
Pudiano	1583
Verde	1590, 1591
Vermelho	1583
puella, Hypoplectrus	1192
unicolor	1192
Plectropoma	1192
puellaris, Cossyphus	1584

	Page.
puellaris, Decodon	1584
Puerco Espino	1745
Pez	1700, 1704
Puffer	1733
Smooth	1728
Southern	1732
Puffers	1726
Sharp-nosed	1740
pugetensis, Artedius	1890
Chitonotus	1890, 1891
Icelus	1891
Puget Sound Pollack	2536
pugetti, Gasterosteus	751
pulchella, Gila	234
Harpe	1584
Moniana	272
pulchelloides, Leuciscus	222
pulchellus, Bodianus	1584
Cheilonemus	222
Cossyphus	1584
Cyclogaster	2127
Cymatogaster	1503
Haplochilus	659
Histiophorus	891
Leuciscus	221
Leucosomus	222
Liparis	2126
Oligocephalus	1089
Squalius	234
Zygonectes	659
pulcher, Eques	1489
Labrus	1585
Neozoarches	2426
Pimelometopon	1585
Semicossyphus	1585
Squalius	234
Trochocopus	1585
pulchra, Harpe	1585
Tigoma	234
pullum, Chondrostoma	206
pullus, Amiurus	141
Cantherines	1713
Monacanthus	1713
Pimelodus	141
pulvereus, Fundulus	652
Zygonectes	652
pulverulentum, Oxygeneum	207
pulverulentus, Myloleucus	246
Pumpkin Seed	1009
Punaru	2397
puncta, Furcaria	1547
punctata, Alutera	1718, 1719
Bairdiella	1434
Gambusia	679
Lamna	48
Muræna	395

	Page.
punctata, Muraenophis	397
Perca	1145, 1146, 1433
. Sciæna	1434
Trigla	2170
punctatissima, Anguilla	348
punctatissimus, Canthigaster	1741
Tetrodon	1741
punctatum, Hyperprosopon argenteum	1502
Myctophum	570
Opisthognathus	2281
. Sicydium	2867
punctatus, Apomotis	997
Balistes	1702
Blennius	2390, 2440
Bodianus	1146
fulvus	1146
Bryttus	998
Caranx	908
Carcharias	41
Ceratacanthus	2860
Clinus	2440
Decapterus	907
Dermatolepis	1168
Diodon	1746
Enneacentrus	1146
Epinephelus	1154, 1146
Eques	1488, 1489
Fundulus	637, 2827
Gadus	2553
Gunnellus	2440
Holocentrus	1153
Hypleurochilus	2390
Ichthælurus	135
Ictalurus	134
Isesthes	2390
Lepomis	998
Micristodus	52
Monacanthus	1713, 1719
Morrhua	2543
Myrophis	371
Opisthognathus	2281
Prionotus	2158, 2164, 2169
Prionurus	1695
Silurus	135
Squalus	26, 43
Stichæus	2439
Tetrodon	1735
Trachinus	1153
Upeneus	859
Xesurus	1694, 1695
puncticeps, Cryptopterus	382
Ophichthus	382
Ophichthys	382
puncticulata, Gambusia	680
Perca marina	1146
	Page.
puncticulatus, Apogonichthys	1111
Arius	131
Chilomycterus	1750
punctifer, Crotalopsis	387
Dionda	215
Ophichthys	387
Hybognathus (Dionda)	215
punctiferus, Bodianus	1147
. Menephorus	1147
punctipinne, Siphostoma	763
punctipinnis, Ayresia	1548
Chromis	1548
Dermatostethus	763
punctulata, Coryphæna	953
Microperca	1104
Uranidea	1949
punctulatum, Boleosoma	1091
Etheostoma	1090
punctulatus, Calliurus	992, 1011
Cottus	1948
bairdi	1950
Gobiesox	2338
Hippocampus	777
Lampugus	953
Minnilus	302
Notropis umbratilis	301
Pimelodus	143
Pœcilichthys	1091
Potamocottus	1949
Pseudoscarus	1646
Scarus	1645
Sicyases	2338
Squalus	26
Puñecas	2195
pungitius, Gasterosteus	745
brachypoda	746
Pygosteus	745
brachypoda	746
punceus, Hyborhynchus	218
Puraque	63
purpurascens, Elops	410
purpuratus, Salmo	492, 499, 2819
bouvieri	496
purpurea, Tigoma	234
purpurescens, Anoplarchus	2423
Lepomis	1006
Salpa variegata	1271
purpureus, Leuciscus	234
Lutjanus	1264
Merlangus	2535
Sebastichthys	1826
Squalius	234
Pursy Minnows	670, 671
pusilla, Aphoristia	2711
Bothrocara	2476
Clupea	426

	Page.
pusilla, Maynea	2476
Perca	1107
Platessa	2647
pusillum, Acanthidium	55
pusillus, Argyrosomus	470
Etmopterus	55
Spinax	55
Symphurus	2710
putaol, Gymnotus	341
putuami, Acipenser	104
Cottogaster	1046
Euchalarodus	2650
Liopsetta	2650
pygmæa, Eucalia inconstans	744
Umbra	624
limi	624
pygmæus, Gadus	2542
Gasterosteus	744
Leuciscus	624
Pygosteus	745
pungitius	745
brachypoda	745
Pylodictis limosus	143
pyramidatus Cyclopterus	2097
pyrrhogaster, Chrosomus	210
pyrrhomelas, Cliola	281
Codoma	281
Notropis	280
Photogenis	281
Pythonichthys	390
sanguineus	390
quadracus, Apeltes	752
Gasterosteus	752
quadrangularis, Selene	1668
quadratus, Porocottus	1998
Zeus	1668
quadricorne, Ostracium	1725
quadricornis, Agonus	2041
Aspidophorus	2041
Cottus	2001
Hypsagonus	2038, 2041
Ostracion	1725
quadridens, Pleuronectes	2654
quadrifasciatus, Chasmodes	2392
Pholis	2392, 2394
quadrifilis, Bathypterois	545
Cottus	1998, 2000
Porocottus	1999, 2000
quadrigutta, Pomacentrus	1570
quadrilateralis, Coregonus	465
quadrilineatum, Hæmulon	1309, 1311
quadriloba, Raia	90
Rhinoptera	90
quadrimaculatus, Diodon	1746
Paradiodon	1746

	Page.
quadriporus, Gobius	2221
quadripunctatus, Scomber	869
quadriremis, Exocœtus	735
quadriscutis, Arius	126
Netuma	126
quadriseriatus, Artedius	1897
Icelinus	1897
Icelus	1897
quadrispinosus, Pseudoscarus	1648
Scarus	1648
quadrituberculata, Platessa	2648
quadrituberculatus, Parophrys	2648
Pleuronectes	2648
Pleuronich- thys	2638
quadrocellata, Anclyopsetta	2635
Platessa	2633
quadrocellatus, Pseudorhombus	2635
quappella, Etheostoma	1804
quartus, Anthias rondeleti	1266
Quasky	514
Quassilabia	198
lacera	199
Quassiremus	380
evionthas	380
nothochir	380
quatuordecimlaminatus, Echeneis	2272
Queenfish	1397
quenseli, Pleuronectes	2654
quercinum, Macrostoma	554
quercinus, Notoscopelus	555
Queriman	810
Querimana	817
gyrans	818
harengus	817
querna, Azevia	2675
Cyclopsetta	2675
Quia-quia	907
Quiebra	898
quiebra, Chorinemus	899
Lichia	899
quiescens, Copelandellus	1100
Etheostoma	1101
Pœcilichthys	1101
Prionotus	2161
Uranidea	1968
Quietula	2251
y-cauda	2251, 2252
Quillback	167
Quill-fishes	2451
Quinnat Salmon	474, 479
quinnat, Oncorhynchus	480
Salmo	480
quinquarius, Pentanemus	828
Polynemus	828
quinqueaculeata, Raja	88

	Page.
quinquecinctus, Pomacanthus	.1680
quinquefasciatus, Epinephelus	1164
Serranus	1164
quinquelineatum, Hæmulon	1311
quiquemaculatus, Centronotus	2430
Opisthocentrus	2430
Quisutsch	480
quoyi, Cestracion	21
Gyropleurodus	21
Rabbit-fish	882, 1748
Rabbit-mouth Sucker	198, 199
Rabdophorus	1672
Rabida	144, 145, 146
Rabirubia	1274, 1275
- de lo Alto	1221
ganizara	1586
inermis	1274
rabirubia, Anthias	1276
Rabirubias	1275
Rabula	390
aquæ-dulcis	390
longicauda	391
marmorea	391
panamensis	391
Raccoon Perch	1023
Rachycentridæ	946
Rachycentron	948
canadus	948
radiale, Diplectrum	1204
radialis, Centropristis	1205
Diplectrum	1205
Serranus	1205
radians, Centropristis	1208
Diplectrum	1208
Labrus	1633
Scarus	1632, 1633
Serranus	1208
Sparisoma	1632
radiata, Raia	69
Raja	69
radiato, Turdus oculo	1703
radiatus, Chærojulis	1591
Halichæres	1591
Iridio	1590
Labrus	1591
Lophius	2738
radiatus, Ogcocephalus	2738
Platyglossus	1591, 1597
Sparus	1596
radiosus, Antennarius	2725
Radulinus	1919
asprellus	1920
boleoides	1919
rafinesquei, Acipenser	106
Collettia	567
Cylindrosteus	111

	Page.
rafinesquei, Myctophus	567
Scaphirhynchus	107
Scopelus	567
Rafinesquiellus	1066, 1068, 1082
Rag Fishes	968
Raia	66
americana	69
birostris	93
chantenay	71
cooperi	73
desmarestia	71
diaphana	71
eglanteria	68, 71
erinacea	68
fimbriata	93
flagellum	88
inornata	73
inermis	73
jordani	73
maclura	87
manatia	93
narinari	88
obtusa	2751
ocellata	69
parmifera	75
quadriloba	90
radiata	69
rhina	72
trachura	76
tuberculata	84
Raiada, Majarra	1561
Raiado, Roncador	1301, 1313
Sargo	1361
Raie tuberculée	84
raii, Brama	958, 959
Sparus	960
Rainbow Darter	1088
Herring	524
Trout	500
Rainwater Fish	665
Raizero	1247, 1251, 1273
Raja	66, 2750
abyssicola	76, 2751
ackleyi	70
aleutica	75, 2751
binoculata	72
bonasus	90
centrura	83
diabolus marinus	93
eglanteria	71
equatorialis	74, 2751
erinacea	69
fyllæ	68
granulata	72
inornata	73
interrupta	2751

	Page.
Raja jamaicensis	81
lævis	71
ocellata	68
ornata	71
parmifera	74
percellens	63
plutonia	69, 70
quinqueaculeata	88
radiata	69
rhina	72
rosispinis	2751
say	86
senta	71
sloani	81
stellulata	75
trachura	75
raji, Brama	960
Rajidæ	66
raleighana, Harriotta	96
rammelsbergii, Mugil	812
ramosum, Pristipoma	1334
ramosus, Pomadasis	1334
Ramularia	2633
dendritica	2633
ransonnetii, Neoditrema	1511
ranula, Careproctus	2134
Liparis	2134
Ranzania	1755
makua	1755
truncata	1755, 1756
Ranzaninæ	1752
rapax, Ptychocheilus	225
raphidoma, Belone	716
Tylosurus	715
raptor, Gadus	2552
Raro	404
rarus, Rhinoscopelus	569
Scopelus	569
Rascacio	1848
rascacio, Scorpæna	1849
Rasciera	1794
Rasher	1794
rashleighanus, Squalus	51
rastralis, Stolephorus	2811
rastrelliger, Physiculus	2549
Sebastichthys	1820
Sebastodes	1819, 1821
Rastrinus	1909
scutiger	1909
Ratfish	95
Rathbunella	2289
hypoplecta	2290
rathbuni, Fundulus	649
Mulloides	857
Paralonchurus	1479
Polycirrhus	1479

	Page.
rathbuni, Pontinus	1857
Upeneus	857
Raton	829
Rat-tail, Common	2583
rauchi, Acipenser	106
raucus, Sargus	1364
rava, Pteroplatea	2754
Raven, Sea	976, 2023
raveneli, Esox	626
ravenelii, Pomotis	1010
Ravens, Sea	2022
Ray, Butterfly	86
California Sting	89
Common Sting	83
Cow-nose	90
Southern Sting	86
Spotted Sting	88
Rays	59
Eagle	87, 89
Electric	76
Round Sting	79
Sting	79, 82
Thick-tailed	60
Whip-tailed	79
Razor-back Buffalo	164
Sucker	184
Razor-fish	1617, 1618
Real, Matajuelo	410
rectangulare, Cichlasoma	1515
rectangularis, Acara	1515
rectifrænum, Eupomacentrus	1553
Pomacentrus	1554
recurvirostris, Sayris	725
Red-bellied Dace	209
Redbreast Bream	1001
reddingi, Orthopristis	1336
Red Drums	1453
Red-eye	990
Little	996
Mullet	814
Red Fallfish	286
Red-fin	281, 298
Redfish	481, 1453, 1760
Bull	1453
California	1585
Little	482
Red Goatfish	858
Grouper	1160
Guativere	1145
Gurnard	2177
Hind	1141, 1158
Redhorse	187
Common	192
Texas	192
redi, Orthragoriscus	1754
Red-mouth Buffalo Fish	163

	Page.
Red-mouth Grunt	1308
Red Parrot Fish	1635
Porgy	1356
Rockfish	1805
Rock Trout	1872
Roncador	1456
Sculpin	1935
Snapper	1264
Sturgeon	106
Sucker	176
Red-sided Shiner	240
Red-spotted Sunfish	1004
Trout	507
Red-tail Snapper	1270
Red-winged Sea-robin	2156
reflexo, Labrus rostro	1677
regale, Cybium	875
Cynoscion	1407
Myctophum	563
Nannobrachium	563
regalis, Cestreus	1407
thalassinus	1408
Cynoscion	1407
Enchelyopus	2553
Johnius	1407
Otolithus	1407
Phycis	2553
Scomber	875
Scomberomorus	875
regis, Atherinops	808
regius, Blennius	2553
Hybognathus	213
Lampris	955
Phycis	2553
Urophycis	2553
Zeus	955
regulus, Psenes	951
Sebastes	1761
Reina	1815
reinhardi, Careproctus	2133, 2134
reinhardti, Corynolophus	2733
Himantolophus	2733
Motella	2559
Oños	2559
Reinhardtius	2610
hippoglossoides	2611
remifer, Archosion	1399
Isopisthus	1399
remiger, Myctophum	573
Ophisurus	384
Remilegia	2270
australis	2270
Remora	2271
albescens	2272
brachyptera	2272
jacobæa	2272

	Page.
Remora remora	2271
remora, Echeneis	2272
Remora	2271
Remoras	2265, 2271
Remorina	2271, 2272
remoroides, Echeneis	2272
Remoropsis	2271, 2272
brachypterus	2272
remotus, Carcharhinus	37
Serranus	1100
Reniceps tiburo	44
repandus, Serranus	1187
Requiem	38
Sharks	27
Requin	38
resplendens, Lampanyctus	555
reticularis, Anchisomus	1735
Antennarius	2719
reticulata, Amia	113
Liparis	2108
Mycteroperca	1187
Pœcilia	2833
Solea	2696
Spatularia	102
Thalassophryne	2325
reticulatum, Bœostoma	2696
reticulatus, Bryttus	998
Catostomus	179
Cestreus	1409
Chilomycterus	1751
Cynoscion	1408
Diodon	1751
Esox	628
Gobiesox	2328
Halieutichthys	2741
Lepadogaster	2328
Lucius	627
Lycodes	2465
Monochir	2696
Notropis	262
Orbis	1750
Otolithus	1409
Pseudoxiphophorus	678
Trisotropis	1187
retifer, Catulus	25
Scylliorhinus	25
retifera, Muræna	401
retiferum, Scyllium	25
retractus, Calliodon	1623
Cryptotomus	1623
retrocurrens, Hæmulon	1297
retropinnis, Catostomus	175, 2791
Microdesmus	2450
Ophichthus	383
Ophichthys	383
retrosella, Amia	1109

	Page.
retrosella, Apogon	1108
retrospinis, Holocentrum	853
Plectrypops	853
retzii, Mola	1754
Orthragoriscus	1754
rex, Catostomus	177
Etheostoma	1026
Percina	1025
rex-mullorum, Apogon	1107
Rey, Peixe	806
Rhacochilus	1507
toxotes	1507
Rhamdella	149, 150, 151
parryi	153
Rhamdia	149, 150
baronis-mulleri	151
brachyptera	151
bransfordi	151
godmani	152
guatemalensis	152
hypselurus	152
laticauda	152
managuensis	153
microptera	153
motaguensis	151
nicaraguensis	152
parryi	153
petenensis	153
polycaulus	153
salvini	152, 2790
wagneri	150, 151
Rhamphoberyx	846
leucopus	847
pœcilopus	847
Rhamphocottidæ	2029
Rhamphocottus	2030
richardsoni	2030
Rhegnopteri	781, 827
Rhencus	1329, 1331
Rheocrypta	1044
copelandi	1046
rhessodon, Arbaciosa	2340
Gobiesox	2340
Rhina	58
squatina	59
rhina, Raia	72
Raja	72
Rhineloricaria	156, 157, 158
Rhinesomus	1721, 1722
rhinichthyoides, Tigoma	312
Rhinichthys	305
arenatus	308
atronasus	307
croceus	308
lunatus	308
meleagris	308

	Page.
Rhinichthys badius	308
cataractæ	306
dulcis	306
dulcis	307
henshavii	312
lunatus	308
luteus	307
marmoratus	306
maxillosus	307
meleagris	308
nasutus	306
(Apocope) nevadensis	311
obtusus	308
ocella	307
simus	307
transmontanus	307
(Apocope) velifer	312
Rhinobatidæ	61
Rhinobatus	61
electricus	63
exasperatus	65
glaucostictus	63
glaucostigma	62
lentiginosus	62, 2750
leucorhynchus	62
marcgravii	63
percellens	63
planiceps	64
productus	63
spinosus	63
stellio	2750
triseriatus	66
undulatus	63
Rhinoberyx	848
chryseus	847
Rhinodontinæ	52
Rhinogobius contractus	2236
Rhinoliparis	2145
barbulifer	2145
Rhinonemus	2560
caudacuta	2560
cimbrius	2561
Rhinoptera	90
bonasus	90
ensenadæ	91
quadriloba	90
steindachneri	91
vespertilio	90
Rhinopterinæ	88, 2753
Rhinoscion	1455
saturnus	1457
Rhinoscopelus	568
andreæ	569
coccoi	568
rarus	569
Rhinoscymnus	56

	Page.
Rhinotriacis	30
henlei	31
rhizophoræ, Fundulus	644
rhodochloris, Sebastichthys	1810
Sebastodes	1809
rhodopus, Trachinotus	941, 943
rhodorus, Ascelichthys	2025
rhodospilus, Gobiesox	2335
rhodoterum, Ditrema	1503
rhodoterus, Holconotus	1502
Rhodymenichthys	2414, 2415, 2416
ruberrimus	2417
rhomalea, Gnathypops	2285
rhomaleus, Opisthognathus	2285
Squalius	233
rhombeus, Gerres	1374
Rhombochirus	2273
osteochir	2273
Rhomboganoidea	108
rhomboidalis, Lebias	672
Otolithus	1404
Turdus	1691
rhomboides, Acanthinion	942
Chætodon	942, 2848
Diplodus	1358
Lagodon	1358
Sargus	1358
Sparus	1358
Trachinotus	2847
Trachynotus	942
Rhomboidichthys	2661
ellipticus	2665
leopardinus	2666
lunatus	2666
lunulatus	2666
maculiferus	2665
ocellatus	2664
spinosus	2663
Rhomboplites	1276
aurorubens	1277, 1278
elegans	1278
Rhombotides	1689
Rhombus	965, 2849
alepidotus	966, 2849
aquosus	2660
aramaca	2626, 2672
argentipinnis	966
bahiensis	2664
crenulatus	966
lævis cornubiensis	2654
longipinnis	966
medius	967
ocellatus	2664
orbicularis	966
palometa	966
paru	965, 2849

	Page.
Rhombus simillimus	967
soleœformis	2672
triacanthus	967
xanthurus	966, 2849
Rhonciscus	1329, 1330, 1333
Rhothæca	1066
blennius	1073
rhothea, Uranidea	1947
rhotheus, Cottus	1946
rhothœcus, Catostomus	181
rhynchæus, Acipenser	106
Rhynchias	2841
septipinnis	2841
Rhynchichthys	847
Rhynchotus	1741
Rhypticus	1229
bicolor	1232
decoratus	1234
maculatus	1234
microps	1232
nigripinnis	1234
nigromaculatus	1233
pituitosus	1234
saponaceus	1232
subbifrenatus	1233
xanti	1231
Rhytidostomus	168
Riado, Sargo	1321
riali, Onus	2530
Ribband Fish	1490
Ribbon Fish	1489
Ribbon Fishes	1485
ricardi, Mesoprion	1273
ricei, Cottus	1952
Uranidea	1953
richardi, Caranx	923
Hemirhamphus	720
Salmo	483
richardsoni, Acipenser	106
Astronesthes	587
Chauliodus	587
Coregonus	465
Corvina	1484
Cottus	1051
Rhamphocottus	2030
Trachidermis	1944
Uranidea	1952
Richardsonius	228, 230, 238
balteatus	239
lateralis	239
rigidus, Pimelodus	155
rijgersmœi, Ocyrus	1276
rim, Scomber	928
rimator, Bathystoma	1308
Hæmulon	1309
Rimicola	2338

	Page.
Rimicola eigenmanni	2339
muscarum	2338
rimiculus, Catostomus	2792
rimosus, Etropus	2688
Ringed Perch	1023
ringens, Atopoclinus	2376
Balistes	1709
Melichthys	1711
Phænodon	586
Stolephorus	449
Sudis	601
Xanthichthys	1709
Rio Grande Chub	233
Trout	495
riparium, Holocentrum	852
risso, Alpismaris	537
rissoanus, Notacanthus	618
rissoi, Trachurus	910
Rissola	2489
marginata	2489
rivalis, Salmo	509, 510
River Chub	322
Drums	1483
Lampreys	10
Perch	1023
Perch of New York	1135
riverendi, Cyprinodon	673
variegatus	673
Trifarcius	673
rivoliana, Seriola	904
rivularis, Salmo	500
rivulatus, Cirrhites	1491
Cirrhitichthys	1492
Diodon	1748
Serranus	1187
Rivulus	662, 2830
cylindraceus	662, 2830
isthmensis	2830
marmoratus	663, 2830
Roach	250
Roaches	243
roanoka, Etheostoma	1036
Hadropterus	1036
Percina	1036
robalito, Centropomus	1123
Robalito de las Aletas Amarillas	1123
Prietas	1119
Robalo	1118
Prieto	1119
Robalos	1117
roberti, Exocœtus	735
Hemirhamphus	721
Hyporhamphus	721
robertsii, Ceratobatis	2756
Siphostoma	2837
Stolephorus	2815

	Page.
Robin, Flying	2183
Round	907
robusta, Gila	227
Perca	1154
robustum, Moxostoma	193
robustus, Benthocometes	2514
Clypeocottus	1938
Exocœtus	736
Fundulus	644
Hippoglossoides	2616
Ictalurus	135
Leuciscus	228
Mylopharodon	219
Neobythites	2515
Ptychostomus	193
Roccus	1131, 1132
chrysops	1132
comes	1407
lineatus	1113, 1132
saxatilis	1133
striatus	1133
rocheanus, Thynnus	868
rochei, Auxis	868
Scomber	867
Rock	1132
Bass	989, 1197
Common	990
Beauty	1684
Cook	1575, 1576
Hind	1152
Salmon	905
Sea Bass	1201
Shellfish	1722
Sturgeon	106
Trout	1866, 1867, 1872
Rockfish	639, 1026, 1132, 1172
Black	1784
Black and yellow	1825
Black-banded	1827
Brown	1817
Flesh-colored	1824
Grass	1819
Orange	1793
Perca	1133
Red	1805
Spotted	1806
Yellow-backed	1822
Yellow-spotted	1826
Yellow-tail	1781
Rock-fishes	1758, 1765
Rocklings, Four-Bearded	2560
Three-Bearded	2557
Rocky Mountain Bullhead	1949
Trout	487
Whitefish	463
rodo, Pristipoma	1323

	Page.		Page.
Rœboides	338	rosæ, Hemirhamphus	722
guatemalensis	338	Hyporhamphus	721
Rogenia	421	Typhlichthys	2835
alba	422	rosarium, Acipenser	106
rogersi, Urolophus	2752, 2753	Rose-back Parrot	1635
Roller, Stone	181, 204	Rose-fish	1760
Rollers, Sand	783, 784	Rose-fishes	1760
Romero	900	roseipinnis, Notropis	298
Roncadina	1461	roseus, Blennius	2420
Roncador	1457	Centronotus	2420
Little	1460	Cryptotomus	1626
Raiado	1301, 1313	Gunnellops	2420
Red	1456	Luxilus	288
Yellow-finned	1467	Minnilus	288
Roncador stearnsi	1457	Notropis	287
roncador, Umbrina	1467	Prionotus	2159
ronchus, Bairdiella	1436	Rosicola	1765, 1775, 1793
Corvina	1436	rosipes, Novaculichthys	1614
Sciæna	1436	Xyrichthys	1615
Ronco	1436	rosispinis, Raja	2751
Amarillo	1303	rossi, Lycodes	2465
Arará	1304	Salvelinus	510
Blanco	1297	rossii, Salmo	510
Carbonero	1300	rostellum, Acipenser	106
Condenado	1306	rostrata, Albula	412
Prieto	1297	Aldrovandia	609
Ronco	1304	Anguilla	348
Roncos	1291	anguilla	348
Roudanin	959	Limanda	2645
rondeleti, Anthias quartus	1266	Loricaria	157
Carcharodon	50	Macdonaldia	617
Scombresox	726	Muræna	348
Xiphias	894	Platessa	2645
Rondeletia	548	rostratum, Cichlasoma	1522
bicolor	548	Holocentrum	852
rondeletii, Exocœtus	733, 734	rostratus, Agonus	2048
Exonautes	2836	Aspidophorus	2048
Orthragoriscus	1754	Brachyopsis	2046, 2048
Sargus	1364	Canthigaster	1741
Rondeletiidæ	547	Gymnothorax	395
Ronquil	2289	Halosaurus	609
Ronquils	2287	Heros	1523
Ronquilus	2289	Heterostichus	2351
jordani	2239	Holocentrus	849
rosacea, Mycteroperca	1184	Lophius	2737
rosaceus, Cymatogaster	1500	Notacanthus	617
Epinephelus	1184	Phycis	2555
Lutjanus	1267	Pseudoscarus	1659
Mesoprion	1267	Scarus	1658
Micrometrus	1500	Sphagebranchus	373
Paraliparis	2142	Squalus	49
Sebastes	1794	Tetrodon	1742
Sebastodes	1808	Zeus	936
Trisotropis	1184	rostro, Labrus reflexo	1677
Zalembius	1500	rotengulus, Leuciscus	221
rosæ, Clevelandia	2255	rotheus, Leucosomus	222

	Page.
rothrocki, Notogrammus	2440
Poroclinus	2434
rotunda, Mola	1754
Round Bass	988
Herring	420
Pámpano	941
Robin	907
Sting Rays	79
Sunfish	988
Whitefish	465
Round-tail	227
rousseau, Siphostoma	767
Syngnathus	767
Roussettes	22
Rovetto	879
Rovetus temminkii	880
rubella, Sciæna	1418
rubellus, Alburnus	293
Leuciscus	293
Minnilus	293
rubens, Centropomus	1107
Labrus tautoga	1579
Trigla tota	2177
ruber, Apogon	1107
Bodianus	1265
fulvus	1146
Caranx	919
Coregonus	538
Dipterodon	1107
Epinephelus	1181
Gadus	2530
Gymnocephalus	1146
Mycteroperca	1180
Rutilus	300
Scomber	919
Sebastes	1818
Sebastodes	1806
Serranus	1181
ruberrima, Mycteroperca olfax	1183
ruberrimus, Gunnellus	2417
Pholis	2417
Rhodymenichthys	2417
Sebastodes	1805, 1806
rubescens, Conger	355
Exocœtus	734
Hypsinotus	1665
Steinegeria	961
rubicauda, Pomotus	1001
rubicunda, Parma	1565
rubicundus, Acipenser	106
Glyphisodon	1565
Hypsypops	1564, 1565
Pomacentrus	1565
rubiginosus, Gobiesox	2337
Sicyases	2337
rubio, Prionotus	2164

	Page.
Rubio Volador	2164
rubra, Liopropoma	1137
Sciæna	849
rubricroceus, Hybopsis	286
Hydrophlox	286
Minnilus	286
Notropis	286
rubrifrons, Alburnellus	295
Alburnus	295
Ceratichthys	320
Fundulus	653
Hybopsis	320
Leuciscus	295
Minnilus	295
Nocomis	320
Notropis	295
Zygonectes	654, 2829
rubripinna, Cliola	281
Cyprinella	281
rubripinne, Sparisoma	1640
rubripinnis, Argyreus	282
Minnilus	298
Scarus	1640
rubrirostris, Catætyx	2505
rubrivinctus, Sebastichthys	1817
Sebastodes	1817
rubropunctatus, Salarias	2396
Scartichthys	2396
rubrum, Chorististium	1136
Rudder-fish	902, 1387
Rudder-fishes	962, 964, 1380
rudis, Abudefduf	1563
Glyphidodon	1563
rufa, Harpe	1583
Morone	1135
Perca	849
Ruffs, Black	962, 963
rufilineatum, Etheostoma	1079
rufilineatus, Nothonotus	1079
Pœcilichthys	1079
rufipinnis, Exocœtus	735
Exonautes	2836
rufolineatum, Etheostoma	1079
rufus, Balistes	1707
Bodianus	1135, 1583
Centropristes	1199
Cossyphus	1583
Holocentrus ascensionis	840
Labrax	1135
Labrus	1583
Onos	2559
Sebastodes	1786
rugifer, Ophichthus	384
rugispinis, Arius	130
Galeichthys	2781
Hexanematichthys	130

	Page.
rugispinis, Tachisurus	130
phrygiatus.	131
rugosus, Platichthys	2652
Runner	898, 906, 921
Runula	2377
azalea	2377
Runulinæ	2346
rupertianus, Acipenser	106
rupestre, Etheostoma	1073
Xiphidion	2426
rupestris, Ambloplites	990
cavifrons	990
Arbaciosa	2341
Bodianus	990
Coryphænoides	2579
Gadus	2541
Gobiesox	2341
Macrourus	2582
Macrurus	2579
Sebastichthys	1813
Sebastodes	1812
Serranus	1174
Sicyaces	2341
Xiphidion	2426
Rupiscartes	2396
atlanticus	2397
rupiscartes, Moxostoma	196
ruppelii, Monacanthus	1713
ruptus, Chærojulis	1593
Platyglossus	1593
Ruscarius	1908
meanyi	1908
Russian Cat	143
Russians, Muksun of the	464
russula, Scorpæna	1851
Rusty Dab	2644
rutila, Moniana	272
Rutilus	243
amblops	321
anomalus	206
bicolor	244
boucardi	247
compressus	282, 296
melanurus	193
olivaceus	244, 2798
plargyrus	282
ruber	300
storerianus	321
symmetricus	245
thalassinus	245
rutilus, Salmo	509
Tetragonopterus	334
rutteri, Neoliparis	2108
Ruvetto	879
Ruvettus	879
pretiosus	879

	Page.
Ryba, Bielaya	480
Chornia	621
Krasnaya	481
Rypticinæ	1131
Rypticus	1229, 1230
arenatus	1232
bicolor	1231
bistrispinus	1233
coriaceus	1233
nigripinis	1234
saponaceus	1232
xanti	1231
Sabalo	409, 414
Sabara, Bacalhao	2230
sabina, Dasibatis	85
Dasyatis	84
Trygon	85
sabinæ, Notropis	262
Sable	889
saburræ, Chasmodes	2392
Sac-à-Lait	638, 987
Saccopharyngidæ	405
Saccopharynx	405, 406
ampullaceus	406
chordatus	406
flagellum	406
Saccostoma	2249
Sacramento Cat	140
Chub	231
Perch	991
Pike	224, 2796
Salmon	479
Sturgeon	104
Sucker	178
sadina, Clupea	420
Etrumeus	420
sagax, Clupea	423
Sagenichthys	1416
ancylodon	1416, 2859
sagitta, Amblyopus	2263
Etheostoma	1080
Pœcilichthys	1081
Tylosurus	711
Tyntlastes	2263
sagittula, Euctenogobius	2229
Gobius	2228
Saibling	508
saida, Boreogadus	2533
Gadus	2534
Saigneur	1691
Sailfish	167, 890
Sailor's Choice	1297, 1338, 1358
Salar	483
lewisi	493
virginalis	495

	Page.
salar, Salmo	486
ouananiche	487
sebago	487
Salariæ	2377
Salarias	2397
atlanticus	2397
chiostictus	2398
margaritaceus	2399
rubropunctatus	2396
textilis	2400
Salariichthys	2400
textilis	2400
Salariinæ	2346
Salarius vomerinus	2400
Sälbling	508
Saleima	1384
aurata	1386
Salema	1358, 1359, 1385
saliens, Chorinemus	899
Oligoplites	899
palometa	899
Scomber	899
salin, Sparus	1360
sallæi, Algansea	212
Ceratichthys	212
Cliola	212
Hudsonius	212
salmarinus, Salmo	509
Salmo	483, 486
adirondacus	505
agassizii	507
alipes	509
alleghaniensis	507
alpinus	509, 514
nivalis	509
amethystinus	505
arabatsch	483
arcticus	521
arcturus	510
argyreus	480
ascanii	509
aurora	493
bairdii	508
brevicauda	493
callaris	508
campbelli	508
canadensis	507
canis	479
carbonarius	509
carinatus	493
clarkii	492, 2819
bouvieri	2819
gibbsii	2819
henshawi	2819
lewisi	2819
macdonaldi	2819

	Page.
Salmo clarkii pleuriticus	2819
spilurus	2819
stomias	2819
tahoensis	2870
virginalis	2819
clupeiformis	466
colii	509
confinis	505
confluentus	480
consuetus	479
cooperi	483
(Coregonus) harengus	469
lucidus	471
tullibee	473
curilus	508, 2223
dermatinus	479
distichus	509
erythrorhynchos	508
fœtens	538
fontinalis	507
gairdneri	497
beardsleei	2819
crescentis	2821
kamloops	499
shasta	502
stonei	503
gibber	478
gibbsii	493
gloveri	487
gorbuscha	478
grayi	509
grœnlandicus	521
hearnei	510
hoodi	505
hoodii	507, 510
hudsonicus	507
immaculatus	507
irideus	500, 2819
agua-bonita	503
gilberti	502
masoni	501
shasta	502
stonei	503
japonensis	479
kennerlyi	483
keta vel kayko	479
killinensis	509
kisutch	481
kündscha	2823
lævigatus	508
lagocephalus	479
lavaretus muchsun	464
leucomænis	2823
lordii	508
lycaodon	483
mackenzii	474

	Page.
Salmo macrostoma	481
(Mallotus) pacificus	521
malma	508
marstoni	516
mas	487
masoni	501
melampterus	485
milktschitch	481
monestichus	509
muikisi	492
muksun	464
mulleri	574
mykiss	487, 492, 2818
agua-bonita	504
bouvieri	496
clarkii	492
gibbsii	493
henshawi	493
lewisi	493
macdonaldi	497
pleuriticus	496
spilurus	495
stomias	497
virginalis	495
myops	533
naresi	515
nerka	483
nigrescens	507
nitidus	509
nummifer	508
omisco maycus	487
oquassa	515
orientalis	480
(Osmerus) olidus	525
pallidus	505
parkei	508
paucidens	483
peushinensis	508. 2819
perisii	509
proteus	478
purpuratus	492, 499, 2819
bouvieri	496
quinnat	480
richardi	483
rivalis	509, 510
rivularis	500
rossii	510
rutilus	509
salar	486
ouananiche	487
sebago	487
salmarinus	509
salvelinus	509
sanguinolentus	481
saurus	537
scouleri	478, 481

	Page.
Salmo siscowet	505
siskawitz	505
socialis	521
spectabilis	508
stagnalis	510
stellatus	493
striatus	481
symmetricus	505
tapdisma	483
(Thymallus) signifer	518
toma	505
trachinus	533
truncatus	499
trutta	487
tschawytscha	480
tschawytschiformis	478
tsuppitch	481, 495
tudes	508
umbla	509
ursinus	505
utah	495
ventricosus	509
warreni	483
willughbii	509
salmoides, Labrus	1012
Micropterus	1012
Salmon	483
Atlantic	486
Blue-back	481
Chinook	479
Coho	480
Columbia	479
Common Atlantic	486
Dog	478
Family	460
Fraser River	481
Hoopid	480
Humpback	478
Jack	1021
Killer	749
King	479
Landlocked	487
Le kai	478
Namaycush	505
Quinnat	474, 479
Rock	905
Sacramento	479
Saw-qui	481
Silver	480
Trout	497, 2818
Tyee	479
White, of the Colorado	225
salmonea, Ericaria	2816
Lepomis	1011
Perca	1021
Salmonete	858

	Page.
Salmonete Amarilla	859
salmoneus, Chanos	415
Esox	538, 627, 629
Mugil	415
Scombrocottus	1862
Salmonidæ	460
Salmonidea	408
Salmoninæ	461
Salmoperca	783
pellucida	784
Salmopercæ	780, 782
Salpa purpurescens variegata	1271
saltans, Chorinemus	899
saltator, Pomatomus	947
Scomberoides	899
Temnodon	947
saltatrix, Gasterosteus	947
Perca	947, 1388
Pomatomus	946
saludana, Oliola	270
saludanus, Alburnops	270
Notropis hudsonius	270
Salvelini	506
Salvelinus	506
alpinus	508, 2822
alipes	509
arcturus	510
aureolus	511
stagnalis	510
bairdii	508
fontinalis	506
agassizii	507
kundscha	2822
malma	507, 2823
namaycush	505
siscowet	505
naresi	515
nitidus	509
oquassa	514, 515
marstoni	515, 2823
naresi	515
parkei	2823
rossi	510
spectabilis	508
stagnalis	509
salvelinus, Salmo	509
salvini, Cotylopus	2208
Heros	1528
Pimelodus	152
Rhamdia	152
Sicydium	2208
Sicyopterus	2208
San Diego Sole	2707
San Pedro Fish	954
sanctæ crucis, Pseudoscarus	1651
Scarus	1651

	Page.
sanctæ-helenæ, Caranx	908
Decapterus	908
Gymnothorax	397
Lycodontis	397
Muræna	397
sanctæ-luciæ, Corvula	1429
sanctæ-marthæ, Vomer	934
sanctæ-petri, Vomer	934
sanctæ-rosæ, Ulvicola	2413
sancti-laurentii, Engyophrys	2668
sancti-pauli, Holocentrus	853
Sand Dab	2614
Darters	1061, 1062
Diver	535
Eel	833
Launces	831, 832, 833
Pike	1022
Rollers	783, 784
Shark	46
Star-gazers	2297
Sucker	1476
Whiting	1474
Sand-fish	1207, 2295
Sanducha	411
sanguifluus, Nothonotus	1077
Pœcilichthys	1077
sanguinea, Muræna	390
sanguineus, Antennarius	2721
Apodichthys	2412
Holocentrus	1761
Pythonichthys	390
sanguinolentus, Oncorhynchus	481
Pomotis	1003
Salmo	481
Sa-peñ-que	487
sapidissima, Alosa	427, 428
sapidissimus, Coregonus	466
Sapo	2314, 2315, 2316, 2321
Bagre	2319
saponaceus, Anthias	1232
Rhypticus	1232
Rypticus	1232
sara, Cybium	877
saragus, Lepodus	960
Saranus apiarius	1142
Sarchirus	109
argenteus	110
vittatus	110
Sarcidium	302
scopiferum	303
Sarcina	887
Sarcura	59, 60
Sarda	871
chiliensis	872
mediterranea	872
sarda	872

	Page.
sarda, Pelamys	872
Scomber	872
Sardina Blanca	332
Bocona	449
de España	423
de Ley	430
Escamuda	431
Machete	433
pseudohispanica	424
sardina, Harengula	430
Menidia	799
Sardinella	430
Sardinæ	864
Sardine, Califorina	423
Sardinella	428, 429, 2811
anchovia	429
apicalis	429
bishopi	430
clupeola	429
humeralis	431
macrophthalmus	430
sardina	430
stolifera	431
thrissina	430
Sardines, Scaled	428
True	422
Sardinia	422
Sargassum Fish	2716
Sargo	1363
Raiado	1321, 1361
sargoides, Chætodon	1562
Sargosomus	1495
fluviatilis	1496
Sargus	1362
ambassis	1346
argenteus	1363
aries	1362
caribæus	1360
caudimacula	1363
flavolineatus	1360
holbrookii	1363
humeri-maculatus	1360
ovis	1361
pourtalesii	1360
raucus	1364
rhomboides	1358
rondoletii	1364
tridens	1364
unimaculatus	1360
variegatus	1364
vitula	1364
sargus, Diplodus	1363
Sparus	1364
Sarothrodus	1672
amplexicollis	1674
atæniatus	1676

	Page.
Sarothrodus capistratus	1678
maculocinctus	1674
nigrirostris	1674
sedentarius	1675
striatus	1677
Sarritor	2072
frenatus	2073
leptorhyncus	2075
sarritor, Prionotus	2169
Satanoperca	1542
crassilabris	1542
satiricus, Neoclinus	2355
saturna, Sciæna	1456
saturnus, Amblodon	1456
Rhinoscion	1457
Saucer-eye Porgy	1349
Sauger	1022
Sault Whitefish	466
Saurels	909, 910
Saurenchelys	364
Sauries	724, 725
Saurus	533
anolis	535
brevirostris	533
fœtens	538
griseus	537
intermedius	535
limbatus	533
longirostris	538
lucioceps	539
mexicanus	538
myops	533
spixianus	538
synodus	536
truncatus	533
varius	536
saurus, Elops	410
Esox	725
Oligoplites	898
Salmo	537
Scomber	898
Scombresox	725
Synodus	537
Trachurus	911
Saury	725
saussurii, Brama	958
Taractes	957
Sauteur	898, 899
Savalle	409
Savanilla	409
savanna, Brachyconger	360
Muræna	360
Murænesox	360
Savola	889
Saw-belly	426
Sawfish, Common	60

	Page.
Saw-kwey	479
Saw-qui Salmon	481
saxatilis, Abudefduf	1561
Chætodon	1562
Glyphidodon	1562
Gymnothorax ocellatus ..	399
Johnius	1475
Lycodontis ocellatus	399
Menticirrhus	1475
Perca	1133
Platessa	2657
Pœcilichthys	1048
Roccus	1133
saxicola, Pleuronectes	2657
Sebastichthys	1799
Sebastodes	1798
say, Dasyatis	86
Raja	86
sayanus, Aphredoderus	786
Scolopsis	787
sayi, Dasibatis	86
Myliobatis	86
Trygon	86
Sayris	725
bimaculatus	725
hians	725
recurvirostris	725
serratus	726
Scabbard Fish	887, 889
scaber, Antennarius	2722
Centropomus	1125
Chironectes	2723
Hexagrammus	1873
scabra, Trinectes	2701
scabriceps, Minnilus	268, 290
Notropis	290
Photogenis	290
scabripinnis, Tetragonopterus	335
Scad	907
Big-eyed	911
Mackerel	907
Scaled Sardines	428
Scaly-fins	1665
Scamp	1184, 1185
Scaphirhyn'chops platyrrhynchus	107
Scaphirhynchus	107
cataphractus	107
platorynchus	107
platyrhynchus	107
rafinesquei	107
scaphiurus, Opisthognathus	2282
scaphopsis, Cœlorhynchus	2590
Macrurus (Cœlorhynchus)	2591
Scaphyrhynchops	107
scapulare, Pristipoma	1321

	Page.
scapularis, Anisotremus	1320
Tylosurus	711
Scaridæ	1572, 1620
Scarinæ	1621
Scartella	2384
microstoma	2384
Scartes	2395
scartes, Fundulus	654
Scartichthys	2395
rubropunctatus	2396
Scarus	1627, 1642, 1643, 1645
abildgaardi	1635
acutus	1652
alternans	1651
amplus	1635
aracanga	1642, 1647, 1648
atomarius	1631
aureoruber	1635
aurofrenatus	1634
bollmani	1646
brachialis	1641
cæruleus	1652, 1654
catesbœi	1638
catesby	1638
chloris	1637, 1640
chrysopterus	1637
circumnotatus	1641
coccineus	1635
cœlestinus	1656
croicensis	1650
cuzamilæ	1648
diadema	1646
distinctus	1636
emarginatum	1641
emblematicus	1654
erythrinoides	1635
evermanni	1651
flavescens	1640
flavomarginatus	1652
frondosus	1636, 1642
gnathodus	1650
guacamaia	1656, 1658
holocyaneos	1654
hoplomystax	1633
humeralis	1641
insulæ-sanctæ-crucis	1651
lacrimosus	1632
lateralis	1637
loro	1654
maschalespilos	1642
melanotis	1638
miniofrenatus	1634
nuchalis	1654
obtusus	1654
oxybrachius	1635
perrico	1659

Page.

Scarus pleianus 1656
    psittacus 1647
    punctulatus 1645
    quadrispinosus 1648
    radians 1632, 1633
    rostratus 1658
    rubripinnis 1640
    sanctæ-crucis............... 1651
    simplex..................... 1656
    spinidens.................... 1637
    squalidus.................... 1640
    strigatus.................... 1639
    superbus 1650
    tæniopterus 1646
    triolabatus 1654
    trispinosus 1648
    truncatus ...:............... 1641
    turchesius 1658
    vetula 1647, 1649
    virens 1640
    virginalis 1647
    viridis 1638
scepticus, Minnilus 296
    Notropis................. 296
    Triglops................. 1925
Schedophilinæ 969
Schedophilopsis 972
    spinosus 972
Schedophilus...................... 970
    enigmaticus.......... 972
    medusophagus 970
Schilbeodes 144, 145, 146
    eleutherus 148
    exilis 147
    funebris 147
    furiosus 149
    gilberti 148
    gyrinus................. 146, 2790
    insignis................ 147
    leptacanthus 146
    miurus 148
    nocturnus 146
Schistorus 1148, 1151
    mystacinus 1151
schlegeli, Orcynus.................... 870
    Priacanthus............. 2858
    Sebastodes 1834
schmidtii, Hoplunnis............... 361
schmittii, Balistes................... 1705
schneideri, Chauliodus............. 585
    Ophichthys 387
    Pœcilia................. 691
schœpfii, Alutera 1718
    Balistes 1718
    Ceratacanthus............ 2860
    Chylomycterus........... 1748

Page.

schœpfii, Diodon.................... 1748
scholaris, Thyrsites 880
schomburgki, Pempheris 978
schoneveldii, Cataphractus 2067
Schoolmaster....................... 1258
schranki, Hæmulon.............. 1302, 1303
schumardi, Etheostoma 1047
Schuylkill Cat...................... 140
Sciadeichthys 119, 120, 122, 2757
    albicans 124, 2760
    emphysetus 122, 2759
    flavescens 123, 2760
    mesops............. 123, 2760
    passany............. 124, 2760
    proops 123, 2760
    temminckianus 122, 2760
    troscheli 122, 2757
Sciades troscheli 122, 2758
sciadicus, Fundulus................. 654
    Haplochilus 654
    Zygonectes 654
Sciæna........................ 1454, 1465
    acuminata 1488
    aluta................... 1438
    amazonica 1419
    caprodes 1027
    chrysoleuca.............. 1439
    chrysura................ 1434
    coro.................... 1324
    croker 1462
    (Corvina) adusta........... 1448
    crouvina 1419
    deliciosa 1455
    edwardi................ 1490
    ericymba................ 1445
    ensifera 1435
    furthi.................. 1441
    fusca................... 1483
    gigas................... 1483
    grisea.................. 1484
    heterolepis.............. 1419
    icistia.................. 1436
    imberbis 1454
    imiceps................. 1451
    jacobi 1457
    lanceolata.............. 1444
    lineata 1133, 1460
    macrops................ 1428
    maculata................ 2198
    magdalenæ.............. 1420
    multifasciata............. 1459
    obliqua................. 1459
    ocellata................. 1454
    opercularis.............. 1461
    ophioscion 1448
    oscitans 1441

	Page.
Sciæna oscula	1484
plumieri	1324
punctata	1434
ronchus	1436
rubella	1418
rubra	849
saturna	1456
sciera	1452
squamosissimus	1418
stellifer	1444
(Stelliferus) stellifer :	1443
surinamensis	1420
typica	1448
undecimalis	1119
vermicularis	1452, 1453
xanthurus	1459
Sciænidæ	1392
Sciænops	1453
ocellatus	1453
sciera, Sciæna	1452
scierum, Etheostoma	1038
scierus, Hadropterus	1037
serrula	1038
Ophioscion	1452
Scirenga	1180
scirenga, Mycteroperca	1181
scituliceps, Synodus	537, 2826
scitulus, Prionotus	2157
sciurus, Diplectrum	1204
Hæmulon	1303
Serranus	1204
Sparus	1304
Sciæninæ	1394
Sclerodermi	781, 1697
Sclerognathus	163
cyprinella	164
elongatus	169
meridionalis	164
urus	165
scofieldi, Stolephorus	2814
Scolecosoma	10
Scoliodon	42
longurio	42, 2748
terræ-novæ	43
scolopacea, Nemichthys	369
scolopaceus, Nemichthys	369
scolopax, Balistes	759
Centriscus	759
Macrorhamphosus	759
Scolopsis sayanus	787
Scomber	865
adscensionis	927
alatunga	871
albacores	870
alleteratus	869
ascensionis	925

	Page.
Scomber balantiophthalmus	911
bisus	867
carangus	920
chloris	938
chrysurus	938
colias	866, 2843
crümenophthalmus	911
crysos	921
dekayi	867
dentex	927
diego	867
fasciatus	904
filamentosus	932
germo	871
gracilis	867
grex	867
guara	927
gunneri	955
beberi	923
hippos	908, 920
kœlreuteri	900
lacertus	867
latus	938
macrophthalmus	867
maculatus	867, 874
mediterraneus	872
niger	948
pelagicus	952
pelamides	869
pelamis	869
pelamitus	872
pelamys	872
plumieri	911
pneumatophorus	867
quadripunctatus	869
regalis	875
rim	928
rochei	867
ruber	919
saliens	899
sarda	872
saurus	898
scombrus	865
sloanei	870
speciosus	928
thazard	867
thynnus	870
trachurus	910
undulatus	867
vernalis	866
zonatus	902
scomberius, Esox	626
Scomberodon	873
Scomberoides, saltator	899
Coryphæna	953
Scomberomorus	873, 4843

	Page.
Scomberomorus caballa	875, 876
concolor	873
maculatus	874, 875, 4843
plumieri	875
regalis	875
sierra	874
Scombresocidæ	724
Scombresox	725
brevirostris	726
camperi	725
equirostrum	726
fosteri	726
rondeleti	726
saurus	725
scutellatum	726
storeri	726
Scombridæ	863
Scombrinæ	864
scombrinus, Caranx	908
Decapterus	908
Scombrocottus	1861
salmoneus	1862
Scómbroidei	781, 860
Scombroides occidentalis	899
Scombroidinæ	896
Scombropinæ	1106
Scombrops	1114
oculatus	1114
scombrus, Scomber	865
Scopelus	569
affinis	571
andreæ	569
arcticus	574
benoiti	573
bonapartii	557
boops	572
borealis	577
caninianus	570
caudispinosus	556
coccoi	569
crassiceps	843
crocodilus	558
elongatus	555
gracilis	572, 574
humboldti	572, 577
hygomii	573
kroyeri	556
madeirensis	557
maurolici	577
mulleri	570, 574
pseudocrocodilus	556
rafinesquei	567
rarus	569
spinosus	575
tenorei	577
scopifer, Notropis	291

	Page.
scopifer, Phenacobius	303
scopiferum, Sarcidium	303
scopiferus, Phenacobius	303
scops, Gnathypops	2283
Scorfanudi Funal	1837
Scorpæna	1839
africana	1833
agassizii	1840, 2860
americana	2023
brasiliensis	1842
bufo	1849
calcarata	1854
capensis	1833
castor'	1856
cristulata	1841
dactyloptera	1837
grandicornis	1850
guttata	1847
histrio	1843, 1846
inermis	1853, 2861
massiliensis	1139
mystes	1849
nematophthalmus	2861
occipitalis	1854
pannosa	1845
plumieri	1848
porca	1839
rascacio	1849
russula	1851
sierra	1860
sonoræ	1852
stearnsi	1843
Scorpænichthyinæ	1880
Scorpænichthys	1889
lateralis	1902
marmoratus	1889
Scorpænidæ	1758
Scorpæninæ	1759
Scorpene	1847
scorpio, Cottus	1973
scorpioides, Cottus	1973
Myoxocephalus	1973
Scorpion	1847
Fishes	1839
Scorpis californiensis	1391
Scorpius virginianus	1976
scorpius, Cottus	1974
Cottus grœnlandicus	1975
scouleri, Oncorhynchus	478
Salmo	478, 481
Scour Fish	879
scovelli, Etheostoma	1082
Siphostoma	769
scripta, Alutera	1719
scriptus, Balistes	1719
Ceratacanthus	2860

	Page.
scriptus, Gymnothorax	398
Monacanthus	1719
Pseudoscopelus	2292
scrutator, Belone	714
Isesthes	2389
scudderi, Diabasis	1300
Hæmulon	1299, 1300
Sculpin	1847
Arctic	1973
Black	1985
Daddy	1974
European	1974
Great	1976
Long-spined	1976
Red	1935
Yellow	1934
Sculpins	1879
Great	1970
Spineless	2025
Stone	1937
Scup, Common	1346
Scuppaug	1346
scutata, Echeneis	2271
scutellatum, Scombresox	726
Scutica	403, 404
scuticaris, Bascanichthys	378
Cœcula	379
Sphagebranchus	379
scutiger, Icelus	1910
Rastrinus	1909
scutum, Achirus	2700
Solea	2700
scylla, Exocœtus	735
Hybopsis	263
Notropis	263
Scylliorhinidæ	22
Scylliorhininæ	22
Scylliorhinus	22
profundorum	22
retifer	25
Scyllium	22
retiferum	25
ventriosum	25
Scymnoid Sharks	56
Scymnus brevipinna	57
Scyphius	774
Scyris	931
Scyris analis	932
Scytalichthys	387
miurus	387
Scytalina	2454
cerdale	2454
Scytalinidæ	2453
Scytaliscus	2454
Scytalophis	381, 384
magnioculis	385

	Page.
Sea Bass	1126
Bats	2736
Catfish	118, 119, 128
Devil	91, 92, 2727
Drums	1482
Mink	1475
Poacher	2031, 2065, 2091
Raven	976, 2622, 2023
Serpent	384
Snail	2105, 2114, 2116
Snipe	714
Trout	1407
Spotted	1409
Sea-horse	775
Common American	777
Sea Robin, Brown-winged	2167
Red-winged	2156
sebago, Salmo salar	487
Sebastapistes	1839
guttatus	1848
Sebastes	1760
auriculatus	1818
capensis	1833
caurinus	1821
dactylopterus	1837, 1838
darwini	1832
elegans	1830
elongatus	1816
fasciatus	1761, 1827
helvomaculatus	1808
imperialis	1837, 1838
inermis	1829
marinus	1760, 1761
viviparus	1761
melanops	1783
nematophthalmus	2861
nigrocinctus	1828
nivosus	1834
norwegicus	1761
oblongus	1830
oculata	1832
paucispinis	1781
regulus	1761
rosaceus	1794
ruber	1818
septentrionalis	1761
steindachneri	1830
taczanowskii	1832
variabilis	1784
ventricosus	1829
viviparus	1761
Sebastichthys	1765, 1777, 1827
atrovirens	1798
aurora	1803
brevispinis	1788
carnatus	1825

	Page.
Sebastichthys chlorostictus	1812
chrysomelas	1826
constellatus	1807
diploproa	1802
entomelas	1786
fasciolaris	1827
flavidus	1782
goodei	1780
introniger	1805
levis	1816
maliger	1823
miniatus	1795
mystinus	1785
nigrocinctus	1828
ovalis	1789
pinniger	1794
proriger	1788, 1793
purpureus	1826
rastrelliger	1820
rhodochloris	1810
rubrivinctus	1817
rupestris	1813
saxicola	1799
serriceps	1827
sinensis	1814
umbrosus	1807
vexillaris	1822
Sebastinæ	1759, 1771
Sebastodes	1765, 1773, 1778
æreus	1807
aleutianus	1795
alutus	1790
atrorubens	1796, 2860
atrovirens	1797, 2860
auriculatus	1817, 1818
dallii	1818
aurora	1802
ayresii	1808
brevispinis	1787
capensis	1833
carnatus	1824
caurinus	1821
chlorostictus	1811
chrysomelas	1825, 1826
ciliatus	1783, 1784
constellatus	1806
crameri	1799, 2860
darwini	1832
diploproa	1801
eigenmanni	1789
elegans	1830
elongatus	1815
entomelas	1785
eos	1810
flavidus	1781
gilberti	1823
gilli	1811

	Page.
Sebastodes glaucus	1777
goodei	1779
hopkinsi	1789, 2860
inermis	1829
introniger	1805
jordani	1778
joyneri	1829
levis	1816
macdonaldi	1786
maliger	1822
matsubaræ	1796, 1833, 2860
melanops	1782, 1783
melanostomus	1803
miniatus	1794
mitzukurii	1831
mystinus	1784, 1785
nebulosus	1826
nigrocinctus	1827
nivosus	1833
oblongus	1830
oculatus	1832
ovalis	1788
paucispinis	1780
pinniger	1793, 1794
proriger	1787, 1792
rastrelliger	1819, 1820
rhodochloris	1809
rosaceus	1808
ruber	1806
ruberrimus	1805, 1806
rubrivinctus	1817
rufus	1786, 2860
rupestris	1812
saxicola	1798
schlegelii	1834
semicinctus	1800
serranoides	1782
serriceps	1827
sinensis	1813
steindachneri	1830
taczanowskii	1831, 2860
trivittatus	1834
umbrosus	1807
ventricosus	1829
vexillaris	1821
vulpes	1835
zacentrus	1814, 2860
Sebastolobus	1761
alascanus	1761
altivelis	1763
macrochir	1763
Sebastomus	1765, 1775, 1805
Sebastoplus	1854
dactylopterus	1837
Sebastopsis	1835
xyris	1835

	Page.
Sebastosomus	1765, 1774, 1781
pinniger	1794
simulans	1783
seco, Luxilus	250
Sectator	1389
ocyurus	1389
sectatrix, Kyphosus	1387
Perca	1388
marina	1388
secundo-dorsalis, Thynnus	870
secundus, Carangops	914
Caranx	914
Hemicaranx	914
sedentarius, Chætodon	1675
Sarothrodus	1675
seemanni, Arius	128, 2772
Galeichthys	2772
Hexanematichthys	128
Tachisurus	129
segaliensis, Brachyopsis	2048
Siphagonus	2048
Syngnathus	2048
Segundo	914
Selache	51
maxima	51
Selachii	15
selachops, Apterichthys	374
Ichthyapus	374
Sphagebranchus	374
Selachostomi	100
Selachus maximus	51
Selanonius	49
walkeri	49
selanonus, Squalus	49
Selar	916, 918
Sclaroides	916
Selenaspis	119
Selenaspis	120, 124, 2760
dowi	125
dowii	2761
herzbergii	124, 2760
luniscutis	125
parkeri	125, 2764
Selene	935
argentea	936
œrstedii	935
quadrangularis	1668
setipinnis	934
vomer	936
selene, Carpiodes	167
Luxilus	269
Minnilus	269
Notropis hudsonius	269
selenops, Hiodon	414
sellaris, Acanthocottus	1998
Porocottus	1996

	Page.
sellicauda, Epinephelus	1155
sellifer, Halichæres	1592
Iridio	1592
sem, Caranx	923
Sema	1498
signifer	1499
semiarmatus, Gasterosteus	747
semicinctus, Gillellus	2298
Halichæres	1593
Iridio	1592
Julis	1593
Platyglossus	1593
Sebastodes	1800
semicoronata, Seriola	904
Semicossyphus pulcher	1585
semifasciatum, Pileoma	1027
Triakis	31
semifasciatus, Serranus	1197
semiloricatus, Gasterosteus	747
semiluna, Sparus	1276
seminolis, Fundulus	647
seminuda, Albula	412
Eleotris	2204
Garmannia	2233
Gila	228
Gymneleotris	2204
seminudus, Gobius	2234
Gymneleotris	2204
Lycodes	2468
semiradiatus, Lepisosteus	110
semiruber, Labrus	1583
semiscaber, Cottopsis	1950
Cottus	1949
semiscabra, Cottus centropleura	1945
Uranidea	1950
semispinosus, Caranx	911
Semitapicis	332
Semotilus	220, 221, 222
atromaculatus	222
thoreauia- nus	223
biguttatus	322
bullaris	222
cephalus	222
corporalis	221, 222
diplemius	222
dorsalis	222
hammondi	222
macrocephalus	222
notatus	659
speciosus	222
thoreauianus	223
senegalensis, Vomer	934
senilis, Gambusia	682
Sennet	826
Señorita	1592, 1601, 2352

	Page.
senta, Raja	71
septentrionalis, Gaidropsarus	2559
Motella	2560
Onos	2560
Perca	1133
Sebastes	1761
septipinnis, Ammodytes	2842
Rhynchias	2841
Septogunnellus gracilis	2436
serena, Dionda	214
Hybognathus	214
Sergeant Fish	947, 948
Major	1561
Seriola	901
argyromelas	950
bipinnulata	907
bonariensis	905
boscii	905
coronata	905
cosmopolita	938, 2847
declivis	905
dorsalis	902
dubia	905
dumerili	903, 904
dussumieri	900
falcata	905
fasciata	904
gigas	903
lalandi	902, 903
ligulata	905
mazatlana	904
picturata	910
pinnulata	907
proxima	904
rivoliana	904
semicomata	904
stearnsii	903
succincta	900
zonata	902
carolinensis	902
Seriolichthys	906
Seriolinæ	896
Seriolophus	895
carangoides	895
Seriphus	1397
politus	1397
serotinus, Acipenser	106
serpens, Gempylus	884
Serpens marinus maculosus	382
Serpent, Sea	384
serpentina, Muræna	348
serpentinus, Blennius	2439
Derichthys	343
Leptoblennius	2439
Serra	597
serra, Alepidosaurus (Caulopus)	597

	Page.
serra, Alepisaurus	597
Gonenion	947
Serran Imperial	1837
Serrana	1489, 1490
Hispana	1488
hispanis	1489
Serranidæ	1126
Serraninæ	1129
Serrano	1207
serranoides, Sebastodes	1782
Serranos	1208
Serranus acanthophorus	1196
acutirostris	1181
æquidens	1211
agassizii	1189
albomaculatus	1197
angustifrons	1159
annularis	1214
apua	1158
arara	1159, 1175
armatus	1165
aspersus	1153
atrarius	1200
auratus	1145
auriga	1221
bivittatus	1205
bonaci	1175
brasiliensis	1221
brunneus	1175
bulleri	1214
calopteryx	1213
capeuna	1311
carauna	1146
cardinalis	1174
camelopardalis	1187
catus	1159
chlorurus	1193
clathratus	1198
colonus	1222
conspersus	1156
coronatus	1142
couchii	1139
courtadei	1152
creolus	1222
cyclopomatus	1175
decimalis	1175
dimidiatus	1179
dubius	1146
emarginatus	1181
erythrogaster	1160
falcatus	1185
fascicularis	1208
felinus	1187
fimbriatus	1154
flaviventris	1221
furcifer	1222

	Page.		Page.
Serranus furvus	1200	Serranus subligarius	1219
fusculus	1211	tabacarius	1215
fuscus	1181	tæniops	1144
galeus	1164	tigrinus	1214
gigas	1154	tigris	1187
guasa	1164	tinca	1181
guativere	1145	trifurcus	1201, 1202
humeralis	1197	undulosus	1181
impetiginosus	1153	unicolor	1192
inermis	1168	varius	1153
interstitialis	1179	Serraria	1028, 1030, 1037
irradians	1208	serrata, Fistularia	758
itaiara	1164	serraticornis, Balistes	1720
jacome	1215	serratogranulata, Perca	1024
labriformis	1155	serratum, Hæmulon	1299
lamprurus	1190	serratus, Gasterosteus	750
latepictus	1175	Naucrates	900
luciopercanus	1216	Sayris	726
lunulatus	1159	serriceps, Sebastichthys	1827
maculatofasciatus	1196	Sebastodes	1827
maculatus	1153	serrifer, Conodon	1324
maculosus	1159	Serrivomer	367
margaritifer	1156	beanii	367
marginatus	1154	serrula, Anisotremus	1323
meutzeli	1154	Chalinura	2576
morio	1160	Hadropterus scierus	1038
mystacinus	1151	Priacanthus	1239
nebulifer	1196	Pristipoma	1324, 1343
nigrescens	1200	Pseudopriacanthus	1239
nigriculus	1153	Seserinus xanthurus	966
nigritus	1161	Sesi de lo Alto	1261
niveatus	1156	sessilicauda, Monolene	2691
oculatus	1283	Setarches	1860
ocyurus	1201	parmatus	1860
olfax	1183	setifer, Argyriosus	936
ongus	1154	Monacanthus	1716
ouatalibi	1146	Stephanolepis	1716
panamensis	1141	setiger, Dasycottus	1991
philadelphicus	1202	setigerus, Lophiomus	2714
phœbe	1212	Lophius	2715
pixanga	1153	setinotus, Chauliodus	585
præstigiator	1214	setipinnis, Argyreiosus	934
psittacinus	1213	Caranx dorsalis	934
quinquefasciatus	1164	gabonensis	935
radialis	1205	Selene	934
radians	1208	Vomer	934
remotus	1160	Zeus	934
repandus	1187	setosus, Mugil	815
rivulatus	1187	Ovoides	1739
ruber	1181	Tetraodon	1740
rupestris	1174	seu conger, Muræna brasiliensis	403
sciurus	1204	sexcornutus, Ostracion	1725
semifasciatus	1197	sexdecemlamellata, Echeneis	2272
stadthouderi	1159	sexfasciatum, Hæmulon	1294
stilbostigma	1217	sexfasciatus, Hæmulon	1295
striatus	1157	sexmaculatus, Diodon	1746

	Page.
sexradiatus, Polynemus	2183
Shad	427
Alabama	2810
American	427
Broad	1372
Common	427
Gulf	2810
Hickory	416, 425
Mackerel	909
Potomac	427
Shad Porgy	1355
Shad-waiter	465
Shads, Gizzard	415
Shark, Bay	37
Bone	51
Bullhead	20
Cat	31
Cow	19
Dusky	35
Elephant	51
Great Blue	33
White	50
Hammer-headed	45
Leopard	31
Long-tail	45
Mackerel	48
North River	427
Nurse	26
Oil	32
Sand	47
Sharp-nosed	43
Shovel-head	44
Shovel-nosed	18
Sleeper	57
Soup-fin	32
Swell	25
Tiger	32
Shark Pilot	902
Sucker	2269
Sharks	15
Angel	58
Basking	50
Blue	33
Bramble	57
Bullhead	19
Cat	22
Cestraciont	19
Cow	17
Cyclospondylous	52
Dog	28
Frilled	16
Hammer-headed	43
Mackerel	47
Man-eater	50
Notidanoid	16
Nurse	25

	Page.
Sharks, Requiem	27
Sand	46
Scymnoid	56
Thresher	45
True	21
Typical	19
Whale	52
Sharp-nosed Flying-fish	728
Puffers	1740
Shark	43
Sharp-tailed Goby	2229
shasta, Cottus	1947
Salmo gairdneri	502
irideus	502
shavianus, Cetorhinus	51
Sheepshead	1358, 1361
Minnow	671
Porgy	1354
Sheepshead, Lake	1484
Shellfish	1723
Shellfish, Rock	1722
Shi Shidai	1665
Shidai, Shi	1665
Shima Soi	1834
Shiner	269, 281
Blunt-nosed	934
Golden	250
Red-sided	240
Spotted	318
Shiners	254
Short-nosed Bat-Fish	2738
Gar	110
Sturgeon	106, 107
shufeldti, Ceratius	2731
Gobius	2221
Mancalias	2730
Typhlopsaras	2731
shumardi, Alburnops	268
Boleosoma	1047
Cottogaster	1046
Hadropterus	1047
Imostoma	1047
Minnilus	268, 269
Notropis	268
shuswap, Agosia	313
falcata	313
sialis, Argentina	526
sialis, Corvula	1428
Noturus	146
sibbaldi, Syngnathus	774
Siboma	228, 231
atraria	233
crassicauda	231
longiceps	233
sicana, Cerna	1162
siccifer, Holocentrus	849

	Page.
sicciferum, Holocentrum	850
sicculum, Chirostoma	806
sicculus, Labidesthes	805
siculus, Lampugus	953
Sicya	2207, 2867
Sicyases	2329, 2330, 2336
carneus	2337
fasciatus	2338
punctulatus	2338
rubiginosus	2337
rupestris	2341
Sicydiinæ	2190
Sicydium	2205
antillarum	2206, 2867
gymnogaster	2208
plumieri	2206, 2867
punctatum	2867
salvini	2208
siragus	2206
vincente	2207
Sicyogaster	2329
Sicyopterus gymnogaster	2208
salvini	2208
Sicyosus	2867
Sidera	392
castanea	396, 2804
chlevastes	399
dovii	397
funebris	396
mordax	396
moringa	395
nigromarginata	400
ocellata	399
panamensis	391
pfeifferi	2805
verrilli	394
vicina	394
siderea, Muræna	2805
siderium, Zophendum	314
siderius, Hyborhynchus	314
sieboldii, Cichlasoma	1516
Heros	1517
Sierra	874, 875
Sierra, Pez	60
sierra, Pontinus	1859
Scomberomorus	874
Scorpæna	1860
Sierrita	713
sierrita, Tylosurus	713
Sigmistes	2863
caulias	2863
Sigmops	581, 582
stigmaticus	583
Sigmurus	1446, 1447, 1452
Signalosa	2809
atchafalayæ	2809
signatus, Bathymaster	2288
Hypoprion	41
Microgobius	2246
signifer, Bryttus	996
Chatoessus	433
Coregonus	518
Monacanthus	1716
Salmo (Thymallus)	518
Sema	1499
Stypodon	220
Thymallus	517, 2871
lewisi	2871
montanus	519
ontariensis	519
tricolor	519
sigolutes, Gilbertina	2028
silenus, Zaprora	2850
Silk Snapper	1262, 2858
Siluridæ	115
Silurus bagre	117
catus	138
clarias	155
cornutus	759
cupreus	140
felis	128
furcatus	140
gyrinus	146
herzbergii	125
limosus	143
lividus	140
marinus	118
melas	141
nebulosus	143
olivaris	143
parkeri	126
punctatus	135
viscosus	143
xanthocephalus	141
Silus	525
ascanii	526
silus, Argentina	526
Silver Chub	221, 320
Hake	2530
Jenny	1370
Salmon	480
Trout	493
Whiting	1477
Silver-fin	278
Silver-fish	409, 795, 889
Silverside	800
Brook	805
Silver-sided Minnow	238
Silversides	788, 796
Silvery Anchovies	439
Lamprey	11
Minnow	213

	Page.
sima, Cliola	267
Eleotris	2198
Simenchelyidæ	348
Simenchelys	349
parasiticus	349
similis, Amphistichus	1504
Fundulus	638
Hæmulon	1304
Hydrargyra	639
simillima, Palometa	2849
simillimus, Poronotus	967
Rhombus	967
Stromateus	967
simotera, Ulocentra	1051
simoterum, Hyostoma	1051
simplex, Aplurus	880
Pseudoscarus	1656
Scarus	1656
Tetragonurus	880
simpsoni, Ictalurus	135
simula, Chalinura	2578
simulans, Enneacanthus	994
Hemioplites	994
Sebastosomus	1783
simulus, Macrurus	2578
Ophioscion	1449
sinus, Alburnellus	627
Careproctus	2131
Menticirrhus	1472
Notropis	267
Rhinichthys	307
sinaloæ, Paralichthys	2872
Siphostoma	2838
Umbrina	1468
sinesis, Sebastichthys	1814
Sebastodes	1813
Singing Fish	2321
sinuatus, Merluccius	2530
Siphagonus	2046
barbatus	2050
segaliensis	2048
Siphateles	243
vittatus	244
Siphostoma	761, 763, 2837
affine	769, 770
albirostre	772
arctum	771, 2838
ascendens	768
auliscus	767
bairdianum	765, 770
barbaræ	765
brachycephalum	769
californiense	764
carinatum	763
cayennense	772
crinigerum	771

	Page.
Siphostoma elucens	768
fistulatum	765
flavirostre	768
floridæ	766
fuscum	770
griseolineatum	764
jonesi	768, 2837
leptorhynchum	764
linea	768
louisianæ	770
mackayi	766
marmoreum	768
pelagicum	767, 2837
picturatum	768
poeyi	766
punctipinne	763
robertsi	2837
rousseau	767, 2837
scovelli	769
sinaloæ	2838
starksi	771, 2838
zatropis	772
siragus, Sicydium	2206
Sirajo	2206
Sirembo guntheri	2523
sisco, Argyrosomus artedi	469
Sisco of Lake Tippecanoe	469
siscowet, Cristivomer namaycush	505
Salmo	505
Salvelinus namaycush	505
siskawitz, Salmo	505
siuslawi, Leuciscus	2797
Skate, Barndoor	71
Big	68, 72
Common	68
Little	68
Skates	15, 66
Skeponopodus	891
guebucu	891
typus	892
skib, Pomatomus	947
Skil	1862
Skil-fishes	1861
Skimback	167
Skipjack	425, 805, 872, 946
Skipper	725
Skowitz	480
Sleeper	2194, 2200, 2216
Shark	57
Slimer	2315
Slippery Dick	1595
Sole	2655
sloanei, Chauliodus	585
Scomber	870
sloani, Leiobatus	81
Raja	81

	Page.
Small Black Lamprey	13
Blindfish	704
Catfish	140, 141
Dolphin	953
Small-mouthed Black Bass	1011
Buffalo	164
Small-scaled Gurnards	2175
Smaragdus	2210
costalesi	2225
stigmaticus	2224
valenciennei	2228
smaragdus, Eleotris	2204
Gobionellus	2228
Gobius	2227
Smaris	1364
lineatus	1378
martinicus	1365
Smear Dab	2653, 2654
Smecticus	1229
bicolor	1232
Smelt, American	523
California	806
Cobessicontic	524
Kodiak	2823
Little	807
Pond	525
Wilton	523
Smelt of the New York Lakes	468
Smelts	519, 522
Surf	524
smiridus, Merluccius	2530
smithi, Carcharodon	50
smithii, Cliola	253
Cyprinus (Abramis?)	413
Smooth Cabezon	2012
Hound	29
Puffer	1728
smyrnensis, Gobius	2118
Snail, Sea	2105, 2114, 2116
Snake Blennies	2435, 2438
Eels	372
Mackerel	883
Snap Mackerel	946
Snapper	1760
Black-fin	1261
Dog	1257
Gray	1255
Lane	1270
Mahogany	1272
Mangrove	1255
Red	1264
Red-tail	1270
Silk	1262, 2858
Snappers	1241, 1247
Snipe Eel	366, 369
Snipe, Sea	714

	Page.
Snipefishes	758
Snip-nose Mullet	964
Snook	1118
Snub-nosed Eels	348
snyderi, Catostomus	2792
Gnathypops	2285
Oligocottus	2871
Soapfish	538, 1229, 1232
Sobaco	1705, 1706
sobaco, Balistes	1706
Canthidermis	1705
Sobacos	1705
sobra, Mesoprion	1266
socialis, Salmo	521
socius, Alburnus	292
Notropis	292
socorroense, Thalassoma	1608, 2859
socorroensis, Chlorichthys	1607
Soft Flounders	2679
sogo, Holocentrus	849
Soi, Aka	1830
Goma	1833
Kuro	1834
Shima	1834
solandri, Acanthocybium	876
Cybium	877
Gempylus	883
solaris, Orthragoriscus	1754
Soldado	848
Soldier Fish	1088
Sole, American	2700
California	2613
Long-finned	2658
Mexican	2698
San Diego	2707
Slippery	2655
Solea	2660
achirus	2702
browni	2701
cynoglossa	2657
fimbriata	2700
fischeri	2700
fonsecensis	2699
gronovii	2696
inscripta	2696
klunzingeri	2697
maculipinnis	2698
mazatlana	2699
panamensis	2702
pilosa	2699
reticulata	2696
scutum	2700
solea, Caranx	927
Soleidæ	2692
Solenostomus	754, 756
soleœformis, Aramaca	2672

	Page.
soleœformis, Hemirhombus	2672
Rhombus	2672
Soleotalpa	2702
unicolor	2703
Soles	2692
American	2693
solis, Lepomis auritus	1001
Pomotis	1001
Somka	2818
Somniosinæ	56
Somniosus	56
brevipinna	57
microcephalus	57
somnolentus, Eleotris	2198
Lobotes	1236
sonntagii, Chironectes	2717
soñoræ, Scopæna	1852
sonoriensis, Girardinus	689
soporator, Gobius	2216
Sorcerers	364
sordida, Muræna	403
Orthopsetta	2680
sordidus, Citharichthys	2679
Psettichthys	2680
Verilus	1284
Soup-fin Shark	32
Southern Flounder	2630
Porgy	1346
Puffer	1732
Sting Ray	86
Striped Gurnard	2168
Spade-fish	101, 1666, 1668
spadicea, Dionda	216
Lampetra	13
spallanzani, Isurus	49
Lamna	49
Leptocephalus	354
Oxyrhina	49
Spanish Flag	1139, 1140, 1817
Hog-fish	1583
Lady-fish	1583
Mackerel	874
of England	866
Sparactodon	946
nalnal	947
Sparada	1498
Sparidæ	1343
Spariform Percoids	1241
Sparinæ	1343
Sparisoma	1625, 1627, 1630
abildgaardi	1635
aracanga	1642
atomarium	1631
aurofrenatum	1634
brachiale	1641
catesbœi	1638

	Page.
Sparisoma catesbyi	1638
chrysopterum	1636, 1637
cyanolene	1633
distinctum	1635, 1636
emarginatum	1641
flavescens	1639, 1640
frondosum	1641, 1642
hoplomystax	1632
lacrimosum	1632
laurito	1637
maschalespilos	1641
niphobles	1633
oxybrachium	1634
radians	1631
rubripinne	1640
strigatum	1639
viride	1638
xystrodon	1630
Sparisomatinæ	1621
sparoides, Labrus	987
Pomoxis	987
Sparopsis	1279
Sparus	1356, 1361
argenteus	1357
argyrops	1346
atlanticus	1153
aureus	1010
bajonado	1352
brachysomus	1353
castaneola	960
caxis	1259
chrysomelanurus	1157
chrysops	1346
chrysurus	1276
cruentatus	1142
falcatus	1585
milneri	1355
niger	960
oblongus	2276
orbitarius	1350
ovicephalus	1361
pagrus	1357
probatocephalus	1361
radiatus	1596
raii	960
rhomboides	1358
salin	1360
sargus	1364
sciurus	1304
semiluna	1276
synagris	1271
tetracanthus	1257
vermicularis	1271
virginicus	1323
vittatus	1323
xanthurus	1346

	Page.
spathula, Polyodon	101
Squalus	102
spatula, Lepisosteus	111
Planirostra	102
Spatularia	101
reticulata	102
Spawn-eater	269
Spearfish	167, 891, 892
Spearing, Ground	533
speciosa, Gambusia	681
speciosus, Caranx	928
Gnathodon	928
Pomotis	1006, 1008
Scomber	928
Semotilus	222
Speck	1047
Speckled Hind	1159
Trout	508
of Lake Crescent	2821
spectabile, Etheostoma cœruleum	1089
spectabilis, Pœcilichthys	1089
Salmo	508
Salvelinus	508
spectrum, Careproctus	2133
Lophius	2723
Osmerus mordax	523
spectruncula, Cliola	265
spectrunculus, Hybopsis	265
Notropis	265
speculiger, Exocœtus	734
Exonautes	2836
speculigera, Lampadena	561
spelæus, Amblyopsis	706
spengleri, Spheroides	1732, 1733
Tetrodon	1733
Spet	826
spet, Esox	826
Sphyræna	826
Sphærina	822
Sphæroides furthi	1737
politus	1736
trichocephalus	1738
tuberculatus	1733
Sphagebranchus	373
anguiformis	374
kendalli	375
rostratus	373
scuticaris	379
selachops	374
teres	379
sphenops, Pœcilia	694
Spheroides	1729, 1731
angusticeps	1731
annulatus	1735
politus	1736
formosus	1736

	Page.
Spheroides furthi	1737
lobatus	1731, 1732
maculatus	1733
marmoratus	1734
pachygaster	1738
spengleri	1732, 1733
testudineus	1734
annulatus	1736
trichocephalus	1737
Sphœrodies	1729
Sphyræna	822
acus	717
argentea	826
aureoviridis	1119
barracuda	2841
becuna	823
borealis	825
ensis	824
forsteri	824
guachancho	824
guntheri	824
lucasana	826
picuda	823
picudilla	824
spet	826
sphyræna	823, 826
viridescens	826
vulgaris	826
sphyræna, Esox	826
Sphyræna	823, 826
sphyrænarum, Echeneis	2268
Sphyrænidæ	822
Sphyrænops	1114
bairdianus	1114
Sphyrna	43, 44, 45
tiburo	44, 2748
tudes	44
zygæna	45
Sphyrnidæ	43
Spicara	1364
martinica	1364, 1365
Spikefish	891
spillmani, Alvordius	1039
spilonotopterus, Exocœtus	740
spilonotus, Monacanthus	1716
spilopterus, Citharichthys	2685, 2686
Leuciscus	279
Photogenis	279
spilopus, Exocœtus	738
spilota, Uranidea	1953, 1962
spilopterygius, Balistes	1702
spilotum, Etheostoma nianguæ	1044
spilotus, Cottus	1961
Hypohomus	1043
spilurum, Cichlasoma	1520
spilurus, Heros	1520

	Page.
spilurus, Pœcilia	697
Salmo clarkii	2819
mykiss	495
Spina, Chopa	1357, 1358
Spinax	55
fabricii	56
hillianus	55
pusillus	55
sucklii	54
Spineless Sculpins	2025
spinescens, Auliscops	754
spinicephalum, Exoglossum	206
Spinicephalus fibulatus	2796
spinidens, Scarus	1637
spinifer, Engraulis	448
Stolephorus	448
spiniger, Icelus	1914
Spinivomer	367
goodei	367
spinosissimus, Agonus	2054
Aspidophorus	2054
Diodon	1746
Leptagonus	2054
spinosus, Calycilepidotus	1937
Centronotus	948
Chilomycterus	1749
Cyclopterus	2099, 2100
Dasyscopelus	575
Diodon	1749
Ehinorhinus	58
Eumicrotremus	2098, 2099
Hemilepidotus	1937
Lumpus	2099
Orthagoriscus	1754
Platophrys	2662
Rhinobatus	63
Rhomboidichthys	2663
Schedophilopsis	972
Scopelus	575
Squalus	58
Trachinotus	942
spinulosus, Gasterosteus	748
Spiny-back Blowfish	1734
Spiny Eels	612
Spiny-rayed Fishes	779
Spirinchus	522
spirlingulus, Leuciscus	282
spixianus, Saurus	538
Synodus	538
spixii, Argyriosus	936
Pimelodus	132
Platysomus	934, 2846
Tachisurus	132
Tachysurus	131, 2783
Vomer	2846
splendens, Beryx	844

	Page.
splendens, Exocœtus	720
splendida, Petenia	1513
spleniatum, Pristipoma	1322
spleniatus, Anisotremus	1321
Split-mouth Sucker	199
Split-tail	223
spongiosa, Halieutæa	2742
Spoon-bill Cat	101
Spot	1458
Spot-tailed Minnow	269
Spotted Cabrilla	1196
Jewfish	1162
Kelptish	2353
Moray	399
Rockfish	1806
Sea Trout	1409
Shiner	318
Sting Ray	88
Suckers	186, 187
Trunk-fish	1723
Weakfish	1409
Spotted-tail Minnow	275
Sprat	432, 450
Spratella	424
Spratelloides bryoporus	422
lamprotænia	419
Springfish	1950
Squalidæ	53, 2749
squalidus, Scarus	1640
squalipeta, Echeneis	2272
Squalius	228
aliciæ	236
ardesiacus	237
atrarius	233
bicolor	232
cæruleus	232
canis	29
conformis	231
conspersus	234
cooperi	236
copei	236
crassus	231
cruoreus	233
egregius	237
elongatus	240
estor	240
funduloides	240
galtiæ	237
gibbosus	231
gula	234
-humboldti	237
hyalope	222
hydrophlox	238
intermedius	235
lemmoni	235
lineatus	233

	Page.
Squalius margaritas	241
modestus	234
montanus	238
niger	235
nigrescens	234
obesus	233
pandora	234
photogenis	296
proriger	240
pulchellus	234
pulcher	234
purpureus	234
rhomaleus	233
squamatus	233
tænia	238
vandoisulus	240
Squalus	53
acanthias	54
acronotus	36
alopecias	46
americanus	47
argus	26
borealis	57
brucus	58
cæruleus	33
carcharias	38, 50
(Carcharias) terræ-novæ	43
cetaceus	47
cirratus	26
cornubicus	49
elephas	51
glacialis	57
glaucus	33
griseus	19
gunnerianus	51
hinnulus	29
hirundinaceus	33
homianus	51
isodus	51
littoralis	47
longimanus	38
macrodus	47
malleus	45
maximus	51
microcephalus	57
monensis	49
nasus	49
norwegianus	57
obscurus	35
obtusus	39
pelegrinus	51
pennanti	49
platyodon	39
punctatus	26, 43
punctulatus	26
rashleighanus	57

	Page.
Squalus rostratus	49
selanonus	49
spathula	102
spinosus	58
squatina	59
sucklii	54, 2749
tiburo	36, 44
vulpes	46
vulpinus	46
zygæna	45
squamata, Tigoma	233
squamatus, Etheostoma (Hadropterus)	1040
Hypohomus	1040
Squalius	233
squamiceps, Etheostoma	1096
squamilentus, Ceratichthys	323
Couesius	323
Paralichthys	2631
Squamipinnes	781, 1665
squamipinnis, Cestreus	1404
Cynoscion	1404, 1405
Gerres	1373
Otolithus	1404
squamosissimus, Diplolepis	1419
Pachyurus	1419
Plagioscion	1418
Sciæna	1418
squamosus, Trachurus	921
squamulosus, Chætodon	1685
Square-mouth	208
Square-tails	975, 976
Squatina	58
angelus	59
californica	59
dumerili	59
fimbriata	59
japonica	59
lævis	59
lewis	59
oculata	59
squatina	58
vulgaris	59
squatina, Rhina	59
Squalus	59
Squatina	58
Squatinidæ	58
Squato	58
Squaw-fish	224
Squeteague	1407
squeteague, Labrus	1407, 1409
Squirrel-fish	845, 847, 1203, 1207
Squirrel Hake	2555
stadthouderi, Serranus	1159
stagnalis, Salmo	510
Salvelinus	509

	Page.
stagnalis, Salvelinus alpinus	510
stahli, Larimus	1423
Monosira	1423
stannii, Cotylis	2332
Star-gazers	2305
Electric	2306
Sand	2297
Star-headed Minnow	656
starksi, Siphostoma	771
Stolephorus	2814
Starksia	2365
cremnobates	2365, 2366
starksii, Siphostoma	2838
Starry Flounders	2651
Stathmonotinæ	2347
Stathmonotus	2408
hemphillii	2408
stearnsi, Blennius	2379
Corvina	1458
Lutianus	1256
Lutjanus	1257
Prionotus	2166
Roncador	1457
Scorpæna	1843
stearnsii, Seriola	903
Steel-backed Chub	205
Steelhead	497
stegophthalmus, Agonus	2036
steindachneri, Chlorichthys	1609
Diabasis	1302
Hæmulon	1301
Ophioblennius	2401
Rhinoptera	91
Sebastes	1830
Sebastodes	1830
Thalassoma	1609, 2859
Steindachnerella	2567
Steindachneria	2567
argentea	2568
Steinegeria	960
rubescens	961
Steinegeriidæ	960
stejnegeri, Stelgistrum	1921
stelgidolepis, Macrourus	2585
Stelgis	2067
vulsus	2067
Stelgistrum	1921
stejnegeri	1921
stelifera, Corvina	1445
Sciæna (Stelliferus)	1443
Xenisma	648
stellata, Platessa	2652
stellatus, Apogonichthys	1110
Caranx	926
Fario	492
Liparis	2118
stellatus, Mustellus	29
Platichthys	2652
Pleuronectes	2652
stellaus, Salmo	493
stelleri, Cottus	1941
Cotylis	2104
Cyclopterichthys	2104
Cyclopterus	2104
Hexagrammos	1871
Labrax	1872
Liparops	2104
Myoxocephalus	1981
Trichodon	2297
Stellerina	2041
xyosterna	2042
Stellicarens	1439, 1440, 1445
Stellifer	1439, 1443
ericymba	1444
furthi	1441
illecebrosus	1442
lanceolatus	1443
microps	1445
minor	1442
oscitans	1440
stellifer	1443
zestocarus	1445
stellifer, Bodianus	1443
Fundulus	648
Sciæna	1444
Stellifer	1443
stelliferoides, Bassogigas	2516
Neobythites	2516
stellio, Perca	1153
Rhinobatus	2750
stellulata, Raja	75
Stenobrachius	561
andreæ	569
coccoi	569
Stenodus	473
mackenzii	474
Stenogobius	2210
Stenotomus	1345
aculeatus	1346
caprinus	1345
chrysops	1346
Stephanoberycidæ	835
Stephanoberyx	836
gillii	836
monæ	836
Stephanolepis	1714
setifer	1716
stephanophrys, Prionotus	2161
Stereolepis	1137
californicus	1138
gigas	1137
Sterletus	103

	Page.
sterletus, Averruncus	2071
Ceratichthys	316
Sternias	1926
xenostethus	1927
Sternoptychidæ	603
Sternoptyx	603
diaphana	603, 2826
gardenii	966
hermanii	603
mediterraneus	604
olfersi	604
Sternopygus humboldti	341
Sternotremia	786
isolepis	787
stevensi, Thaleichthys	521
Stichæinæ	2349
Stichæus	2439
aculeatus	2433
anguillaris	2436
enneagrammus	2441
hexagrammus	2441
islandicus	2439
lumpenus	2438
maculatus	2433
medius	2436
nubilus	2438
punctatus	2439
unimaculatus	2441
Stickleback, Alaska	749
Brook	744
California	751
Common Eastern	748
European	747
Nine-spined	745
Two-spined	748
Sticklebacks	742, 746
stigma, Gymnelis	2477
stigmæa, Ulocentra	1047
stigmæum, Boleosoma	1048
Etheostoma	1048
stigmæus, Citharichthys	2681
stigmatias, Paralichthys	2636
stigmaticus, Ceratichthys	323
Gobionellus	2224
Gobius	2224
Sigmiops	583
Smaragdus	2224
stigmatisticum, Lepophidium	2483, 2484
Stigmatogobius	2210
stigmatura, Bollmannia	2239
Cliola	275
Codoma	275
stigmaturus, Gobius	2220
Notropis	275
Photogenis	275
Stilbe	249, 250

	Page.
Stilbe americana	250
stilbe, Zalocys	2848
Stilbiscinæ	359
Stilbiscus	363
edwardsi	363
Stilbius	249, 250
stilbius, Leuroglossus	527
Minnilus	293
Notropis	293
stilbostigma, Prionodes	1216
Serranus	1217
stimpsoni, Triglopsis	2005
Sting Rays	79, 82
Stingaree	83
stipes, Atherina	790
Joturus	821
Stipvisch	1702
stirurus, Chloroscombrus	938
Stit-tse	499
Stizostedion	1020, 1021
canadense	1022
boreum	1022
griseum	1022
vitreum	1021
Stizostethium	1020
Stoasodon	88
narinari	88
Stolephorus	439
argyrophanus	441
astilbe	2815
brownii	443
chœrostomus	444
clupeoïdes	447
compressus	447
cubanus	442
cultratus	443
curtus	445
delicatissimus	444
engymen	2815
eurystole	445
exiguus	442
hiulcus	443
ischanus	442
lucidus	446, 2811
macrolepidotus	449
miarchus	441
mitchilli	446
mundeolus	2812
naso	2813
opercularis	445
panamensis	448
perfasciatus	441, 445
perthecatus	442
poeyi	445, 2811
productus	447
rastralis	2811

	Page.
Stolephorus ringens	449
robertsi	2815
scofieldi	2814
spinifer	448, 2814
starksi	2813
surinamensis	447
stolifera, Clupea	432
Dussumieria	419
Jenkinsia	419
Sardinella	431
stolzmanni, Belone	713
Cestreus	1412
Cynoscion	1412
Otolithus	1412
Tylosurus	713
stomata, Hippoglossina	2620
Stomias	588
affinis	588
ferox	588
fieldii	586
stomias, Atherestes	2609
Bathytroctes	454
Chasmistes	2794
Esox	585
Platysomatichthys	2610
Salmo clarkii	2819
mykiss	497
Trisotropis	1178
Stomiatidæ	587
Stomiatinæ	587
Stomodou	2529
bilinearis	2531
Stone Bass	1139
Cat	143, 144
Lugger	181, 205
Roller	181, 204
Sculpins	1937
Sturgeon	106
stonei, Salmo gairdneri	503
irideus	503
storeri, Acipenser	105
Aledon	1754
Atherina	807
Cetomimus	550
Leuciscus	222
Scombresox	726
storeria, Cichla	987
storeriana, Cliola	270
storerianus, Hybopsis	270, 321
Leuciscus	270
Rutilus	321
stouti, Bdellostoma	6
Polistotrema	6
strabo, Icelinus	1897
Ophioscion	1448
straminea, Oliola	262

	Page.
stramineus, Hybognathus	262
Hybopsis	262
stratus, Monacanthus	1713
Strawberry Bass	987
Straw-colored Minnow	261
striata, Alausa	431
Argentina	526
Perca	1311
striatulus, Brycon	337
Chalcinopsis	337
striatum, Bathystoma	1310
Hæmulon	1311
striatus, Anthias	1157
Blennius	2388
Bodianus	1259
Centropristes	1199
Chætodon	1677
Epinephelus	1157, 1208
Holocentrus	849
Hypsoblennius	2388, 2392
Isesthes	2388
Labrus	1200
Roccus	1133
Salmo	481
Sarothrodus	1677
Serranus	1157
stricticassis, Arius	126
Netuma	126, 2765
Tachisurus grandicassis	126
strigata, Trigla	2167
strigatum, Sparisoma	1639
strigatus, Antennarius	2720
Gobius	2228
Holacanthus	1683
Prionotus	2167
Scarus	1639
strigilata, Loricaria	158
strigosus, Anarrhichas	2447
Striped Anchovy	443
Bass	1131, 1132
Grunt	1296, 1313
Gurnard	2167
Mullet	811
Surf-fish	1505
strœmii, Zeus	955
Stromateidæ	964
Stromateus alepidotus	966
cryptosus	968
gardenii	966
longipinnis	966
palometa	967
paru	966
simillimus	967
triacanthus	968
stromii, Macrourus	2579

	Page.
strumosus, Gobiesox	2333
Studfish	648
Sturgeon, Common	105
Green	104
Lake	106
Ohio	106
Oregon	104
Red	106
Rock	106
Sacramento	104
Short-nosed	106
Stone	106
White	104, 107
Sturgeons	102
Shovelnose	107
Sturio	103
vulgaris	105
sturio, Acipenser	105
sturioides, Paragonus	2063
Podothecus	2063
Sturisoma	156, 157
Stygicola	2500
dentata	2500
dentatus	2500
Stylephoridæ	2601
Stylephorus	2601
chordatus	2601
stylifer, Hippocampus	778
Stypodon	220
signifer	220
suareus, Caranx	908
suavis, Cliola	272
Cyprinella	272
subæqualis, Corvilla	1429
Corvina	1429
subarcuata, Zygæna	45
subarcuatum, Hæmulon	1306
subarmatus, Acanthurus	1691
Subatka	595
subbifrenatus, Rhypticus	1233
subbifurcata, Pholis	2440
Ulvaria	2440
subbifurcatus, Eumesogrammus	2440
subcærulea, Amia	113
subfuscus, Labrus	1578
subligarius, Centropristis	1219
Dules	1218
Serranus	1219
suborbitalis, Holocentrus	850
Macrurus (Nematonurus)	2573
Nematonurus	2572
Plectromus	841
subrotundus, Ostracion ventre glabro	1749
subterraneus, Lucifuga	2501

	Page.
subterraneus, Typhlichthys	704
subtruncata, Belone	711
subtruncatus, Homoprion	1434
Tylosurus	711
subulatus, Orcynus	871
succincta, Seriola	900
sucetta, Cyprinus	186
Erimyzon	185, 186
oblongus	186
Sucker, Blue-headed	171
Brook	178
Carp	166
Columbia River	178
Common	178
Eastern Carp	168
Fine-scaled	178
Flannel-mouthed	174
Gourd-seed	168
Hare-lip	199
Hog	181
Hump-backed	184
June, of Utah Lake	183
Large-scaled	192
Long-nosed	176
Lump	2096
May	199
Missouri	168
Northern	176
Pea-lip	199
Rabbit-mouth	198, 199
Razor-back	184
Red	176
Sacramento	178
Sand	1476
Split-mouth	199
Tahoe	177
Webug	180
White	178, 192
Winter	187
Suckerel	168
Sucker-mouthed Buffalo	164
Suckers	161
Carp	165
Chub	185
Fine-scaled	173
Lump	2094
Mountain	169, 170
Suckers Spotted	186, 187
White-nosed	190
Suck-fish	2328
Sucking-fish	2269
sucklii, Acanthias	54
Catostomus	179
Spinax	54
Squalus	54, 2749
Sudis	599

	Page.
Sudis borealis	601
coruscans	602
intermedius	600
ringens	601
suensonii, Chilorhinus	372
sucuri, Coryphæna	953
Cyprinus (Catostomus)	195
sufflamen, Balistes	1706
Canthidermis	1706
sugillatus, Ophisura	387
Suillus	1580
suillus, Lachnolaimus	1580
sujef, Murænoides	2419
sulcatus, Herpetoichthys	382
Trachonurus	2591
sumichrasti, Citharichthys	2686
Summer Flounder	2629
Herring	426
Sunapee Trout	511
Sunfish	931, 1753
Black-banded	995
Blue	1005
Blue-spotted	996
Common	1009
Green	996
Long-eared	1002
Mud	989
Red-spotted	1004
Round	988
Sunfishes	984, 999
Banded	994
Pigmy	981
Sunny	1009
suœrii, Meletta	425
superbus, Pseudoscarus	1650
Scarus	1650
superciliosus, Aspidophorus	2036
Hexagrammos	1872
Hippocephalus	2036
Hyborhynchus	218
Labrax	1873
Surf Smelts	524
Whiting	1477
Surf-fish	1503
Common	1504
Striped	1505
Wall-eyed	1493, 1501
White	1506
Surgeon, Blue	1691
Common	1691
Surgeon-fishes	1688
surinamense, Pristipoma	1319
surinamensis, Anisotremus	1318, 1319
interruptus	1319
Arius	130

	Page.
surinamensis, Batrachoides	2314
Batrachus	2314
Engraulis	447
Galeichthys	2780
Hexanematichthys	129
Holocentrus	1236
Lobotes	1235, 2858
Lutjanus	1319
Plagioscion	1419
Pleuronectes	2666
Pœcilia	691
Pseudosciæna	1420
Sciæna	1420
Stolephorus	447
Tachisurus	130
Surmullets	855, 856
susanæ, Boleosoma	1059
sutor, Blepharis	932
Caranx	932
swaini, Notropis	290
Pœcilichthys	1086
Swainia	1039, 1040
Swallowers, Black	2291
swampina, Hydrargira	641
Hydrargyra	645
Fundulus	645
swanii, Bothragonus	2086, 2088
Hypsagonus	2088
swannanoa, Etheostoma	1070
Swellfish	1729, 1748
Swell Shark	25
Toad	1732, 1733, 1748
Swingle Tail	45
Swordfish, Common	894
Swordfishes	893
Syacium	2670
latifrons	2673
micrurum	2672
ovale	2674
papillosum	2671
Symbranchia	341
Symbranchidæ	342
Symbranchoid Eels	342
Symbranchus	342
immaculatus	342
marmoratus	342
vittatus	342
symmetricus, Apomotis	998
Caranx	910
Ceratichthys	246
Lepomis	999
Leucosomus	246
Pogonichthys	246
Rutilus	245
Salmo	505
Trachurus	910

	Page.
Symmetrurus	223
argyreiosus	224
Symphurus	2704, 2705
atramentatus	2706
atricaudus	2707
diomedeanus	2711
elongatus	2707
fasciolaris	2707
lcei	2708
marginatus	2706
nebulosus	2712
piger	2705
plagiusa	2710
plagusia	2709
pusillus	2710
williamsi	2711
Synagris	1288
macronemus	1289
synagris, Lutjanus	1271
Neomænis	1270
Sparus	1271
Synaphobranchidæ	350
Synaphobranchus	351
affinis	351
bathybius	352
infernalis	352
kaupii	351
pinnatus	351
Synapteretmus	545
Synauceia cerous	1941
Synbranchus fuliginosus	342
transversalis	342
Synchirinæ	1883
Synchirus	2023
gilli	2024
Synecoglanis	133
beadlei	135
Synentognathi	707
Synentognathous Fishes	707
Syngnathi	760
Syngnathidæ	760
Syngnathinæ	760
Syngnathus	761, 774
abboti	764
æquoreus	774, 2839
affinis	769
albirostris	772
arundinaceus	765
ascendens	768
bairdianus	770
brachycephalus	769
brevirostris	765
californiensis	764
cayennensis	773
dekayi	771
dimidiatus	765

	Page.
Syngnathus elucens	768
ethon	767
fasciatus	771
fistulatus	765
flavirostris	768
fuscus	770
griseolineatus	764
heckeli	2839
hippocampus	775
jonesi	768
leptorhynchus	765
linea	768
louisianæ	770
marmoreus	768
milbertianus	771
peckianus	771
peckii	770
pelagicus	770
picturatus	7(8
rousseau	767
segaliensis	2048
sibbaldi	774
tenuis	766
viridescens	771
Synodontidæ	532
Synodus	533, 2807
argenteus	411
evermanni	535
fasciatus	536
fœtens	538
intermedius	535, 536
jenkinsi	537, 2826
lacerta	537
lacertinus	536
lucioceps	539
myops	533
poeyi	536
saurus	537
scituliceps	537, 2826
spixianus	538
synodus	536
synodus, Esox	536
Saurus	536
Synodus	536
Sypterus	946
Syrrhina	61
exasperata	65
syrtensium, Argentina	526
tabacaria, Fistularia	757, 758
Haliperca	1215
tabacarius, Centropristes	1215
Prionodes	1215
Serranus	1215
tachete, Le Diodon	1746
Tachisurus albicans	124

	Page.		Page.
Tachisurus brandti	122	tæniotus, Tetragonopterus	334
dowi	125	Tæniophis	392
dubius	127	westphali	396
fissus	131	tæniops Bodianus	1144
flavescens	123	Enneacentrus	1144
furthii	132	Epinephelus	1144
grandicassis	126	Serranus	1144
stricticas-		tæniopterus, Balistes	1702
sis	126	Cottus	1979, 1988
gulosus	133	Perissias	2667
jordani	129	Platophrys	2668
kessleri	127	Pseudoscarus	1646
longicephalus	130	Scarus	1646
luniscutis	125	Tæniosomi	782
mesops	123	Tæniotoca	1505
nuchalis	131	lateralis	1505, 1506
oscula	127	Tahoe Chub	2798
platypogon	127	Lake Trout	493, 2870
proops	124	Sucker	177
rugispinis	130	tahoensis, Catostomus	177
phrygiatus	131	Salmo clarkii	2870
seemani	129	taiasica, Awaous	2236
spixii	132	Chonophorus	2237
surinamensis	130	Gobius	2236
temminckianus	123	Tail, Hard	921
variolosus	132	Tailor Herring	425
Tachysurinæ	115, 2757	Tails, Square	975, 976
Tachysurus	119, 121, 131, 2782	Talismania	455
emmelane	2785	æquatoris	456
emphysetus	122	antillarum	455
fissus	131, 2782	Tally-wag	1199
furthii	132, 2787	Tambor	1732, 1734, 1805
herzbergii	125	Tang	1691
lentiginosus	122	Blue	1691
liropus	2784	Ocean	1693
melanopus	132, 2784	tang, Mugil	812
multiradiatus	132, 2788	Tangbrosme	2438
nuchalis	2782	tanneri, Hyperchoristus	589
nuchalus	131	tapdisma, Salmo	483
passany	124	tapeinosoma, Auxis	868
peruvianus	122	Taractes	957
spixii	131, 2783	saussurii	957
variolosus	132, 2788	Tarandichthys	1891
taczanowskii, Centronotus	2416	cavifrons	1891
Pholis	2416	filamentosus	1892
Sebastodes	1831	tarascorum, Algansea	2796
tænia, Bassozetus	2510	Tarentola	537
Bathyonus	2510	Tarletonbeania	575
Blennius	2418	crenularis	575
Clinostomus	238	tenua	575
Murænoides	2418	Tarpon	409
Squalius	238	atlanticus	409
tæniatum, Hæmulon	1308	Tarpons	408
tæniatus, Anisotremus	1322	Tarpum	409
Chirocentrodon	435	Tates, Tom	1308
Evoxymetopon	886	tau, Batrachus	2316

	Page.
Tau, Batrachoides	2314
Batrachus beta	2316
pardus	2317
Gadus	2316
Opsánus	2315
Tauridea	1942, 1943, 1952
taurina, Chrysophrys	1354
taurinus, Calamus	1354
taurocephala, Cliola	253
taurocephalus, Alburnops	253
taurus, Abudefduf	1563
Carpiodes	165
Glyphidodon	1563
Tautoga	1577
americana	1579
cærulea	1577
niger	1577
onitis	1578, 1579
tessellata	1579
tautoga, Labrus	1579
fusca	1579
rubens	1579
Tautogolabrus	1576
adspersus	1577
Tautogs	1577, 1578
taylori, Chilara	2489
Ophidium	2489
Tchaviche	479
Tectospondyli	53, 58
Teipalcate	2698
Teleostei	113
Teleostomi	97, 1241
Telescops	1111
telescopus, Leuciscus	292
Minnilus	292
Notropis	292
arcansanus	292
Photogenis	292
Telestes	228
Telipomis	995
temminckianus, Bagrus	123
Sciadeichthys	122, 2760
Tachisurus	123
temminckii, Arius	123
Ditrema	1510, 1511
temminkii, Acanthoderma	880
Rovetus	880
Temnistia	1934
ventricosa	1936
Temnodon	946
saltator	947
tenebrosus, Alepocephalus	453
Antennarius	2719
Chironectes	2719
tenellus, Fundulus	659
Hyborhynchus	218

	Page.
tenorei, Scopelus	577
Ten-pounder	410
tentabunda, Trigla	2183
tentaculatus, Cottus	2000
Porocottus	2000
tenua, Tarletonbeania	575
tenue, Moxostoma	186
tenuifilis, Antennarius	2721
tenuirostris, Anguilla	348
tenuis, Atherinopsis	802
Gadus	2555
Icelinus	1894
Leuresthes	802
Opisthocentrus	2430
Phycis	2555
Syngnathus	766
Uranidea	1966
Urophycis	2555
teres, Alosa	420
Catostomus	179
Cœcula	379
Cyprinus	179
Etrumeus	420
Sphagebranchus	379
Teretulus	187
cervinus	197
teretulus, Phenacobius	303
liosternus	303
tergisus, Hiodon	413
terræ-novæ, Carcharias	43
Lycodes	2466
Scoliodon	43
Squalus (Carcharias)	43
tessellata, Plagusia	2709
Tautoga	1579
Tessellated Darter	1057
tessellatum, Boleosoma	1046, 1057
Etheostoma	1078
tessellatus, Hadropterus	1070
Labrus	1578
Nothonotus	1078
Testar	2332
testar, Lepadogaster	2332
testudineus, Sphæroides annulatus	1736
Spheroides	1734
Tetraodon	1735
Tetrodon	1735
Tétard	2332
tetard, Guavina	2200
Tête-de-roche	1323
Tetrabranchus	342
tetracanthus, Acara	1540
Centrarchus	1540
Gasterosteus	748
Heros	1539
Sparus	1257

	Page.
tetradens, Zipotheca	887
Tetradon erethizon	1739
ornatus	1742
Tetragonopterinæ	331
Tetragonopterus	333
æneus	333
argentatus	336
brevimanus	335
fasciatus	334
fischeri	334
fuscoauratus	334
humilis	335
mexicanus	335
microphthalmus	334
microstoma	334
œrstedii	334
panamensis	334
petenensis	335
rutilus	334
scabripinnis	335
tæniatus	334
Tetragonoptrus	1672
Tetragonuridæ	975
Tetragonurus	975
atlanticus	976
cuvieri	976
simplex	880
tetranemus, Hybopsis	315
Tetraodón setosus	1740
testudineus	1735
tetraodon, Ostracion	1740
Tetraodontidæ	1726, 1727
Tetraodontinæ	1727
Tetrapterus amplus	892
imperator	892, 2844
indicus	892
tetrapturorum, Echeneis	2273
Tetrapturus	891
albidus	892
belone	892
georgii	892
herschelii	892
lessoni	892
tetraspilus, Upeneus	860
Tetrodon	1727
ammocryptus	1735
angusticeps	1731
annulatus	1736
capistratus	1742
caudicinctus	1742
(Cheilichthys) pachygaster	1738
curvus	1728
formosus	1737
furthi	1737
geometricus	1735, 1736
heraldi	1736

	Page.
Tetrodon hispidus	1733
lævigatus	1728
lineolatus	1728
lune	1754
mathematicus	1728
mola	1754
nephelus	1733
oxyrhynchus	1741
pachycephalus	1729
pachygaster	1738
plumieri	1733
politus	1736
psittacus	1740
punctatissimus	1741
punctatus	1735
rostratus	1742
spengleri	1733
testudineus	1735
trichocephalus	1738
truncatus	1756
turgidus	1733
nephelus	1733
Tetronarce	77
californica	77
occidentalis	77
tetrophthalma, Lioglossina	2622
Tetroras	51
Teuthididæ	1688
Teuthis	1689
aliala	1693
australis	1691
bahianus	1693
cœruleus	1691
crestonis	1692
hepatus	1692
tractus	1693
triostegus	1690
Teuthys	1689
texana, Anguilla	348
Cyprinella	274
texanus, Catostomus	192
Notropis	274
Texas Redhorse	192
texensis, Dionda	215
textilis, Salarias	2400
Salariichthys	2400
Thærondotis	392
thalassinum, Etheostoma	1071
Moxostoma	191
Nothonotus	1072
thalassinus, Cestreus regalis	1408
Cynoscion	1407
Doratonotus	1612
Gobius	2245
Lepidogobius	2245
Microgobius	2245

	Page.		Page.
thalassinus, Myloleucus	245	Thrissa	422
Otolithus	1408	thrissa Clupea	432
Ptychostomus	192	thrissina, Clupea	431
Rutilus	245	Sardinella	430
Thalassoma	2859	thrissoides, Megalops	409
bifasciatum	1610, 2859	Thumb, Miller's	1941, 1950
bifasciatus	1610	Thunder-pumper	1484
grammaticum	1610, 2859	thunnia, Orcynus	869
lucasanum	1607, 2859	Thynnichthys	869
nitidissimum	2859	Thynnus	869
nitidum	1608, 2859	Thunnus	869
socorroense	1608, 2859	thynnus	870
steindachneri	1609, 2859	Thymallidæ	517
virens	1611, 2859	thymalloides, Coregonus	518
Thalassophryne	2323	Thymallus	517
dowi	2326	lewisi	2871
maculosa	2324	ontariensis	518
reticulata	2325	montanus	519
Thaleichthys	521	signifer	517, 2871
pacificus	521	montanus	519
stevensi	521	ontariensis	519
thaleichthys, Osmerus	522	tricolor	519, 2871
thazard, Auxis	867	monta-	
Scomber	867	nus	2871
Theragra	2535	tricolor	519
chalcogramma	2535	Thynnichthys	868
fucensis	2536	brevipinnis	869
Theraps	1540	thunnia	869
irregularis	1540	thynnoides, Auxis	868
thermalis, Poecilia	693	Thynnus	868, 869
theta, Diaphnus	565	affinis	869
Diaphus	564	argenti-vittatus	871
Theuthis	1689	atlanticus	871
Thick-tailed Rays	60	balteatus	871
Thimble-eyed Mackerel	866	brachypterus	870
thoburni, Alcidea	1887	brasiliensis	869
Mugil	813	brevipinnis	869
Tholichthys	1672	coretta	870
thomponi, Acipenser	105	leachianus	869
Amia	113	macropterus	871
Carpiodes	167	mediterraneus	870
Podothecus	2060	pacificus	871
Ptyonotus	2005	pelamys	869
Triglopsis	2005	pompilus	900
thomsonii, Cottunculus	1993	rocheanus	868
Cottus	1994	secundo-dorsalis	870
thoreauianus, Semotilus	223	thunnia	869
atromacu-		vulgaris	870
latus	223	thynnus, Albacora	870
Threadfins	827	Orcynus	870
Threadfishes	931	Scomber	870
Thread Herring	432	Thunnus	870
Three-angled Trunk-fishes	1721	Thyrina	803, 2840
Three-bearded Rocklings	2557	crystallina	804
Thresher	45	evermanni	804
Sharks	45	Thyris	2690

	Page.
Thyris pellucida	2691
Thyrsites acanthoderma	880
niger	879
pretiosus	880
scholaris	880
Thyrsitinæ	877
thyrsitoides, Lemnisoma	884
Thyrsitops violaceus	879
Thyrsoidea aterrima	396
concolor	396
cormura	394
maculipinnis	394
marginata	394
miliaris	398
Tiaroga	305
cobitis	305
tiburo, Reniceps	44
Sphyrna	44, 2748
Squalus	36, 44
Zygæna	44
Tiburon	39
Ticky-ticky	659
Tiger Shark	32
Tigoma	228, 230, 231
bicolor	232
conformis	231
conspersa	234
crassa	231
egregia	237
gibbosa	235
gracilis	236
humboldti	237
intermedia	235
lineata	233
nigrescens	234
obesa	233
pulchra	234
purpurea	234
rhinichthyoides	312
squamata	233
tigrinus, Galeocerdo	32
Holocentrus	1214
Myrichthys	376
Prionodes	1214
Serranus	1214
tigris, Antennarius	2723
Carcharias	49
Chironectes	2723
Epinephelus	1187
Mycteroperca	1187
camelopardalis	1187
Serranus	1187
Trisotropis	1187
Tigrone	32, 39
Tilefish	2278
Tilesia	2537

	Page.
Tilesia gracilis	2538
tilesii, Hemilepidotus	1936
timpauogensis, Hybopsis	233
Minnilus	233
Timucu	711, 715
timucu, Belone	715
Esox	711
tinca, Serranus	1181
tincella, Algansea	211, 2796
Leuciscus	211
Leucus	211
Tinker Mackerel	866
Tiñosa	924
tippecanoe, Etheostoma	1090
Tippecanoe Sisco	469
Tirantes	885
Tiru	537
Tirus	533
marmoratus	537
Toad, Swell	1732, 1733
Toadfish	1733, 1748, 2313, 2315
Toad-fishes, Poison	2323
Tobacco Box	68
Toeroe	1403
toeroe, Otolithus	1404
Togue	504
Tokenoko me waru	1829
Tom, Mad	144, 147
toma, Salmo	505
Tomcod, California	2539
tomcod, Gadus	2540
Microgadus	2540
Tomcods	2538, 2540
tomcodus, Gadus	2540
fuscus	2540
luteus	2540
mixtus	2540
Tomicodon	2329
Tomtate	1308
Tondo, Pesce	48
Tongue Fish	2704, 2710
Toothed Herring	413
Top Minnow	659, 680
topeka, Notropis	266
Topes	31
Toro	920, 1724
torpedinus, Trygonobatus	81
Urolophus	81
Torpedo	77
bancrofti	78
brasiliensis	78
californica	78
occidentalis	77
pictus	78
torpedo, California	77
torquatus, Labrus	1609

	Page.
Torrentaria	1066, 1063, 1080
torridus, Lutjanus	1264
torsk, Blennius	2561
Gadus	2561
torvus, Cottunculus	1994
tota, Perca maculis	1153
Toter	181
toto, Cyprinus	415
Totuava	1411
townsendi, Gobius	2250
Lampanyctus	558
Myctophum	558
toxotes, Ditrema	1508
Rhacochilus	1507
Trachelocirrus	950
Trachichthyidæ	836
Trachichthys pretiosus	837
Trachidermis richardsoni	1944
Trachinocephalus	533
myops	533
Trachinoid Fishes	2273
Trachinoidea	2273
Trachinoidei	781
Trachinotinæ	897
Trachinotus	939
argenteus	944
carolinus	944
cayennensis	945
culveri	942
cupreus	944
falcatus	941, 2847
fuscus	942
glaucus	940
goodei	943
kennedyi	942
paloma	945, 2848
rhodopus	941, 943
rhomboides	2847
spinosus	942
Trachinus adsencionis	1153
cirrhosus	2019
gasteropelecus	2297
osbeck	1153
punctatus	1153
trichodon	2296
trachinus, Malacanthus	2276
Salmo	533
Trachisurus parkeri	126
Trachonurus	2591
sulcatus	2591
trachura, Raia	76
Raja	75
Trachurops	911
brachychirus	911
crumenophthalmus	911
plumieri	912

	Page.
Trachurus	909
aliciolus	904
boops	922
cuvieri	910
europæus	911
fallax	910
fasciatus	904
imperialis	927
linnæi	911
picturatus	909, 2844
rissoi	910
saurus	911
squamosus	921
symmetricus	910, 2844
trachurus	910
trachurus, Caranx	910
Cottus	1936
Gasterosteus	747
Hemilepidotus	1936
Scomber	910
Trachurus	910
Trachynotus argenteus	944
carolinus	944
cupreus	944
fasciatus	941
glaucoides	941
glaucus	941
goreensis	943
nasutus	941
ovatus	942
pampanus	944
rhomboides	942
trachypoma, Myripristis	846
Trachypterus	2870
trachyurus	2601
Trachyrhamphus	761, 2568
Trachyrinchinæ	2562
Trachyrincus	2568
Helolepis	2568
trachyurus, Trachypterus	2601
tractus, Acanthurus	1693
Teuthis	1693
Trahiras	330
tranquebar, Aspidophoroides	2092
transmontana, Columbia	784
transmontanus, Acipenser	104
Rhinichthys	307
transversalis, Synbranchus	342
transversum, Pœcilosoma	1089
traski, Hysterocarpus	1496
treculii, Dioplites nuecensis	1012
Treefish	1827
Trematopsis	1753
willugbei	1754
Triacanthidæ	1697
Triacanthodinæ	1697

	Page.
triacanthus, Argyriosus	936
Nauclerus	900
Poronotus	2849
Rhombus	967
Stromateus	968
Xenochirus	2084
Triacis henlei	31
triagramma, Heros	1529
Triakis	31
semifasciatum	31
triangularis, Ostracion tuberculus	1723
Pachynathus	1705
triangulum, Lutjanus	1454
Tribe, Flounder	2607
Halibut	2605
Turbot	2608
tribulus, Prionotus	2171, 2172
Trigla	2172
Trichidion	828
approximans	829
octofilis	830
octonemus	830
opercularis	831
plumieri	830
Trichiuridæ	888
Trichiurus	889
argenteus	889
caudatus	887
ensiformis	887
gladius	887
lepturus	889, 2844
trichocephalus, Sphæroides	1738
Tetrodon	1738
Trichocyclus	1743, 1744
erinaceus	1744
Trichoderma	1714
Trichodiodon	1743, 1744
pilosus	1743, 1744
Trichodon	2295
japonicus	2297
lineatus	2297
stelleri	2297
trichodon	2295
trichodon, Drachinus	2297
Mugil	816
Trachinus	2296
Trichodon	2295
Trichodontidæ	2295
Trichonotus	409
Trichopsetta	2669
ventralis	2669
trichopterus, Pteraclis	956
trichroistia, Oliola	276
Codoma	276
trichroistius, Notropis	275
tricocephalus, Spheroides	1737

	Page.
tricolor, Chætodon	1684
Engraulis	443
Genicanthus	1684
Holacanthus	1684
Pomacanthus	1684
Thymallus	519
signifer	519
Tricopterus	915, 917, 920
tricornis, Lactophrys	1724
Ostracion	1725
tricuspidatus, Hyporhamphus	720
tricuspis, Cottus	2009
Gymnocanthus	2008
Phobetor	2009
tridecemlineatus, Esox	628
trideus, Acropœcilia	690
Archosargus	1360
Centropristis	1202
Lutjanus	1202
Pœcilia (Acropœcilia)	690
Sargus	1361
tridentata, Lampetra	12
tridentatus, Ammocœtes	12
Entosphenus	12
Ichthyomyzon	12
Petromyzon	12
tridentiger, Gambusia	2833
tridigitatus, Dactyloscopus	2301
Polynemus	2177
Trifarcius	670
felicianus	676
riverendi	673
trifasciata, Hydrargyra	639
Lepomis	1011
trifurca, Centropristes	1202
Perca	1202
trifurcens, Anthias	1202
Centropristis	1202
Serranus	1201, 1202
trigammus, Chirus	1872
Trigger Fishes	1698, 1699
Trigla	2176, 2867
carolina	2156, 2172
cuculus	2177
digitis vicensis palmatis	2183
evolans	2169
fasciata	2183
lineata	2167
palmipes	2156
pictipinnis	2176
pini	2177
punctata	2170
strigata	2167
tentabunda	2183
tota rubens	2177
tribulus	2172

	Page.
Trigla volitans	2183
Triglidæ	2147
Triglochis	46
Triglops	1923
beani	1924
pingeli	1923
pleurostictus	1923
scepticus	1925
xenostethus	1927
Triglopsis	2005
stimpsoni	2005
thompsoni	2005
Trigonobatus	82
trigonum, Ostracium	1724
trigonus, Lactophrys	1723, 1724
Ostracium	1724
trilineatum, Pristipoma	1320
trilineatus, Anisotremus	1320
trilobatus, Scarus	1654
Triloburus	1198, 1199, 1201
trilobus, Blepsias	2019
Lutjanus	1200
trimaculatus, Heros	1529
Trinectes	2693
scabra	2701
triostegus, Acanthurus	1691
Chætodon	1691
Teuthis	1690
tripes, Nealotus	881
Triple-tail	1235
Tripteronotus	461
Tripterygium carminale	2350
tripterygius, Cottus	2023
tripunctulatus, Maurolicus	578
Valenciennellus	578
triqueter, Lactophrys	1722
Ostracion	1723
triserialis, Murænopsis	384
Ophichthus	384
Ophichthys	384
triseriata, Platyrhina	66
triseriatus, Platyrhinoidis	65, 66
Rhinobatus	66
trisignatum, Moxostoma	179
Trisotropis	1169, 1172
aguaji	1175
bonaci	1175
brunneus	1175
calliurus	1186
camelopardalis	1187
cardinalis	1174
chlorostomus	1179
dimidiatus	1179
falcatus	1185
interstitialis	1179
microlepis	1178

	Page.
Trisotropis petrosus	1172
reticulatus	1187
rosaceus	1184
stomias	1178
tigris	1187
trispinosa, Corvina	1443
trispinosus, Odontopyxis	2085
Pseudoscarus	1648
Scarus	1648
trispinous, Odontopyxis	2086
tristis, Moniana	272
tristœchus, Esox	111
Lepisosteus	111
Litholepis	111
triurus, Bodianus	1236
trivittatus, Diabasis	1311
Grammistes	1311
Sebastodes	1834
Trochocopus darwinii	1586
pulcher	1585
Trompa	1653
Lija	171
Trompetero	754, 757
tropica, Echeneis	2268
tropicus, Atractosteus	111
Lepisosteus	111
Tropidichthys	1741
Tropidinius	1278
arnillo	1279
dentatus	1279
Tropidodus	20
troscheli, Heros	1537
Sciadeichthys	122, 2757
Sciades	122, 2758
troschelli, Glyphidodon	1562
Trout	483
Black-spotted	487
Blue-back	514, 2819
Brook	506
Bull	507
Coast Range	500
Colorado River	496
Columbia River	492
Cut-throat	487, 492, 493
Dolly Varden	507
Dublin Pound	507
Gila	226
Golden of Mount Whitney	503
Great Lake	504
Green-back	497
Kamchatka Salmon	2818
Kamloops	499
Kern River	502
Lac de Marbre	515
Lake Tahoe	493, 2870
Mackinaw	504

	Page.
Trout McCloud River Rainbow	502
Nissue.........................	503
No-she........................	503
of Utah Lake.................	495
Oquassa......................	514
Rainbow......................	500
Red Rock.....................	1872
Red-spotted	507
Rio Grande...................	495
Rock 1866, 1867	
Rocky Mountain	487
Salmon	497
Sea	1407
Silver	493
Speckled	506
of Lake Crescent ..	2821
Sunapee	511
Truckee	493
Waha Lake...................	496
Yellow-fin	496
Yellowstone	493
Trout Perch 782, 784	
trowbridgii, Abeona 1497	
Holconotus	1497
Homalopomus	2531
Trucha........................	819
Truckee Trout	493
truculentus, Chorophthalmus	542
True Eels	346
Fishes	97
Sardines....................	422
Sharks	21
Trumpet-fish.......... 754, 756, 759	
truncata, Belone 714, 715	
Malthæa	2738
Ranzania............ 1755, 1756	
truncatus, Blennius	2381
Dinectus	106
Orthagoriscus...........	1756
Salmo	499
Saurus	533
Scarus	1641
Tetrodon	1756
Trunk-fish............ 1720, 1721, 1722, 1723	
Spotted	1723
Three-angled...........	1721
Trutta 483, 486, 487	
trutta, Salmo	487
Truttæ	483
Trycherodon	247
megalops	249
Trygon..........................	82
gymnura	84
hastata	84
osteosticta.................	84
sabina	85

	Page.
Trygon sayi.........................	86
tuberculata.................	84
Trygonobatus torpedinus	81
Trygonorhina alveata	65
Tschawytscha.....................	479
tschawytscha, Oncorhynchus	479
Salmo	480
tschawytschiformis, Salmo	478
tsiltcoosensis, Catostomus.........	2793
tsuppitch, Fario....................	493
Oncorhynchus	481
Salmo 481, 495	
tuberculata; Dasibatis.............	84
Raia....................	84
Trygon................	84
tuberculatus, Catostomus	186
Sphæroides	1733
tuberculé, Le Sphéroide............	1733
tuberculée, Raie....................	84
tudes, Gobiesox	2333
Salmo	508
Sphyrna	44
Zygæna	44
tuditana, Cliola.....................	253
tuditanis, Hybopsis	253
tuditanus, Hypargurus	253
Leuciscus..............	253
Tullibee.....:	473
tullibee, Argyrosomus	473
bisselli	473
Coregonus	473
bisselli.........	473
Salmo (Coregonus)	473
tumidus, Carpiodes	167
Chironectes	2717
Lophius..................	2716
Tuna...........................	870
tunicata, Liparis 2121, 2128	
tunicatus, Liparis	2120
Tunnies........................	869
Little	868
Tunny	870
Little......................	869
Turbot	1701
Ocean	1706
Tribe......................	2608
turchesius, Pseudoscarus	1659
Scarus	1658
Turdus cauda convexa..............	1145
cinereus peltatus	1373
flavus	1583
oculo radiato 1591, 1703	
pinnis branchialibus	1257
rhomboidalis	1691
turgidus, Tetrodon................	1733
nephelus.......	1733

	Page.
turneri, Lycodalepis	2469
Lycodes	2469
tuscumbia, Etheostoma	1100
Psychromaster	1100
Two-spined Stickelback	748
Tyee Salmon	479
Tylosurus	708
acus	716
almeida	715
angusticeps	712
ardeola	713
caribbæus	717
contrainii	717
crassus	716
diplotænia	712
euryops	711
exilis	714
fodiator	715
galeatus	716
gladius	716
hians	718
longirostris	714
marinus	714
microps	712
notatus	710, 711
pacificus	716
raphidoma	715
sagitta	711
scapularis	711
sierrita	713
stolzmanni	713
subtruncatus	711
timucu	711
Tympanomium	1753
planci	1754
Tyntlastes	2262
brevis	2262
sagitta	2263
Typhlichthys	704, 2835
rosæ	2835
subterraneus	704, 2835
Typhlogobius	2261
californiensis	2262
Typhlopsaras	2729
shufeldti	2731
typica, Sciæna	1448
Typical Sharks	19
typicus, Ophioscion	1448
Plargyrus	283
typus, Achirophichthys	388
Skeponopodus	892
tyrannus, Anguilla	348
Brevoortia	433
aurea	434
brevicaudata	434
patronus	434

	Page.
tyrannus, Clupea	434
Tyrant Fish	886
Ugly Fish	137
uhleri, Citharichthys	2684
Ulæma	1371
lefroyi	1371
Ulca	2021
marmorata	2021
Ulcina	2088
Ulka	1974
Ulke	1974
Ulocentra	1047, 2851
davisonii	1049
gilberti	1049, 2852
histrio	1050, 1051
meadiæ	2852
phlox	1052
simotera	1051, 2853
stigmæa	1047
verecunda	1049
ulochir, Paraliparis	2144
ulvæ, Xiphidion	2424
Xiphistes	2423
Ulvaria	2440
subbifurcata	2440
Ulvicola	2413, 2860
sanctæ-rosæ	2413
umatilla, Agosia	313
Umbla	506
minor marina	823
umbla, Salmo	509
Umbra	623, 2807
delicatissima	621
limi	623
pygmæa	624
pygmæa	624
umbratilis, Alburnellus	299
Alburnus	299
Minnilus	299
Notropis	298
ardens	301
atripes	300
cyanocephalus	300
fasciolaris	301
lythrurus	300
matutinus	301
punctulatus	301
umbratilis	299
Umbridæ	622
umbrifer, Notropis	274
Urolophus	2752
Umbrina	1465
alburnus	1475
analis	1468
arenata	1474

	Page.
Umbrina broussonetii	1466
coroides	1466
dorsalis	1469
elongata	1476
furnieri	1463
galapagorum	1468
gracilis	1474
januaria	1474
littoralis	1477
martinicensis	1474
nasus	1473
nebulosa	1475
panamensis	1473
phalæna	1475
roncador	1467
sinaloæ	1468
undulata	1467, 1476
xanti	1467, 1468
umbrosa, Cliola	273
Cyprinella	273
Lepidopsetta	2642
Narcine	78
umbrosus, Eques acuminatus	1487
Esox	627
Gymnothorax	390
Platichthys	2643
Pleuronectes	2643
Sebastichthys	1807
Sebastodes	1807
Umbrula	1469, 1471, 1476
Unbarana	411
uncinatus, Artediellus	1905, 1906
Centridermichthys	1906
Cottus	1906
Icelus	1906
uncompahgre, Xyrauchen	184
undecimale, Hæmulon	1300
undecimalis, Centropomus	1118
Sciæna	1119
undecimradiatus, Centropomus	1119
undulata, Murænophis	403
Perca	1462
Umbrina	1467, 1476
undulatus, Menticirrhus	1476
Micropogon	1461
Ostracion	1724
Rhinobatus	63
Scomber	867
undulosus, Serranus	1181
unerarak, Ophidium	2477
Unibranchapertura	342
grisea	342
immaculata	342
lineata	342
marmorata	342
unicolor, Ammocœtes	10

	Page.
unicolor, Bryttus	1001
Chætodon	1676
Holocentrus	1192
Hypoplectrus	1190, 1192
aberrans	1193
accensus	1193
affinis	1193
bovinus	1193
chlorurus	1193
crocotus	1192
gummigutta	1192
guttavarius	1192
indigo	1193
nigricans	1192
pinnivarius	1192
puella	1192
vitulinus	1192
Perca	1192
Petromyzon marinus	10
Serranus	1192
Solea	2702
Soleotalpa	2703
Unicorn Fish	1719
unicornis, Citharichthys	2683
Unicornu pisces bahamensis	1719
unicornus, Balistes	1720
unifasciatus, Hemirhamphus	720, 721
Hyporhamphus	720
unimaculata, Perca	1360
unimaculatus, Archosargus	1359, 1360
Argyreiosus	934
Clinus	2441
Grammistes	1360
Sargus	1360
Stichæus	2441
uninotata, Heterandria	687
uninotatus, Ctenlabrus	1577
Girardinus	687
Lutjanus	1271
Mesoprion	1271
unionensis Centropomus	1122
univittatus, Apodichthys	2412
Upeneus	857, 2843
balteatus	860
dentatus	859
flavovittatus	860
grandisquamis	860
maculatus	858
martinicus	859
parvus	859
punctatus	859
rathbuni	857
tetraspilus	860
xanthogrammus	860
Upselonphorus	2306
guttatus	2310

	Page.
ve ifer **Agosia**	212
ve ox, **ircharhinus**	2747, 2748
iola	253
deptorhamphus	724
venadom, **Mycteroperca**	1186
Venefi	365
procera	365
veneno, Mycteroperca	1172
apua	1173, 1174
guttata	1174
marina	1172
ve enos, Epinephelus	1172
ve osideletta	426
venosi **Aluterus**	1719
ventra, **Arnoglossus**	2670
Brosmophycis	2503
Citharichthys	2670
Cottus	2008, 2009
Dinematicthys	2503
Novaculichthys	1615
Ogilbia	2503
Trichopsetta	2669
Xyrichthys	1616
ven Ostracion subrotundus	
glabro	1749
ven rich, **Apocope**	309
Temnistia	1936
ven rieus, Blepsias	1936
Ceratichthys	309
Cotylis	2104
Cyclopterichthys	2104
Cyclopterus	2104
Salmo	509
Sebastes	1829
Sebastodes	1829
ven om, Scyllium	25
ven Cliola	274
Cyprinella	274
Emmeekia	1602
Limia	665
Lucania	665
ven Dinemus	854
Notropis	274, 275
Pseudojulis	1602
Xyrichthys	1619
rany ybium	877
Verasp	2618
moseri	2619
Verde huny	1195
ojarra	1538
orena	396
idiano	1590, 1591
abrillas	1194
rcula, Ulocentra	1049
rcumm, Etheostoma	1050
Lepidion	2543

	Page.
Upselonphorus y-græcum	2308
uracantha, Loricaria	158
Uraleptus	2545
maraldi	2545
Uranichthys	381
brachycephalus	382
havannensis	382
Uranidea	1963
aspera	1944
bendirei	1964
boleoides	1968
cognata	1955
formosa	1969
franklini	1967
gobioides	1968
gracilis	1968
greenei	1965
gulosa	1945
hoyi	1969
kumlieni	1967
marginata	1965
microstoma	1958
pollicaris	1954
punctulata	1949
quiescens	1968
rhothea	1947
ricci	1953
richardsoni	1952
semiscabra	1950
spilota	1953, 1962
tenuis	1966
vheeleri	1950
viscosa	1968
uranidea, Cottogaster	1044
Etheostoma	1045
uranops, Phenacobius	304
Uranoscopidæ	2305
Uranoscopinæ	2306
Uranoscopus anoplos	2308
y-græcum	2308
uranoscopus, Ceratias	2730
Mancalias	2729
Uraptera	66
binoculuta	73
Uraspis	916, 918, 926
Uriphæton	1143
Urocentrus	2414, 2415
pictus	2416
Uroconger	358
vicinus	358
Urolophinæ	79
Urolophus	79
aspidurus	81
asterias	82, 2752
goodei	81
halleri	80

	Page.
Urolophus jamaicensis	81
mundus	81
nebulosus	80, 2752
rogersi	2752, 2753
torpedinus	81
umbrifer	2752
Uronectes parii	2478
urophthalmus, Heros	1537
Urophycis	2552
chesteri	2556
chuss	2555
cirratus	2553
earlli	2554
floridanus	2554
regius	2553
tenuis	2555
Uropsetta	2624
californica	2626
Uropterygius	403
necturus	404
urostigma, Oliola	275
Urotrygon	80
mundus	81
Uroxis	82
ursinus, Salmo	505
urus, Bubalichthys	164
Carpiodes	164
Ictiobus	164
Sclerognathus	165
ustus, Callyodon	1624
Cryptotomus	1624
Utah Lake Chub	232
Mullet	179
Trout	495
utah, Salmo	495
utawana, Catostomus	179
uter, Catulus	25
Uwo Aka	1833
Vaca	1190
Vacas	1189
vacca, Capriodes	168
Damalichthys	1510
Ditrema	1510
Vacuocua	1427
vafer, Myrophis	372
vagrans, Chirostoma	795
Kirtlandia	794
laciniata	2840
Menidia	795
laciniata	795
vahlii, Lycodes	2463
Vaillantia	1054, 1060
camura	1060
chlorosoma	1060
valenciennei, Smaragdus	2228
Valenciennellus	577

	Page.
Valenciennellus tripunctulatus.....	578
valenciennesi, Arius	124
Bagrus	124
Erotelis	2204
Moxostoma..........	190
vampyrus, Cephalopterus	93
Ceratoptera.............	93
Vandellius	886
lusitanicus	887
vandepolli, Girardinus	2834
Pœcilia..............	696
arubensis	2834
vaudoisula, Gila....................	240
vandoisulus, Leuciscus	239
Squalius..............	240
Vaqueta de dos Colores	1684
varia, Perca......................	1200
variabilis, Acanthocottus	1975
Dioplites	1012
Perca ....:... 1784, 1796	
Pomacentrus	1552
Sebastes.................	1784
variatum, Boleosoma...............	1070
Etheostoma..............	1069
variatus, Alvordius	1034
Characodon	669
Hadropterus..............	1070
variegata, Lampetra	2745
Loricaria......,......	159
Muræna.................	2805
Salpa purpurescens......	1271
variegatus, Batrachoides	2316
Cyprinodon	671
riverendi ..	673
Limnurgus	666
Orbis lævis..............	1735
Pachylabrus	1508
Sargus	1364
variolosus, Arius	132
Tachisurus.............	132
Tachysurus 132, 2788	
varius, Cephalus	1756
Malacoctenus	2357
Monacanthus	1716
Myxodes	2357
Saurus	536
Serranus	1153
velata, Chalisoma	1703
velatum, Moxostoma...............	190
velatus, Ptychostomus.............	190
veliana, Atherina	790
velifer, Carpiodes	167
Catostomus.................	167
Ictiobus...................	167
Letharcus.................	375
Rhinichthys (Apocope).....	312

	Page.
velifera, Agosia.....................	212
velox, Carcharhinus.............. 2747, 2748	
Cliola	253
Euleptorhamphus............	724
venadorum, Mycteroperca..........	1186
Venetica	365
procera	365
venenosa, Mycteroperca......,.....	1172
apua 1173, 1174	
guttata....	1174
marina	1172
venenosus, Epinephelus............	1172
venosa, Meletta	426
venosus, Aluterus	1719
ventralis, Arnoglossus..............	2670
Brosmophycis ....:...	2503
Citharichthys	2670
Cottus................. 2008, 2009	
Dinematichthys	2503
Novaculichthys..........	1615
Ogilbia	2503
Trichopsetta	2669
Xyrichthys..............	1616
ventre, Ostracion subrotundus glabro	1749
ventricosa, Apocope..............	309
Temnistia...............	1936
ventricosus, Blepsias..............	1936
Ceratichthys	309
Cotylis	2104
Cyclopterichthys	2104
Cyclopterus...........	2104
Salmo	509
Sebastes	1829
Sebastodes	1829
ventriosum, Scyllium	25
venusta, Cliola	274
Cyprinella	274
Emmeekia	1602
Limia	665
Lucania.................	665
venustus, Dinemus.................	854
Notropis................. 274, 275	
Pseudojulis...............	1602
Xyrichthys..............	1619
verany, Cybium	877
Verasper	2618
moseri	2619
Verde, Johnny	1195
Mojarra...............	1538
Morena....................	396
Pudiano 1590, 1591	
Verdes, Cabrillas..................	1194
verecunda, Ulocentra..............	1049
verecundum, Etheostoma	1050
Lepidion	2543

	Page.
Upselonphorus y-græcum	2308
uracantha, Loricaria	158
Uraleptus	2545
maraldi	2545
Uranichthys	381
brachycephalus	382
havannensis	382
Uranidea	1963
aspera	1944
bendirei	1964
boleoides	1968
cognata	1955
formosa	1969
franklini	1967
gobioides	1968
gracilis	1968
greenei	1965
gulosa	1945
hoyi	1969
kumlieni	1967
marginata	1965
microstoma	1958
pollicaris	1954
punctulata	1949
quiescens	1968
rhothea	1947
ricci	1953
richardsoni	1952
semiscabra	1950
spilota	1953, 1962
tenuis	1966
wheeleri	1950
viscosa	1968
uranidea, Cottogaster	1044
Etheostoma	1045
uranops, Phenacobius	304
Uranoscopidæ	2305
Uranoscopinæ	2306
Uranoscopus anoplos	2308
y-græcum	2308
uranoscopus, Ceratias	2730
Mancalias	2729
Uraptera	66
binoculata	73
Uraspis	916, 918, 926
Uriphæton	1143
Urocentrus	2414, 2415
pictus	2416
Uroconger	358
vicinus	358
Urolophinæ	79
Urolophus	79
aspidurus	81
asterias	82, 2752
goodei	81
halleri	80

	Page.
Urolophus jamaicensis	81
mundus	81
nebulosus	80, 2752
rogersi	2752, 2753
torpedinus	81
umbrifer	2752
Uronectes parii	2478
urophthalmus, Heros	1537
Urophycis	2552
chesteri	2556
chuss	2555
cirratus	2553
earlli	2554
floridanus	2554
regius	2553
tenuis	2555
Uropsetta	2624
californica	2626
Uropterygius	403
necturus	404
urostigma, Cliola	275
Urotrygon	80
mundus	81
Uroxis	82
ursinus, Salmo	505
urus, Bubalichthys	164
Carpiodes	164
Ictiobus	164
Sclerognathus	165
ustus, Callyodon	1624
Cryptotomus	1624
Utah Lake Chub	232
Mullet	179
Trout	495
utah, Salmo	495
utawana, Catostomus	179
uter, Catulus	25
Uwo Aka	1833
Vaca	1190
Vacas	1189
vacca, Capriodes	168
Damalichthys	1510
Ditrema	1510
Vacuocua	1427
vafer, Myrophis	372
vagrans, Chirostoma	795
Kirtlandia	794
laciniata	2840
Menidia	795
laciniata	795
vahlii, Lycodes	2463
Vaillantia	1054, 1060
camura	1060
chlorosoma	1060
valenciennei, Smaragdus	2228
Valenciennellus	577

	Page.
Valenciennellus tripunctulatus.....	578
valenciennesi, Arius	124
Bagrus	124
Erotelis	2204
Moxostoma..........	190
vampyrus, Cephalopterus	93
Ceratoptera..............	93
Vandellius	886
lusitanicus	887
vandepolli, Girardinus	2834
Pœcilia..............	696
arubensis	2834
vandoisula, Gila....................	240
vandoisulus, Leuciscus	239
Squalius..............	240
Vaqueta de dos Colores	1684
varia, Perca......................	1200
variabilis, Acanthocottus	1975
Dioplites	1012
Perca 1784, 1796
Pomacentrus	1552
Sebastes.................	1784
variatum, Boleosoma...............	1070
Etheostoma..............	1069
variatus, Alvordius	1034
Characodon	669
Hadropterus..............	1070
variegata, Lampetra	2745
Loricaria	159
Muræna.................	2805
Salpa purpurescens......	1271
variegatus, Batrachoides	2316
Cyprinodon	671
riverendi ..	673
Limnurgus.........	666
Orbis lævis..............	1735
Pachylabrus	1508
Sargus	1364
variolosus, Arius	132
Tachisurus..............	132
Tachysurus 132, 2788
varius, Cephalus	1756
Malacoctenus	2357
Monacanthus	1716
Myxodes	2357
Saurus	536
Serranus	1153
velata, Chalisoma	1703
velatum, Moxostoma..............	190
velatus, Ptychostomus.............	190
veliana, Atherina	790
velifer, Carpiodes	167
Catostomus.................	167
Ictiobus..................	167
Letharcus..............	375
Rhinichthys (Apocope).....	312

	Page.
velifera, Agosia.....	212
velox, Carcharhinus.............. 2747, 2748
Čliola	253
Euleptorhamphus...........	724
venadorum, Mycteroperca..........	1186
Venetica	365
procera	365
venenosa, Mycteroperca......,.....	1172
apua 1173, 1174
guttata....	1174
marina	1172
venenosus, Epinephelus.............	1172
venosa, Meletta	426
venosus, Aluterus	1719
ventralis, Arnoglossus..............	2670
Brosmophycis	2503
Citharichthys	2670
Cottus................. 2008, 2009
Dinematichthys	2503
Novaculichthys..........	1615
Ogilbia	2503
Trichopsetta	2669
Xyrichthys..............	1616
ventre, Ostracion subrotundus glabro	1749
ventricosa, Apocope.............	309
Temnistia..............	1936
ventricosus, Blepsias..............	1936
Ceratichthys	309
Cotylis	2104
Cyclopterichthys	2104
Cyclopterus	2104
Salmo	509
Sebastes	1829
Sebastodes	1829
ventriosum, Scyllium	25
venusta, Oliola	274
Cyprinella	274
Emmeekia	1602
Limia	665
Lucania.................	665
venustus, Dinemus.................	854
Notropis................. 274, 275
Pseudojulis..............	1602
Xyrichthys..............	1619
verany, Cybium	877
Verasper............................	2618
moseri	2619
Verde, Johnny	1195
Mojarra.....................	1538
Morena.....................	396
Pudiano 1590, 1591
Verdes, Cabrillas..................	1194
verecunda, Ulocentra..............	1049
verecundum, Etheostoma	1050
Lepidion	2543

	Page.		Page.
Verilus	1283	vetula, Pseudoscarus	1649
sordidus	1284	Scarus	1642,1649
Verma	374	vetulus, Parophrys	2649
kendalli	375	Pleuronectes	2641
Yermelho, Pudiano	1583	vexillare, Boleosoma nigrum	1096
vermicularis, Corvina	1453	Etheostoma nigrum	1096
Ophioscion	1452	vexillaris, Sebastichthys	1822
Sciæna	1452,1453	Sebastodes	1821
Sparus	1271	vexillarium, Holocentrum	852
vermiculatus, Esox	627	vexillarius, Holocentrus	852
Exocœtus	740	Vexillifer	2445
Lucius	627,2827	wheeleri, Uranidea	1950
Xyrichthys	1619	Viajaca	1600
vermiformis, Neoconger	362	vicensis, Trigla digitis	2108
Vermiglia, Pesca	1811	vicina, Muræna	394
vernalis, Clupea	426	Murænophis	394
Gobio	321	Sidera	394
Hydrargyra	639	vicinalis, Icelus	1916
Hyodon	413	vicinus, Gymnothorax	394
Pomolobus	426	Lycodontis	394
Scomber	866	Uroconger	365
vernullas, Batrachoides	2316	victoriæ, Moxostoma	197
verreauxi, Conger	355	Vicuda	824
verrea. Cossyphus	1583	Vieja	1635,1636,1649
Lutjanus	1583	Colorada	1639
verrilli, Gymnothorax	294	Muger	1639
Lycenchelys	2470,2471	Viejas	1637
Lycodes	2471	vigil. Ioa	1645
Lycodontis	393	vigilax. Ceratichthys	253
Sidera	394	Ciliola	253
verrucosa, Occa	2043	Villarius	2799
verrucosus, Brachyopsis	2044	dugesii	2799
Cottus	1980	pricei	2799
Diodon	1749	villosa, Clupea	521
Myoxocephalus	1979	villosus Cottus	2023
Verrugata	1476	Mallotus	520
Verrugato	1462,1463	vincente, Sicydium	2297
versicolor, Abramis	250	vinciguerræ, Exocœtus	734
Girardinus	689	Exonautes	2636
Heterandria	688	Vinciguerria	577
Labrus	1346	attenuata	577
Malacocterus	2359	vinctipes, Nanostoma	1475
Myxodes	2359	vinctus. Blennius	2282
Pœcilichthys	1089	Caranx	915
versipunctatus, Gymnothorax	394	Fundulus	687,2827
vertagus, Gadus	2541	vinolentus, Lethotremus	2101
verticalis, Echeneis	2270	viola. Antimora	2544
verticallis, Pleuronichthys	2638	Eques	1466
verus. Carcharias	50	Haloporphyrus	2544
Conger	355	violacea, Bipinnula	578
vespertilio, Holorhinus	90	violaceus, Apodichthys	2427
Malthæa	2737	Cebedichthys	2427
Ogcocephalus	2737	Escolar	4043
Rhinoptera	90	Thyrsitops	870
veternus, Podothecus	2063,2064	Viper-fishes	578,584
vetula, Balistes	1703	virens, Chlorichthys	1619

	Page.
Verilus	1283
sordidus	1284
Verma	374
kendalli	375
Vermelho, Pudiano	1583
vermicularis, Corvina	1453
Ophioscion	1452
Sciæna	1452, 1453
Sparus	1271
verrniculatus, Esox	627
Exocœtus	740
Lucius	627, 2827
Xyrichthys	1619
vermiformis, Neoconger	362
Vermiglia, Pesca	1811
vernalis, Clupea	426
Gobio	321
Hydrargyra	639
Hyodon	413
Pomolobus	426
Scomber	866
vernullas, Batrachoides	2316
verreauxi, Conger	355
verres, Cossyphus	1583
Lutjanus	1583
verrilli, Gymnothorax	394
Lycenchelys	2470, 2471
Lycodes	2471
Lycodontis	393
Sidera	394
verrucosa, Occa	2043
verrucosus, Brachyopsis	2044
Cottus	1980
Diodon	1749
Myoxocephalus	1979
Verrugata	1476
Verrugato	1462, 1463
versicolor, Abramis	250
Girardinus	689
Heterandria	688
Labrus	1346
Malacocterus	2359
Myxodes	2359
Pœcilichthys	1089
versipunctatus, Gymnothorax	394
vertagus, Gadus	2541
verticalis, Echeneis	2270
verticallis, Pleuronichthys	2638
verus, Carcharias	50
Conger	355
vespertilio, Holorhinus	90
Malthæa	2737
Ogcocephalus	2737
Rhinoptera	90
veternus, Podothecus	2063, 2064
vetula, Balistes	1703

	Page.
vetula, Pseudoscarus	1650
Scarus	1647, 1649
vetulus, Parophrys	2640
Pleuronectes	2641
vexillare, Boleosoma nigrum	1058
Etheostoma nigrum	1058
vexillaris, Sebastichthys	1822
Sebastodes	1821
vexillarium, Holocentrum	852
vexillarius, Holocentrus	852
Vexillifer	2495
vheeleri, Uranidea	1950
Viajaca	1539
vicensis, Trigla digitis	2183
vicina, Muræna	394
Mu/ /ænophis	394
Sidera	394
vicinalis, Icelus	1916
vicinus, Gymnothorax	394
Lycodontis	394
Uroconger	358
victoriæ, Moxostoma	187
Vicuda	824
Vieja	1635, 1636, 1649
Colorada	1639
Muger	1639
Viejas	1627
vigil, Ioa	1065
vigilax, Ceratichthys	253
Cliola	253
Villarius	2789
dugesii	2789
pricei	2790
villosa, Clupea	521
villosus Cottus	2022
Mallotus	520
vincente, Sicydium	2207
vinciguerræ, Exocœtus	734
Exonautes	2836
Vinciguerria	577
attenuata	577
vinctipes, Nanostoma	1075
vinctus, Blennius	2382
Caranx	918
Fundulus	637, 2827
vinolentus, Lethotremus	2101
viola, Antimora	2344
Eques	1486
Haloporphyrus	2544
violacea, Bipinnula	878
violaceus, Apodichthys	2427
Cebedichthys	2427
Escolar	4843
Thyrsitops	879
Viper-fishes	578, 584
virens, Chlorichthys	1610

	Page.
virens, Gadus	2534
Pollachius	2534
Scarus	1640
Thalassoma	1611, 2859
virescens, Apodichthys	2412
Clupea	426
Cynoscion	1415
Gallus	932
Gymnothorax	394
Lycodontis	394
Otolithus	1415
Pantosteus	171, 172
virgata, Coryphæna	953
virgatulus, Gobiesox	2333
virgatum, Etheostoma	1093
virgatus, Delolepis	2442
Pœcilichthys	1093
virginalis, Salar	495
Salmo clarkii	2819
mykiss	495
Scarus	1647
virginianus, Acanthocottus	1976
Cottus	1976
Scorpius	1076
virginicum, Pristipoma	1323
virginicus, Anisotremus	1322, 1323
Polydactylus	829
Polynemus	830
Pomadasys	1323
Sparus	1323
viride, Ophidium	2477
Sparisoma	1648
viridescens, Atherina	800
Fundulus	641
Osmerus	523
Sphyræna	826
Syngnathus	771
viridipallidus, Gobius	2259
viridis, Amia	113
Centrarchus	992
Centropomus	1118
Diacope	1246
Esox	110
Evoplites	1246
Genyoroge	1246
Gymnelis	2477
Lepidosteus	111
Lutjanus	1246
Piscis bahamensis	1638
Scarus	1638
viscosa, Uranidea	1968
viscosus, Cottus	1968
Silurus	143
vitrea, Ammocrypta	1065
Ioa	1064
Lucioperca	1022
vitrea, Perca	1021
vitreum, Stizostedion	1021
vitreus, Pœcilichthys	1065
vitta, Xyrichthys	1617
vittata, Argyrotænia	833
Channomuræna	404
Oliola	258
Clupea	421
Codoma	258
Echeneis	2269
Gymnomuræna	404
Hemitremia	242
Inermia	1366
Lepidomeda	328
Pœcilia	692
vittatum, Exoglossum	327
vittatus, Ammodytes	833
Cyprinus	307
Eummelichthys	1365, 1366
Erythrichthys	1366
Esox	628
Ichthyophis	404
Leuciscus	282
Proceros	102
Sarchirus	110
Siphateles	244
Sparus	1323
Symbranchus	342
vitula, Sargus	1364
vitulinum, Plectropoma	1192
vitulinus, Hypoplectrus unicolor	1192
vitulus, Carpiodes	165
Viuva	1788
vivanet, Bodianus	1257
vivanus, Anthias	1224
Hemianthias	1223
Lutjanus	1264, 1265
Mesoprion	1263
Neomænis	1262
Pronotogrammus	1224
vivax, Ammocrypta pellucida	1063
Cliola	253
vivipara, Pœcilia	691
Viviparous Perch	1498
viviparus, Lophius	2715
Sebastes	1761
vlamingii, Coryphæna	953
Voilier	891
Volador	740, 2183
volador, Exocœtus	733
Rubio	2164
Voladora, Aguja	891
Volantin	914
volitans, Cephalacanthus	2183
Dactylopterus	2183
Exocœtus	734, 736

	Page.
volitans, Trigla	2183
volucellus, Hybognathus	263
Hybopsis	263
Leuciscus	263
Notropis	263
Vomer	933
brasiliensis	934
browni	934, 2846
cayennensis	934
columbianus	934
cubæ	934
curtus	934
dominicensis	934
dorsalis	934
gabonensis	934, 2846
goreensis	934
martinicensis	934
noveboracensis	934
sanctæ marthæ	934
sanctæ petri	934
senegalensis	934
setipinnis	934, 2846
spixii	2846
vomer, Argyreiosus	936
Selene	936
Zeus	936
vomerina, Atherina	793
vomerinus, Anarrhichas	2447
Salarius	2400
vorax, Lucius	628
Mesoprion	1281
Platyinius	1281
Pseudorhombus	2626
Ptychocheilus	227
Voraz	1280
vulgaris, Acanthias	54
Ameiurus	140
Amiurus	140
Auxis	868
Brosmius	2561
Conger	355
Gunnellus	2419
Hippoglossus	2612
Liparis	2118
Lumpus	2097
Merluccius	2530
Molva	2552
Mustelus	29
Pagrus	1357
Pimelodus	140
Pomotis	1010
Sphyræna	826
Squatina	59
Sturio	105
Thynnus	870
vulnerata, Apocope	312

	Page.
vulneratum, Etheostoma	1077
vulneratus, Nothonotus	1077
Pœcilichthys	1077
vulpeculus, Pimelodus	141
Vulpes bahamensis	411
vulpes, Albula	411
Alopecias	46
Alopias	45
Esox	411
Pimelodus	135
Sebastodes	1835
Squalus	46
vulpinus, Squalus	46
Vulsiculus	2181
imberbis	2181
vulsus, Agonus	2068
Podothecus	2068
Stelgis	2067
wabashensis, Anguilla	348
Wachna Cod	2537
wagneri, Pimelodus	151
Rhamdia	150, 151
Waha Lake Trout	496
Wahoo	876
walkeri, Selanonius	49
Wall-eyed Herring	426
Pike	1021
Pollack	2536
Surf-fishes	1501
Warmouths	991, 992
warreni, Boleichthys	1103
Pœcilichthys	1103
Salmo	483
Waru, Me	1829
watauga, Hybopsis	319
Weakfish, Bastard	1406
Common	1407
Spotted	1409
webbi, Embiotoca	1505
webbii, Blennophis	2401
Ophioblennius	2401
Webug Sucker	180
welaka, Notropis	2799
Welshman	848
Wench, Old	1703
West Indian Lancelet	3
westphali, Tæniophis	396
Whale Sharks	52
wheatlandi, Gasterosteus	749
Whiffs	2678
whipplei, Boleosoma	1096
Oliola	279
Etheostoma alabamæ	1095
whipplii, Boleichthys	1096
Cyprinella	279

Page.

whipplii, Etheostoma 1095
    Notropis 278
Whip-tailed Rays................. 79
Whirligig Mullet 818
White Bass 1132
    Cat 134, 138
    Croaker...................... 1397
    Grunt........................ 1310
    Hake 2555
    Lake Bass 1132
    Mullet 189, 813
    Perch...... 1133, 1134, 1484, 1501, 1509
    Salmon of the Colorado ..... ·225
    Sea Bass of California 1413
    Sturgeon..................... 104, 107
    Sucker....................... 178, 192
    Surf-fish 1506
White-bill......................... 431
White-bone Porgy 1353
White-eye 1021
Whitefish.................... 433, 461, 2276
    Broad 464
    Common....................... 465
    Humpback 466
    Menominee.................... 465
    Mongrel 473
    Musquaw River 466
    Rocky Mountain 463
    Round........................ 465
    Sault 466
White-mouthed Drummer.......... 1462
White-nosed Suckers............,... 190
Whiting 2530
    California................... 1477
    Carolina 1474
    Northern 1475
    Sand......................... 1474
    Silver....................... 1477
    Surf 1477
Whiting of Lake Winnipiseogee ... 466
Widow-fish......................... 1788
Wife, Old.......................... 940, 1703
wilderi, Lampetra 13, 2745
Will, Black........................ 1139
williamsi, Symphurus.............. 2711
williamsoni, Coregonus............. 463
            cismontanus 463
            Gasterosteus......... 750
            Gasterosteus micro-
cephalus 751
willoughbyi, Acrotus 973
Willow Cat 2788
willughbeii, Canthidermis 1707
            Salmo................ 509
            Trematopsis 1754
wilsoni, Cottus 1952

Page.

Wilton Smelt....................... 523
winchelli, Hybopsis................ 321
Wind-fish 221
Window Panes................... 2659, 2660
Winninish 487
Winnipiscogee, Whiting........... 466
Winter Flounders................... 2646
    Sucker 187
Wolf Eel........................... 2448
Wolf-fish.................... 595, 2445, 2447
woolmani, Paralichthys 2628
Worm Eels.......................... 370
Wrasse-fishes ..,.................. 1571
Wreckfish 1138, 1139
wrighti, Etheostoma 1047
Wry-mouth 2442, 2443
wurdemanni, Gobius 2225

xænocephalus, Hybopsis 289
    Minnilus............. 289
    Notropis............. 289
xænura, Oliola 280
    Codoma 280
xænurus, Minnilus................. 280
    Notropis................. 280
xaninrus, Catulus................. 24
Xanthichthys....................... 1708
    cicatricosus 1709
    mento 1760
    ringens............. 1709
xanthocephalus, Ameiurus 141
    Silurus 141
xanthogrammus, Upeneus 860
xanthomelas, Anguilla............. 348
xanthops, Odontoscion 1427
xanthopteron, Hæmulon 1307
xanthopterum, Hæmulon 1307
xanthopus, Catostomus 181
xanthopygus, Caranx 921
xanthosticta, Mycteroperca bonaci. 1176
xanthostigmus, Citharichthys...... 2680
xanthulum, Cestreus............... 1411
xanthulus, Cynoscion 1410
xanthurus, Homoprion............. 1434, 1459
    Leiostomus......... 1458
    Pomacentrus.......... 1557
    Rhombus.............. 966, 2849
    Sciæna............... 1459
    Seserinus .... ·966
    Sparus 1346
xanti, Labrisomus 2362, 2363
    Rhypticus.................. 1231
    Rypticus.................. 1231
    Umbrina 1467, 1468
    Xenichthys 1288
xantusi, Lepidopus 2844, 4843

	Page.
xenarcha, Mycteroperca	1180
Xenarchi	780, 785
xenachus, Epinephelus	1180
xenauchen, Galeichthys	2777
Xenichthyinæ	1244
Xenichthys	1287
agassizii	1287
californiensis	1286
xanti	1288
xenops	1288
xenurus	1015
xenicus, Fundulus	662
Xenisma	633, 635, 648
catenata	648
stellifer	648
xenisma, Prionotus	2154
Xenistius	1286
californiensis	1286
Xenochirus	2079
alascanus	2081
latifrons	2082
pentacanthus	2080
triacanthus	2084
Xenocys	1285
jessiæ	1285
xenodon, Calotomus	1626
Xenomi	620
Xenomystax	360
atrarius	361
xenops, Xenichthys	1288
xenopterus, Cypsilurus	2836
Xenopterygii	782, 2326
xenostethus, Sternias	1927
Triglops	1927
Xenotis	999, 1000, 1002
aureolus	1003
lythrochloris	1003
xenura, Kuhlia	1015
xenurus, Xenichthys	1015
Xererpes	2413
fucorum	2413
Xesurus	1694
clarionis	1695
laticlavius	1695
punctatus	1694, 1695
Xiphias	893
gladius	894
imperator	892
makaira	891
rondeleti	894
Xiphidiinæ	2348
Xiphidion	2424
cruoreum	2425
mucosum	2425
rupestre	2426
ulvæ	2424

	Page.
Xiphiidæ	893
Xiphister	2424
chirus	2424
mucosum	2425
rupestris	2426
Xiphistes	2423
chirus	2424
ulvæ	2423
Xiphophorus	701
bimaculatus	678
gillii	692
gracilis	683
guntheri	702
helleri	701, 702
Xurel	909, 923
del Castilla	937
xyosterna, Stellerina	2042
xyosternus, Brachyopsis	2043
Xyrauchen	184
cypho	184
uncompahgre	184
Xyrichthyinæ	1575
Xyrichthys	1617
cultratus	1619
infirmus	1616
jessiæ	1613
lineatus	1619
martinicensis	1617
modestus	1619
mundiceps	1618
novacula	1619
psittacus	1618, 1619
rosipes	1615
ventralis	1616
venustus	1619
vermiculatus	1619
vitta	1617
xyris, Sebastopsis	1835
Xyrula	1612
jessiæ	1612, 1613
Xystæma	1372
cinereum	1372
Xyster	1384
xyster, Zapteryx	65
Xystes	2076
axinophrys	2076
Xystophorus	900
Xystreurys	2623
lioplepis	2623
xystrodon, Sparisoma	1630
Xystroperca	1169, 1170, 1181
Xystroplites	1006
gillii	1007
longimanus	1008
xysturus, Myrichthys	2802
Ophichthys	2802

	Page.
xysturus, Ophisurus	376
Pisodonophis	376
yalei, Ostracion	1724
Yanagi Nomai	1830
Yarrella	583
blackfordi	584
yarrelli, Acipenser	105
yarrowi, Agosia	309
y-canda, Gillichthys	2252
Quietula	2251, 2252
Yellow-backed Rockfish	1822
Yellow Bass	1134
Belly	1001
Cat	139, 143
Yellow-finned Grouper	1155, 1172
Roncador	1467
Trout	497
Yellow Fish	1144
Goatfish	859
Grouper	1183
Grunt	1303
Jack	919
Mackerel	921
Perch	1023
Pike	1021
Sculpin	1934
Yellow-spotted Rockfish	1826
Yellowstone Trout	493
Yellow-tail	902, 906, 1275, 1433
Croaker	1467
Rockfish	1781
y-græcum, Astroscopus	2307
Upsilonphorus	2308
Uranoscopus	2308
Yuriria	314, 315, 321
zacentrus, Sebastichthys	1815
zachirus, Glyptocephalus	2658
Zaclemus	1477, 1478, 1480
zadocki, Cylindrosteus	111
Zalembius	1499
rosaceus	1500
Zalieutes	2738
elater	2738
Zalocys	2848
stilbe	2848
Zalopyr	1795, 2680
Zalypnus	2246
cyclolepis	2246
emblematicus	2247
Zanclurus	890
Zanclus	1687
canescens	1688
centrognathus	1688
cornutus	1687, 1688

	Page.
zanclus, Dermatolepis	2854
zanderi, Argyrocottus	1995
zanemus, Ceratichthys	319
Zaniolepidinæ	2862
Zaniolepis	1876
frenatus	1877
Latipinnis	1876
Zapaters	898
Zaphotias	2826
pedaliotus	2826
Zaprora	2850
silenus	2850
Zaproridæ	2849
Zapteryx	64
exasperatus	64
xyster	65
zatropis, Siphostoma	772
zebra, Acanthurus	1691
Arbaciosa	2341
Asproperca	1027
Chætodon	1691
Fundulus	641, 647
Gerres	1373
Gobiesox	2342
Gobius	2226
Hydrargyra	647
Percina caprodes	1027
Pileoma	1028
Psychrolutes	2027
zebrinus, Fundulus	646
Zeidæ	1659
Zeinæ	1660
zelotes, Dactyloscopus	2303
Hermicaranx	2845
Zenion	1661
hololepis	1661
Zenopsis	1660
ocellatus	1660
zenopterus, Exocœtus	738
Zeoidea	1559
zephyreus, Astroscopus	2309
Pristis	2749
Zesticelus	1990
profundorum	1990
Zestidium	1439, 1442
Zestis	1439, 1440
zestocarus, Stellifer	1445
zestus, Nebris	1417
Zeus capillaris	936
ciliaris	932
crinitus	932
gallus	936
geometricus	936
guttatus	955
imperialis	955
luna	955

	Page.
Zeus niger	936
ocellatus	1661
opah	955
quadratus	1668
/ regius	955
rostratus	936
setipinnis	934
stræmii	955
vomer	936
zeylonicus, Leuciscus	415
Ziphotheca	887
tetradens	887
Zoarces	2456
anguillaris	2457
elongatus	2457
fimbriatus	2457
gronovii	2457
labrosus	2457
polaris	2469
Zoarchidæ	2455
Zoarchus	2456
zoarchus, Lycodes	2464
Zoarcidæ	2455
Zoarcinæ	2455
zonale, Etheostoma	1075
arcansanum	1075
Nanostoma	1075
zonalis, Astatichthys	1075
Pœcilichthys	1075
zonata, Seriola carolinensis	902
zonatum, Elassoma	982
zonatus, Alburnus	285
Chætodipterus	1668
Ephippus	1669
Esox	639
Fundulus	657
Notropis	285
Scomber	902
Zygonectes	659
Zonichthys	901, 904
bosci	905
coronatus	905
declivis	905
gigas	903
zonifer, Clinus	2359
Erilepis	1863
zonifer, Myriolepis	1863
Zygonectes	657
zonipectus, Pomacanthodes	1682
Pomacanthus	1681
zonistius, Luxilus	285
Minnilus	285
Notropis	277, 285
Zonogobius	2210
zonope, Jordania	1884
Zonoscion	1477, 1478, 1479

	Page.
zonurus, Malacocottus	1994
Zophendum	308
australe	212
plumbeum	216
siderium	314
zopherus, Potamocottus	1952
zophochir, Ophichthus	385
Ophichthys	385
zosteræ, Hippocampus	778
zosterura, Evermannia	2256
zosterurum, Gobiosoma	2257
zunnensis, Leuciscus	227
Zygæna	43
lewini	45
malleus	45
subarcuata	45
tiburo	44
tudes	44
zygæna, Sphyrna	45
Squalus	45
Zygobatis	90
Zygonectes	650, 2828
atrilatus	682
auroguttatus	654, 2829
brachypterus	682
chrysotus	656
cingulatus	655, 656
craticula	657
dispar	659
escambiæ	658
floripinnis	651
funduloides	650
guttatus	658
henshalli	653, 2829
hieroglyphicus	658
inurus	682
jenkinsi	652
lateralis	659
lineatus	649
lineolatus	657
luciæ	655
macdonaldi	651
mannii	664
melanops	682
notatus	659
nottii	657
pulchellus	659
pulvereus	652
rubrifrons	654, 2829
sciadicus	654
zonatus	659
Zygonectes zonifer	657
Zygonectus	633, 635
zyopterus, Galeorhinus	32
Zyphothyca	883

i

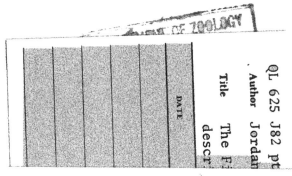

9 780331 660524